In Defense of the World's Most Despised Species

Some animals and plants injure or kill millions of people annually, and others cause trillions of dollars in property damage and loss. Such harmful species are understandably hated. However, the vast majority of the planet's millions of species are disliked simply because of how they look and act. This bias is endangering numerous species that play important roles in maintaining both the natural ecosystems and the human economies of the world. *In Defense of the World's Most Despised Species* examines the psychological motivations that lead people to make judgments about the attractiveness of species, noting the overwhelming importance of visual cues. It describes in considerable detail the physical and behavioral traits of species that lead us to love or hate them. Full color illustrations throughout present beautiful, charming animals and plants, species that seem loathsome, behavior of people in relation to such divergent species and their characteristics, and numerous explanatory diagrams of relevant biological and psychological phenomena. The aim of this book is to give readers insights into how we humans arrive at biased judgments and to promote the welfare of valuable, albeit sometimes unlovable animals and plants that consequently suffer from discrimination. Many of the ugliest, most disgusting, and feared species, such as vultures, toads, hyenas, sharks, spiders, and even the vast majority of cockroaches, in reality are some of our most valuable friends.

Features

- Theme of the book – human preferences for and against species – is novel, scarcely examined to date.
- Multidisciplinary analysis, especially psychology, biological conservation science, and ecology, as well as philosophy, agriculture, urban planning, human health, and law.
- Text is accessible, user-friendly, concise, and well-organized, making numerous complex topics comprehensible, readable not only by specialists but also by students and the educated layperson.
- Includes over 2,000 high-quality, entertaining, and informative color figures.

In Defense of the World's Most Despised Species

Why We Love Some Species but Hate Most, and Why It Matters

Ernest Small

CRC Press is an imprint of the
Taylor & Francis Group, an **informa** business

Designed cover image: Ernest Small

First edition published 2024
by CRC Press
6000 Broken Sound Parkway NW, Suite 300, Boca Raton, FL 33487-2742

and by CRC Press
4 Park Square, Milton Park, Abingdon, Oxon, OX14 4RN

CRC Press is an imprint of Taylor & Francis Group, LLC

ISBN: 9781032525013 (hbk)
ISBN: 9781032536453 (pbk)
ISBN: 9781003412946 (ebk)

DOI: 10.1201/9781003412946

Typeset in Times
by codeMantra

Contents

Executive summary

FIGURE 0.1 Human preference for beauty, evidenced by animals likely to be chosen to survive on Noah's Ark. Attractive animals, as shown here, are given much more priority for conservation than ugly species. Painting by Jessica Hsiung.

FIGURE 0.2 Human prejudice against ugliness, evidenced by animals unlikely to be chosen to survive on Noah's Ark. Unattractive animals, as shown here, are much less likely to be conserved than attractive species. Painting by Jessica Hsiung.

Biodiversity is often defined as a combination of species (living things), habitats (their homes), and ecosystems (the interactions of environment and constituent species in given regions). Countless publications and thousands of concerned organizations have raised the alarm that all three components of biodiversity are in crisis. Habitats and ecosystems are rapidly being degraded, and species are undergoing extinction faster than at any time since humans came into existence. It is widely recognized that humans are both the cause and the potential cure of the world's ecological problems. This book addresses a very poorly appreciated and largely ignored problem that is key for humans to, literally, save the living components of the world. The issue is ferocious, largely ingrained preferences and prejudices that determine human behavior toward non-human species. Most people hate most species, and are indifferent at best to their survival, while not appreciating that they are indispensable for our welfare.

This book describes in considerable detail the physical and behavioral traits of species, which lead us to like or dislike them. Just as prejudice based on appearance is widespread among humans, there is parallel bias directed against non-human species, primarily because of the way they look. Similarly, favoritism toward good-looking people is matched by strong preferences for a very select number of attractive species. Astonishingly, many of the physical features that elicit prejudices and preferences toward people are the same that elicit prejudices and preferences toward animals, and sometimes even plants. This reflects the instinctive or hard-wired determinants of many prejudices and preferences, which are difficult to modify. Nothing is as attractive as a human baby, so it is hardly surprising that baby features retained in adult humans and even non-humans elicit profound admiration. Unfortunately, the vast majority of non-human species have been judged by most people to be so ugly that they are unworthy of consideration.

The attractive species people favor are evident from zoos, botanical gardens, and indeed from the biota we maintain around our own homes. Size is, by a considerable margin, the most admired of all features. The giants of the world (such as whales, elephants, and colossal trees) always demand respect; tiny 'bugs' – making up the majority of the world's species – are fortunate not to be deliberately crushed underfoot. The 'charismatic megafauna' include big glamorous animals such as pandas and tigers, which have been adopted as icons of environmental and biodiversity organizations. They receive the lion's share of public and private financial support, publicity, research, and protective legislation. Such species usefully increase public awareness of the environmental and biodiversity crises and assist in mobilizing financial support for conservation. They are widely touted as critical to the cause of conservation, not just symbolically, but also because preservation of their habitats, it has been claimed, can simultaneously preserve other species. However, there are only limited benefits to other species and habitats. Indeed, by monopolizing conservation funding and biasing support toward the welfare of the glamorous species, support is being diverted from the most degraded ecosystems and the majority of species in danger of extinction. Our natural emotional preference for a small number of species is beneficial for conservation, but unless it is tempered with a rational program of support for all habitats and species, the world is destined to become a much poorer place.

Just as the most charismatic and useful species generate compassion for biodiversity, the most despised species generate hatred and indeed destructiveness toward the natural world. Human attitudes to species are of course partly determined by rational considerations, such as whether they are dangerous or valuable. We are harmed physically and economically by many animals and plants, which understandably have often led to hatred against them. However, many species are so ugly and their lifestyles so disgusting that people seem unable to tolerate them. Indeed, the physical appearance and behavior of the world's species are critical determinants of how well or badly they are treated by people. While there are moral and ethical reasons why all living things deserve respect, far more critical from a human perspective is that our negative emotional reactions to most of the world's species are preventing appreciation of their indispensable values for our welfare – indeed, for our very survival in the perspective of current ecological crises threatening the planet. This book examines in detail the world's most despised creatures (listed below), illustrating that for the most part they are victims of prejudice, have significant compensating values, and their harm potential is better controlled by an improved understanding of their biology rather than blind attempts at extermination.

Principal despised species of the world.

'Bugs'	**'Lower' Vertebrate Animals**	**Toxic Plants**	**Environmental Weeds**
Bedbugs	Sharks	Opium poppy	Cactus pear
Clothes moths	Frogs and toads	Tobacco	Common reed
Cockroaches	Snakes	Sugar cane	English ivy
Fleas	Vultures	Poison ivy and relatives	Purple loosestrife
Houseflies	Feral pigeon	Ragweed	Water hyacinth
Leeches			
Lice	**Mammals**	**Agricultural Weeds**	**Urban Weeds**
Locusts	Bats	Barnyard grass	Crabgrass
Mosquitoes	House mouse	Bermuda grass	Dandelion
Spiders	Hyenas	Lambsquarters	Japanese knotweed
Termites	Rats	Nutsedges	Kudzu
Ticks	Skunks	Water hyacinth	Canada thistle

WHY ARE THERE SO MANY PICTURES OF ANIMALS, PLANTS, AND PEOPLE IN THIS BOOK?

Humans have understood the testimonial value of pictures of species since prehistory. The earliest crude paintings and engravings on rocks and in caves showed the animals of greatest significance, including bison, horses, mammoths, and deer that were hunted for food, and lions, bears, and wolves that were dangerous predators and competitors. (Analyses of cave paintings made from 3000 to 8000 B.C.E. in Europe showed that only animals were depicted [Leroi-Gouram 1982; Ruspoli 1987].) These primitive drawings reflect simple economic motivations for liking or disliking species. By contrast, consider which images of species are most often exhibited today: photos of other people (loved ones, but also admired media stars and athletes), pets, and charismatic animals (especially big mammals and colorful birds). For women, floral images are universal – on clothing, bedding, and even wallpaper. Conversely, aside from butterflies, how often does one observe framed photos of insects (which happen to make up most of the world's species)? Clearly, we have strong preferences for and prejudices against the visual appearance of both other people and other species.

The familiar maxim 'What you see is what you get' reflects how extremely reliant most of us are on our vision. But as many eventually learn, looks can be misleading, and even our perception of what we see is conditioned by what we have been told (remember the expression 'The Emperor has no clothes'). Nevertheless, as detailed in this book, we humans are programmed to accept good looks as the basis of merit, and bad appearance as the basis for disapproval. This is of course superficial, and results in bias and prejudice that have led to preferential favoring of good-looking individuals (both human and non-human) and disrespectful behavior toward less fortunate individuals (including the majority of species). This book analyzes how quite parallel perception of the visual beauty or ugliness of people is the key consideration that determines how well or badly non-human species are treated by people. The result is not merely academic: we humans are conserving hundreds of the most beautiful species while millions of unattractive ones that are at least as valuable for human welfare are being exterminated.

This book presents more than 2,000 illustrations in color (arranged in about 800 plates), and this is not merely for entertainment. They serve to demonstrate how pervasive and decisive aesthetic appearance of species is to determining our behavior toward them, and therefore that it is essential to appreciate and control our natural visual biases.

Academic volumes generally have few if any illustrations, and given the inevitable technical complexity of subjects and language, most such books are typically challenging to read and, frankly, tedious or even boring. Animals and plants are not boring – they are fascinating, and the best way of presenting them is often pictorially. Art lovers will notice that many of the paintings reproduced here to illustrate various themes are often by history's greatest artists, and in some cases by masterful cartoonists and caricaturists whose works are also featured in art museums. Moreover, some of the figures are provocative: sardonic, sad, or funny, consistent with the aim of this book to provoke and analyze emotional responses toward the welfare of the species under discussion. All of the many figures included in this book are beautiful, but in some cases, this requires thought. 'Beauty will save the world' is the famous utterance of a character in *The Idiot*, one of the novels of Russian author F. M. Dostoevsky. Analysts of the quotation have pointed out that Dostoevsky's idiot was unconsciously pointing out the profound observation that true beauty underlies nature but often takes on the appearance of ugliness, just as evil often hides under the disguise of beauty. Readers of this book are invited to explore and judge these issues for themselves.

Preface

Modern science is extensively compartmentalized into very specialized disciplines, many of which are so arcane that even other scientists, let alone the public, often find incomprehensible. Science benefits greatly from interactions among different kinds of researchers, but some disciplines seem not to combine well, like ketchup and ice cream. This book addresses two apparently disparate fields, conservation biology, and psychology. In fact, as explained in the text, there is a budding field of 'conservation psychology' with the general goal of understanding and guiding the behavior of humans toward the natural world for the benefit of both. One cannot help but notice (since so many scientists say so) that the natural world is in crisis, and immediate remediation is vital to prevent imminent disaster. Although the public has focused almost exclusively on the deterioration of the climate (it's hard not to notice how erratic the weather has become), in fact humankind is also significantly endangered by deteriorating biodiversity (living things and their habitats). This book addresses value judgments, based particularly on the physical appearance of species, that determine how people behave toward them. Just as many people are victims of biases and prejudices, most species are similarly misjudged and mistreated. Indeed, most species are disliked, even detested. This is facilitating the current disastrous disregard for the world's ecosystems and the associated extinction of countless species. Because the welfare of humans depends on these species, their conservation needs to be promoted. Understanding the underlying psychology of concerns (inevitably biased) for different species is critical to alleviating impending threats not just to non-human species but to ourselves.

The fields of science dealing with living things have tended to separate research concerning humans from research into all other species. For example, the economic relationships of humans with each other and with the environment fall under 'Economics,' while the economic relationships of non-human animals with each other and with the environment tend to fall under 'Ecology.' This book stresses that controlling human behavior toward species is critical ecologically for human survival and critical economically for human welfare. This is why this book includes chapters on ecology, agriculture, and urbanization, the primary areas in which humans are endangering themselves by endangering the biota of the planet.

Brain-controlled behavior of humans is the subject matter of 'Psychology' while the counterpart in non-human animals is 'Ethology.' Although the two fields are traditionally separated, as reviewed in this book there are numerous parallels between humans and non-humans that explain how we think and behave. As will be stressed, a substantial portion of our behavior is instinctive, just as in so-called 'lower animals,' and this often accounts for biases, prejudice, and counterproductive activities.

The central topic of this book, human behavior toward species, demands examination of information from the disciplines of ecology, the behavioral sciences, and related fields. This information is voluminous, as reflected by the very extensive references (over 2,000) listed. Authors are often advised to refer to primary (i.e., earliest) papers dealing with scientific findings, but in the main recent reviews are cited here, since this allows interested readers to quickly locate background information should they wish.

Psychology, indeed the human social sciences in general, can rarely predict human behavior with the mathematical precision of the physical sciences. Similarly, ethology (animal psychology) deals with complex, variable phenomena that are difficult to analyze. In this book, claims are examined based mainly on the most recent authoritative scientific reviews. The attempt is made consistently to reflect majority scientific opinion, but many aspects are controversial, speculative, and conflicting, and need more study. Researchers in the fields of behavioral sciences are sometimes prone to drawing grand conclusions based on scant evidence (which is often all that is available), and likely reviewers of this book will find that some overgeneralizations have been repeated. Readers are advised to be skeptical.

As emphasized in the previous section, this book is rather lavishly illustrated because visual criteria strongly determine how well or how badly we treat other species. Most people are pleased by photos and paintings of beautiful and interesting species and their interactions with people. But this book is also about ugliness. Many species that are hostile to humans (particularly disease-causing microorganisms) cause horrific symptoms in humans and animals. Illustrations of such pathology are appropriate for medical and veterinary presentations, but not here. In illustrating non-photogenic species, and more particularly in showing cruel and painful behavior toward them, I have been restrained, simply because of the need not to offend the heterogeneous audience for which this book is addressed. I have frequently employed paintings and drawings, since by nature they can diffuse the harshness that is more evident in photos. I have also often included humorous cartoons in criticizing exploitative segments of society, since once again this is less shocking than the associated pain or damage that could be more harshly illustrated.

This book includes information on both animal conservation (keeping them and their supporting habitats from vanishing in nature) and 'animal welfare' (preventing cruelty and pain). As described in detail in the book, animal (along with plant and habitat) conservation is vital for the future welfare of humans because of the indispensable associated supplies of materials, as well as the indispensable support of the health of our environment. We can and should save as much as possible of the natural world, including as many species as possible, simply for our own (selfish) survival. Species that harm us and/or our interests need to be controlled, but judiciously, because of the potential of harming ourselves. All these are *pragmatic* reasons for supporting conservation. Although there are also moral reasons for conservation of nature the presentation in the book is primarily pragmatic. The animal welfare aspects, which are mostly ethical in nature, are examined to a much lesser degree, simply because, for most people, pocketbook and other human benefits trump ethical issues. For example, most of us prefer to eat animals and accept that rodents should continue to be employed in medical research, although these activities generate pain. References are provided for the viewpoint that pain to sentient beings that we exploit needs to be minimized, with the long-term objective of total elimination. The epilogue to this book makes basic recommendations on how we should address the welfare of non-human species that are harmful to us.

This work presents extensive information gathered from the literature, and some error, omission, and misinterpretation are inevitably incorporated into compilations of this type. Moreover, scientific knowledge concerning the material is rapidly evolving. Liability arising directly or indirectly from the use of any of the information is specifically disclaimed.

Acknowledgments

Brenda Brookes, my invaluable colleague and friend for many years, skillfully assembled and enhanced most of the more than 2,000 illustrations in this book and also prepared dozens of original figures. I also thank Jessica Hsiung, our staff illustrator, who prepared more than a dozen outstanding plates. Illustrations purchased from the commercial provider Shutterstock are cited according to the required format. Sources of other illustrations are given in their accompanying captions. Many of the figures are reproduced consistently with the following Creative Commons Licenses: CC BY 1.0 (Attribution 1.0 Generic): https://creativecommons.org/licenses/by/1.0/; CC BY 2.0 (Attribution 2.0 Generic): http://creativecommons.org/licenses/by/2.0/; CC BY 3.0 (Attribution 3.0 Unported): http://creativecommons.org/licenses/by/3.0/. CC BY 4.0 (Creative Commons Attribution 4.0 International license): https://creativecommons.org/licenses/by/4.0/deed.en. CC BY ND 2.0 (Attribution NoDerivs 2.0 Generic): https://creativecommons.org/licenses/by-nd/2.0/; CC BY SA 1.0 (Attribution ShareAlike 1.0 Generic): https://creativecommons.org/licenses/by-sa/1.0/; CC BY SA 2.0 (Attribution ShareAlike 2.0 Generic): https://creativecommons.org/licenses/by-sa/2.0/. CC BY SA 2.5 (Attribution Share Alike 2.5 Generic): https://creativecommons.org/licenses/by-sa/2.5/; CC BY SA 3.0 (Attribution ShareAlike 3.0 Unported): http://creativecommons.org/licenses/by-sa/3.0/. CC BY SA 4.0 (Attribution ShareAlike 4.0 International): https://creativecommons.org/licenses/by-sa/4.0/deed.en. CC0 1.0 (Universal Public Domain Dedication): https://creativecommons.org/publicdomain/zero/1.0/deed.en.

Some of the material for this book is adapted from several of my previous publications in the journal *Biodiversity*, including Small 2011c, 2011d, 2012a, 2012c, 2016a, 2019a, 2019b, 2021a, and 2021b (see Literature cited).

About the Author

Dr. Ernest Small received a doctorate from the University of California at Los Angeles in 1969. He has since been employed with Agriculture and Agri-Food Canada, the country's national department of agriculture, where he presently holds the status of Principal Research Scientist. He specializes in the evolution and classification of economically important plants, dealing particularly with food, forage, biodiversity, and medicinal species. The species *Trigonella smallii* (Small's sweetclover) was named in his honor, and he himself has named dozens of new species. He is the author of 15 previous books, 6 of which received or were nominated for major awards. He has also authored over 400 scientific publications, mostly on economically important plants. Dr. Small's career has included dozens of appearances as an expert witness in court cases, acting as an adviser to national governments, presenting numerous invited university and professional association lectures, participating in international societies and committees, journal editing, and media interviews. He has been an adjunct professor at numerous universities and continues to supervise doctoral candidates. Dr. Small has received several professional honors, including: election as a Fellow of the Linnean Society of London; the G.M. Cooley Prize of the American Association of Plant Taxonomists for work on the marijuana plant; the Agcellence Award for distinguished contributions to agriculture; the Queen Elizabeth Diamond Jubilee Medal for contributions to science; the George Lawson Medal, the most prestigious award of the Canadian Botanical Association, for lifetime contributions to botany; the Lane Anderson Award, a $10,000.00 prize for science popularization; the Industry Leadership Award of the Canadian Hemp Trade Association (subsequently renamed in his honor); the Outstanding Paper in Plant Genetic Resources Award of the Crop Science Society of America; and appointment to the Order of Canada, the nation's highest recognition of achievements.

1 Introduction and Chapter Summaries

INTRODUCTION TO THE BOOK

Bias and prejudice based particularly on appearance are major problems for society. We humans are obsessed with our own appearance and that of others (note Figure 1.1), resulting in widespread unfair treatment. This book documents the harmful results of astonishingly parallel bias and prejudice directed against the majority of non-human species, primarily because of the way they look.

Bias and prejudice are negative beliefs, attitudes, and behaviors based on incomplete, usually mistaken information. Frequently the negativism is acquired because of erroneous information provided by others (particularly parents and peers). Also frequently, the negativism is acquired by a single painful experience (like a dog bite, leading to a lifelong fear of dogs). Both sources of bias and prejudice are significant, but as discussed in great detail in this book, there is another determinator: our innate nature. Our species is hardwired to react instinctively to many features of other living things. The initial reactions are emotions, like fear, pleasure, and disgust, which inevitably lead to attitudes (positive or negative) and behaviors. For example, the first experience with the smell of a skunk leads to suspicion about all small black and white mammals, while smelling a perfumed flower leads to trying to smell many other flowers.

Scientists argue over whether we are born with particular instinctive behaviors, or have predispositions to acquire the behaviors with early experience, but the result is the same. Obviously (at least, it is clear to most people) that sexual 'preference' is fixed very early in human development, and so each of us is hardwired to admire specific visual characteristics of the opposite sex. As this book documents, humans also appear programmed to approve of a wide range of visual features, and conversely to disapprove of contrasting features. The result is that we have innate evaluative tendencies that significantly bias our attitudes and behaviors toward not only other people but toward other species.

It is clear (documentation is provided in this book) that across human cultures (i.e., wherever one lives in the world), similar biases develop, proving that we humans have innate perceptual tendencies, toward both humans and other species. Perhaps the strongest evidence for innate biases against other species is furnished by animal phobias – why should fear of spiders, snakes, and frogs be common, and why are women much more frequently afraid?

FIGURE 1.1 'Beauty and the beast' is a fairy tale, encountered in various versions, tracing back thousands of years. The yarn relates how a young attractive girl meets a hideous beast, who eventually transforms into a handsome prince. The story points out the preoccupation of people with beautiful appearance, and the reluctance to accept ugliness. This painting (public domain) is from: Crane, W. 1874. Beauty and the Beast. London, George Routledge and Sons.

DOI: 10.1201/9781003412946-1

Remarkably, many, indeed the majority of physical traits that are employed as criteria to (mis)judge other humans are paralleled by similar criteria used to evaluate other species. A simple example: think of how we react to an elephant and an ant. We immediately respect large living things (human or non-human) but are relatively indifferent to small creatures. Clearly, humans are hardwired to judge both other humans and other animals, and while such innate judgments are (or were in historical times) important for survival, in modern times biased judgments can be destructive to human interests.

Indeed, prejudice against other species is highly counterproductive for human welfare. As will be described, most people are extremely biased against most other species (this is obvious since most species are insects, and aside from entomologists, bugs elicit little admiration). The fact is, however, most species contribute usefully, indeed indispensably, to the welfare of the planet in general and the well-being of humans in particular. Unfortunately, the majority of species are quite unattractive by human standards and are accordingly viewed with disdain. The resulting hostility to most species has facilitated an alarming explosion of extinctions. As will be explained, while most people aren't that concerned, they should be, because the result is a planet that is deteriorating, endangering human welfare.

Living things (technically 'organisms,' colloquially 'creatures') exist in discernibly different groups called 'species,' of which the most familiar is our own, humans (*Homo sapiens*). The subject of this book is human attitudes (particularly 'preferences' and 'prejudices') toward and consequent treatment of all the other millions of species. As will be noted, we spontaneously love and respect a minority of creatures with attractive physical characteristics or economic values, but despise and fear a much larger number with opposing characteristics (Figure 1.2). The consequences are enormous – for numerous other species since they are progressively being eliminated – but also for people, because the species we are eradicating have ecological and resource values contributing significantly to human welfare (Table 1.1).

There are different ways of examining human attitudes and behaviors toward other species. Indeed, several fields of inquiry are concerned specifically with human values and attitudes. Economics is sometimes defined as the study of wealth or finance but more broadly deals with how people use resources. Aesthetics (esthetics) is a branch of philosophy judging beauty, taste, and sentiment, especially with regard to art and culture, but it also addresses nature. Ethics, another branch of philosophy, deals with morals (right and wrong, good and bad).

Neuroaesthetics, a young science, is 'the study of the neural processes that underlie aesthetic behavior' (Skov and Vartanian 2009) and attempts to relate brain functions to perceptions of beauty in many fields, including biology. Psychology is the study of how the mind determines attitudes and behavior, including emotional reactions and value judgments. There have been attempts to establish a field of 'conservation psychology' (Clayton and Myers 2015) with the associated hope that psychology can be employed to

FIGURE 1.2 Extreme examples on a preference or prejudice scale ranking the likability of species. The cockroach is frequently cited as the most objectionable and unwelcome of creatures. By contrast, humans almost universally regard themselves as the most magnificent of species, and human babies as the supreme examples of perfection.

TABLE 1.1
A Concise Summary of How Human Preferences and Prejudices Influence the Survival of All Other Species and the Ecological Welfare of the World. Detailed information is provided in subsequent chapters.

Human Judgments, Responses ↓	Despicable Species (Especially 'Pests,' 'Vermin,' and 'Weeds,' But by Extension Most Invertebrates, Consistent with the Saying 'The Only Good Bug Is a Dead Bug')	Admirable Species (Especially Domesticated Livestock and Pets, Cultivated Plants, Harvested Lumber and Seafood Species, Birds, and Popular Mammals)
Perception	Destructive, dangerous, harmful, inferior, ugly, useless	Beautiful, benign, likable, impressive, superior, valuable
Emotions induced	Anger, disgust, fear, hate, pain	Admiration, love, pleasure, respect, satisfaction
Actions	Pest control, with unintended associated reduction of biodiversity; indifference to the majority of species and their habitats as they are degraded by human activities	Breeding of animals and plants so that domesticated forms now dominate the planet; highly inadequate wild species preservation, concentration on saving charismatic mammals
Consequences	Species extinctions, ecosystems degradation, environmental pollution, planetary resources crisis, threats to human survival	Selective species conservation, increase of beautiful and valuable species, conversion of wildlands and natural habitats into cities, farmland, and wasteland

save biodiversity (Saunders et al. 2006; Blumberg 2015). All of these fields bear on human preferences and prejudices. Prejudice is usually conceived as an opinion or judgment that is 'unfair' (a moral and legal issue) or more importantly based on insufficient knowledge (an education issue). However, as will be examined, to at least some extent some of our likes and dislikes are hardwired into our brains. This is not to excuse bias, but to point out that to promote rational behavior, in our own interests we must comprehend why and when our attitudes are not entirely appropriate. Much of this book is concerned with the psychological determinants of our likes and dislikes of other species and the consequences.

Most of the world's species are animals, these are extremely diverse, and very few are attractive by conventional standards. Human beauty contests (admittedly increasingly viewed as sexist and socially damaging) provide an instructive model of how most people judge the appearance of animals: most species are 'losers' while only a select few are beautiful (compare Figures 1.3 and 1.4). Indeed, some animals are celebrities, admired like entertainment stars, because they exhibit certain physical and behavioral features which are widely considered to be attractive. Animals that have achieved public star status are almost completely represented by a very small number of wild mammals and birds, and a few butterflies. These are the animals that are featured in zoos, in the visual and print media, and indeed even in the names of sports teams. The appeal of these species is emotional, due mostly to how well they conform to human concepts of beauty and power.

Plants are so different in appearance and behavior from the majority of animals that it is hard to imagine that they are the victims of prejudice. But they are. One only has to consider attitudes toward plants in lawns. Only the primary turf grass (commonly Kentucky bluegrass) is welcome, and homeowners go to great lengths to eradicate every other plant, contemptuously termed 'weeds.' A few dozen animals (such as dogs, cats, canaries, and goldfish) have been 'beautified' into ornamental breeds by humans, showing their dissatisfaction with the appearance of their wild progenitors. But plants have suffered this fate to a much greater extent: thanks to genetic selection, thousands of species have been 'prettified' – most often their flowers have been enormously expanded, strikingly colorized, and made to last for long periods (Figure 1.5). Curiously, the world of wild species is also being beautified by virtue of conservation efforts being heavily directed on the basis of aesthetic standards: attractive species are the primary beneficiaries of protection from extinction, while the plight of the overwhelming majority of endangered species is receiving limited attention.

Almost everyone at some point in their lives is the victim of unfair treatment by society because of the ways people frequently instinctively think or have learned to think. Various aspects of physiological status or physical appearance have long been the basis of prejudicial treatment. Sexism – usually the preferential treatment of males – is perhaps the most ancient and widespread bias-based 'ism.' People with pathological conditions caused by disease, genetics, or accident have long suffered from neglect and indifference. Age, mental disability, and sexual orientation are other common physiological bases for discrimination. Legislation now prevents the worst abuses, but it is safe to say that no form of human bias toward other humans has been eliminated. However, bias toward other species has scarcely been addressed. The importance of understanding and controlling our biases concerning other species is the subject of this book. Most of the world's species are currently viewed as insignificant and inconsequential, if not contemptible. In reality, they need to be respected and conserved because without the invaluable

FIGURE 1.3 Beauty contests. Source: Shutterstock. Contributors: (a) BNP Design Studio; (b) Viktoriia_P.; (c) Guingm.

materials and indispensable ecological services they provide the welfare of humans is at severe risk.

Millions of years ago, when the human species was in its infancy, we were an experiment by nature with little prospect of survival, since we lacked fangs, claws, horns, and other weapons possessed by many of the other dominant species. Modern humans have existed on Earth for perhaps hundreds of thousands of years, but during most of our history, we were merely one of millions of species that shared the planet. Through technology and population increase, much of the world has been usurped and modified to satisfy human needs and tastes. Unintentionally, we have made much of the planet less safe for ourselves. Our artificial habitats suit us and a minority of other species, although not the majority. Until recently, the survival of species was decided by nature, but increasingly it is being determined by humans. Thanks to our intelligence and luck, we have come to rule the world. In effect, mankind has taken on the responsibilities of God and Nature. We seem destined to progressively usurp from nature every remaining square cm that can be made hospitable by technology. Unfortunately, mankind (womankind to a lesser degree) is extremely destructive, and we have reached the point where continuing ruination of our environment (both the living and inanimate parts) threatens our continued existence.

SUMMARY OF CHAPTER 1 (INTRODUCTION AND CHAPTER SUMMARIES)

This chapter (1) as stated above, introduces the basic issue addressed in this book: how bias and prejudice directed at most of the world's species, largely and superficially based on appearance, is so harmful to human welfare that it has resulted in a planetary existential crisis; (2) as presented in the following, provides brief summaries of all of the remaining chapters, which detail various aspects bearing on the causes and extent of bias against species; and (3) concludes that in parallel with discriminatory behavior unjustly directed against human beings, the key solution is education about the beauty, joy, and values of non-human beings.

FIGURE 1.4 Animal beauty contests. Source: Shutterstock. (a) Fish beauty contest. Notice that the fish at left is jealous of the winner at right. (Contributor: Natas). (b) Cow beauty contest. (Contributor: Alexcoolok). (c) Dog beauty contest. (Contributor: Mikhail Sedov).

FIGURE 1.5 Competitive exhibitions of flower varieties result in continuing selection of more beautiful plants. (a) Tulips at the Keukenhof Garden in the Netherlands. Photo by Krysi@ (CC BY 3.0). (b) Flower show at the Royal Hospital Chelsea, London. Photo by Kent Wang (CC BY SA 2.0).

SUMMARY OF CHAPTER 2 (CRUEL & COMPASSIONATE SIDES OF HUMANS)

At the heart of some of the most intractable problems faced by people is human nature – we have both a compassionate side (which is why we are kind to some species) and a cruel side (leading to damage to other life forms and their living quarters). Chapter 2 explores the cruel and compassionate sides of human nature (note Figure 1.6). On the cruel side, we have some very nasty behaviors toward other people, such as bullying, racism, slavery, and war. On the compassionate side, we are a social species, often exhibiting kindness, at least to our kin. Similarly, we are nice to some species, particularly toward those that seem to represent kinship (albeit, relatively distant). But most humans are, at best, indifferent to most species, and quite hostile to most.

A critical contrast is made in Chapter 2 with regard to human motivation affecting sentient animals. The distinction concerns the difference between 'animal conservation' and 'animal welfare' – the former concerned with promoting species survival, the latter with reducing or eliminating cruelty (i.e., subjecting animals to pain). The central theme of this book is that we need to protect all (or at least nearly all) species (which is addressed by conservation) simply because they are valuable either directly to human welfare or contribute indirectly by maintaining planetary ecology upon which humans depend. However, aside from these pragmatic justifications, there are also ethical and spiritual issues connected with animal conservation, since preservation or elimination of biodiversity is also a moral question. By contrast, the need to provide humane treatment to sentient animals that are exploited is basically a moral issue (a topic very extensively examined in the literature, largely from a philosophical perspective). As discussed in this book, both with regard to animal conservation and animal welfare, sympathy for other species is heavily determined by our innate preferences and prejudices.

SUMMARY OF CHAPTER 3 (SPECIESISM)

There are innumerable forms of prejudice in the human sphere (those mentioned in this book include ableism, ageism, favoritism, lookism, racism, sexism, and sizeism). Chapter 3

FIGURE 1.6 Good and bad behavior. (a) Good behavior toward humans: 'The Good Samaritan' by Austrian painter Josef Arnold the Elder (?–1879), a fresco of the parish church of Lajen (South Tyrol). Photo by Wolfgang Moroder (CC BY 3.0). (b) Bad behavior toward humans: A pickpocket scene (public domain) by English illustrator Isaac Robert Cruikshank (1789–1856). Source: Woodward, G. M. 1819. The caricature Magazine 5: Folio 75. Credit: Lewis Walpole Library. (c) Good behavior toward animals: Painting entitled 'Suspense' (public domain) by English artist Charles Burton Barber (1845–1894). (d) Bad behavior toward animals: Fox hunt. Source (public domain): Surtees, R. S. and J. Leech. 1892. Handley Cross; or, Mr. Jorrock's Hunt. London, Bradbury, Agnew and Company. Credit: University of North Carolina at Chapel Hill.

FIGURE 1.7 Genocidal prejudice. (a) Shooting African Americans on May 2, 1866, during a race riot in Memphis. Illustration (public domain) by Alfred Rudolph Waud in Harper's Weekly, 26 May 1866. Source Tennessee State Library and Archive: Tennessee Virtual Archive. (b) Passenger pigeon (*Ectopistes migratorius*), driven to extinction in the early 20th century. Source (public domain): Whitman, C.O. 1920. Orthogenetic Evolution in the Pigeons. 3 vols. Credit: K. Kayashi. (c) Dutch sailors pursuing dodos (*Raphus cucullatus*). This flightless bird was endemic to the island of Mauritius, east of Madagascar in the Indian Ocean. It became extinct during the mid-to-late 17th century because of persecution by humans. Source (public domain): Johnston, H. H. 1914. Pioneers in South Africa, London, Blackie (figure by Walter Paget).

addresses 'speciesism,' a term which summarizes the biases and prejudices that humans manifest toward other species. Many animals and some plants suffer from negative stereotypes. A propaganda war is being waged by the pest control and pesticide industries, targeted against 'bugs,' 'vermin,' and 'weeds' but affecting many innocent species, including people. The lack of concern for both humans and non-human species sometimes amounts to genocide (Figure 1.7). (The word 'genocide' traditionally refers to extermination activities directed at humans. Its use here in reference to other species, and similar use in this book of the phrase 'a holocaust' (not *The* Holocaust) is by extension, with apologies to those whose value systems do not accept that the terms are also appropriate for non-human species.)

SUMMARY OF CHAPTERS 4–6 (DETERMINANTS OF HUMAN PREJUDICE AGAINST & PREFERENCE FOR SPECIES)

These three chapters provide quite detailed reviews of physical (Chapters 4 and 5) and behavioral (Chapter 6) characteristics of animals and plants that are employed as criteria to judge their attractiveness and merit. The aesthetic and utilitarian features that we humans value in animals are summarized in Table 1.2, and those we value in plants are presented in Table 1.3. The physical beauty features detailed in Table 1.2 are, to a considerable extent, what people employ to base their biases and prejudices toward *both* other people and species.

TABLE 1.2
Features That Make an Animal Attractive or Unattractive to Humans

Attractive Features	Unattractive Features
Utilitarian Considerations	
Source of resources (food, clothing, medicine, etc.)	Competitor for resources or preys on humans, livestock, or damages crops (predators, parasites, pests, vermin)
Safe (friendly, limited harm potential, non-toxic)	Dangerous (aggressive, threatening, venomous, or pathogenic)
Appearance	
Phylogenetic (morphological) proximity to humans: a mammal or at least a vertebrate	Quite unlike human form: 'creepy-crawlies,' 'bugs,' 'worms,' snakes
Possession of infantile human features that suggest vulnerability and evoke protective sentiments (large head, large eyes, large forehead, flat face, rotundness, soft contours, clear complexion, small mouth, small limbs)	Possession of mature or aged human-like features (e.g., baldness, missing teeth, wrinkled or sagging skin); aggressive or powerful appearance
Absence of natural weapons that threaten people (horns and antlers are dangerous, but not intended for humans)	Possession of prominent killing weapons (large jaws, teeth, claws) (Additionally, many species have much less evident but dangerous weapons – e.g., snakes that inject toxin or asphyxiate by constriction; electric eels)
Healthy appearance: body organs present and conform to norms of size and appearance; lack of evident physical damage	Possession of human-like features that appear exaggerated, unhealthy, or diseased and provoke urge for avoidance (e.g., large proboscis, large stomach, large mouth, bow legs, warts, irregular teeth or tusks, wrinkled or sagging skin; missing body parts; missing teeth; scars, wounds, discolorations, pathological growths)
Beautiful, spectacular, distinctive, humorous, or entertaining features	Conventional, uninteresting
Large size	Tiny or microscopic
Luxuriant, soft, fluffy, or fuzzy, colorful covering (hair, fur, feathers); ornamentation (horns, antlers)	Naked, or covered with scales, slime, warts, wrinkles; spines (sometimes considered ornamental)
Behavioral, Ecological	
Similar to admirable and complex behavior of humans: intelligence, maternal and social instincts, monogamy, gregariousness, communication, cleanliness (grooming), use of tools; apparent capacities for feeling, thought, pain, emotion, compassion	Solitary life style, simple instinctual behavior, hostility, cannibalism, polygamy
Dominant: a powerful predator or extremely resistant to predators	Vulnerable prey
Athletic and graceful; locomotory virtuosity and speed, on land, in the air, in water, or in trees	Clumsy, slow or erratic movements, sedentary or sessile
'Honorable' or 'clean' food source: ideally a carnivorous hunter, less preferably a herbivore or omnivore	'Dishonorable' or 'filthy' food source: parasites (including cuckoos and bloodsuckers) or eats carrion (scavengers) or rotting organic matter (including feces)
Pleasant, sounds (e.g., bird song; mimics of human speech)	Unpleasant, persistent sounds (screams, squeals, rasps), such as by howler monkeys, cicadas, and crickets
Pleasant or neutral smell	Bad smell (skunks, etc.)
Constructs domicile above ground	Lives in earth, mud, or inside other species
Viewable	Not viewable (shy, nocturnal, 'sneaky,' or lives below ground or in the sea, or simply microscopic)
Homeothermy (= warm-blooded; temperature maintained more or less constant at a desired temperature; 'heterothermic' animals exhibit both homeothermy and poikilothermy depending on time, environment, or season)	Poikilothermy (= cold-blooded; temperature determined mostly by environment)

Summary of Chapter 4 (Size, the Most Important Determinant of Human Prejudice against and Preference for Species)

The most universally admired physical characteristic is size, and simply because of its dominating importance, the topic is separated from other visual considerations (which are discussed in detail in Chapter 5). Huge creatures elicit great respect (Figure 1.8), whereas the majority of species, which are small, tend to be ignored or disdained.

Summary of Chapter 5 (Visual/Beauty Determinants of Human Prejudices against and Preferences for Species)

People evaluate species primarily by aesthetic considerations – usually just relative visual beauty or ugliness, discussed in considerable detail in Chapter 5. Most of the more than a dozen physical assessment attributes employed are the same or at least reminiscent of features that we admire or dislike in other people. Glamorous

TABLE 1.3
Features That Make Plants Attractive or Unattractive to Humans

Attractive Features	Unattractive Features
Utilitarian Considerations	
Useful for materials (food, fodder, fiber, medicine, etc.), groundcover, ornament, etc.	Harmful: Aggressive weeds or toxic (orally, by skin contact, or pollen causes hayfever), or spines cause injuries
Attracts entertaining animals (birds, butterflies, bees, squirrels)	Attracts problem animals (bears, deer, rabbits, moles, groundhogs)
Appearance	
Impressive size, long-life (usually large trees, but dwarfed trees, i.e., bonsai, are also attractive)	Short-lived, evanescent
Large, colorful, long-lasting floral display	Small or no flowers
Decorative fruit (large, numerous, colorful, and/or persistent)	Fruit small or absent, or messy and staining (e.g., mulberry, *Morus* species)
Unique, decorative foliage (dissected or colorful)	Foliage indistinct, unattractive shade of green or yellow
Unique stems and habit (succulent, colorful bark, unusual branching, interesting shapes, vines)	Scraggly, leafless much of year, or climbs upon or among, and thereby reduces appearance of, more attractive plants
Miscellaneous	
Pleasant smell	Repugnant smell (e.g., Bradford Pear, *Pyrus calleryana* cultivar Bradford; female ginkgo trees, *Ginkgo biloba*; tree-of-heaven, *Ailanthus altissima*)

FIGURE 1.8 The importance of size (public domain figures). In the famous book *Gulliver's Travels* by Anglo-Irish writer Jonathan Swift (1667–1745), the central character Gulliver visits exotic lands, including one in which he was huge compared to the inhabitants, and another in which he was tiny by comparison. The contrast illustrates how relatively large size results in respect, whereas smaller size is associated with disdain. (a) Gulliver as a giant in a land of dwarfs. Source: Louis Rhead's cover illustration of a 1913 edition of Gulliver's Travels. (b) Gulliver as a dwarf in a land of giants. Source: A 1797 Spanish cartoon from Digital Bodleian. (c) In a similar vein, giant people and animals are the star attractants of movies, as shown by this poster by Reynold Brown (1917–1991) for the 1958 film *Attack of the 50 Foot Woman.*

appearance of animals is critical for sympathetic attention, but bizarre or ferocious appearance, if entertaining, can also be important. Vertebrates (including fish, amphibians, snakes, birds, and mammals) include a wide spectrum of species that humans consider ugly, and others that are judged to be beautiful (Figure 1.9). As will be detailed, we are hardwired to admire many mammals, provided that they have features indicative of beauty, health, intelligence, and/or power in humans, or are 'cute and cuddly' as indicated by many physical attributes of human babies. Most bird species also possess admired traits. Fish are very variable, some impressively ugly or pretty, but mostly neither. Frogs and snakes are decidedly handicapped by both their appearance and behavior. Animals distantly related to humans, particularly invertebrates, usually have few features considered attractive. Most of the world's creatures are invertebrate animals, and extremely few of these conform to human concepts of beauty. The majority of all species are

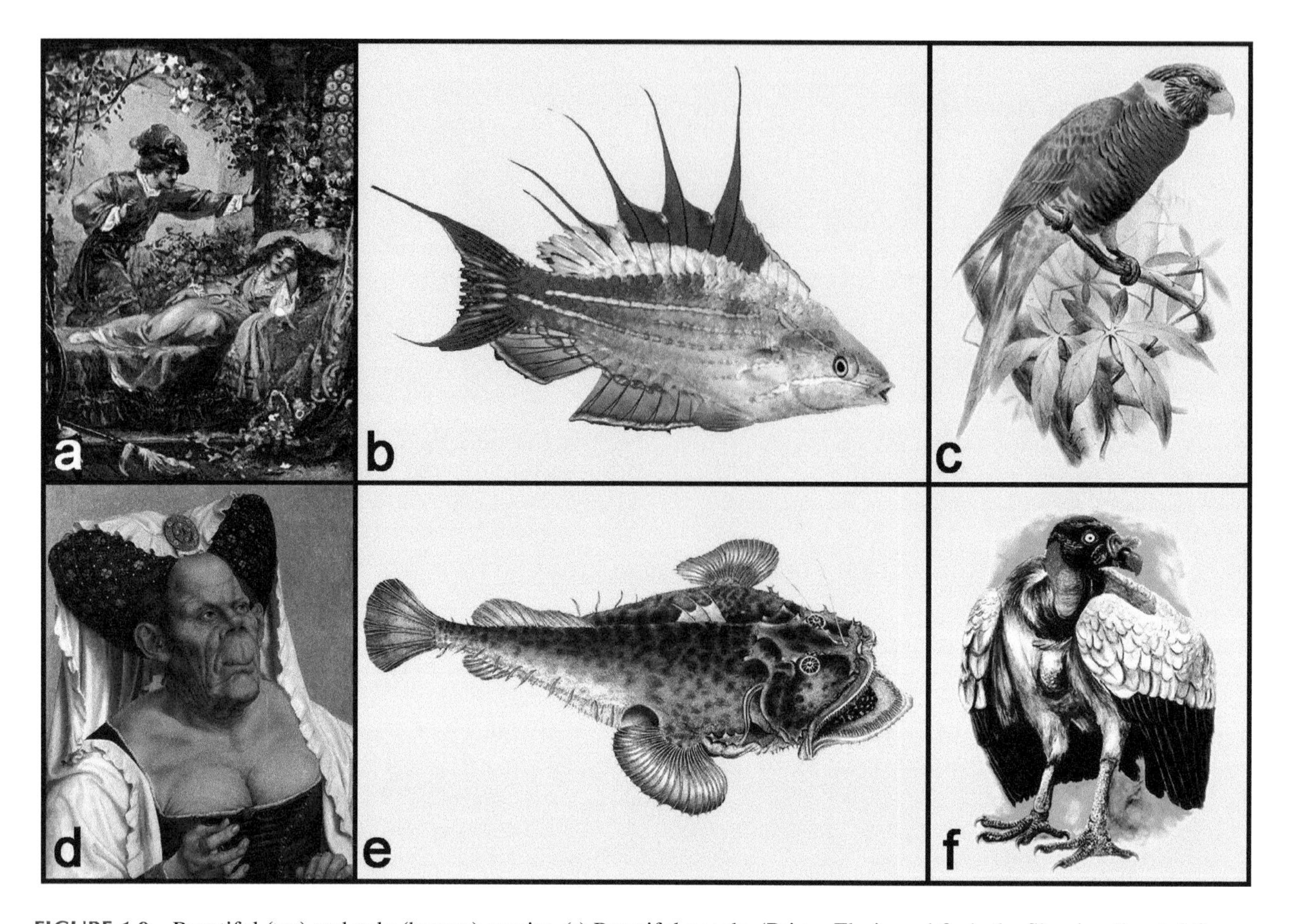

FIGURE 1.9 Beautiful (top) and ugly (bottom) species. (a) Beautiful people. 'Prince Florimund finds the Sleeping Beauty.' Source (public domain): Forbush, W. B., E. V. Hale, and H. W. Mabie. 1916. Childhood's Favorites and Fairy Stories. The Young Folks Treasury, Volume 1. University Society. (b) Paine's flasher wrasse (*Paracheilinus paineorum*) Photo by Rickard Zerpe (CC BY 2.0). (c) Rainbow lorikeet (*Trichoglossus moluccanus*). Source (public domain): Smit, J. 1881. Report on the Scientific Results of the Voyage of H.M.S. Challenger during the Years 1873–76. (d) 'The Ugly Duchess,' a satirical portrait by Flemish artist Quentin Matsys (1466–1530). (e) Fishing frog or angler (*Lophius Piscatorius*). Source (public domain): Donovan, E. 1802. The Natural History of British Fishes. Digitally enhanced by Rawpixel (CC BY SA 4.0). (f) King vulture (*Sarcoramphus papa*), painting by Pavel Galván (CC BY SA 3.0).

insects, but except for butterflies and bees, most are usually perceived very negatively. Ornamental plants and many of our pets have undergone very intensive selection to conform to human concepts of beauty.

Summary of Chapter 6 (Non-visual Determinants of Human Prejudices against and Preferences for Species)

Chapter 6 analyses several non-visual determinants of human preferences for and prejudices against species. Once again, considerations that are reminiscent of features that people admire or dislike in other people are key to our attitudes. We dislike obnoxious features like bad smells and disgusting behaviors like drunken carousing (Figure 1.10) and living in unsanitary conditions, but admire athleticism, caring parents, and lifestyles reminiscent of our own.

SUMMARY OF CHAPTER 7 (SYMBOLIC/REPRESENTATIONAL CREATURES AS REFLECTIONS OF HUMAN PREJUDICES AGAINST AND PREFERENCES FOR SPECIES)

Much of human attitudes toward other species is odd, indeed often irrational. Human culture has generated a mythology in which many species have come to epitomize the best and worst of human behaviors, and accordingly are respectively idolized or despised. Such species are valued or hated in religion, commerce, literary tradition, entertainment media, and other aspects of society. The separateness of humans and other animals has even become obscured in some circumstances, with various beings conceived as semi-human hybrids or blends with other species. Favored species are commonly humanized and depicted conducting themselves much like people (Figure 1.11). Moreover, many have anthropomorphized their pets as child surrogates. These developments, explored in Chapter 7, provide key

FIGURE 1.10 Obnoxious human and non-human animals. (a) A gang of drunken sailors. Old colored etching by W. Elmes. Source: Wellcome Library, London (CC BY 4.0). (b) A gang of party animals. Sign at a bar in Valencia. Photo by J. Bizzie (CC BY 2.0). (c) A gang of carousing bugs. Source: Historia de un grillo sabio (public domain; part of a vintage collection of chromolithographs published in Barcelona).

FIGURE 1.11 Anthropomorphic animals (all figures are public domain). (a) Funeral for a horse, by artist J. S. Pughe (1870–1909). Credit: Library of Congress. (b) Elephants from a children's book illustration by British artist Harry B. Neilson (1861–1941). (c) Rollerskating dogs by German artist Carl Robert Arthur Thiele (1860–1936). (d) Singing cats, also by C. R A. Thiele. (e) A family of hedgehogs. Source: Mein erstes Märchenbuch, Stuttgart, Verlag Wilh. Effenberger; a late 19th-century German fairy tale book illustrated by Heinrich Leutemann and Carl Offterdinger. Photo credit: Harke. (f) Lions on the cover of an 1898 comic book, The Funny Household, published by McLaughlin Bros, New York.

indications of how humans think about other species and reflect many of our preferences and prejudices.

SUMMARY OF CHAPTER 8 (INDISPENSABLE VALUES OF SPECIES FOR HUMAN WELFARE)

Species have extensive and varied utilitarian values. As noted in Chapter 8, non-human animals (1) provide humans with food, clothing (e.g., leather, wool), construction materials, manure, and industrial materials; (2) assist in the maintenance of ecosystems upon which humans are dependent; (3) are experimental subjects of biomedical research and toxicity testing; (4) furnish labor (as draught/towing and riding animals); (5) provide assistance (as guard and herding dogs, and guides for the blind); (6) contribute companionship; and (7) act as entertainers. Plants are much more important than animals for (1) and (2), and additionally are more important for providing medicines and beautifying the living areas of people. In addition to physical and behavioral attractiveness, people employ such economic values to evaluate the merits of animals and plants. Under these circumstances, practical value trumps appearance.

As emphasized in Chapter 8, almost all species are wild. Species that are harvested from nature (notably marine animals, especially fish, and timber trees) provide such essential resources that humans appreciate their value regardless of their external appeal. (Domesticated livestock, service animals, and crops are so indispensable that they too are not handicapped by their appearance.) Wild fish and seafood stocks are invaluable, although lobsters and indeed most other edible marine creatures are not emotionally attractive to humans. Even stinging bees are forgiven for being insects, if they furnish honey. Most thinking people, even if they dislike the appearance and/or the behavior of most animals, recognize that certain wild species supply invaluable products (Figure 1.12) and so wilderness areas of the world need to be maintained at least to sustain these species.

However, as discussed in Chapter 8, the public is hardly aware that the millions of seemingly unimportant species are in fact indispensable for human welfare, specifically for maintenance of ecosystem services (Figure 1.13). Unfortunately, the current holocaust eliminating many of these species represents an existential threat to the future of humans. Chapter 8 strives to explain that species, including the human species, are inseparable from their habitats, ecosystems, and the welfare of the planet. Ecosystems regulate key cycles and processes (such as soil formation, atmospheric maintenance, and purification of fresh water), and species are necessary contributors. Society has become acutely aware of how human welfare is being negatively influenced by climate change and global warming associated with degradation of the atmosphere, but this is just one component of the 'environmental crisis.' The public scarcely recognizes that countless species indispensable for maintenance of ecosystem functions, and accordingly for human welfare, are under severe threat. The overwhelming majority of the world's species are unattractive by human standards, and as emphasized in this book, this represents a critical roadblock to providing the necessary support for their conservation.

SUMMARY OF CHAPTER 9 (EXTINCTION – HOW BIASED ELIMINATION OF SPECIES ENDANGERS HUMANS)

Chapter 9 examines the extent of the extinction threat to species, which is widely misconceived by the public as simply a situation requiring selective preservation of charismatic animals such as pandas and whales. Indeed, as emphasized in this chapter, biased favoring of particular species oversimplifies and leads to inadequately addressing the ecological threats. An additional confusion concerns 'de-extinction' as a solution, but technology just can't regenerate real dinosaurs or resurrect other spectacular extinct animals. Still another misconception is that extinction is just 'normal'

FIGURE 1.12 Exemplary harvests from the wild. All images are public domain. (a) Lumber from forests. Source: Whipple, W. 1915. The story of young Abraham Lincoln. Philadelphia, Henry Altemus. (b) Fishing. Source U.S. National Oceanic and Atmospheric Administration. (c) Preparing sugar maple syrup. Painting by Dutch-Canadian artist Cornelius Krieghoff (1815–1872). Photo credit: Library and Archives Canada.

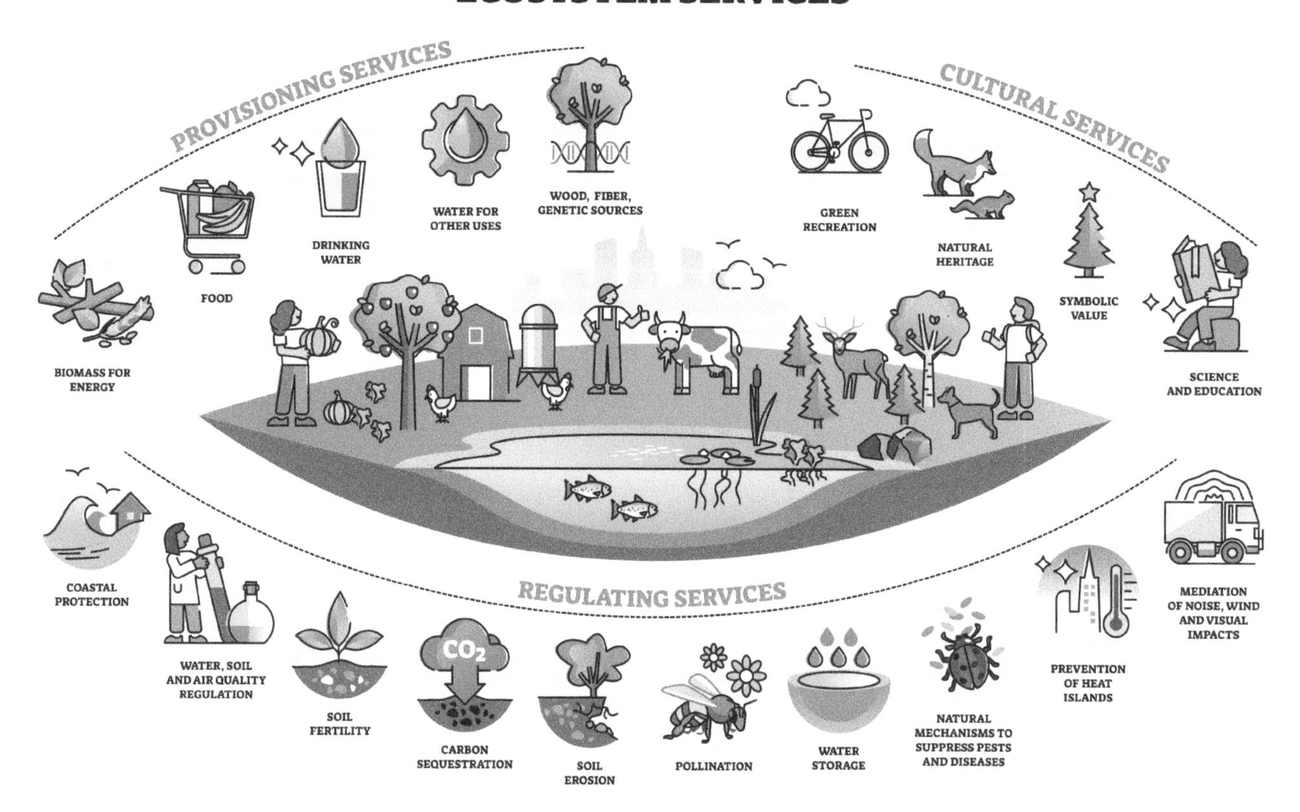

FIGURE 1.13 Ecosystem services. Source: Shutterstock (contributor: VectorMine).

(so is death, but not when the survival of the human species is at risk). The extent of reduction of the living world compared to past times is staggering (Figures 1.14 and 1.15). The biodiversity/bioresources crisis has two interrelated symptoms: species reduction/extinction and ecosystem/habitat degeneration. Human expansionism – population growth and associated destructive exploitation of the life systems of the planet – are the drivers. Overpopulation is the result of normal reproductive behavior, and is difficult to control, although efforts are underway. Overexploitation of the planet's limited resources, also caused by normal human behavior, is also difficult to control. The frequent failure of people to protect ecosystems independently exploited by different interests (a phenomenon termed 'the tragedy of the commons') is examined in this chapter as a key problem to be addressed. The key organizations and

FIGURE 1.14 Overfishing resulting in a huge decrease of the natural supply. Central panel prepared by Hans Hillewaert (CC BY SA 4.0), remainder by B. Brookes.

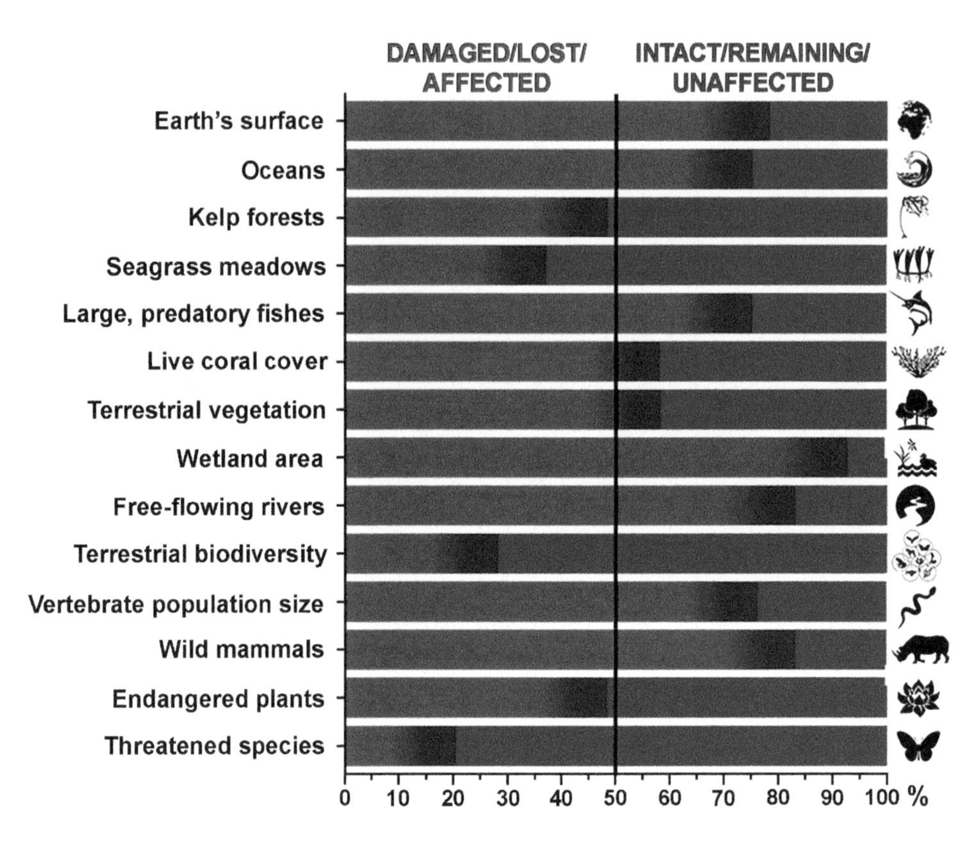

FIGURE 1.15 Damage caused by humans to environmental categories, compared to pre-human influence. Figure from Bradshaw et al. (2021; CC BY 4.0), slightly altered by B. Brookes.

international agreements addressing extinction threats are identified. It is clear that current international attempts to address impending ecological disasters are inadequate.

SUMMARY OF CHAPTER 10 (BIAS AND PREJUDICE IN SPECIES CONSERVATION)

'Conservation' of species and their supporting habitats concerns the reduction of the threats to species and ecosystems, and attempts at remediation. Unfortunately, as discussed in Chapter 10, there is considerable disagreement on which species, which habitats and ecosystems (indeed, which geographical regions), and what approaches are appropriate. Moreover, there is very strong bias and prejudice against most species coupled with strong preferences for a minority of species. This chapter examines the extent of bias in the world of biological conservation and makes recommendations on how to control this.

Conservation, remediation, and restoration of at least a portion of remaining wildlands and their biodiversity depends on convincing the public, philanthropic organizations, and political leaders to provide the considerable funding required. A principal technique that conservation-oriented organizations have found effective for obtaining financial support is to highlight the plight of the most valued species that are endangered. In theory, species could be evaluated on the basis of economics, ecological importance, scientific interest, or cultural significance, but in practice aesthetics trumps other considerations. Not surprisingly, conservationists individually and as organizations show very strong favoritism for exactly the same charismatic, attractive species as does the general public (Figure 1.16). A chief issue is whether it is preferable to concentrate on the welfare of particular ecosystems or the welfare of certain species (because of their attractiveness in raising public support and funding, or because they play key ecological roles supporting ecosystems, or because they are associated with many other species that indirectly benefit). This chapter examines the advantages, disadvantages, limitations, and misunderstandings associated with the different approaches. Controversial solutions include species triage (saving some, disregarding others) and abandoning given species in nature while maintaining them only in parks or zoos. Given that all proposed solutions involve preferential treatment of species, it is important to involve the diversity of society in making decisions. As reviewed in the chapter, legislators, educators, religious leaders, corporations, hunters, artists, and others, in addition to professional conservationists, all can contribute to minimize bias and prejudice.

SUMMARY OF CHAPTER 11 (DEALING WITH DANGEROUS SPECIES)

Some animals inspire terror and are greatly feared although in reality they very rarely kill people (sharks are the best example), others, notably dogs, are quite lovable, but they frequently kill people. Albeit a very small proportion of all

FIGURE 1.16 Favoritism for funding of conservation of attractive species penalizes ecologically essential but ugly animals. Prepared by B. Brookes.

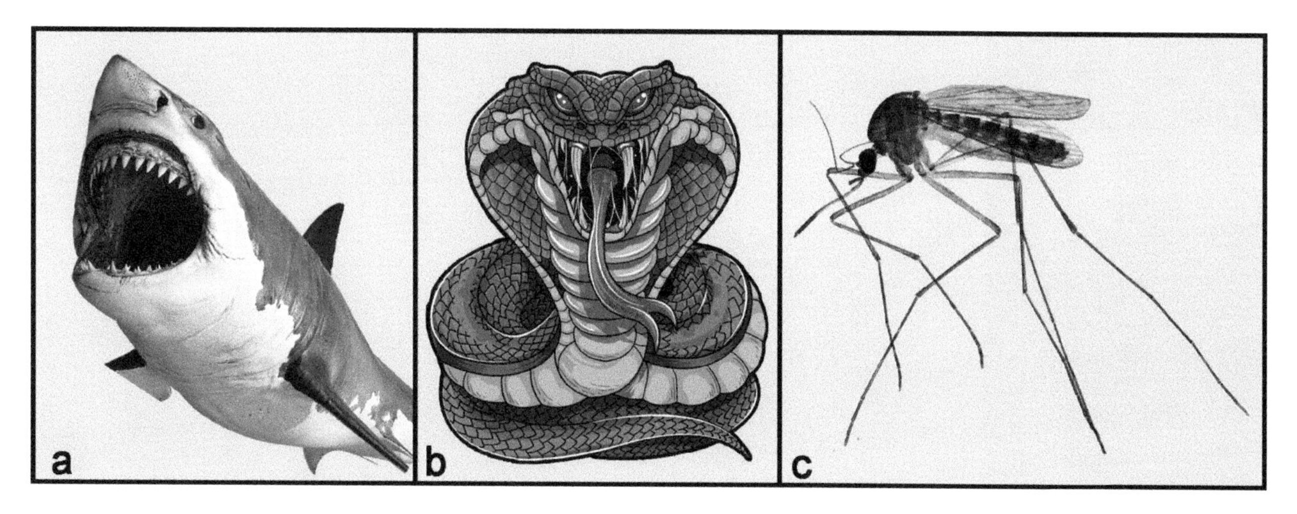

FIGURE 1.17 Examples of greatly feared animals that people have tried to exterminate. (a) White shark. Source: Shutterstock (contributor: ARTYuSTUDIO). (b) Cobra. Source: Shutterstock (contributor: DEYASA_346). (c) Female *Culex* mosquito Photo by Alan R Walker (CC BY SA 3.0).

species, many animals are harmful to humans, threatening the welfare of ourselves, our livestock, pets, and crops. Some are merely annoying, but others are quite dangerous. Our traditional response has been to kill off species that appear to be threatening (Figure 1.17), but as discussed in Chapter 11, in many cases there are less drastic solutions that allow co-existence.

SUMMARY OF CHAPTERS 12 AND 13 (REFORMING AGRICULTURE AND URBANIZATION – THE GREATEST THREATS TO SPECIES)

The most acute problem of modern times is the threat of environmental degradation ruining the planetary ecosystems upon which world's species depend. The world's biota and biosphere are in catastrophic decline because the indispensable habitats and ecosystems in which species live are being degraded by human activities or converted to human uses. Humans have created two principal artificial environments – cultivated lands and cities – and these are the chief sites of habitat elimination, and hence the principal causes of species reduction and extinction. The agricultural use of land is the most common cause of habitat decline, but urban spread is also significant. Agriculture is discussed in Chapter 12, urbanization in Chapter 13. Both agriculture and urbanization fall under the headings 'land intensification' (Lindenmayer et al. 2012), which is hostile to co-existence with species that live in the same or adjacent areas. Both agriculture and urbanization are also expanding, their 'footprints,' taking over much of the Earth's terrestrial areas, and thereby simply physically excluding most other species.

Both in cities and farms, humans are at war with the specialized pests that harm us or our possessions (including property, domesticated animals, and crops). Enormous quantities of 'pesticides' are being employed to kill harmful animals, and enormous quantities of 'herbicides' are similarly used to eliminate harmful plants. (Some authorities include all chemicals designed to eliminate unwanted plants and animals under the term 'pesticide.'). Unfortunately, in attacking these species, we are causing them to become more resistant to the chemicals, and the environment is being contaminated, harming innocent species including ourselves. Moreover, since these detrimental wild species receive extensive negative publicity, the public's view of wild species in general is being degraded. As discussed in Chapters 12 and 13, it is possible to live much more harmoniously with wild creatures by conducting agriculture in more sustainable ways and by designing cities with species-friendly green spaces (Figure 1.18).

FIGURE 1.18 Sustainable vs. unsustainable occupation of land for agriculture (top row) and cities (bottom row). The key to biodiversity sustainability is breaking up the landscape with permanent green spaces that provide habitats for wildlife. (a) Example of a sustainable farm near Klingerstown, Pennsylvania, the area separated by woodlots that provide habitats. Photo (public domain) by Scott Bauer, U.S. Department of Agriculture. (b) A wheat monoculture being harvested. Monocultures are highly productive and efficient but very substantially exclude wildlife. Photo by Scott Bauer, USDA (public domain). (c) A wonderful city green space promoting habitats for wildlife: Central Park, New York. Photo by David Shankbone (CC BY SA 3.0). (d) Extreme urbanization leaving little habitat for wildlife: The urban jungle of Hong Kong. Photo by Cattan2011 (CC BY 2.0).

Summary of Chapter 12 (Reforming Agriculture – The Greatest Threat to Species)

Historians argue over whether agriculture was 'invented' or 'discovered.' Regardless, agriculture has been the greatest innovation in the history of humankind, because (as discussed in Chapter 12) it has saved billions of people from starvation. On the negative side, agricultural practices have been the principal causes of the current environmental crisis and the accompanying decimation of the world's species. Several challenges are facing agriculture (Figure 1.19), making it imperative to adopt more sustainable practices. The need to feed the world's population (which hopefully will stop growing) dictates that agriculture cannot be abandoned. However, as discussed in Chapter 12, there are steps that should be taken to limit harm to the environment and biodiversity. Animal welfare is an allied issue. Livestock constitute an extremely small percentage of the world's species but represent the majority of opportunities for humans to demonstrate their humanity toward other sentient creatures.

Summary of Chapter 13 (Reforming Urbanization – The Second Major Threat to Species)

The great irony of cities, which now provide housing for the majority of the human species, is that they do so at the cost of eviction of most other species. The second irony is that cities not only have been constructed for the comfort of humans, but they have proven to also provide ideal habitats for some species of wildlife (Figure 1.20) and weeds, including the majority of the world's most despised pests. As discussed in Chapter 13, priority should be given to designing cities that are more compatible with biodiversity, but also there should be greater tolerance of species that are merely annoying or pose quite limited threats.

FIGURE 1.19 Some key issues facing agriculture. (a) Methane emissions from cattle contribute to greenhouse gas pollution of the atmosphere. Source: Shutterstock (contributor: BlueRingMedia). (b) Honeybees, essential for crop pollination, are dying off. Source: Shutterstock (contributor: Laurie Barr). (c) Agriculture uses more water than any other sector, but the world is running out of fresh water, and drought is endangering food crop production. Source: Shutterstock (contributor: Elegant Solution). (d) Small, sustainable farms such as this are disappearing and being replaced by ecologically damaging factory farms and monocultural crops. 'A Milking We Will Go,' painting (public domain) by American artist Edward Mason Eggleston (1882–1941).

FIGURE 1.20 Urban pests. Source: Shutterstock. (a) Rats looking for food from trashcans (contributor: GraphicsRF.com). (b) Raccoon raiding a trashcan (contributor: bilha golan). (c) Homeowner looking at a mole damaging his lawn (contributor: Aleutie). (d) Pigeons pooping on a car (contributor: Vector Micro Master).

SUMMARY OF CHAPTER 14 (ADVANCING TECHNOLOGIES AND THE FATE OF THE WORLD'S SPECIES)

Chapter 14 deals with advancing technologies that offer possibilities of drastically altering the relationships of humans and other species. Technology may progress to the level that genuine species may no longer be needed, and artificial machines meeting human needs are produced (Figure 1.21). In theory, people may be able to manipulate genes so extensively that revolutionary inter-species creations may come to dominate the planet. Eliminating undesirable species could become a simple exercise. The majority of the world's wild species find refuge in sites that are as yet too hostile for human exploitation, but with new technologies virtually the whole planet could be exploited, excluding most species. One can expect that human taste and values (or biases) will dictate what life forms are permitted to exist in the future. Doubtless there is both promise and peril.

SUMMARY OF CHAPTERS 15–21 (THE WORLD'S MOST DESPISED SPECIES)

The remaining chapters examine 22 groups of animal species and 20 of plant species that are (with some exceptions) extremely damaging to human health, natural ecosystems, agriculture, and urban property. It will be emphasized that although they are the world's most despised, dangerous, and destructive of all creatures, they have beneficial properties so a better understanding of the biology of these seemingly villainous species is key to managing both their adverse and advantageous potentials.

Chapters 15–17 deal respectively with the world's most despised species of animals, respectively of invertebrates, non-mammalian ('lower') vertebrates, and mammals. These species (often called 'pests,' vermin,' and 'parasites') are detrimental, despised, and persecuted, but often excessively. Frequently, such pariah species have redeeming values. Numerous innocent species are often confused with similar injurious ones – reminiscent of harmful stereotyping among people.

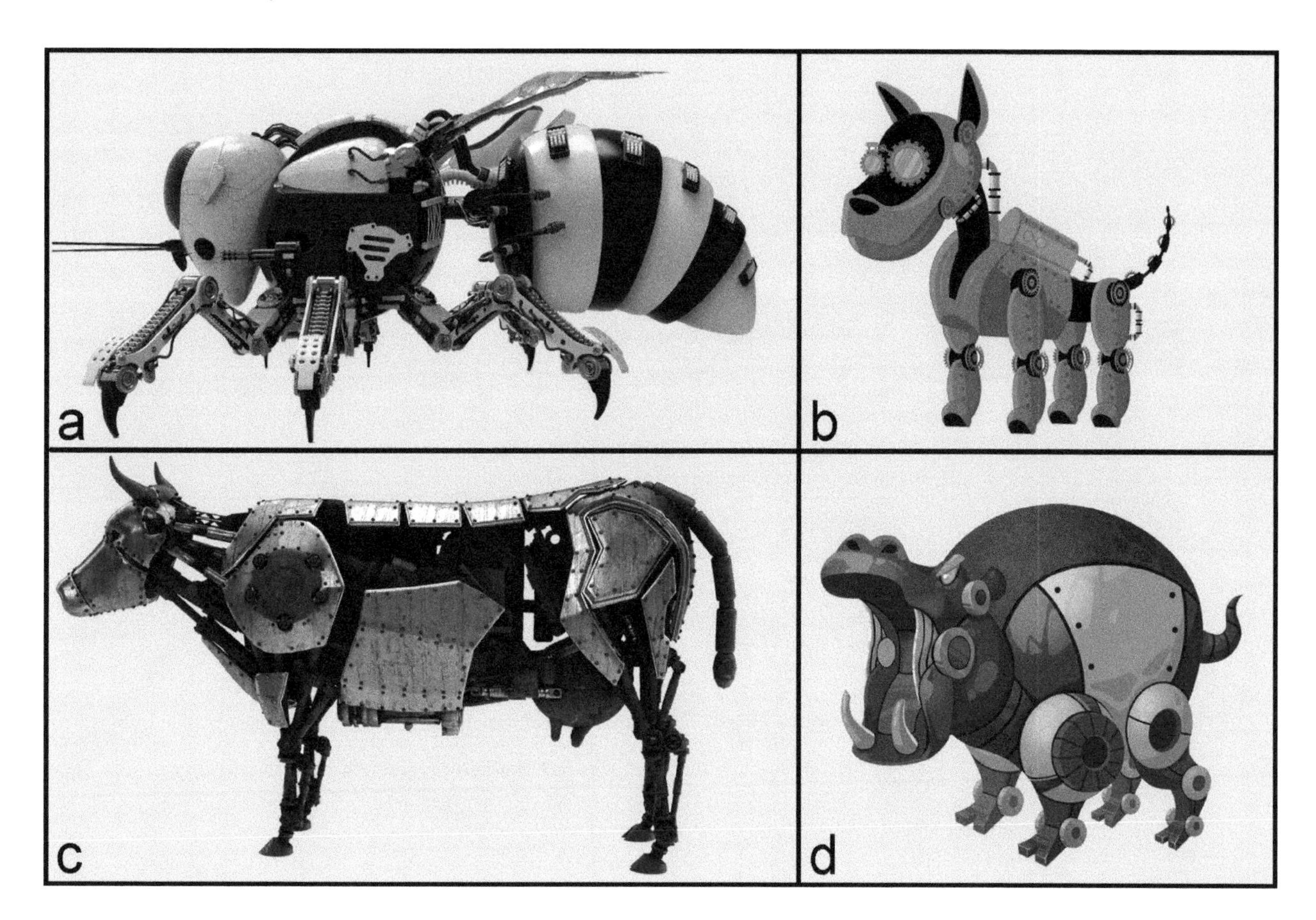

FIGURE 1.21 Robotic interpretations of animals that conceivably advancing technology could substitute for the real versions. Source: Shutterstock. (a) Bee (contributor: Josh McCann). (b) Dog (contributor: Photo-kot). (c) Cow (contributor: Vitalii Gaidukov). (d) Hippo (contributor: VHar).

Chapters 18–21 examine the world's most despised species of plants, including those that are directly toxic to humans and animals (Chapter 18), or are harmful weeds of agriculture (Chapter 19), natural ecosystems (Chapter 20), and urban areas (Chapter 21).

Summary of Chapter 15 (In Defense of the World's Most Despised Invertebrate 'Bugs')

Chapter 15 deals with invertebrate pests. The majority of the most strongly despised species are insects. Along with equally hated spiders and ticks (which are correctly 'arachnids,' not insects) these are often derisively termed 'bugs.' Invertebrates are remarkably diversified, but only a minority cause serious problems (Figure 1.22). The true insects include bedbugs, clothes moths, cockroaches, fleas, houseflies, lice, locusts, and mosquitoes. Leeches are also included in this chapter, although they are 'worms.' This set of 'least-wanted' villains (most of which are parasites of humans), in the minds of most people, are the worst of the worst, deserving merciless extermination. Nevertheless, as will be explained, they have some desirable features, and limiting their negative effects requires careful risk/benefit balancing of the effects of control measures. Unfortunately, these harmful animals have caused almost all insects to be painted with the same stereotyping brush. Bugs, although making up the majority of the world's species, are widely considered inferior, insignificant, disgusting, harmful, and deservedly killed. Changing this perception is essential.

Summary of Chapter 16 (In Defense of the World's Most Despised 'Lower' Vertebrate Animals)

Chapter 16 concerns 'lower vertebrates' (exemplified by sharks, frogs and toads, snakes, vultures, and pigeons). Except for poisonous snakes (which kill as many 100,000 people in some years), there is little justification for the very negative attitude people have for these species, all of which provide critical ecological services contributing to the health of the environment. The deep prejudice against the appearance of many of these species is illustrated by the traditional tale of a princess transforming an ugly frog into a handsome prince (Figure 1.23).

Summary of Chapter 17 (In Defense of the World's Most Despised Mammals)

Chapter 17 discusses the most despised mammals: bats, mice, hyenas, rats, and skunks. The presence of mice and rats in buildings (Figure 1.24) triggers intense disgust and

FIGURE 1.22 Examples of pest insects. Source: Shutterstock. (a) Ants stealing a sandwich (contributor: Teguh Mujiono). (b) Housefly alarming a girl (contributor: Piscary). (c) Mosquitoes chasing a man (contributor: Ron Leishman). (d) A flea-infested dog (contributor: Teguh Mujiono).

FIGURE 1.23 Illustrations (public domain) of the familiar fairy tale relating how a princess transforms an ugly frog into a handsome prince by kissing it. (a and c) Source: Crane, W. 1874. The Frog Prince. New York, George Routledge and Sons. (b) Source: Pixabay, modified by B. Brookes.

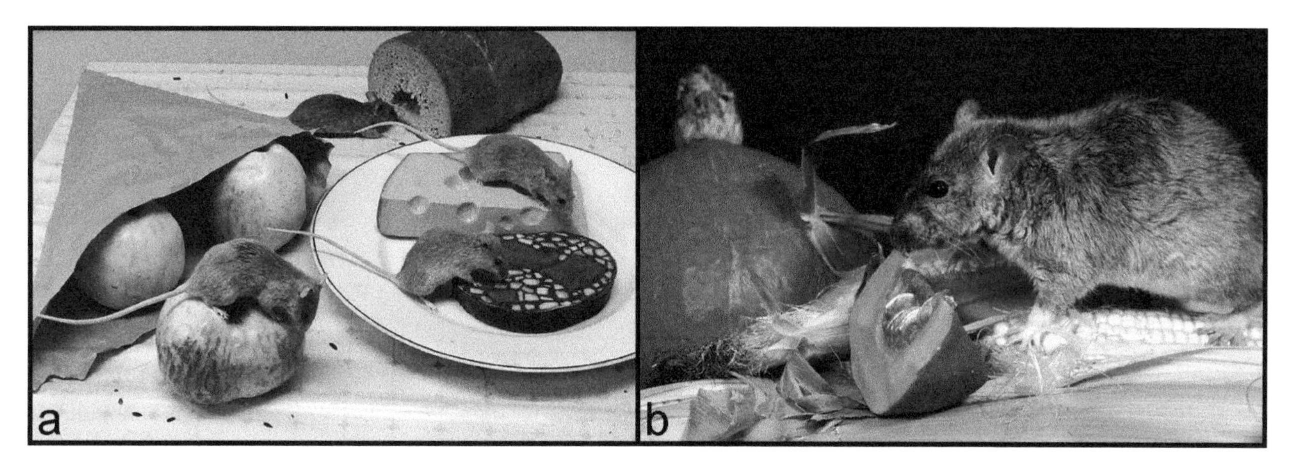

FIGURE 1.24 Rodent infestations. (a) House mouse (*Mus musculus*) exhibit at the Staatliches Museum für Naturkunde Karlsruhe, Germany. Photo by H. Zell (CC BY SA 3.0). (b) Norway rat (*Rattus norvegicus*) in a pantry. Source: Shutterstock (contributor: Torook).

alarm. These mammals are significant vectors of diseases, but ironically, as will be noted, they have contributed immeasurably, as experimental research subjects, to cures for diseases. Like the lower vertebrates discussed above, the repulsiveness felt for them by most people is mostly the result of their offensive appearance and assassination of their reputations. Remarkably, several of these despised species have been domesticated, and attractive, friendly breeds have been selected which make good pets.

Summary of Chapter 18 (In Defense of the World's Most Despised Toxic Plants)

Chapter 18 deals with the world's most toxic plants. Many plant species are dangerous, either when consumed or touched, and some are deadly (Figure 1.25). Poisonous weeds extensively harm domesticated livestock (wild animals instinctively avoid most toxic plants), and some toxic garden ornamentals and houseplants are a potential health threat,

FIGURE 1.25 Socrates and poison hemlock – the most famous example of plant poisoning. Accused by the Athenian government of denying the gods and corrupting the young through his teachings, Socrates (469–399 B.C.E.) was offered the choice of renouncing his beliefs or being sentenced to death by drinking a poison hemlock potion. (a) "The Death of Socrates" (1787) by French artist Jacques-Louis David (1748–1825). Public domain image (credit: New York Metropolitan Museum of Art). (b) Poison hemlock (*Conium maculatum*). All parts of the plant are extremely toxic. Source: public domain image from Köhler, F. E. 1887. Medizinal-Pflanzen, Volume 2. Berlin, Germany: Gera-Untermhaus.

mostly to children and pets. However, unlike many pathogenic microorganisms and some dangerous animals which kill many people, few humans are killed by being attacked by plants. Accidental or unwise consumption of toxic or tainted plant materials does occur but is relatively avoidable. By contrast, involuntary physical contact with some noxious plants is very widespread and often serious. The species that are particularly infamous for causing misery are the skin-irritating poison ivy and the allergenic-pollen-producing ragweed. The pain produced by these plants generates hostility toward the natural world and so reduces respect for biodiversity. The deadliest species, however, are the dangerously addictive opium poppy, tobacco, and sugar cane, each of which prematurely ends the lives of millions of people annually. Coincidentally, these are major crops with some detrimental effects on ecosystems. Of course, it is unfair to blame these plants for self-destructive human behavior, but it is equally unfair to condemn all of the other problematical species that harm people, which came into existence long before humans appeared and changed the balance of nature. As noted in Chapter 18, all of these species have very valuable redeeming features, and their harm can be greatly alleviated by appropriate management.

Summary of Chapter 19 (In Defense of the World's Most Despised Agricultural Weeds)

Chapter 19 examines weeds that threaten agriculture, the most indispensable activity contributing to human existence. Indeed, agriculture occupies about 38% of Earth's terrestrial surface and represents the largest use of land on the planet (https://ec.europa.eu/environment/integration/research/newsalert/pdf/sustainable_food_IR8_en.pdf). Agricultural weeds (e.g., Figure 1.26) are a grave threat to people because they injure the crops that sustain human life. They also indirectly degrade biodiversity by provoking environmentally unsustainable agricultural practices to control the weeds. Indeed, there is an inextricable destructive link between agriculture and nature: the former usurps land from the latter, and agricultural activities represent the greatest threat to biodiversity and natural ecosystems. Because agricultural weeds greatly reduce agricultural productivity, they increase demand for converting natural lands and also increase the need for biodiversity-damaging pesticides and ecosystem-damaging overuse of fossil fuels and water. The world's worst agricultural weeds include barnyard grass, Bermuda grass, lambsquarters, the nutsedges, and weedy rice, discussed in detail, but there are hundreds more. These will not be as familiar to most people as the other pests, animals, and plants highlighted in this book, but they are detested by farmers. While the worst weeds are responsible for devastating damage to both agriculture and the environment, as will be described they possess remarkably important useful properties.

Summary of Chapter 20 (In Defense of the World's Most Despised Environmental Weeds)

Chapter 20 is concerned with weeds that degrade ecosystems of the natural environment. Weeds that specialize in natural habitats (e.g., Figure 1.27) are a grave threat to biodiversity because they can degrade fragile natural ecosystems and eliminate susceptible indigenous species. Cactus pear, common reed, English ivy, purple loosestrife, and water hyacinth, among the world's leading weeds harming natural habitats, are highlighted.

Summary of Chapter 21 (In Defense of the World's Most Despised Urban Weeds)

Chapter 21 deals with annoying urban weeds like crabgrass, dandelion, and Canadian thistle. In most poorer/developing (so-called 'Third World') nations, a large majority of the population works directly or indirectly in

FIGURE 1.26 Eliminating agricultural weeds, the principal plants that directly harm human economic welfare. (a) A Kenyan farmer removing weeds by hand, the main method of controlling weeds in the Third World. Photo by Neil Palmer, CIAT (CC BY SA 2.0). (b) A tractor applying herbicides and other pesticides to a crop. Although efficient, pesticides inevitably affect non-targeted wild species and natural ecosystems, and so agricultural weeds also negatively affect biodiversity. Photo by Aqua Mechanical (CC BY 2.0).

FIGURE 1.27 Love vine (*Cassytha filiformis*), an example of a very serious environmental weed (even where it is native). Love vine is an apparently leafless (actually, leaves are reduced to scales), climbing, twining, parasite which infests coastal woody plants throughout the tropics worldwide. It also can infest orchards and gardens. Love vine weakens and often kills plants, not just by lowering the light available for photosynthesis, but also by piercing the host with haustoria (root-like extensions) through which it withdraws nutrients. (Love vine has very little chlorophyll and cannot photosynthesize sufficiently for itself.) (a) Love vine draped over a tree in Florida. Photo by Forest & Kim Starr (CC BY 3.0). (b) Volunteers removing love vine by hand from wildland in Florida. Photo by U.S. Department of the Interior, Bureau of Land Management, Eastern States (CC BY 2.0).

agricultural activities, and so weeds of crops and grazing lands are of predominant concern. By contrast, in rich/developed countries, agriculture occupies far fewer people (e.g., in the U.S., farmers and ranchers make up just 1.3% of the labor force), and so agricultural weeds are of little concern for most. Similarly, although people increasingly conduct leisure activities in wildlands, they rarely become acquainted with damaging environmental weeds. However, weeds that occur in urbanized areas – such as those found associated with residences, commercial areas, roadways, and parks – are viewed constantly by most people and so have acquired exaggerated importance. As a result, although weeds of commercial crops and natural wildlands cause much more economic and environmental damage, urban weeds generate much more hostility to nature. A few species, like Japanese knotweed and kudzu (also discussed in detail in Chapter 21), are indeed capable of physically damaging property, but by a considerable margin urban weeds are merely an aesthetic problem (Figure 1.28). Eliminating weeds from lawns, examined in detail in this chapter, exemplifies the considerable biodiversity-harmful consequences of overzealous persecution of unwanted but basically harmless plants. The discussion concludes that much greater use of sustainable groundcovers is desirable to reduce the excessive ecological problems associated with lawns. Moreover, the socially responsible trend of simply tolerating weed growth on public lands should be encouraged.

KEY MESSAGES OF THE BOOK

As humans, we are prisoners of our unconscious reflexive positive and negative reactions to the superficial appearance of species, regardless of whether they are in fact friends or foes. But we should recognize that all species have virtues, which are often indispensable, and we need to temper our hostility to the majority of living things with knowledge of their values. We have preferences for and prejudices against most of our fellow creatures. This is having momentous effects on conservation of biodiversity and, consequently, will inevitably also affect the ecosystems that sustain life on our planet. Aside from the moral aspects of eliminating numerous species, we must realize that our indifference to most of the world's species is short-sighted. The appearance and even the economic usefulness of species for the most part are independent of the vital ecological roles they play in sustaining the ecosystems vital to our own lives. What is important is not that we suppress our natural admiration for certain life forms, but that we moderate our prejudices with understanding of the inapparent values of all species, for the long-term welfare of ourselves and our planet.

This is a big volume, with lots of complex information, so it is important to emphasize a simple, key practical recommendation, which in fact is mentioned throughout the book. Biases and prejudices that have been acquired by most adults are very difficult to escape, so it is critical to educate our children about the beauty, joy, and value of species (Figure 1.29).

FIGURE 1.28 A 'perfect' lawn (but very hostile to biodiversity) and a perfectly awful lawn. (a) Lawn seed advertisement (ca 1915). Credit: Henry G. Gilbert Nursery and Seed Trade Catalog Collection; Peter Henderson & Co., Biodiversity Heritage Library (CC BY 2.0). (b) A dandelion-infested lawn. Photo by Mike Mozart (CC by 2.0).

FIGURE 1.29 Children being educated to value and respect species. Source: Shutterstock. (a) Contributor: BlueRingMedia (b) Contributor: White Space Illustrations.

2 The Cruel and Compassionate Sides of Human Nature

INTRODUCTION

Human behavior is complex, with numerous examples of cruelty and kindness to other people (Figure 2.1). The brilliant English poet Alexander Pope (1688–1744) wrote 'The proper study of Mankind is Man.' Indeed, human nature is of predominant importance for many issues because it controls what people do, and the ability of people to alter the world is becoming God-like. This book is about the constructive and destructive interrelationships between humankind and all other species, upon which, as will be explained, depend on the ultimate survival of both. Humans have the choice of living harmoniously within the constraints of nature or continuing to engage in exploitation-for-the-short-term, suicidal ruination of the planet's life-support systems. The latter alternative is rapidly leading to degradation of the planet and endangering the future. A key factor, which has scarcely been examined to date, is how human psychology leads to how we behave antagonistically toward other living things, which are frequently invaluable because they generate indispensable resources and services. In particular, as will be presented, most of society is inclined to hostility toward most species, albeit we are also programmed to favor a minority, with the result that we are naturally destructive to most of the living world. As noted in the following, humans appear to have intrinsic

FIGURE 2.1 'The Kosovo Maiden' by Serbian artist Uroš Predić (1857–1953), a painting (public domain) illustrating both compassion and cruelty. The scene illustrates a famous poem in which a young beauty fruitlessly searches the battlefield for her fiancé after the Battle of Kosovo in 1389 between Serbia and the Ottoman Empire. She is shown helping wounded Serbian warriors by providing water, wine, and bread.

DOI: 10.1201/9781003412946-2

TEXTBOX 2.1 COMPASSION FOR ALL LIVING THINGS

Sympathy beyond the confines of man, that is, humanity to the lower animals, seems to be one of the latest moral acquisitions.... This virtue, one of the noblest with which man is endowed, seems to arise incidentally from our sympathies becoming more tender and more widely diffused, until they are extended to all sentient beings.

—*Charles Darwin (1809–1882, foremost student of evolution; in Darwin 1871)*

We must fight against the spirit of unconscious cruelty with which we treat the animals. Animals suffer as much as we do. True humanity does not allow us to impose such sufferings on them. It is our duty to make the whole world recognize it. Until we extend our circle of compassion to all living things, humanity will not find peace.

—*Albert Schweitzer (1875–1965; German-born humanitarian, philosopher, and physician, medical missionary in Africa, Nobel Laureate; in Schweitzer 1959)*

If only we can overcome cruelty, to human and animal, with love and compassion we shall stand at the threshold of a new era in human moral and spiritual evolution – and realize, at last, our most unique quality: humanity.

—*Jane Goodall (1934–; English primatologist and anthropologist, viewed as the foremost expert on the behavior of wild chimpanzees, environmental activist; in Goodall 1999)*

tendencies to behave savagely and pitilessly, although civilization tends to moderate our worst behaviors, and some of the world's best minds advocate a greatly increased benevolence toward animals (Textbox 2.1).

This paragraph represents a brief excursion into philosophy, which the reader may choose to ignore since most people find the subject to be abstruse. 'Altruism' has been defined as 'uncalculated consideration of, regard for, or devotion to others' interests sometimes in accordance with an ethical principle' (Webster's Third New International Dictionary 1993), a definition which stresses unselfishness. There are sophisticated analyses of the biological nature of altruism (Dawkins 1989; Stevens 2004; Batson 2011) suggesting that most if not all generosity toward others is just reflective of the survival value of such behavior, not of 'goodness' or 'badness.' In the context of this book, this does not invalidate advocating for the conservation of all species because this is good for humans, but it represents a philosophical challenge for doing so for moral reasons. But there are also numerous analyses of the moral nature of good and evil (e.g. Shermer 2004). Psychologically, some humans are saints, and others are devils (Kaufman et al. 2019). Paulus and Williams (2002) defined a socially aversive set of personality traits as 'The Dark Triad' (Figure 2.2). These include narcissism (self-importance), Machiavellianism (strategic exploitation and deceit), and psychopathy (callousness and cynicism). Probably all of us have at least a trace of these traits. However, 'strongman' personalities in leadership positions (dictators, despots, totalitarians, tycoons, tyrants, and other autocrats – notably Vladimir Putin and, some would contend, Donald Trump) 'score significantly higher than non-autocrats on the Dark Triad' (Nai and Toros 2020). To control our barbaric tendencies, it is critical that we understand both our personal cruel and compassionate nature, and the obsessive destructive nature of some of our influential leaders.

CRUELTY

Background

The concept of cruelty is vague (it covers a range of behaviors) and ambiguous (it may be defined differently, especially in law and philosophy). Cruelty by people typically involves indifference or even pleasure while inflicting suffering on sentient creatures, including both humans and non-human animals (Figure 2.3). Sociopathy (or psychopathy) is an inability to perceive or at least to sympathize with pain in others. Sadism is an extreme, deviant, compulsive form of cruelty. Psychological inability to empathize may free one from moral responsibility, but not from legal consequences. As noted in this chapter, there is a questionable but well-established tradition of employing intelligence of animals as a guide to their capacity to feel pain, and accordingly to judge the degree, if any, of cruelty toward them. The principal concern in the following discussion is with how human tendencies to harm living things (including other humans) requires understanding and control simply as a practical matter in order to promote the health of both human society and the natural world which supports the human species. For many, however, the ethical aspects of cruelty by humans are their principal concern.

FIGURE 2.2 The 'Dark Triad,' a model based on three pathological psychological traits thought to contribute substantially to harmful, destructive personalities. Prepared by B. Brookes.

The words 'human' and 'inhuman' are often employed to mean kind and brutal, respectively, but humans are by no means always kind. In practice, humans frequently behave viciously toward other living things. To be a human being necessarily includes having an animal nature with potential for bestial behavior. Understanding and controlling our instinctive tendencies to harm our own and other species is critical because, as discussed in this book, our welfare is dependent on that of countless other organisms. Unfortunately, as also discussed, it is altogether too easy to hate and victimize the majority of living things. This chapter addresses the hostile, exploitative capabilities of the human species with a view to relevance to deal constructively with other species.

Evolution favors behavior that enhances survival of one's genes (Dawkins 1989), so selfishness for oneself and altruism, empathy, and compassion for one's close relatives (who share the same genes) are natural, while indifference or hostility to distantly related individuals is also somewhat 'natural.' Complex species, including humans, tend to have an instinctive (genetically programmed) side resulting in kindness to beneficial organisms, tolerance to harmless ones, and hostility to species perceived as dangerous or competitive. In fighting for limited resources, it is often the case that closely related organisms – sometimes even parents and siblings – become chief competitors, and this often generates ferocious combat. Humans are ingenious and often seemingly merciless competitors toward each other. History reveals that widespread homicide was practiced, frequently to usurp resources or power. As well, torture was commonly employed in order to coerce others to adopt the conventions and values (especially religious dogma) of those in power. Slavery, rape, and other atrocities that are exploitative have been widespread phenomena dating back to pre-history and regrettably still occur.

Racial Hostility among Humans as a Model of Hostility to Other Species

Perceived differences among humans matter, as evidenced by how many people view and behave toward the discernibly different groups of our fellow human beings (Figure 2.4a).

FIGURE 2.3 Caricatures (public domain) of cruelty to humans (top three figures) and non-human animals (bottom three figures). (a) Early 19th-century print showing a schoolmaster flogging a student. Credit: British Museum. (b) Late 18th-century print (public domain) showing two women in a pillory. Credit: British Museum. (c) 'The abolition of the slave trade,' an illustration of inhuman treatment by Scottish caricaturist Isaac Cruikshank (1764–1811). Credit: Royal Museum Greenwich. (d) Early 19th-century print showing an excessively heavy man (dressed as Napoleon) weighing down an unfortunate horse. Credit: Bodleian Libraries, Oxford University. (e) Spanish-style bullfight by British caricaturist James Gillray (1756–1815). Credit: British Museum. (f) 'The Death of the Fox' by English caricaturist Thomas Rowlandson (1756–1827). Credit: Yale Center for British Art.

Embarrassingly, there is a long history of human bigotry, savagery, and cruelty to groups of other humans that are racially different (Figure 2.4b). Indeed, many readers will have been the victims of abusive prejudice. At the root of much of human conflict is 'intraspecific competition,' in which members of the same species compete for exactly the same limited resources. When tribes or nations compete, even minor racial differences may serve to identify the 'enemy,' and stereotypes can be established that needlessly maintain hostilities for generations. Occasionally, conflict between groups is initiated when members of one group conclude or imagine that they have been harmed by members of another group. The result can be the overgeneralization that all members of the other group are equally capable of harm – i.e., *guilt by association*. Such stereotyping is not only a major contributor to racism, but it is also basic to why humans are very hostile to most other species ('speciesism,' as discussed in Chapter 3). Moreover, as will be presented, we humans may even be programmed so that we are instinctively hostile to many creatures. Unfortunately, we also seem to be susceptible to believe the worst about many animals – notably insects, spiders, snakes, and rodents – when in fact the vast majority are not only harmless, they indispensably contribute to our welfare.

Are Humans Natural Killers?

'Murder,' the unlawful or unethical killing of another human, is a social construct; the concern here is simply deliberate killing of animals (human or not). Killing other humans is 'homicide' (see review by Daly and Wilson 1988, and the epilogue to this book). Among humans, killing occasionally reflects psychopathy (i.e., abnormal psychology) or sometimes results from extreme overcrowding or stress, but killing for gain or revenge occurs in every society (i.e., in a statistical sense, killing is a 'normal' associate of intraspecific competition). Because we are a social species, most killing is the result of war between groups. Possibly most combat killing and mass genocides are based on the human social predisposition to obey dominant male leaders (leading to the excuse 'I was only following orders'). Traditionally, killing for personal gain or satisfaction is considered unjustified. In Judeo-Christian tradition as related in Genesis, Cain kills his brother Abel,

FIGURE 2.4 Racial variation in humans as a basis for inter-group conflict. (a) 'Races of mankind,' illustrating the extensive biological variation among humans. From Roe and Leonard-Stuart (1911), photo by Sue Clark (CC BY 2.0). (b) A painting (public domain) commemorating war in 1824 between the British and the Ashanti Empire in the region of the Gold Coast of Africa (now Ghana). Source: https://commons.wikimedia.org/wiki/File:Aschanti_Gefecht_11_july_1824_300dpi.jpg.

and mankind inherits the 'mark of Cain,' suggesting that humans are inherently murderous (Figure 2.5). Perhaps we are, at least potentially.

Certain species are natural killers at birth: some newborn chicks kill off their younger and smaller nestmates ('siblicide'), so only the firstborn survives. Some sharks, such as sand tiger sharks, cannibalistically consume their brothers and sisters while still in the uterus. Some salamanders are induced to become cannibals when food becomes scarce (Wakahara 1995). However, these examples are debatably comparable to humans, since we are social by nature, i.e., we thrive in groups, and families produce so few offspring that all individuals are valued. 'Comparative ethology' is concerned with the behavior of social groups of animals (Crook and Goss-Custard 1972). Aside from clarifying the nature of the animals themselves (such as how they organize hierarchically), the study of social animals is valuable as clues to the nature of humans (Wilson 2000).

FIGURE 2.5 People killing animals and each other. (a) Diorama entitled 'Primitive Man Hunting Animals' at the Museum of Vietnamese History. Photo by HappyMidnight (CC BY SA 3.0). (b) Roman gladiators fighting to the death at the Colosseum, a painting (public domain) by French artist Jean-Léon Gérôme (1824–1904). Photo credit: phxart.org. (c) Aztec human sacrifice depicted in the 16th century Ignote, codex (public domain; source: http://www.latinamericanstudies.org/aztec-human-sacrifice.htm). (d) Painting (public domain) of 'Custer's last stand' by Edgar Samuel Paxson (1852–1919). Photo credit: Whitney Gallery of Western Art.

Diet is key to how 'bloodthirsty' particular animals are. Social animals that are predators, such as lions and wolves, are indeed natural killers: as obligative carnivores, they rely on killing other animals for food. Social animals that are herbivores, such as elephants, horses, and bison, do not seek to destroy other species unless attacked (although males may compete against each other for dominance). Humans are omnivorous, physiologically adapted to both animal and plant foods, but most individuals have a distinct preference for flesh. Historically, humans have been hunter-gatherers, eating whatever was locally available. In modern times, we culture enormous quantities of livestock and crops for food, but most sustenance that is still harvested from nature is animal rather than plant, especially from the seas. Harvesting animals inevitably is accompanied by pain. Many humans are concerned with humane treatment of animals, but cruelty remains widespread. As a philosophical principle, the majority of people are comfortable with killing animals for economic purposes (although few would choose to visit an abattoir). However, for religious or moral reasons, many are completely opposed to killing animals.

Territoriality is a second key to killer instinct in animals. Social predators often kill competitors in their territory, particularly intruders of the same species. *Sex* is a third key, as exemplified by male lions taking over another pride and killing the young in order to bring the lionesses into sexual receptivity so that they can sire their own cubs. Whether the motivation is food, territory, or sex, the fundamental driver is competition for *resources*.

Although most animal-based food is generated by agriculture, numerous people continue to hunt and fish, often recreationally, as a leisure activity. While obtaining food and enjoying the outdoors may be motives, it is clear that, for some, a principal impetus is *enjoyment* associated with the act of killing (Oven 2019), often euphemistically termed 'harvesting' since admitting that killing is pleasurable seems deviant. The pleasure associated with hunting and fishing frequently is directly proportional to how big and/or how dangerous the prey is (Child and Darimont 2015; Figure 2.6). As discussed later, the most respected animals are the 'charismatic megafauna,' and many of the biggest and most dangerous of these have also been traditionally admired as prey ('big game'). Increasingly, these animals have become endangered, and there is now a strong element of shame in killing them for 'sport.' In the Western world, the shame element markedly increased in 2015, when a male African lion named Cecil (Figure 2.7), under scientific study in Zimbabwe, was killed by arrows from an American recreational big-game trophy hunter. International media outrage ensued, and the U.S. Fish and Wildlife Service added

FIGURE 2.6 Hunting scenes (public domain paintings) reflecting the traditional view that big game recreational hunting is heroic. (a) 'The lion hunt' by French artist Horace Vernet (1789–1863). (b) Hunting bear in Russia, a 1931 advertising poster of Intourist. (c) Tiger hunt by British artist Briton Rivière (1840–1920). Photo by Lot-tissimo.

FIGURE 2.7 Cecil the lion, a protected research lion in Africa killed in 2015 by a recreational hunter, resulting in widespread protests. Source: Shutterstock (contributor: Paula French).

African and Indian lions to their endangered species list. It is notable that the 'blood lust' of trophy hunters has been moderated by the perception of society that hunting the world's most admired animals to extinction is unacceptable. Nevertheless, hunting and fishing remain extremely popular recreational activities (Figure 2.8), and killing (not 'catch-and-release' or simply photographing) is an essential element. For exemplary articles regarding the range of views on the ethical and environmental aspects of recreational hunting, see Gunn (2001), Dickson (2009), Cohen (2014), Von Essen and Hansen (2016), and Di Minin et al. (2021a).

Humans are primates, and so comparison is appropriately with other social primates. Tarsiers (including about 20 species in three genera), which are native to Indonesia, Philippines, and Malaysia, are the only extant completely carnivorous primates, consuming insects and small vertebrates. Most non-human primates eat meat, mostly vertebrates, but only as a small part of their diet, and some species are entirely herbivorous (Watts 2020). The closest social relatives of humans are chimpanzees (*Pan troglodytes*) and bonobos (*Pan paniscus*), followed by gorillas (*Gorilla* species). Chimpanzees consume more meat than any other primate except humans, and bonobos also have a definite taste for meat. Gorillas, the largest living primates, are mainly vegetarians, feeding on stems, foliage, and fruits, although western lowland gorillas have been observed eating termites and ants. Orangutans (*Pongo* species) are solitary, not social, and are mostly vegetarians although they do eat insects, bird eggs, and small vertebrates. Humans are omnivorous like our closest relatives but eat far more meat than any other

FIGURE 2.8 Recreational (sport) hunting paintings (public domain). (a) 'Battue d'Automne' by French artist Bernard Boutet de Monvel (1881–1949); credit: rawpixel.com (CC BY 4.0). (b) Woman with a dead turkey, by American artist L. M. Glackens (1866–1933), published by J. Ottmann Lith. Co., NY in 1904. (c) Illustration of fox hunting by English artist Henry Alken (1784–1851). Credit: rawpixel.com (CC BY 4.0).

primate. Humans and chimpanzees, which prefer meat, are considered to be relatively prone to violence, while bonobos and gorillas, which prefer plants, are not, unless provoked.

It has been proposed that the roots of violence in humans originate in our primate ancestors (Ferguson and Beaver 2009; Gómez et al. 2016), at least as reflected by chimpanzees, which seem to be violent because their aggression results in increased resources (Wilson et al. 2014). Wrangham and Peterson (1996) specifically trace human male violence to chimpanzee male violence. From an evolutionary viewpoint, if killing competitors increase resources and leads to leaving more progeny, then such behavior should be favored. Of course, competitors may respond from an evolutionary perspective by evolving countermeasures, including also becoming more violent. Peaceful co-existence is also a potential strategy for survival, and it is comforting to consider the possibility that just as we may have inherited a tendency for violence from our chimp relatives, our bonobo cousins may reflect a counter-tendency for pacifism (Figure 2.9).

Much research and speculation have been advanced regarding the possible inclination of humans (especially males) to be brutal toward others. Violent tendencies toward other humans could predispose people to also be vicious against other species. Anthropologists are sharply divided on this question (Gabbatiss 2017). Our anatomy and intellect show that we are quite capable of harming other people and other creatures, albeit with the aid of tools. Our history is one of very extensive warfare against each other and destruction of other species. Modern man is capable of eradicating all life with modern weaponry, and wars have become a major threat to biodiversity (Lawrence et al. 2015; Hanson 2018). Clearly, it is important to understand our aggressive behaviors for the future survival of our own, let alone the other species of the world.

FIGURE 2.9 The closest relatives of humans. (a) A snarling, aggressive chimpanzee (*Pan troglodytes*). Source of photo: Shutterstock, contributor: Edwin Butter. (b) Peaceful bonobos (*Pan paniscus*). Photo by Wcalvin (CC BY SA 4.0).

'Bloodsports' are defined as entertainment or sporting events in which blood is shed, but how much blood is unclear, so it is arbitrary whether certain human vs. human sports (such as hockey, football, boxing, and mixed martial arts) are included. Bloodsports include setting animals against each other (dog fights, cock fights, bull baiting, bear baiting, etc.), human-non-human animal fights (primarily bullfighting and several other rodeo events), and recreational hunting and fishing (occupational harvesting is usually excluded). 'Bloodthirsty' means eager to observe or participate in violence and killing. In past times, human sacrifice was not uncommon, gladiatorial combat occurred, and public executions and punishment were sometimes considered to be entertainment. In modern times, the worst combat and sacrificial practices have been banned or discouraged, often as a result of lobbying by groups dedicated to the protection of humans and other animals. Nevertheless, a large proportion of people are strongly entertained by violence and associated harm to living things, as manifested in fictional entertainment media (Weaver 2011). Spectator contact sports are commonly said to be a means of diffusing natural warmongering tendencies of humans, which may be true, but it frequently happens that riotous behavior is a consequence of the home team either winning or losing. Indeed, it is debatable whether violent sports, recreational hunting, and even violent computer games – all of which are often vicarious psychological substitutes for killing other humans – make homicide more likely or less so (Rothmund et al. 2018).

Are Humans Natural Slavemasters?

Slavery in its ultimate form is the complete ownership and control of one person by another, including the right to exploit, restrict, and damage the slave in any manner. The word 'slavery' is also employed to designate extremely abusive and unfair interhuman relationships of certain kinds, particularly economic and sexual exploitation, which are significant modern concerns. Although slavery was widespread historically and geographically, a particularly large industry that thrived until the 19th century involved kidnapping and enslaving Africans to work in the American colonies to produce crops such as tobacco and cotton (Figure 2.10). Cruel forms of slavery were widespread in the past but are condemned and vigorously prosecuted in

FIGURE 2.10 Slavery. (a) Dutch enslavement of indigenous people of Brazil between 1630 and 1654. Painting (public domain) prepared by Dutch artist Frans Post (1612–1680). (b) Slaves working on a cotton plantation beside the Mississippi River. Currier & Ives 1884 print (public domain). Credit: U.S. Library of Congress. (c) Lucy Higgs Nichols (1838–1915), an African American escaped slave and a nurse for the Union Army during the American Civil War. She was accepted as the only honorary female member of the army of the U.S. Republic. This painting by Lucy Grant (CC BY SA 3.0) shows Nichols and her daughter escaping slavery in Tennessee in 1862.

modern times. The benefits of slavery to slave owners are great, so it is not surprising that the phenomenon was so common in the past, when survival was much more challenging for people than it is today.

Among wild species, slave-like relationships are rare, but do occur, especially among ants, which particularly enslave other ant species (Wilson 1975; Buschinger 2009) and other insects (Figure 2.11a). However, there does not appear to be a good example of any animal species, other than humans, enslaving other members of its own species. Among some social insects (like bees and termites), individuals develop into castes (such as workers, soldiers, and sexual forms), which might seem to be unfairly discriminatory in human terms. One might (with considerable justification) argue that the relative burdens of human mothers and fathers are not only unequal but also unfair. Criticizing Mother Nature or Evolution is pointless; criticizing lazy fathers is not.

Human relationships with the species we have domesticated are at least partially comparable to a master-slave association (Kendrick 2018) although to some extent domesticated animals receive benefits (Figure 2.11b and c). However, the benefits tend to be one-sided. Farmers literally own livestock, and despite welfare regulations, they are often subjected to extremely hostile conditions. Beasts of burden typically have lifestyles that would be intolerable to people. Medical experimentation causes pain to countless animals, once again despite regulations and oversight. Many undomesticated animals have been captured and confined to lives of captivity in zoos and aquaria for the entertainment of humans. Most of our pets are far more privileged but nevertheless are the property of their owners.

A 'symbiotic' relationship is one in which both participants benefit. The relationship between dogs and humans (and perhaps to a lesser degree between other pet species and their owners) exemplifies a symbiotic association which is simultaneously a slave relationship (at least technically). However, among humans, slavery is always exploitative, harming some people for the benefit of other people, and many pets are treated as kindly as most children. *Animal Liberation*, a pivotal book by Singer (1975) argued that while animals do not have 'rights' comparable to those of humans, limiting the pain of animals is just as important as limiting the pain of humans. Some radical proponents of the 'Animal Liberation Movement' (a phrase overlapping with the 'Animal Rights' movement or front) have illegally freed captive animals (both domesticated and wild) from their confines. However, numerous domesticated animals and plants (indeed, all of our major crops) are so changed from their wild ancestors that should they be 'liberated' to the wild, they could no longer survive. Indeed, because their natural habitats have been so reduced, the best chance for survival of some wild animals is in zoos. Although we humans exercise absolute control over many other species that is in some respects comparable to slavery, in many cases we cannot abandon the relationships without causing more harm than good.

Bullying

Pain deliberately inflicted on other humans can vary from simple annoyance to torture. Volk et al. (2014) defined bullying as 'aggressive goal-directed behavior that harms another individual within the context of a power imbalance.' Bullying is usually harmful, intimidating, or coercive action toward someone perceived as vulnerable. It is generally less harmful than homicide and slavery discussed above but can have devastating effects on victims. In the human sphere, homicide, slavery, and bullying are so 'evil' that they are controlled by legislation and tradition, but in the context of evolution, harming or exploiting competitors may increase resources and so improve the genetic fitness of those who engage in such behaviors (Figure 2.12). Dominance behavior in social animals is common and is a frequent form of bullying. Social animals frequently benefit

FIGURE 2.11 Slave-like inter-species relationships. The extent to which the seemingly exploited partner is a 'slave' is debatable. (a) A 'farmer' ant tending her 'cows' (aphids). In this symbiotic relationship, the ants benefit by obtaining honeydew secreted by the aphids, which in turn are protected from predators by the ants. Photo by Stuart Williams (CC BY 2.0). (b) Milking a cow. In this and the next example, the exploited animal at least receives a regular supply of food. Painting (public domain) by American illustrator Clara Miller Burd (1873–1933). (c) Raising rabbits for meat. Painting (public domain) by Polish artist Antoni Kozakiewicz (1841–1929). Photo credit: www.muzeum.leszno.pl.

FIGURE 2.12 Scenes showing bullying behavior in humans and other animals. (a) Bullying between children. Cartoon by Oscarioval (CC BY SA 4.0). (b) Gull about to steal a fish from a penguin (such theft has been termed 'kleptoparasitism' by animal behaviorists). Source: Pixabay (public domain). (c) The mobbing of a long-eared owl by other birds. Painting (public domain) by Transylvanian-born artist Tobias Stranover (1684–1756). Mobbing is harassment of dangerous species practiced by their potential victims.

from having a strong leader (effectively the biggest bully in the group) and from denying resources to the weakest members. Bullying is widespread among humans, almost invariably developing among children, requiring adult supervision, suggesting that it is 'natural,' and not a form of abnormal psychology (although some bullies are clearly psychopathic). Competitive sports frequently involve intimidation – a form of bullying. Individuals often join gangs to protect themselves against bullying, while facilitating the bullying of others.

The aforementioned discussion is based on intraspecific bullying, whether in humans or in other animal species. What about inter-species bullying? Larger and/or more aggressive animals frequently chase away other species from a food or habitat resource. Parasitism is a different phenomenon, but like bullying it represents one species harming but not necessarily killing another species. Predators frequently injure young prey animals and present them to their young to train them to hunt. Equally cruel, pet cats often play with their injured prey. When placing fish species in an aquarium, some species must not be placed together because one may pester another incessantly until it dies. Clearly, inter-species bullying occurs naturally.

The possibility arises that humans may also have intrinsic drives to bully other species. As discussed above, many people (hunters) do seem to take pleasure in killing animals, and we clearly benefit from enslaving animals, so it seems reasonable that at least some people naturally would bully animals. Domesticated livestock – particularly working animals – are frequently driven to near exhaustion for the benefit of people, which falls within the definition of bullying. Some rodeo events and other forms of animal-based entertainment similarly stress and sometimes endanger the animals. Traditional British foxhunting – really more aptly described as fox bullying than hunting since the event can last 3 hours or more – has been deservedly condemned as cruel.

COMPASSION

The Distinction between Animal Welfare and Animal Conservation

'Animal rights' is based on the view that there are moral constraints limiting the exploitation of animals by humans, especially cruelty and inflicted suffering. The established phrase 'animal welfare' is somewhat misleading since it does not address conservation aspects, which are vital to the 'welfare' of species. Animal welfare is mostly concerned with livestock and pets (both mostly domesticated) but also wild animals that are being harvested or employed for various purposes, including entertainment. Prior to the 1950s, limited attention was paid to the welfare and protection of animals, although compassion for animals is an old tradition in several religions (Figure 2.13) and the cause of animal welfare has been championed by some for centuries. However, animal welfare is increasingly a matter of global concern, and there are now tens of thousands of societies dedicated to the humane protection of animals (Sneddon et al. 2021). For the most part, the movement addresses cruelty to pets, livestock, and animals employed experimentally, largely for medical and/or health purposes. Animal protection organizations are supported largely by charitable donations from individual donors and accordingly are much more common in rich than in poor countries.

Unique among species, humans are (sometimes) concerned with the feelings of sentient species especially those that they exploit. The motive may be pragmatic – an unhappy animal may be unproductive – but many have empathy for the suffering of all creatures. Nature is quite savage and cruel by human standards, but it is perhaps in the area of limiting the infliction of pain on other species that humans can potentially claim superiority. The ethical issues are controversial, and the following reviews are recommended: Linzey 2009; Ferguson et al. 2013; Office of

FIGURE 2.13 Francis of Assisi (1181/1182–1226). St. Francis was an Italian Catholic friar, commemorated as the patron of animals and ecology. He epitomizes compassion for animals and indeed for all species. (a) Painting (public domain) 'St. Francis and the Animals' by 17th-century Flemish painters Lambert de Hondt and Willem van Herp. (b) Painting (public domain) showing St. Francis, entitled 'Sermon to the birds,' by Italian artist Giotto (1266–1337). Credit: Musée du Louvre. (c) Blessing a pet dog, a tradition in some churches on October 4, the feast day of Francis of Assisi. Photo by Siena College (CC BY 2.0).

Laboratory Animal Welfare 2015; Robertson 2015; Pacelle 2016; Paul et al. 2016; McWilliams 2017; Butterworth 2018; Feldstein 2018; Jabr 2018; Conte et al. 2021.

Animal welfare is different from animal conservation, which is primarily concerned with the preservation of wild animals and their habitats, but also addresses the conservation of animal genetic resources, primarily livestock breeds. Animal welfare and animal conservation are motivated by different objectives but are united in their desire to increase respect for and the well-being of nonhuman animals. Both movements are handicapped by the predominant view that most animals simply do not merit much, if any protection. Moreover, the degree of protection (whether for prevention of cruelty, or conservation in nature) is influenced very strongly in both cases by human prejudicial perception that ugly, primitive, dumb creatures deserve less (indeed, far less) consideration than beautiful, intelligent animals. This book stresses the importance of such prejudice especially in the pragmatic context of species conservation, because the issue is vital to the economic well-being, indeed survival of people. Animal welfare, however, is fundamentally an ethical issue and bears on the moral well-being of humans.

The public is often confused by the motives of self-appointed defenders of animals. When Greenpeace ships protect whales, is it to keep these magnificent animals from being physically damaged, or to keep the species from going extinct? When movie stars criticize the wearing of fur, is it to protest cruel trapping methods, or to maintain the species in nature? Such uncertainty does not serve the goals of either animal welfare or animal conservation. However, both movements are founded on respect, indeed love of animals, and both reflect moral convictions about human responsibilities. There should be opportunities for cooperation. For example, wild birds are under severe danger from mirror-glass-clad buildings into which they inadvertently fly, and (independently) they are also being decimated by cats. The result is threat to the survival of bird species (an animal conservation issue), but also numerous crippled birds requiring rehabilitation (an animal welfare issue).

Genetic Similarity to Humans as a Principal Determinant of Compassion for Animal Species

Although racism is basically irrational prejudice, in a twisted way it might serve to advance one's own genetic interest group (i.e., race) at the expense of others. Racism amounts to judging the merit of others simply on the basis of how closely they are (or appear to be) genetically related to one's own race. (This assumes that racists are knowledgeable about the biology of race, which is challengeable.) We humans not only tend to favor other people who are closely related to us, we also tend to judge all other species on the basis of relatively closeness to us (Tisdell et al. 2006; Miralles et al. 2019). Modern research has clarified the ancestry of the major lineages of living things with considerable precision. Species genetically closer to humans tend to be judged worthier of better treatment. There is in effect a 'ladder of life' with mankind at the top and the 'higher,' more deserving forms located on the higher rungs (Figures 2.14–2.16). Empathy for animal species is determined to a considerable extent by how closely ('phylogenetically') they are related to humans, or at least are reminiscent physically, behaviorally, or in intelligence. Thus primates are ranked very high, so are mammals in general, less support is given to reptiles, invertebrates receive very little support, and so forth. However, there are exceptions: birds, for example, receive extensive sympathy, although they are closest to reptiles and quite distant from humans. Arluke and Sanders (1996) referred to the spectrum of human preferences for animals as 'the 'sociozoologic scale,' and Pizzi (2009) termed it 'taxonomic chauvinism.'

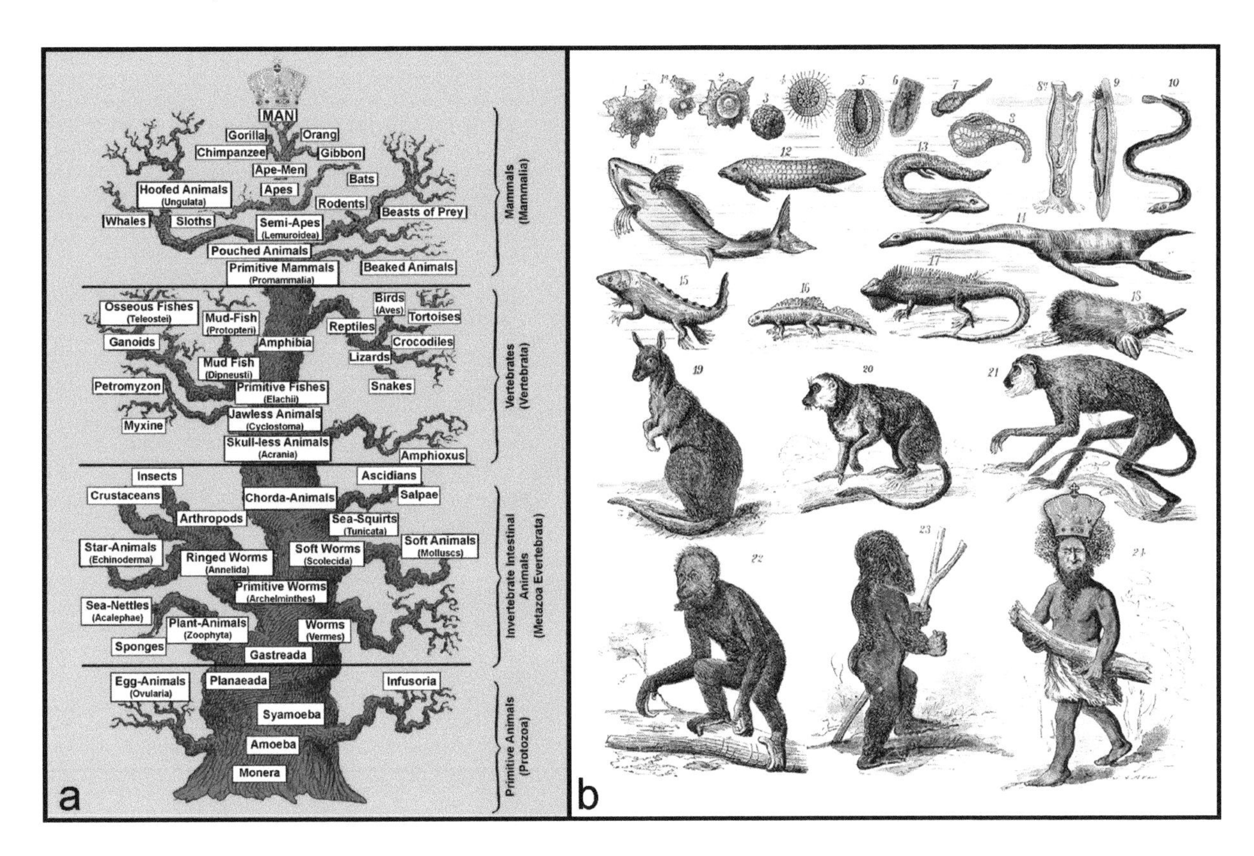

FIGURE 2.14 The traditional view that humans (note crowns) are the ultimate accomplishment of nature, interpreted as the ascendant apex of a pedigree tree (a) or the descendent bottom of a set of increasingly complex species (b). (a) Illustration entitled 'Pedigree of Man' or 'Tree of Life' from: Haeckel, E. 1879. *The Evolution of Man: A Popular Exposition of the Principal Points of Human Ontogeny and Phylogeny*. New York, Appleton. (b) Illustration of the 'Descent of Man' from: Avery, G. 1874. A critique of Haeckel, in *Scientific American*. March 11, p. 167. (The animals are identified at https://commons.wikimedia.org/wiki/File:Human_pidegree.jpg.)

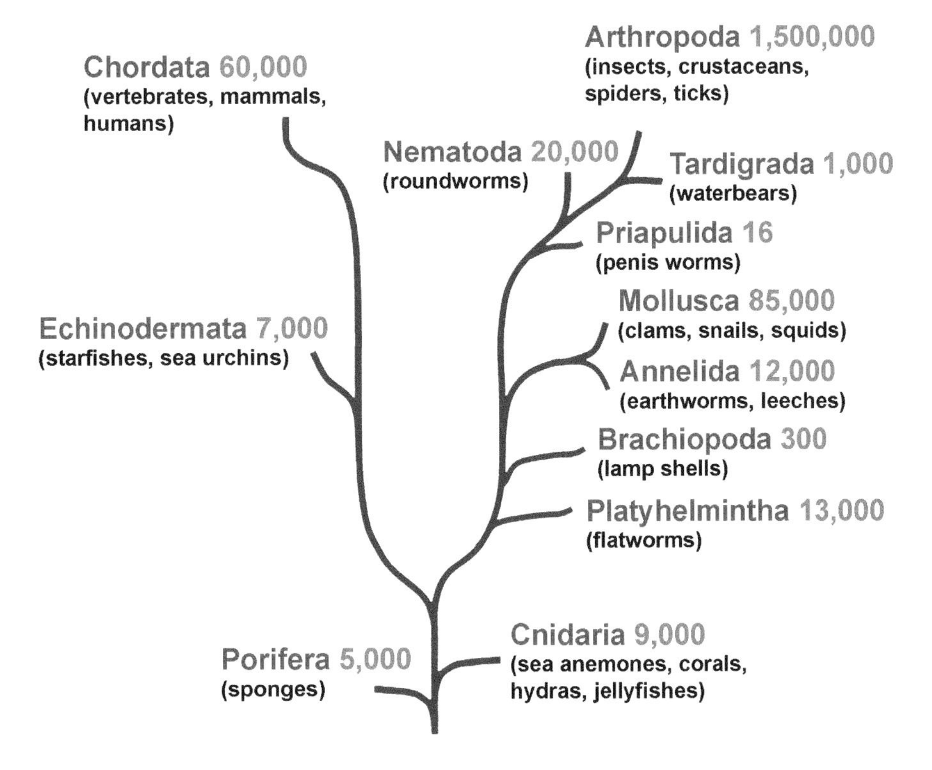

FIGURE 2.15 A simplified phylogenetic tree of multicellular animals, showing major groupings, the numbers of species, and examples of the kinds of species included in the groupings. After Li et al. 2016 (CC BY SA 4.0).

FIGURE 2.16 A simplified phylogenetic tree of the major groups of vertebrate animals. Prepared by B. Brookes.

Intelligence as a Principal Determinant of Compassion for Animal Species

Humans greatly value intelligent people as well as intelligent animals that can be trained for service or entertainment (Figure 2.17). Indeed, it appears that the brains of several (e.g., cat, dog, goat, pig) have been altered by domestication so that they exhibit forms of behavior considered to be intelligent. Not surprisingly, the mental abilities of animals are important to human perception of their merit. On the basis of a survey, Callahan et al. (2021) found that decreasing intellectual and emotional abilities are ascribed respectively to mammals, birds, reptiles, amphibians, and fish. However, the measurement of intellectual capacities in non-human animals is problematical (Van Schaik 2006; Pouydebat 2017; Pearce 2019; Boesch 2020), and brain size in particular has not proven to be a simple indicator of relative intelligence (Logan et al. 2018; Smaers et al. 2021).

Most adults are humane and empathetic and respect the need to put boundaries on pain caused to animals that we exploit, particularly for food and for medical research. However, most people also apply this compassion very unevenly to animal species. In practice, invertebrates are rarely included in animal welfare regulations. For example, silk production involves boiling silkworms alive in their cocoons, and several other invertebrates are consumed alive – most commonly oysters, which are eaten raw. (Bates [1967] wrote that 'if oysters shrieked as they were pried open, or squealed when jabbed with a fork, I doubt whether they would be eaten alive.') The so-called 'lower' or 'cold-blooded' vertebrates (fish, amphibians, and reptiles), if mentioned in legislation, are usually only given limited protection (Zacchariah 2012; Crook 2013; Horvath et al. 2013; Figure 2.18). This only leaves mammals and birds, and protection for them corresponds to the widespread conviction that they (or at least many of them) are exceptionally intelligent (the same is true for squids and octopuses, rare examples of sometimes protected invertebrates). Several dozen countries now protect the other species of the 'Great Apes' besides humans (orangutans, gorillas, chimpanzees, and bonobos), all of which are exceptionally intelligent. As noted earlier, intelligence and ability to experience pain are not necessarily correlated, although this is a common assumption.

The apparent intelligence of animals is a very strong determinant of how much concern people have for the

FIGURE 2.17 Animals admired for their intelligence. (a) Chimpanzee (*Pan troglodytes*). Photo by William Warby (CC BY 2.0). (b) Bornean orangutan (*Pongo pygmaeus*). Photo by William Warby (CC BY 2.0). (c) African savanna elephant (*Loxodonta africana*). Photo by Felix Andrews (CCBY SA 3.0). (d) Red fox (*Vulpes fulva*). Public domain image from Pixabay.com. (e) Octopus (*Octopus vulgaris*). Photo by H. Zell (CC BY SA 3.0). (f) Bottlenose Dolphin (*Tursiops truncatus*). Public domain image from Pixabay.com. (g) Barred owl (*Strix varia*). Public domain image from Pixabay.com. (h) Scarlet macaw (*Ara macao*), alleged to have the intelligence of a 4–8 year old human. Photo by Matthew Romack (CC BY 2.0). (i) American crow (*Corvus brachyrhynchos*). Photo by Trougnouf (CC BY SA 4.0). (j) Pigeon (rock dove; *Columba livia*). Public domain image from Pixabay.com.

suffering of animals. In discussing animal welfare, Rollins (2011) commented 'it seems reasonable to give priority to those animals which seem to be able to experience positive and negative mental states similar to our own and display intelligence and innovative problem-solving abilities.' A frequent rejoinder to this position that intellectually superiority entitles species to more humane treatment is the scenario that one day aliens of superior intelligence could attack Earth and exploit us just as we treat most other species. If extraterrestrial beings are indeed much more advanced than us, we may be lunch to them (Figure 2.19), but would this be justified?

Welfare of Wild Animals

The world of living things is constantly at war, and is a very dangerous place for most organisms, the vast majority dying long before they mature. Most species directly exploit other species for gain (e.g., as food) or indirectly harm other species (e.g., by compromising their environment). Evolution (or Mother Nature) selects only a very small number of the most adapted individuals for survival, leaving the rest to perish. In addition, regardless of how well-adapted or how weak an individual is, many survive or expire simply by chance. The scale of carnage in nature

FIGURE 2.18 Discrimination on the basis of perceived intelligence. The perception is common that 'higher' or intelligent animals deserve protection from pain but 'lower' animals do not because of their alleged lack of intelligence and inability to feel pain. (a) Painting (public domain) by P. Mathews ca. 1838 of the 'Trial of Bill Burns,' a man prosecuted for beating his donkey, one of the earliest cases of a court case regarding animal cruelty in Great Britain. (b) Cartoon of a lobster protesting being cooked. The possibility that lobsters feel pain has resulted in the practice of boiling them alive being banned in some jurisdictions. (c) An unhappy worm. Impaling a worm on a fishing hook is widely believed to be essentially painless or simply of insignificant importance. (b and c) Prepared by B. Brookes.

FIGURE 2.19 A scene of big-brain extraterrestrials eating humans. Does superior intelligence justify harming less intelligent animals, when humans are the inferior species? Prepared by B. Brookes.

FIGURE 2.20 Examples of humane treatment of wild animals. (a) Elephant orphanage in Sri Lanka. Photo by Bernard Gagnon (CC BY SA 3.0). (b) Orphaned baby kangaroo at The Kangaroo Shelter, Alice Springs, Australia. Credits: UI International Programs/Riley O'Day/University of Newcastle, (CC BY ND 2.0). (c) Pelican contaminated by an oil spill in Louisiana in 2010 being cleaned. Photo by Brian Epstein (CC BY 2.0). (d) Attempt to rescue a 12 ton juvenile female northern bottlenose whale in the River Thames in central London in 2006. Photo (public domain) by John Hyde. (e) Bird house. Public domain photo from Pixabay.com. (f) Feeding an orphan squirrel. Public domain photo from Pixabay.com.

is so vast that humans are basically incapable of reducing the pain experienced by most wild species. But there are efforts. Because there is extremely uneven empathy for the 'lower' and 'higher' animals, almost no one would call for the rescue of wounded frogs, snakes, fish, or invertebrates, but many wild bird and mammal species are often rehabilitated when injured. Orphaned young of the charismatic megafauna such as bear cubs and elephant babies are especially appealing candidates for rescue. There are many examples of humane treatment of wild animals, mostly involving attractive mammals and birds (Figure 2.20).

Welfare of Livestock

Most livestock today are employed as food, and while food animals are generally treated reasonably (at least in some countries), too much respect generates guilt when they are killed and eaten, a phenomenon termed 'the meat paradox' (Wang and Basso 2019). A poignant example develops when a calf or piglet raised as a pet by a farmer's son or daughter becomes a candidate for slaughter (Figure 2.21). Of course, we humans do not seek advice from our livestock in regard to our dietary habits (Figure 2.22), but ending the lives of sentient creatures is distinct from consuming cereals, vegetables, and fruits. In Western society, women are more likely than men to become vegetarians ('salad eaters') for ethical and emotional reasons (Johnson et al. 2021), and men tend to have been persuaded that eating meat (and taking charge of the barbecue) is a reflection of their masculinity (Ruby and Heine 2011, 2012; Ruby et al. 2013). Livestock are mostly mammals and birds, and their discomfort is often not easy to ignore. A substantial minority of people are vegetarians because of the associated suffering (albeit minimized by regulation) of food animals.

Welfare of Work and Exhibition Animals

'Beasts of burden,' especially horses, donkeys, and mules, have been enormous contributors to the welfare of humans. Unfortunately, they have often been treated cruelly in historical times and frequently continue to have difficult lives in many third-world countries where humane treatment is considered unaffordable. Animals used for entertainment – especially circuses – also have often been abused historically, although today the popularity of animal exhibitions is diminishing. Rodeos sometimes appear to be quite cruel, but the animals need to be in very good physical condition, and they generally receive much better treatment than livestock. Regardless, most commercial animals end up as food, commercial products, and fertilizer. Most farmers simply cannot afford to be sentimental. A very small minority is afforded the luxury of retirement to a life of leisure, and these are usually mammals of special appeal to people (Figure 2.23).

Welfare of Experimental Animals

Animals are sacrificed experimentally for many purposes. Aside from exploiting animals for food, the predominant deliberate infliction of harm on animals is

FIGURE 2.21 'The Meat Paradox' illustrated by reluctance to slaughter a pet piglet. Most people are sensitive to the welfare of animals, but nevertheless choose to consume meat, necessitating their sacrifice. Prepared by B. Brookes.

for medical research for the benefit of humans. Some 'privileged' animals subjected to pioneering research have even been named (Figure 2.24) but most remain anonymous. Vast numbers of rodents and much smaller samples of other mammals are the primary subjects/victims. The associated ethical issues are briefly explored in Chapter 17, where the ethics and justification of employing experimentally harmed animals (especially rodents) are examined. The benefits to the health of humans from sacrificing animals have been enormous, but the suffering of laboratory animals should not be regarded as necessary in perpetuity for human welfare, given the continuing advances of technology. This book is concerned in part with the unfairness of human treatment to other humans and to other species, and most perceptive people intuitively appreciate that causing pain to creatures with nervous systems isn't quite right, even if the cause is noble (cf. Figure 2.25).

FIGURE 2.22 Suggestions for dinner from the potential main dishes. Prepared by B. Brookes.

FIGURE 2.23 Retired animals. (a) One of the 'Presidential Turkeys' at the Willard Hotel in Washington, DC in Nov. 2020, in advance of the 'National Thanksgiving Turkey Pardoning Ceremony' at the White House. Turkeys pardoned by the President are not sacrificed but retire to a life of leisure. Official White House photo (public domain) by Andrea Hanks. (b) Chimpanzee at 'Chimp Haven,' the world's largest chimpanzee sanctuary, a nonprofit facility in Keithville, Louisiana, housing more than 300 chimpanzees retired from laboratory research. Photo by ChimpHavenOfficial (CC BY SA 4.0). (c) Retired thoroughbred race horses at 'Old Friends,' a nonprofit equine retirement facility in Georgetown, Kentucky. More than 100 horses are present. Photo (2015) by Jlvsclrk (CC BY SA 4.0).

FIGURE 2.24 Animals famous for their involuntary contributions to science as experimental guinea pigs. (a) 'Herman the Bull' (1990–2004), the first non-human mammal modified with human DNA (with the human gene coding for lactoferrin, a protein in colostrum, the first type of breast milk produced after a baby is born). Photo by Peter Maas (CC BY SA 3.0). (b) Romanian stamp (public domain) from 1959 showing 'Laika the Space Dog' (1954–1957), launched into low orbit in 1957 on Soviet spacecraft Sputnik 2, where she died. (c) 'Dolly the Sheep,' the first mammal cloned from an adult cell. Born in 1996, she died in 2003 at age 6, about half the life span typical for sheep. Photo by Toni Barros (CC BY SA 2.0).

FIGURE 2.25 Humans being treated as laboratory animals by other species. (a) 'Lab humans.' Prepared by B. Brookes. (b) 'Vivisection of a human,' satirical cartoon (public domain) published in Lustige Blatter (Berlin), ca. 1910. Credit: Wellcome Foundation (CC BY 4.0).

Welfare of Pets

Most pets ('companion animals') are extremely attractive to humans and often become psychological surrogates, substituting for children and friends. 'Anthrozoology' as a term was purported to be an analysis of the evolution and relationships of pets and people (Bradshaw 2017a) but has been more generally understood to be a subset of ethnobiology, dealing with interactions between humans and other animals (note Freeman et al. 2011). Ethnobiology includes ethnozoology and ethnobotany. 'Ethnozoology' is the study of the past and present interrelationships between human cultures and the animals in their environment (Alves and Albuquerque 2017). 'Ethnobotany' is similar but addresses plants, not animals. The most loved pets are birds and mammals, and the warmest treatment is reserved for dogs and cats (Figures 2.26 and 2.27). Most animal shelters are dedicated to cats and dogs, which are the most frequent long-term pets (aquarium fish have been claimed to be more popular but are treated like gift flowering plants, often discarded when they decline). The use of dogs as experimental animals is especially carefully regulated, if not forbidden in given jurisdictions. (The use of dogs as food in some countries exemplifies just how different attitudes can be toward given species.) In Western countries, dogs and to a lesser degree cats are the flagship species used in fund appeals by animal welfare groups. Unfortunately, pets are sometimes treated savagely by pathological individuals. As with human charitable campaigns, mistreated individuals are often employed in advertising to elicit sympathy.

Most pets are domesticated, but some are wild, only slightly domesticated, or domesticated as livestock. These are sometimes collectively referred to as 'exotic,'

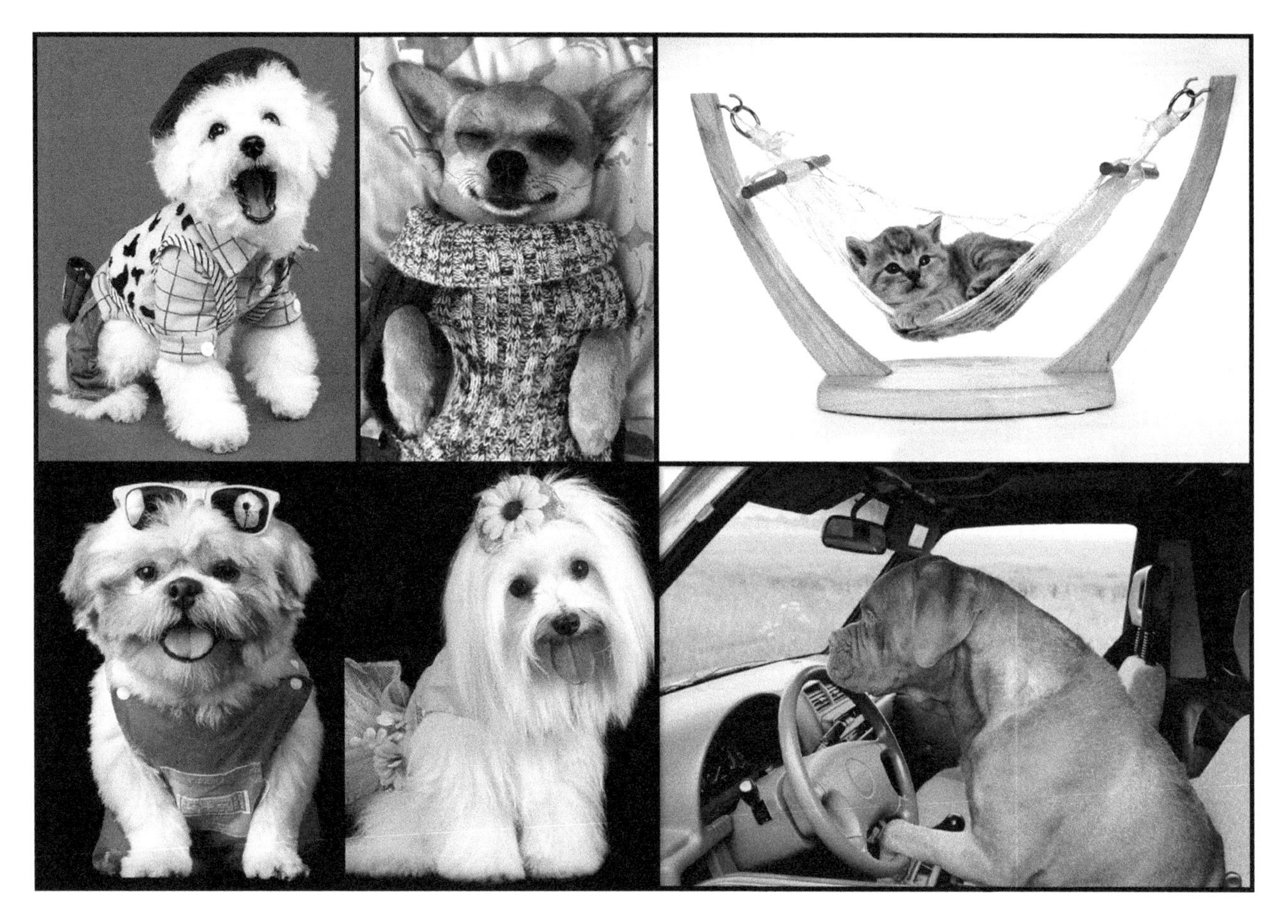

FIGURE 2.26 Pampered pets. Photo (public domain) from Pixabay.com.

FIGURE 2.27 Compassionate canine care. (a) A boy scout treating a wounded dog. Painting (public domain) by American artist Norman Rockwell (1894–1978). (b) A girl scout treating a wounded dog. Illustration (public domain) published on the cover of the Saturday Evening Post, 25 Oct 1924. (c) A sick dog being examined in a veterinary clinic. Source: Shutterstock (contributor: DeepGreen).

'non-traditional,' or 'unconventional.' Examples are shown in Figure 2.28. Because their dietary and environmental needs are often much less understood than conventional pets or they simply require much more care than apparent, they often are simply unsuitable as pets for most owners (Schuppli et al. 2014; Whitehead and Vaughn-Jones 2015). These unusual pets – especially those that are marketed in retail outlets – are useful in indicating the preferences of people. Cage birds are extremely popular and often receive as much love from humans as mammals. Aquarium fish are also popular (as noted above, it is sometimes claimed they are the world's most numerous pets)

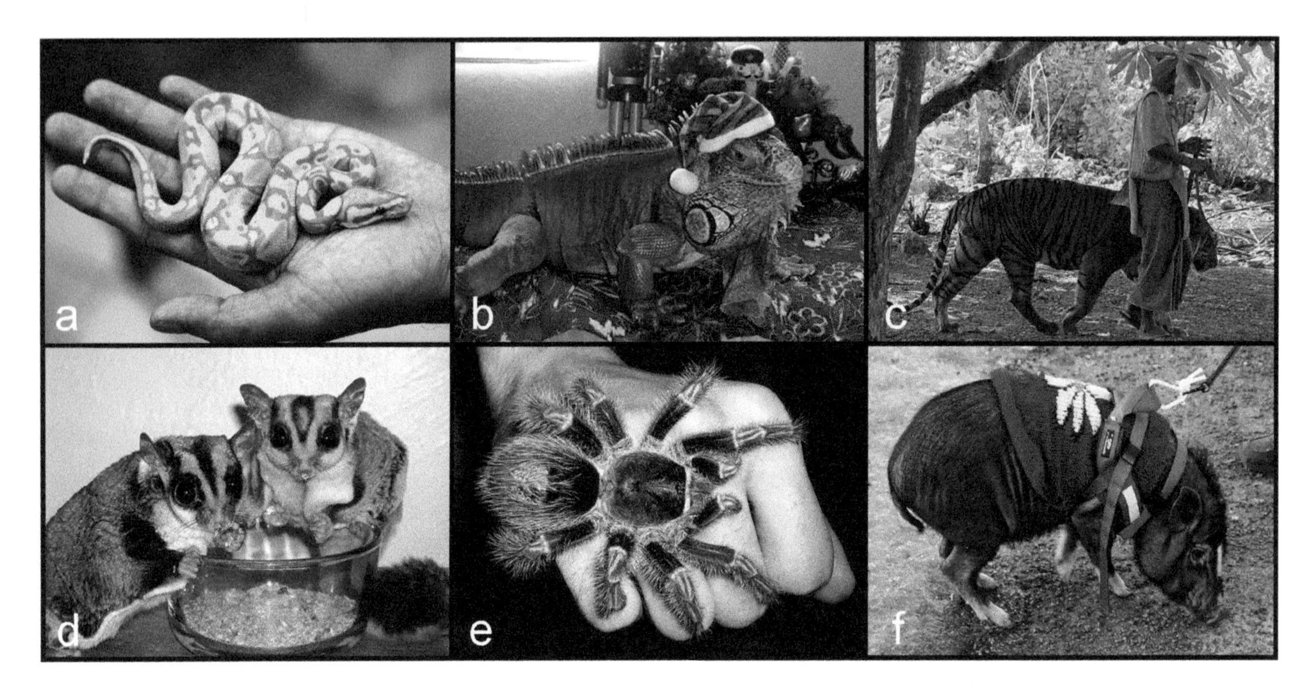

FIGURE 2.28 Examples of 'exotic' pets. (a) Ball python (*Python regius*). Photo by Tris T7 (CC BY SA 4.0). (b) Green iguana (*Iguana iguana*). Photo (public domain) by Pineapple-Cove. (c) Monk walking a male tiger on a leash at Tiger Temple, Thailand. Photo by Michael Janich (CC BY SA 3.0). (d) Male and female sugar gliders (*Petaurus breviceps*, a marsupial) eating meal-worms. Photo by OberonNightSeer (CC BY SA 3.0). (e) Chaco golden knee female tarantula (*Grammostola pulchripes*). Photo (public domain) by PavelSI. (f) Pet pig. Photo by Eirik Newth (CC BY 2.0).

but veterinarian care for them is scarcely available and their lives are often short. Large carnivores such as big cats, bears, wolves, big snakes, and crocodiles are dangerous and usually not allowed in most jurisdictions but seem to be appealing to a minority as status symbols. Reptiles (especially snakes, lizards, and turtles) are also appealing to a minority. Animals that most people find scary, such as tarantulas, piranhas, and skunks, are maintained by a few. Although frogs are not attractive to most people, there is a substantial pet trade in wild amphibians, often resulting in the introductions of alien invasives (Measey et al. 2019). Because they are so reminiscent of humans, some people find the great apes (gorillas, chimpanzees, bonobos, and orangutans) and monkeys attractive, but primates are highly unsuitable as pets (Soulsbury et al. 2009).

Biophilia and Domesticated Animals

Wilson (1984, 1993) defined 'biophilia' (based on Greek, literally meaning love of life or of living things) as 'an innate positive emotional response of human beings to other living organisms,' and it is not difficult to find examples (Figure 2.29). 'Zoophilia' has been defined as human sexual excitement through stroking or fondling animals (Adams et al. 2010), which is a quite unrelated concept. Biophilia represents the positive polar end of a wide spectrum of human likes and dislikes of species (Dhont and Hodson 2019). Most people no longer hunt or fish, so domesticated animals are the principal ones that influence human attitudes to the animal kingdom. As will be noted (Chapter 12), domesticated animals are of enormous economic importance to humankind. Numerous people have acquired considerable respect, indeed a deep love for their domesticated animals, and frequently spend considerable money on their welfare. The widespread concern for humane treatment of animals is different from the issue of animal conservation, but the emotional connection that people have for their own pets ('companion animals') carries over into sympathy for wild animals. Pets may function as ambassadors for livestock and wild animals (Serpell 1995, 2000; Serpell and Paul 1994). 'Studies have shown that affection for pets goes hand-in-hand with concern for the natural world' (Bradshaw 2017b). It is probably true that dog owners are especially predisposed to support wolf conservation, cat owners to support big cat conservation, horse owners to support the conservation of equids, and cage bird owners to support bird conservation. Farmers also have a natural interest in certain animal groups. Rare breeds are a form of biodiversity and are quite deserving of conservation efforts, albeit this issue is divorced from conservation of natural ecosystems. However, rare breed and livestock societies should be interested in the conservation of wild animal habitats related to their specialities. Very

FIGURE 2.29 Examples of biophilia (love of other species). (a) A tender moment with an orangutan. Photo by Gautsch (CC BY 2.0). (b) Young boy with a python. Photo by Shafika Tahir (CC BY SA 4.0). (c) Young girl with ducklings. Photo by Henry Burrows (CC BY SA 2.0). (d) Young girl and puppy. Photo by David~O (CC BY 2.0). (e) Hugging a pot-bellied pig. Photo by LadyDragonfly (CC BY 2.0). (f) A girl with her cat. Photo by Niels Kliim (CC BY 2.0).

few animals are fortunate or unfortunate enough to have been domesticated, but at least their wild relatives may benefit from the emotional predisposition of people to promote their welfare.

PARALLELS OF PREJUDICES AND PREFERENCES OF INTEREST GROUPS IN HUMANS AND NON-HUMAN ANIMALS

Individuals have preferences and prejudices, but so do groups. The statements were made earlier in this chapter 'At the root of much of human conflict is 'intraspecific competition,' in which members of the same species compete for exactly the same limited resources. When tribes or nations compete, even minor racial differences may serve to identify the 'enemy,' and stereotypes can be established that needlessly maintain hostilities for generations.' Humans seem to be naturally suspicious of groups different from their own (Öhman 2005). Many animals are social, some forming permanent groups, others associating together temporarily. Such groups may be basically genetically based and long-term (especially families and tribes) or very short-term and composed of unrelated individuals (such as vegetarian animals drinking collectively at a waterhole because this increases safety from ambush by predators). Groups exist because there are advantages for the members, usually at the cost of disadvantages for others. Geographically based gang formation among youth provides protection from attack by one's fellow members in the neighborhood, but non-members are subject to abuse when they visit neighborhoods controlled by other gangs. Many social animals claim ownership of regions much like humans (Figure 2.30). Wolf packs and lion prides share a geographical area with member of their own group but will viciously attack intruders. Among humans, political groups have the purpose of protecting their members, but this may be associated with quite vicious behavior toward others. All forms of supremacism (which could be based on age, gender, race, ethnicity, religion, sexual orientation, language, social class, ideology, national origin, or culture) reflect a degree of bias, although the primary intent may be quite legitimate. The liberal-conservative spectrum is based on a combination of economic

FIGURE 2.30 Examples of geographical gang warfare in humans and animals. (a) Painting (public domain) of an annual fistfight on a bridge between two neighborhoods in Venice, Italy, by German artist Joseph Heintz the Younger (1600–1678). Photo credit: Anagoria. (b) Imaginative painting (public domain) of wolf packs fighting by British artist George Bouverie Goddard (1832–1886). Photo credit: Art UK.

power and social constraint, which necessarily rewards some at the cost of others. Many political groups are overtly biased, reflecting: jingoism (extreme patriotism or nationalism associated with aggressive or warlike foreign policy), xenophobia (fear and hatred of strangers or foreigners), or populism (resentment against alleged elitists and privileged people in favor of common or ordinary people).

Much (possibly most) of the behavior of animals, including humans, is instinctive (genetically programmed), designed by nature and evolution to benefit individuals, and this inevitably means that the result is often harmful to others. The human concepts of bias, prejudice, and bigotry, and indeed all moral/ethical terminology, scarcely if ever apply to non-human species. However, the conduct of individuals and groups of humans, which amount to bias, prejudice, and bigotry, in considerable measure is controlled by the same biological impulses. This book emphasizes that to alleviate the associated harms it is critical to appreciate the underlying drives that direct and constrain human behavior.

SIN	VIRTUE
Envy	Kindness
Lust	Chastity
Gluttony	Temperance
Greed	Charity
Pride	Humility
Sloth	Diligence
Wrath	Patience

FIGURE 2.31 Contrast of the 'Seven Virtues' (Seven Capital Virtues) and the 'Seven Sins' (Seven Capital Sins), which trace to ancient and medieval Christian tradition.

HOPE FOR THE FUTURE

Whether we humans are morally superior to other species has been argued (De Waal et al. 2006). Regardless, we alone appear capable of collectively modifying our individual behaviors for the long-term good of our own species. The simplistic contrast of 'sins' and 'virtues' (Figure 2.31) reflects recognition that the key to successful living with others is simply controlling our natural selfish urges. The emerging field of social psychology of environmentalism and conservation has potential for improving human relationships with other species (St John et al. 2010; Wauters et al. 2017; Campbell 2019).

3 Human Prejudice against Other Species (Speciesism)

INTRODUCTION

Living things provoke emotional responses, but the sentiments span a very wide range from love and respect to nausea and hatred. Disgust and fear far exceed approval and concern. Most people are: (1) not just ignorant of but indifferent to almost all of the species on the planet; (2) slightly to extremely negative toward the majority of species they encounter; and (3) strongly positive toward certain species that are valuable or simply have characteristics appreciated by the human psyche. This chapter concentrates on negativity toward species. There are several terms for dislike of other species, but these are not well known. 'Misothery' is aversion, hatred, or contempt for non-human animals. Rarely, 'biophobia' is employed to mean fear or aversion of living things (e.g. Olivos-Jara et al. 2020).

Most humans are completely unaware of the hundreds of millions of species in the world, but on the basis of experience with a few familiar animals and plants, people have nevertheless developed unflattering stereotypes for various life forms, most infamously 'bugs,' which include the majority of the world's species. We rely on a relatively small number of species for most of our physical resources, particularly crops, livestock, forest trees used as lumber, and wild fish. Of course, these are esteemed, but since most food and other goods are obtained from stores, people tend not to appreciate that the ultimate source is nature. In a similar vein, we love our pets, but these are treated as members of the household, quite different from the natural, wild world. A small number of wild species are greatly admired – notably the icons of environmental organizations such as whales, pandas, and tigers – but almost all people only see most living wild species in zoos and on video screens, and species that are either very attractive or entertaining are predominantly shown. By contrast, most people live in urban situations, and aside from cultivated flowering plants viewed in gardens, the largest number of species encountered are despised pests, vermin, and weeds, giving an unfortunate biased perspective of the world of living things.

DISTINCTIONS BETWEEN PREJUDICES, PREFERENCES, AND INSTINCTS WITH RESPECT TO ATTITUDES TOWARD SPECIES

This book argues that prejudices and preferences that humans exhibit for and against each other are key analogues of attitudes toward animals and plants. What is the difference between 'preferences' and 'prejudices'? In theory, a preference can be based on valid, logical examination of alternatives resulting in choice or choices of one or more of the alternatives. The 'pre' in 'prejudice' indicates that an attitude toward something has been acquired ('pre-judged') on the basis of insufficient evidence. However, many of our behaviors, likes, and dislikes are instinctive (i.e., biologically inherited), and although instincts have evolved because they are adaptive (or were for our ancestors), some may be inappropriate in modern society. Science by nature insists that we must be prepared to change our beliefs when evidence that they are not appropriate becomes available, and this credo is particularly applicable to prejudices. All of us are products of our particular histories, and it is not possible to know everything, or even enough about most subjects to be certain, so necessarily we adopt much of the 'common knowledge' (including misinformation) about most subjects. Once incorrect information is acquired, it is difficult to correct because of a phenomenon termed 'confirmation bias' – the human tendency to selectively recall and interpret information to affirm prior beliefs. Much misinformation is harmless or relatively so, but incorrect knowledge that leads to the maltreatment of living things is dangerous. Racism, the most harmful manifestation of prejudice, is fundamentally based on misinformation. As will be discussed, the substantial prejudice that people have acquired for most of the world's species is parallel to racism among people. For our future welfare, it is important that we learn the importance and values of species. However, as will be described, our attitudes toward non-human species are based to a considerable extent on instinctive ways we evaluate physical and behavioral features of living things. As a result, we are partially programmed to like or dislike other species, depending on what they look like and how they behave. Phrased alternatively, we instinctively like or at least respect species that we perceive as beautiful, impressive, powerful, and reminiscent of humans, and dislike species that we perceive as ugly, unimpressive, weak, and quite unlike humans. This intuitive thinking greatly distorts the true values of most other species, which as will be explained play indispensable ecological roles that sustain the world in a habitable state for people. Hopefully, the information presented in this book will persuade many of the necessity of friendlier relationships with all life forms.

Wilson (2000) famously concluded that human belief and behavior are mostly deterministic – i.e., is scarcely modifiable, and it does seem that to a great extent changing human behavior is very challenging. In a 1912 essay, the founder of psychoanalysis, Sigmund Freud, wrote 'anatomy is destiny' (the English translation of the original German

DOI: 10.1201/9781003412946-3

can be found in volume 12 of Strachey 1953–1974). The phrase has been very controversial because it has been applied to sexual orientation, and it is now clear that male and female anatomy does not always determine gender identification or preference. Moreover, this biological issue is related to the ethical issue of fairness. Undeniably, however, sexual anatomy has a very large influence on preferences and behavior. In a similar vein, this book points out that we humans are substantially programmed to like or dislike much of the anatomy (i.e., the appearance) of species (especially of other humans), that this in turn substantially determines our behavior toward them, which is often unfair.

CLINICAL ZOOPHOBIA

Clinically serious phobias affect about 10% of people. Zoophobia (fear of animals) is the most common class of phobia, with a lifetime prevalence (affecting people at some time in their lives) estimated at 3.3–5.7% (Eaton et al. 2018). Animal phobias collectively have been stated to be the world's most frequent mental illness (Steel et al. 2014). Curtis et al. (1998) concluded that the most prevalent specific fears were of animals among women, and of heights among men. The animals most frequently feared are dogs, snakes, rodents, insects, and spiders, but other species are also causes (Figure 3.1). Frequently, zoophobia

FIGURE 3.1 Paintings reflecting seriously phobic hatred of certain animals. (a) Biblical depiction of Adam, Eve, and the snake in the Garden of Eden. Painting (public domain) by Italian artist Marcantonio Franceschini (1648–1729). The stereotype of snakes as untrustworthy has contributed to their bad image in Western societies. (b) 'Little Miss Muffet,' public domain illustration, by English artist Arthur Rackham (1867–1939), based on the Mother Goose story of a girl being frightened by a harmless spider. (c) Fear of a mouse. Painting (public domain) by Spanish artist Cecilio Pla (1860–1934). (d) Jonah and the Whale, public domain painting by Dutch artist Pieter Lastman (1583–1633). Fear of whales is termed cetaphobia. (e) Illustration of the Biblical Plague of Frogs in Egypt, by Sweet Media (CC BY SA 3.0), reflecting prejudice against these usually innocuous creatures.

FIGURE 3.2 Cartoon depiction of 'speciesism' in a segregated lunch counter, reflecting human preferential treatment for 'higher' animals and prejudice against 'lower' species. Prepared by B. Brookes.

develops during childhood and resolves by adulthood. It may be noted that some psychologists have contended that humans (and other animals) are born with innate fears, or at least the predisposition to acquire fear very early in life, of some apparently threatening creatures, most notably snakes (LoBue and DeLoache 2008; DeLoache and LoBue 2009; Polák et al. 2020; see Wilson's 1984 chapter 'The Serpent,' which discusses the impact of snakes on the human psyche). While it may be rational to dislike species that have harm potential, it must be remembered that, by definition, phobias are irrational fears, and so species that are the subjects of phobias are unfairly stigmatized. Unfortunately, convincing patients of particular phobias, let alone most of the public, to sympathize with such species, is difficult.

SPECIESISM

The term 'speciesism,' by analogy with racism, was coined by British psychologist R. D. Ryder in 1970 (Ryder 2010; cf. Singer 1975) to refer to prejudice against non-human species (Figure 3.2). It has been contended that humans and non-human animals are so different that any comparison of human traits, values, volition, and behaviors with those of animals is simply invalid, but this position is simply ignorant (Textbox 3.1). Many philosophers have argued that human prejudices toward animals are indeed fundamentally similar to prejudices against humans (Everett et al. 2019). The concept has been used particularly by animal rights advocates, but some have generalized it beyond

TEXTBOX 3.1 COMPARISONS OF HUMANS WITH OTHER ANIMALS ARE VALID BECAUSE WE ARE FUNDAMENTALLY PART OF THE SAME GENETIC POOL THAT DETERMINES OUR NATURE

It is a widespread notion that humans differ fundamentally from all other animals and so much that comparisons are invalid. It is also a widespread belief that somewhere in the world it is possible to find a culture where people live in harmonious, non-competitive, altruistic bliss with each other, and were it not for the existence of Western culture we would be able to achieve this ideal state. Both claims are erroneous. Humans carry an incredibly large baggage of evolutionary history, and the mere fact that our DNA sequences are similar to those of our nearest relatives among the great apes by as much as 99% makes it a highly unlikely claim that we could just step out of ape dress. Human nature is to a large extent universal. This includes certain beauty standards, and the ways in which males and females interact.

—*Grammer et al. (2003)*

sentient beings to apply to all forms of life (sometimes even to inanimate, inorganic objects such as rocks). The moral justification or rejection of speciesism has been discussed extensively in the context of philosophy and religion (Horta 2010; Ryder 2011; Horta and Albersmeier 2020).

'LOOKISM' AND DISLIKE OF UGLY NON-HUMAN SPECIES

In the human world, physical appearance is a strong determinant of how we perceive and behave toward other people. Good-looking people are comparatively successful in mating and other social interactions, and in achievements in numerous occupations where personal appearance is a factor, by comparison with unattractive individuals, often crudely termed 'losers' (Hamermesh 2011). Most obviously, individuals in the entertainment industries (especially motion pictures and television) tend to be exceptionally good-looking (although unattractive people are often assigned to play villains and menial roles). 'Lookism,' a relatively uncommon term that has nevertheless been employed for decades (Saiki et al. 2017), is negative perception and behavior toward unattractive people, and like most 'isms' is prejudicial and unfair, and increasingly the subject of human rights employment legislation (Cavico et al. 2013; Mason 2021). Racism, sexism, and ageism (discrimination on the basis of age) overlap with lookism and are allied forms of prejudice. Good looks are often unmentioned criteria for employment in occupations requiring extensive interaction with the public (Figure 3.3a–d). Attractive animals, even when they are pests, are tolerated much more than unattractive animals (Figure 3.3e and f). While we humans are obsessed with the attractiveness of our personal looks, animals in general are unconcerned (Figure 3.4). Making fun of ugly people, as exemplified by 'freak shows' of past times, is no longer acceptable, but there are many websites dedicated to identifying the world's ugliest animals, such as the naked mole rat (Figure 3.5, Textbox 3.2).

FIGURE 3.3 Looks matter. Top row: Occupations in which appearance is known to be a tacit criterion for employment. (a) Cheerleader. Photo by Palmount45 (CC BY SA 4.0). (b) Waitress. Photo by Brianfagan (CC by 2.0). (c) Stewardess. Photo by Fabio (CC BY 2.0). (d) Ceremonial guard at Buckingham Palace, London. Photo (public domain) from Pixabay.com. Bottom row: Different treatment of attractive and unattractive rodents. (e) A beautiful red squirrel being fed. Photo (public domain) by Artem Beliaikin. (f) A rat killed in a trap. Photo by Glogger (CC BY SA 4.0).

FIGURE 3.4 Self-acceptance of their appearance by animals. All figures are from Shutterstock. Contributors: (a) Vitasunny, (b) Iliukhinao777, (c) Tadienna, (d) Kejt19801, (e) Nuvolanevicata, (f) Oleg Zhevelev.

FIGURE 3.5 Naked mole rat (*Heterocephalus glaber*). Photo by TxanTxunai (CC BY SA 4.0).

TEXTBOX 3.2 DON'T JUDGE A NAKED MOLE RAT BY ITS COVER

The naked mole rat (*Heterocephalus glaber*), a quite unique burrowing East African rodent weighing approximately 35 g, is almost always included in presentations of the world's ugliest animals. Its offensive characteristics if present in humans would indicate disease, injury, extremely old age, or very abnormal development (which is why these features seem unsightly to us); however, they are all just special adaptations to underground existence. 'Beady eyes' in humans are considered unattractive, but the small eyes and corresponding poor vision of naked mole rats simply reflect a limited need for sight. The large projecting 'buck teeth' are superb instruments for digging tunnels, and the splayed toes for moving earth. The absence of all but a few hairs facilitates movement in the tunnels, as does the wrinkled skin, which allows easy motion of the torso in tight quarters. The pinkish or yellowish skin, which in humans could indicate disease, simply reflects lack of a need for skin pigments as protection against sunlight. Bowlegs are also a sign of disease in humans, but in naked mole rats, the curved legs are ideal for movement in tunnels (the animals can move backward as fast as forward). Compounding the aesthetic problems of this little beast is its depressing name (something like 'sociable desert hampster' would have been preferable). Fortunately, the naked mole rat is not considered endangered, for who would champion its conservation? But it would be extremely shortsighted to judge it (or any other animal) merely by appearance. This longest-lived rodent (up to 28 years) is mostly immune to cancer, and clarification of its genetic mechanism of resistance could be instrumental for improving human health.

PREJUDICIAL LANGUAGE AND THE DISTINCTION OF HUMANS AND NON-HUMAN ANIMALS

In biology, an 'animal' is a member of the animal 'kingdom.' In common language, an animal is frequently defined as 'one of the lower animals: a brute or beast, as distinguished from man: any creature except a human being' (Webster's Third New International Dictionary 1993). In philosophy and allied disciplines, the concern has been not so much biological relationship, but what is uniquely 'human' in the context of social, political, literary, or ethical frames of reference (Mathäs 2014). Humans generally consider themselves highly superior to all other animals, and this attitude is reinforced by derogative terms such as brute and beast. Indeed, the word animal is often employed pejoratively (Figure 3.6). Moreover, numerous animal-based terms (such as bitch, chicken, cockroach, cow, pig,

FIGURE 3.6 The Elephant Man Joseph Carey Merrick ('John Merrick,' 1862–1890), a severely deformed Englishman exhibited at a freak show as the 'Elephant Man,' was the subject of a famous 1980 film in which he uttered 'I am not an animal! I am a human being!' (a) Wax model of John Merrick, the Elephant Man, in the Museum of the Weird, Austin, Texas. Photo by Cory Doctorow (CC BY SA 2.0). (b) Model in the Brussels Museum of the Strangest. Photo by Miguel Discart (CC BY SA 2.0). (c) Cap and hood worn by Merrick to hide his deformities. Photo (public domain) by Jack 1956.

rat, skunk, and snake) are employed as slurs (Egan 2021), a phenomenon termed 'semantic derogation' (Fontecha and Catalán 2003). Plant terms are occasionally employed to characterize the merit of people (Sommer 1988; for example, peach and tomato refer to a beautiful woman, prune to an ugly woman, and nut to a crazy person). People extensively, often cruelly, exploit other animals (to say nothing of other humans), and downplaying the importance and feelings of the victims by using pejorative language seems to be a way of avoiding or at least reducing feelings of guilt. For additional information, see the discussion of slavery in Chapter 2.

COMPARATIVE PREFERENCES AND PREJUDICES OF MEN AND WOMEN FOR ANIMAL SPECIES

For simplicity, the present discussion is limited to males and females regardless of how defined (i.e., biologically or heritably determined sex vs. gender identification or preference is not addressed).

A popular nursery rhyme holds that little boys are made of frogs, snakes, snails, and puppy-dogs' tails while little girls are made of sugar and spice and everything nice. This likely reflects traditional attitudes to the roles of males and females, but there is also a biological, especially a hormonal, basis of differences in compassion or empathy between the sexes. Is it possible that there is a correlated preference or bias with respect to non-human animals? Maternal behavior in humans and indeed many other animals often is far more caring for offspring by comparison with paternal behavior. Women additionally seem to be intrinsically more sympathetic to other people than men, so they may also be more caring for animals. Indeed, although there is considerable overlap, 'Women, on average, show higher levels of positive behaviors and attitudes toward animals (e.g., attitudes toward their use, involvement in animal protection), whereas men typically have higher levels of negative attitudes and behaviors (e.g., hunting, animal abuse, less favorable attitudes toward animal protection)' (Herzog 2007).

Oxytocin facilitates childbirth and nursing for women, and is associated with increased caregiving and empathy, even toward other species (Handlin et al. 2012; Odendaal and Meintjes 2003).

There is a much greater degree of antihunting and antitrapping sentiment among women compared to men (Kellert and berry 1987). Although women are more concerned about the welfare of animals, they are more timid and reluctant to touch them (Herzog et al. 1991). Women are also more fearful of predators compared with men (Alves et al. 2014; de Pinho et al. 2014; Kaltenborn et al. 2006; Prokop and Fančovičová 2013). This is consistent with the 'parental investment hypothesis,' which postulates that since females take more care of children they need to be more sensitive to danger in order to protect them (Røskaft et al. 2003). It is also consistent with the observation, discussed previously in this chapter, that animal phobias are significantly more common among women than men.

As child-bearers, it seems plausible that women are especially favorably biased toward cute, huggable, small animals compared to men. There is ample evidence of this (Bjerke et al. 2001; Lindemann-Matthies 2005; Prokop and Tunnicliffe 2010; de Pinho et al. 2014). Conversely, men tend to be more tolerant of 'creepy-crawlies' than women (Boren 2015). Borgi and Cirulli (2015) found that fear and dislike of invertebrate animals were notably higher among young girls than boys.

Sexism may be defined as ideological belief that one sex is superior to the other, with consequent biased behavior. In practice, harmful discrimination based on sex is overwhelmingly males behaving badly toward females. Women are naturally more sensitive to sexism and aggressive sexual behavior, which by human standards is common in the animal kingdom. One might expect that a given sex would tend to be more sympathetic to the same sex of other species, and vice versa. It is clear, as discussed in this book, that humans have much more regard for mammals and birds than they do for other animals, so people are unlikely to care more about males or females of snakes, frogs, fish, and invertebrates (although female mosquitoes are far more annoying than the males). Indeed, most people would be unable to identify males and females of these groups. In mammals and birds, males and females are produced approximately in a 50:50 ratio, although there is appreciable deviation, sometimes related to environmental circumstances (Clutton-Brock 1986; Clutton-Brock and Iason 1986).

In a few species (notably some birds and frogs), the males assume all of the duties of child care after eggs are laid. In seahorses, the female deposits her eggs into the male's pouch where he fertilizes and incubates them for weeks until they emerge (an example of so-called 'male pregnancy'). However, in most species, males usually shirk parental care, and so animal mothers with their young may particularly resonate with human females (Figure 3.7). Conversely, males may be turned off just thinking about how much work would be involved if they had to assume the female's responsibilities.

In humans, sexism is exacerbated by the advantages of greater size and aggression in men, and the biological burdens of motherhood and associated maternal behavior in women. The pattern of larger males than females occurs in most vertebrates. However, females are probably larger than males in most species of invertebrates (notably in honey bees and spiders), and this is also the case in many species of fish, in some species of amphibians and reptiles, and in just a few mammals such as hyenas (Ralls 1976; Figure 3.8). Students in introductory biology classes usually find the idea that females can dominate males challenging to their assumptions.

Polygyny is a form of social polygamy in which a male maintains multiple females (often polygamy and polygyny are treated as synonyms). The reverse, domination of many males by a female, occurs primarily in social insects, such as bees, termites, and ants. Males of species such as lions, gorillas, and seals take charge of a harem of females by

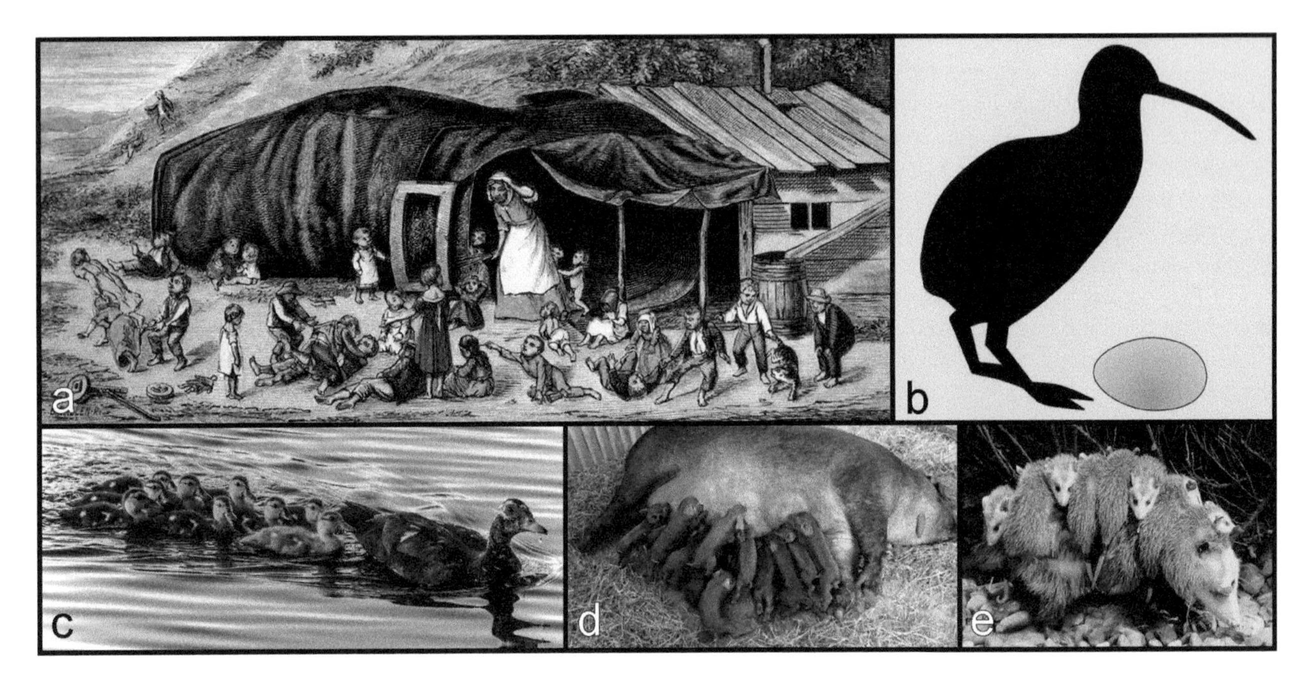

FIGURE 3.7 The burdens of motherhood. (a) Nineteenth century advertising card (public domain) illustrating an old poem: 'There was an old woman who lived in a shoe, She had so many children she didn't know what to do; She gave them some broth without any bread, She whipped them all soundly and sent them to bed.' Photo credit Boston Public Library (CC BY 2.0). (b) Comparison of an egg of kiwi (flightless bird of the genus *Apteryx* endemic to New Zealand) and the mother. The egg is one of the largest relative to the size of any mother bird, ranging up to 20% of her weight, so giving birth is especially laborious. Prepared by Shyamal (CC BY 3.0). (c) Muscovy duck (*Cairina moschata*) mother and her many ducklings. Photo by Charles Patrick Ewing (CC BY 2.0). (d) Sow and her litter of 13 piglets. Photo by Mike Ohlhausen (CC BY SA 3.0). (e) Virginia opossum (*Didelphis virginiana*) of North and Central America. This mother is carrying nine young on her back. Photo by SpecialJake (CC BY SA 3.0).

FIGURE 3.8 Examples of species in which females are much larger than males. (a) A large female golden orb-weaver spider (*Nephila pilipes*) and its tiny male suitor. Males are typically smaller than females in spiders. Photo by Graham Winterflood (CC BY SA 2.0). (b) A large female common toad (*Bufo bufo*) with a smaller male on its back. In most frogs and toads, males are smaller than females, although there is usually overlap in size. Photo (public domain) by Bernie. (c) Two small males attempting to mate with a large female garter snake (*Thamnophis sirtalis*). Females are much larger than the males in garter snake species. Photo (public domain) by Miles Frank, U.S. Fish and Wildlife Service.

besting other males in fights and excluding them from access to the females (Figure 3.9). These dominant males tend to be much larger, aggressive, and dangerous compared to the females, often possessing huge antlers, teeth, or tusks. Unfortunately for them, they attract trophy hunters. Also to their misfortune, because they are very impressive animals, they are prized as zoo and water park exhibits. Polygyny was once common in humans, and powerful men in leadership roles often maintained very large harems (Figure 3.9a). One might expect that men and women tend to differ in their attitudes to such unequal behavior in other species, the former tending to admire aggressive animals capable of acquiring harems.

In most bird species, males are more colorful or ornamented than females (Figure 3.10), and so tend to attract bird watchers and visitors to zoos and aviaries. Peacocks,

FIGURE 3.9 Male-dominant species and their harems. (a) 'Harem dancers' painting (public domain) by Italian artist Fabbi Fabbio (1861–1946). (b) A northern fur seal (*Callorhinus ursinus*) and his harem. Photo (public domain) by M. Boylan. (c) A red deer (*Cervus elaphus*) stag and his does. Photo by Carolyn Granycone (CC BY SA 2.0). (d) A rooster and his hens. Photo by Christine (CC BY SA 2.0). (e) A silverback gorilla (*Gorilla gorilla*) and his clan. Prepared by Blanca Martí de Ahumada (CC BY SA 3.0). (f) A lion and two of his lionesses. Photo by GinaFranchi (CC BY SA 4.0).

for example, notably outshine peahens. However, both sexes of pet birds tend to be equally attractive, and indeed their sex is sometimes difficult to identify.

Males of the major livestock species are usually viewed very negatively because they are more dangerous and/or less useful. Of course, bulls and roosters are necessary for breeding, but cows produce milk and chickens lay eggs. And, naturally, it is the females that produce young, and only a small number of males are necessary to inseminate a large number of females. Young animals of the unwanted sex can be eliminated simply by employing them as meat. Male sheep, cattle, goats, and pigs are routinely castrated (surgically by removal of the testes, or chemically by hormones). Castration lowers testosterone levels, reducing aggressiveness and potential harm to humans and other animals. Castration may also increase the market qualities of the meat. Male horses are often castrated (becoming geldings) to make them docile. In industrial production, different breeds of chickens are employed for meat and eggs. For egg production, males are culled as soon as the chicks are recognized (on a world basis, billions are killed annually, and there is concern about associated lack of humaneness). For meat production, so-called broiler strains are raised. Half of these are males, half females, and they are harvested before they reach sexual maturity. In the past, it was common to reserve roosters for meat production, often castrating them (to become capons) to eliminate their aggressiveness and improve the quality and quantity of meat produced. In the production of foie gras using ducks and geese, the female (instead of the male) chicks are culled because the males can put on more weight.

The situation for crop plants is only slightly comparable to animals. Well over 90% of plant species bear bisexual flowers or both male and female unisexual flowers on the same plant, and only a few, notably marijuana (*Cannabis sativa*), hop (*Humulus lupulus*, used for beer), and date palm (*Phoenix dactylifera*) are divided into male plants and female plants. Once again it is the females that are mostly grown, and when males appear, they are usually eliminated.

Male racing horses (stallions, geldings, or colts) far outnumber females (fillies). Nevertheless, the females occasionally win prestigious horse races, such as the Kentucky Derby. Champion females can be bred once a year, while the semen of champion males can be used to breed numerous offspring, and so males can be of much greater value. Once again, the preference for males is simply an economic decision. Male greyhound dogs tend to be slightly faster than the females (Entin 2007), but the sex difference is less significant than in racehorses.

Comparative preference for male and female pet dogs has been extensively explored. Preferences seem to differ internationally, but on the whole, the two sexes are acquired as pets in equal numbers. Breeders of purebred dogs sometimes charge more for female puppies, because they have the potential of producing more puppies for sale. Vodičková et al. (2019) studied adoptions of 955 dogs from a shelter in the Czech Republic and found no significant differences between men and women with respect to whether the dogs

FIGURE 3.10 Male and female of bird species in which the male is much more colorful. (a) Indian peacock and peahen (*Pavo cristatus*). Photo by Michelle Kinsey Bruns (CC BY SA 2.0). (b) Mallard ducks (*Anas platyrhynchos*). Photo by Richard Bartz (CC BY SA 2.5). (c) Mandarin ducks (*Aix galericulata*). Photo by Francis C. Franklin (CC BY SA 3.0). (d) Chinese monal (*Lophophorus lhuysi*). Source (public domain): Gould, J. 1850–1883. Birds of Asia. London, Taylor and Francis. (e) Greater bird of paradise (*Paradisaea apoda*). (f) Goldie's bird of paradise (*Paradisaea decora*). (e and f) Source: Gould, J. and W. M. Hart. 1875–1888. *The Birds of New Guinea and the Adjacent Papuan Islands*. London, Henry Southeran.

were male or female. In a survey of 877 respondents in Australia, men preferred female dogs and women preferred male dogs (King et al. 2009). In a survey of 770 respondents in Italy, men preferred male dogs and women preferred female dogs (Diverio et al. 2016). These three studies all produced different results, suggesting that if there are different preferences for male and female dogs, between men and women, they are not substantial. Miller et al. (2019) studied the adoption of over 2,600 cats from an Australian shelter and found no significant difference between rates at which male cats and female cats were adopted. Thus the most common mammalian pets apparently do not suffer generally from human sexist attitudes. However, male and female cats and dogs differ in a number of behavioral characteristics (especially in certain breeds), and many individuals gravitate much more to one sex than the other.

In all of the above examples of domesticated species, the human preferences for females or males of other species are based on economic considerations and/or behavior or appearance. The degree of consequent prejudicial treatment of one or the other sex of other species is simply not comparable to the harm associated with sexism within the human species.

In medical research, there has been a tendency by many researchers to employ male rather than female laboratory animals, principally mice and rats (Zucker and Beery 2010). The prejudice against using female animals has been justified on the contention that they are more variable than males because of cyclical reproductive hormones. Unfortunately, studies using only male animals are often significantly less applicable to human females than to males, so women do not receive equal benefit from the research. For example, pain signals transmitted in the spine have been found to differ between male and female rats, and similarly between male and female humans, and given that women report pain much more than men, research based only on male rats penalizes women (Dedek et al. 2022).

UNFAIR PUBLIC IMAGES OF SOME SPECIES

In this information age, with most people trying to present a favorable online appearance, unfair public images have become a new concern for many. However, unfair reputations and stereotypes are an old problem, and many animals are victims. So entrenched are the reputations of some

FIGURE 3.11 Views of animals that contradict their pejorative images. (a) A clean pig (by Hermandesign 2015). (b) A friendly shark (by Catalyst Labs). (c) A pleasant-smelling skunk (by Maria Bell). Source of figures: Shutterstock.

animals that when depicted in a more favorable light they appear comical (Figure 3.11).

Defamation of Species in Literature and the Media

As humans, we admire heroes but fear and despise villains. Species that exhibit behavior in nature that in human terms is deceitful, disgusting, greedy, murderous, or sly are often considered villainous. This includes aspects of the behavior of many highly intelligent carnivorous animals, notably foxes, hyenas, rats, skunks, vultures, weasels, and wolves. Plants and animals are common subjects of stories, sometimes religious, and over time some species have been portrayed as heroic, others as despicable. Species that have been portrayed positively in literature tend to be treated well subsequently (Małecki et al. 2020). However, folklore has often bestowed bad reputations on species that biologists know is undeserved (Figure 3.12). Snakes, for example, are feared or loathed in most cultures but provide essential roles in controlling rodents. Bats are also widely feared, although they too contribute to natural pest control, and most are absolutely harmless. The European magpie has the folkloric reputation of being the leading pilferer of the bird kingdom (Figure 3.13), claimed to compulsively line its nest with shiny things. However, it has been shown that the bird avoids opportunities to steal new objects it encounters (Shephard et al. 2014). Sharks have suffered more than any other animals as a result of their fearsome image, and the public continues to be terrified by frequent, gory reports of attacks (although lightning strikes people about 50 times as often).

Bugs: Unfair Stereotyping of the Majority of Living Things

Chapter 15 deals in detail with 'bugs,' and only some general remarks will be made here. 'True bugs' are a group of perhaps 80,000 species of insects called Hemiptera, most of which feed on plants, but it also includes the bedbugs. However, the colloquial word bug is much more comprehensive, usually including not just insects but also spiders and numerous other small, usually terrestrial invertebrates, especially 'creepy' pests. Often the word bug is applied to diseases caused by microorganisms ('I caught a bug') and to computer viruses and is often employed as a pejorative catch-all term (e.g., as a fault in a machine). Calling something a bug is not a compliment, and because the word is applicable to most of the world's species, the very word conveys contempt for the majority of living things. Relentless advertising for eliminating bugs from residences and gardens serves to reinforce this hostility. Insects, the largest group of species, are incredibly diverse (Figure 3.14) and mostly harmless and beneficial, although often depicted as threatening (Figure 3.15). Rehabilitating the reputations of bugs is a formidable undertaking, but, inasmuch as most are indispensable for human welfare, it is important to persuade the public to be more tolerant.

E. O. Wilson (1929–2021), one of the world's leading entomologists, championed the view that humans innately love animals (Wilson 1984, 1993), although in fact most people strongly dislike insects (see citations in Fukano and Soga 2021; Joye and De Block 2011 provided an extensive rejection of the concept). Many insects are passionately despised (Figure 3.16), and it is challenging to encourage respect for them (Simaika and Samways 2018). In Chapter 15, see the quotation from Lockwood (2013), suggesting additional reasons why people may be especially fearful of insects. Fukano and Soga (2021) provided three reasons why dislike of insects matters: (1) Since insects make up most animal biodiversity, their negative image is a major impediment to global conservation. (2) Insects are major triggers for specific animal phobias, and reducing fear of them could decrease associated psychological, social, and economic burdens. (3) Insects represent a huge, environmentally sustainable potential human food source, but entomophagy (consumption of insects) is rare because of the negative perception of insects.

FIGURE 3.12 Examples of vilification of animals in literature. (a) The fox as an untrustworthy and deceitful. This figure shows the classic Aesop fairy tale of a fox tricking a crow to drop a piece of cheese. Modified from a 1908 public domain figure in the journal Punch by J. S. Pughe, from U.S. Library of Congress. (b) A princess contemplating a frog. The familiar story reflects the traditional stereotype that frogs and toads are as so ugly and disgusting that they can be admired only if transformed into a handsome prince or princess. Illustration (public domain) of 'The Frog Prince' by Anne Anderson. Source: http://www.artsycraftsy.com/anderson_prints.html. (c) The wolf as a vicious predator of humans. Illustration of *Little Red Riding Hood* by Jessie Willcox Smith from the 1911 book (public domain) *A Child's Book of Stories*. Source: https://socialistreadinggroup.wordpress.com/2018/07/08/little-red-cap-and-brier-rose/. (d) Lurid cover of a Moby Dick novel, portraying the whale as a dangerous animal hostile to humans. Source: https://commons.wikimedia.org/wiki/File:Moby_Dick_for_Wikicommons.jpg (CC BY 4.0). (e) The ostrich as an animal so dumb that it buries its head in the sand to avoid predators – a myth that has been widely accepted. Source (public domain): openclipart.org. (f) The snake as treacherous. 'The Fall of Man,' showing the biblical depiction of a serpent betraying mankind in the Garden of Eden. Public domain painting by German artist Lucas Cranach the Younger (1515–1586).

Vermin: Unfair Stereotyping of Many Invaluable Animals

'Vermin' denotes harmful or pest animals, especially (but not restricted to) pests of crops and livestock and disease-transmitting pests of cities. This often includes harmful 'bugs,' small and medium-sized mammals, and birds (Figure 3.17). Before humans evolved, all of these occupied relatively restricted habitats and natural distribution ranges, where they survived on local plants and/or animals. When humans created urban and agricultural areas, the species found them to their liking. Some of these pests originated in foreign locations and followed civilization to many areas, but often native species just added local farms and residences to their shopping sites. The degree of damage to human interests varies enormously, depending on species, and in some cases, their net effect is actually neutral or beneficial. Chapters 16 and 17 analyze the most serious vermin species in detail.

FIGURE 3.13 The European magpie (*Pica pica*) has an undeserved reputation for stealing valuable objects, reinforced by Gioachino Rossini's 1817 opera *The Thieving Magpie*. The plot is based on a maid who almost goes to the gallows for stealing silver, before it is discovered that the culprit was a magpie. The public domain images shown here are based on the story. (a) 'A Thief of a Magpie,' painted by English artist William James Webbe (1830–1904). Photo by Irina (CC BY 2.0). (b) An educational late 19th-century card. (c) Source: Pocock, I., L. C. Caigniez, and M. d'Aubigny. 1815. *The Magpie or the Maid?* London, John Miller.

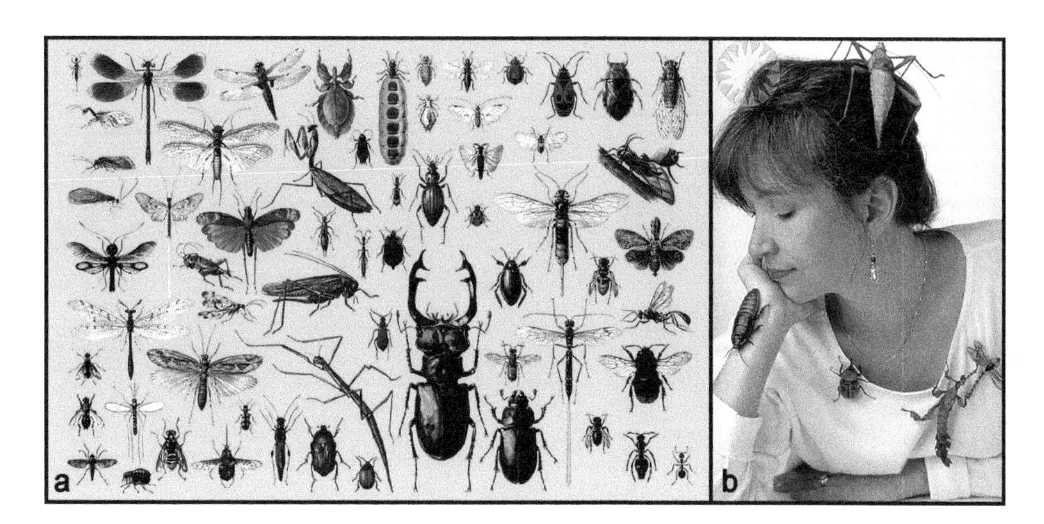

FIGURE 3.14 Insects. (a) Variation among insects. Source (public domain): *Der Neue Brockhaus, vol. 2* (1937). F. A. Brockhaus, Leipzig. Image scanned by Julo. Identifications are available at https://commons.wikimedia.org/wiki/File:Insects_in_Brockhaus_1937.jpg. (b) Sally Love, former director of the Insect Zoo at the U.S. National Museum of Natural History, wearing a selection of living insects. Credit: Laurie Minor-Penland, Smithsonian Institution (CC BY 2.0).

FIGURE 3.15 Some pest insects shown as extremely threatening. Source of images: Shutterstock. (a) Cockroach (contributor: ayelet-keshet). (b) Hornet (contributor: Teguh Mujiono). (c) Mosquito (contributor: Teguh Mujiono).

FIGURE 3.16 A 1910 postcard (public domain) showing a woman contemptuously kicking a mosquito.

FIGURE 3.17 Examples of 'vermin.' (a) American beaver (*Castor canadensis*). Photo by Steve (CC BY SA 20). (b) Groundhog (*Marmota monax*). Photo (public domain) by N. Lewis, U.S. National Park Service. (c) Pigeon (rock pigeon, *Columba livia*). Photo by Diego Delso (CC BY SA 3.0). (d) Coyote (*Canis latrans*). Photo by Rebecca Richardson (CC BY 2.0).

Weeds: Unfair Stereotyping of Many Invaluable Plants

Weeds have been defined simply as plants growing where they aren't wanted, but serious weeds cause considerable economic damage to agriculture (Figure 3.18b), urban landscapes (Figure 3.18a), and natural ecosystems (Figure 3.18c). Like vermin, described above, before humans altered nature, weeds were confined to natural habitats but found the changed environments caused by humans to be to their liking. All weeds have very significant redeeming values, which sometimes more than compensates for the damage that they cause. The worst weeds are examined in Chapters 18–21.

The War on 'Narcotic' Drug Species: Unfair Persecution of Medicinal and Spiritual Plants and Animals

The 'War on Drugs' refers to a U.S.-led global campaign started in 1971 against psychoactive drugs considered dangerous and designated as illegal. The initiative has been marked by extremely vigorous military, police, and incarceration campaigns. At various times in the past, and sometimes to the present, individuals convicted of criminal offences involving intoxicant drug use have been treated very harshly, even put to death. Large numbers from racial minorities have been prejudicially jailed. Nevertheless, illicit drug industries (often based on chemical synthesis and not on extraction from species) supplying recreational black markets have grown enormously, generating great suffering and economic losses, and widespread condemnation. Inevitably, the reputations of species with intoxicant potential have suffered, and this penalizes the search for legitimate applications (Small 2004). Fortunately, progressively more humane approaches are being adopted. Increasingly, education and treatment are receiving more support than legal prosecution, and the unfair reputation of some drug species is being rehabilitated.

Numerous species contain drugs that affect human behavior, consciousness, cognition, mood, or perception, and many of these have been employed deliberately to induce pleasure,

FIGURE 3.18 Damaging weeds. (a) Damage to an urban area. A house overrun by weeds. Photo by Infrogmation of New Orleans (CC BY SA 4.0). (b) Damage to agriculture. Manual weeding of an agricultural field in India. Photo by Deeptrivia (CC BY SA 3.0). (c) Damage to a natural ecosystem. Purple loosestrife (*Lythrum salicaria*), an invasive Old Word alien that has taken over a natural marsh in North America, causing considerable damage to local plants and animals. Photo by Liz West (CC BY 2.0).

reduce pain, or produce spiritual trances. Such uses can be very beneficial or extremely harmful. Frequently when employed illegally or when dangerously addictive, the drugs are termed hallucinogens, intoxicants, narcotics, or psychedelics, and when used spiritually, they are called entheogens (see Small 2016b for a presentation of psychopharmacological terminology; the general term 'intoxicant' is employed here, and 'psychoactive' is also often encountered). Species with consciousness-altering chemicals generally employ these as toxins to repel other species from harming them. In practice, plants furnish most mind-altering preparations used by people, but some fungi and animals are also sources (Schultes and Hofmann 1979; Orsolini et al. 2018; Figure 3.19). In Chapter 18, the most dangerous addictive plants are discussed in detail. Often dating back thousands of years, in most of the world people have adopted local mind-modifying species, mostly for use in spiritual ceremonies, but also as pain relievers and recreational drugs. Alcoholic beverages, which are mostly based on yeast digesting the sugars in plants, represent humankind's oldest and most familiar drug, and also illustrate the ambiguity with which intoxicating drugs have been regarded. Some religions accept the use of alcohol, others reject this, but everyone recognizes that there are both pleasurable and adverse potentials. Drugs from some plants like opium poppy have been both enormously harmful when employed recreationally and hugely helpful medically. Sometimes a drug species accepted in the past as harmless has proved otherwise, and vice versa, and further research is likely to reveal both useful and harmful aspects of other narcotic species.

TABOO FOOD SPECIES

Taboos are social prohibitions, usually unwritten, against specific behaviors. They are often associated with religions or cultures. Some taboos are obviously meant to discourage harmful behavior or maintain respect for customary values, and other taboos have no obvious purpose. Many taboos affect how humans view and treat species. Species that have positive religious importance – notably sacred cows – are admired and often protected (see Chapter 7). Numerous taboos are concerned with species employed as food. The only near-universal prohibition against eating species seems to be against consuming human flesh. Almost all human societies and most religions specify that certain species (mostly animals but sometimes plants) are unfit for human consumption. In Judaism, for example, camels, hares, pigs, bears, shellfish, eagles, vultures, falcons, frogs, ravens, owls, most insects (except some locusts), and others are 'unclean' (although by tradition all animals on Noah's Ark were considered to be available for food). Islam prohibits all predatory terrestrial animals. Most Hindus, especially Brahmins, are vegetarians, completely abstaining from eating meat, while those who do will not consume beef. Pork is the world's most widely eaten meat, but consuming pigs is forbidden in Islam, Judaism, and some Christian denominations, such as Seventh-day Adventists. In Western culture, dogs and cats are pets, and as such considered inappropriate as human food, but they are commonly consumed in Asia.

There is debate among anthropologists regarding the rationale, if any, of particular taboos (Meyer-Rochow 2009). One of the most curious taboos was practiced by the Pythagoreans of antiquity – the followers of the sixth-century BCE Greek philosopher. They abstained from eating the common fava or faba bean (*Vicia faba*), a dietary staple of the time. It has been speculated that this was due to susceptibility to favism, common in the Mediterranean region. This is an inherited potentially deadly enzyme deficiency (not an allergy) triggered by fava bean that results in blood destruction. It has been proposed that some taboos arose to promote the protection or efficient utilization of a resource (Meyer-Rochow 2009). A taboo that prevents harvesting of a wild species represents an effective means of protecting it. Ironically, such protected species may be regarded contemptuously as unworthy. Conversely, a strong preference for certain wild species endangers them, although it also makes them more likely to be domesticated. Plants and animals that represent health threats or simply taste bad are sometimes among taboo foods. Vultures, for example, often can carry diseases from the carrion they consume and are avoided by humans for this reason.

FIGURE 3.19 Intoxicant (psychoactive) species. (a) Colorado River toad (*Incilius alvarius*). This native of northern Mexico and the southwestern United States exudes toxic hallucinogenic compounds from its parotid glands (located on the back, neck, and shoulder). Its toxin can kill dogs but nevertheless has been employed as a street drug. Photo by kuhnmi (CC BY 2.0). (b) Psilocybin mushrooms (*Psilocybe weilii*), also called magic mushrooms and shrooms. Several genera of fungi are called psilocybin mushrooms. Psilocybin, the main chemical of most hallucinogenic mushrooms, has been employed in indigenous New World spiritual ceremonies and as a street drug. Photo (public domain) by GumbyDude. (c) Fly agaric (*Amanita muscaria*). This common mushroom's active agents are muscimol and ibotenic acid. It also has been used as a spiritual and recreational hallucinogen. Photo by Thomas Pruß (CC BY SA 3.0). (d) Marijuana plant (*Cannabis sativa*). Photo by Plantlady223 (CC BY SA 4.0). (e) Cocaine shrub (*Erythroxylum coca*). In its native South America, the leaves are commonly chewed for energy (similar to how coffee is consumed), but extracted cocaine is misused in Western countries and is much more dangerous. Photo by Darina (CC BY SA 3.0). (f) Peyote (*Lophophora williamsii*). This native of Mexico and southwestern Texas is a small, spineless cactus with psychoactive alkaloids, particularly mescaline. It has been used by indigenous North Americans as a spiritual drug for centuries. Photo by Maja Dumat (CC BY 2.0).

SUPERSTITIOUS BELIEFS CONCERNING SPECIES

There are innumerable superstitions concerning plants and animals, many negative and inducing fear, many others positive and suggesting benefits. (See the introduction to Chapter 7 for references on both entirely mythological species and real species with alleged magical properties.) For the most part, the superstitions are entirely irrational, not based on fact, but sometimes they trace to some evidence (for example, that a plant employed medicinally actually has curative properties). Regardless, species often unjustly acquire reputations, which amount to prejudice. Discouraging negative superstitions that endanger vulnerable species is important for their conservation (Holmes et al. 2018). On the positive side, some species and ecologically important sites have religious or spiritual values and are accordingly protected (as discussed in Chapter 7).

Example 3.1: British Fairy Trees

In many parts of the world, trees acquire magical reputations. The British Isles are notorious for folklore, and one of the most interesting entrenched beliefs concerns hawthorns (species of *Crataegus*), which typically are small, thorny trees (Figure 3.20), or shrubs which produce small, apple-like fruit. Specimens that stand isolated in a field are often considered to be 'fairy trees,' reputedly the homesites of fairies or elves (Brunsdon 2019). Such trees can bring good luck, and so people often leave offerings at their base, or in their branches. A frequent practice is to tie ribbons or colorful rags to the branches (Figure 3.20c) while making wishes for improved health or for blessings, in which case the trees are often called 'wishing trees.' However, harming such trees is thought to invite disaster and bad luck. In 1968, a road had to be realigned to pass a hawthorn that was on a new road being built from Ballintra to Rossnowlaught. In 1982, workers in the De Lorean car plant in Northern Ireland protested that the site was cursed because a hawthorn had been destroyed during construction of the plant. To calm everyone, management ceremoniously planted another fairy tree on the property. On the grounds of Belfast University, a center of level-headed modern thought, people refused to trim a hawthorn for fear of offending the fairies, and it is said that when new buildings were planned, they had to be placed so that they would not disturb the tree. Hawthorns, of course, benefit from the superstitious fear of harming them.

Example 3.2: The Aye-Aye

The aye-aye (*Daubentonia madagascariensis*; Figure 3.21) is a highly endangered lemur, indigenous to Madagascar. Adults weigh about 2 kg and have a head+body length of about 40 cm and a tail about 60 cm long. This primate taps rapidly along the surfaces of dead trees, listening to reverberations to identify wood-boring insects and larvae inside the trunks. It then uses its rodent-like teeth to create holes and then employs a special long slim hooked middle finger to pull grubs out. In its native area, the species is often considered to be a harbinger of evil and death and is frequently killed on sight (Simons and Meyers 2001). Some believe it employs its middle finger to kill people. In fact, the aye-aye is quite harmless to humans, so it must be its quite strange appearance that led to the pronounced bias against it. Despite its bad reputation, there is some support for its conservation in Madagascar (Randimbiharinirina et al. 2021).

FIGURE 3.20 British hawthorn fairy trees. (a) A hawthorn (*Crataegus monogyna*) in flower in England. Photo by Nilfanion (CC BY SA 3.0). (b) A vintage illustration of a hawthorn fairy tree in fruit. Source (public domain): Pixabay.com. (c) Ribbons attached to a hawthorn fairy tree in England, to make wishes, traditionally for cures to illness. Photo by Jim Champion (CCC BY SA 4.0).

FIGURE 3.21 An aye-aye (*Daubentonia madagascariensis*). Photo by Klaus Rassinger, Museum Wiesbaden (CC BY SA 4.0).

4 Size, the Most Important Determinant of Human Prejudice against and Preference for Species

INTRODUCTION

The size of organisms is determined by physical, environmental, ecological, and genetic constraints (LaBarbera 1989; Calder 1996). Species with smaller body sizes tend to be much more abundant than species with larger bodies, but the general relationship between frequency and size has proven difficult to assess (Ulrich 2008). The concern here is not why some creatures are big or little, but how the perception of size of other species affects human judgments of their merit.

The most important physical characteristic that humans admire in other species is size (examples are shown in Figure 4.1). Like the proverbial 800 pound gorilla, huge creatures, whether animal or plant, always get respect. Visit any museum of natural history, and the exhibits that attract the most fascination are the largest dinosaurs. As a rule, the biggest animals are also the star attractions in zoos and animal preserves (Frynta et al. 2013; Moss and Esson 2010; Colléony et al. 2017; Skibins et al. 2017), Ward et al. (1998) found that preference for exhibits of larger animals in zoos increases directly with size. The largest animals are also the subjects of extensive scientific study (Rosenthal et al. 2017), and of considerable media coverage, for example, dominating the covers of nature magazines (Clucas et al. 2008). Mythological giant humans, non-human animals, and plants also fascinate people and reflect the considerable impression they make on the human psyche (Figure 4.2).

Larger size is correlated with brute strength, which enables larger individuals to more easily alter their physical environment to suit their needs. More importantly, bigger organisms can usually exploit or outcompete smaller competitors – both other species and other individuals in the same species, as illustrated in Figure 4.3. In infancy, we are confronted by our parents and other adults, all of whom appear to be enormous, and indeed who exercise great control over our lives. As we mature, we become aware that the size of others, particularly their height, remains an important determinant of respect from others. Although intelligence is currently the key determinant of leadership, in more primitive times the biggest (presumably the most powerful and aggressive) male was likely the leader of human groups, as is the case in numerous social animals. Taller men have more reproductive success (Pawlowski et al. 2000) and indeed benefit from many of the advantages described in Chapter 5 for attractive people.

Large size of both humans and animals conveys the power that can facilitate law and order (Figure 4.4). Large men are commonly employed for security, especially as bouncers and bodyguards. (Unfortunately, criminals similarly employ large men as enforcers to extract resources from victims.) Large police officers and personnel are valuable for crowd control, as are large horses. The sight of either or both is usually sufficient to inhibit trouble-makers. The presence of large dogs also tends to reduce criminal activity.

There are many techniques and tools that people use to visually increase their height: sitting on a high throne or a high object (note how often royalty is shown atop horses), high heels, elevator shoes, tall hats (Figure 4.5), bouffant hair, and clothing with vertical stripes. In a similar vein, four-footed animals sometimes stand up on two legs, the tallness signaling dominance.

THE CHARISMATIC MEGAFAUNA (ATTRACTIVE LARGE ANIMALS)

The 'charismatic megafauna,' by definition, are large (often 'majestic'), glamorous animals that attract public attention. The species are usually at least the size of a large dog, and generally bigger than a large man (Figure 4.6). Most members of the charismatic megafauna are very photogenic (most notably the giant panda), although attractiveness and charisma are somewhat subjective, and the basis of 'attraction' could be horrible or bizarre appearance or terrifying behavior. Most species in this select group are mammals. The endangered blue whale (*Balaenoptera musculus*), which may weigh 180 tons and grow to 30 m in length, is the largest animal ever to evolve on Earth. Whales now epitomize the cause of conservation of sea life. Other mammals of the charismatic megafauna include bears, elephants, dolphins, giraffes, great apes, lions, rhinoceroses, seals, tigers, wolves, and zebras, giving rise to the criticism that humans are 'mammal-centric.' However, some of the largest birds (including ostriches, condors, eagles, penguins, and whooping cranes), fish (sharks, rays), and reptiles (cobras, pythons, crocodiles, alligators, tortoises, Komodo dragon) are also considered to be members of the charismatic megafauna. If *Tyrannosaurus rex* were not extinct, it would certainly qualify as a member, as indeed would many extinct huge creatures. Cephalopods are a class of mollusks (which include snails and bivalves), the biggest

DOI: 10.1201/9781003412946-4

FIGURE 4.1 Examples of large individuals and large species commanding respect and admiration. (a) Hand-colored etching (public domain) 'The Surprising Irish Giant of St. James's Street' by British artist Thomas Rowlandson (1757–1827), housed in the Metropolitan Museum of Art, New York. (b) William Bradley (1787–1820), the Yorkshire Giant, claimed to be the tallest recorded British man that ever lived, measuring 2.36 m in height. Published in London by S. W. Fores in 1810. Credit: Wellcome collection (CC BY 4.0). (c) Dinosaur exhibit in California. Photo by Catchpenny (CC BY ND 2.0). (d) Another dinosaur exhibit in California. Photo by Loren Javier (CC BY ND 2.0). (e) Sperm whale (*Physeter macrocephalus*), the largest toothed predator. Source: Craig, H. ed. 1880. Johnson's Household Book of Nature, New York, H. J. Johnson. Credit: Biodiversity Heritage Library (CC BY 2.0). (f) Big Bingo, a giant two-story high elephant exhibited by Ringling Bros. Lithograph published by The Strobridge Litho. Co., Cincinnati & New York in 1916. Credit: Boston Public Library (CC BY 2.0). (g) A giant pumpkin. Record-setting pumpkins weigh well in excess of 1 t. Photo by Mike Mozart (CC BY 2.0). (h) Giant sequoia (*Sequoiadendron giganteum*), the world's largest tree species in terms of mass (up to 1900 tons). Source (public domain): Flore des serres et des jardins de l'Europe, vol. 9 (1853–1854), Gand, Belgium, L. van Houtte.

FIGURE 4.2 Famous mythological giants. (a) King Kong, from a 1933 Swedish movie poster (public domain) published by J Olséns Litografiska Anstalt, Stockholm. (b) A statue of the Jolly Green Giant (mascot of B&G Food), 17 m tall, in Blue Earth, Minnesota. Photo by Wallace Parry (CC BY SA 3.0). (c) Depiction of the giant and the giant beanstalk in the fairy tale 'Jack and the beanstalk.' Source: Shutterstock (contributor: Delcarmat). (d) The legendary giant lumberjack Paul Bunyan and his giant blue ox 'Babe.' Source: Shutterstock (contributor: John Kaestner). (e) The giant Goliath, defeated by David in the Bible story. Source: Shutterstock (contributor: Anton Brand).

(and most intelligent) of which are octopuses and squids (Figure 4.7). These largest invertebrates should be included in the charismatic megafauna but are not often thought of in this way.

Substantial size of animals can be disadvantageous, as many large species have been overharvested, and some that have been competitive with people have been reduced or eliminated. By definition, 'big-game' species are large animals that merit being hunted, and usually also being made into trophies. Most large animals have been hunted – as sources of meat and other products, for sport, or because they were thought to be dangerous or detrimental – and if not critically endangered in the wild, their populations are being alarmingly reduced. Large animals are also powerful,

FIGURE 4.3 Examples of advantages of larger size. All figures are from Shutterstock. (a) Furniture mover, an occupation in which size and associate strength is advantageous (contributor: Vector Vision). (b) A bully using his size and strength to intimidate and extract concessions from an associate (contributor: Pretty Vectors). (c) An athletic contest, in which the larger man is advantaged by his size (contributor: Pretty Vectors). (d) Bigger fish eating smaller ones, reflecting the advantages of size to predators (contributor: Peter Hermes Furian). (e) A big fat cat, metaphorically representing a predatory businessman who has exploited smaller, more vulnerable individuals, represented by the small mouse (contributor: IDraw). (f) A huge draft horse and a much smaller pony, the large size of the former required to move very heavy loads (contributor: Sammy33). (g) Cartoon of a vintage, horse-drawn double-decker bus, requiring large powerful draft animals (contributor: Iralu).

and so subject to being enslaved as beasts of burden or as exhibits for entertainment (Figure 4.8).

GIANT ANIMAL CLONES

Some social animals often collect together into impressively large groups even when the individual is quite small (e.g., a hive of bees, a swarm of grasshoppers, and a colony of ants). However, the individuals of such species are separate, easily identified as such, and if they are naturally small, they are not notable for impressive size (unlike elephants, which are not only huge but also live in large groups). In contrast, some simple invertebrate animals (often sessile, i.e., attached to a site) adhere to each other, and as the number of individuals increases (often predominantly asexually), the collective colony in some species appears to grow like a bush or tree (and so often appear plant-like), or in other strange forms (Jackson and

FIGURE 4.4 Examples of size contributing to security. Illustrations from Shutterstock. (a) A big bouncer responsible for crowd control at a night club (contributor: Filthydanus). (b) A king and his big bodyguards (contributor: Askib). (c) A London, England policewoman, whose stature and apparent authority are greatly increased by riding a large horse (contributor: Elmm). (d) A big guard dog providing security for property (contributor: ONYXprj).

Coates 1986; Simpson 2021). Such clones can grow to large sizes, appear collectively to be single individuals, and so are given far more respect than the tiny component organisms. The champion in this respect is coral reefs. Coral animals secrete external skeletons of calcium carbonate which adhere like bricks in a wall, and the inert mineral forms most of the reef (Figure 4.9).

THE CHARISMATIC MEGAFLORA (ATTRACTIVE GIANT TREES)

By analogy with the charismatic megafauna, a 'charismatic megaflora' is recognizable (Figure 4.10). As with the animals, they are from a select taxonomic group – vascular plants (flowering plants and conifers) – although many 'algae,' especially the 'seaweeds,' also attain huge size. However, seaweeds by definition are submersed, and so apparent only to divers. The most obvious large plants are species of trees that grow to gigantic size (height and/or girth), and because they are (or were) sources of timber, and deforestation is a worldwide issue, such species are frequently the focus of conservation campaigns. Very large trees can evoke astonishingly protective feelings in people, as evidenced by the opposition that develops when it becomes known that land developers or city planners intend to cut down very old ('old growth') trees or groves within a neighborhood. Of the approximately 60,000 species of trees in the world (Beech et al. 2017), the giants such as the redwoods of California have a special status that is well suited to convincing the public and politicians of the value of conservation. The respect that huge trees are afforded is ironic, since they are mostly dead wood (just as coral reefs are mostly inert minerals).

AMBIVALENCE TOWARD TREES

By their sheer size, trees naturally dominate much of the terrestrial world. Dense growth of trees characterizes forests

FIGURE 4.5 Tall headgear employed to heighten the visual authority of powerful men. (a) Painting of early 19th-century uniforms of the Russian Pavlovsky Grenadier Regiment. The higher status of the officers is indicated by their taller hats. Source (public domain): Charlemagne, A. J. 1890. History of the Pavlovsky Life Guards Regiment 1790–1890. (b) Pope Benedict XVI in 2008. Photo by Rvin88 (CC BY 3.0). (c) Painting (public domain) of a member of the British parliament, Fitzstephen French, by English artist George Hayter (1792–1871). Credit: Museum of Fine Arts, Houston. (d) Toyotomi Hideyoshi (1537–1598), a Japanese samurai who was one of the most influential men in Japan. He was known for his unique helmet. Public domain painting by an unknown artist. Source: https://commons.wikimedia.org/wiki/File:Toyotomi_Hideyoshi_on_his_horse.jpg. (e) Portrait (public domain) of field-marshal Manuel Inácio Martins Pamplona Corte Real (1762–1830), Portuguese nobleman and politician. Painted in the 19th century by an unknown artist. Source: Cabral Moncada Leilões – Leilão 1228 (Leilão Online de Antiguidades e Objectos de Decoração). (f) U.S. military recruitment poster (public domain) of Uncle Sam, painted by James Montgomery Flagg in 1916–1917.

and jungles. 'Open habitats' are places lacking dense tree cover, such as plains, tundra, polar barrens, grasslands, and deserts. Our primate ancestors were adapted to living in and eating the fruits of trees, but humans thrive in open habitats. We can't climb nearly as well as monkeys and apes, but we are much superior walkers and runners. Any continuous tracts of tall vegetation can hide predators and make movement difficult. Hart and Sussman (2008) hypothesized that the human predilection for 'beautiful scenery' evolved from our ancestors scanning the African grasslands for danger. So it seems logical that humans naturally prefer open landscapes, and accordingly perhaps are more comfortable with animals that are easily seen because they are not hidden among trees. Although giant cities are sometimes called 'concrete jungles,' in fact they are designed to facilitate movement while keeping people relatively safe. Although the trees in cities are sometimes called 'urban forests,' in fact their location is strictly regulated to not interfere with human activities. One of the first concerns of many settlers in the past was simply to clearcut an area so they could establish farms. While people today genuinely love trees, they have limited tolerance for the damage and danger they pose when they are too close to a residence. Unfortunately, the forests and jungles of much of the world are being converted to agricultural lands, which are more profitable.

FIGURE 4.6 Some prominent members of the charismatic megafauna, painted by T. Knepp, U.S. Fish and Wildlife Service (public domain images). (a) African elephant (*Loxodonta africana*). (b) Bengal tiger (*Panthera tigris tigris*). (c) Black rhinoceros (*Diceros bicornis*). (d) Great hammerhead shark (*Sphyrna mokarran*).

FIGURE 4.7 Giant cephalopods. (a) Colossal squid (*Mesonychoteuthis hamiltoni*). Invertebrate animals, which constitute the majority of known species, are generally very small. A striking exception is the colossal squid, the largest known invertebrate. A specimen collected in Antarctic waters in 2007 was 10 m long and weighed 450 kg, but it is thought that much larger animals exist. The size of the human for scale has been disputed. Figure by Citron (CC BY 3.0). (b) A mythological 'kraken' (giant octopus-like cephalopod) attacking a ship. Source (public domain): Pierre Dénys de Montfort, 1801. (c) A 'giant squid' fighting with sailors. Source (public domain): N. C. Wyeth (1882–1945) in The Adventure of the Giant Squid, 1939.

FIGURE 4.8 Examples of human domination of large animals for transportation and entertainment. (a) Bull riding at the Calgary Stampede. Photo by Chuck Szmurlo (CC BY SA 3.0). (b) Lion taming. Photo by HaleyDara (CC BY ND 2.0). (c) Girl on horse. Photo by Patrick Doheny (CC BY 2.0). (d) Boy on elephant. Photo by Sheikh Rafi (CC BY 2.0). (e) Trainer standing on 'Shamu the killer whale.' Photo (2009) by Milan Boers (CC BY 2.0). (f) Dolphin show in Valencia, Spain. Photo by Javier Yaya Tur (CC BY 2.0).

FIGURE 4.9 Coral reefs. These are constructed underwater mostly by colonies of corals, which are tiny colonial animals (usually no larger than 1.5 cm in diameter) held together by their surrounding calcium carbonate shells (exoskeletons). The mineral makes up the stony framework of reef systems, which command human respect because they can be gigantic (the largest, the Great Barrier Reef, extends for over 2,300 km, covering 345,000 km^2 on the northeastern coast of Australia). Coral reefs are home to at least one-quarter of marine species and are the source of considerable harvested natural resources and ecosystem services. The reefs are under grave threat because habitat destruction, particularly from climate change, is destroying them. (a) Great Barrier Reef coral species. Source (public domain): Saville-Kent, W. 1893. The Great Barrier Reef of Australia. London, Riddle and Couchman. Photo credit: Rawpixel (CC by SA 4.0). (b) A coral reef ecosystem. Photo by Pedro Fernandes (CC BY 2.0).

FIGURE 4.10 Some prominent members of the charismatic megaflora (attractive giant trees). (a) African baobab (*Adansonia digitata*), considered to be the world's largest tree species in terms of individual trunk girth (circumference up to 16 m). Photo by F. Reus (CC BY SA 2.0). (b) 'The Hundred-Horse Chestnut' painted by Jean-Pierre Louis Laurent Houël (1735–1813). This is the largest and oldest known chestnut (*Castanea sativa*) in the world. Located on the eastern slope of Mount Etna in Sicily, it is 2,000 to 4,000 years old. Guinness World Records listed it for the record of 'Greatest Tree Girth,' with a circumference of 57.9 m when measured in 1780. Above ground, the tree has split into multiple large trunks. The tree's name originated from a legend in which a queen of Aragon and her company of 100 knights supposedly took shelter under the tree. (c) Saguaro cactus (*Carnegiea gigantea*) in Saguaro National Monument, Arizona. This native of the Sonoran Desert of extreme southeastern California, southern Arizona, and adjoining northwestern Mexico is the largest cactus in the United States, and probably the most recognized cactus in the world, commonly shown as background in Wild West movies. The plant grows to a height of 5–20 m, but very slowly, often only 2.5 cm in a year. Photo (public domain) from U.S. Department of Agriculture, Forest Service.

RELATIVE UNATTRACTIVENESS OF SMALL CREATURES

Prejudice against small size in humans has been termed sizeism (sizism), but this term is also applied to prejudice against fat people. Prejudice against short people is called heightism and is considerable for males (Ulinski 2018). People have comparable prejudices against small animals.

Large animals require large amounts of energy, but there simply isn't enough energy in many environments to support a large number of large individuals, and conversely, it is much easier for environments to support large numbers of small individuals (van Valen 1972). In nature, most organisms are small, and typically habitats and ecosystems support progressively larger numbers of smaller individuals (or smaller species). As observed by Hutchinson and MacArthur (1959), this relationship usually holds not just collectively for the entire world, but in large geographical areas (a relationship termed the body size-species richness distribution pattern). A related consideration is that the world's living biomass of animals is predominantly in small species, which are therefore of great importance.

However, the relative appeal of small species is a different issue. Conservation scientists have difficulties in obtaining research support for most small invertebrates. However attractive or important species are, if they are not readily visible it is difficult for most people to sympathize with their need for conservation. As a practical matter, it is much easier to demonstrate that funds dedicated to the preservation or restoration of large species have *visibly* led to success than is the case for small creatures. Most small creatures are invertebrates, but larger invertebrates such as lobsters and cephalopods do get funding for conservation (Cardoso et al. 2012).

Microorganisms by definition are tiny. These include bacteria, protozoa, most 'algae,' and most 'fungi.' Viruses, which are often not considered to be 'alive,' are sometimes included. However, some fungi (notably 'mushrooms') and some algae (notably 'seaweeds') are quite large. There has been extremely limited interest in preventing the extinction of microorganisms. However, efforts are underway to stimulate the conservation community to pay attention to fungi that are endangered (Mueller et al. 2014), even the pathogenic species, which are potential sources of pharmaceuticals (May 2005).

Domesticated small animals (dogs, cats, hamsters, etc.) are very appealing, but their features have been selected by humans. Most urban wildlife are small animals, and most are considered to be pests. The main exceptions are birds, and the popularity of birdwatching is significant, although most species have not risen to the iconic status of the charismatic large mammals. The hobby culture of small aquarium fish is also very popular. There are animal fanciers who like to share their living quarters with pet turtles, lizards, snakes, crocodilians, and frogs, but they are a minority. Most really small animals are insects, and on the whole, these have very little visual appeal (except of course to entomologists and naturalists). The conspicuous exception is butterflies. Their charms are obvious: they are like flying flowers, exhibiting not just wonderful colors, patterns, and forms, but also the power of flight. Unlike the myriads of annoying and harmful insects that most people regularly confront (and would probably vote for their extinction), butterflies are harmless and well-behaved, seemingly with no

FIGURE 4.11 Contrast of beautiful and ugly phases of a butterfly. (a and b) Peacock Butterfly (*Aglais io*). (a) Beautiful adult. Photo by Didier Descouens (CC BY SA 4.0). (b) Ugly caterpillar. Photo by Ragnar1904 (CC BY SA 4.0). (c) An ugly version (public domain) of the caterpillar in Alice's Adventures in Wonderland by English painter Arthur Rackham (1867–1939).

other interest but sipping nectar from blossoms. Were butterflies judged by their much less attractive larval (caterpillar) stage and the habit of some larvae of feeding on garden plants, they would be much less popular (Figure 4.11).

Much of the world of life presents astonishing architectural beauty that is rarely seen because of inaccessibility (Figure 4.12). The majority of species are tiny, usually out of sight and so also out of mind. Species that live in the sea (where approximately 90% of the planet's biomass occurs) or underground (where most terrestrial biomass occurs), or are simply adept at not being seen, are also at a tremendous disadvantage in eliciting public interest, although many are stunningly attractive. A toadstool, *Amanita ostyae*, in the Malheur National Forest of Oregon, is known to cover an area of almost $9\,km^2$, arguably making it the largest creature on Earth, at least on an area basis (Minter 2010); however, most of this fungus is underground. Species that are visible but very small also have very limited attractiveness just because they are hard to see. This means that only an extremely small percentage of the world's species are sufficiently visible to be of substantial concern to most people.

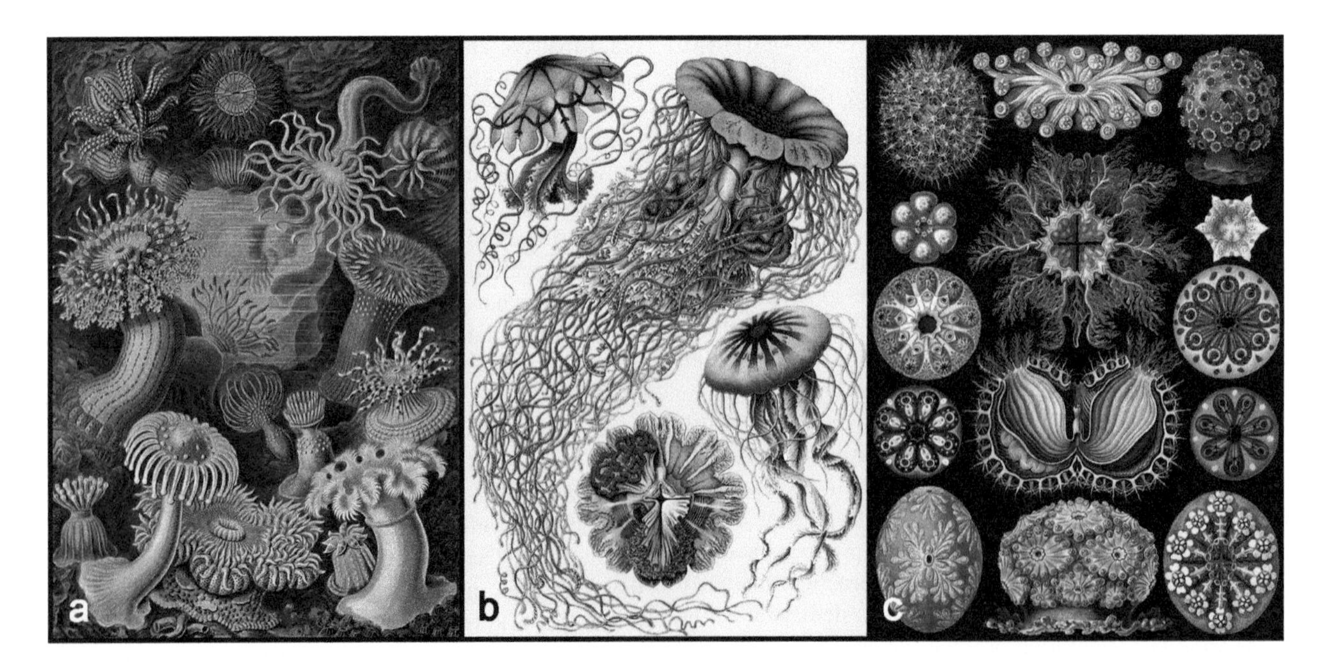

FIGURE 4.12 Examples of beautiful sea animals that most people never view. Source (public domain): Haeckel (1899–1904). (a) Sea anemones (plate 49). (b) Jellyfish (plate 8). (c) Sea squirts (plate 85).

LARGE GROUP SIZE AS COMPENSATION FOR SMALL SIZE OF INDIVIDUALS

Aside from a few people who choose to live as hermits, virtually all humans spend their lives in the company of others. Indeed, the collective force of groups of people, combined with our ability to pass down acquired knowledge to future generations, is the basis of civilization (Wilson and Hölldobler 2005). As discussed in this section, numerous animals are also social, but their collective power is sometimes a threat to humans, and so perceived with alarm. This may be the case for ant armies and swarms of locusts and bees. In the plant world, algae often develop impressive 'blooms' (algae do not have flowers) although the component individuals may be microscopic. Sometimes large groups of animals are also much more impressive, indeed more aesthetic than the single organisms, and sometimes individual animals that interact cooperatively with their fellows are perceived quite favorably by humans.

Animals commonly congregate together, in accord with the maxim 'there is strength in numbers.' This is true for both predators and prey, the former being more able to catch the latter, and the latter being more able to detect and/or fight the former (Figure 4.13). Flocks of birds can be mesmerizing (although their droppings are not appreciated). The same is true for bats, but since they fly mainly at night this isn't usually evident. Herds of mammals, especially the species that migrate in great numbers, similarly inspire awe.

The power of human groups is not simply in absolute numbers, but also in how required activities are partitioned among individuals who have special talents, knowledge, or training. This enormously increases efficiency. In groups of animals, there are often simple divisions of labor based on sex (females almost always are the primary caregivers of the young) and some combination of maturity, size, and ferocity (characteristics usually exhibited by leaders). Some species have evolved to produce specialist 'castes' with different functions (sexual reproducers, sterile workers, sterile soldiers). The best-known examples are ants, bees, wasps, and termites. The queens of these insects are always the largest animal in the colony, and 'soldiers' are often notably large. Other invertebrates with this life style include some shrimps. Mole rats are the only known vertebrates with such division into functional groups. Ants (with over 20,000 species) are extraordinarily successful and have been estimated to make up 15%–20% of all terrestrial biomass (Schultz 2000). The popularity of ant-farm ('formicarium') kits reflects admiration for the extraordinary cooperation that ants exhibit.

FIGURE 4.13 Social animals. Assembling in groups makes predators more efficient and prey more protected. Large groups also elicit more respect (fear or admiration) from humans. (a) A colony of emperor penguins. Photo (public domain) by Denis Luyten. (b) A herd of African elephants. Photo by Vaughan Leiberum (CC BY 2.0). (c) A flock of James's flamingo (*Phoenicoparrus jamesi*) in Bolivia. Photo by Pedro Szekely (CC BY SA 2.0). (d) A swarm of Pacific Sea nettles (*Chrysaora fuscescens*). These carnivorous animals catch their prey by their stinging tentacles, which can harm humans, although usually not lethally. Photo credit (public domain): Chad King, U.S. National Oceanic and Atmospheric Administration. (e) A pack of wolves hunting a herd of bison in Yellowstone National Park, USA. Photo by D. R. MacNulty, A. Tallian, D. R. Stahler, and D. W. Smith (CC BY 4.0). (f) A swarm of ants attacking a lizard in India (species not identified). Photo by Rajeshkunnoth (CC BY SA 1.0). (g) A swarm of red-bellied piranha (*Pygocentrus nattereri*). Image by Damian Soltys (CC BY SA 3.0).

Since prehistorical times, swarms or plagues of insects have represented threats. Fire ants, killer (Africanized) bees, and hornets can damage and even kill, locusts can wipe out crops, termites can ruin houses, and biting insects such as mosquitoes can annoy and carry diseases. Even solitary insects can set off alarms: it is said that if you see one bed bug or cockroach, there are probably thousands hiding. As pointed out elsewhere, the overwhelming majority of insect species are harmless to people, but unfortunately large assemblages are alarming and sometimes dangerous.

5 Visual ('Beauty') Determinants of Human Prejudices against and Preferences for Species

INTRODUCTION

Human attitudes to species are complex. Serpell (2004) observed that factors influencing opinions about animals include education, occupation, relationships with pets and livestock, religious tradition, and age. However, two considerations – pragmatic value of species (examined in Chapter 8) and emotional perception of them – primarily influence how they are evaluated and treated by people. Emotional reactions are mainly determined by appearance (reviewed in detail in this chapter) but also by non-visual considerations (reviewed in Chapter 6). Because of its importance, size, the most important visual determinant was dealt with separately, in Chapter 4. Clearly, people have very strong preferences regarding human appearance, and many of the biases that have been developed are reinforced by ceaseless advertising for commercial services and products. This book addresses parallel preferences and prejudices toward species by humans leading to distorted judgments about them – why this is important and what corrective actions are advisable.

Notwithstanding the old maxim that 'beauty is only skin deep,' our behavior is strongly determined by sight. Of the various senses, sight is the most valued (Enoch et al. 2019). Specific areas of the human brain address different functions, such as the auditory cortex for hearing and the olfactory cortex for smell. Of these functional areas, the single largest is the visual cortex, devoted to vision. Additionally, other areas of the brain are also involved in sight, processing information from the visual cortex. Estimates of the proportion of the brain devoted to vision range up to well over 50%, although there is no unanimity about this. But it is clear that human behavior is substantially determined by the appearance of things. This is the most extensively illustrated chapter in this book because visual appearance is so important.

How we look to each other influences how we behave toward each other, especially with respect to faces, since emotions like trust, pleasure, anger, fear, and disgust show up in faces and trigger responses that can enhance survival and reproduction or lead to violence (Darwin 1872; Ohman et al. 2001a, b; Ohman 2009; LoBue and Rakison 2013; Vashi 2015). As detailed in this chapter, humans are hard-wired to like human features, but also certain visual characteristics of other species that seem beautiful or decorative (like flowers, fur, feathers, wings, shells, and antlers). Conversely, most humans dislike or are neutral with respect to numerous other features of most species. This chapter examines visual characteristics employed as criteria to judge the beauty of species. There are few published comprehensive analyses of this topic (a chapter by Prokop and Randler 2017 is notable). Tables 1.2 and 1.3 in Chapter 1 summarize all considerations used to evaluate species.

HUMAN BEAUTY AS A DETERMINANT OF SUCCESS

'Evolutionary aesthetics' deals with how and why we humans have come to perceive some things as ugly, other things as beautiful. Some scientists have linked aesthetic responses to objects to their positive or negative value to the survival of our ancestors (Voland & Grammer 2003). Objects, environments, and situations that increased chances of survival and reproductive success have become fixed in the human psyche as beautiful, whereas those that decreased survival are perceived as ugly (Ruso et al. 2003; Thornhill 2003; Ulrich 1986). In theory, emotional responses to beauty and ugliness, therefore, should represent appropriate adaptive reactions to potentially beneficial or harmful situations. In reality, many of our ingrained responses amount to biases and prejudices that have misled us into making incorrect evaluations.

We humans evaluate the appeal of just about all visible material things (such as buildings, cars, clothing, clouds, food, furniture, and tools) as well as situations that affect us (like the price of fuel, the weather, and how noisy it is). The key considerations about beauty and ugliness are that the former generates pleasure tending to result in positive behavior and the latter produces dissatisfaction, tending to result in negative behavior. In evolutionary terms, our preference for beauty and our aversion to ugliness enhance our survival (Voland and Grammer 2003; Rusch and Voland 2013). Suggestive of the importance of the perception of beauty, part of our brain – the prefrontal cortex – is activated when we observe beautiful things (Cela-Conde et al. 2004).

Beauty is a major limiter or promoter of achievement for humans (Versluys and Skylark 2017; Elmer and Houran 2020). More physically attractive people tend to get preferential treatment in employment, are more likely to be elected to office, often receive better judicial decisions, more frequently earn higher salaries ('beauty premium' refers to the monetary advantage), have longer and more

DOI: 10.1201/9781003412946-5

FIGURE 5.1 Cosmetic enhancement of appearance. (a) Relative percentage of serious cosmetic surgical procedures (based on 1.1 million operations in the US in 2018, as reported in American Society of Plastic Surgeons 2019). Figure prepared by B. Brookes. (b) A woman at her dressing table applying cosmetics. Painting (public domain) by American artist Frederick Carl Frieseke (1874–1939). (c) Displays of cosmetics in a store. Photo by Pear285, released into the public domain.

stable marriages, and produce more children (Jokela 2009; Prokop and Fedor 2011). They are also judged to be more healthy, intelligent, honest, and sociable (the so-called 'halo effect' or 'physical attractiveness stereotype') – evaluations that are almost always advantageous. This phenomenon has also been summarized in the phrase 'What is beautiful is good' (Dion et al. 1972). Conversely, people who are not judged to be attractive are discriminated against (see the discussion of 'lookism' in Chapter 3).

However, there is some evidence that, at least for women, under some circumstances people associate good lucks with *lack* of competence (Braun et al. 2012; Lee et al. 2015). Fidrmuc et al. (2017) described an exception to the advantage of good looks in men: the probability of winning a Nobel Prize in Physics, Chemistry, Medicine, or Economics, which proved to be significantly negatively correlated with beauty (most candidates for these honors are men).

The advantages of good looks are widely appreciated, so people strive to improve their appearance. In the U.S. alone, over 18 million purely cosmetic procedures (including about 16 million minimally invasive procedures such as application of Botox® and laser hair removal) are carried out annually by plastic surgeons – over 90% on women patients, particularly on breasts and faces (American Society of Plastic Surgeons 2019). Illustrative of how sensitive people are to their appearance, Pressler et al. (2022) found that close-up selfies (30–45 cm) with smartphone cameras make noses appear larger, resulting in increased rhinoplasties (nose jobs) among the young. In modern society, beautiful appearance has become such an expectation for women that an enormous cosmetics industry has developed (Figure 5.1), and females of all ages devote substantial time to beautification.

While the pressure on women to conform to beauty standards is enormous, males are also not immune from societal stress to improve their looks, particularly with respect to muscle development and hair loss (Figure 5.2). Looking at a male gorilla (Figure 5.3) might produce jealousy for those who don't measure up in these respects.

Myostatin is a growth factor controlling muscle development. Mutations decreasing it can cause 'double muscling' – greatly increased growth of muscles, and the appearance of extreme muscularity (Figure 5.4). This is well known in some breeds of cattle, such as Belgian blue, and is sometimes considered desirable, since muscular beef increases quality. Indeed, livestock breeders frequently select for increased carcass muscularity by influencing myostatin levels (Michel 2010; Mirhoseini and Zare 2012). Unfortunately, in addition to other physiological problems, the calves can be so big that births are painful, requiring cesarean section. Whippet dogs that are heterozygous for the mutation (i.e., having inherited the defective gene from just one parent) are not only relatively muscular but are faster competitive runners. However, the double mutant animal, while massively muscular, is slower. The mutation is known in several other mammals, and in humans (first discovered in a young boy with the muscularity of a professional body builder). Livestock breeders are exploring the economic value of double muscularity, while animal welfare groups are concerned about the associated pain to the mutated animals. Media reports (Griffin 2015) indicated that Chinese scientists created double-muscled myostatin-deficient beagles, suggesting that double-muscled pet dogs could be available in the future. There is clearly a market for such dogs, both for appearance and strength, but it is concerning that the animals could be destined for lives of discomfort.

IS BEAUTY ONLY SKIN-DEEP?

There is no definition of 'beauty' that satisfies everyone. English poet John Keats (1795–1821) wrote that 'Beauty is truth, truth beauty – that is all ye know on Earth and all ye need to know,' but even this famous quotation is disputed (Gardner 2007). To a considerable degree, 'beauty' is the basis we use to judge the value of other species.

FIGURE 5.2 Illustrations of stress on men to improve their appearance. (a and b) Concern over muscularity. (a) Muscular male superhero created by Mitch Hallock/ComiCONNMitch, released into the public domain. (b) Update (prepared by B. Brookes) of the famous Charles Atlas comic advertisement showing a thin man insulted about his appearance at a beach, who subsequently becomes muscular and attractive to the opposite sex. (c and d) Concern over hair loss. (c) Postcard circulated in 1910 (public domain) of a man fretting over his baldness. (d) Nineteenth-century baldness cure advertisement (public domain).

Beauty can be defined as the degree of psychological gratification associated with the appearance of things. Physical appearance (primarily by sight but also by other senses such as smell, hearing, and touch) of objects is necessarily how we perceive them, at least initially. Some beautiful things produce pain, some ugly things produce pleasure, so with experience and familiarity, perception of their beauty may change. Our mothers become more beautiful with age, irrespective of physical changes that occur. Nevertheless, it is clear that humans are genetically programmed to admire some physical characteristics and to dislike others. As with many aspects of living things, individuals differ to at least some extent in their responses and can be conditioned to alter their perceptions. It is well known that cultural factors alter standards of beauty, but it is also clear that humans, like other animals, have innate and mostly inalterable tendencies to judge some objects as attractive, others as repellent. Of course, the query 'is beauty only skin-deep' does not address the scientific facts regarding human perception of attractiveness and indeed is merely a rhetorical question. It is really a statement that physical beauty is or at least should be insignificant for many issues of importance to humans (Figure 5.5). However, it is important to understand how perception of beauty warps rational judgments, in order that we behave well.

FIGURE 5.3 Impressive musculature and hairiness of gorillas – characteristics valued in men. (a) Gorilla (*Gorilla gorilla*). Photo by Anagoria (CC BY 3.0). (b) King Kong, modified by B. Brookes from a 1933 poster (public domain).

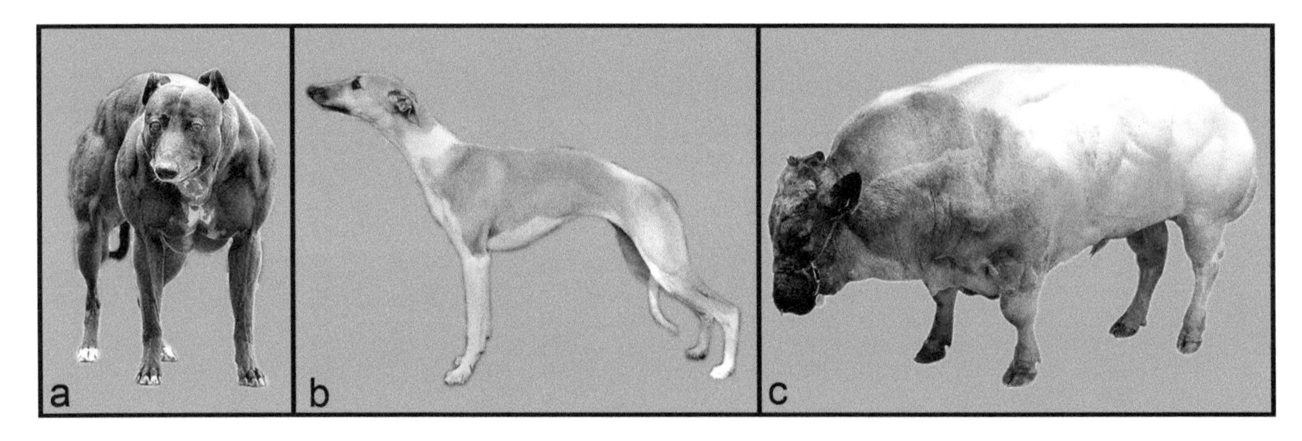

FIGURE 5.4 Double-muscled (mutant myostatin-deficient) animals. (a) A double-muscled female whippet. Photo by Charles Williams (CC BY 2.0). (b) A normal female whippet. Photo by Thomas Barregren (CC BY ND 2.0). (c) A Belgian blue double-muscled bull. Photo by Roby (CC BY SA 3.0).

BEAUTY AS A CURSE

Ironically, some impressive species are harvested specifically for their possession of beautiful fur, feathers, or ivory (Figure 5.6), in which case their beauty is a curse. Human harvesting or hunting of a wild species (animal or plant) frequently is associated with the possibilities that it will become dangerously rare or even extinct, but sometimes targeting a species for a given characteristic can result in natural selection of less attractiveness of that feature. For example, in Canada, hunting for the silver variation of fox fur strongly eliminated wild foxes with this desired color variation from the wild (Haldane 1942). Also in Canada, trophy hunting for bighorn sheep led to a decrease in horn size (Pigeon et al. 2016). Similarly, in Africa, intensive poaching for elephant ivory resulted in the evolution of tuskless elephants (Campbell-Staton et al. 2021).

Despite the phrase 'beauty is in the eyes of the beholder,' we humans consistently favor physical traits reflecting good health, fertility, and conformation to ideals of appearance. Hans Christian Andersen's ugly duckling eventually was accepted because it transformed into a beautiful swan (Figure 5.7), leaving unanswered the awkward question of its fate had it grown into an ugly duck. Consciously and unconsciously (and fairly and unfairly) appearance determines in considerable measure the successes and failures of individuals and groups. In a similar vein, the welfare of all other species on the planet is being determined to a large extent by human evaluation of their beauty (Figure 5.8).

FIGURE 5.5 'After all, beauty is only skin deep!' Cartoon (public domain) published in 1914 by German artist Heinrich Kley (1863–1945) in *Puck* magazine, vol. 75, no. 1940, May 9.

The main goal of this chapter is to examine the psychological determinants of human perception of the beauty of all other species. Many of the physical and behavioral traits valued in judging humans are also strongly valued in judging other species. Unfortunately, the vast majority of nonhuman species do not conform to human concepts of beauty, and accordingly tend not to be sympathetically treated. Aside from ethical considerations (O'Neill 1997), these species are vital to the future well-being of the world's ecosystems, and accordingly to the welfare of humans.

A small number of despised species not only are viewed as ugly and disgusting but also are considered very dangerous to humans. These hated species, the subject of many of the chapters of this book, have a dramatically negative effect on human attitudes to nature in general, and to the hundreds of millions of other, quite innocent, wild species in the natural world. This book defends the most maligned species, which as will be explained, almost always have very significant redeeming features, and often suffer from undeserved bad reputations.

FIGURE 5.6 Species victimized for their beautiful fashion products. (a) Leopard. Photo by www.walkers.sk (CC BY 3.0). (b) Leopard coat painting (public domain) by Harrison Fisher (1875–1934), photo (public domain) by Halloween HJB. (c) Peacock (*Pavo cristatus*), illustration prepared by Mahesh Iyer (CC BY SA 3.0). (d) Indian dancer in peacock costume. Photo by Arian Zwegers (CC BY 2.0). (e) Crocodile. Source: Danske Billeder (Danish Pictures), 1892, published by Alfred Jacobsen, Denmark. Credit: Dansk Skolemuseum (CC BY SA 4.0). (f) Alligator bag. Photo (public domain) by Ashley Van Haeften. Credit: Los Angeles County Museum of Art. (g) Indian Elephant (*Elephas maximus indicus*) confronting Bengal Tigers (*Panthera tigris tigris*). Painting (public domain) from Brockhaus, F. A. 1910. *Kleines Konversations - Lexikon (Small Conversation Lexicon)*, 5th edition. Leipzig, F. A. Brockhaus. (h) Ivory seahorse brooch with gemstone eye. Public domain photo. Credit: Hiart, Honolulu Museum of Art.

FIGURE 5.7 The Ugly Duckling – a fairy tale about the unfairness of judging only on the basis of appearance. (a) Illustration (public domain) by Theo van Hoytema, from the book by Hans Christian Andersen, published in 1893 by C. M. van Gogh, Amsterdam. (b and c) Figures (public domain) from Mabel Lucie Attwell's drawings in a 1914 version of Hans Andersen's Fairy Tales. Illustrations modified by B. Brookes.

FIGURE 5.8 A beauty contest, celebrating species considered to be attractive by humans, while the rejected ugly creatures have been banished to the audience. The scene is a metaphor for how humans honor and protect a minority of glamorous, charismatic animal celebrities, while most beasts and bugs are disdained and mistreated. Prepared by Jessica Hsiung.

In the following presentation, characteristics that people use to make aesthetic judgments about other species are examined. Many of these characteristics are extensions of how people judge other people. Frequently these features serve as sexual attractants and/or indicators of health. Sociologists and psychologists have made observations about the importance of features that serve to attract the sexes to each other, such as breast size, height, body symmetry, muscularity, hair color and length, facial divergence from normal, skin tone, waist-to-hip ratio, and ratio of limbs to body (Frederick et al. 2010). Many of these are simply not applicable or doubtfully relevant to how we judge animals (almost none are relevant to how plants are judged), but some are, as will be discussed.

FACIAL FEATURES

Importance of Human Facial Beauty

The idiom 'don't judge a book by its cover' expresses the wisdom of not judging the value of something by its outward appearance. But this is basically what we do with faces. Features of the face are key to judgment of human beauty (Rhodes 2006; Muñoz-Reyes et al. 2015; Todorov 2017). The word 'physiognomy' may simply mean the general form or appearance of something, but in past times it referred to facial features or expression, employed pseudo-scientifically as indicative of character (Figure 5.9). Nevertheless, in modern times, indeed probably since humans first set foot on Earth, people have employed human faces as principal indicators of whether they like or dislike, and trust or fear others. Phrenology is an allied pseudoscience, which attempted to correlate personality and character to head shape (Figure 5.9d). This section highlights physically attractive features of humans in order to address possible parallel evaluation of animals.

FIGURE 5.9 Studies of physiognomy (the discredited practice of evaluating moral character from the shape of faces). (a) 'Dublures of Characters,' public domain, by English caricaturist James Gillray (died 1815). 'Dublure' or 'doubloure' is French for doubled, and the artist has portrayed his interpretation of the hidden characteristics of the people shown (mostly politicians) by the grotesque doubles in the background. Credit: United States Library of Congress. (b) Another depiction of physiognomy. Artist unknown (early 19th-century France). Source: Wellcome collection (CC BY 4.0). (c) A physiognomic analogization of human and nonhuman faces. Source: Le Brun, C. 1809. Dissertation sur un traité de Charles le Brun, concernant le rapport de la physionomie humaine avec celle des animaux. Paris. Credit: Wellcome collection (CC BY 4.0). (d) Caricature entitled 'Bumpology' prepared by George Cruikshank in 1826. Phrenology (head shape analysis) was a pseudoscience allied to physiognomy. Credit: Wellcome collection (CC BY 4.0).

The Attractiveness of Human Infantile Features with Special Reference to Faces

Most people intuitive employ the word 'cute' without realizing its profound biological importance. 'Cuteness' ultimately is based on a special syndrome of baby features, especially facial characteristics. Konrad Lorenz (1903–1989), a famous Austrian ethologist (ethology is the study of animal behavior) and Nobel Laureate, hypothesized (Lorenz 1943) that humans have a 'cute response' to a set of immature features characteristic of babies (Kindchenschema, German for 'baby schema'). These features included: head large relative to body size, rounded, with a large protruding forehead; eyes large relative to face and below midline of head; cheeks rounded, plump; body rounded, chubby (with soft contours, soft surface body elasticity); movements awkward, clumsy (Figure 5.10). Crying, chuckling, giggling, and

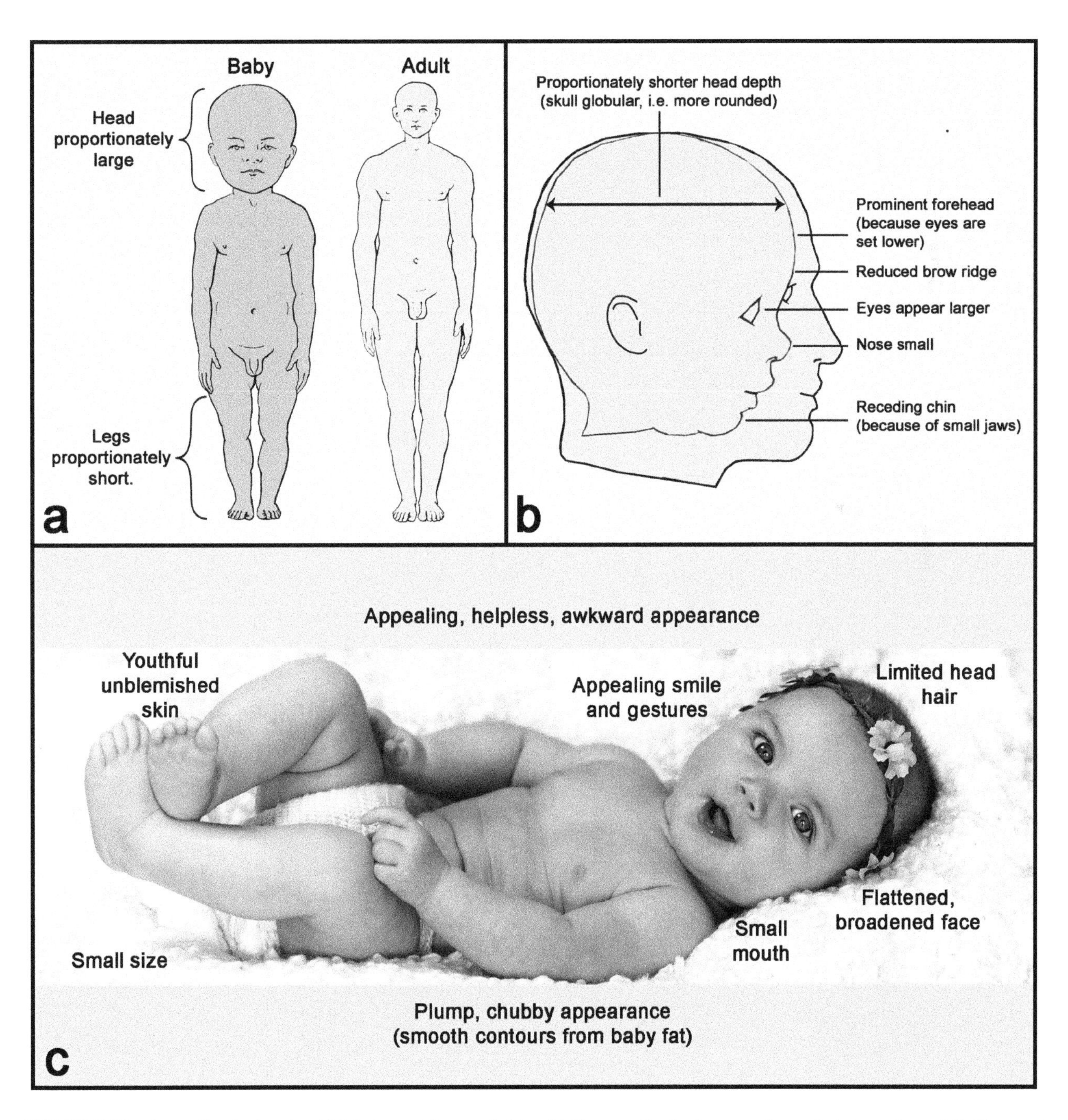

FIGURE 5.10 Appealing (cute, adorable) features of human babies. (a) Comparative body proportions of a baby and an adult. The most obvious adult maturation features are that the human head becomes proportionately smaller, the legs become proportionately longer, and the overall size increases greatly. Figure modified from Choulant's Anatomic illustrations (Geschichte und Bibliographie der anatomischen Abbildung; Leipzig, 1852), credit: Wellcome Images (CC BY 4.0). (b) Comparative head proportions of a baby and an adult. Key differences are indicated. Prepared by B. Brookes. (c) Appealing features of a baby. Photo (public domain) from Pixabay.com.

other behaviors, with associated facial contortions, also contribute to the set of baby signals. Humans, like many other mammals, respond to the immature appearance of their babies with caring, parental, protective behavior (Hildebrandt and Fitzgerald 1979; Karraker and Stern 1990; Morreall 1991; Sprengelmeyer et al. 2009).

The Value of Infantile Features in Public Relation Campaigns

Babies and infants are the strongest elicitors of empathy in humans. Of course, they represent the most helpless, fragile, and vulnerable period of human life, so evolution has made numerous human infant characteristics extremely attractive ('cute') and evocative of protective instincts ('cuddlesome'), as discussed in this chapter. Humanitarian charity appeals to help underprivileged victims of diseases, injuries, or disasters in a country that is too poor to look after its own usually feature very photogenic but sad-looking children in obvious distress. This works because tugging at the heartstrings loosens the purse strings. Many baby animals (especially mammals) possess infantile characteristics remarkably comparable to those of humans. Some baby animals are incredibly cute and cuddly; they are star attractions in zoos, movies, and television, and are models for toys (most notably the teddy bear). Such animals make excellent focal points for appeals to conserve biodiversity (Textbox 5.1: Figure 5.11).

Neoteny (Retention of Infantile Features) with Special Reference to Female Beauty

The attractiveness of female faces (and men's to a significantly lesser degree) is a sensitive topic because it is the basis of extensive discrimination and associated pain. Natural biological variation dictates that only a minority have the genetic background to develop an 'ideal' appearance. More importantly, societal pressure to strive for

FIGURE 5.11 The attractiveness of baby harp seals. (a) A living baby harp seal. Photo by M. Godbout (CC BY 3.0). (b) Cartoon emphasizing cuteness of a baby harp seal. Source: Shutterstock (contributor: Andryuha1981). (c) A baby boy playing with a toy seal. Note the comparable cuteness. Source: Shutterstock (contributor: Calek). (d) Anti-sealing cartoon. Source: Shutterstock (contributor: ArtbyAnny).

TEXTBOX 5.1 WHY BABY HARP SEALS AROUSE MORE CONCERN THAN MILLIONS OF OTHER SPECIES

Baby harp seals (*Phoca groenlandica*) are white from the age of 3 to about 15 days (the pelt color changes subsequently), while they lie helpless on ice flows as their mothers go off to harvest fish (Figure 5.11). The harp seal is prey for sharks, polar bears, and killer whales, and is also harvested annually by hunters in northern nations for food, but more particularly for its beautiful fur. The sight of adorable baby whitecoats with big pleading eyes being cruelly clubbed and bleeding to death (whitecoats are not harvested in Canada, where the hunt of maturer seals is supervised by veterinarians) has been the basis of extensive, fanatical, international activism to ban seal hunting. Famous entertainers rail against the hunt, and young women strip naked in public (to demonstrate that they do not need fur to be beautiful?). Arguments have often been framed in the context of cruelty (i.e., whether humane or not), conservation (whether seal populations are endangered), economic necessity (e.g., whether too many seals reduce the cod harvest), or morality (do people really need to wear fur?). Of course, all of these contentions are secondary to a simple reality: in a world in which almost all baby animals perish cruelly, usually not by the hand of humans, whitecoat harp seals have succeeded above all other (possibly over 100 million) species in generating sympathy, simply by virtue of their overwhelming cuddly, fuzzy, snuggly cuteness, which triggers the powerful response to protect human helpless, vulnerable babies. It is a serious mistake to dismiss the campaign against seal hunting as merely irrational emotionalism of animal rights and animal welfare groups because it reflects fundamental human perception of what kinds of living beings merit protection. From a conservation viewpoint, what is required is to constructively channel emotional attachment to species so that there will be tangible benefits for the ecology and biodiversity of the world.

FIGURE 5.12 Facial neoteny (retention of infantile features) in women. The attractive facial features of babies are evident in the faces of the women. (a) Photo by Heather Katsoulis (CC BY SA 4.0). (b) Photo by Jim Hammer (CC BY SA 2.0). (c) Photo by tec_estromberg (CC BY 2.0.). (d and e) Photos by fastweightlossmotivation99 (public domain). (f) Photo by Eyes_Master (CC BY 2.0).

unachievable levels of physical beauty places unreasonable stress on many (indeed, probably the majority). Fortunately, there are now movements that focus on the importance of self-acceptance and respect for others.

This book argues that it is unfair to judge animals by their appearance, but certainly people can have and do have different tastes. Nevertheless, it is safe to say that from the perspective of beauty, the external features of female

FIGURE 5.13 Care and enhancement of the female face. Illustrations from Shutterstock. (a) Contributor: Mentalmind. (b) Contributor: SkyPics Studio. (c) Contributor: Ansty. (d) Contributor: Takzany Thippon. (e) Contributor: Elysart.

humans have been judged to be more attractive by most people at least since the beginning of recorded history. The phenomenon of neoteny – the retention into adulthood of features of babies that humans are programmed to admire – has been advanced as a major contributing factor of what makes women attractive, See Figure 5.12. Babies are small and helpless, and while it would be insulting to suggest that this applies to women, the larger size, muscularity, low-pitched voices, body hair, and aggressiveness of men indicate their considerable divergence from how babies look and act. Women are encouraged to 'show more skin' because, like babies, the body's surfaces are attractively smooth and plump, the result of accumulated subcutaneous fat, and also (sometimes with the aid of shaving) hairlessness. Women's faces can be particularly neotenic: rounded or oval, with a small nose and small chin, full cheeks, full lips, and large eyes. Facial makeup for women is very strongly intended to accentuate the youthful babyface syndrome. Babies tend to have perfect complexions (often with rosy cheeks), and the chief aim of cosmetics is to replicate this, hiding the effects of aging and exposure to the elements. The eyes of baby humans appear enlarged, occupying a larger amount of the face, and this appearance is accentuated by long lashes and arched brows. To emulate this, women are expected to become skillful in the artistic application of eye cosmetics, including eye liner, eye shadow, mascara, brow makeup, and artificial eyelashes, and to take extraordinary care of their faces (Figure 5.13). Moreover, just like a baby's facial expressions are key to communication (at least with mothers), women's painted faces are often skillfully employed to indicate mood and intentions. Note that in addition to emphasizing (i.e., exaggerating) youthfulness, facial cosmetic for women additionally appear to emphasize sexual dimorphism, i.e., differences from men (Jones et al. 2015).

Comparative Sensitivity of Women and Men to Infantile Features of Animals

Men and women demonstrably tend to differ in attitudes to species. This is obviously the case for plants: as will be noted later in this chapter, ornamental flowering plants are of hugely greater importance to women than to men. As observed in Chapter 3, women are considerably more likely to develop phobias to animals. Nevertheless, as one would expect, women respond more emphatically to infantile features of animals than men, tending to exhibit

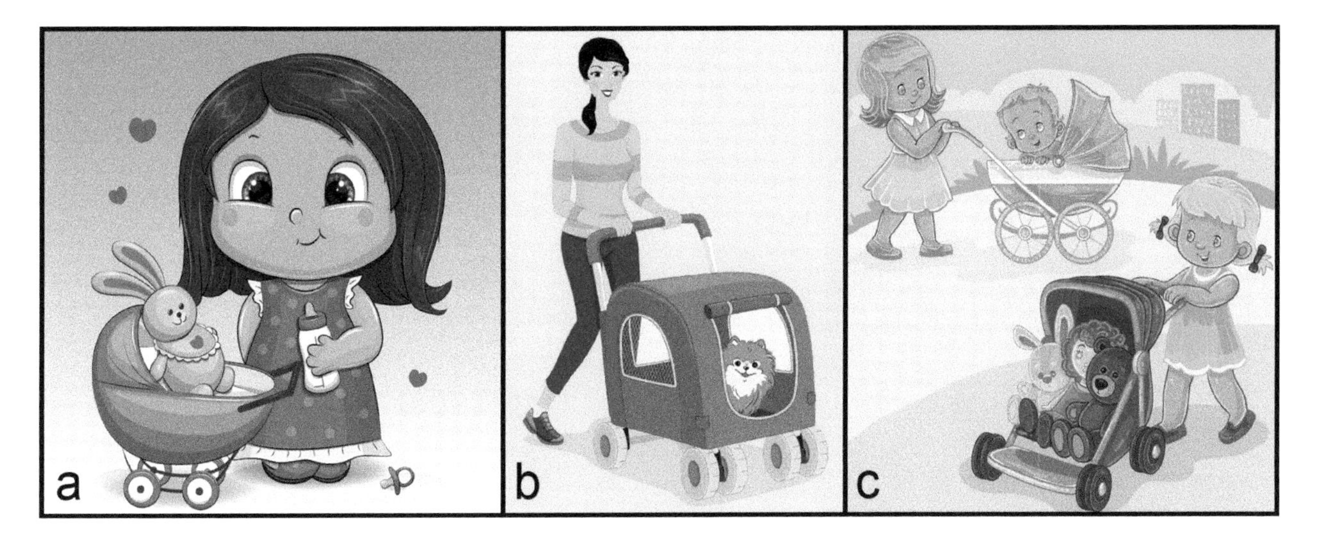

FIGURE 5.14 Transfer of human maternal behavior toward cute animals. Source: Shutterstock. (a) Contributor: Vectorpocket. (b) Contributor: BNP Design Studio. (c) Contributor: Vectorpocket.

similar behavior toward animals as they would to human babies (Figure 5.14). Maternal (nurturing) behavior, which is elicited by human baby features, is of course much more developed in women than in men, and women respond more strongly than men to neotenic characteristics of human babies (Sternglanz et al. 1977; Hildebrandt and Fitzgerald 1978; Koyama et al. 2006; Lobmaier et al. 2010; Sprengelmeyer et al. 2013). Serpell (2004) concluded that with respect to evaluating animals, women, compared to men, tend to more strongly emphasize emotional, rather than pragmatic, considerations, favor 'cute' animals, and especially favor vulnerable baby animals.

Mothers (also fathers to a much lesser degree) characteristically speak in an elevated pitch (timbre) to their infants, apparently because it promotes development (Piazza et al. 2017). The same gender-based behavior also is evident when people 'speak' to puppies and kittens. Women also are much more alerted than men to the high-pitched sound of babies and infants, and probably also to similar high-timbre vocalizations of young animals.

Men's Faces: Development of Aggressive Appearance

As they mature, men acquire beards (often interpreted by evolutionary anthropologists as a threat display, warning competitors), which obscure their faces (of course, if unshaven). Although some males, like most females, retain babyface features, most do not. The facial appearance of men seems to be part of a syndrome of testosterone-driven features that reflect aggression. As noted in the next paragraph, men's head proportions may be one of these characteristics suggesting ferocity. Muscle development, combined with much less subcutaneous fat than in females, and a much greater tendency to body hairiness, makes the male's surfaces bumpier, and thereby more threatening. The voice also deepens (and the larynx protrudes awkwardly from the neck), additionally considered to be a threat display feature.

'Aspect ratios' of rectangular objects such as video screens, paintings, photographs, books, and flags are measured as the ratio of their width to their height, and for some industries, standard formats have been adopted (which often change over time). In anthropology, the human face has often also been evaluated for width-to-height ratio. For this purpose, researchers usually define face height as 'upper face,' i.e., between the highest point of the upper lip and the highest point of the eyelids, or sometimes the eyebrows. Face width is approximately between the ears (strictly, 'bi-zygomatic breadth,' measured between the zygomatic bones, i.e., the cheekbones). Measured in this way, not only female faces but also male faces can be readily evaluated even if bearded, but sometimes not if moustached. Ratios usually range between 1.5 and 2.2. There is evidence that higher testosterone levels in men produce faces with a higher 'width-to-height ratio,' so that faces (excluding the forehead and the lower jaw) appear 'squatter' (Geniole et al. 2015; Třebický et al. 2015; Figure 5.15). Moreover, the higher testosterone has also been linked (somewhat controversially) to a greater level of aggressive dominance-related behaviors, including being less trustworthy, more self-centered and deceptive, prone to infidelity, and more psychopathic (Anderl et al. 2016; Noser et al. 2018; cf. Kordsmeyer et al. 2019 and Wang et al. 2019 for contrary evidence). People intuitively tend to perceive such faces as authoritative (hence warranting respect) but relatively dominating and alarming, and so not necessarily more attractive (see the analysis of the faces of U.S. presidents by Lewis et al. 2012). Early humans had faces that were extremely wide, with very large width-to-height ratios, perhaps reflecting the need to intimidate others and to be very aggressive in

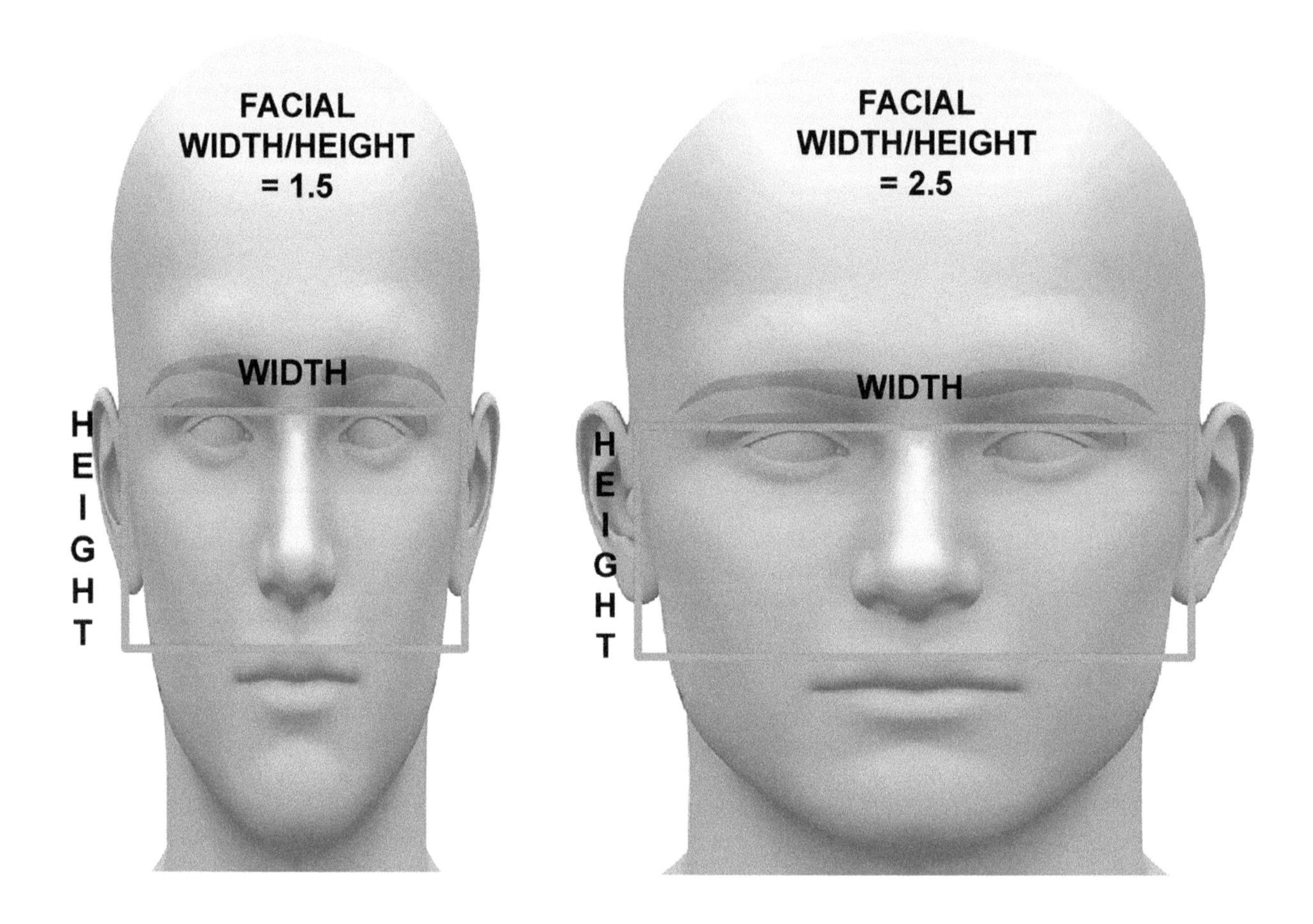

FIGURE 5.15 Model male faces demonstrating a very low and a very high facial width-to-height ratio. Prepared by B. Brookes.

order to survive (Figure 5.16). While there is evidence that wider facial ratios differentiate some men, it is less clear that such faces also tend to distinguish men from women (Lefevre et al. 2013).

Parallel studies with nonhuman primates support the observation that facial width-to-height ratios are important to how animals are perceived. Mature male orangutans (all three species) have large floppy cheek-pads called flanges, which make their faces look very wide (Figure 5.17a). The flanges, which are very attractive to females, develop in response to testosterone, and males with poor testosterone levels may not produce them at all (Marty et al. 2015). The 'Great Apes' include humans (genus *Homo*) and three other primate genera (*Pan*, Gorilla, and *Pongo*). The males of the latter two genera have impressively wide faces, and indeed these inspire both respect and fear in people (Figure 5.15). Wilson et al. (2020) noted a positive relationship between face width-to-height ratio and dominance in female chimpanzees (*Pan troglodytes verus*). Martin et al. (2019) found the ratio was associated with dominance in both male and female bonobos (*Pan paniscus*). Lefevre et al. (2014) found that facial width-to-height ratio was positively associated with alpha status and assertive personality in both male and female capuchin monkeys (*Sapajus* spp.). Borgi and Majolo (2016) observed a similar relationship in macaques (species of *Macaca*).

Readers are reminded of the earlier discussion in this chapter of the pseudoscience of physiognomy (evaluating moral character from the shape of faces). Facial width-to-height ratio, albeit controversial, has appreciable validity, but this does not justify interpreting the measurement of a particular individual as an accurate evaluation of his moral attributes, since there is simply too much variation for such a conclusion.

Consistent with the above evidence regarding wide faces, it appears that frontal ('face') views of sports cars with high width-to-height ratios are very appealing to potential buyers who want to demonstrate their dominance (Maeng and Aggarwal 2016; Figure 5.18). All of the above information regarding perception of faces refers to the situation within a given species, so it is speculative, but perhaps not unreasonable to suggest that humans may also use facial features of other species (particularly other great apes) as criteria for liking or disliking them.

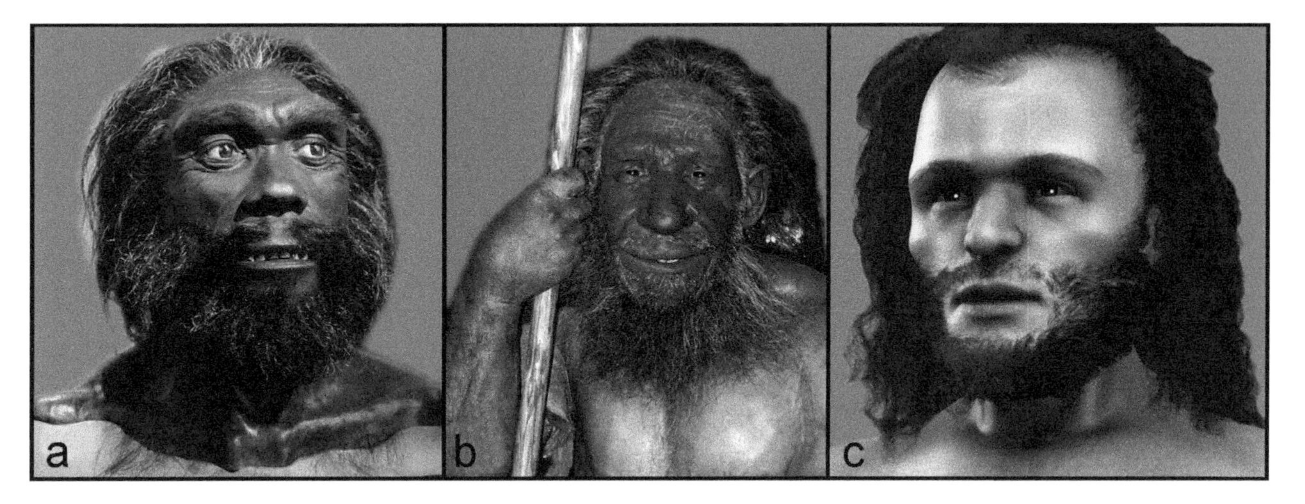

FIGURE 5.16 Early members of the genus *Homo*. Note the very broad (horizontal) heads and short (vertical) upper faces, so that the facial width-to-height ratios are at the extreme upper range of most present-day humans. Such faces, as discussed in the text, are decidedly intimidating and could reflect aggressive behavior that was necessary for survival in the past. (a) *Homo heidelbergensis* (*Homo sapiens heidelbergensis*). The group occupied northern Eurasia until it went extinct about 200,000 BP. Photo by Tim Evanson (CC BY 2.0). (b) Neanderthal Man (*Homo neanderthalensis* or *Homo sapiens neanderthalensis*). He lived in Eurasia and contributed a small number of genes to modern humans before going extinct about 40,000 years ago. Photo credit: Neanderthal-Museum, Mettmann (CC BY SA 4.0). (c) Early 'anatomically modern human' (Cro-Magnon man). Thriving in Europe after 40,000 BP, this was the main ancestor of modern humans. Photo by Cicero Moraes (CC BY SA 4.0).

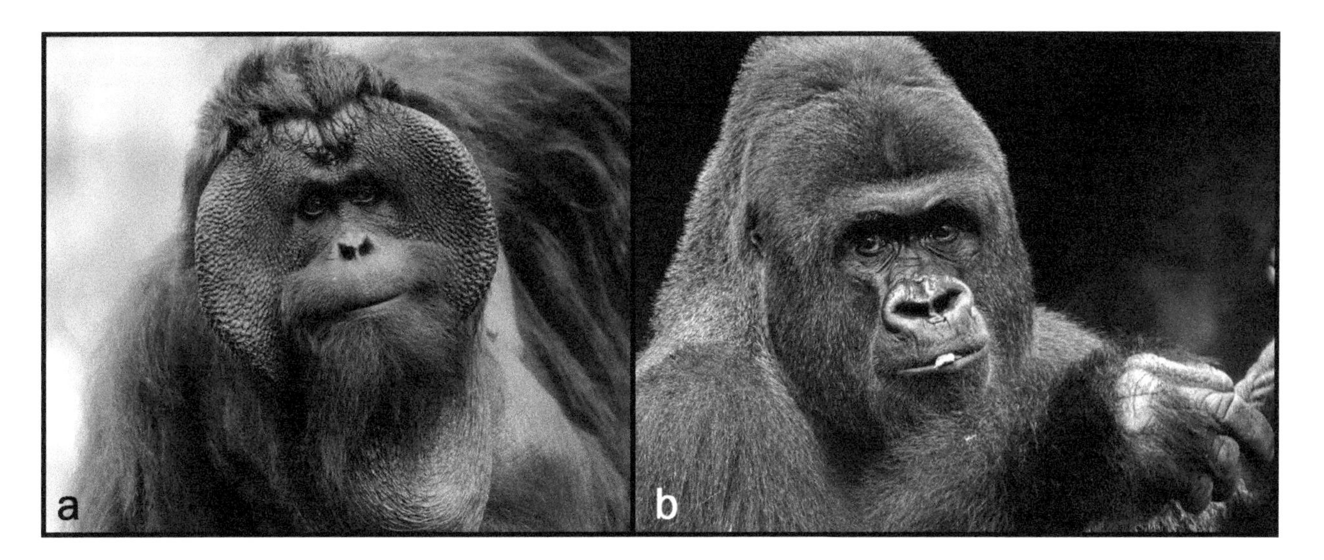

FIGURE 5.17 Impressively wide faces of two male primates closely related to humans. (a) Bornean orangutan (*Pongo pygmaeus*), with big cheeks. Photo by Eric Kilby (CC BY SA 2.0). (b) Gorilla (silverback, *Gorilla gorilla*). Source: Pixabay.com (public domain).

FIGURE 5.18 Relatively wide-faced men, dogs, and vehicles all seem to generate fear and respect. Prepared by B. Brookes.

Judging Animal Faces

Face features are very important in assessing the attractiveness of animals. Many animals are considered to be ugly by human standards (Figure 5.19).

Some young animals show similar immature features to those of human babies (Figures 5.20 and 5.21), and indeed some animals (including female humans as noted above) retain these juvenile features. Cuteness features have been selected by humans in a variety of breeds of pet dogs and cats (Lorenz 1971; Tuan 1984; Serpell 1996; Figure 5.22). Archer and Monton (2011) observed preferences for infant facial features in pet dogs and cats. Golle et al. (2015) found similar positive responses to cute infants and cute puppies. There is evidence that humans unconsciously use the baby features not just to judge the attractiveness of other humans but also of wild animals (Borgi et al. 2014; Markwell 2021). Skilled cartoonists wishing to make animals look attractive draw them with pronounced juvenile features (Gould 1979, 1980; Etcoff 1999; Figure 5.23), or with aggressive features to make them appear dangerous (Figure 5.24). Estren (2012) cautioned that animals lacking cuteness are in danger of receiving poor treatment.

Many nonhuman animals produce facial expressions which sometimes resemble human emotive facial expressions (Darwin 1872; Waller and Micheletta 2013). Most animals do not have faces that can be varied, and their constant appearance can give them a strange, emotionless aspect that is difficult to sympathize with. However, many, especially mammals, produce a variety of facial expressions, and often these function as communications to other individuals within social species and serve to maintain group harmony. Deliberate

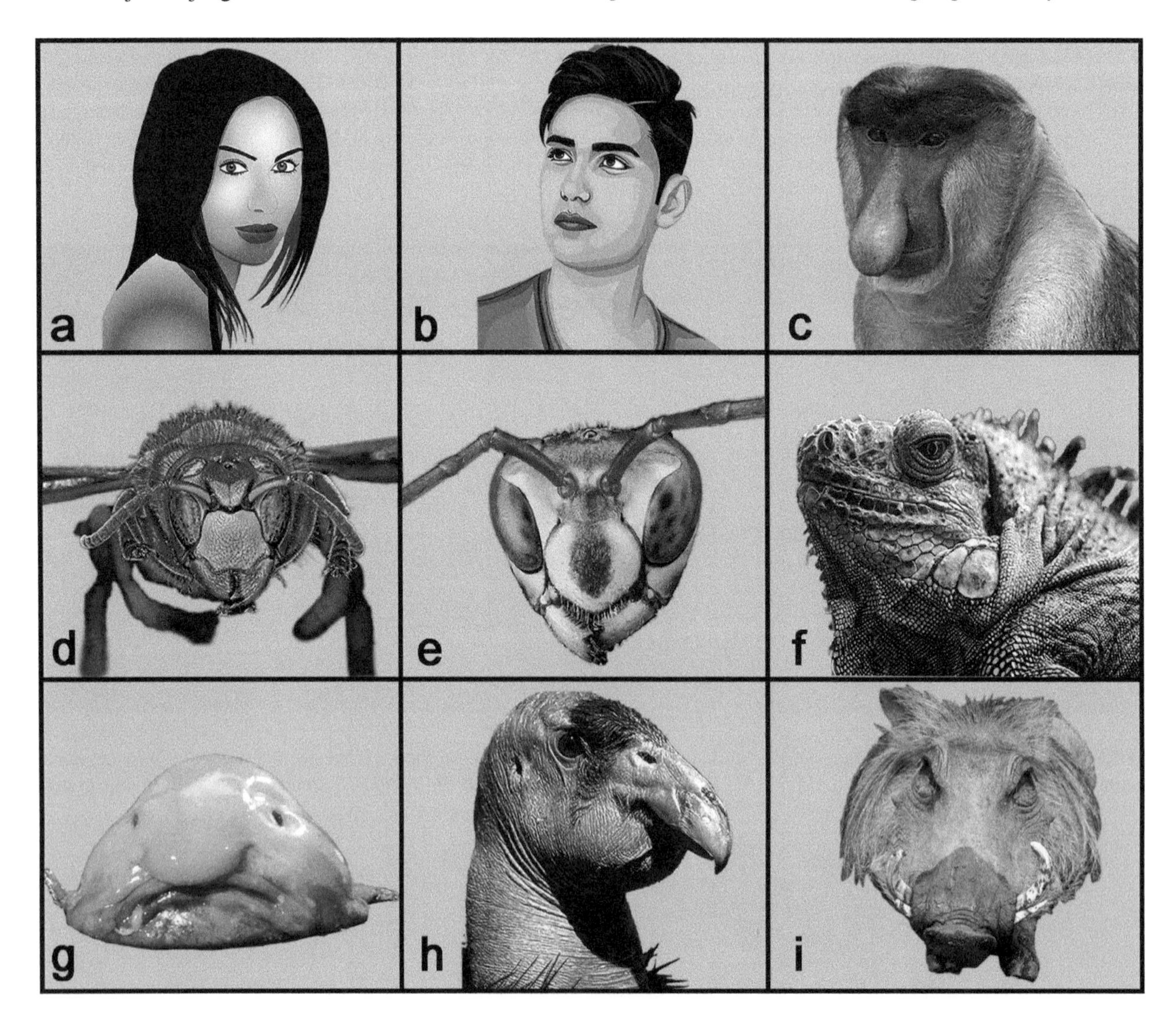

FIGURE 5.19 Ugly animal faces, judged by deviation from the standards of ideal human faces. (a) Human female. Public domain image from Pixabay.com. (b) Human male. Public domain image from Pixabay.com. (c) Proboscis monkey (*Nasalis larvatus*). Photo by Charles J Sharp (CC BY SA 4.0). (d) European hornet (*Vespa crabro*). Photo by Llez (CC BY SA 3.0). (e) Guinea paper wasp (*Polistes exclamans*). Photo (public domain) by Sam Droege, U.S. Geological Survey. (f) Green iguana (*Iguana iguana*). Photo by Tambako the Jaguar (CC BY ND 2.0). (g) Blob fish (*Psychrolutes marcidus*). Note: Blobfish live at depths of 600–1,200 m and are adapted to strong atmospheric pressures. When brought to the surface, their appearance distorts, making them appear much odder than they are normally. In profile, they do not look markedly different from typical fish. Photo by James Jowel (CC BY ND 2.0). (h) California condor (*Gymnogyps californianus*). Photo (public domain) by U.S. Fish and Wildlife Service. (i) Warthog (*Phacochoerus africanus*). Photo by Jim Bowen (CC BY 2.0).

FIGURE 5.20 Comparative appearance of a puppy and an adult golden retriever. Note small body size, comparatively large head size, plumpness, short legs, and flat face of the puppy – all features that are also appealing in human babies (compare Fig. 5.10). (a) Source: Pixabay.com (public domain). (b) Photo by Barras (CC BY SA 3.0).

FIGURE 5.21 Cute and cuddly baby animals are strong elicitors of protective instincts in humans. (a) Koala cub (*Phascolarctos cinereus*). Photo by E. Veland (CC BY SA 3.0). (b) Tiger kitten (*Panthera tigris*). Photo from Pixabay.com (public domain). (c) Polar bear cub (*Ursus maritimus*). Photo by Jensk369 (CC BY SA 2.0). (d) White-tailed deer fawn (*Odocoileus virginianus*). Photo from Pixabay.com (public domain). (e) Muscovy duckling (*Cairina moschata*). Photo by Luis Miguel Bugallo Sánchez (CC BY SA 3.0). (f) Baby meerkat (*Suricata suricatta*). Photo from Pixabay.org (public domain).

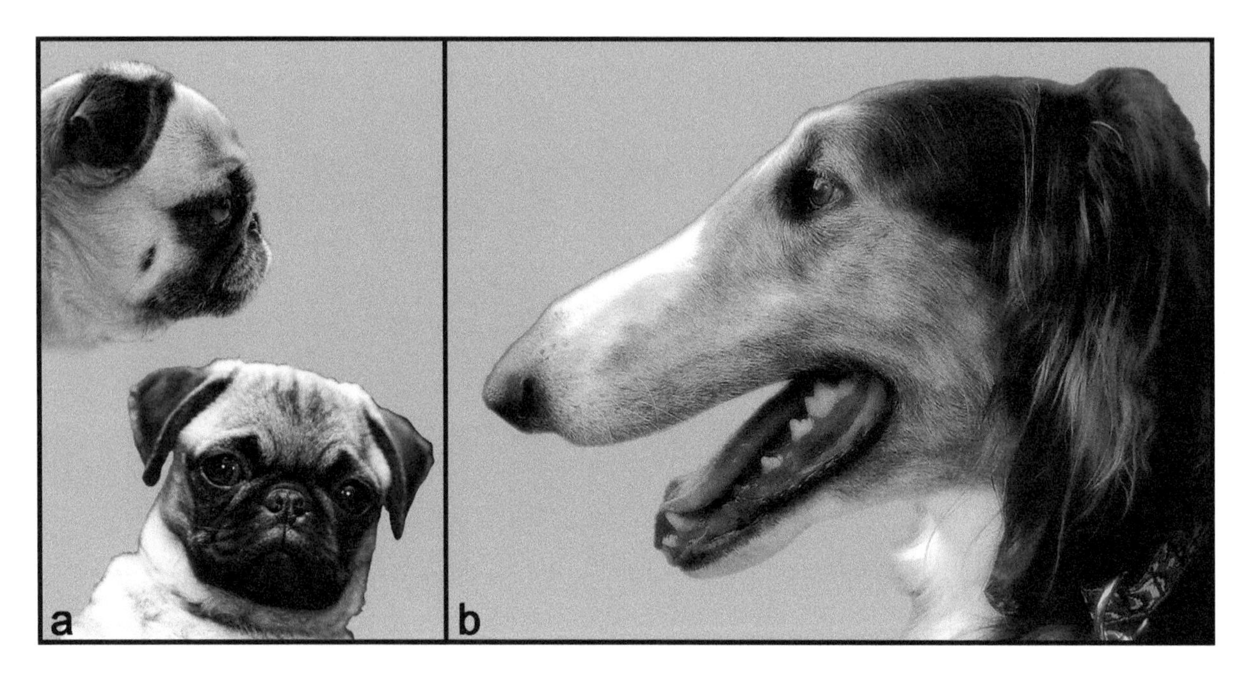

FIGURE 5.22 Comparison of a flat-faced ('brachycephalic') dog, the pug, with a long-snout ('long-nosed,' 'dolichocephalic') dog, the borzoi (also sometimes called the Russian wolf hound). Flat-faced dogs have been selected for their cute baby faces. Unfortunately, this is often correlated with more restricted breathing passages leading to health problems. The formidable jaws of the borzoi, while not as attractive, suit this sitehound to hunting large prey. (a) Upper photo of pug by DaPuglet (CC BY SA 2.0). Lower photo of pug by Pharaoh Hound (CC BY 2.0). (b) Photo of borzoi by Autangelist (CC BY SA 3.0).

FIGURE 5.23 Cartoon animals with infantile features that make them appealing. Illustration released into the public domain by Shutterstock.

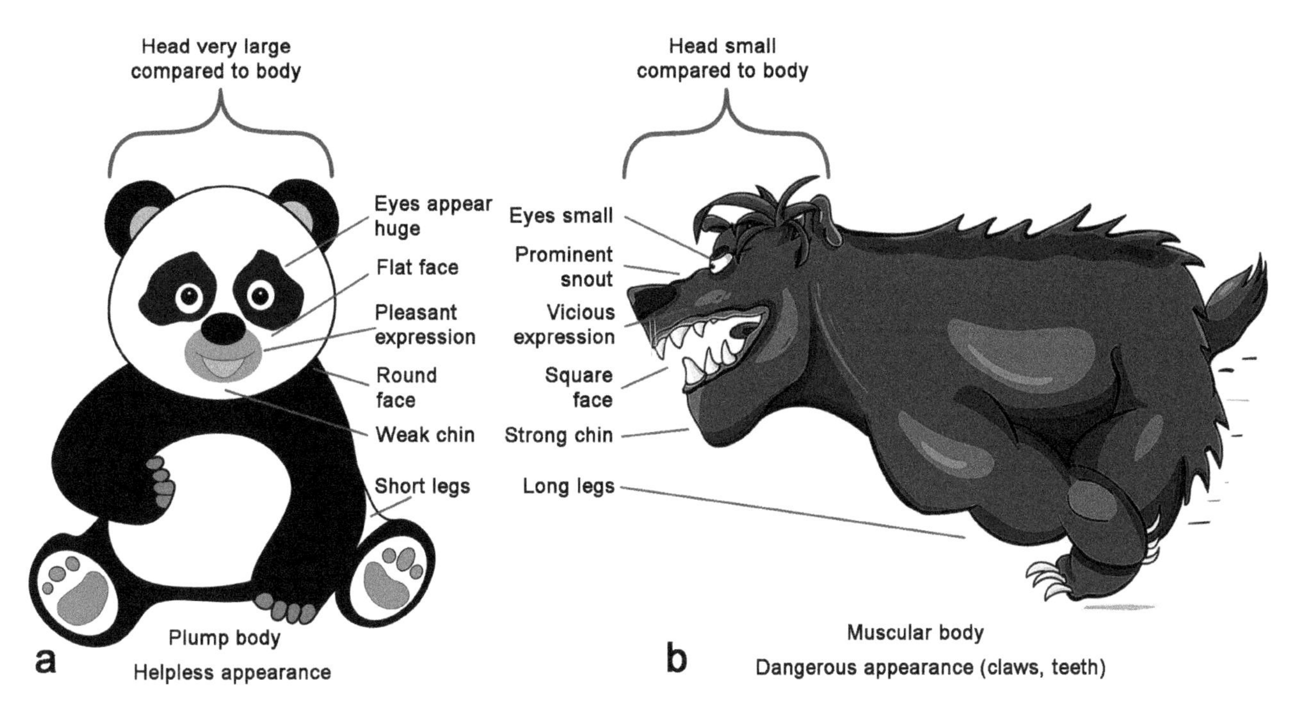

FIGURE 5.24 Contrast of (a) attractive infantile features and (b) frightening mature features in cartoon bears. Prepared by B. Brookes.

TEXTBOX 5.2 FACIAL EXPRESSIONS AS KEYS TO COMMUNICATING FRIENDSHIP

Within the visual realm, the face is the primary communication tool for terrestrial mammals, and this is where facial expressions are generated. Many mammals also have facial fur of contrasting patterns and colors that are themselves used as a kind of signaling in social interactions. Human (and most mammalian) faces make facial expressions using the numerous mimetic muscles found deep to and attached to the skin of the face. These facial expressions help individuals with cohesion of social and kin groups, maintaining relationships, and potentially signal the intent and emotional state of the sender… accumulating evidence is demonstrating that humans and dogs are uniquely attuned to one another's facial expressions to the exclusion of other domestic mammals. Not only are humans good at reading facial expressions of dogs, but dogs are good at reading facial expressions of humans… Dogs also engage in mutual gaze with humans to the exclusion of any other species. During this mutual gaze, oxytocin is released in both species, which strengthens the care-giving response that many humans feel toward dogs. It has even been hypothesized that during the process of domestication, dogs 'hijacked' human emotions that resulted in the intense care-giving response that many humans display toward dogs. During the domestication process, humans may have preferentially associated with gray wolves that displayed affiliative, non-aggressive behaviors that elicited these care-giving responses, resulting in contemporary dog breeds that generally display neotenic facial features relative to gray wolves.

—Burrows et al. (2021)

communication between humans and other species is mostly between people and their pets, but also with livestock (Burrows et al. 2021). During the evolution of dogs from gray wolves, dogs have acquired quite astonishing abilities to communicate with humans through facial expressions (Burrows and Omstead 2022). Most people with pet dogs understand how dogs manipulate their ears, mouth, lips, and eyes to express emotions. For example, submission is indicated by lowered head, flattened ears, and lips retracted horizontally, aggression by opened mouth, erect ears, and forward staring, and playfulness by lips drawn back horizontally, teeth exposed, ears more or less erected, and eyes partly closed.

However, no other animal has acquired as much ability to communicate with humans as have dogs. Students of animal behavior sometimes homologize nonhuman expressions with emotive expressions in people, but clearly there are many species-specific expressions. Primate species closely related to humans particularly may be judged to be more or less attractive depending on their facial expressions. Chimpanzees appear to be grinning in response to a variety of situations, but this is not always equivalent to smiling in humans (Parr and Waller 2006). Nevertheless, among mammals and many other species, a gaping mouth with lips retracted to emphasize the size of fangs is probably a warning ignored at one's peril. Regardless of the actual mood of an animal, whether its face seems threatening or friendly (Textbox 5.3; Figure 5.25) can influence its attractiveness to people.

TEXTBOX 5.3 THE APPEAL OF SMILING FACES IN HUMANS AND OTHER ANIMALS

Domesticated animals have kinder, more human-like, or human-desired facial features, analogous to some of the features humans desire in themselves. This is particularly evident looking at the dog versus the wolf. The shape of the dog's mouth is sometimes upturned, even at rest, almost in a smiling position, whereas the wolf's mouth is flatter. In human facial aesthetics upturned oral commissures (corners of the mouth) are considered more desirable. As people age and facial soft tissues become lax, the corners of their mouth turn down. This oral shape is commonly perceived by others as unpleasant or angry, regardless of the person's actual temperament. Consequently, in the world of aesthetic plastic surgery, this is valued as an undesirable appearance, perceived by others as being unpleasant or grumpy. Among other things, aging humans often seek out procedures to correct downturned corners of their mouth (e.g., Botox to the depressor anguli oris muscle, fillers to raise the oral commissure, face lifts, etc.) to give them a friendlier, smiling appearance. Perhaps humans selected dogs to have this oral shape because they perceive it as friendly and desirable in a human-like way?

—*Burrows et al. (2021)*

FIGURE 5.25 Animals that appear to be smiling tend to be appealing. All of these animals have mouths that naturally form a smile. (a) Samoyed dog. Photo by Michael Neufeld (CC BY SA 4.0). (b) Chimpanzee. Photo by Tambako the Jaguar (CC BY ND 2.0). (c) Piglet. Photo by Дино4ка228 (CC BY SA 4.0). (d) Orangutan. Source (public domain): Publicdomainpictures.net. (e) Dolphin. Photo by Matthias (CC BY 3.0).

AGEISM

Among humans, ageism (sometimes spelled agism) is discrimination based on age. Ageism is usually a subcategory of lookism (discussed in Chapter 3). Although stereotypes and prejudices affect the young, serious discrimination against older individuals is common. Ageism against the old is ethically and legally unacceptable, but there is a partial biological basis. Health deteriorates with age and, in parallel, so does performance and survival (which is not to deny that many older individuals can outperform younger ones, or that experience and accumulated knowledge often make age irrelevant). The proverb 'Youth will be served' reflects the simple truth that some aspects of life, such as athletic abilities, decrease with age. Of course, external appearance – such as skin wrinkling and hair color – rather than birth certificates are usually employed to judge age. Mating preferences are keys to reproduction, and the strong appeal of younger women (compared to similarly aged males) to the opposite sex reflects their relatively shorter period of fecundity.

Domesticated animals are subject to age discrimination, but the motivation is often simply economic. Livestock raised for food are harvested at the age when doing so is most efficient, although frequently preference is given to younger animals because they are tastier. Livestock raised for athletic performance – such as for transportation and racing – similarly reach a 'best before' age (Figure 5.26). Champion male racehorses are especially privileged since they can supply semen for years after retirement from competition. Although most people dearly love their aged pets, some do not, and it is much easier to find new homes for younger animals.

As discussed in this book, most wild animals (which are predominantly insects) are not attractive to people, so their age is scarcely relevant. Moreover, some animals (notably turtles and tortoises) hardly show external signs of age deterioration. Nevertheless, hunters, photographers, and zoos are quite concerned with the appearance of animals and seek out specimens in their prime, which usually does not include the very old. Nature is merciless in eliminating individuals once their athletic prowess deteriorates, so unlike senior citizens in human societies relatively few animals reach old age. ('Ableism' is discrimination against people with disabilities. Nature's elimination of the old is probably better characterized as ableism than as ageism.)

Numerous plants are annuals, or at least the above-ground parts live for only 1 year, so the traditional concept of ageism scarcely applies. However, woody plants (shrubs and trees) do eventually deteriorate, and so cultivated specimens are cut down when they become unsightly or dangerous. Extremely old trees are often very picturesque (Figure 5.27), and judging their aesthetic value can be difficult.

FIGURE 5.26 Ageism applied to horses. (a) Chromolithograph (public domain) showing young, vigorous horses, marketed by the print vending company Currier & Ives in 1886. Source: U.S. Library of Congress. (b) Old, worn-out horses about to be reduced to industrial byproducts. Prepared by B. Brookes.

FIGURE 5.27 Examples of very old trees that have become picturesque with age, albeit scraggly. (a) Mostly dead Great Basin bristlecone pines (*Pinus longaeva*) in the White Mountains of California. These trees are believed to be the oldest living things on the planet, many over 4,000 years old (the result of clonal reproduction, i.e., genetically identical young trees bud off from the old ones). Photo by Rick Goldwaser (CC BY 2.0). (b) An old Benjamin fig (*Ficus benjamina*) in the Botanical Gardens, Kandy, Sri Lanka. Photo by Ronald Saunders (CC BY SA 2.0). (c) An ancient olive tree (*Olea europaea*) in Lebanon. Photo by Serge Melki (CC BY 2.0).

BODY COVERINGS

Huggability

From birth, physical contact with one's mother (clinging, suckling, resting) is essential for normal human social and emotional development (Cascio et al. 2019). For almost everyone, hugging, cuddling, or just holding hands is part of mating, and often also of social bonding.

Physical contact results in the release of the hormone oxytocin (Chen et al. 2020), resulting in a sense of well-being, reducing fear, and promoting emotional connections (explaining why shaking hands on first contact has been so prevalent). Physical touch also increases levels of dopamine and serotonin, two neurotransmitters that regulate mood and relieve stress and anxiety; increases endorphins, the natural pain relievers which trigger feelings of relaxation; and reduces cortisol, which is a key associate of stress (Young et al. 2020). The immune system is also hypothesized to benefit from touch (Suomi 1995). Huggable animals promote these benefits (Young et al. 2020), which is presumably why therapy dogs ('emotional support animals') are so popular (Figure 5.28), and why petting zoos/farms (Figure 5.29) are similarly common. Animals that have abrasive, spiky, slimy, or even toxic surfaces do not offer these psychological benefits to people, and so have lowered attractiveness (Figure 5.30d–f). Humans lack appreciable body hair ('fur'), but at least contact with smooth skin is quite sensuous, and presumably contributes to the satisfaction of babies (Figure 5.30a). Furry toys and pets (Figure 5.30b and c), and even soft blankets, can provide infants with pleasing softness.

'Tree-hugger,' a pejorative term for an environmental activist, evokes the image of conservationists embracing large trees to prevent them from being felled. It is sad that some people have not experienced the emotional satisfaction that trees can provide, but many do appreciate them (Figure 5.31). As noted in Chapter 4, some trees represent the largest, oldest creatures on Earth, and deserve respect. The profound spiritual love for some groves of trees, and tree worship, is examined in Chapter 7.

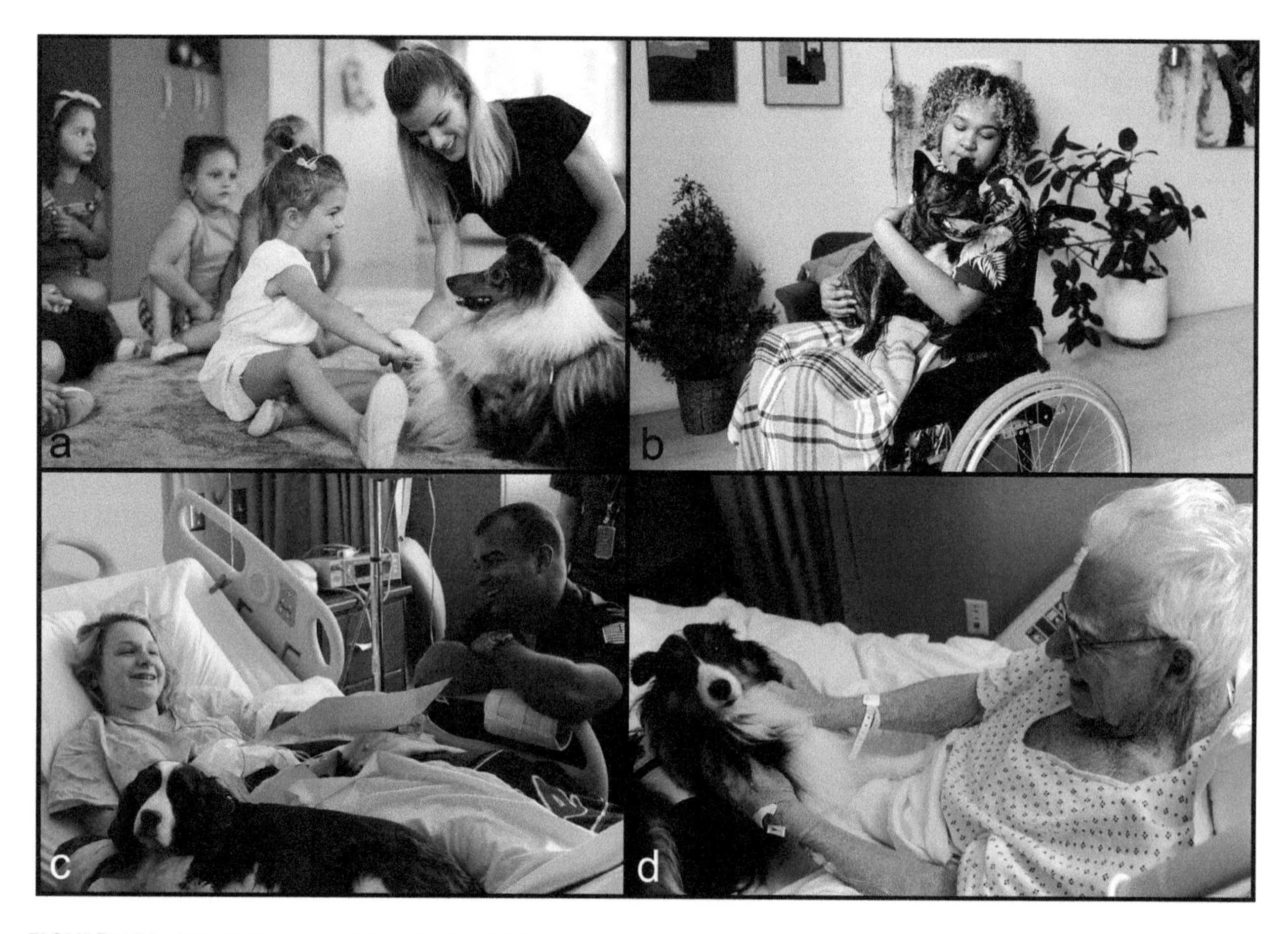

FIGURE 5.28 Physical contact with professional 'therapy animals' such as these dogs provides psychological benefits and so such species are greatly valued. (a) Collie visiting a class of young students. Source: Shutterstock, contributor: Gpointstudio. (b) French bulldog visiting a hospital. Source: Shutterstock, contributor: AnnaStills. (c) Quinny, a cocker spaniel, visiting a patient at Le Bonheur Children's Hospital, Memphis, Tennessee. Photo (public domain) by Amanda Rae Sullivan, U.S. Navy. (d) Molly, a Sheltie, visiting Langley Air Force Base Hospital in Virginia. Photo (public domain) by Zachary Wolf, U.S. Air Force.

FIGURE 5.29 Physical contact with furry, fluffy, or fuzzy animals at 'petting farms' ('petting zoos') provides psychological benefits, and so such species are greatly valued. Source of photos: Shutterstock. (a) Pony. Contributor: Roman Rybaleov. (b) Duckling. Contributor: Mak_Jen. (c) Piglets. Contributor: Galitsin. (d) Goat. Contributor: MNStudio. (e) Lamb. Contributor: Critterbiz. (f) Bunny. Contributor: Serhiy Kobyakov.

Hair and Fur

Fur is a thick layer of hair covering the surface of animals, mainly mammals. Thick hair (more than long hair) serves for thermoregulation (insulation against heat loss in cold environments, insulation against heat acquisition in hot environments) and camouflage. A covering of hair also is protective against ultraviolet radiation from the sun, which is presumably why, when humans evolved to become 'naked apes,' head hair was retained to reduce the damaging effects of overhead light. Hair is protective against many tiny parasites but is a natural habitat for others. Lengthy, luxuriant hair in humans may be the result of good health and youthfulness, and so is a component of what makes people attractive. However, cultural and religious traditions often determine how long hair is allowed to grow or even displayed in public, and fashion has varied considerably over the centuries (Thanikachalam et al. 2019). A variety of evolutionary hypotheses has been debated regarding why there are hair differences among the races and sexes of humans (Robbins 2012; Frost 2015).

Head hair for women is usually kept long, and great efforts are made to fashionably display their 'crowning glory.' Body hair is a sexually dimorphic characteristic of humans, so underarm and other body hair (usually excepting pubic hair), in accord with the conventional expectations of women, is shaved off. Attitudes toward masculine body hair are variable: some consider it sexy (especially chest hair), others judge it unattractive. In men, head hair is usually kept short and often lost with age. (It was once fashionable for high-status men, and sometimes also women, to wear elaborate wigs.) All things considered, in modern times the beauty of one's head hair is much more valued by women than men (Figure 5.32), and the much greater appreciation of fur coats by women compared to men is consistent with this preference. It seems reasonable to predict that not just the fur but the animals bearing attractive fur would be more appreciated by women than by men.

Hair 'texture' is a complex topic. Principal aspects concern curl pattern (such as wavy or frizzy), strand thickness, and density. Tightly coiled hair is characteristic of some (but by no means all) people of African origin, and some African-Americans wear an 'Afro' hairstyle as an expression of natural beauty, not conforming to Eurocentric (white-centric) beauty norms. Tightly coiled hair has been termed 'kinky,' which also has pejorative meanings, so the term 'Afro-textured' often replaces it. Numerous black women use 'relaxers' to straighten their hair to conform to predominant societal styles. Many of these chemicals are carcinogenic (Brinton et al. 2018; Eberle et al. 2020). Regrettably, black women often suffer considerable prejudice because of their naturally curly hair and so are pressured to alter it (Prince 2010). In 2022, The U.S. House of Representatives voted to prohibit discrimination on the basis of hair texture and hairstyles that are tightly coiled, curled, or worn in locs, cornrows, twists, braids, Bantu knots, or Afros (Amri 2022). Such prejudice does not seem to be carried over to the animal kingdom although wild animals very rarely have curly hair or fur (our relative the mountain gorilla is curly-haired). However, several domesticated animals have curly hair. There are numerous curly-haired dogs (like poodles and Airedale terriers) and a few cats (such as the Selkirk rex and LaPerm) which are much loved. Sheep hair is usually naturally curly/kinky, which is what makes

FIGURE 5.30 Huggability and non-huggability as factors in human judgment of the attractiveness of species. (a–c) Examples of huggability. (a) Madonna and Child, painting by Italian artist Raphael (1483–1520). Photo by José Luiz Bernardes Ribeiro (CC BY SA 4.0). (b) A girl and her Teddy Bear. Source (public domain): Pixabay.com. (c) A girl and her cat. Source (public domain): Pixabay.com. (d–f) Examples of animals that are too spiny to hug. (d) Thorny devil (*Moloch horridus*), an Australian lizard. Photo by Steve Shattuck (CC BY 2.0). (e) Regal horned lizard (*Phrynosoma solare*), native to Mexico and the Southwestern U.S. Photo by Room237 (CC BY SA 3.0). (f) Indian crested porcupine (*Histrix cristata*), from Gray, J. E. 1830–1834. Illustrations of Indian zoology. Digitally enhanced by Rawpixel (CC BY 4.0). (g–i) Examples of animals with toxic surfaces. (g) Hooded Pitohui (*Pitohui dichrous*) of New Guinea, a bird that secretes a neurotoxin causing numbness and burning when touched. Photo by Benjamin Freeman (CC BY SA 3.0). (h) Cane toad (*Rhinella marina*), a native of South and Central America, with skin secretions so toxic that many dogs have been killed. Photo by MyFWC Florida Fish and Wildlife (CC BY ND 2.0). (i) Red lion fish (*Pterois volitans*), a native of South Pacific reef ecosystems, with venomous fin spines. Photo by Michael Gäbler (CC BY SA 3.0).

FIGURE 5.31 Tree-huggers. (a) Endangered Bengal tiger (*Panthera tigris tigris*) in India. Photo by Abhishek Chikile (CC BY SA 4.0). (b) Smokey Bear and friends. Painting (public domain image) by Rudy Wendelin. Credit: U.S. National Agricultural Library. (c) Students hugging a giant redwood (*Sequoia sempervirens*). Public domain photo. Credit: Redwood National Park, California.

FIGURE 5.32 Women's hair: a possible beauty criterion applied to judging the value of fur. (All figures are public domain.) (a) Eighteenth-century painting of Marie Antoinette, Queen of France, by an unknown artist. Note her elaborate hair style. (b) Illustration by Paul Hey of the Brothers Grimm fairy tale in which Rapunzel is locked in a high tower until her rescue by a prince who climbs up to her using her long hair as a rope. Source: https://www.childstories.org/vi/rapunzen-1832.html. (c) Long-haired lady. Nineteenth-century advertising card for a hair preparation published by Dr. J. C. Ayers & Co.

woolen sweaters so cozy. Alpacas also have curly hair that resembles the wool of sheep and similarly is employed for warm clothes. The same is true for Angora goats, their curly wool known as mohair. Other domesticates with curly hair include Mangalica pigs, frillback pigeons, Texel Guinea pigs, Sebastopol geese, and curly horses (Leary 2020). As evidenced by this extensive list, curly-haired animals have often been selected by humans, indicating that the appearance of strongly curled hair is beautiful (Figure 5.33).

Bakmazian (2014) reviewed psychological research on beards and concluded that facial hair influences the perception of physical characteristics of men, such as age, sexual maturation, and attractiveness. From an evolutionary viewpoint, facial ornamentation in animals may be a threat display (warding off other species and competitors of the same species) or a sex attractant (a hypothesis advanced by Charles Darwin in 1871), and this may be the case for human males. Facial hair could also insulate against cold. It has been

FIGURE 5.33 The cuteness of naturally curly-textured hair. (a) A bichon fris dog. Photo by Heike Andres (CC BY SA 3.0). (b) An attractive African-origin female. Photo (public domain) by Le Dinh Thuy. (c) Another attractive African-origin female. Photo (public domain) by Kimberly, Pixabay.com. (d) A lamb. Photo by size4riggerboots (CC BY ND 2.0).

claimed that a lion's mane serves to protect vital areas like the throat and jaw from lethal attacks (Beseris et al. 2020), and in a similar vein, it has been hypothesized that facial hair cushions blows during combat between men (Beseris et al. 2020 found support for this, Dixson et al. 2018 did not). The classical Romans, for a period, allegedly discouraged long hair and beards as either could be grabbed by opponents while fighting (many historians contend the Roman preference for shaving was mostly a fashion trend). During the First World War, beards wouldn't fit under respirators, and a gas attack could be deadly, so soldiers had to shave. Police and military organizations often discourage long hair, but as women and religious groups are progressively conscripted this is changing. Several religions forbid cutting of hair. Conversely, some occupations associated with hazardous equipment or wearing protective masks make long hair dangerous. Some men refuse to shade off facial hair because they oppose modern expectations of appearance or because they are just lazy. Today, beards, moustaches, and sideburns are fashion statements, and often particular styles become trends.

'Pogonophilia' is the cultivation and admiration of beards, while 'pogonophobia' is its opposite. It may be assumed that men with facial hair think it makes them more attractive, particularly to women (indeed, in some cases, facial hair hides blemishes and scars). There is some evidence that women prefer men with facial hair (Miller 1998; Skamel 2003; Dixson et al. 2005; Dixson and Rantala 2016). However, Bakmazian (2014) found that facial hair not only tends to make men less attractive to women, men with beards are rated as less intelligent, and less trustworthy. Jach and Moroń (2020) reviewed the literature and concluded that women were ambiguous in their preferences for beards on men, while men with beards preferred them for themselves but disliked beards on other men. Because the growth of facial hair is dependent on testosterone levels which is correlated with aggression, facial hair does appear to be physiologically reflective of potential hostility. Facial hair makes men appear more forceful and aggressive (consistent with the interpretation that beards are a threat display). Not surprisingly, men with beards or even fast-growing facial stubble are viewed as being more dominant (Etcoff 1999; Puts 2010; Bakmazian 2014). In earlier times in Europe, monarchs such as Henry VIII in England and Peter the Great in Russia placed taxes on beards, so only the rich could afford to demonstrate their high status with facial hair.

Of concern in this section is whether humans find facial hair (or hair more generally distributed) in animals attractive, and whether such judgments could be related to preferences for or against human beards. For humans, facial hair appears to vary from being attractive to looking quite eccentric, and the same seems to be the case for animals (Figure 5.34). The facial hair (essentially the mane) of lions is unquestionably handsome on the animal (Figure 5.34b), but it reflects dangerous power, and similarly so much hair on a human appears threatening (Figure 5.34a); a goatee looks odd both on goats (Figure 5.34d) and on people (Figure 5.34c), although it has been associated with talent and status. Moustaches, representing masculinity but not hiding the face, are considered relatively sexy by women and manly as reflected by frequent association with the military (Figure 5.34e). However, a moustache on an animal may appear quite odd (Figure 5.34f).

As noted in the previous paragraph, people are extremely concerned about the appearance of their own hair, so one might well expect that they regard the fur of animals with approval, indeed possibly with envy. In fact, it has been shown that many have a strong preference for nonhuman mammals with luxuriant fur, especially if it is patterned with spots, rings, or variegations (pied or piebald) (Landová et al. 2018). Animals have been killed for their fur since prehistorical times, and the insulation value of fur allowed people to colonize cold climates. Humans have less of their body covered with hair than virtually any other mammal

FIGURE 5.34 Parallels of facial hair between people and other mammals. (a) Painting (public domain), 'Study of an old man with a beard,' by Antoine van Dyck (1599–1641), in the Metropolitan Museum of Art. (b) Lion. Photo by Tambako the Jaguar (CC BY ND 2.0). (c) Drawing (in 1920; public domain) of Uncle Sam (icon representing presidency of the U.S.) with a goatee, by artist James Montgomery Flagg. (d) Goat with a goatee. Source (public domain): Pixabay.com. (e) Portrait of Soviet dictator Joseph Stalin (1878–1953) with a walrus moustache. Photo by Ephraim Stillberg (CC BY SA 3.0). (f) Walrus, public domain photo. Credit: Megapixie.

(but not the naked mole rat), so fur was a great gift from nature. Fur is one of the softest durable materials that can be employed for clothing and other objects that people touch, and it is very pleasant to the feel. Pet mammals are huggable substantially because their fur is so soft. Fur coats remain fashionable in many countries (Figure 5.35), but for several decades campaigns against cruelty (especially associated with trapping) to fur-bearing animals, and for conservation of the rarer mammalian species, have made genuine fur a pariah item in some countries. Fake fur looks like fur and satisfies the human bias that favors the look of fur.

FIGURE 5.35 Mammalian fur, so admired that it is made into decorative rugs or fur coats. (a) Bearskin rug, Photo by Thomas Quine (CC BY 2.0). (b) Raccoon coat. Photo by Robert Sheie (CC BY 2.0). (c) Lady in fur. Credit: Bea Serendipity/Ghee (CC BY 2.0). (d) Tigerskin rug. Painting entitled 'Scherzando' (public domain) by Ignaz Gaugengigl (1853–1932). Photo credit: https://collections.mfa.org/objects/32126/scherzando. (e) Domestic cat (*Felis catus*) pelts for the fur industry. Photo by Mickey Bohnacker, Presse-Fotograf, Frankfurt/Main, released into the public domain. (f) Timberwolf jacket. Photo by Kuerschner, released into the public domain.

Prejudice against 'Unsightly' Complexions

Humans have been described as 'naked apes,' but shortly after being born naked, we spend the majority of our lives in clothing. However, in most of the world today, females are expected to be much less clothed ('show more skin'). Enormous industries are concerned with skin beautification. Dermatologists are doctors who specialize in treating thousands of conditions involving the skin, hair, and nails. Many of these conditions, like skin cancer, are serious, but many others simply represent humankinds' search for beauty (a statement that also applies to plastic surgeons). The word 'aesthetician' is usually applied to professionals licensed to provide cosmetic skin care treatments (such as facials, hair removal, and makeup application), but depending on local regulations these specialists sometimes employ techniques requiring medical training. Cosmetics have been employed for thousands of years, but in recent times their use has been expanded enormously. Pressured by ceaseless advertising, women (and girls) have become persuaded to compulsively beautify their skin by employing preparations such as 'foundations' and 'moisturizers' to improve the health and/or appearance of the skin. Some products (lipstick, rouge, bronzer, eye shadow) simply paint over the normal appearance of the skin, and others ('concealers') hide or minimize imperfections, scars, and sores. Some products whiten/lighten the skin, reflecting prejudice based simply on natural skin color (see the discussion of color presented later in this chapter). During maturation, progressive change to skin is normal, but camouflaging evidence of aging has become one of the primary purposes of cosmetics.

The prevailing ideal of youthful female beauty calls for a smooth, wrinkle-free, uniformly unblemished complexion. Of course, skin that is covered by clothes is of less concern, and similarly animals that are covered by fur or feathers usually don't appear unattractive. However, animals with bare skin that is wrinkled, bumpy, otherwise uneven, or apparently discolored (Figure 5.36) are at a definite disadvantage insofar as beauty is concerned. Most terrestrial mammals are covered by fur and most birds by feathers, but there are exceptions. Vultures often have naked (feather-free) 'bald' heads, to avoid picking up pathogenic bacteria. This contributes considerably to their often ugly appearance. Animals that build and live most of their lives underground (e.g., earthworms, naked mole rats) may be naked

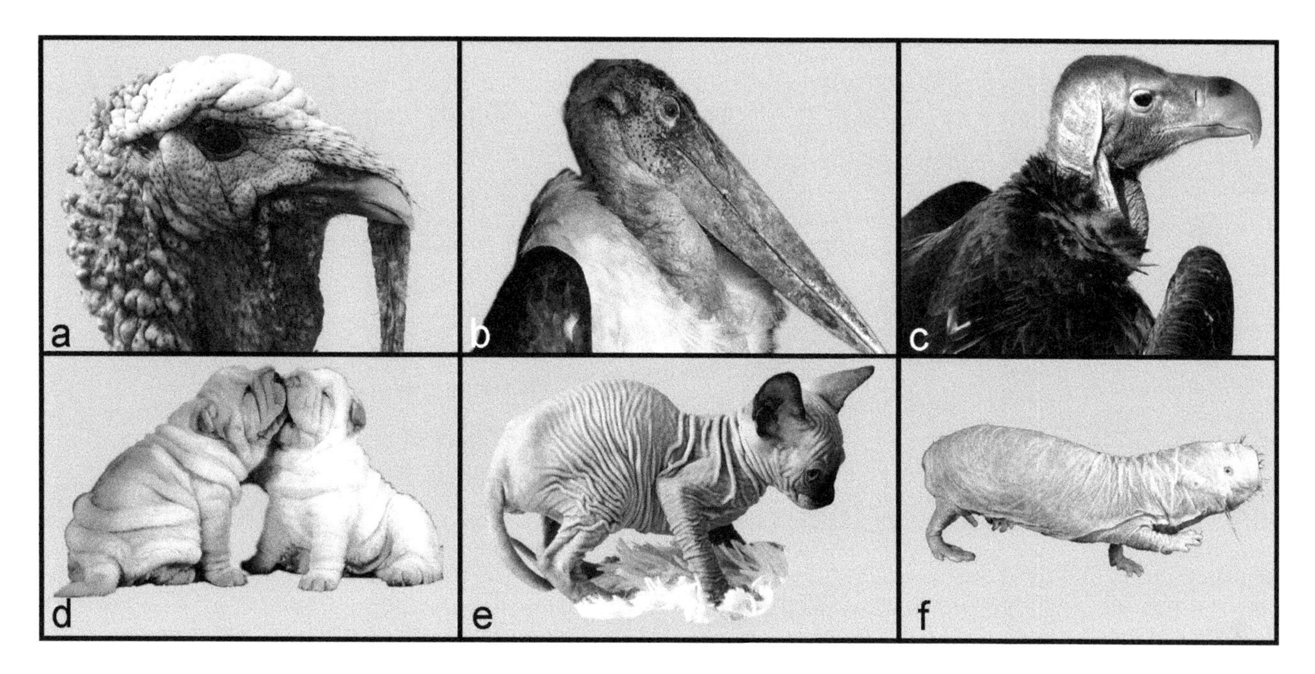

FIGURE 5.36 Animals with skin appearance often considered undesirable. (a) Wild turkey (*Meleagris gallopavo*). Photo by Russ (CC BY 2.0). (b) Marabou stork (*Leptoptilos crumenifer*). Photo by Puffin11uk (CC BY SA 2.0). (c) Lappet-faced vulture (*Torgos tracheliotos*). Photo by Belgian chocolate (CC BY 2.0). (d–f) Animals with very wrinkled skin. (d) Sharpei dogs. Photo by Kitty.green (CC BY SA 2.0). (e) Sphynx kitten. Photo by Dmitry Makeev (CC BY SA 4.0). (f) Naked mole rat (*Heterocephalus glaber*). Photo credit (public domain): Mehgan Murphy, Smithsonian's National Zoo & Conservation Biology Institute.

because this minimizes friction while they are burrowing. Some domesticated animals, particularly certain dog and cat breeds, are quite naked and wrinkly, indicating the wide variety of beauty perception among humans, but nevertheless also reflecting a minority view.

SHAPE

Attempts have been made to 'explain' beautiful shapes mathematically (e.g., Gielis 2017). Regardless, the overall shape of organisms (their 'gestalt') as well as the shapes of individual parts clearly contributes to how humans perceive beauty.

Sexual Dimorphism

Sexual dimorphism refers to the different appearances developed by males and females of many species. The phrase includes differences in reproductive organs but especially emphasizes other ('secondary') features. Female choices of preferred features in males and vice versa are enormously important in evolution (Prum 2017). While external visually apparent features are usually of principal interest, the differences may also be behavioral. Of course, sexual preference (and prejudice) in humans for other humans is very strongly determined by biological sex or by sexual orientation. People strongly respect more impressive species, and when this differs between the sexes of another species (for example, if the males are more visually impressive), there is a corresponding bias.

Sexual Attraction to Other Species

We humans spend much of our lives in the pursuit of sex and judge each other to a very considerable extent by our sexual attractiveness, so the question arises of the extent to which our intrinsic sexual values similarly influence our judgment of other species. A major purpose of sexually dimorphic features of one sex that are attractive to the other sex of a species is to promote sexual reproduction. Of course, for this goal there is no advantage of members of one species being sexually attracted to a different species that is sexually incompatible. Some animals mimic characteristics that serve as sexual attractants in other species in order to lure them close enough to trap them for food (Figure 5.37c). Such features may include physical mimicry, mating calls, and pheromones (chemicals affecting behavior) distributed over an area. Sex pheromones are commonly used to trap pest insects. Some animal-pollinated orchids cleverly modify their flowers to look like female insects to attract pollinators (a phenomenon termed 'sexual deception;' Figure 5.37a). The resulting 'pseudocopulation' can be energetically costly for the pollinator should he ejaculate (Gaskett et al. 2008).

FIGURE 5.37 Dangerous, deceptive, sexual lures attracting one species to a different one. (a) King-in-his-carriage (*Drakaea glyptodon*). This 'hammer orchid' of Australia has one petal (in purple) modified to look like a female wasp. When a male wasp attempts to mate with it, the hinged stalk on which it sits carries the bee against the part of the flower with the stigma and pollen, resulting in pollination. Photo by Mark Brundrett (CC BY SA 3.0). (b) Painting (public domain) of 'sirens' by German artist Hans Thoma (1839–1924). In Greek mythology, sirens were dangerous hybrids of women and birds, who employed enchanting music to lure sailors to become shipwrecked on rocky coasts. Photo credit: Auction House Kaupp, Sulzberg, via ARCADJA auction. (c) Femme fatale firefly (*Photuris lucicrescens*). Females (and sometimes males) of the firefly (beetle) genus *Photuris* hunt species of another firefly genus, *Photinus*, and so are called firefly 'femmes fatales.' Male fireflies flash characteristic, species-specific patterns as they search for stationary females at night, and the females usually respond with single flashes after a specific interval. *Photuris* females lure *Photinus* males by faking the latter's flash signal, and when they arrive expecting sex they are eaten. Photo by Bruce Marlin (CC BY SA 2.5).

Among humans, women widely employ the beauty of flowers (which are essentially sexual in nature) and the scent chemicals of plants and animals (which are also frequently sex attractants) to make themselves more alluring to men. However, these practices do not amount to sexual activity between humans and other species.

'Zoophilia' is sexual attraction of a human toward a nonhuman animal (some authors also include sexual activity). One definition of 'bestiality' (often misspelled 'beastiality') is sexual intercourse between a person and a nonhuman animal ('sodomy' and 'buggery' are sometimes used synonymously, but these terms also have other meanings). Such activity (which is of course non-consensual) is widely considered psychologically deviant and is subject to moral censure and criminal penalties. Domesticated mammals are the chief subjects (or victims), and so bestiality is a concern of animal welfare advocates. 'Human sexual relations with animals has been part of the human race throughout history, in every place and culture in the world' (Podberscek and Beetz 2005). Bestiality porn is widespread on the internet, but the extent to which this reflects frequency in the general population is uncertain (Grebowicz 2010; Holoyda et al. 2018). In literature, fictional hybrids between nonhuman animals and people are sometimes known for sexual prowess, notably goat-human creatures (see satyrs in Chapter 7), and sexually attractive nonhuman creatures are popular in literature (Figure 5.37b).

Preference for Bilateral Symmetry

Numerous simpler organisms exhibit 'spherical symmetry' (roundness; Figure 5.38a) or 'radial symmetry,' a repeating pattern around a central axis, like a pie, with a top and bottom (or a front and back), but without left and right sides (Figure 5.38b). However, most animal species are 'bilaterally symmetrical' (at least externally) with a single plane dividing them into two approximately mirror image left and right halves (Figure 5.38c). Bilateral symmetry is advantageous in allowing the parts of one side to compensate for damage to the corresponding parts of the opposite side (for example, should an eye or an arm be damaged, the other eye or arm would still be functional).

Humans and some other animals find symmetrical patterns to be more attractive than asymmetrical ones (Enquist and Arak 1994). In particular, people find bilateral facial symmetry of other humans to be particularly attractive (Thornhill and Gangestad 1999), a phenomenon that has been explained as due to symmetry reflecting developmental stability and therefore desirability in mate choice (Rhodes et al. 1998). Most discussions of facial beauty cite symmetry as a key determinant. However, overall judgment of the beauty of human faces is complex, and various studies have found that asymmetrical faces are often judged to be as attractive as symmetrical faces (Zaidel and Hessamian 2010). Zaidel et al. (2005) suggested that people judge facial symmetry as a stronger

FIGURE 5.38 Symmetry patterns. (a) Covid19 (=COVID19) virus, exemplifying spherical symmetry. Public domain image from Centers for Disease Control and Prevention, United States Department of Health and Human Services. (b) Necklace starfish or Indian red sea star (*Fromia monilis*), exemplifying radial symmetry. This species of the Indian Ocean and Western Pacific grows to 30 cm in diameter. Photo by Frédéric Ducarme (CC BY SA 4.0). (c) Cinnabar moth (*Tyria jacobaeae*), a widespread Old World species, exemplifying bilateral symmetry. Source (public domain): Nemos, F. 1895. Europas bekannteste Schmetterlinge. Beschreibung der wichtigsten Arten und Anleitung zur Kenntnis und zum Sammeln der Schmetterlinge und Raupen. Berlin, Oestergaard Verlag.

indication of health than of beauty (although there is often strong overlap of health and beauty).

Almost all animals with teeth have these arranged bilaterally symmetrical (i.e., similarly on both sides of the face), but upper and lower teeth may differ dramatically, and this may also be the case from the front of the mouth to the back. Dental symmetry and uniformity of teeth are key indicators of both human attractiveness and health (Figure 5.39a–c). Human teeth are specialized, but compared to many other animals, appear to be relatively uniform and indeed are sometimes compared to Chicklets™ gum. Many animals have sets of teeth within which there are dramatically different kinds specialized for feeding, killing, or defense, and often these appear odd, diseased, or threatening (Figure 5.39d–f).

The extent to which we humans employ symmetry or asymmetry of other species as a clue to their attractiveness

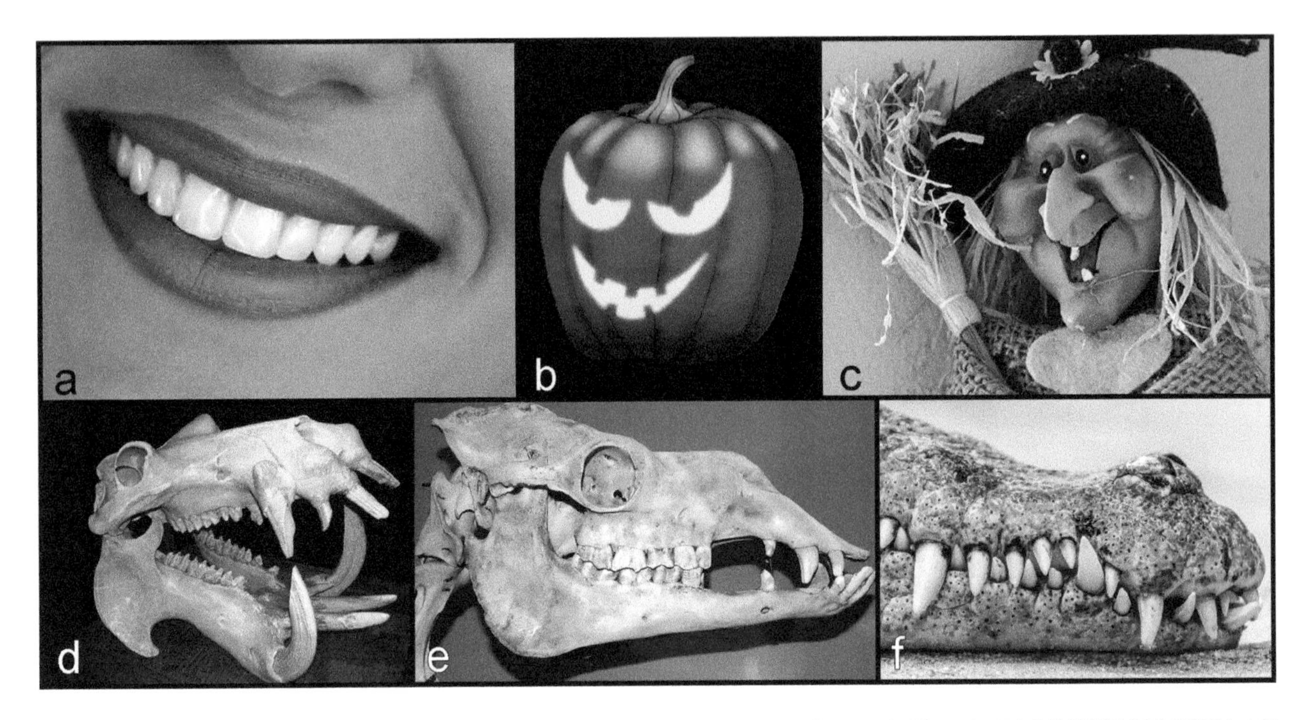

FIGURE 5.39 The contribution of symmetrical, uniform teeth to beauty. (a) Excellent teeth. Photo by Liz20151222 (CC BY SA 4.0). (b) Jack-o'lantern and (c) a witch with scattered teeth contributing to eccentric appearance. Source: Pixabay.com (public domain). (d) Skull of a hippo (*Hippopotamus amphibius*) showing odd arrangement of specialized teeth. Photo by Raul654 (CC BY SA 3.0). (e) Skull of a camel (*Camelus dromedarius*) showing odd specialized teeth. Photo by David J. Stang (CC BY SA 4.0). (f) Snout of a Nile crocodile (*Crocodylus niloticus*) showing irregular teeth. Photo by Ermell (CC BY 3.0).

has not been extensively documented. Some animals appear odd because of their asymmetrical features (Figure 5.40), but their asymmetries are highly adaptive. Pronounced asymmetry in an individual that is unusual for its species tends to indicate disease, injury, or pathological development, and this may trigger hostile behavior from others, both predators and other members of the same species. Bonsai (artistic dwarfing of woody plants) is frequently deliberately asymmetrical (reflecting how plants grow naturally, bending in response to environmental stresses) and seems quite beautiful (Figure 5.41).

Preference for 'Averageness'

For well over a century, it has been known that faces that are artificially made up to represent average features of a group of people are usually more beautiful than any of the individuals (Breyer 2000). (Such so-called 'composite portraiture' should not be confused with 'facial composites' prepared by artists employed by the police to represent the average of several eyewitnesses' memories of a criminal's face.) In an 1877 letter to his close relative Charles Darwin (https://galton.org/essays/1870-1879/galton-1878-nature-composite.pdf), Francis Galton, the pioneer in this area, noted 'I find by taking two ordinary carte-de-visite photos of two different persons' faces, the portraits being about the same sizes and looking about the same direction, and placing them in a stereoscope, the faces blend into one in a most remarkable manner, producing in the case of some ladies' portraits in every instance *a decided improvement* in beauty.' It seems counterintuitive that people with the most beautiful faces have faces that are 'average' in a sense. However, humans have been found to favor features that are indeed average for our species (it's necessary to distinguish an 'average face' from a face representing averages, i.e., the statistical means, of features). People with most or all of their features exhibiting averageness indeed seem beautiful but are rare. It is thought that the phenomenon is biologically based: people apparently perceive that average physical features represent healthy development (Rhodes et al. 2001; Zaidel and Hessamian 2010).

The extent to which humans prefer 'averageness' in other species is uncertain. Judged on the basis of admired paintings and photos of animals and indeed many plants, as well as of living species exhibited in zoos, it is clear that exceptional representatives of many species are often preferred (e.g., a deer with extraordinarily large antlers), but at the same time features that are clearly detrimental are disliked (e.g., a deer with a misshapen leg). Domesticated animals and plants have been selected to be highly divergent from their wild ancestors and indeed have acquired exaggerated features that are desired by humans. Within many domesticated breeds, however, there is often a 'standard' (a set of characteristics each with a specified acceptable range) that breeders are expected to satisfy and perpetuate (breed standards sometimes differ according to organization). Such standards are especially well developed for dogs (Figure 5.42). Clearly,

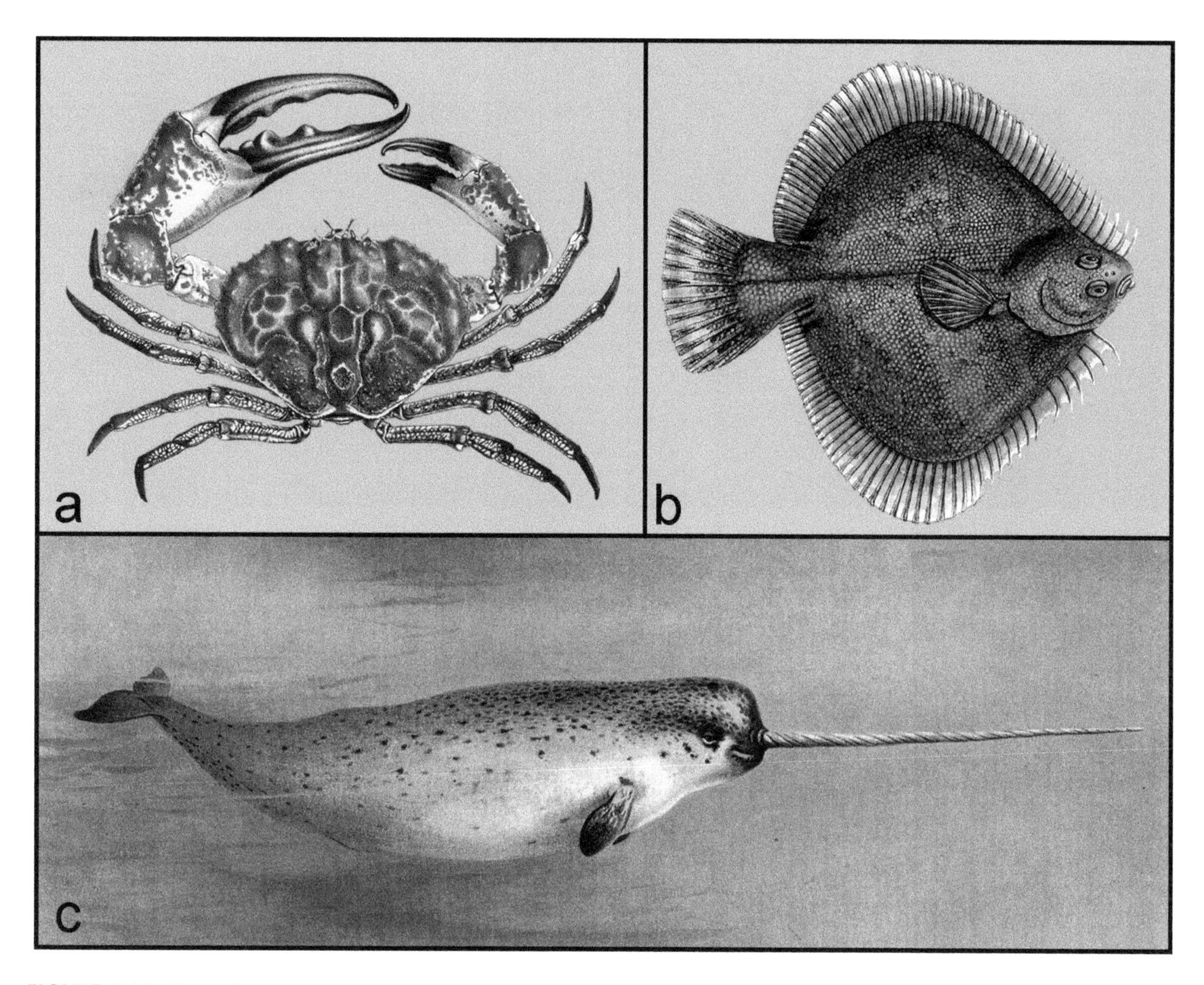

FIGURE 5.40 Examples of asymmetrical animals. (a) Tasmanian giant crab (*Pseudocarcinus gigas*), illustrating the remarkable asymmetrical development of one of the claws of some male crabs. Males can reach 18 kg in weight. The species occurs in the oceans of southern Australia. Lithograph (public domain) by John James Wild (1824–1900). (b) Sand flounder (*Rhombosolea plebeia*) a flatfish of New Zealand coastal waters (note eyes on same side of the head, an adaptation that allows the fish to lie inconspicuously on the ocean floor). Source (public domain): South Pacific fishes by Frank Edward Clarke (1849–1899). Photo credit rawpixel.com. (c) Narwhal (*Monodon monoceros*). Males have a long helical tusk arising from an upper left canine tooth. The species is a resident of Arctic waters. Unlike the previous examples, the asymmetry is not visible externally. Source (public domain): Thorburn, A. 1920. British Mammals.

FIGURE 5.41 Asymmetrically grown beautiful bonsai. (a) Japanese black pine (*Pinus thunbergii*). Photo by Ragesoss (CC BY SA 4.0). (b) Blue atlas cedar (*Cedrus libani* var. *atlantica*). Photo by Ragesoss (CC BY SA 3.0). (c) Indian azalea (*Rhododendron indicum*). Photo by Marufish (CC BY SA 2.0).

FIGURE 5.42 Purebred dog shows and their 'beauty criteria' – standards, which are sets of specifications for acceptable ranges of characteristics. (a) Poster for the 2013 European Dog Show. Photo by Palexpo Genève (CC BY ND 2.0). (b) Sporting Group at the 2010 Westminster Kennel Club Dog show. Photo by Audrey_sel (CC BY 2.0). (c) Diagram showing relative body proportions of boxers to assist in judging conformity to their breed standards. Prepared by Loudenvier, based on a drawing by Pearson Scott Foresman (CC BY SA 3.0). (d) Rottweiler anatomical features included in judging. Prepared by Caronna (CC BY SA 3.0).

since dog breeds are spectacularly divergent and many have very odd features, the attractiveness of dog breeds is a matter of considerable subjective taste. Notably, the 'best in show' may harbor serious but hidden genetic diseases, a bad temperament, and lack of intelligence, indicating that beauty is indeed often just skin-deep.

Prejudice against Fat Animals

Fat people are currently the victims of prejudice, which is acquired during early childhood (Feldman et al. 1988). Curiously, during the Middle Ages and the Renaissance in Europe, obesity was often a reflection of wealth and was relatively common among the elite. The great Flemish painter Peter Paul Rubens (1577–1640) was well known for depicting 'full-figured' naked women, leading to the term 'Rubenesque' (Figure 5.43a). Obesity is sometimes considered to be the most pressing medical problem of modern times, so there is reason to avoid becoming overweight. On the other hand, for genetic reasons, a large proportion of people are simply stout and/or physiologically unable to maintain low weights, and in recent decades, there have been movements advocating acceptance of body image. 'Steatopygia' is the accumulation of substantial tissues on the buttocks and thighs, and is associated with increased adipose tissue in the buttock region. This condition is noted among some women (to a lesser degree in men) of Sub-Saharan Africa, notably among the Khoisan of Southern Africa and Pygmies of Central Africa. Such steatopygia has been associated with racist theories and speculation about its possible adaptive values (Hudson et al. 2008). Animals with what appears to be excessive weight are not viewed as attractive, although fat accumulation may be adaptive. Animals store fat in generalized body areas and often additionally in specific sites. For example, the humps of camels are centralized sites of fat storage. Some mammals and reptiles store

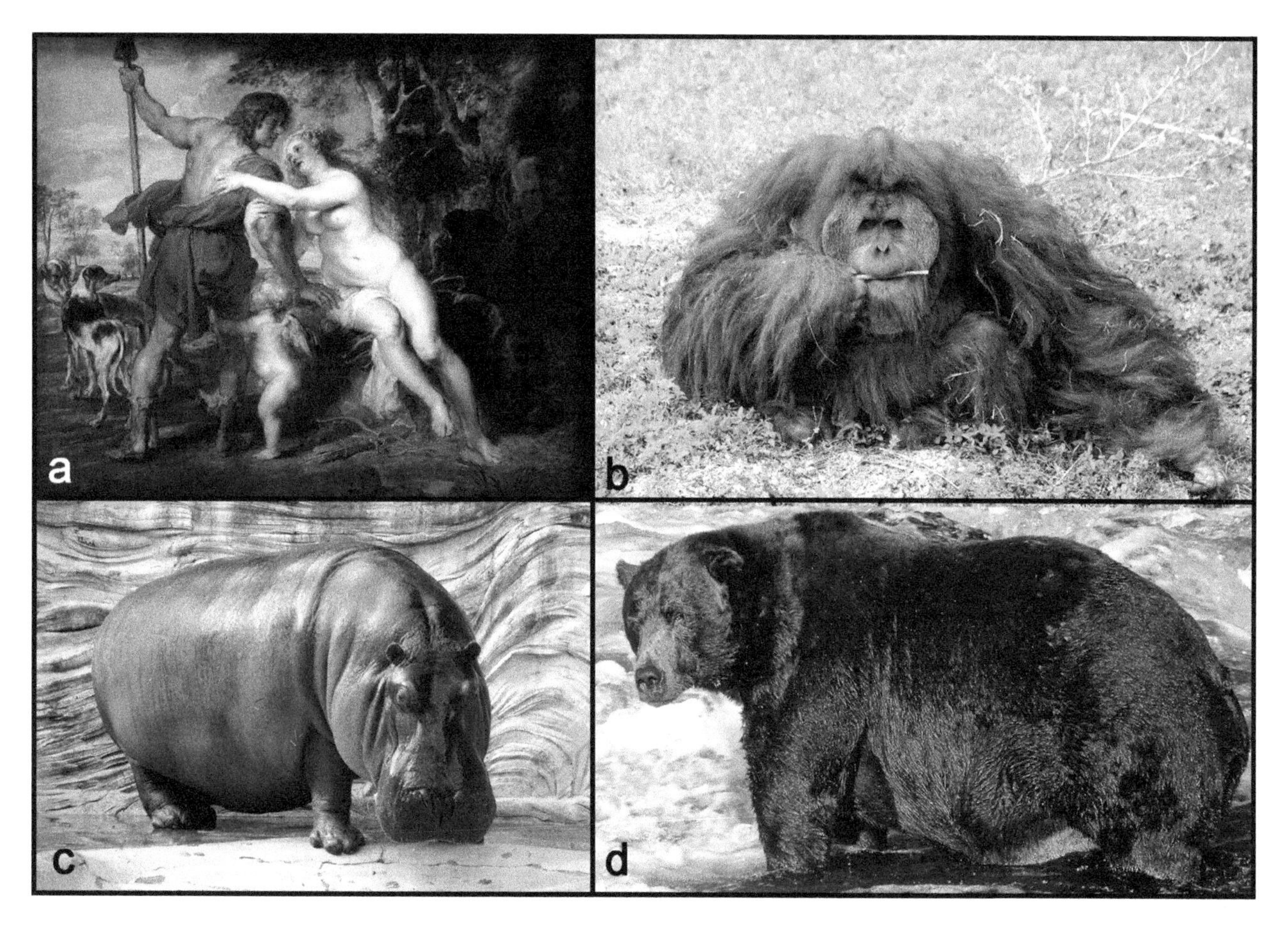

FIGURE 5.43 Animals that are considered less than attractive when they have accumulated fat. (a) 'Venus and Adonis,' painting (public domain) by Peter Paul Rubens in the first half of the 17th century. Housed in the New York Metropolitan Museum of Art. (b) Fat orangutan (*Pongo pygmaeus*). Photo by Amada44 (CC BY 2.0). (c) Fat hippo (*Hippopotamus amphibious*). Photo by Kabacchi (CC BY 2.0). (d) Fat brown bear (*Ursus arctos*) in Katmai National Park and Preserve, Alaska. Photo by N. Boak, U.S. National Park Service (public domain).

FIGURE 5.44 Fat livestock, considered very attractive by the farmers who raised them (19th-century public domain lithographs). (a) Pig; source: British Art Museum. (b) Turkey; published by Jos. Schulz in Germany. (c) Cattle; source: U.S. Library of Congress.

fat in their tails. Bornean orangutans (*Pongo pygmaeus*; Figure 5.43b) consume low-protein fruit to survive, so they store fat by becoming fat when food is abundant in the wild in order to survive periods of famine, probably like early humans (Vogel et al. 2012). Similarly, bears that hibernate in the winter must accumulate large stores of fat by becoming fat in order to survive (Figure 5.43d). Many herbivorous animals must have large stomachs to house large vegetation meals which are slowly converted to nutrients by cooperating bacteria (Figure 5.43c). Although fat livestock may not be attractive to most people (Figure 5.44), they are of course considered very attractive to farmers. However,

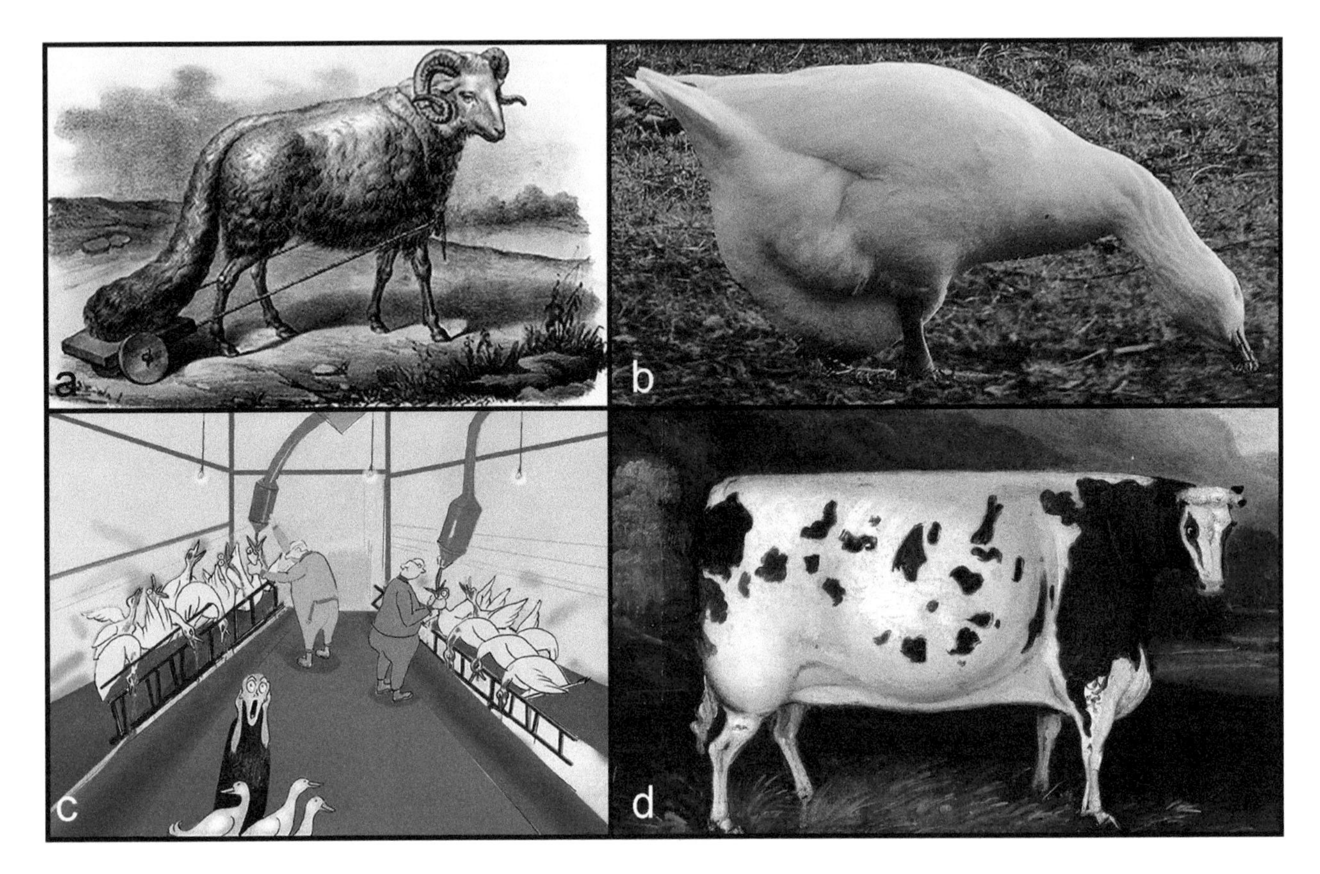

FIGURE 5.45 Livestock that are overfattened to the point of causing pain. (a) Fat-tailed sheep, a breed bred for large production of tail meat. The excessive size of the tail can make movement difficult. Lithograph (public domain) by Raimundo Petraroja (1863–1900). (b) An overfed domestic goose. Photo by Stanzebla (CC BY SA 2.0). (c) A Spanish anti-foie gras poster (public domain), protesting the cruel force-feeding of geese in order to harvest their hypertrophied liver. Revised by B. Brookes. Photo credit: Paco Catalán. (d) An overfed cow, the 'Craven Heifer,' weighing about 2 metric tons, and claimed to be the heaviest known cow ever recorded in England. Painting (public domain) by an unknown artist in 1811.

overfattening becomes an issue when it causes unnecessary pain. So-called Oriental sheep are breeds that are popular in the Mediterranean area. Some of these develop exceptionally large tails valued for their fatty meat, but the tails of some animals become excessively burdensome (Figure 5.45a). 'Foie gras' is a specialty food made from duck or goose liver. Although occasionally produced by natural feeding, it is mostly the product of 'gavage' – force-feeding, mostly in France, which fattens the bird (Figure 5.45b), particularly enlarging the liver by as much as ten times. Because of the cruelty (Figure 5.45c), force-feeding in general and foie gras in particular are widely banned or shunned.

Preference for Slimness

In the youthful phase of development, the bodies of humans and many other animals (and even numerous plants) elongate more than widen, with the result that young bodies are often more slender than mature ones. Thus among people, slimness is correlated with youth, which, as noted earlier in this chapter, is strongly attractive. In the preceding discussion, the modern phobia of obesity is discussed; here, a parallel mania for slim bodies is addressed. In modern times, an obsession with slimness has developed, demonstrated particularly by the fashion industry's employment of female models so skinny they often appear emaciated and suffering from anorexia (Sobolevskaya 2015). Young girls have been particularly affected. As noted in Chapter 3, females are larger than males in many animals. However, in humans, females tend to be smaller and slimmer. Female slimness (Figure 5.46) likely appeals to men as a reflection of youthfulness and so fecundity. Of course, when pregnant, women are much less slim, signaling that they are mated and so sexually unavailable. Accordingly, there may be a biological basis for the male preference for slimness in women. In people generally, slimness also tends to be related to health, since it reflects limited accumulation of fat and a healthy lifestyle and/or good genetics.

Humans are natural 'endurance runners.' Although we are much slower than many four-footed animals, we are (or at least our ancestors were) capable of pursuing prey for hours (like wolves) until the victims drop from exhaustion. This appears to have been a critical adaptation that allowed humans to abandon the arboreal lifestyle characterizing our primate relatives (Bramble and Lieberman 2004; Lieberman et al. 2009). Slimness is a great advantage to human distance runners, since it minimizes the weight carried, so it could be that we have inherited an admiration

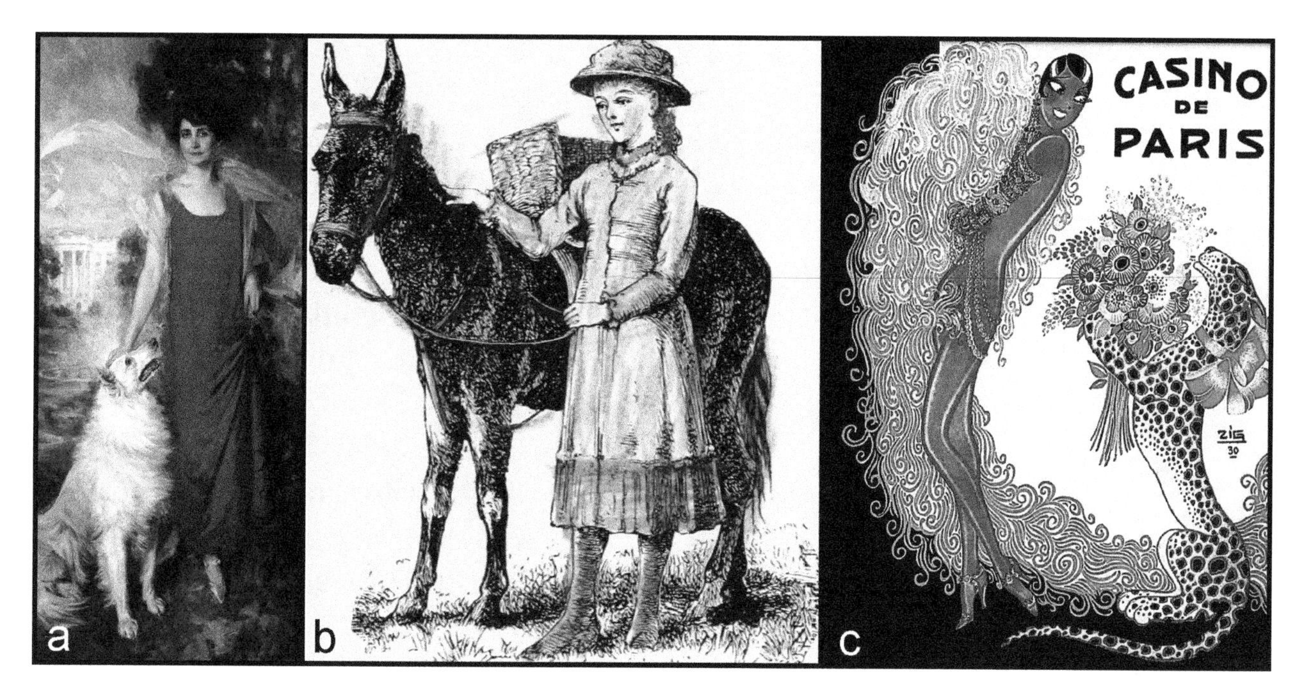

FIGURE 5.46 Slim people and their slim animals (public domain illustrations). (a) Official portrait (public domain) of First Lady Grace Coolidge (1879–1957) by American artist Howard Chandler Christy (1873–1952). Source: U.S. White House. (b) Slender girl and her slender donkey. Source: Johonnot, J. 1885. *Book of Cats and Dogs, and Other Friends, for Little Folks*. New York, D. Appleton and Company. (c) Dancer Josephine Baker (1906–1975) and an admiring cheetah, drawn by French poster artist Zig (Louis Gaudin, 1882–1936).

for this trait. The preference for slimness in humans may generalize to other animals that are also distance runners. The best known of these are racehorses, greyhounds, and cheetahs, which are indeed quite lean (Figure 5.46) and greatly admired.

Attitudes in Favor of and against Species with Unconventional Shapes

The size and shape of animals are limited by physical constraints such as gravity, the need to absorb and distribute oxygen to cells remote from the animal's surface, and limitations of heat loss or absorption from the surface (Smithsonian Institution Press 1987; Sharma et al. 2014). These environmental constraints result in animal sizes and shapes being confined within limits. Flying birds, for example, conform to an easily recognizable pattern, shown by most of the perhaps 18,000 species (Barrowclough et al. 2016). Similarly, most of the more than 30,000 species of fish also are easily recognizable by their shape. Relative constancy of shape applies not just to animals, but to numerous items encountered in daily life – hammers, chairs, spoons, etc., obviously because their functions demand certain shapes. Animals that do not conform to expected conventional shapes often will appear to be odd or unbalanced and may seem objectionable (Figure 5.47a–c). For the most part, exaggerated features of animals adapt them to their habitats. Some animals with very unconventional physical dimensions are admired. Others are subject to ridicule. Elephants with their elongated noses (trunks), giraffes with their elongated necks, penguins with their very short legs, eagles with their aquiline beaks, and birds of paradise with phenomenally long tail feathers are admired. Proboscis monkeys, with their inflated noses, vultures with wattles, camels with humps, and sloths with their huge fingernails, are ridiculed. Most invertebrates, fish, snakes, and amphibians are so different in appearance from humans that their physical proportions are not admired.

Weirdly shaped plants (Figure 5.47d–f), in contrast to weirdly proportioned animals, are viewed much more sympathetically. While plants are mostly admired for their flowers, oddly proportioned and shaped leaves, branches and trunks are also widely appreciated. Indeed, for centuries, European plant explorers searched for 'curiosities' to bring back to exhibit in public botanical collections. To this day, public conservatories (very large greenhouses with collections of plants from throughout the world) feature the oddest plants on the planet. Of course, plants are so different from humans that their features are unlikely to elicit comparisons with human features, as is the case with animals.

However, fruits and vegetables that do not conform to expected shapes because of asymmetry or size, such as spotted apples or curved cucumbers, are often discarded simply for cosmetic reasons (i.e., they fail to meet current beauty standards). So-called 'ugly food' refers to vegetables and fruits that are oddly shaped, colored, or sized. Such produce may appear blemished, deformed, twisted, or look like it is covered by cancerous growths (Figure 5.48).

FIGURE 5.47 Weirdly proportioned species. (a) Aardvark (*Orycteropus afer*) of Africa, with a long snout specialized for eating termites. Photo by Chris Bartnik (CC BY ND 2.0). (b) Platypus (*Ornithorhynchus anatinus*) of Australia, noted for its duck-like snout and other unusual features. Photo by Brisbane City Council (CC BY 2.0). (c) Star-nosed mole (*Condylura cristata*) of northern North America, the nose modified into touch-sensitive appendages. Photo by gordonramsaysubmissions (CC BY 2.0). (d) Dragon blood tree (*Dracaena cinnabari*) of the Socotra archipelago, with its unusual umbrella-like crown, is named for its red sap. Photo by Gerry & Bonni (CC BY 2.0). (e) Grandidier's baobab (*Adansonia grandidieri*) of Madagascar. Baobabs have unusual barrel-like trunks that can store water. Photo by Bernard Gagnon (CC BY SA 3.0). (f) Veitch's pitcher plant (*Nepenthes veitchii*) of Borneo. The leaves of pitcher plants are modified into insect-trapping pitchers. Source (public domain): The garden, an illustrated weekly journal of horticulture in all its branches, edited by William Robinson, Vol. 17, 1880.

In fact, such fruits and vegetables are merely the result of natural growth variation. Estimates of associated wastage are of the order of one-quarter to one-third (Mookerjee et al. 2021). Because of the resulting very large loss of perfectly edible but 'imperfect' or 'suboptimal' food, there are movements in various countries to widen the public's acceptance of 'ugly fruits and vegetables' (Yuan et al. 2019; Xu et al. 2021). This isn't easy, as people seem to assume that good-looking food tastes better (Prokop and Fančovičová 2012). Indeed, culinary competitions frequently assign half the points to taste and half to appearance.

Preference for Long Necks

Some cultures interpret long necks, at least long slim necks in women, as attractive (Figure 5.49a and b). Populational studies comparing neck dimensions of men and women are limited, but it does seem that women tend to have proportionately slimmer necks than men, that this is considered attractive by many men, and that accordingly media columns to women provide advice on how to accentuate their neck slimness. Obesity can reduce neck slimness, so to some extent neck slimness may be a health indictor (Selvan et al. 2016). Barbie, the American toy company Mattel's fashion doll, is noted for her outrageously long slim neck (to say nothing of her other proportions). Heavy brass rings are worn around the neck by Kayan (Padaung) women of Myanmar (Shimoda and Ohsawa 2017; Figure 5.49a), and the myth is widespread that this lengthens their necks. In fact, the rings push down the collarbone and upper ribs, creating the illusion that the necks appear longer. Male animals (including football players) that engage in competitive head-butting exercises benefit from muscular, thick, relatively short necks. (Giraffes batter each other using their long necks but do not attack each other head-on.) Women suffer whiplash injuries much more frequently than men do in car accidents, and this is generally attributed to be significantly related to their different neck structure (Mordaka and Gentle 2003). Long necks are advantageous in some animals for foraging (Wilkinson and Ruxton 2012; Martinez 2015; Wilson et al. 2017; Figure 5.49c–f). The necks of the extinct sauropod dinosaurs reached 15 m in length, adult bull giraffes can attain a neck of 2.4 m, and ostrich necks can exceed 1 m (Taylor and Wedel 2013). Long necks contribute to height of animals, and as noted previously, height is very strongly admired. Generally, vertebrate animals with long necks tend to have an attractive or at least impressive appearance.

FIGURE 5.48 Misshapen fruits and vegetables that are generally discarded because of their objectionable appearance. (a) Twined apples, photo by Zonk43 (CC BY SA 3.0). (b) Octopus-like strawberry, photo by Silverije (CC BY SA 3.0). (c) Duck-shaped zucchini, photo by Alex Gee (CC BY SA 2.0). (d) Heart-shaped potato, photo by Mylenos (CC BY SA 4.0). (e) Wrinkled tomato, photo by Joe Shlabotnik (CC BY 2.0). (f) Branched carrot, photo by Bnilsen (CC BY SA 2.0). (g) Lobed red pepper, photo by Sara Jane (CC BY SA 2.0). (h) Cherry with tumor-like outgrowths, photo by Silveridge (CC BY SA 3.0). (i) Furrowed orange, photo by Forest and Kim Starr (CC BY 2.0).

Attitudes Regarding Long Leg-to-Body Ratios

The four limbs of humans are divided into two pairs: arms for grasping and legs for walking (we are bipedal). Studies have shown that while limbs can be perceived as too long or too short (i.e. there is a range of favorability), there are sex-related preferences for longer legs. (Arm:torso length has not been nearly as fully examined; Versluys and Skylark 2017.) Men prefer women with longer leg-to-body ratios, but women prefer the reverse in men (Swami et al. 2006, 2007; Sorokowski 2010; Kiire 2016), which might partly explain the popularity of high heels (as noted earlier, height per se is admired in humans). In fact, leg-to-body ratio is a sexually dimorphic character: women, in general, tend to have a higher ratio than men (Swami et al. 2006). How this divergence in male and female preference for each other affects evaluation of other animals is unknown. As noted previously, very short legs (i.e., small leg:torso ratio) is a highly attractive feature in babies, so animals that stimulate this feeling might be considered attractive. However, relatively short legs can be the result of poor nutrition (Bogin and Varela-Silva 2010), so humans may be conditioned to regard this as undesirable in adults.

Contrasting limbs of humans and other animals is problematical although other mammals (Figure 5.50c) and birds (Figure 5.50d) can be reasonably compared. A 'limb' can be defined as a jointed or prehensile appendage of an animal body. However, some biologists restrict the term to legs,

FIGURE 5.49 Human females and other animals with long necks. (a) A young Karen (Padaung) woman wearing neck rings for long, slim appearance. Photo by Dave Bezaire (CC BY 2.0). (b) Bust of Nefertiti (1352–1332 B.C.E.), wife of Pharaoh Akhenaten, in Neues Museum (Berlin, Germany). Photo by Philip Pikart (CC BY SA 3.0). (c) Restoration of the extinct *Barosaurus lentus*, one of the largest dinosaurs, defending itself from a pair of *Allosaurus fragilis*. The largest specimens may have exceeded 48 m in length, 66 tons in weight, with a neck length of at least 15 m. Photo by Fred Wierum (CC BY SA 4.0). (d) Giraffe (*Giraffa camelopardalis*). Public domain photo from Pixabay.com. (e) Ostrich (*Struthio camelus*). Photo by Erick Kilby (CC BY SA 2.0). (f) American Flamingo (*Phoenicopterus ruber*). Photo by Charles J. Sharp (CC BY SA 4.0).

arms, wings (of birds), and fins (of fish) but exclude tails. All terrestrial vertebrates, including amphibians, reptiles, birds, and mammals, had (or at least their ancestors had) four limbs (and so are 'tetrapods'). Some, particularly snakes, have lost their limbs (although vestigial legs may persist in some species). The fins of fish were the precursors of legs but are so different that visually fish, like snakes, are legless. Invertebrates have developed many limb patterns. They are often legless, sometimes with numerous legs, but most tend to have six or eight similar limbs. Insects have six legs, spiders and their relatives have eight, octopuses have six arms and two legs (not eight tentacles), squids have eight uniform legs and two longer tentacles with sucker pads for grabbing prey, and there are other patterns (note the lion's mane jellyfish, Figure 5.50b, whose 'tentacles' are not true limbs but can be interpreted visually as such). Humans use two relatively large limbs for walking, so the natural tendency is to make comparisons with animals that also walk erectly or hop on two limbs (like some other bipedal mammals, birds, and frogs). Mammals and birds with very long limbs in relation to their torsos do look curious but not really objectionable. Invertebrates with unusually long legs (Figure 5.50e–g) are so different from humans (regardless of their limbs) that it is difficult for people to identify with them.

Foot Preferences

Feet (as well as hands) are sexually dimorphic characteristics in humans: they are significantly larger in men, presumably an adaptation for a more physically aggressive life. Like some other differences between the sexes, they are visual signals that serve as sexual attractants. In practice, this has meant that small feet are valued in females, with the unfortunate result today that many women force their feet into painfully small shoeware. The traditional tale of Cinderella having beautifully small feet allowing her to wear

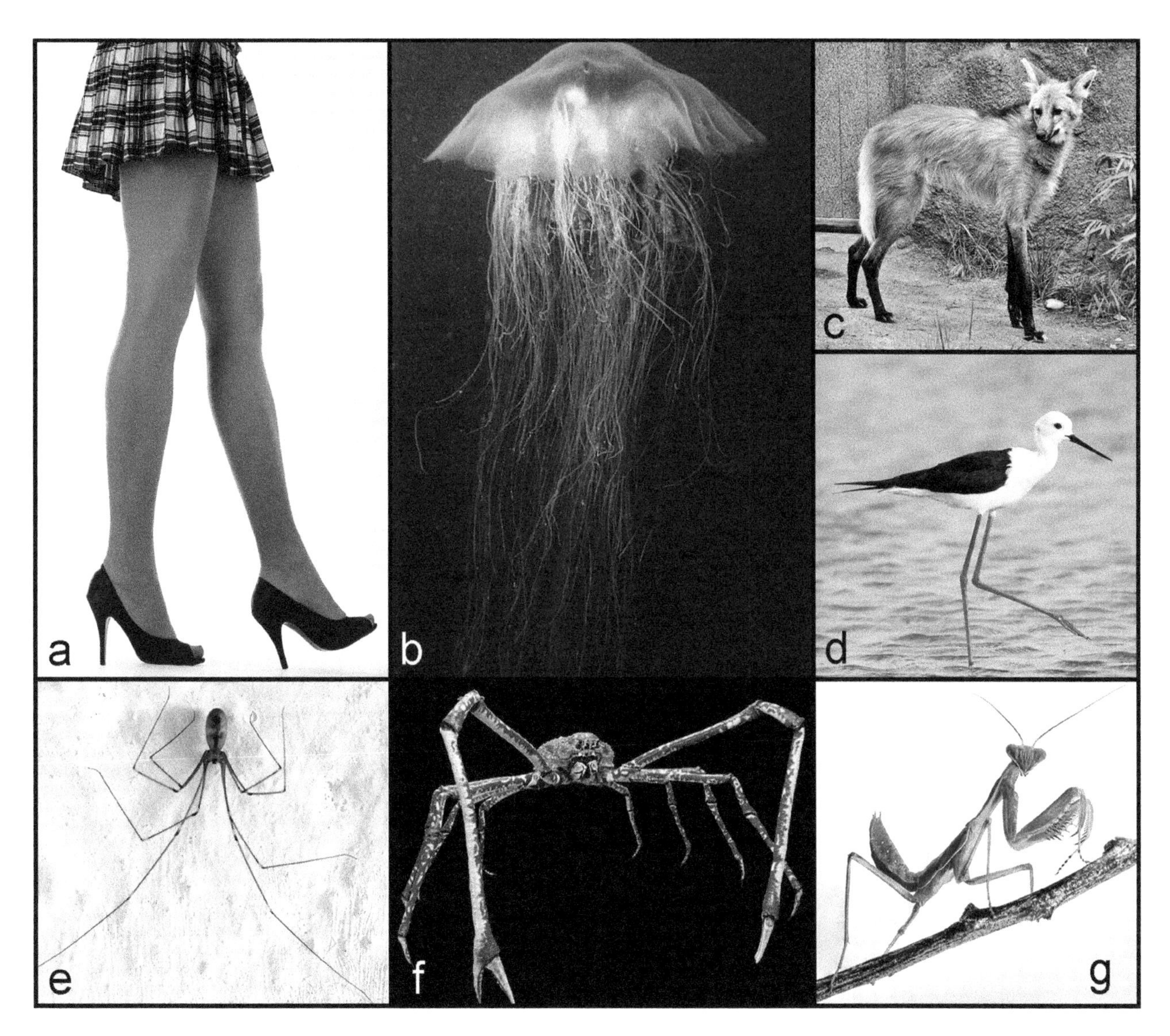

FIGURE 5.50 Examples of animals with large limb to body ratios. (a) A long-legged woman. Photo (public domain) by Pixabay.com. (b) Lion's mane jellyfish (giant jellyfish, *Cyanea capillata*), the largest jellyfish, and possibly the longest animal. Its 'bell' is up to 2.5 m across, and its tentacles can exceed 35 m, longer than the blue whale. It occurs in the world's cold northern seas and uses its stinging tentacles to capture prey. Photo by Derek Keats (CC BY 2.0). (c) Maned wolf (*Chrysocyon brachyurus*), of South America. Its 1 m height allows it to see over grass while hunting. Photo by Tomás Del Coro (CC BY SA 2.0). (d) Black-winged stilt (*Himantopus himantopus*). Such wading bird species frequently have very long legs. Photo by JJ Harrison (CC BY SA 3.0). (e) Daddy long-legs spider (*Pholcus phalangioides*). Photo by André Karwath (CC BY SA 2.5). (f) Japanese spider crab (*Macrocheira kaempferi*). This marine crab lives in the sea around Japan. It has the largest leg span of any arthropod: up to 3.7 m, and can weigh up to 19 kg. Photo by Lycaon (Hans Hillewaert; CC BY SA 3.0). (g) European mantis (*Mantis religiosa*). There are over 2,500 species called mantises, also known as 'praying mantis' for the distinctive posture of the first pair of greatly enlarged forelegs, which are spiked and adapted to grasp prey. Photo by Ángel M. Felicísimo (CC BY SA 2.0).

a tiny glass slipper (Figure 5.51a) illustrates how entrenched is the bias against large feet in women. In the same vein, the feet of young Chinese upperclass women were once bound so that they would be tiny, although this simply distorted the feet and crippled the women (Figure 5.51b). By contrast, clowns traditionally wear huge shoes, enhancing their image as buffoons (Figure 5.51e). The feet of animals are extraordinarily diverse (Figure 5.51c, d and f), reflecting very different functions and adaptations. Probably most people do not take particular notice of the feet of most animals, except as curiosities. Some domesticates (such as horses, and pet dogs in winter) are shoed, indicating that sometimes people are concerned with the feet of animals. Humans are members of the 'great apes' (Hominidae), but unlike the other members (gorillas, orangutans, the chimpanzee, and the bonobo), we are bipedal and lack prehensile feet like the more arboreal members which are adapted to climbing trees (either full- or part-time). Essentially the feet

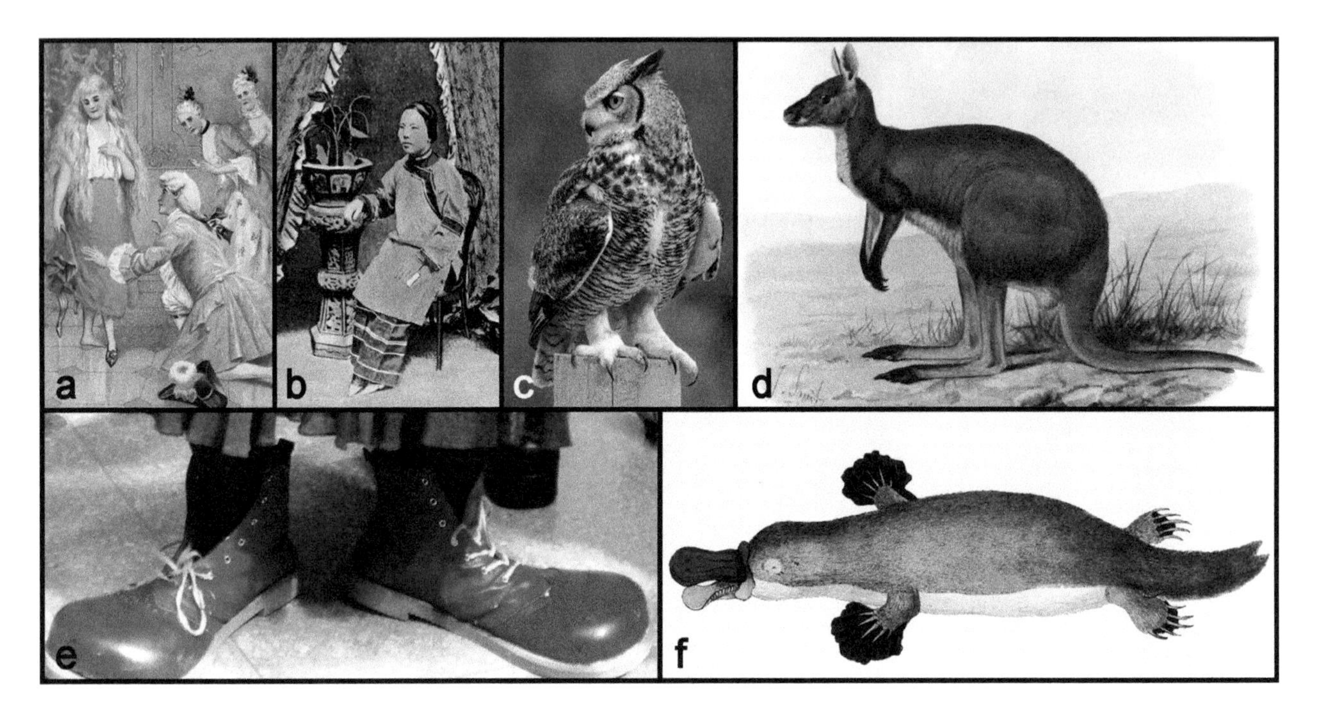

FIGURE 5.51 Extreme feet. (a) Cinderella successfully wearing the tiny slipper that demonstrates that she is worthy of being the prince's true love. Painting (public domain) by artist Jenny Nysrøm (1854–1946). Credit: National Library of Norway. (b) Postcard (public domain) entitled 'A high class Chinese lady with small feet.' Presumably she was a victim of the practice of binding feet to produce the tiny feet valued in early China. Published in 1907 by M. Sternberg. Scanned by the New York Public Library. (c) Great horned owl (*Bubo virginianus*) with killing talons. Photo by Peter K Burian (CC BY SA 4.0). (d) Common wallaroo (*Macropus hagenbecki*) with jumping feet. Wallaroos include two species of kangaroo-like mammals native to Australia. Painting (public domain) by Joseph Smit (1836–1929), published in 1907 in Novitates Zoologicae, vol. 14. (e) A pair of huge clown shoes being worn. Photo by In pastel (CC BY 2.0). (f) Paddle feet of a duck-billed platypus (*Ornithorhynchus anatinus*). Illustration (public domain) by Frederick Polydore Nodder in Shaw, G. 1799. *Naturalist's Miscellany*. London, Nodder & Co.

of the other great apes (and indeed many other primates) are rather hand-like, and having four hands seems like something to be greatly admired.

The Attractiveness of Posteriors

In anatomical biology, the word 'posterior' often designates (1) the rear (back, dorsal, hind) side or (2) the lower end of an animal. The following discussion is based on the first meaning and basically refers to part of an animal that is viewable when it is retreating. In common language, posterior often means buttocks (backside, derrière). Often, there are notable features of the posterior of an animal. The 'butt' of humans is comparatively much better developed than in related species, partly due to our upright style of walking ('bipedality') and the associated musculature (especially the gluteus maximus) and perhaps also as a site for energy (fat) storage (Lieberman et al. 2005; Shapiro 2020). Expanded female buttocks are linked to a wider pelvis and associated ease of childbirth, and not surprisingly are an extremely strong attractant to males (male buttocks are a weaker attractant to females). Havelock (1927) hypothesized that corsets and bustles are intended particularly to draw attention specifically to the buttocks of human females. At least judged by past fashion trends, clothing has been employed as an attention-getter to the back of people, especially females (Figure 5.52a–c). The practice of attaching fluffy tails to the backside of Playboy Bunnies was in this tradition. In a similar vein, the back parts of some animals are colored to serve as sex attractants (e.g. Figure 5.52d and f), and since humans like color, this would also be viewed as attractive to people.

The only thing strikingly reminiscent of a human posterior in the plant kingdom is the fruit of coco de mer (*Lodoicea maldivica*), a palm tree endemic to the Seychelles (an island archipelago nation northeast of Madagascar). The fruit weighs up to 32 kg and contains the largest seed in the plant kingdom. The seed is enveloped in a huge bi-lobed fruit, called the double coconut, which is highly reminiscent of a woman's derrière (Small and Catling 2003; Figure 5.53). Because of this resemblance, the fruits were valued in the past as an aphrodisiac. Today, they are sometimes sold as curiosities and frequently exhibited in museums, where they are a source of amusement.

Evaluating the attractiveness of the posterior of animals to people is complicated by the existence of 'tails.' Tails have been defined in a restricted way as distinctive, flexible posterior appendages of the torso. In numerous animals, the tail functions to aid in balance. In some animals (such as many New World monkeys), it is prehensile, aiding in

FIGURE 5.52 Attractiveness of the posterior of animals: (a–c) clothing that emphasizes the posterior; (d–f) spectacular tails or rear segments. (a) Dress coat with tails. Source (public domain): Vanity Fair, 14 March 1885. (b) Woman's dress with a fishtail train. Painted by the Mademoiselles Giroux (France, circa 1880). Public domain photo released by the Los Angeles County Museum of Art. (c) Advertisement of women wearing bustles in the 1880s. Photo by C. Perrien (public domain). (d) Male superb lyrebird (*Menura superba*) of Australia. Painting (public domain) by Thomas Davies, published in 1810 in Transactions of the Linnean Society of London, Volume 6. (e) Butterfly tail goldfish (*Carassius auratus*). Photo by Syberspace (CC BY 3.0). (f) Male peacock spider (*Maratus volans*) in a courtship pose. Photo by KDS444 (CC BY SA 3.0).

climbing trees. In dogs, tails serve notably as communication devices. Some animals, like red kangaroos and scorpions, use their tails as weapons. Fish and other aquatic animals such as whales and crocodiles employ tails for propulsion. Some animals use their big tails as parasols for protection from overheating by the sun, or conversely as a blanket for warmth. Still other animals, such as some lizards, store fat in their tails. Curiously, human embryos have vestigial tails, and tails sometimes are produced in people as developmental abnormalities. However, humans and other great apes are tailless as adults (facetiously, we have been accused of having 'tail envy'). In general, tails (unlike faces) are rather uncomplicated and are not negatively viewed by people. Indeed, some domesticated animals have been selected in part for their attractive tails (Figure 5.52e). Tails usually hide the anus, which most people do not wish to view (it has been said that if people had tails, pants that hide the buttocks would not be necessary). Mermaids – female humans in which the bottom half of the body has been replaced by a fishtail – are quite beautiful, albeit mythical, indicating that tails are not unattractive. Ponytails are particularly attractive on ponies and women (Figure 5.54a and b), perhaps because trailing ponytails in motion convey health and athleticism (Figure 5.54c).

FIGURE 5.53 Coco de mer (*Lodoicea maldivica*). (a) Fruit, resembling a woman's posterior. Photo by Didier Descouens (CC BY SA 4.0). (b) A Seychellois boy, his right hand on a coco de mer fruit, his left hand on an elongated cluster of male flowers. Photo (public domain) from a 1977 album by Maxime Fayon, photo credit: Dino Sassi & Marcel Fayon, Eden Ltd. (c) Painting (public domain) of coco de mer trees in the Seychelles, by British artist Marianne North (1830–1890).

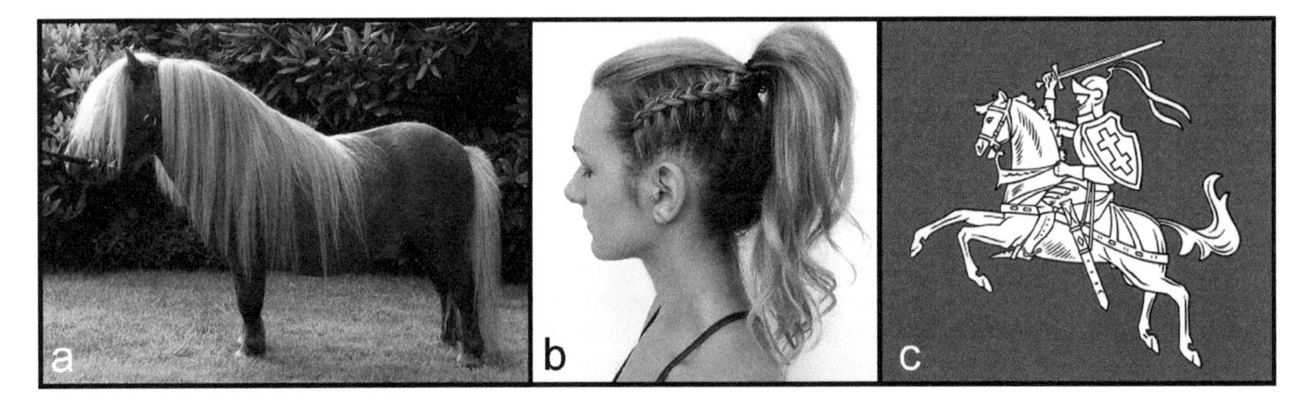

FIGURE 5.54 Attractive tails. (a) A Shetland pony with an enormous tail. Photo by Frederik De Graeve (CC BY SA 3.0). (b) A woman with a ponytail. Photo (public domain) by Ivan Sudorenko. (c) State flag of the Republic of Lithuania until 1940. Note the athletic impression conveyed by the flowing tail and the ribbon attached to the knight's helmet. Public domain photo.

The Attractiveness of Narrow Waists

Clothing that constricts the waist is an unsubtle way of emphasizing both the female pelvis and bust. Corsets are garments worn to train and hold the torso into a desired shape, particularly producing a smaller waist, sometimes also enhancing the bust and/or bottom. Women's girdles are garments that enclose the lower torso, extending below the hips. Fashion trends over the centuries have promoted (often painfully) cinched waistlines producing 'wasp-waisted' female silhouettes (Figure 5.55a–c). Extremely narrow waists are indeed characteristic of wasps, which are any of hundreds of thousands of narrow-waisted insects of the insect order Hymenoptera. The idealized hourglass figure of women is based on sexual attraction to men, but the somewhat similar architecture of wasps, and indeed of the related ants (which evolved from wasps), has no such appeal (Figure 5.55d–f). However, there is a certain similarity of directing the eye to three areas: head, upper body, and lower body, albeit normally arranged horizontally rather than vertically as in humans. Children easily learn to draw wasps and ants, so at the very least their body plan intuitively is familiar, unlike the quite strange appearance of many invertebrates. Although wasps and ants are considered to be pests, at least one can appreciate the symmetry and impressiveness of their bodies.

Narrow waists contrast with big bellies, which in past times, when food was often scarce, were viewed as a reflection of prosperity. Today, especially for men, a large stomach

FIGURE 5.55 Narrow waists. (a) Queen Elizabeth I (1533–1603) in 1592, painting at the National Portrait Gallery in London. Photo by HarshLight (CC BY 2.0). (b) Lady Elizabeth Vernon in 1576, Countess of Southampton, Maid of Honor and Lady in Waiting to Elizabeth I of England. Photo (public domain) by Peter K. Levy. (c) Late 18th-century painting (public domain) by U.S. artist John Collet entitled 'Fashion Before Ease.' (d) Asian mud-dauber wasp (*Sceliphron curvatum*). Photo by Gubin Olexander (CC BY 4.0). (e) German wasp (*Vespula germanica*). Photo (public domain) by Bernie. (f) African thief ant (*Carebara vidua*), wingless queen. Photo by Bernard Dupont (CC BY SA 2.0).

is a sign of obesity and health issues (related to harmful abdominal fat) and is considered unattractive. Indeed, a torso characterized by 'six-pack abs' has become a male beauty standard. In women, a swollen midsection often indicates pregnancy – a condition that may signal to men (except the father) that the woman is not a potential mate, but to women that she is not a competitor. Herbivorous mammals may have quite expanded bellies that act as fermentation chambers in which symbiotic bacteria convert otherwise indigestible plant materials to nutrients for their host. However, mammals with such 'pot bellies' may appear to be comical (Figure 5.56). For additional information, see the previous discussion concerning prejudice against fat animals.

POSTURE

The work ethic is so firmly engrained in modern society that people at rest during what are normally working hours seem to be lazy. Accordingly, animals that spend daylight hours lying down appear lazy. Lions typically spend 20 hours a day sleeping (often hunting at night), but they are so majestic their apparent laziness seems not to bother observers. Pigs, on the other hand, lying down, often seem to be indolent. More serious are poses assumed by animals that make them seem to be threatening, which have been shown to induce negative emotions in viewers (Prokop and Fančovičová 2017). A skunk

FIGURE 5.56 Species that tend to develop pot bellies. (a) Caricature by James Gillray (died 1815) of George IV, King of the United Kingdom, 1820–1830, known for being overweight. Photo credit: U.S. Library of Congress. (b) A domestic pig (*Sus scrofa*) with a pot belly. Photo by Thomas Au (CC BY 2.0). (c) Proboscis monkey (*Nasalis larvatus*). Photo by Bernard Dupont (CC BY SA 2.0). (d) White-spotted puffer (*Arothron hispidus*). Pufferfish have stretchy stomachs which they enlarge for protection by rapidly pumping water into them. Source (public domain): Jordan, D. S. and B. W. Evermann. 1903. Bulletin of the United States Fish Commission, Vol. 23. (e) Central bearded dragon (*Pogona vitticeps*), an Australian species often kept as a pet. Photo by Lahm12 (CC BY SA 4.0). (f) Big-belly seahorse (*Hippocampus abdominalis*). Photo by Zureks (CC BY SA 3.0).

aiming its back end should not be challenged. Of course, an animal that looks like it may attack is dangerous, but as repeatedly pointed out in this book, looks can be deceiving. A coiled snake could be ready to strike, or merely bluffing. In the discussion of hyenas in Chapter 17, it is noted that their shorter hind legs contribute to an appearance that has been described as 'skulking' and therefore scary. Morris (1969) contended that the popularity of penguins, primates, bears, pandas, and dogs was increased by their ability to stand vertically, a posture resembling the appearance of humans. Prokop et al. (2021) showed that bipedal posture contributes to the perceived cuteness of small animals, but when large animals like bears stand up they can be perceived as threatening. Note in Figure 5.57 how small animals standing up seem especially cute, but the larger animals standing on their hind legs appear to be especially dangerous. Cartoonists commonly show animal characters intended for children in poses resembling people standing up, because this makes them more attractive (Figure 5.58).

COLOR

Color has significant effects on human psychology (Elliot and Maier 2014; Elliot 2015). 'Color' is defined in physics as the range of the light spectrum visible to humans (when reflected by objects), a definition that excludes black (which is due to absorption of all wavelengths; i.e., black is absence of light). White is due to reflection of all wavelengths and has been arbitrarily considered to be or not to be a color. 'Hue' technically refers to colors excluding black, white, and gray, but in common language, it is a synonym of color. 'Gray' has been termed a 'shade' of white. ('Shades' are degrees of darkness of colors; 'tone' is variously defined, such as 'degree of lightness of a color,' and for this discussion, it means the same thing.) Psychologically, black and white are colors (a black dress is a black dress, not a colorless dress), but it is convenient to treat the black-white continuum as a separate component of the subject, since it has special significance to bias, bigotry, discrimination, prejudice, and racism.

FIGURE 5.57 Relative human perception of small and large animals that stand erect. Notice that the cuteness of the small animals in the top row is enhanced, whereas the large animals in the bottom row appear more threatening. (a) Bengal (domestic) cat. Photo by Lightburst (CC BY SA 4.0). (b) Dachshund. Source: Shutterstock (contributor: Maxtimofeev). (c) Long-tailed weasel (*Neogale frenata*), Photo by USFWS Mountain-Prairie (CC BY 2.0). (d) Meerkat. Photo by Anthony Kelly (CC BY 2.0). (e) Circus animal trainer confronting a tiger. Source (public domain): Pixabay.com. (f) Two Siberian (Amur) tigers (*Panthera tigris altaica*) fighting in Siberian Tiger Park, Harbin, China. Source: Shutterstock (contributor: Gudkov Andrey). (g) Komodo dragon in Komodo National Park, Indonesia. Source: Shutterstock (contributor: Gudkov Andrey). (h) Polar bear. Photo of a display in the Children's Museum of Indianapolis by Daniel Schwen (CC BY SA 4.0).

In nature, color is highly adaptive. Particular colors or color combination patterns often attract, deter, or hide animals. Most animals do not wish to be noticed most of the time, and so are drab or camouflaged. Color is usually easy to notice, and so a differently colored animal in a crowd is easy to pick out, and so can be a target for predators (Figure 5.59). Similarly, a colored minority in a region can also be easily recognized and subjected to unequal treatment. One or both sexes of some animals are prominently colored (at least during the reproductive period) in order to attract a mate. Highly venomous animals sometimes avoid predators by developing prominent warning coloration, and many harmless animals often mimic them to take advantage of the scary colors. In plants, flower colors serve as attractants to pollinators, and color changes are important in advertising to animals that fruits are immature or are ripe enough to consume.

Preferences for and Prejudices against Black and White

Many animals have 'monochromatic vision' – they cannot see colors other than black, white, and gray. These include some nocturnal mammals such as various bats and raccoons, and some rodents. Marine mammals with monochromatic vision include seals, sea lions, walruses, and cetaceans such as dolphins and whales. Sharks also have monochromatic vision, but most fish have good color vision. For animals with monochromatic vision, shades or combinations of black, white, and gray are especially important, for example for recognizing prey and avoiding predators. Humans have excellent color vision, and while a variety of colors are appreciated, black and white also figure importantly in determining preferences for other species, as discussed in this section.

FIGURE 5.58 Anthropomorphic animals, their attractive human appearance enhanced by standing upright on their hind legs. (a) Vintage (before 1917; public domain) old Russian Easter postcard showing rabbits delivering a giant egg. Source: http://fotki.yandex.ru/users/koziuck-vladimir/album/158391/. (b) Comic postcard featuring a monkey mailman delivering a letter to a female monkey in a tree; advertisement for Sander's Sinclair Station near Lindberg. Source (public domain): https://collections.carli.illinois.edu/cdm/compoundobject/collection/nby_teich/id/435051. (c) Kitten caught pilfering jam by its mother. A cartoon (public domain) entitled 'Caught in the Act,' from an 1880 book, Pets at Play, published by McLoughlin Bros. (d) Dog from a poster by Konrad-Adenauer-Stiftung, a German political foundation (CC BY SA 3.0). (e) Vintage trade card (public domain) showing a fishing frog, published by J. D. Larkin & Co. Source: digital.lib.muohio.edu/u?/tradecards, 2441. (f) Elephant acting as a janitor. Cartoon by Auguste Vimar (1851–1916) published in the book Le Boy de Marius Bouillabès d'Auguste Vimar, Paris, H. Laurens. (g) Poster (public domain) for Bock Beer featuring a goat, published by Calvert Lithographing Co., Detroit. Source: Library of Congress.

FIGURE 5.59 Individuals that stand out because of coloration are potential victims. Source: Shutterstock. Contributors: (a) Volkanakmese, (b) TaiChesco, (c) Nuvolanevicata.

The word 'colorism' is thought to have been coined in 1982 to mean 'prejudicial or preferential treatment of same-race people based solely on their color' (Norwood 2015). The word shadeism has acquired this meaning, and the word colorism is now also used to designate discrimination among people based on hue (white, black, yellow, red). This discussion deals with the subject of relative darkness along the white-black continuum. In humans, skin darkness has been selected as a compromise between avoiding damage from UVB radiation at low latitudes and manufacturing vitamin D in short-season environments. Dark appearance can also be adaptive in other animals. In modern societies, people with dark pigmentation frequently experience discrimination based on gradations of skin color (Chen and Francis-Tan 2021). Such skin tone bias overlaps with racial discrimination (Norwood 2014). 'Pigmentocracy' refers to the superior status afforded to people with lighter compared to darker skin tone. The issue is complicated by sexual dimorphism in facial coloration: women tend to have lighter skin than men (Nestor and Tarr 2008), a phenomenon which is consistent across racial and ethnic groups (Frost 2005). Of course, men have beards, which are usually dark in younger individuals, contributing to facial darkness. The issue of skin color preference/prejudice is complicated by parallel gender preference differences for shade between men and women, the latter preferring men who are darker (Carrito et al. 2017), as reflected by the expression 'tall, dark, and handsome.' In Asia, there is a pronounced tendency for women to whiten their skin in the belief that this makes them more beautiful (Paramita and Winahjoe 2014). In China, men prefer women with lighter skin, and vice versa, according to Wang and Cao (2020). In India, Nagar (2018) found that both men and women prefer lighter-skinned people of the opposite sex. Similar attitudes have been found in the U.S. with respect to degree of lightness among black people (Crivens 1999). Duckitt et al. (1999) found that young (3- and 5-year-old) children in South Africa showed strong anti-black, pro-white color bias toward humans, which they suggested was the result of early environmental conditioning. Ioannis et al. (2014) found that consumers (whether black or white) preferred goods packaged in white rather than black. The issue of whether there is an innate preference for white or black or whether this is learned is unsettled, but, regardless of origin, there is extensive prejudicial treatment with regard to human skin tone. American civil rights activist and Baptist minister, Martin Luther King Jr., in 1963 memorably stated 'I have a dream that my four little children will one day live in a nation where they will not be judged by the color of their skin but by the content of their character.'

The degree to which human prejudice based on skin pigmentation leads to similar prejudice against animal species is unclear. The superstitious prejudice against black cats is well known (Figure 5.60). Some studies suggest that people adopt black cats and dogs in animal shelters at significantly lower rates than when these animals are otherwise colored (Porter 2016). 'Black dog syndrome' or 'big black dog syndrome' is the alleged preference for lighter-colored animals over black dogs (Sinski et al. 2016). However, Miller et al. (2019) found that black and white cats in an Australian shelter were adopted at the same rate. Aside from pet dogs and cats, wild animals that are entirely white or black are uncommon but frequently are perceived as beautiful and are often highly valued as zoo exhibits (Figures 5.61 and 5.62). White-furred buffalo (i.e., American bison, *Bison bison*) are very rare and are considered sacred in several Native American religions (Figure 5.63b). Curiously, rare white mutants in normally black species and black mutants in normally white species are sometimes rejected by others within their species (Binkley 2001), but these are examples of failure of recognition, not of genuine prejudice in wild animals.

Some mutations produce unusually black ('melanistic') variations of animals, or individuals that are much blacker than normal. The word 'panther' is ambiguous, for example defined as 'A large wild cat such as a leopard or jaguar, especially in a color form with black fur' or 'any of various big cats with black fur.' In Asia and Africa, so-called black panthers are leopards – *Panthera pardus* (Figure 5.61b); in Latin America, they are jaguars – *Panthera onca*

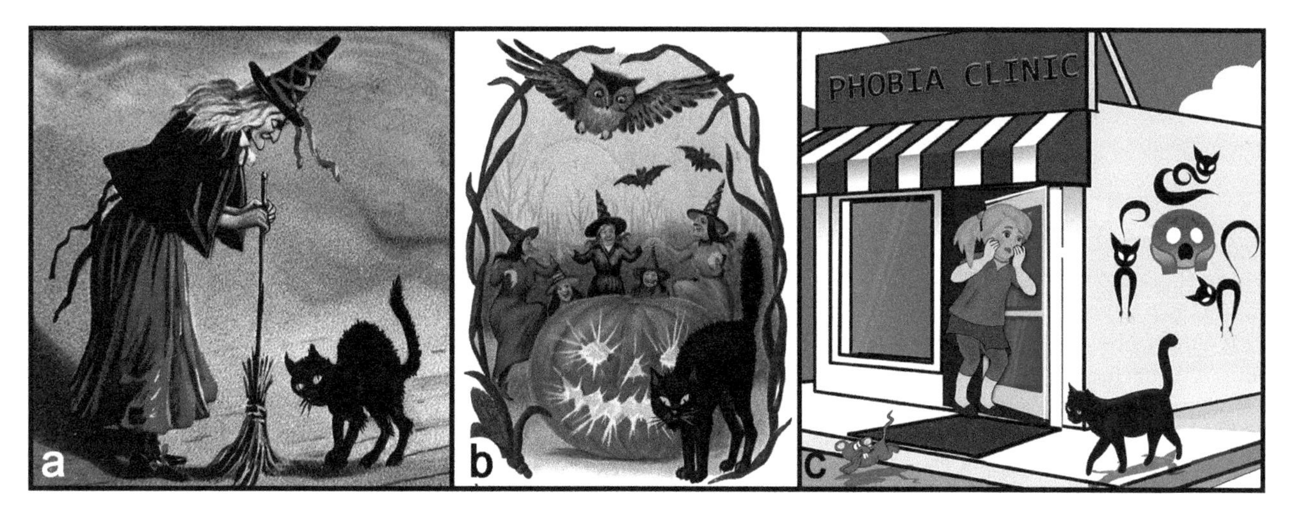

FIGURE 5.60 Black cats have long suffered from prejudice and superstitious beliefs, although they are honored in some cultures. (a and b) Early 20th-century postcards (public domain) showing association of black cats and witchcraft, especially at Halloween. (c) Illustration of the widespread folklore that a black cat crossing one's path is unlucky. Prepared by B. Brookes.

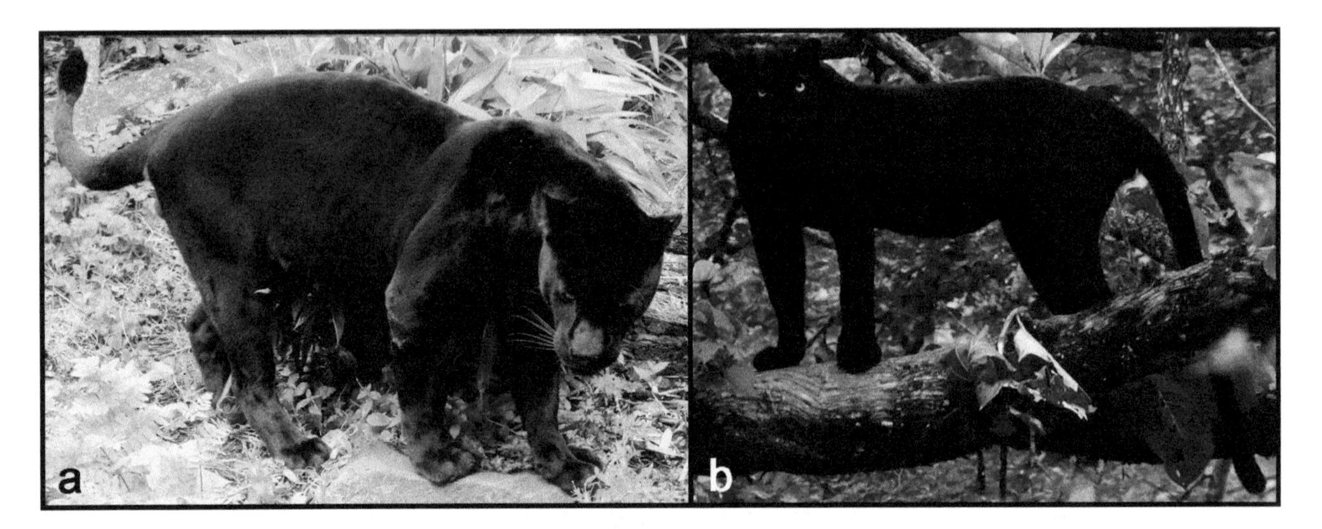

FIGURE 5.61 'Black panthers.' (a) Black jaguar (*Panthera onca*) from South America. Photo by Bardrock (CC BY 3.0). (b) Black leopard (*Panthera pardus*) from India. Photo by Davidvraju (CC BY SA 4.0).

FIGURE 5.62 Color variants of wolves (*Canis lupus*). (a) Black wolf. Photo by Bruce McKay (CC BY 2.0). (b) White wolf. Source: Pixabay (public domain).

(Figure 5.61a). Black wolves (Figure 5.62a) and white wolves (Figure 5.62b) are melanistic variants of the gray wolf, *Canis lupus*. Black forms of other vertebrates also are regularly observed. The extent to which black coats are adaptive is unclear. They would be advantageous to predators hunting at night. In a hot environment, a black coat could lead to overheating from the sun's rays, but in a cold area this could be beneficial. Male lions (*Panthera leo*) with darker manes are very attractive to lionesses and feared by male rivals (Cuthill et al. 2017).

Albinism, whiteness due to the unusual absence of pigmentation, is the opposite of melanism. It is known to affect all vertebrates (note examples in Figure 5.63) and can influence skin, hair, scales, feathers, and eyes. The degree of absence of pigment can vary. In humans, albinism can be a serious disease, and (similar to other disease conditions) people have shown serious social discrimination against albinos. In nature, a white coat could be beneficial to avoid absorbing heat in hot environments but could attract predators (except in a snowy environment). White skin is in danger of damage from ultraviolet radiation, but animals living in perpetual darkness (as in caves) often are white.

Gray, a compromise between black and white, is relatively indistinctive and seems adaptive for animals, especially

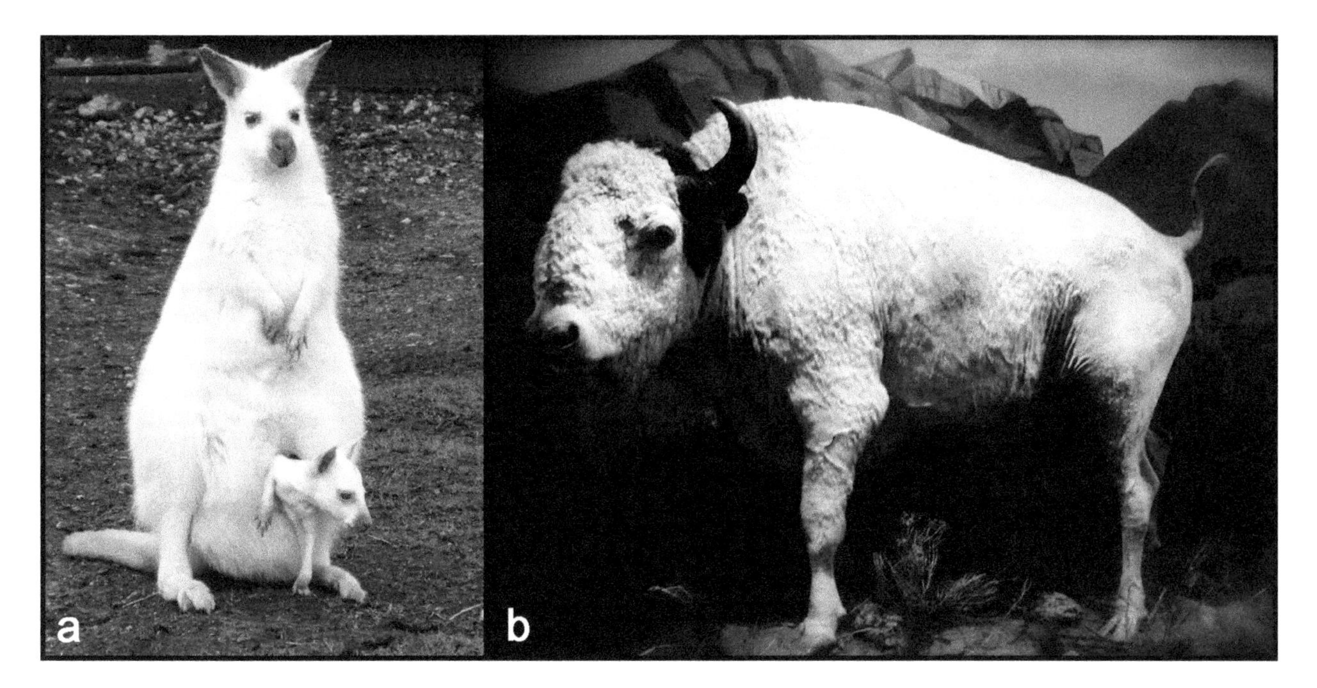

FIGURE 5.63 Examples of valued white animal mutants. (a) Bennett's wallaby (*Macropus rufogriseus*) with a joey in her pouch at an animal breeding farm in the Netherlands. Photo (public domain) by Apdency. (b) 'Big Medicine,' a taxidermy-preserved sacred white 'buffalo' (American bison, *Bison bison*) that lived from 1933 to 1959 and is on permanent display at the Montana Historical Society. Photo by Los Paseos (CC BY 2.0).

those that are nocturnal, such as moths. Curiously, some gray (so-called 'silver') hair is considered to be an attractive feature of mature male entertainment stars (often affectionately termed 'silver foxes') and in male gorillas ('silverbacks'), but usually not for the hair of female humans (an example of ageism).

Mixed black and white patterning is remarkably common among animals, and this is clearly because it is adaptive (especially at night, when many animals have no or poor color vision). In skunks, the pattern is a warning sign. In some animals, it may represent camouflage in snowy or shady environments (in 'disruptive coloration' the change from dark to light can draw the eye away from the true outline of the animal). Darkness around the eyes can reduce glare from the sun. All of these explanations have been advanced as the reason why giant pandas are black and white (Caro et al. 2017; Figure 5.64a). Humans appear to be extremely fond of black and white patterning, as evidenced by tuxedo apparel (Figure 5.64b), and indeed black and white mammals are common exhibits in zoos.

Plant foliage is green because of the presence of chlorophyll, which is necessary for photosynthesis (a few parasitic plants have no or limited chlorophyll, and are white or whitish). Horticulturalists have selected varieties of ornamental plants in which the leaves are variegated – i.e., portions of the leaves are not green, usually white (Figure 5.65), and variegated cultivars are extremely popular. White areas mostly lack chlorophyll, and so the plants' ability to produce energy for growth is reduced. Yellowish and purplish areas have reduced chlorophyll, and also are less productive, albeit some photosynthesis is possible. Such mutations are quickly eliminated in nature, but they are valued by people, and so are very common in gardens.

Preferences for Other Colors

As individuals, people have different color preferences (Bakker et al. 2015). Color preferences of humans appear to be partly innate, selected as survival responses during evolution because they are particularly associated with either pleasant or unpleasant objects, and partly learned. Early researchers on color preferences in humans suggested that in general people prefer, in decreasing order, blue, red, green, violet, orange, and yellow (a sequence that leaves out white flowers, which are widely admired). Blue has been shown to be the most pleasant color to the majority of people (perhaps explaining why baby boys are traditionally honored with blue), possibly because a blue sky indicates calm weather. By contrast, dark yellow and orange are the colors of urine, feces, vomit, and rotting food, and there is a general lack of preference for these hues in all human cultures (Palmer and Schloss 2010). We are all born with an innate attraction to reddish nipples to obtain nourishment from breasts, perhaps accounting for why red is such an attention-getter and explaining the advertising slogan 'red sells.' Another hypothetical explanation is that red is exciting because it is the color of blood. There have been claims that women prefer red slightly more than men do (supposedly why little girls are dressed in pink), speculatively explained as useful for our ancestors who relied on

FIGURE 5.64 Black and white patterning. (a) Giant panda in snow, illustrating possible camouflagic value of the black and white pattern. Photo by Ninara (CC BY 2.0). (b) Black and white formal wear, the standard for men's suits. Lithograph made in 1906. Credit: Albert and Victoria Museum (CC BY 4.0).

women gatherers to spot red berries against green foliage, while the men hunted. Increasing the redness of women's faces using cosmetics makes their faces more attractive to men (Pazda et al. 2016). The psychological effects of red have been extensively studied but remain controversial. Certainly red is an extraordinarily attracting stimulus and is the most popular of colors insofar as ornamental flowering plants is concerned.

Learning plays a very large role in color preferences, and it is important to be aware of ethnic, religious, and social traditions with respect to color. Hinduism, Buddhism, and Confucianism have special respect for yellow and orange; Muslims have considered green to be a sacred color; and Christianity has suggested special salvation-related meaning for red, white, and black. (In addition, regardless of color, some particular species of ornamental plants often have special significance for given religions.) In modern times, social constraints limit men from displaying pink and purple, and this has extended to recommendations for giving flowers to men (although the reverse gifting is still predominant). Sexist stereotypes related to women's hair color have become common: blondes are attractive but dumb, brunettes are studious and competent, and redheads are smart but temperamental. Men develop darker head hair than women, which grays with age, stimulating some males to tint their hair darker to retain a youthful image.

FIGURE 5.65 Examples of ornamental plants with white/green variegated foliage. Such mutations are attractive but penalize the plants because the white portions do not photosynthesize. (a) *Dieffenbachia bowmannii* cultivar Camilla. Photo by LucaLuca (CC BY SA 3.0). (b) African violet (*Saintpaulia ionantha*). Photo by A. Benedito (CC BY SA 4.0). (c) Hosta cultivar Hanky Panky. Photo by Jean Jones (CC BY 2.0).

Women also tint their hair to avoid gray, but in contrast to choosing to darken their hair like men, many women tint their hair blonde, perhaps an unconscious recognition that lighter hair is a sexually dimorphic character and so emphasizes femininity. (Accordingly, the saying 'blondes have more fun' may have a biological rationale.) Jiang and Galm (2014) found blonde hair increased tips for waitresses (Figure 5.66a–c). Price (2008) observed that blonde female fund raisers obtained more donations for charitable causes than brunettes. Blonde women tend to earn higher wages than women with hair of other colors (Johnson 2010). Extrapolating human preference for hair color to wild animals is difficult, since few are blonde (lions and tigers are blondish; Figure 5.66d). The most blonde of dogs is the golden retriever (Figure 5.66e), which is one of the most popular of canine breeds.

Attractively Colored Animals

With the popularization of cell phone cameras, on-line photographs of animals provide a guide to which species are considered worth recording. Clearly, people greatly value brilliantly or spectacularly colored animals, as reflected in the examples shown in Figure 5.67. Curtin and Papworth (2020) found that multi-colored animals are especially appealing.

Novelty or rarity of the color of an individual of a familiar species can greatly elevate its status among humans. Familiar commercial lobsters (*Homarus americanus*, the American lobster and *H. gammarus*, the European lobster) are bluish green or greenish brown, turning bright red when cooked. One out of millions of freshly caught specimens occurs in quite different colors, such as blue, orange, red, yellow, or white (Figure 5.68), and usually such lobsters are treated with great deference, either being released back to the ocean or donated to an aquarium. Ironically, such color mutations are normally extremely deleterious in nature, attracting predators, but when caught by humans such lobsters are usually treasured and saved. Rarer color morphs of given species may sometimes be ecologically adaptive in nature, but this is not well understood (Nijhawan et al. 2019).

Attractively Colored Ornamental Flowers

Ornamental plants are valued for a variety of characteristics, such as trees for shade, turf grasses for lawns, and vines for hiding imperfections of buildings and fences. However, the ornamental plant industry is predominantly based on the attractiveness of flowers, and this is due mostly to displays of color (Figure 5.69). In nature, flowers have characteristics that are attractive to pollinators, and of these color is particularly important because some animals prefer certain flower colors. For example, hummingbird flowers are frequently red. Some butterflies prefer yellow or white, while others prefer blue (although most do not seem to have a strong color preference). Many flies prefer white or dull colors, while numerous social insects and birds prefer bright colors. For humans, however, flower color does not reflect a natural, symbiotic relationship. Hůla and Flegr (2016), based on a survey of photos of flowers of Czech native plants, concluded that flower shape is more important for human preferences than color, but this does not seem accurate for ornamental species, which normally provide masses of color, obscuring floral shape.

FIGURE 5.66 Blonde attractiveness. Top row: waitresses, an occupation that favors blond females. Source: Shutterstock, contributors: (a) Alexy Grigorev, (b) Nicoletaa Ionescu, (c) Mahno. (d) 'Daniel in the Lions' Den,' painting by Flemish artist Peter Paul Rubens (1577–1640), in the U.S. National Gallery of Art. (e) American golden retriever. Photo by Newyrker10021 (CC BY SA 4.0).

Purchasers of commercial cut-flower bouquets have been found to prefer reddish flowers (especially when bought by men) and avoid yellowish flowers (Yue and Behe 2010). On the whole, the color preferences of people for flowers seem to be determined mostly by individual taste (although green flowers are obviously not desired because plants are already overwhelmingly green). Moreover, a variety of colors is usually considered attractive, both in bouquets and in gardens (although as a rule, it is recommended that bedding plants of a particular color should be planted close together to accentuate the display). In foliage plants, variegated leaves have been selected in numerous cultivars, which also demonstrate a pattern preference for variation in color.

New flower colors are continually appearing for the leading ornamental plants, but there is a tendency for traditionally available colors to remain especially popular. In the case of roses, for example, the most popular colors are red, pink, white, yellow, lavender, and purple; and for carnations the leading colors are red, pink, and white. Today, numerous ornamental plant species are available in a wide range of flower colors, and often the chief concern of many is simply to coordinate flower color with dress or decor.

Of all possible colors, black is the most unnatural for flowers. The phrase 'black flowers' is used loosely to refer to very dark flowers of some ornamental species. Usually such flowers are a deep purple, and almost always a close look reveals that the flowers are not truly black. Orchids, tulips, carnations, calla lilies, roses, and pansies are among the species for which black-flowered variants have been selected. Black is associated with evil and death, which limits the use of black flowers in many situations. Nevertheless, the rarity and novelty of black as a floral color has stimulated breeders to select very dark-flowered varieties, and they add considerable interest to a garden or bouquet.

The Extraordinary Importance of Colorful Plants to Women

In a survey of the preferences of children in grades 1–12 for animals and plants, the girls ranked 'flowers' first, while the boys ranked them last (Badaracco 1973). As noted in the next paragraph, females also show greater preferences for colorful appearance. As discussed in the following (and as is common knowledge), women are far more attracted to colorful flowers than men.

Differences between male and female humans may be due to genetics or social constraints (especially learned behaviors and values). Regardless of why, women appear to place much more value on color than men do.

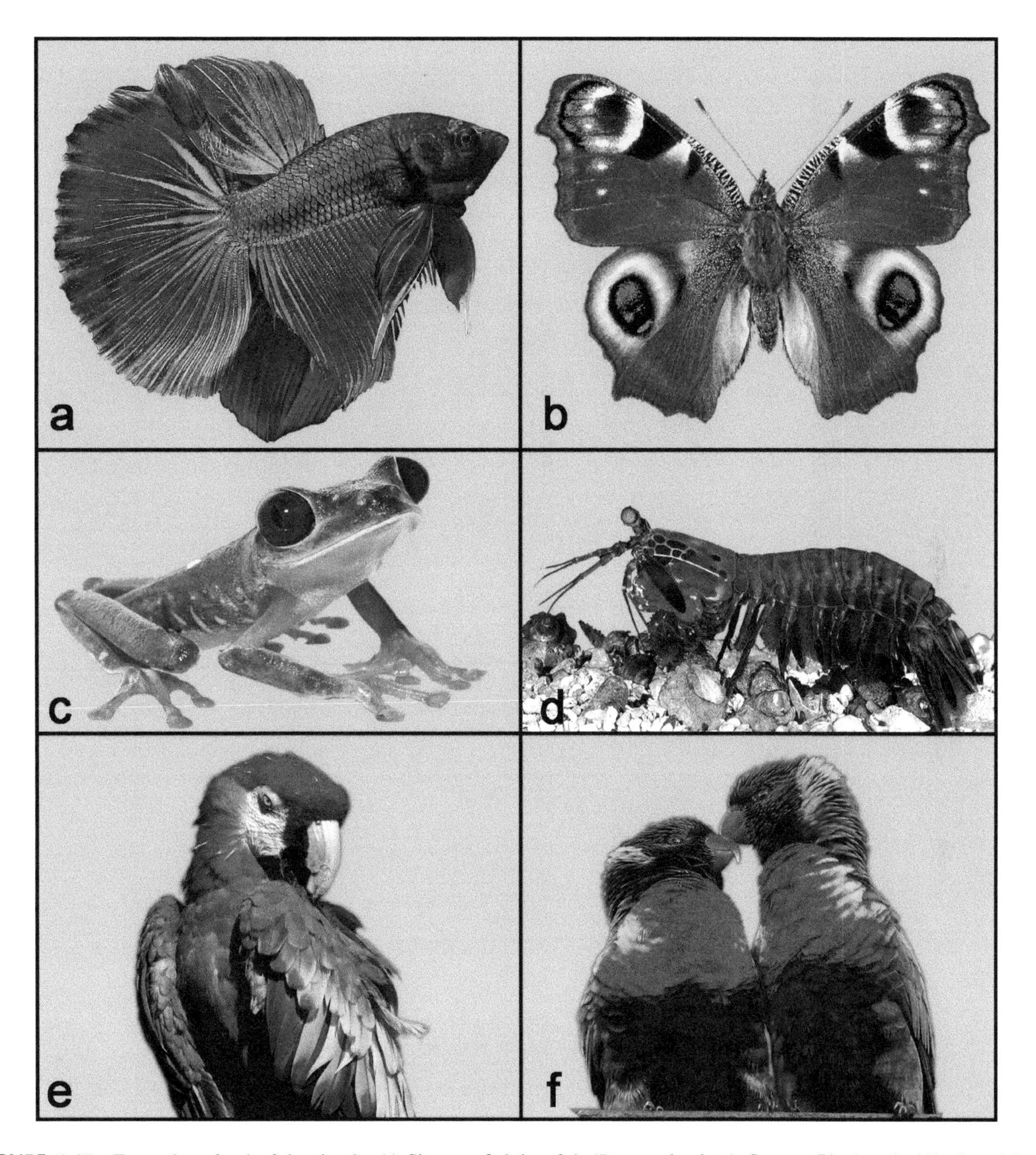

FIGURE 5.67 Examples of colorful animals. (a) Siamese fighting fish (*Betta splendens*). Source: Pixabay (public domain). (b) Peacock butterfly (*Aglais io*). Photo by Didier Descouens (CC BY SA 4.0). (c) Red-eyed tree frog (*Agalychnis callidryas*). Photo by Brian Gratwicke (CC BY 2.0). (d) Mantis shrimp (*Odontodactylus scyllarus*). Photo by Roy L. Caldwell (public domain). (e) Scarlet macaw (*Ara macao*). Photo by Russ May (CC BY 2.0). (f) Rainbow lorikeet (*Trichoglossus haemotodus*). Photo by John (CC BY SA 2.0).

Notwithstanding the popularity of the 'little black dress,' in most modern societies, women dress far more colorfully than men (Figure 5.70). Women also develop immensely more appreciation of attractive colorful plants than men. Women are responsible for most purchases of ornamental plants (Figure 5.71a), and floral motifs widely decorate gifts intended for women (Figure 5.71b and c). Women appear to be instinctively aware that their beauty and attractiveness is accentuated when associated with beautiful flowers (Figures 5.72 and 5.73). In turn, men widely recognize the courtship value of flowers for women (Figure 5.74).

It has been speculated that the human attraction to flowers traces to their value in indicating to our ancient ancestors that a habitat had good potential for resources. Wildflower species come into flower at particular times, so they are indeed indicators of when they and other species could be harvested (Orians and Heerwagen 1992). Honey in particular, produced by bees, would become available

FIGURE 5.68 Very rare colors in commercial lobsters: American lobster (*Homarus americanus*) and European lobster (*H. gammarus*). Most of these were photographed in aquaria, where they were preserved. (a) Red American lobster (lobsters turn red when cooked but are normally camouflaged darkly bluish green or greenish brown, blending with the ocean floor). Photo by U.S. National Oceanic and Atmospheric Administration (public domain). (b) Split-color American lobster. Photo by Tv. Rule (CC BY SA 4.0). (c) Albino/light blue American lobster. Photo by Syrio (CC BY SA 4.0). (d) Blue American lobster. Photo by Bryan Harvey (CC BY SA 2.0). (e) Blue European lobster. Photo by Bart Braun (public domain). (f) Blue, white-speckled European lobster. Photo by Albarubescens (CC BY SA 4.0).

FIGURE 5.69 Beautiful gardens. (a) Azalea garden, Tokyo. Photo by Peachbird (CC BY SA 3.0). (b) Butchart Gardens, British Columbia. Photo by MasterShake (public domain). (c) Tulips at the Keukenhof in the Netherlands, one of the world's most famous gardens. Photo by Luu (public domain). (d) Japanese Garden in Portland, USA (featuring a Japanese Maple, *Acer palmatum*). Photo by John Fowler (CC BY 2.0).

FIGURE 5.70 Examples of the importance of colorful dress to women. (a) Indian dancers. Photo by Ramnath Bhat (CC BY 2.0). (b) Chinese dancer. Photo by Chen Wen (CC BY 2.0).

FIGURE 5.71 Illustrations showing the relationship of flowers and women. (a) 'The Flower Seller' (1891; public domain), Avenue de L'Opera, Paris, by French artist Louis Marie de Schryver (1862–1942). (b) 'Mother's Day.' Photo by Quinn Dombrowski (CC BY SA 2.0). (c) Victorian valentine card (dated 1899; public domain) decorated with flowers and girls. Photo by DiethartK (CC BY SA 4.0).

FIGURE 5.72 Paintings of young women by French artist Émile Vernon (1872–1920), who characteristically showed beautiful females associated with beautiful flowers. All figures are public domain.

when flowers are abundant ((Heerwagen and Orians 1995). Women, traditionally gatherers of resources from nature (while men were primarily hunters), may have evolved special sensitivity to flowers. For additional information, see Hůla and Flegr (2021).

BEAUTIFICATION OF PETS BY BREEDING

As discussed earlier, numerous plants have been domesticated not just for food and various other necessities, but for ornament, i.e., for their beauty. Thousands of plant species have been domesticated for their ornamental value, for the most part for their flowers, which are attractive largely because they are colorful. Plants are so different from humans that they offer little that can be compared with the parts of people that are considered attractive or ugly.

Animals have been domesticated primarily for resources (such as meat, milk, eggs, and various byproducts such as leather), as transporters of people and goods, and as aids in hunting and security. However, several dozen animals have also been domesticated at least substantially for their ornamental value. The most popular pets are dog, cats, and various other small mammals, birds, and fish (Figure 5.75). (The contrived politically correct phrase 'companion animals' is imprecise, since (1) it is not applied to livestock, which are in fact our most indispensable companion animals, and (2) logically, human friends are also companion animals.)

FIGURE 5.73 'Helen of Troy,' painting (public domain) by English artist Evelyn De Morgan (1855–1919). In Greek mythology, Helen was said to be the most beautiful woman in the world. The artist has embellished her beauty with roses, claimed to be the most beautiful of flowers.

FIGURE 5.74 Flowers play prominent roles in courtship. (a) 'Bouquet de primevères' (1893) by French artist Alfred Guillou (1844–1926). Photo by Grégory Lejeune (public domain). (b) Nineteenth-century print (public domain) showing a romantic couple, the young man dressed in the style of Napoleon, giving a bouquet of violets to his companion. (c) Hindu deity Lord Krishna courting his beautiful consort Radha, goddess of love, tenderness, compassion, and devotion, by adorning her with a garland of flowers. Illustration from Infinite Eyes (public domain).

The oldest domesticated animal, the dog, is derived from the gray wolf, *Canis lupus* (Figure 5.76a and b). Initially dog breeds were selected primarily for their pragmatic values: As hunting aids, shepherding assistants, security guards, weapons of war, controllers of rodents and other pests, for pulling or transporting people and goods, and (most regrettably) for exhibition fighting against other dogs, bulls, bears, and others. To this day, numerous dog breeds retain

FIGURE 5.75 Great paintings (public domain) showing the most popular family pets: small mammals, birds, and fish. (a) 'Family of cats and a dog in the salon' by French artist Louis Eugène Lambert (1825–1900). (b) 'A treat for her pet' by French painter Guillaume Dubufe (1853–1909). (c) 'Admiring the goldfish' by Italian painter Vittorio Reggianini (1858–1938).

behavioral and physical characteristics suiting them to their past or present trades. In recent times, most dogs (whether purchased as purebred or mongrels adopted as rescues) are chosen partly for their presumed behavioral traits, but primarily for their appearance. Very large dogs, very muscular dogs with powerful jaws, and very athletic dogs appeal to some, reflecting human respect for these physical features. Conversely, very small dogs, especially with short muzzles, seem to elicit maternal instincts and serve as baby substitutes. Long or impressively luxuriant fur is widely admired (although shedding is not), as is patterned coloration, and indeed these features reflect tastes in fur coats and accessories worn by humans. Some features that are potentially comical – floppy ears, greatly elongated body, underbite, curious expressions – seem to serve as fear-reducing signals to humans, eliciting compassion.

The wild rock dove ('pigeon,' *Columbia livia*, Figure 5.76c) is the ancestor of 'fancy pigeons' – breeds selected simply for their ornamental value (Figure 5.76d). There are over 1,000 breeds, differing in size, shape, color, and behavior, and many of these are exhibited at pigeon shows (see the discussion of feral pigeons in Chapter 16).

Choices of cat breeds reflect preferences parallel to those for small dog breeds, although the range is much smaller, reflecting much more limited domestication. Although there is considerable variability among domesticated breeds of dogs and cats, for the most part they show evidence of retention of neotenic (infantile) features admired by people. Curiously, silver foxes (*Vulpes vulpes*) domesticated for their fur also appear to have undergone similar selection for cute features (Elia 2013). In parallel, aquarium fish (especially goldfish, as shown in Figure 5.77) have also been selected, not for feathers but for colorful, spectacular fins. Butterflies have not been domesticated (although the silkworm, which is one of the few domesticated insects, is a kind of butterfly), but they similarly exhibit wings that are attractively colored. Clearly, humans greatly value diverse species – ornamental flowering plants, birds, fish, and butterflies – for much the same reason: very impressive and diversely displayed coloration.

DO PETS RESEMBLE THEIR OWNERS?

Sometimes by chance people have some resemblance to animals (note Figure 5.34), but popular wisdom holds that pets often look like their masters, which if true would not be a random occurrence. People tend to overestimate their physical attractiveness (Greitemeyer 2020), so it would seem that they would at least tend not be deterred from choosing pets that they indeed resemble. Some research indicates that in fact people are attracted to dogs that have some of their features (Figure 5.78). Coren (1999) found that women with long hair preferred long-haired springer spaniels and beagles, while women with short hair preferred shorter-haired basenjis and huskies. Observers tasked with matching photos of people and photos of dogs significantly correctly identified which dogs were owned by which people (Roy and Christenfeld 2004; Nakajima et al. 2009). It is apparent that large, muscular men accompanied by tiny dogs look odd, and it does seem that males who want to project a hypermasculine appearance are attracted to dogs with an aggressive appearance (note Figure 5.20).

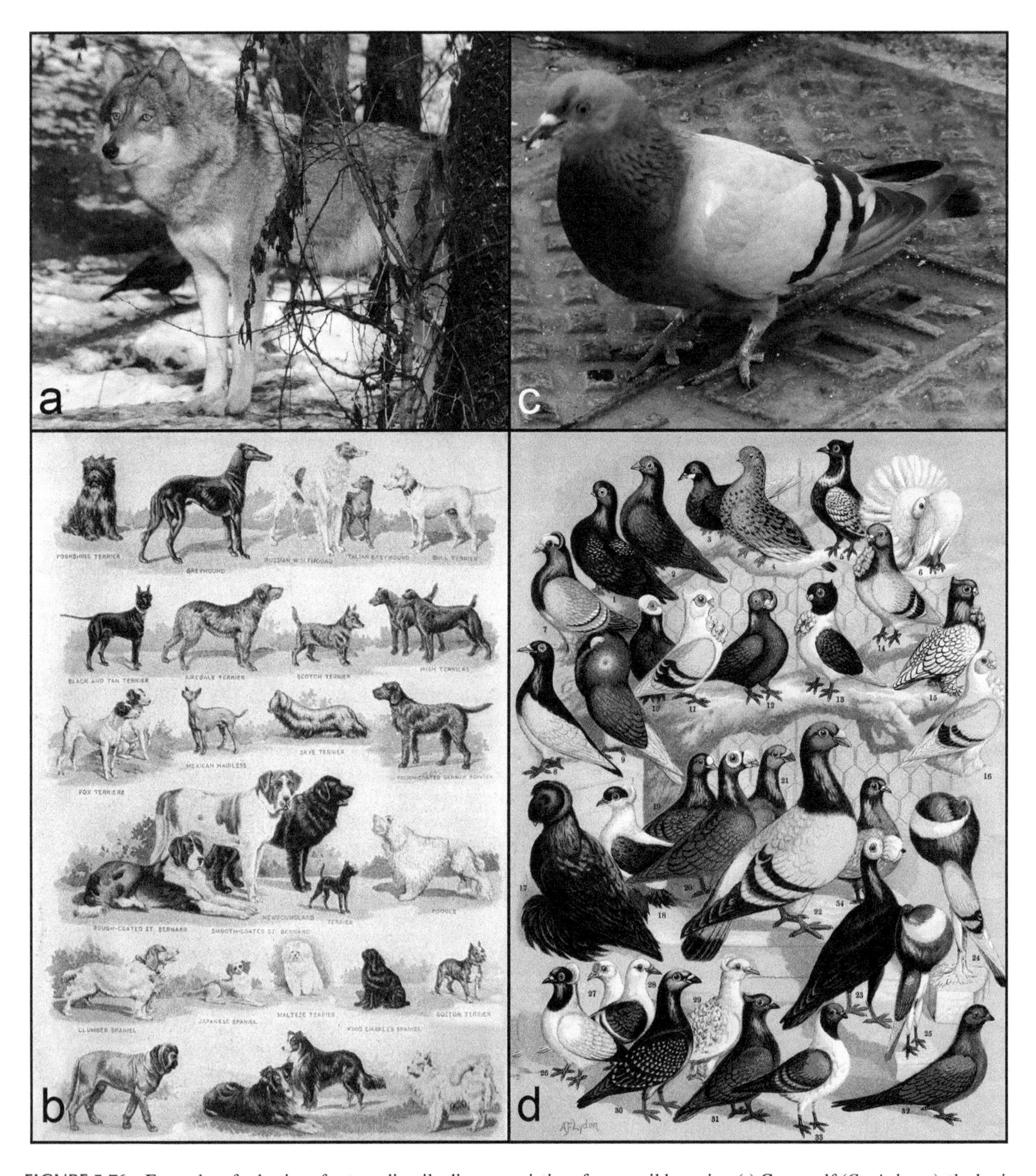

FIGURE 5.76 Examples of selection of extraordinarily diverse variations from a wild species. (a) Gray wolf (*Canis lupus*), the basic ancestor of dogs. Photo by Alois Staudacher (CC BY 2.0). (b) A selection of dog breeds (*Canis lupus familiaris*). Public domain figure from Roe and Leonard-Stuart (1911). (c) Common pigeon (rock dove, *Columbia livia*), the basic ancestor of fancy pigeons. Photo by Sean MacEntee (CC BY 2.0). (d) Fancy pigeons. From an English poster (public domain) showing Victorian breeds, published in 1891.

As noted earlier in this chapter, women's faces are quite neotenic and women respond more strongly than men to neotenic characteristics not just of human babies but also of animals, so as a result it would seem that women's faces in general would tend to resemble the faces of many pets that they have chosen.

THE PUBLIC RELATIONS VALUE OF ANIMALS BEING PERCEIVED AS BEAUTIFUL

Advertisers are keenly aware of the persuasion value of beauty, especially the physical attractiveness of celebrity endorsers (Saad 2007), and the same principle dictates that

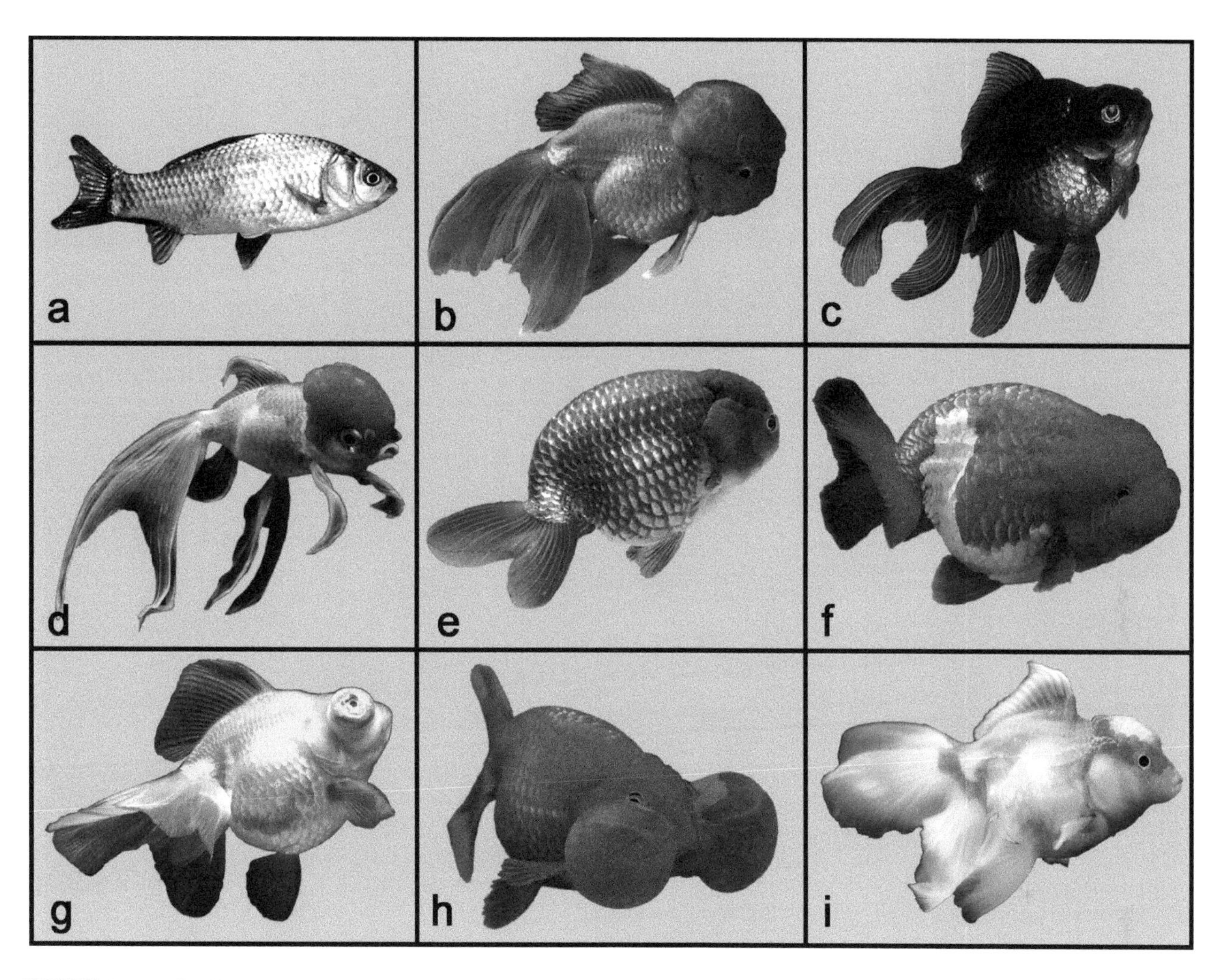

FIGURE 5.77 Goldfish (*Carassius auratus*) varieties, illustrating human preferences. (a) Wild fish from China, representative of the ancestor from which domesticated (fancy) forms originated. Photo by DeborahY (public domain). (b) Photo by Lawrencekhoo (CC BY SA 4.0). (c) Photo by KD Blackmore (public domain). (d) Photo by Vassil (public domain). (e) Photo by Lawrencekhoo (CC BY SA 4.0). (f) Photo by Melanochromis (public domain). (g) Photo by Lawrencekhoo (CC BY SA 4.0). (h) Photo by Lerdsuwa (CC BY SA 3.0). (i) Source: Petanos.com (public domain).

FIGURE 5.78 Pets that look like their owners. (a) Dogs reminiscent of their masters (contributor: Good Stock). (b) A girl who resembles her new pet goldfish (contributor: Alina Ermokhina). Source of photos: Shutterstock.

FIGURE 5.79 Charismatic megafauna (glamorous large animals) employed in advertising. (a) An 1890s advertising poster (public domain) showing an attractive woman drinking Coca-Cola. (b) 1882 advertisement (public domain) showing circus elephant Jumbo (1860–1885) feeding a laxative called Castoria to a baby elephant. (c) 1890s French advertisement (public domain) for a stomach medicine called Vin Bravais. Photo credit: Halloween HJB. (d) Wine advertisement prepared by Robin Hutton (CC BY 2.0). (e) 1910 advertisement (public domain) for the fur coat company P. Rückmar, by artist Ernest Montaut (1879–1909). Photo credit: Susanlenox. (f) Nineteenth-century soap advertisement (public domain) for Kendall M'f'g. Co. Credit: Boston Public Library.

animals employed to advertise products should also generally be beautiful (Figure 5.79). 'Beauty is in the eye of the beholder,' but while there is wide variation in what people find beautiful, there is also wide agreement. To biologists and naturalists, every living creature has intrinsic beauty, but in the court of public opinion species differ spectacularly in their natural appeal. Beauty is often sufficient for public backing, and lack of beauty often prevents public support. Advocating the conservation of beautiful creatures is likely to be much more successful than advocating for most species in need of conservation. Support usually goes toward the welfare of habitats of the attractive species, which necessarily benefits other species in the same habitats, often including some threatened species. Unfortunately, this approach frequently leaves out threatened species that are unattractive and do not live in habitats of attractive species judged to merit protection.

CLOTHING AND COSTUMES

Clothing serves various purposes, such as thermoregulation, modesty, body armor, and indicators of status and group membership (Buckner 2021). Most animals have natural body adaptations that serve these functions, although some adopt external materials for these purposes. For example, some species protect themselves from attack by coating their bodies with objects. However, people are unique in the extent to which they cover their bodies. Most humans dress themselves most of the time, although sex, age, tradition, and climate influence how much clothing is worn. Aside from practical values, there are artistic aspects of clothing that reveal our preferences and prejudices. Clothing serves to exaggerate desired features and to disguise undesired ones. The padded shoulders of men's suits and the narrowly cinched waists of female attire obviously emphasize idealized sexual features. Much of modern society demands that very conventional, and frankly rather boring, costumes be worn – most obviously the civilian business suit and the military uniform. It is during celebrations that encourage people to 'dress up,' such as Halloween, that human aspirations for appearance tend to be most dramatically revealed, especially with respect to animal costumes (Figure 5.80). As noted in Chapter 1, the most favored animals are mammals and birds, and this is reflected in costume preferences. 'Fursuits' are animal costumes custom-made by enthusiasts called 'furries'

FIGURE 5.80 Popular animal costumes. (a) Fursuiters at a furry convention. Photo by Murphy (CC BY SA 4.0). (b) Boy in a Halloween lion suit. Contributor: B. Calkins. (c) Dancing cat and bunny in San Antonio, Texas. Contributor: Jennifer_Crowder_artist. (d) Two-man pantomime elephant in Myanmar. Contributor: LiteChoices. (e) Chicken man sitting down to a fried chicken dinner. Contributor: Lisa F. Young. Photos b to e are from Shutterstock.

(Figure 5.80a), a trend that exploded in popularity in the 1990s. The movement clearly reveals a preference for mammals, which of course are the main furry animals. Batman and Spiderman have become movie icons and have also inspired animal costumes for bats and 'bugs,' which nevertheless remain strongly disliked, and similarly shark and dinosaur costumes are popular, but because they inspire terror rather than because these animals are liked.

ENHANCING THE APPEARANCE OF SPECIES

As discussed earlier, breeding of ornamental plants and domesticated animals has greatly changed their appearance in ways considered attractive to people. Not content with the natural appearance of flowers, the cut-flower industry sometimes dyes flowers to enhance their appearance, packages them in brilliant foil, and sells them in colorful pots. Expensive fruits are sometimes sold in very attractive containers. However, domesticated animals are much more commonly 'dressed up' to make them more attractive. Some pet owners clothe their animals like children. For celebrations and holidays, livestock are frequently elaborately clothed or painted, especially in Asia (Figure 5.81). Pet grooming, including services such as bathing, brushing, and nail clipping, has become a multi-billion-dollar industry in the world, catering especially to dogs (Figure 5.82).

HAPPINESS IS A WARM PUPPY

This chapter has detailed how we humans are obsessed with appearance, often painfully so. We should conclude with a happier note. The well-known quotation 'Happiness

FIGURE 5.81 Elaborately decorated livestock reflecting the high respect they are given. (a) Indian elephant in Jaipur, Rajasthan. Photo by Faraz Usmani (CC BY 2.0). (b) Camel in Karachi, Pakistan. Photo (public domain) by A. Savin. (c) Yak in Yunnan Province, China. Photo by Randy Levine (CC BY SA 4.0). (d) Mounted knights at a jousting fair in England. Photo by Bradley Howard (CC BY 2.0). (e) Ox in India. Photo by Bhaskaranaidu (CC BY SA 3.0). (f) Water buffalo racing in Bali. Photo by Wiaskara (CC BY SA 4.0).

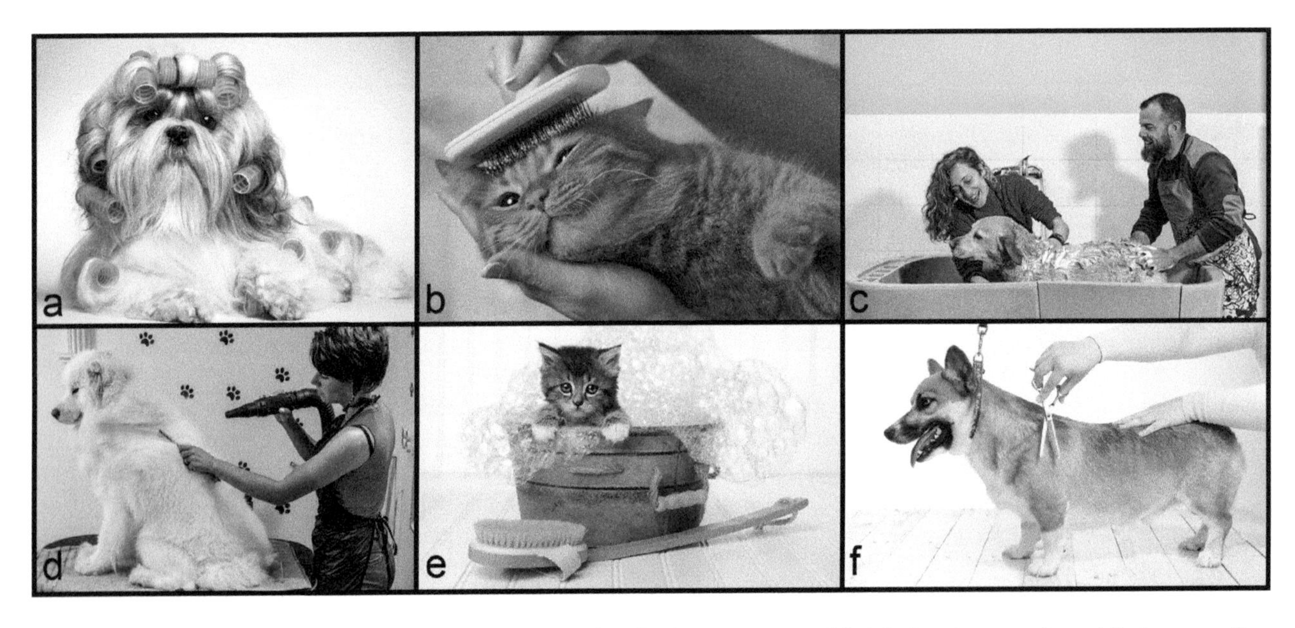

FIGURE 5.82 Pet grooming, illustrating the value people place in the appearance of their beloved companions. All photos are from Shutterstock. (a) Shih tzu with curlers. Contributor: Chaoss. (b) Brushing a cat. Contributor: Koltsov. (c) Bathing a Labrador retriever. Contributor: Contributor Enadan. (d) Blowdrying a samoyed. Contributor: AF-Photography. (e) Kitten in a bubble bath. Contributor: Katrina Brown. (f) Clipping a corgi. Contributor: Makistock.

FIGURE 5.83 Paintings (public domain) illustrating the theme 'Happiness is a warm puppy.' (a) 'A Young Girl and Her Dog' by British artist Joshua Reynolds (1723–1792). Credit: Tokyo Fuji Art Museum. (b) 'Girl with a Dog in Her Hands' by French artist Jean-Baptiste Greuze (1725–1805). Credit: National Museum Warsaw. (c) 'The Actress Rejane and Her Dog' by Italian artist Giovanni Boldini (1842–1931). Credit: http://www.wikiart.org/en/giovanni-boldini/the-actress-rejane-and-her-dog.

is a warm puppy' came from the newspaper comic Peanuts, by Charles M. Schulz (in April 25, 1960). Lucy, the neighborhood 'crabby' person, hugged Snoopy, Charlie Brown's dog, and discovered the joy inherent in other species. This is a metaphor for how we can find happiness by being less concerned with our own problems and more attuned to the everyday simple marvels of life. Millions have learned this lesson, at least temporarily (Figure 5.83).

6 Non-visual Determinants of Human Prejudices against and Preferences for Species

INTRODUCTION

Although visual appearance provides the key features people use to judge other humans and non-humans, some other considerations can be important. Humans especially value behaviors that promote the welfare of society (such as maternal instinct, cooperativeness, and cleanliness) and despise characteristics that are inadaptive (such as thievery, sociopathy, laziness, and extreme aggressiveness). Of course, we perceive these behaviors in animals mostly or entirely with our eyes, but the concern of this chapter includes human perceptions exclusive of the visual appearance of the animals themselves, particularly behavior. The considerations that determine societal class among people – especially the impressiveness of one's home and the quality of food – are important to human judgments of others. This chapter examines how we judge the merit of animals on similar bases.

Aside from visual criteria, other sensory perceptions can influence our judgment of species. Smell and sound can also be sensory criteria: we tend not to appreciate malodorous, noisy people, and the same is often true for how we evaluate animals. The importance of odor and sound in judging species is examined in this chapter. In Chapter 5, the importance of touch was discussed under the heading 'body coverings,' where it was pointed out that the softness of fur is greatly liked but spiny coverings are not (this topic could have been discussed here, but texture of species is mainly perceived visually). Some creatures are naturally wet or slimy (snails, for example), and many find this objectionable. Taste, another sense, is of course a key criterion for judging whether we like food species. And, pain perception caused by harmful species dramatically indicates to us that such creatures are to be avoided. We can also easily distinguish temperature, but whether a species is hot or cold rarely factors into our opinion of them.

THE APPEAL OF LOVING ANIMAL PARENTS

Sexual reproduction of most animals requires proximity of males and females, at least for the transfer of sperm to ova. Once birth occurs, there is a wide range of care of progeny among animals. Many (especially 'lower animals') provide no care for their offspring, in which case huge numbers are usually produced to compensate for their very high mortality. This behavior is foreign to humans and therefore is not considered admirable. Some animal species living in resource-poor environments cannot afford to share the limited supplies, either with competing species or even with other members of the same species. Bears, for example, usually are solitary (except when females are raising cubs) because individuals require large areas to sustain themselves. Such species are naturally aggressive, indeed often foul-tempered, and even sometimes engage in cannibalism or infanticide of members of their species. At least in these respects animals with hermit lifestyles are not perceived as attractive.

Humans, however, invest massive amounts of care in just a few children, and animals that do the same tend to be admired. In most animals (indeed, also in most plants), maternal investment in the success of offspring exceeds (usually by a very wide margin) paternal investment. The sight of mothers protectively close to their young is highly appealing, even if the animals are vicious and quite dangerous (Figure 6.1). The mother-child image is so valued that many wildlife photographers have risked death to capture a perfect pose (or less courageously have simply visited a zoo). The uplifting images of baby animals and their mothers are so appealing that their use in conservation campaigns is almost universal. Humans, unlike most animals, are largely long-term male-female pair bonders, and species that exhibit this behavior evoke special admiration (Figure 6.2). The highest approval goes to species in which the fathers participate in child-rearing.

ODOR

Preference for Perfumed Species

The sense of smell (i.e., perception of volatile chemicals in the air) plays a major role in human evaluation of many things, including other species. Human odor detection is considered to be inferior compared to many other animals (Tronson 2001), but nevertheless is good. Species perceived as pleasant-smelling are usually viewed positively, while those that are bad-smelling tend to generate disgust. Malodorous smells may be termed 'stink' or 'stench,' while pleasant smells are often termed 'scent,' 'aroma,' and 'fragrance.' The word 'odor' tends to refer to scents in general

DOI: 10.1201/9781003412946-6

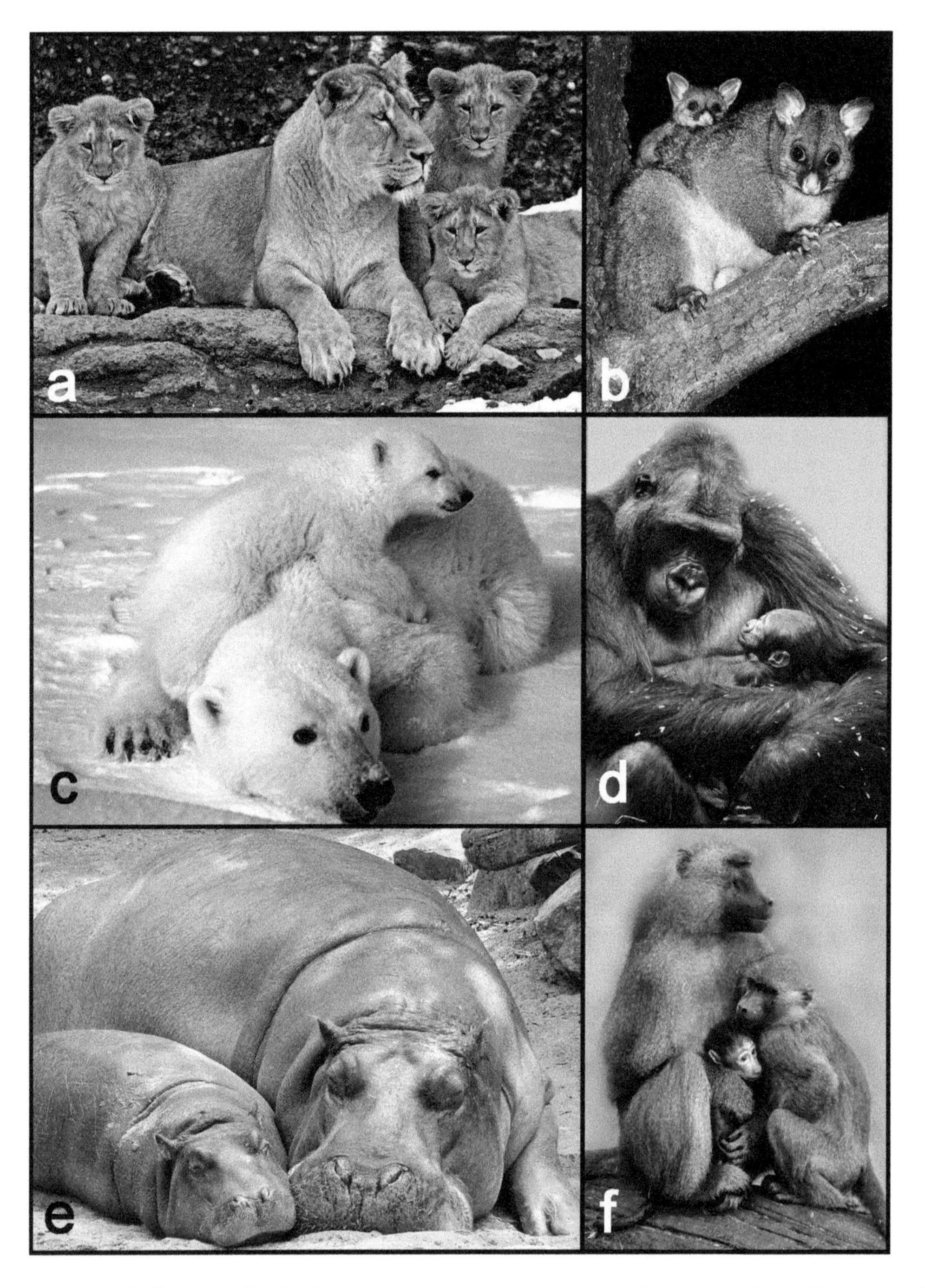

FIGURE 6.1 Mother-and-child images of animals are extremely appealing to the public because they reflect human family values. (a) Lions (*Panthera leo*). Photo by Tambako the Jaguar (CC BY ND 2.0). (b) Common brushtail possums (*Trichosurus vulpecula*). Photo by JJ Harrison (CC BY SA 2.5). (c) Polar bears (*Ursus maritimus*). Photo (public domain) by S. Schliebe, U.S. Fish and Wildlife Service. (d) Gorillas (*Gorilla gorilla*). Photo by Tambako the Jaguar (CC BY ND 2.0). (e) Hippos (*Hippopotamus amphibius*). Photo by Tambako the Jaguar (CC BY ND 2.0). (f) Olive baboons (*Papio anubis*). Photo from Pixabay (public domain).

in Commonwealth, English-speaking nations, but often to unpleasant scents in the U.S. and elsewhere. Animal species have evolved odor detection capabilities because some smells indicate potential rewards, especially to find food and locate or attract mates, while other smells are associated with danger (especially predators and disease).

Foul odors are particularly generated by decay and disease, and often reflect unhygienic, infective conditions, and so are usually signs of danger. Unpleasant odors are related to the disgust avoidance response, which has been selected by evolution to protect us (Kelly 2011). (There are various kinds of disgust [Prokop and Randler 2017], such as moral disgust; in the present context, the issue is 'pathogen disgust.')

Smell is determined by chemicals in the air, but 'aroma' is considered to represent a combination of smell and taste – i.e., taste detectors modify the perception. Generally, bad-smelling species are strongly avoided

FIGURE 6.2 Lovebirds (any of nine, mostly African, parrot species of the genus *Agapornis*). (a) Masked lovebird (yellow-collared lovebird, *Agapornis personatus*). Photo by Mertie (CC BY 2.0). (b) Fischer's lovebird (*Agapornis fischeri*). Photo by Peter Békési (CC BY SA 2.0). (c) Rosy-faced lovebird (*Agapornis roseicollis roseicollis*). Photo by Charles J. Sharp (CC BY SA 4.0). (d) 'Lovebirds,' a composition by Joseph Christian Leyendecker, in The House of Kuppenheimer Style Book, Fall/Winter 1918–1919. Photo by Susi Pator (public domain). (e) Vintage storybook illustrations titled 'Love Birds.' Source (public domain): Burnside, H. M. M. A. Hoyer, M. Dickens, and E. Nisbet. 1893. *Told by the Sunbeams and Me*. London, Tuck & Sons.

by humans. However, de-sensitization from prolonged exposure can occur. In the Middle Ages in Europe, peasants commonly kept livestock inside their huts (Figure 6.3), and indeed so-called 'house barns' are a more sophisticated variation in which there is separation of animals and humans within the same building. Farmers who regularly contact manure often come to be indifferent to its smell. During Medieval times, copious amounts of dung from livestock, as well as the waste products of various trades, contaminated cities, but people became quite indifferent. When smoking was popular, non-smokers were forced to develop some tolerance to the odor of tobacco.

Frequently, odors attractive to particular animals are detested by humans. Animals that smell pleasant to humans are uncommon, although pungent extracts from some animals are major components of commercial perfumes. By comparison, numerous plant species produce pleasant perfumes that are well known to attract pollinators. Why these are predominantly enjoyable for humans, who are not pollinators, is unclear (Raguso 2004). Certainly, one of the motivations of humans to assemble gardens is the pleasing perfume of many of the species (Figure 6.4).

Ambivalence Regarding Stinky Plants

Many ethnic plant foods have strong aromas, which are perceived as attractive by those who have become habituated to them (e.g., durian fruit, Figure 6.5a). However, there is considerable genetically based variation among humans in regard to their reactions to the aroma of foods, and so our preferences/dislikes of many foods is often partially genetically based (Tepper and Ullrich 2002). At the same time, whether the aroma of particular foods is associated with good or bad experiences (i.e., environmental conditioning) also determines whether given aromas become liked or disliked (Schloss et al. 2015).

Some plants are naturally intensely stinky, and because of this are not welcome in the neighborhood. Trees with particularly repugnant smells include Bradford pear (*Pyrus calleryana* cultivar Bradford), female ginkgo trees (*Ginkgo biloba*), and tree-of-heaven (*Ailanthus altissima*). Curiously,

FIGURE 6.3 A peasant house in Medieval Europe. Livestock were often kept indoors, sometimes to protect them from wild animals, but more often to prevent their theft. Prepared by Jessica Hsiung.

FIGURE 6.4 Ornamental flowering plants with pleasant scents are widely valued, as reflected in these great paintings (all are public domain). (a) 'Sweet flowers' by Belgian artist Jean-François Portaels (1818–1895). (b) 'Little gardener sniffs flowers' by French artist Jean-Paul Haag (1854–1906). (c) 'The soul of the rose' by English artist John William Waterhouse (1849–1917).

FIGURE 6.5 Some of the world's stinkiest plants. (a) Durian (from *Durio zibethinus*), widely considered to be the world's most foul-smelling (although delicious) fruit. This street art scene from Singapore shows a merchant who has given a piece of durian to a child. Photo by Sasha India (CC BY SA 2.0). (b) The bloom of rafflesia (*Rafflesia arnoldii*), found in the rainforests of Borneo and Sumatra. Famous as the largest flower in the world, it emits a stench similar to that of decaying meat to attract the flies that pollinate the plant. Photo by Rahmat Andriansa (CC BY SA 4.0). (c) Corpse plant or titan arum (*Amorphophallus titanum*). The huge phallic spike is made up of numerous tiny flowers, and the whole structure is commonly misinterpreted as a single flower. Like (b), it produces an odor like rotting meat to attract pollinating insects. Photo by Lothar Grünz (public domain).

some very rare plants produce gigantic very stinky flowers that attract insect pollinators, and when exhibited, they attract huge audiences (Figure 6.5b and c).

Prejudice against Stinky Animals

Relatively few animals have bathing facilities available, so it should not be surprising that they do not meet modern human standards of hygiene. Some animals, as noted below, use foul-smelling chemicals as defensive weapons against predators, and many animals are attracted by smells that humans dislike. Many species employ feces for their repellent or attracting smell. Some vultures deliberately coat themselves with their own excrement in order to dissuade predators (see Chapter 16). Similarly, some beetle larvae coat themselves with 'fecal shields' for protection (Nogueira-de-Sá and Trigo 2002). 'Scent-rolling' in the poop of other species, by wolves and other animals (including many dogs), is also common and has been hypothesized as a way of concealing their natural scent and so not alerting prey. Notably stinky species are shown in Figure 6.6.

JUDGING ANIMALS ON THEIR SOUNDS

Many animals are well known for their vocalizations (Figure 6.7). The sounds produced by both humans and non-humans can be pleasant and reassuring, or unpleasant and disturbing, and accordingly, they reflect well or badly on those responsible. The very word 'songbird' reflects human perception that the chirping of numerous birds is musical and enjoyable. Some birds additionally can mimic the human voice, eliciting admiration of their apparent intelligence. The cacophony of howler monkeys is highly annoying, as is the incessant droning whistle of frogs, buzzing of crickets, and screams of cats that disturb sleep at night. In most of these cases, the intent of the sounds is communication, especially to declare ownership of territory and attract mates.

One of the most annoying of all animal sounds is the high-pitched buzz of female mosquitoes, caused by rapid beating of their wings, foreshadowing a coming bite. A common misconception is that the iconic noise is caused by the wings slapping the air, but the whine is actually produced by an organ at the base of the mosquito's wings that scrapes when the wings beat. The mosquitoes are not trying to warn people that they are about to be bitten – they can't help it. The sound has been shown to be critical to attracting the males for sex (Mosquito Reviews 2022).

Many animals, often especially males, employ yells and screams as threats to warn off competitors and express dominance. The roars of large predators such as lions and the collective howls of wolf packs can be frightening, but sometimes we humans find them entertaining. In a similar vein, snarling guard dogs are valued to protect property. Men have deeper voices than women, clearly part of their testosterone-determined arsenal of threat displays. Interestingly, there is evidence that both men and women use a deeper voice to indicate dominance in respect to their audience, and a higher voice to suggest respect. Moreover, as women take on power roles, they seem to be speaking with deeper voices (Robson 2018) – not just to other humans, but also to pets!

FIGURE 6.6 Some of the world's stinkiest animals. (a) Musk ox (muskox; *Ovibos moschatus*), a specialist of Arctic regions. Males dribble pungent urine over themselves and are famous for their musk scent. Photo by Quartl (CC BY SA 3.0). (b) Wolverine (*Gulo gulo*). Found in the northern parts of the Northern Hemisphere, like many other mustelids (such as weasels, badgers, otters, ferrets, martens, and minks), it has potent anal scent glands used for marking territory and sexual signaling. Its pungent odor has given rise to the nickname 'skunk bear.' Photo by MatthiasKabel (CC BY SA 3.0). (c) Tasmanian devil (*Sarcophilus harrisii*). Native to Tasmania, this ferocious marsupial produces a strong odor when under stress. Photo by JJ Harrison (CC BY SA 3.0). (d) Striped skunk (*Mephitis mephitis*), North America's principal skunk. Photo (public domain) from Needpix. (e) Striped polecat (*Ictonyx striatus*). Native to Africa, like skunks, it sprays predators. Photo by U.Name.Me (CC BY SA 4.0). (f) Stink bird or hoatzin (*Opisthocomus hoazin*), a tropical species of South America. Uniquely for birds, it mostly eats leaves, which are fermented in its enlarged crop, contributing to its manure-like odor. Photo by Bill Bouton (CC BY SA 2.0). (g) Brown marmorated stink bug (*Halyomorpha halys*). Many stink bug species secrete foul-smelling, often toxic substances to repel attackers or anesthetize prey. Stink bugs can invade residences, especially to overwinter, and sometimes represent a significant pest control problem. Photo by Chris Hedstrom (CC BY SA 2.0).

JUDGING ANIMALS ON THEIR ATHLETICISM

The inexact terms 'sessile' (sometimes defined as attached) and 'sedentary' (sometimes defined as confined to a small area) describe animals with very limited, if any, motility (Welch 1967). Many species, such as oysters, clams, and mussels, are essentially immobile for most or all of their lives, and so are excluded from being admired for athletic abilities. In relation to their small size, some insects exhibit extraordinary lifting, pulling, or jumping powers. This is largely a matter of the physics of scaling: the strength of muscles is roughly proportional to their cross-sections, but weight is roughly proportional to volume, so as 'weight' (mass, size) increases, strength decreases. In theory, shrinking a man to 1/100th normal size would increase his strength-to-weight ratio by 100 times. This accounts for why ants and dung beetles are able to carry much more than their own weights, and why fleas can jump many times their height. It also helps explain why the monarch butterfly is able to fly thousands of kilometers despite its tiny size. These feats of strength and endurance are admirable, but of limited interest to most people except as curiosities of nature. Of much greater interest to most is the athletic ability of animals at least as large as birds. Also, there is interest in animal abilities comparable to those of superior human athletes, especially with regard to speed, power (ability to

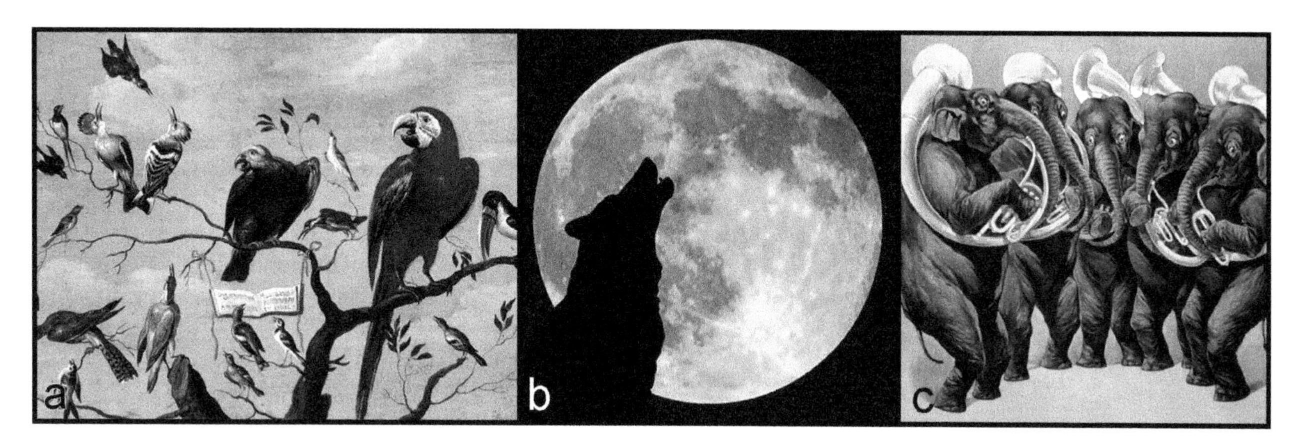

FIGURE 6.7 Respected animal calls (public domain illustrations). (a) Painting entitled 'A chorus of birds' by Flemish artist Jan van Kessel the Elder (1626–1679). (b) Icon (public domain) of a wolf howling at the moon. Credit: U.S. Fish and Wildlife Service. (c) An 1899 Ringling Bros. Circus poster showing trumpeting elephants. The calls of elephants can reach up to 2 km. Source: U.S. Library of Congress.

FIGURE 6.8 Slow race. Cartoon by French caricaturist J. J. Grandville, published in Hetzel, P.-J. 1842. Scénes de la vie privée et publique des animaux. Paris. Colorized by B. Brookes.

push, pull, or lift), balance, flexibility (as in gymnastics), and endurance. In terms of raw power, the blue whale, the largest animal that ever existed, would necessarily be the champion (simply moving its weight, often over 100 tons, is impressive), but among land animals, the African elephant, sometimes weighing over 6 tons, is capable of hauling more than its own weight (and with its trunk, lifting objects weighing several hundred kilograms). Flying birds are naturally athletic, notably the peregrine falcon, which sometimes achieves a diving speed of 400 km/h. Most fish are also naturally athletic, and sport fishing for the large species is challenging. Many arboreal mammals are spectacular gymnasts (e.g., apes, monkeys, squirrels) and are fascinating to watch. Predators naturally need to be athletic to catch and kill their prey, and people are often attracted to nature television shows featuring rather gory hunting spectacles. For millennia, we humans have depended on killing animals for survival, and we may be genetically programmed to admire animals that hunt, such as powerful lions and fast cheetahs. Unfortunately, we may also be programmed to be contemptuous of very slow creatures, such as sloths and slugs, and the poster child of slowness, the snail (Figure 6.8).

Humans have taken advantage of the athletic abilities of some other species by domesticating them to provide services. 'Draft' animals ('draught' in British English) such as oxen, mules, and elephants have been enormously important historically to compensate for the relative weakness of humans to pull heavy weights. (Similarly, pack-bearing and 'mounted' animals are employed for transport and riding, respectively.) In tribute to the remarkable athletic abilities of horses (Figure 6.9), the term 'horsepower' was adopted in the late 18th century by Scottish engineer James Watt, based on the power of draft (heavy) horses. Horses can typically drag one-tenth of their body weight or pull wheeled loads weighing 1.5 times their body weight. However, some strong draft breeds can drag 10–15 times their body weight over short distances, with one horse having pulled almost 30 tons. The biggest horses can weigh over 1 ton (the largest weight recorded was over 1.5 tons). Other notable domesticated working athletic animals include dogs (for various purposes, as shown in Figure 6.10), elephants for logging, cats as mousers, and pigeons for sending messages long distances. The athletic abilities of wild (undomesticated) animals have also been employed: hawks, diving birds, and cheetahs for hunting.

In parallel to human combat events, for entertainment purposes, animals have been forced to fight. Bull-baiting

FIGURE 6.9 The three athletic functions of domesticated horses. (a) Riding horse. 'Napoleon crossing the Alps,' a painting (public domain) by French artist Jacques-Louis David (1748–1825). (b) Draft horses. Photo of heavy horses by Airwolfhound (CC BY SA 2.0). (c) Racing horse. Diamond Jubilee (1897–1923), a British thoroughbred champion race horse. Source (public domain) Richardson, C. 1911. *The New Book of the Horse*. London, Waverley.

FIGURE 6.10 Athletic dogs. (a) Border collie jumping for a frisbee. Photo by Lucie Schönová (CC BY SA 3.0). (b) Border collie herding sheep. Photo by SheltieBoy (CC BY 2.0). (c) A sled dog team of yakut laika, an ancient working dog breed that originated in the Arctic seashore of the Yakutia (Sakha) Republic of Russia. Photo by Ajarvarlamov (CC BY SA 3.0). (d) Two merchants selling milk from a dogcart. Late 19th-century painting (public domain) from Brussels, Belgium. Credit: U.S. Library of Congress. (e & f) Paintings (public domain) of St. Bernard dogs saving children. Source: McLoughlin Bros. 1870. *Dog of St. Bernard*. New York, McLoughlin Bros.

and the similarly cruel bear baiting have been banned, dogfighting and cockfighting have been largely but not entirely eliminated, bullfighting (really, bull slaughter) persists in some countries, and in Asia staged fights persist for Siamese fighting fish and crickets. Rodeos and race events (featuring horses, dogs, or camels) are still quite popular. All animals, including those as noted here that perform tasks for people, are admired, at least for their services (although the opinion of the animals concerned might be less charitable).

JUDGING ANIMALS ON THEIR PHYSICAL AGGRESSIVENESS

Animals compete for resources, with other species and with other members of their own species. Physical fighting (combat) is a choice that can often be made to increase an individual's access to resources, but this requires assessment of the possible benefits weighed against the risks involved. Often, competitors are so much larger or dangerous that the risk is inadvisable (sometimes it is better to be a live coward instead of a dead hero). On the other hand, simply a threat (even if it amounts to a bluff) might suffice to scare away competitors. 'Agonistic behavior' designates competitive conduct in the presence of others (the same or different species). Predator-prey interactions are generally excluded. Agonistic behaviors could include aggression, submission, or retreat. Survival is often at stake, potentially involving deadly or at least harmful attacks while fighting for food, shelter, territory, and/or sexual partners. Often, violence pays (Georgiev et al. 2013). Animals, especially males, frequently threaten in the hope of avoiding combat (Figure 6.11). Threats may include flaunting of one's size, making loud noises, and engaging in postures that look like an attack is imminent. Dogs bare their teeth, growl, and the hair on their back stands up, with the result that a weaker opponent will turn with its tail between its legs and run off. Silverback gorillas pound their chests and grunt. Male mandrils yawn to display their huge canine teeth.

These considerations apply to human as well as non-human animals, so people naturally have perspectives of other people's degree of aggressiveness or passivity. We are often contemptuous of passive people, who may seem to be cowards. On the other hand, we tend to admire aggressive people, at least when they exhibit courageous behavior, although 'adrenalin junkies' who are natural risk-takers may seem to be foolish. In the military and police spheres, the ability to project controlled aggressiveness is essential. In the world of sports, aggressive behavior is highly admired (note the discussion of 'bloodsports' in Chapter 2). Sports teams' animal mascots are usually quite aggressive animals (Chapter 7).

Aggressive behavior associated with physical combat is particularly driven by testosterone in males. 'Toxic masculinity' refers to the notion that 'manliness' requires domination and aggression. Regardless, aggression in male animals, including humans, is clearly advantageous in increasing evolutionary fitness (i.e., producing more progeny) and is valued to at least some degree by both men and women.

Animals frequently use threats to usurp the acquired materials of others (e.g., lions claiming the recent kills of cheetahs). In the human sphere, such behavior seemingly amounts to 'theft' (a moral and legal judgment). Nevertheless, people have often been comfortable acquiring the 'spoils of war,' reflecting admiration of aggression. However, within a society overly aggressive individuals are frequently incarcerated.

It is important to distinguish physical aggression from aggressive tactics that do not employ actual violence. In modern times, the most famous 'pacifists' were Mahatma Gandhi (1869–1948) and Martin Luther King (1929–1968). Their passivity involved a refusal to engage in physical

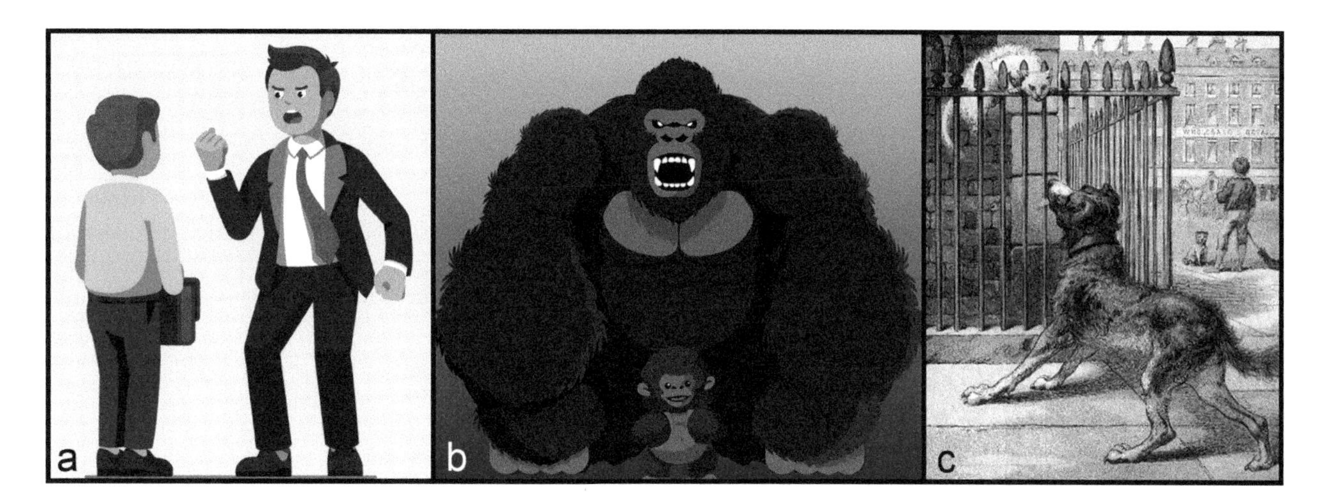

FIGURE 6.11 Threat displays. (a) A display of intimidation between humans. Source: Shutterstock (contributor: Iconic Bestiary). (b) Gorilla protecting a youngster. Source: Shutterstock (contributor: Australier). (c) Confrontation between a cat and a dog. Source (public domain): Valentine, D. 1865. *Aunt Louisa's Birthday Gift*. Scribner, Welford and Co. Credit: https://www.reusableart.com/white-cats-03.html.

FIGURE 6.12 Animal symbols representing passivity and aggressiveness. Public domain images. (a) A dove carrying an olive branch, both representing peace. Photo credit: Elembis, Sammy pompon and Mark Miller. (b) 'Don't tread on me,' the 'Gadsden flag' (created by C. Gadsden in 1775) featuring a coiled rattlesnake, symbolic of the aggressive struggle for American independence from British rule, and recently associated with far-right political groups. (c) The greater coat of arms of the U.S. The olive branch and the arrows held in the eagle's talons denote the power of peace and war. Photo by Ssolbergj (CC BY SA 3.0).

combat, but their so-called 'passive resistance' in reality was an extreme form of aggression intended to challenge and revolutionize norms.

In psychiatry, 'passive-aggressive personality disorder' is a deviant or pathological negativistic bullying form of behavior by a perpetrator intended to hurt a victim (bullying is discussed in Chapter 2). Defining the condition has been contentious, and the 'Diagnostic and Statistical Manual of Mental Disorders' of the American Psychiatric Association discontinued use of the phrase in 2013. The condition, which is not relevant to this discussion, is mentioned here only to avoid possible confusion.

How do the perceptions that people have of aggressiveness and passivity apply to non-human animals? Some domesticated animals have been selected for belligerence, especially for combat in front of spectators (bullfighting, dogfighting, cockfighting, and the like). Most domesticated animals (both livestock and pets) have usually been selected for passivity, of course because animals that attack their owners or masters are dangerous. In livestock species with extremely aggressive males responsible for many females (bull/cows, rooster/chickens, etc.), a minimum number of males are maintained (Chapter 3). Recreational hunters and fishers, in contrast, value wild aggressive animals – for the sporting experience of catching them (Chapter 2). At least in the safety of their living rooms, people often admire very aggressive animals shown on television programs.

The leading national animal emblems (lions, tigers, eagles, and wolves) are extremely aggressive, reflecting respect for animals that are natural killers. By contrast, doves (usually white and often carrying olive branches), which are much less aggressive, are frequently symbolically associated with peace and pacifism. (Doves, as discussed in Chapter 16, are not clearly distinguished from pigeons. They are any of hundreds of species in the bird family Columbidae.) Well-known animal symbols symbolic of aggression and passivity are shown in Figure 6.12.

JUDGING SPECIES ON THE ATTRACTIVENESS OF THEIR HABITAT

Ecologists, somewhat arbitrarily, split up the parts of the Earth where species live into units. Just as humans have pronounced preferences and prejudices for species, they also have different preferences for these units. 'Ecosystems,' the most comprehensive units, are relatively large portions of the planet where a set of species interact with their physical environment and with each other. Echeverri et al. (2017) found that people prefer forest and ocean ecosystems over grassland and tundra ecosystems. Cronon (1995) noted 'Most of us, I suspect, still follow the conventions of the romantic sublime in finding the mountaintop more glorious than the plains, the ancient forest nobler than the grasslands, the mighty canyon more inspiring than the humble marsh.' There have been attempts to prioritize the needs for conservation of ecosystems (Keith et al. 2013; Bland et al. 2017, 2019). See Chapter 9 for information on the IUCN Red List of Ecosystems, which evaluates risk to the world's ecosystems. 'Habitats,' the most important localized ecological units, are essentially the home areas (neighborhoods) that supply the resources necessary for survival and reproduction. (Ecosystems intergrade with habitats and sometimes are the same.)

'Wastelands' are arbitrarily defined on the basis of being scarcely habitable or exploitable by humans – either because of being naturally stressful environments, or being despoiled by human activities. 'Badlands' are extensive tracts of heavily eroded, uncultivable land with little vegetation (once again, the terminology reflects human judgment). 'Deserts' are large areas of land, often in hot regions, where there is little vegetation because

FIGURE 6.13 Stressful habitats with limited biodiversity, and limited appeal to most people. (a) Mojave Desert near Las Vegas Nevada. Photo by Fred Morledge (CC BY SA 2.5). (b) Tundra in Nunavut, northern Canada. Photo by ADialla (CC BY 2.0). (c) Alpine zone of Mount Madison, New Hampshire. Photo by Ken Gallager (public domain). (d) Sphagnum moss peat bog in Parc national de Frontenac, Quebec, Canada. Photo by Boréal (CC BY SA 3.0). (e) Salt flat in Bolivia. Photo by Yoann Supertramp (CC BY 3.0). (f) Eroded badlands in the Four Corners region of New Mexico. Photo (public domain) by U.S. Bureau of Land Management.

of aridity, and consequently where human occupation is limited. High mountainous areas, frigid zones, and several kinds of soils (bogs and other substrates with levels of elements that are too toxic or too limited for most plant survival) also tend to be very hostile for human occupation. The animals and plants capable of living in such regions tend to be small (because resources are very limited) and often possess very unique (often odd) features and behaviors that adapt them to survive. The sparsity of vegetation and animals and the natural climatic and landscape stresses make such habitats relatively unattractive to most humans (Figure 6.13). Visiting such sites (often designated as national parks or reserves) does attract tourists (viewing the odd biota and landscapes is entertaining), but these areas are nevertheless quite foreign and somewhat unsettling compared to the comfort of the places where most people live. By contrast, those who grow up in or near these special sites tend to acquire a deep respect for them, suggesting that human biases and preferences for habitats once established tend to persist.

JUDGING ANIMALS ON THEIR CONSTRUCTED DOMICILES

We humans are highly class-conscious and judge each other by indicators of wealth such as automobiles, clothing, jewelry, and the prestige of professions. The biggest investment of most people's lives is their residence, and accordingly, the opulence of one's domicile (Figure 6.14) is a principal way we judge each other. Most animals are adapted to living without the protection of housing or with very simple burrows in soil, vegetation, or even inside other animals. Some species, however, fabricate housing that is so complex it produces admiration for them. This includes termites that construct gigantic communal mounds, social bees and their hives, weaver birds and their astonishing nests (sometimes communal), and beavers and their constructions (Figure 6.14).

Male bowerbirds include about 27 species of the family Ptilonorhynchidae, centered in the tropical regions of New Guinea and northern Australia. They are extraordinary in building nesting structures in front of which are beds of sticks and brightly colored objects in order to attract mates (Figure 6.15).

JUDGING ANIMALS ON THEIR 'GARMENTS'

Unlike other creatures, humans wear clothes, and while in past times clothing was employed mostly as protection from the weather, today, at least in wealthy nations, clothing is primarily chosen to enhance appearance (often at the cost of comfort). Indeed, for most people clothes are the primary way their image is enhanced, and the fashion industry is immense. (See 'clothing and costumes' in Chapter 5.)

Although animals do not wear clothes like humans, some do analogously adopt materials in their environments to cover themselves, usually for physical protection or camouflage (Figure 6.16). Many animals, including elephants and pigs, coat themselves with mud or dust as protection against sunburn – resulting in an appearance that humans do not find attractive. Caddisfly and stonefly larvae build portable protective cases for themselves using silk that they produce along with stones, sticks, sand, and other materials (wearing this

FIGURE 6.14 Comparative impressiveness of constructed living quarters determines attitudes toward the occupants. (a) Biltmore Estate in Asheville, North Carolina. Photo by Kolin Toney (CC BY SA 2.0). (b) A shack. Photo (public domain) by Anne and David. (c) Male baya weaver (*Ploceus philippinus*) of India at the entrance tunnel of its hanging woven nest. Weaver birds include over 100 species, mostly of tropical Africa and Asia, the males of which construct elaborately woven nests out of vegetation. Photo by Kamble Vidhin Sangola (CC BY SA 4.0). (d) Termite mound in Australia. Photo by Thinboyfatter (CC BY 2.0). (e) Canadian beavers (*Castor canadensis*) and their lodge. Beavers have been called the greatest of wild animal architects. Source (public domain): Craig, H. 1880. *Johnson's Household Book of Nature*. New York, H. H. Johnson.

armor, they move to different areas seeking food). Similarly, hermit crabs move about while wearing shells discarded by others. Some octopuses also wear shells for protection. Decorator crabs stick materials on their bodies for camouflage, and sometimes even attach noxious organisms to ward off predators. Although these behaviors are instinctive, they appear quite clever and often evoke admiration from people.

JUDGING ANIMALS ON THEIR DIET

'Food snobbism' ('wine snobbism' is perhaps the best-known subcategory) is common among humans. Religions and cultural traditions are associated with food taboos (discussed in Chapter 3) that confer low status on those who consume forbidden foods. 'Carnism' (coined by Joy in 2001 and popularized in Joy 2009), referring to the use and consumption of animal products, especially meat, is the predominant ideology today. (Mahlke (2014) considers carnism to be a form of speciesism, discussed in Chapter 3.) However, in recent times, vegetarianism and a variety of specialized dietary fads have acquired elevated status in certain quarters. For meat-eaters, the different parts of livestock have different status (and price), and of course, restaurants are judged on the quality (and expense) of their

FIGURE 6.15 A human marriage bower and bower birds. Just as the male human has prepared an impressive display for his bride, the male birds prepared a display of collected objects in front of a lodge constructed of vegetation. Public domain illustrations. (a) 'The Nuptial-Bower' by British caricaturist James Gilray (1756–1815). Credit: U.S. Library of Congress. (b) Great bowerbird (*Chlamydera nuchalis*) by British animal painter William Matthew Hart (1830–1908), published in: Sharpe, R. D. 1891–1898. *Monograph of the Paradiseidae, or Birds of Paradise, and Ptilonorhynchidae, or Bower-birds.* Vol. 2. London, Henry Sothern. (c) Spotted bowerbird (*Ptilonorynchus maculatus*) by Elizabeth and John Gould, British ornithologists and artists, from their book *The Birds of Australia*, Vol. 4, published by the authors in 1848 in London.

offerings. Indeed, the expense (and even the quantity) of food is one of the chief indicators of social class among people. The foods that animals consume and how they eat also at least partly determine human attitudes toward various species (Figure 6.17). Of course, animals that eat us – parasites and predators – get no respect for such culinary behavior. Animals that consume decaying food, especially scavengers of corpses, evoke revulsion. Eating discarded food from refuse bins is regrettably a sign of poverty among humans, but many animals likely regard garbage containers and dumps as sources of the finest cuisine.

Humans are descended from arboreal primates with particular fondness for sweet fruits, and we have retained our instinct for this preference. We also are adapted to consuming calorie-rich plant storage organs such as seeds, nuts, and tubers, but we are not particularly foliage and stem eaters, despite the dietary advantages of salads, sometimes derisively termed 'rabbit food.' Chapter 12 provides criticism of the need to reduce the ecological harm from raising livestock by eating more plants, but the human preference for meat is undeniable. Predators that hunt and eat the same animals that people eat get some respect for having the same tastes, but competitors of humans tend to be persecuted or even eliminated. The same is true for animals that consume our crops, garden plants, and houses (Figure 6.18).

Humans have a natural aversion to excrement, so animals that appear to like and even eat feces are viewed with disgust. This includes some that are otherwise considered 'cute,' notably pandas, horses, and dogs. Coprophagy (consumption of feces) may be a temporary, adaptive behavior in some species. Sometimes the young of animals eat their mother's dung to obtain their symbiotic intestinal bacteria. Similarly, termites eat the feces of their nestmates to obtain the gut protists which digest the wood that they eat. Mother dogs lick away excrement from their young puppies for the benefit of the latter. However, animals that are adapted to coprophagy as a means of obtaining basic nutrition are also common (Soave and Brand 1991), most famously dung beetles and flies, both of which include thousands of species. In some herbivores, notably rabbits, eating their own feces is a way of extracting more nutrients (this allows the animal to have a smaller digestive system, so it weighs less and can run away faster from predators). Despite their way of life, scarab dung beetle adults have received considerable respect since ancient times (Simmons and Ridsdill-Smith 2011; Figure 6.19).

'Insectivores' are insect-eating species, which includes a very large number of animals and some plant species. 'Entomophagy' is eating of insects. Chapter 12 discusses consumption of insects by humans, which is widespread, has huge potential for more efficiently feeding the world and reducing environmental damage due to current livestock production but is viewed with disgust by most people in Western culture. Most insectivores are small animals – especially small vertebrates and birds – while large carnivores generally rely on larger prey for food, and as noted in Chapter 4, humans have a very strong preference for large animals that often outweighs other considerations.

The sight of animals consuming other live animals is not appealing for most people, and few pets are fed live animals. Indeed, except for certain species, humans do not eat animals while they are still alive and tend to disapprove of unnecessary pain (see Chapter 2). In nature, of course, survival, not cruelty, is the predominant concern.

The vernacular term 'grub' refers to a stubby worm-shaped larva of some insects, especially scarab beetles. Such grubs frequently feed on the roots of turf and pasture grasses, which are often further damaged by skunks, raccoons, birds, and other small animals rooting out the grubs to eat. The non-technical term 'maggot' similarly refers to limbless grubs, especially of the insect order Diptera (flies). Maggots are frequently found in refuse, and both decaying

FIGURE 6.16 Animal apparel. (a) Painting (public domain) of ostentatiously dressed William IV of the United Kingdom by artist Martin Archer Shee (1769–1850). (b) Painting (public domain) of ostentatiously dressed Queen Victoria of the United Kingdom by artist Franz Xaver Winterhalter (1805–1873). (c) Soldier wearing vegetation camouflage. Photo (public domain) by Anonymous. (d) African bush elephants coating themselves with mud. Photo by Mgiganteus (CC BY SA 3.0). (e) Caddis fly larva in its protective case. Photo by Devon Buchanan (CC BY 2.0). (f) A purple hermit crab (*Coenobita brevimanus*) wearing an old soup can. Photo by Naturalselections (CC BY SA 4.0). (g) Coconut octopus (*Amphioctopus marginatus*) wearing shells. Photo by Nick Hobgood (CC BY SA 3.0). (h) Kono's carrier snail (*Xenophora konoi*), which coats itself with stones, shells, and whatever else it can find, for camouflage. Photo by James St. John (CC BY 2.0). (i) Spider crab (*Achaeus spinosus*), a species of 'decorator crab' which camouflages itself with bits of local material. Photo by Christian Gloor (CC BY 2.0).

and living flesh, and so are viewed with disdain. Both words have quite pejorative meanings when applied to people, and this reflects the very negative perception of the lifestyle of such creatures. Nevertheless, maggots have some use in medical 'debridement therapy' (cleaning dead flesh from wounds), removing flesh from skeletons for museum display (using dermestid beetles), and as a farmed food for livestock (grubs are grown in waste flesh that would otherwise have to be discarded).

Plants grow over scattered areas, which tend to separate foraging herbivores spatially, reducing conflict among individuals. Very small prey animals, such as insects, also are scattered, and this reduces competition. However, animals, whether normally social or solitary, often congregate at a food source resulting in competitive food frenzies (Figure 6.20). Large prey, even if captured communally, may not be shared democratically, and generally, there is a 'pecking order' determined by size, sex, and aggressiveness.

FIGURE 6.17 Judging species by their foods. (a) Dog with bone. Attribution: Guy Gilroy, www.myguysmoving.com (CC BY SA 2.0). (b) Raccoon at a garbage can, prepared by Knuth, Virginia State Park staff (CC BY 2.0). (c) 'Der Gourmet,' public domain painting by British artist Alexander Austen (1891–1909).

FIGURE 6.18 Despised animals that eat our crops, garden plants, and houses. Source: Shutterstock. (a) Contributor: Kseni Now. (b) Contributor: Bassarida. (c) Contributor: Refluo.

FIGURE 6.19 Dung beetles. (a) Sacred scarab (*Scarabaeus sacer*). Most scarabs are dung beetles. Public domain painting from Fabré, J. H. 1921. *Fabre's Book of Insects*, New York, Dodd, Mead and Company. (b) Male rainbow scarab (*Phanaeus vindex*). Public domain photo. Credit: A. Santillana, Insects Unlocked project, University of Texas at Austin. (c) Pectoral (breast plate) of Tutankhamun (Egyptian pharaoh 'King Tut,' c.1341–c.1323 B.C.E.), with a winged scarab beetle (top center) and semi-precious stones. They were worshipped in ancient Egypt. Photo attribution: https://www.flickr.com/photos/dalbera/

FIGURE 6.20 Food frenzies. (a) Caricature (public domain) of a dining club by English artist Thomas Rowlandson (1757–1827). Credit: The Elisha Whittelsey Collection, Metropolitan Museum of Art. (b) Kodiak bears (*Ursus arctos middendorffi*). Photo (public domain) by Lisa Hupp/USFWS. (c) Koi (*Cyprinus rubrofuscus*). Photo by Don Dexter Antonio Photography (CC BY ND 2.0). (d) Black skimmers (*Rynchops niger*). Photo by Charles Patrick Ewing (CC BY 2.0).

The sight of large carnivores fighting over their share of a meal is disturbing or entertaining, depending on people's sensitivities. Modern dinner table etiquette is so refined that any departure of behavior by animals may seem inappropriate, although gorging, drooling, grunting, spitting, and similar behaviors were likely once normal (Figure 6.20a).

Social animals tend to eat communally. In most social animals, individuals are usually concerned only with their own access to food, not their neighbors, who are merely tolerated (parents, of course, tend to ensure that their young are fed). In humans and some other species (notably other primates), apparently altruistic behavior occurs, individuals deliberately sharing food, either just after it has been acquired by foraging or hunting, or at meals (Jaeggi and van Schaik 2011). This has been explained in evolutionary terms as adaptive because the behavior promotes group solidarity and/or genetic fitness (or alternatively, as simply avoiding harassment from hungry members of the group; Stevens 2004). Regardless of motivation, the sight of animals and indeed other people cooperatively feeding rather than squabbling tends to elicit human approval (Figure 6.21).

DOG PEOPLE VS. CAT PEOPLE

Human personality is complex and, of course, influences preferences and prejudices for and against pets (Kidd et al. 1984; Podberscek and Gosling, 2000). An interesting contrast exists between dog owners and cat owners, reflecting the more interactive and dependent nature of dogs and the more independent and solitary nature of cats (Amiot and Bastian 2015; Figure 6.22). It has been said that dogs regard their owners as their bosses, while cats treat their owners as their employees. People who self-identify as 'dog people' have been found to tend to be relatively extroverted, agreeable, and conscientious, but less neurotic and less open to experiences than those who label themselves as 'cat people' (Gosling et al. 2010). Perrine and Osbourne (1998) concluded that dog people score higher for masculinity, independence, and athleticism. Children in homes with dogs have been rated higher for empathy than those in homes with cats (Daly and Morton 2003). Cat owners have been found to avoid their cats more than dog owners avoid their dogs (Zilcha-Mano et al. 2012). Photos of people shown with dogs are rated more positively than the same people in photos without dogs (Rossbach and Wilson 1992). Also, people accompanied by dogs in photos are rated more positively than those with cats (Geries-Johnson and Kennedy 1995).

JUDGING ANIMALS ON THEIR STATUS

'Respect' may be based on admiration or simply fear of the power possessed by someone or something. In the human sphere, people often are respected for some particular natural talent (such as singing) or acquired

FIGURE 6.21 Social animals eating communally, a life style attractive to people. (a) A pack of lions (*Panthera leo*) feeding on an African buffalo (*Syncerus caffer caffer*) in Botswana. Photo by Diego Delso (CC BY SA 4.0). (b) Orangutans (*Pongo pygmaeus*) feeding cooperatively at a rehabilitation center in Borneo (normally orangutans are solitary). Photo by Nino Verde (CC BY SA 3.0). (c) Painting (public domain) entitled 'Freedom from Want' by American artist Norman Rockwell (1894–1978). Credit: U.S. National Archives and Records Administration. (d) Painting (public domain) entitled 'Christmas Eve' by Swedish artist Carl Larsson (1853–1919). Photo credit: Carl Larsson.

competence (like medical doctors), or for being rich, or having political power, or merely for possessing a gun (Figure 6.23a). Whatever the basis, people who are respected by the majority (regardless of our own opinion) gain benefits from society. Among animal communities, dominant species are respected by other animals primarily for being dangerous, but also just for being large. But for this discussion, the key issue is not how animal species respect each other, but how humans view the status of animals. In the oceans, whales clearly are the most respected simply for their size, and sharks because of their fearsome reputation. Similarly on land the largest herbivores such as elephants, rhinos and hippos are admired, and so are the most fearsome predators. However, of all animals the lion has gained the most universal respect, as reflected in its title as the 'king of beasts' (Figure 6.23b). Lions are the world's most popular national symbol (as noted in Chapter 7) and seem to have made a greater impression on the human psyche than any other animal.

FIGURE 6.22 Dog vs. cat people. Illustrations from Shutterstock. (a) Contributor: Iconic Bestiary. (b) Contributor: StockSmartStart. (c) Contributor: Iconic Bestiary.

FIGURE 6.23 Importance of status. (a) 'The Pyramid of Capitalist System,' an American cartoon caricature (public domain) critical of capitalism, copied from a Russian flyer published in the 1911 edition of *Industrial Worker* by The International Publishing Co., Cleveland, OH. In this hierarchy, a wealthy few have dominant status while the impoverished masses are at the bottom. (b) Lions guarding a waterhole in Etosha National Park, Nambia, intimidating other animals from approaching. Photo by Buiobuione (CC BY 4.0).

7 Symbolic (Representational) Creatures

Reflections of Human Prejudices against and Preferences for Species

INTRODUCTION

Species are (or were, if extinct) real, living organisms, but we humans have embellished the nature of some to represent important aspects of cultures and have created artificial, imaginative species-like creatures for literary and psychological purposes. These may be religious, spiritual, mythological, allegorical, or emblematic, and often there is an anthropomorphic aspect (i.e., illustrating human traits, emotions, values, or intentions). Cryptozoology and cryptobotany are pseudosciences respectively addressing animals and plants, generally extremely unusual, alleged to exist but for which there is no reasonable evidence. However, there are sometimes valid motives for the 'existence' of physically unreal creatures (consider Santa Claus, the Tooth Fairy, the Easter Bunny, and Smokey Bear). Many imaginary beings represent characteristics, features, events, people, etc. that a culture or group thought to be sufficiently important to warrant recognition. Typically, such constructs reflect valuable or attractive matters, but also occasionally harmful or fearful phenomena. There is an extensive folklore associated with species of plants (Lehner and Lehner 1960; De Cleene and Lejeune 2003) and animals (Allen 2016), but relatively few of these acquire much representational significance or importance for people. But some do, and since their symbolism is embodied in species or species-like entities, these often reveal the characteristics that people especially value or dislike in species. The minority of genuine species that acquire widespread prestige, especially for religious significance or as representatives of political regions, are very sympathetically treated when they become endangered, since mistreating them is like insulting a religion or a country. Additionally, many purely symbolic species are important icons for supporting species conservation.

TOTEMS

A 'totem' is an animal, plant, or other natural object serving emblematically among certain traditional people. (The basic concept has been enlarged or modified by various authors.) Animals are much more frequently employed as totems than plants (Sommer 1988). Adopting natural objects as symbols (totemism) has been a major topic of research by psychologists and anthropologists, since it seems to indicate how humans think and evolve values (Lee et al. 2018b). Highly artistic totemic representations of animals have been made by indigenous peoples in North America (Kramer 2008; Stewart 2009) and Taiwan (Peng and Chung 2017). The vertical sequence of figures on totem poles sometimes reflects historical events. The figures often blend human and animal features (Figure 7.1) and often indicate great respect for local animals and plants. In colonial North America, Christian missionaries were highly intolerant of totem worship, which they considered idolatrous. Most totem poles were destroyed or moved as exhibits to museums and entertainment parks. In recent decades, there has been a resurrection of carving totem poles by indigenous people.

SPECIES AND RELIGION

Species have been extensively incorporated into the mythology,[1] dogma, and traditions of various religions (Alves et al. 2017). 'Zoolatry' is the worship of animals, which was remarkably extensive among the ancient Egyptians, who deified (and mummified) numerous species. Species that acquire religious significance have to be memorable – they may be big, beautiful (or quite ugly), useful, or dangerous, or they may have some behavior that is significant. Turtles are exemplary: their slowness, longevity, and impressively protective shell preadapted them to being incorporated into mythology. They have often been subjects of creation myths, and many cultures honor sacred turtles (Figure 7.2). In Hindu mythology, Akupara is a tortoise (termed a 'World Turtle' and a 'Cosmic Turtle') supporting on his back elephants which in turn uphold the world (Figure 7.2a). Many Indigenous groups of North America similarly conceive of a turtle that carries the Earth upon its back (Figure 7.2b).

The Abrahamic religions (Christianity, Islam, and Judaism) all relate the story of Noah's Ark, in which every living species on Earth is saved on God's order. As reflected by the classical Christian view of animals in paradise (Figure 7.3), the species that people like most are pretty birds and impressive mammals, particularly those that have been domesticated. Various plants and animals have biblical or Koranic significance, and some are used in sacred ceremonies, but none has acquired great status. Islam is

[1] The word 'myth' can be pejorative, indicating falsehood, which would be insulting to those who interpret religious narratives literally. A more reasonable interpretation is that spiritual mythology is a creative, poetic, symbolic, allegorical expression of fundamental truths.

DOI: 10.1201/9781003412946-7

FIGURE 7.1 Totem poles and costumes. (a) An old totem pole in Alaska. Photo by Leonard Kaplan (CC BY SA 3.0). (b) Bella Coola (Nuxálk) people of British Columbia conducting a religious ceremony in which totemic costumes representing animals are worn. Also note the totem poles. Painting (public domain) from Sievers, S. 1894. Amerika, Eine allgemeine Landeskunde. Leipzig, Bibliographisches Institut. (c) Totem poles in Formosan Aboriginal Culture Village, Taiwan. Photo by Bernard Gagnon (CC BY SA 3.0).

'aniconic' with respect to the employment of illustrations of animals (and humans) in a religious context, based on the tradition that this is a form of idolatry (however, flowers, often stylized, are popular). In general, species with particular religious significance for advancing the cause of biodiversity preservation in the Western World are hard to identify. Moreover, very conservative practitioners tend to emphasize the schism between humans and all other species, to the disadvantage of the latter (Amiot and Bastian 2015). A survey of conservative Protestants revealed low support for animal rights (DeLeeuw et al. 2007). However, there are recent efforts by the Abrahamic religions to support environmentalism (e.g., Tirosh-Samuelson 2015; DeLong-Bas 2018; McKim 2020). Indeed, in recent decades, representatives of the monotheistic religions have become enthusiastic and influential supporters of biodiversity conservation.

Eastern (Asian) religions emphatically support a profound respect for species, both plants (Figure 7.4) and animals (Figure 7.5). Buddhism and Hinduism are the most widespread non-Abrahamic religions. Buddhism advocates non-violence toward all species, and similarly Hinduism teaches respect for every living thing. Many plants and animals are prominent in religious texts and traditions, and their considerable divine significance has served to protect them from harm. Such highly respected species are especially useful icons for biodiversity conservation in Asia. The ginkgo tree owes its survival to having been preserved beside religious shrines. In India, species with religious significance include tigers, lions, elephants, peacocks, owls, swans, mangoes, coconuts, tamarinds, sacred basil, and many others (Sinha 1995; Sinha et al. 2005; Sood et al. 2005).

Aboriginal peoples invariably have a spiritual dimension that is deeply respectful of many species. In most of Africa, the Americas, Australia, and other parts of the world, indigenous religions have been supplanted by Christianity and/or Islam. Nevertheless, indigenous people have retained much of their traditional respect for biodiversity in general, and for some species in particular.

'Buddhist environmentalism' is a movement dating to the 1970s that has become widespread throughout the world (Sponsel and Natadecha-Sponsel 2017; Darlington 2019; Tu

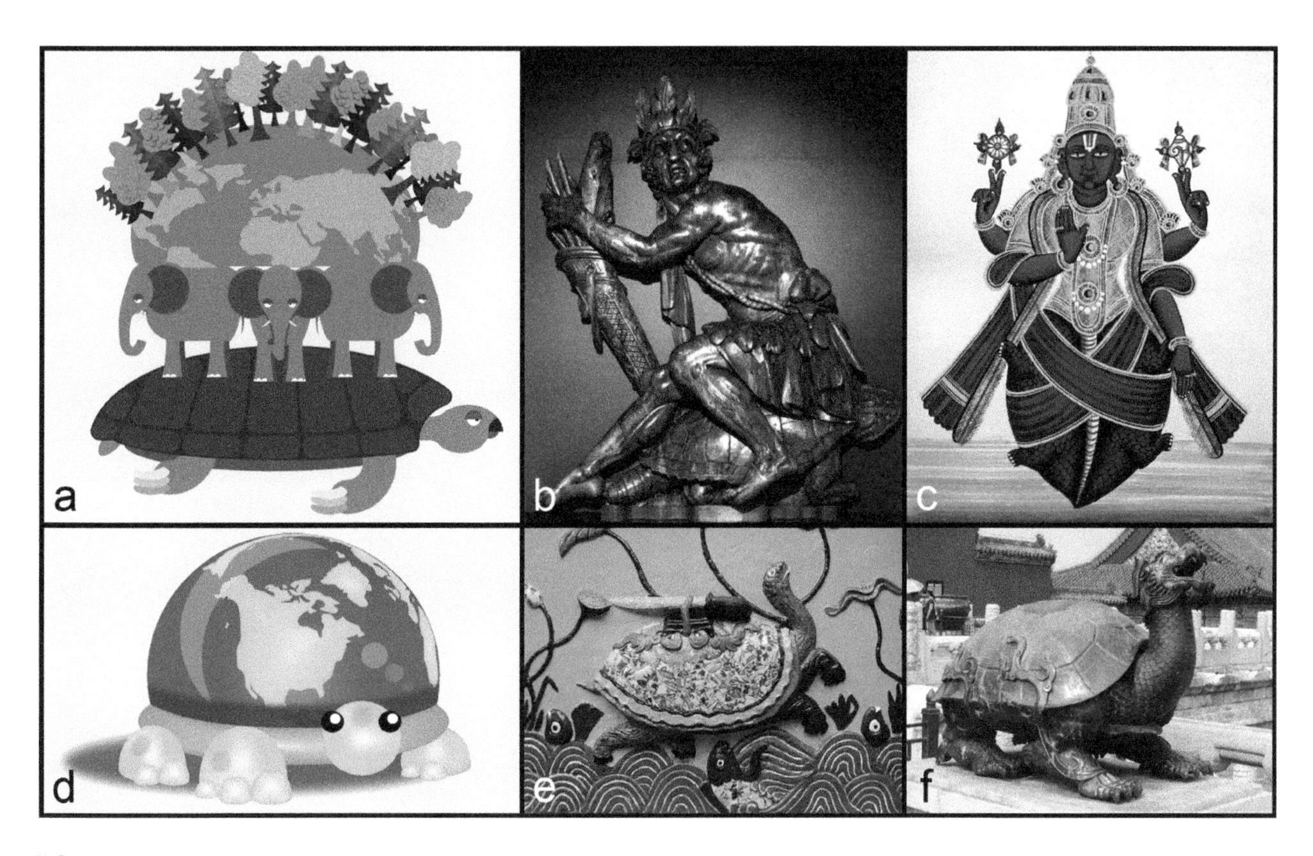

FIGURE 7.2 Sacred turtles. (a) The 'World Turtle' Akupara, of Hindu mythology, supporting elephants which in turn uphold the world. Source: Shutterstock (contributor: Top Vector Studio). (b) Nineteenth century carving of an Iroquois Indian sitting on a turtle, in reference to the Great Turtle that carries the Earth in Iroquois mythology. Source: Naval Museum, Brest, France (CC BY SA 2.0). (c) Painting (1850, public domain) of the Hindu God Krishna, transforming into Kurma, his turtle form. He is shown here as half-turtle half-human. Source: https://www.britishmuseum.org/collection/object/A_1993-0806-0-6. (d) A clever interpretation of the World Turtle shown above. Created by Jerry Paffendorf (CC BY 2.0). (e) Vietmam's Golden Turtle God (Kim Quy) with a sword on his back named Heaven's Will that the turtle loaned to a 15th-century emperor. Photo of a mural in the Temple of the Jade Mountain, by Rdavout (CC BY SA 3.0). (f) Dragon turtle, one of the four Chinese mythological spiritual creatures (Sì Shòu), each guarding a direction of the compass. This one is the protector of the North. Photo in the Forbidden City by Jonathan O'Donnell (CC BY 2.0).

FIGURE 7.3 Paradise, as conceived by 16th-century European artists (public domain images). Note that the animals are all mammals and birds. (a) Painting by Flemish artist Jan Brueghel the Elder (1568–1625). (b) Copy of an original work by Flemish artist Pieter Brueghel the Elder (1526/1530–1569) housed in Museo Nacional del Prado.

FIGURE 7.4 Examples of sacred Asian plants. (a) Offerings at a Shinto shrine in Japan. In the foreground are branches from the evergreen tree sakaki (*Cleyera japonica*), a standard oblation to the spirits. Photo by Katorisi (CC BY SA 3.0). (b) Banyan fig (*Ficus benghalensis*), known as sacred fig and bo-tree, it has considerable religious significance in Asia and is therefore useful there as a conservation icon. Photo by D. Alpern (CC BY SA 3.0). (c) Holy basil (*Ocimum tenuiflorum*; also known as *O. sanctum*), called tulasi and tulsi in Asia, it is perhaps the holiest plant in Hinduism. Shown here is a decorated plant at a religious ceremony in India. Photo by Antrabhardwaj2 (CC BY SA 4.0). (d) The date palm (*Phoenix dactylifera*) was sacred in Mesopotamian religions, and in ancient Egypt represented immortality. Palm branches were subsequently adopted as religious symbols in the Abrahamic religions. Photo by Shehzad.abbasi (CC BY SA 4.0). (e) Sacred lotus (*Nelumbo nucifera*), a native of Asia and Australia, is symbolically important in Buddhism and Hinduism. Photo by T. Voekler (CC BY SA 3.0). (f) The sacred fig or bodhi tree (*Ficus religiosa*) is native to the Indian subcontinent and has religious significance in Buddhism, Hinduism, and Jainism. Photo by Ji-Elle (CC BY SA 3.0).

and Thien 2019). It combines Buddhist concepts and principles with environmental issues, particularly biodiversity conservation. A striking example of its success was the rescue of the Asian open-billed stork (*Anastomus oscitans*; Figure 7.5b), which had been extirpated from most of its natural Asian range. The only remaining colony in Thailand was near a Buddhist temple, *Wat Phai Lom*, north of Bangkok. The temple was recognized as a bird sanctuary in 1970, and in a decade, the number of birds in Thailand alone increased dramatically and came to represent half of the entire world population.

In regions in which local plants or animals have sacred status, the use of these species as symbolic icons can be very effective for the conservation of natural resources, species, and ecosystems (Colding and Folke 1997; Negi 2005; Bhagwat et al. 2011; Saini et al. 2011; Niroula and Singh 2015). Religion-based respect for particular species can contribute importantly to their protection. For example, among Hindus, several species are associated in scripture with different gods and goddesses (Figure 7.6) and their sacred status protects them against harvesting.

SACRED GROVES: COLLECTIVE-SPECIES ICONS

Humans instinctively respect largeness (see Chapter 4), and so in past times, mountains, seas, and celestial bodies (especially the sun) were often worshipped as living entities. Some trees are the largest living things on Earth, and standing beside the base of a huge, centuries-old tree is awe-inspiring (Figure 7.7). Tree worship has been remarkably widespread in human history, and sometimes certain species or given trees had sacred status (Omura 2004; Dafni 2006; Parthasarathy

FIGURE 7.5 Examples of Asian animals protected by their religious affiliation. (a) Axolotl (*Ambystoma mexicanum*), a Mexican salamander almost exterminated from its freshwater habitat by pollution and over-fishing. Its traditional Aztec religious significance, as well as a recent effort by Christian nuns, is leading to its rehabilitation. This photo (public domain, courtesy of Pixabay.com) shows an albino form, popular in the exotic pet trade. (b) Open-billed stork (*Anastomus oscitans*), a species saved in India from extreme population decline because of the religious importance of storks. Photo by A. Staudacher (CC BY SA 3.0). (c) Snow leopard (*Panthera uncia*), native to the mountain ranges of Central and South Asia, where its numbers have been declining alarmingly. Its sacred status in parts of Asia is contributing to efforts to conserve the species. Photo by Bernard Landgraf (CC BY SA 3.0).

FIGURE 7.6 Artwork showing Hindu deities in poses depicting several of the most sacred animals in India. (a) Ayyappan, a popular deity in the South Indian state of Kerala, frequently shown riding a tigress. Public domain photo. Source: https://www.columbia.edu. (b) The elephant-headed Ganesha (Lord Ganesha, Ganesh). Photo by Adityamadhav83 (CC BY SA 3.0). (c) The god Krishna and goddess Radha with some animal companions, notably a cow. Photo by Infinite Eyes (public domain).

and Naveen Babu 2019). Tree worship came to be regarded in much of Eurasia as a pagan practice and was suppressed, but there still are regions where trees have religious importance. For example, the African baobab is widely protected because of its spiritual importance.

Forests, even if composed of relatively small trees, are very effective in evoking deference and awe. Being in a dense grove of trees, encircled by trunks and overtopped by the canopy, perhaps elicits a primitive, satisfying memory of protected existence in the womb. Walking in a forest often produces a remarkable sensation of spiritual closeness to nature, which has been compared to praying in a tall, very quiet, dimly lit cathedral. Forested areas are ecosystems, which are collections of interacting species, not really 'individuals,' but nevertheless they have widely been conceptualized as discrete living icons. So-called 'sacred

FIGURE 7.7 Young ladies dwarfed by awesome redwoods (*Sequoiadendron giganteum*) in Sequoia National Park, California. Source of photos: Shutterstock. Contributors: (a) My Good Images. (b) Ivanova Ksenia.

FIGURE 7.8 Examples of sacred groves. (a) Gethsemane, an ancient olive grove at the foot of the Mount of Olives in Jerusalem, the site of Jesus just before his crucifixion. Photo by Tango7174 (CC BY SA 4.0). (b) Sacred grove in Kerala, South India. Photo by Renjusplace (CC BY SA 4.0). (c) The Kitano Tenmangu Shrine in Japan, built in 947 in honor of Sugawara no Michizane, the 'God of Agriculture.' Note the spectacular flowering plum trees. The vegetation surrounding such Shinto shrines has sacred status. Photo by Japanexperterna.se (CC BY SA 3.0). (d) The Sacred Wood, Blida, Algeria. A photochrome print from 'Views of People and Sites in Algeria' from the 1905 catalog of the Detroit Publishing Company, where it is noted that 'two picturesque tombs of saints are shaded by superb groups of Aleppo pines, araucarias, and olive-trees.' Photo by Ashley Van Haeften (CC BY 2.0).

groves' are religiously protected smaller or larger habitat patches dominated by trees, and throughout the world, there are examples of such reserves, often representing the only surviving examples of ecosystems (Chandran and Hughes 2000; Ntiamoa-Baidu 2008; Figure 7.8). Sacred groves have provided significant protection for forest habitats and undoubtedly have limited biodiversity loss for millennia. In a similar vein, there are thousands of 'sacred natural sites' around the world (Shinde and Olsen 2020) that by their nature are substantially protected against damage. Many of these encompass landscapes (ecosystems and species) that consequently are safeguarded.

SPECIES AS NATIONAL AND REGIONAL EMBLEMS

Numerous countries and often also political subdivisions of nations have official representative plants and animals, which are almost always very attractive or very useful, and in many cases occur only or predominantly in the region they represent. These are like ambassadors, calling attention to the importance of preserving the region's natural biological heritage – for practical sustainable harvesting in many cases – or simply in order to prevent their beauty from disappearing from nature. Emblematic species almost always have been chosen because interest groups, naturalists, or scientists lobbied or persuaded politicians of their merit. Because of their prominence on flags, seals, official stationary, stamps, coins, and the like, the status of emblematic species provides them with a measure of protection. It is embarrassing to those in charge of a political region should their emblematic species be at risk and not receiving sufficient protection (a frequent situation). The American bald eagle (*Haliaeetus leucocephalus*; Figure 7.9c) provides an instructive example. It is the national animal (and the national bird) of the United States and has been used as a major representative symbol of the country at least since 1782. By the mid-1950s, DDT and poaching had reduced the population in the lower 48 states to about 400 breeding pairs. Conservation measures were enacted, and today, there are over 10,000 pairs in the contiguous U.S. The iconic status of the species contributed substantially to its rehabilitation from extreme decline. A similar example is the brown pelican (*Pelecanus occidentalis*). In 2010, an explosion on the Deepwater Horizon oil rig located in the Gulf of Mexico, 66 km off the coast of Louisiana, resulted in the largest marine oil spill in history. Over

FIGURE 7.9 The leading national animal emblems, all of which are apex predators in some locations. (a) Lion (*Panthera leo*), the world's most popular national symbol. Photo by Yunaila (CC BY ND 2.0). (b) Tiger (*Panthera tigris*), a national emblem of Bangladesh, India, and Malaysia. Photo (public domain) by J. and K. Hollingsworth, U.S. Fish and Wildlife Service. (c) American bald eagle (*Haliaeetus leucocephalus*), official symbol of the United States. Eagles (various species) are the world's second most popular national symbol. Photo (public domain) by M. Lockhart, U.S. Fish and Wildlife Service. (d) Wolf (*Canis lupus*), a national symbol of Estonia, Italy, Serbia, Portugal, and Turkey. Photo by MrT HK (CC BY 2.0).

FIGURE 7.10 Rescue of the brown pelican (*Pelecanus occidentalis*) from the Deepwater Horizon oil rig Gulf of Mexico oil spill in 2010. The iconic status of the bird was key to its rehabilitation. Illustrations from Shutterstock. (a) Flag of Louisiana, featuring the brown pelican. Contributor: Norsworthy. (b) Cleaning an oil-soaked pelican. Contributor: Breck P. Kent. (c) Cartoon of the oil rig polluting the ocean. Contributor: Vectorpouch. (d) Cartoon of an oil-contaminated pelican protesting pollution. Contributor: Bogadeva1983.

50,000 birds, including many brown pelicans, were killed, and many others damaged by the oil (Figures 7.10c and d). Because the species is the state bird of Louisiana (Figure 7.10a), it received extensive restoration and rehabilitation (Figure 7.10b). A less constructive solution was made by the Canadian province Prince Edward Island, in which the showy lady's slipper (*Cypripedium reginae*) was once the floral emblem. However, in the early 1960s, it was realized that the species had become alarmingly rare, and in 1965 another species, the much more common stemless lady's slipper (*C. acaule*) was substituted as the official flower of the province.

DRAGONS: METAPHORICAL PERSONIFICATIONS OF HUMAN FEAR OF DANGEROUS ANIMALS

Dragon-like beasts are part of the mythology of numerous civilizations. Dragons aren't real, but like nightmares they likely were generated by the genuine fears that people accumulated during their daily lives. It is doubtfully a coincidence that the human psyche has so often condensed these fears into the form of dangerous beasts with serpentine bodies, scales, substantial size, menacing teeth, and (at least in Western culture) abilities to project fire from their mouths and to fly. The extremely widespread ancient popularity in various cultures of legendary heroes (always men and often rescuing helpless women) killing these mythological beasts (Figure 7.11) may indicate that humans are programmed to fear the features of dragons and accordingly to enjoy stories of humankind triumphing over these symbolic monsters. The Swiss psychiatrist Carl Jung theorized that all humans share innate unconscious psychological drives, which he termed archetypes (Jung 1959). He hypothesized that similarities between the myths of different cultures reveal the existence of these universal archetypes. The features of dragons are reminiscent of those of snakes, and as discussed in Chapter 16, it is possible that humans have acquired instincts to avoid

FIGURE 7.11 Paintings showing the slaying of dragons. (a) 'Roger freeing Angelica,' painting (public domain) by Swiss artist Arnold Böcklin (1827–1901). The scene is based on a 16th-century epic poem in which the knight Roger rescues Angelica who was left as a human sacrifice to a sea monster. (b) 'Bahram Gur killing the dragon,' painted in Persia in 1370. In the Shahnameh, Iran's national epic, Bahram Gur eliminates a gargantuan dragon. Source: Chah-namah, Topkapı Palace collection, Folio 203. Photo credit: Michel Bakni (CC BY SA 4.0). (c) 'Archangel Michael fights the dragon and rebel angels,' a dome fresco in Abbey Church, Austria, based on the *Book of Revelation XII*, 1–17, painted by Austrian artist Paul Troger (1698–1762). Photo credit: Wolfgang Sauber (CC BY SA 4.0). (d) 'Perseus freeing Andromeda,' painted by Italian artist Paolo Caliari (1528–1588). Based on Greek mythology, the scene shows the hero Perseus flying through the air in combat to kill the sea dragon who has come to kill Andromeda, chained to a cliff by the sea shore. (e) Persian legendary hero Rostam killing a dragon. Image by Adel Adili (CC BY SA 3.0). (f) 'Saint George defeating the Dragon,' painting by German artist Johann König (1586–1642). The often-illustrated legend of Saint George and the Dragon tells of Saint George (died 303) slaying a dragon that demanded human sacrifices.

such often dangerous animals. The feature of dragons are also reminiscent of large carnivorous dinosaurs (which are also reptiles), and although extinct, they generate awe and fear when viewed in museums and movies. Dragons have unattractive features, and as such they seem to fulfill the role of villains. In medieval Europe, people were fascinated by representations of devilish grotesque monsters (Eco 2005), and perhaps the stresses of the period demanded such physical incarnations as representatives of evil that could be overcome by heroes. Gargoyles commonly decorated buildings (Figure 7.12), partly to indicate that such threatening monsters protected the sites, but also to suggest that they were under the control of humans.

OGRES: METAPHORICAL TRANSFERENCE OF FEARED ANIMAL FEATURES TO HUMANS

Ogres are mythological human-like (sub-human or quasi-human) creatures that are subjects of folklore in many cultures. The form they take is variable (Figure 7.13), but they are usually powerful giants, ugly (even hideous), and very nasty (bestial, devilish, odious). In many European fairy tales, ogres consume people, especially children. As with dragons discussed above, ogre-like figures appear in the mythology of so many cultures that it cannot be due to chance alone (Jobling 2001; Cross 2016). Some ogres are green, like the contemporary fictional movie personality Shrek, which is curiously reminiscent of green Martians and green dragons, suggesting the color has acquired a connotation of strangeness. Some ogres have fangs and combined with their large size, aggressive nature, and cannibalistic diet, these features clearly reflect human fear of dangerous, carnivorous predators. The repulsiveness of ogres is a reflection of traditional association of evil and ugliness in literature. However, Shrek is an admirable anti-hero and is the subject of attempts to teach children about prejudice (Melchiori and Mallett 2015).

FIGURE 7.12 Gargoyles. Especially in gothic architecture, these were carved stone grotesque figures often combining human and non-human features, situated to project from roof gutters to direct rainwater away from the building. Gargoyles also symbolically protected the building from evil spirits. More generally, any monstrous statue associated with a building may be termed a gargoyle. (a) Gargoyle on the Cathedral Notre-Dame de Paris. Photo by Sharon Mollerus (CC BY 2.0). b) A griffon (mythical beast) sculpted by François Gilbert on the Palais Longchamp in Marseille, France. Photo by Rvalette (CC BY SA 4.0). (c) Gargoyle at the Union Banking Company Building, Douglas, Georgia, U.S. Photo by Bubba73/Jud McCranie (CC BY SA 4.0).

FIGURE 7.13 Traditional interpretations (public domain images) of ogres. These mythical creatures combine human appearance and terrifying features of beasts. (a) Painting (public domain) of an ogre by the Japanese artist Soga Shōhaku (1730–1781). (b) Satirical painting (public domain) entitled 'The big German ogre' by Italian artist Alphonse Lévy (1843–1918). Source: Marseille, MuCEM, Musée des Civilisations de l'Europe et de la Méditerranée. (c) Caricature entitled 'The ogre and the fairy' by French painter Jean Veber (1864–1928). Photo credit: Le Salon de la Mappemonde (CC BY ND 2.0).

HUMANOIDS: THEORETICAL AND ARTIFICIAL HUMAN-LIKE CREATURES

We humans are so enamored of the appearance and nature of our own species that we find it difficult to conceive of advanced forms of life that are not based on our own physique. Particularly favored human features include bilateral symmetry; upright bipedality with two manipulative arms (not like bipeds such as kangaroos with a balancing tail); and high intelligence, reflected by large complex brains (at least, big foreheads). Conway Morris (2005) argued that evolution inevitably generates dominating intelligent life forms (but this may require a very long time). Convergent evolution (independent species developing similar characteristics in response to similar pressures, such as long necks to reach food in high places) is well known, so it is possible that some highly adaptive human characteristics (such as opposable thumbs that facilitate tool use) could evolve in other animals. But it is a stretch to think that humans are the ultimate ideal animal structure for all

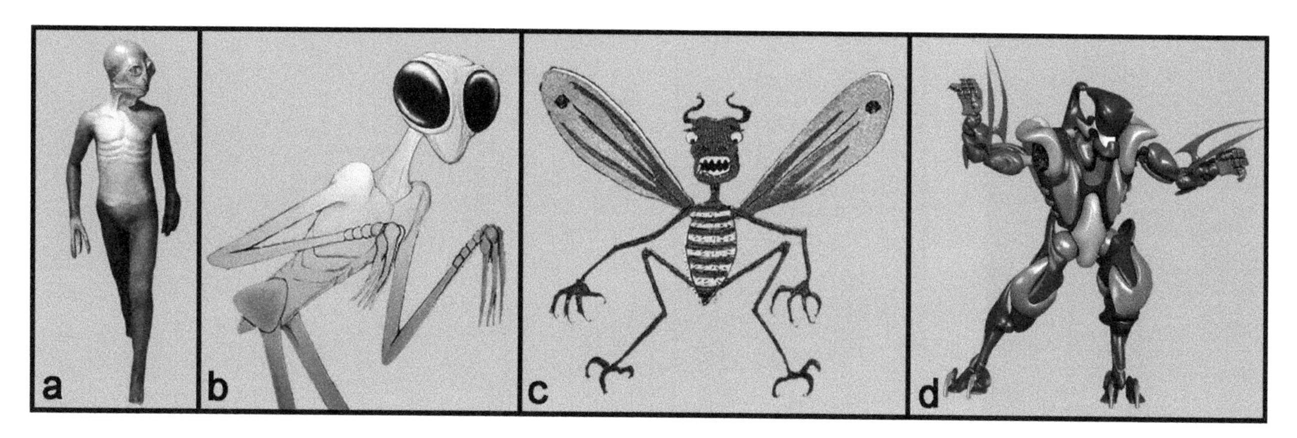

FIGURE 7.14 Examples of humanoids (theoretical human-like creatures or creations). (a) 'Dinosauroid,' a model in The Dinosaur Museum, Dorchester, Dorset, England, postulated to be a potential product of evolution (see text). Photo by Jim Linwood (CC BY 2.0). (b) An insectoid resembling a praying mantis. Credit: FolsomNatural (CC BY 2.0). (c) An insectoid resembling a fly. Source: A late-19th century public domain cartoon. Credit: Schenking van de heer F. G. Waller, Amsterdam. (d) An android robot. Source (public domain): Pixabay.com.

environments (see discussion in Gould 1989). The term 'humanoid' denotes a non-human creature or being with human form or characteristics (Figure 7.14). Humanoids may include 'androids,' i.e., robots. Curiously, in some respects, humanoid androids reduce genuine human contact (e.g., automated answering machines; supermarket check-out systems; automated computer figures such as Siri and Alexa; self-driving automobiles). Postulated advanced extraterrestrial life forms are most often reminiscent of people, often with some monstrous differences. Oddly, invading aliens are usually shown as green in color (see the preceding discussion of dragons, which reflect dangerous greenish reptiles). The word 'insectoid' denotes creatures or objects with bodies or traits like insects (arachnids, notably spiders, are also sometimes included). The term 'reptiloid' similarly refers to reptile-like beings or things. It has been hypothesized that a 'dinosauroid' comparable to humans could have evolved were it not for the demise of dinosaurs in past times (Russell and Séguin 1982; Naish 2012; Figure 7.14a).

MYTHICAL PART-HUMAN HYBRIDS

This section is concerned with a unique subcategory of imaginary anthropomorphic animals. In its most general sense, the word 'hybrid' means a combination of two things. In biology, the word is employed conventionally simply to designate organisms derived from sexual breeding between two parents. Such hybrids often exhibit various degrees of intermediacy of all or most body parts between the characteristics of the parents. In the mythology of ancient and medieval peoples, the word hybrid has more often designated imaginary creatures that combine whole body parts of different kinds of creatures, often between humans and other animals or sometimes plants (Figure 7.15). ('Chimera' is employed in biology to indicate combinations of different living tissues, but the word has other meanings and is too ambiguous and unfamiliar to employ here.) Dozens of part-human beings were conceived by different cultures, normally involving vertebrate animals, especially mammals, but also fish, birds, snakes, and even plants. This is a special form of anthropomorphism because it reflects an extremely close perceptual kinship and respect (if not necessarily admiration) for non-human species. Depending on culture, these hybrids often had (or continue to have) the status of gods or associates of gods, and so from the perspective of people they embody great power. Not surprisingly, the animals chosen for such status are strong (such as the bull, lion, and horse), have special skills (birds with the ability to fly, fish that can swim), or simply have notable importance (such as the scarab beetle for the ancient Egyptians). Satyrs – tracing to ancient Greek and Roman mythology – perhaps reflect a less respectful view of humans, animals, or both. The Romans conceived of them as figures (usually male) with goat ears, tails, hooves, and horns (highly reminiscent of features attributed to devils), often drunk, and carousing. Numerous classical European paintings show them with an exaggerated erection, harassing women (satyriasis is uncontrollable or excessive sexual desire in a man).

ANTHROPOMORPHIZATION OF SPECIES

'Anthropomorphism' is the ascribing of human characteristics to non-human things. Some definitions (accepted in this book) restrict the word to living things; others include inanimate objects (e.g., fashioning the feet of bathtubs and furniture to look like lions' feet). 'Zoomorphism' is a similar but also ambiguous term. Some authorities restrict the word zoomorphism to projection of animal features onto humans (e.g., comparing someone to a pig). There is a large body of literature dealing academically with anthropomorphism in a very broad sense (see, for example, Libell 2014; Lee 2019), which can include not just animals but also naming and talking to one's god(s), plants, car, guitar, computer, or even personal body parts.

FIGURE 7.15 Examples of mythological hybrids between humans and other species. (a) Mermaid (fish-human hybrid) Figure entitled 'The Little Mermaid' from: *Stories from Hans Anderson* illustrated by E.S. Hardy. Boston, DeWolfe, Fiske & Co. (1890). (b) A centaur (horse-human hybrid), from a 1920 lithograph advertising Centaure tonic wine. Source: Wellcome collection (CC BY 4.0). (c) Faun (goat-human hybrid), a creature of Roman mythology. Painting (public domain) by Swiss artist Arnold Boecklin (1827–1901). Photo credit: Hajotthu. (d) Painting showing harpies (bird-human hybrids) by Austrian artist Johann Jakob Zeiller (1708–1783). Photo by Wolfgang Sauber (CC BY SA 4.0). (e) A mandrake (human-plant hybrid), tracing to stories of the ancient Greeks, Romans, and Jews. Medieval Europeans came to believe it had a highly medicinal root but was very dangerous to approach. Source: Meydenbach, J. 1491. Hortus sanitatis. Image (public domain) from the U.S. National Library of Medicine. (f) Damballah, the Haitian Voodoo serpent god. Painting (public domain) by Haitian artist Hector Hyppolite.

Anthropomorphization of animals is of particular importance, because the chosen species reflect how people have viewed nature in the past, and probably also how people acquire prejudices about the animals. At least since recorded history, we humans have been entertaining ourselves by conjuring up creatures – mostly animals but also plants – that are human-like (Figure 7.16). Human-like cartoon animals, which are now daily fare for almost all children, are important because they are very influential in establishing preferences and prejudices.

Anthropomorphization of Pets

The most intense expression of anthropomorphism relates to our pets, which for some people become human surrogates – substitutes for friends and children. As most dedicated pet owners know, love for pets can become as deep as for family members (Archer 1997). Divorce is often an occasion for battles over ownership of pets that rival concerns over custody of children. It is often difficult for many to empathize with the depth of these feelings, or to sympathize with the prioritization that is often afforded to pets at times when many humans are suffering from lack of resources. Conversely, those who passionately love their pets tend to have little tolerance for those who don't have similar feelings. There does seem to be an instinctual tendency to humanize pets, and this may simply be comforting, which is likely why we do it. It probably is a good thing, making us more sensitive to the protection of domesticated animals (Webster 2011; Butterfield et al. 2012) and the welfare of wild animals and nature in general (Tam 2019; Manfredo et al. 2020; Rottman et al. 2021; Williams et al.

FIGURE 7.16 Animals in anthropomorphic poses. (a) Cats in a restaurant. Source: Wellcome Images (CC BY 4.0). (b) Elephants. Image (public domain) from 1988 book *The Circus Procession. A Lyrical Parade of Colorful Circus Characters* published by McLoughlin Bros, New York. Credit: U.S. Library of Congress. (c) Chromolithograph entitled *The Dog in the Manger* from an 1880 book for children published by McLoughlin Brothers, New York. (d) Goats. Source (public domain image): Pixabay.com. (e) Pelican fishing (public domain). Source: 1880 Pearl Series book published by McLoughlin Bros.

2021). Some pet specialists claim that animals really crave discipline, not pampering, and some psychologists warn of potential harm from distorting reality. Most of the time relating to pets as if they were humans is likely harmless, even if it annoys them when overdone, and appears silly and eccentric to one's neighbors (Figure 7.17).

Anthropomorphization of Pests

Harmful species (particularly identified as pathogens, germs, insects, bacteria, or vermin) are often depicted as dangerous, disgusting, villainous, and extremely ugly, in order to sell products that will control or eliminate the offending organisms (Huang et al. 2019; Figure 7.18). Such very negative advertising particularly reinforces traditional stereotypes of insects and rodents as meritless creatures deserving merciless, painful eradication.

SPECIES AS COMMERCIAL SYMBOLS

Animals, and sometimes plants, are frequently referenced in the names of organizations and/or their logos. The resulting phrases and figures are usually copyrighted or otherwise protected as intellectual property. Nearly one-third of major league teams in North America base their names on animals. Automobile companies are fond of naming their models after horses, bulls, cobras, and big cats because they suggest grace, precision, power, and speed (note Figure 7.19b). Owl symbols, suggesting wisdom, are frequently employed by educational industries. Exotic animals such as peacocks, pheasants, and ostriches may suggest luxury. Sometimes a product benefits from branding that associates it with a memorable animal like the gorilla simply because it is easy to remember. Much of advertising employs pleasantly anthropomorphized animals (e.g., Charlie the Tuna, Tony the Tiger; Figure 7.19a and c) and plants (e.g., Mr. Peanut, the California Raisins) to sell their products (Brown and Ponsonby-McCabe 2014). So long as the product remains respected, the reputation of the associated species isn't harmed and may even be improved.

SPECIES AS MASCOTS

Animals are commonly adopted as mascots, especially by sports teams, and these tend to reflect stereotypes of

FIGURE 7.17 Examples of excessive/obsessive anthropomorphization of pets. These costumed animals are, nevertheless, cute. Photos: (a–c) Dogs and (d–f) cats are by Petful (CC BY 2.0). Photos: (g and h) Pigs are by Sheilapic76 (CC BY 2.0). Photo: (i) Orangutan is by Joel Ormsby (CC BY 2.0).

aggressiveness, toughness, strength, and stamina. Animals that are not admired are rarely adopted as mascots, since their function is to complement people. ('Wildwing,' the anthropomorphic duck mascot of the Anaheim Ducks hockey team, is an unusual choice. So is the vulture, the mascot of Flamengo, a soccer team in Brazil.) Most mascots are cartoonish figures drawn by artists, but a few are real (Figure 7.20), and because these require considerable care, they reflect species that are highly valued. Some mascots, especially for universities, are living animals and are displayed at public events. For the most part, these are relatively docile domesticated animals, but a few are wild and quite dangerous, and so do not represent wise choices to exhibit at sports events (as is sometimes done).

ANIMALS AND PLANTS AS CODED MESSAGES

The use of plants as codes traces to ancient times but became highly developed in 19th-century England (Greenaway 1884). The 'language of flowers' (floriography) was a Victorian-era system of communicating private emotions by mentioning plants or flowers in letters or placing them in small bouquets called tussie-mussies which were very fashionable at the time. For example: red roses meant passionate, romantic love; white roses: virtue and chastity; yellow roses: friendship or devotion; gerbera: innocence or purity; daffodil: regard; and geranium: gentility. Most messages were positive or neutral, but some were negative, for example lobelia meant malevolence;

FIGURE 7.18 Advertising (all public domain figures) emphasizing the ugly, dangerous aspects of certain species. (a) Spraying herbicide against giant caterpillars. Original from: Washington State Apple Commission. 1917. Better Fruit. Better Fruit Pub. Co., Hood River, OR. Modified and colorized by B. Brookes. (b) A late 19th-century advertising postcard by the insecticide company Zacherl (Vienna, Austria). The scene shows a bee-like insect spraying and killing off bad roaches to the delight of the other bees. (c) A mosquito, from a U.S. government anti-malaria poster. (d) A rabid fox. Image courtesy of the National Library of Medicine Digital Collection, from 'The Fox Can Transmit Rabies,' published in 1951 by the Health Publications Institute in cooperation with the U.S. Public Health Service. (e) A rat made to look vicious, from a Chicago anti-rat poster.

FIGURE 7.19 Commercial animal logos (all images are public domain). (a) 'Elsie the cow,' advertising character developed in 1936 as a mascot for the Borden Dairy Company (demised). (b) A ram hood ornament on a 1990 Dodge Ram truck. Photo by Christopher Ziemnowicz. (c) Early 20th century pack of Camel cigarettes.

FIGURE 7.20 Live animal mascots, representing universities, athletics, and the U.S. military. Note that each of these animals is noted for strength and/or athleticism. (a) Mike VI, tiger mascot of Louisiana State University. Public domain photo by Nowhereman86. (b) Zan the goat, Shepherd University (West Virginia) mascot. Photo by Ron Cogswell (CC BY 2.0). (c) Paladin at West Point Academy, one of the mules that has served as the U.S. Army's mascot. U.S. Army photo (public domain) by John Pellino. (d) Chesty XIV, English bulldog, mascot of the U.S. Marine Corps. Photo by Adrian R. Rowan (public domain). (e) Apollo the falcon (a gyr-peregrine hybrid), one of several U.S. Air Force Academy mascots. Photo (public domain) by Mike Kaplan. (f) Thunder III, mascot of the Denver Broncos football team. Photo by Jeffrey Beall (CC BY 4.0).

hydrangea: heartlessness; begonia: beware; and lime blossoms: illicit sex.

One of the most interesting examples of the use of flowers to transmit messages concerns violets (Selin 2020). Napoleon Bonaparte (1769–1821) was fond of violets, perhaps because they reminded him of his childhood in the woods of Corsica. He met his future wife Josephine (1763–1814) at a ball, where she wore a coronet of violets and carried a bouquet of violets, which she threw to him from her carriage while departing. In memory of this, she wore a wedding gown embroidered with violets, and he gave her violets on each anniversary of their wedding. When he was sentenced to live out his days on the island of Elba, he told his followers that he would return in the spring, with violets. Napoleon's supporters sometimes determined people's political views by asking 'Do you like violets?' In 1815, Napoleon returned to the royal palace in Paris, showered by violets. Accordingly, he was called by his followers *Caporal Violette* ('Corporal Violet') and *Papa-Père la Violette* ('Daddy Violet'), and the violet was adopted as the emblem of the Imperial Napoleonic party. After Josephine's death, Napoleon had her grave covered with violets. He kept some of them along with a lock of Josephine's hair in a locket near his breast until he died. When the French monarchy was restored, the wearing of violets was banned, and until 1874 French governments forbade any reproduction showing a violet, the symbol of Bonaparte supporters. Nevertheless, the violet flourished in the early days of the French Republic. France was flooded with postcards picturing innocent-looking violets with Napoleon's portrait cleverly hidden among the flowers. (Figure 7.21a2; cf. Figure 5.74b).

Perhaps the most frequent romantic gift of animals expressing love is a pair of European turtle doves (*Streptopelia turtur*, Figure 7.21b), especially when offered during Valentine's Day. A much less attractive example of coded symbolism in the form of an animal gift is based on the phrase 'To sleep with the fishes,' meaning to be murdered, the body hidden from the public. The expression is popular in media with regard to American Mafia or organized crime sending a gift of fish to competitors to inform them that they have eliminated one of their associates (Figure 7.21c).

FIGURE 7.21 Examples of symbolic messages associated with plants and animals (see text). (a) 'Violettes du 20 mars 1815' by Jean-Dominique-Étienne Canu. In this widely circulated 19th-century drawing, the profiles of Napoleon (2, under the green leaf on the right, which resembles Napoleon's bicorne hat), his second wife Marie Louise (1, facing Napoleon, on the far left, under the second violet), and their son, Napoléon François Joseph Charles Bonaparte (3, on the right of the central stems, next to the lower violet) are hidden (it requires some study before their faces become apparent). (b) A pair of turtle doves, expressing long-lasting love. Source: Shutterstock, contributor: Dneprstock. (c) Gangsters preparing a victim to 'sleep with the fishes' by embedding his legs in cement. Source: Shutterstock, contributor: Viki1984.

FLOWERS AND MASCULINITY

As discussed in Chapter 5 (and is common knowledge), women are immensely more appreciative of ornamental flowers than men. Delicate, extremely attractive flowers naturally represent characteristics considered to be feminine. By the Victorian era, violets represented the qualities of an ideal wife: humility, faithfulness, and modesty (Small 2021c). Boutonnières once represented the only masculine recognition of floral beauty (Figure 7.22a), but even this concession has been generally abandoned. Prejudice against male homosexuals has often been expressed by pejorative slang terms based on flowers, reflecting the perceived lack of masculinity associated with them. In particular, 'pansy' (*Viola ×wittrockiana*) refers to a weak or effeminate male, or a male homosexual, meanings which appear to have become prominent about a century ago. Other flowers, notably daisies, lavender, and violets, also have been

FIGURE 7.22 Flowers and perceptions of masculinity. Source of photos: Shutterstock. (a) A gentleman wearing a boutonnière on his tuxedo, an obsolescent practice reflective of the reluctance of most males to associate with flowers. Contributor: Mary981. (b) A flowerbed planted in the sequence of rainbow colors (red, yellow, green, blue, violet) at an annual LGBTQ+ event in Hamburg, Germany. Contributor: Uellue.

FIGURE 7.23 Flower-class naval military vessels ('corvettes'). Although these ships belong to the flower class, they were built in Canada, which declined the British practice of assigning floral names. (a) HMCS Regina. Photo (circa 1942–1943; public domain) by Canada National Defence. (b) 'Pom-pom' naval guns (40 mm British cannons used against aircraft), from the flower-class corvette HMCS Kamloops. Additional guns were employed to launch depth charges against submarines. Photo (of a display in the Canadian War Museum) by JustSomePics (CC BY SA 3.0).

associated with male homosexuality. The bright colors of flowers are now represented in the spectral rainbow that has become symbolic of LGBTQ+ gender rights (Figure 7.22b).

Given the lack of male enthusiasm for ornamental blooms, it comes as a surprise to many that during the World Wars, hundreds of combat vessels were named after flowers (Figure 7.23). In particular, the British Royal Navy constructed about 300 so-called corvettes, many designed for minesweeping and escort operations near ports (they could navigate in waters less than 4 m deep), especially during the Second World War. These were called 'flower-class' ships because all of them were assigned unique flower names like buttercup, geranium, pansy, petunia, poppy, and rosemary (for a list of some of these warships see https://military-history.fandom.com/wiki/Arabis-class_sloop). Canadians, the US, France, and other allies were assigned many of the ships, renaming them because floral labels seemed insufficiently aggressive.

8 Indispensable Values of Species for Human Welfare

INTRODUCTION

Table 8.1 summarizes eight different beneficial values that species have for humans. These are important in determining the practical and emotional attitudes that people have for species. Of course, negative aspects of species, which are analyzed in other chapters, are also critical. Of the eight values of species, the first two – resource materials and ecological services – are indisputably indispensable to human welfare and indeed human survival, and so are highlighted in this chapter. The remaining values are important, indeed indispensable to many, and are mentioned in other chapters. As noted previously, 'non-economic' support for biodiversity conservation is decidedly determined by aesthetics (cf. Martín-López et al. 2007).

'Resources' or 'materials' are quantifiable monetarily, and so their economic value is relatively easily determined. Domesticated animals (especially livestock) and plants (especially crops) are the most 'economically' important species – they provide the bulk of the world's food and generate more money and commercial activity than all other species combined. Their relationship with humans is detailed in Chapter 12. However, almost all species are wild, and this chapter is concerned with the economic aspects of wild species and people. Humans still partly depend on wild species, mostly from the oceans (for seafood) and forests (for lumber). Much of the world's living animal biomass is in the oceans, which remain the primary source of wild food, but this resource is now threatened by pollution, overharvesting, and other unsustainable practices. Similarly, most plant biomass occurs in trees, but logging is degrading this resource.

In addition to tangible materials that come directly from species, the living world also participates in many cyclical processes that renew and maintain the planet in a livable form. The phrase 'ecosystem services' has come to designate these benefits to humans. Unfortunately, as discussed in the following, the phrase is vague and ambiguous, and what is included is often unclear. Nevertheless, it is indisputable that living things and the ecosystems they occupy provide indispensable services that maintain the functioning of the planet's life-supporting systems. These services are essential for the welfare of humans but only in recent times have economists attempted to quantify their dollar values (note Dasgupta 2021). Species are perceived by most economists to be valuable to humans principally as harvested resources, but this is short-sighted because it ignores ecological services. Some economists advocate what have been termed 'sustainable growth,' 'green growth,' and 'sustainable development' (Völker et al. 2020). The vaguely defined associated concept 'circular economy' (Rizos et al. 2017 analyzed over 100 definitions) generated over 10,000 publications from 2018 to 2022. See Corvellec et al. (2022) for a critique. In his 1849 novel David Copperfield, Charles Dickens wrote 'Annual income twenty pounds, annual expenditure nineteen and six, result happiness. Annual income twenty pounds, annual expenditure twenty pounds ought and six, result misery.' The lesson is simply to limit consumption, and perhaps this is more of a human psychology challenge than an economic issue.

TABLE 8.1
Chief Positive Values of Species to Humans[a]

Objective (existential) values

1. Materials (e.g., food, medicine, clothing, fuels, and building materials)
2. Ecological services (for ecosystem functions; e.g., wild pollinators, maintenance of groundwater, and carbon sequestration)
3. Commercial services (e.g., animals employed for transportation, protection, and sport; plants used for preventing erosion or adding nitrogen to soil)
4. Subjects of study for scientific knowledge, medicinal research
5. Indicators of imminent environmental threats (like the miner's canary)

Subjective (mental, emotional) values

6. Aesthetic enjoyment and entertainment (e.g., beauty, oddity, impressiveness, or behavior of animals in zoos, birdwatching, rodeos, beauty and growth cycles of ornamental plants)
7. Compassionate satisfaction (transference of love and respect for other humans to love and respect for other species, most notably for pet animals and selected wild species)
8. Spiritual (moral and religious dogma) satisfaction (from continued existence of other species)

[a] A variety of schemes accounting for values of biodiversity or species have been devised; this system emphasizes that some values are 'material' or 'tangible,' and others are 'emotional.'

DOI: 10.1201/9781003412946-8

FIGURE 8.1 Examples (public domain) of harvesting wild species resources. (a) Fishing: 'Catching a Brook Trout,' 1854 painting by Arthur Fitzwilliam Tait, published by Nathan Currier. (b) Hunting: Fox hunting, painting by English artist George Wright (1860–1944). Courtesy of the Walker Gallery Liverpool via ArtUK. (c) Harvesting wild berries, painting by German artist Wilhelm Schütze (1840–1898). (d) Harvesting wild honey. The majority of crops depend on insects (mostly wild bees and managed honey bees) for pollination. 'Bee-Farmer' by English artist Francis D. Bedford (1864–1934). (e) Ranching: Roundup scene. Grazing lands, mostly public, are the basis of much of the world's livestock industries. Source: Siringo, C. 1886. *A Texas Cowboy; or Fifteen Years on the Hurricane Deck of a Spanish Pony.* (f) Logging: Advertisement for 1948 Autocar Trucks. Photo by Alden Jewell (CC BY 2.0).

WILD ANIMALS AND WILD PLANTS AS RESOURCES

In past times, people were intimately acquainted with the animals and plants growing in their locality, which they relied on for food, shelter, medicines, and tools. Numerous species were not only respected but were afforded totemic status – i.e., they were regarded as honored spiritual or symbolic representations (see Chapter 7). Wild species still provide significant resources (Figure 8.1), although domesticated animals and plants are of more importance economically. Therefore, it is not surprising that most people today have very limited knowledge of wild species. Perhaps the chief source of information about wild species for most people is broadcast nature shows, which are decidedly biased toward large animals that are entertaining and photogenic but hardly represent the range of important services for people.

WHY PLANTS ARE INDISPENSABLE FOR ALL FORMS OF BIODIVERSITY

All living things construct their bodies and carry on their activities (such as growth, digestion, and reproduction) using energy. The ultimate source of energy for virtually all organisms is sunlight. (The exceptions include sulfur bacteria which obtain their energy by oxidizing hydrogen sulfide (H_2S) and iron bacteria which get their energy by oxidizing iron compounds.) Only species capable of photosynthesis (plants, algae, certain bacteria, and certain protists) can capture solar energy directly (these are termed 'primary producers'). (Note that some animals are symbiotic hosts of simple photosynthetic plants, and some animals have even acquired chloroplasts.) Other classes of living things get their energy by (1) eating plants (herbivores or omnivores and some parasites, these are 'primary consumers'); (2) eating animals that have eaten plants (carnivores, omnivores, some parasites, and scavengers, these are 'secondary consumers'); (3) eating animals that have eaten animals that have eaten plants (these are mostly top predators, including apex predators, and are termed 'tertiary consumers'); or (4) absorbing energy from tissues of decomposing or decaying animals or plants (ultimately carried out by microorganisms, either free-living or associated mutualistically with soil-dwelling invertebrates; in soil ecology, the terms 'decomposer' and 'detritivore' are often applied both to bacteria and animals employing bacteria to break down organic tissues). The complex relationships among these feeding classes are based on 'trophic' (feeding) aspects, and the phrases 'feeding chain' and 'feeding network' are employed. The interrelationships involve energy flow, ecological connections, and ecosystem structure.

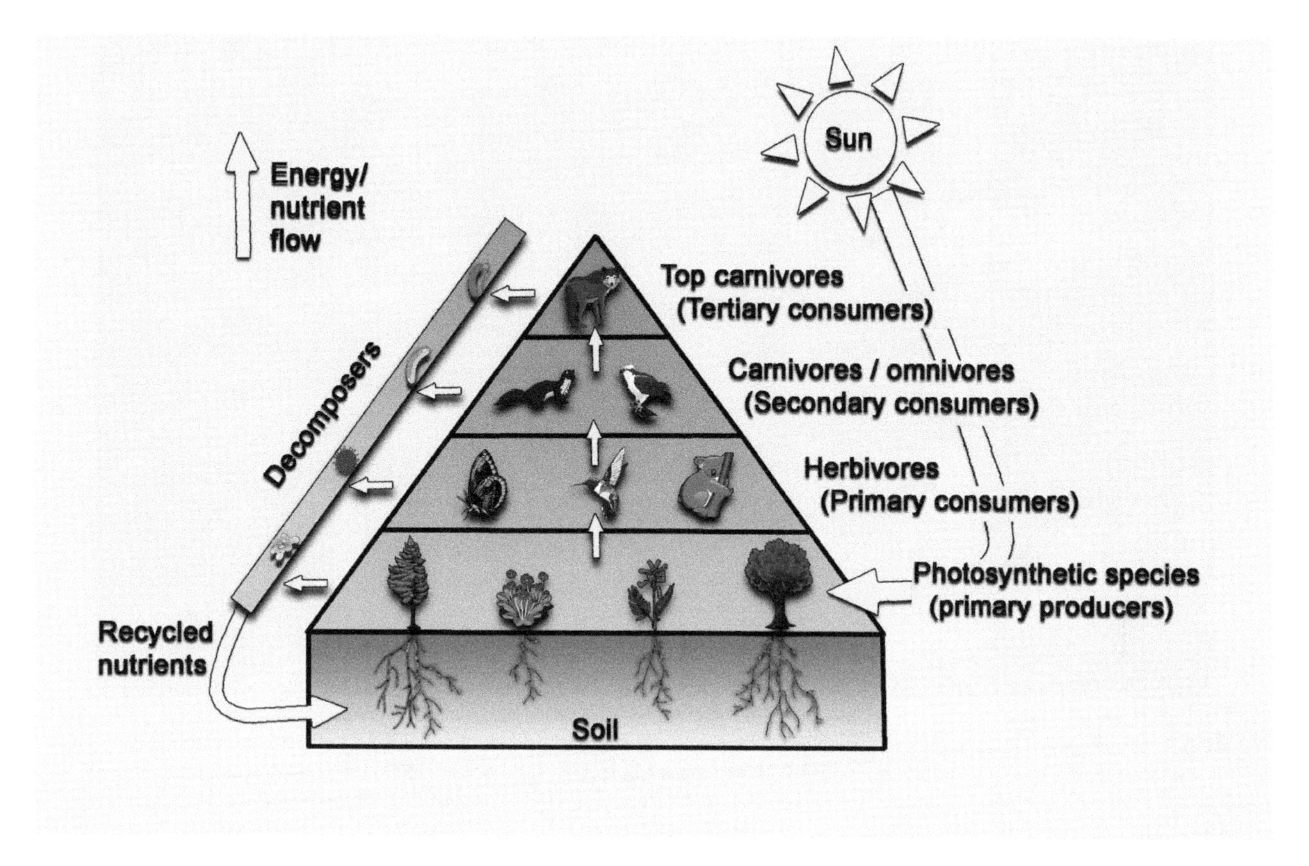

FIGURE 8.2 Trophic (feeding)–ecological–energy pyramid, illustrating indispensable relationship of plants to other forms of biodiversity. Prepared by B. Brookes.

All these aspects of biodiversity are traditionally modeled as a pyramid, in which the base is made up of plants (Figure 8.2). Indeed, plants are appropriately conceived as the *foundation* of biodiversity. Although the world of living things is truly wonderful, the plant kingdom is especially admirable for its invaluable contributions, without which most biodiversity could not exist. Nevertheless, some plant species are decidedly harmful to other life forms, especially humans, and destructive of ecosystems. How the very worst of such plants should be regarded and treated is the subject of Chapters 18–21.

WHY PLANTS ARE INDISPENSABLE FOR HUMANS

Although we humans are omnivorous, most of our food energy is based on directly eating plants (Figure 8.3), and the remainder from eating animals that have directly or indirectly fed on plants. We also get our fossil fuels and petrochemical industrial materials from ancient plants that have been turned to coal, oil, and natural gas. Most of our textile fiber comes from plants, and most of the world's lumber (a particularly valuable form of fiber) is harvested from wild trees (Figure 8.4). Unfortunately, much of the world's forests are being sacrificed for urban and agricultural expansion (Figure 8.4c). The majority of the world's medicines also originate from plants. And, since gardening has become the world's leading recreational pastime, one should not forget the beauty of ornamental plants (Figure 8.5).

THE ROLE OF WILD RESOURCES IN PERSUADING THE PRIVATE SECTOR TO SUPPORT BIODIVERSITY CONSERVATION

Biodiversity as a source of essential harvestable materials is very easy to understand, has appeal to a wide audience, and is a basis for economic justification for biodiversity preservation. Contributing to the maintenance of ecosystem functions, discussed next, is also of vital economic importance but is much harder to use as the basis of a simple, comprehensible, persuasive message. The private sector is mostly responsible for harvesting the world's wild living resources, and so should understand the need to promote long-term continuing supplies.

Some wild plants and animals are so important as commercial resources that the need for their preservation is beyond question (Figure 8.6). However, this is not the case for most wild species. Nevertheless, an enlightened private sector should be in the forefront of biodiversity protection. The world's wild biodiversity is a huge source of materials that are necessary for commerce, and protecting

FIGURE 8.3 Food plants. (a) A variety of food plants. Photo (public domain) by Agricultural Research Service, United States department of Agriculture. (b) Painting (public domain) by Italian artist Giuseppe Arcimboldo (1527–1593), who was famous for showing human heads as vegetables, flowers, and fruits. This work depicts Rudolf II, Holy Roman Emperor, as Vertumnus, the Roman God of the seasons. The original of this 1591 painting is in Skokloster Castle, Stockholm. This allegoric interpretation reflects the saying 'You are what you eat,' which traces to several authors in recent centuries.

FIGURE 8.4 Harvesting timber. Forests represent a key natural biodiversity resource requiring sustainable management. (a) Photo of logging in a tropical rainforest (Sabah, Borneo) by T. R. Shankar Raman (CC BY SA 4.0). (b) A 1907 postcard (public domain) showing an elephant in Ceylon transporting a log. Credit: New York Public Library. (c) An area of illegal deforestation of native vegetation of the Brazilian Amazon forest. Source: Shutterstock, contributor: Tarcisio Schnaider.

FIGURE 8.5 Ornamental gardens. (a) Floral art at the Garden of Versailles (Paris), illustrating the use of plants as ornamentals. Photo by Daniel Stockman (CC BY SA 2.0). (b) Royal Pavilion, Chiang Mai, Thailand. Source: Shutterstock, contributor: Think4photop. (c) Tulips at the Beijing Botanical Garden. Source: Shutterstock, contributor: Haines. (d) The Sunken Garden, Kensington, London. Source: Shutterstock, contributor: Geert Van Keymolen.

FIGURE 8.6 Examples of wild species that are harvested commercially in very large quantities. (a) Atlantic bluefin tuna (*Thunnus thynnus*), the basis of much commercial fishing, is critically endangered in parts of its range. Painting (public domain) by T. Knepp, U.S. Fish and Wildlife Service. (b) Teak tree (*Tectona grandis*), a large deciduous tree harvested for lumber. One of the world's most valuable timber trees, it is grown in plantations in many countries, but illegal logging persists in the large forests of Myanmar. Photo by PJeganathan (CC BY SA 4.0). (c) Base of a huge teak tree. Photo by Prashanth Dotcompals (CC BY 2.0). (d) Sockeye salmon (*Oncorhynchus nerka*), the most important commercial species of North Pacific salmon. Painting (public domain) from: Evermann, B. W., E. L. Goldsborough, and E. Lee. 1907. *The Fishes of Alaska*. Washington, DC, Department of Commerce and Labor Bureau of Fisheries. (e) Brazil nut (*Bertholletia excelsa*), a giant evergreen South American tree, the source of Brazil nuts. This is the only internationally traded nut collected almost entirely from the wild. Photo by Mauroguanandi (CC BY 2.0). (f) Bowl of Brazil nuts. Photo by Nick Fullerton (CC BY 2.0).

FIGURE 8.7 'Greenwashing': the practice by corporations of pretending to be environmentally friendly to hide their polluting, contaminating activities. Prepared by B. Brookes.

this fountain of wealth is surely essential for sustainable economic progress. However, there are serious caveats. First, not all business people are enlightened. Second, not all businesses are as concerned with long-term sustainability as they are for profits from short-term exploitation. Third, some businesses by their exploitive nature are simply incompatible with the conservation of specific habitats, areas, or species. Fourth, economies are complex, involving chains and webs of relationships, and this sometimes poses insurmountable challenges for partitioning costs and responsibilities for supporting biodiversity.

Nevertheless, business interests, especially large corporations, currently provide significant support to biodiversity conservation, particularly to some high-profile conservation organizations. While the motivation may be altruistic in some cases, improving public image is obviously the main goal. 'Greenwashing' is disinformation by an organization to present an environmentally responsible public image, especially by corporations (de Freitas Netto et al. 2020). Some corporations with terrible environmental records especially tend to greenwash their images by supporting conservation projects (Figure 8.7). Conservation resources are so limited that corporate contributions almost always need to be accepted. Such funding will tend to be directed toward high-profile ('cute and cuddly') species, projects in inspiring habitats (forests of towering trees), and scenic circumstances (eagles on mountains) that are visibly and emotionally appealing, since media reports and commercials based on these images make the sponsors look good. In general, private sector support for public-good activities is very beneficial to society. The principal difficulty is that private-sector support tends to obligate the beneficiaries to use the funds in directed ways that may not be entirely in the best interest of biodiversity.

THE INDISPENSABLE ECOSYSTEM SERVICES BY LIVING THINGS FOR HUMAN WELFARE

The preceding discussion focused on species as the operational units of concern. However, it is also useful to examine the importance of ecosystems, of which living things are components. Analyses from the viewpoint mainly of species or mainly from ecosystems are necessarily overlapping but are valuable in providing different perspectives of the threats to nature and what should be done. The following information summarizes the specific values of ecosystems.

Ecology is the science that addresses the relationships among organisms and the environment. All living things, including humans, exist because they have access to essential life-sustaining materials (e.g., food for energy, water for metabolism, oxygen for breathing) and an environment in which the physical conditions (e.g., temperature, pressure, irradiation) are suitable. Interactions among species with each other and with the environment are fragile and extremely complex, but nature tends to create relatively stable associations of interacting species and environments, known as ecosystems.

Just what constitutes an ecosystem is subject to interpretation (Boitani et al. 2015). Where one ecosystem begins and another ends can be arbitrary. The oceans collectively can be conceived as one ecosystem, or they can be considered as several. Areas of the world have been drastically modified by humans, and some of these areas (particularly cities and agricultural lands) have been conceived to constitute ecosystems (e.g., 'agroecosystems'), albeit artificial ones. Such artificial ecosystems are discussed later – Chapter 14 – under the heading 'Advancing Technology and the Fate of the World's Biodiversity.' While not completely incompatible with other species, landscapes that have been very extensively modified by humans mostly contribute to rather than alleviate the biodiversity and environmental crises.

Natural ecosystems are usually conceived of as being made up of both inanimate and living portions. The inanimate part is made up of the physical environment (particularly soil and atmosphere), and the animate part is the living species. There is regular interchange between the two, and natural cyclical geophysical processes maintain ecosystems. The best known ecological cycles are shown in Figure 8.8. Species are essential contributors to these feedback cycles, which are necessary for the maintenance of life on Earth. Species, including humans, benefit in various

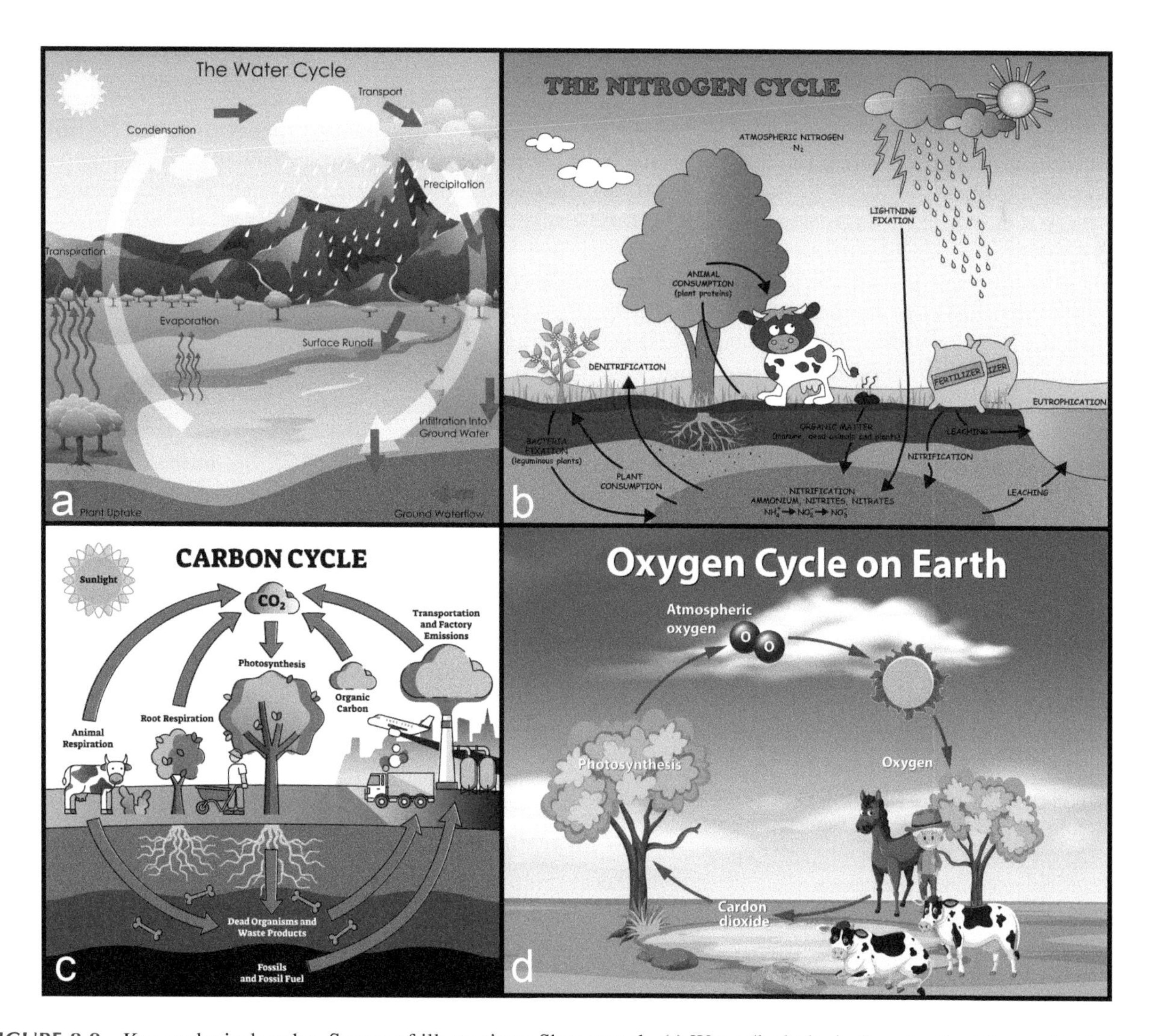

FIGURE 8.8 Key ecological cycles. Source of illustrations: Shutterstock. (a) Water (hydrological) cycle. Contributor: Stockshoppe. (b) Nitrogen cycle. Contributor: Milena Moiola. (c) Carbon cycle. Contributor: VectorMine. (d) Oxygen cycle. Contributor: Brgfx/.

FIGURE 8.9 Examples of key decomposers, responsible for recycling the energy in waste. Source of figures: Shutterstock. (a) A mushroom. The thread-like underground mycelium absorbs organic compounds from dead materials and, when sufficient nutrients have been accumulated, sends up mushrooms to disperse spores. Contributor: Amadeu Blasco. (b) Earthworms eat and therefore recycle dead organic materials but also will digest living microorganisms that are taken in. Contributor: Leka Leck.

ways by being components of ecosystems. Such benefits may be only from the inanimate part (e.g., the majority of commercial minerals) but most originate entirely (e.g., edible plants) or partially (e.g., nitrogen fixed from the atmosphere by bacteria that serves as fertilizer) from the living part. What have been termed 'resources' (such as food and lumber) are a principal ecosystem service, and examples were presented earlier.

All living things produce waste, which is an unpleasant topic that most people would rather not consider, although the need to handle such material can't be ignored for long. Aside from recycling efforts, humans flush body wastes down their toilets and put their garbage outdoors for collection, or eventually their living quarters would be drowning in waste products. Living things also produce waste products like excrement and dead bodies. Nature recycles all such organic wastes, employing species termed decomposers. Decay organisms (primarily fungi and bacteria) benefit by extracting energy, and the breakdown products (simple compounds, particularly nutritional chemicals) are employed by plants and some other species. Most decomposers are microscopic, but some fungi (mushrooms) are large (Figure 8.9a), and many invertebrates (sometimes called detritivores, indicating that they eat rather than absorb nutrients through their external surfaces), including earthworms (Figure 8.9b), termites, and millipedes, also contribute (often they house microorganisms that assist in digesting waste). Species that participate in waste recycling are among those that are least respected by people, but it is well to remember that without them our world would be covered by excrement, carcasses, plant litter, and other forms of undesirable matter.

The 'Millennium Ecosystem Assessment' (Reid 2005) was a major examination of the human impact on the environment, sponsored by the United Nations. In evaluating the state of the planet's ecosystems, the authors concluded that 'any progress achieved in addressing the Millennium Development Goals of poverty and hunger eradication, improved health, and environmental sustainability is unlikely to be sustained if most of the ecosystem services on which humanity relies continue to be degraded.' The publication popularized the phrase 'ecosystem services,' meaning the benefits to humans from ecosystems. Benefits were divided into four categories termed provisioning (valuable materials), cultural (psychological and educational benefits), supporting (specifically nutrient recycling, soil formation, and 'primary production', i.e., photosynthesis), and regulation (specifically of climate, water, and diseases). Since the original publication of the work, ecologists have modified the analysis and presented their interpretations in diagrams, some very complex and often differing in what is included in the four categories, and indeed within which category a given benefit should be included (see, for example, Value of Nature to Canadians Study Taskforce 2017). Clearly the phrase ecosystem services is open to interpretation, but it is now entrenched. The concept of ecosystem services is useful in emphasizing that there are indispensable benefits associated with the welfare of the world's ecosystems. A simplified analysis giving examples of benefits is presented in Figure 8.10. Regardless of how categorized, 'nature's goods and services are the ultimate foundations of life and health' (Corvalán et al. 2005). Examples of the 'ecocidal' consequences that can occur when human activity degrades an ecosystem are shown in Figures 8.11–8.13.

THE POLLINATOR CRISIS

'Pollinator services' are often considered a subsection of 'ecological services' but are so important and so threatened that the topic deserves to be highlighted. Close to 90% of flowering plants are at least partly dependent on animal pollination for optimum seed set (Ollerton et al. 2011) while almost 20% use other means, primarily wind pollination (Ackerman 2000). Hundreds of thousands of species

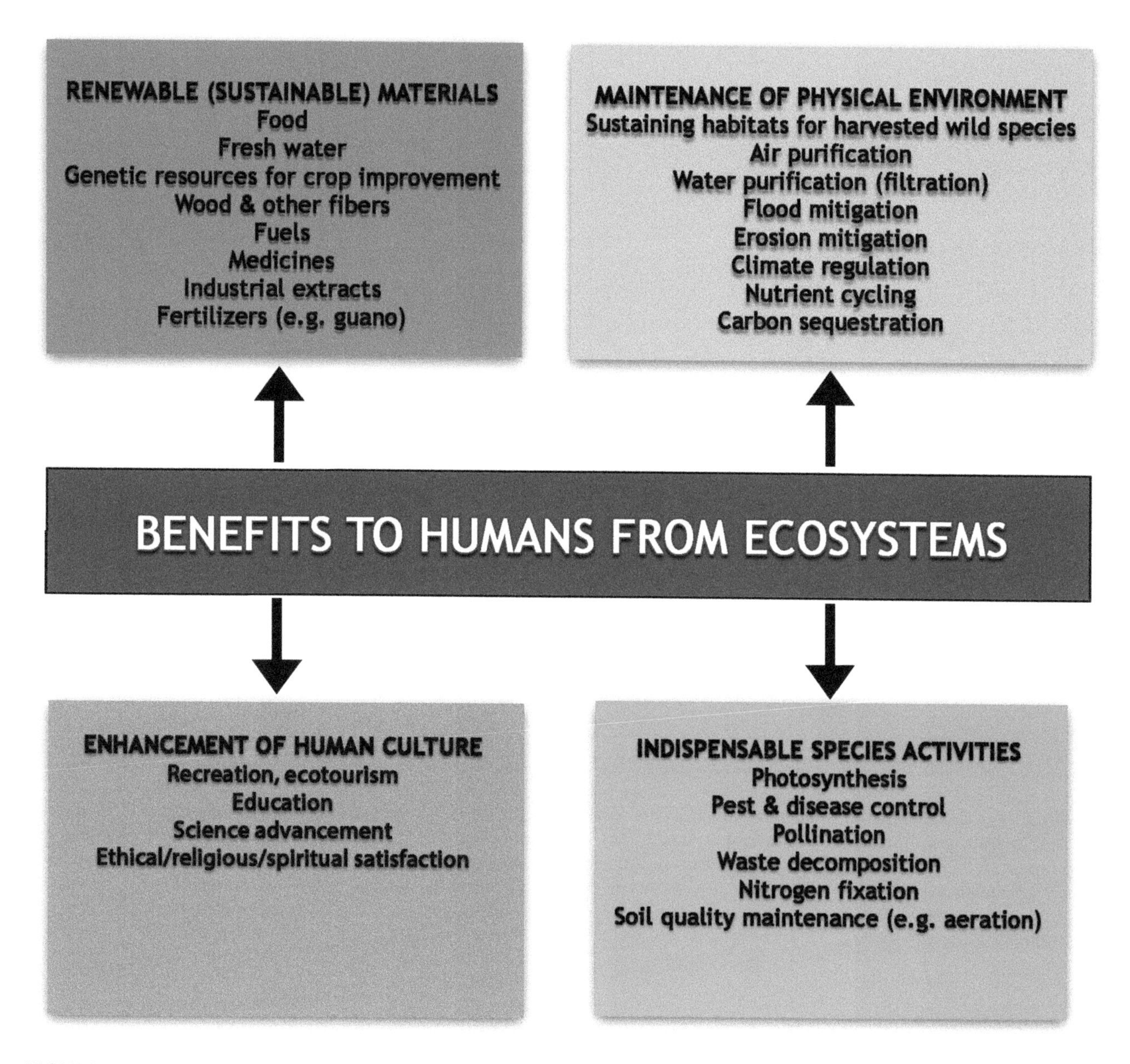

FIGURE 8.10 Examples of ecosystem services (all benefits are provided by species). Prepared by B. Brookes. Compare Figure 1.13.

of animals, mostly insects but also some mammals and birds, are pollinators (Abrol 2012). Pollination is essential for the productivity of cross-pollinated fruit and seed crops. Pollinators, especially native bees such as bumblebees, are needed for one-third of the crop species that humans consume (Ollerton et al. 2011) although the major cereals are wind-pollinated. Of course, most wild plants also cannot afford to lose their pollinators. In recent years, there has been a worldwide decline in wild pollinators, attributed to the loss of suitable habitats, introduced harmful species, and use of pesticides (Zwarun and Camilo 2021). Moreover, managed bees (mostly honeybees) are also in decline (see Chapter 12). The current catastrophic global decline in pollinators is known as the 'Pollinator Crisis' (Shivanna et al. 2020; van der Sluijs et al. 2021) and has generated great concern (Figure 8.14).

SPECIES AND THE ENERGY CRISIS

Introduction

This section relies particularly on Small and Catling (2006a). Energy sources are classified as 'renewable' (also termed sustainable or green) and 'non-renewable' or finite (Figure 8.15). Most of the world's energy is currently sourced from fossil fuels, particularly crude oil, natural gas, and coal, which are non-renewable (i.e., eventually the sources will be exhausted). Soaring oil prices, political instability of several major sources, and environmental concerns about the issues of climate change and global warming (related to the Kyoto Protocol's curbs on emissions of carbon dioxide and other greenhouse gases) have stimulated a movement to reduce reliance on fossil fuels by developing 'biofuel' technologies. Biofuels are fuels

FIGURE 8.11 The Aral Sea ecosystem catastrophe. This scene shows the former floor of the Aral Sea in Uzbekistan, Central Asia. The Aral Sea was once the world's fourth largest inland body of water but has shrunk by over 70%. It is in a desert area and only exists because it is fed by rivers. The abandoned ships rest in a toxically polluted wasteland, caused by draining rivers feeding the sea in order to irrigate cotton and overusing pesticides and fertilizers. This is one of the world's most tragic examples of the destructive effects of unsustainable agricultural practices and is one of the greatest environmental catastrophes ever recorded. Photo taken in 2011 by Sebastian Kluger (CC BY 3.0).

derived from biomass, i.e., from recently living organisms or their metabolic products (such as cow manure). By contrast, fossil fuels (oil, coal, natural gas, peat), although also of biological origin, are the remains of ancient plants and animals which decayed and transformed over geological time. Technically, fossil fuels originated from species. Both biofuels and fossil fuels contain stored energy that was captured by photosynthesis.

A number of disadvantages of fossil fuels compared to biofuels have been identified:

1. Burning fossil fuels releases carbon dioxide that would otherwise not be released to the atmosphere. Burning biofuels does not, because the CO_2 released in this case is part of the currently circulating CO_2 rather than the CO_2 that has been unavailable for millions of years as a result of being bound in rocks and sediments.
2. Depending on the technology used, the combustion of fossil fuels can release emissions that are harmful to the environment and species.
3. Obtaining fossil fuels (often in fragile landscapes) and accidental spills may also degrade ecosystems and harm biodiversity.
4. Fossil fuels are non-renewable, and their increasing scarcity is associated with political instability. Conflict and war lead to destruction of natural ecosystems.

Biofuels are thought to have potential for alleviating all of these problems associated with fossil fuels, but the extent to which this is possible is currently being debated. For example, burning a liter of ethanol biofuel instead of a liter of gasoline has been claimed to reduce the accumulation of atmospheric CO_2 by as much as 80%, although some studies suggest much smaller percentages.

Curiously, automobile technology in its infancy was closely associated with biofuels (Sharma 2020). Otto von Nikolous designed his newly invented four-stroke combustion engine in 1878 to run on ethanol. Similarly, Rudolf Diesel designed his 1892 invention, the diesel engine, to run on peanut oil. The Ford Model T, an American staple

FIGURE 8.12 The Dust Bowl: America's worst ecosystem collapse. The photo (public domain) shows a dust storm approaching Stratford, Texas in 1935. Semi-arid portions of the western prairies of the U.S. and adjacent Canada used to be covered by native grasses with extensive root systems penetrating as deep as 2 m to reach water during drought periods. Early in the 20th century, the U.S. Congress encouraged expanded farming in areas that had been lightly used, including parts of Colorado, New Mexico, Kansas, and the panhandles of Oklahoma and Texas. The prairie grasses were replaced with wheat, the roots of which are too shallow to stabilize soil efficiently. During the 1930s, severe drought resulted in crop failures. Much of the area is subject to very strong winds, and this resulted in phenomenal wind erosion of the poorly anchored topsoil, which turned to dust blowing away in huge clouds. More than three-quarters of the topsoil was blown away in some regions. Tens of thousands of farms were abandoned, hundreds of thousands became homeless and indigent, and extensive financial ruin resulted. Photo credit: NOAA George E. Marsh, U.S. National Oceanic and Atmospheric Administration.

FIGURE 8.13 Burning and deforestation of the Amazon Rainforest to make grazing lands. The Amazon Rainforest extends over nine nations of South America, about 60% in Brazil, and is also home to numerous Indigenous people. It has the largest concentration of animals and plants of any comparable region. It includes 10% of the world's species, and doubtless some of these are potential sources of new medicinal and industrial compounds. The Amazon Rainforest has been called 'The Lungs of the Planet' because it produces more than 20% of the world's oxygen, and it is considered to be a critical world resource for keeping the climate of the world healthy (Strand et al. 2017; Peng et al. 2020). Since the 1960s, deforestation in the name of economic development has been devastating the rainforest. During the last half century, more than 15% of the forest has been eliminated, particularly for cattle ranching. Modeling studies suggest that should deforestation exceed 40% of the forest area, a tipping point will be passed that does not allow significant remediation of this invaluable ecosystem (Nombre et al. 2016). Photos by Amazônia Real (CC BY 2.0).

FIGURE 8.14 The importance of wild pollinators. Wild pollinators, especially bees, are indispensable for the reproduction of sexually reproducing wild plants and cultivated crops. (a) A U.S. Government poster illustrating important bee pollinators and both wild plants and crops that require their services. Artist: Steve Buchanan; for additional acknowledgments, see https://www.fs.usda.gov/wildflowers/features/posters/BountyOfBees.pdf (b). A demonstration to save wild bee pollinators. Photo by Bogusz Bilewski/ Greenpeace PL (CC BY ND 2.0).

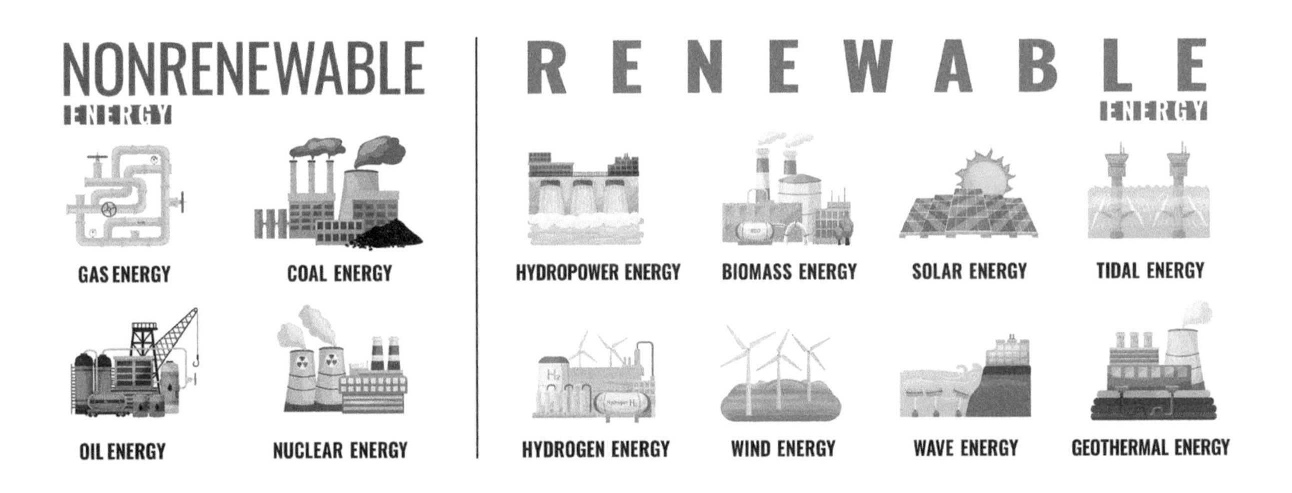

FIGURE 8.15 A contrast of the principal non-renewable and renewable types of energy. Source: Shutterstock, contributor: Double Brain.

between 1903 and 1926, was originally designed to run on ethyl alcohol (the growing availability of cheap petroleum made gasoline the fuel of choice) and was optionally available with a modified carburetor so that farmers could use the vehicle with ethanol they could make themselves. (Prohibition lasted from 1920 to 1933, making it illegal for people to produce their own 'moonshine.')

Petroleum is not only used for fuel but also to manufacture thousands of industrial products ('petrochemicals' are converted petroleum components). The current emphasis on biofuels is part of a more general movement to utilize renewable raw material for industrial processes and products, rather than non-renewable petroleum. This growing movement to replace non-renewable fossil fuels with renewable materials of biological origin has spawned a new vocabulary. Recent terms in addition to 'biofuels' include 'biomaterials,' 'bioenergy,' 'green energy,' 'biopower,' 'biorefinery,' and 'agroenergy.'

Wood harvested from woody plants is a traditional source of fuel but is becoming increasingly more expensive. Herbaceous (i.e., non-woody) plants or parts of plants are also being grown deliberately as 'biomass' for energy extraction, either simply to be burned for heat and electricity or to be processed into petrochemical fuels. Maize (corn), sugar fruit, palm fruit, and oilseeds are prominent domesticated crops being grown for energy. In temperate regions, the starch from corn is the principal raw material for the production of bioethanol, which in tropical areas sugar from sugar cane is the main source. The U.S. corn lobby exerts considerable influence in promoting corn-generated ethanol for adding to gasoline. In Europe, biodiesel has become the principal biofuel choice for automobiles because diesel-powered vehicles have long been popular there. Biodiesel is produced from oil, which may be from animal fats, and any oilseed (Singh et al. 2021). In practice, biodiesel is obtained mainly from soybean, canola/rapeseed (canola is a class of low-erucic acid Canadian-bred rapeseed), and sunflower in temperate regions, and in tropical regions from jatropha (*Jatropha curcas*, a poisonous plant, grown in India), oil palm (notably in Malaysia), castor bean (notably in Brazil), and other species. 'Microalgae' (microscopic, unicellular photosynthetic 'plants' or 'microorganisms') also have potential for producing biodiesel (Small 2011b; Konur 2021; Mizik and Gyarmati 2021; Figure 8.16). Microalgae are promising because many of the species are rich in lipids suitable for conversion to biodiesel, they can grow very rapidly (consuming large quantities of carbon dioxide which are accordingly removed as climate change atmospheric gases), and they do not require arable land (they are cultured in tanks or ponds) like conventional crops.

'Cellulosic' ethanol production is in its infancy, but this technology is very promising (Machineni 2020; Chen et al. 2021). Most plants are composed of cellulose (fibrous material making up cell walls), hemicellulose (a wrapping for the cellulose), and lignin (which acts as cement to hold plant cells together). The cellulose and hemicellulose are collectively termed cellulosic material. The most promising technique for producing cellulosic ethanol is the use of enzymes to separate the lignins; once the separation is achieved, the lignin can be burned as an energy source (typically used to fuel the production of more ethanol). The cellulosic components are hydrolyzed (split) into their component sugars (cellulose and hemicellulose are polymers, i.e., chemical chains, with the links made up of simple sugars). Paper pulp and municipal solid waste can be used, but usually raw plant materials are employed, particularly

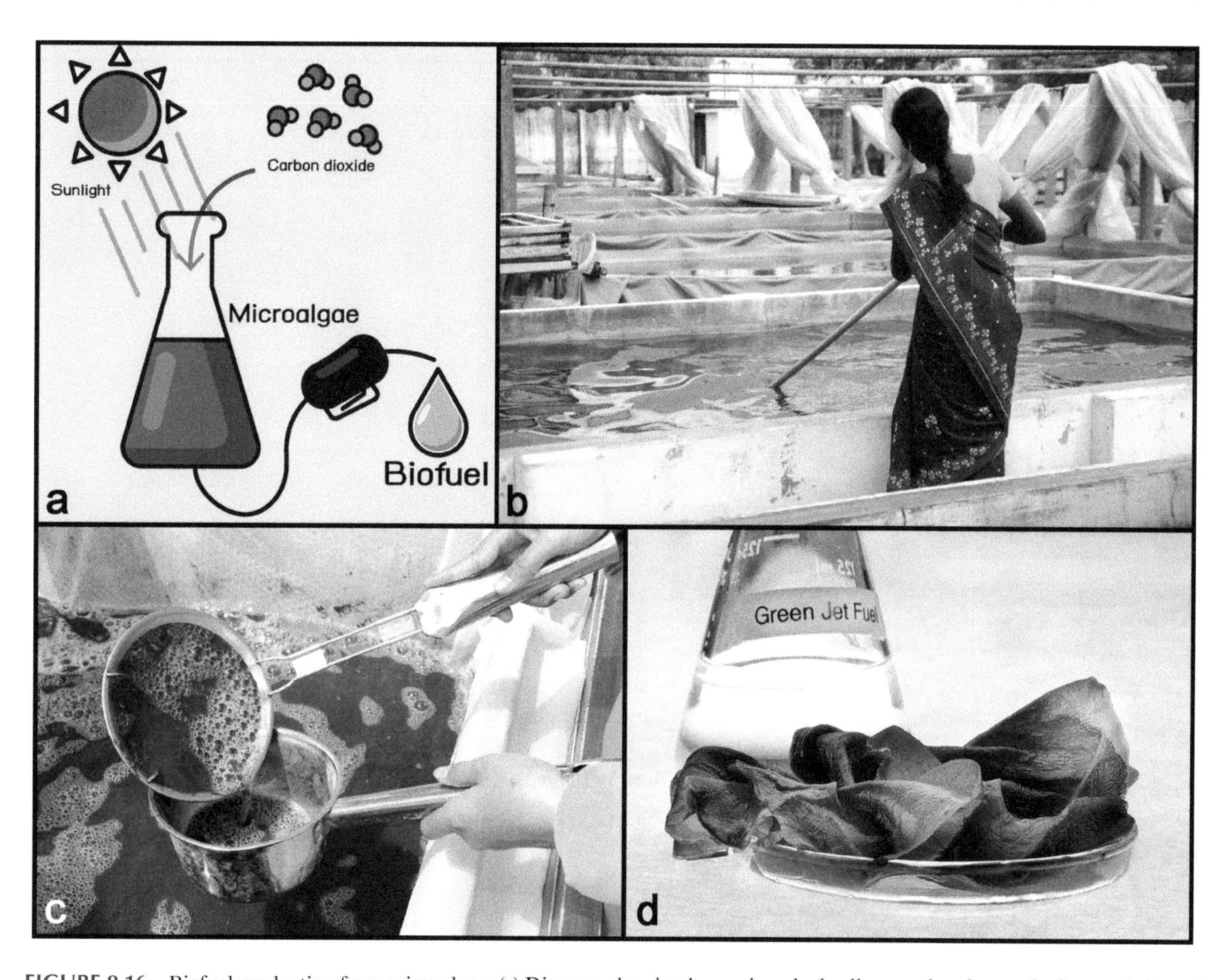

FIGURE 8.16 Biofuel production from microalgae. (a) Diagram showing how microalgal cells reproduce by employing sunlight and carbon dioxide, and the resulting biomass can be converted to biofuel. Source: Shutterstock, contributor: Thurakit T. (b) A worker in India caring for a commercial vat of the microalga *Spirulina* intended for production of biofuel. Photo by PWRDF (CC BY 2.0). (c) *Spirulina* being filtered out on a commercial microalga production plant. Source: Shutterstock, contributor: Mintra Chumpoosueb. (d) Photo (public domain) of dried micro-algal-based biofuel produced by the Honeywell Corporation.

parts of crop plants that remain after harvest, such as straw. Some plants are being developed specifically for use as cellulose sources for fuel alcohol, most notably switchgrass, *Panicum virgatum* (Jacot et al. 2021), but also miscanthus, hemp, and fast-growing poplars, willows, and eucalyptus. There are numerous trees and grasses that could also be used in tropical regions. To date, such materials have been simply burned to produce heat and/or electricity. The comparative benefits of using such cellulosic sources for biofuels for vehicles, compared to sugar/starch sources, depend on improved technologies for converting the cellulosic materials to ethanol (Haberzettl et al. 2021).

In addition to transportation fuels, energy use in general is being given close scrutiny. Plants can not only furnish fuel for vehicles (Figure 8.17a and b), but they can also be burned in lieu of fossil energy sources to supply heat and electricity (Figure 8.17c). Waste materials from livestock and humans are also useful for the production of biogas (Figure 8.17d). The economic justification for such wide use of 'energy crops' is unclear, but the technology is potentially worthwhile as a means of reducing ecological problems, compared to current alternatives, such as hydroelectric power, nuclear, and coal-based power, all of which can lead to extensive environmental damage (although some technological improvements are possible). Biofuels may also replace the use of gasoline in the huge array of gas-powered equipment (lawnmowers, snowblowers, chain saws, etc.) that currently exist and will continue to require conventional gasoline for many years.

Are Biofuels Environmentally Friendly?

Burning fossil fuels adds to the current carbon dioxide concentration of the planet, while burning biofuels does not (as explained above). Ethanol, the primary biofuel to date, is generally non-toxic directly to the environment and is also highly biodegradable. However, the preceding observations do not take into account the hidden environmental costs of production. Fossil fuels are currently used for machinery and agricultural inputs required to produce biofuel crops. Moreover, cultivating these crops produces pollution because employing fertilizers and pesticides result in release of nutrients (particularly nitrogen and phosphorus, causing eutrophication of lakes and rivers) and

FIGURE 8.17 Biomass employed as transportation biofuels (a, bioethanol; b, biodiesel), fuel stock for production of electrical power (c), and biogas (d). (a) Source: Shutterstock, contributor: Vectorfair. (b) A car fueled by biodiesel produced from hemp oilseed, prepared by B. Brookes. (c) Electrical power produced from biomass. Source: Shutterstock, contributor: Double Brain. (d) Production of biogas from cow manure. Source: Shutterstock, contributor: Poul Carlsen.

pesticide residues, which result in loss of biodiversity. Hill et al. (2006) showed that (when all such factors are considered, and based on U.S. production): (1) ethanol (from corn) and biodiesel (from soybean) produce more energy than required to make them (contrary to some earlier assertions); (2) relative to fossil fuels, greenhouse gases are reduced 12% by the production and combustion of ethanol, and 41% by biodiesel; (3) biodiesel production is associated with far less release of nitrogen, phosphorus, and harmful pesticides to the environment (this applies to U.S. production systems; the authors did not attempt a comparison with environmental problems associated with fossil fuels); (4) supplementing gasoline with relatively small amounts (15%) of ethanol lowers emission of carbon monoxide, volatile organic compounds, and small particles compared to using pure gasoline but, on a life cycle basis, increases five major air pollutants, including carbon monoxide, volatile organic compounds, oxides of sulfur, and oxides of nitrogen (biodiesel was often superior in these respects). It is important to emphasize that the study of Hill et al. is based on U.S. production conditions for corn and soybean, and similar analyses for biofuel production for other crops in other countries need to be carried out before generalizing on the benefits of biofuels vs. fossil fuels.

Are Biofuels Bad for Biodiversity?

Manufacturing biofuels is associated with at least some destruction of biodiversity, but the extent is not clear at present. There is concern that as food crops are progressively used for biofuel, demand for new crop land will result in sacrifice of natural habitats. Also, as trees are increasingly used for biofuel, natural forests will also be sacrificed. In both cases, the land that once supported natural ecosystems is replaced by monocrops, resulting in substantial decline in biodiversity.

Can Biofuels Replace Fossil Fuels?

Biofuels, particularly ethanol, are becoming more popular, but current production methods and conversion technologies may be capable of producing only modest environmental benefits. 'In the US case, the use of ethanol would require enormous areas of corn agriculture, and the accompanying environmental impacts outweighs its benefits. Ethanol cannot alleviate the United States' dependence on petroleum' (Dias de Oliveira et al. 2005). Aside from the current state of technology, the greatest hurdle to biofuel production in the long term is land availability and the growing cost of agriculture. Rising temperatures, decreasing supplies of water, increasing demands for space (urban sprawl) and resources associated with population growth, and rising expectations are harbingers of greatly increasing demand for fuel, which is unlikely to be met just by increasing the production of biofuels.

Biofuels vs. Food

When food crops are used for fuel instead of for food, there is a potential impact on food prices and availability, particularly in developing countries. An increasing proportion of grains are being used for fuel alcohol production, and world grain supplies have fallen to alarmingly low levels. It has been estimated that the corn used to produce 95 L of ethanol, required to fill the tank of some SUVs, could be used to feed a person for 1 year in poor countries. The issue of diverting crops currently used as human and livestock food to biofuel production has become a major consideration (Muscat et al. 2020). As the price of fuel increases, it can become profitable to sell crops for automobile fuel rather than as food. It has been estimated (Hill et al. 2006) that global demand for food will double within the coming 50 years, but that global demand for transportation fuels will increase even more rapidly. Biofuels that are based on food plants, therefore, cannot be expected to solve the developing energy crisis.

Prospects

From an environmental viewpoint, biofuels as substitutes for fossil fuels are useful for improving air quality, reducing greenhouse gases, and lowering toxicity resulting from accidental spills. Politically, they can decrease energy dependence on foreign imports. Socially, biofuels are labor intensive, and so can create employment. Unfortunately, biofuels have the potential of replacing only a small portion of the vanishing supply of fossil fuels. Given the fact that new arable land is very hard to find, it is unlikely in the long term that food crops can continue to be used to produce biofuels. The much more abundant cellulosic materials in agricultural and forest residues have more potential. 'Marginal' lands can be increasingly used in the future for cellulosic biofuel production, but these lands often contain significant biodiversity and supply essential ecological services. Alternative sources of energy are urgently required. Biofuels represent only a small part of the needed fix. Possible additional solutions include ways of harnessing energy from the sun (solar power), the movement of the oceans and tides, the planet's heat (geothermal and ocean gradients in temperature can be exploited), wind, fuel cells, and additional technologies, as pointed out earlier (Figure 8.15).

9 Extinction
How Biased Elimination of Species Endangers Humans

INTRODUCTION

Individual living things die, which is sad, but far sadder is the ending of entire species. Today, a frightening number of species are on their deathbeds. This chapter examines how the current extinction crisis is a threat not just to the world's wild species but also to the future of humans.

NATURAL EXTINCTION

Very long time periods are defined in geology by the intervals between planetary-scale catastrophes in which a substantial portion of living species were exterminated, and life was forced to survive very difficult conditions. Barnosky et al. (2011) defined a mass extinction as an episode in which the Earth lost 75% of its species in a geologically short time. There have been at least five such episodes due to catastrophic geological events in the last half billion years (Figure 9.1). Most familiar is the Cretaceous-Paleogene transition 66 million years ago, generally acknowledged to have been caused by an asteroid which slammed into Earth. This resulted in a mass extinction, famously of most dinosaurs (Figure 9.2). Generally, however, conditions on the planet and the world's biota change slowly, for example due to drift of continents over billions of years to different part of the Earth, related to plate tectonics. Over the 3.5 billion years of life on Earth, 4 billion species are estimated to have gone extinct, representing 99% of all that ever existed (Barnosky et al. 2011). Most of the time, extinction has been balanced by the evolution of new species (the average species lasting of the order of 10 million years).

WHAT HUMAN (DIS)INTEREST IN EXTINCT SPECIES REVEALS ABOUT PREFERENCES AND PREJUDICES

As noted above, 99% of all the species that ever existed are extinct. Most of these are completely unknown to science, but there are numerous fossils representing the various kinds of species that nature generated. Many thousands of extinct species have been described, illustrated, and displayed in museums. However, the public's interest in extinct species is extremely limited – to a small minority of animals that are very large, especially predators. As noted in Chapter 4 with respect to living species, people are most attracted by large, charismatic mammals, but also to smaller beautiful animals (especially birds and butterflies) and beautiful flowering plants. The skeletal remains of extinct animals and other fossilized material from living things are not especially attractive, and dead objects simply don't elicit empathetic feelings in people in ways that living things do. Nevertheless, reconstructions of very large animals do seem to command respect (or fear). It should not be surprising that huge dinosaurs (Figure 9.3) dominate the interest of people in extinct species, followed by large mammals (Figure 9.4) and big birds (Figure 9.5), but this does reflect the inherent prejudice and biases of people against most life forms. While mammals dominate the extant 'charismatic megafauna,' the extinct charismatic megafauna is dominated by reptiles.

ANTHROPOGENIC EXTINCTION

Recent domination of Earth by humans is dramatically degrading the natural environment (hence the environmental crisis, of which global warming is the most publicized aspect) as well as damaging both ecosystems and their constituent species (hence the biodiversity crisis). Unlike previous geological cataclysms, the ongoing catastrophic degradation of the Earth has been termed 'anthropogenic' (caused by humans). 'Anthropogenic extinction' is human-caused extinction of living things. There may have been occasions in the past when people deliberately eliminated a competing or dangerous species, but extinction due to human activities is almost always unintended. Over 1,000,000 species of plants and animals are now considered to be in danger of extinction (United Nations 2020). Over 500 species of vertebrates are known to have less than 1,000 individuals left alive (Ceballos et al. 2020). The general extinction rate is very difficult to estimate (He and Hubbell 2011), possibly 30,000 species per year or three species per hour; perhaps a thousand times as much as would be occurring in the absence of humans. There is suspicion that the alarmingly high current rate of human-driven extinction of species is of the same magnitude as happened during the five recognized episodes of mass extinction that occurred in the past. This has led some to refer to the present period of anthropogenic extinction as the sixth mass extinction, and the Holocene extinction, and it has been proposed that it should be recognized as a formal geological epoch (time period) termed the Anthropocene. The chief problem associated with anthropogenic extinction is not necessarily whether the planet can survive periods of catastrophic reductions of biodiversity – it

DOI: 10.1201/9781003412946-9

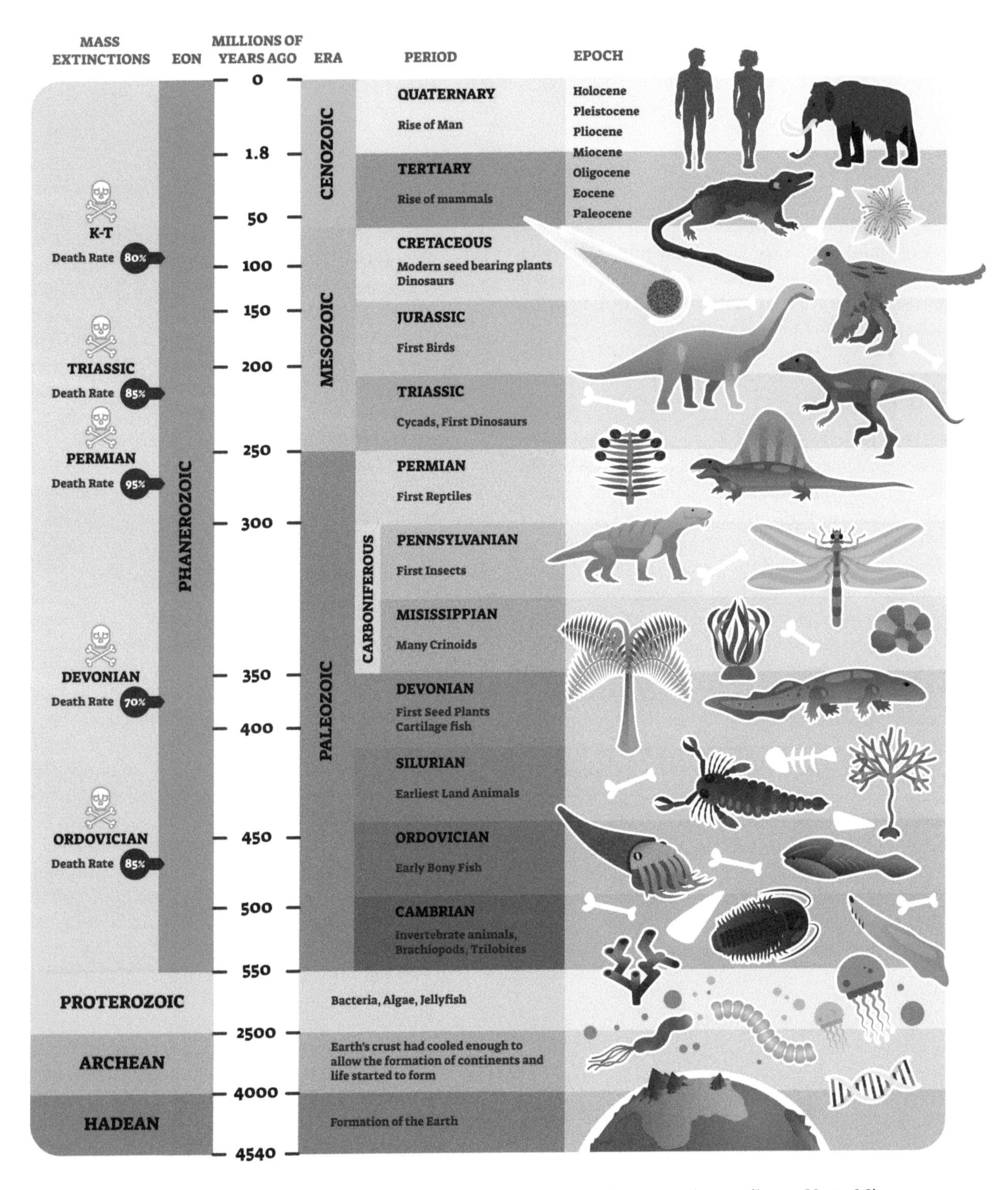

FIGURE 9.1 Geological timeline showing history of mass extinctions. Source: Shutterstock, contributor: VectorMine.

FIGURE 9.2 An artist's conception of the Cretaceous/Paleocene mass extinction that occurred 65 million years ago, when an asteroid is hypothesized to have collided with Earth, killing off about three-quarters of all living species, including dinosaurs. Source (public domain) Pixabay.com.

FIGURE 9.3 Giant extinct reptiles of the Mesozoic era (252–66 million years ago), known as the 'Age of Reptiles' and 'the Age of Dinosaurs.' (a) Diagram showing the Age of Dinosaurs with respect to the geological time scale. Source: Shutterstock, contributor: Captainz. (b) Visitors to a museum featuring reconstructed dinosaurs. Source: Shutterstock, contributor: Denis Cristo. (c) Dinosaurs of the Cretaceous period (145–66 million years ago) of the Mesozoic era. Source: Shutterstock, contributor: Liliya Butenko. (d) Pterosaurs are extinct flying reptiles dating from the Mesozoic to the end of the Cretaceous (228–66 million years ago). They are often referred to as 'flying dinosaurs' but are a separate lineage. Shown here is an artist impression of the giant pterosaur *Quetzalcoatlus northropi* foraging on a prairie. A juvenile titanosaur has been caught by one pterosaur. The height of the animal may have ranged up to 10 m when standing, with a wing span exceeding 11 m. The weight is uncertain, some estimates ranging up to 250 kg. Source: Witton, M. P. and D. Naish. 2008. "A Reappraisal of Azhdarchid Pterosaur Functional Morphology and Paleoecology." *PLoS One* 3(5): e2271. doi:10.1371/journal.pone.0002271. (e) Dinosaurs of the Jurassic period (200 to 145 million years ago) of the Mesozoic era. Source: Shutterstock, contributor: Liliya Butenko.

FIGURE 9.4 Large extinct mammals. (a) The extinct genus *Megatherium* includes species called ground sloths endemic to South America that lived from the Early Pliocene through the end of the Pleistocene (about 35 million to 11,000 years ago). *Megatherium americanum*, the giant ground sloth, was native to the Pampas through southern Bolivia. It was as large as an elephant, weighing up to 4 t and standing over 4 m in height. Source: Shutterstock, contributor: Esteban De Armas. (b) Public domain painting of *Smilodon populator* by artist Charles R. Knight, 1903, from the American Museum of Natural History. Specimens ranged in weight to over 400 kg and in height to over 1.2 m at the shoulder. The largest and best known of the saber-toothed cats, it was not closely related to modern cats. *Smilodon* species lived in the Americas during the Pleistocene epoch (2.5 million years to 10,000 years ago). (c) The Entelodon or hell pig, an extinct omnivorous mammal that lived during the Late Eocene to Middle Oligocene (37–27 million years ago) on the plains of Eurasia. It was about 3 m long, 2 m tall, and weighed about 500 kg. It is not closely related to modern pigs and is thought to be related to hippopotamuses. Painting (public domain) by German landscape artist Heinrich Harder (1858–1935).

FIGURE 9.5 Giant extinct birds. (a) The largest extinct bird in the world, the flightless elephant bird (*Vorombe titan*), endemic to Madagascar, where it became extinct about 1,000 years ago. It was about 3 m tall and could weigh over 700 kg. Its closest living relatives are the much smaller kiwis. Source: Shutterstock, Contributor: YuRi Photolife. (b) The giant moa (*Dinornis novaezealandiae*) of New Zealand. Moas were ostrich-like flightless birds standing as tall as 3 m and weighing as much as 250 kg. Source (public domain): Rothschild, L. W. 1937. Extinct birds. Drawn by F. W. Frohawk in 1907. (c) The terror bird (*Titanis walleri*). This flightless carnivorous North American bird was up to 2 m tall, weighed about 150 kg, and is thought to have become extinct several million years ago. Note the formidable dinosaur-like head. Painted by Russian artist Dmitry Bogdanov (CC BY SA 3.0).

has done so repeatedly. The problem is that most species do not survive such periods, and it is possible that humans could be subjected to extremely inhospitable conditions in the short term and could even become extinct.

HUMANS AND ANIMAL EXTINCTION

There are three basic ways that humans endanger other animals: unintentionally (primarily by habitat destruction), deliberate elimination of species that are perceived as competitors or sources of danger, and overharvesting for resources (especially for food). Vertebrates, especially fish and the larger mammals, are natural prey of humans, and so have suffered greatly. As humans increased in numbers in past times, hunting caused the extinction of numerous vertebrates, especially mammals (Figure 9.6), with numerous others driven to the brink (Figure 9.7).

HUMANS AND PLANT EXTINCTION

Possibly 500,000 species are photosynthetic. Of these, the flowering plants (about 400,000 species) are much better known and utilized than all other categories of plants. In sharp contrast to the animal world, for which only a few dozen species have been domesticated (changed genetically to better serve humans) and are maintained as livestock and pets, thousands of plant species have been domesticated, and possibly as many as 100,000 are cultivated (whether or not they have been domesticated), at least in a limited way. Agricultural crops are the most important of all plants. Millions of seed collections of these have been preserved for breeding improved cultivars (Stolton et al. 2006). Of course, most plants are far easier to maintain outside of their natural habitats than are most animals. From a conservation perspective, this is both good and bad. It is good because

FIGURE 9.6 Species of animals that went extinct in recent times because of hunting by humans. (a) Paleo-Indians (the earliest known settlers of the Americas) hunting a glyptodont (an early South American relative of armadillos). These armored herbivores were hunted to extinction at the end of the last ice age, 2 millennia after humans arrived in South America. Painting by Heinrich Harder (1858–1935). (b) Dodo (*Raphus cucullatus*), a flightless bird which was endemic to the island of Mauritius, east of Madagascar in the Indian Ocean. It became extinct during the mid-to-late 17th century because of habitat destruction, hunting, and predation by introduced mammals. This model is in the Museum National d'Histoire Naturelle in Paris. Photo (public domain) by Jebulon. (c) Woolly rhinoceros (*Coelodonta antiquitatis*) being hunted. It lived on the northern steppes of Eurasia, surviving the last ice age. Exhibit, Horniman Museum, London. Photo by Jim Linwood (CC BY 2.0). (d) Scene showing the hunting of a woolly mammoth (*Mammuthus primigenius*). The size of an African elephant, it was found in northern Eurasia and northern North America. The species went extinct about 5,000 years ago. (e) Passenger pigeon (*Ectopistes migratorius*), driven to extinction in the early 20th century. Source (public domain): Whitman, C. O. 1920. *Orthogenetic Evolution in the Pigeons*. 3 vols. Credit: K. Kayashi. (f) Tasmanian tiger (*Thylacinus cynocephalus*, also known as the Tasmanian wolf and the thylacine). Although it looked like a striped dog, like other marsupials the females gave birth to immature young that they raised in their pouch. The animal weighed 15–30 kg and was the largest marsupial predator. The Tasmanian tiger became extinct on the Australian mainland about 3,000 years ago but made a last stand in Tasmania. English colonists offered a bounty because the tigers allegedly killed many sheep. Thousands were slaughtered from 1830 to 1909, and the last living individual died in a zoo in 1936. Source (public domain): Gould, J., and H. C. Richter. 1841. *Mammals of Australia*, vol. I, plate 54. London, J. Gould & H. C. Richter.

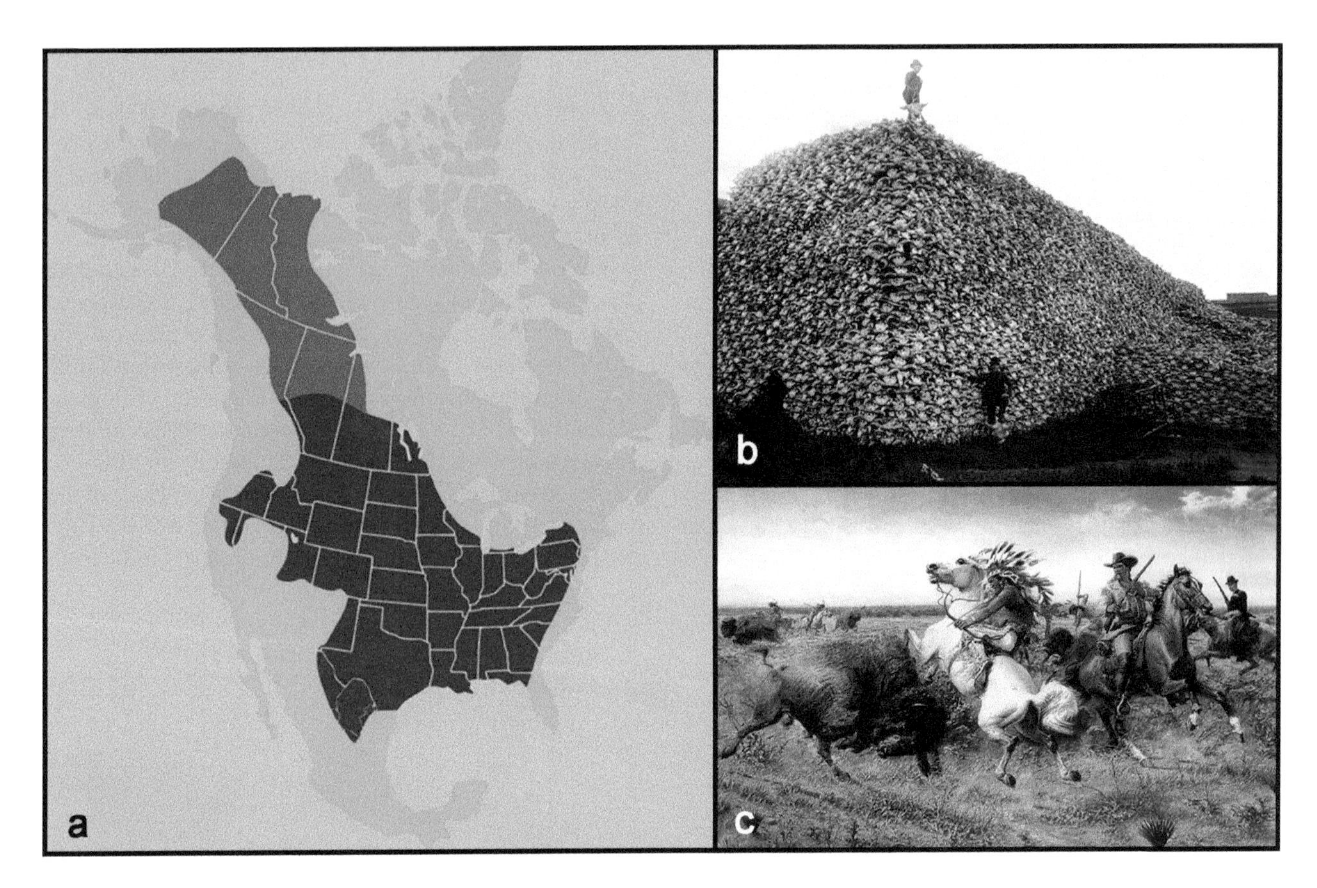

FIGURE 9.7 The American bison ('buffalo,' *Bison bison*), which almost went extinct. Unregulated hunting and diseases from domestic cattle reduced the population from over 60 million in the late 18th century to less than 600 by 1889. Conservation efforts have resulted in the production of over 30,000 individuals today, mostly in national parks and reserves. (a) The species once ranged from Alaska to Mexico. The original distribution of plains bison (*Bison bison bison*) is shown by the upper area in orange. The original distribution of the wood bison (*Bison bison athabascae*) is shown by the lower area in brown. Prepared by Cephas (CC BY SA 3.0). (b) Photo (public domain), ca. 1892, of a pile of American bison skulls waiting to be ground for fertilizer. Source: Burton Historical Collection, Detroit Public Library. (c) 'The Great Royal Buffalo Hunt' (public domain), ca. 1895, by German-American lithographer Louis Maurer (1832–1932). The illustration is based on an event staged in 1872 for Grand Duke Alexei Alexandrovich of Russia. Shown are the famous Buffalo Bill Cody (who claimed to have killed thousands of buffalo) and warriors of different Sioux tribes, hired for the hunt.

numerous species are being protected from extinction. It is bad because only a limited sample of the range of variation can be maintained outside of nature, and this leaves much of their genetic diversity unprotected. Moreover, since plants are so widely maintained outside of their natural habitats, some have mistakenly concluded that it is not essential to protect most species within their habitats.

More plants are cultivated as ornamentals than for any other purpose. Unlike animals, plants are rarely considered ugly (many are bizarre, but this usually makes them charming). Plants with especially attractive flowers have become superstars as cultivated ornamentals (Figure 9.8), but there are thousands whose blooms are sufficiently pretty that they are maintained in horticulture. While it is very disappointing that thousands of less attractive plant species are being extinguished in nature, at least a huge number that are naturally attractive to humans are being maintained in cultivation.

Unfortunately, the fact remains that thousands of flowering plants are endangered. Big trees like big animals are quite attractive to conservationists. However, like the majority of small animals, most herbaceous plants aren't particularly attractive to most people – for the most part simply because they don't produce conspicuous flowering displays like most ornamentals. Often, a given plant species is essential to the welfare of animal species, so that should the plant go extinct, the animals will follow suite.

The Danger of Potentially Valuable Medicinal Plants Becoming Extinct before They Are Discovered

Medicinal plant conservation is an important issue in many Third World Nations where medicine is based predominantly on herbal medicine (Chen et al. 2016; Kumar and Jnanesha 2016). As noted in Figure 9.9, some of the world's most valuable medicinal plants have been threatened, sometimes eliminated from the wild, but fortunately often saved from extinction. Several of the most useful pharmacological compounds from plants were discovered only recently, illustrating the importance of conserving as many species as possible to allow future research to find more invaluable medicines for humankind. As with hunting of animals, excessive harvesting is a threat to wild

FIGURE 9.8 Vintage commercial flower catalogs (public domain) showing popular ornamental plants. The attractiveness of these species protects them from the threat of extinction. Photo credit: U.S. Department of Agriculture, National Agricultural Library. (a) Dingee & Conard 1898 "Our new guide to rose culture" catalog. (b) Childs' spring catalog of seeds, bulbs and plants for 1910, featuring sweet pea. (c) Miss C.H. Lippincott 1897 catalog, featuring hollyhocks. (d) Farquhar's 1902 seed catalog, featuring hollyhocks. (e) Childs' combination catalog for 1910, featuring pansies. (f) Peony (*Paeonia moutan*), in 1900 catalog of The Yokohama Nursery Co., Ltd. Japan.

plant resources. However, destruction or degradation of habitats has been the main way that people have been causing extinction of species. The pharmacology of most wild species has scarcely been examined, and there is considerable danger of the extinction of many plant species with invaluable constituents before their value has been discovered. Plants produce thousands of medicinal constituents, not to alleviate human illness but usually to serve as toxins to protect themselves from herbivores. Sedentary animals that, like plants can't run away from their enemies, probably also often produce valuable protective chemicals, and their preservation similarly needs to be considered.

Animals Used Medicinally in Relation to Their Conservation

'Zootherapy' is the treatment of human ailments with remedies made from animals and their products (Alves and Alves 2011). Plants and plant-derived materials constitute most ingredients employed in medicine, but materials from animals are also employed. While some animal ingredients are employed in modern Western medicine, animal-based therapies are mostly employed in Third-World countries, the medicinal systems often termed traditional or herbal (a term usually reserved for plants). When employed in Western countries, the use of animal therapies is often termed alternative or complementary medicine. Numerous cultures have employed animal ingredients in medicine, especially the Chinese, who employ about three dozen animal species substantially (Call 2006). Thousands of animal species have been used in folk therapies, the majority likely of no benefit (aside from possible placebo effects), and some actually harmful to the patients. This is a sensitive area, since major users of animal-based therapies are mostly indigenous peoples and ethnic groups, and no one (especially those in the privileged Western World) wishes to insult them by challenging their entrenched cultural traditions and beliefs. However, in some cases endangered

FIGURE 9.9 Examples of valuable medicinal plants that are threatened or extinct in the wild and have fortunately been conserved in cultivation. (a) Ginkgo (*Ginkgo biloba*), a large tree originating from China, where it is probably extinct in the wild. Ginkgo trees, which are significant in Buddhism, Confucianism, and Taoism, owe their survival to having been preserved beside religious shrines. In recent decades, Ginkgo extracts have been explored for various medicinal uses, notably for dementia and other memory decline problems, and there is continuing exploration of other promising applications. Photo by J.-P. Grandmont (CC BY SA 3.0). (b) Madagascar periwinkle (*Catharanthus roseus*), an endemic herb of Madagascar, where it is endangered because of extensive habitat destruction. It furnishes the alkaloids vincristine and vinblastine which are important in the treatment of leukemia and lymphoma. Photo by Jane Wong S.K. (CC BY SA 3.0). (c) Pacific yew (*Taxus brevifolia*) a small coniferous tree confined to the Pacific Northwest of North America, which for years was cavalierly destroyed so more lucrative trees would grow. In the late 20th century, it became the chief source of the chemotherapy drug paclitaxel (the commercial preparation Taxol), widely used to treat breast, ovarian, and lung cancer, although now other *Taxus* species are also being used as sources. Photo (public domain) by Glacier NPS.

animal species are being harvested for animal-based products (notably tiger bone, sea turtle shell, and bear bile), so there is urgency in addressing the problem. Curiously, one of the most publicized examples – the use of rhinoceros horn as an aphrodisiac (a discredited practice) – appears to be a very minor practice, while the illicit trade in rhino horn is primarily for expensive curios (Hsu 2017).

STATE OF KNOWLEDGE OF EXTINCTION RISK TO KNOWN WILD SPECIES OF THE WORLD

The International Union for Conservation of Nature (IUCN) is perhaps the leading organization dedicated to conservation of nature and allied projects throughout the world. It is best known for its Red List of Threatened Species (IUCN 2021), which is the world's most comprehensive inventory of the conservation status of species and the leading guide to information on species in need of conservation. (This should not be confused with the IUCN Red List of Ecosystems, which assesses risks to ecosystems.) Extinction risk of species is assessed in nine categories, as indicated in Figure 9.10. 'Threatened' species are those listed as Critically Endangered, Endangered, and Vulnerable. The project is continuing, with updated lists published online regularly. (Chapman 2009 provides a more comprehensive, albeit dated analysis based not only on the IUCN 2009 data but also on additional references; Christenhusz and Byng 2019 provide supplementary information for plants.)

Less than 6% of known (i.e., described) species have been assessed (Table 9.1). Moreover, extensive data are available only for mammals, birds, and flowering plants. The reported numbers of threatened species shown in Table 9.1 are highly biased in favor of the larger, more charismatic plants and animals, notably vertebrate animals and vascular plants. Very small numbers of species of other groups have been analyzed so reliable estimates are not available of the percentage of species that are threatened. As indicated in Chapter 1, conservation concern to date has been overwhelmingly centered on vertebrates and higher plants. There can be no doubt that extinction is far more common than currently known for the other life forms, which are accordingly in dire need of conservation attention.

A KEY MISCONCEPTION ABOUT THE NEED TO LIMIT EXTINCTION

The quotation in Textbox 9.1 is probably representative of the current misunderstanding of much of the general public – that the principal goal of conservation is to save the world's most beautiful animals from extinction. This is the inevitable result of relentless well-meaning campaigns by wildlife organizations soliciting funding to 'save' giant pandas, tigers, elephants, and other glamorous animals. Such emotional appeals based on charisma detract from the far more important and pragmatic information that can appeal to the self-interest of most people: the survival of both biodiversity in general and humans in particular is dependent on the well-being of the world's habitats and ecosystems. Educating the public about the practical harvested

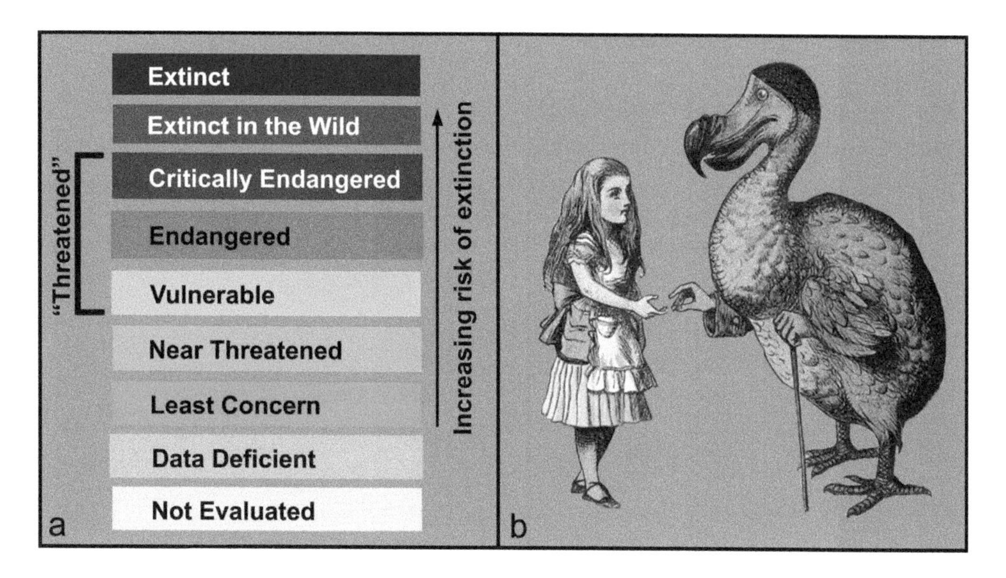

FIGURE 9.10 Extinction risk. (a) International Union for Conservation of Nature (IUCN) Red List of Threatened Species categories. Prepared by B. Brookes. (b) Alice and the dodo, symbolizing the possibility that humans could become extinct like the dodo unless the environmental threats to the planet are reduced. Source (public domain): Tenniel, J., 1890. *The Nursery Alice* (based on Alice's Adventures in Wonderland). Modified by B. Brookes.

TABLE 9.1

International Union for Conservation of Nature 2021 Red List (Version 2021-1) Data for 'Threatened Species' (Critically Endangered, Endangered, and Vulnerable, Collectively) for the Major Groups of Organisms

	Estimated Number of Described Species	Number of Species Evaluated	Percent of Described Species Evaluated	Number of Threatened Species	Best Estimate of Threatened spp. as % of Extant Data (for Sufficiently Evaluated Species)
Vertebrates					
Mammals	6,513	5,940	91	1,323	26
Birds	11,158	11,158	100	14,481	14
Reptiles	11,341	8,492	75	1,458	
Amphibians	8,309	7,212	87	2,442	41
Fishes	35,797	22,005	61	3,210	
Subtotal	73,118	54,807	75	9,914	
Invertebrates					
Insects	1,053,578	10,865	1.0	1,926	
Molluscs	81,719	8,881	11	2,305	
Crustaceans	80,122	3,189	4	743	
Corals	2,175	864	40	237	
Arachnids	110,615	393	0.36	218	
Velvet worms	227	11	5	9	
Horseshoe crabs	4	4	100	2	100
Others	151,801	844	0.56	148	
Subtotal	1,480,241	25,051	2	5,588	
Plants					
Mosses	21,925	282	1.3	165	
Ferns and allies	11,800	678	6	266	
Gymnosperms	1,113	1,016	91	403	41
Flowering plants	369,000	52,077	14	20,883	
Green Algae	11,616	16	0.1	0	
Red algae	7,291	58	0.8	9	
Subtotal	422,745	54,127	13	21,726	

(*Continued*)

TABLE 9.1 (*Continued*)
International Union for Conservation of Nature 2021 Red List (Version 2021-1) Data for 'Threatened Species' (Critically Endangered, Endangered, and Vulnerable, Collectively) for the Major Groups of Organisms

	Estimated Number of Described Species	Number of Species Evaluated	Percent of Described Species Evaluated	Number of Threatened Species	Best Estimate of Threatened spp. as % of Extant Data (for Sufficiently Evaluated Species)
Fungi and Protists					
Lichens	17,000	57	0.3	50	
Mushrooms, etc.	120,000	368	0.3	196	
Brown algae	4,317	15	0.3	6	
Subtotal	141,317	440	0.3	252	
TOTAL	2,117,421	134,425	6	37,480	

Prepared from data for 2001 from the IUCN red List of Threatened Species (https://www.iucnredlist.org/).

TEXTBOX 9.1 A COMMON MISCONCEPTION: BIODIVERSITY EXTINCTION IS HARMFUL PRIMARILY BECAUSE IT ELIMINATES LIKEABLE SPECIES

Will the world and humankind be very much the poorer if we lose a thousand or so species?... I passionately believe in saving the whale, the tiger, the orangutan, the sea turtle and many other specifically identified species. What I do not accept is the general principle that all species alive today should carry on existing for ever... it isn't a tragedy if we lose quite a few along the way.

—*Marcel Berlins, The Guardian, http://www.guardian.co.uk/commentisfree/2008/oct/08/features.comment*

materials from, and the pragmatic ecological contributions of biodiversity, is challenging but necessary.

NATURAL BUT HARMFUL DESTRUCTIVE HUMAN BEHAVIOR: THE ULTIMATE BASIS OF EXTINCTION THREATS TO ECOSYSTEMS AND SPECIES

All species, including humans, compete for resources, and so it is inevitable that we are sometimes hostile to other people, indeed other living things. There are many different combinations of beneficial/harmful interaction among species (positive for one or both, positive for one and negative for the other, neutral for one but negative for the other, etc.), but for the most part human technology has enabled humans to dominate and thereby exploit all other life forms. Unfortunately, humans are often so greedy that they are willfully or blindly destructive. In response to exaggerated emotions – hatred, bigotry, prejudice, fear – we are the most dangerous species on the planet – to each other and to all other creatures. Indeed, it has been hypothesized that we (*Homo sapiens*) eliminated the other human (i.e., *Homo*) species that existed in the past (Adams 2009). Our poorly controlled behavior, unfortunately, is now threatening the welfare of the entire world (Figure 9.11) and hence our future existence.

The biodiversity/bioresources crisis came into being because of the perpetual human tendencies to reproduce, consume, exploit the environment, and develop technologies to more efficiently carry out these objectives. These intrinsic behaviors were adaptive in the distant past but now are threatening the welfare of our planet and hence our species. A large part of the problem is simply that our destructive hard-wired predispositions, which amount to personal and collective greed, are extremely difficult to moderate. The solution is to control our reckless behavior toward the planet and its other inhabitants. As noted in the following, national and international remediation is required, but concerned citizens can be helpful in their neighborhood. Local protection of habitats – for example, a pond or bog where rare and endangered species occur – is important, especially for plants and animals with extremely limited distribution areas.

HUMAN OVERPOPULATION: A CRITICAL ISSUE

Many biologists (e.g., Mora and Sale 2011; Cafaro and Crist 2012; Aukema et al. 2017; Crist et al. 2017; Marques et al. 2019) have suggested that what is probably of greatest importance to address the ecological health of the planet is to stabilize human population growth (Figure 9.12), reduce consumption of resources, and increase efficiency

FIGURE 9.11 Examples of human damage to the planet. (a) 'Battle of Courcelette' (1918), public domain painting by Canadian war artist Louis Whirter (1873–1932), illustrating destruction from warfare. (b–d) Illustrations of pollution and ecosystem damage. Public domain images from Pixabay.com.

of production technologies now supporting civilization. On the positive side, while population growth is not projected to stabilize, at least the growth rate is decreasing. A key determinant controlling human reproductive behavior today is health or 'well-being' – because where economic conditions are harsh and survival is relatively low, people have more babies – and the situation in these respects is most distressing in Africa (Murdoch et al. 2018). Discouragingly, consumption levels are increasing, and the traditional economic model based on unending growth has not changed. (For analyses of the incompatibility of infinite growth economics and the ecological welfare of the world, see Global Footprint Network 2011; Martin et al. 2016; and Fressoz and Bonneuil 2017. Also note the comments regarding 'circular economy' in the Introduction to Chapter 8.) Efficiency of production technologies is increasing, but at the same time other technologies are facilitating the increased agricultural exploitation of marginal lands that are the chief refuge of biodiversity.

'THE TRAGEDY OF THE COMMONS' AS A DRIVER OF EXTINCTION

Achieving agreement of all or most stakeholders on an issue is difficult, sometimes impossible, because individuals may differ in the perception of the relative cost/benefit to themselves, and to their competitors. Additionally, personality, greed, and circumstances of the parties involved may differ significantly. Wars are fought over resources and territory, and such areas can be seriously degraded by some of the users. It is often to the benefit of most or all to have agreements on sharing. This applies particularly to areas that are employed to harvest resources (such as bodies of water and forests) and grazing areas (for ranching or herding), which are easily degraded to the misfortune of all, including biodiversity and habitats. 'The tragedy of the commons' is such a situation, in which individual stakeholders have access to a shared valuable resource and can act independently in their own selfish interests even if they degrade that resource to the

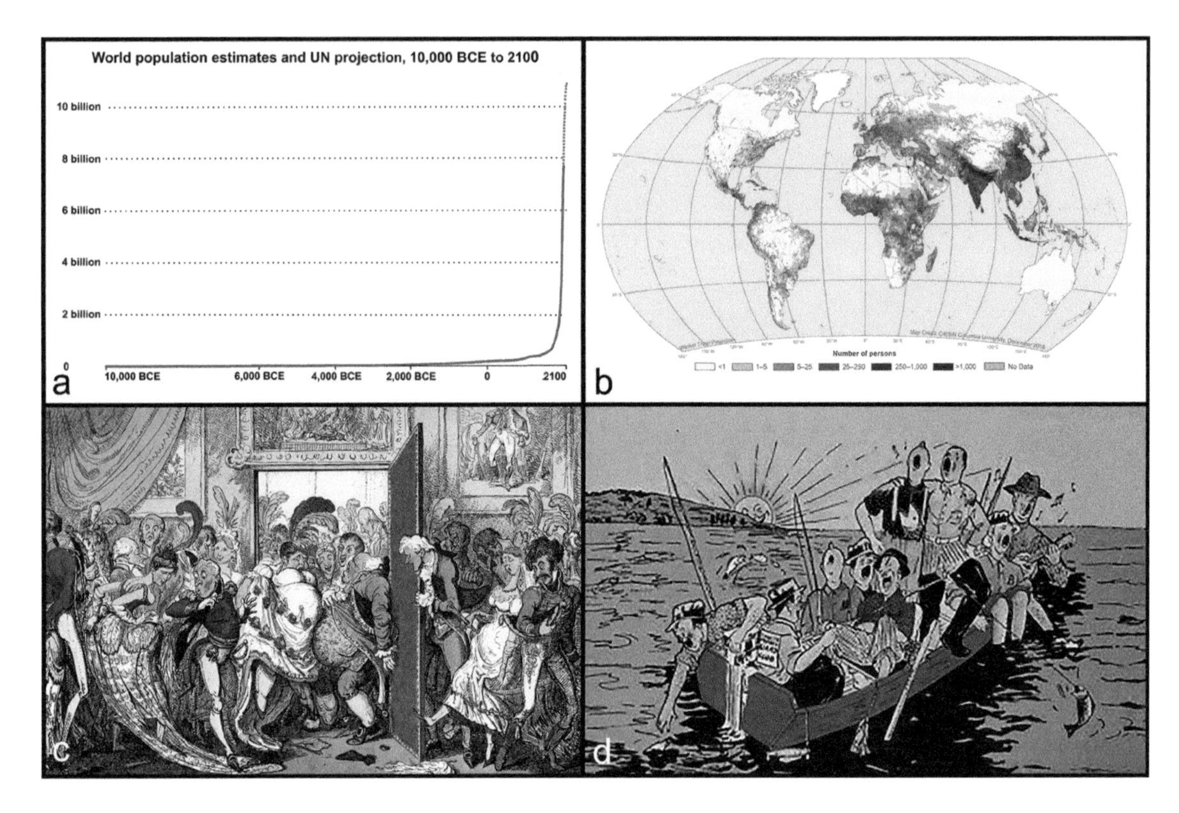

FIGURE 9.12 Scenes summarizing human overpopulation. (a) World human population since 10,000 B.C.E. Source: OurWorldinData.org (CC BY 4.0). (b) Estimates of human population density (numbers of persons per square kilometer) as of 2020. Credit: Trustees Columbia University (New York; CC BY 4.0). (c) An overcrowded drawing room (1818) by British caricaturist George Cruikshank (1792–1878), illustrating the human tendency to overpopulate a limited space. (d) Vintage postcard (public domain) showing an overcrowded fishing boat – a frequent metaphor for how people are overpopulating the planet.

detriment of everyone (Figure 9.13). The phrase was originated by ecologist Garrett Hardin (1968) and has since been applied in economics, ecology, and other fields. Political scientist Elinor Ostrom demonstrated that the tragedy of the commons often was avoided by locals finding sustainable equitable solutions (Aligica 2010). Nevertheless, overexploitation of shared resources is a widespread problem that can only be solved if those concerned honorably agree to limit usage to sustainable levels. In essence, the current ecological crises facing planet Earth – for climate, biodiversity, and other threatened resources – are tragedies of the commons on a grand scale, which urgently require workable international solutions. The human tendencies to postpone or minimize substantial personal sacrifice and to gamble that others will work toward solutions are major roadblocks to progress. Unfortunately in international relations, passing the buck – refusing to confront a growing threat in the hope that another player will – is all too frequent.

THE STATE OF INTERNATIONAL AGREEMENTS TO PROTECT HARVESTED SPECIES

Illegal overharvesting of wild species is a principal threat to their continued existence. National and international efforts are critical since species and their supporting ecosystems often occur in several nations, and those involved in the illicit wildlife trade have no respect for borders. The oceans are largely open to all users, and so are especially difficult to regulate. Among nations, regulations vary according to local policies. Enforcement of control measures also depends on availability of resources, which are often limited in countries where the needs are most pressing. In rich countries, where illicit exports are predominantly directed, enforcement of import regulations is strong (Figure 9.14). In poor countries, poaching of animals to supply the illicit wildlife and bushmeat trades, as well as excessive logging and fishing, is a very serious threat to the survival of many species (Figure 9.15). Similarly illicit harvesting of plants, fungi, and invertebrates also endangers biodiversity. In the following sections, see the information on the CITES organization which addresses trade in endangered species, and note the comments regarding difficulties in reaching agreements on regulating harvesting marine species.

INTERNATIONAL ENVIRONMENTAL AGREEMENTS BEARING ON BIODIVERSITY CONSERVATION

There are numerous international agreements ('conventions' or 'treaties') addressing conservation of the world's species. Several deal with ecosystems (e.g., the United Nations Forum on Forests; the Convention on Wetlands

FIGURE 9.13 A contrast of a resource (a) being employed sustainably and (b) being overexploited. Prepared by B. Brookes.

of International Importance) and climate issues (e.g., the United Nations Framework Convention on Climate Change; the United Nations Convention to Combat Desertification), which bear indirectly on species. Also, some other international agreements conserve sites that support species (e.g., the World Heritage Convention). Often agreements have subsidiary economic motives, particularly protecting resources for sustainable harvesting and sharing the benefits. Agreements are usually at least partly administered by organizations, and sometimes their conservation activities are more important than the agreements. Compacts to control fishing on the High Seas have had limited success (see Oda 2003 for analysis of the 'Convention on Fishing and Conservation of Living Resources of the High Seas' and Bianca et al. 2021 regarding the newest initiative). The most important agreements directly regulating the welfare of species are noted in the following.

Convention on Biological Diversity

This agreement was established in Nairobi in 1992. It is governed by the 'Conference of the Parties' (COP; meetings occur approximately every 2 years), which includes all of the approximately 200 country signatories. In addition to addressing conservation of biological diversity, there are two additional goals: the sustainable use of its components, and the fair and equitable sharing of benefits arising from genetic resources. The COP is a complex organization with

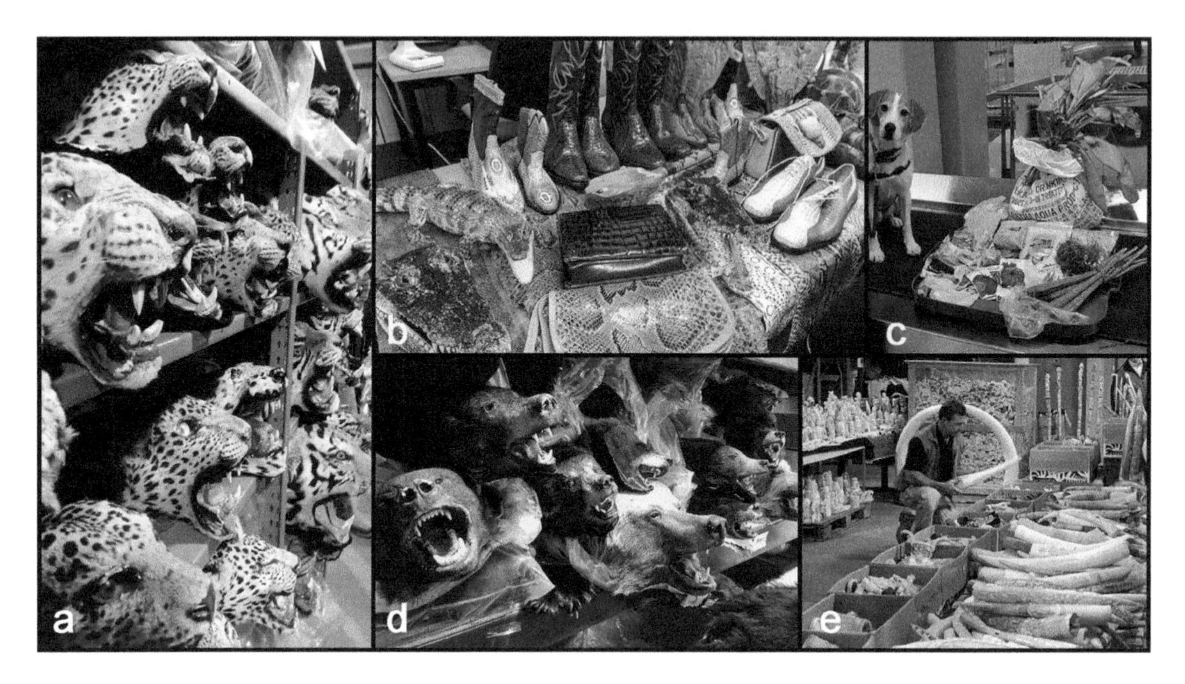

FIGURE 9.14 Illegally imported species seized by customs. (a) Illicit leopard heads seized by U.S. Customs. Photo Credit: Ryan Moehring, U.S. Fish and Wildlife Service. (b) Crocodile and alligator hide products confiscated by British Revenue & Customs. Photo by The Wub (CC BY SA 4.0). (c) Illegal plant items revealed by a detection dog. Photo (public domain) by U.S. Customs and Border Protection. (d) Bear heads seized by U.S. Customs. Photo Credit: Ryan Moehring/U.S. Fish and Wildlife Service. (e) Ivory seized by U.S. Customs. Photo credit (public domain): Gavin Shire, U.S. Fish and Wildlife Service.

FIGURE 9.15 Examples of unjustified harvesting of species. (a) White-fronted brown lemur (*Eulemur albifrons*) killed in northeast Madagascar for bushmeat. Photo by 'Does not wish to be named for safety reasons' (CC BY SA 3.0). (b) Illegal logging in the Amazon basin of Brazil. Photo by Brazilian Environmental and Renewable Natural Resources Institute (CC BY 2.0). (c) Overfishing: an estimated 360 t of Chilean jack mackerel (*Trachurus murphyi*) in a Chilean purse seiner near Peru. Photo (public domain) by C. Ortiz Rojas, U.S. National Oceanic and Atmospheric Administration.

very extensive activities. Countries are required to prepare 'National Biodiversity Strategies' for the conservation and sustainable use of biological diversity, and most participants have done so. Supplementary 'protocols' expand the activities of the convention – for example, the 2010 'Nagoya Protocol on Access to Genetic Resources and the Fair and Equitable Sharing of Benefits Arising from their Utilization to the Convention on Biological Diversity.' (Secretariat of the Convention on Biological Diversity 2011). Information source: https://www.cbd.int/

Convention on International Trade in Endangered Species of Wild Fauna and Flora (CITES)

This agreement protects endangered wild plants and animals by regulating international trade of wild fauna and flora through a system of permits and certificates. More than 180 'parties' (nations and regional organizations) have agreed to be bound by the convention. The Convention entered into force in 1975. Information source: https://cites.org/eng

Convention on the Conservation of Migratory Species of Wild Animals

Migratory species are those that move between breeding and non-breeding areas, and are difficult to manage because they often roam among countries (Miller et al. 2018). This accord aims to conserve terrestrial, aquatic, and avian migratory species throughout their range. The organization is based within the United Nations. It entered into force in 1982 and now has over 130 members. Information source: https://www.cms.int/

International Treaty on Plant Genetic Resources for Food and Agriculture

The treaty aims to conserve, exchange, and sustainably use the world's plant genetic resources related to food and agriculture. Other aims include promotion of fair and equitable benefit sharing, and recognition of farmers' rights. The treaty is sponsored by the Food and Agriculture Organization of the United Nations. It entered into force in 2004 and now has about 150 parties. Information source: https://www.fao.org/plant-treaty/en/

International Plant Protection Convention

This agreement is intended to limit the introduction and spread of pests of plants and plant products. The Convention includes protection of natural flora, although stress is on plants and plant products moving in international trade. It is overseen by the United Nations Food and Agriculture Organization. It was established in 1951 and now has over 180 parties. Information source: https://www.ippc.int/en/structure/

KEY ORGANIZATIONS SUPPORTING SPECIES CONSERVATION

There are numerous organizations concerned with species conservation and more generally with environmental preservation. These include charities, trusts, government organizations, and non-governmental organizations, and can be local, regional, national, or global. Many groups specialize in protests (e.g., Figure 9.16). The most significant international organizations that actually conduct conservation work are cited in the following (there are dozens more providing valuable efforts, but too many to include). For discussion on the nature of organizations supporting biodiversity and the criteria they employ, see Gordon et al. (2005) and Goud Collins (2016). Criticism of these organizations frequently is based on the biasing potential of accepting funding from corporations; on the bias associated with funding originating mostly from rich Western nations; and on the criteria employed for choosing and funding projects.

FIGURE 9.16 Demonstrations protesting extinction risk by Extinction Rebellion. This global environmental organization, named for the Anthropocene Extinction, was founded in 2018, and addresses risks to climate, biodiversity, and ecology. (a) Placard warning of potential human extinction. Photo (public domain) by Sebastian Dooris. (b) Parade in Melbourne, Australia. Photo by John Englart (CC BY SA 2.0). (c) Demonstration titled 'Dance to Extinction' protesting lumber harvest in old-growth forests of Victoria, Australia. Photo by Matt Hrkac (CC BY 2.0).

International Union for Conservation of Nature and Natural Resources (IUCN)

The IUCN is a non-governmental organization, dating to 1948, and is the world's oldest global environmental organization. It is centered in Gland, Switzerland, with members representing governments and civil society organizations. The goals include nature conservation and promotion of sustainable use of natural resources. Its policy regarding how business interests should respect biodiversity is at https://www.iucn.org/theme/business-and-biodiversity/our-work/business-key-areas-work. With the consultation of thousands of experts (including your author), it prepares the IUCN Red List of Threatened Species (https://www.iucn.org/resources/conservation-tools/iucn-red-list-threatened-species), the world's most comprehensive inventory of the conservation status of species. IUCN also is involved in habitat conservation and produces the 'IUCN Red List of Ecosystems,' an evaluation system for the state of ecosystems that is parallel to the Red List for Threatened Species. Other goals include expanding national parks, other protected areas, and oceans and marine habitats. IUCN cooperates with many other conservation organizations and also with some corporations. Information source: https://www.iucn.org/

Commission on Genetic Resources for Food and Agriculture

This organization is part of the Food and Agriculture Organization of the United Nations. It is concerned with the management of biodiversity of relevance to food and agriculture. It was established in 1983. The membership includes about 180 countries and the European Union. Information source: https://www.fao.org/cgrfa/en/

World Wide Fund for Nature

This NGO was formerly named the World Wildlife Fund, which remains its name in Canada and the U.S. It addresses biodiversity loss, unsustainable use of natural resources, and allied issues. It was founded in 1961 and is headquartered in Gland, Switzerland. It is the world's largest conservation organization, with millions of supporters worldwide, and thousands of projects underway. The allied funding foundation is supported financially by individuals, governments, and corporations. Information source: https://help.worldwildlife.org/hc/en-us/articles/360008012153-World-Wide-Fund-for-Nature

'DE-EXTINCTION' – AN UNREALISTIC SOLUTION

'De-extinction' is the process of resurrecting extinct species, the term sometimes expanded to include creating an organism which greatly resembles an extinct species. Sometimes re-introduction to a natural ecosystem is also considered part of the process (Novak 2018). Attention to the concept expanded widely with Michael Crichton's 1990 science fiction novel 'Jurassic Park.' The goal of de-extinction is to utilize genetic engineering (or 'synthetic biology') to recreate species that have gone extinct. The process depends on using DNA or RNA (at least fragments) that is in good condition, which usually requires the availability of frozen specimens such as from the wooly mammoth (Figure 9.17), or from museum-quality specimens that went

FIGURE 9.17 Woolly mammoth (*Mammuthus primigenius*), the species most frequently considered for de-extinction. (a) A mural showing a herd of woolly mammoths. Illustration (public domain) painted in 1916 by wildlife artist Charles R. Knight (1874–1953). (b) A baby wooly mammoth preserved for 35,000 years in a frozen lake in Russia and maintained as shown in this photo in a cold exhibition chamber. Well-preserved tissues in such specimens provide DNA that might be used to produce a modern wooly mammoth clone. Photo by Cyclonaut (CC BY SA 4.0). (c) Diagram showing how to resurrect a wooly mammoth. Illustration by France-Presse (CC BY SA 4.0).

extinct very recently, such as the passenger pigeon and the Tasmanian devil. De-extinction for the most part is not yet possible because of current technological limitations and presents critical ethical and conservation issues (Seddon and King 2019; Preston 2021). Currently, it appears that 'facsimiles,' 'proxies,' or 'functional replacements' rather than exact copies of extinct species is usually the best that can be expected (Shapiro 2017). In the context of this book, de-extinction is important because many naively conclude that so long as species can be brought back from the dead their current welfare is unimportant (Banks and Hochuli 2017). Also, the species that are being considered for de-extinction – usually animals that are charismatic, mammalian (Adams 2017), terrestrial, predatory, and/or ones that are culturally important (Turner 2017) – dramatically reveal the preferences toward species that are the antithesis of those being defended in this volume. However, it is possible to prioritize species to be resurrected on the basis of their potential ecological/conservation values (Iacona et al. 2017). For example, keystone species might be capable of restoring ecosystems and therefore should be leading candidates. Selbach et al. (2018) argue that parasites of resurrected species should also be resurrected, but such a prospect seems unlikely given the prejudice against them.

There are (at least theoretical) moral issues associated with de-extinction. Are we obligated to bring back species that humans drove to extinction? Could resurrected 'necrofauna' or 'necroflora' be patented? Should extinct humans, such as Neanderthals, be brought back? Most importantly, should individuals or corporations be allowed to 'tinker' so profoundly with nature?

WHY ISN'T THE PUBLIC ALARMED ABOUT BIODIVERSITY EXTINCTION?

The most pressing issue facing humankind is the expanding degradation of our planet and the consequent growing threat to the welfare of the world's species, including ourselves (Food and Agriculture Organization 2019a, b; IPBES 2019). Human-caused climate change ('global warming'), perhaps the most obvious symptom of the problem, has captured the interest of many because it demonstrably has the potential to harm health and economic welfare. This is good for biodiversity (the animate part of the environment), because concern for the inanimate environment is likely to improve the habitats that sustain living things, and climate change is one of the great threats to biodiversity. But there are other, debatably greater threats to biodiversity, particularly habitat degradation and elimination, and overexploitation of natural resources (Textbox 9.2). Conservationists have bemoaned the fact that while climate remediation has been identified as an issue of predominant concern that has attracted international political and financial support, by contrast support for species conservation is alarmingly limited (Textbox 9.3). Numerous scientists have analyzed the frightening rate of extinction of species (IPBES 2019), and the substantial associated costs to human society, but quite unlike the climate change threat, both public and

TEXTBOX 9.2 THE IMPORTANCE OF CLIMATE CHANGE FOR BIODIVERSITY SHOULD NOT SUBORDINATE THE RELEVANCE OF OTHER MAJOR DETERMINANTS

During the past 25 years, more than 85% of published scientific studies that have explored the responses of biodiversity to future environmental changes – i.e., the so-called 'biodiversity scenarios' – have relied on simulations that only project changes in climatic conditions. Although we do not intend to downplay the future impacts of climate change on biodiversity, we are concerned that too strong a focus on climate change in biodiversity scenarios may discourage scientists from examining the effects of other, at least equally important issues. For instance, the destruction and modification of natural habitats resulting from land-use change are among the most important and immediate threats to biodiversity.

—*Titeux et al. (2016)*

TEXTBOX 9.3 SUPPORT FOR SPECIES CONSERVATION IS ALARMINGLY LIMITED

The negative impact of biodiversity loss on ecosystem functioning and services, and ultimately on human well-being, has been unequivocally established; however, despite all efforts, biodiversity is still declining worldwide. It is widely accepted that biodiversity awareness is crucial for its conservation. Nevertheless, after many initiatives to alert society about the consequences of losing biodiversity, biodiversity loss is still perceived as a minor environmental risk compared to others such as climate change.

—*Arroz et al. 2016*

governmental support for biodiversity conservation have been disappointing (Legagneux et al. 2018). Moreover, what support there is, is decidedly slanted toward a small, privileged set of species, especially the 'charismatic megafauna' including such stars of the animal kingdom as whales, giant pandas, elephants, lions, and tigers.

On the whole, the majority of the public is currently only mildly concerned with and largely indifferent to biodiversity loss (Platt 2019a). Just as 'climate change denial' (and 'global warming denial') is based on ignorance and the narrow self-interest of parties (including politicians) benefiting from the status quo, so 'biodiversity extinction denial' has appeared (Platt 2019b; Lees et al. 2020; cf. Sinatra and Hofe 2021; Figure 9.18). There are several factors making it difficult to convince the public of the seriousness of species losses. People predominantly live in human-modified landscapes and so are disconnected from nature and its values (Louv 2005). The biodiversity/bioresources crisis is apocalyptic in nature, but not cataclysmic, i.e., a huge change is occurring, but not suddenly. Habitats and species are disappearing incrementally, but out of sight of most people, so it is easy for many to deny that there is a significant issue, and to give priority to other problems. The biodiversity problem can be alleviated somewhat by relatively inexpensive measures, but on the whole massive expenditures are needed, and neither the public nor politicians are currently prepared for the required sacrifices. Environmentally conscious people now realize that remediation and conservation of ecosystems is necessary, as well as a future based on sustainable use of resources. But delivering this message to the public is very challenging.

WHY IS THE CLIMATE CHANGE DISASTER BEING MUCH MORE FORCEFULLY ADDRESSED THAN THE BIODIVERSITY EXTINCTION DISASTER?

Human-caused climate change is a serious development justly receiving public attention (Figure 9.19), but so is biodiversity extinction. Both climate change and the biodiversity/bioresources crisis are international issues, requiring governments to agree on shared funding, and this is challenging. Nevertheless, the climate change issue is currently receiving extensive domestic and international attention for two reasons: (1) extreme weather events in recent times are causing disastrous damage to property and people, making it evident that remedial actions are essential; (2) concerned individuals and organizations have educated the public about its seriousness. Damage from extreme weather is evident to everyone, but damage to wild species is evident primarily to those who harvest sea food stocks and timber resources, and they represent a small minority. The ecological services provided by species are essential to maintenance of the planet but are scarcely appreciated by most people. Most economists term these contributions as 'non-market' values (Gowdy 1973), whereas the detrimental costs of weather disasters are obviously financial. Accordingly, the public is much less concerned with biodiversity extinction than with climate change. Educational initiatives are clearly required to counter the impending disastrous consequences of biodiversity loss to the planet and ourselves.

FIGURE 9.18 Biodiversity extinction denial. Prepared by B. Brookes.

FIGURE 9.19 Portrayals of climate change and global warming. (a) A collage of typical climate and weather-related concerns (left to right): drought, hurricane, wildfire, glacial ice melting. Credit: U.S. National Oceanic and Atmospheric Administration (public domain). (b–d) Shutterstock images; contributors: (b) SquishyDoom, (c) Tote, (d) YummyBuum.

10 Bias and Prejudice in Species Conservation

INTRODUCTION

'Biodiversity' is sometimes defined as having three components: genetic variation within species, species, and 'landscapes' (the physical environments occupied). Habitats are components of ecosystems (an ecosystem is a combination of interacting species in a shared environment), which often overlap, and are often somewhat arbitrarily defined. These distinctions are rather academic and can obscure the purpose of nature conservation. Biological conservation addresses the intertwined issues of (1) reduction and elimination of species and (2) degradation of their supporting environment (including their habit, ecosystem, and landscape). The main cause of biodiversity loss is destruction of the environment of species, so the two issues are usually inseparable. The ideal solution is restoration of the supporting habitat of threatened species in nature, but if not feasible it may be necessary to maintain them in artificial settings.

Conservation science is a complex field, and there are differences of opinion, outlined in this chapter, on what criteria are appropriate for deciding on which species and habitats should receive priority. However, the viewpoint is documented here that not science, but preferences and/or prejudices chiefly determine conservation priorities. As phrased by Castillo-Huitrón et al. (2020), understanding human emotions is key to conservation support for wildlife management (cf. Steg and de Groot 2018). As phrased by Fančovičová et al. (2022), 'Aesthetic judgment plays a role in conservation science because it significantly influences the willingness to protect animals by means of emotional processes. Species with low aesthetic value generally activate the emotion of fear or disgust and receive low conservation support. In contrast, species with high aesthetic value receive high conservation support from the public and are more frequently kept in zoos than less attractive species. In addition, there are considerable cross-cultural similarities in aesthetic preferences for animals, suggesting that the psychological processes underlying aesthetic preference are universal and findings from one culture can be generalized to other cultures.'

EXTINCTION RISK AS A CRITERION FOR THE RELATIVE IMPORTANCE OF CONSERVATION OF SPECIES

If all species were considered equally deserving, the IUCN Red List (described in Chapter 9), based on extinction risk, could be used to prioritize relative allocation of conservation resources (Keller and Bollman 2004). However, the degree of threat to most taxonomic groups of the world's approximately 2 million named species (there may be as many as 100 million unnamed, undiscovered species) is very imperfectly known. As noted in Chapter 9, only a small percentage of species – primarily vertebrates and flowering plants – have been evaluated.

There are also philosophical disagreements about the relative attention that should be paid to species that are currently endangered, naturally rare in nature, or endemic (confined to limited geographical areas, often to specific habitats). Some have reasoned that most such species are 'evolutionary dead ends' which are naturally relatively unsuccessful, therefore not deserving of special attention. By contrast, it has been argued, attention should be put on common (high frequency) species suffering population losses, since they represent the majority of biodiversity (at least with respect to numbers of individuals), and are usually the organisms of primary significance to both humans and the ecology of the planet. Moreover, the traditional emphasis on rare species underemphasizes the importance of preserving the considerable genetic variation within many species, which while not in danger of extinction are undergoing drastic reduction.

Assuming that uncommon species should be the focus of conservation (the majority opinion), it is unclear which species deserve attention. 'Rarity and endemism are often incorporated uncritically as components of conservation status, but do not necessarily equate to vulnerability or threat of extinction' (Stewart and New 2007). Indeed, there are so many rare species that evidence of population decline or threat is increasingly being viewed as a necessary condition for investment in conservation measures.

BIOMASS AS A CRITERION FOR THE RELATIVE IMPORTANCE OF CONSERVATION OF THE TAXONOMIC GROUPS OF LIVING THINGS

The most comprehensive analysis of the relative amounts of living biomass of the world's major groups of species is Bar-On and Phillips (2018; Figure 10.1). Plants make up over 80% of the world's living biomass, bacteria almost 13%, animals collectively constitute only about 0.4%, and humans a mere 0.01%. If rarity is employed as an indicator of the need for conservation, it would seem that priority should be given to animals. However, plants are indispensable for animal survival, and microorganisms are indispensable for energy recycling. Photosynthetic creatures are responsible for almost all energy used by other species,

DOI: 10.1201/9781003412946-10

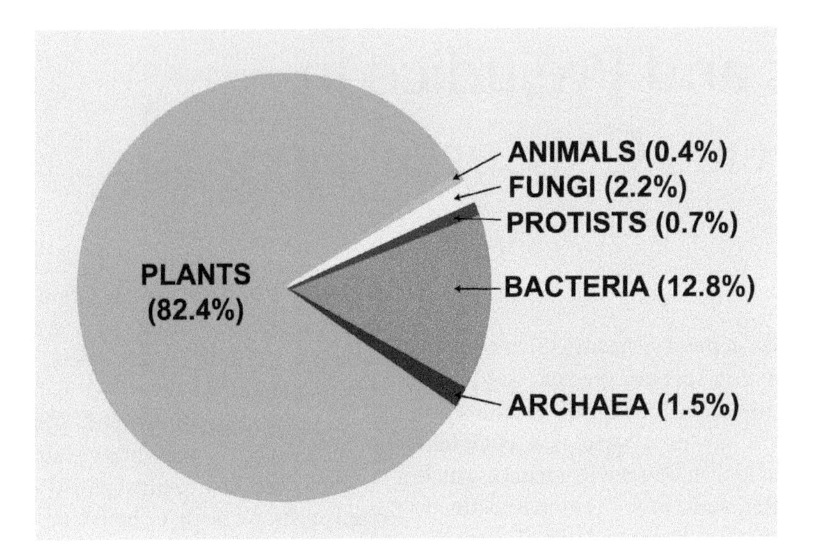

FIGURE 10.1 Relative percentages of the major taxonomic groups that make up the world's biomass. Based on data in Bar-On and Phillips 2018.

and without organisms responsible for energy recycling by decay (fungi, bacteria, etc.), organic material would simply accumulate. The non-animal world represents the absolutely indispensable living things on Earth, because, in the absence of animals in general and humans in particular, other life forms could continue. Biomass is too crude a measure to guide decisions regarding conservation, but it does point out that animals, the form of life of chief concern to most people, represent a tiny proportion of the living world. Moreover, as discussed in this book, although there are millions of animal species, the public is focused on an extremely limited (i.e., biased) selection.

GEOGRAPHY AS A GUIDE TO SPECIES CONSERVATION

Conservation Hotspots

It is generally agreed that biodiversity conservation necessarily considers both species and their habitats, since for most purposes they are inseparable (although many species can be maintained outside of their natural areas). Local ecology is a critical determinant of species adaptation and survival (see World Wildlife Fund's presentation of the planet's ecoregions, at http://www.worldwildlife.org/science/ecoregions/item1847.html). The density of species increases from north to south but is highly localized in some regions of the world. Species at risk are concentrated in some locations, particularly in tropical regions, and since very little of these areas are conserved, it has seemed especially desirable to concentrate on them.

Geographical schemes have been created to prioritize conservation areas (Brooks et al., 2006). Several major conservation organizations, including the three largest non-governmental ones (Conservation International, The Nature Conservancy, and the World Wildlife Fund), use these systems at least in part to allocate funding. The most widely employed biogeographical conservation scheme recognizes 36 'biodiversity hotspots' (Figure 10.2), which collectively comprise just 2.5% of Earth's land surfaces, but are thought to contain more than half of the world's species (Myers et al. 2000; Lamoreux et al. 2006; Pimm et al. 2014; Habel et al. 2019; Trew and Maclean 2021). Most of the hotspots are located in tropical developing countries challenged by overpopulation and poverty, and less than 10% of the original vegetation survives.

Conservation Coldspots

Analyses of the world's areas of high concentrations of endemic species are the best basis currently available for the identification of conservation priorities but have been criticized for being biased in species representation or unrepresentative of particularly unique or important species, or of many species that are simply restricted in distribution (Shrestha et al. 2019). In particular, it has been argued (Kareiva and Marvier 2003) that attention primarily to biodiversity-rich hotspots in mostly tropical regions is at the cost of reducing conservation of species in other regions, colorfully labeled 'coldspots' (the polar bear is a good example of a species not in a hotspot but deserving attention). Moreover, biodiversity-rich hotspots are often in countries where poor governance and corruption waste funds intended for conservation (Eklund et al. 2011).

Biased Conservation of Species Resulting from the Geography of Funding Sources and Targets

'Conservation geography' is 'the subfield of conservation science that studies where, when, and what conservation actions should be implemented in order to mitigate threats and promote sustainable people–nature interactions' (Di

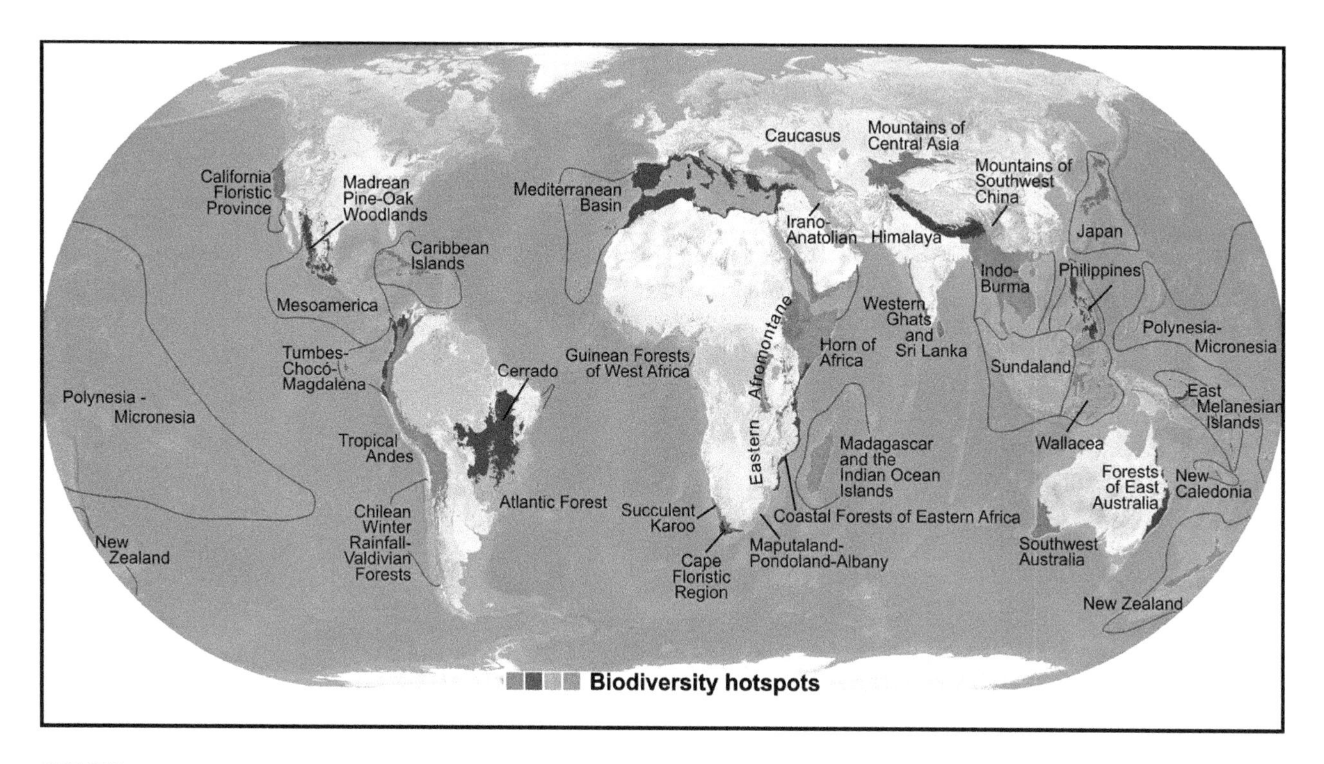

FIGURE 10.2 'Biodiversity hotspots,' areas with very high concentrations of species, are the only natural homes of numerous plant and animal species and are heavily threatened by habitat loss and degradation. The perimeters of the hotspot areas are shown as red lines. (The areas often include small islands that don't show up on the map but are the residences of rare species.) Source: Conservation International (conservation.org; CC BY SA 4.0).

Minin et al. 2021b). Habitats and biodiversity are most threatened in developing countries, largely in tropical and subtropical areas of the world (Murray et al. 1996). About 80% of the world's biodiversity and 80% of the world's peoples are in developing nations. However, about 85% of the world's money and 95% of the world's scientists and engineers are in the industrialized nations (World Book 1998). Ninety percent of conservation funding is spent by nations within their own boundaries (Brooks et al. 2006), so funding is extremely disproportionate to areas of greatest need, with much more funding directed to temperate areas relatively low in biodiversity than to tropical countries much richer in biodiversity (McClanahan and Rankin 2016; Reed et al. 2020). Poverty, population growth, and international trade policies restrict the resources that poor and developing countries are able to dedicate to conservation, and in some areas, progress is blocked by political instability or political priorities that are inimical to conservation. Capitalism, industrialization, colonization, and other historical developments all share blame, but arguments over who is responsible (i.e., guilty) for the present unsatisfactory state of affairs are not helpful. Necessarily, countries of the temperate region and Western World must assume a disproportionately large share of responsibility for conservation everywhere simply because they have most of the resources necessary for remediation.

The greatest conservation challenges are posed by marine habitats, since the seas are largely an internationally exploited resource that is relatively poorly regulated, and the status of most species is undetermined. Identifying geographical areas deserving protection is difficult (Astudillo-Scalia and de Albuquerque 2020). To date, marine conservation has been very inadequately financed (Douvere 2016). Whales and other marine mammals make very attractive conservation icons, but the problem of slowing the deterioration of the seas requires very large sacrifices, suggesting that pragmatic conservation goals are best phrased in terms of sustaining stocks of commercial fish and other edible seafood.

SPECIES OF SPECIAL CONSERVATION IMPORTANCE

It is a credo of democratic societies that all people are equal. Nevertheless, for certain purposes, some people are much more important ('more equal') than most. Similarly, in the context of biodiversity conservation, there are two general purposes for which some species have very special significance: (1) certain species are extraordinarily attractive or useful, and so they are very persuasive for obtaining public and political support for conservation (Figure 10.3a and b); (2) certain species have (or at least are alleged to have) extraordinary importance ecologically – either as dominant controllers of the orderly functioning of ecosystems – i.e., their extinction would result in far greater damage than the extinction of most other species; (Figure 10.3c and d). Some species also seem especially useful as indicators of the

FIGURE 10.3 Examples of how particular species are especially important for the conservation of many other species. (a) A flagship species important for fund raising. Because pandas are so attractive, they are widely employed to generate contributions for conservation. (b) Tourism attraction. Whale watching is the most popular form of biodiversity tourism and generates sympathy for conservation of marine environments and their many species. (c) Community protection by a species with flagship and keystone functions. Like other large predators, tigers require extensive hunting areas to survive, and accordingly tiger preserves protect the many other cohabiting creatures (even if some are eaten by the tigers). (d) Food and habitat provision. Oak trees, like the central tree here, provide food (especially acorns) for thousands of other species as well as habitat (for nesting) for many others.

health of ecosystems (like canaries in a coal mine). The term 'surrogate species' has frequently been applied to all of these species because they are special representatives of their ecosystems. Some ecologists have complained that money-generating species and ecologically influential species have been confused in the past, leading to misunderstanding. However, as noted in Table 10.1 (part C), some surrogate species concepts deliberately combine the two kinds.

FUNDRAISING ALTERNATIVES FOR SPECIES CONSERVATION: COMMERCIAL VALUES VS. ECOLOGICAL VALUES VS. POPULAR APPEAL

Table 8.1 presented a list of eight pragmatic, emotional, and philosophical values that have been attributed to biodiversity, and these represent criteria for crafting a public relations/fundraising campaign on behalf of species conservation. There have been attempts to quantify the value of preserving particular species on the basis of perceived (subjective) importance, measured economic importance, and ecological importance (e.g., Bulte and Van Kooten 1999), but this theoretical approach has had little practical application to date. Financial support can come from government, philanthropists, non-government organizations, the private sector, and the public. In practice, and in common with marketing techniques generally, two kinds of values have been emphasized to date in species conservation campaigns: Rational, economic considerations based on utility (particularly harvestable materials and ecological services), and relatively subjective satisfaction based on human values, prejudices, instincts, or sensations (which may be termed aesthetic, moral, spiritual, empathetic, and/or compassionate). Both categories are useful, but on the whole are applicable to different audiences.

TABLE 10.1
Terminology for Species with Special Status for Conservation Purposes.

A. Sociopolitical: Conservation and Fundraising Promotion or Management

CHARISMATIC SPECIES: Species that attract attention, normally positive because they possess traits that are admired or at least respected (sometimes feared). Some have argued that species that attract negative attention (because they are repellent, disgusting, harmful, or at least reputedly so) are also charismatic (for the purpose of their control), but only attractively charismatic species are useful as conservation icons.

EMBLEMATIC SPECIES: A relatively infrequent phrase that can be used to refer to species employed as emblems of political regions or causes.

FLAGSHIP SPECIES: A species selected as a persuasive representative to attract attention and funding for a given habitat or environmental issue, campaign, or cause.

ICONIC SPECIES: A well-known species for which there is widespread concern for its conservation or at least its welfare.

MANAGEMENT INDICATOR SPECIES: 'Any species, groups of species, or species habitat elements selected to focus management attention for the purpose of resource production, population recovery, or maintenance of population viability of ecosystem diversity' (United States Department of Agriculture Forest Service 1984).

MARQUEE SPECIES: (by extension from a definition of 'marquee plant' as defined by the EarthScholars Research Group at the 2001 annual meeting of the Botanical Society of America): A species that strongly attracts the public's attention and invites its direct observation, that during some or all of its life cycle is capable of drawing a crowd at a public exhibition, and that may serve as a portal to public understanding of species.

POSTER SPECIES: A relatively informal phrase, referring to a species that is very attractive or interesting and so is useful for attracting attention to a subject.

B. Important Ecological or Indicator Role

APEX SPECIES: Usually a reference to animals: the top consuming animal or animals ('apex consumers,' including herbivores, omnivores, and/or carnivores) or the top predator or predators ('apex predators': Carnivores or sometimes omnivores) in an ecosystem; the phrase has also been used more generally to refer to dominant species (plants ('apex producers'), animals, fungi, or microorganisms), i.e., those that are predominant in an area.

DOMINANT SPECIES: The most frequent species in an area, or the species that constitutes the most biomass in an area. This concept is confusable with food chain rank among carnivores, where the apex predator may be uncommon but nevertheless is the 'king of the jungle.' Relatively infrequent but large herbivores (like elephants) may also 'dominate' an area in the sense that they exert tremendous control over other species.

FOCAL SPECIES: A species targeted for analysis because it is especially sensitive to a habitat's or ecosystem's environmental conditions that are being altered or threatened. Several focal species better represent an ecosystem than a single species (see Lambeck 1997).

FOUNDATION SPECIES: A dominant, primary producer in an ecosystem (provides much or most of the food, and perhaps also habitat, for animals).

INDICATOR SPECIES: a species that indicates the degree of normal functioning of an ecosystem. Good indicator species are sensitive to environmental changes (like a canary in a coal mine) or are reflective of the presence of many associated species.

KEYSTONE SPECIES: A native species that plays an essential role in a habitat or ecosystem, without which the functioning, structure, or productivity of the habitat or ecosystem would be degraded with consequent harm to other species. (A keystone in an arch is just one of many stones, but without it the arch collapses. By analogy, keystone species are claimed to be much more important than their limited biomass would suggest.) Human beings have been characterized as the ultimate keystone species (O'Neill and Kahn 2000). Sometimes introduced but established alien invaders have also been considered to be keystone species (which modifies the concept of a keystone species from one that prevents 'degradation' of an ecosystem to one that (for good or evil) is simply a principal determinant of an ecosystem's function).

UMBRELLA SPECIES: A wide-ranging species (usually a large vertebrate) that represents the ecological requirements of many other species in its range, and so when protected, protection is also provided to the other species.

C. Combination of A and B, above

PRIORITY SPECIES: (World Wildlife Federation definition) A flagship or keystone species chosen to represent and important to the welfare of an ecological region.

SURROGATE SPECIES[a]: A species chosen to represent (1) the ecological welfare of all of the species in an ecosystem, and/or (2) biodiversity conservation issues.

[a] Surrogate species have been defined in various ways, and considerable overlap is possible among concepts; see Caro and O'Doherty (1999), Caro and Girling (2010), and Barua (2011) for critiques of several of these surrogate species definitions. There are no rules or registration procedures for ranking a species as a surrogate, so anyone can advocate this status for any species. A curious byproduct of the growing recognition of surrogate species over the last several decades is the attempt by some biologists to have the particular species that they are studying (sometimes a species that has just been recognized) elevated to the status of surrogate species, since this inflates the importance of both the species in question and the scientist.

Important commercial resource species such as the leading harvested seafood species contribute greatly to the economy of many countries. Similarly, important recreational species such as game ducks are vital to some outdoor industries. In many countries, harvesting of wild plants and animals remains quite significant. In these circumstances there is substantial business and political support for targeted conservation of the economic species, and since this requires maintenance of their habitats, associated species gain protection.

As noted in Chapter 7, species and sites of symbolic, religious, spiritual, or political importance in particular jurisdictions are protected, at least locally.

As emphasized in this book, most people are indifferent, indeed ignorant of the majority of species. The argument that these species and their supporting ecosystems are important to the health of the world is at best received politely by most of the general public, albeit it is accepted by academia, many NGOs, and informed decision makers in government. But focus on the natural appeal of glamorous species has proven to be key to enlisting public support. Our natural empathy and compassion for other people extend very strongly to a select group of animals, and this is the basis for conservation appeals based on these particular 'poster animals.' The underlying motive is frequently to preserve the habitat of the targeted animals, so that numerous other species within the same habitats will also benefit. However, the other animals usually do not have much appeal, so featuring them tends to be counterproductive. The values and limitations of this approach are examined in the following subsections.

Public Relations and Fundraising Value of Focus on Selected Species

In practice, advertising posters for conservation causes typically display highly attractive, charismatic, species that generate sympathy and financial support (Figure 10.4). Home et al. (2009) noted: 'The use of a flagship species brings at least two advantages to conservation organizations. Firstly, it is easier to present the organization as a specialized and coherent organization. Secondly, it creates fewer mental barriers when carrying out fundraising activities, because donors can attach their support to a tangible subject that gives substance to the conservation concept; the simpler the message, the higher the willingness to donate.'

Indeed, in these times of very limited resources, it is important to demonstrate that funds dedicated to biodiversity conservation are being used effectively. Regardless of whether or not support for a given high-priority species simultaneously improves habitats and the lot of other species, it is critical that the public sees short-term results, and it is relatively easy to increase the size of a selected

FIGURE 10.4 Posters for iconic species that generate sympathy for conservation. (a–e) Source: Shutterstock. Contributors: (a) Tatyana Komtsyan, (b) Vecter Art, (c) Kudryashka, (d) Charactoon design, (e) Vector Tradition. (f): BOSFoundation (CC BY SA 4.0). Modified by B. Brookes.

population. This provides validation of single-species support in the eyes of sponsoring governments, organizations, and the public, although many scientists believe the approach is short-sighted.

Appeals for funds to conserve foreign species are necessarily concentrated in wealthy Western nations. For this purpose, the species that are brought to the attention of the Western public must of course be familiar and admired. Threatened exotic mammals of Africa and Asia, such as lions, tigers, and elephants, are so well-known and respected that they are ideal representatives for the purpose. Using threatened exotic plants to exemplify the need for conservation is a more difficult exercise because most people do not appreciate the geographical origins of plants and, except for large trees, high emotional attachment is not present.

Ecological Values of Focus on Selected Species

The World Wildlife Fund Global Species Programme justified focus on selected species in its definition of flagship species: Those that 'act as ambassadors for a natural habitat, issue, campaign, or environmental cause. By focusing on, and achieving conservation of that species, the status of many other species which share its habitat – or which are threatened by the same threats – may also be improved.' McGowan et al. (2020), for example, contend that flagship species provide substantial protection to neighboring species (i.e., in the ecosystem).

Contradictory Evidence of the Ecological Value of Focus on Selected Species

Large animals occupy large areas, and this fact alone tends to make them useful for conservation of large areas. The degree to which the large charismatic animals actually reflect the ecological requirements of associated species, however, seems to be quite variable, and hence these animals are often not good representatives of other threatened species (e.g., Beazley and Cardinal 2004; Ray 2005). The domination of the conservation world by the charismatic megafauna has led to complaints that other groups of species are being ignored, for example plants (Marinelli 2005), fungi (Moore et al. 2001), and insects (Dunn 2005).

Emphasizing ecological and conservation concerns for single species (the 'fine-filter' approach) contrasts with 'coarse-filter' emphasis on all of the species and the detailed ecology of habitats and communities. Scientifically, the latter strategy is preferable because it directly addresses all possible issues of habitats and species conservation. However, resource limitations severely restrict detailed studies of all species in given areas. Study of a single species of course provides invaluable information for that species, but how representative it is of other species and their ecology is open to question. It is often alleged that highlighting certain species is justified because they are 'surrogates,' fairly representing the ecological and conservation needs of other species. There is variable support for this and the associated conclusion that benefits 'trickle down' from the species of focus to other species, some scientists concluding that these ideas are simply erroneous (e.g., Bond 2001; Caro et al. 2004), undetermined (e.g., Simberloff 1998; Lu et al. 2001), valid (e.g., Fleishman et al. 2000), or that there are sometimes benefits from studying single or a few species in an ecosystem, the results sometimes indicating important information for some situations, but limited benefits in other situations (e.g. Andelman and Fagan 2000; Lindenmayer et al. 2002; Roberge and Angelstam 2004; Branton and Richardson 2010; Isasi-Catala 2011). What is particularly disturbing is that 'conservation networks built around surrogate species may fail to capture rare, endangered, or endemic species' (Favreau et al. 2006). Severns and Moldenke (2010) provide an example of how stressing conservation of one focal species actually harmed another threatened species. Sitas et al. (2009) examined the allocation of conservation resources to about 700 threatened mammals and 100 critically endangered amphibian species and provided a particularly discouraging conclusion: 'Our results provide evidence of the strong biases in global conservation attention. We find that most threatened species receive little or no conservation, and that the small number receiving substantial attention is extremely biased. Species most likely to receive conservation attention are those which are well-studied, charismatic and that live in the developed world. Conservation status and evolutionary distinctiveness appear to have little importance in conservation decision-making at the global scale. Most species inhabit the tropics and are both poorly known and uncharismatic. Therefore, the majority of biodiversity is being ignored by current conservation action.'

EXAMPLES OF ANIMAL SUPERSTAR SPECIES THAT HAVE SERVED AS THE STIMULUS TO ESTABLISH CONSERVATION AREAS

National governments have played major roles in establishing parks and reserves in their own countries. These areas are frequently very important for conservation, but the regions designated are (1) often chosen for multiple uses rather than ecological or biodiversity considerations, and (2) usually selected for picturesque and scenic properties, not for habitat significance. Accordingly the conservation importance and degree of protection of both habitats and species are often limited. However, when high-profile animals and plants have been the basis for creating significant conservation areas, the motivation is generally primarily for conservation. Some examples follow.

With no disrespect to lions, tigers have become the kings of conservation reserves. Save-the-tiger campaigns are popular and have attracted considerable funds (Hollywood's highest paid star, Leonardo di Caprio, announced a personal gift of $1,000,000.00 in 2010). In 2011, the governments of the 13 countries that contain tiger populations signed a

FIGURE 10.5 Tiger reserves in India. (a) Idealistic painting of the Pilibhit Tiger Reserve in India. Photo credit: Voiceofpilibhit (CC BY SA 4.0). (b) Tigress and three cubs in Tadoba Andhari Tiger Reserve. Photo by Ajinkya Vishwekar (CC BY SA 4.0). (c) Tiger in Kanha Tiger Reserve. Photo by Davidvraju (CC BY SA 4.0).

declaration of intent to make tiger conservation a top priority and to double tiger numbers by 2022. There certainly is a need for conservation of the world's estimated 3,200 remaining tigers, thought to have numbered about 100,000 a century ago. Tigers are under severe threat by deforestation and poaching for use of body parts in Asian folk medicine. More than half of the world's tigers are in India, where there are several dozen tiger reserves (Figure 10.5).

The monarch butterfly (*Danaus plexippus*; Figure 10.6), probably the world's most familiar butterfly, is native from southern Canada to Central and South America, and is the subject of international conservation efforts. Although many populations are permanently resident in the southern part of the range, most participate in portions of a summer breeding residency in Canada and/or the United States, migrating to, overwintering, and additionally breeding in the southern U.S. (especially California) and Mexico, and returning to the northern areas in the spring. Where the migration route is not extensive, some butterflies make the two-way trip, but most fly only or mostly one way, their progeny completing the return trip. The larvae feed only on the foliage of milkweeds (more than 100 *Asclepias* species in North America), the adults on nectar from various flowers. Populations are under threat particularly because of habitat reduction along the flyways (many milkweed plants have been exterminated by herbicides) and deforestation in the overwintering habitat. The monarch butterfly's migration was recognized by the International Union for the Conservation of Nature as an 'Endangered Phenomenon' in 1983, the first such designation for a biological phenomenon as opposed to a species in the history of international conservation. In 2022, the International Union for the Conservation of Nature added the butterfly to its Red List of threatened species and categorized it as 'endangered.' The Monarch Butterfly Biosphere Reserve in the states of Michoacán and México west of Mexico City was founded in 2006 and is a key site where millions of butterflies overwinter, congregating in giant clusters on fir trees. Illegal logging in the area needs to be controlled for the future welfare of the butterfly. The region is a magnet for butterfly tourism, which may be excessive and has been accused of threatening the insect. The monarch butterfly is the subject of conservation activities by several governmental and non-governmental organizations, a variety of protective legislation, and considerable research. It is a favorite biology subject for schoolchildren, and indeed effectively serves to indoctrinate respect for insects in a way that probably no other species can. It also can be credited with teaching children about nature in a very pleasant way and alerting the public about threats to biodiversity. And it also dramatically illustrates how the beauty and attractive behavior of a species can generate huge backing and consequent conservation activities, which may not be the most effective use of limited funding but nevertheless does represent support that otherwise probably would not have materialized.

FIGURE 10.6 Monarch butterfly (*Danaus plexippus*). (a) Top and bottom views of a male. Photo by Didier Descouens (CC BY SA 4.0). (b) Larva (caterpillar). The head end, with longer tentacles, is at the top. Photo by Ryan Hodnett (CC BY SA 2.0). (c) A swarm in Michigan. Photo (public domain) by G. Niemenen, U.S. Fish and Wildlife Service. (d) A tree in Mexico completely covered by monarch butterflies. Photo by Carlos Adampol Galindo (CC BY SA 2.0).

FIGURE 10.7 Komodo dragon (*Varanus komodoensis*). (a) Photo by T. Vickers, released into the public domain. (b) Tourist group, photo by Dion Hinchcliffe (CC BY SA 2.0). (c) Photo by Wikilianto (CC BY SA 4.0).

The Komodo dragon (*Varanus komodoensis*) is confined to the islands of Komodo National Park and nearby Flores, in Eastern Indonesia. The national park was founded in 1980 specifically to protect the dragon, the world's largest lizard, which grows up to 3 m in length and up to 70 kg in weight. Although established to protect just one species, the area also serves to conserve a variety of terrestrial and marine species and their habitats. The Komodo dragon is a ferocious predator with a frightening reputation – like a kind of terrestrial alligator – and although it occasionally attacks people and livestock, it is greatly valued for attracting a large volume of ecotourism (Figure 10.7).

EXAMPLES OF PLANT SUPERSTAR SPECIES THAT HAVE SERVED AS THE STIMULUS TO ESTABLISH CONSERVATION AREAS

Saguaro cactus (*Carnegiea gigantea*) is native to the Sonoran Desert of extreme southeastern California, southern Arizona, and adjoining northwestern Mexico. This plant is spectacular, the largest cactus in the United States, and probably the most recognized cactus in the world, commonly shown as background in Wild West movies and Roadrunner/Coyote cartoons. The plant grows up to 20 m in height, but very slowly, often only 2.5 cm in a

FIGURE 10.8 Saguaro National Park. (a) Photo by John Fowler (CC BY 2.0). (b) Picturesque sunset. Photo by Jeanmimi 2000 (CC BY SA 2.0). (c) Vintage photo (public domain). Photo credit: Oregon State University special collections.

year. Habitat destruction, vandalism, and 'cactus rustling' for sale of the plant as an ornamental have taken a toll on the iconic plant. To protect it, in 1933 the 337 km^2 Saguaro National Monument (upgraded to a National Park in 1994; Figure 10.8) was established in southeastern Arizona, and it has become a major tourist area. The cactus is protected by the Arizona Native Plant Law. No part of the cactus may be harvested without a permit on public land. On private property, the plants can be destroyed although some cities impose tougher requirements. Special exemption applies to the indigenous Native Americans of the region, who are permitted to carry on their tradition of utilizing the fruit.

Modern domesticated Corn (*Zea mays*) has several wild relatives popularly called 'teosinte' (Figure 10.9). These include *Z. diploperennis* and *Z. perennis*. The latter may well be the most important species ever discovered for improving crops and is estimated to have contributed billions of dollars of value annually to corn (Tyack et al. 2020). With the specific purpose of conserving endangered teosinte in the wild habitat, the 139,000 ha Sierra de Manantlán Biosphere Reserve was created in 1987, one of several preserves in the world established specifically to maintain wild germplasm related to crops (Stolton et al. 2006). Although created to preserve wild relatives of corn, the reserve is a haven for many other species that are endemic to Mexico, including over a thousand plant species and over a dozen mammals. (For additional information, see Small and Cayouette 1992.)

The Coco de mer or double coconut palm (*Lodoicea maldivica*; Figure 10.10) is native to two small islands (of the approximately 115) in the Seychelles. The 'coconuts' of this palm tree are in fact fruits with just one seed, the largest seeds of any plant, weighing up to 32 kg. The plant has proven to be almost impossible to grow anywhere else in the world, and it has become a major tourist attraction of the Seychelles. Fortunately, Seychelles has an excellent record of conservation activities, with almost 50% of the land area devoted to conservation. Notably, Aldabra, the world's largest raised coral atoll, and the unique and endemic Coco de Mer Palm Forest of the Vallée-de-Mai on Praslin Island, are two UNESCO World Heritage Sites. (For additional information, see Figure 5.53 and the associated information.)

SPECIES TRIAGE: SAVING ONLY SELECTED SPECIES FROM EXTINCTION

The limited willingness of society to stem the loss of biodiversity has stimulated the view that hard choices will have to be made about what to salvage. The issue of what species to maintain has been referred to as 'the Noah's

FIGURE 10.9 Wild corn (teosinte), the basis of a biodiversity conservation area, the Sierra de Manantlán reserve, Jalisco, Mexico. (a) Comparison of a teosinte (wild ancestral form of corn) and modern corn. Modern corn is much less branched with much fewer ears (clusters of kernels), which are much larger and lack a hard case around the kernel. Public domain image. Credit: Nicolle Rager Fuller, National Science Foundation. (b) Part of the Sierra de Manantlán reserve, Jalisco, Mexico. Photo by Sheys Peich (CC BY SA 4.0). (c) Comparison of a 'cob' of teosinte and modern corn. Photo by John Doebley (CC BY 2.0). (d) Diorama showing ancient cultivation of corn by Iroquois. Photo (public domain) by New York State Museum.

FIGURE 10.10 Coco de mer (*Lodoicea maldivica*). (a) Painting (public domain) by English artist Marianne North (1830–1890) of coco de mer palm trees growing in a gorge in the island of Praslin, Seychelles. (b) A tree bearing many coco de mer fruits. Photo by Z Thomas (CC BY SA 4.0). (c) Coat of Arms of the Republic of Seychelles showing the exceptional respect to the native biota of the country. Center: A coco de mer tree above a giant tortoise (*Testudo gigantea*). Top: A white tailed tropic bird (*Phaeton lepturus lepturus*). Sides: sail fish (*Istiophorus gladius*).

Ark problem' (Weitzman 1998). Conservation triage (note Figure 10.11) is methodology to most efficiently utilize inadequate resources to maximize species and/or habitat conservation, according to various criteria that have been proposed, usually with risk of extinction being a principal criterion. (Bottrill et al. 2008 is widely considered to be a landmark paper on the subject.) Several other criteria have been proposed, many of these discussed previously (including economic value, ecological value, relative cost, and political feasibility). Another criterion that is commonly included is taxonomic uniqueness, i.e., phylogenetically isolated species, with no near relatives, should be preferentially preserved (Isaac et al. 2007; Faith 2008; Tucker et al. 2012). Based mainly on mammals, some ecologists employ

FIGURE 10.11 'Triage Restaurant.' Conservation triage is the different assignment of effort to conserve species on the basis of one or more criteria. In this case, the pandas are receiving first class service, the frogs are being tolerated, and the skunks are being ignored. Prepared by B. Brookes.

a 'minimum viable population size' of 5,000 adult individuals, below which a species is claimed to be past the point of no return and is doomed to extinction, at least in nature. See Sanderson (2006) for discussion of target numbers for conservation purposes. Flather et al. (2011) have demonstrated that the concept of minimum viable population is of questionable validity. In other words, where there's life there's hope.

The triage concept in principle would seem to be a reasonable or at least realistic way of approaching the biodiversity crisis, but there are difficulties that should be understood in the light of the chief thesis of this review, that conservation is being guided principally by human bias in favor of species with desired physical characteristics.

First, since the concept of conservation triage arose by analogy with medical triage, some clarification of the latter is in order. Emergency medical triage arose to meet the needs of treating injured soldiers on battlefields, where resources are inadequate to treat everyone. In its initial form, triage consisted of grouping patients into those who seem too severely injured to waste resources on; those with relatively minor injuries, and those whose treatment can be postponed for an extended interval; and those in the middle (in terms of gravity of injuries), with treatment given first (and perhaps only) to those in the middle, based on the philosophy that this will maximize benefit to the majority. In modern times, a variety of more sophisticated medical triage systems has been created, sometimes with more than three categories, sometimes with emphasis simply on creating an order of patient treatment priority with de-emphasis on categories.

Inherent in medical triage is a recognition that scarce resources should not be wasted on those who are so severely injured that they are unlikely to recover. The implication that the same philosophy should be applied to species extinction has met considerable opposition, those opposed holding that it is defeatist. Those in favor contend that allowing some species to expire is merely realistic and reflective of current practice, which is mainly accurate, but not true for iconic species as discussed in this presentation. In any event, when a medical patient cannot be saved, the issue is simply that technology is not available for the purpose. By contrast, technology is available to save virtually every species at risk. The analogy with medical triage has succeeded in emphasizing the distasteful ethical conclusion that present limited funding can conserve only a minority of species at risk, and choices need to be made.

In practice, and totally unlike medical triage, there is not one universal or even several sets of criteria that are being widely employed to allocate resources to address the problem of species extinction. Various well-meaning systems have been proposed, but these are necessarily based on mammal, birds, and higher plants, since the majority of species haven't even been discovered. The transfer of the concept of medical triage to conservation has somewhat muddied the simple-to-understand goal of prioritization of attention to species at risk, but it has been very useful in stressing that decision-making is inevitably based on criteria. But in the final analysis, as emphasized in this book, the criteria are mainly reflective of human biases and choices. Consider whether the whooping crane, panda, tiger, or other beloved iconic species, so close to extinction, should be allowed to go extinct, the same way that battlefield triage calls for the most severely injured soldiers to go untreated. Obviously not, and obviously human bias overwhelms all other considerations.

PERMANENT CAPTIVITY AS A MEANS OF SAVING SPECIES DESTINED FOR EXTINCTION IN NATURE

Conservation of species in nature (*in situ*, i.e., in natural sites) is always preferable, but it seems that only a fraction of the world's species can be saved, and *ex situ* (offsite) options need to be emphasized for many species, which obviously will be selected on the bases and biases discussed in this review. Many animals are critically endangered (Figure 10.12), and measures to maintain them either in nature or in zoos or parks are usually difficult and expensive. On the whole, terrestrial plants can be maintained relatively easily in cultivation, and botanical gardens and arboreta are extremely important for conservation of critically endangered species. On a practical basis, the wild relatives of economically important domesticated species (particularly crops) can furnish genes of immense value and clearly merit protection. Seed collections have been made of numerous crops and many wild plants (see Chapter 12). Unfortunately, the seeds of many tropical plants are 'recalcitrant' – extremely difficult to maintain under the very low temperatures that can preserve temperate region seeds in a dormant condition for up to hundreds of years. Regrettably, the technology of tissue, gamete (sperm and egg), and embryo preservation (cryoconservation) of most animals is far less advanced and much more costly than is the case for plants. Cryoconservation is being practiced to save rare livestock breeds (FAO 2010) and, at least in theory, offers the possibility of conserving animals on the brink of extinction much as plants can be conserved by seed banks.

Because they are so large, the terrestrial charismatic megafauna generally require huge territories to exist in nature. In all probability, most of these will be progressively confined to restricted areas, and it is a semantic issue whether small conserves are *in situ* or *ex situ* (a very small nature conserve is often hardly different from an artificial animal park). Megafauna have appreciable ecotourist value, but it is questionable whether the poor nations in which most occur can afford to maintain them without much more external funding than currently available. Polar bears, arguably the world's largest land predators (polar bears and brown bears [*Ursus arctos*] have both exceeded 1 t), are highly threatened by global warning, and in the long-term, preservation in zoos may be the only option. The continuing pollution of the seas is very threatening for all marine species, notably for whales which cannot be reasonably maintained in aquaria.

The giant panda is a particularly poignant case in point. Its dependence on bamboo makes it very vulnerable in nature because bamboo populations regularly crash (in cycles of 15–100 years, plants of given bamboo species simultaneously produce flowers and seeds, then die in large numbers, making the food source scarce in certain regions). Moreover, human population growth has reduced most of the natural areas (the species was once present in much of China and even in Burma but has been reduced to the mountains around Chengdu). Fortunately, the phenomenal attractiveness of pandas ensures their survival in zoos and in designated (doubtfully 'natural') conservation areas. Also fortunately, China and hence the panda are not constrained by funding, and the Chinese have been very successful in establishing panda reserves and captive breeding programs (which have now produced hundreds of bears) in China (Figure 10.13).

BIASED CONSERVATION OF SPECIES RESULTING FROM SPECIESISM BY CONSERVATIONISTS AND GRANTING AGENCIES

Science is a search for truth, which is incompatible with bias. However, with regard to which species deserve attention, there is a kind of openly acknowledged bias that is justified: numerous species are critical to the welfare of humans, so it is natural that they should receive extra attention. Many scientists choose to study economically important species, either because they are motivated to help mankind or help themselves to the relatively lucrative employment and funding available. Like the general public, scientists also find certain high-profile species especially attractive and interesting (whether or not they are important economically) and choose to work on them simply to satisfy their curiosity or to gain a scientific reputation that is much harder to achieve when one works on obscure species of limited or no interest to the public. The wild species that naturally dominate public attention are to a considerable extent the same ones that attract many scientists and receive very strong research funding, and consequently for which there are large numbers of research papers, extensive protective legislation, and a highly biased effort at conservation compared to the millions of other species.

Non-governmental organizations, foundations, and land trusts may have a considerable degree of independence from reliance on public funding, especially if financially endowed. Additionally, organizations primarily dedicated

FIGURE 10.12 Some critically endangered animal species. As of 2021, of the more than 120,000 species tracked by the International Union for Conservation of Nature, over 8,000 were ranked as critically endangered. (a) Himalayan brown bear (*Ursus arctos isabellinus*), a critically endangered subspecies of the brown bear, found in northern Afghanistan, northern Pakistan, northern India, west China, and Nepal. This animal is thought by some to be the source of the legend of the Yeti (Abominable Snowman). Photo by Shahzaib Damn Cruze (CC BY SA 4.0). (b) The Scottish wildcat (*Felis silvestris silvestris*), a European wildcat subspecies in Scotland, once widely distributed in Great Britain. Source (public domain): Thorburn, A. 1920. *British Mammals*. London, Longmans, Green & Co. (c) The golden-bellied capuchin (*Sapajus xanthosternos*), also known as the yellow-breasted or buff-headed capuchin, is a species of New World monkey, currently restricted to the Atlantic forest of Southeastern Bahia, Brazil. Photo by Miguelrangeljr (CC BY SA 3.0). (d) Siamese crocodile (*Crocodylus siamensis*), a medium-sized freshwater crocodile native to Southeast Asia. Photo by Rlevse (released into the public domain). (e) Black-and-white ruffed lemur (*Varecia variegata*), endemic to the eastern rainforest of Madagascar. Source (public domain): Buffon, G. L. L. 1782. De algemeene en byzondere natuurlyke historie, vol. XIII, plate XXVIII. Amsterdam, S. H. Schneider. (f) Geometric tortoise (*Psammobates geometricus*) native to a very small section in the South-Western Cape of South Africa. Source (public domain): Sotheran, H. 1872. *Tortoises, Terrapins, and Turtles*. London, J. Baer & Co. (g) European sea sturgeon (*Acipenser sturio*), also known as the Atlantic sturgeon and common sturgeon. It was once abundant in coastal habitats all over Europe. Source (public domain): Bloch, M. E., 1795–1797. Illustrations de ichtyologie ou histoire naturelle générale et particulière des poisons. Moritz. (h) Cotton top tamarin (*Saguinus oedipus*), native to Northwest Colombia. Photo by Cuatrok77 (CC BY 2.0). (i) Addax or white antelope (*Addax nasomaculatus*), found in the Sahara Desert. Photo by MathKnight and Dr. Zachi Evenor (CC BY SA 4.0). (j) Male and female Javan blue-banded kingfishers (*Alcedo euryzona*), endemic to Java, Indonesia. Source (public domain): Keulemans, J. G. 1868–1871. *A Monograph of the Alcedinidae: or, Family of Kingfishers*. London, published by the Author.

FIGURE 10.13 Conservation breeding of the giant panda (*Ailuropoda melanoleuca*), threatened with extermination by destruction of its natural habitat. More effort has been made to prevent its extinction than for any other species. The panda reproduces extremely slowly (because its bamboo diet is nutrient-poor), so extraordinary care is required. It has been saved because its overwhelming cuddly cuteness is greatly valued by people. (a) Mother and child. Photo by Angela (CCC BY 2.0). (b) A 1-week-old cub at Chengdu's Giant Panda Breeding Research Base. At birth, the panda is only 90–180g in weight, but at maturity it may weigh over 100kg. Photo by Colegota (CC BY SA 2.5). (c) Young pandas gathered for a bamboo meal. Source: Shutterstock, contributor: Hung Chung Chih. (d) A 3-month old baby panda. Source: Shutterstock, contributor: Flysnowfly.

to conserving selected tracts of land are particularly deserving of support, since they have chosen an approach that is much less appealing to the public than 'Save-the-Tiger' types of initiatives. The world's largest independent conservation organization, the World Wide Fund for Nature ('World Wildlife Fund' in North America), stresses conservation of both selected species and selected habitats. It currently focuses on about three dozen high-profile species (including bigleaf mahogany, dolphins, elephants, porpoises, tunas, and whales) and also on about three dozen ecoregions. The WWF's 28 flagship species (shown at http://www.worldwildlife.org/species/specieslist.html) are almost all charismatic megafauna. In North America, the Nature Conservancy and the Audubon Society have protected huge acreages. Directors of such organizations are usually extremely dedicated and informed, playing critical roles in guiding the appropriate selection of habitats to be conserved. Probably no one is free from preferences for species and habitats, but when independent people decide on priorities there is a greater likelihood that biodiversity will be better protected than when it is necessary to rely heavily on public contributions.

Reminiscent of racism, speciesism has a shameful quality, so scientists and scientific agencies keep it hidden, not from evil intentions, but because it is embarrassing (the elephant in the room syndrome). Lorimer (2006) described the situation as follows: 'There is a clear reluctance in official circles to acknowledge the emotional underpinnings to natural history for fear of jeopardizing its political power… the official rationale advanced for biodiversity conservation in

FIGURE 10.14 Biased funding by granting agencies for conservation. Glamorous, charismatic animals receive the bulk of support. Prepared by B. Brookes.

the nonhuman realm is reduced to a collection of resources – potential cures for cancer and ecological support systems… While this is no doubt a strategic intervention in the realms of policy it poorly reflects the real motivations of those carrying out nature conservation on the ground. By effacing these emotional and subjective attachments this official approach is at heart dishonest.'

Harrison et al. (2021) noted that 'Ultimately, conservation decisions are influenced by many factors – political, economic, scientific, and social.' Whether the considerations are termed 'values,' 'preferences,' or 'prejudices,' human psychology plays a critical role in guiding species conservation. Whether justified or not, most if not all humans have very strong biases that determine preferences for or against many other life forms. There is good reason to believe that these biases are the principal determinant of how funding is and will continue to be provided for biodiversity conservation (Harris 2014; Figure 10.14). This may be a cynical view, but appropriate management of human biases in support of the more likeable life forms may well be the most realistic strategy for the preservation of biodiversity in general.

ACADEMIC (THEORETICAL) VS. PRACTICAL (APPLIED) BIAS

The current species extinction issue is a worsening crisis, and like many issues facing humankind, it is important to both scientifically analyze the nature of the problem and to address it with practical remedial measures. The relative allocation of limited resources to theoretical and applied conservation currently appears to be biased toward the former. As noted by Sutherland et al. (2009), 'there is a widely acknowledged mismatch between the priorities of academic researchers and the needs of practitioners.' One of the definitions of 'academic' is 'very learned but inexperienced in or unable to cope with the world of practical reality' (Webster's Third New International Dictionary 1993). Given the urgency of environmental issues, the present need for action seems greater than the need for more analysis.

A more subtle distinction was pointed out by Martín-López et al. (2007). Environmental scientists tend to be motivated by factual considerations based on the ecological needs of both individual species and collections of species, whereas 'nature lovers' are basically motivated by emotional factors.

SPECIES CONSERVATION FROM THE PERSPECTIVE OF SOCIETAL FUNDING PRIORITIES

Biodiversity conservation is one of the major existential issues of modern times, but it requires major investment. Huge public expenditures should of course be made rationally, on such bases as risk/benefit analysis and moral principles. Frequently, however, there is also a substantial emotional determinant. For example, although space exploration has been justified on the basis of military advantages and scientific discoveries of potential benefit, it is clear that the exciting images of astronauts in space suits being launched on gigantic rockets, carrying out their tasks, and conversing with earthlings has been a major public relations success for the initiative. Obviously the romantic image of space travel, not an urge for public service, explains why so many children now aspire to be astronauts. Medical research to cure illness hardly needs justification, but support for particular diseases varies greatly, not just with frequency and seriousness of the illnesses, but with several emotional determinants, e.g., how deserving the victims seem to be (breast cancer receives huge support, lung cancer from smoking is not viewed sympathetically) or how embarrassing the disease is (note the distaste for diseases of the lower alimentary canal). Humanitarian aid is still another example of the importance of emotional factors; there are pronounced racial, ethnic, and class preferences that influence international aid. Nothing in this paragraph should be interpreted as support for prejudice or bias; what is important is to understand that human emotion, even if irrational or unfair, is often critical to human action and needs to be countered or channeled constructively.

An informed public provided with accurate scientific information about the deteriorating state of the planet's habitats, ecosystems, and biodiversity would seem sufficient to correct the biodiversity crisis. After all, human welfare is contingent on the welfare of the environment upon which biodiversity depends, and biodiversity is by far the most important source of materials and ecological services without which humans cannot exist. Public education on the issues, as recommended in the next section, is vital to addressing the environmental and biodiversity crises. Unfortunately, important as they are, the environment and biodiversity are among a crowd of issues that are also of vital importance to most people: deteriorating economies, social justice, wars, famine, poverty, health, and overwhelming technological innovations, to name a few. Moreover, lucrative revenue-producing activities sponsored by government and business are often the greatest threats to biodiversity, and it is difficult to challenge such powerful interests. An objective examination of the budgetary allocations of the richest countries indicates that only a tiny percentage of funding is targeted to environmental and biodiversity issues. Democratic governments reflect the priorities of the public, and accordingly the most pressing need for addressing the world's declining biodiversity is to persuade the public that the issue deserves a higher priority. Adelaja et al. (2008) examined motivations to spend on conservation and environmental issues in the U.S. and concluded that complex socioeconomic and political considerations, often different among the states, were important. Maxwell and Miller (2013) suggest that pre-existing biases and ideologies need to be taken into account when attempting to persuade a given audience to support environmental issues. Similarly, Saunders et al. (2006) argue that conservation of other species requires a better understanding of our own species' psychology. Indeed, these are goals of this book.

RECOMMENDATIONS TO INCREASE RESPECT FOR MALIGNED SPECIES

In the human world, prejudice against groups is addressed primarily by policies, education, legislation, and enlisting the support of key sectors of society. As noted in the following, these initiatives are also useful for alleviating prejudice against other species and promoting rational, harmonious coexistence with the natural world.

Policies to Increase the Welfare of Species

Much of this book provides background information on the nature of prejudice against species, the associated harm, and how to address the issue. Throughout, it is emphasized that human attitudes to the millions of other species in the world are determined not only by practical considerations by also by emotional determinants, which often distort our perceptions. We are largely hard-wired to instinctively like or dislike numerous external characteristics of other species. Just as we idolize the physical beauty of a small minority of entertainment stars, we similarly greatly admire a minority of other species, which are accordingly given preference. Conversely, most other species are perceived as quite unattractive and are treated with considerable prejudice. Indeed, most people would be tempted to crush underfoot any of the world's other species that crossed their path! The threat to the world's species has become an acknowledged crisis, endangering our own survival. For our own welfare, we need to understand and control our unconscious biases against other living things. In civilized nations, policies have been enacted to control prejudice and limit discrimination against other members of the human species. At the very least, there needs to be greater understanding and tolerance of other living things.

The goal of this book is to promote an improved understanding and tolerance of other living things. Chapter 8 reviewed the importance of wild species as sources of materials and indispensable ecological services, the two basic pragmatic foundations on which species conservation policies need to be based. Chapter 11 discusses methods of co-existing harmoniously with the dangerous species of the world. Chapter 12 provides ways of reforming agriculture, which more than any other area has been responsible for damage to the world's ecosystems and harm to species. Similarly, Chapter 13 provides ways of reducing the harms associated with urbanization, the next most harmful threat to species. And Chapter 14 addresses the dangers posed by advancing technologies to the world's species.

Public Education to Increase the Welfare of Species

Conservation organizations have become the principal promoters of the welfare of wild species and their habitats. As discussed earlier, they rely heavily on funding campaigns based on extremely attractive, charismatic animals. As also noted, this is an approach with limited validity, masking the primary need to maintain the ecosystems of the world, not just selected species. But at the level of the general public, the latter tactic generates considerable funding and protects some species and their habitats. Although unattractive species can benefit from general appeals to prevent their extinction (Veríssimo et al. 2017), support is limited. At best, many people are simply indifferent to the majority of species. However, most species are insects and suffer from stereotypes and stigmas caused by hated human parasites such as cockroaches, mosquitoes, bedbugs, and fleas. A substantial proportion of the public is still unaware of conservation issues, so continuing the approach of highlighting the fate of glamorous whales, pandas, and the like remains a useful approach.

The alternative to emphasizing the conservation of individual species is to highlight the welfare of the ecology of the world's ecosystems, and indeed this is the primary approach of many conservation organizations. However, compared to the global climate crisis, convincing the public to support remedial conservation measures has had limited success. The climate change issue has captured public interest because recent extreme weather events have made it evident that human welfare is at stake. There is a parallel need to convince and alarm the public that the welfare of biodiversity is similarly in desperate need of protection because the millions of threatened species, not just those celebrated in the media, are essential for our quality of life. Educating the public about biological conservation issues is critical since their support is indispensable both for governmental legislation and for private funding.

Many organizations dealing with species, such as zoos, museums, and parks, which attract the general public, need to justify their expenditures with evidence of attendance and often also income. Inevitably, there needs to be a partition of effort between entertainment and education. Superficial but attractive exhibits and events generate finances, but responsible administrators have opportunities to also persuade the public to support the current priorities of the living world (Figure 10.15). Balmford et al. (1996) criticized the tendency of zoos to spend inordinately on breeding large charismatic animals, which albeit generate public interest, are much more expensive to support than many equally threatened but less attractive species.

FIGURE 10.15 Public biodiversity education presentations. (a) *Hall of Biodiversity*, American Museum of Natural History (New York). Photo by Anagoria (CC BY 3.0). (b) Natural History Museum in Leiden, Netherlands. Photo by Henk Caspers/Naturalis Museum (CC BY SA 3.0). (c) Public lecture on biodiversity. Photo by Åge Hojem / NTNU University Museum (CC BT SA 2.0).

Student Education to Increase Respect for All Species

Although we have innate tendencies to react positively and negatively to species, as discussed in this book, experience is also determinative of how we treat other forms of life (Kos et al. 2021). Education is critical in forming children's perceptions of and behaviors toward living creatures. Prejudice is mostly learned, but so is tolerance and respect, which are most efficiently implanted in the young. Love and respect for nature and its species is best initiated in early childhood (Figure 10.16). The difference between swatting a mosquito and deliberately stepping on an ant (Figure 10.17) is a useful topic for exploration. Young students are remarkably receptive to the values of nature (Keith et al. 2022). Teaching methods and guidelines for educating students about biodiversity are reviewed by Trombulak et al. (2004), Jeronen et al. (2017), and Yli-Panula et al. (2018). A survey of zoos found that educating visitors, including school children, is the highest priority activity from the point of view of both zoo managers and the public (Roe et al. 2014). Children need to be taught that charismatic animals are only part of the problem, and local species also deserve consideration (Textbox 10.1). Students tend to sympathize with local species that are endangered (Lindemann-Matthies; Liles et al. 2021), and the importance of teachers having personal knowledge of local species was emphasized by Wolff and Skarstein (2020). Cartoon depictions of animals in movies, comics, books, games, and toys are important to conditioning young children to accept animals currently often shown as ugly villains (Small 2016a; Figure 10.18).

Legislation to Increase Species Conservation

Legislative protection of wild animals, indeed other wild species, has largely been based simply on the need to sustain them for human exploitation, not out of ethical concerns. Nevertheless, in recent times, numerous species in danger of extinction are being afforded protection (Figure 10.19) simply because there is now an international consensus that the species selected for this status merit continued existence. Principal international agreements governing conservation protection of wild species were noted in Chapter 9. Countries, political regions (states, provinces), and local areas (especially cities) often have their own legislation concerning species, and invariably these reflect local prejudices and preferences, as well as the practical values, associated with species. Chapter 11 discusses the need for conservation legislation to be respectful of local property owners to avoid backlash. Attempts to elicit interest in safeguarding 'lower' animals have often been met with derision, making attempt at legislative protection very challenging.

FIGURE 10.16 Cartoon depiction of children being educated about the merits of animals usually considered to be ugly, disgusting, and/or dangerous. Prepared by B. Brookes.

FIGURE 10.17 Avoiding stepping on ants – a moral and conservation lesson. Prepared by B. Brookes.

TEXTBOX 10.1 EDUCATING CHILDREN ABOUT LOCAL BIOTA IS AS IMPORTANT AS EMPHASIS ON CHARISMATIC SPECIES

While highlighting charismatic megafauna is effective at getting kids interested in wildlife and conservation efforts, using these species tends to provide children with the perception that nature is located in far-away, exotic lands such as Africa, the circumpolar North, tropical areas, and Australia. Some have argued that the overemphasis on distant charismatic megafauna results in a disconnection between children and their local plant, insect, and animal populations and environment. Suggested solutions to the problem of overemphasizing the charismatic include engaging young minds in an appreciation of local species and biodiversity.

—*Hund (2012)*

FIGURE 10.18 Ugly (left) and attractive (right) cartoon animals, reflecting how they are illustrated contributes to undesirable or desirable stereotypes. Based on public domain figures from Pixabay.com.

FIGURE 10.19 Legislation to protect species from harm is essential for both their conservation and sustainable harvesting. Prepared by B. Brookes.

Religion and the Promotion of Species Conservation

Chapter 7 discussed the importance of species in relation to religious and spiritual traditions. In the past, species were viewed mostly as exploitable resources, particularly in the Abrahamic religions. However, followers of the non-monotheistic belief systems have a long tradition of respect and promotion of the welfare of species. In the light of increasing threats to the welfare of the planet, all spiritual movements need to be increasingly concerned with nature conservation. Affluent nations, particularly in the Western World, have the majority of financial resources required to remediate the environmental crises. Since Christianity is the dominant Western religion, it is especially positioned to play a large role in conservation of the natural world (Hessel and Ruether 2000; O'Brien 2010; Blanchard and O'Brien 2014; Nita 2016). 'Green Christianity' refers to Christian-based efforts to protect the environment and its living constituents, which in fact has biblical justification (Figure 10.20).

Corporations and the Promotion of Biological Conservation

Corporations control much of the wealth of the world and are a large source of economic support for various causes. As discussed in Chapter 8, ensuring the welfare of habitats and species is in the interest of businesses dependent on natural resources, which accordingly should be supportive of biological conservation. As also discussed, some polluting industries are motivated merely to 'greenwash' their reputation, in which case they may nevertheless also be potential contributors of funds for conservation. As discussed in Chapter 9, corporations are subsidizing the world's major conservation organizations (ironically, themselves with corporate structures), which is concerning since the interests of the former may conflict with the welfare of species and habitats (Robinson 2012; Figure 10.21). An additional concern is the 'continuous expansion/growth' model that corporations (indeed, most economists) follow, which is incompatible with long-term sustainability. Corporations are not altruistic philanthropies, and while their contributions to environmental issues can be valuable, they may also be Faustian (Pulver and Manski 2021).

Recreational Hunting and Fishing and the Promotion of Conservation

Contrary to the intuition of many, valuable allies of biodiversity and nature conservation can be those who regard the world of non-human species primarily as objects to be killed for recreation. In the past, and continuing to the present,

FIGURE 10.20 Paintings (public domain) of Noah's Ark, the featured ship of a biblical narrative in Genesis in which God spares Noah and examples of all the world's animals from a world-scale flood. The story is recognized in all Abrahamic religions, but is particularly highlighted in Christianity. The account is widely recognized today as the basis for mandatory religious respect of all biodiversity. (a) Painting by Italian artist Aurelio Luini (1530–1593). Photo by Pierre 5098 (CC BY SA 4.0). (b) Painting by Greek artist Theodore Poulakis (1622–1692). Credit: Digitized Archive of the Hellenic Institute of Venice. (c) 'Noah's Ark and the Deluge,' 19th century painting. Source: U.S. Library of Congress. Photo credit: Frank Zimmerman.

FIGURE 10.21 Corporations, although potential funding sources for conservation, can be harmful to species. Prepared by B. Brookes.

hunting and fishing were not conducted sustainably. Today, it is ironic that those who support conservation efforts in order to have animals to hunt and kill just for fun provide vital support for the welfare of these very animals. This makes sense since conservation of the habitats of game fish, birds, and mammals is essential in order to sustain the supply. Theodore Roosevelt (1858–1919; Figure 10.22), president of the U.S. from 1901 to 1909, was an avowed naturalist and hunter who understood that conservation was critical the future of outdoor recreation, hunting, and resource management. He accomplished much more legislatively on behalf of conservation than any other North American politician.

FIGURE 10.22 Cartoons (public domain) depicting Theodore Roosevelt's love of nature. (a) Roosevelt refusing to kill a baby black bear during a hunting trip because he considered it dishonorable. This cartoon (public domain) by Clifford Berryman appeared in The Washington Post in 1902 and is considered to represent the story that led to stuffed 'Teddy bears' becoming the favorite toy of children. Colorized by B. Brookes. (b) Roosevelt initiating nature preserves. As president, he set aside 51 million ha of forest in the U.S. and also instituted hundreds of protected areas (National Forests, National Monuments, National Parks, bird reserves, and game preserves). Published April 4, 1903, in Des Moines Register and Leader. Colorized by B. Brookes. (c) 'A Thanksgiving Truce' lithograph by J. S. Pughe (1870–1909) showing Roosevelt sharing a feast with many wild animals sitting around a large banquet table in the wilderness. Courtesy of U.S. Library of Congress Prints and Photographs Division.

FIGURE 10.23 Species-sympathetic art in high-traffic public places. (a) Huge wall mural on the 'Blind Walls Gallery' in Breda, Netherlands, which presents dozens of outdoor murals. Painted by Mantra (=Youri Casell), titled 'The most beautiful butterflies of the city,' showing (from top to bottom): a peacock butterfly, (*Aglais io*), painted lady (*Vanessa cardui*), and a large blue (*Phengaris arion*). Photo by ReneeWrites (CC BY SA 4.0). (b) Mural on the wall of a parking structure in Downtown Juneau, commissioned by Public Art Works Juneau. Photo (public domain) by Bernard Spragg. (c) Barren ground caribou mural by Joyce Wieland, on the wall of a Toronto subway station. Photo (public domain) by Gniw. (d) Provocative street art in Benicarló, Spain, showing trees as sentient creatures. Photo by Gordito1869 (CC BY SA 4.0). (e) Photograph (public domain) of Maman the spider sculpted by Louise Bourgeois, exhibited at Bürkliplatz, Zürich, Switzerland.

Art and the Promotion of Appreciation for Species

The problem highlighted in this book is that we humans, albeit often unconsciously and instinctively, are prejudiced against most species because they are unattractive. But viewed through an appropriate lens, and in an appropriate context, all species in fact are beautiful. Living things are marvels of architectural construction, albeit often on a small scale, and ecosystems are filled with natural wonders superior to everything built by humankind. Art is capable of revealing this beauty, and inspiring empathy for species and their habitats, and so artists can play important roles in promoting the welfare of biodiversity. Some examples are shown in Figure 10.23. Also note the online presentation 'Artists on the Front Line of Biodiversity' at https://www.whatcommuseum.org/exhibition/endangered-species/

11 Dealing with Dangerous Species

INTRODUCTION

Humans, like every other living thing, are subject to illness and destruction caused by numerous other organisms. Some malevolent species (most notably microorganisms and their vectors such as mosquitoes and ticks) are responsible for immense suffering, and great efforts have been made to control or even eradicate them (as noted in Table 11.1, 18% of human deaths are due to infectious and parasitic diseases). Additionally, many species attack crops, livestock, pets, and wild species that are harvested, and accordingly are very harmful to the interests of humankind. Many of these simply compete for the same resources as people. Examples of harmful species in the animal kingdom include vermin such as rats and coyotes, and toxic species such as some snakes, spiders, scorpions, and jellyfish. However, as noted in Table 11.1, attacks by animals cause much less than 1% of all human deaths. Plants that are hostile include toxic plants such as ragweed (the cause of hayfever; see Chapter 18) and poison ivy (a cause of dermatitis; also see Chapter 18), and weeds (which cause enormous agricultural losses; see Chapter 19–21). Many species that seem very harmful to humans likely play useful roles in nature, and some probably have potential values that could be discovered in the future. Nevertheless, it is probably futile to ask humans to sympathize with species that are very harmful to people.

FEAR OF DANGEROUS ANIMALS AS MOTIVATION FOR THEIR NEGATIVE PERCEPTION

Some animals are so dangerous that they can kill people very quickly. Most animal species, including humans, are in danger from predators that want to eat them and territorial species that can kill them simply for getting too close. Such potentially deadly animals – especially those that are large and aggressive like crocodiles, lions, and bears – generate fear. Today, most people live in relatively safe urban settings and are rarely threatened by wild animals. In fact, humans are now the world's dominant super-predator. Nevertheless, our reactions to potential animal attackers are instinctively negative. Ohman and Mineka (2001) hypothesized that such innate responses are embedded in specific neural structures and circuits of the mammalian brain, the so-called 'fear module.'

TABLE 11.1
Causes of Death of Humans

Category	%
Non-communicable diseases (heart disease, cancer, etc.)	73
Infectious and parasitic diseases (tuberculosis, malaria, etc.)[a]	18
Accidental injuries (automobile accidents, drowning, etc.)	6
Violence (war, suicide, etc.)	2
Animal attack (80% due to poisonous snakes)[a]	0.15

Source: After *International Statistical Classification of Diseases and Related Health Problems*, https://www.who.int/standards/classifications/classification-of-diseases.

[a] Due to species.

Venomous animals defending themselves from humans kill or disable far more people than large animals that prey on humans for food. Most venomous animals (including virtually all snakes) employ their toxins to hunt small animals, but many also use their venom for protection. Regardless, humans have justifiable fear of the thousands of species that possess venom (Figure 11.1).

Of all the ways that non-human species represent a threat to humans, it is probably safe to say that being eaten by huge animals with menacing teeth is the most frightening. This is rather hypocritical of us, since we consume more species than any other organism. Paradoxically, we are fascinated, indeed awed by the sight of animals with threatening teeth (Figure 11.2). This ambivalence works both for and against the reputations that dangerous animals generate: The sight of gaping mouths with enormous fangs is highly entertaining, so people often support their maintenance in conservation areas or at least in zoos. However, when such animals intrude into areas frequented by humans, their elimination is likely.

Probably because powerful people, particularly leaders, have been critical to the success of human societies, human beings seem to have an instinctual regard for power in other animals. This is likely why large creatures inspire respect, and so do all forms of offensive animal weaponry, including teeth and poisonous fangs, claws and talons, constrictor muscles of snakes, huge beaks, etc. To a lesser extent, we are also fascinated by defensive animal weaponry, including horns, antlers, porcupine quills, and the squirted toxins of various animals. A large proportion of 'nature films,' seemingly dedicated to the protection of animals, depicts savage predators viciously attacking prey (reminiscent of the media motto: 'if it bleeds it leads'). Disturbingly, the human species is spellbound and entertained by dangerous animals tearing other animals to pieces. The most prominent of the charismatic megafauna are in fact dangerous predators. Humans are very ambivalent about these animals, very willing to provide support for their conservation, especially in foreign lands, but not when they endanger their own families and property. In

DOI: 10.1201/9781003412946-11

FIGURE 11.1 Venomous animals. (a–c) Examples of animals that inject venom. (a) Mexican scorpion (*Centruroides vittatus*), one of the world's most poisonous scorpions. Of the 2,000 scorpion species, only about 25 are capable of killing a human, and this is a rare occurrence. Photo by Tomascastelazo (CC BY SA 3.0). (b) King cobra (*Ophiophagus hannah*), the world's longest venomous snake (up to 5.6m), and one of Asia's most feared reptiles, although how many humans are killed is uncertain. Snakes kill about 100,000 people annually and permanently disable about 400,000 more. Painting (public domain) by T. Knepp, U.S. Fish and Wildlife Service. (c) Mexican redknee tarantula (*Brachypelma smithi*). Tarantulas are a group of more than 1,000 species of spiders which are often large and can deliver painful (but almost never deadly) bites. Photo byFir0002 (CC BY SA 3.0). (d–f) Examples of animals with skin-secreting toxins. (d) Hooded pitohui (*Pitohui dichrous*), of New Guinea, one of very few birds that is poisonous. It produces batrachotoxin compounds in its skin and feathers, the same kind of toxins in many poisonous dart frogs. Photo by Benjamin Freeman (CC BY 4.0). (e) Common lion fish (*Pterois miles*), native to the Indian Ocean and introduced elsewhere, its highly venomous fin spines when contacted can cause extreme pain, even death. Photo by Michael Gäbler (CC BY 3.0). (f) Cane toad (*Rhinella marina*). This large, invasive South American toad with very poisonous skin is killing native animals (especially in Australia) and pets that mistakenly assume it is edible.

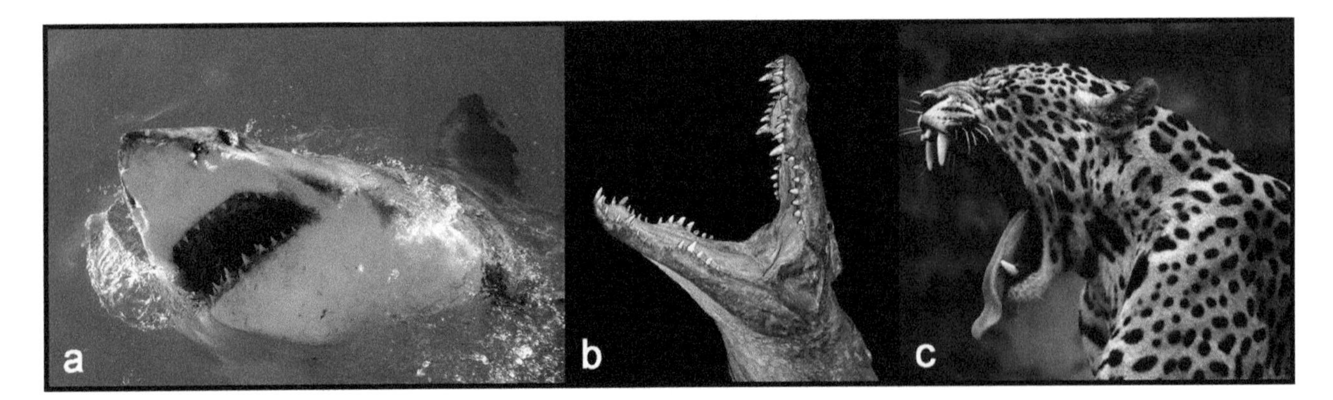

FIGURE 11.2 Examples of dangerous predators with terrifying teeth, capable of killing humans. (a) Great white shark (*Carcharodon carcharias*), the largest and most feared of all sharks. Less than 100 shark attacks (all species collectively) are normally reported worldwide annually, with usually less than 10 fatalities. Photo by Olga Ernst (CC BY SA 4.0). (b) Nile crocodile (*Crocodylus niloticus*). The species kills hundreds of people annually. Photo by Jean-Pol Grandmont (CC BY 4.0). (c) Jaguar (*Panthera onca*). Records of jaguars killing humans are rare. Photo by MarcusObal (CC BY SA 3.0).

the case of some Third World people whose children and pets are still being killed by wild predators, it is understandable that they are often hostile to them, no matter how impressive or endangered they may be. However, charismatic predators are often valuable attractions for ecotourism, in which case the animals are more likely to be tolerated.

FEAR OF INFECTION, CONTAMINATION, OR DISEASE FROM SOME SPECIES GENERATES DISGUST AND REPULSION

The intolerant way that lepers were treated in past times (Figure 11.3a) tragically illustrates how fearful people were of acquiring the condition from others suffering from

FIGURE 11.3 Illustrations of harsh treatment in the past of people suffering from pathology. (a) Two lepers are denied entry into the city, one with a rattle to announce his coming. Drawing (14th century, public domain) by Vinzenz von Beauvais (in France). (b) An old Barnum & Bailey circus poster (public domain) illustrating a 'freak show' exhibition of abnormally and pathologically developed people, a practice that has been banned.

the disease. People with pathological conditions that distort their appearance – whether from disease, genetics, or accident – have long suffered from prejudice and ridicule (Figure 11.3b). Disease-causing species are responsible for much more human pain than the much bigger, dangerous animals discussed in the above section. Unlike large predators that kill quickly, many species kill or at least seriously debilitate people slowly. The majority of species producing diseases (both infectious conditions and debilitation) are parasites (Figure 11.4). Parasites, which are species that

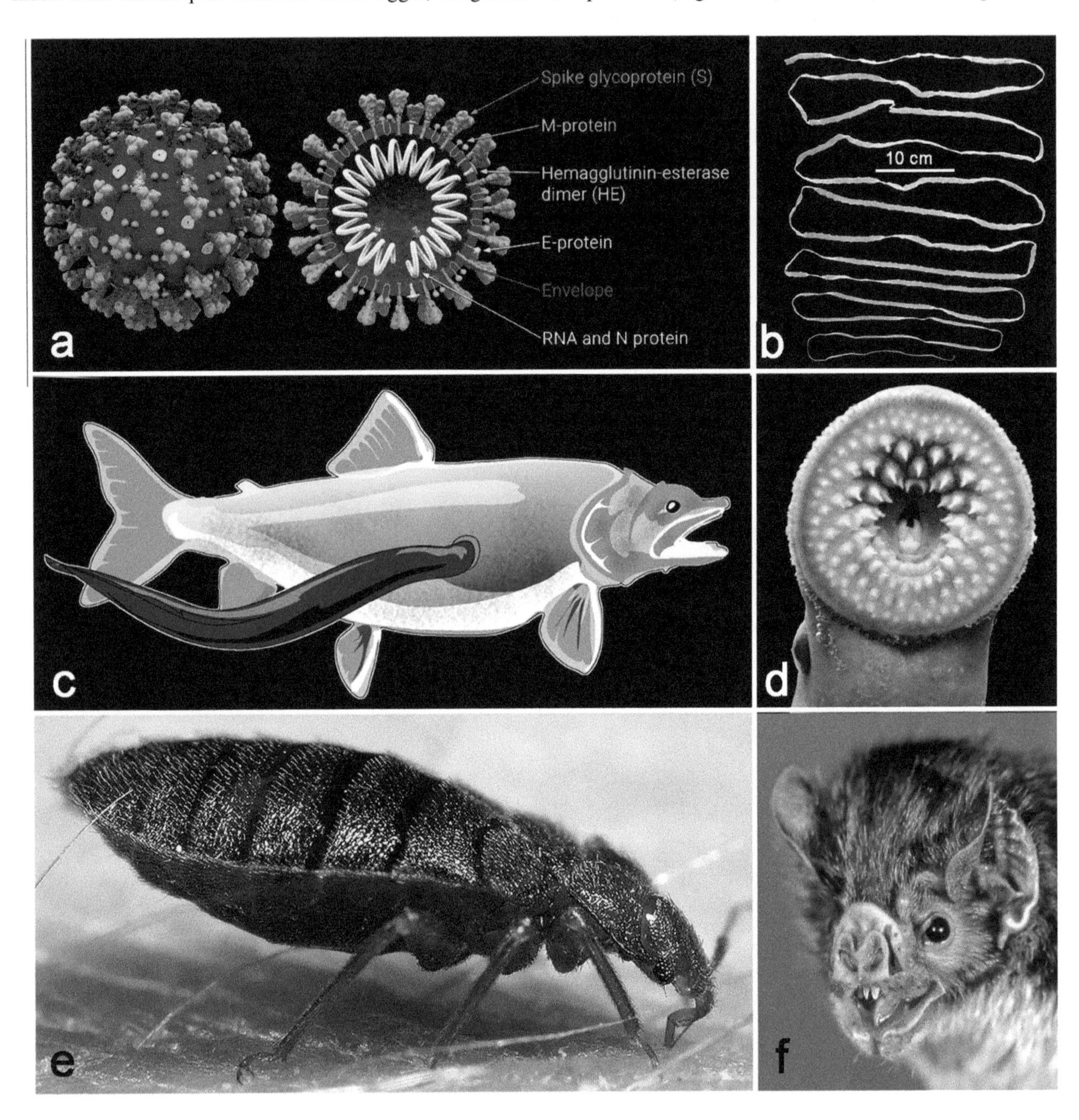

FIGURE 11.4 Examples of parasites, a lifestyle that people find repugnant. It has been estimated that about half of all species are parasitic. (a) Coronavirus. Many biologists do not consider viruses to be living organisms, in which case they are not considered to be parasites in the conventional sense, although the distinction is semantic. Source: Wikimedia (public domain). (b) Beef tapeworm (*Taenia saginata*), one of thousands of species of intestinal tapeworms. Specimens of this species over 22 m in length have been recorded. Humans become infected by eating undercooked infected meat, and the worm can survive for years in the small intestine. Photo (public domain) by U.S. Centers for Disease Control. (c) Diagram of a lamprey attached to a fish. Prepared by LadyofHats (public domain). (d) Mouth of a sea lamprey (*Petromyzon marinus*). The mouth functions like a suction cup, and the attachment is reinforced by the gripping teeth. This native of the Atlantic Ocean has become a serious invader of the Great Lakes of North America. Photo (public domain) by Joanna Gilkeson, U.S. Fish and Wildlife Service. (e) Bedbug (*Cimex lectularius*) ingesting a blood meal from a human. Photo (public domain) by U.S. Centers for Disease Control and Prevention. (f) Common vampire bat (*Desmodus rotundus*), thought to be responsible for the death of about 100,000 domestic cattle annually in South and Central America, by transmitting rabies. Photo by Uwe Schmidt (CC BY SA 4.0).

FIGURE 11.5 Anti-household-pest posters. (a) 'Pest Control Around the House,' by Insightpest (CC BY SA 2.0). (b) Attack by a giant bedbug. Extract from 'Halloween Magic Lantern Show' (public domain), photo by Terry Borton (CC BY 2.0). (c) 'Rats in a Kitchen Threaten Human Health.' Credit: https://www.vecteezy.com/free-vector/vector. Vectors by Vecteezy. (d) 'Flies infecting food' by A. Games. Credit: Wellcome Collection (CC BY 4.0).

live in or on other living species (the hosts), deriving nutrients from them, are discussed in Chapter 15. Most parasites are small (they especially include microorganisms which can't be seen, and numerous invertebrates which are usually inconspicuous). They play critical roles in maintaining the balance of species in nature and, in some cases, provide some benefits to their hosts (see the discussion of coexisting with parasites later in this chapter). People of course have learned which parasites are most harmful, and what animals transmit them, and tend to develop quite negative emotions toward both (Prokop et al. 2010).

Most parasites, indeed most disease-causing organisms, are microscopic, but they are frequently transported by larger creatures, especially rodents and insects, which consequently receive the lion's share of blame. 'Disgust' is a psychological adaptation to produce pathogen-avoidance behavior (Fukano and Soga 2021), protecting the individual against disease and contamination (Curtis et al. 2011). Disgust appears to be particularly generated by threats to health, especially sanitary conditions (Tucker and Bond 1997). This emotion is often triggered by the small animals that are known or suspected of transmitting diseases and thrive in dwellings (Figure 11.5), but also by the unconventional or unsettling appearance of species or individuals of some species (Figure 11.6). Excrement is held in especially low esteem (probably because it often harbors diseases), so that animals that rely on it to reproduce (such as some flies and beetles) are considered to be disgusting. Most people cannot stomach the concept of eating insects (probably a disgust reaction associated with insect presence on spoiled food), although they represent a huge potential sustainable food source. 'Repulsion' (which is basically a strong form of disgust) is often related to perception that other species have characteristics that in humans reflect pathology (disease or abnormal development; note Figure 11.3). While we may be programmed to dislike certain features of living things, it needs to be remembered that stereotypes represent exaggerated and unjustified evaluations.

COEXISTING WITH WILD PREDATORS DANGEROUS TO PEOPLE

As humans overpopulated the terrestrial world in historical times, animals that treated people as prey were usually killed, and often driven to extinction. Inevitably, as civilization expanded and natural habitats were destroyed, conflict with predators increased. Aside from microorganisms, large carnivorous predators have been the most fear-provoking

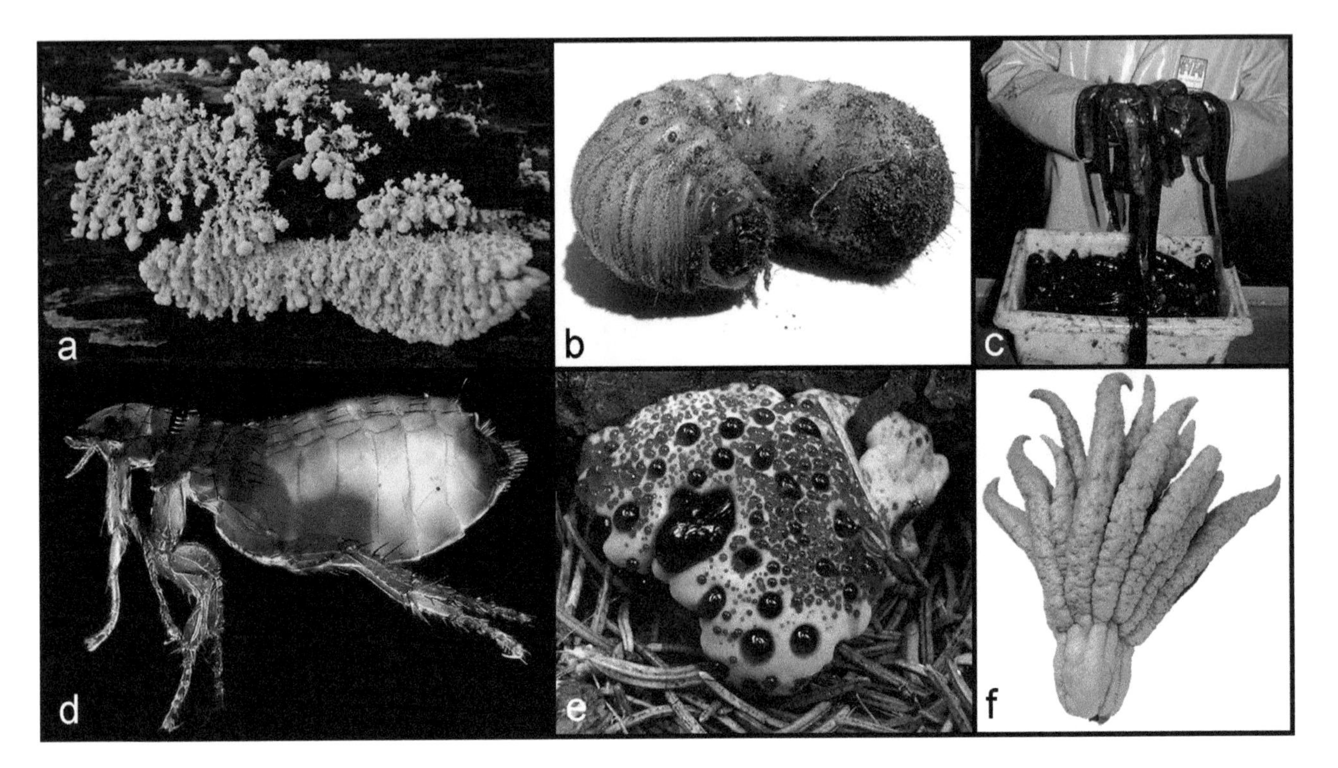

FIGURE 11.6 Species whose appearance evokes disgust. (a) Dog vomit slime mold (*Fuligo septica*) Photo by Henk Monster (CC BY 3.0). (b) Beetle grub. Photo by Toby Hudson (CC BY SA 3.0). (c) Black hagfishes (also known as slime eels; *Eptatretus deani*). This species has been termed 'the most disgusting of all sea creatures.' Public domain photo by U.S. National Oceanic and Atmospheric Administration. (d) Cat flea (*Ctenocephalides felis*). Photo by Andrei Savitsky (CC BY 4.0). (e) Devil's tooth fungus (*Hydnellum peckii*). Photo by B. Baldassari (CC BY SA 3.0). (f) Fruit of Buddha's hand (also called the fingered citron; *Citrus medica* var. *sarcodactylis*). The diseased appearance of this piquant citrus is normal. Photo (public domain) by Kaldari.

species, and there has been concern about limiting their damage (Treves and Naughton-Treves 1999; Røskaft 2004; Treves and Palmqvist 2007; Staňková et al. 2021). The need is to adopt control techniques and approaches that do not significantly harm our environment, including innocent species and ourselves. Only recently has it become a priority to conserve very large and dangerous predators (Hovardas 2018). However, favoring such potential killers is mostly by people in safe cities where only occasionally does a bear, puma, or crocodile accidentally wander into urban areas. Most large predators in need of conservation are in poor third-world nations, typically with rapidly expanding populations who view conversion of wildlands as necessary for their economic development. Ordering locals not to harm tigers and other great cats that are eating their children is unlikely to be constructively received, although the conservation of big cats is widely supported (Athreya et al. 2013; Schulz et al. 2017; Holland et al. 2018). The issue of conservation of dangerous predators and their habitats is to a considerable degree based on a combination of values and economics: both motivation and resources are needed. The situation of conservation of polar bears in Churchill, northern Manitoba, Canada, is exemplary (Figure 11.7). Canada is home to two-thirds of the global population of polar bears, which congregate in large numbers in Churchill. The area is managed as a tourist site for viewing the bears from the safety of large trucks (Archibald 2017). Because the bears regularly come into the city, officers need to be posted as guards to protect the population, and problem animals are tranquilized and transported to a safe area. Inasmuch as Churchill is remote from other populated areas, costs are necessarily incurred to support the industry (Dawson et al. 2010; see Brandt and Buckley 2018 for cautions regarding the ecological downside of ecotourism).

COEXISTING WITH WILD PREDATORS DANGEROUS TO LIVESTOCK

Large carnivores are perceived throughout the world as threats not just to lives but also to livelihoods. Assaults by carnivorous animals on humans are relatively uncommon today, but attacks on livestock are very common, leading to hatred of the offending species, retaliatory killing, and opposition toward conservation efforts. As agriculture continues to expand, conflicts with large carnivores are also expected to increase (Robertson et al. 2020). In Africa, a wide variety of large predators (particularly spotted hyenas but also leopards, baboons, lions, jackals, and others) kill livestock and as a result generate hostility to conservation (Holmern et al. 2007). In Europe, most livestock predation is attributed to wolves (*Canis lupus*), which are the most widespread large carnivore on the continent, although free-ranging dogs are sometime responsible. In Australia, dingoes (*Canis lupus dingo*) are the major predator (Van Eeden

FIGURE 11.7 Managing polar bears as a tourist enterprise in Churchill, northern Manitoba, Canada. (a) A 'tundra buggy' employed to safely observe polar bears in Churchill, Manitoba. Photo by Studiogirl54 (CC BY SA 4.0). (b) A polar bear visiting a tundra buggy in Wapusk National Park, south of Churchill. Photo by Ansgar Walk (CC BY SA 3.0). (c) A tranquilized polar bear in Churchill being transported out of town. Photo by Emma (CC BY 2.0).

FIGURE 11.8 Exaggerated presentations of wolf attacks on humans by European artists. (a) Painting (public domain) by Flemish artist Pieter Breughel the Younger (1564–1638) entitled 'The Good Shepherd' housed at Musées royaux des Beaux-Arts de Belgique, Brussels. Photo by Rama (CC BY SA 2.0). (b) Painting (public domain) by Polish artist Józef Chełmoński (1849–1914) entitled 'Wolves attack,' Source: Polish Army Museum. (c) Illustration (public domain) from the French illustrated Weekly, *Le Petit Journal* (25 January 1914), depicting a wolf snatching a child. Source: Bibliothèque Nationale de France.

et al. 2020). Coyotes (*Canis latrans*) are the principal livestock predator in North America, especially on sheep, goats, and cattle (Mitchell et al. 2004). Falsely reported predations are sometimes declared in error or simply to obtain compensation (Caniglia et al. 2013). Most species of large carnivore are in decline globally (Berger 2006). As so often happens when stereotyping occurs, some predators acquire reputations for damaging livestock that far exceeds their real harm (Suryawanshi et al. 2013). There is a long tradition in Europe of considering wolves as bloodthirsty, not just to domesticated animals but also to humans (Figure 11.8), the result of unfair stereotyping (Rutherford 2022).

FIGURE 11.9 Illustrations of the predator control problem for livestock. (a) Great Pyrenees sheep dog guarding a flock. Photo by Don DeBold (CC BY 2.0). (b) Llama guarding sheep in Wisconsin. Photo by Jereme Rauckman (CC BY 2.0). (c) A goat boma in Kenya. A 'boma' is an (often crude) woody enclosure used in many parts of Africa to protect livestock and often also families against dangerous predators. Photo by Regina Hart (CC BY 2.0). (d) A coyote attacking a lamb. Photo (public domain) by U.S. Department of Agriculture. (e) Killed dingoes strung in a tree in Australia. Public domain photo from Pixabay.com. (f) Wolf hunters in Russia displaying their catch. Photo attribution: www.volganet.ru (CC BY SA 3.0).

Efforts to reintroduce large predators such as wolves into ranges where they have been exterminated have often led to opposition on the contention that the animals are extremely dangerous to livestock. Such restoration projects need to be based on skillful evaluation of the sociology and psychology of the stakeholders (Van Eeden et al. 2020). Typically, lobbyists representing game-keepers, farmers, and ranchers oppose such efforts (Janeiro-Otero et al. 2020). Generally, one should expect at least some predation of livestock so provisions for adequate monitoring and compensation to the people affected are essential. Throughout the world, government-backed programs of predator elimination, such as bounties, and classification as vermin, have often been effective at reducing populations. However, frequently control techniques do not work well (Eklund et al. 2017), and sometimes unexpected disturbances to the ecology of a region have resulted. Given that humans have been dealing with livestock predation for thousands of years, a variety of ways of protecting valued animals have been adopted (Figure 11.9), which have usually proven to be partially effective. Unfortunately accepting the loss of some livestock to predators may be a necessary compromise. Lute and Carter (2020) on the basis of coyotes, wolves, and grizzly bears in the American West suggested that the key to coexistence with these carnivores is equitable distribution of costs and benefits among the stakeholders.

Conserving large carnivores is important because they are important for wider biodiversity protection. As discussed previously, they are among the keystone, umbrella and flagship species of special ecological significance.

COEXISTING WITH PARASITES

Parasites are creatures that extract benefits from other individuals while harming them relatively slowly, often over long periods, during which (in the case of animal hosts) there may be considerable prolonged pain. In contrast, carnivores kill their prey quickly, which from a human perspective seems preferable as it limits suffering. Also, most carnivores are hunters, a lifestyle that appeals to many humans. Perhaps over half of all species are parasites (Windsor 1998; most disease-causing species, for example, are parasites). Almost every free-living species is host to parasites, so by default parasites outnumber their hosts. Parasites, like predators, play a critical role in regulating the success of species in nature, and ecologists frequently argue that this is 'good' because it maintains a 'desirable' balance of species and resources so that ecosystems remain 'healthy.' Biological control of pests often involves employing their parasites to reduce their damage. Arguably (see Chapter 15), some parasites even provide some direct health benefits to their hosts, such as stimulating the immune system, despite being basically detrimental. In theory, some parasites may even be essential to the welfare or even the survival of their host species, but there do not seem to be known examples (and in any event this situation may more resemble symbiosis than parasitism). 'Most parasite species remain understudied, underfunded, and underappreciated' (Dougherty et al. 2016; compare Wood and Johnson 2015).

Although there are many kinds of parasitism, the public image of parasites gravitates to tiny creatures that extract

FIGURE 11.10 Street demonstration on behalf of parasite conservation. Prepared by B. Brookes.

nourishment by attaching long term to skin or occupying the body internally, as exemplified by leeches and tapeworms, respectively. Biting parasitic insects like mosquitoes, bedbugs, and ticks additionally reinforce the gruesome reputation of parasites. Such species are considered repulsive and disgusting, and as discussed earlier, the disgust emotion in humans appears to be instinctive – selected to be genetically fixed in our psyches in order to avoid diseases. Almost all members of the public intensely despise parasites, and indeed even in the profession of biology, only a few parasitologists advocate for the conservation of parasites (Carlson et al. 2020). The thieving, pain-inducing lifestyle of parasites is repugnant to people since comparable behavior in humans is considered immoral (Windsor 2021). The parasites that harm human beings are responsible for great suffering, and there are continuing efforts to control them and even drive them to extinction. Modern medicine is constantly improving techniques to address diseases. Unfortunately, microorganisms are principal causes of disease, and they constantly mutate, requiring new treatments. Many animals carry and spread disease-causing microorganisms, and new viruses affecting people evolve from wild animals, and so there is an ongoing conflict not just with microorganisms but also with their animal vectors. In addition, there are innumerable parasites that are hazardous to animals valued by humans. It seems futile to attempt to generate respect for parasites (Figure 11.10) when so many other species are currently in desperate need of conservation. The key issue regarding parasites is how to balance hatred for them with the need to avoid control measures that would damage ecosystems to the point of harming ourselves.

COEXISTING WITH INSECT PESTS

The most hated insect pests (such as those illustrated in Figure 11.11) are examined in Chapter 15. Insect pests are responsible for transmitting considerable disease, substantially decreasing agricultural and forestry production, and damaging buildings. The public has no tolerance, indeed considerable hatred of insect pests, and they are considered appropriate targets for merciless extermination by any means. However, the key to appropriate control measures is to 'reduce the pest impact to a level where the marginal cost of further measures would exceed the marginal revenue to be gained' (May 1976; cf. Uspensky 1992). The

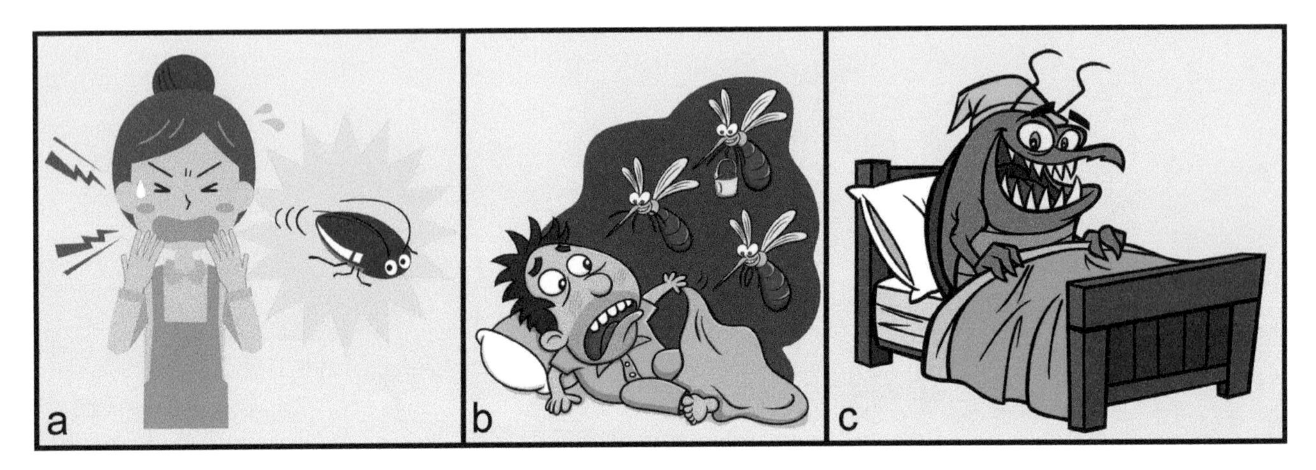

FIGURE 11.11 Most of the world's species are insects, and while most are harmless, there are so many that are harmful and difficult to eradicate that they are especially hated. Three of the most despised are illustrated here: (a) the cockroach (contributor: Mono_Abe), (b) the mosquito (contributor: Chompoo), and (c) the bed bug (contributor Larry-Rains). Source: Shutterstock.

predominant means of insect control is insecticides, which pose dangers to the ecology of the world and to the health of people. As also observed by May (1976), 'we must in the future be more selective in our use of insecticides, and more efficient in their application, than we have been in the past.'

COEXISTING WITH VERTEBRATE PESTS

The most hated vertebrate pests are examined in Chapters 16 and 17. Some vertebrates cause considerable damage to agriculture, property, human health and safety, and natural resources (Witmer 2007; Figure 11.12), and inevitably there are demands that they be eliminated. As for insect pests discussed above, the key is to minimize control measures at least to the point where more harm than good results to people. The most serious vertebrate pests are mammals and birds although snakes kill many more people than all other vertebrates combined. Because people have much higher regard for them than for 'lower creatures,' many pest species (such as feral dogs, cats, and horses) and urban wildlife (such as squirrels, rabbits, foxes, and raccoons) are tolerated to a considerable degree and are often even afforded legislative protection (Broom 1999; Littin et al. 2004). As a result, in advanced countries such as the U.S., some vertebrate pests such as deer, geese, and blackbirds have enjoyed population increases (Dolbeer 1999). In some tropical and subtropical countries, aggressive vertebrate pests such as wild monkeys and even leopards are tolerated in cities.

COEXISTING WITH WEEDS

Weeds and other harmful plants are discussed in Chapters 18–21. Weeds, like animal pests, are quite detrimental to agriculture, and similarly, the chief control measure has been the excessive application of chemicals, i.e., herbicides, which have negative consequences for the environment, biodiversity, and human health (Nimal 2014; Figure 11.13). Weeds of course are plants, and the public is generally very fond of plants, with the exception of lawn and garden weeds. In Western countries, farmers constitute a small fraction of

FIGURE 11.12 Leading vertebrate pests. (a) Snakes are the most dangerous vertebrates, killing about 100,000 people annually, and provoking more fear than any other animal. Source: Shutterstock, contributor: Sogno Lucido. (b) Rats and mice cause enormous economic damage. Source: Shutterstock, contributor: BlueRingMedia.

FIGURE 11.13 'Progress' in weed control technology. Weeds of agriculture are the most harmful of plant pests. In the past, weeding was done manually, which was labor intensive but safe for people and the environment. Today, chemical control is much more efficient but is associated with health risk and environmental pollution. (a) Removing weeds by hand from a flax field, painting (public domain) by Belgian artist Emile Claus (1849–1924), housed in the Royal Museum of Fine Arts Antwerp. Photo credit: Ophelia2. (b) Tractor spraying herbicide on a field. Photo by Aqua Mechanical (CC BY 2.0).

the population so most people have little appreciation of the damage caused by agricultural weeds. The most damaging weeds of natural environments are ornamental plants that have escaped from gardens. They usually have beautiful flowers, and so the public has much less sympathy for campaigns to eliminate them than for comparable efforts to eradicate mosquitoes and other insect pests. Additionally, many weeds of native environments are wild invasive aliens, usually imported unintentionally from foreign locations. Unlike animal pests, very few plants seem ugly, disgusting, and dangerous, although many are extremely damaging to native species and ecosystems. Most people are, at worst, indifferent to most plants but have very negative views of numerous animals. The resulting much greater public tolerance for plant pests compared to animals has promoted the spread of weeds.

COEXISTING WITH HARMFUL SPECIES: FINDING A *MODUS VIVENDI*

Singh and Rahman (2012) stated 'Modern man with his thirst and hunger for economic growth and blinded by science and technology has strayed from a life of coexistence with nature.' However, restoring peaceful coexistence can be challenging since an enormous number of species now threaten our health and economic welfare. Tolerance is an ideal (or at least idealistic) value, but it's difficult to coexist peacefully with species that don't have our best interests at heart (Figure 11.14). Indeed for the entire recorded history of people we have been fighting competing species for our own survival (Winston 1997). Making matters worse, the risk from many noxious pests is growing as a result of increasing globalization and environmental change (Spence et al. 2020).

'*Modus vivendi*' means a feasible arrangement or a practical compromise. Of course, we need not be kind to predators, parasites, and pests of humans and our supporting animals. Practicality and common sense, as well as ethical/moral considerations, should govern when and how we respond to the dangers presented by other species. The commonsense view that is advocated in this book is that control measures need to be based on objective analyses of harm vs. benefit, with due consideration not only for commercial interests but also society, biodiversity, and the planet. And, we need to be mindful that prejudice against many species results in excessively harsh treatment.

COEXISTING WITH PROTECTED SPECIES THAT LIMIT PROPERTY RIGHTS

Well-meaning conservation efforts have frequently limited certain rights of people by forbidding them to harm protected species or their habitats when they occur on their property. Such species may actually be dangerous to health (for example, in Asia villagers are often forbidden from injuring big cats that regularly attack children). In Western nations, however, protected species are merely a threat to the economic welfare of some people. Nevertheless, nothing is as counterproductive to the purpose of conservation as angering the people who live in critical conservation areas (Bowen-Jones and Entwistle 2002; Walpole and Leader-Williams 2002). The presence of a rare species which prevents the free use of land may outrage the landowners or those accustomed to access (Billé et al. 2012; Figure 11.15). Even where it is clear to the local population that habitat destruction and excessive harvesting of biodiversity is harmful in the long term, very poor people may have little choice but to engage in unsustainable activities. Clearly in such circumstances conservation needs to be carried out in ways that respect the sensitivities of local people and the potential economic harm they might incur.

At times, conservation of the habitat of a rare species leads to conflict with business interests, who wish to use the area occupied for economic development. The conservation

FIGURE 11.14 Peaceful coexistence with harmful species. Prepared by B. Brookes.

achieved in such cases is often at the cost of antagonizing powerful political and commercial forces, indeed sometimes also a large population that would have benefited from the area that has been reserved for maintaining a threatened species (Beattie 1995). There are several particularly controversial examples. Furbish's louse-wort (*Pedicularis furbishiae*; Figure 11.16a) is an endangered, endemic perennial herb confined to the shores of the St. John River in Canada and the United States. A hydroelectric project worth over $200 million was canceled by the U.S. Congress in 1986 because it would have reduced the species' habitat, leading to criticism that important economic development was being sacrificed for a 'lousy plant.' The northern spotted owl (*Strix occidentalis caurina*; Figure 11.16b) is a threatened species found from southern British Columbia to California. Protection of the owl's forest habitat became a cause célèbre in the latter half of the 20th century, pitting environmentalists against loggers and sawmill operators. The preservation of species such as these represents victories on behalf of biodiversity. However, this has been at the significant cost of discouraging some business interests and politicians from supporting conservation. Perversely, it may be that some landowners are motivated to destroy habitat before it becomes 'infested' with an endangered species, as has been alleged to have occurred for the golden cheeked warbler (*Setophaga chrysoparia*), an endangered native of Texas (Parenteau 1998; Figure 11.16c).

Recovery/reintroduction programs require fair treatment of people negatively affected by new limitations on the uses of their property, but unfortunately it is sometimes not in the nature of governments to provide adequate financial compensation. In any event, in this situation, balancing ecological sustainability and social justice is difficult (Pinkerton et al. 2019).

PHYSICAL SEGREGATION OF DANGEROUS SPECIES FROM HUMANS

Territorial Separation

Strategically, enemies can be controlled or eliminated by hostile measures or accommodated by tactics which reduce their harm to tolerable levels. Nations warring over territory have frequently decided that the costs of continuing warfare in relation to the potential benefits are best reduced by compromising – usually by dividing the disputed land. A key consideration is that territorial separation of an enemy reduces his threat. Pests are often highly invasive, and not easily excluded from one's homeland, but there are continuing attempts. Nations maintain border inspection stations

FIGURE 11.15 Example of frustration with laws protecting endangered species at the cost of obstructing people. Prepared by B. Brookes.

FIGURE 11.16 Examples of initiatives to preserve endangered species and their habitats that resulted in severe backlash from economic interests determined to exploit the land occupied. The photos are public domain from the U.S. Fish and Wildlife Service. (a) Furbish's louse-wort (*Pedicularis furbishiae*). (b) Northern spotted owl (*Strix occidentalis caurina*). Photo credit: J. and K. Hollingsworth (c) Golden cheeked warbler (*Setophaga chrysoparia*).

FIGURE 11.17 Control of dingoes in Australia. (a) Dingoes. Source (public domain): Gould, J. 1863: *The Mammals of Australia, Vol. 3*. London, J. Gould. (b) Dingoes on the hunt. Source (public domain): Fuertes, L. A., and E. H. Baynes. 1919. *The Book of Dogs; An Intimate Study of Mankind's Best Friend*. Washington, DC, National Geographic Society. (c) Map showing location of dingo fence in Australia. The Dingo Fence (also called Dog Fence) is a barrier that extends over 5,600 km to exclude predatory dingoes from south-eastern Australia where sheep are widely raised. Based on a map by Roke~commonswiki (CC BY SA 3.0).

to exclude pests that might accompany returning travelers. Within countries, there are sometimes similar large-scale attempts to exclude pests, such as rat exclusion in Canada's province of Alberta (Bourne 1998) and the 'Dingo Fence' in Australia (Figure 11.17).

In many countries, the richest people often live in prestigious enclaves, sometimes 'gated' and guarded to prevent unauthorized people from entering (Figure 11.18b and c). This is the modern equivalent of walled villages and walled quarters within cities, which similarly served for security (Figure 11.18a). While not concerned with military attack, as fortresses were, nevertheless security is the apparent chief motive. This may be innocent, for example reflecting a need to have exclusive access to facilities such as swimming pools and tennis courts for which the inhabitants have paid hefty fees. However, in some countries, prejudicial social exclusion of minorities is associated with gated communities. Gated communities do facilitate control of the natural ecology of an area, for example to establish desired landscape plantings, control weeds, and exclude pest animals such as rats. Such areas are usually immaculately maintained, and as such, they exclude much or most of local natural plants and animals. Moreover, since wealth is associated with political power, adjacent areas that are sources of undesirable smells and pests such as mosquitoes are likely to be 'sanitized.' The conspicuous consumption of the rich inhabitants is usually associated with excessive use of fossil energy and water, and generation of pollutants, so there is an ecological cost that belies the artificial beauty of the buildings and landscapes. Moreover, there is a disconcerting parallelism of excluding 'undesirable' people and other undesirable species (Figure 11.18d).

Living on Earth in Biodiversity-Free Bubbles

At least theoretically, one can consider generalizing the concept of private gated communities to living in even more exclusionary 'bubbles' that keep out all, or at least most of the other species on the planet.

There is extensive information on the ability of humans to exclude microorganisms such as viruses, bacteria, and numerous species of fungi (some of these are large, although most are microscopic). These are ubiquitous on Earth, and even occur within the bodies of living humans, so it is virtually impossible to completely escape the presence of other organisms. Indeed, humans and other species have evolved immune systems to fight off the inevitable challenges from microorganisms. Because of medical susceptibility to some 'germs,' a few unfortunate individuals need to isolate (sometimes in so-called 'bubbles'). Biological laboratories, manufacturing facilities, and research facilities often also need to maintain localized sterile sites. The reverse of keeping microorganism out of a location is keeping them confined in a location. All of these endeavors to isolate microorganisms are extremely challenging. The world's deadliest viruses are maintained for research in Containment Level 4 labs, but these are very elaborately engineered facilities and are expensive to operate.

Completely preventing microorganisms from contacting humans or valued property is impossible at present. However, countries maintain regulations and border inspections to keep harmful animals and plants from entering, and it is possible to reduce pests that have become established, albeit controlling them is a never-ending costly battle. Chapters 14 and 15 address new technological solutions under consideration for eliminating pests.

Space Colonization: Which Species Should Accompany People?

A few humans prefer to live as hermits, but generally just to remain apart from other humans, not necessarily from other species. Some humans have established secure underground bunkers in the hope of surviving nuclear catastrophes, which would likely greatly reduce other species, at least locally and temporarily. Perhaps the most extreme attempts to live in isolation from other species have been carried out as experimental scientific projects, frequently to determine just how few species are essential to human

FIGURE 11.18 Gated communities. (a) Damascus Gate, Jerusalem, illustrating a classical city with strong walls for security. Photo by twiga_swala (CC BY SA 2.0). (b) Entrance to a modern gated community in Tempe, Arizona. Photo by Nick Bastian (CC BY ND 2.0). (c) London's One Hyde Park, the most expensive gated community in the world. Some of the 86 units in this apartment complex have sold for over $200 million. Photo by Ell Brown (CC BY SA 2.0). (d) A modern gated community provides opportunities to exclude pests, much as it also restricts entry of undesired humans. Prepared by B. Brookes.

survival in order to colonize other planets. Such systems have been constructed on Earth, the best known example being Biosphere II, a research facility in Arizona prepared in the 1980s with the goal of researching the operation of a mini-ecosystem in a completely enclosed environment that could support humans. Many unanticipated challenges occurred, numerous of the species died, and the project collapsed (Rand et al. 2020; Biosphere II was taken over by the University of Arizona and is current employed for various research and teaching projects).

In the 1970s, the U.S. National Aeronautics and Space Administration/Ames Research Center commissioned a number of artistic conceptions of potential space colonization (see https://space.nss.org/settlement/nasa/70sArtHiRes/70sArt/art.html. These were based on reproducing portions of Earth – including buildings, forests, farms, and even rivers along with bridges in gigantic space colonies. Today, such a scenario – simply scooping up species, landscapes, and cities – seems utopian. Indeed, as discussed next, it is likely that very few species will be allowed to leave our planet.

Humans are destined to attempt to colonize areas outside of the Earth, variously termed space colonization, space settlement, and extraterrestrial colonization. While the possibility theoretically exists that there are planets that already are habitable, it is presumed that people will have to employ advanced technologies in order to survive. To remain away from the Earth for lengthy periods, it will be necessary to employ 'controlled (closed) ecological life-support systems.' These are self-supporting systems to provide food, reusable water, and oxygen for space stations and colonies and are based on a very limited number of plants and microorganisms (Ferl et al. 2002; Arena et al. 2021). Human spaceflights and the operation of an orbiting space

FIGURE 11.19 Plants grown on the International Space Station. (a) Dwarf wheat growing in the Advanced Plant Habitat (APH), an experimental facility on the Earth-orbiting International Space Station (ISS). Photo (public domain) by U.S. National Aeronautics and Space Administration (NASA) in 2018. (b) Mizuna (Japanese mustard greens; *Brassica rapa* var. *niposinica*) in the APH on the ISS. Photo credit: NASA.

outpost have been achieved. Experimental growth of plants has been underway for some years on the International Space Station (Figure 11.19), and there are conceptions of crops being grown in gigantic space stations (Figure 11.20) and on the moon and planets in greenhouses (Figure 11.21). Clearly, only an extremely restricted number of plants and indeed possibly animals as well would be included in such colonies. Obviously the pragmatic values of these species would be the principal basis of choosing species to include, but just as obviously human preferences and prejudices would also inevitably become important. On Earth, millions of species interact to maintain natural ecosystems, but as described in this book, most of these are disliked, and probably exporting species to space would require some sort of permit. As noted, the millions of species on Earth provide indispensable and reliable ecological services, but space colonies are simply not natural, and experience has shown that transporting species to foreign places often is disastrous, so preventing species from traveling outside of the planet may actually be desirable. In the very long term, it would also be necessary to confront the sensitive question of which races of humans would be chosen to reproduce people in space. Much of biology is concerned with nature as it is found on Earth, but the artificial, extremely simplified systems that are likely to be established elsewhere in the universe are likely to present quite different concerns to scientists.

BIOLOGICAL WARFARE

People have employed large domesticated animals such as dogs, horses, and elephants as military weapons for millennia, but, ironically, the potential damage they can cause is much more limited than the use of tiny species. 'Biological warfare' is the employment of biological agents (live organisms or their products) to harm an enemy ('bioterrorism' may be a synonym but emphasizes threats against civilian populations). This may be based on the use of toxic chemical extracts such as ricin (a toxin from the castor bean plant, *Riccinus communis*) to physically harm people. (However, the use of chemicals is alternatively placed in the field of 'chemical warfare.') When infectious agents such as bacteria, viruses, insects, or fungi are employed to harm humans or their valued crops and livestock, the field is often referred to as 'germ warfare.' Crude employment of biological warfare (such as introducing individuals suffering from a contagious disease into an enemy population) date back to antiquity (Riedel 2004; Barras and Greub 2014; Mojoodi et al. 2020). British troops in conflict with native Americans are alleged to have deliberately infected Native Americans with smallpox (Figure 11.22a; there is no doubt that European diseases introduced by colonists caused pandemics which decimated the native population by about 90%). Biological and chemical warfare are banned under international regulations but are occasionally employed surreptitiously by unscrupulous regimes. While some sociopathic, immoral powers may favor the weaponization of infectious species, this is surely an area beyond issues of prejudicial attitudes. From a military viewpoint, the use of biological agents may be unwise, because once a contagion is established among the enemy it may spread to one's own homeland. Also, like nuclear weaponry, the results can be so catastrophic that one wants to avoid similar usage by the enemy. The destructive effects would likely extend to the natural world and harm biodiversity (Dudley and Woodford 2002). The COVID (SARS-CoV-2) pandemic illustrates how difficult it is to control a biological agent once it has been released. Most scientists have concluded that it probably arose in nature from an animal and was transmitted to humans. However, some suspect that the virus leaked from the secretive Institute of Virology (Figure 11.22b) in Wuhan, China, where the first COVID-19 cases were reported (Maxmen and Mallapaty 2021). The future possibilities of creating 'designer' biological agents are terrifying, and it is disturbing that considerable technology is being tested without public scrutiny.

FIGURE 11.20 Conceptions (public domain) of giant space colonies prepared in the 1970s by artists commissioned in the 1970s by the U.S. National Aeronautics and Space Administration. Note that ecosystems of Earth have been essentially reproduced in space. (a) Giant doughnut-like ring 1.8 km in diameter, which rotates once per minute to provide artificial gravity on the inside of the ring via centrifugal force. The population would be similar to a dense suburb, with part of the ring dedicated to agriculture and part to housing. Artist: Don Davis. (b) View inside the ring. Artist: Don Davis. (c) A double cylinder colony. Each of the giant cylinders was designed to hold up to a million people, essentially reproducing the ecosystem of Earth. Art work: Rick Guidice. (d) View inside one of the cylinders. Artist: Don Davis.

FIGURE 11.21 Figure 11.21 Artist conceptions of controlled environment space farms. (a) Greenhouses on Mars. Photo (public domain) released by NASA in 2004. (b) 'Lunar farm.' Photo credit: Space Studies Institute (CC BY 3.0).

FIGURE 11.22 Pandemic potential of viruses. (a) Scene showing Pontiac, an Ottawa Indian chief, meeting Colonel Henry Bouquet (1719–1765), a leader of British forces, at Muskingum, Ohio. Bouquet is alleged to have authorized his troops to spread smallpox among Native Americans by deliberately infecting blankets after peace talks. Painting (public domain), in 1766, by Benjamin West (1738–1820), American-British artist. Source: U.S. Library of Congress. (b) Wuhan Institute of Virology, China. Photo by Ureem2805 (CC BY SA 4.0). (c) International symbol (public domain) for a biohazard (toxin of biological origin or a microorganism).

DANGEROUS PETS

Wild Animals

Wild, undomesticated animals dangerous to humans (such as shown in Figure 11.23) generally do not belong in private households, not just because they can harm the occupants but because they can escape and injure neighbors. In many jurisdictions (ranging from entire countries to cities), keeping wild animals is simply illegal. In additional to harm for people, there is potential harm for the animals, since

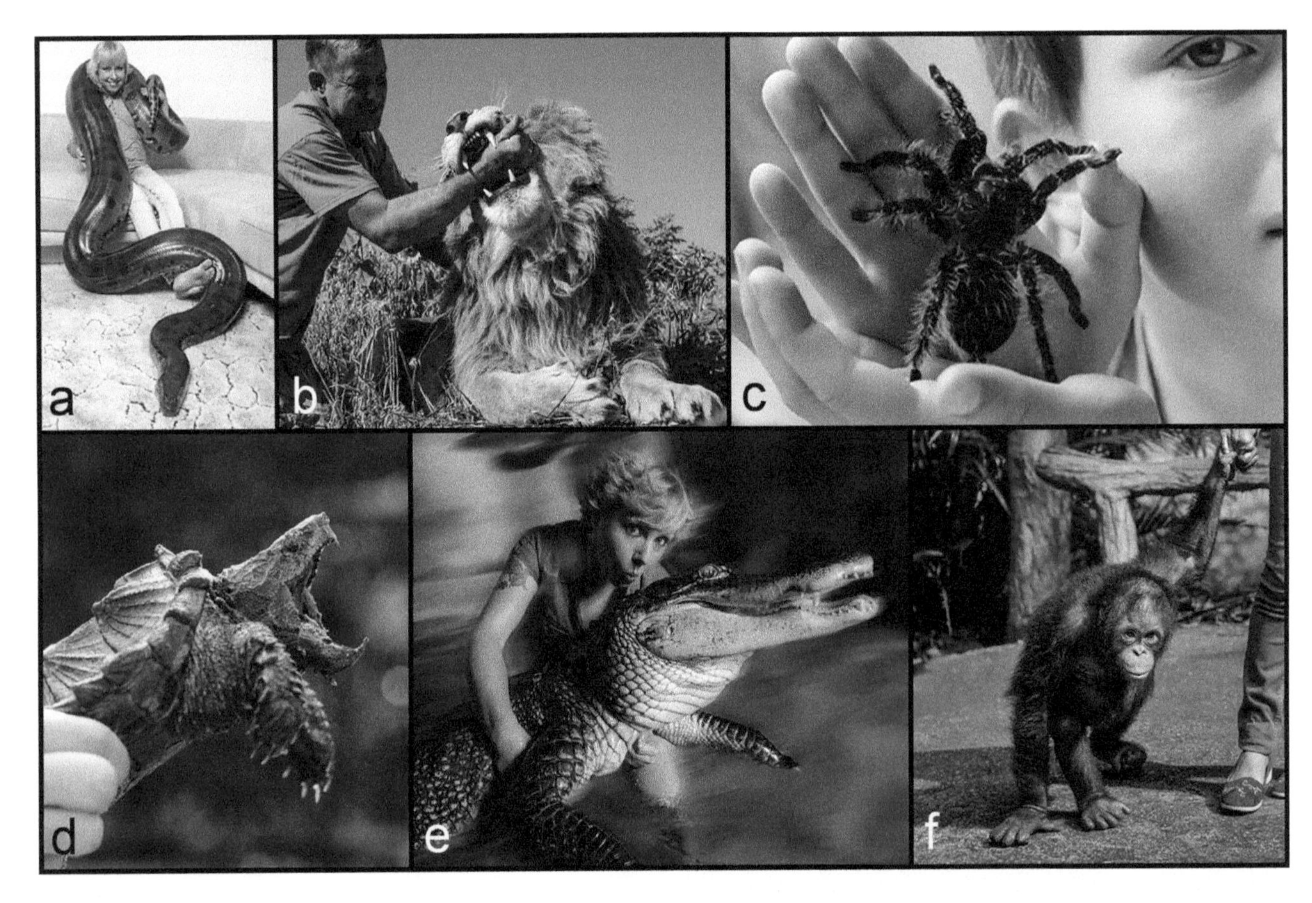

FIGURE 11.23 Dangerous wild pets. Source of illustrations: Shutterstock. (a) Anaconda. Contributor: Holger Kirk. (b) Lion. Contributor: Kit Korzun. (c) Tarantula. Contributor: Lipatova Maryna. (d) Alligator snapping turtle. Contributor: Sista Vongjintanaruks. (e) Crocodile. Contributor: Holger Kirk. (f) Orangutan. Contributor: Yatra4289.

TEXTBOX 11.1 WHY DO PEOPLE KEEP WILD ANIMALS AS PETS?

Some animal studies scholars see an analogy between economic and cultural subordination of the foreign, exotic peoples by the European colonial powers in the sixteenth and seventeenth centuries and the contemporary desire to possess and display an exotic animal in one's home. In the past, only the kings and princes could afford to maintain zoos or menageries which testified to their wealth and power. Even today exotic animals are much rarer, harder to obtain, and more difficult to maintain so tigers, bears or apes kept in the house point to the high social status of their owners. Moreover, the wild animal may be attractive for some people precisely because of its wildness... The owners also often want to prove that they have power over it and to shock people. It is noteworthy that gender may also influence exotic pet preferences. While male owners prefer dangerous predators, women are more likely to adopt 'cute' creatures.

—*Rutkowska (2017)*

few have the expertise and facilities to maintain them safely in good health. Moreover, few veterinarians will deal with such risky animals, so their welfare simply can't be met in captivity, outside of zoos (Loeb 2020). The motivations for keeping dangerous wild animals vary (Textbox 11.1). Some may have scientific or conservation reasons for keeping dangerous wild animals. Frequently, abandoned or injured animals are rescued as cute babies. Primates (monkeys, gorillas, orangutans) are so similar to humans that they have natural appeal to people, especially when young. As noted in Chapter 4, size of animals, and their associated power, very strongly elicit human respect, and so big, aggressive, carnivores such as the big cats (especially lions and tigers), constrictor snakes (especially pythons), and crocodilians (alligators, crocodiles, and others) are especially appealing. Some people enjoy possessing dangerously venomous animals, such as snakes and spiders, often presumably because they like to have unique possessions that impressive others.

There is a large, legal, quasi-legal, and illicit trade in so-called 'exotic' animals originating from foreign lands. While most (especially fish and birds) have little likelihood of physically harming people, they have the potential of escaping and becoming invasive aliens. Animals acquired from the wild can vector dangerous diseases like rabies, and although the possibility is remote, it is conceivable that viral diseases that they harbor could mutate into new significant infections of people (like COVID).

Domesticated Animals

Dogs are the leading cause of injuries from pets. Countless people are bitten by dogs, not just breeds considered to be dangerous (Textbox 11.2). In some countries, dogs are the leading animal responsible for killing people, often because they are the chief transmitter of rabies in the region. Dog bites are a serious public health problem but are largely preventable, requiring dog owners to manage their pets responsibly (Figure 11.24). It has been said that 'there are no bad dogs, only bad owners.'

TEXTBOX 11.2 DANGEROUS DOGS

There is no credible scientific evidence that any one breed of dog is more likely to bite than another. All breeds and sizes of dogs have the potential to bite – from American pit bull terriers and rottweilers to chihuahuas and Yorkshire terriers. Despite this, large dogs are capable of delivering more than 450 pounds of pressure per square inch in a single bite – enough to penetrate light sheet metal – making the bite of a large dog more capable of delivering injury, compared with the bite of a small dog... The vast majority of animal bites to people involve dogs (85%–90%), followed by cats (5%–10%).

—*Christian (2011)*

Recent media interest regarding dangerous dogs has made the dogs the focus of attention. The focus should be on their owners. Any dog will bite if not socialized, and it is inadequate/irresponsible owners that have tended to be the major factor, not the dogs themselves. However, the more powerful the dog, the more disastrous the consequences can be, particularly with children and other vulnerable individuals. Irresponsible owners can be divided roughly into two groups. The first lacks knowledge on how to train dogs, or doesn't care, while the second... uses the dogs for nefarious purposes, for example protecting criminal assets, intimidation, or attacking people or other animals... Dogs used for protection and intimidation are invariably larger and/or more powerful... In the right hands, with good care and training, none of these dogs is any more likely to bite or attack a human than any other breed or type of dog. However, the injuries caused by pit bulls trained to be aggressive can be particularly serious, as this breed has an intensity and duration of attack not seen in other dogs. This leads to severe injuries and, in some cases, death of the victim (human or animal). This trait has been created by cruel individuals who want animals that are 'game' for dog fighting.

—*Grant (2011)*

FIGURE 11.24 Bad dogs. Source of illustrations: Shutterstock. (a) Contributor: Khabarushka. (b) Contributor: Guingm.

12 Reforming Agriculture
The Greatest Threat to Species

INTRODUCTION

Agriculture, which now provides the bulk of the world's food, has become the chief threat to the ecology of the planet (Dudley and Alexander 2017; Bélanger and Pilling 2019; Gustafson et al. 2020). Because agriculture generates more harm to the world than any other activity, making it more environmentally friendly can be of greater benefit to the world's species, including humans, than perhaps any other endeavor. The intimate link between agriculture and biodiversity is examined in publications by the Food and Agriculture Organization of the United Nations (2019a, b).

Previous chapters in this book have examined in detail our preferences for the relative physical attractiveness of species. In agriculture, appearance is much less important than utilitarian values. Nevertheless, there is a disturbing parallelism between the short-sighted aesthetic evaluation of species (including humans) simply on the basis of their looks and the equally short-sighted market evaluation of species simply on the basis of their utilitarian values. As discussed in this chapter, our adulation of a very small set of crop and livestock species has enslaved us to rely on them so excessively that our planet and its life support systems are being degraded. As pointed out previously, bias against less attractive wild species obscures their considerable economic and ecological values. In the same vein, bias against many of the agricultural species currently of lesser market value also obscures their considerable economic potential and ecological values.

OUR ANCESTRAL WAY OF LIFE: DIFFICULT BUT SUSTAINABLE

Humans are animals, and like all other animals, we require food for survival. Animals may be herbivores (plant eaters), carnivores (flesh eaters), or omnivores (consumers of both plants and animals). Animals may also differ in the range of species consumed, some restricted to few, others to many. Some species obligately feed primarily or entirely on an extremely narrow range of other species. For example, the giant panda subsists almost exclusively on bamboo species. Others, such as pigs and the common rats, will eat almost anything edible. Animals that depend on very few species for food are constantly in danger of starving, even going extinct, if their primary food sources become unavailable. On the other hand, animals capable of eating numerous other species are less likely to starve. Sometimes the food source species are endangered by a population explosion of the species that consume them. Many animals are mobile, and when food becomes unavailable locally, they may be able to migrate to a more hospitable area. Sometimes, however, an imbalance between consumers and what they eat may significantly disturb an ecosystem, resulting in the loss of one or more species. Such pressures can result in the evolution of new species or even new ecosystems, so that on balance nature remains healthy. Humans are fortunate in being omnivores, and today, we consume a wider range of foods than any other species. In pre-agricultural times, people were nomadic hunter-gatherers, typically occupying an area until food resources became scarce, and setting up temporary camps elsewhere (Figure 12.1). This limits local damage to ecosystems from harvesting of wild animals and plants, allowing time for recovery. In a word, this ancient lifestyle was *sustainable*.

HOW HUMANS CREATED A DANGEROUSLY FRAGILE OVERDEPENDENCE ON JUST A FEW KEY FOOD SPECIES

Like many other animal species, humans expanded their populations until the food supply from their territories became limiting. This set the stage for the invention of agriculture (Cohen 1977). To support growing populations and prevent starvation, it became necessary to cultivate crops and raise livestock in order to provide sufficient food (Bender 1975; Harlan et al. 2012). Agriculture and other technological innovations created a new relationship with nature. By domesticating selected animals and plants and raising them on a huge scale in artificial ecosystems, we are now, to a considerable degree, free of the need to obtain food from wild species. But this was part of a Faustian bargain, as humans are now dependent on domesticated species.

DOMESTICATED ANIMALS AND PLANTS: AN INESCAPABLE UNION WITH HUMANS

In common language, the word 'domesticated' often describes an animal that is 'tame' or 'tamed.' However, domestication in the scientific sense specifically refers to the process of genetically altering a group of living things so that their characteristics have become more adapted to surviving in association with humans. In the past, most such selection was probably unconscious (i.e., people were unaware or at least not clearly aware that the plants and animals under their care were changing in desirable ways). Today, professional breeders deliberately select features so

DOI: 10.1201/9781003412946-12

FIGURE 12.1 Scenes of former hunter-gatherer societies. (a) Crow Indians hunting bison. Diorama in the Milwaukee Public Museum. Photo by Evan Howard (CC BY SA 2.0). (b) Stone-age gatherers, a diorama in Hong Kong Museum of History. Photo by Musetress (CC BY 3.0). (c) Diorama of Indians fishing. Credit: Royal Alberta Museum. Photo by Jason Woodhead (CC BY 2.0). (d) Indian hunters and wild rice gatherers. Painting (public domain) by Margaret Martin in 1939, housed in the Smithsonian American Art Museum. (e) Hunter-gatherer nomadic encampment of Crow Indians. Public domain painting by American artist Charles M. Russell (1864–1926). Photo by mark6mauno (CC BY 2.0). (f) Paleolithic era hunting scene (public domain). Credit: MothsART.

that the resulting plants or animals (and some fungi and bacteria) are superior as sources of food and other useful products.

Agriculture is the controlled production and harvest of animals and plants, and almost all of these species are either domesticated livestock or domesticated crops. Most livestock are raised for food (and also used secondarily for other products, such as hides). Most crops are also raised for food (for humans and livestock) but are also employed for fiber, industrial products, and production of ornamental plants. Especially valued crops are sometimes assigned titles like 'King of Vegetables' or 'Queen of Cereals' (Table 12.1). Some undomesticated wild animals, such as ostrich and bison, are raised to a minor extent, and similarly so are some undomesticated plants, such as ginseng (for medicine) and milkweed (the floss attached to the seeds used for fiber). Most agriculture, especially for crops, occurs on intensively managed, confined property. In ranching (which is a part of agriculture), herds are managed for livestock production on natural pastures or at least grazing lands. Shepherding is the same, but an individual (the shepherd) provides regular, often daily supervision of the animals, often because they are especially susceptible to predators. The carrying capacity of the world in pre-agricultural (pre-Neolithic) times is speculative (estimates from 10 million to over 1 billion have been suggested), but it is clear that without domesticated species the world would be unable to support the present human population, let alone the projected expansion to 10 billion in 2050 (United Nations 2019; Ortiz et al. 2021).

Hunting, fishing, and foraging, which sustained humans in pre-agricultural times, cannot supply enough food to maintain more than a fraction of the world's people. While it is conceivable that technical innovations could one day liberate humankind from dependence on domesticated plants and animals (Figures 12.2 and 12.3), at least for the foreseeable future most people would simply starve without them. Indeed, today four crops (wheat, maize/corn, rice, and potato) provide over half of plant-based calories in the human diet, while about a dozen species of animals (especially cattle, chickens, and pigs) provide 90% of the animal protein consumed globally (Chiarelli and Annese 2009). Unfortunately, extremely

TABLE 12.1
'Royalty' Designations of Highly Valued Crops

Crop Category	King	Queen
Beverages	Coffee	Tea
Cereals	Rice, Wheat	Corn (maize)
Fiber	Cotton	Flax (linen); also silk (produced by silkworms eating mulberry)
Flowers	Rose	Rose
Forage/fodder	Berseem clover	Alfalfa (lucerne)
Fruit	Apple, Mango	Mangosteen
Herbs	Basil, Tarragon	Holy basil (tulsi)
Oilseeds	Soybean, Peanut, Rapeseed	Sesame
Nuts	Walnut	Pecan
Pulses	Chickpea, Lentil	Pea
Spices	Black pepper	Cardamom
Vegetables	Potato	Okra

These informal designations differ according to local popularity.

FIGURE 12.2 Several of the world's most important domesticated livestock and crop species. (a) A cattle herd. Photo (public domain) by U.S. Department of Agriculture, NRCS Texas. (b) Chickens. Photo (public domain) by Bruce Dupree, Alabama Extension. (c) Pigs. Photo by CSIRO, Australia (CC BY 3.0). (d) Rice (upland) harvest in Texas. Photo (public domain) by Lance Cheung, U.S. Department of Agriculture. (e) Potato harvest. Photo by GRIMME Group (CC BY 2.0). (f) Wheat. Photo (public domain) from Pixabay.com.

FIGURE 12.3 European honeybee (also called Western honeybee; *Apis mellifera*). It is, by a considerable margin, the most important domesticated insect species and is a critical contributor to agriculture. The honeybee is responsible for pollinating crops that generate about a third of the food that people eat. It is currently under attack by the parasitic varroa mite, and since wild pollinators are declining, the future of food production is under threat. (a) Honeybee collecting pollen on a flower. Photo by Andreas Trepte, www.avi-fauna.info (CC BY SA 2.5). (b) Honey bee hives in a field of rapeseed (*Brassica napus* subsp. *napus*). The resulting pollination increases crop yield, and the bees additionally produce honey. Photo (public domain) credit: Zachtleven, Pixabay.com. (c) Painting (public domain) of a little girl asking a bee for some honey. Prepared by A.L.O.E. (1821–1893), published in the 1871 book Favorite Picture Book for *The Nursery*.

extensive ecological and environmental damage has been a side effect, since the culture of domesticated livestock and crops has not only usurped vast amounts of land and water previously used by wild species but has also been associated with degradation of adjacent ecosystems and catastrophic reduction of numerous wild species. The meat industry of the world is responsible for almost a third of the total global anthropogenic greenhouse gases (United Nations 2012).

Most people obtain their food from supermarkets and indeed give little thought to the farmers who produced it. Even less appreciation is usually given to the livestock and crop varieties upon which we now depend. However, when driving through farm country, fields of corn, wheat, and other staples do command respect, as they should. Probably most people encounter livestock not in working farms, but at fairs and petting zoos, where they are often admired, albeit the animals have been prettified. Farm animals are very familiar animals, often stereotyped, but in no danger of becoming extinct (although older breeds are often not maintained). The most important farm animals are birds (mostly chickens) and mammals (mostly cows and pigs). The wild ancestors of crops and livestock harbor invaluable genes and are the subject of intensive conservation activities.

UNSUSTAINABLE INTENSIVE AGRICULTURE

Only a few domesticated crops and livestock species provide most of the food consumed by people. Without the current extremely efficient production of food by 'intensive agriculture' on a massive scale, the present world population could not be fed. Unfortunately, the current principal crops are grown as huge monocultures, which by their nature exclude almost all other species and replace natural ecosystems (Figure 12.4). Moreover, heavy employment of synthetic fertilizers, fossil fuels, and pesticides pollutes the soil and atmosphere. At least as damaging to the natural world, the current principal livestock species are raised by 'factory farming' (Figure 12.5) – extremely concentrated (albeit efficient) heavily mechanized production, which like monocultural crops results in alarming environmental contamination. Our food production and harvesting technologies have so severely degraded the planet that we are endangering our survival. Changing these high-efficiency practices to make agriculture more sustainable is very challenging, but fortunately there are some measures that can be taken to alleviate the crisis.

FIGURE 12.4 A contrast of a huge cereal monoculture and a small vegetable farm. (a) A gigantic rice monoculture in China. Photo (public domain) by Sasin Tipchai, Pixabay.com. (b) A small vegetable farm in China. Photo (public domain) by Anna Frodesiak.

FIGURE 12.5 Contrast of small-scale and factory livestock farming. (a) A smallholder farm in Kenya. Photo by Enochkp (CC BY SA 4.0). (b) A factory chicken farm in India. Photo by Matthew T. Rader (CC BY SA 4.0).

THE NEED TO GROW A GREATER VARIETY OF SMALL-SCALE CROPS

Simply increasing the number of crops grown can be advantageous for biodiversity, sometimes with little or no decrease in food productivity (Lanz et al. 2018). Smaller farms and smaller areas devoted to given crops are beneficial because this breaks up the landscape, providing refuges in the intervening and transitional areas where wildlife can survive. Ideal in this respect are 'market gardens' or 'truck farms,' which are small-scale operations producing fruits, vegetables, or flowers as cash crops, often sold directly to consumers, restaurants, and local retail stores Expanding the number of crops grown on a given acreage is also a way of increasing food security, since dependence on very few crops could be disastrous should one of them be threatened by a disease. However, the current trend of large corporations monopolizing huge tracts of land in order to grow gigantic monocultures makes this difficult.

Wise agricultural policies could reduce dependence on the small number of major crops and livestock species by encouraging the production and consumption of a greater variety of sustainable food species. 'Polyculture' is the planting of several crops in a relatively small area (Figure 12.6), a practice which is relatively friendly to the environment compared to monocultures. 'Market farmers,' who supply produce to small, local outlets, often grow vegetables using polyculture. There is expanding consumer interest in ethnic foods, which is supportive of growing a greater variety of vegetables, although in the commercial marketplace these are usually comparatively costly. In rich nations, personal food gardens are more of a hobby than a necessity, but they also contribute to sustainability, and should be encouraged.

FIGURE 12.6 Outstanding examples of small-scale food production. (a) The vegetable garden at the Château de Villandry, Francc. Photo by Marcok (CC BY SA 4.0). (b) A U.S. Government Printing Office poster (public domain), printed 1895–1972, entitled 'Uncle Sam Says, Garden to Cut Food Costs.'

THE NEED TO REVERSE THE TREND OF DISAPPEARING SUSTAINABLE SMALL FARMS

The small family farm (Figure 12.7) is an endangered species in many advanced countries, a victim of large-scale agribusiness (also called corporate agriculture and 'Big Agriculture'). This is a reflection of short-sighted agricultural policy, as small farms strongly promote both agricultural and ecological diversity and welfare (Boyce 2006; Bisht 2013). In poor countries, 'smallholders' are small-scale farms typically managed by families (in essence, they are family farms, but usually very small, and intergrade with what are termed gardens in Western countries). They are extremely numerous and often represent most of the agricultural production of countries, as well as the principal employment. Such small farms may be employed to produce exportable crops like coffee, rubber, and palm oil, but they are a critical source of food for personal use or for sale in a local market. They typically produce vegetables, chickens, and pigs. Often, manure rather than synthetic fertilizers is employed, to the benefit of the soil and the atmosphere. Such restricted operations are physically separated, providing habitat in the intervening zones for many wild plants and animals. While not as efficient as crop monocultures and factory farming, they are far more environmentally friendly. They usually are highly sustainable and compatible with biodiversity, but unfortunately they are being outcompeted by large, industrial-scale agri-businesses (Hazell and Rahman 2014).

PROMOTION OF EMPATHY FOR LIVESTOCK BY SMALL FARMS

The majority of long-term interactions of people and animals occur on farms, where humans may be humane or inhuman in their behavior. Small farming operations facilitate, indeed often depend on an intimate relationship between farmers and livestock (Figure 12.8). Often the farmer knows the personality and needs of each animal, and exercises concern for each, not just because his income depends on the welfare of his animals, but because genuine caring relationships develop. Children growing up on farms especially acquire a love of, and respect for, livestock. In urban areas where small livestock, especially chickens and rabbits, are allowed in backyards, the animals are often treated as pets. Even amateur or hobbyist producers of honey come to greatly value their bees. It probably is not unreasonable to believe that such

FIGURE 12.7 Examples of small farms. (a) 'Evening on the Farm' by O. C. Fisher (1885–1974). (b) 'Rural Free Delivery' by Criss Glasell (1898–1971). Both murals are public domain paintings by American artists, prepared for the Iowa Post Office, and are the property of the Smithsonian American Art Museum.

FIGURE 12.8 Empathetic treatment of farm animals. (CC BY 2.0). (a) A Quechua girl and her llama in Peru. Photo by Thomas Quine (CC BY 2.0). (b) Cambodian farmer and his Asian buffalo. Public domain photo from Pixabay.com. (c) A girl with a chicken in Afghanistan. Public domain photo, credit: M. Lueders/USAID.

sustainable production, based on small-scale operations, produces or at least promotes citizens who care about other people, other sentient species, and the welfare of nature. By contrast, factory farming, as described above, treats animals as collective commodities to be produced with as little humane care as required legally. It probably is not unreasonable to believe that such production coarsens people, and as noted above, industrial-scale agriculture tends to degrade the environment.

THE NEED TO PROMOTE URBAN AGRICULTURE

'Urban agriculture' is an approach to managing cities that emphasizes the use of the local residential, business, and public landscapes, insofar as possible, for food production (Lin et al. 2015; Clucas et al. 2018). This can involve utilization of whatever public spaces are available for small community farms or orchards (Figure 12.9b) and encouraging

FIGURE 12.9 Forms of urban agriculture. (a) Rooftop garden in Rotterdam Netherlands. Photo by Wim de Jong (CC BY 2.0). (b) Community vegetable gardens in the middle of a city (Shanlong) in Taiwan. Photo by Chen Huang (CC BY 3.0). (c) 'Vertical farm' (indoor facility with several levels illuminated by artificial light) in Finland. Photo by ifarm.fi (CC BY 2.0). (d) A tiny streetside vegetable garden on public property in Newport England. Photo by Jaggery (CC BY SA 2.0).

the planting of edible landscaping (often termed 'food-scaping'). In southern areas of the world with long, warm climates, rooftop gardens (Figure 12.9a) are common, and these are beneficial by reducing pressure to convert wild-lands to agriculture. 'Green roofs' include all roofs planted with vegetation, and they all tend to benefit biodiversity (Williams et al. 2014). An important component of urban agriculture is so-called 'vertical agriculture' – multi-story commercial production on a site, especially employing soil-less/hydroponic techniques (Demirbas 2019; Chole et al. 2021; Nijwala and Sandhu 2021; Figure 12.9c). 'Guerrilla gardening' is a special kind of urban agriculture in which edible or ornamental plants are established, often without permission, on public property (Figure 12.9d) or land that someone else owns. This may be simply to beautify land that is uncared for or to make food plants available in neighborhoods where food is scarce.

THE NEED TO REPLACE RED MEAT CONSUMPTION WITH PLANT FOODS

Meat-based vs. plant-based diets has become both a health and an ecological issue with monumental consequences for both people and the planet. Both forms of nourishment are appealing (Figure 12.10). The disadvantages to human health of consuming large amounts of 'red meat' (primarily beef, veal, and pork), combined with the associated generation of waste gases polluting the environment, are persuasive arguments for eating alternative foods. It has long been known that, expressed in units of protein production on a land area or on a cost basis, harvesting crops is much more energy efficient than traditional livestock harvesting (Pimentel and Pimentel 2003). Alarmingly, animal farming is far more damaging to the environment than raising crops. World meat production based on the cow (*Bos taurus*), at least based on market reports, is less than for poultry and pigs (FAO 2021), but cows are valued also for milk. Nevertheless, beef is the largest contributor to global warming of all marketed animal products (de Vries and de Boer 2010). Breeding cows that produce less gas and employing management techniques for this goal (Bell et al. 2012; Lahart et al. 2021) are under exploration, but the potential is unclear. Harvesting protein (in the form of meat) from a ruminant, such as a cow or sheep, produces about 250 times more emissions than a parallel amount of protein from a legume, such as alfalfa (Tilman and Clark 2014).

FIGURE 12.10 Consumption of red meat vs. vegetables: A critical issue for the welfare of the world. (a) A display of vegetables. Photo (public domain) by www.Pixel.la Free Stock Photos. (b) An attractive salad. Photo (public domain) by Tandocjared, Pixabay.com. (c) A display of meat. Photo by Phototram (CC BY 2.0). (d) An attractive hamburger. Photo (public domain) by Shutterbug75, Pixabay.com.

Poultry and sea foods are considered to be healthier choices, but 'plant-based meat alternatives' (meat-like foods manufactured from true [i.e., flowering] plants, algae, or fungi) have potential to be generated more economically and compatibly with the ecology of the world. Health is a persuasive motivation for changing diet, but for most people the taste of food is decisive, and the majority of humans crave the taste and texture of meat. The food industry is attempting to create meat alternatives prepared from non-animal proteins, but with similar appearance, mouthfeel, and aroma to traditional meat (He et al. 2020). 'In vitro meat' (grown from tissue culture, and also known as cultured, cell-based, lab-grown, clean, and cultivated meat) is currently being explored, but the potential value of such 'cellular agriculture' remains to be demonstrated (Bhat et al. 2017). The most promising development is the improving technology for producing plant-based meat analogues (D'Silva and McKenna 2018), until recently a market primarily of interest to vegetarians. Remarkable mimics of hamburgers, sausages, nuggets, and other popular meat products are now being produced (Rubio et al. 2020). Nevertheless, conversion from a predominantly meat diet to a plant-based diet is very challenging for most people. Moreover, it appears that people in different regions of the world have evolved different degrees of relative tolerance or dependence on meat (Buckley et al. 2017). For example, Southern Europeans appear quite tolerant of high-plant diets, while the Inuit of Greenland are better optimized (physiologically, not just by cultural adaptation) to process considerable meat and animal fat (their traditional food).

All species employed in agriculture affect the planet, usually negatively but sometimes positively. The damage or benefit depends on various factors, particularly on where and how the species are raised. Profitability, market demand, competitiveness, labor availability, and similar practical economic considerations are currently employed to determine which species are grown. However, in deciding on the choice of species to be grown, it is possible to also consider the ecological consequences of growing particular species (Montford and Small 1999a, b).

INSECTS AS A CRITICAL INCREASED SOURCE OF PROTEIN

As stressed in this book, most humans are disrespectful of, indeed quite hostile to, insects, which make up most of the world's species. This is extremely short-sighted because they represent the most underutilized large potential food resource (Dossey et al. 2016; Mitsuhashi 2016; Sogari et al. 2019; van Huis et al. 2021). Entomophagy by humans refers to eating insects, although the term is often expanded to include spiders (Evans et al. 2015). Harvesting of insects for food by humans is widespread in some cultures (Figure 12.11). Indeed, insects are a portion of the traditional diets of at least 2 billion people (Doi et al. 2021). Insects are also sometimes employed as feed for animals which are subsequently employed as humans' food. Mealworms, crickets, and locusts are being raised on an industrial scale by 'insect farms,' and to avoid the widespread disgust in Western nations by people at the prospect of consuming recognizable insects, they are typically processed into a flour. The product is nutritious and represents a far more sustainable source of protein for human consumption than currently grown mammalian livestock (Ayensu et al. 2019; Tang et al. 2019). 'Entomophagy can make a significant contribution to insect conservation if they are sustainably harvested in conjunction with appropriate habitat management' (Yen 2009). Compared to traditional sources of meat, insects are thought to be useful for mitigating greenhouse gas emissions, cutting land uses and polluted water, and reducing environmental contamination (Raheem et al. 2019). Videbæk and Grunert (2020) found that younger consumers and males are more positive toward entomophagy than older people and females. Regardless, we all need to become more open-minded to the value of insects.

THE NEED TO REDUCE FOOD WASTE

Thousands of scientific publications are being produced annually on food waste, indicating the importance of the

FIGURE 12.11 Entomophagy (consumption of insects). (a) Roasted crickets in a market near Mexico City. Photo by Meutia Chaerani/Indradi Soemardjan (CC BY 2.5). (b) A food stall in Bangkok, Thailand, selling deep-fried insects. Counter-clockwise, from the back-left to the front: locusts, bamboo-worms, moth chrysalis, crickets, scorpions, diving beetles, and giant water beetles. Photo by Takoradee (CC BY SA 3.0). (c) Centipedes on a stick in Beijing, China. Photo by Denise Chan (CC BY SA 2.0).

topic. Over one billion t of food is wasted every year globally, while almost one billion people go undernourished and another 800 million go hungry (Grosso and Falasconi 2018). Food production is associated with consumption of limited resources (freshwater, fossil fuels), deforestation, pollution (especially greenhouse gas emission), and loss of habitats, so wasting food exacerbates all of the problems associated with agriculture and biodiversity (Feldstein 2017). Food wastage also generates waste, adding to the burden of garbage pollution of the natural world. Waste can occur at every stage of the food chain, from production through processing and distribution, but the leading occurrence is by people in households – estimated at over 40% (Bilska et al. 2020; Figure 12.12a). Discarded food can be employed as animal food (Figure 12.12c) or composted (Figure 12.12b), but a greater respect needs to be generated for food utilization. As noted in Chapter 5, there is considerable waste of perfectly edible deformed vegetables and fruits, termed 'ugly food' (see Figure 5.48). An associated problem with food waste is that the edible garbage produced maintains large populations of pests (Figure 12.13).

THE NEED TO CONSERVE AGRICULTURAL GENETIC RESOURCES

As noted in the earlier part of this chapter, of all the millions of species in the world, humans have become dependent on crops and livestock. Our welfare now rests on their welfare. Crop and livestock species originated from wild species, which often still exist in nature on the brink of extinction. Because domesticated species have been changed genetically, they invariably lack some of the genes still present in the wild, which can be useful for improving crop varieties and livestock breeds. The impending extinction of wild relatives of agricultural species is a huge threat to the future welfare of society. Additionally, old crop varieties and livestock breeds, which are no longer maintained by farmers, similarly frequently retain valuable genes that have been lost in the currently raised kinds, but once again are often nearly extinct. The conservation of old crop varieties and livestock breeds is not a 'sexy' or 'hot' area in the world of biological conservation, and the general public is scarcely aware of its importance. Unlike the conservation of wild species, the public is

FIGURE 12.12 Food waste. (a) Discarded surplus food in a dumpster. Photo by OpenIDUser2 (GNU FDL). (b) Composting food waste. Photo (public domain) by Trish Walker. (c) Backyard heritage chickens (Rhode Island reds) eating kitchen food waste. Photo by Rbreidbrown (CC BY SA 4.0).

FIGURE 12.13 Pests maintained in urban areas by food waste in garbage. Source of illustrations: Shutterstock; (a) by YummyBuum, (b) by GraphicsRF.com.

not a significant source of funding support. Governmental agriculture departments, of course, strongly support the conservation of such 'genetic resources' that are essential to improvement of crops and livestock. Unfortunately, government funding is always insufficient in this area, if for no other reason than politicians receive much more credit for giving priority to issues that the public identifies as important. The private sector also is involved in breeding better crops and animal breeds, but since it is profit-motivated, genetic resources are not shared and are only maintained for the duration of particular breeding projects.

In Situ vs. *Ex Situ* Conservation

The diversity of wild species is ideally preserved by allowing them to grow in their natural undisturbed habitats, where the often considerable range of genic variation can continue to exist (Ingram et al. 2017). Unfortunately, human domination of the planet is degrading or exterminating habitats that support wild species. In some cases, wild areas are reserved to allow the organisms to continue to survive (see Figure 10.9, which shows a wildland dedicated to wild corn/maize conservation). Maintenance of genetic resources by allowing or facilitating the species to persist in the wild is termed '*in situ*' conservation.

Alternatively, selected individuals of species which are highly threatened with extinction are sometimes cared for by people in non-natural circumstances. Such preservation may be in the form of living specimens allowed to reproduce conventionally (in parks, institutional gardens, zoos, and the like), or as conserved reproductive material (such as seeds or frozen tissues). Preservation outside of natural habitats is categorized as *ex situ* conservation. *In situ* conservation is usually far less costly and is capable of maintaining much more genic diversity than is possible with *ex situ* conservation, which necessarily is based only on selected samples. The normally small samples of material kept in germplasm collections are subject to mutations and accidental hybridization during periodic replication, and loss of alleles of genes that occurs naturally when reproduction occurs in small populations, so over time *ex situ* collections tend to become less representative of the original wild population. Species on the brink of extinction may necessitate *in situ* conservation, at least unless and until sufficient numbers are generated that they can be used to repopulate a habitat. For the most part, species that are raised by people are maintained *ex situ* by or for government agriculture and forestry departments. Non-economic species may also be protected *ex situ* in certain very large government institutions, often supported internationally. Many zoos, whether backed by governments or not, participate in efforts to save selected animal species from extinction.

Conservation of Invaluable Crop Germplasm

Crop germplasm – the breeding materials that are indispensable for creating improved cultivars –has been in decline for some time (Ebel et al. 2021). The conservation of plant germplasm has been extensively and formally covered by international agreements. The Food and Agriculture Organization (FAO) of the United Nations is the most important international agency dealing with genetic resources. At a conference in Rome in 1983, it was resolved in an agreement called 'The FAO International Undertaking on Plant Genetic Resources' that 'plant genetic resources of economic and/or social interest, particularly for agriculture, will be explored, preserved, evaluated and made available for plant breeding and scientific purposes.' In 1993, a multilateral treaty dealing with biodiversity, 'The Convention on Biological Diversity,' was agreed to after meetings in Rio de Janeiro. This called for conservation, sustainable use, and equitable sharing of biological diversity; the sustainable use of its components; and the fair and equitable sharing of benefits arising from genetic resources. In 2001 in Madrid an expanded agreement (consistent with the two previous agreements) called 'The International Treaty on Plant Genetic Resources for Food and Agriculture' was initiated (coming into force in 2004). The history of these international treaties is outlined by Culet (2003).

The seeds of some species can remain viable under ambient conditions for decades, rarely for centuries, but those of most plants will not germinate after a few years, unless preserved under conditions of low temperature and/or humidity. Under cold storage, the otherwise short-lived seeds of some species will last decades, other for hundreds of years, lower temperatures tending to prolong viability. Many seeds can be maintained in liquid nitrogen at its boiling point of −196°C. Millions of collections of seeds of crop varieties and crop wild relative are maintained under long-term cold storage in the world's major plant genebanks. Some tropical species have seeds which do not tolerate freezing and require more complicated methods for conservation. Plant breeders and researchers employ seeds for plant breeding. The most important crops, especially the cereals, have the largest number of accessions. The Svalbard Global Seed Vault (Figure 12.14) is particularly concerned with crops and their wild relatives and is a backup facility for other institutions. England's Millennium Seed Bank (Figure 12.15) is dedicated to collecting all plant species and, in this respect, is unlike most long-term seed collections which are usually dedicated to crops and their wild relatives.

Some very valuable root or tuber crops (e.g., cassava, potato, sweet potato, taro, yam), fruit crops (e.g., apple, banana, cranberry, date, hop, orange, pear, strawberry), nut crops (e.g., hazelnut, hickory nut, macadamia, walnut), and indeed many others are mostly propagated vegetatively, and living cloned plants are maintained in special gardens.

Living materials are sometimes preserved as continuously propagated cell cultures (i.e., they are maintained as single cells or proliferating cells not organized into tissues) or as tissue cultures, which can be employed to grow innumerable identical plantlets (Figure 12.16). As noted in the next section, animal tissues, principally semen, ova

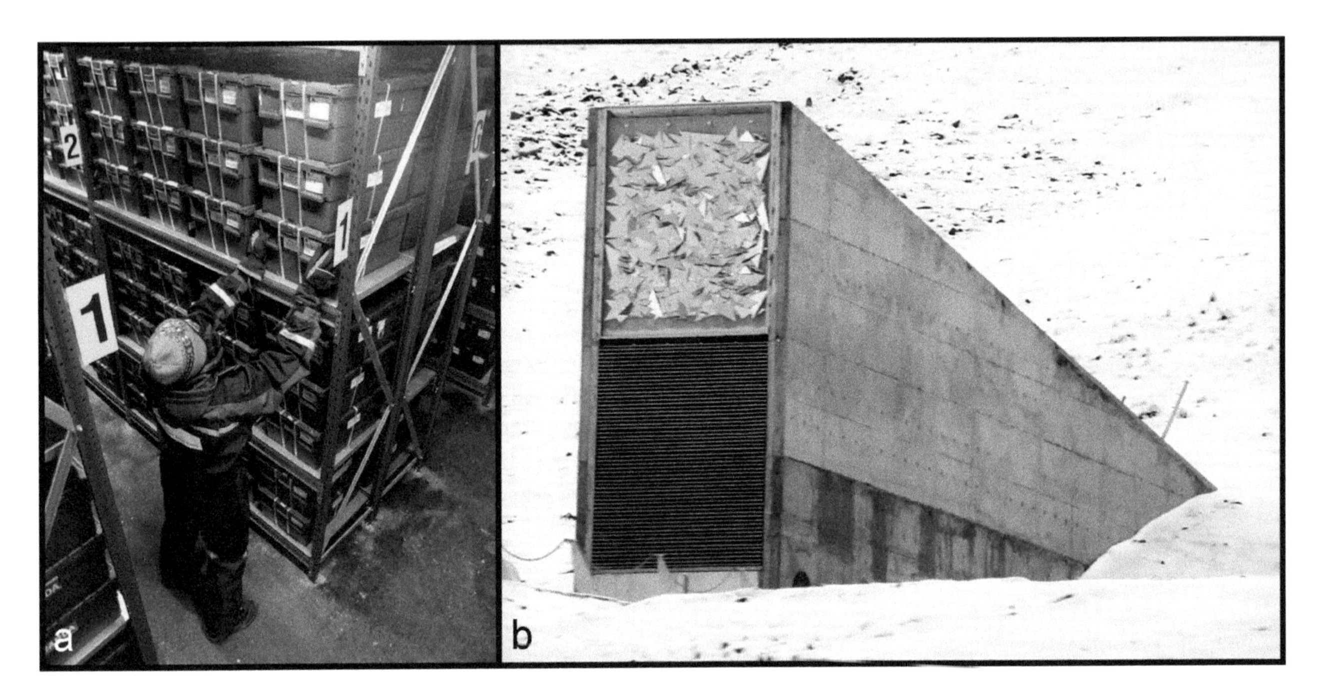

FIGURE 12.14 The Svalbard Global Seed Vault, a secure seed collection in an abandoned coal mine on the Norwegian island of Spitsbergen in the remote Arctic, a site providing a continuously frozen environment. Seeds of crops are stored at −18°C, and if the electrical supply fails, the permafrost keeps temperatures at no higher than −3.5°C. Almost 900,000 seed samples of more than 5,000 different species have been deposited. The facility is expected to grow to become the world's largest collection of seeds of crops, with a capacity to store about 5 million samples. Known as 'the doomsday vault,' the purpose is to back up important crop germplasm in other seedbanks which could be destroyed by disaster. Unlike conventional seed banks which exchange seeds with numerous individuals and other institutions, only the depositor seedbanks have access to their materials. (a) Containers of seeds stored on shelves. Public domain photo by NordGen/Dag Terje Filip Endresen. (b) Entrance. Photo by Bjoertvedt (CC BY 3.0).

FIGURE 12.15 The Millennium seed bank building in Wakehurst Place Garden, West Sussex, England. The associated Millennium Seed Bank project is an international effort which has collected 2.4 billion seeds of over 39,000 plant species, representing over 13% of the world's plants. A particular effort is made to collect seeds of species in danger of extinction. Seeds are sometimes available by negotiation. Photo by Patche99z (CC BY 3.0).

FIGURE 12.16 Plant tissue cultures being grown for germplasm preservation by the U.S. Department of Agriculture. Photo by Lance Cheung (public domain photo).

(unfertilized eggs), and embryos, of very valuable livestock are now often stored cryogenically in liquid nitrogen, and some plant gene banks today similarly conserve apical meristems (growing points or 'buds') of some species. Pollen grains also can be maintained long-term under cold storage for breeding purposes, although they are unsuitable for directly growing into new plants.

Conservation of Invaluable Animal Germplasm

Animal genetic resources are maintained 'in vivo' (as living animal populations; also called '*in situ*' as for plants) and 'in vitro' (as cell cultures, primarily reproductive cells; also called '*ex situ*' as for plants). Cryopreservation (frozen samples, normally of semen and embryos), as for plants, is the basic method of maintaining animal genetic resources (Figure 12.17a), since animals do not reproduce by structures comparable to seeds. Animals are generally more expensive to maintain, either as living captive populations or as cultures, and so most such conservation is for terrestrial mammalian and avian livestock. Very few of the millions of wild animals are represented in gene banks. An international framework for the management of animal genetic resources was provided by Scherf and Pilling (2015). International collections comparable to those developed for plant germplasm are not available although there are regional sharing agreements. A global site that backs up animal gene banks, similar to the Svalbard gene bank noted previously, has not been established. There are numerous national animal germplasm genebanks, which currently maintain a world total of several million in vitro samples, primarily of the principal livestock species (De Paiva et al. 2016). The private sector concerned with animal breeding has its own private gene banks. The semen from champion thoroughbred stallions is claimed to be the world's most expensive liquid (a typical ejaculation of 50 ml or more can impregnate more than a dozen mares, and stud fees can be in the millions).

Because conserving animal germplasm in a frozen state is very expensive, very few species or breeds can be chosen. For the private sector, motivated by profit, the basis of choice is simply an issue of short-term risk vs. reward. However, the long-term value of germplasm depends on mostly unpredictable future events, so it is wise for governments to conserve a large, number of different kinds, even if they are of limited current market value. Genebank managers often exclude potential additions that are genetically repetitive of their holdings, but the major animal breeds are remarkably diverse, so the choice often becomes subjective. As reviewed in Chapter 5, people are heavily influenced by appearance, and even scientists find it difficult to exclude their biases. Candidates for animal germplasm conservation can look quite odd (Figure 12.17), and it probably is the case that unconscious prejudices are playing a role in choosing those destined for survival. Private individuals and non-governmental conservation organizations are currently maintaining selected rare breeds, and it is likely that decisions on which lines to protect sometimes were made on the emotional appeal of their perceived beauty.

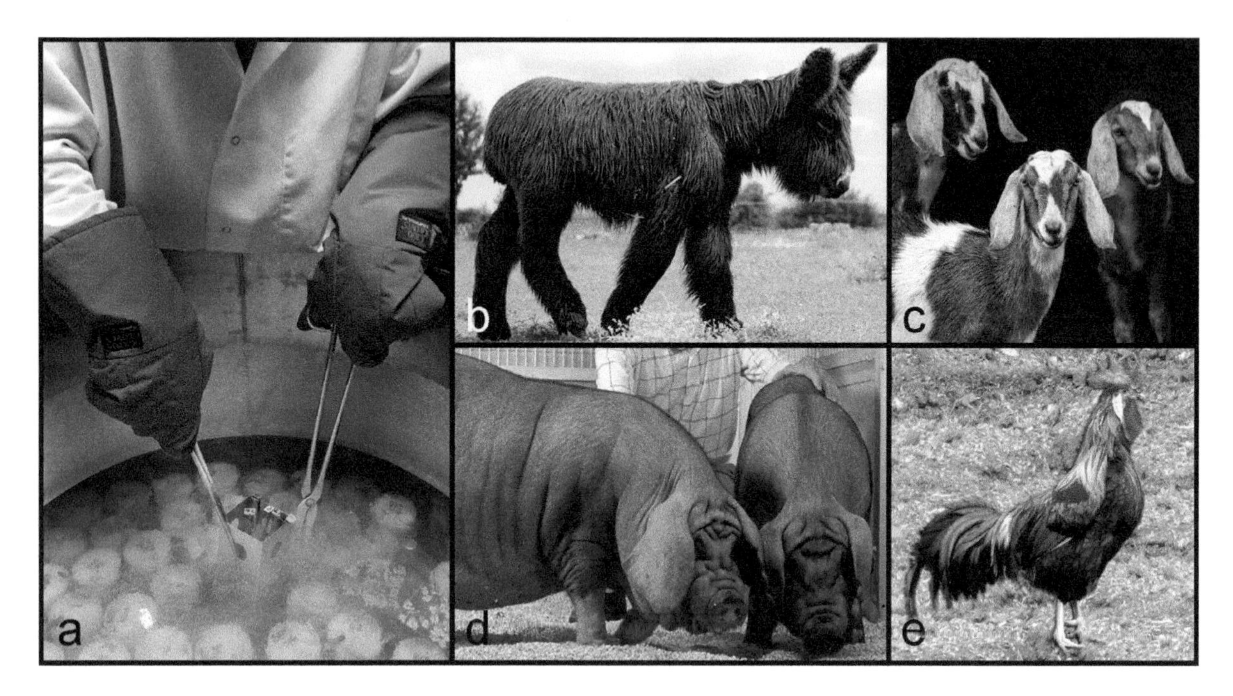

FIGURE 12.17 Rare breeds that are candidates for conservation of germplasm. (a) Livestock germplasm being frozen in liquid nitrogen for long-term ultra-cold cryostorage. Photo (public domain) by United States Department of Agriculture Gene Bank. (b) The Baudet du Poitou or Poitou donkey, a French breed. In 1977 only 44 specimens were found worldwide, but conservation efforts have now produced hundreds. Photo by Sudorculus (CC BY SA 3.0). (c) African Nubian goats. Their long ears are a cooling adaptation for hot dry environments. Photo (public domain) by Lance Cheung, United States Department of Agriculture. (d) Meishan pigs. In 2018 the breed was declared to be critically endangered worldwide. Photo (public domain) by Keith Weller, United States Department of Agriculture. (e) A brown leghorn rooster. Photo by Fernando de Sousa (CC BY SA 2.0).

WOODLOTS: SHARING FARMLAND WITH WILD BIODIVERSITY

In the next chapter, the advisability of reserving land in and around cities for biodiversity is examined. Here, the parallel concept of reserving some areas of farmlands for biodiversity is discussed.

'Woodlots' are plots of land maintained for small-scale logging, fuelwood, maple syrup production, or recreation. Woodlots were traditionally maintained on farms in colonial times for firewood and lumber, and as sites for hunting. Sometimes trees on the least suitable portions of a farm for crops were allowed to survive. With the conversion of many farms to cities, large woodlots have sometimes been maintained as public lands serving to provide green spaces or to divide neighborhoods. Wooded areas bordering farms are still common in rural areas (Figure 12.18) but increasingly are being converted for crops or pastures. Some wise

FIGURE 12.18 Woodlots bordering farms. (a) A farm in Finland. Photo by Janne Järvinen (CC BY SA 3.0). (b) A farm in Vermont. Photo by Putneypics (CC BY 2.0).

governments subsidize the maintenance of woodlots on farmlands in recognition that they conserve biodiversity (e.g., wild pollinators) and maintain the local ecology (e.g., stabilize soil water reserves, prevent soil erosion). This is such a simple measure, benefiting farmers, society, and nature, that it deserves to be much more widely adopted.

WILD HORSES VS. TAME COWS

Horses were once native to North America but became extinct on the continent about 10,000 years ago. Modern horses (*Equus caballus* or *Equus ferus caballus*) and burros (*Equus asinuus* or *Equus africanus asinus*) were introduced to North America in post-Columbian times by explorers and missionaries, especially by the Spanish. Many escaped or were released to the wild, and their descendants are established in parts (especially the West) of the U.S. and Canada (Figure 12.19). Feral horses are termed 'wild horses.' Feral burros are termed 'wild burros,' although in the U.S. the word burro is frequently restricted to feral animals while tamed animals are termed donkeys. There are more than 100,000 wild horses and perhaps 20,000 burros in the U.S. wandering on federal and state lands, including tribal, public, and private property. Free-roaming horses and burros receive some legal protection, especially on public lands, but are considered to be a serious threat to the ecology in some regions, and a danger to the economic welfare of ranchers, and there are attempts to reduce the population (about 200,000 animals have been removed in recent decades). Measures have included fertility control, offering the public the opportunity to adopt the animals, euthanasia, and sale for slaughter for pet food. Particularly with respect to the cattle industry, the situation represents a clash of how different segments of society value wild horses in relation to livestock production. Wild horses have considerable emotional value to people as symbols of beauty and freedom, and cows of course are sources of economic well-being. The situation is difficult because many are sensitive to harming much-loved horses and their possible slaughter for food contradicts societal taboos. (Ironically, the horses are sacred cows.) The situation is also complicated because the feral animals are not indigenous (at least in the last millennium) and their range is the result of human introductions. Under these circumstances, there does not seem to be a solution that can please everyone, or perhaps anyone. By no means is the issue of troublesome feral horses confined to North America – it also occurs in Africa, Asia, Argentina, Australia, and New Zealand. For analyses, see Beever (2003), Kincaid (2008), Danvir (2018), Frey and Thacke (2018), and Hurwitt (2018).

FIGURE 12.19 Wild (feral) horses and burros in the Western U.S. (a) Wild horses in Idaho. Photo by Peter Robbins (CC BY 2.0). (b) Wild stallions fighting in Shackleford Banks, North Carolina. Photo (public domain) by C. Wasley, U.S. National Park Service. (c) Captured wild horses awaiting adoption at the Bureau of Land Management corral in Hines, Oregon, in 2018. Photo by Tara Thissell, BLM (CC BY 2.0). (d) A wild burro in Nevada. Photo by Tomás Del Coro (CC BY SA 2.0). (e) Wild burros in the Spring Mountains, Nevada. Photo by Andrew (CC BY ND 2.0). (f) Captured wild burros awaiting adoption at the Bureau of Land Management corral in Nevada. Photo (public domain) by Beth Freniere, BLM.

13 Reforming Urbanization
The Second Major Threat to Species

INTRODUCTION

'Urbanization' is usually understood as population shift from rural to urban areas, but it may also mean the creation and expansion of cities (see Moll et al. 2019 for examination of what urbanization means in the context of biodiversity). Urbanization began thousands of years ago with the shift of hunter-gatherers into permanent settlements. In 1950, half of the world's people lived in cities, and this is expected to increase to two-thirds by 2050 (United Nations 2018). Most cities grow, often vertically (resulting in greater density of people) but also horizontally (resulting in greater areas of occupation). Population increase throughout the world, increasingly in urban areas, is the indirect cause of agricultural expansion with its associated serious harm to species and ecosystems, as discussed in the previous chapter. In parallel, population increase is the chief driver of urban expansion and consequent harm to species and ecosystems (Figure 13.1). The extremely high concentrations of people in the 'great cities' of the world strip water, other materials, and energy from surrounding natural areas, while contaminating them with large amounts of sanitary waste, garbage, and atmospheric pollution. Most major cities are located near large rivers, lakes, or seas, so their polluting effluents can spread widely. 'Green urbanism' refers to promoting communities that are not only beneficial to humans but also to the environment (Lehmann 2010). Everyone, let alone city planners, instinctively knows that beautiful cities are green cities (Figure 13.2). 'Green belts' (greenbelts), 'green spaces,' or 'urban growth boundaries' that limit development around and often inside cities are simple, highly effective promoters of sustainable urbanization (Figure 13.3). Urban architecture can be designed compatibly with nature in ways that satisfy the needs for human comfort in relation to environments (Browning et al. 2014).

URBAN BIODIVERSITY: WELCOME AND UNWELCOME WILD SPECIES LIVING IN CITIES

Cities directly harm the living world, both by physically displacing wild species and by degrading their habitats (McKinney 2002; Ricketts and Imhoff 2003). However, not all wild species find cities to be hostile environments. Occasionally, cities provide survival refuges for wild species being eliminated from their natural habitats (Elmqvist et al. 2013, 2016). Of course, most large land animals, especially the carnivores, can be dangerous to humans and they are not welcome. Nevertheless, city landscapes are heterogeneous patchworks of natural, seminatural, and modified habitats, and are capable of benefiting many species, some attractive, many undesired. The word pest is pejorative and is appropriate for given species only when their presence or activities are objectionable. Unfortunately, cities provide ideal conditions for some species to become problems (Johnson and Munshi-South 2017). What people call 'garbage' is an essential food and habitat resource for large populations of pests in cities. Cities also provide considerable light at night, facilitating the lifestyles of many wild animals (Leveau 2020). Cities are home to many of humankind's most hated pests (examples are shown in Figure 13.4), including invasive and exotic species, none of which is in danger of extinction. Some are harmful such as rats, mice, bedbugs, houseflies, lice, and cockroaches, and some are mostly just annoying, such as pigeons and raccoons. Controlling these 'vermin' is extremely costly. Curiously, many attractive ornamental plants introduced into urban gardens because of their beauty have escaped to the wild where they now are often major threats to native ecosystems. In parallel, feral pet dogs and cats have become significant problems worldwide. As argued throughout this book, it is important to increase tolerance to despised species because their inevitable presence among biodiversity in general should not be allowed to compromise efforts to conserve the other species that accompany them.

Urban wildlife management has become an important part of applied biology (McCleery et al. 2014; Angelici and Ross 2020). However, only a very small proportion of city-dwelling species are judged worthy of analysis. As pointed out by Egerer and Buchholz (2021), the species highlighted in studies of urban wildlife are mostly high-profile birds and mammals that attract public interest, while other forms of urban life, including most plants and invertebrates, are often largely ignored. Even among scientists, different forms of biodiversity are far from equal. Insects are, by a considerable margin, the most vilified and understudied of all species.

Fukano and Soga (2021) hypothesized that urbanization accentuates hostility to insects in two ways (Textbox 13.1). Their first point verifies the proverb 'familiarity breeds contempt' insofar as increasing numbers of pest insects in houses increases dislike. Their second point is that unfamiliarity also breeds contempt, insofar as decreasing knowledge of pest insects also increases dislike (but of all unfamiliar insects).

Not all landscape management tools appropriate for wildlife in wildlands are suitable for urban areas. 'Green bridges' and 'wildlife corridors' which allow movement across (usually under) transportation routes are often vital

DOI: 10.1201/9781003412946-13

FIGURE 13.1 Ecosystem destruction and endangerment of wild species resulting from urban expansion. Prepared by Jessica Hsiung.

for big animals to facilitate their usage of the large territories that they require (Riley et al. 2014). However, there are good reasons to exclude bears and deer from cities. Also, to a much greater extent than in the natural environment, wild animals can be quite hazardous to the health of people, and this needs to be considered (Wookey 2022).

For most people, the majority of living species viewed is in their home city. Humans have been increasingly losing contact with most life forms, but plants and animals that they observe can play important roles in reconnecting people with nature. Serenity is difficult to find in cities, which are centers of stress, but the sight of living things can be reassuring. Unfortunately, as emphasized in this book, not all species are admired. The pests in cities contribute greatly to an overwhelmingly negative view of most small creatures encountered, especially insects and small

FIGURE 13.2 Two of the top ten most beautiful cities of the world, illustrating the importance of the 'urban forest.' (a) Paris: Eiffel Tower region. Photo by David McSpadden (CC BY 2.0). (b) Vancouver: Main mall of the University of British Columbia. Photo by University of BC (CC BY SA 4.0).

FIGURE 13.3 Green belts. (a) Location of green belts in England. The belt around London is in red. Prepared by Hellerick (CC BY SA 3.0). (b) The Barton Creek Greenbelt of Austin, the state capital of Texas. This reserve contains over 20km of trails. Source: Shutterstock, contributor: Trong Nguyen. (c) Aerial panorama of Oxford, England, which is at the center of the Oxford Green Belt. Photo by Chensiyuan (CC BY SA 4.0).

FIGURE 13.4 Common urban pests. (a) Raccoon (*Procyon lotor*) and skunk (*Mephitis mephitis*) feasting on cat food. Photo by Piepie & GeeAlice (CC BY SA 2.5). (b) Norway rat (*Rattus norvegicus*). Photo by Reg Mckenna (CC BY 2.0). (c) Pigeons (*Columba livia*). Photo by Myloismylife (CC BY SA 3.0).

TEXTBOX 13.1 URBANIZATION MAY INCREASE DISLIKE OF INSECTS

There are two pathways by which urbanization increases the intensity and breadth of feelings of disgust toward insects: (1) urbanization increases the extent to which people see insects indoors, and insects that are seen more often indoors induce stronger feelings of disgust than is induced by insects seen outdoors; and (2) urbanization reduces people's natural history knowledge about insects, and decreased knowledge results in a broader range of insects eliciting feelings of disgust.

—*Fukano and Soga (2021)*

terrestrial vertebrates, which accordingly are often persecuted. Conversely, a small number of species possessing features admired by humans, including our pets, are protected. These prejudicial views of despised animals by the public tend to become generalized to the biodiversity of the world, limiting support for conservation of the world's species and the ecosystems that support them and us.

FEEDING URBAN PESTS: CHOOSING SIDES

Almost no one views the most serious urban pests (like rats and bedbugs) as welcome, but the more attractive ones (usually birds and mammals) appeal to a minority. In Christian theology, charity is one of the seven virtues. For many, providing charity to animal pests in the form of food offerings seems admirable (Figure 13.5), but for others it simply enlarges the urban pest problem. Pigeons, gulls, ducks, and geese are chief beneficiaries among birds, to the annoyance of many. Some (especially women) seem to be compulsive feral cat feeders. In Britain, where most large wildlife was exterminated long ago, foxes are privileged guests (except when they are victims of traditional fox hunting). Squirrels are not often considered to be significant pests, except at bird feeders, in which case baffles may be adopted. Hummingbird feeders may be fitted with guards, to prevent bees and wasps from the feeding port, but saucer-shaped feeders with top ports are better because the syrup level is usually too low for the insects to reach. Ants can be discouraged from hummingbird feeders by using dripless feeders. Some thoughtless people in highly populated areas deliberately feed local wildlife such as bears, deer, coyotes, and raccoons, but this can result in harm to both people and the animals.

Conflict between urban cats and wild birds especially illustrates how human preferences influence the survival of given species. There are many rural cats, but most domesticated felines are found in urban areas. Free-roaming cats (both feral and 'housecats' that are unconfined) are responsible for the death of many birds (Blancher 2013; Figure 13.6). Domestic cats kill more than a million birds every day in Australia (Woinarski et al. 2017), about 5 million a day in the United States according to Harris (2000), and more than a billion annually in the U.S. according to Nico and Cooper (2009). This represents an interesting city wildlife dynamic and an equally interesting example of human preferences. Most people love most wild birds (many maintain bird houses and feeders), and most also love most cats (some go to great lengths to feed strays). Cats often lurk near feeders and birdbaths awaiting a chance to ambush their prey. However, few cat lovers are willing to confine their pets in order to preserve birds.

FIGURE 13.5 Feeding urban pests. (a) Feeding feral pigeons. Photo by Laura Hadden (CC BY 2.0). (b) Feral cats at lunch. Photo by Scott Granneman (CC BY SA 2.0). (c) Treats for feral rhesus macaques (*Macaca mulatta*) in Myanmar. Photo by Jakub Hałun (CC BYSA 4.0). (d) Feeding Canada geese. Photo by Liz West (CC BY 2.0).

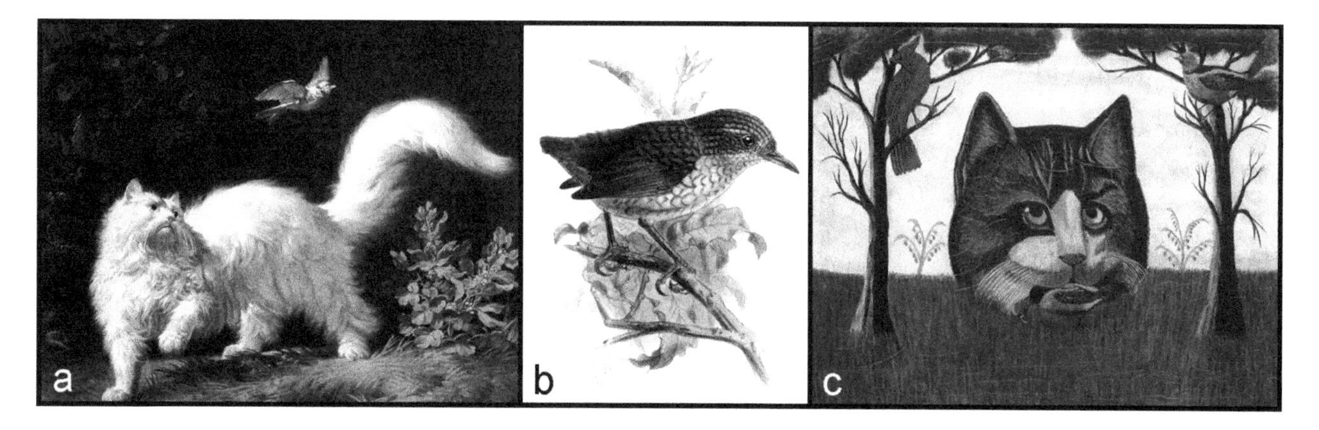

FIGURE 13.6 Paintings (public domain) bearing on the harm to birds by cats. (a) 'An Angora Cat,' painting by French artist Jean-Jacques Bachelier (1724–1806). (b) Lyall's wren or the Stephens Island wren (*Traversia lyalli*), a small flightless New Zealand bird, driven to extinction by feral cats in 1925. This painting was prepared in 1895 by Dutch ornithologist John Gerrard Keulemans (1842–1912). (c) 'The Cat,' a 19th-century anonymous painting in the National Gallery of Art, Washington, DC.

THE IMPORTANCE OF GREEN SPACES FOR URBAN SPECIES

In the previous chapter, it was pointed out that the damage from agriculture is greatest from monocultures which totally monopolize huge areas, compared to situations in which crops are grown in separated places, allowing species to survive in the intervening regions. A parallel consideration applies to cities. 'Green spaces' that act as 'habitat patches' allow many species to coexist with humans in urban areas. Wise urban planning allows for this (Apfelbeck et al. 2020), but almost all cities have expanded with little

FIGURE 13.7 Use of greenery in cities to provide urban habitats for animals. (a) A tall building in the city of Milan, Italy, the balconies generously decorated with ornamental plants including trees. Photo by Greta Cadei (CC BY 3.0). (b) Urban wildlife (a mule deer) in Baker City, Oregon. Photo by Baker County Tourism (CC BY ND 2.0). (c) National Botanic Gardens in Glasnevin, Ireland. Photo by William Murphy (CC BY SA 2.0).

or no forethought. Megacities likely have far more damaging effects on biodiversity and ecosystems than areas with comparable populations but divided up into multiple small cities. This is due to the high cost of land in huge cities, promoting development of every available location, and so limiting unused spaces that can support biodiversity.

There are various ways of maintaining the welfare of wild species in cities (Goddard et al. 2010; Panlasigui et al. 2021). An important method is to maintain and increase vegetation along streets and in parks and residential areas, including on the exterior of large buildings (Figure 13.7). Even weedy areas can be beneficial, although they may provide habitats for animal pests, and it is hard to persuade the public that plants perceived as weeds should not be eradicated. However, some weeds support butterflies (as is the case for milkweeds, which are essential for monarch butterflies) or pollinating bees, so many environmentally friendly citizens will plant them in their gardens. The expression 'urban forest' is usually interpreted as referring just to trees, but all plants growing in cities can provide habitats that promote biodiversity (Angold et al. 2006; Alvey 2006). 'Urban jungle,' sometimes equated to urban forest, refers to an urban area characterized by ruthless struggle and competition among humans, but the concept applies to the ecology of wild species trying to survive in cities (cf. Read 2012).

BROWNFIELDS AND CONSERVATION WASTELANDS: ABANDONED AND NEGLECTED URBAN LANDS AS REFUGES FOR SPECIES

Areas that have been developed by people for cities, agriculture, or other purposes need maintenance, rehabilitation, or renewal, or eventually they will fall into disrepair. Almost everywhere, there are sites that have been abandoned or allowed to degenerate, with the result that they have been recolonized by plants and animals that could not compete with humans inside the cities. Often such areas have been contaminated. Ionizing radiation from the nuclear accidents of Chernobyl (Ukraine, 1986) and Fukushima (Japan, 2011) has been harmful to local animal species (Mousseau and Møller 2016), but in the absence of humans in the contaminated area, many wild species are nevertheless thriving (Deryabina et al. 2015; Lyons et al. 2020; Figure 13.8). The United States Environmental Protection Agency (2021) defines 'brownfields' as 'any land in the United States that is abandoned, idled or under used because redevelopment and/or expansion is complicated by environmental contamination that is either real or perceived.' In Great Britain and sometimes elsewhere, 'brownfields' refers to all abandoned or derelict sites, regardless of possible toxicity (Alker et al. 2000). Despite the name, so-called brownfields are usually green with plant life, and they are often islands in cities where biodiversity can thrive, albeit often temporarily (Buglife n.d.; Kattwinkel et al. 2011; Hunter 2014).

Abandoned and neglected urban areas tend to be very negatively perceived by the public because they are unkempt, and some of their constituent species are significant pests (Macadam and Bairner 2012). Such sites (Figure 13.9) are another example of human perception of ugliness not corresponding to what is in the best interests of other species. The expression 'conservation wasteland' is sometimes applied to areas whose habitats and/or ecosystems are so seriously degraded that their ecological rehabilitation does not seem feasible or worthwhile. Kareiva and Kareiva (2017) proposed that such degraded lands might be considered for future sites for human needs that are hostile to biodiversity, such as wind farms, solar energy facilities, and oil palm plantations. In contrast, Bonthoux et al. (2014) view degraded lands as opportunities to promote biodiversity. The term 'rewilding' (re-wilding) refers to deliberately allowing an area to return to nature, without significant human efforts to guide the resulting ecology (Soulé and Noss 1998).

FIGURE 13.8 Scenes at the Chernobyl, Ukraine, nuclear disaster radioactive exclusion area (30 km around the power plant blast in 1986; the exclusion area extends north into Belarus). (a) New confinement sarcophagus (the building with the curved roof) constructed in 1986 over the power plant that exploded. 'Sarcophagus' usually means a stone coffin, and by extension, it is also employed for the giant concrete covering shown here intended to reduce radiation. Note how vegetation is taking over the site. Source: Shutterstock (contributor: Leshiy985). (b) A tawny brown owl chick (*Strix aluco*) behind a window in an abandoned house. Source: Shutterstock (contributor: Romm). (c) Wolves running among abandoned houses. By the mid-1990s, wolves were so common they were considered to be a nuisance to farmers. Source: Shutterstock (contributor: Film Studio Aves). (d) Przewalski's horse, an endangered wild Asian horse (photo in 2006). The rare horse was deliberately introduced into the exclusion zone after the accident. Photo by Xopc (CC BY SA 2.5).

FIGURE 13.9 Abandoned and neglected areas of a city, although detested by the public, provide abundant habitats for biodiversity. (a) An abandoned house in Russia. Photo (public domain) by Dmitriy Protsenko. (b) An abandoned house in Detroit. Photo (public domain) by Notorious4life. (c) An automobile junkyard in Great Britain. Photo by Ian S (CC BY SA 2.0).

GARBAGE DUMPS AND SANITARY LANDFILLS AS HABITATS

Garbage dumps or, euphemistically, 'sanitary landfills,' accompany urbanization, albeit they are usually situated out of sight (sometimes even out of the country). They are often ecological nightmares, filled with toxic waste, but nevertheless can be quite supportive of wildlife (including pests). It is a sad commentary on society that many humans as well as animals are reduced to surviving on garbage for food (Figure 13.10).

FIGURE 13.10 Animals feeding on garbage. (a) A herd of several dozen wild elephants that have become dependent on a garbage dump in east Sri Lanka. Several of the animals died from consumption of toxic materials. Photo taken in 2020 by Tharmapalan Tilaxan (CC BY SA 4.0). (b) Birds congregating at a garbage dump in Yucatan, Mexico. Photo by Katja Schulz (CC BY 2.0). (c) Polar bears consuming garbage at Churchill, Manitoba. Source: Shutterstock, contributor: Keith Levit. (d) A herd of Przewalski's horse, an endangered wild Asian horse, consuming garbage. Source: Shutterstock, contributor: Stockphoto mania.

BIODIVERSITY OFFSETTING

Hunting of endangered animals in some countries is sometimes justified with the argument that the money raised goes toward their conservation ('kill them to save them'). In some jurisdictions, companies or industries that pollute are able to purchase 'carbon credits,' the money paid ostensibly dedicated to reducing an equal amount of pollution elsewhere. 'Biodiversity offsetting' is a similar trading scheme that allows the loss of biodiversity (both organisms and habitat) in one place provided that this is compensated for by efforts to preserve or restore biodiversity at an ecologically equivalent site elsewhere (Maron et al. 2012). There are many such offsetting schemes employed globally (Madsen et al. 2011). As reviewed by Tierney et al. (2017), there are significant theoretical and practical difficulties with these efforts. The goal of biodiversity offsets is 'no net loss,' but in practice this is often murky (Pope et al. 2021). Such compensation programs aren't the same as triage, which is the deliberate abandonment of efforts to conserve some species (as discussed in Chapter 10), but they also represent attempts to minimize damage to the planet where funding is limited. However, biodiversity offsetting may be more useful for the public image of land developers than for the welfare of ecosystems and biodiversity.

COMPARATIVE BIODIVERSITY OF AFFLUENT AND IMPOVERISHED NEIGHBORHOODS

Social and economic inequalities characteristically result in urban areas being segregated into separate regions differing dramatically (Figure 13.11). In rich neighborhoods, both private properties and public streets will be very attractively landscaped, and parks are probably interspersed, where entertaining birds, squirrels, and other harmless animals are found. Such areas will be well-policed, allowing well-dressed inhabitants to walk their pedigreed pooches in safety. Poor neighborhoods, however, tend to be in a state of disrepair, perhaps with abandoned cars, junk, and garbage strewn about, which provide habitats and food for various less reputable animals. The absence of maintenance of the grounds separating buildings often results in weeds becoming dominant. Moreover, city maintenance services tend to be less available to rectify problems. Serious pests like flies, bedbugs,

FIGURE 13.11 Contrast of urban biodiversity in affluent and poverty-stricken neighborhoods. Prepared by Jessica Hsiung.

cockroaches, and rats are typically concentrated in the poorest urban areas (Biehler 2013), compromising the welfare of the locals, who tend to have health problems that they cannot afford to get treated. Because of the lack of planted trees and the consequent canopies that shade and cool the environment, heat builds up in the summer, sometimes killing those who cannot afford air conditioning. Clearly, the rich benefit from living in healthier ecosystems, and conversely, the poor suffer from being forced to survive in unhealthier habitats (Huynh et al. 2022). It is an uncomfortable truth that society often views both the human and non-human inhabitants of the poorest neighborhoods as unworthy of much care, and even as subjects deserving to be subjected to control measures.

14 Advancing Technologies and the Fate of the World's Species

INTRODUCTION

There have been various answers to the question 'What separates humans from other animals,' but one common response is our ability to create 'tools' to solve complex challenges. Today, biotechnological tools are in preparation that are capable of more profoundly modifying life than anything previously available in the entire history of humankind. This will allow people to exercise their preferences and prejudices regarding other species in ways that may be extremely wise or exceedingly foolish and destructive.

'Biotechnology' is applied science that employs aspects of organisms to achieve practical goals. It may deal with parts or chemicals of organisms; physiological or biochemical processes within individuals; relationships among individuals, populations, or ecosystems; and even artificial or synthetic analogues (see the discussion of 'synthetic biology' in Fernau et al. 2020). This is an extremely broad field and has produced millions of publications in recent decades. In particular, thousands of analyses have been published examining the benefits and disadvantages of biotechnology specifically in regard to humans, biodiversity, and the major ecosystems of the world. Predictably, industry and governments tend to emphasize the positive aspects for human health and economic welfare, environmental groups express alarm about the harm to biodiversity and the health of the planet, and a varied set of humanitarians are concerned about issues such as: the ethics of playing the role of God in altering genes; exploitative aspects such as the pain inflicted on animals; and the growing domination of major corporations. Recent comprehensive analyses include: Sherlock and Morrey (2002); Oksanen and Siipi (2014); Gonzalez (2015); Chaurasia et al. (2020); and Thompson (2020).

As detailed in Chapter 8, most humans primarily value wild species as bioresources, so as science reduces this economic dependence by creating new life forms, there is danger that wild biodiversity will be viewed less sympathetically. The following information is concerned with the considerable potential of biotechnology to influence biodiversity with particular regard to the theme of this book: the important role of human preferences and prejudices in determining the fate of species.

ROBOTS WITH ARTIFICIAL INTELLIGENCE: CHALLENGES TO THE WORLD OF LIVING THINGS, INCLUDING HUMANS

A frequent theme in science fiction is the escape of intelligent robots (Figure 14.1) from human control, with the resulting enslavement or even elimination of humankind. The late theoretical physicist Stephen Hawking warned that 'The development of full artificial intelligence could spell the end of the human race' (Cellan-Jones 2014). The scenario that is feared is that computerized machines could become super-intelligent, reprogram themselves, and outcompete

FIGURE 14.1 Conceptions of thinking human-like all-machine robots with frightening potential to dominate humans. (a) Credit: Mikemacmarketing via www.vpnsrus.com (CC BY 2.0). (b) Prepared by Chitra Sancheti (CC BY SA 4.0). (c) Prepared by Lothar Dieterich, Pixabay (public domain).

DOI: 10.1201/9781003412946-14

humans and indeed all life forms in battles for resources and self-preservation. Whether such 'thinking machines' would actually have human-like consciousness is debatable, but more critical is the unpredictability of anything comparable to ethics, morality, and values accompanying their potential enormous power to alter the world. While many consider the dangers to be hypothetical and negligible, others regard the threats as serious (Bostrom 2014; Kemp et al. 2016; Gouveia 2020). Just as humans continue to modify the planet and its inhabitants, it is conceivable that artificially intelligent machines would do the same, but perhaps even more drastically.

CYBORG: COMBINATIONS OF HUMANS AND MACHINES

'Transhumanism' is a loosely defined philosophical movement that in part endeavors to improve humans, including potential biotechnological advances that could make people more intelligent and robust. A 'cyborg' or 'cybernetic organism' is a hybrid of human and machine that enhances human capabilities. In principle, the contributions of human and machine could differ considerably, so relatively minor contributions, such as hip replacements, mechanical heart valves, and artificial teeth might suffice to constitute cyborg (note Figure 14.2c). Conceivably, and more drastically (note Figure 14.2b), the human brain could be grafted to a machine (i.e., a robot) that is more supportive than the human body (the Robocop movie series is based on this idea). In this situation, the robot portion would be closely controlled by a human brain, which is much less threatening than intelligent robots discussed above. In theory, at least, one's brain could be kept alive forever, which could quickly lead to overpopulation (although conceivably brains could be merely filed away in some sort of library). More drastically, all of the internal parts of a cyborg could be a machine, as in the Terminator movie series (cf. Figure 14.2a).

CHIMERAS: COMBINING SPECIES BY BIOTECHNOLOGY

In the natural world, there are examples of very different species that maintain their separateness while remaining closely attached to other species. One of the most intimate of such associations is a lichen. Lichens are a symbiotic combination of a fungus and an alga or a photosynthetic bacterium, the latter providing the benefits of photosynthesis and the former furnishing a home (in principle, humans and their crops share a comparable relationship). As such, the approximately 20,000 'species' of lichens are 'composite organisms,' although they are named for their fungal component.

In the ancient Greek epic tale the Iliad, the poet Homer described a fearsome beast known as a chimera which inhabited Asia Minor. It had the head of a lion, the body of a goat, and the tail of a serpent, and breathed fire. This usage of the word chimera reflects an old meaning, 'monster.' Chapter 7 dealt with mythical part-human hybrids termed chimeras that are based simply on combinations of parts of humans and other animals. The word chimera is also employed in still another sense to indicate an organism that originated as a fertilized egg but subsequently differentiated (e.g., by mutation) into zones (this is often the case in ornamental plants in which the leaves are variegated, i.e., with both green and white areas, as shown in Figure 5.65).

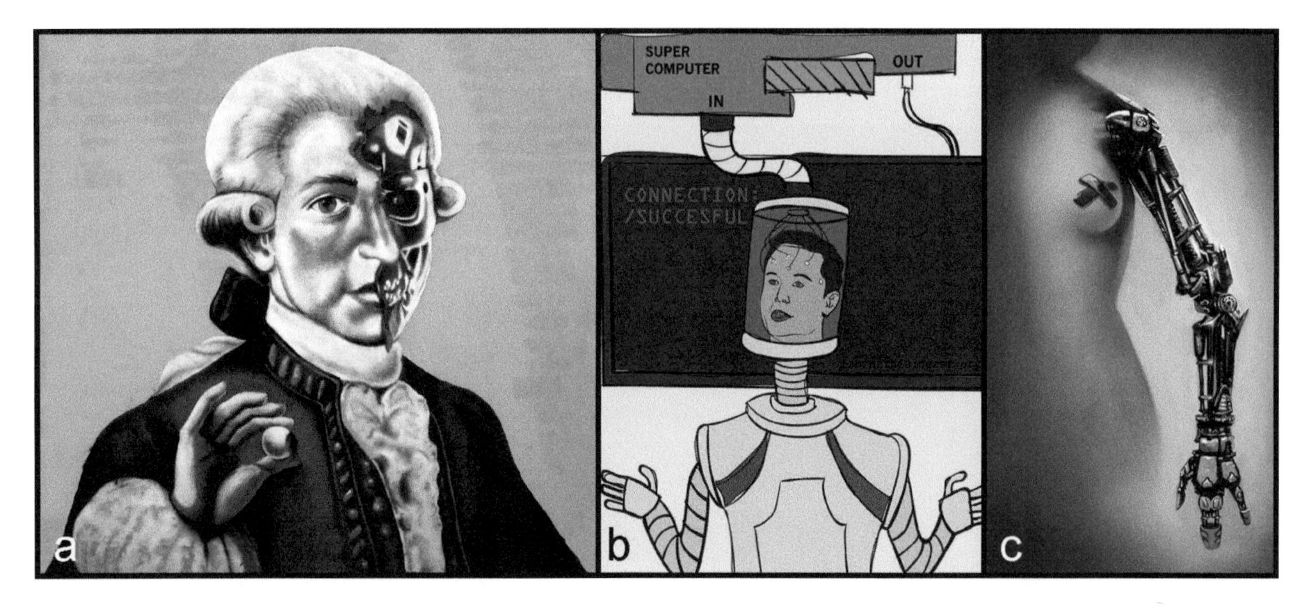

FIGURE 14.2 Conceptions of cyborg (part human – part machine). (a) Graffiti on a wall in Graz, Austria, depicting Wolfgang Amadeus Mozart as a cyborg. Photo by Clemens Stockner (CC BY SA 4.0). (b) Conception of entrepreneur Elon Musk's head as part of a cyborg – a tribute to his expressed concern that Artificial Intelligence could make humans obsolete. Drawing (public domain) by tamingtheaibeast.org. (c) Robotic arm grafted to a human. Created by David Revoy (CC BY 3.0).

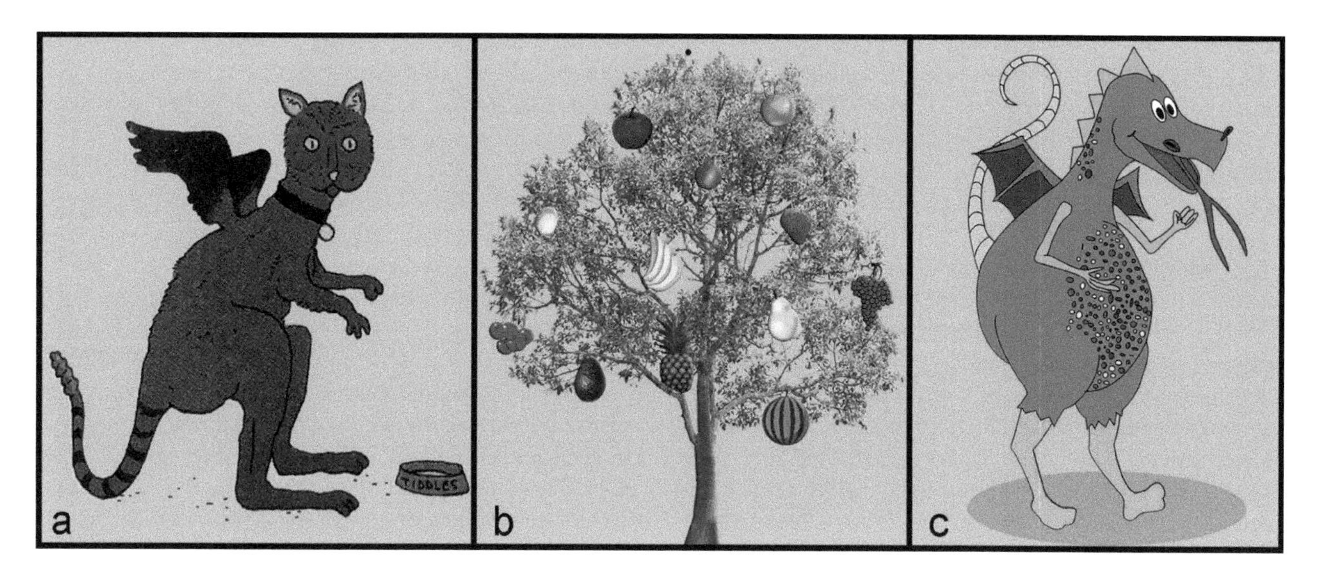

FIGURE 14.3 Some theoretical chimeras (organisms made of a combination of species). (a) Pet animal. Prepared by Gwydion M. Williams (CC BY 2.0). (b) Multi-fruit tree. Prepared by B. Brookes. (c) Another weird combination of species. Source (public domain): Pandanna Imagen, Pixabay.com.

In humans, one form of chimera occurs when a person has one blue eye and one brown eye. Chimeras are known to arise in humans as a result of cells from a fraternal twin in the uterus being absorbed by his brother. Such patterns are sometimes referred to as 'mosaics' in zoology. In general, the word chimera refers to a 'blend' or 'combination' of at least two species (note Figure 14.3). There are degrees of conceivable intimacy of such blending. The various concepts of chimeras theoretically challenge the concepts of individuality and species, and with advances in biotechnology, chimeras may bear more realistically on the relationships of humans and other species.

A 'cellular chimera' has been defined as an organism that has two or more genetically different cell populations that come from different zygotes (the point to note is that the resulting creature is not genetically uniform and may even represent more than one species). Cellular chimeras are created by inserting cells from one animal (of the same or a different species) inside another, and in most cases, the resulting 'organism' isn't really that different – it's just an individual from one species with a few living cells of a different individual. Researchers have created cellular chimeras of different animals, including combinations of humans and monkeys (Tan et al. 2021), quails and chickens, and mice and rats. Needless to say, many regard this area of research and development as Frankenstein-ish and fraught with ethical concerns (Nezerith and Wareham 2020; Rollins 2020).

Xenotransplantation is another form of combining species. It is the grafting of organs between animals and represents a potential means of replacing diseased organs in humans. Grafting is much easier in plants than in animals, and numerous cultivated plants are being grown in this fashion (most cultivated rose varieties, for example, are grown on roots of different varieties). The nomenclature of plant 'graft-chimeras' (graft hybrids) is governed by an international code (Brickell et al. 2016). Graft-chimeras are often given new genus names, combining the names of the two graft-parents' genera (e.g., *+Laburnocytisus* 'Adamii' (broom laburnum) is a graft-chimera of *Cytisus purpureus* (purple broom) and *Laburnum anagroides* (golden chain tree). Alternatively, the graft-chimera may be identified as the combination of the two species (e.g., *Cytisus purpureus+Laburnum anagroides*). Aside from the nomenclatural difficulties associated with creatures generated by humans that are at least partly composed of more than one species, there will be issues of how much priority they should be afforded for conservation and other purposes. In the future, there will likely be a bewildering variety of unusual chimeric creatures.

Chimeras: Combining Humans and Non-human Species by Biotechnology

Grafts between genetically different organisms are a form of chimera. Chimera medical research combining human and non-human animal tissues is very controversial and tightly controlled (often prohibited) in most Western nations. The issue of living beings with mixed human and animal elements is a growing concern of bioethics (Bokota 2021). Although it is challenging, surgeons often replace failing body organs using contributions from living donors or cadavers. 'Xenotransplantation' is inter-species organ exchange, and there have been some achievements with transfers between non-human animals (Lu et al. 2020). However, the possibility of replacing diseased organs of humans with organs of animals, or at least tissues that could reinvigorate the organs, is far from reality. Growing human organs in vitro (Latin for 'in glass,' i.e., in the laboratory) has been unsuccessful to date. Growing human organs parasitically on animals has been considered. In 1996, pictures were widely circulated of a laboratory mouse with what appeared to be a human ear grown on its back. Many thought that this was the beginning of using genetically engineered animals to grow organs for transplanting to humans. However, the 'ear' was the result of mechanically forcing the growth of the skin into ear shape and had nothing to do with genetic engineering (Hugo 2017). The modern field of 'tissue engineering' often is involved in exploring tissues cultured on scaffolds to replace body parts such as ears for medical applications. Several biotechnology companies are attempting to genetically engineer pigs to make their organs more compatible with the human body so that they can be transplanted (Kawai et al. 2019; Wolf et al. 2019; Figure 14.4). In particular, it has been proposed that large livestock could become incubators for human hearts, kidneys, livers, lungs, and other organs (this has been termed 'organ farming'). In 2021, surgeons in New York maintained a pig kidney transplanted into a brain-dead human for 3 days. In 2022 in Maryland, a heart from a genetically modified pig was transplanted into a human, who lived for two more months. Controversial, cutting-edge research is underway to implant genetically modified animal embryos with human stem cells that might be directed to form human organs or at least their precursors. Possibly, stem cells salvaged from patients might be injected into chosen animals where a large supply of these cells could be grown and re-implanted in the patient in hopes of regenerating failing organs. Many people accept the idea of transplanting animal organs into humans for therapeutic reasons, but transplanting brains or brain tissue is an area of particular controversy. At least in fiction, human heads of people with dying bodies have been grafted onto animals to prevent their imminent death, and there have been laboratory attempts (mostly unsuccessful or short-lived) at head transplants employing non-humans (Lamba et al. 2016). The resulting experimental combinations look grotesque, painful, and cruel. Whole brain transplants are not technically possible at present, and the transfer of brains tissues or grafts to cerebral regions within the human species is currently an experimental field with the objective of curing diseases. Brain tissue xenotransplantation involving humans is controversial. One remote possibility that has troubled some ethicists is the possibility that transferring human brain cells into animals could 'humanize' them intellectually. Could such chimeras possess at least elementary human-like intelligence, have a sense of moral injustice, and be capable of intellectualizing their own suffering? Indeed, human brain cells have been transplanted into mice brains (e.g., Xu et al. 2019), and the issue of the degree of humanness of such mice has been examined (Loike 2015; Figure 14.5).

BIOLOGICAL CONTROL OF PESTS USING THEIR NATURAL ENEMIES

Measures to combat the most unwanted animals and plants are usually only partially effective. Everywhere in the world there are both native and introduced species that are harmful to people and property, as well as to natural ecosystems

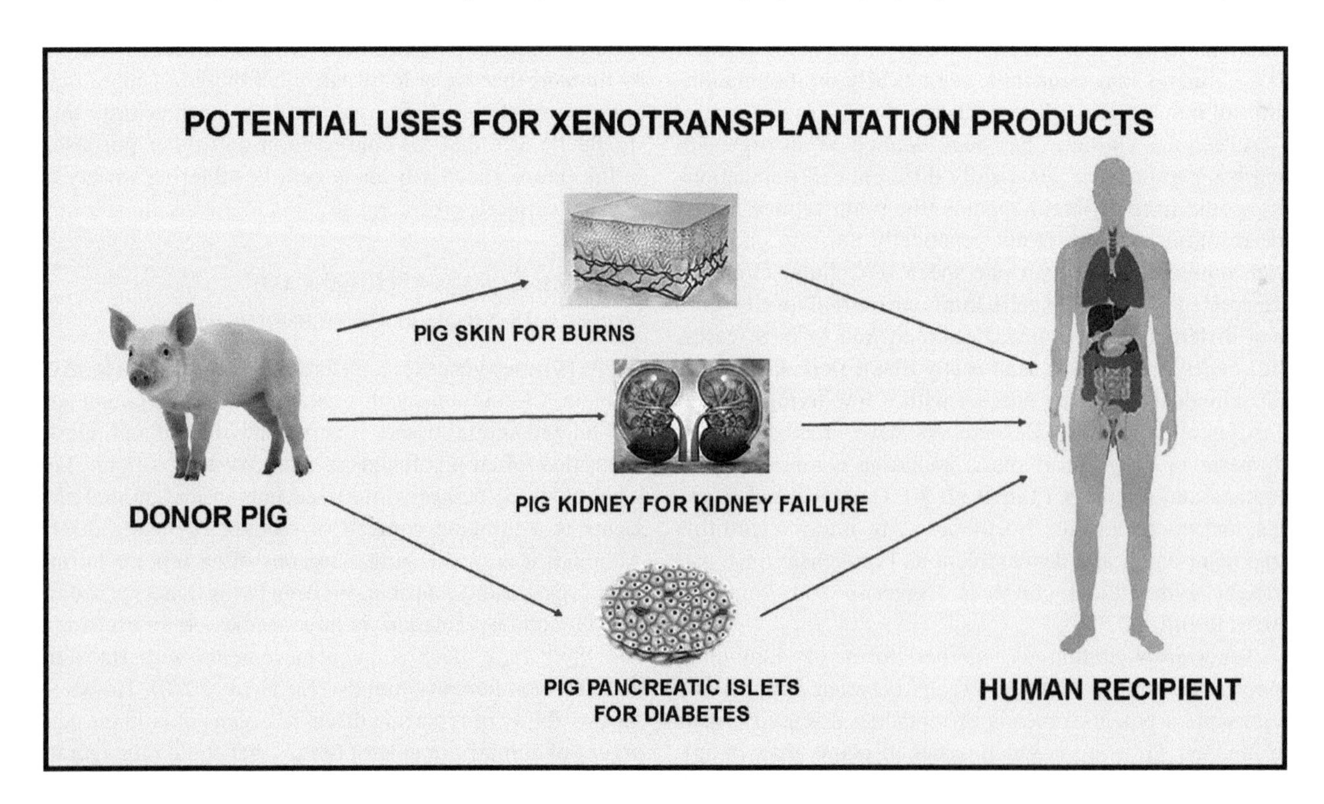

FIGURE 14.4 Potential medical benefits of transplanting tissues and organs from pigs to humans. Source: U.S. Food and Drug Administration (public domain).

FIGURE 14.5 The possible creation of intelligent mice by transfer of human brain cells to their brains. Prepared by B. Brookes.

and their native species. Controlling these harmful species is costly. Animal and plant pests are largely addressed today with chemical measures (particularly herbicides and pesticides), which can harm people and the environment. 'Biological control' (biocontrol) – the deliberate introduction of species that harm the harmful species – has proven to have considerable merit but requires extensive studies to ensure that introduced agents will harm only the intended targets. 'Classical biological control' is the intentional importation into a domestic area of foreign specialist natural enemies of invasive alien species to establish permanent populations which will reduce their harm on a continuing basis. Other kinds of control may involve biopesticides; the regular release of artificially reared biocontrol species into a limited area, such as a greenhouse; and various methods that conserve established biocontrol species to sustain natural ecosystems. The fable of the 'old lady who swallowed a fly' illustrates the unpredictability of using one creature to control another (Figure 14.6). Biocontrol agents for destroying or controlling a given plant species are first identified by systematically searching for the natural parasites of the species as well as its closest relatives, and then extensive tests are necessary to ensure only target species are affected. 'Biological warfare' employing bacteria, viruses, insects, or fungi to harm humans or their valued crops and animals is an act of war (see Chapter 11). Biological control, which similarly employs species to control species harmful to humans, is legitimate. The distinction between biological warfare and biological control is that in the latter case the punishment to targeted species is appropriate and limited. Biological control is useful but requires so much investment in research and development that it is applicable only to some of the numerous species that are harmful to the health of people and their property. Moreover, as discussed in this book, most harmful species have redeeming features, and sometimes harming a species that is dangerous to one interest group is bad for a second interest group. Examples follow.

Cactus pear (*Opuntia ficus-indica*), discussed in detail in Chapter 20, is a widely cultivated fruit crop in numerous countries and is ideal for dry regions because it requires little water. It is also an invasive weed in many regions. To control the weed, insects were introduced as biological control agents, which have indeed reduced the harm caused by the weedy plants but have been harmful to cactus pear fruit growers.

As detailed in Small (2016b), research was conducted in recent decades to find insects and fungi, and even to create genetically engineered microorganisms, that could destroy plants of marijuana (*Cannabis sativa*), until recently considered by governments to be a major cause of health and social problems. Fortunately, the attempts were unsuccessful, because the cannabis plant has been transformed into a major crop for food, fiber, and health products. Had pathogens been introduced into nature to kill off plants grown illicitly for the black market, they could now be decimating legitimate crops. Indeed, the possibility of employing microorganisms to control pests remains a potentially hazardous undertaking, because microbes are exceptionally difficult to control in nature. They can evolve much faster than higher organisms like insects (which are the main biocontrol agents), and once released, they are virtually impossible to recapture. Eliminating an entire species (and perhaps its relatives) from a region can require very widespread distribution of the control agent and represents an experiment with natural ecosystems that is dangerous.

FIGURE 14.6 Possible unintended consequences from biological control (using one species to kill another) as illustrated by the classic fable 'The Old Lady Who Swallowed a Fly.' Prepared by B. Brookes.

There was an old lady who swallowed a fly;
She swallowed a spider to catch the fly;
She swallowed a bird to catch the spider;
She swallowed a cat to catch the bird;
She swallowed a dog to catch the cat;
She swallowed a goat to catch the dog;
She swallowed a cow to catch the goat;
She swallowed a horse; she's dead, of course!

GENETIC ENGINEERING

Genetic engineering has been variously defined but basically includes biotechnological techniques that directly modify the genetic makeup of organisms – adding or subtracting genes, or altering gene expression. This may involve transfer of natural genes within and between species, synthesizing genes, or just changing (editing) the gene complement of an organism to modify development. The field is revolutionizing much of biology and is developing so rapidly that the possibilities for altering biodiversity are only beginning to be appreciated. Some of the more challenging issues are examined in the following.

Deliberate Weakening or Even Eradication of Unwanted Species by Contaminating Their Genes

Eradicating microorganism-caused diseases of humans is not a goal that many would disagree with, although completely killing off such pathological species is challenging. Smallpox caused by variola virus and rinderpest (a viral disease of cattle) have apparently been eliminated except for laboratory collections. Indifference to the consequences of overhunting or control measures for perceived pests has led to the unintended extinction of species in the past, but today the issue of deliberately eradicating species is receiving cost-benefit examination (Figure 14.7).

In recent decades, the makeup and control mechanisms of particular genes have been clarified, and it now seems possible to deliberately exterminate harmful species by several techniques. 'Gene drive' is a technique that 'enhances' the inheritance of a characteristic – even if it is harmful to the survival of a species. Accordingly, a target species can be 'infected' with an engineered gene that is harmful to that entire species (not just a few individuals initially infected). It now seems possible to use the technique to reduce the harm caused by pest species, for example by lowering their infectivity or pathogenicity. More dramatically, harmful species could be completely exterminated, for example by introducing a gene that greatly lowers reproduction. Gene drive technology has the potential to eliminate invasive predators, pests, and disease vectors, not only for the benefit of humans, but potentially also to prevent endangered species from extinction (Sandler 2020). Since insects are among the

FIGURE 14.7 Theoretical illustration of a genetic drive technique to eliminate mosquitoes. Male mosquitoes are shown here, engineered to spread harmful genes (represented by dynamite sticks) to the females. Prepared by B. Brookes.

leading pests, they have been identified as principal targets. Mosquitoes in particular, since they transmit several of the world's most devastating diseases, are among the current experimental targets (see Chapter 15). There are proposals to modify the insects with genes intended to spread through populations and either completely exterminate the species or simply eliminate their ability to carry diseases. Toward this end, in 2021 male GMO mosquitoes bearing genes designed to kill off the population after mating were released for the first time in the U.S. (Waltz 2021). This technology has also been proposed to control vertebrate pests (Thresher 2007). However, there is fear among many scientists that such technologies are potentially hazardous, both to humans and to the ecology of the planet (ETC Group and Heinrich Böll Foundation 2018; see Chapter 15). The difficulty is that once genes in nature have been modified, they may disseminate and mutate in unexpectedly dangerous ways. Additionally, there is concern that nature is unpredictable, and alternative problems could arise after a given species is eliminated. In the long term, our collective attitudes to other species (currently extremely negative regarding insects) are likely to determine how extensively humankind will decide to eradicate other species.

Favoring Particular Wild Species by Strengthening Their Genes

It is obvious that controlling, possibly even eradicating species that harm humans and our domesticated animals and plants is a goal meriting consideration, but what about similarly controlling or eradicating diseases of wild species judged to be important? Certainly people are strongly motivated to control species that harm important wild economic species such as lumber trees and seafood animals, and employing the genetic techniques discussed in the previous section are being considered. As reviewed by Reynolds (2021), well-meaning conservationists are contemplating employing genetic engineering techniques to assist selected threatened species to survive. For example, the American chestnut tree, which has been nearly extinguished by a fungal blight (Figure 14.8), could be made more resistant to it (Kosch et al. 2022); and corals suffering in Australia's Great Barrier Reef might be genetically modified to withstand warmer and more acidic seawater caused by elevated atmospheric concentrations of greenhouse gases (Mankad et al. 2021). (Coral animals require associations with photosynthesizing algal species to survive. The Great Barrier Reef, illustrated in Figure 4.9, has over 400 species of coral animals, so the problem is not easily addressed.)

Exploiting Wild Species by Genetically Engineering Them to Be More Useful in the Wild

As discussed in Chapter 12, as the human population expanded, wild species provided insufficient food and other resources, and so crops and livestock were *invented* (cf. distinction of invention and discovery in Chapter 12). Most domesticated species have been so profoundly altered from their wild ancestors that they can no longer survive in nature and require considerable care. Some wild species continue

FIGURE 14.8 Example of a nearly exterminated wild species that might benefit from genetic engineering. The American chestnut (*Castanea dentata*), a dominant tree of the eastern U.S. and southern Ontario until the early 20th century, after which it was virtually exterminated by chestnut blight, an Asian fungus that first appeared in the U.S. about 1904. (a) Collecting nuts in the later 19th century. This wood engraving (public domain) was prepared by American artist Winslow Homer (1836–1910) in 1870. Credit: Brooklyn Museum. (b) An 1864 lithograph (public domain) entitled 'The Village Blacksmith' published by the poster company Currier and Ives. 'The Village Blacksmith' is a poem by American poet Henry Wadsworth Longfellow, first published in 1840. It begins 'Under a spreading chestnut-tree, The village smithy stands.' Longfellow once stated that the tree that inspired his poem was a horse chestnut (*Aesculus hippocastanum*) a native of Europe widely planted as an ornamental, not the American chestnut. However, the artist of this painting would not have known this when he drew the generic tree shown here, probably meant to be an American chestnut. (c) Photo (public domain) of gigantic American chestnut trees in the early 20th century. Source: U.S. Department of the Interior, Office of Surface Mining Reclamation and Enforcement.

FIGURE 14.9 Examples of wild species (maple sugar trees and pearl-producing molluscs) that might be genetically engineered to be more productive in their natural habitats. (a) Lithographic print (public domain) of the painting 'Maple Sugaring' published by Currier & Ives in 1872. (b) A pearl diver in Japan. Photo (public domain) by Fg2. (c) A black pearl and its shell. Photo by Brocken Inaglory (CC BY SA 3.0).

to be harvested, but most do not furnish products of value, but as emphasized in this book, they do provide invaluable ecological services. To date, only domesticated (cultivated) species have been changed genetically (either by traditional breeding or more modern techniques). Although it could significantly change ecosystems, the possibility exists of altering the genetics of wild-growing species so that they will provide useful products while still growing in the wild. For example, most molluscs with shells can produce pearls (Figure 14.9b and c), but rarely do, so gem-quality pearls are mostly obtained from farmed oysters that need to be artificially stimulated to generate pearls. If appropriate genes were introduced to wild molluscs, conceivably the supply of pearls would be hugely increased and much cheaper. Another example is the commercial use of natural stands of maple trees (*Acer* species, especially *A. saccharum*; Figure 14.9a). Sugar concentration of the sap is less than 3% on average, but possibly by introducing gene changes this could be raised considerably. These scenarios are like conventional domestication in altering the genetics of species for the benefit of people but allow the species to remain in the wild without human assistance. While most people are opposed to deliberate release of GMOs to establish in the wild, society (or at least some countries or private interests) may choose to deliberately alter ecosystems in the hope of increasing the direct benefits (regardless of the associated ecological perils).

The Problem of Confining Engineered Genes

Once organisms have been created by genetic engineering, whether in domesticated or wild species, the issues arise of the possible uncontrolled spread of modified genes with unpredictable consequences (Figure 14.10). There are two undesirable possibilities. First, transgenic individuals could simply escape and establish in the wild and pose problems to particular species and to ecosystems. Second, transgenes could be transferred from a GMO to related organisms by hybridization. Transgenic plants could be transported to the wild by seeds. Recent decades of research with genetically modified crops have demonstrated that transgenes are often transferred (usually by pollen) to the same species (other cultivated varieties or wild forms) but also (less frequently) to wild-related species (Rizwan et al. 2019). In effect, this is a form of 'genetic contamination' of wild species although there has been very little demonstration of actual harm as a result.

Most countries where genetic engineering is occurring have very strict regulations to prevent GMOs and transgenes from accidental release. However, some nations are less concerned, and some idiosyncratic or disturbed individuals could deliberately contaminate nature. This is evidenced by the release of many exotic pets which consequently have become serious invasive pests. In the main, GMOs to date have been created for economic reasons, but in the future, it is very likely that aesthetic criteria will be added. Just as the survival of species today is being affected by human perception of their beauty and their usefulness to humans, so human judgment is likely to determine which engineered genes will be released in the future. However, human perception favoring the survival of species on the criteria of beauty and usefulness is a collective process (billions of people contribute to the judgments), whereas engineered genes are being released in some countries simply on the whims of individuals, sometimes with the possibility of establishing irretrievably in nature.

FIGURE 14.10 The issue of unintended gene release. Genetically engineered genes, once released, may have unpredictable consequences. Prepared by B. Brookes.

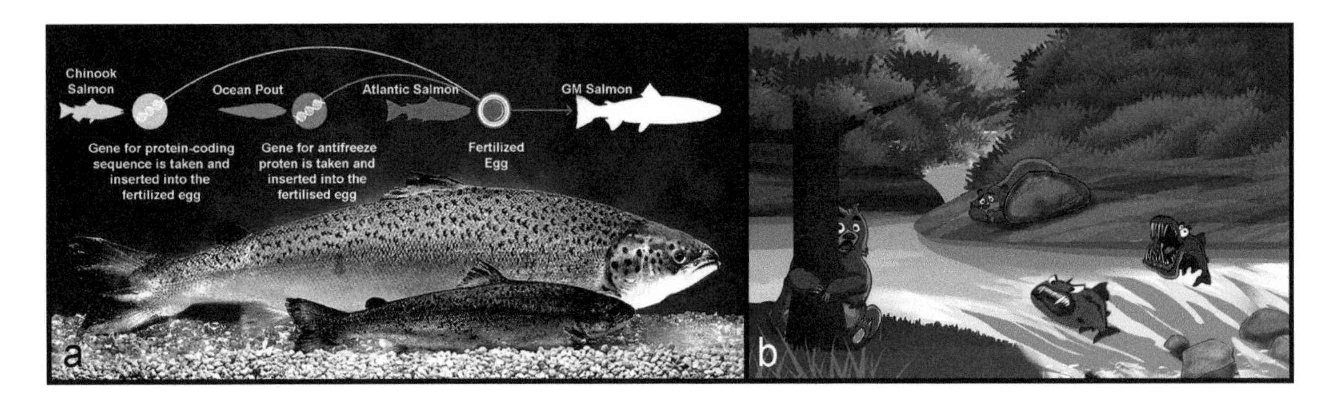

FIGURE 14.11 Genetically modified salmon. (a) Figure illustrating origin of GMO salmon. The GMO salmon in the background is the same age as the Atlantic salmon in the foreground. Prepared by J Levin W (CC BY SA 4.0). (b) Illustration of escaped GMO salmon altering the balance of nature. Prepared by B. Brookes.

Transgenic fish are a particular concern, since once they escape to large bodies of water it is virtually impossible to eliminate them, and they may outcompete their wild relatives (Dunham 2009). The majority of the world's consumed salmon is farmed Atlantic salmon, not wild-caught, mostly raised in large cages anchored offshore. Because of the importance of the species, it has been a prime target for genetic engineering. Transgenic salmon illustrate the ecological difficulties associated with the prospect of purposely establishing GMOs in nature. In 1989, AquAdvantage® salmon, a transgenic GMO salmon, was created (Figure 14.11a). The transfer of a growth hormone-regulating Pacific Chinook salmon gene surrounded by promotor and terminator regions of the anti-freeze gene of the eel-like ocean pout (Arctic pout) enabled the new fish to grow continuously under cold conditions and reach market size faster than wild fish, with less food. (The idea that the fish becomes much larger than normal salmon is an urban legend.) This was the first GMO animal approved for food, at least in North America. The fish eggs are treated with pressure, which causes most of them to acquire an extra set of chromosomes and become sterile females. The fish (like most other farmed salmon) are raised in captivity (in principle, remote from waters to which they could escape). These two containment measures are reminiscent of those taken in Jurassic Park to prevent dinosaur escape to the wild. Nevertheless, concern has been expressed that captive genetically engineered fish will escape, breed with wild salmon, and so irreversibly altering the wild species and its ecosystems (Figure 14.11b). (Aside from GMOs escaping, there is also concern about non-GMO domesticated salmon escaping, since they too can affect wild fish; see Wacker et al. 2021.) The ecological and commercial issues are controversial and hotly debated (see, for example, Grossman 2016; Van Slyck 2017; Tramper 2019; Callegari and Mikhailova 2021).

The Creation of Unprecedented Novel Life Forms by Genetic Engineering

Most genetically modified organisms to date have been created either for experimental purposes or to alter their physiology in ways that are desired by people. For the most part, the goals of genetic engineering have been quite pragmatic: to improve the health of humans and the productivity of livestock and crops (Cotter et al. 2020). GMOs are most familiar in the public sphere in the context of 'Frankenfood' – allegedly toxic food from genetically modified food species (Figure 14.12). There are fears that genetic engineering has harmed or could harm people, other species, and the ecology of the world, and there are associated ethical concerns (see for example, Vermeersch 2000; Kotze 2018; Samaddar 2020). Important as these aspects are, they are peripheral to the concerns of this book. The emphasis in this chapter is how human preferences for and prejudices against other species could affect them in the context of new technologies. From this perspective, the potential of genetic engineering is very much in its infancy. Species that are already perceived as useful will likely be candidates for improvement by genetic engineering, while harmful species will be candidates for reduction or even elimination. In principle, this is no different from traditional human reaction to useful and harmful species, although there is potential for profoundly altering individual species and the balance of nature.

There is every reason to expect that new, powerful genetic modification techniques that can alter other species will be applied at least partly in relation to preferences and prejudices. Genetic engineering, like most technologies, will of course be exercised to meet basic consumer needs (food, health, etc.) but also could be guided by desires for entertainment and other aesthetic considerations. One should expect that the innate prejudices and preferences described in Chapters 4–6 will be the basis for modifying familiar domesticates but also for creating new domesticates.

As documented in Chapter 4, size is, by a considerable margin, the most important characteristic admired by people. The resurrected giant dinosaurs of Jurassic Park captivated audiences and prophetically indicate how the entertainment industry could exploit genetic engineering by creating spectacular animal attractions (Whittle et al. 2015).

FIGURE 14.12 Anti-GMO food images. 'Frankenfood' – a highly pejorative term applied to food produced from genetically modified organisms – has generated undue fear in the public that such edible products are toxic. There are, however, legitimate concerns about some ecological and social issues associated with GMOs, and as noted in the text, they have enormous potential to modify biodiversity. Prepared by B. Brookes.

Supersizing animals is limited by physical and physiological limitations that are associated with the way the organs of bodies balance their activities, so the prospects of generating giant gorillas, for example, are limited (Pak 2021). A more likely scenario is to start with animals that are already very large, and modify them, as illustrated by the following possibility. In recent times, zoos and aquaria have been under great pressure by animals rights groups to cease exhibiting large animals (most notably elephants and whales) because of the psychological cruelty to the animals, but with genetic engineering one might produce animals that are 'naturally' passive, like livestock. This may seem far-fetched, but there have been proposals to engineer farm animals to be relatively non-sentient, without the 'stress genes' that could produce suffering during their lives on industrial factory farms (Perzigian 2003). Domesticated cats are a favorite pet, but some people keep lions and tigers as house pets (often illegally), which is highly inadvisable because they are so dangerous, but possibly their behavior could be modified genetically so that they might indeed become safe companions for humans. Another possibility is genetically engineering very large animals so that they are quite small, and accordingly much more manageable as pets. Standard farm pigs can weigh well over 100 kg, and even the so-called 'miniature pigs' like the Vietnamese potbellied pig sometimes grow to over 50 kg. In 2015, a storm of controversy arose over the possibility that gene-edited 'micropigs' created by a Chinese institute would be sold as pets (Cyranoski 2015). At maturity, the animals weigh about 15 kg, similar to a medium-sized dog, and it was suggested they would be available in various color patterns to be produced by additional genetic engineering.

As documented in Chapter 5, to a considerable degree, humans have often domesticated breeds of animals and varieties of plants for their ornamental values, with emphasis on color and exaggerated physical features. Genetic engineering has revolutionary prospects of creating unprecedented designer pets and radically modified livestock and plants (Figure 14.13).

Bioluminescence ('cold light') is the phenomenon of organisms emitting visible light (Hyde 2014). This makes the species glow in the dark. It is best known in the flashing lights of fireflies (also called lightning bugs and glowworms) and has been documented in protists, bacteria, fungi, insects, and many marine invertebrates (Meyer-Rochow 2001; 2004; Figure 14.14a and b). Most marine animals, including many fish, produce their own light (or contain bacteria that do it for them), and not surprisingly usually occur in the dark deep zones of oceans. Bioluminescence is very widely confused with biofluorescence, which is absorption of light followed by re-emission as lower-energy light (some species are both bioluminescent and fluorescent). Fluorescent protein genes have been transferred to experimental animals (including dogs and cats) so that when ultraviolet light is shone on them they glow in the dark, but this is not bioluminescence (although frequently represented as such on the Internet). 'GloFish' refers to a patented and trademarked brand of different species of genetically engineered fluorescent fish (Figure 14.14c). Reminiscent of the ancient fable of mice belling a cat to provide warnings of its presence, it has been proposed that vampire bats (see Chapter 17) be genetically altered so they glow in the dark (Marks 2019).

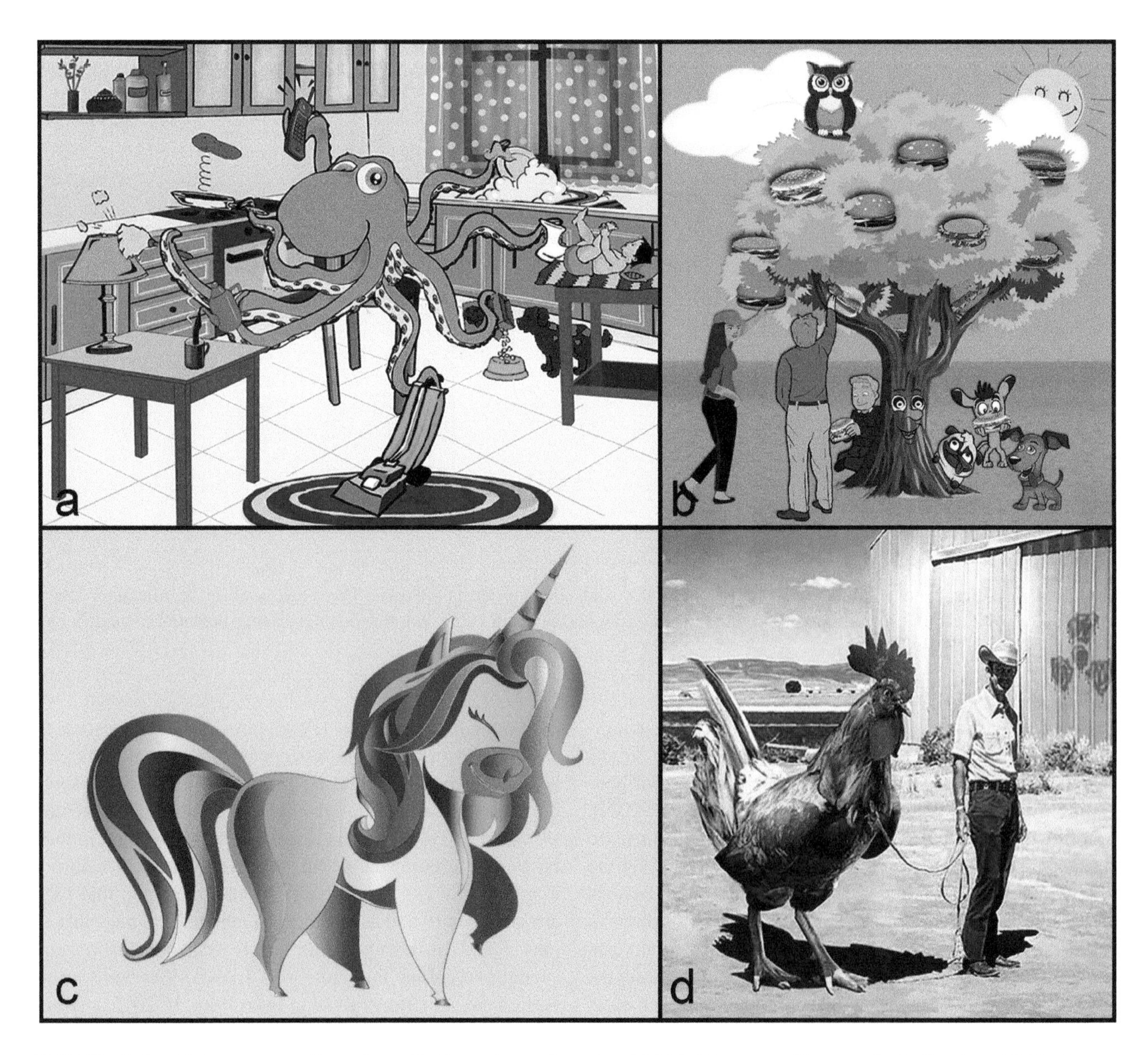

FIGURE 14.13 Extreme conceptions of genetically engineered species. (a) Helpful octopus. Prepared by B. Brookes. (b) Hamburger tree. Prepared by B. Brookes. (c) Colorful unicorn. Public domain image by Gordon Johnson, Pixabay.com. (d) Giant cock. Photo prepared by Lucyfrench123 (CC BY 2.0).

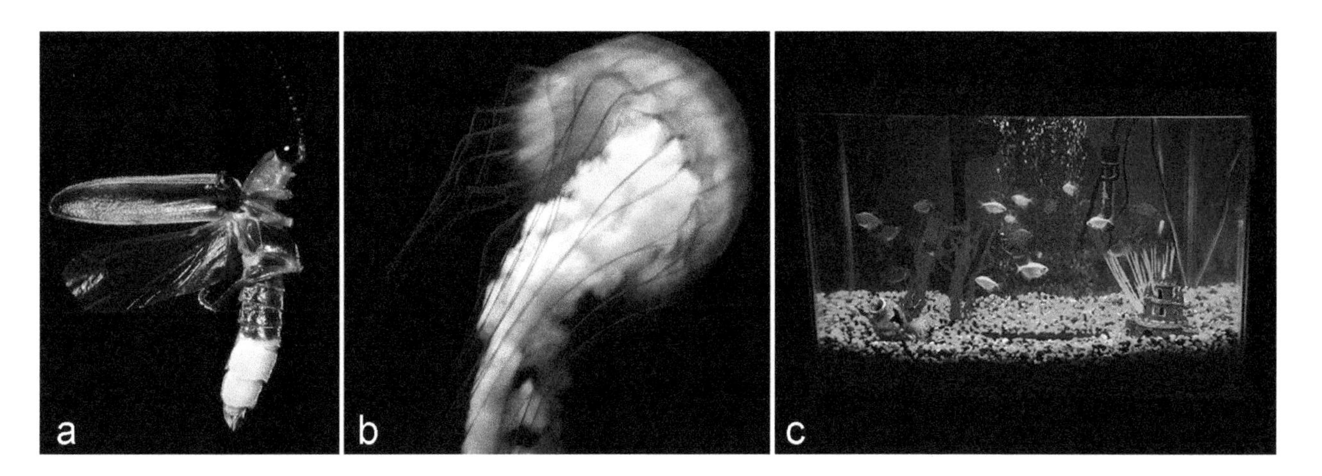

FIGURE 14.14 Animals that glow in the dark. (a) Firefly (*Photinus pyralis*), a beetle with a bioluminescent flashing light organ on the ventral side of its abdomen that serves to attract males to the females. Photo by Art Farmer (CC BY SA 2.0). (b) A jellyfish glowing by bioluminescence. Photo by Morrisayoder (CC BY SA 3.0). (c) GloFish®: A genetically engineered biofluorescent tetra fish that glows under black light. Photo by Matt (CC BY 2.0).

Estéve (2007) recommended making urban trees bioluminescent to replace street lighting. In principle, any species could be made bioluminescent, and some conceivably could be useful as flashlights. Species could be made bioluminescent simply for ornamental purposes (Fleiss and Sarkisyan 2019), so it is possible that in the future there will be many pets and houseplants that glow in the dark. One can imagine that some bioluminescent pets would eventually escape to the wild, and perhaps light up the environment.

Human Gene Interchange with Other Species

As noted above, the creative imagination of humankind to alter the genetics of other species is great, but of course human decisions are often made on the basis of preferences and prejudices. The special issue of possible transfer of genes from other species to humans, and vice versa, is the subject of this section. Transfer of human genes to other animals and even plants has been achieved although the science is in its infancy. Futurists speculating on the prospects of genetic engineering have been most concerned with how humans will alter our own genes, and certainly there will be opportunities to modify our physiology to improve health, which should not be very controversial. More troubling are possibilities of changing cosmetic, sexual, athletic, and intelligence aspects – which challenge moral precepts (Metzl 2019; Figure 14.15). Gene editing and gene transfer represent a means of profoundly altering the genomes of individuals and populations. 'Gene therapy' may refer to curative genetic changes to the somatic (non-reproductive) cells, in

FIGURE 14.15 Genetically modified humans. (a) A designer baby gene store. (b) A genetically modified basketball specialist. Figures prepared by B. Brookes.

which case the changes are not heritable. Gene therapy may also affect the reproductive cells (sperm, eggs), which can lead to passing on altered genes to children and so potentially changing gene frequencies in human populations.

The informal phrase 'designer baby' refers to a hypothetical human baby whose genetic makeup has been constructed by genetic engineering. Just how extensively such customizing should be is debatable. The primary justification for modifying the genetics of humans is to prevent or alleviate susceptibility to inherited diseases. In principle, advanced genetic modification techniques applied to date to animals could be applied to humans, but ethical considerations are preventing most use of living humans as experimental subjects. A scandalous, unauthorized experiment initiated in 2017 to edit the genome of human embryos leading to birth apparently succeeded but led to widespread condemnation and the jailing of the scientist (Greely 2019).

Eugenics: Old and New

Eugenics refers to deliberate efforts to alter the genetics ('genome' collectively for an individual or a population) of a population of humans in ways that are considered desirable. The term is highly pejorative because it reflects racial discrimination, contempt for the mentally or physically challenged, and highly questionable science in the 19th and 20th centuries, most notably by the German National Socialist regime (the Nazis). In the past, the key methodology was selective breeding – encouraging reproduction by individuals with favored traits and discouraging reproduction (often employing forced sterilization) by those with less favored characteristics. One of the most memorable efforts was a project to distribute the sperm of Nobel Prize laureates to raise the intelligence of people (Plotz 2006). Sperm banks today often include donors who have been screened for 'socially desirable characteristics' such as height and robustness of physique, and to a considerable degree, donors are white university students (Almeling 2007). In effect, the 'assisted reproduction' industry has commodified human genes and is engaging at least indirectly in selective breeding (Figure 14.16).

Domestication of animals and plants by humans is quite comparable to eugenics in concept although ethical considerations make the comparison odious. A simple but widely

FIGURE 14.16 Cartoon illustrating the eugenic nature of commercial sperm banks. Prepared by B. Brookes.

accepted definition of evolution of a population is the altering of its gene (allele) frequencies, so in effect eugenics amounts to deliberate direction of the evolution of the human species.

Classical eugenics has largely been abandoned, at least with respect to the coercive and overtly biased racial aspects. The phrases 'new eugenics' and 'liberal eugenics' have been employed in many recent publications, often wholly or partly based on the new genetic engineering technologies applied to people. Regardless of terminology, genetic engineering has reinvigorated the possibilities of fundamentally altering the genetics of humans – not just of individuals (i.e., locally), but of the world population. The basic subject of this book is how human preferences and prejudices are altering all of the other species of the planet. The parallel subject of how we may alter ourselves deserves mention because our intrinsic nature is already changing biodiversity and modifying ourselves may exacerbate the current already dire plight of most other species. As noted above, ethicists have extensively examined the morality of the subject, but one may expect that human preferences and prejudices will play determinative roles, regardless. The identification of the sex of fetuses by modern technology has often led to abortion of females, demonstrating the ugly reality of sexism. Women employing non-partner sperm donors to become pregnant also exercise preferences and prejudices based on available information about the contributors (Whyte 2015).

Probably the most universally welcome aspect of genetic engineering is the likelihood of improving the health of the world's humans by altering their genetics. However, people have other aspirations in addition to good health. Human preferences for achievement span a wide variety of fields such as athletics, science, literature, art, music, dance, and entertainment media, and it may not be easy to identify genetic factors that determine the relevant talents. Intelligence and athletic ability are widely admired and have substantial genetic determination, and so conceivably one day genes will be available to enhance performance. However, as presented in Chapter 5, physical appearance dominates how most people judge others, and indeed themselves. Prejudices against and preferences for body features have enormous consequences for individuals in human societies, and inevitably to the extent possible, people will choose to acquire features that increase the success and happiness of themselves and their children, while similarly suppressing features that are perceived as undesirable. How might this affect human attitudes to other species? Future technologies may narrow the range of people developing physically undesirable features, and possibly in parallel this will be accompanied by accentuation of prejudices against animals with physically undesirable features.

One scenario is currently too radical to consider seriously for the foreseeable future. This is the possibility of cloning humans (Figure 14.17). A clone is a group of genetically uniform individuals – like identical twins, but with many copies available. A doppelgänger (doppelganger) is a biologically unrelated double, of a living person, but clones not only look alike, they are genetically alike. Cloning is an asexual means of reproduction. Clones are easily generated in plants by vegetative reproduction (typically a twig is cut off and rooted in soil), and many crops are propagated as clones (e.g., potatoes, apples, and marijuana strains). Some lower animals reproduce clonally, and the technology of cloning vertebrates, especially livestock, is advancing (Avise 2015; see 'Dolly the Sheep' in Figure 2.23). Human clones already exist in a restricted sense: as tissue cultures. For example, 'Hela' is a standard cell line used in medical research, taken from cervical cancer cells from **He**nrietta **La**cks, who died of cancer in 1951. Humans usually choose

FIGURE 14.17 An orchestra of cloned musicians. Prepared by B. Brookes.

outstanding individual animals and plants to clone because they demonstrate exceptional desirability, and once cloned, a constant supply of predictably uniform material will be available. For some purposes, clones may be the best of the best, and one can imagine a world populated by a restricted set of elite humans. Presumably, they would look like movie stars and champion athletes, have the intelligence and talents of Nobel Prize laureates, and be impervious to all diseases. While such science fiction is imaginary, it is useful in examining the prospects of reducing variation of the world's living things, which is what we humans are already doing.

ARTIFICIAL REALITY

Real creatures, even if quite admirable, can be quite troublesome and expensive to maintain, and human ingenuity is often capable of creating substitutes. Statues of lions are ubiquitous, but lions are not. Of course, an inanimate statue barely conveys the charm of the genuine beast. But consider Christmas trees: increasingly, artificial trees are replacing real trees, and the best models appear remarkably life-like. The key to the acceptance of artificial substitutes is perception of their reality. Technology is rapidly improving the ease of convincing people that unreal things are actual.

'Virtual reality' is technology-generated experience, a topic which is greatly increasing in sophistication with advances in digital technologies. The experiences can simulate the real world or be quite fictional. To date, emphasis has been primarily on visual perception, but auditory, odor, tactile, and other kinds of simulation of sensory stimuli are in early development. Depending on the accuracy of simulation, additional phrases such as augmented reality, mixed reality, and extended reality are encountered. The field is particularly useful for entertainment (especially video games), training, education, and commerce.

In the 1990 film Total Recall starring Arnold Schwarzenegger as Douglas Quaid, implanted memories involving extremely elaborate experiences were commercially available. Indeed, Quaid had the option of actually visiting the planet Mars or having a fake but extremely entertaining memory of doing so. Imagine a similar scenario wherein people had the option of actually visiting the world's premier animal reserves, parklands, and forests, or experiencing simulations of Jurassic Park and comparable mind-blowing adventures. For most people, the synthetic alternative might be chosen, and the result could be that support for maintenance of natural environments and their biota would diminish (Figure 14.18). Of great concern is the fact that young people already are so immersed in

FIGURE 14.18 Potential use of virtual reality to ignore the world's problems, including the biodiversity crisis. Prepared by B. Brookes.

FIGURE 14.19 'Nature Deficit Disorder': Estrangement from the real world and its wonderful life forms. Prepared by Jessica Hsiung.

their digital devices that they scarcely have time to explore the real world, resulting in the condition 'Nature Deficit Disorder' characterized by alienation from nature (Louv 2005; Figure 14.19).

SCIENCE AND PUBLIC INTEREST

The scenarios described in this chapter are to some extent hypothetical, but it is clear that technological advances are currently changing the world so rapidly that there is (or should be) concern about the wisdom of future changes and the ability of society to control these changes. The complexity of science is a roadblock to getting the public interested in developments and issues, and unfortunately, scientists themselves sometimes seem aloof from basic issues of concern to most citizens. Many scientists in industry and government are restricted in their ability to communicate with the public, and there are still areas of life in which research

FIGURE 14.20 Tampering with nature. Source of photos: Shutterstock. (a) Creating a giant tomato. Contributor: Nuvolanevicata. (b) Creating a tiny dinosaur. Contributor: ProStockStudio.

is difficult to undertake. This chapter, and indeed this book attempts to outline some challenging, indeed threatening scenarios, but it is necessary for the world of science and the many scientists with special knowledge to make a greater effort to evaluate and communicate concerns to the public. Nature belongs to everyone, and private interests need to respect this principle before altering this shared resource (Figure 14.20).

15 In Defense of the World's Most Reviled Invertebrate 'Bugs'

INTRODUCTION

Species of invertebrate animals, notably insects, are undergoing an alarmingly high rate of extinction, coupled with minimal support for their protection, even from the world's leading conservation organizations. This is intolerable, as invertebrates constitute over 95% of the world's species, have indispensable economic values, and provide ecological services without which life on Earth would virtually cease. Much of the lack of public and governmental support for invertebrate conservation is due to the abhorrent tiny pests that have persuaded most people that 'bugs' are bad, and consequently, the only species worthy of support are the charismatic superstar mammals like pandas and tigers that currently are the mainstays of biodiversity fundraising. Just as these respected, highly attractive icons are effective ambassadors of biodiversity conservation, so certain detested pests have poisoned the public image of invertebrates and indeed have made it seem to many that most wildlife is hostile. The 'dirty dozen' bugs that particularly are a hindrance to improving public investment in biodiversity are bedbugs, clothes moths, cockroaches, fleas, houseflies, leeches, lice, locusts, mosquitoes, spiders, termites, and ticks. Except for spiders, these species, admittedly, are responsible for enormous damage to health and economic welfare. Nevertheless, this chapter shows that most have at least some compensating values, their harm has often been exaggerated, and all have related species that are good citizens by human standards. Six of the dozen 'least wanted' invertebrates highlighted are blood parasites of people, and these 'bad apples' are very hard to defend since parasitism seems abhorrent. Remarkably, however, at least half of the world's tens of millions of species are also parasites, and without them, most ecosystems would be in danger of collapse. To improve invertebrate conservation, it is advisable that efforts be made to educate the public regarding their importance. Since prejudices against 'bugs' are primarily acquired during childhood, special attention is needed to persuade the young that most invertebrates are harmless, valuable, and entertaining. Recent advances in genetic engineering ('synthetic biology,' 'genetic drives'), as will be presented, have led to very serious consideration of deliberately eliminating the world's worst pests of humans. While these extermination technologies could greatly increase support for invertebrate conservation by annihilating their most despised representatives, the dangers of unforeseen damage to ecosystems and hence to biodiversity are substantial. These issues are examined in detail in this chapter.

Conservationists are keenly aware that the key strategy to improve support for biodiversity conservation, just as has been the case for the climate change issue, is the generation of greater public and governmental concern. Toward this goal, the most successful tactic has been the adoption of charismatic iconic species – giant pandas, whales, spectacularly gorgeous butterflies (Figure 15.1), and towering trees – whose survival people spontaneously support (Bennett et al. 2015; Jepson and Barua 2015; Thomas-Walters and Raihan 2017; Macdonald et al. 2017; Albert et al. 2018b). The public relations value of these 'flagship,' 'surrogate,' or 'ambassador' species has proven to be immense, and happily their conservation is necessarily achieved by preserving their habitats, which coincidentally conserves many other species (Yamaura et al. 2018). However, the approach often fails to adequately support many other partially associated species (e.g., Fourcade et al. 2017; Kramer et al. 2019) and of course ignores the vast majority of species that occur outside the geographical ranges and habitats of conservation icons. Moreover, there is often disagreement about which charismatic species to feature in conservation campaigns (Smith et al. 2012). These issues are detailed in Chapter 10.

In the noble quest to address the biodiversity extinction crisis by highlighting the world's most attractive, lovable, and charming species, a key consideration has been tacitly ignored: the world's most harmful, offensive, disgusting, and despised species, which regrettably have been allowed to greatly lower respect for the living world. Advocating on behalf of biodiversity's most reviled species is not an easy exercise – how do you defend mosquitoes, bedbugs, cockroaches, fleas, vermin and other harmful, ugly, offensive pests? A simple statistic illustrates the magnitude of the problem: more than half of the world's tens of millions of species are parasites (Morand 2015). Parasitism is a life form that is repugnant to most people – even most biologists and conservationists – who have limited if any sympathy for such species. Nevertheless, as will be reviewed, parasites are indispensable components of biodiversity and indeed have values that deserve to be identified. Most importantly, the widespread fear and disgust associated with the world's most despised species need to be tempered with a realistic understanding of and respect for their useful roles in nature, in order to minimize the disrespect for biodiversity that they generate.

The animal kingdom is classified dissimilarly by different authorities. Some recognize about three dozen comprehensive groupings termed phyla, the majority of which are completely unknown to most people. Vertebrates

DOI: 10.1201/9781003412946-15

FIGURE 15.1 Beautiful butterflies, the only invertebrates that receive universal approval. Source: publicdomainpictures.net.

(a subgroup of the phylum Chordata) are animals with a spinal column or backbone and include mammals, birds, amphibians, reptiles, and fish. Invertebrates (an artificial grouping defined by lack of a spinal column or backbone) are all other animals. Most of the noxious animals examined in this chapter are classified into two subdivisions of the phylum Arthropoda: Insecta (the insects) and Arachnida (including spiders, ticks, and others). The leeches are in the phylum Annelida (best known for earthworms).

Over 95% of the world's animals are invertebrates (Scanes 2018; note Figure 15.2). Of the tens of millions of animal species, the average person is particularly aware of the 12 'bugs' highlighted here. These are the antithesis of the charismatic, photogenic species that are employed as conservation icons to evoke sympathy for biodiversity. Unlike the very large 'megafauna' such as pandas, tigers, and elephants, almost all are miniature animals occurring in vast numbers. Most alarmingly, with the exception of spiders, they negatively affect huge numbers of people. Most significantly in the context of biodiversity, these are the creatures most responsible for generating hostility to nature and the living world, because they are so ubiquitous, familiar, hated, and the subjects of constant vilification.

So what can be done? As reviewed in the following, efforts at rehabilitating the reputations of the principal pests are required. The worst of the worst – parasites and pests – need to be correctly identified and not mistaken for the millions of their innocent relatives. Moreover, their destructiveness should be clearly understood but not exaggerated, and control measures adopted that are targeted but not harmful to nature and biodiversity in general. Our invertebrate enemies have remarkable behavioral and anatomical adaptations that make them efficient competitors with humans, and while it is difficult to admire society's worst biological foes, a measure of respect is desirable, since most play significant positive roles in nature, which are emphasized in this chapter. Most people are unaware that the vast majority of life forms are harmless invertebrates, particularly insects, and that they contribute in critical ways to the orderly functioning of ecosystems, and indeed to human welfare. The principal goals of this chapter are to correct the widespread impression that the invertebrate world is validly characterized by its most offensive representatives and to point out that even the most despicable have redeeming values and that there are innumerable related species that play indispensable roles in nature.

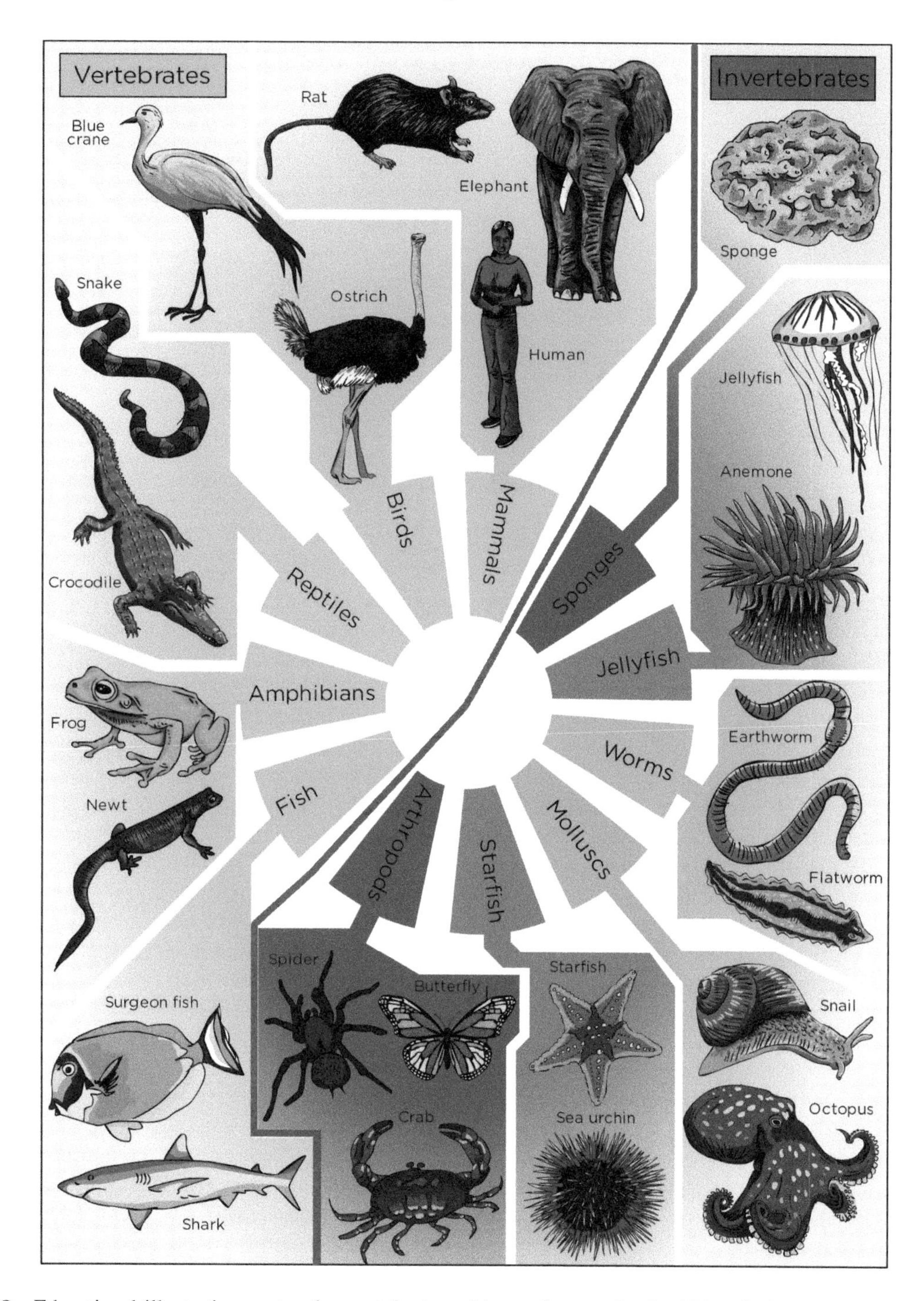

FIGURE 15.2 Educational illustration contrasting vertebrate and invertebrate animals. Although the two groups are given equal prominence in this plate, in fact invertebrates constitute more than 95% of all animal species and are much more diverse than indicated here. Source: Siyavula Education (CC BY 2.0).

Importance of Size of 'Bugs' for How Humans View Them

As detailed in this book, human psychology determines what species are likable or detestable. The vernacular derogatory, pejorative noun 'bug' when applied to living things may denote both pathogenic microorganisms ('germs') and small, relatively unattractive animals (mostly insects and spiders). However, attitudes toward the vast array of so-called bugs differ. A minimum size is an essential criterion because if something can't be seen, it's hard to conceptualize it as a living being (Figure 15.3). Pathogenic microorganisms that are invisible to the naked eye, notably viruses, bacteria, and protozoans, generate diseases that are greatly feared (excessive, unreasonable fear is 'germophobia'), but the causal species per se are not really detested like the invertebrates discussed here. Most of the especially despised species examined in this chapter

FIGURE 15.3 Humorous illustrations of the coronavirus pandemic reveal intense public fear and dislike of the resulting disease but limited recognition of the causal agent as a hateful species because it is invisible. (a) Love in the COVID era, by Hilda Chaulot (CC BY SA 4.0). (b) Infected anti-vaxxer, by Welleman (CC BY SA 4.0). (c) Cartoon (public domain) by Ray Shrewsberry/Pixabay.com.

are terrestrial (leeches are more often aquatic, and so are mosquito larvae), reflecting the fact that the majority of human encounters with biodiversity are on land. Almost all terrestrial invertebrates are small, but at least large enough to be noticed, especially insects. Possibly the most detestable of small animals are internal parasites, such as tapeworms and botflies (see Scanes and Toukhsati 2018 for a review). These are not discussed here because, at least in Western countries, they rarely occur today in humans (although a considerable problem in Developing Countries and for livestock) and can be managed by hygiene and modern medicine. Many invertebrates are serious agricultural pests. These are not discussed here because, in Western countries where biodiversity conservation requires popular public support, agriculture is carried out by a remarkably small proportion of the population. There are many invertebrates whose stings are occasionally fatal, notably certain ants, wasps, and scorpions, but these are relatively rarely encountered. (Ants in residences, gardens, and picnic areas are greatly disliked, but on the whole they are viewed by the public as hard-working, respectable super-athletes.) Remarkable invertebrate diversity occurs in the marine environment, but because the majority of these species are infrequently viewed and most do not threaten humans (jellyfish are an exception), they are largely ignored by people. Indeed, the vast majority of invertebrate species are hidden from human view, and in accord with the old proverb 'out of sight, out of mind,' they are not of concern to most people. Unfortunately, it is the common pests that are, by a considerable margin, most evident and influential.

BEDBUGS

Introduction

The Hemiptera or 'true bugs' may contain as many as 80,000 species, most of which feed on plants. The bedbugs (also spelled bed bugs and bed-bugs) represent only a little more than 100 of the described Hemiptera. Bedbugs are ectoparasites – parasites feeding on the outer surface of animals. Recent literature assigns them to two families, the Cimicidae, which live on birds and mammals, including humans, and the Polyctenidae, which live on bats and are called bat bugs (Reinhardt and Siva-Jothy 2007). However, 'the taxonomy of bat-associated bugs is currently in a state of transition' (Hornok et al. 2017), and all bedbugs are often included in the Cimicidae. Bedbugs are small nocturnal insects which feed on the blood of humans and other warm-blooded hosts, primarily mammals (often bats) and birds (including poultry). The common bedbug, *Cimex lectularius* (Figure 15.4), occurs in temperate climates throughout the world. For the most part, the unqualified word 'bedbug' refers to this species. *Cimex hemipterus* (also called 'bedbug') is a more significant pest of humans in tropical regions. Bedbugs are one of the most familiar of insects, as reflected by the popular saying 'Good night. Sleep tight. Don't let the bedbugs bite,' the origin of which is obscure (Bologna 2018). Bedbugs may be familiar, but one of the reasons that infestations rapidly expand before they can be eliminated is that most people cannot identify the insect (Romero et al. 2017). Adult bedbugs (Figure 15.4a) are seemingly wingless (actually their wings are vestigial, represented by short, non-functional wing pads) and cannot fly. They are reddish brown (more reddish after a blood meal), flattened, oval, banded and about 4–5 mm long. As noted by Berenbaum (2010), 'their sexual practices are bizarre even by insect standards: because the female bedbug has no genital opening, the male inseminates her by using his hardened, sharpened genitalia to punch a hole through her abdomen. With no elaborate courtship ritual, males in a frenzied pursuit of sexual congress often blunder into and puncture the bodies of other males, occasionally inflicting fatal wounds.' Bedbugs usually feed at night, piercing skin with two hollow tubes (Figure 15.4b). The saliva, containing anticoagulants to facilitate blood flow, is injected through one tube, while blood is withdrawn through the

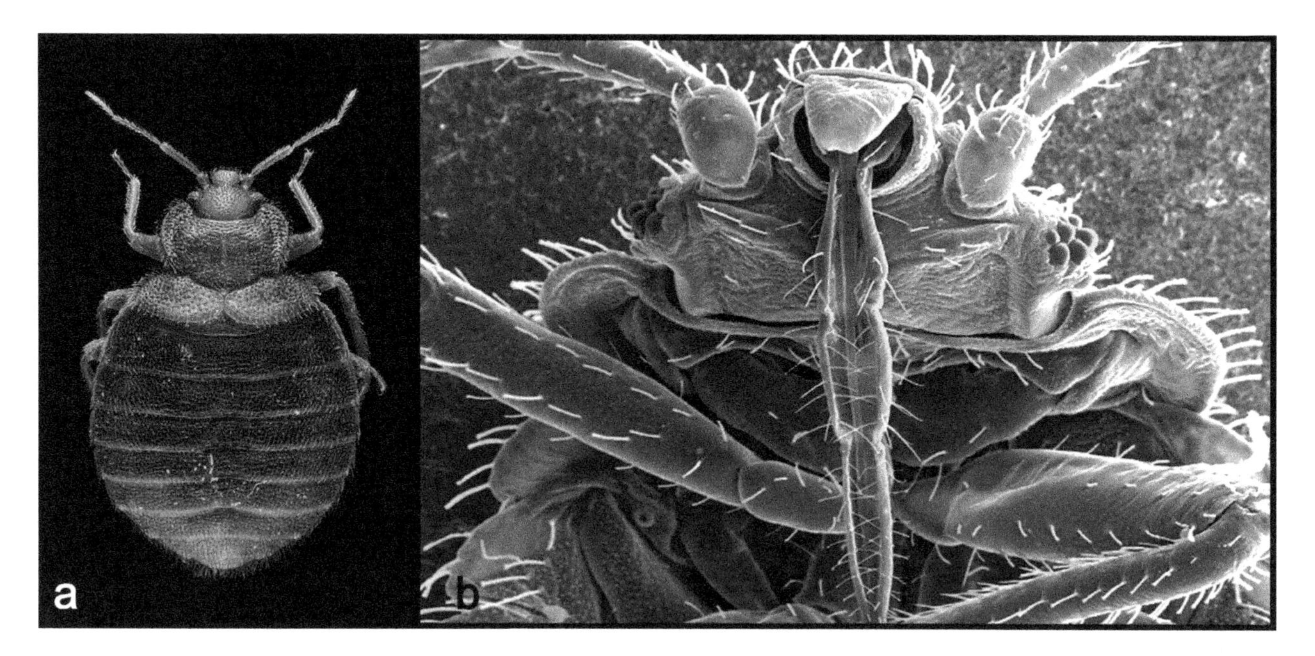

FIGURE 15.4 Bedbug (*Cimex lectularius*). (a) Adult female. Photo (public domain) by Piotr Naskrecki, U.S. Centers for Disease Control and Prevention. (b). Digitally colorized scanning electron micrograph of the lower (ventral) surface. The mouthparts employed to pierce skin and suck up blood are shown in purple. Photo (public domain) by Janice Harney Carr, U.S. Center for Disease Control.

other tube (Figure 15.4b). Victims of these tiny vampires are unaware that they are donating blood because of anesthetics in the bedbug saliva. Often bedbugs produce a distinctive linear group of three bites (macabrely referred to as 'breakfast, lunch and dinner'). Another disturbing clue to their presence is that bedbug feces stains bedding black while blood from bites results in brown stains. Flea bites are mainly around the ankles, but bedbug bites may occur on any area of skin that was exposed while sleeping. Another distinction is that flea bites tend to have a red spot in the center.

Common bedbugs (*C. lectularius*) are extremely well adapted to cohabit with humans and are thought to have parasitized people since ancient times. It has been believed that they transferred from cave-living bats to humans, possibly in the Pleistocene epoch (Talbot et al. 2019), but recent evidence suggests that bedbugs evolved much earlier (Roth et al. 2019). Bedbugs are small and flat, capable of hiding in tiny crevices, and also avoiding light which would reveal them to people. They often take refuge in mattresses, furniture, carpets, baseboards, and clutter. Bedbugs prefer humans but will also parasitize pets and rodents in homes. When blood is unavailable, they can remain dormant for months. As with mosquitoes, a blood meal is necessary for egg production, and females can produce hundreds of eggs in their lifetime. Bedbugs can travel short distances but rely on humans to move between buildings. These insects are exceptional hitchhikers, traveling on clothing, bedding, luggage, furniture, and boxes. For centuries, people have made efforts to remove or kill bedbugs in their sleeping quarters before retiring (Figures 15.5 and 15.6). The use of the insecticide DDT for pest control in the 1940s and 1950s reduced bedbug frequency, but subsequent increased international travel and exchange of goods have made them very widespread (Borel 2016). Bedbugs are extensively found in apartment complexes, dormitories, hotels, cruise ships, trains, and buses. Crowded living quarters where control measures are difficult, such as homeless shelters and refugee camps, tend to attract bedbugs. Although often thought to be associated with lack of sanitation, bedbugs show up in immaculate homes.

Harmful Aspects

With the recent worldwide increase in bedbugs, the cost of their removal from commercial and private buildings has become enormous (Scarpino and Althouse 2019). Although bedbugs can harbor pathogens in their bodies, such as plague and hepatitis B, they have not been linked to the transmission of any disease and are not regarded as a particularly dangerous medical threat (Goddard and deShazo 2009; Delaunay et al. 2011; Pospischil 2015; Lai et al. 2016). However, for sensitive individuals, the welts and swelling produced are more itchy and longer-lasting than mosquito bites. Additionally, some people develop skin infections and scars from scratching bites. Frequently, the discovery of bedbugs produces considerable stress, and sometimes even a paranoia ('delusional parasitosis') that bedbugs are present when they are not. The misguided stigma that bedbugs indicate uncleanliness also causes psychological stress for many. Occasionally, respiratory symptoms, dermatitis, and allergic reactions result from secretions produced by bedbugs. Sometimes serious poisoning has occurred as a result of incompetent usage of pesticides to control bedbugs. Unfortunately, resistance has developed to common pesticides (Romero and Anderson 2016). All bedbugs are

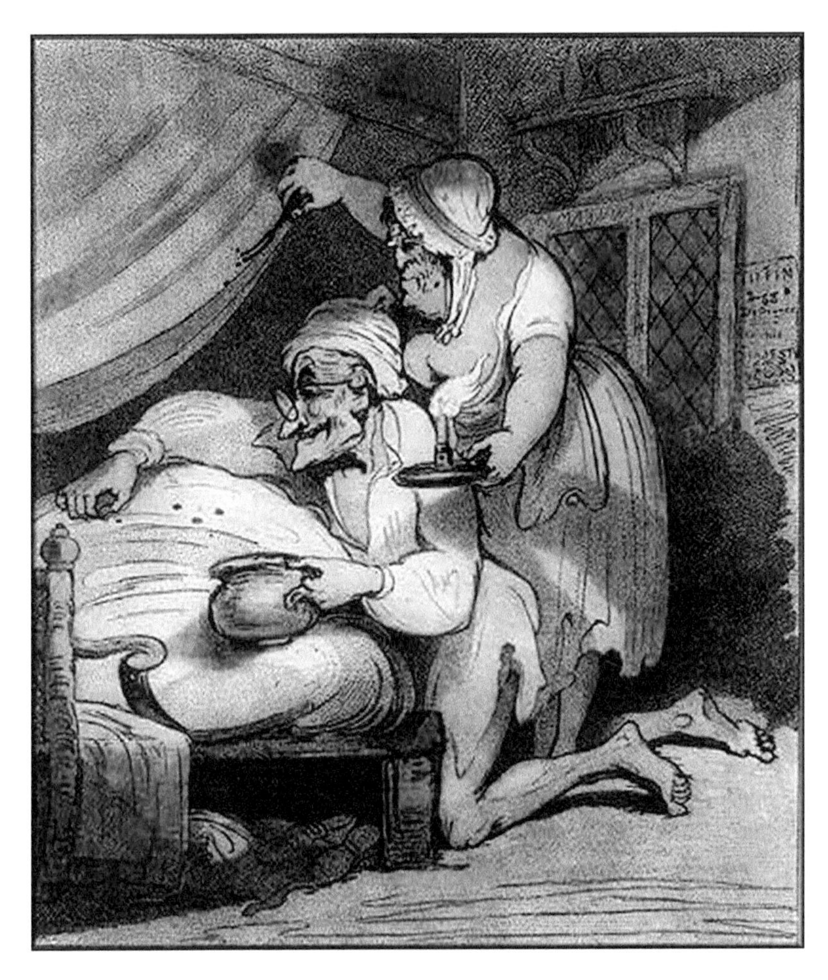

FIGURE 15.5 A British cartoon (public domain) published in 1872 showing an old couple catching bedbugs. Source: U.S. Library of Congress.

FIGURE 15.6 Early 20th-century American advertisement (public domain) for bedbug insecticide. Druggists commonly prepared and advertised their own concoctions, the safety of which was undetermined.

FIGURE 15.7 Bedbug cartoon (public domain) from Clipart Panda.

blood parasites, and as such are scorned by virtually all people (Figure 15.7).

Beneficial Aspects

What can be said in defense of bedbugs? Doggett et al. (2012) suggested the possibilities that extracts from bedbugs might be useful to inhibit the growth of bacteria and prevent germination of pathogenic fungal spores, and that they might be a potential resource for the discovery of new drugs. Most of their relatives in the huge insect order Hemiptera do not harm people although admittedly some are significant pests of crops. Bedbugs are food for some predators, notably spiders. They provide employment for professional pest control companies since most infestations need to be treated by experts. Bedbugs have become the basis of million-dollar lawsuits, providing work for lawyers (another form of parasite, some would unkindly remark). Bedbugs also stimulate research, for example into their genome, which was intended specifically to find ways of killing them (Nield 2016). Bedbugs also are useful subjects of ethical debate, for example regarding why Noah brought them aboard his ark.

Conservation Aspects

Bedbugs, like most parasites, generally do not threaten the very existence of host species, which have usually evolved adaptations to survive the attacks. However, when the population of a host species has become reduced to the point of endangerment, parasites can be the straw that broke the camel's back. For example, bedbugs appear to threaten the survival of the crowned eagle (*Harpyhaliaetus coronatus*), a large raptor of southern South America, currently considered a globally endangered species, with a world population under 1,000 (Santillán et al. 2009). The bedbug species specializing on humans have not been found to have beneficial properties (at least to date), and their extinction would likely not trouble conservationists (*The Guardian* 2013).

CLOTHES MOTHS

Introduction

The Lepidoptera constitute one of the largest orders of insects, with about 200,000 species. The larvae ('caterpillars') of many species are significant pests of crops, and some are toxic or allergenic (Mullen and Zaspel 2018). On the positive side, many Lepidoptera are important pollinators, and most serve as food for birds, mammals, amphibians, other insects, and spiders (Perveen and Khan 2018). Some moth species are specialist feeders on weeds, which makes them good biological control agents for harmful plants. Silk cocoons prepared by larvae of the domesticated silkworm moth (*Bombyx mori*) are the chief source of silk, perhaps the most familiar economic product of the Lepidoptera. The larvae of numerous wild lepidopterans, as well as of the domesticated silkworm moth, are widely employed as food in some countries.

Butterflies and moths are somewhat arbitrarily distinguished groups. Butterflies usually have thin antennae with small balls or clubs at the ends, while moths lack the club ends. Most butterflies are diurnal (active during daylight, inactive at night), while most moths are nocturnal (active during nighttime, inactive during daylight) or crepuscular (active at dawn and/or dusk). Most butterflies and many of the moths that fly during daylight have brightly colored wings, while nocturnal moths are usually brown, gray, white, or black, and drab compared to most butterflies. Because colorful Lepidoptera are much more attractive than plain species, and the dull ones are reminiscent of harmful clothes moths, there are very dramatic preferences for most butterflies and distaste for moths (Figure 15.8). Only about 10% of lepidopteran species are considered to be butterflies, the rest are moths, so most adult Lepidoptera are not viewed favorably, and almost all immature stages (especially the caterpillars) of all species are considered suspicious at best.

Clothes moths are just a few species of the large moth family Tineidae. The most unwelcome species is *Tineola bisselliella* (Figure 15.9), best known as the common clothes moth, but also often called cloth moth, clothing moth, and webbing cloth moth. Adults have a body length of 6–7 mm and a wing span of 9–16 mm. The larvae are creamy-white caterpillars up to 15 mm long. The common clothes moth has been speculated to have originated in Africa (Plarre and Krüger-Carstensen 2011) but has been distributed to many countries by humans. The larvae (caterpillars) are feeding machines, but the adult moths do not feed. Several other pest moth species are also economically significant 'clothes moths,' most notably *Tinea pellionella*, the case-making clothes moth or case-bearing clothes moth (Figure 15.10), so-named because the larvae feed and move about in a silken case. Adult common clothes moths are a uniform, buff-color, with a small tuft of reddish hairs on the top of the head (less evident reddish hairs may also be on the casemaking clothes moth). The two moths are similar, but case-making clothes moths have dark specks on the wings (less evident in stored museum specimens). Curiously, clothes moths develop more slowly or not at all on clean wool, because they require Vitamin B and various salts as essential nutrients, and these are lacking in well-cleaned wool. However, perspiration and other kinds of fabric soiling can provide vitamin B and salts, and the larvae focus their feeding on patches of cloth soiled with sweat, oil, urine, beverages, or other materials. Clothes moths are rarely observed because they avoid light. Because clothes moths are frequently in heated buildings where conditions are favorable, they can reproduce year-round, typically producing two generations annually.

Harmful Aspects

True to their name, clothes moths are notorious for damaging garments (Figure 15.11). The larvae feed mostly on natural fibers employed by humans, particularly animal hair such as wool, silk, and fur, but also plant fabrics like cotton and linen, and synthetic fibers blended with natural fibers. Apparel, bedding, blankets, carpets, curtains, drapes, feathers, leathers, piano felts, rugs, upholstery, animal bristle brushes, and many other valued objects are in danger (Hodgson et al. 2008). Even paper and wallpaper may be consumed. Expensive and irreplaceable materials (such as tapestries and stuffed animals maintained in museums)

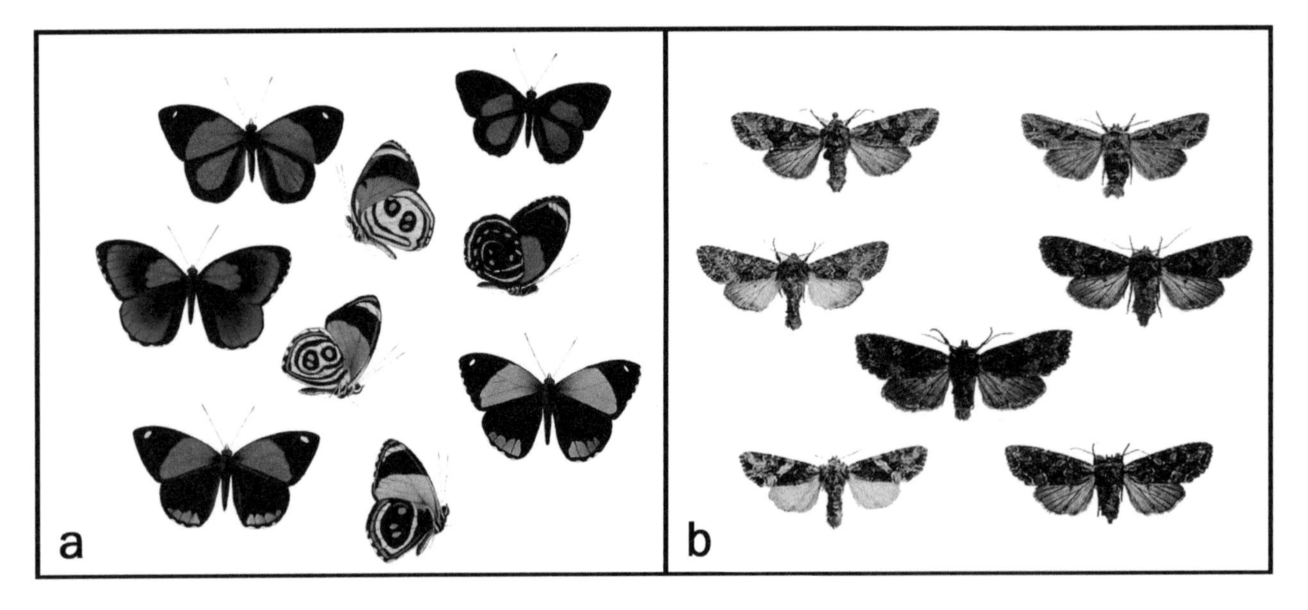

FIGURE 15.8 Contrast of attractive 'butterflies' and drab 'moths' (public domain plates). For better or worse, much of human judgment is based on appearance, and flying insects lacking impressive coloration are disadvantaged. (a) Butterflies. Source: Hewitson (1856–1876). Credit: biodiversitylibrary.org/page/12839744. (b) Moths. Source: South (1907).

FIGURE 15.9 Common clothes moth (*Tineola bisselliella*). (a) Adults. Photo by Frdel (CC BY SA 4.0). (b) Close-up of upper parts. Photo by Dack9 (CC BY SA 4.0).

FIGURE 15.10 Casemaking clothes moth (*Tinea pellionella*). (a) Adult. Photo by Janet Graham (CC BY 2.0). (b) Lateral view of larva. Photo by Caroline Harding, MAF Plant Health & Environment Laboratory (CC BY 3.0).

that are stored for long periods are often damaged beyond repair. The larvae can also feed on other dry protein-rich animal materials, including fishmeal, dried meat extracts, drugs containing albumin, and insect remains (Cox and Pinniger 2007). Occasionally the larvae damage stored grains and seeds, ground seeds like soya bean meal, spices, and tobacco (Stejskal and Horák 1999), but other moths are more harmful agricultural pests. Adult clothes moths do not feed. Sometimes carpet beetles (*Anthrenus* species) are responsible for the damage attributed to moths. Chemical control measures have been extensively employed to control moths. Mothballs or moth flakes (with naphthalene or para-dichlorobenzene) are toxic to humans and pets, and their use is now discouraged. The traditional storage of garments in cedar-lined closets and chests to deter clothes moths has proven to have limited value.

FIGURE 15.11 A woman examining moth damage to her stored garments. From a public domain illustration published in *The Ladies' Home Journal*, March, 1926.

Beneficial Aspects

In nature, clothes moths contribute significantly to the recycling of dead animals, concentrating on the epidermal tissues that most other species find to be indigestible. The larvae are specialist feeders on the exterior of carcasses of birds and mammals. They are adapted to consuming animal fibers, especially fur, silk, feathers, and leather. These materials contain the protein keratin, which cannot be digested by many other animals, but is to the taste of the larvae (Cox and Pinniger 2007). Pests that cause considerable economic damage, like the clothes moth, attract research funds, not just for control measures, but to understand their basic biology, and the resultant knowledge is of great scientific and practical value for utilizing and protecting biodiversity.

Conservation Aspects

As noted by New (2004), 'Moths have long been regarded as the "poor cousins" of butterflies in Lepidoptera conservation, and have lagged well behind in popularity and in the attention given to their conservation status and needs. Only rarely do they gain greater prominence, despite the enormous taxonomic and biological variety they display.' Unfortunately, the harmful clothes moth does little to promote the cause of moth conservation.

COCKROACHES

Introduction

There are about 4,400 species of cockroaches (Brenner and Kramer 2018; Carlson et al. 2018), which are now considered to be allied with the termites (Djernæs 2018). The main household pests, shown in Figure 15.12, are the German cockroach (*Blattella germanica*) and the American cockroach (*Periplaneta americana*), both actually from Africa (the former is far more prevalent in buildings). Pest cockroaches are reputed to have extraordinary

FIGURE 15.12 Major cockroach pests. (a) German cockroach (*Blattella germanica*). Photo by Lmbuga (CC BY SA 3.0). (b) American cockroach (*Periplaneta americana*). Photo by Gary Alpert (CC BY 2.5).

survival skills, and a study of the DNA of the American cockroach suggested that this is because of the acquisition of genes that enhance the insect's abilities to detect bitter (toxic) substances and to detoxify poisonous substances that are ingested (Li et al. 2018; Yin 2018). Cockroaches were so common in the Carboniferous (late Paleozoic era) that this geological period is sometimes referred to as 'The Age of Cockroaches.'

Harmful Aspects

Ants, termites, fleas, houseflies, and bedbugs are among the candidates for the title of 'worst household pest,' but most surveys give the title to cockroaches. According to a Neapolitan folk saying, 'Every cockroach is beautiful to its mother' (Arruda and Pomés 2013), but most people find them repellent, if not extremely ugly. They carry bacteria (and sometimes protozoa and viruses) which can contaminate food and spread disease, especially gastroenteritis (such as food poisoning, dysentery, and diarrhea). They can also serve as intermediate hosts of parasites that affect domestic animals and potentially humans (Brenner and Kramer 2018). Cockroaches produce protein fecal matter and decaying moulted skeletons, releasing foul odors, and causing allergic rhinitis and asthma. Roaches like starchy food like cereals, sugary substances, and commercial meat preparations, all of which can be contaminated by their fecal droppings. They also feed on books and their bindings and on paper. Cockroaches can even chew through electrical wiring and short-circuit electronics. With considerable justification, cockroaches have come to epitomize poor sanitation (Figure 15.13). The mere appearance of cockroaches causes considerable emotional distress in some individuals, and some develop an irrational fear of them (katsaridaphobia). Cockroaches are the most common pests of indoor dwellings (Lai 2017), causing damage to reputations when found in cruise ships, hotels, restaurants, and supermarkets. Broad-spectrum insecticides employed to control cockroaches are increasingly ineffective because of the development of insecticide resistance, necessitating increased dosages which in turn can harm humans, pets, and livestock (Fardisi et al. 2019).

Beneficial Aspects

As pointed out by Bell et al. (2007), more than 99% of the thousands of cockroach species have never set foot in a kitchen! Indeed, as with mosquitoes, most cockroaches are not troublesome. Only about 30 species are associated with human habitats, and only about four are significant pests. Many cockroaches are important members of ecosystems as recyclers of decaying plant and animal matter, upon which they feed. They are thought to particularly contribute to recycling nitrogen in forests, making this normally scarce element available to the trees. Some cockroaches are kept as pets, particularly the Madagascar hissing cockroach, *Gromphadorhina portentosa* (Figure 15.14). Cockroaches are a source of food for many birds and small insectivorous

FIGURE 15.13 Most people find cockroaches loathsome and extremely unsanitary, as this illustration indicates. Drawing by Brenda Brookes.

FIGURE 15.14 Madagascar hissing cockroach (*Gromphadorhina portentosa*). Photo by Husond (CC BY SA 3.0).

FIGURE 15.15 Roasted cockroaches on sale in a Chinese market in Burma. Photo (public domain) from Pixabay.

mammals such as mice and rats, and some parasitic wasps specialize on cockroach eggs. A few cockroaches are also used as food for humans or livestock (Figure 15.15). Some cockroaches pollinate plants (Suetsugu 2019; Vlasáková et al. 2019). Because cockroaches are adapted to filthy conditions, they may have the ability to produce antimicrobials of potential value to humans (Mosaheb et al. 2018). Cockroaches serve as laboratory models for scientific studies of metabolism, insecticide resistance, chemical communication, and neurobiology.

Conservation Aspects

The 1997 science-fiction movie *Men in Black* featured a struggle between one of the heroes and a gigantic extra-terrestrial cockroach who acquired the name Edgar. During a confrontation, the hero remorselessly crushed cockroaches underfoot in order to attract Edgar's attention. In the film, Edgar demonstrated admirable respect for Earth's insects by killing a bug exterminator for doing his job, and a clerk who was swatting flies. However, the indifference of audiences of the film for the welfare of cockroaches demonstrates how difficult it is to seek financial support for their conservation, even the innocent species. Indeed, several cockroach species are threatened with extinction (e.g., Carlile et al. 2018). Another conservation issue that involves cockroaches is the fact that they are an important food source for some rare species. For example, roaches make up half the diet of the endangered red-cockaded woodpecker (*Leuconotopicus borealis*; Figure 15.16) of the Southeastern U.S. (*The Guardian* 2013).

As with mosquitoes, most informed entomologists and ecologists are strongly opposed to the prospect of driving all cockroaches to extinction, since the vast majority are harmless to humans and serve important roles in sustaining ecosystems. However, also as with mosquitoes, the idea of totally eliminating the chief pest species is not unattractive to many, on the theory that the possible harm to the ecology of the world would pale in comparison to the benefits for people. Nevertheless, there are many for whom deliberate extermination of even species as objectionable as pest cockroaches is wrong in principle (Textbox 15.1).

FLEAS

Introduction

Fleas are any of about 2,500 species of small, flightless insects classified in the order Siphonaptera. (Some so-called

FIGURE 15.16 The endangered red-cockaded woodpecker (*Leuconotopicus borealis*), which feeds extensively on cockroaches. Photo by Sam D. Hamilton, U.S. Fish and Wildlife Service (CC BY 2.0).

TEXTBOX 15.1 A VIEW OPPOSING ALL SPECICIDE, EVEN OF PEST COCKROACHES

You should never wish for the extinction of an entire species, especially insects as they always play an important role in the food chains of an ecosystem, typically being at the bottom of the food chain. Even if a creature is a 'pest' to us humans, it still had reasons to evolve the way they have and a niche within the environment they evolved in. A lot of larger species prey on insects and while they may annoy you, even cockroaches have their place in the environment. Wiping them out would mean a decline in population for any species that prey on them. Cockroaches also feed on dying/dead organic and plant material and release nitrogen through their feces, which helps fertilize soil and plants to grow.

—*Hurricane Matthew (2015)*

fleas are not true insects but have jumpy movements like those of fleas; 'water fleas' are tiny crustaceans, mostly found in fresh water; 'snow fleas' are springtails, which typically feed on decaying matter in moist soil.) Fleas inhabit all continents and a very broad range of habitats. They are external parasites, adapted to suck blood from their hosts, about 95% of which are mammals, the remainder birds (Durden and Hinkle 2018). Adult fleas are 1–8 mm long, usually brown, with flattish bodies facilitating movement through their host's fur or feathers. These insects possess strong claws to hold onto their hosts and well-developed hind legs for jumping. Fleas often bite in a cluster or line of three wounds (reminiscent of bedbugs). Although many fleas occur predominantly on one host species, most flea species are capable of living off a variety of hosts. The cat flea, *Ctenocephalides felis*, is the most common flea on both cats and dogs, as well as numerous other mammals, including humans and several species of livestock (Rust 2018). The dog flea *C. canis* (Figure 15.17b) is particularly fond of dogs but also occurs on other species. The human flea, *Pulex irritans*, parasitizes a range of mammal species as well as humans (Ziegler 2016). Figure 15.18 suggests that there are fleas parasitizing fleas, but there is no known case of this (although fleas have their own parasites).

Harmful Aspects

Flea bites are usually just a nuisance to people but can remain itchy and inflamed for weeks, which in turn causes both humans and other animals to scratch, sometimes damaging the skin. Scratching can result in hair loss in wild and domestic animals. Blood loss from large infestations may result in anemia. Occasionally an allergic skin reaction occurs. More seriously, fleas are vectors of diseases caused by viruses, bacteria, and rickettsias (a bacterial group), as well as protozoan and helminth parasites. The most potentially serious conditions include murine typhus, tularemia, cat scratch disease, and bubonic plague. The chigoe flea or jigger (*Tunga penetrans*) causes tungiasis, a skin disease which is widespread in tropical countries. (Jiggers should not be confused with chiggers, which are mites [members of the Arachnida, which also includes spiders and ticks], a group of tiny parasites attaching and feeding on the skin of animals, including humans, causing itching.) The Oriental rat flea, *Xenopsylla cheopis* (Figure 15.17a), is credited with transmitting *Yersinia pestis*, the bacterium causing the bubonic plague, to rodents, which in turn spread the disease to humans, resulting in devastating plagues, notably the black death (ca. 1350). During World War II, the Japanese army employed the insects as a biological weapon, dropping plague-infested fleas in China. At least in most industrialized countries, flea-borne diseases are a limited threat today. However, fleas are a constant threat to household pets. Veterinary care to prevent or cure flea infestations represents a considerable cost. While mature adults are short-lived (typically 2 or 3 months), other stages of flea development (young adults, nymphs, eggs) have been known to survive in a dormant state for a year in a house.

Beneficial Aspects

Unlike the adults, flea larvae are not parasites. After emerging from their eggs, the larvae feed on organic materials such as dead insects, feces, and vegetable matter, and so contribute to organic recycling. While it is hard to accept that the spreading of infective parasites to vulnerable hosts by adult fleas is a good thing, this does benefit the parasites, and also regulates populations of the hosts, and so keeps them in balance with the remainder of the ecosystem (Textbox 15.2).

Although one of the world's best known pests, fleas are sometimes treated with respect. They are a common subject of fairy tales and fables, and in the past, they were shown in paintings as a routine part of life (Figure 15.19).

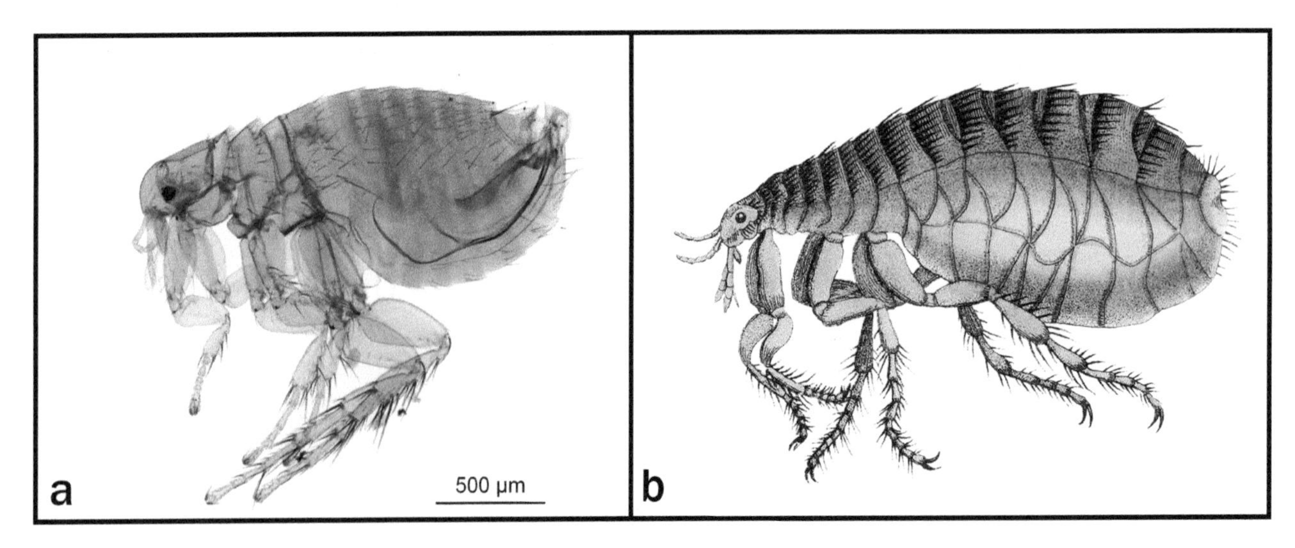

FIGURE 15.17 Examples of flea species. (a) Oriental rat flea (*Xenopsylla cheopis*). Photo by Olha Schedrina, The Natural History Museum (CC BY 4.0). (b) Dog flea (*Ctenocephalides canis*). Source: Shaw (1813). Photo enhanced by rawpixel.com (CC BY 4.0).

FIGURE 15.18 A hierarchy of flea parasitism, conceived by Anglo-Irish satirist Jonathan Swift (1667–1745). Illustration by B. Brookes.

The vermin only teaze and pinch
Their foes superior by an inch.
So, naturalists observe, a flea
Has smaller fleas that on him prey;
And these have smaller still to bite 'em,
And so proceed ad infinitum.

—**Jonathan Swift, 1733,** *On Poetry: A Rhapsody*

Because of their extraordinary jumping and athletic abilities, fleas came to be popular in flea circuses from the 19th century onward (Figure 15.20). Craftspeople dressed fleas as humans or prepared them to tow miniature carts. The well-known expression 'flea market,' referring to a small marketplace, is not considered pejorative.

Conservation Aspects

Fleas that are specialists on endangered animals are in danger of going extinct when their host does so (Platt 2018). Indeed, numerous species of fleas are known to have become extinct, and many others may do so (Kwak 2018). However, most biologists, like most other people, are not sympathetic to fleas. The black-footed ferret (*Mustela nigripes*), once common on the Great Plains of North America, is highly endangered. Prairie dogs constitute over 90% of their diet, and the ferrets often occupy their burrows. Conservation organizations protect the prairie dogs from outbreaks of lethal sylvatic plague (which kills both the prairie dogs and the ferrets) by eliminating fleas that carry the plague bacterium (World Wildlife Fund 2014; Eads et al. 2021).

HOUSEFLIES

Introduction

'True flies' are insects of the order Diptera, which probably contains about 1 million species, although only about 150,000 have been described to date (Gerhardt and Hribar 2018). Many insects with 'fly' in their name, such as butterflies, mayflies, stoneflies, and whiteflies, are not true flies. Mosquitoes, discussed later, are the most harmful of all dipterans to humans, followed by the common housefly, examined here. Many dipteran species are important pollinators, assist in organic decomposition, and serve as food for other animals, and some make excellent subjects for genetic studies. Forensic post-mortem pathologists sometimes employ certain flies – mostly flesh-flies (sarcophagids) and blowflies (calliphorids) – to estimate the time of death (Harvey et al. 2016b). These insects are quick to visit a fresh corpse. The eggs they lay hatch between 8 and 24 hours later, and their subsequent development provides an estimate of how long the body has been dead.

TEXTBOX 15.2 WHAT ARE FLEAS FOOD FOR?

If you think about the benefits of fleas regarding the food chain, environment, and so on, it could be argued that fleas (like all species) are merely filling an available ecological niche. All organisms are part of the food chain; whether they are consumed by animals, microorganisms, or fungi, fleas help keep nutrients flowing through the system of life. You could (controversially) argue that bloodsucking parasites help to re-balance populations that are out of control by being vectors for disease. That's all well and good until it's your species that's under the threat of disease and death!

—Bungay (2018)

FIGURE 15.19 Humorous painting entitled 'A monkey physician examining a cat patient for fleas.' Source: Wellcome Collection (CC BY 4.0).

FIGURE 15.20 Advertising poster (public domain) for 'Roloff's Floh-Circus' (German for Roloff's flea circus). Originally published by Adolf Friedlaender, Hamburg, Germany, in 1906.

Unfortunately, numerous true flies are significant pests of livestock and crops, and 'no other group of insects has as much impact on human and animal health' (Gerhardt and Hribar 2018). 'Of all the insect groups, the flies (Diptera) most frequently play negative roles in human symbolism. Flies typically represent evil, pestilence, torment, disease and all things dirty. This association is likely a result of the fact that those flies most familiar to people have a close association with filth' (Hogue 2009).

Although several species of flies commonly occur in houses, the predominant housefly (also spelled house fly and house-fly) is *Musca domestica* (Figures 15.21 and 15.22), thought to have originated from the savannahs of central Asia before it spread throughout the world (Khamesipour et al. 2018). Adults are usually 6–7 mm long. Females lay batches of about 100 eggs on decaying organic matter such as food waste, carrion, and feces. The eggs hatch into legless, white maggots up to 12 mm in length (Figure 15.22). These transform into adults in 2–6 weeks. Adults usually live 2–4 weeks but can hibernate overwinter. Adult houseflies cannot bite humans and are adapted to a moist diet. They feed on liquid and semiliquid substances, and solid materials softened by their saliva. Houseflies are attracted to garbage, manure, sewage, compost, and other wastes. The housefly is closely associated with humans, and it has been claimed that it is the most widely distributed insect in the world. Certainly, it is the most widespread dipteran in buildings. Houseflies are truly at home in homes. They can walk on vertical window panes or hang upside down on a ceiling, because of the surface-tension properties of a secretion produced by their feet. Houseflies rest at night but are active when illumination is good and humans are also awake. Houseflies have extremely rapid responses to movement and so can usually avoid being swatted.

Harmful Aspects

Houseflies transport pathogens on and in their bodies, constantly defecating and salivating on food, thereby depositing bacteria, viruses, and other disease-causing organisms (Iqbal et al. 2014; Figure 15.23). The contaminated food may in turn result in the transfer of serious illnesses such as anthrax, cholera, diarrhea, dysentery, eye infections (conjunctivitis, trachoma), poliomyelitis, salmonellosis, tuberculosis, skin infections (yaws, leprosy), and typhoid fever (Malik et al. 2007; Khamesipour et al. 2018). Houseflies can occur in large numbers and are often annoying, distracting, and a significant nuisance during work and leisure as they buzz around faces. Fly defecation soils human belongings. Accordingly, the presence of houseflies is viewed with justification as an indicator of poor sanitation and unhygienic conditions, and this can have a negative psychological impact. To date, pesticides have been the most effective control measure, but houseflies have acquired pesticide resistance (e.g., to DDT), and the use of chemical controls remains ecologically and medically problematical.

Beneficial Aspects

What good are houseflies? They feed many predators, including amphibians, birds, insects, reptiles, and spiders. Houseflies (especially the maggots) contain large amounts of protein and are beneficial to the many birds, reptiles, and other insects that prey on them. Maggots are even used as a source of protein in commercial fish and livestock feed, called 'magmeal,' which has as much as 60% protein. By feeding on decaying matter, maggots hasten the recycling of elements that eventually become incorporated in plants and nourish the animals that feed on them. 'Maggot debridement therapy' is the use of maggots (especially of

FIGURE 15.21 Housefly (*Musca domestica*). (a) Adult. Public domain photo by U.S. Department of Agriculture. (b) Close-up of head. Photo by Sanjay Acharya (CC BY SA 4.0).

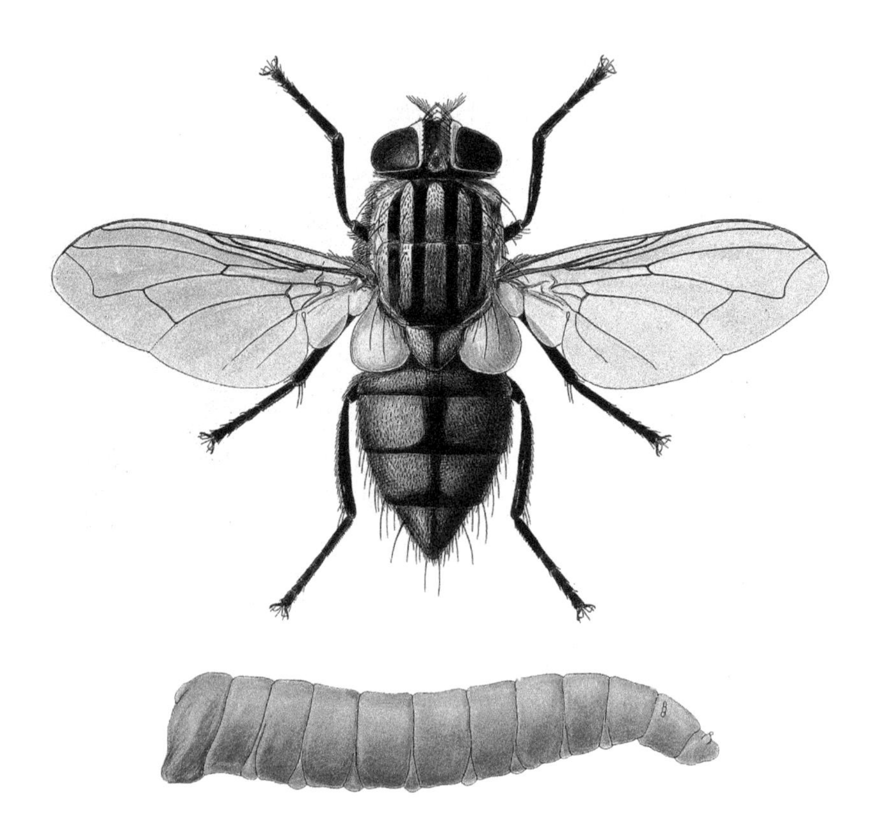

FIGURE 15.22 Adult of a female housefly (*Musca domestica*) and a larva. Drawn by Amedeo John Engel, colored by A. J. E. Terzi. Credit: Wellcome Collection gallery (CC BY 4.0).

FIGURE 15.23 Health posters (public domain) concerning danger from houseflies. (a) Anti-fly poster prepared in 1944 by the U.S. War Department. Source: U.S. National Library of Medicine. (b) Poster highlighting the disease transmission of houseflies. Source: Archives, New Zealand Department of Health.

the housefly) to clean wounds and bone infections. The maggots consume and remove the dead tissue and disinfect wounds. Houseflies contribute to commerce, providing work for pest control specialists, and manufacturers of fly-swatters, fly sprays, flypaper, and electrocuting grids.

Conservation Aspects

Diptera are among the species that are especially endangered (Sánchez-Bayo and Wyckhuys 2019). Unfortunately, houseflies have so sullied the reputation of this large group of insects that conservation attempts specifically targeted at endangered flies (or any species that has 'fly' in their name) are likely to face ridicule (Textbox 15.3). As is obvious, 'Houseflies are highly abundant and not threatened or endangered' (Doctor 2013).

In addition to houseflies, there are other, highly undesirable Diptera. Tsetse flies (genus *Glossina*) are large biting flies of tropical Africa that feed on the blood of vertebrates. These 'flies of death' transmit trypanosomes (pathogenic protozoans), which cause human sleeping sickness and animal trypanosomiasis. Because of the enormous health and economic effects of tsetse flies, their deliberate extinction has been considered (Textbox 15.4). However, tsetse flies protect many of the native large mammals in Africa (which have evolved with these flies) by preventing the encroachment of humans and their invasive cattle. The eradication of tsetse flies could potentially doom the last remaining strongholds of large African mammals.

LEECHES

Introduction

Leeches are segmented worms, classified as subclass Hirudinea of the phylum Annelida. The most familiar annelid is the earthworm (one of the very few invertebrates that people welcome on their land). There are about 500 freshwater, 100 marine, and 100 terrestrial species of leeches. Most have a sucker at each end, the anterior one with the head used for both feeding and anchorage, the posterior one to assist in anchorage. About 25% of leech species are predators of small invertebrates, usually feeding on worms, snails, and insect larvae (Aloto and Eticha 2018). Some leeches are scavengers. The diverse kinds of leeches have jaws adapted to different feeding strategies (Aloto and Eticha 2018; Kuo and La 2019). However, about three-quarters of leeches are blood suckers. Many absorb

TEXTBOX 15.3 THE DIFFICULTY OF CONSERVING ENDANGERED FLIES

Maybe some endangered species should be endangered. Take, for instance, *Rhaphiomidas terminatus abdominalis*, the scientific name for a great big fly that calls some 45 acres of California sand dunes home. This fly is a pest in more ways than one. Already, builders of a hospital have lost about $4 million because part of their construction was going to encroach on the fly's habitat. The federal government required a change in plans, and it is also holding up a project in the area that some believe could lead to 20,000 new jobs over time. The fly is getting this kind of special treatment, of course, because it is thought to be the last of its kind. Under the government's interpretation of the Endangered Species Act, this creature, which no one even knew about until a relatively few years ago, is due privileges few human beings can claim. But what would its extinction cost anyone or anything? Some sadness on the part of a few entomologists, maybe. Nothing else. Its preservation, meanwhile, is costing the Endangered Species Act its credibility and a number of Californians economic opportunities. Our view is that this is an ugly, stupid, worthless bug. Squash it.

—*Scripps Howard News Service (1997)*

(Note: This species, the Delhi Sands Flower-loving fly, is one of very few fly species on the U.S. Endangered Species List, and its protection has been extremely controversial.)

TEXTBOX 15.4 SHOULD THE 'FLY OF DEATH' BE EXTERMINATED?

Humans have developed the technical capacities to purposefully eradicate undesirable species, such as insect vectors of a variety of pathogens. Policymakers are now tasked to determine whether and to what ends such technologies should be used. The moral seriousness of this decision cannot be overstated... We suggest a full ethical evaluation of tsetse fly elimination is essential to support sound decision-making around the usage of new technologies, to assess whether or under what conditions elimination of an endemic species harmful to humans might be justified... We suggest there is a good case to be made against the global eradication of tsetse fly species.

—*Bouyer et al. (2019)*

blood from their hosts through their anterior (head-end) sucker (Figures 15.24 and 15.25), within which is a mouth equipped with a set of razor-sharp jaws. After the host is wounded, a muscular pharynx (part of the gut behind the mouth) pumps the blood into the leech's digestive tract. A group of 'jawless' leeches employs a proboscis to pierce skin and extract blood. Leeches inject anticoagulants to prevent blood clotting and are suspected of also anesthetizing the wound so that they can remain undetected while feeding. Most leeches feed on a variety of hosts but may specialize on certain groups, such as frogs, fish, or birds (Sawler 1981). The majority of leeches stay on their hosts for short times while feeding, but some marine leeches remain attached for long periods.

FIGURE 15.24 Medicinal leech (*Hirudo medicinalis*) sucking blood from a human subject. The anterior (head) sucker is at right, and the posterior (tail) sucker is barely visible at left. Photo by GlebK (CC BY SA 3.0).

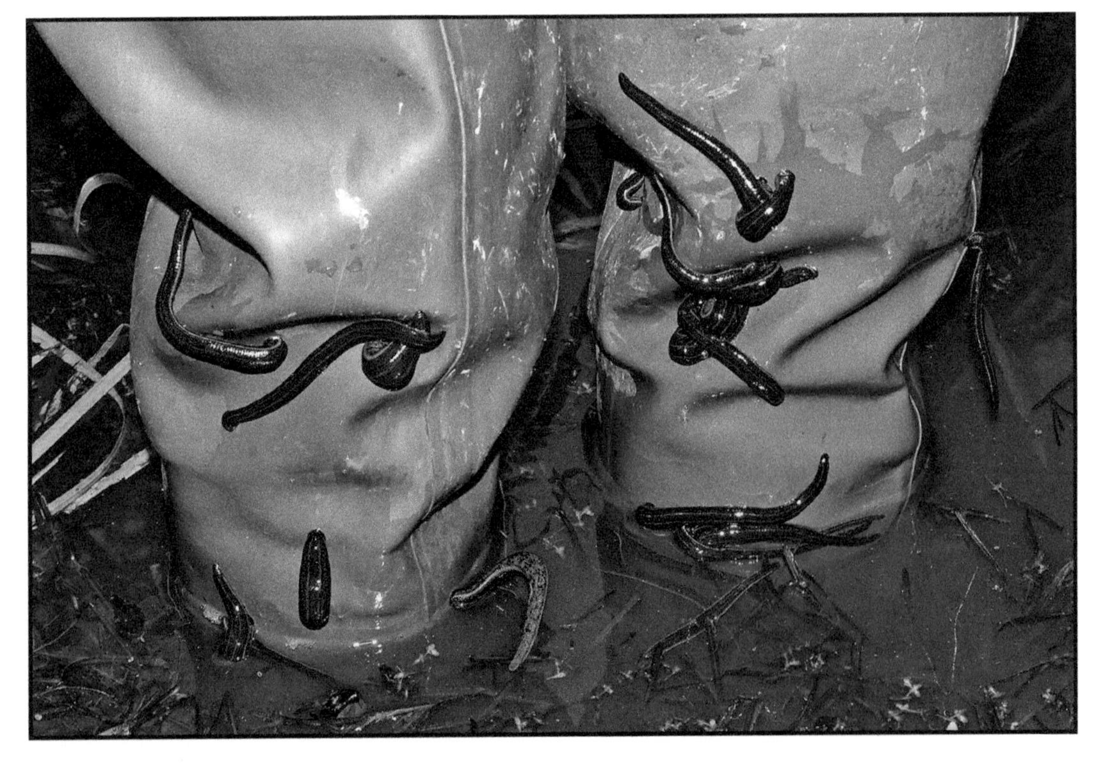

FIGURE 15.25 Attack on rubber fishing pants by medicinal leeches (*Hirudo medicinalis*), while wading across a small pond in Germany. Photo by Christian Fischer (CC BY SA 3.0).

Harmful Aspects

Rarely, leech bites produce significant allergic or anaphylactic reactions. Leeches can transmit viruses and bacteria, but there are very few examples of disease transfer to humans (Al-Khleif et al. 2011). Very rarely, leeches can get inside eyes, ears, noses, digestive systems, and urinary organs, and cause significant harm. Leeches also damage livestock. An attack by 100 or more leeches can produce significant blood loss. Many people avoid outdoor activities, especially swimming, because of fear of leeches. Some individuals become quite hysterical when they find a single leech on their body, although the blood loss is trivial. Of the many parasites attacking humans, the leech has been adopted in the phrase 'social leech' as representing this reprehensible character flaw (note Figure 15.26). However, the widespread fear and loathing of leeches is far out of proportion to their potential to harm humans.

Beneficial Aspects

The best known leech species is the medicinal leech, *Hirudo medicinalis*, but at least five other species are often employed medically (Saglam 2018), and it appears that in recent times these other species are the primary ones employed (Sket and Trontelj 2008). The practice of bloodletting using leeches was common in medicine from ancient times to the 19th century (Figure 15.27). Leeches also used to be widely employed to draw blood from areas swollen by poisonous stings and bites. Today, leeches are sometimes used to treat joint and vein diseases, black eyes, and as a substitute for stiches in microsurgery (Whitaker et al. 2004; Porshinsky et al. 2011; Zaidi et al. 2011). Leeches are also employed to maintain circulation after microsurgery, such as after the re-attachment of ears (Soucacos et al. 1994). The anticoagulant hirudin from leeches is employed as a drug to treat some blood-clotting disorders, and leeches have been found to have many other pharmacologically significant compounds (Sig et al. 2017). Fish and birds widely consume leeches, which are a natural component of food webs.

Conservation Aspects

'Leeches are indispensable organisms of the aquatic ecosystem,' but 'they are largely ignored in many ecological and environmental research projects' (Saglam 2018). Regrettably, 'Leeches are most affected by agricultural activity, excessive collection, dangerous chemical compounds in water, increasing urbanization and global climate change' (Saglam 2018). Analysis of DNA accumulated in blood-feeding leeches has potential to monitor the conservation status of wild vertebrates that the leeches feed on (Kampmann et al. 2017; Tessler et al. 2018).

In the past, leech collection for medical usage was a popular trade (Figure 15.28). The medicinal leech has been overcollected for therapeutic usage in some countries, leading to shortages and the need to import other leech species (Elliott and Kutschera 2011). The traditional medicinal leech, *H. medicinalis*, is now protected and/or listed as endangered in many European countries (Sket and Trontelj 2008). Leeches are a popular live fishing lure, and some concern has also been expressed about the over-harvesting of local wild populations for bait. There are several quite rare leech species that are of conservation concern.

FIGURE 15.26 'Social leech' (always a guest, never a host), prepared by B. Brookes.

FIGURE 15.27 The use of leeches in bloodletting. (a) Vase for medicinal leeches. Photo by Paulo O (CC BY 2.0). (b) Humorous painting with three physicians in the form of leeches recommending bloodletting to treat a grasshopper patient. Photo of 19th-century painting from Wellcome Collection (CC BY 4.0).

FIGURE 15.28 Old painting, 'Three women wading in a stream gathering leeches.' Leech collectors were usually poor country women who used their bare legs as bait. Colored aquatint by R. Havell, 1814, original by G. Walker. Credit: Wellcome Collection (CC BY 4.0).

For example, the European land leech, *Xerobdella lecomtei*, discovered in 1868, is one of the rarest animals on Earth and is badly in need of protection (Kutschera et al. 2007). Unfortunately for leeches, they are particularly despised.

LICE

Introduction

Lice include about 5,000 wingless species of the insect order Phthiraptera, about 4,000 of which are external parasites of birds, while about 800 species parasitize mammals. 'Chewing lice' (also known as biting lice) occupy hair or feathers, feeding on skin and debris on the body of the host, while 'sucking lice' absorb blood and sebaceous secretions. Less than 1,000 species are classified as sucking lice, all of which are specialists on placental mammals, and these are the principal species of economic significance to humans (Durden 2018). 'Most species of mammals and birds are infested by at least one but up to six species of lice' (Barker 1994). Lice paste their eggs, called nits, to hairs or feathers. True lice should not be confused with a variety of other so-called lice. For example, 'plant lice,' better known as aphids, are sap-sucking small insects that cause considerable damage to cultivated plants, and 'sea lice' are parasitic crustaceans which damage fish.

Three kinds of lice are found on humans (Bonilla et al. 2013; Figure 15.29): the human head louse (*Pediculus humanus capitis*), the body louse (*Pediculus humanus humanus*, commonly called the 'cootie' and also known as the clothes louse), and the pubic louse (*Pthirus pubis*, also known as the crab louse). The head louse and body louse can interbreed, but rarely do so (Durden 2018). These three groups are usually found only on people, although sometimes companion animals such as dogs acquire short-lived infestations (Durden 2018). While body lice feed on the human body, they live on clothing, which they must have access to in order to survive. Clothing lice evolved from head lice ancestors (Leo and Barker 2005), after modern humans adopted clothing, perhaps about 170,000 years ago, which coincidentally corresponds to the rapid onset of an ice age when garments would have been necessary for survival (Toups et al. 2011). Pets such as dogs and cats are not responsible for transmitting lice among humans. Lice can only crawl and cannot jump, hop, or fly. In the past, people accepted lice as a problem to be addressed regularly (Figure 15.30).

Harmful Aspects

The derogatory word 'lousy,' applied in many ways, reflects human disdain for these tiny insects, and learning that one is

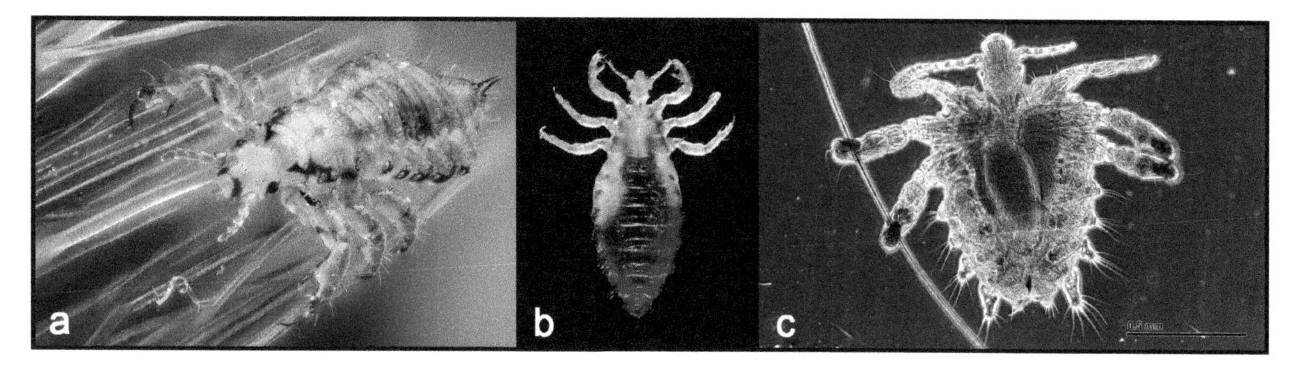

FIGURE 15.29 Lice found on humans. (a) Human head louse (male, *Pediculus humanus capitis*). Photo by Gilles San Martin (CC BY SA 2.0). (b) Human body louse (male, *Pediculus humanus humanus*). Public domain image by U.S. Centers for Disease Control and Prevention. (c) Pubic louse (*Pthirus pubis*). Photo by Josef Reischig (CC BY SA 3.0).

FIGURE 15.30 An 1812 colored etching showing a Spanish family picking lice from each other's heads. Source: Wellcome Trust (CC BY 4.0).

infested usually results in fear and embarrassment. An even less charming word is 'pediculosis,' meaning an infestation of hairy body parts or of clothing with the eggs, larvae, or adults of lice. Lice are capable of vectoring microbial diseases and helminths (parasitic worms), but most spend their lives on a single animal, limiting their capacity to infect many individuals. Blood-sucking lice inject saliva into their host to prevent blood from clotting, and this can result in an allergic reaction. Occasionally, scratching lice bites result in secondary bacterial infections. Lice are very significant parasites of livestock, pets, and wild animals (Durden 2018). They cause considerable anemia in cattle in the northern United States (Peek and Buczinski 2017). Some dangerous chemicals have been employed in the past to get rid of lice infesting humans, and some authorities question the safety of some of the preparations conventionally applied today.

The human body louse is the chief health hazard to humans. It is a transmitter of typhus, trench fever, louse-borne relapsing fever, and other diseases. In the past, the body louse was a constant companion of people, but today it is chiefly found in less developed nations and in unsanitary conditions such as associated with war, famine, natural disasters, and homelessness. Severe infestations of more than 30,000 lice on some people have been recorded.

Head lice infestations are a worldwide problem with prevalence between 1% and 3% in elementary school-aged children (Sneath and Toole 2010). Head lice are far more likely to be encountered in Western nations than body lice. Head-to-head contact among people is considered the most common way that head lice are transmitted, with girls being infected much more often than boys. Lice on the scalp are often not detected until itching starts, 2 or 3 months following colonization. Lice nits (eggs; Figure 15.31) historically were removed with lice combs (Figure 15.32a), now considered obsolescent (Mumcuoglu 2008). Medical shampoos are now usually used to kill lice, although genetic resistance to some of the treatment chemicals is developing. Head lice are detested, especially when found on children, and their presence normally results in frenzied attempts to get rid of them. However, unlike body lice, they (as well as pubic lice) are not known to vector infectious diseases, although they can transmit bacteria causing skin infections (Durden 2018). Large infestations of head lice can result in severe irritation and consequent scratching leading to secondary infections. Head lice are not considered to be a significant health hazard or an indication of poor hygiene (a widespread misconception).

The crab louse or pubic louse is mostly attached to pubic hair but may occur on coarse hair elsewhere, such as eyebrows, eyelashes, beards, chest, and armpits. It typically transfers during human copulation, and in France, crab lice are sometimes charmingly described by the ambiguous phrase 'papillons d'amour' (butterflies of love). Transfer from bed linens, sofas, and toilet seats can occur, but crab lice can survive only a few hours off the host. Crab lice cause intense itching.

Beneficial Aspects

It has been suggested that infestations by head lice might promote a natural immune-increasing response that defends against the more dangerous body louse (Rozsa and Apari 2012). Conversely, it has been demonstrated that lice can exert an immune-suppressive effect on their hosts, a phenomenon which may have practical application for treating auto-immune diseases (Jackson et al. 2009). Because it has the smallest known insect genome (Kirkness et al. 2010), the body louse has proven to be useful as a model research

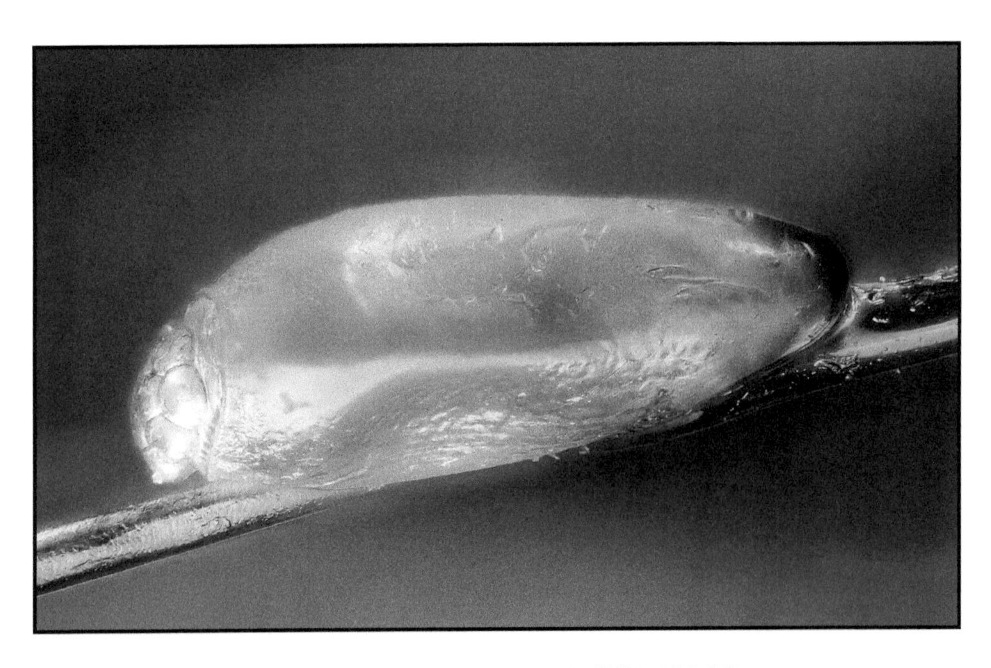

FIGURE 15.31 Nit (egg of human head louse). Photo by Gilles San Martin (CC BY SA 2.0).

FIGURE 15.32 Lice removal with a lice comb. (a) A double-sided lice comb from 19th-century India, housed in the London Science Museum. Source: Wellcome Trust (CC BY 4.0). (b) A woman combing lice out of her daughter's hair. From a public domain, old French postcard.

organism. Moreover, studies of the comparative genomes of lice and their hosts are useful in assessing their evolutionary and ecological evolution (Toon and Hughes 2008).

Conservation Aspects

'Co-extinction' occurs when a species that is indispensable to another species (such as in obligate parasitism and mutualism) goes extinct, causing the dependent species to also become extinct. When the species essential to the other is endangered, the dependent species is 'co-endangered.' Several species of lice are known to have died out because their hosts went extinct, and dozens of other lice species have been identified that are in danger of going extinct because their hosts are in danger of going extinct (Rózsa and Vas 2015). The case for saving endangered lice species is made in Textbox 15.5.

The louse species *Colpocephalum californici*, associated only with the California condor, *Gymnogyps californianus*, is thought to have gone extinct as a result of veterinary delousing routines during the captive-breeding attempts at the Los Angeles Zoo, to keep the bird from extinction (Dunn 2009). It is probably safe to say that most conservationists would not have opted to maintain the louse species, since it was a threat to the future existence of the much-admired condor. As pointed out by Windsor (1995) 'when trying to save the last remaining individuals of a species, a parasitic burden may not be appropriate.'

TEXTBOX 15.5 WHY SHOULD LICE SPECIES IN DANGER OF EXTINCTION BE PROTECTED?

There are several reasons why conservationists should care about threatened parasites. They not only constitute a large proportion of global biodiversity but also exert selective pressures to increase host diversity, and therefore, harboring a unique parasitic fauna can increase the conservation value of the host. Furthermore, parasites carry phylogenetic and population genetic information about the evolutionary past of their hosts. On the other hand, the preservation of parasite species that pose considerable medical or veterinary threats would not be widely accepted. Not all parasites are equally important. For example, the critically co-endangered gorilla louse *Pthirus gorillae* is of particular value because it is closely related to the human pubic louse *Pthirus pubis*, thus its loss would deprive us of a unique possibility to study the evolution and ecology of a human pathogen... The potential costs and benefits of reintroducing infested vs. non-infested animals are open to debate. As far as we are aware, no practical work has been carried out to conserve any species of louse.

—*Rózsa and Vas (2015)*

LOCUSTS

Introduction

Locusts and grasshoppers are not assigned to different taxonomic groups. Locusts are simply certain species of grasshopper that have a swarming phase. The insect family Acrididae has over 10,000 species, of which about 500 (both grasshoppers and locusts) can cause damage to pastures and crops; about 50 of these are considered major pests (Zhang et al. 2019; Figure 15.33). The word 'locust' derives from the Latin *locusta*, meaning grasshopper (certain plant species of the pea family [Fabaceae] are also known as locusts). Normally, locusts are solitary, occur in low numbers, and are innocuous. However, when exposed to drought followed by abundant availability of vegetation, they reproduce rapidly and swarm as winged adults. A large swarm can number in the billions, and densities on the ground can exceed thousands per square meter (Zhang et al. 2019). Locusts can strip crops, both as the immature stage (when they are called hoppers) and during their winged phase.

Harmful Aspects

Locust plagues (Figure 15.34) have occurred since prehistory and have produced extreme damage on all continents where crops are grown (Latchininsky 2013). They have caused famines and human migrations. The use of insecticides to control them has had some deleterious effects on biodiversity, but in recent times better monitoring, biological control, and improved agricultural practices have decreased the significance of locusts. Nevertheless, large locust outbreaks, often promoted by climate change and shifts in land usage, continue to be a major, albeit sporadic problem in many parts of the world (Lomer et al. 2001;

FIGURE 15.33 Some locust species. (a) Garden locust (*Acanthacris ruficornis*) in Ghana. Photo by Charles J. Sharop. (CC BY SA 4.0). (b) Migratory locust (*Locusta migratoria*) in Germany. Photo by H. Crisp (CC BY 3.0).

FIGURE 15.34 Locust swarms. (a) Painting of a locust swarm, by the German artist Wilhelm Friedrich Kuhnert (1865–1926). Prints (public domain) first published in 1904. (b) Plague locusts in Australia. Photo by CSIRO (The Commonwealth Scientific and Industrial Research Organisation) (CC BY 3.0).

Lecoq 2010; Zhang et al. 2019). Locust swarms often devastate natural vegetation cover, and this can result in soil erosion, increased runoff, and habitat alteration that is deleterious to native plants and animals (albeit, such cycles have been occurring for millennia, and native species have survival adaptations).

Grasshoppers, including locusts, have been regarded since the dawn of agriculture as crop pests. The historical disrespect for these insects probably is the basis for well-known fables in which a dislikable grasshopper spends the summer making music while ants work industriously to store up food for winter (Figure 15.35). When winter arrives, the grasshopper starves, providing the lesson that idleness is sinful.

Beneficial Aspects

In nature, locusts and grasshoppers are, except in years of great abundance, desirable components of healthy ecosystems, controlling plant densities and participating in nutrient cycling and the food chain, both as consumers and consumables (Latchininsky et al. 2011; Figure 15.36).

FIGURE 15.35 A depiction (public domain) of the lazy grasshopper and the industrious ants, in a 1919 version of Aesop's fairy tale illustrated by Milo Winter.

FIGURE 15.36 Locust being consumed by a cotton-top tamarin (*Saguinus oedipus*), a critically endangered South American monkey. Photo by Mickey Samuni-Blank (CC BY SA 3.0).

FIGURE 15.37 Fried locusts on sale by a street vendor in Chad. Photo by Keithk (CC BY SA 4.0).

TEXTBOX 15.6 THE CRITICAL CONSERVATION ISSUE IN LOCUST CONTROL: BOMBARDING NATURAL HABITATS WITH INSECTICIDES

In contrast to pests developing in close association with a particular host crop, locusts and grasshoppers are often controlled in natural or semi-natural landscapes, exposing structurally and functionally diverse communities to agrochemicals, chemicals to which they are not adapted. This suggests that insecticide-induced perturbations may be severe... Few insect taxa raise such controversial, and often irrational, views and feelings as acridids. In Central Europe, many grasshoppers are nowadays considered as indicators of biodiversity and ecosystem quality, and sometimes even as flagship species, i.e., 'popular, charismatic species that serve as symbols to stimulate conservation awareness and action.' In other parts of the world, however, they remain feared pests of crops and pastures and may trigger control campaigns of considerable scale and intensity. While the status of both locusts and grasshoppers as pests of rangeland may presently be questioned, status as pests of crops is certainly not. The potential to bring havoc to crops is generally acknowledged, and the debate is rather on control strategies than on whether acridids are worth controlling.

—*Peveling (2001)*

From the human perspective, locusts are edible insects, and indeed considered a delicacy by many cultures, especially in African, Middle Eastern, and Asian countries (Figure 15.37). Their consumption is permitted both in Judaic and Islamic dietary practice, unlike many other animal foods. As a source of animal protein, locusts are far less damaging to the planet than conventional livestock and represent a potential alternative. Locusts are an important source of food for many animals.

Conservation Aspects

Most major agricultural crops are grown in large, concentrated monocultures, and while this has deleterious consequences for biodiversity, at least the application of pest control measures, particularly insecticides, can be localized to the cultivated area. However, locust swarms are mobile, migrating among crops and natural areas such as deserts or semi-deserts, steppes, savannas, and grassland biomes, and so pesticides are often applied in very large amounts to extensive natural areas, potentially endangering ecosystems and biodiversity (Textbox 15.6). There is obviously a need for guidelines to avoid damaging natural areas (Wiktelius et al. 2003). 'Chemical pesticides applied for locust control represent a risk for humans, terrestrial non-target fauna and aquatic ecosystems' (Everts and Ba 1997).

Locusts typically occur in such vast numbers that it is difficult to contemplate their possible extinction. Nevertheless, this happened to the Rocky Mountain grasshopper (or locust), *Melanoplus spretus*, which once was widespread over the Great Plains from Canada to Texas, periodically devastating the crops of homesteaders and farmers. By 1902, it had completely disappeared without any apparent cause (Lockwood 2004).

MOSQUITOES

Introduction

Mosquitoes (also spelled mosquitoes) are members of the fly family Culicidae, of which about 3,500 species have been described, and many more await discovery (Foster and Walker 2018). Mosquitoes begin life as eggs, laid in water. Hatched eggs develop into motile larvae (Figure 15.38), which feed on algae and detritus. Adults – both males and females – feed on plants, particularly on nectar. Adult males (and sometimes females as well) feed only on plant juices. Some female adults eat other insects and do not suck blood. However, in order to produce eggs, female adults of many species require a blood meal, which they obtain by extracting blood from the surface of animals by employing their proboscis (tube-like mouthpart), as shown in Figure 15.39.

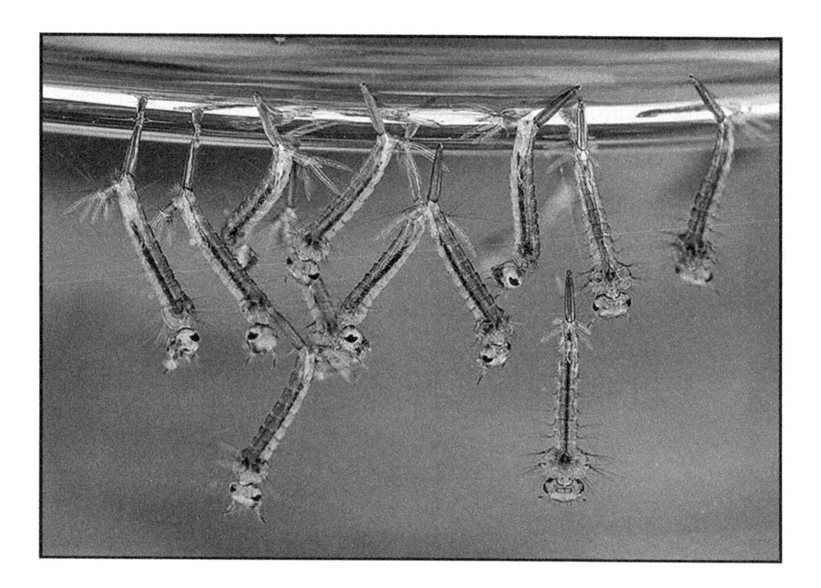

FIGURE 15.38 Mosquito larvae (genus *Culex*) hanging upside down at the surface of water. Photo (public domain) by James Gathany, U.S. Centers for Disease Control and Prevention.

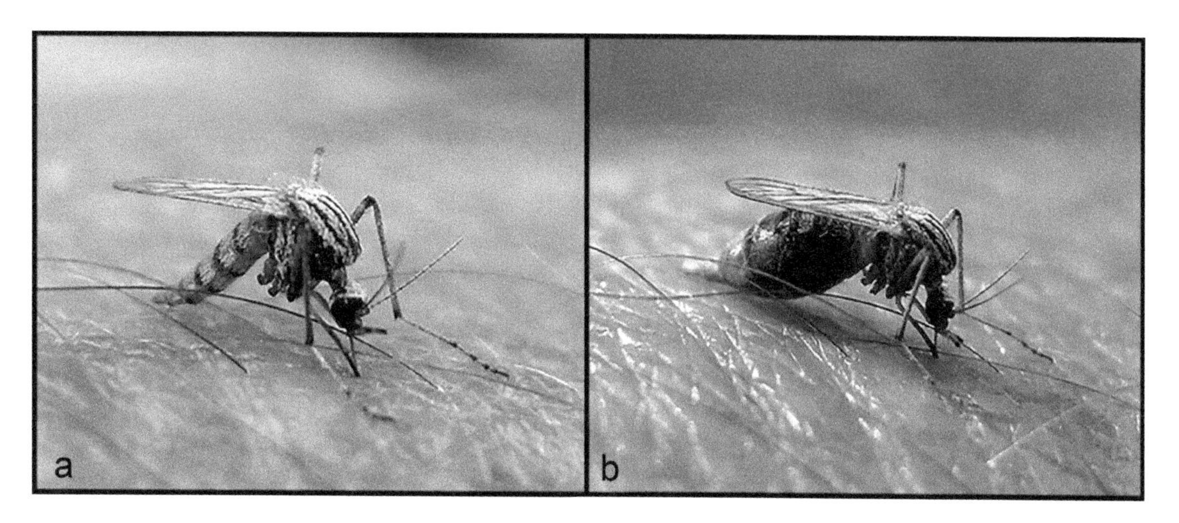

FIGURE 15.39 Progress of a mosquito-sucking blood. (a) Mosquito identified as 'probably *Ochlerotatus vittiger*' of NSW Australia, just beginning to extract blood. (b) Same mosquito at left, after several minutes of feeding. Photos by John Tann (CC BY 2.0).

Mosquito saliva has anticoagulants to keep blood flowing freely, so the proboscis does not become clogged with blood clots. Females of blood-sucking mosquito species feed on the blood of all classes of vertebrates (even fish if they are temporarily above water) and some invertebrates. Some species specialize on certain hosts, and others are less selective. Adults rarely live for more than 2 weeks, and the males often less than 1 week. Most blood-sucking mosquitoes are crepuscular (feeding at dawn or dusk). Mosquitoes are attracted to human hosts by CO_2 and other compounds expelled during breathing, and by sight. They exhibit preferences to sweat and other body odors, and body heat, so some individuals are much more attractive than others to the insects.

Harmful Aspects

Mosquitoes are responsible for annoying, itchy bites, which are usually a mild irritant (Figure 15.40). However, people differ in their responses, and sometimes some individuals experience life-threatening allergic reactions. The chief threat from mosquitoes is that they transmit deadly diseases such as malaria, chikungunya, West Nile virus, yellow fever, Zika fever (named for the Ziika or Zika Forest of Uganda), dengue fever, various forms of encephalitis, and filariasis. These diseases kill millions annually and debilitate many more, especially in developing tropical countries. Malaria, the world's most significant infectious disease, harms over 250 million people worldwide and may kill more than 1 million annually (World Health Organization 2014). Because they are extraordinarily efficient at removing blood from one animal and transmitting it along with pathogenic microbes into another, mosquitoes are widely considered to be the most dangerous of all animals (Winegard 2019). 'Half of the global population is at risk of a mosquito-borne disease. They have had an untold impact on human misery' (F. Hawkes, cited in Bates 2016). Mosquitoes also 'strain the resources of health services and reduce human productivity, thereby perpetuating economic hardship' (Foster and Walker 2018). The average mosquito weighs only about 2.5 mg and its meal amounts to just approximately 5-millionths of a liter of blood, but these insects are capable of generating huge numbers and have proven extremely difficult to control. Not only are mosquitoes threats to people, they also harm livestock, pets, and wildlife. Swarms in Alaska have been thick enough to asphyxiate caribou (Fang 2010). Very extensive efforts have been made to eliminate mosquitoes, and in the past, DDT has been a principal weapon (Figure 15.41).

'Blood sucker' is an insulting phrase reflecting the disgust that humans have for 'sanguivores' (animals that feed on drawn blood). Numerous species, such as ticks, fleas, lice, mites, bedbugs, lampreys, leeches, and some bats and birds, engage in this vampire life style. Of all of these, mosquitoes are the most hated creatures everywhere, with few

FIGURE 15.40 'The American Mosquito,' painting (public domain image) by José Guadalupe Posada (Mexican, 1851–1913), housed in The Metropolitan Museum of Art.

FIGURE 15.41 Second World War poster (public domain), advocating the use of pesticide (predominantly DDT) to eliminate malarial mosquitoes. Source: U.S. National Archives and Records Administration.

exceptions such as Antarctica and Iceland which are just too cold to support significant populations.

Beneficial Aspects

Despite their blood-thirsty reputation, some species of mosquito do not feed on blood at all. Of the several thousand species, only about 200 bite or bother humans, and only about half of these are serious vectors of human disease. Mosquitoes are vegetarians most or all of the time (Figure 15.42). Although females of most species need blood protein as a prerequisite to develop eggs, at other times, like the males, they employ their long proboscis to feed on the nectar of flowers or juice from fruit or stems. Some orchids (notably *Platanthera* species) rely on mosquitoes for pollination. Other plants, such as goldenrods (*Solidago* species), and some aquatic or semi-aquatic plants also benefit from pollination by mosquitoes that reproduce in nearby waters. Mosquitoes are a valuable source of food for many species, especially fish, amphibians, birds, bats, and spiders. Some mosquito larvae (notably of *Toxorhynchites*, 'elephant mosquitoes,' so named for their large size) even eat the larvae of other mosquito species, and so have been considered as potential control agents of harmful mosquitoes (Focks 2007). Adults of *T. speciosus* (Figure 15.43) are believed to be the world's largest mosquitoes, reaching a length of over 3.4 cm; fortunately, the adults are entirely herbivorous. Aquatic larvae of some insects, such as dragonflies and damselfly nymphs, also feed on mosquitoes. Mosquitofish (*Gambusia* species) are particularly effective predators of mosquito larvae and are often introduced into ponds and pools to control mosquitoes. 'Wiping out a species of mosquito could leave a predator without prey, or a plant without a pollinator' (Fang 2010). However, not all species that benefit from mosquitoes evoke sympathy from humans. 'The botfly depends on mosquitoes to carry its larvae to hosts. The botfly lays eggs on the mosquito, which hatch and jump off when the mosquito lands on a mammal. The botfly larvae then burrow into the mammal and develop there' (Shelomi 2016a).

FIGURE 15.42 Male mosquito (*Culex salinarius*) feeding on a bramble (*Rubus*) flower. Some mosquito species are essential pollinators for certain plant species. Some species and all male adults are strictly vegetarians. Photo by Beatriz Moisset (CC BY SA 4.0).

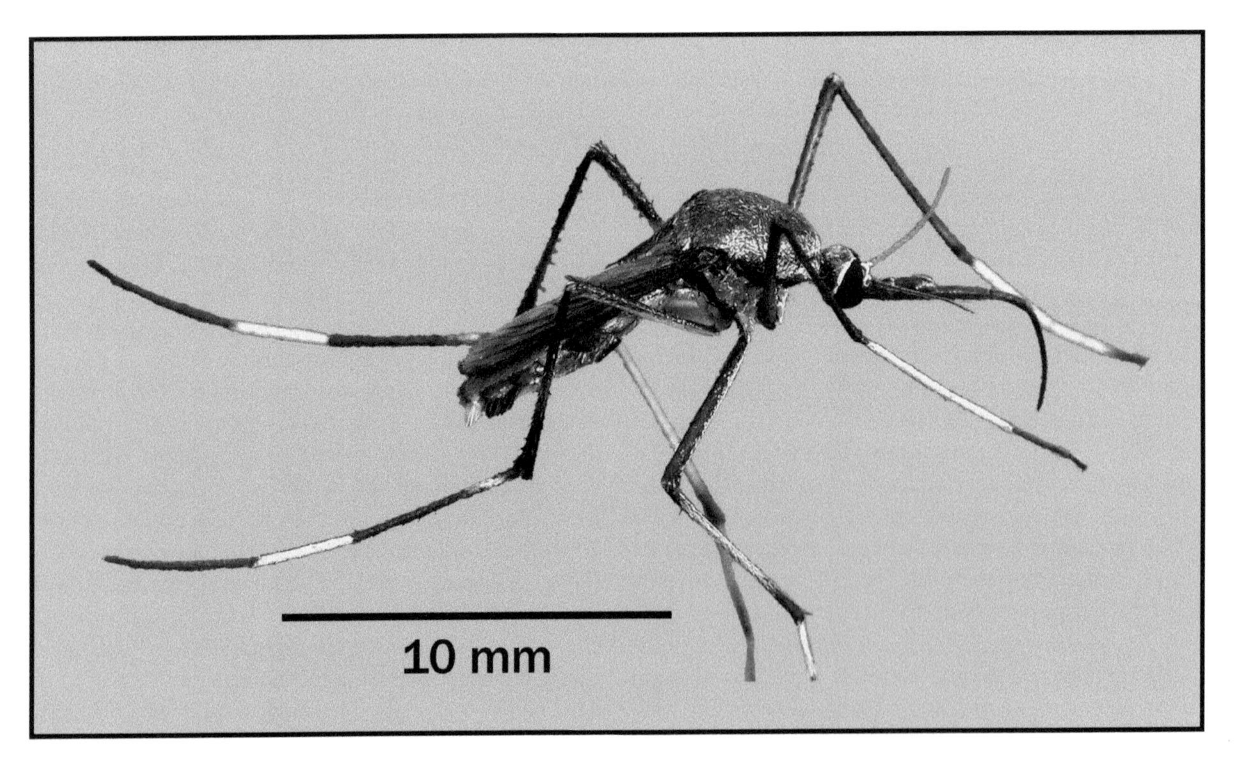

FIGURE 15.43 The Australian elephant mosquito (*Toxorhynchites speciosus*), reputedly the world's largest mosquito. Photo by Summerdrought (CC BY SA 4.0).

Conservation Aspects

'The romantic notion of every creature having a vital place in nature may not be enough to plead the mosquito's case' (Fang 2010). Not all arguments in favor of maintaining mosquitoes in nature are convincing. Some of the microbes that cause diseases would go extinct were it not for mosquitoes transmitting them to hosts, although few would mourn such losses of biodiversity. Another defense on behalf of mosquitoes is that fear of them has slowed the destruction of tropical rainforests, where they can make life intolerable for people. In a similar vein, malaria and other mosquito-borne diseases have protected the people of some malaria-endemic regions from military and cultural colonization. For example, in 1892, after 11 years of effort to construct the Panama Canal, the French abandoned the project after one-third of their workforce died from yellow fever. Mosquitoes have killed many people, thereby limiting the growth of the world's population, and so indirectly have limited the damage caused to the planet and its inhabitants. And, controlling mosquitoes and treating their destructive effects stimulate employment, research, and technological developments. It's also satisfying to know that many parasitic microorganisms only infect and cause diseases in mosquitoes. However, mosquitoes that get sick are often the most effective transmitters of diseases, because numerous microbes have reproduced in their bodies and are available for injection into their victims. The preceding arguments are, of course, insufficient to allay concerns about the harm that mosquitoes cause. The quotations in Textboxes 15.7 and 15.8 are representative of the common view that at least selective elimination of the most deadly mosquito species is warranted and would not result in disastrous harm to ecosystems. There are proposals to employ genetic engineering to infiltrate wild mosquitoes with 'extinction genes' (LaFrance 2016; Plummer 2016; Kyrou et al. 2018; Shrivastava et al. 2019). Mosquitoes are the chief target of attempts to eliminate harmful species employing genetic modification, and because of concern about the potential harmful effects on biodiversity, guidelines for experimentation have been developed (World Health Organization 2014). Unfortunately, there are potential unforeseen consequences. For example, 'The removal of an entire species, such as a mosquito, could have effects on other organisms in the ecosystem, which could in turn lead to unwanted changes, such as an increase in the population of another insect disease vector as it fills the ecological niche opened by suppression of mosquito populations' (National Academies of Sciences, Engineering, and Medicine 2016).

SPIDERS

Introduction

The word 'bug' to most people means an insect or other similarly small animal, and this would include spiders. However, spiders and insects are distinctly different although both have jointed legs and a hard external skeleton. Unlike insects which have a three-part body (head, thorax, and abdomen) and six legs, spiders (and their relatives, ticks, discussed next) have a two-part body (cephalothorax and abdomen) and eight legs. Spiders are *arthropods* ('phylum'

TEXTBOX 15.7 THE VIEWPOINT THAT ERADICATING DEADLY MOSQUITO SPECIES IS JUSTIFIED

Say we eliminate *Aedes aegypti* and a salamander species and an orchid are eliminated along with it: that is a trade we can live with, and by 'we' I mean the millions who will no longer die from yellow fever. The other extinctions will be a tragedy, yes, but the loss of yellow fever will be a triumph worthy of the Nobel Peace Prize. Compared to the losses of the dodo and the Tasmanian tiger, which came with no benefit to society and are thus completely unfortunate, the benefits of the loss of *A. aegypti* [an important vector of dengue, yellow fever, Zika, chikungunya, West Nile, and La Crosse viruses] or *Anopheles gambiae* [the chief transmitter of malaria] would outweigh even the most pessimistic estimates of costs.

—*Shelomi (2016b)*

TEXTBOX 15.8 WHAT WOULD HAPPEN IF ALL MOSQUITOES WERE ELIMINATED?

This is where the opinions of many scientists differ. Some say that it would actually be alright because another species would simply take their place in the ecosystem. (We just don't know if that other species would be better or worse for us than mosquitoes.) Others say that every animal plays an important role in the ecosystem, so removing them entirely could have a lot of negative effects. Mosquitoes are a source of food for a lot of birds, fish, and even plants, so many animals would actually suffer from their extinction. So currently, scientists are working on ways to kill only the species that carry deadly diseases like malaria – or genetically modify them, so that they either can't carry the disease anymore or die before they can infect humans with it.

—*Sirwinchester (2016)*

Arthropoda), a very large group of invertebrate animals (almost two-thirds of all species that have been described are arthropods), which includes insects, daddy-long-legs, crabs, lobsters, shrimp, and others. Spiders belong to the arachnids ('class' Arachnida), a subgroup of arthropods with over 100,000 named species, but which may prove to include over 500,000 species. Mites, ticks, scorpions, and other creatures are also placed within the arachnids. Spiders are often classified as a subgroup of the Arachnids, the 'order' *Araneae*, with over 45,000 known species (possibly another 75,000 species remain to be cataloged). Spiders are extremely diverse, occupying many habitats, and occurring in much of the world (Marusik and Koponen 2000). Except for *Bagheera kiplingi*, which is mainly vegetarian, all other known spiders are predators (Meehan et al. 2009). Spiders mostly feed on insects and other spiders, but a few large species can capture small or immature vertebrates such as birds, lizards, snakes, and mice. Most species of spiders are relatively small, about 2–10 mm in body length. Tarantulas (a thousand or more species of the family Theraphosidae) are the largest and heaviest spiders, the bodies of some as long as 9 cm, and despite their scary appearance, they are frequently kept as pets. Like most arthropods, there is much research needed to identify the many species not yet discovered (Coddington 1991).

Harmful Aspects

Only a few dozen spider species, mostly in the Tropics, are considered to be dangerous to humans (Mullen and Vetter 2018; Figure 15.44). Diaz (2004) reported that in the entire 20th century, only about 100 deaths from spider bites were reliably reported. In discussing the world's most dangerous spiders, Kularatne and Senanayake (2014) wrote: 'About 12

FIGURE 15.44 Examples of extremely toxic spiders. (a) The Sydney funnel web spider (*Atrax robustus*) is a native of eastern Australia. Males (above) wander more than females (below) and so tend to be involved more in biting humans. Deaths of humans have not been reported for decades, thanks to the use of anti-venom since 1981. Photo of a display at the Australian Museum, Sydney by Sputniktilt (CC BY SA 3.0). (b) The Brazilian wandering spider (*Phoneutria nigriventer*) is native to South America. Several other species of *Phoneutria* are also called Brazilian wandering spider, and the name 'banana spider' is also used because they are often found in shipments of bananas. These large, aggressive spiders are responsible for serious poisonings in South America but also in other countries when found in exported bunches of bananas. Photo by João P. Burini (CC BY SA 3.0). (c) The western black widow (*Latrodectus hesperus*) of western North America. Photo by Marshal Hedin (CC BY 2.0). Spiders of the genus *Latrodectus* occur on several continents and are known as black widow spiders and brown widow spiders ('widow' in the name is based on the female's habit of often eating the males following sex). Female widow spiders have relatively large venom glands, and unlike the males, their bite can be harmful to humans although death or serious complications are rare. The three North American black widow species are considered to be the most poisonous spiders on the continent.

species of spiders stand out as clinically important. These include widow spiders, recluse spiders, banana spiders and Australian funnel web spiders. The most dangerous are the female widow spiders of the genus *Latrodectus*, e.g., black widow (*L. mactans*), gray widow (*L. geometricus*), American *Loxosceles* causing cytotoxin-mediated local cutaneous damage, and the world's most toxic, Australian funnel web spiders. In Brazil, three groups of spiders are found: Banana spiders (genus *Phoneutria*; phoneutrism), recluse or violin spiders (genus *Loxosceles*; loxoscelism) and widow spiders (genus *Latrodectus*; lactrodectism). Widow spiders are widely distributed in all continents including Australia.'

'Chelicerae' of spiders are mouthparts made up of a basal portion containing venom glands and hollow fangs that inject venom (chelicerae are often called fangs or jaws). Almost all spiders use their fangs to inject venom, and while most cannot bite through human skin, some can inject dangerously toxic doses into people (also into pets and livestock), and some people are allergic to small amounts. Some scientists hypothesize that harmful spider bites during the evolution of humans may have resulted in genes favoring instinctive fear of spiders (see Zvaríková et al. 2021 for analysis). A similar situation may exist regarding fear of snakes. The amygdala region of the brain appears to play a role in the fear response of mammals, and experiences with dangerous spiders and snakes may have sensitized this region to fear these animals (Spitzer 2005; LeDoux 2012).

Many people find spiders to be spooky, but for a minority they are terrifying. Significant fear of spiders is widely known as arachnophobia, but some authorities employ the term araneophobia for fear of spiders while employing the word arachnophobia to also include related arachnids such as scorpions, ticks, and mites. Fear of spiders ranks with fears of snakes, heights, and public speaking as the leading phobias. Studies of responses to different classes of arthropods have shown that people show relatively high fear of spiders (Gerdes et al. 2009). Women have been found to have a greater fear of spiders than men (Cornelius and Averill 1983), consistent with the same observation for fear of snakes. A survey in Sweden reported that 5.5% of adults and children have snake phobias and 3.5% have spider phobias, and that these phobias are suffered by about four times more women than men (Fredrickson et al. 1996; Figure 15.45). This might indicate that women have superior genetic adaptation to avoid dangerous animals that could harm their relatively susceptible infants, or alternatively, that men are adapted to greater risk behavior (Bower 2009). It is important to understand that arachnophobia and

FIGURE 15.45 Miss Muffet public domain illustration by Hope Dunlap, from a 1908 children's book, illustrating the familiar nursery rhyme (of obscure origin, but at least two centuries old). The fear exhibited by the girl reflects the widespread but invalid conviction that spiders are a significant danger. Photo by Pearlmatic (CC BY 2.0).

Little Miss Muffet
Sat on a tuffet,
Eating her curds and whey;
Along came a spider
Who sat down beside her
And frightened Miss Muffet away.
(Tuffet = low seat; curds and whey =
moist soft cheese and liquid remaining
after milk has been curdled and strained)

indeed other fears of animals are psychological illnesses that can be treated (Bouchard et al. 2014), and those suffering from such medical conditions should not be condemned as haters of biodiversity. Irrational fear of spider bites has been known since ancient times. In the Middle Ages, a hysterical condition called tarantism (for the Italian province of Taranto, where the mania was first reported), involving mass frenzied dancing, was claimed to arise from spider bites, although the cause is unclear (Donaldson et al. 1997).

Dislike of spiders is likely not only because a small minority of species are actually dangerous. In terms of human preferences, most people find spiders unattractive physically. However, some spiders are quite beautiful, rivaling butterflies in their brilliant combination of color (Figure 15.46). Nevertheless, the behavior of spiders does not endear them to people. Spiders are literally 'creepy' in the way they move. And, their hunting and feeding style – ambushing and/or laying sticky traps, paralyzing their prey, and sucking them dry – are gruesome in human terms. Cobwebs in houses, sometimes with the remains of insect corpses, need to be cleaned away regularly, generating additional resentment of spiders.

Beneficial Aspects

Spiders do not deserve their bad public reputation as dangerous. Frequently, bites blamed on spiders came from mosquitoes, biting flies, or fleas, and indeed undiagnosed skin irritations are often termed 'spider bites' (spider bites may show two puncture marks from the fangs). Almost all spiders, including venomous species, are timid, avoiding confrontation with humans and biting only in self-defense, typically when handled, cornered, or injured. Most spider bites are less painful than a bee sting. Spiders are not transmitters of diseases.

Spiders play a critical role in controlling insect pests (Maloney et al. 2003; Ndava et al. 2018), their primary food, although they occasionally eat plant material (Nyffeler et al. 2016). In turn, spiders are food for some insects and for many vertebrates, including birds, reptiles, amphibians, and mammals. Fried spiders are a delicacy in Cambodia (Figure 15.47), and the practice of consuming spiders ('arachnophagy') by indigenous peoples is well known (Meyer-Rochow 2004). Spiders are deliberately employed as biological control agents to reduce pest insects (Riechert and Lockley 1984; Young and Edwards 1990; Marc et al. 1999; Sunderland 1999). Spider venom is being explored for its potential medicinal properties (Matavel et al. 2016; Ting et al. 2019) and pesticide possibilities (Ikonomopoulou and King 2013; King and Hardy 2013). Spider silk also has possible medical applications (Harvey et al. 2016a), and it has been speculated that spider silk genes could be inserted into bacteria, other insects, and plants to transform them into silk sources (Liu et al. 2011). Spiders are key components of many ecosystems and are potentially useful indictors of ecosystem health (New 1999; Textbox 15.9).

FIGURE 15.46 Examples of colorful spiders (all are males; a–c are 'jumping spiders'). (a) *Habronattus coecatus*. Photo by Ryan Kaldari (public domain). (b) *Pelegrina pervaga*. Photo by Pete Carmichael (CC BY 2.0). (c) *Maevia inclemens*. Photo by Opoterser (CC BY 3.0). (d) *Epeus flavobilineatus*. Photo by Guido Bohne (CC BY SA 2.0).

FIGURE 15.47 Spiders as food. (a) The edible spider (*Haplopelma albostriatum*) of Southeast Asia. Photo source: https://commons.wikimedia.org/wiki/File:HAPLOPELMA_ALBOSTRIATUM_HEMBRA_ADULTA.jpg (CC BY SA 2.5). This palm-sized tarantula is quite venomous, but the toxin is believed to be neutralized by frying. The spider occupies burrows in the ground, from which it is collected, but it is also often reared in holes in the soil for later harvest and marketing. (b) Fried spiders for sale at a market in Skuon, Cambodia. Photo by Mat Connolley (Matnkat) (CC BY SA 3.0).

TEXTBOX 15.9 SPIDERS AS CLIMATIC INDICATORS

When the French chemist Quatremere-Disjonval was imprisoned in Utrecht after being banished from France by Napoleon, he reputedly learned to predict the weather by observing spiders and their webbing behavior. Consequently, he was able to predict a period of freezing weather when Utrecht was under siege by the French in 1795; his message smuggled to the French general was instrumental in the successful capture of the city. Quatremere-Disjonval, released in gratitude, was perhaps the first person to utilize spiders directly as a climatic indicator.

—*New (1999)*

Conservation Aspects

'The IUCN Red List of Threatened Species is the most widely used information source on the extinction risk of species... Spiders currently comprise over 47,000 species described at the global level. Of these, only 200 species (0.4%) have been assessed' (Seppälä et al. 2018). Unfortunately, the negative view of spiders by most people makes their conservation a challenging endeavor (New 1999; Skerl 1999; Textbox 15.10), and their image is in need of rehabilitation. Most illustrations of spiders make them seem quite threatening (but note Figure 15.48).

TERMITES

Introduction

Termites are a group of about 2,300 insects closely related to cockroaches (Inward et al. 2007). Most are 4–15mm in length, although queens of *Macrotermes bellicosus* are over 10cm. Termites are very successful, occurring on all continents except Antarctica. There are about 1,000 species in Africa and about 400 in each of Asia, South America, and Australia. Relatively few species are in North America and Europe, as termites are generally not cold-adapted. Termites divide labor among reproductive males ('kings') and females ('queens'), and sterile 'workers' and 'soldiers' (Figure 15.49a). This social structure is reminiscent of the organization found in some bees and wasps, and especially in ants. Termites are not ants although they are sometimes called 'white ants.' The kings and queens have a brief winged phase (at which time they are called 'alates'; Figure 15.49b), while the sterile forms are wingless and are mostly blind. Worker termites are usually the type found in infested wood, as in many species they have the role of digesting cellulose in wood and feeding the other castes. Some termites are remarkable. The queens of some species live for as long as 50 years, a record among insects. Some termites build huge, elaborate mounds (Figure 15.50), considered to be the greatest architectural achievement in the insect world. Most termites construct nests, often underground (Eggleton 2011).

Harmful Aspects

Several hundred termite species significantly damage wooden buildings (Su and Scheffrahn 2000; Figure 15.51)

TEXTBOX 15.10 CONSERVATION CHALLENGES FACING SPIDERS

Spiders, like many invertebrates, receive little attention from the conservation community. This may be due to fear and dislike of their appearance, behavior, or venomous nature; the fact that most spiders are probably widely dispersed and not presumed to be threatened; or because relatively little is known about the distribution and abundance of these creatures… There are many reasons to conserve spiders, even without considering that all species have intrinsic value in and of themselves. Spiders are clearly an integral part of global biodiversity since they play many important roles in ecosystems as predators and sources of food for other creatures. Spiders also have utilitarian value. For many years, spiders have been model organisms for research in ecology, behavior, and communication. They may also be important as biological control agents in agro-ecosystems, providers of silk for materials science, and suppliers of venom for both medical and insecticide research… Many threats to spider diversity have been documented. The primary threat is habitat loss and degradation, as with many other elements of biodiversity. More specifically, some spiders have become imperilled due to urban development, land-use management techniques, air and ground water pollution by pesticides and fertilizers, the introduction of alien species, and in some cases, collection and trafficking due to the pet trade. For a few species, these threats have pushed them to the threshold of extinction.

—*Uniyal (2004)*

FIGURE 15.48 'Friendly spider,' a late 19th-century advertising image (public domain) for the Merrick Thread Company, which suggested their thread was as strong as spider silk.

FIGURE 15.49 Formosan subterranean termites (*Coptotermes formosanus*). (a) Most of the termites are workers; the soldiers present have larger (orange) heads with large jaws. (b) Winged (alate) stage. Public domain photos by Scott Bauer, Agricultural Research Service, U.S. Department of Agriculture.

FIGURE 15.50 African elephants next to a termite mound (constructed by *Macrotermes* sp.) in Tanzania. Photo by Whitney Cranshaw, Colorado State University/Bugwood.org (CC BY 3.0).

FIGURE 15.51 Termite damage (from *Reticulitermes* sp.) to a house in Mississippi that was less than 5 years old. Photo by Wood Products Insect Lab, USDA Forest Service, Bugwood.org (CC BY 3.0).

and crops, both herbs and trees (Rouland-Lefèvre 2011). In some countries, damage to buildings amounts to billions of dollars annually. Extremely toxic pesticides are often used to control termites, and if not applied professionally, damage to biodiversity and humans may result. 'They are the most problematic pest threatening agriculture and the urban environment… especially in the semi-arid and sub-humid tropics' (Verma et al. 2009). Termites are thought to be responsible for producing as much as 11% (some estimates are as high as 40%) of the greenhouse atmospheric gas methane as a result of the digestion of cellulose (Zimmerman et al. 1982; Thakur et al. 2003; Ritter 2006). While termites can be beneficial to nature, ranchers often find them objectionable.

More than two dozen termite species are considered to be invasive (Evans et al. 2013). The Formosan subterranean termite (*Coptotermes formosanus*, Figure 15.49), actually a native of China, may be the most destructive of all termites. It is often called the super-termite because it is invasive in many areas of the world, producing huge colonies (some with millions of insects) and consuming wood at an alarming rate (sometimes an individual eats 400 g/day). It is one of North America's most serious insect pests, estimated to cause over 1 billion dollars in damage annually in the southern U.S. Several other termite species are also considered extremely destructive (Govorushko 2019).

Beneficial Aspects

Termites primarily consume dead plant material, such as wood, litter, and animal dung. Their recycling of organic matter is of huge ecological significance in subtropical and tropical areas, particularly improving tropical soils (Donovan et al. 2001; Mokossesse et al. 2012). They also play an important constructive role in temperate forests (Maynard et al. 2015). Termites are food for numerous predators, including insects, especially ants, also spiders, lizards, frogs, toads, many birds, and numerous mammals including some specialists such as aardvarks, aardwolves, anteaters (Figure 15.52), and pangolins. Termites are eaten and employed as folk medicine by some ethnic groups (Figueirêdo et al. 2015).

Conservation Aspects

Termites play a surprisingly significant role in supporting ecosystems and maintaining habitat for many other animals. Their mounds provide living quarters for many species, especially ants (Dejean and Durand 1996; Dejean et al. 1997). Termite mounds also appear to play a valuable role in stabilizing climates and biodiversity (Bonachela et al. 2015; Textbox 15.11).

Ambivalence about the Values of Termites

As for many of the invertebrates discussed in this chapter, from a human perspective, termites may be both harmful and useful (Ibrahim and Adebote 2012; Govorushko 2019; Textbox 15.12). 'Termites are viewed as both beneficial and destructive in many ecosystems. In some instances they are thought to be important in nutrient cycling and soil formation, whereas in others they are regarded as major factors in range deterioration and soil erosion' (Bodine and Ueckert 1975).

TICKS

Introduction

Ticks include about 1,000 species of small (usually 3–5 mm long) relatives of spiders, discussed previously (Sonenshine and Roe 2014; note Figure 15.53). They are distributed worldwide, from the Arctic to tropical regions, occurring especially in countries with warm, humid climates. Ticks have been assigned to subclass Acari of the insect class Arachnida. There are two major families of ticks, about 80% of species in the Ixodidae (hard ticks, so-called because of their hard dorsal shield) and 20% in the Argasidae (soft ticks, with a flexible leathery cuticle). A third family the Nuttalliellidae has just one species. The classification of ticks requires study (Mans et al. 2019). Ticks are blood

FIGURE 15.52 Old painting of giant anteaters (*Myrmecophaga tridactyla*) of South America, feeding on termites. Source: Wellcome Trust (CC BY 4.0).

TEXTBOX 15.11 THE VALUE OF TERMITES FOR STABILIZING CLIMATE, BIODIVERSITY AND FOOD PRODUCTION

Termites might not top the list of humanity's favorite insects, but new research suggests that their large dirt mounds are crucial to stopping the spread of deserts into semi-arid ecosystems and agricultural lands... In the parched grasslands and savannas, or drylands, of Africa, South America, and Asia, termite mounds store nutrients and moisture, and – via internal tunnels – allow water to better penetrate the soil. As a result, vegetation flourishes on and near termite mounds in ecosystems that are otherwise highly vulnerable to 'desertification,' or the environment's collapse into desert.

— *Kelly (2015)*

Worms and termites are not likely to win hearts and minds, but they, along with lichens and microbes, are vital to food security... Worms, termites, lichens, and soil microbes may well be the heroes of food production as without these species land-based biodiversity would collapse and food production cease... Safeguarding the underlying ecological foundations that support food production, including biodiversity, will be central to feeding 7 billion inhabitants, climbing to over nine billion by 2050.

— *Jena (2012)*

TEXTBOX 15.12 AMBIVALENCE ABOUT THE VALUES OF TERMITES

Although termites are excellent decomposers and have a positive impact on numerous ecological functions, they become serious issues when they attack crops and constructions. Consequently, the positive roles played by termites are often overshadowed by their status as pests threatening agriculture in the tropics where billions of US$ are annually spent on their prevention and extermination... more effort has been spent on eradicating termites than on understanding their environmental impacts, and/or how to use their impacts for improving specific ecological functions in agro-ecosystems. Hence, it appears that the role of termites, as ecosystem service providers, is clearly underappreciated and that more research is needed to better evaluate the importance of termite activity and diversity in tropical agro-ecosystems.

—*Jouquet et al. (2018)*

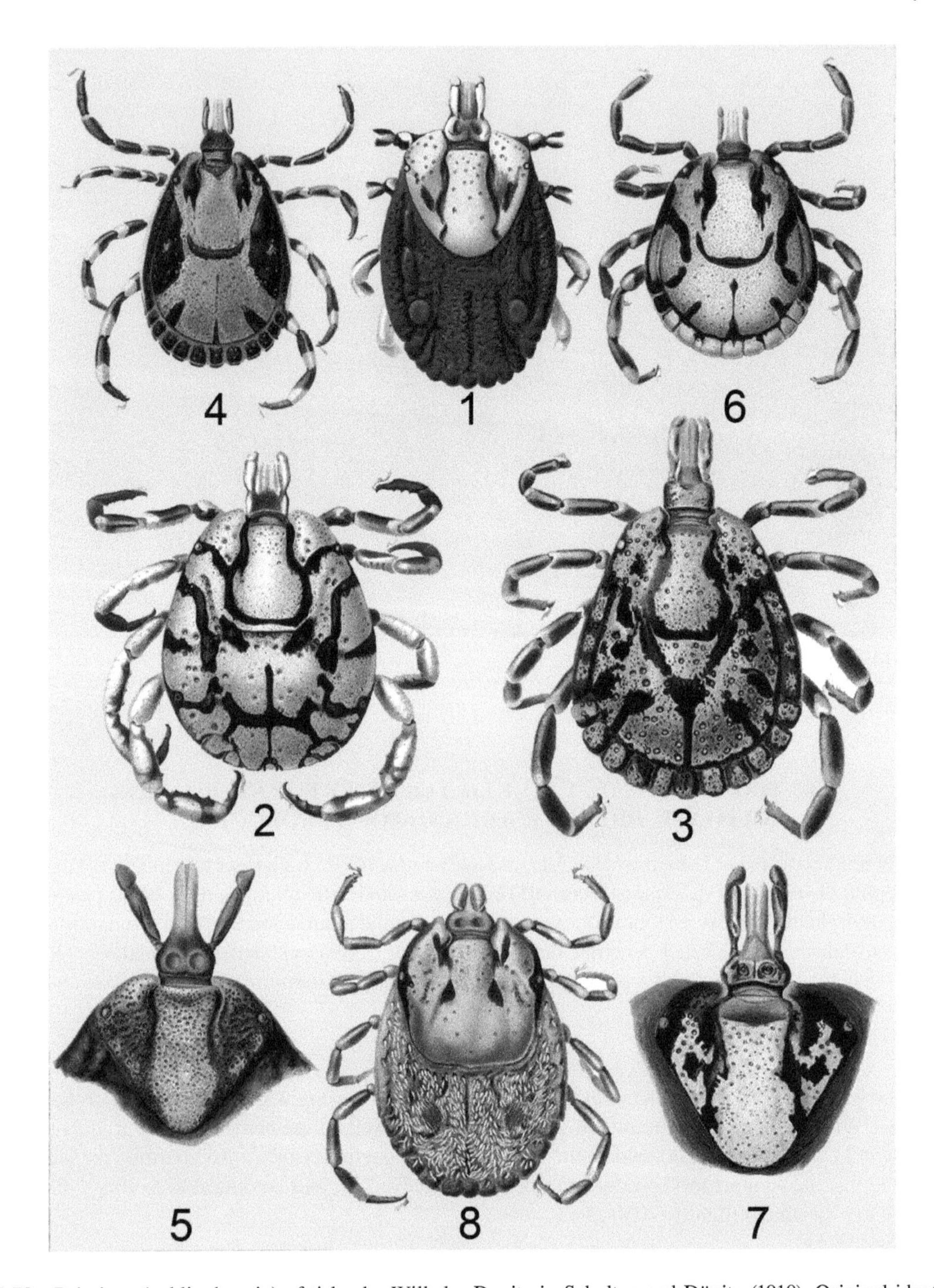

FIGURE 15.53 Paintings (public domain) of ticks, by Wilhelm Donitz in Schultze and Dönitz (1910). Original identifications: 1, *Hyalomma hippopotamense* (female); 2, *Hyalomma hippopotamense* (male); 3, *Amblyomma marmoreum* (male); 4, *Amblyomma variegatum* (male); 5, *Amblyomma variegatum* (female); 6, *Amblyomma hebraeum* (male); 7, *Amblyomma hebraeum* (female); 8, *Dermacentor rhinocerinus.*

parasites, feeding on mammals, birds, and to a much lesser extent on reptiles and amphibians. Ticks cannot fly or jump. They crawl up on low vegetation, awaiting animals to which they adhere. They feed by cutting a hole in the skin of their hosts, attaching by a harpoon-like barbed structure (the hypostome) near the mouthparts, while injecting an anticoagulant or clotting inhibitor to keep the host's blood flowing. Males and females tend to be distinctive (Figure 15.54a and b). Some ticks are capable of consuming 100 times their weight in blood, and they can become quite swollen after feeding for some time (Figure 15.54c). Migrating birds can carry ticks, along with the diseases they harbor, for long distances (Cohen et al. 2015).

Harmful Aspects

Ticks transmit diseases among animals, including humans, livestock, pets, and wild species (Magnarelli 2009; Asebe

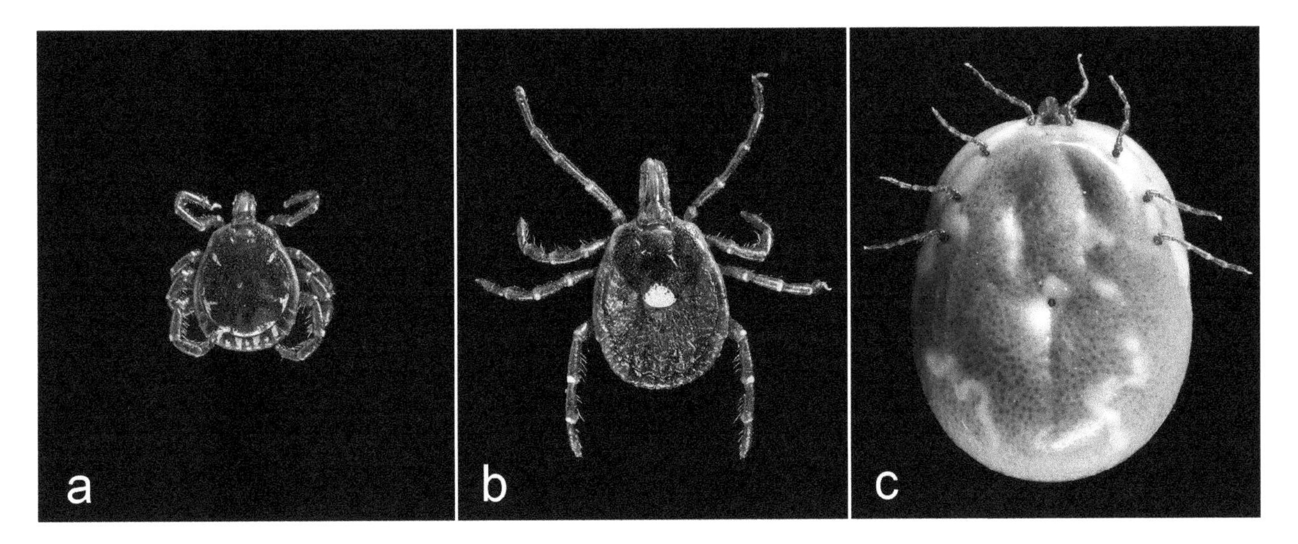

FIGURE 15.54 Lone star tick, *Amblyomma americanum*, the most common tick of the Southeastern U.S. (a) Dorsal (top) view of male. (b) Dorsal (top) view of female. Females, but not males, have a white or yellow spot on their back. (c) Ventral (bottom) view of an engorged female. Females average 4–6 mm in length when unfed and up to 16 mm or larger when fully fed. Males averages 2–5 mm regardless of feeding state. Public domain photos; credit: James Gathany, U.S. Centers for Disease Control and Prevention.

et al. 2016). 'Among arthropod vectors of disease, ticks transmit the most diverse array of infectious agents and ticks are the most important arthropod vectors, globally, of pathogens to humans and domestic animals' (Kikel 2018). Blood-borne infectious agents may be bacteria, viruses, and protozoa. The bacterium *Rickettsia* can cause typhus, African tick bite fever, Rocky Mountain spotted fever, and other serious conditions. Additional diseases carried by ticks include Colorado tick fever, tularemia, Lyme disease, and many more (CDC 2018). The venomous Australian paralysis tick can, true to its name, cause paralysis. Many other ticks that attach and suck blood near the spinal cord can also produce a condition called 'tick paralysis' in humans and other animals. 'Tick toxicosis' is a condition that causes sickness and death in animals (Nicholson et al. 2018). Livestock production around the world suffers enormous losses due to ticks (Narladkar 2018). Tick-caused wounds promote secondary microbial infections and reduce the value of the hides of livestock. As blood parasites, ticks in large number can cause anemia. Climate change is affecting the distribution of ticks, sometimes warmer, more humid conditions extending the range of diseases (Dantas-Torres 2015; Oliveira et al. 2017). Enormous research efforts are underway to control ticks, but some pessimism has been expressed about the prospects of success (Pfeiffer 2018).

Lyme disease (named after Old Lyme, Connecticut, where the disease was first diagnosed), caused by bacteria of the genus *Borrelia*, is the most common and the most widely feared tick-transmitted disease in the Northern Hemisphere (Figure 15.55). It has been estimated that about 300,000 new cases of infection are occurring annually in the U.S. (Kuehn 2013). Lyme disease is carried by ticks belonging to the genus *Ixodes*, notably the deer tick (*I. scapularis*) of eastern North America, the western black-legged tick (*I. pacificus*) of the West Coast of the U.S., the sheep ick (*I. ricinus*) of Europe, and the Taiga tick (*I. persulcatus*) of China.

Fear of ticks, unfortunately, discourages people from enjoying the outdoors. It encourages the use of pesticides. It also raises distrust of wild animals, especially birds, deer, and rodents, which may be carriers of tick-borne diseases.

Beneficial Aspects

Ticks are consumed by birds (Figure 15.56), reptiles, amphibians, insects, nematodes, and mites, and so contribute to the orderly cycling of food energy in ecosystems. Like all parasites, ticks regulate the population sizes of their hosts, preventing excessive expansion that might disturb ecosystem functioning. There is evidence that tick diversity can serve as an index to local animal biodiversity (Esser et al. 2019).

Conservation Aspects

Ticks that are obligate parasites of a given host species are of course dependent for survival on that species. For example, the sambar deer (*Rusa unicolor*) has four specifically associated ticks, which would become extinct if the deer were extinguished. In fact, dozens of ticks are in danger of extinction (Durden and Keirans 1996; Mihalca et al. 2011), and ticks have been identified as being especially subject to extinction because of climate change (Carlson et al. 2017a). Because these blood-sucking parasites are despised and can transmit viruses, bacteria, and pathogenic protozoans to a variety of hosts, it is difficult (perhaps virtually impossible) to promote their welfare.

FIGURE 15.55 A 1992 poster (public domain) warning of risk of Lyme disease caused by ticks. Credit: U.S. Army Environmental Hygiene Agency.

FIGURE 15.56 Red-billed oxpeckers (*Buphagus erythrorhynchus*) feeding on ticks infesting Cape buffalo (*Syncerus caffer*) at Kruger National Park, South Africa. Photo by Derek Keats (CC BY 2.0).

SPECIAL CONSIDERATIONS IN THE DEFENSE OF INVERTEBRATES

Six chapters follow this one, each a defense of the worst representatives of the major classes of vertebrate animals and plants. However, invertebrates, particularly insects (which constitute most of the world's species), are by a considerable margin the most disliked and disrespected of all living things. Accordingly, the remainder of this chapter addresses issues related to the intense intolerance toward 'bugs.'

What Are Invertebrates Good for?

This chapter has examined the most disliked of invertebrates, which have contributed to the very poor image of invertebrates in general. Although invertebrates make up as much as 97% of all animal species, they receive far less biodiversity research than vertebrates (Titley et al. 2017). 'Despite their high diversity and importance for humankind, invertebrates have largely been neglected in conservation studies and policies worldwide' (Cardoso et al. 2011). It is important to educate the public about the importance of invertebrates.

The following contributions to human welfare have been advanced as their chief values. (1) The principal importance of invertebrates is their indispensable values in contributing to ecosystem functions (New and Yen 1995; Kellert 1996; Cock et al. 2012). Soil invertebrates (including arthropods, mollusks [mollusc is the preferred spelling outside of North America], nematodes, protozoa, and especially worms) are essential to the health of soil. Invertebrates are also necessary to decompose waste and recycle organic matter, and indeed without their contributions, we would be living in a world of dung. (2) Most pollination is conducted by insects, without which agriculture and indeed civilization would be in dire straits (Basu and Cetzal-Ix 2018; Figure 15.57). (3) Invertebrates directly or indirectly provide food for humans, livestock, and harvested animals. Wild game animals, particularly fish, have frequently fed on invertebrates, so indirectly food is made available to people. Invertebrates also furnish food directly to humans (e.g., shrimp, clams, crabs, lobsters, oysters, honey). (4) Invertebrates provide a variety of non-food economic products (e.g., pearls, dyes, shellac, silk; Figure 15.58). (5) Invertebrates are occasionally employed medicinally (e.g., leeches for stitching wounds) and have great potential as sources of pharmaceuticals (Textbox 15.13). (6) Invertebrates are also widely employed in laboratory research. (7) So-called 'beneficial insects' are predators of other insect species that are detrimental to humans, especially agricultural pests. Some invertebrates are useful as biocontrol agents because they feed on pests (either animals or weeds). (8) Some invertebrates provide habitats for other, economic species. Coral polyps build coral reefs, the home of a quarter of all ocean species, including many consumed by humans. (9) Invertebrates are often displayed in educational and amusement exhibits (such as aquaria and insectaria). (10) Invertebrates are excellent bioindicators of the health of habitats (McGeoch et al. 2011).

FIGURE 15.57 Bee visiting a rose. Pollinators, particularly insects, are essential for many crops and for reproduction of numerous wild plants. Photo by Debivort (CC BY SA 3.0).

FIGURE 15.58 A woman wearing a silk dress, the product of the silkworm moth. Public domain photo from Pixabay.

'Animal Rights' as Applied to Invertebrates

One of the significant uses of invertebrates is as experimental animals. Fruit flies and nematodes, for example, have been mainstays of genetic studies. Generally, regulations demand humane treatment of vertebrates in laboratories, farms, zoos, and in nature, a reflection of how similar humans are to other vertebrates, and how we consequently have compassion, empathy and pity for our nearest animal relatives. In contrast, except occasionally for the largest cephalopods (particularly octopuses), invertebrates are not protected from infliction of pain or tissue damage of any kind by humans. Key to this 'animal rights' philosophical issue is whether invertebrates are capable of experiencing pain and suffering, or even if they experience consciousness and stress in ways comparable to humans (Horvath et al. 2013). If 'lower animals' are no more capable of experiencing pain than plants, then humane treatment of them seems irrelevant to many. This is a much-discussed and unsettled issue for invertebrates, and indeed, there are ongoing debates regarding the humane treatment of all animals.

Most people, however much they strongly object to causing pain to vertebrate animals, are much less generous to invertebrates, and it does seem that, at least in Western societies, the majority are substantially or completely indifferent to the possibility of invertebrate suffering (Figure 15.59). This lack of concern is likely due in considerable part because the most familiar invertebrates are mainly reviled pests. Some Asian religions, notably Jainism, accept even the tiniest of invertebrates as genuinely sentient creatures meriting considerable human respect and protection (Rankin 2018). Ancient Jain monks had such respect for all forms of life that as they walked they would sweep the ground in front of them to avoid stepping on even the humblest of insects (Figure 15.59f). Even some modern Jains wear cloth over their nose to avoid accidentally inhaling small flying insects like gnats! Spiritual and religious movements have important roles to play in persuading society that invertebrates deserve much more consideration than they currently are afforded.

What Are Parasites Good for?

Half of the invertebrates discussed in this chapter are blood parasites (bedbugs, fleas, leeches, lice, mosquitoes, and

TEXTBOX 15.13 THE POTENTIAL VALUE OF INVERTEBRATES AS SOURCES OF INVALUABLE MEDICINALS

Infectious diseases pose a serious threat to humankind, accounting for an estimated 17 million deaths. These statistics comprise a daily toll of 50,000 men, women, and children dying, despite advances in antimicrobial chemotherapy. Contrary to the wide belief that infectious diseases have been largely alleviated, malaria, cholera, and tuberculosis alone remain significant threats, while HIV/AIDS, Ebola, dengue, and Zika pose a major risk to human health… Hence, there is an urgent need to discover novel and effective antibiotics. This has sparked the antibiotic hunt from natural sources. Plants and marine algae have often been acclaimed for their beneficial antimicrobial properties. However, instead of focusing on an extinguishable antimicrobial source from the flora, it would be interesting to consider the fauna as well. Since invertebrates represent a staggering 95% of the fauna and have existed for millions of years in hazardous environments, they are promising candidates. These creatures are believed to have developed antimicrobials to protect themselves from the pathogenic microbes.

—*Mosaheb et al. (2018)*

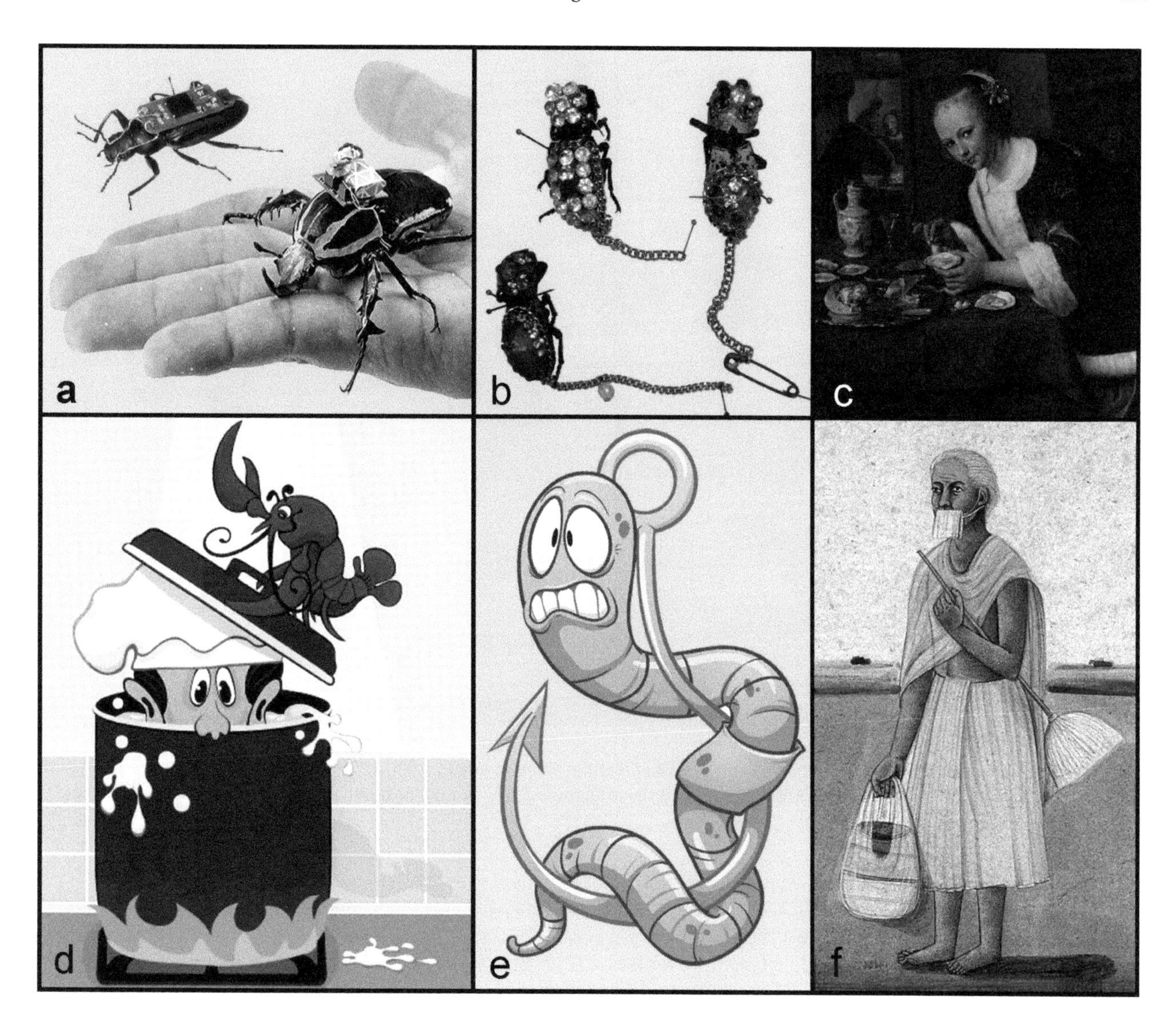

FIGURE 15.59 Questionable ethical treatment of invertebrates. (a) Beetles (*Zophobas morio* at top, *Mecynorrhina torquata* in hand) fitted with electronic backpacks and batteries for experimental usage as flying cyborgs. Photo by Beetleman26 at Biological Machine Laboratory, NTU, Singapore, in 2017 (CC BY SA 4.0). (b) Bejeweled Zopheridae (a family of beetles), the living insects worn as a fashion accessory. Photo (CC BY SA 2.5) by Shawn Hanrahan at the Texas A&M University Insect Collection in College Station, Texas. (c) 'Girl eating oysters,' public domain painting by Dutch artist Jan Steen (1625/1626–1679). Oysters are commonly eaten alive. Photo credit: The Yorck Project (2002). (d) Lobster boiling a chef. Pain experienced by lobsters being boiled alive has long been debated. Source: Shutterstock (contributor: Kittasgraphics). (e) Worm on a fishing hook, also debated regarding the pain inflicted. Source: Shutterstock (contributor: Memo Angeles). (f) A Jain monk carrying a broom to move insects without harming them and wearing a mask to avoid accidentally inhaling flying insects. Painting by a 19th-century Indian monk. Source: Wellcome collection (CC BY 4.0).

ticks), which are perhaps the most despised of all reviled creatures responsible for huge investments to eliminate them (e.g., Figure 15.60). Parasites are by no means limited to invertebrates: They also occur in vertebrates and flowering plants, and especially in fungi and microorganisms. Indeed, it is commonly estimated that about 50% or more of all species are parasitic (Jones et al. 2015). Nevertheless, the very high incidence of blood parasites of humans has contributed to the very poor image of invertebrates in general. It is important to educate the public about the importance of parasites, which as explained in the following is mainly as major regulators of animal populations.

Parasites are organisms that live in or on individuals of another species (the hosts), deriving nutrients at the hosts' expense. (A more expansive definition of biological parasite would broaden the meaning to include reliance of one species on another species for existence or support without adequate compensation. For example, 'brood parasites' are species that rely on others to raise their young, as illustrated by cuckoo species.) As noted earlier, to be judged as 'despicable' species need to be visible, and 'microparasites' such as viruses and some bacteria are too small to evoke this judgment, while 'macroparasites' that are at least the size of lice and ticks are the ones that evoke hatred.

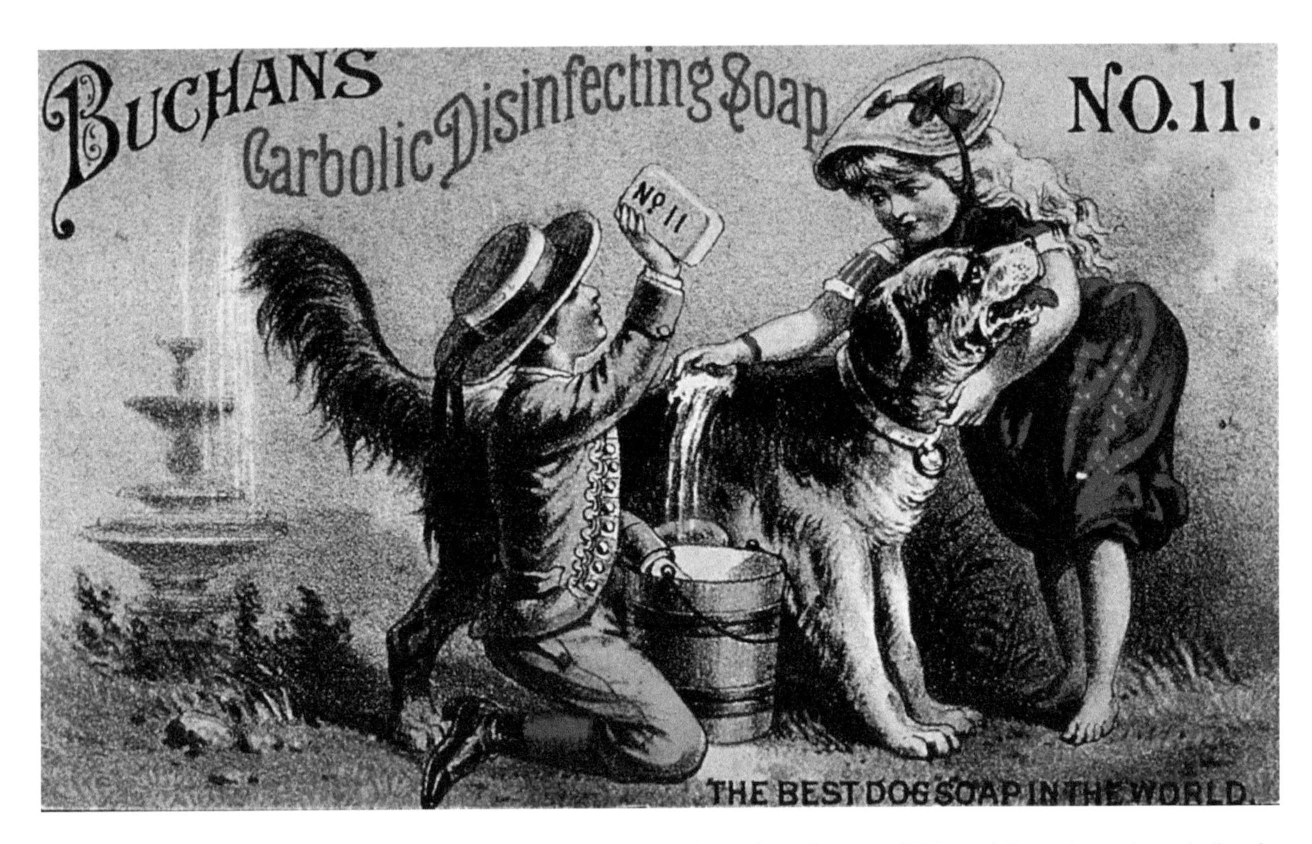

FIGURE 15.60 Advertisement for Buchan's Carbolic Disinfecting Soap, featuring two children giving a large dog a bath using the soap. Product information claimed that the medicine 'kills all parasitic life on man or beast.' Public domain photo from the U.S. National Library of Medicine.

Parasites keep populations of their hosts in check and stimulate their hosts to evolve more efficient protective adaptations. Almost certainly the world would have far less interesting biodiversity were it not for parasites. As people, our human values and personal experiences with parasites lead us condemn their life styles on moral ('parasites are evil') and aesthetic ('parasites are repulsive') grounds, but nature makes no such judgments, and those who advocate on behalf of biodiversity need to strive to keep an open mind. Many of mankind's most hated parasites have some significant medical uses, as pointed out earlier. Numerous parasites are used in 'biological control' as natural predators to eliminate introduced pests of natural and agricultural landscapes (illustrating the maxim 'the enemy of my enemy is my friend'). Particularly important in this regard are 'parasitoids,' species that live on or in a host, eventually killing it (Figure 15.61). About 10% of described insect species are parasitoids, mostly in the Order Hymenoptera (which includes bees, wasps, and ants).

The 'hygiene hypothesis' holds that 'decreased exposure to infectious agents early in life increases susceptibility to allergy (and perhaps autoimmune diseases) by limiting immune system development' (Kerksiek 2008). Intrinsic in this idea is that many parasites are actually beneficial to the immune system of some humans and have therapeutic potential (Capron 2011; Dhingra et al. 2013; Tunnessen and Hsieh 2018). Autoimmune diseases such as arthritis, asthma, diabetes, and multiple sclerosis are caused by the body excessively reacting to its own constituents, especially T-cells attacking proteins. But what may be an excessively active immune system today could have been adaptive for our ancestors, who had to endure a far greater load of parasites than today's people. Some physicians have reasoned that deliberately infecting autoimmune patients with parasites may serve to channel the immune system's attack toward the parasites and away from the patients' bodies. Another hypothesis deals with intestinal parasites that remain for very long periods within the body. In this case, it has been argued that such parasites deliberately weaken the immune system so they will not be attacked, and if so, this may benefit autoimmune patients by also weakening the attack on their bodies. Although it seems disgusting, some autoimmune patients today are deliberately swallowing intestinal parasites to improve their overall health. It has been claimed that some people in the early 19th century consumed tapeworm eggs so that they resulting tapeworms would cause them to lose weight; this extreme form of dieting appears to be mythical and is certainly dangerous. There is some interest in exploring the medical application of parasites to relieve some conditions (Sobotková et al. 2019). It also needs to be noted that while parasites take advantage of other species, they themselves are subject to very extensive predation by other species and thereby contribute to food chains and webs (Orlofske et al. 2015).

FIGURE 15.61 Parasitoid wasp (*Aleiodes indiscretus*) ovipositing in its host, a gypsy moth caterpillar. United States Department of Agriculture photo (public domain) by Scott Bauer.

When all is said that can be found in defense of parasites (Textbox 15.14), they are perhaps the hardest creatures to defend (Textbox 15.15).

Insect Conservation Challenges

Bedbugs, clothes moths, cockroaches, fleas, houseflies, lice, locusts, mosquitoes, and termites, examined in detail in this paper, are insects ('class' Insecta). There are more insect species on Earth than any other form of life (over 1 million have been cataloged, and there are believed to be tens of millions of undescribed species). The huge number of insect species masks the fact that many of these are facing extinction (Forister et al. 2019; Textbox 15.16).

Unfortunately, apart from butterflies and some bees, there are very few examples of concern for the survival of insects, and indeed most are viewed negatively (Kellert 1993; Barua et al. 2012; Textboxes 15.17 and 15.18). 'Unless it's a "pretty" bug like a butterfly, dragonfly, or ladybug, people tend to shriek in fear or look down in disgust at these little critters. They're offered no sympathy. They're ignored. Feared. Hated. Killed.' (Harr 2016). Indeed, even the limited efforts at insect conservation show evidence of bias toward more attractive species (Leandro et al. 2017). This chapter argues that much of the fear, disgust, and hatred is due directly to the least-liked but unavoidable species examined in this presentation, and that education and improved public relations are key ways to improving insect conservation.

Hatred of Insects Caused by Commercial Campaigns against Household Pests

In times of war, exaggerated, unrealistic, and ugly visual stereotypes and vicious propaganda have been commonly created to eliminate respect for the enemy, so that they can be eliminated without guilt. Traits that are emphasized include dangerousness, ugliness, prevalence, and horrible, disgusting behavior. Indeed, ethnic groups were commonly labeled as the insect groups discussed here (Rafles

TEXTBOX 15.14 IN DEFENSE OF CONSERVATION OF PARASITES

A growing body of work has shown that parasites are a critical part of ecosystems, acting as regulators of food webs and host populations, and serving an important role in energy flow through trophic levels. The increasingly apparent benefits of parasites make a case for their recognition as an important neglected target for conservation, especially given that parasitic life cycles are already known to be particularly extinction-prone due to cascading co-extinctions with hosts… While institutions such as the IUCN have spent decades developing centralized frameworks for prioritizing the conservation of free-living biodiversity, parasites are rarely included in mainstream assessments; for example, only two animal macroparasites are listed on the IUCN Red List (*Hematopinus oliveri*, the pygmy hog louse, and *Hirudo medicinalis*, the medicinal leech). The under-representation of parasites speaks to… the comparative bias against parasites in conservation.

—Carlson et al. (2017b)

Parasites, particularly the host-specific species, are perhaps the most imperilled group of organisms on Earth... specialized parasites are at the mercy of their host species, which actually makes them more prone to extinction than their hosts… this cryptic loss effect means many parasite species are at risk of extinction, even if they are not currently recognized as such. For every threatened vertebrate species listed on IUCN Red List, there is a number of unrecognized co-threatened parasites waiting to go extinct should their host decline… Why should we worry about conserving parasites? It turns out that despite the negative connotations of the word, most parasites don't kill their hosts, and they play some important roles. We certainly know that parasites play key roles in food chains, nutrient cycling, and in helping their host's immune system stay strong and effective.

—Platt 2018

TEXTBOX 15.15 EQUAL RIGHTS FOR PARASITES?

An important issue in conservation biology is lying dormant. The term biodiversity seems to be used almost entirely for free-living animals and plants. Parasites seem to be ignored or regarded as a threat to the conservation of endangered species… Parasitology is usually taught from a medical or veterinary perspective in which parasites are nasty critters to be eliminated. Informing most people, even most biologists, that parasites are going extinct is sure to bring a response such as 'good riddance.'… Parasites are part of our biosphere, and we, as biologists, must accord them the same respect we exhibit for their hosts. If we truly appreciate biological diversity, we must advocate that all species are precious, even parasites… My concern for this matter was expressed in my slogan 'Equal Rights for Parasites!' Apparently, this slogan is not as catchy as I had hoped, because it has not caught on.

—Windsor (1995)

TEXTBOX 15.16 CHALLENGES TO THE CONSERVATION OF INSECTS

An estimated 11,200 species [of insects] have gone extinct since the year 1600. Some estimates are that half a million insects may go extinct in the next 300 years, while some projections suggest that perhaps a quarter of all insect species are under threat of imminent extinction… Only about 10% of all insects have scientific names, with many taxonomic revisions still required… Describing all unknown species before they become extinct is the taxonomic challenge... Another great challenge for insect conservation is the perception challenge. Even among some general conservation practitioners, insects are often considered insignificant or given scant attention. This lack of appreciation of insects can reach major proportions among some sectors of human society, who may only recognize the dirty cockroach and the nuisance fly.

—Samways (2007)

Biodiversity loss has become a major global issue, and the current rates of species decline – which could progress into extinction – are unprecedented. Yet, until recently, most scientific and public attention has focused on charismatic vertebrates, particularly on mammals and birds, whereas insects were routinely underrepresented in biodiversity and conservation studies in spite of their paramount importance to the overall functioning and stability of ecosystems worldwide … At present, about a third of all insect species are threatened with extinction… The pace of modern insect extinctions surpasses that of vertebrates by a large margin… Because insects constitute the world's most abundant and speciose animal group and provide critical services within ecosystems, such… cannot be ignored and should prompt decisive action to avert a catastrophic collapse of nature's ecosystems.

—Sánchez-Bayo and Wyckhuys (2019)

TEXTBOX 15.17 'THE ONLY GOOD INSECT IS A DEAD INSECT'?

In recent decades, there is no mistaking the fact that Western attitudes toward the natural world have changed markedly. Protecting endangered species, preserving wilderness areas, moderating global warming, and sustaining fragile ecosystems are common themes of modern life… Unfortunately, however, this apparent newfound love for and connection to the natural world has not pervaded the insect world. How is it that we have come to embrace the preservation of and a deep affinity with whales, wolves, polar bears, cats, dogs, and gerbils, to name just a few, while at the same time kill without hesitation any and all creeping, crawling things that have the misfortune of crossing our paths? For all the progress the human species has made in recent decades toward finding its way back to nature, the language of war, loathing, and eradication still informs our attitude toward insects. We seem locked in an engrained specicide that sanctions the wholesale extermination of insects – as if by doing so the world would be a much better and safer place for everyone. When it comes to insects, it seems our mantra is 'Kill often, kill on sight, and kill mercilessly.'

—Ryan (2014)

TEXTBOX 15.18 PREJUDICE AGAINST BUGS

'What sort of insects do you rejoice in, where you come from?' the gnat inquired.

'I don't rejoice in insects at all,' Alice explained, 'because I'm rather afraid of them.'

—Lewis Carroll in 'Through the Looking Glass,' (1872)

Most of the general public and farmers view invertebrates with indifference, aversion, and disdain. Most in our study expressed dislike of bugs, beetles, ants, crabs, spiders, ticks, and cockroaches; a strong aversion to insects in the home; extreme dislike of biting and stinging insects; a desire to eliminate mosquitoes, cockroaches, spiders, fleas, and moths altogether. A large majority viewed invertebrates as lacking in the ability to experience pain, suffering, or consciousness. Most viewed invertebrates as incapable of feeling or emotion, affect or individuality, intellect or rational decision-making. Few supported making significant expenditures or economic sacrifices on behalf of protecting endangered invertebrates.

—Kellert (1996)

The human reaction to insects is neither purely biological nor simply cultural. And no one reacts to insects with indifference. Insects frighten, disgust, and fascinate us. Our emotional response to insects on our bodies and in our homes is not merely a modern, socially constructed phenomenon. Rather, it is a vital part of being human. Our perception of insects is deeply rooted in our species' evolutionary past… Some scientists argue that people are disgusted by bugs as this behavior evolved to help us stay away from toxic, poisonous substances. However, people might have never been exposed to harmful bugs and yet are still afraid or dislike them. This reaction seems already part of collective unconsciousness, even though kids are not naturally afraid and even act curious toward bugs. Different cultures even view different bugs differently, such as delicacies or even pets. Our fear for insects may be explained rationally as we know that they transfer diseases or poison through their invading, biting, or stinging. However, fear often has an irrational nature. For most humans, insects are largely mysterious and alien.

—Oudejans (2015)

2010). In modern times, a propaganda and chemical war is being waged against invertebrates that harm the property of humans (Figure 15.62). The chief exposure to invertebrates by most people is the ceaseless advertisements for insecticide sprays and repellents, which depict 'bugs' as maliciously as possible. 'Insect bombs' and electric 'zappers' often boast of killing everything in the vicinity, omitting any mention of harm to beneficial species. The fear of disease factor is typically highlighted, emphasizing for example that mosquitoes carry Zika virus and ticks transmit Lyme disease. Additionally, housekeepers are often warned that their domicile won't be 'clean' or smell good if insects are present. Often, it is suggested that a chemical barrier needs to be laid down to prevent surreptitious entry into the home of any form of 'bug' (Figure 15.63). And, since household pests usually remain hidden, pest control companies offer inspections to relieve homeowners' anxieties. The result is not just that most people do not hesitate to kill invertebrates that appear in their homes and gardens, but that they view almost all insects and spiders that they encounter with suspicion, if not indeed with contempt. The harm that this does to the cause of biodiversity preservation is immense, because public attitudes and values are the ultimate drivers of investment in biological conservation.

The Psychology of Entomophobia

Wilson (1993) defined 'biophilia' as 'the innately emotional affiliation of human beings to other living organisms' (cf. definition in Wilson 1984) and suggested that most people instinctively like most other life forms. In reality, most humans are biophobic,' i.e., dislike the majority of species they encounter. Earlier, arachnophobia (defined either as fear of spiders or, more generally, fear of all arachnids) was discussed. 'Entomophobia' has been defined as fear of insects (all or just particular groups), more generally as fear of arthropods, or more vaguely as fear of bugs (Figure 15.64). There has been much academic and some experimental analysis of whether phobias to animals are learned or instinctive (see earlier discussion of arachnophobia). Lockwood's (2013) book (available online), *The Infested Mind: Why Humans Fear, Loathe, and Love Insects*, is perhaps the best analysis of why people dislike bugs. He wrote:

> Entomophobia is rooted in six 'fear-evoking perceptual properties.' Insects can: (1) invade our homes and bodies; (2) evade us through quick, unpredictable movements, to which it might be added that the furtive skittering of a cockroach, for example, with its head lowered as if slinking out of the room, evokes a sense that the creature is guilty or ashamed; (3) undergo rapid population growth and reach staggeringly large numbers, threatening our sense of individuality; (4) harm us both directly (biting and stinging) and indirectly (transmitting disease as well as destroying woodwork, carpets, book bindings, electrical wiring and food stores); (5) instil a disturbing sense of otherness with their alien bodies – they are real-world monsters associated with madness (e.g., 'going bugs'); and (6) defy our will and control through a kind of radical mindless or amoral autonomy.

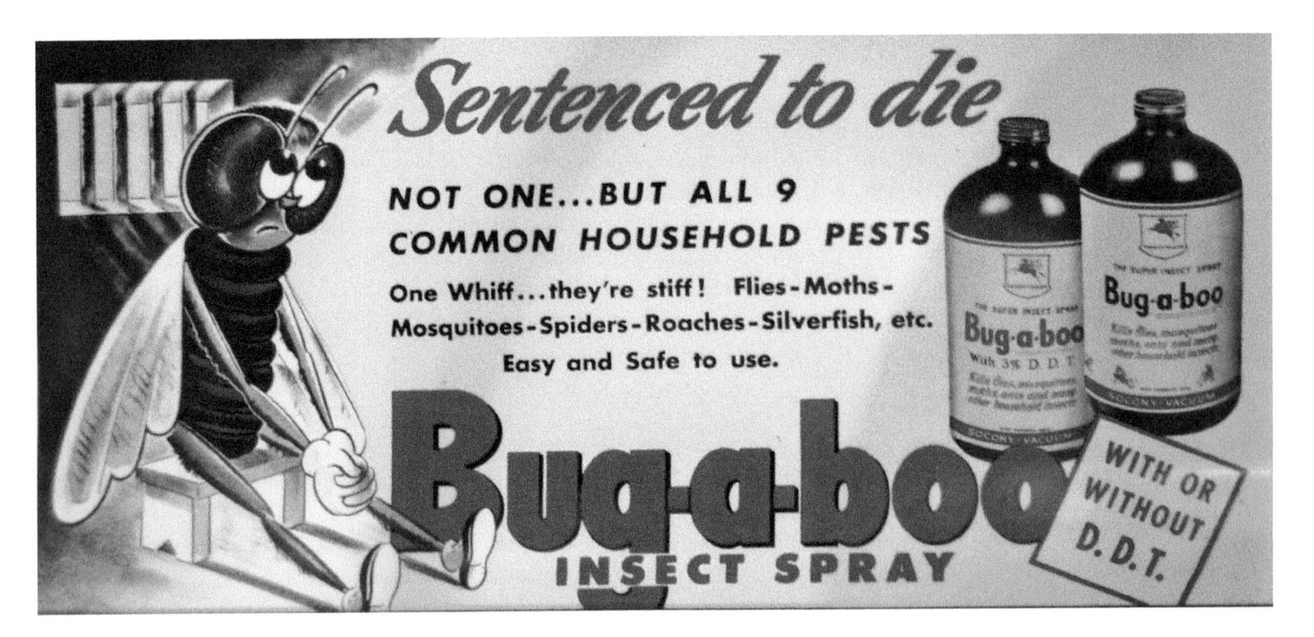

FIGURE 15.62 Vintage pesticide poster. Photo by Kevin Krejci (CC BY 2.0).

FIGURE 15.63 Advertising cartoons designed to scare homeowners into hiring pest removal firms. Source: Shutterstock. (a) Scary microbiological 'bugs' in a dirty toilet (contributor Akkachai thothubthai). (b) A couple discovering bed bugs (contributor: Ivector). (c) A woman in a panic over bugs in the room (contributor: Vector Micro Master). (d) A shocked couple finding a giant cockroach in their house (contributor: Aleutie).

FIGURE 15.64 Fear of bugs. (a) Fear of a cockroach. Source: Shutterstock (contributor: Ayelet-keshet). (b) Fear of spiders. Source: Shutterstock (contributor: Sylverarts Vectors).

Improving the Public Image of Bugs on Behalf of Invertebrate Conservation

Guiney and Oberhauser (2008) recommended that conservation efforts for insects be focused on charismatic insects, endangered insects, and insects that provide important ecological services. Unfortunately, there are very few charismatic insects (notably bees and butterflies), and very many endangered insects, and those most in danger of extinction often make their last stands in foreign, hostile, or isolated locations of little relevance to most people. A survey of the use of local charismatic insects to promote conservation among children had mixed results (Schlegel et al. 2015).

New (2007), noting the difficulties of garnering support for conservation of insect species specifically, recommended that efforts be directed to campaigns that support communities of organisms, so that invertebrates present would also benefit. In a similar vein, Simaika and Samways (2018) also observed the lack of support for insect conservation and recommended that this could be improved by efforts to educate the public about the values of nature in general. These approaches, at least, could contribute to the cause of invertebrate conservation.

Education of the public regarding ecological services is certainly in order. Since health has become a major concern of most people, it is important to emphasize the value of insects as sources of natural compounds of medicinal value (Cherniack 2010, 2011), and the fact that since most insect species have not yet even been discovered, their extinction could greatly compromise human welfare. People need to be made aware that 'Of all insect species, over 97% of those usually seen in the home landscape are either beneficial or are innocent bystanders' (Stewart and Coverstone n.d.).

Children are frequently naturally curious and sympathetic to many insects (Lockwood 2013), but fear of insects and spiders usually begins in childhood (Hardy 1988), so it is clear that early education is a critical period to generate appreciation of, and diminish fear of, bugs (Albo et al. 2021). Illustrated children's books, prepared by nature-friendly authors, are invaluable (Figure 15.65). Libraries should have such a category prominently displayed. Since children increasingly pass their time in front of electronic screens, carefully prepared digital presentations are needed. Most importantly, contact with living insects is critical. Ideally, teachers and parents should introduce their children to the local fauna, but given the widespread negative attitudes to insects, it may be preferable to leave this task to specialists. Exhibits by museums and zoos, featuring living invertebrates explained by skilled teachers and guides, are likely to be very impressionable on children. Religious professionals also can be extremely helpful, stressing that living things are spiritual creations meriting respect. Regardless of theological tradition, teaching children to respect all species is helpful in generating respect for all people.

Historically, most artists have depicted insects as hostile, ugly, alien, and utterly lacking in relatable human qualities. The best children's books anthropomorphize insects, giving them charm, character, and human qualities that serve to emphasize that they are valued citizens of the world (Figure 15.65). In the cause of increasing support for biodiversity, one cannot underestimate the persuasive value of art featuring animals (Small 2016a).

FIGURE 15.65 Art from children's stories, sympathetically illustrating insects. (a) The hookah-smoking caterpillar from Lewis Carroll's 'Alice's Adventures in Wonderland,' drawn by John Tenniel, published in 1865. Colored by MrWalletPants (CC BY 2.0). (b) Cover of 'The Butterfly Ball and the Grasshopper's Feast,' a vintage children's book published in 1800 in Great Britain. Illustrator unknown. Photo by Paul K (CC BY 2.0). (c) Beetle carrying a grasshopper, dragonflies and a gnat flying above, from 'The Butterfly Ball and the Grasshopper's Feast.'

The word 'insectarium' (plural insectariums, insectaria) is employed primarily to mean an institution exhibiting insects and other small, primarily terrestrial invertebrates (especially other arthropods), for public education and entertainment. These may be museums displaying mounted insects or zoos with living insects (Figure 15.66). Some retail stores selling preserved butterflies and similar curios also sometimes advertise themselves as insectaria. By far, the most popular kinds of insectaria are 'butterfly houses' (also known as lepidopteraria) specializing in captive live butterflies and moths. 'Butterfly gardens,' by contrast, are open facilities that attract free-living indigenous Lepidoptera. Insectaria are present in many of the world's largest cities and are superb places for the public to become more sympathetic to insects.

Synthetic Biology and the Future of Undesirable Bugs

'Synthetic biology,' a catchall term for which a universally accepted definition does not exist, refers to genetic engineering science that attempts, much more substantially than in previous times, to modify the genetic makeup and biology of species, populations, and organisms; alter the components and metabolism of natural living systems; and even create artificial life-like systems, all for practical purposes (for background, see Secretariat of the Convention on Biological Diversity 2015; European Commission 2016). The relevance of the diverse synthetic biology possibilities to biodiversity has been recently extensively reviewed (National Academies of Sciences, Engineering, and Medicine 2016; Redford et al. 2019), and analysts foresee both extremely promising advantages and very dangerous hazards for conservation. In the context of this paper, synthetic biology offers possibilities of modifying or even totally eliminating parasites, organisms that produce or vector diseases, and indeed any species considered deleterious, employing a technique termed 'gene drive.' A gene drive is a 'system of biased inheritance that enhances the ability of a genetic element to pass from an organism to its offspring through sexual reproduction' (National Academies of Sciences, Engineering, and Medicine 2016). This technology allows engineered genes to very rapidly infiltrate a population or species, even if the altered genes are detrimental or deadly. (The use of such technology as bioterrorism is a concern. However, not all species are easily susceptible.)

Earlier, the objectives of eradicating cockroaches, mosquitoes, and tsetse flies were discussed, and certainly these and some other 'bugs' are among the principal candidates under consideration for deliberate extinction (Leftwich et al. 2016). This issue is of course a slippery slope, with potentially disastrous unanticipated consequences for ecosystems and biodiversity. The 'precautionary principle' (when human activities may lead to unacceptable or irreversible harm, proceed only if or when risk is established to be minimal) has been discussed in relation to synthetic biology (Holm 2019), and the U.N. recently recommended cautions (COPD 2018; note Textbox 15.19). However, there are urgent social and economic pressures to reduce the harm of the world's major pests, and experimentation to eliminate them is proceeding (e.g. McFarlane et al. 2018; Moro et al. 2018).

FIGURE 15.66 Insectariums. (a) Butterfly house in Vienna, Austria. Source: Shutterstock, contributor: Lals Stock. (b) Exhibition of preserved insects at the Newfoundland Insectarium. Photo by Ricksearle (CC BY ND 2.0). (c) Preserved insects on display in the insectarium portion of the Museum of the Sea, Uruguay. Photo by Museo del Mar (CC BY SA 3.0). (d) Butterflies feeding on fruit in a butterfly house in Mindo, Ecuador. Source: Shutterstock, contributor: Matyas Rehak. (e) Butterflies feeding on fruit in the butterfly house, Mannheim, Germany. Photo by Dierk Schaefer (CC BY 2.0). (f) Butterflies feeding at a zoo in Emmen, Netherlands. Photo by Sandra Vos (CC BY 2.0).

TEXTBOX 15.19 GENETIC TECHNOLOGIES TO ELIMINATE UNWANTED SPECIES ARE PREMATURE BECAUSE THERE ARE UNCERTAINTIES AND UNEVALUATED POTENTIAL DANGERS

Molecular biology has developed at an increasing speed in the last years and now offers the possibility to generate artificial gene drives (GD) to alter or eliminate wild populations. There are now concrete research projects seeking to utilize GD as a 'silver bullet' to combat invasive alien species… We consider the technology not to be fit for practical use at present. Because of the potential of GD to alter ecosystems and to eliminate species, we recommend adjusting the legal framework and increasing biosafety research, especially on ecological consequences. In addition, ethical and societal questions need to be addressed before the application of GD technology.

—Simon et al. (2018)

Research on gene drive systems is rapidly advancing. Many proposed applications of gene drive research aim to solve environmental and public health challenges, including the reduction of poverty and the burden of vector-borne diseases, such as malaria and dengue, which disproportionately impact low- and middle-income countries. However, due to their intrinsic qualities of rapid spread and irreversibility, gene drive systems raise many questions with respect to their safety relative to public and environmental health. Because gene drive systems are designed to alter the environments we share in ways that will be hard to anticipate and impossible to completely roll back, questions about the ethics surrounding the use of this research are complex and will require very careful exploration.

— National Academies of Sciences, Engineering, and Medicine (2016)

Most humans respect nature, but when our welfare is at stake, we modify the world to suit our needs. Indeed, far exceeding all other species, we have created our own artificial habitats and ecosystems while sacrificing other species, their habitats, and the world's natural resources. Environmentalism and the goal of sustainability offer hope of preserving biodiversity, but, as argued in this book, human attitude to other species is a key determinant that has been insufficiently addressed. As pointed out earlier, more than half of the world's species are parasites, a life style that we humans detest, and most species are insects, another life form that most humans have learned to disrespect. Most conservationists, indeed most people who love nature, understand that the ecology of the world is complex, fragile, and altered at our peril. Recent history and indeed current events shows that the human species is capable of genocide of its own populations and annihilation of other species, albeit to date by relatively crude, inefficient, and slow methods. Today, 'gene drives could be a tool for modifying wild species to suit human needs, perhaps to bring about their extinction, perhaps to alter them to suit aesthetic preferences' (National Academies of Sciences, Engineering, and Medicine 2016). The prospect of living in a world without pests, or even just without the most annoying and harmful insects, is seductively attractive, even to dedicated environmentalists and biologists. The public would be much more sympathetic to insects and indeed to nature in general if only the worst bugs were eliminated. The case can be made that driving the greatest enemies of humans to extinction would be beneficial not just for people but for biodiversity (Redford et al. 2019).

Nevertheless, the invention of revolutionary genetic-based extermination biotechnologies is frightening, even if there are desirable potentials because we humans have demonstrated that we are prone to risky and destructive behaviors. In previous times, the most comprehensive scientific effort to eliminate the world's pests centered on the insecticide DDT. The Swiss chemist Paul Hermann Müller was awarded the Nobel Prize in Medicine or Physiology in 1948 for developing the chemical as a pesticide. In 1945, Müller announced that DDT could 'send malaria mosquitoes, typhus lice and other disease-carrying insects to join the dodo and the dinosaur in the limbo of extinct species, thereby ending these particular plagues for all time' (Russell 2001). DDT, of course, became infamous for its disastrous environmental and health impacts, but it is important to appreciate from the DDT story how we humans are altogether much too eager to adopt technologies that seem capable of eliminating our major invertebrate enemies (Figure 15.67). This chapter has stressed that most people are too prejudiced against the majority of invertebrate animals to appreciate that most are harmless, useful, and rather admirable. Indeed, numerous 'bugs' that provide indispensable services would be crushed underfoot by many, ignorant of the importance of their preservation. Now that technologies are being created that could efficiently eliminate animals considered undesirable, it is critical that efforts be made to highlight the merits of all species, not just the narrow selection that the public currently admires. In particular, the war against the most reviled invertebrate pests needs to be tempered with appreciation of their ecological roles and contributions to the overall welfare of biodiversity.

FIGURE 15.67 Illustrations (public domain) from the 1940s showing misplaced enthusiasm to employ DDT for exterminating the world's pests. (a) Second World War poster, U.S. Department of Agriculture, U.S. National Archives. (b) Detail from an advertisement for Pennsalt DDT products that appeared in *Time Magazine*, June 30, 1947. Credit: Science History Institute. Philadelphia.

16 In Defense of the World's Most Despised 'Lower' Vertebrate Animals

INTRODUCTION

As emphasized in this book, all biodiversity is vital to the welfare and survival of humans – even pests, some of which admittedly are responsible for significant damage to health and economic welfare. 'Vermin' refers to animals that are perceived as being so detrimental to human health or welfare that they warrant extermination. Nevertheless, as explained in this book, such species often play important ecological roles and have compensating economic values. Moreover, their harm has often been exaggerated, and their very negative public images are undeserved. One of the goals of this book is to generate understanding of the economic values and useful roles the world's most well-known vertebrate pests play in nature. Chapter 15 examined the conservation aspects of reviled invertebrates. This and the next chapter are concerned with conservation aspects of vertebrate animals that are detested and/or dangerous, and therefore least likely to receive support. Lack of sympathy, indeed blind hatred for such species, sometimes has some justification, but it lessens respect for biodiversity in general and control measures often endanger thousands of innocent species. This chapter deals with the most reviled 'lower' vertebrate species: sharks (representing fish); frogs and toads (representing amphibians); snakes (representing reptiles); and vultures (representing birds). Chapter 17 will deal with mammals.

Invertebrates, the subject of the last chapter, include all animals except vertebrates. Vertebrates are animals with a spinal cord surrounded by cartilage or bone. About 70,000 species have been named, including fish, amphibians, reptiles, birds, and mammals. Many are highly threatened (Ripple et al. 2017; Table 16.1). Vertebrates make up less than 5% of the world's documented animal species but are viewed far more sympathetically than invertebrates, because humans are empathetic with the appearance and behavior of many of them, particularly, as discussed in earlier chapters, the charismatic superstar mammals like pandas and tigers, and beautiful birds, that currently are the mainstays of biodiversity fundraising (Figures 16.1 and 16.2). Unfortunately for invertebrates, aside from butterflies and bees there are very few that are charismatic, so they have limited ability to attract public sympathy. Vertebrate animals receive much more targeted biodiversity research and conservation than do invertebrates (Donaldson et al. 2016; Titleyet al. 2017), but are nevertheless in desperate need of attention (Ceballos et al. 2017). The main drivers of biodiversity loss in vertebrates were identified as agricultural expansion, logging, overexploitation, and invasive alien species by Hoffmann et al. (2010), while Ducatez and Shine (2017) examined the harmful effects of habitat alteration, invasive species, climate change, and overexploitation. Although vertebrates are greatly preferred by people, the beauty criteria discussed in earlier chapters dominate which species are preferred (note Figure 16.3). Although many vertebrates are charismatic, their level of support depends on just how charismatic they are (Bellon 2019). 'Charismatic wildlife species are likely to obtain

TABLE 16.1
Comparison of Key Economic and Psychological Considerations Bearing on Human Relationships with the Groups of Vertebrates

Group	Fish	Amphibians	Reptiles	Birds	Mammals
IUCN species assessed[a]	19,239	6,794	7,829	11,147	5,850
IUCN species assessed as 'Threatened'[a]	2,674	2,200	1,409	1,486	1,244
Percentage assessed as 'Threatened'[a]	13.9	32.4	18.0	13.3	21.3
Domesticated pets ('companions')	Popular	Limited popularity	Limited popularity	Very popular	Extremely popular
Domesticated livestock	Few but large potential	Few and limited importance	Few and limited importance	Few but high importance of chickens	Many and of high importance
Harvested value (wild+ cultured)	Huge	Low	Low	High	Very high
Charisma	Low	Very low	Very low	Very high	Very high

[a]*Source:* IUCN (International Union for Conservation of Nature): Table 3a: Status category summary by major taxonomic group (animals) – https://nc.iucnredlist.org/redlist/content/attachment_files/2019_3_RL_Stats_Table_3a.pdf; this Table also shows numbers of extinct species. 'Threatened' includes species assessed as 'Critically Endangered,' 'Endangered,' and 'Vulnerable.'

DOI: 10.1201/9781003412946-16

FIGURE 16.1 Popular art, reflecting how people prefer large mammals and beautiful birds. Source: Shutterstock. (a) Noah's Ark (contributor: Brgfx). (b) 'Welcome to the Zoo' (contributor: GraphicsRF.com).

FIGURE 16.2 Cute cuddly charismatic megafauna mammals whose conservation is enthusiastically supported by the public and biodiversity organizations. (a) Giant panda cub. Photo by Playlight55 (CC BY 2.0). (b) Bengal tiger cub. Public domain photo from Anthony (U.K.), Pixabay.

more public funding for efforts to conserve them than less charismatic ones' (Tisdell and Nantha 2006).

This chapter deals with the most vilified 'lower' vertebrate species, i.e., those that aren't mammals but represent fish, amphibians, snakes, and birds (Figure 16.4). Chapter 17 deals with the most vilified mammals, i.e., the remaining vertebrate species. Fish, amphibians, snakes, and sometimes also birds are often considered 'lower' because they are evolutionarily older, or simply because they aren't mammals, reflecting human arrogance that we belong to the 'highest' group of animals (Figure 16.5). Cooper (1977) noted that some zoologists consider birds and mammals (both of which are warm-blooded) as 'upper vertebrates,' but he restricted the term to include only mammalian vertebrates, while employing the phrase 'lower vertebrates' as in this paper, to cover birds, reptiles, amphibians, and fish. The Collins English Dictionary defines 'lower vertebrates' as 'relatively simple and primitive vertebrates,' giving birds and reptiles as examples, and excluding mammals (https://www.collinsdictionary.com/dictionary/english/the-lower-vertebrates). Most people are indifferent to fish, except as food and aquarium pets, but sharks are widely hated. Most people find amphibians, particularly frogs, to be unattractive and, of all vertebrates, amphibians are particularly imperilled (Table 16.1). Most people are afraid of reptiles, particularly snakes. Most birds (now known to be derived from reptiles) are widely admired, but vultures are considered repulsive and some species, most notably pigeons, are ranked as extreme pests by many. The predominantly cold-blooded (exothermic) vertebrates (fish, amphibians, and reptiles) have much lower public appeal than birds and mammals, and are relatively neglected in conservation programs (Textbox 16.1). Conversely, it is much easier to enact legislation eradicating cold-blooded pests than warm-blooded (endothermic) pests (Fenoglio et al. 2018). Citing George Orwell's 'All animals are equal, but some animals are more equal than others,' Mather (2019) commented 'There is a huge bias in favor of mammals, which is not consistent with

FIGURE 16.3 Street art (graffiti) showing imaginary, hostile-appearing monsters, reflecting how features foreign to human shape, as exhibited by many lower vertebrates, tend to appear unattractive, albeit entertaining. (a) Painting by Ces53, Photo by Janisorlando (public domain). (b) Photo by Kotzian (CC BY 3.0). (c) Photo by Tim Bartel (CC BY SA 2.0).

their frequency… 0.2% of those on the planet.' Reasons for these biases were discussed in earlier chapters.

SHARKS

Introduction

'Fish' is the conventional plural for a group of specimens or individuals all belonging to the same species (although employing 'fishes' in this situation is debatably a misspelling – e.g., 'Jesus fed the crowd with five loaves and two fishes'). However, 'fishes' is appropriate to referring to a group of more than one species, and this is frequent in scientific publications (although employing 'fish' would not usually be considered an error). When referring to fish(es) in general in expressions like 'all the fish(es) in the sea' and 'he sleeps with the fish(es),' the shorter form of the plural tends to be preferred.

Fishes are classified in various ways, and the following is just one commonly recognized system. Twenty-seven thousand living species of fish have a skeleton that is made of bone and are classified in class Osteichthyes. By contrast, only a little more than 1,000 species have a skeleton made of cartilage (the flexible, light material prominent in human noses and ears), and they constitute class Chondrichthyes (Carrier et al. 2012). The term 'shark' is sometimes used in a broad sense to refer to all of the chondrichthyan fishes, but more commonly the word shark refers to a smaller subgroup (superorder or subdivision Selachii) of about 500 species (Team Ocean Portal 2018). Within class Chondrichthyes, the group of 500 or so sharks along with about 650 species of skates and rays make up the subclass Elasmobranchii or 'plate-gilled' fish.

Some sharks are huge. The whale shark (*Rhincodon typus*; Figure 16.6) is the largest living shark, usually less than 10 m in length although one individual grew to over 18 m. It is a filter feeder, surviving on zooplankton and small fish. The extinct megalodon (*Carcharocles megalodon*; Figure 16.7), the largest known shark, is thought to have sometimes reached 18 m or more in length (Figure 16.8).

FIGURE 16.4 The most vilified 'lower' vertebrate pests (sharks, frogs and toads, snakes, and vultures). Painting by Jessica Hsiung.

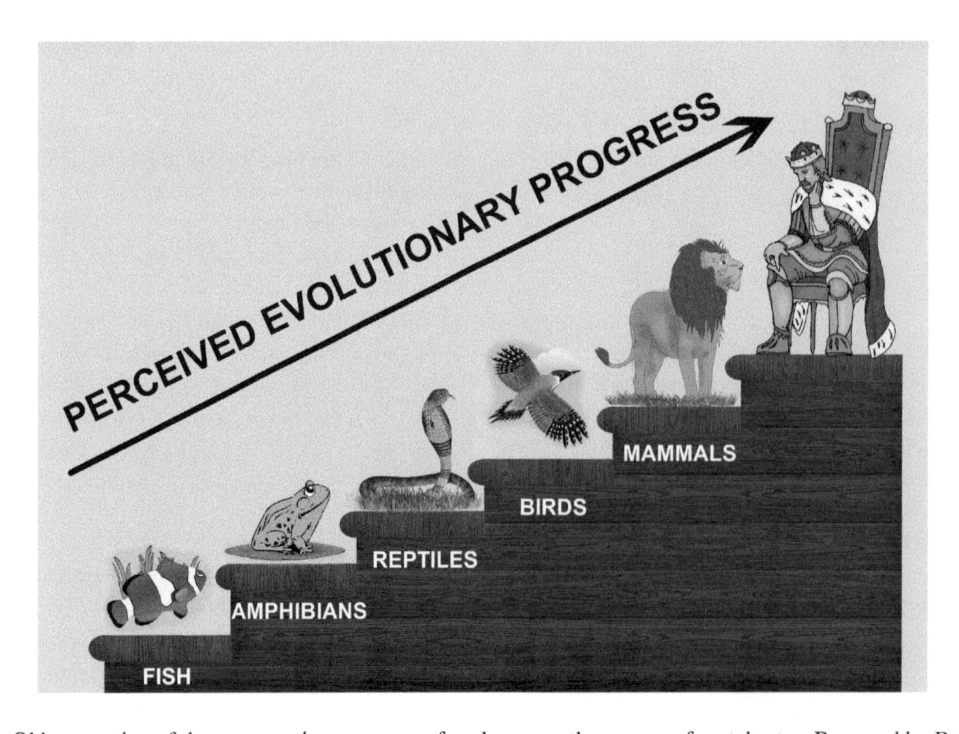

FIGURE 16.5 Old perception of the comparative sequence of rank among the groups of vertebrates. Prepared by B. Brookes.

TEXTBOX 16.1 CHARISMATIC SPECIES DOMINATE CONSERVATION FUNDING TO THE DETRIMENT OF MOST ENDANGERED VERTEBRATES

The interrelationship between public interest in endangered species and the attention they receive from the conservation community is the 'flywheel' driving much effort to abate global extinction rates. Yet big international conservation non-governmental organizations have typically focused on the plight of a handful of appealing endangered species, while the public remains largely unaware of the majority... interest was higher for mammals and birds at greater risk of extinction, but this was not so for fish, reptiles, and amphibians. Our analysis reveals a global bias in popular interest toward vertebrates that is undermining incentives to invest financial capital in thousands of species threatened with extinction. Raising the popular profile of these lesser known endangered and critically endangered species will generate clearer political and financial incentives for their protection.

—*Davies et al. (2018)*

FIGURE 16.6 A 1912 postcard (public domain) showing a harvested whale shark (*Rhincodon typus*) in Miami, Florida, claimed to weigh 13.5 metric tons and to be 14 m long.

The dwarf lantern shark (*Etmopterus perryi*) from the Caribbean matures at 16–19 cm and is considered to be the world's smallest shark. Sharks are carnivorous, feeding on a wide range of prey. Most have several rows of teeth in their jaws. The teeth are constantly shed and replaced, and thousands of teeth are developed in a shark's lifetime. Most sharks are cold-blooded (their temperature matching their environment), but several, including the great white, can generate some heat. Most sharks are marine, occurring in all oceans, but a few can also survive in fresh water. The Greenland shark (*Somniosus microcephalus*) is known to have reached an age of about 400 years (Nielsen et al. 2016), a record for vertebrates.

Harmful Aspects

Sharks well-known for deadly attacks on humans are shown in Figure 16.9. Curtis et al. (2012), provide the following key information. 'Unprovoked attacks by sharks on humans are infrequent, but they can be extremely traumatic events. In general, the risk of shark attack is exceptionally low when compared with other dangers potentially encountered by beachgoers (e.g., drowning, rip currents, surfboard accidents, stingrays, jellyfish, etc.). However, similar to other animal attacks, they draw a disproportionate amount of public and media attention because of their dramatic circumstances. Millions of people engage

FIGURE 16.7 Artist impression of a megalodon pursuing two extinct *Eobalaenoptera* baleen whales. Prepared by Karen Carr (public domain).

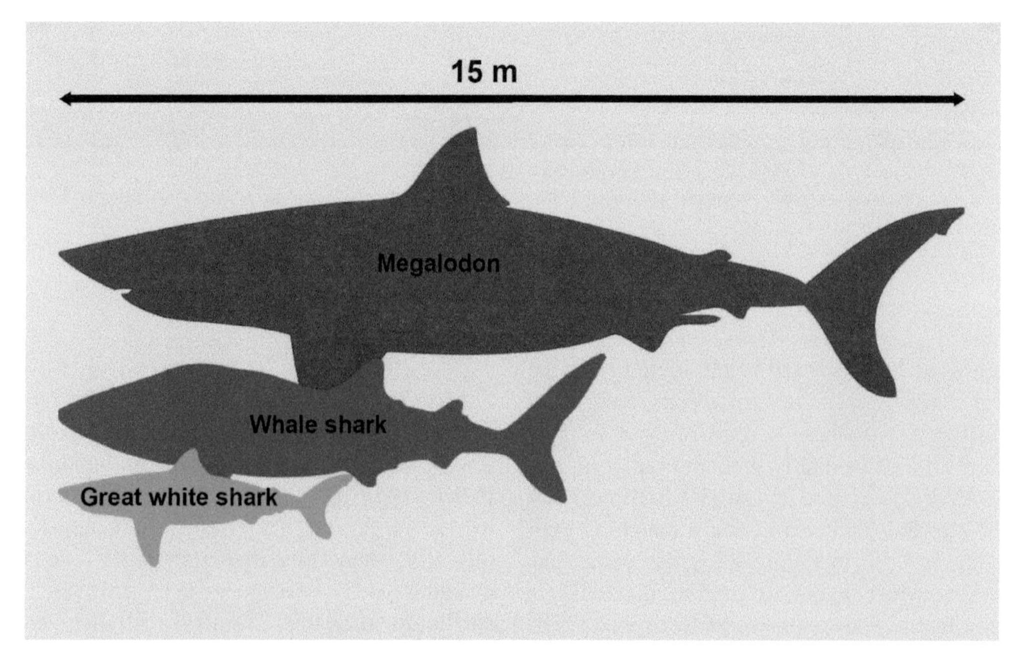

FIGURE 16.8 Comparative sizes of the extinct megalodon, the whale shark, and the great white shark. Prepared by Scarlet 23 (CC BY SA 3.0).

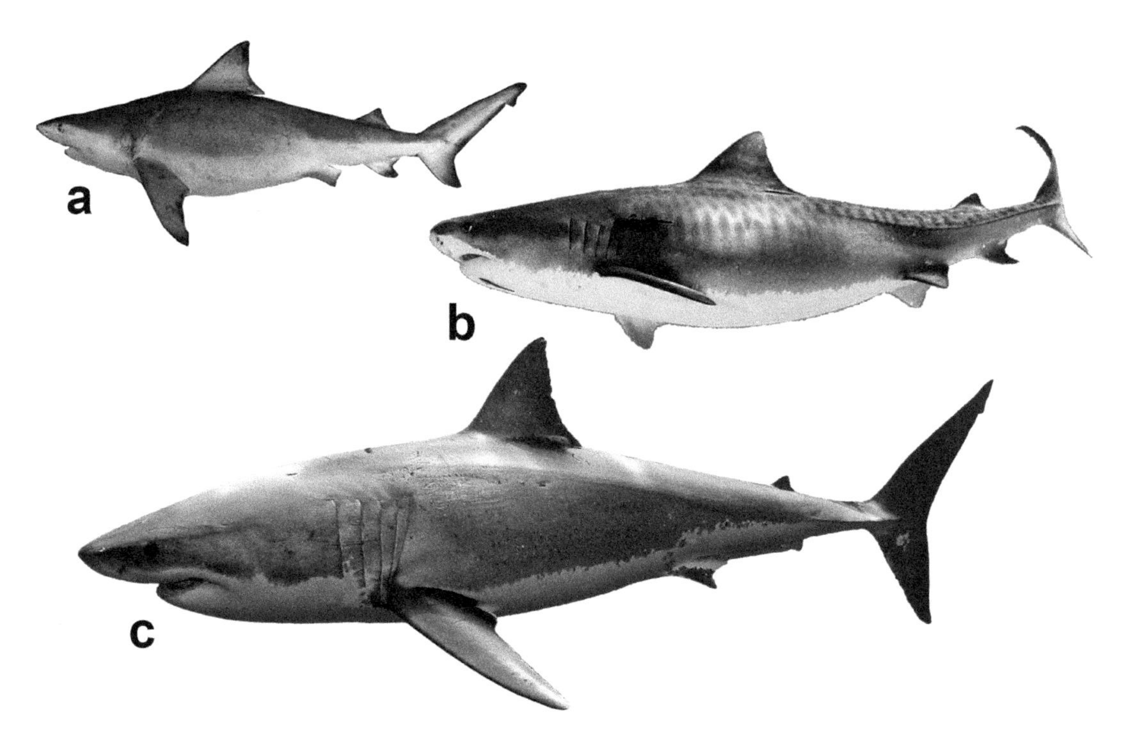

FIGURE 16.9 Leading shark attackers of humans (at relative scale). (a) Bull shark (*Carcharhinus leucas*). Photo (public domain) by Brenda Bowling. (b) Tiger shark (*Galeocerdo cuvier*). Photo by Albert Kok (CC BY SA 3.0). (c) Great white shark (*Carcharodon carcharias*), considered to be the most dangerous of sharks. Photo by Elias Levy (CC BY 2.0).

in swimming, surfing, boating, snorkeling, or scuba diving in the ocean each year, providing billions of dollars in revenues to coastal communities worldwide. Repeated shark attacks within a certain area that result in injuries or deaths are not only extremely traumatic to those involved but can also lead to adverse economic impacts on coastal communities in close proximity to attack locations. This may result in considerable public pressure to take action to reduce the risk of shark attacks in such areas. There are over five hundred shark species in the world's oceans, yet only about thirty species have been documented to attack humans. White sharks have been implicated in a total of three hundred forty-six unprovoked attacks on humans worldwide since 1839, including 102 fatalities. the white shark is most frequently cited as responsible where the identity of the attacking species is ascertained.' The bull shark (*Carcharhinus leucas*) and tiger shark (*Galeocerdo cuvier*) are also believed to be leading attackers of humans (Midway et al. 2019). Six global shark bite 'hotspots' have been identified: the West Coast of the United States, South Africa, Australia, Brazil, Reunion Island, and the Bahamas (Chapman and McPhee 2016). Shark bites of humans are claimed to have been increasing in recent years (see discussion in Midway et al. 2019, who point out that attacks seem to be decreasing in some areas). Increase in attacks has been attributed to several factors, including the rise in human population, habitat destruction, and climate change (Chapman and McPhee 2016). Forty percent of the world's population lives within 100km of a coast, contributing to an increased likelihood of shark-human meetings.

Beneficial Aspects

Like all species, sharks are components of ecosystems in which many species have evolved natural balances. The large sharks that are of particular interest to the public are apex predators, which have been especially influential in shaping and maintaining marine ecosystems (Heupel et al. 2014).

It is highly ironic that sharks are widely regarded as 'man-eaters' (Neff and Hueter 2013), whereas in truth humans are 'shark-eaters.' Sharks (as well as their chondrichthyan relatives) are commercially important for meat (Figure 16.10), and for shark fin soup (Figure 16.11b), a traditional and usually expensive (often over $100.00/bowl) delicacy in some Asian countries, especially China. 'Finning' frequently involves slicing off a shark's fins (Figure 16.11a) and discarding the body at sea, and bans on this barbaric, wasteful practice have been instituted in some regions.

Biodiversity ecotourism – non-harmful observation of wild animals in their natural habitats – is potentially valuable for biological conservation efforts. 'Shark ecotourism' is the commercial practice of bringing tourists as close as possible to sharks in the wild without resulting in harm to either people or sharks (Klimley and Curtis 2006; Gallagher and Hammerschlag 2011). Like whale watching, it can be conducted responsibly, but in reality, there is an element of danger. There is also the possibility of harm to natural ecosystems. Shark tourism has been touted as a means of reducing shark fishing in some regions and promoting shark conservation (Cisneros-Montemayor and Sumaila 2014; Gallagher et al. 2015). Viewing of white sharks underwater, wearing diving gear and from the safety of a

FIGURE 16.10 Harvested shortfin mako sharks (*Isurus oxyrinchus*) on sale in the port of Vigo, Spain. The flesh taste is reminiscent of tuna. Photo by José Antonio Gil Martínez (CC BY 2.0).

FIGURE 16.11 Shark fins for food. (a) Harvested shark fins drying on a rack in Japan. Photo by Takonomakura (CC BY SA 3.0). (b) Shark fin soup. Photo by Cedric Seow (CC BY SA 2.0).

shark cage, has become particularly popular (Figure 16.12). However, the sharks are typically attracted with 'chum' (a mixture of fish body fluids, oils, and macerated tissues), and even sometimes fed by the tourists, and there is concern that this practice conditions the sharks to approach boats and people (Bezeredi 2014).

Sharks have unusually robust physiology (they are resistant to diseases and heal quickly), and research on their biology has the potential to assist in developing medications and treatments for humans. However, claims that shark tissues, especially cartilage, can cure human cancers are unfounded (Posadzki 2011).

FIGURE 16.12 Sharks for ecotourism: People inside shark-proof cage, with sharks swimming outside in Hawaii. Photo by Kalanz (CC BY SA 2.0).

Conservation Aspects

A key website for information on the conservation of sharks is provided by the International Union for the Conservation of Nature Shark Specialist Group: https://www.iucnssg.org/. Sharks include the largest predatory fish in the oceans, and the largest are in turn hunted by very few natural marine predators. However, their natural environment has become a victim of human activities, like so much of marine life (Textbox 16.2). Sharks are threatened by deliberate overfishing (especially for fins), incidental or accidental catches ('by-catch' from becoming tangled in nets or caught on bait hooks not intended for them), pollution, habitat loss, and climate change (Dulvy et al. 2014). They are especially vulnerable because of their slow growth, late sexual maturity, and low reproductive output (Walker 1998). Perhaps 90% of large sharks have been eliminated. As top marine predators, long-lived sharks are significant bioaccumulators of pollutants such as mercury (Camhi et al. 1998). The reduction of sharks has deleterious economic and ecological consequences (Ferretti et al. 2010; Chapman 2017; Textbox 16.3).

Sharks are popular subjects for public and hobby aquaria (Morris et al. 2010; Grassmann et al. 2017; Figure 16.14), but like zoos, keeping animals in captivity may be beneficial or harmful to biodiversity (Lück 2007). Most species are unsuitable for aquaria, and hobbyists need to be cautious about offerings in the trade which may have been acquired from shrinking wild resources.

People are fascinated by large sharks that can eat humans, but this morbid interest is often unhelpful.

TEXTBOX 16.2 HUMANITY'S DESTRUCTION OF ITS MATERNAL AQUATIC ECOSYSTEM

It is a curious situation that the sea, from which life first arose, should now be threatened by the activities of one form of that life.

—Rachel Carson (1951; Figure 16.13)

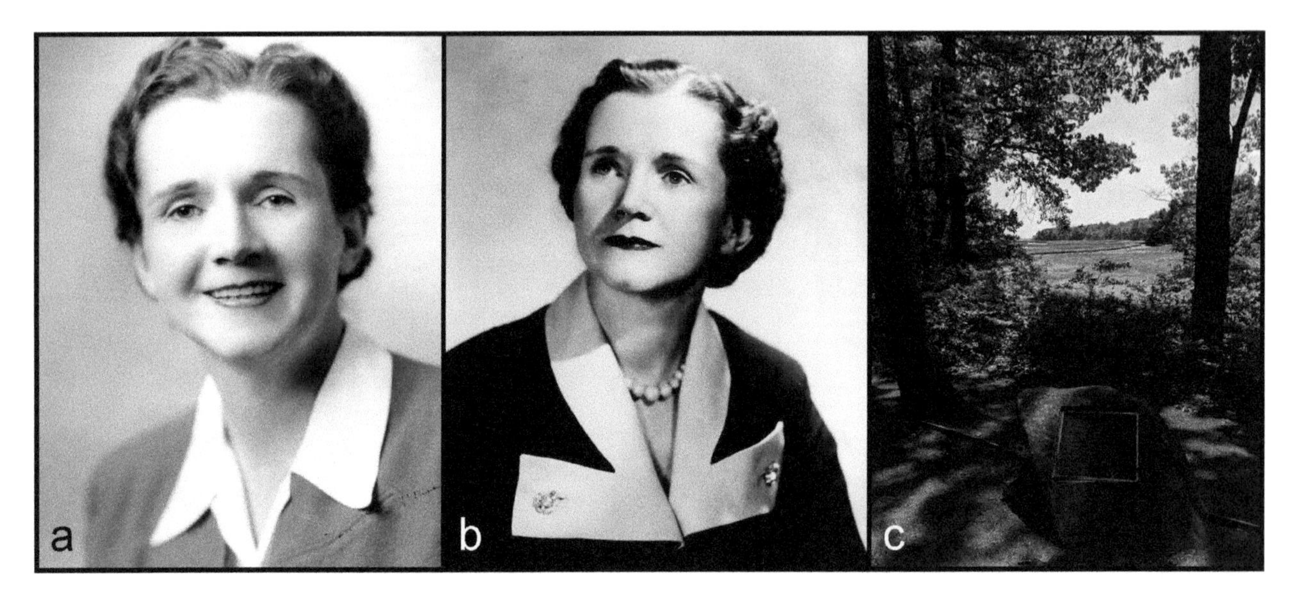

FIGURE 16.13 Rachel Louise Carson (1907–1964), renowned American science writer and conservationist. She was a professional marine biologist who raised concerns about the ecology of the oceans, with such popular books as *The Sea Around Us* (1951). Her award-winning bestseller *Silent Spring* (1962) alerted the public about the dangers of synthetic pesticides for wildlife and human health. (a) 1944 photo (credit: U.S. Fish and Wildlife Service). (b) 1961 photo (credit: Smithsonian Institution). (c) Rachel Carson National Wildlife Refuge, established in 1966 in the state of Maine, where she resided from 1953 until her death. The sanctuary protects more than 2,000ha of salt marshes and estuaries for migratory birds along the Maine coastline. Photo by Harry Cutts (CC BY SA 4.0).

TEXTBOX 16.3 ECONOMIC AND ECOLOGICAL HARM ASSOCIATED WITH DECLINE OF SHARKS

Shark and ray populations in many parts of the world's oceans are in decline. These populations face a variety of threats, most notably from fishing, habitat degradation, pollution, and climate change. These declines are exacerbated by their life history – slow growth, late maturity, and small numbers of young relative to most other aquatic taxa – and as a result, populations have less potential to sustain fishing or to recover from depletion than do most teleost fish or invertebrates. Although no species of shark or ray are known to have become extinct in the wild, several species have been extirpated from large parts of their range, and 67 species are currently listed as Critically Endangered or Endangered on the IUCN Red List... The loss of some shark and ray populations from aquatic ecosystems has socioeconomic and ecological consequences. First, sharks provide a source of protein, as well as a variety of other products (e.g., leather, fins, cartilage, liver oil), that are important to communities in both developing and developed nations. Whereas some fished populations are managed within sustainable limits (e.g., gummy shark, *Mustelus antarcticus*, in southern Australia), most are fished without knowledge of their sustainability or at levels above scientifically recommended limits. The lack of sustainable fishing practices for shark and ray populations will mean that this source of protein will need to be replaced by other sources, most of which are already at or above sustainable limits, or consumption will need to decline. Second, the decline of shark and ray populations has ecosystem consequences. The role of some shark species as top predators exerts top-down effects on ecosystems, and their loss or decline may have important direct and indirect effects on populations that can cascade through marine ecosystems. The loss of sharks may result in substantial changes to ecosystems that affect other organisms and the industries and human communities that rely on them.

—*Simpfendorfer et al. (2011)*

For centuries, sharks have been depicted as extremely dangerous (Figures 16.15 and 16.16). Sensationalistic and often misinformed reports of gory shark attacks make it hard to view sharks sympathetically (Pearce 2015; Textbox 16.4), and to motivate the public to support conservation of sharks (Friedrich et al. 2014). The expression 'loan shark' reflects the widespread villainous and heartless image of sharks. Similarly, a 'pool shark' is a skilful billiards player who habitually takes advantage of competitors. Regrettably for sharks, they are almost never pictured sympathetically as in Figure 16.17. The 1975 movie *Jaws* particularly reinforced fear of sharks, and indeed

FIGURE 16.14 Whale shark (*Rhincodon typus*) in the Georgia (U.S.) Aquarium. Photo by Zac Wolf (CC BY SA 2.5).

FIGURE 16.15 A typical historical depiction of sharks as extremely dangerous. 'Watson and the Shark,' an oil painting (public domain image) by the American painter John Singleton Copley (1738–1815), showing a shark attack.

FIGURE 16.16 Cartoon showing widespread conception that sharks are dangerous bloodthirsty predators of humans. Prepared by B. Brookes.

TEXTBOX 16.4 LACK OF ACCURATE AND RELEVANT KNOWLEDGE HARMS SHARK CONSERVATION

With the cultural shift toward environmentalism, public opinion about sharks has been changing. Popular culture now emphasizes that these animals are not a marine menace but are instead stewards of the environment that play a critical role at the apex of marine food chains. The public is told in plaintive tones that sharks are misunderstood and that there is a need to study them because fisheries are depleting their numbers to such an extent that there will likely be dire consequences for marine ecosystems. As a result, a wave of instant 'shark researchers' has sprung up all over the world. Most seem to be associated with tourist diving operations using protective cages. These businesses often cater to a sense of adventure while offering clients the impression that they are contributing to a scientific understanding of these animals. Unfortunately, despite considerable press coverage and shifting public awareness, this has done little to promote much needed basic research.

—Naylor and Aschliman (2013)

Gaining support for shark conservation has been extremely difficult due to the negative preconceived notions the general public holds toward sharks... The media... can play a significant role in promoting conservation, but unfortunately media coverage of sharks has been controversial recently with the airing of several non-factual, fake documentaries. To promote shark conservation the media's message has to be unbiased, non-sensationalized, and accurate to ensure people are receiving the information necessary to build strong pro-shark conservation behaviors.

—O'Bryhim and Parsons (2015)

FIGURE 16.17 An unusual, sympathetic, humorous, anthropomorphic depiction of a shark befriending a sailor. Source: Holder and Jordan 1909. Photo credit: Freshwater and Marine Image Bank, University of Washington Libraries.

FIGURE 16.18 A display in favor of shark conservation. Photo by Socheid (CC BY SA 4.0).

stimulated sports fishermen to deliberately eliminate them (Francis 2012). Nevertheless, the massive star of the film, the great white, hardly represents the majority of sharks, which are no threat whatsoever to people. The widespread image of sharks as violent, bloodthirsty, unfeeling 'eating machines' has led to their persecution and senseless slaughter. 'Despite its relative rarity, shark attack is a cultural phenomenon that draws intense public interest in the popular media with myths and misconceptions routinely perpetuated on television, in magazines and newspapers, and in the social media' (Midway et al. 2019). Notable in this regard is 'Shark Week,' an annual, TV block (often week-long) which began in 1988 at the Discovery TV Channel, much criticized for sacrificing scientific accuracy for sensationalistic aspects related to sharks. Similarly, the 'Sharknado' TV series, starting in 2013, has dealt with waterspouts that lift sharks out of the ocean and deposit them on people with terrifying consequences. In fact, sharks are an insignificant hazard to humans: worldwide, about a dozen people are killed by sharks every year, while millions of sharks are killed by humans annually. Culling of sharks, often organized by governments, has been found to be ineffective, but in areas where sharks are known to pose risks, beach barriers and management techniques do reduce shark-human interactions (Curtis et al. 2012).

While there is conservation concern for sharks (Figure 16.18), to date international agreements to prevent their overfishing have been very limited (Shiffman and Hammerschlag 2016). Many sharks are threatened, but for the most part, researchers are concentrating on relatively common charismatic sharks and not the smaller, less interesting, but endangered sharks, which are much rarer (Momigliana and Hartcourt 2014). The three most dangerous sharks are in fact relatively protected in legislation: the white shark is classified by the IUCN (International Union for Conservation of Nature) Red List as 'vulnerable,' while bull and tiger sharks as 'Near Threatened.'

FROGS AND TOADS

Introduction

There are approximately 7,000 living species in the vertebrate class Amphibia (the amphibians), which in addition to frogs and toads includes salamanders, newts, mudpuppies, and other groups (Pyron and Wiens 2011). The Chinese giant salamander (*Andrias davidianus*), the world's largest amphibian, sometimes reaches 2 m in length (Turvey et al. 2019). Frogs inhabit all continents except Antarctica, and while a few reach the Arctic Circle, the greatest diversity occurs in the Tropics. They occupy most terrestrial and aquatic habitats but avoid salt water. Frogs (and toads) mostly lay their eggs in water, and the eggs develop into swimming tadpoles and then transform into frogs (or toads), which may spend their adulthood mostly in water, on land, or in both (Figure 16.19). The classification of frogs and toads is unsettled. There are about 6200 species that are called frogs and toads (Vitt and Caldwell 2013). Commonly, the 'true frogs' are placed in the family Ranidae. There are over 600 species of frogs found worldwide that are in this family. Other families that are often recognized include Bufonidae (the true toads), Ascaphidae (the tailed frogs), Pelobatidae (the spade foots), and Hylidae (the tree frogs) (Figure 16.20).

The goliath frog (*Conraua goliath*) from Africa is the largest living frog, growing up to 33 cm in length and weighing up to 3.25 kg. The extinct 'devil frog' (*Beelzebufo ampinga*) form the Late Cretaceous period of Madagascar may be the largest frog that ever lived. It grew to 41 cm in length and weighed about 4.5 kg. *Paedophryne amauensis* from Papua New Guinea may be less than 8 mm in length and is considered the world's smallest known vertebrate (Rittmeyer et al. 2012).

The distinction between what should be included in 'frogs' in its most comprehensive sense and 'toads' in its comprehensive sense is debatable. In practice, frogs have smooth, moist skin (which may appear slimy), a narrow body with round bulging eyes, and long hind legs designed for lengthy, high jumps. Frogs usually live near and are often found in water, and have many predators. By contrast, toads have rough, dry, bumpy skin, a wide body with oval eyes which do not bulge as much as the eyes of frogs, and short hind legs designed for small hops rather than jumps. Frogs usually have teeth, toads usually do not. Toads can live in dry situations away from water, and their bitter-tasting skin deters predators. These differences do not hold for all species called frogs and toads. Moreover, the word 'frog' is often used in a comprehensive sense to refer to both frogs and toads. Regardless of technical classifications, frogs and toads share an unmistakable appearance (Figure 16.21), characterized by short, tailless bodies, broad, flat heads, big mouths, and long, muscular hind legs.

Harmful Aspects

Toxins on the outer parts of frogs often protect them from predators. The skin of many frogs (indeed, of numerous amphibians) may contain tiny glands that secrete a variety of toxic chemicals (Daly 1995). Many toads and some frogs also have large glands on the side of their heads behind the eyes that manufacture defensive toxins which are secreted externally. Various terms have been applied to these glands (Tyler et al. 2001). They are best called parotoid glands (or alternatively, paratoid glands). However, they have also frequently been termed parotid glands (even in technical zoological literature), which is confusing because the latter phrase is also applied to a different kind of gland that occurs in mammals. In mammals, parotid glands excrete saliva within the mouth.

Many frogs accumulate toxic alkaloids from the animals they consume, and some are able to synthesize their own toxins (Saporito et al. 2012; Santos et al. 2016). The toxins of 'poison dart frogs' are employed in South America to produce poison that is added to darts (Figure 16.22), and such species are so poisonous that they merit avoidance. Some are so toxic

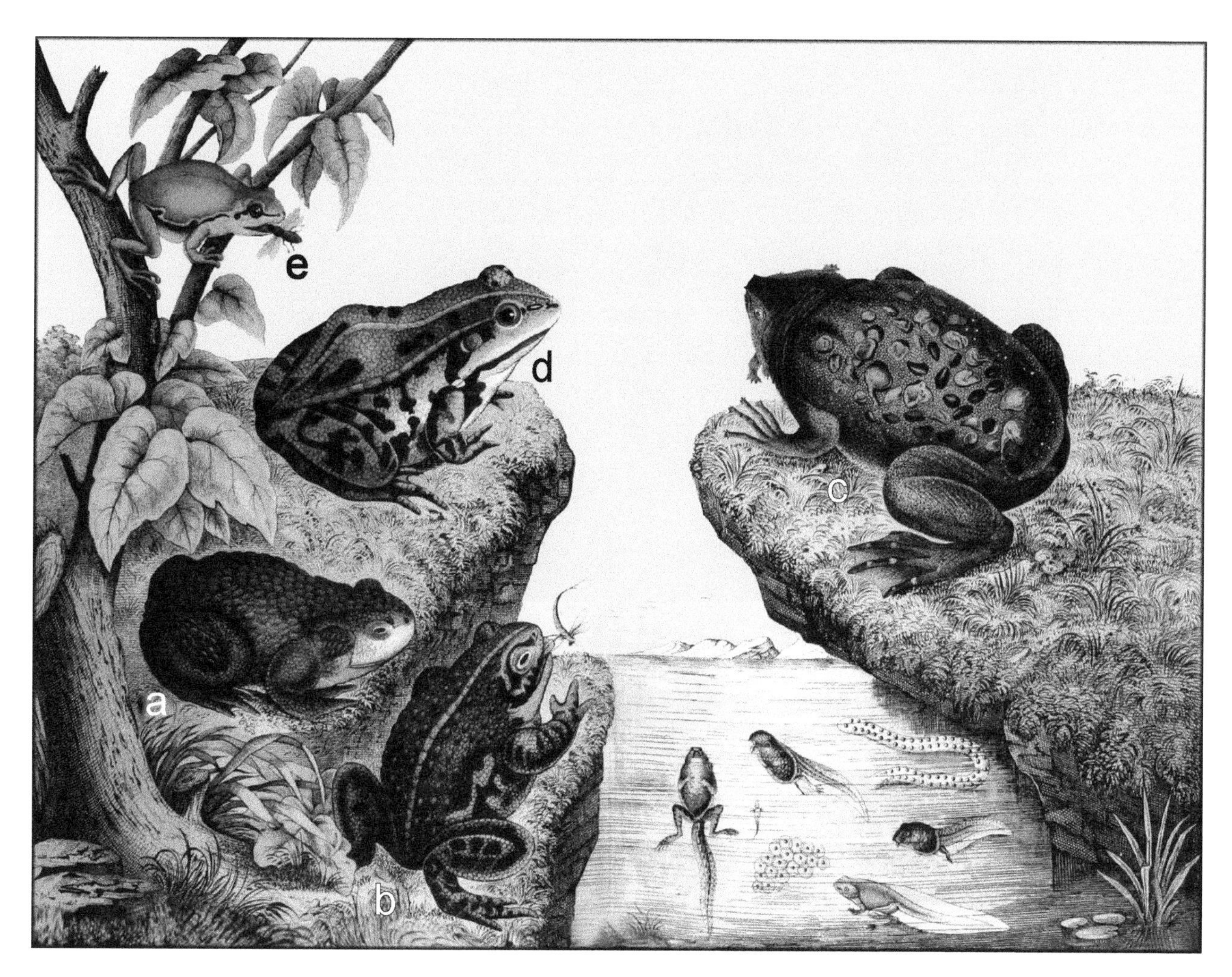

FIGURE 16.19 Plate from a public domain 19th-century encyclopedia (Kirby and Schubert 1889) showing frogs and toads in their natural habitat. a: Common toad (*Bufo bufo*); b: Natterjack toad (*Epidalea calamita*); c: Surinam toad (*Pipa pipa*); d: Edible frog (*Pelophylax esculentus*); e: Green tree-frog (*Hyla arborea*).

FIGURE 16.20 The world's largest and smallest frogs. (a) The extinct 'devil frog' (*Beelzebufo ampinga*) of Madagascar, thought to be the largest frog that ever lived. Prepared by Nobu Tamura (CC BY 3.0). (b) *Paedophryne amauensis* from Papua New Guinea, considered the world's smallest vertebrate. Source: Shutterstock, contributor: Alex_Gor.

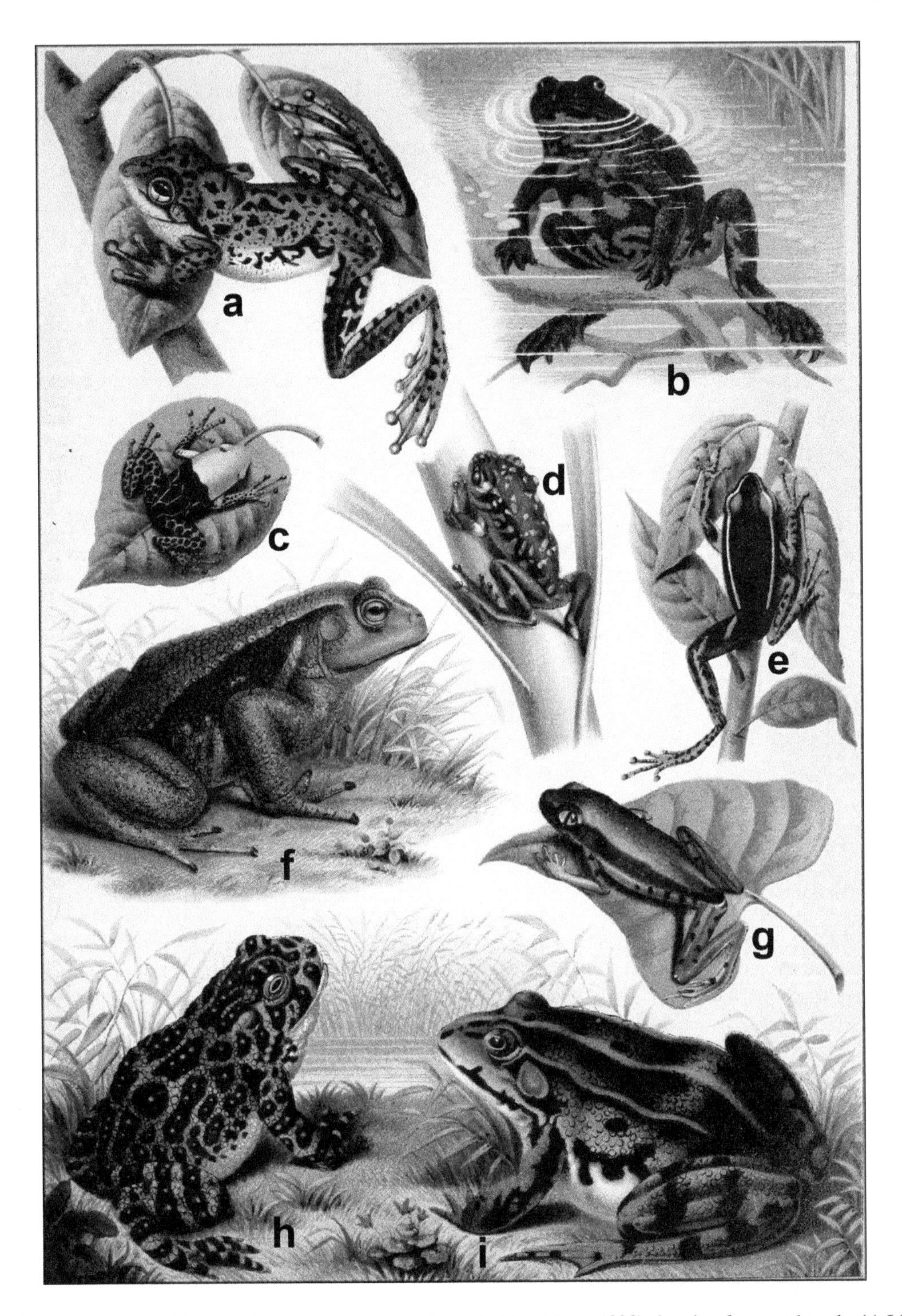

FIGURE 16.21 Plate from a public domain 19th-century encyclopedia (Brockhaus 1892) showing frogs and toads. (a) *Litoria peronei* (Australia). (b) *Bombina bombina* (Europe). (c) *Ranitomeya fantastica* (Peru). (d) *Leptopelis flavomaculatus* (Africa). (e) *Allobates femoralis* (South America). (f) *Bufo japonicus* (Japan). (g) *Phyllomedusa hypochondrialis* (South America). (h) *Bufo balearicus* (Italy). (i) *Pelophylax lessonae* (Europe).

that one drop of their skin secretions can kill an adult human. Many toxic frogs are brilliantly colored, to warn away predators (Figure 16.23). Some frog secretions are hallucinogenic and are occasionally unwisely employed by people to get high.

The cane toad (*Rhinella marina*; Figure 16.24) is the world's largest 'toad.' This native of South and mainland Central America has been introduced to islands throughout Oceania and the Caribbean, and Northern Australia. Imported in some places to control agricultural pests (especially in sugarcane, hence 'cane' in the name), it has become an invasive pest and a serious threat to native species (Tingley et al. 2017). The skin of cane toads is very

FIGURE 16.22 Amazonian Indians with blowguns employed to shoot darts coated with poison frog secretions. (a) Yagua in Peru. Photo by Chany Crystal (CC BY ND 2.0). (b) Native Columbians. Photo by Yves Picq (CC BY SA 3.0).

FIGURE 16.23 Poison dart frogs. (a) Strawberry poison frog (*Oophaga pumilio*). Photo by Marshal Hedin (CC BY 2.0). (b) Amazonian poison dart frog (*Ranitomeya amazonica*). Photo by V2 (CC BY SA 3.0). (c) Blue poison dart frog (*Dendrobates azureus*). Photo by Michael Gäbler (CC BY 3.0). (d) Doris' poison frog (*Ranitomeya dorisswansonae*). Photo by Mauricio Rivera Correa (CC BY SA 2.5). (e) Green and black poison dart frog (*Dendrobates auratus*). Photo by Adrian Pingstone (released into the public domain). (f) Yellow-banded poison dart frog (*Dendrobates leucomelas*). Photo by H. Krisp (CC BY 3.0).

toxic, and dogs attacking them have died. Cane toads are reviled in Australia where they seem to be unstoppable despite merciless programs of extermination (Shine 2018).

Beneficial Aspects

Frogs and toads in their natural ecosystems play important roles as predators and prey. They consume agricultural pests and disease carriers and are an important food source for birds, snakes, and other animals. Vast numbers of insects are destroyed, some species known to eliminate thousands of mosquitoes in a single night. Tadpoles are usually aquatic, and those consuming algae assist in keeping waterways clean. In parts of the world, frogs are eaten and provide a significant source of protein. Where there are very large supplies of frogs or toads, their skins are sometimes harvested for leather. Frogs are also employed in traditional medicine to treat many conditions. Frogs

FIGURE 16.24 Cane toad (*Rhinella marina*). (a) Photo by Dropthepuck88 (CC BY SA 3.0). (b) Photo by Sam Fraser-Smith (CC BY SA 2.0).

are also used to a degree in Western medicine. They are common subjects of laboratory experimentation, including manipulations by schoolchildren (a subject that raises ethical issues, and usage is decreasing due to animal welfare concerns). The African clawed frog (*Xenopus laevis*) was extensively employed in the past in pregnancy testing (urine from a pregnant woman injected into a female frog causes it to lay eggs). The skin of amphibians often contains defensive biological compounds, which have potential to contribute to pharmaceuticals against infective microorganisms. Sometimes frogs are kept as pets, but it should be understood that amphibians often carry salmonella bacteria, so handling them could produce intestinal discomfort.

Conservation Aspects

Amphibians worldwide are facing numerous threats to their existence (Heatwole and Wilkinson 2012; Tapley et al. 2015; Textbox 16.5; Figure 16.25). The Amphibia are proportionally the most threatened group of vertebrates. A very high percentage of frogs and toads – 30% – is classified as 'severely threatened' (Morrison et al. 2012; IUCN 2017); by comparison, 22% of mammals and 13% of birds have this status. About 150 species of frogs are believed to have gone extinct in recent history (Vitt and Caldwell 2013). The skin of frogs and toads is thin and permeable to water ('semi-permeable' to infiltration of many compounds in water), facilitating entry of harmful pollutants. Many wild populations are in serious decline due to harmful chemicals in their environment. Indeed, because of their sensitivity to pollution, frogs have been likened to the canary in the coal mine that indicates environmental danger. Frogs are considered by ecologists to be a particularly good indicator of the health of ecosystems, and the current rapid reduction in their numbers coupled with the frequent development of physical malformations (such as extra legs) is viewed with alarm. Many breeding ponds are being

TEXTBOX 16.5 THE THREAT TO SURVIVAL OF FROGS AND TOADS

Amphibians appear particularly vulnerable to global change as the world enters its sixth mass extinction event. Because of their roles as important prey and predators, susceptibility to water-soluble toxins through permeable skin, and a life history that straddles aquatic and terrestrial environments, amphibians are good indicators for environmental degradation and community stability in the face of the major drivers of species loss. Although the causes of amphibian declines are diverse and interactive, individual mechanisms include habitat loss, environmental contamination, global climate change, disease and pathogens, spread of invasive species, and overharvesting.

—*Warkentin et al. (2009)*

FIGURE 16.25 Stubfoot toads (*Atelopus* of Central and South America). These are highly diminished by habitat loss, pollution, introduced species, and disease. Source (public domain): Franz et al. (1918).

filled in because of housing development or polluted by agricultural pesticides. Large numbers are killed on roads, especially toads. Some species are threatened by collection from nature for the trade in living animals. The introduction of more competitive foreign species is another factor, and so is climate change. As if amphibians didn't have enough problems, a deadly parasitic chytrid (*Batrachochytrium dendrobatidis*; chytrids are usually interpreted as fungi) is rapidly eradicating amphibians throughout the world (Skerratt et al. 2007; Bacigalupe et al. 2017).

Frog-legs (frogs' legs) are a prominent dish in many national cuisines, particularly in France and China (Figure 16.26). Although frogs are farmed in China, Vietnam, and Taiwan, the vast majority that end up on a plate are harvested from the wild. Many countries have severely restricted local collection, but huge quantities are exported from Asia, mostly from Indonesia. Currently, about a billion frogs are exported annually to the U.S., about twice as many to Europe (Warner 2011). Many environmentalists are opposed to consumption of frog-legs in view of the general decline of amphibians (Warkentin et al. 2009). While it has been assumed that a very small number of species are being harvested from nature, it appears that in fact a broad range of frogs are being taken from the wild (Ohler and Nicolas 2017). The mountain chicken (*Leptodactylus fallax*), a native of the Caribbean islands of Dominica and Montserrat, has become critically endangered because of overcollection for its legs which are said to taste like chicken (Schuessler 2016).

Attitudes by the public toward frogs and toads are quite polarized, and this is critical in determining support for conservation. Many people think frogs and toads are just ugly and disgusting. A plague of frogs was one of God's punishments in the Old Testament (Figure 16.27). On the whole, historical attitudes to frogs and toads have been rather mixed, many fairy tales depicting amphibians favorably, others not. According to a particularly common myth, touching warty toads (or any frog) gives one warts. 'The Frog Prince' fairy tale (especially a version by the Brothers Grimm) involves a Princess kissing a frog which causes it to transform into a handsome prince (Figure 16.28). Fortunately, many people find frogs to be cute, and there are many illustrations of frogs and toads appearing to behave like humans (Figure 16.29).

On the other hand, some people are pathologically afraid of frogs and toads. Phobic fear of frogs, toads, or other amphibians is called batrachophobia or ranidaphobia (the latter term sometimes restricted just to frogs). Extreme fear of toads has been termed bufonophobia. Fear of frogs is basically irrational, but it is real, and people with this condition need sympathetic, professional guidance.

There are efforts being made to save frogs and toads. Many schools now teach that these, as well as snakes, deserve to be protected. Almost 7% of all amphibians are represented in zoos, where they receive a measure of protection against extinction (Murphy and Gratwicke 2017). There are also international efforts to limit the trade in frogs collected from the wild (Figure 16.30).

FIGURE 16.26 A dish of frog's legs. Photo by Anagoria (CC BY 3.0).

FIGURE 16.27 Depictions of the second of the ten biblical (Exodus 8:1–4) plagues of ancient Egypt, a plague of frogs. (a) A serious illustration: An 18th-century etching. Source: Wellcome Foundation (CC BY 4.0. (b) A humorous illustration: Source: Shutterstock, contributor: Askib.

FIGURE 16.28 'The Frog Prince,' public domain painting by Paul Meyerheim (1842–1915).

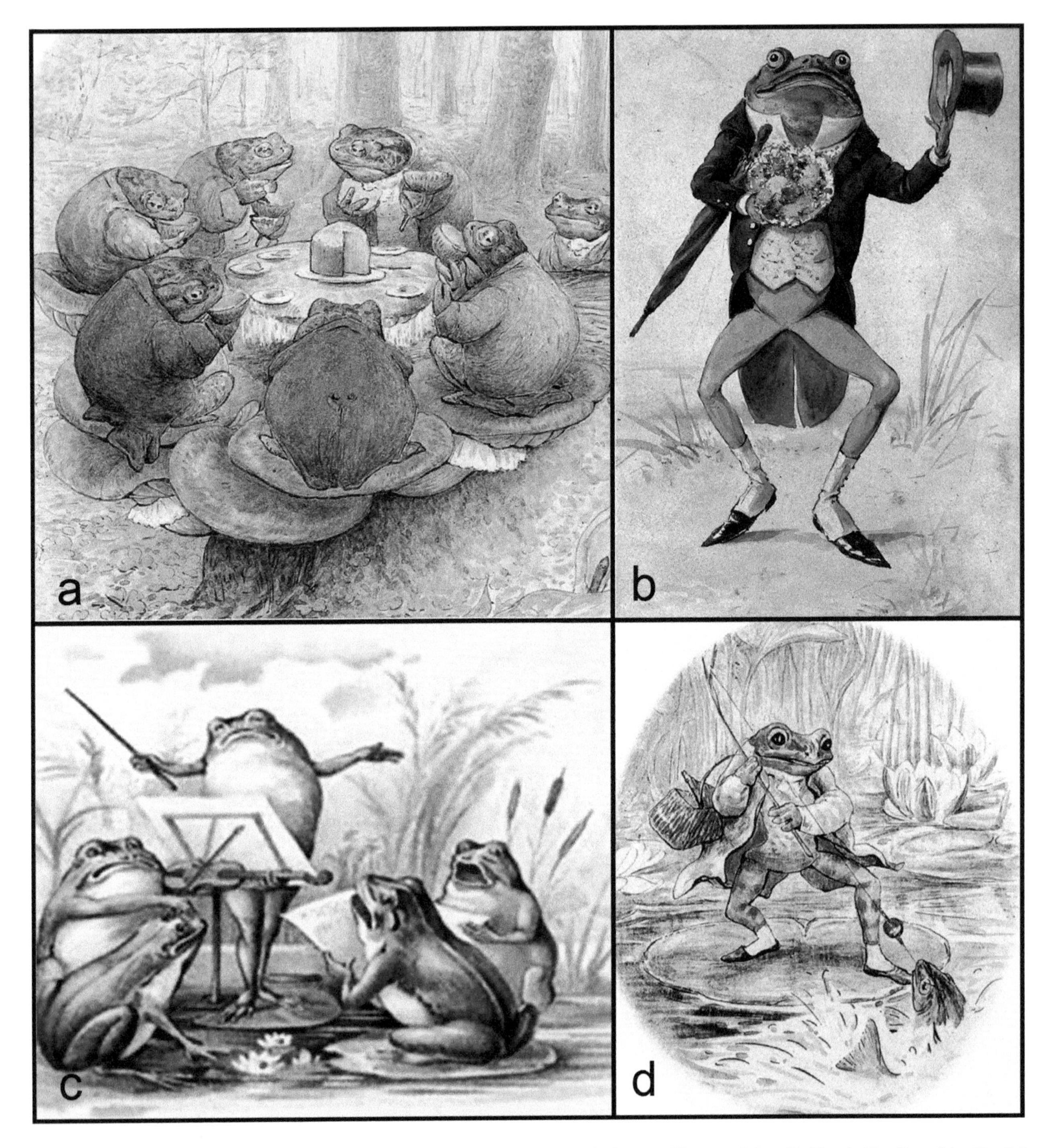

FIGURE 16.29 Anthropomorphic depictions of frogs and toads – shown in human-like activities. (a) 'Toads Tea-Party' by English author, illustrator, mycologist, and conservationist Helen Beatrix Potter (1866–1943; her published works entered the public domain in 2014). (b) A toad in morning dress, holding an umbrella and a bunch of flowers. Drawing by G. Hope Tait, ca. 1900. Credit: Wellcome Collection (CC BY 4.0). (c) 'Chorus of frogs,' a vintage postcard (public domain) from the early 20th century. (d) Frog fishing, from Beatrix Potter's 'The Tale of Mr. Jeremy Fisher.'

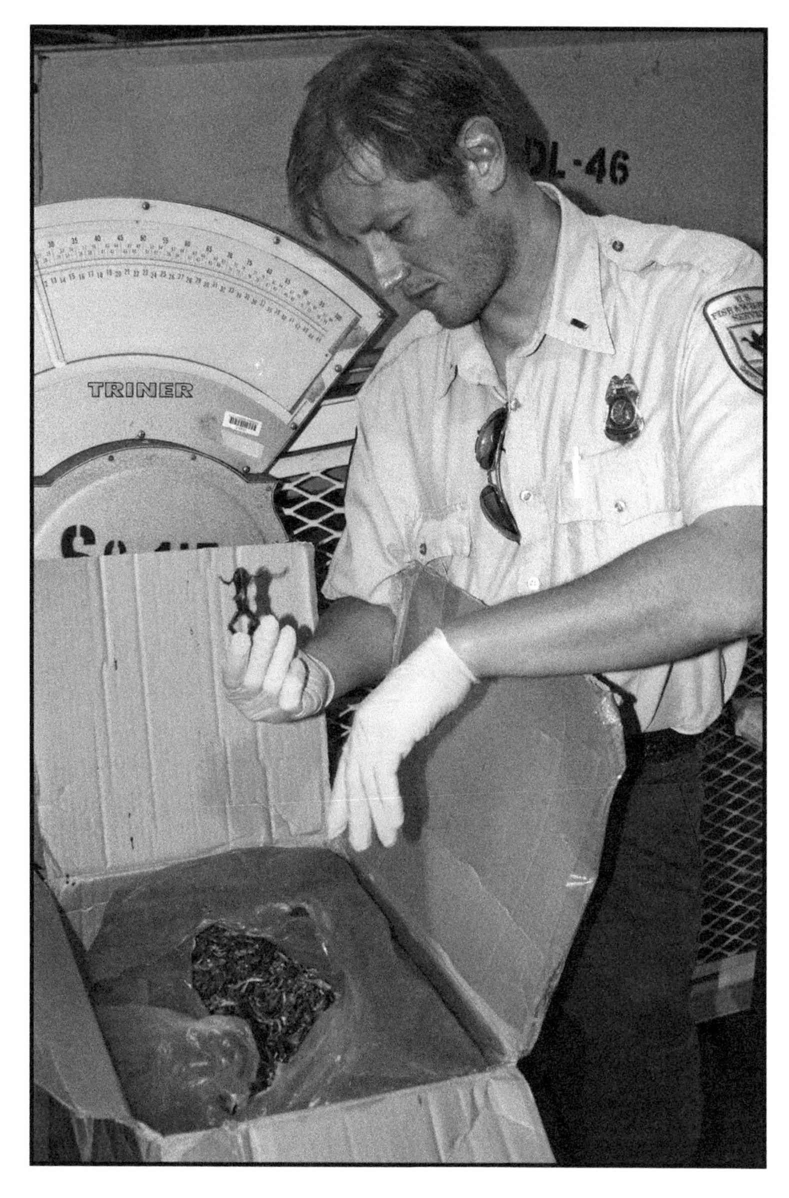

FIGURE 16.30 United States border inspector checking dried frog shipment for adherence to regulations. Photo (public domain) credit: Bill Butcher, United States Fish and Wildlife Service.

SNAKES

Introduction

The vertebrate class Reptilia, with about 10,000 species, is classified in different ways by different authorities. In addition to snakes, other creatures accepted as reptiles include turtles, lizards, crocodiles, and others (Pincheira-Donoso et al. 2013). Some authorities include birds, which in fact are known to be derived from reptiles (Modesto and Anderson 2004). Snakes are assigned to suborder Serpentes (the word serpent is synonymous with snake). There are about 3600 species of snakes living today. Snakes occur on every continent except Antarctica but are absent from some islands, notably Greenland, Iceland, Ireland, and New Zealand (although sea snakes often visit coastal areas). Most snakes are terrestrial, but about 70 marine species, which are extremely venomous although docile, live in the Indian and Pacific oceans. So-called thread snakes – blind, burrowing, worm-like animals – are as short as 10 cm in length. The longest snake is the reticulated python growing to about 7 m in length (the extinct *Titanoboa cerrejonensis* possibly grew to 14 m; Figure 16.31). The green anaconda is not the longest, growing to over 5 m, but is believed to be the heaviest of living snakes, approaching 100 kg. Snakes are 'cold-blooded' (poikilothermic), i.e., unable to generate their own heat metabolically, and in freezing climates they overwinter in a dormant condition in sheltered, above-freezing situations, sometimes denning communally with hundreds, even thousands of other individuals. Snakes are scaly (covered by overlapping scales) but, contrary to frequent belief, are not slimy. Snakes are of considerable interest to science, and there are many monographs (e.g. Mehrtens 1987; Mattison 2007; Lilywhite 2014; O'Shea 2018). Many snakes are quite beautiful (Figure 16.32).

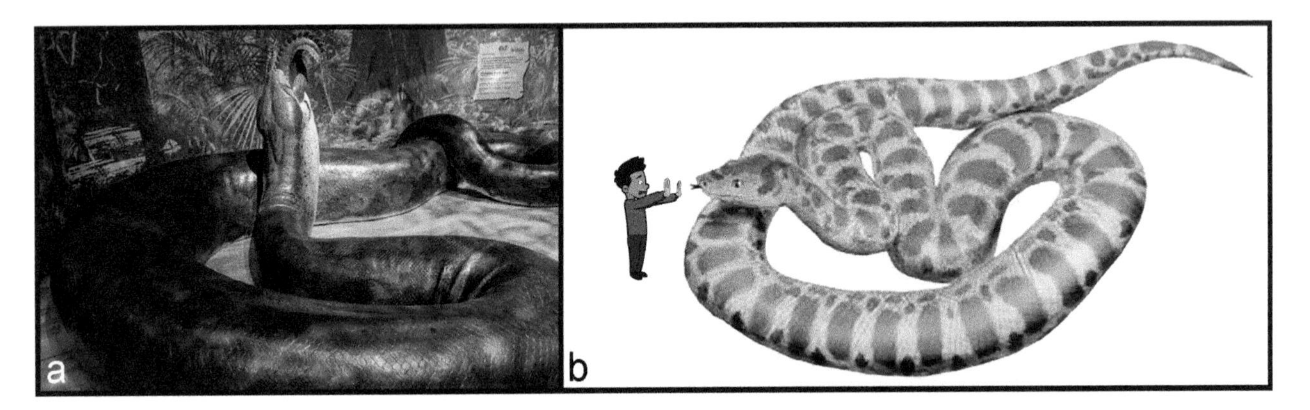

FIGURE 16.31 *Titanoboa cerrejonensis*, an extinct snake of northeastern Colombia. The largest snake that ever existed, it may have exceeded 14 m in length and weighed over 1 t. It lived during the Middle to Late Paleocene epoch (about 60 to 58 million years ago) following the extinction of the dinosaurs. (a) Exhibit of the snake swallowing a crocodile in the Florida Museum of Natural History, Gainesville. Photo by Kelly Verdeck (CC BY ND 2.0). (b) Prepared by Nobu Tamura (CC BY 3.0), model human added by B. Brookes.

Harmful Aspects

Some biologists arbitrarily state that no snake is 'poisonous' when a poison is defined as a substance that is ingested or inhaled – which excludes injection. In any case, the venom (which is based on modified saliva) is usually a combination of many chemicals and can be quite toxic. The compounds present may include proteolytic enzymes (which break down proteins), myotoxins (which destroy muscle tissue), neurotoxins (which block or destroy nervous system tissues), and hemorrhagic toxins (which attack the circulatory system and prevent blood clotting). The principal function of venom for snakes is prey subjugation, but venom also serves as a deterrent to predators (Jackson and Fry 2016; Jackson et al. 2019). Most snakes are nonvenomous (at least, they do not have significantly toxic saliva), but about 725 species are venomous, and of these about 250 can kill a human with a single bite. The internet abounds with lists allegedly representing the world's most venomous or dangerous snakes, but there is considerable disagreement about how to make such judgments (Silva 2013; Walls 2013), and snake identification is sometimes erroneous, leading to false records. Examples of particularly poisonous snakes are shown in Figure 16.33. Some snakes, such as many of the sea snakes (Figure 16.34), have extremely toxic venom – i.e., very little is required to kill – but they very rarely attack humans. Aggressive venomous snakes that are common and likely to encounter people are responsible for most snakebites. Walking barefoot in snake-infested regions is frequent in developing countries, particularly in parts of Asia and Africa, where casualties are highest. Due to lack of treatment facilities coupled with occurrence of venomous snakes, it has been estimated that about 93 million people are at greatest risk, including many in sub-Saharan countries, Indonesia, and other parts of southeast Asia (Longbottom et al. 2018). 'Snakes with major clinical importance belong to the families Elapidae (African and Asian cobras, Asian kraits, African mambas, American coral snakes, Australian and New Guinean venomous snakes, and sea snakes) and Viperidae (Old World vipers, American rattlesnakes and pit vipers, and Asian pit vipers)' (Gómez-Betancur et al. 2019). A detailed geographical guide to the principal venomous snakes is found in World Health Organization (2016). Approximately 100,000 people die annually from snake bites, and every year, 400,000 suffer permanent disabilities such as amputations and permanent neurological damage (Williams et al. 2019a; Textbox 16.6). Antivenom reduces mortality from venomous snakes by 90% but is frequently unavailable in poor countries, accounting for the large number of deaths. By comparison, hippos, the deadliest land mammal, kill an estimated 500 people annually. Shark bites attract far more attention but claim as little as five deaths worldwide yearly. Snakes are natural predators of wild animals, but they also sometimes victimize livestock and pets. Occasionally, very large nonvenomous snakes also kill humans and domesticated animals.

Some introduced snakes are causing considerable harm to natural ecosystems (Kraus 2015). The most widely cited example is the brown snake introduced to Guam, where it climbs trees and decimates native birds not expecting

TEXTBOX 16.6 SNAKEBITE HARM OCCURS MOSTLY IN POOR, TROPICAL COUNTRIES

Snakebite envenoming (SBE) affects as many as 2.7 million people every year, most of whom live in some of the world's most remote, poorly developed, and politically marginalized tropical communities. With annual mortality of 81,000 to 138,000 and 400,000 surviving victims suffering permanent physical and psychological disabilities, SBE is a disease in urgent need of attention. Like many diseases of poverty, SBE has failed to attract requisite public health policy inclusion and investment for driving sustainable efforts to reduce the medical and societal burden.

—*Williams et al. (2019a)*

FIGURE 16.32 Beautiful snakes. (a) Blue striped garter snake (*Thamnophis sirtalis similis*), slightly venomous, a Florida subspecies of the common garter snake. Photo by Glenn Bartolotti (CC BY SA 4.0). (b) Texas coral snake (*Micrurus tener*), venomous, endemic to the southern United States and northeastern and central Mexico. Photo by L. A. Dawson (CC BY SA 2.5). (c) Rainbow snake (*Farancia erytrogramma*), nonvenomous, endemic to coastal plains of the southeastern United States. Photo by Charles Baker (CC BY SA 4.0). (d) Texas long-nosed snake (*Rhinocheilus lecontei tessellatus*), nonvenomous, endemic to the western United States and northern Mexico Photo by L. A. Dawson (CC BY SA 2.5). (e) Coast garter snake (*Thamnophis elegans terrestris*), slightly venomous, one of numerous subspecies of this species of North American garter snakes. Photo by Steve Jurvetson (CC BY 2.0). (f) Mud snake (*Farancia abacura*), nonvenomous, endemic to the southeastern United States. Photo by U.S. Fish and Wildlife Service (public domain). (g) Southern ringneck snake (*Diadophis punctatus*), slightly venomous, southern phase of a species found in much of the U.S., central Mexico, and southeastern Canada. Photo by Glenn Bartolotti (CC BY SA 3.0).

it (Textbox 16.7). The Burmese python (*Python bivitattus* or *Python molurus bivittatus*) is currently invading the Florida Everglades with widespread reductions of native animals (Dorcas et al. 2017; Willson 2017; Figure 16.35).

Beneficial Aspects

Most snakes are beneficial because they control rodent and other pests. Without snakes, increase in rodent populations would significantly result in property damage, spread of infectious diseases, and destruction of agricultural products. Species such as garter snakes also feed on typical garden pests such as slugs and snails, and some snakes consume insects like grasshoppers and crickets that damage crops. Snakes are prey for larger animals such as hawks, owls, and herons. Snakes are a key component contributing to the maintenance of ecosystems.

Snakes are exploited in several ways by humans, especially as food, pets, and (notably in Afro-Brazilian religions) in traditional medicine and magic/religious ritual (Alves

TEXTBOX 16.7 INVASIVE SNAKES CONTRIBUTE TO BIODIVERSITY LOSS

Invasive species are considered to be among the major causes of biodiversity loss… Increasing urbanization creates a desire for contact with nature for people living in towns and cities, and keeping pets is one way of fulfilling this need. The popularity of reptiles as pets has been growing steadily since the second half of the 20th century. Unfortunately, reptiles can have enormous negative ecological impacts, e.g., invasion of the brown tree snake (*Boiga irregularis*) on Guam island caused the extinction of 77% of the island's native birds and 75% of its native lizards. Also, other direct impacts on humans (e.g., (venomous) snakes or power outages caused by these snakes).

—*Kopecký et al. (2019)*

and Filho 2007). Snake skin, especially from large snakes such as pythons, is employed to make a premium leather (Figure 16.36). Snake venom is under investigation as a potential source of pharmaceuticals (Estevão-Costa et al. 2018; Munawar et al. 2018). The venom of particularly toxic snakes is harvested in order to produce anti-venoms to treat snakebite (Figure 16.37). Antivenom is produced by injecting small amounts of venom into a domestic animal, and after the animal's blood has produced antibodies, collecting and purifying them.

Aust et al. (2017) reviewed the status of snake farming in China and Vietnam, where the reptiles are raised for meat, hides, and medicine (Figure 16.38). China and Viet Nam are believed to be the largest and most important producers of, and markets for, snake meat. Aust et al. (2017) estimated that there are 4,000 farms producing several million snakes of at least 15 taxa, including the monocled cobra (*Naja kaouthia*), the Chinese cobra (*Naja atra*), the oriental rat snake (*Ptyas mucosus*), and the king cobra (*Ophiophagus hannah*), which are raised especially for meat. Aust et al. (2017) noted that 'The livestock industry often attracts the attention of animal welfare groups because it advocates the intensive production of higher-order vertebrates' but that this is less of a problem because 'Snakes display markedly inferior cognitive abilities compared to endothermic species such as poultry and pigs.'

Despite the fear and loathing of many for snakes, they are favorite subjects of zoological displays. A 'serpentarium' (also known as an ophidiarum) is a zoo dedicated to snakes. A herpetarium is a zoo specializing in both reptiles in general as well as amphibians. Usually, snakes are exhibited as sections of zoos and museums. Snakes are sometimes included in entertainment acts (Figure 16.39),

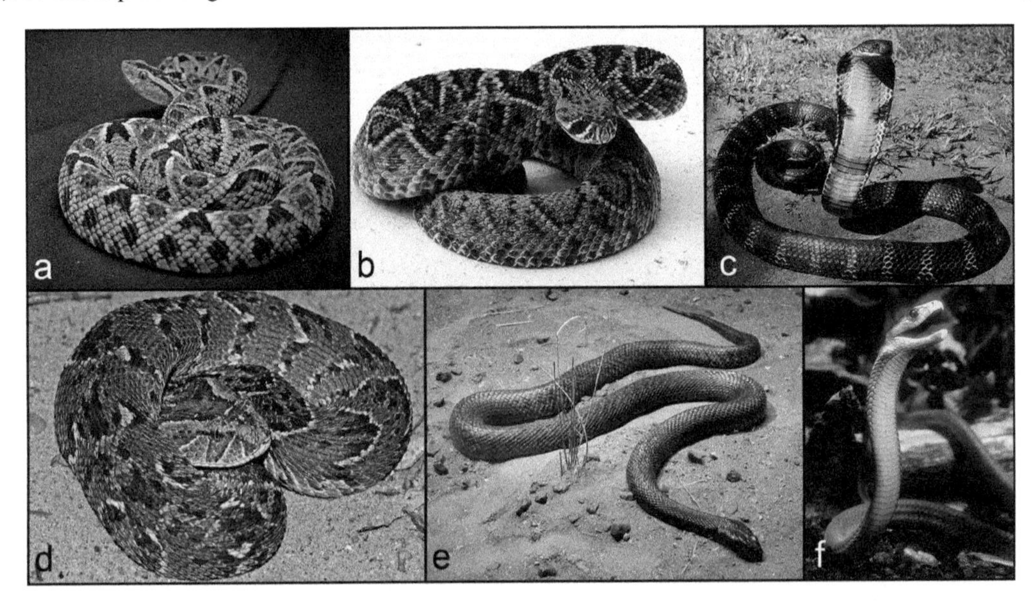

FIGURE 16.33 Some of the world's most dangerous and feared venomous snakes. (a) Fer-de-lance (*Bothrops asper*), the most dangerous snake of Central and South America. Photo by Robert Wedderburn (CC BY SA 3.0). (b) Eastern diamondback (*Crotalus adamanteus*) of the Southeastern U.S., considered the most dangerous snake of North America. Photo by Bladerunner8u (CC BY SA 3.0). (c) King cobra (*Ophiophagus hannah*) of India and Southeast Asia, the most feared snake of Asia, and the world's longest venomous snake, sometimes almost 6 m in length. Photo by Anand Titus and G. N. Pereira (CC BY SA 2.0). (d) Puff adder (*Bitis arietans*), which kills more people in Africa than any other snake. Photo by Bernard Dupont (CC BY SA 2.0). (e) Inland taipan (fierce snake; *Oxyuranus microlepidotus*) of Australia, the venom of which has been claimed to represent the most toxic land snake on Earth. Photo by XLerate (CC BY SA 3.0). (f) Black mamba (*Dendroaspis polylepis*), considered to be the most dangerous and deadly snake in the world. Very aggressive and alleged to be the fastest of venomous snakes, this is Africa's longest venomous snake, reaching to over 4 m. Photo by Bill Love (CC BY SA 3.0).

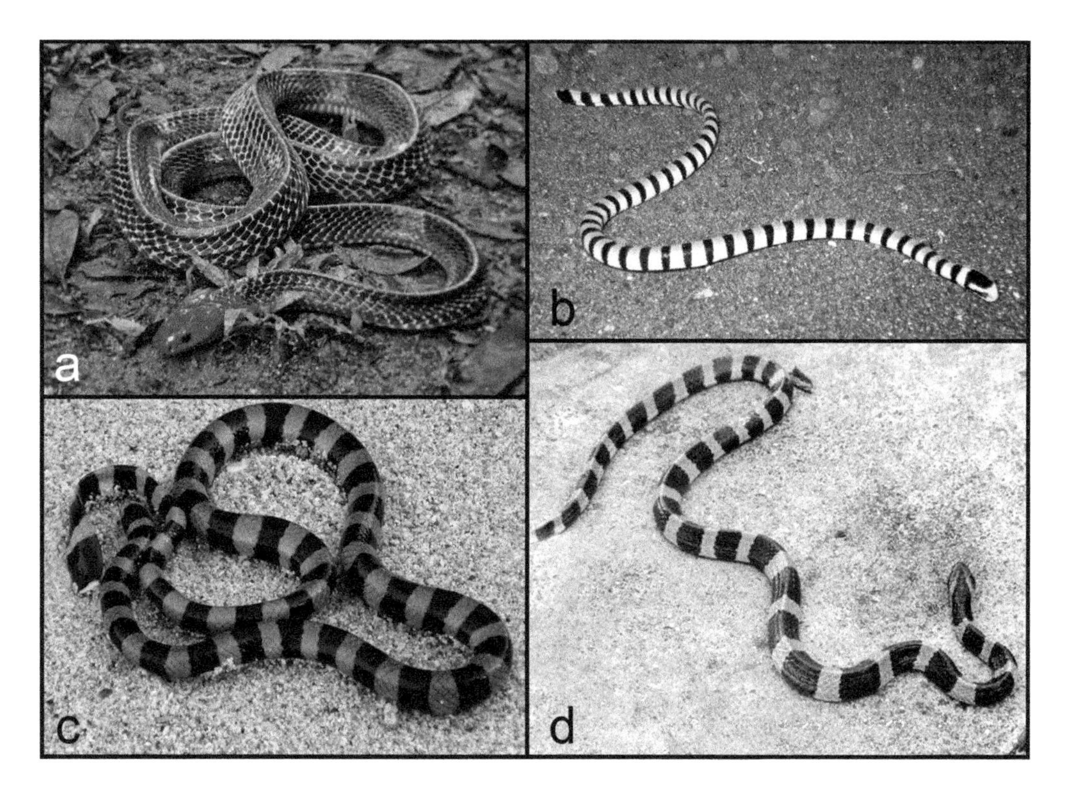

FIGURE 16.34 Sea snakes: dangerously venomous but rarely attack humans. (a) Red-headed krait (*Bungarus flaviceps*). Photo by Touchthestove (CC BY 4.0). (b) Banded sea krait (yellow-lipped sea krait; *Laticauda colubrina*). Photo by Jens Petersen (CC BY SA 3.0). (c) Blue-lipped sea krait (*Laticauda laticaudata*). Photo by Bramadi Arya (CC BY SA 4.0). (d) Banded krait (*Bungarus fasciatus*). Photo by AshLin (CC BY SA 2.5).

FIGURE 16.35 Burmese Python (*Python bivittatus*) locked in a struggle with an American alligator (*Alligator mississippiensis*) in Everglades National Park, Florida (2009 photo). Credit: Lori Oberhofer, U.S. National Park Service (public domain).

FIGURE 16.36 Boots made with snakeskin leather. Photo by Greg O'Beirne (CC BY SA 3.0).

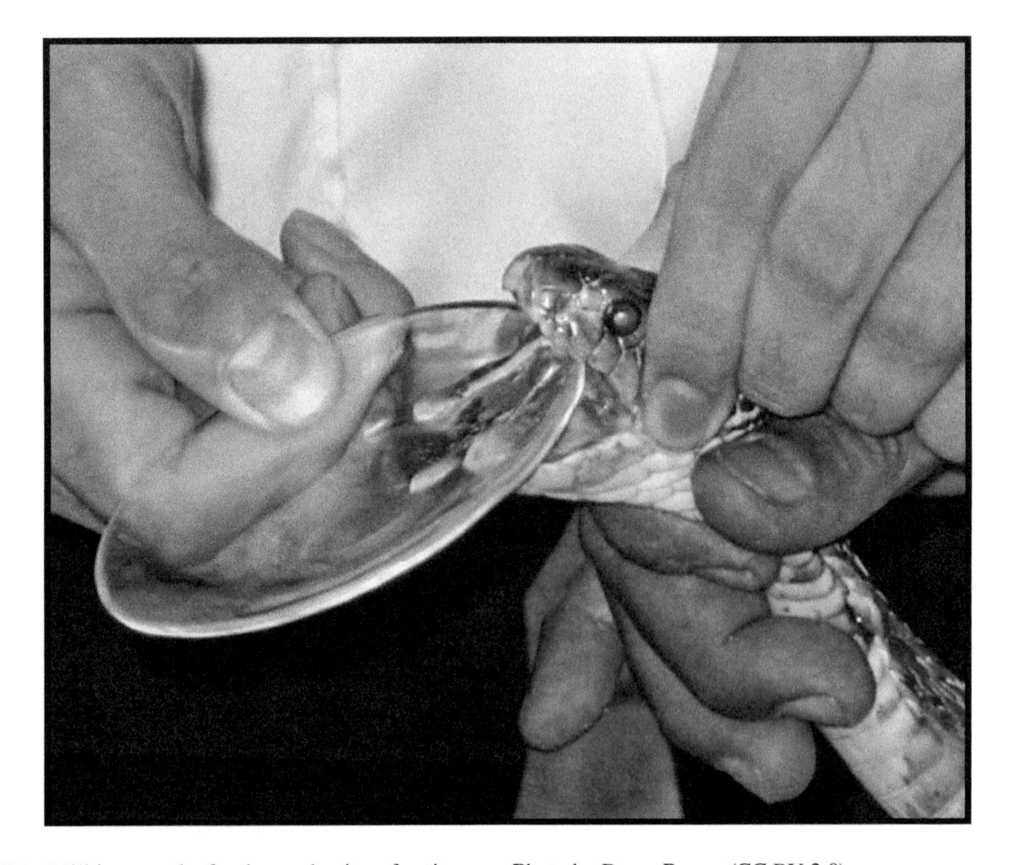

FIGURE 16.37 Milking a snake for the production of antivenom. Photo by Barry Rogge (CC BY 2.0).

FIGURE 16.38 Snakes as human food. Source of photos: Shutterstock. (a) Snake farm in Vietnam. Contributor: Renatok. (b) Grilled snakes on sale in a food market in China. Contributor: WeStudio.

FIGURE 16.39 Vintage print (public domain) showing circus act featuring snakes.

which are sometimes unseemly, as noted in the following. Snake charming, most common in India, involves playing and waving around a wind instrument called a pungi in front of a snake (typically a cobra), which the audience is persuaded to believe becomes hypnotized (Figure 16.40). Strip-tease acts often involve snakes, a reflection of the association of the sexuality of women with the interpretation of snakes as phallic symbols of fertility (Foubister 2003). Some American Christian sects handle poisonous snakes, inspired by a passage of Scripture, Mark 16:15–18, which allegedly instructs followers of Christ to handle poisonous 'serpents' as a part of their worship, as a confirmation of their faith (an illegal practice in some states).

Conservation Aspects

Snakes are the victims of habitat loss, particularly caused by urban sprawl, pollution, and deforestation (Mullin and Seigel 2009). About 100 species are listed by the IUCN Red List (https://www.iucnredlist.org/) as endangered. About 15% of sea snakes are either threatened with extinction or 'Near Threatened' (Elfes et al. 2013).

FIGURE 16.40 Snake charming, a chromolithograph by Alfred Brehm, ca 1883. Public domain image.

Some rare species are imperilled by overharvesting (Alves and Filho 2007). Because snakes are predators, they are susceptible to the accumulation of environmental contaminants absorbed by their prey (Campbell and Campbell 2001).

There has been concern that snake farms may contribute to the overexploitation of wild species harvested from nature but represented as legitimately raised farm snakes (Lyons and Natusch 2011; Kasterine et al. 2012). Wild amphibians and rodents are often fed to farmed snakes (also recycled low-value waste protein from the livestock and fishery industries), so there also needs to be attentive to their conservation.

Snakes are heavily penalized by their bad reputations which have resulted in their being considered to be principal protagonists of humans (Figure 16.41). Snakes are among the most persecuted of animals (Textbox 16.8). In addition to venom, snakes use constriction (Figure 16.35), or simply swallowing to kill their prey. All of these methods are repulsive to the human psyche. Moreover, numerous snakes are ambush predators, and so are by nature 'sneaky,' further deteriorating the way humans perceive snakes. One of the dictionary definitions of snake is 'a treacherous or deceitful person.' This view that snakes are wicked traces back at least to the story in the first book of the Bible of a serpent in the Garden of Eden persuading Eve to consume the forbidden fruit from the Tree of Knowledge (Figure 16.42). Western stories have often portrayed snakes as untrustworthy (Foubister 2003). On the other hand, snakes generally are respected creatures in most religions and in mythology (Stanley 2008; Sasaki et al. 2010; Figure 16.43).

'Herpetophobia' is fear of reptiles, typically snakes and lizards, but sometimes also animals like turtles, alligators, and crocodiles. 'Ophidiophobia' refers to an irrational or extreme fear of snakes. However, it is not clear that fearing snakes is 'abnormal.' Perhaps a third of people exhibit some fear of them, more commonly women (Rakison 2009). Some have theorized that humans, and indeed many other animals, have an innate (instinctive) aversion to the sight of snakes acquired during evolution, as an adaptation to survive encounters with dangerous snakes (Kawai 2019). There is even an indication that we are programmed to become alert when we hear snakes hissing (Erlich et al. 2013). On the other hand, life experiences clearly condition some people to fear snakes. A more sophisticated view is that we have an inherent capacity to learn about characteristics that signal that an animal is dangerous, and that snakes trigger a lifelong fear response, perhaps as early as infancy (Souchet and Aubret 2016; Baynes-Rock 2017). It is clear that some animals are genetically programmed either to fear their predators from birth or to selectively become afraid of snakes (Cook

FIGURE 16.41 Conflict between snakes and humans, exemplified by painting (public domain) by Emile Doepler (1855–1922) entitled 'Thor and the Midgard Serpent' illustrating Norse mythology.

Source: Doepler and Ranisch (1903).

TEXTBOX 16.8 FEAR AND IGNORANCE AS OBSTACLES TO SNAKE CONSERVATION

A surprisingly large number of persons in our society either cannot, or will not, accept the truth about one of our most useful forms of animal life – our harmless common snakes. This unfortunate attitude exists because it seems to be a characteristic of human beings to avoid learning any good about something which they prefer to consider as harmful. They refuse to learn how to tell the harmless from the poisonous ones.

—*Nelson (1960)*

Due to an irrational, albeit subconscious, fear of snakes, many snakes are killed needlessly. This could be avoided with a little understanding and education. Fear of snakes can be overcome regardless of its innate origins.... However, overcoming this fear will first require an attitude of caring about our environment and its inhabitants followed by education about the reality of these misunderstood reptiles.

—*Stanley (2008)*

Though threatened by many of the same issues that affect other wildlife, including habitat loss, climate change and disease, negative attitudes may be the biggest barrier to snake conservation because it often impedes efforts to address other threats. For example, public outcry based on fear and misinformation recently halted a scientifically sound conservation plan for timber rattlesnakes. A similar project at the same location that involved releasing eagles was embraced by the community. Rattlesnakes are no less iconic or important to the ecosystem than eagles. In fact, they may help reduce the incidence of Lyme disease, which affects tens of thousands of people in the United States each year, by reducing the number of rodents that harbor this disease. But facts often play second fiddle to emotions where snakes are concerned. Snakes are important components of biodiversity, serving as both predators and prey in nearly every ecosystem on Earth.

—*Amarello (2017)*

FIGURE 16.42 Adam, Eve, and a snake in the Garden of Eden. (a) Painting (public domain) showing Eve receiving a forbidden apple from the snake and giving one to Adam. Unlike the evil snake, all the other animals are innocent. Prepared (ca. 1615) by Jan Brueghel de Oude and Peter Paul Rubens, housed in the Mauritshuis Museum, Netherlands. (b) A charming, if much simpler, illustration of the story. Source: Shutterstock, contributor: Ottoflick.

FIGURE 16.43 Respectful views of snakes in Hindu tradition. (a) Vishnu flanked by two wives resting on Shesa, the serpent on the waters. In Hindu belief, the god Vishnu is watched over by Shesha. Chromolithograph by Ravi Varma (1848–1906). (b) Painting showing Manasā, Indian goddess (Hindu deity) of snakes and protection against snakebite. Source: Wellcome Trust (CC BY 4.0).

and Mineka 1990). In studies of human infant responses, it can be difficult to tell whether snake-like stimuli are evoking a 'fear' response or simply 'interest' (Thrasher and LoBue 2016).

There are dozens of books written for children that sympathetically discuss snakes (e.g., Thompson and Wildlife Conservation Society 2006). Unfortunately, movies today often reinforce fear of snakes. The 1963 American epic movie *Cleopatra* included the suicide of the young queen of Egypt by a poisonous asp. (According to popular belief, Cleopatra committed suicide in 30 B.C.E. by applying an 'asp' – actually an Egyptian cobra, *Naja haje* – to her breast.

Note Figure 16.44.) The memorable line 'Snakes, why did it have to be snakes' was uttered by Indiana Jones in the 1981 film *Raiders of the Lost Ark* when he discovered that the Well of Souls was crawling with them. And the 2006 horror film *Snakes on a Plane* garnered a large cult following.

An outstanding defense of reptiles (and amphibians as well) was made in 'Life in Cold Blood,' a BBC nature documentary series written and presented by David Attenborough (Figure 16.45), and first broadcast in the United Kingdom in 2008. The series was associated with the following comments:

> From a purely biological standpoint, cold-blooded amphibians and reptiles long have commanded the respect of scientists for their sheer resilience, if nothing else. These animals dominated the world for an estimated 200 million years, and today, more than 14,000 species of them still thrive... Reptiles and amphibians are sometimes seen as simple, primitive creatures. That's a long way from the truth. The fact that they are solar-powered means that their bodies require only 10% of the energy that mammals of a similar size require. At a time when we ourselves are becoming increasingly concerned about the way in which we get our energy from the environment and the wasteful way in which we use it, maybe there are things that we can learn.

VULTURES

INTRODUCTION

Vultures are large birds that entirely or predominantly scavenge the bodies of dead vertebrate animals. They occur worldwide, except in Antarctica and Australia. There are 16 species of Old World vultures (examples are shown in Figure 16.46). There are also seven species of New World vultures, including the Andean Condor and the endangered California Condor. Vultures are sometimes locally called buzzards, an ambiguous name that is also applied to some species that are not vultures. The Old World species (thought

FIGURE 16.44 The suicide by snake of Queen Cleopatra of Egypt (69–30 B.C.) (a) 'The Death of Cleopatra,' public domain painting by British artist Reginald Arthur (1871–1934). Photo credit: Roy Miles Gallery, London. (b) Cleopatra and a snake, public domain painting by Dutch artist Jan Berends (born about 1679, died after 1742). Source: https://www.sothebys.com/en/buy/auction/2021/master-paintings-part-ii/cleopatra-and-the-asp-in-an-elegant-interior. (c) The extremely poisonous Egyptian cobra (*Naja haje*), the so-called 'asp' that reputedly killed Cleopatra. Photo by Ghorayr (CC BY SA 4.0).

FIGURE 16.45 Sir David Attenborough (1926–), renowned English broadcaster, biologist, natural historian, and author, accompanied by two of the many species named in his honor. (a) Life restoration of *Attenborosaurus conybeari* by Nobu Tamura (CC BY SA 4.0). *Attenborosaurus* is an extinct genus of marine reptiles from the Early Jurassic Period of England. (b) Photo of David Attenborough at the Great Barrier Reef, circa 2015. Attribution: Department of Foreign Affairs and Trade website – www.dfat.gov.au. (c) Attenborough's fan-throated lizard, *Sitana attenboroughii*, of India. Photo by Kalesh sadasivan (CC BY SA 4.0).

FIGURE 16.46 Some Old World vultures. All illustrations are from Wellcome Trust (CC BY 4.0). (a) Palm-nut vulture (Angolan vulture; *Gyphierax angolensis*). Color lithograph, ca. 1875. (b) Bearded vulture (Lammergeier; *Gypaetus barbatus*). Chromolithograph by W. Greve after A. Thorburn, ca. 1885. (c) Cinereous vulture (*Vultur monachus*). Chromolithograph by W. Greve, after A. Thorburn, ca. 1885.

to have eagle ancestors) are not significantly related to the New World species (thought to have stork ancestors), so it is clear that their life style of feeding almost entirely by scavenging dead bodies evolved independently.

Scavengers are species that include in their diet dead animal or plant material that they did not kill or collect, such as animals that died naturally, or the remains of prey that was abandoned. While numerous scavengers will spend at least some of their time hunting and killing live animals or foraging on living plants, vultures are the only vertebrates that are obligate (full-time) scavengers, although sometimes they kill wounded, dying, newborn, or small and weak animals. There are also several obligate invertebrate scavengers, such as blowflies, and various deep-sea creatures, including some fish, that are thought to rely mainly if not exclusively on scavenging. 'Necrophagy' refers to feeding on corpses (dead bodies) or carrion (bodies that are decaying, but not too extensively), and such animal tissues almost always provide high-quality food.

The biology of vultures has been studied extensively (see Campbell 2015 for an excellent review). Dead animals putrefy (similarly dead plant materials decay), and the causal microorganisms frequently produce toxins. Vultures specialize on corpses (some occasionally consume rotting fruit), and their bodies are adapted to such feeding and to withstand the associated toxic germs. The 'bald' head and often the neck of vultures is mostly or entirely bare of feathers to facilitate thrusting their heads into the carcasses while avoiding contaminated flesh and blood accumulating on head feathers. An alternative, somewhat speculative interpretation is that the lack of feathers facilitates cooling in hot environments (Ward et al. 2008). The feet of New World vultures are adapted to walking on ground, rather than grasping and killing prey. (Old-World vultures have talons, like their eagle and hawk relatives.) Large wings allow soaring with minimal expenditure of energy, and along with keen vision, this facilitates the location of dead bodies. Sharp, hooked beaks are designed to rip meat apart. Vultures of the genus *Cathartes* (including turkey vulture, greater yellow-headed vulture, and yellow-headed vulture) also have a well-developed sense of smell that aids in the detection of carrion as far away as 2 km. Stomachs of vultures are capable of withstanding extremely corrosive acidity, which is necessary to kill the large intake of bacteria. Strong immune systems further serve to survive infections from contaminated meat. The Egyptian vulture has learned to drop stones on ostrich eggs to open them. Bearded (lammergeier) vultures (Figure 16.46b) drop bones on rocks to break them open and use their specialized tongues to extract the marrow.

Harmful Aspects

The two most common species of vultures in North America (the distributions of both extend to South America) are the turkey vulture (*Cathartes aura*; Figure 16.47b) and the black vulture (*Coragyps atratus*; Figure 16.47a), and both are significant pests, their sharp nails, strong beaks, and corrosive droppings capable of considerable damage (Textbox 16.9; Figure 16.48).

As noted in the next section, vultures play important roles in limiting the spread of diseases from carcasses. However, 'while some bacteria are destroyed in a vulture's digestive tract, some bacteria coulfd be transmitted on the bird's feathers and feet. Thus, while vultures can reduce the spread of most infections in a locality, they also could introduce infections into new areas' (Avery and Lowney 2016). In fact, Old World vultures have recently been demonstrated to be carriers of disease (Meng et al. 2017), including histoplasmosis and salmonella. However, the risk of disease transmission to humans seems low (Marin et al. 2014).

Beneficial Aspects

Despite the bad reputation of scavengers in general and vultures in particular, they are very beneficial to ecosystems (Wilson and Wolkovich 2011; Ogada et al. 2012; Devault et

TEXTBOX 16.9 PEST PROBLEMS CAUSED BY THE TWO MOST COMMON NORTH AMERICAN VULTURES

Black and turkey vultures cause problems in several ways. The most common problems associated with vultures are structural damage, loss of aesthetic value and property use related to offensive odors and appearance, depredation to livestock and pets, and air traffic safety... Livestock losses to black vultures are a major concern for many producers. Black vulture depredation of livestock involves killing or injuring animals that are sick, weak, or otherwise unable to defend themselves. This usually involves newborn calves, piglets or lambs, and the associated heifers, sows, and ewes... Property damage, especially from black vultures, includes tearing and removing window caulking, screen enclosures, roof shingles, vinyl seat covers from boats and tractors, windshield wipers and door seals on cars, and plastic flowers at cemeteries. Droppings of turkey and black vultures create nuisance conditions, especially when the birds loaf on roofs of houses, office buildings, communication towers, and electrical transmission structures. The accumulation of droppings on electrical transmission towers causes arcing and power outages. Vultures pose hazards to aircraft, especially when landfills, roosts, or other congregating sites are located near approaching or departing flight paths... In addition, vultures can cause human health and safety problems by contaminating water sources with their droppings. Contamination has occurred when coliform bacteria from droppings entered water towers or springs from which residences drew water... Many people consider vultures a nuisance because of the white-wash effect their droppings leave on trees and structures at roost sites, the ammonia odor emanating from roost sites, and a general feeling of doom when vultures congregate nearby... Black vultures often plunder dumpsters and garbage cans, and they frequent waste transfer stations, zoos, and any place where food scraps are regularly available.

—*Avery and Lowney (2016)*

FIGURE 16.47 The two most common New World vultures, painted by J. J. Audubon (1785–1851) for his 1832 four-volume work *The Birds of North America.* (Audubon 1827–1838). (a) Black vultures (*Coragyps atratus*; plate 106) feeding on a mule deer. (b) Turkey vulture (*Cathartes aura*; plate 151).

FIGURE 16.48 Black vultures (*Coragyps atratus*) damaging a car in Everglades National Park, Homestead, Florida. The birds rip off windshield wipers and weather-stripping. Photo by Judy Gallagher (CC BY 2.0).

al. 2016; Figure 16.49). Of particular ecological importance is the role of vultures in controlling diseases that can be transferred from carrion (Markandya et al. 2008; Santangeli et al. 2019). Carcasses may be sources of many diseases, including rabies, anthrax, bubonic plague, mad cow disease, and foot and mouth disease. Vultures have extremely acidic stomachs (pH = 1) where very few viruses and bacteria can survive. Bacteria contaminating the surfaces of vultures have likely served to protect them from predators, which would likely become sick if they ate them. In ranching areas of North America, vultures were once killed in the belief that they carried and transmitted anthrax and hog cholera, but these disease organisms do not survive passage through a vulture's digestive system.

Reductions of vultures in areas of Asia have resulted in increased accumulation of carcasses. Feral dogs have

FIGURE 16.49 Painting (public domain) by Thomas Baines (1820–1875) showing a dead buffalo being approached by vultures. Baines, an English artist, produced many paintings showing landscapes of southern Africa.

increased dramatically in parts of Asia because of the increased supply of corpses that serve to feed the canines instead of the birds. The presence of vulture suppresses this problem, as well as the associated danger of rabies (Ogada et al. 2016).

The practice of leaving the dead in designated places for vultures to eat (termed 'air burial' and 'sky burial') traces to the Zoroastrianism religion, probably beginning in the first millennium BCE in ancient Persia. This still occurs in certain parts of Asia, but sometimes there just aren't enough vultures available to dispose of the bodies (MaMing et al. 2016).

Conservation Aspects

About 70% of vultures are listed as threatened or near-threatened by the International Union for Conservation of Nature (Ogada et al. 2016). Vultures have declined catastrophically over the last three decades, especially in Asia and Africa, and are now the most threatened group of birds in the world (Buechley and Sekercioglu 2016a, b).

Vultures are threatened by a variety of factors, some of them common to many birds (Campbell 2015; Ogada et al. 2016; Textbox 16.10). Collisions with electrical lines, wind turbines, windows, and aircraft occur. Sanitation measures, such as burying dead animals, while beneficial for human health, have not been helpful for vultures in some places. (Conversely, increases in road kill have been good for vultures.) Agricultural chemicals are another source of harm. DDT is still used in some regions, and new organochlorine pesticides are dangerous for vultures. Plaza et al. (2019) state that the chief current threat to vultures is pesticides. By the early 2000s, more than 95% of India's vultures had vanished, due particularly to the use of diclofenac in cattle. This veterinary anti-inflammatory drug has been widely employed to treat pain and swelling in sacred cattle but is very toxic to vultures, causing kidney failure. The drug is now banned in much of Asia, and vulture populations appear to have stabilized. Lead in ammunition (currently banned in some regions) has induced toxicity (the strong acid in the digestive systems of vultures dissolves lead to a greater extent than in most other birds). When carcasses are poisoned to kill livestock predators, vultures sometimes become innocent victims. In Sub-Saharan Africa, poisoning of carcasses to eliminate pests such as hyenas and jackals has greatly decreased vulture populations. Many animals are attracted to carcasses or dying animals by the presence of vultures hovering above, and wildlife officers are similarly made aware of protected animals killed illegally. To avoid attracting game wardens, poachers sometimes poison carcasses to eliminate vultures (University of Utah 2016). Safford et al. (2019) argue that poisoning (from all sources) is the chief cause of vulture

TEXTBOX 16.10 THE ECOLOGICAL ROLE AND CONSERVATION PLIGHT OF VULTURES

Vultures are nature's most successful scavengers, and they provide an array of ecological, economic, and cultural services. As the only known obligate scavengers, vultures are uniquely adapted to a scavenging lifestyle. Vultures' unique adaptations include soaring flight, keen eyesight, and extremely low pH levels in their stomachs. Presently, 14 of 23 (61%) vulture species worldwide are threatened with extinction, and the most rapid declines have occurred in the vulture-rich regions of Asia and Africa. The reasons for the population declines are varied, but poisoning or human persecution, or both, feature in the list of nearly every declining species. Deliberate poisoning of carnivores is likely the most widespread cause of vulture poisoning. In Asia, *Gyps* vultures have declined by >95% due to poisoning by the veterinary drug diclofenac, which was banned by regional governments in 2006. Human persecution of vultures has occurred for centuries, and shooting and deliberate poisoning are the most widely practiced activities. Ecological consequences of vulture declines include changes in community composition of scavengers at carcasses and an increased potential for disease transmission between mammalian scavengers at carcasses. There have been cultural and economic costs of vulture declines as well, particularly in Asia. In the wake of catastrophic vulture declines in Asia, regional governments, the international scientific and donor communities, and the media have given the crisis substantial attention.

—*Ogada et al. (2012)*

decline. Other factors harming vulture survival include trade in traditional medicines made from vultures and killing them for food ('bushmeat'). The slow reproductive rates of many vultures make them susceptible to population decrease. Climate change may benefit some species and harm others such as the hooded vulture in Africa where desertification, in combination with urbanization and habitat alteration, is decreasing the supply of corpses.

Among vultures, condors (Figure 16.50) have a privileged position in the eyes of most people. The simple fact that condors do not have the word 'vulture' in their names strongly reduces the very negative image conjured up by this label. Another significant advantage is their size: they are massive, the wing spans sometimes exceeding 3 m. Condors are the largest flying land birds in the Western Hemisphere, and among the largest birds in the world that are able to fly (but see Figure 16.51). It is a truism that large animals very strongly tend to be idolized by humans (Chapter 4). The Andean Condor (Figure 16.50b) has been a respected icon for millennia among indigenous peoples of South America and is currently a national symbol of six countries throughout its range. Similarly, the California Condor (Figure 16.50a) is a traditional esteemed symbol among California native peoples. The enormous wingspans of condors greatly help in keeping them aloft but limit them to windy areas where they can glide on air currents with little effort. Andean condors are confined to Andean mountains, coasts that also supply strong ocean breezes, and deserts with strong thermal air currents. In parallel, the California condor is currently restricted to the western coastal mountains of the U.S. and Mexico and the northern desert mountains of Arizona (the species once ranged from Mexico to Canada). Persecution by hunters and farmers greatly reduced the number of condors (Haemig 2012). By 1982, there were only 22 birds of the California Condor left alive. Intensive conservation efforts led to the species rebounding, and there are now hundreds in California, Arizona, Utah, and Baja California (Mexico). The Andean Condor, currently considered 'Near Threatened,' has similarly benefited from conservation and re-introduction policies.

Vultures, often referred to as 'nature's clean-up crew,' are like garbage collectors (indeed, they are highly attracted to garbage) and undertakers – indispensable to health, but nevertheless viewed with discomfort, even revulsion. Eating unsanitary decaying corpses is disgusting from a human perspective. The birds have been interpreted as bad luck omens. They are stereotypically associated with impending disaster, such as in cartoons depicting circling vultures as a harbinger of death. Vultures are often added to horror movies, and scenes showing human carnage shortly after a war. Language also indicates the lack of esteem in which vultures, indeed scavengers in general, are viewed. The word 'scavenger' is often applied to humans in a very pejorative sense to refer to a person who searches refuse for food, so all scavengers tend to be held in low esteem. The word 'vulture,' as well as 'buzzard' often used as a synonym, has some even more pejorative meanings, such as 'a contemptible or rapacious person,' 'a mean or cantankerous person,' and 'a contemptible person who preys on or exploits others.' Expressions like 'vulture capital,' 'vulture investment,' and 'legal vulture' reflect these predatory meanings. 'Vulture funds' are private investment funds that trade in the defaulted or soon-to-default debts usually issued by the world's most heavily indebted poor countries, providing creditors opportunities to extract exorbitant fees to extend payment schedules.

Historically, attitudes to vultures have ranged between respect and hatred. Some spiritual traditions were quite respectful of vultures (Kushwaha 2016), but vultures have also been associated with witchcraft and demons. Often, vultures have been considered to be quite undesirable and have been persecuted. Some vultures defend themselves by vomiting foul-smelling material on attackers (a practice that also reduces the weight of the birds facilitating

FIGURE 16.50 The two species of New World condors. (a) An old male California condor (*Gymnogyps californianus*), painted by J. J. Audubon (1785–1851). Source (public domain): Audubon, J. J. 1838. Vol. 4, *The Birds of America*, vol. 4, plate 426. (b) Andean Condor (*Vultur gryphus*), shown attacking a lamb on a rock. Males like this can easily be identified by the large comb (caruncle) on the head and the wattle on the neck that is formed by folds of skin. Colored etching prepared by W. Panormo (1796–1867) after a drawing by H. Smith (active 1858). Source: Wellcome Trust (CC BY 4.0).

FIGURE 16.51 Giant teratorn (*Argentavis magnificens*), an extinct (as of 6 million years ago) predatory bird of Argentina, believed to be a relative of the condors. Among the largest flying birds ever to exist, it is speculated to have had a wing span reaching 7 m and a weight that exceeded 70 kg. Credit: Stanton F. Fink (CC BY SA 3.0). Outline of woman added for scale.

escape). Some vultures prevent overheating by 'urinating' (i.e., emptying their cloaca) on their legs, with subsequent evaporative cooling (a behavior which seems less repugnant under the technical term 'urohidrosis'). These habits do not serve to endear the birds to humans.

Most people are very sympathetic to birds and find their appearance and life styles to be very attractive. However, most vultures are the complete opposite of the 'cute and cuddly' charismatics species that easily attract conservation support. Even Charles Darwin, perhaps the greatest student of biodiversity, upon observing a turkey vulture from the deck of the Beagle in 1835, called it a 'disgusting bird' whose bald head was 'formed to wallow in putridity' (Royet 2016). Many vultures have been described as having faces that only a mother could love. The king vulture, at least, is exceptionally colorful and is quite impressive in flight (Figure 16.52). Moreover, some artists have succeeded in depicting vultures in quite attractive poses (Figure 16.53). Cartoonists tend to emphasize the ugly, disgusting aspects of vultures (Figure 16.54a), but they can be shown to appear quite charming (Figure 16.54b).

Despite their morbid reputation and appearance, vultures have managed to achieve some help, at least

FIGURE 16.52 King vulture (*Sarcoramphus papa*) of Central and South America. (a) Photo by Renato Augusto Martins (CC BY SA 4.0). (b) Photo by Eric Kilby (CC BY SA 2.0). (c) Photo by Weltvogelpark Walsrode (CC BY SA 3.0).

FIGURE 16.53 Painting (public domain) titled 'Four Vultures of Different Species' by Flemish artist Philipp Ferdinand de Hamilton (ca. 1664–1750), housed in Belvedere Museum, Vienna.

FIGURE 16.54 Contrasting illustrations of life styles of vultures. Source: Shutterstock. (a) A vulture family shown as disgusting scavengers. Contributor: Oriol San Julian. (b) A vulture family shown as loving companions. Contributor: Cernecka Natalja.

in Western countries, where they have legislative protection (Balmford 2013). Shooting vultures in North America and the European Union is now illegal. In the U.S., the Migratory Bird Treaty Act protects vultures, which are managed by the federal government and can be killed only with a Migratory Bird Depredation Permit from the U.S. Fish and Wildlife Service (Avery and Lowney 2016). Farmers have frequently been the cause of vulture decline, and it is particularly important to educate the farming and ranching communities that on balance vultures are highly beneficial (Cailly Arnulphi et al. 2017).

FERAL PIGEONS

Introduction

What is the most hated bird? Depending on geography, people identify various species, such as starlings, crows, seagulls, Canada geese, house sparrows, and cormorants, as the most undesirable avian pest. But the most universally encountered annoying bird is the 'common pigeon,' also known as the domestic pigeon, city pigeon, street pigeon, and simply as the pigeon (Blechman 2007; Figure 16.55). Occasionally, it is also referred to as a dove (especially if it is white, or in the context of religion, literature, or art). The 'white dove' emblemizes peace and goodwill, but ironically it is often the same species as the despised pigeon. In 2004, British and American ornithologists officially renamed the feral pigeon as the 'rock pigeon,' but the phrase has traditionally been employed to designate the wild phase of the species, and in any event is mostly encountered in scientific circles. The names pigeon and dove are applied to over 300 species in the bird family Columbidae, but *Columba livia*, including wild, domesticated, and escaped birds is the species of dominant interest. The different kinds of birds within the species can hybridize, and interbreeding between wild and domesticated kinds has produced a spectrum of intermediate kinds.

Feral pigeons are derived from the rock dove (also known as the rock pigeon), *Columba livia*. The ancestral wild birds were native to Europe, North Africa, and western Asia. The wild birds are adapted to open and semi-open environments. They occupy cliffs and rock ledges, and

FIGURE 16.55 Feral pigeon (*Columba livia*). (a) Typical appearance. Photo by Elke Wetzig (Elya) (CC BY SA 3.0). (b) Pigeon in flight. Photo by Hari K Patibanda (CC BY 2.0). (c) A white pigeon. Photo by Optimistic pushpendra2 (CC BY SA 4.0). (d) Nine-day old nestlings ('squabs'), normally born as a pair. Photo by Ciell (CC BY SA 2.5).

since big buildings and large bridges provide comparable habitats, escapes from the domesticated forms have found cities to be ideal. Feral pigeons now live in urban environments around the world, surviving largely on human food waste (Capoccia et al. 2018; Figure 16.56).

Pigeon mate for life and can breed up to eight times a year, typically producing two eggs at a time. Both parents feed the young with a special regurgitated 'pigeon milk.' Pigeons characteristically bob their heads, because their side-mounted eyes function better with stationary images, and the jerking motions while they walk keep their heads stationary briefly between steps as the rest of their bodies are moving continuously. The mechanisms underlying the ability of pigeons to navigate are not well understood.

In Judeo-Christian tradition, the pigeon (or dove) is a symbol of peace and of the Holy Spirit (Figure 16.57a and c). In Asian religions, including Muslims, Hindus, and Sikhs, pigeons are often afforded sacred status and are protected. In southern Asia, huge flocks are fed daily at temples (Figure 16.57b and d). However, pigeons are sometimes sacrificed in religious ceremonies.

Pigeons (not just the common pigeon) were often shot in the hundreds as targets, but this blood sport has become unpopular in most of the world. Clay pigeons are mostly employed today, but the practice is still legal in some places. Pigeons were also once often used as quarry in competitive falconry.

Harmful Aspects

In cities, most people enjoy watching and feeding pigeons. Nevertheless, they are viewed as pests, particularly because their corrosive droppings require cleaning of buildings, statutes, and vehicles (Spennemann and Watson 2017;

FIGURE 16.56 Feral pigeons in urban centers. (a) On the Empire State Building, New York City. Photo by ZeroOne (CC BY SA 2.0). (b) In Trafalgar Square, London. Photo by ProhibitOnions (CC BY SA 3.0). (c) At St. Mark's Basilica, Venice. Source (public domain): Kirk Fisher/Pixabay.com. (d) Pigeons covering temples and pagodas in Durbar Square, Kathmandu. Source: Shutterstock, contributor: Doctor J.

FIGURE 16.57 Respect for pigeons (in the form of white doves) in religion. (a) Stained glass window in a church in Boynton Beach, Florida, showing the Dove of Peace overlapping a cross. Created by Tony Villante, photo by Deisenbe (CC BY SA 4.0). (b) Feeding sacred pigeons in Bangladesh. Photo by David Stanley (CC BY 2.0). (c) Stained glass window of Irish missionary Saint Columbanus (543–615) in the Abbey of Bobbio in Italy. Columbanus (the Latinized form of Columbán, meaning the white dove), and usually pictured with a dove, is venerated by the Eastern Orthodox Church and the Roman Catholic Church. Photo by Trebbia (CC BY SA 3.0). (d) A child feeding pigeons at a Jain temple in Calcutta. Photo by Sushovanbasak (CC BY SA 3.0).

Textbox 16.11; Figure 16.58). Fifty birds can produce over 1 t of 'poop' in a year. Pigeons are widely thought to harbor diseases, and so they generate fear. Feral pigeons indeed can carry a variety of pathogenic agents (Cano-Terriza et al. 2015; Mia et al. 2022). While serious disease transmission is possible to humans and ito livestock, this is infrequent. Pigeons are communal, often associating in colonies of hundreds of individuals, and to some, such large numbers are an eyesore. At night, their billing and cooing, and wing flapping, can disturb sleep. Indicative of the contempt they are often shown, pigeons have been called 'flying rats' and 'rats with feathers' (Jerolmack 2008). An accessible guide to controlling pest pigeons is at https://www.pigeoncontrolresourcecentre.org/.

Beneficial Aspects

As for dogs, cats, and other animals employed as pets, pigeon enthusiasts have selected hundreds of ornamental forms called fancy pigeons. These are often exhibited in competitions (Wendell 1977; Figure 16.59). Specimens are particularly valued for luxuriant, colorful plumage. Tumbler pigeons are varieties selected for their ability to tumble or roll over backward in flight.

TEXTBOX 16.11 THE HARMFUL SIDE OF PIGEONS

The rock dove… is regarded as a common pest species. This is probably due to their defecation on landmarks, sidewalks, statues, and buildings. They negatively affect other species; for example, the breeding success of white-tailed tropicbirds *Phaethon lepturus* in Bermuda is compromised through nest competition. They impact the economy negatively as they depredate crops. For example, they cause damage to maize, seeds, legumes (soybeans and chickpeas), and sunflower crops. In the USA alone, the damage associated with feral pigeons in urban areas amounts to around US$1.1 billion annually. The rock dove carries pathogens, diseases, and parasites that are harmful to humans and other animals.

—*Shivambu et al. (2020)*

FIGURE 16.58 The problem of pigeon droppings in urban areas. Source of photos: Shutterstock. (a) A statue of Diana the Huntress by Louis Auguste Leveque with a pigeon on top, at Tuileries Garden in Paris, France. Contributor: Devis M. (b) A park bench stained with pigeon droppings. Contributor: Denklim. (c) A bronze statue with a pigeon on its head. Contributor: Paolo Paradiso.

FIGURE 16.59 Examples of fancy pigeons. (a) A fantail pigeon. Source (public domain): Keulemans, J. G. 1869. *Onze vogels in huis en tuin*, vol. 1. Leyden, P.W.M. Trap. (b) Suabians. (c) Blue owls. Source of (b and c): Tegetmeier, W. B. and H. Weir. 1868. *Pigeons: Their Structure, Varieties, Habits, and Management.* London, G. Routledge and Sons. (d) Vogtland Whitehead Trumpeter. (e) Maltese. (f) Pouters. Source of (d–f): Schachtzabel, E. 1906. *Illustriertes Prachtwerk sämtlicher Taubenrassen.* Würzburg, Königl. Universitätsdruckerei H. Stürtz a.g.

Pigeons bred for special purposes are provided with different kinds of shelters. 'Coop' is a general term meaning a cage or pen for confining poultry. Dovecotes (dovecots) are shelters housing pigeons or doves. The term columbarium refers to a building with compartments for the storage of funerary urns with cremated remains but in the past was sometimes also applied to dovecotes. Dovecotes are located outdoors and typically contain pigeonholes for the birds. Lofts are shelters on top of or inside building where pigeons are kept. In practice, the words dovecote and loft are often employed interchangeably. Shelters for pigeons for meat and guano production were often huge in the past. Elaborate shelters were often a status symbol for the rich in the past, especially in Europe. Representative pigeon shelters are shown in Figure 16.60.

Ironically, although pigeon droppings are now considered to be a major health and hygiene issue, they were an invaluable resource from the 16th to the 18th centuries in Europe. Pigeon guano was a prized fertilizer for crops. In England in the 16th century, it was the only known source of saltpeter, an essential ingredient of gunpowder.

The Old Testament story of Noah releasing a dove from his Ark demonstrates that humankind has been familiar with the bird's orientation ability for millennia. 'Messenger pigeons' (also called homing pigeons and mail pigeons), by virtue of their remarkable guidance instincts, were employed in the past to transmit communications from distant locations (Figure 16.61), especially by the military. A pigeon named G. I. Joe is credited with saving the lives of a thousand soldiers in World War II after British troops had established a position within an Italian town that mistakenly was due to

FIGURE 16.60 Shelters for pigeons. (a) A special-built, overlapping-hearts cage housing a pair of doves symbolizing love (pigeons mate for life) at a marriage ceremony. Source: Shutterstock, contributor: Kovalchynskyy Mykola. (b) A simple garden shelter ('dovecote'). Source: Shutterstock, contributor: Widyasto. (c) Pigeon ranch, Los Angeles, California (between 1907 and 1914) – a gigantic commercial shelter for pigeons. The main product was young birds (squabs) for meat. Source (public domain): Loyola Marymount University Digital Collections. (d) An old-fashioned special-built stone dovecote. Source: Shutterstock, contributor: Mark Paterson.

FIGURE 16.61 Navigation abilities of common pigeons. (a) A 12th-century mosaic (public domain) in Basilica di San Marco, Venice, depicting Noah sending a dove from his ark to determine whether land is nearby. (b) Homing pigeons in flight. Photo by Spanishguitar101 (CC BY SA 4.0). (c) An example of the communication usage of pigeons: 'The messenger of love' by Leonard Straszyński (?–1879).

FIGURE 16.62 Heroic military pigeons. (a) A monument to military pigeons in Brussels, Belgium. Photo by EmDee (CC BY SA 3.0). (b) Cher Ami. In 1918, a carrier pigeon named Cher Ami (assumed to be a male, although this has been disputed), shot through the breast and leg by enemy fire, managed to return to his loft with a message capsule dangling from his wounded leg. The message saved the lives of about 200 allied soldiers in France. Cher Ami was awarded the French Croix de Guerre and was inducted into the Racing Pigeon Hall of Fame in recognition of his extraordinary service during World War I. Cher Ami died in 1919 from his wounds. His preserved body is on display at the Smithsonian National Museum of American History in Washington, DC. Photo (public domain), credit Line: War Department. U.S. Signal Corps. https://www.si.edu/object/cher-ami%3Anmah_425415. (c) Monument in the Argonne Forest, France, to Cher Ami. Photo (public domain) by Wilson44691.

be bombed by allied planes. Commando, a pigeon with the British armed forces during World War II, was awarded a medal for transporting crucial intelligence. A pigeon named Cher Ami (French for dear friend) is celebrated for saving allied soldiers in France in World War I (Figure 16.62).

In modern times, homing pigeons are employed in the sport of pigeon racing. They have been recorded flying up to 1,800 km (Shivambu et al. 202). Additionally, they are often released in ceremonies as flying white doves.

Pigeons (often termed 'utility pigeons' and 'King pigeons' [with a capital K]) have been bred for meat (Figure 16.63). Large breast muscles are common in utility pigeons. Young birds, called squabs, are preferred. In numerous countries, raising squabs for food has been and remains a major

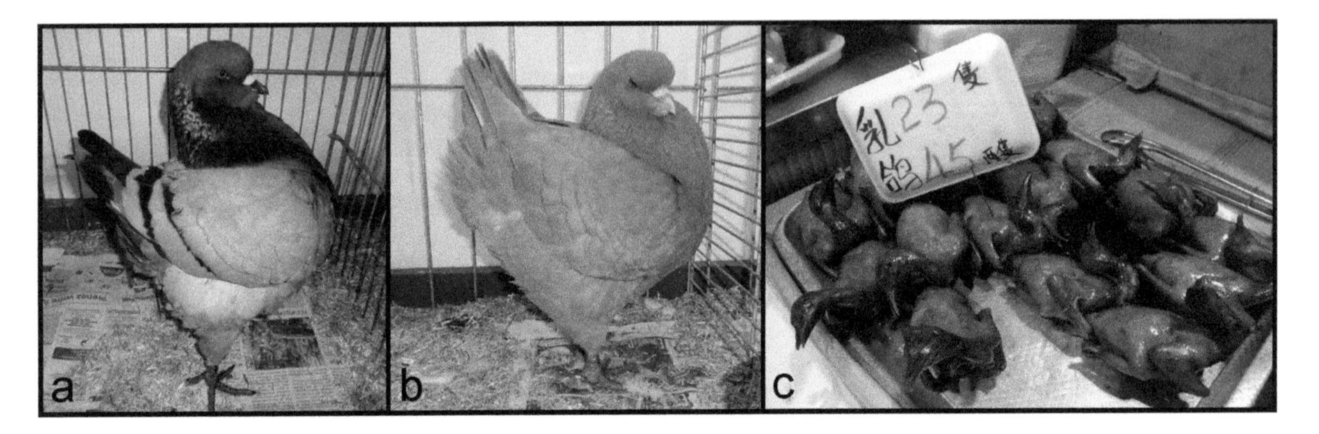

FIGURE 16.63 Meat pigeons. (a and b) Photos by Uikitireza (CC BY SA 4.0). (c) Roasted squab (young pigeons) for sale in Hong Kong. Photo by HomJeovasaA (CC BY SA 3.0).

industry. Chicken produced on an industrial scale is cheaper, but in some nations, squab is considered as haute cuisine.

Domestic pigeons are used in laboratory experiments in biology and medicine. They are considered to be one of the most intelligent of birds and are often employed in behavioral studies requiring brainpower.

As with all major pests, pigeons support employment and industries to research, prevent, and repair the damages they cause.

Conservation Aspects

Feral pigeons are invasive in many parts of the world, harming indigenous species and disturbing ecosystems (Shivambu et al. 2020). They can be a problem for solar panel installations on roofs, requiring regular removal of droppings.

Some pigeon owners have been known to illegally trap hawks and falcons threatening their pigeons. Indeed, raptors (i.e., birds of prey, such as hawks, owls, eagles, and falcons) native to many regions have turned to introduced pigeons for some of their food, sometimes as their primary prey (Figure 16.64a). The peregrine falcon is the champion natural predator of pigeons and is often deliberately employed as a control measure.

Attempts to eliminate or at least reduce feral pigeons from cities are often controversial and met with emotional opposition from much of the public. Many people enjoy feeding pigeons, which of course contributes to generating greater numbers. Poisoning, shooting, and trapping (and euthanizing) are rarely realistic options in urban areas, but these practices are allowed in many countries in rural/agricultural areas (note Figure 16.64b). Since the birds are non-migratory and are pests, they are usually not included in protective legislation. Contraceptive chemical fertility control is an option but difficult to apply to large populations. Enticing the birds to nest in dovecotes, then collecting the eggs is practiced, but is labor-intensive. As noted above, feral pigeons have been called 'flying rats,' but their emotional appeal to humans is their saving grace, while rats are exterminated with little sympathy.

FIGURE 16.64 The principal predators of pigeons. (a) A raptor attacking a pigeon. Source: Shutterstock, contributor: Mustafa Kocabas. (b) A 1905 post card (public domain) lampooning the bravery of hunters.

17 In Defense of the World's Most Despised Mammals

INTRODUCTION

The theme of this book is human preferences for and prejudices against species. Of all the major groupings of animals, the mammals receive the most sympathetic attention. We humans arrogantly consider mammals to be the 'highest' group of animals because we are the predominant member. Nevertheless, just as we express both our love for, and hatred against certain groups within our own species, other mammals receive very selective admiration or detestation. Mammals such as those included in the 'charismatic megafauna' (big and attractive as shown in Figure 17.1) and the cute and cuddly species (Figure 17.2) are admired and make ideal aids for enlisting conservation support. In parallel, as noted in this chapter, certain species of mammals are passionately hated and are often the subjects of extermination campaigns. These abhorred species are major pests, and their intrusions into the artificial habitats of people provoke quite negative sentiments against the genuine world of wildlife from which they originate. As noted in previous chapters, the most reviled species of invertebrate animals and 'lower vertebrates' significantly but irrationally prejudice the public against wildlife and militate against efforts to rehabilitate and conserve biodiversity. The same is true for mammalian pests. We humans wage war against our enemies, and our impulse is to remove the objectionable species not just from our urbanized world but everywhere. However, as will be noted, the most offensive pest mammals are superlative competitors and survivors, and attempts to eliminate them require such extensive measures that inevitably many other species are endangered. What is required is to find ways of living with these pests that minimize their harmful effects, and a key first step is to learn to understand and respect their needs. Toward this goal, this chapter presents both the negative and positive aspects of our most detested mammalian pests, emphasizing that on balance their benefits to humans exceeds their harm.

FIGURE 17.1 Examples of charismatic megafauna mammals that strongly attract conservation support (all photos are public domain). Credits: (a) Elephant: Hưng Lê from Pixabay. (b) Orca: Clker-Free-Vector-Images from Pixabay. (c) Giant panda: Dušan Smetana from Pixabay. (d) Giraffe: PublicDomainPictures from Pixabay. (e) Tiger: George Desipris, https://www.pexels.com/photo/close-up-photography-of-tiger-792381/.

DOI: 10.1201/9781003412946-17

FIGURE 17.2 Examples of cute and cuddly mammals that strongly attract conservation support. Credits: (a) Meerkat: Public domain photo by Skeeze from Pixabay. (b) Koala: Public domain photo by Skeeze from Pixabay. (c) Harp seal (baby whitecoat phase): Photo by Lysogeny (CC BY SA 4.0). (d) Ring-tailed lemurs: Photo by Alexandra (CC BY SA 3.0). (e) Long-tailed chinchilla: Photo by Matteo De Stefano/MUSE (CC BY SA 3.0). (f) Red panda: Public domain photo by Mathias Appel.

Burgin et al. (2018) recognized 6,400 species of living mammals, which represent less than 10% of recognized vertebrate species, and less than 0.5% of all animal species. Nevertheless, mammals overwhelmingly dominate conservation initiatives. By a considerable margin, the public supports conservation and rehabilitation of certain 'charismatic' mammals much more than any other species (Table 17.1).

In relation to their relatively small number of species, mammals play disproportionately large economic roles in human existence. Domesticated mammals, and in some regions wild mammals, provide food and hides. Livestock mammals furnish most of the world's meat, milk, leather, and wool (as noted by Thornton 2019, there are currently 1.5 billion cows, 1 billion sheep, and 1 billion pigs in the world). Some species are important beasts of burden for riding, hauling, and plowing. The dung and urine of livestock provide an agricultural fertilizer that is superior to today's synthetic versions. Dogs and cats have become the world's major pets, and dogs are also invaluable working assistants. (Unfortunately, most of the approximately 1 billion dogs [Atitwa 2018] and about 600,000 cats [Migiro 2018] in the world are 'free-range' or feral, causing enormous ecological problems.) Recreational hunting is primarily for mammals and birds. Rats and mice are the principal vertebrates employed in medical research and scientific studies. (As pointed out later, on a world basis about 100 million experimental mice and a smaller but substantial number of rats may be sacrificed annually in laboratories.)

In nature, mammals are also often disproportionately important – as members of food chains and food webs, as grazers, and as predators. Mammals are often keystone species – critical for maintaining services and functions associated with sustaining a balanced ecosystem. On the negative side, some invasive mammals are causing extensive damage to ecosystems and biodiversity.

The attention that we humans pay to the welfare of wild mammals is also highly disproportionate to their relatively small number compared to other animals. 'The International Union for the Conservation of Nature (IUCN) is the world's oldest and largest global environmental network that aims to help the world find pragmatic solutions to our most pressing environment and development challenges by supporting scientific research' (from its website: https://www.iucnssg.org/who-we-are.html). The IUCN organizes 'Specialist Groups' which disband and reform every 4 years (https://www.iucn.org/commissions/ssc-groups; e.g. Shark Specialist Group, Duck Specialist Group). An analysis of the specialist groups devoted to animals shows that mammals are the most studied species (Table 17.2).

TABLE 17.1
The Most Charismatic Animals ('Deemed Charismatic Mainly because They Were Regarded as Beautiful, Impressive, or Endangered') as Determined and Ordered by Albert et al. (2018). All except the shark and the crocodile are mammals.

1. Tiger	6. Panda	11. Chimpanzee	16. Dolphin
2. Lion	7. Cheetah	12. Zebra	17. Rhinoceros
3. Elephant	8. Polar Bear	13. Hippo	18. Bear
4. Giraffe	9. Wolf	14. Shark	19. Koala
5. Leopard	10. Gorilla	15. Crocodile	20. Blue Whale

TABLE 17.2
Number of International Union for the Conservation of Nature Specialist Groups

Vertebrates	Number of Groups	Invertebrates	Number of groups
Mammals	35	Insects	4
Birds	16	Other arthropods	3
Reptiles and amphibians	12	Coral	1
Fish	10	Mollusks	1
Total	73	Total	9

Conversely to the useful roles described above, some mammals are decidedly harmful to humans. Some rodents destroy crops growing in fields and stored in buildings and also ruin property. Several mammals transmit deadly infectious diseases. A few wild carnivorous mammals prey on domestic animals and occasionally on people. Some, most notably skunks, are simply repellent to human senses, while others, particularly hyenas, seem to represent the worst of human character defects. The species examined here are reputedly the worst of the worst.

The most despised mammals include bats, hyenas, mice, rats, and skunks (Figure 17.3). Notably, most of these disreputable mammals are no larger than a housecat, whereas the most respected mammals are usually huge. Size is one of the characteristics that strongly determine whether a species is liked or disliked by humans (Chapter 4), although smaller species are at least as vital to the welfare of the world as the giants. A goal of this chapter is to generate understanding of the economic values and useful ecological roles of the world's most disliked mammals, in order to minimize the disrespect for biodiversity that they generate. Indeed, as will be presented, the harm of these pests has often been exaggerated, and their very negative public images are undeserved.

BATS

Introduction

Bats make up the mammalian order Chiroptera, with over 1,300 species (Taylor and Tuttle 2018; examples are shown in Figure 17.4). The traditional taxonomic division into two suborders, the Megachiroptera (the Old World fruit bats or flying foxes, shown in Figure 17.5, which are exclusively herbivores) and the more numerous and diverse Microchiroptera, has been losing support because of recent genetic studies. Bats occur throughout the world except the Antarctic. There are twice as many species in the Old World compared to the New World, but the greatest diversity occurs in South America. Unlike some other air-borne mammals, bats are capable of true (self-powered) flight, not just gliding. The wings are folds of skin stretched between elongated finger bones and various parts of the body. Some species are small, like the bumblebee bat (*Craseonycteris thonglongyai*) weighing as little as 1.5 g, others exceed 1.5 kg and have a wingspan as wide as 1.7 m. The smaller bats ('microbats') usually roost in caves or other similar protected areas, while many large bats ('megabats') often overnight in the open. Bats are mostly nocturnal, usually spending days in caves, trees, or other refuges. Most bats are rather drab, reflecting their usual active period during nights. However, the coloration of a few is rather attractive (e.g., Figure 17.6). In cold seasons, bats may hibernate in dens, especially caves. While at rest, they hang upside-down. Some species are solitary, others are social, sometimes congregating in the millions. Some bats eat fruits, but about 70% are insectivores (insect eaters), and some species are large enough to hunt fish, frogs, lizards, birds, and mammals – even other bats. True 'vampire bats' include only three species, ranging from Central to South America (occasionally in Mexico), which feed exclusively on blood (they are the only mammals that do so entirely). Bats are hunted by some animals, especially birds of prey and snakes. There are numerous scientific books on bats (e.g., Hill and Smith 1984; Altringham 2011; Adams and Pedersen 2013; Fenton and Simmons 2015; Voigt and

FIGURE 17.3 The most vilified mammals (bats, hyenas, mice, rats, and skunks). Painting by Jessica Hsiung.

Kingston 2016; Taylor and Tuttle 2018), which have proven to be appealing to scientists, if not to most of the public.

Compared to other mammals, flying is the most notable adaptation of bats, but echolocation (sonar radar) is also an extraordinary talent. Many bats produce ultrasonic sounds (which humans can't hear) and interpret their reflections in order to locate objects, either to avoid them in flight or to hunt prey while flying in the dark. Meat-eating bats tend to use echolocation to hunt their prey, but fruit eaters do not need this adaptation. (Several marine mammals and a few terrestrial insect-eating mammals have also evolved echolocation, but the bats are the experts in this regard.) There are no blind bats (contrary to the expression 'as blind as a bat'), but this idea is so entrenched that a bat is called a 'blind mouse' in some languages (murciélago in Spanish, slijepi miš in Bosnian). However, some bats that are specialized for night activity have relatively poor vision.

FIGURE 17.4 Painting of bats (public domain) from Craig (1897, plate 4). Identifications provided: 1: Common bat; 2: Vampire Bat (Old World 'vampire bats' are misnamed); 3: Horseshoe bat; 4: Dog-headed bat; 5: Fruit-eating bat.

Harmful Aspects

Bats carry several deadly viruses such as ebola, different kinds of coronavirus (such as SARS, MERS, possibly COVID-19), henipavirus, and astrovirus, which can cause lethal infections in other mammals. Bats appear to be a continuing source of new viruses infecting humans (Wong et al. 2007). Besides rabies, none of the other viruses seem to produce any symptoms in bats (the diseases are said to be 'subclinical'), which appear to be extraordinarily resistant, and accordingly are efficient vectors of many diseases (Klimpel and Melhorn 2013; Wang and Cowled 2015). It has been speculated that the high heat produced by bats while flying somehow immunizes them against most of the effects of viruses and may even stimulate the viruses to become more virulent (Gross 2020). While bats are known to carry over 200 viruses, the extent to which they are responsible for particular outbreaks is in need of study (Moratelli and Calisher 2015), especially with respect to COVID-19 (Tuttle 2020).

In the United States and Canada, the most recently documented cases of rabies have been attributed to bat rabies

FIGURE 17.5 Fruit bats. Source (public domain): Lydekker (1895).

viruses (Dato et al. 2016). 'Bat transmission of rabies virus to humans occurs only in the Americas and involves hematophagous [blood-eating], frugivorous [fruit-eating] and insectivorous [insect-eating] bat groups' (Beltz 2018). Most bat species do not carry the rabies virus, but some do. Bat bites of humans are rare, and rabies in humans from such bites is extremely rare.

Flying bats can easily carry deadly viruses long distances. Many bats, including fruit-eating bats, discard less-digestible portions of their food along with infective agents, which can be ingested by other animals, spreading diseases (Jennings 2019). Fruits chewed by bats can be harvested and eaten by people, infecting them. Indeed consuming any of a variety of food items contaminated by saliva, feces, or urine of bats can infect livestock and people (Schneeberger and Voigt 2016). An outbreak of Nipah virus, which causes symptoms such as fatal encephalitis (brain inflammation), was traced to harvested juice from date palm contaminated by bats (Hunt 2020).

Horseshoe bats are so named for their large horseshoe-shaped nose-leaf, which acts like a parabolic reflector, helping focus echolocation. Severe acute respiratory syndrome

FIGURE 17.6 Honduran white bat (*Ectophylla alba*), a unique species of 'leaf-nosed bat,' found in Honduras, Nicaragua, Costa Rica, and western Panama. The nose leaves of at least some bats are thought to serve roles in modifying and directing echolocation calls (the sonar calls are directed through their nostrils, rather than through the mouth as in many bats with small noses). This tiny species (3.7–4.7 cm long) has snow white fur (one of the very few white bats), a yellow nose, and yellow ears. The bats construct delicate roosting tents from the leaves of *Heliconia* plants by nibbling at the side veins. They mostly eat figs (*Ficus colubrinae*). (a) Bat captured in Costa Rica. Photo by Geoff Gallice (CC BY 2.0). (b) Costa Rican stamp (public domain image) depicting a group of bats.

coronavirus-like viruses have long been known to occur in horseshoe bats (Woo et al. 2006). Zhou et al. (2020) found that the genome of the COVID-19 virus was 96% identical to that of horseshoe bats (identified as *Rhinolophus affinis*) in China. It has been speculated that the COVID-19 pandemic traces to the consumption of horseshoe bats in China, in so-called 'wet markets' where close contact between humans and food animals has resulted in the transmission of many microbes from animals to humans (Gross 2020). (A wet market sells perishable goods such as food and flowers; particularly outside of the Western world, wet markets often sell living animals for food.)

The three species of true vampire bats (which are mostly confined to tropical and subtropical regions of the New World) are the only exclusively blood-feeding mammals and have captured the imagination of many. The concept of vampires traces to Eastern European Middle Ages legends of a corpse returning to life and sucking blood from the neck of human victims (Figure 17.7). The vampire bat was named for the legend rather than the reverse.

False vampire bats are any of several Old World bat genera: *Megaderma*, *Cardioderma*, and *Macroderma* (family Megadermatidae) and the New World genera *Vampyrum* and *Chrotopterus* (family Phyllostomatidae). These large bats used to be believed to feed on blood, like the true vampire bats, but in fact they prey mostly on small vertebrates.

True vampire bats are New World species which feed mainly on the blood of cattle, horses, and wild mammals, and rarely bite humans. They extract relatively small quantities of blood but expose livestock to secondary infections, parasites, and the transmission of viral-borne diseases such as rabies. The common vampire bat (Figure 17.8) became a serious pest when Europeans introduced mammalian livestock into the New World Tropics. In Latin America, paralytic rabies transmitted by vampire bats (almost completely the common vampire bat) is a major cause of mortality in cattle. The common vampire bat is thought to be responsible for the death of about 100,000 domestic cattle annually in South and Central America (Altringham 2011).

Bats widely employ human structures such as buildings and bridges as den sites, and they are usually unwelcome. Bat excrement in attics can pose serious health risks to home owners, including diseases such as histoplasmosis (a fungal infection that can be acquired by inhaling spores) and the introduction of ticks and fleas. Professional removal may be necessary.

Collisions with birds are a well-known hazard to flying aircraft. Although aircraft have been damaged by large flocks of bats (Figure 17.9), this is quite rare (Hill and Smith 1984).

Beneficial Aspects

The most important positive contribution of bats is their enormous consumption of insect pests, particularly benefitting agriculture (Boyles et al. 2011; Riccucci and Benedetto 2014; Taylor et al. 2018; Textbox 17.1). Even small bats can consume thousands of insects in a night, often amounting to half their body weight (Taylor and Tuttle 2018). Some species of fruit-eating bats can

FIGURE 17.7 Postage stamp issued in 1997, featuring the fictional vampire Dracula (from the Bram Stoker 1897 novel of the same name), who could transform into a bat. © Canada Post.

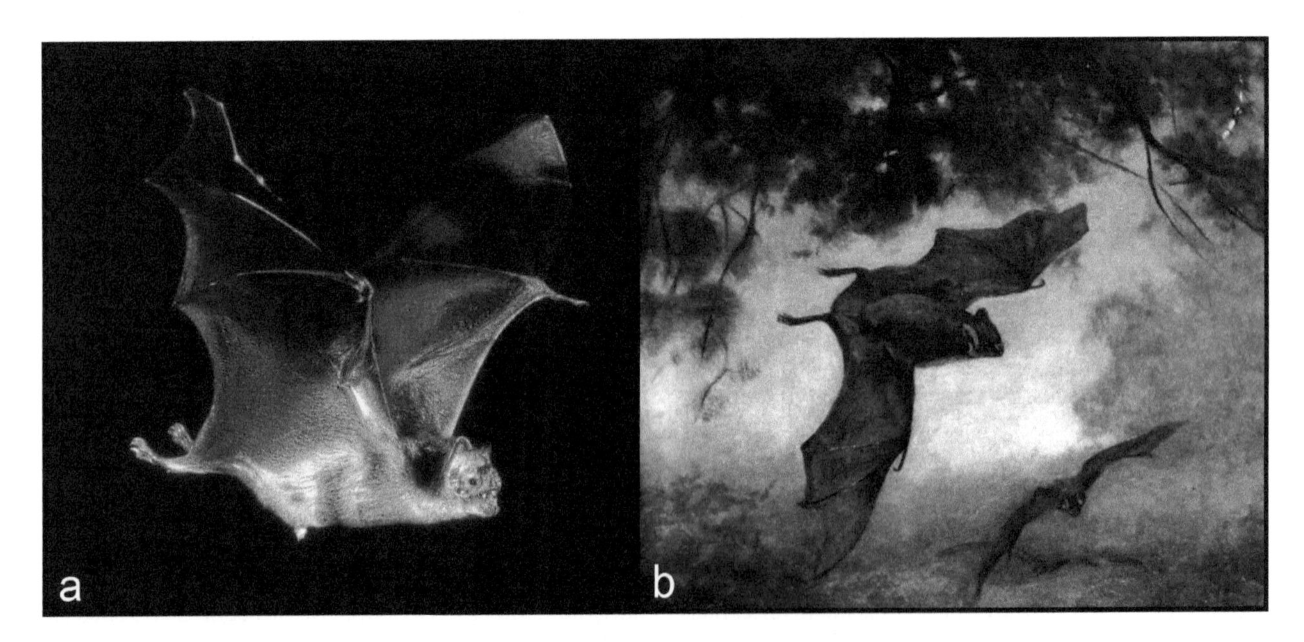

FIGURE 17.8 The common vampire bat (*Desmodus rotundus*) in flight. (a) Photo by Uwe Schmidt (CC BY SA 4.0). (b) Painting (public domain) from Finn (1909).

FIGURE 17.9 Mexican (or Brazilian) free-tailed bats (*Tadarida brasiliensis*), native from the southern half of the U.S. to South America. This bat has been claimed to fly horizontally (as opposed to diving) faster than any other animal, reaching speeds over 160 km/h. The species tends to congregate in the millions at just a few locations, making it vulnerable to habitat destruction. Photo (public domain) by Ann Froschauer, U.S. Fish and Wildlife Service.

TEXTBOX 17.1 IMPORTANCE OF BATS

Bats are often critical to ecosystems. Bats are often considered 'keystone species' that are essential to some tropical and desert ecosystems. Without bats' pollination and seed-dispersing services, local ecosystems could gradually collapse as plants fail to provide food and cover for wildlife species near the base of the food chain. Bats are also critical to the ecosystem through their pest-control services. Throughout the United States, scientists estimate, bats are worth more than $3.7 billion a year in reduced crop damage and pesticide use. And that, of course, means fewer pesticides enter the ecosystem.

—*Weebly.com (2015)*

damage crops, but this pales by comparison with the enormous pest-control contributions that bats make to agriculture.

Guano is accumulated excrement of seabirds and bats. Bat dung is deposited in and near caves housing bats. It is an invaluable fertilizer for plants because of its very high nitrogen content. (Bat guano is a source of saltpeter – sodium nitrate – once employed in the production of gunpowder and other explosives.) Some bats are pollinators, indispensable for hundreds of species of tropical plants. Plants that rely on nectar- or pollen-eating bats for pollination usually open their flowers at night. Notably, bats pollinate agave plants (*Agave* species), from which tequila is made; the durian (*Durio* species), a stinky but popular fruit of Asia; and the African baobab (*Adansonia digitata*), a charismatic tree with an enormously wide trunk (Figure 4.10). Fruit-eating bats distribute seeds. Despite the negative perception of bats by many, they have some potential for ecotourism (Pennisi et al. 2004).

Bats are widely consumed by people in Asia, the Pacific Rim, and Africa, and to a lesser extent in Central and South America (Mildenstein et al. 2016). Eating bats is rare in Western culture, in part because most temperate region bats are relatively small. To a much lesser extent, bats are also sometimes hunted for their perceived medicinal values. Food use of the large-bodied fruit bats of the Old World tropics is especially common. Consumption of bats is contrary to Jewish and Islamic dietary laws. In West Africa, sale of bat bushmeat is sometimes discouraged because of the danger of disease transmission.

Anticoagulants from the common vampire bat (*Desmodus rotundatus*), especially an anti-clotting agent called 'Draculin' or 'Desmoteplase,' might have potential for treating patients with blood-clotting problems (Taylor and Tuttle 2018).

Conservation Aspects

Bats have a severe image problem that has resulted in their persecution. Since the dawn of humanity, people have been frightened by bats. Their nocturnal nature predisposed them to causing fear, since people often naturally are scared of the dark and indeed of most creatures of the night, especially those like bats that are mostly unseen and indeed unheard because of their silent flight. In many cultures, bats were viewed as bad omens and symbols of fear and death. Bats are portrayed as usual companions of witches (Figure 17.10), and vampires (Figure 17.7) traditionally transform themselves into bats. In many cultures, bats figure prominently in folklore and are often associated with evil, bad luck, and death. Demons are often shown with bat-like wings. In Greek mythology, bats were associated with the underworld, and in Mayan culture, the bat was a god of death. There are various persistent myths about bats, including the notion that they attack the hair of humans. Another misconception is that they are dirty, but in fact bats are very clean animals and groom themselves and their young much as cats do.

Occasionally, bats are depicted in a positive way. In the ancient kingdom of Macedonia, bats were considered to bring good luck, a belief that persists today in the Gypsy world. Similarly, in Poland and Arabic culture, bats are occasionally viewed as bringers of good luck. In China, bats are sometimes interpreted as symbols of longevity and happiness. Bats are also sometimes used as heraldic emblems. The well-known comic-book and movie character Batman is a superhero, fighting rogues on behalf of society. Three U.S. states have an 'official state bat': Texas and Oklahoma have named the Mexican free-tailed bat their state bat, and Virginia has dubbed the Virginia big-eared bat as their state bat.

FIGURE 17.10 Creepy Halloween vintage (public domain) postcards featuring witches associating with bats. (a) Circulated in 1917. (b) Circulated in 1910.

'Zoonotic transfer' of infectious agents may occur directly to humans from a primary source such as bats, or through a secondary food animal source infected by the primary source. The outbreak of COVID-19 is thought to trace to the Huanan Seafood Market city in Wuhan, the capital of the Hubei province of China. There, numerous wild animals such as bats, marmots, venomous snakes, and deer are on sale. It has been hypothesized that the novel coronavirus from bats infected uninspected animals and subsequently spread to humans. Critically endangered Malayan pangolins (*Manis javanica*) were shown to harbor the virus, and pangolins in general have been accused of being intermediate carriers of the novel coronavirus between bats and humans (Gross 2020). The COVID-19 pandemic has posed a severe public relations problem for bats (MacFarlane and Rocha 2020). As previous experience has demonstrated, once suspected of being the source of a disease, the response often has been to systematically eliminate the animals considered guilty, regardless of how threatened the species are in nature. Bats are not really to blame for zoonotic diseases such as COVID-19 – humans are (Walsh and Cotovio 2020). The responsible approach is to reduce or eliminate intensive harvest and marketing of diminishing wild animal species from nature, where alternative food sources are available. Regardless of campaigns to eliminate bats suspected of posing an immediate disease threat to people, the use of bats as food represents a survival risk for the rarer bat species (Mickleburgh et al. 2009; Mildenstein et al. 2016).

The potential of bats to transmit diseases represents a significant source of danger to humans, but current reporting tends to be alarmist, endangering conservation efforts (Textbox 17.2). Responsible reporting is particularly important in regard to rabies transmission by bats (Lu et al. 2016, 2017). Bats do pose dangers to humans, but humans are far more dangerous to bats.

Bats in North America are the victims of white-nose syndrome, a skin disease caused by the fungus *Pseudogymnoascus destructans*. 'White-nose syndrome is the most devastating wildlife disease of mammals in recorded history' (Bure and Moore 2019). The fungus grows about the muzzles and wings of bats while they hibernate. The emerging disease was first discovered in New York State in 2006, initially spread over eastern North America, and is currently expanding in North America (Hoyt et al. 2020). The fungus was introduced to North America from Europe, where its potential for expansion is undetermined (it does seem that many Old World bats are resistant). White-nose syndrome has killed millions of bats in North America, wiping out most of the bats of some species in some regions. Three endangered species and one threatened species have been affected by the disease. People visiting bat caves may pick up the fungus on their clothes, and spread it.

There is evidence that bat fatalities occur at wind turbines because of 'barotrauma' – lung tissue damage (embolism) caused by rapid air-pressure reduction near moving turbine blades (Baerwald et al. 2008). 'Collisions with wind turbines and white-nose syndrome are now the leading causes of reported multiple mortality events in bats' (O'Shea et al. 2016).

About two dozen bat species have been evaluated as Critically Endangered (the highest level of threat defined by the International Union for the Conservation of Nature), about 50 as Endangered, 100 Vulnerable, and 80 as Near Threatened, but more than half have not yet been evaluated (Taylor and Tuttle 2018).

Despite the dislike of many for bats, there are substantial numbers of people and organizations involved in bat conservation, at least in Western nations (Figure 17.11). 'Bat gates'

TEXTBOX 17.2 ALARMIST REPORTS ABOUT VIRUS-CARRYING BATS ENDANGERS THEIR CONSERVATION

The recent upsurge in bat-borne virus research has attracted substantial news coverage worldwide. A systematic review of virological literature revealed that bats were described as a major concern for public health in half of all studies (51%), and that their key role in delivering ecosystem services was disregarded in almost all studies (96%). Although research on zoonoses is of the utmost importance, biased framings of bats can undermine decades of conservation efforts. We urge researchers and science communicators to consider the conservation impacts of how research findings are presented to the public carefully, and, whenever possible, to highlight the ecological significance of bats, their dire conservation situation, and their importance for human well-being.

—*López-Baucells et al. (2018)*

There is a well-established body of evidence indicating that people seldom protect, and often despise, or even kill, animals they fear. This makes bats exceptionally vulnerable. Throughout history, they have been objects of fear and hostility across many cultures, arguably due to their nocturnal and elusive behavior. Also, biased media coverage has framed bats as exceptionally dangerous virus reservoirs, generating frightening headlines worldwide, that are jeopardizing decades of conservation progress.

—*Tuttle (2017)*

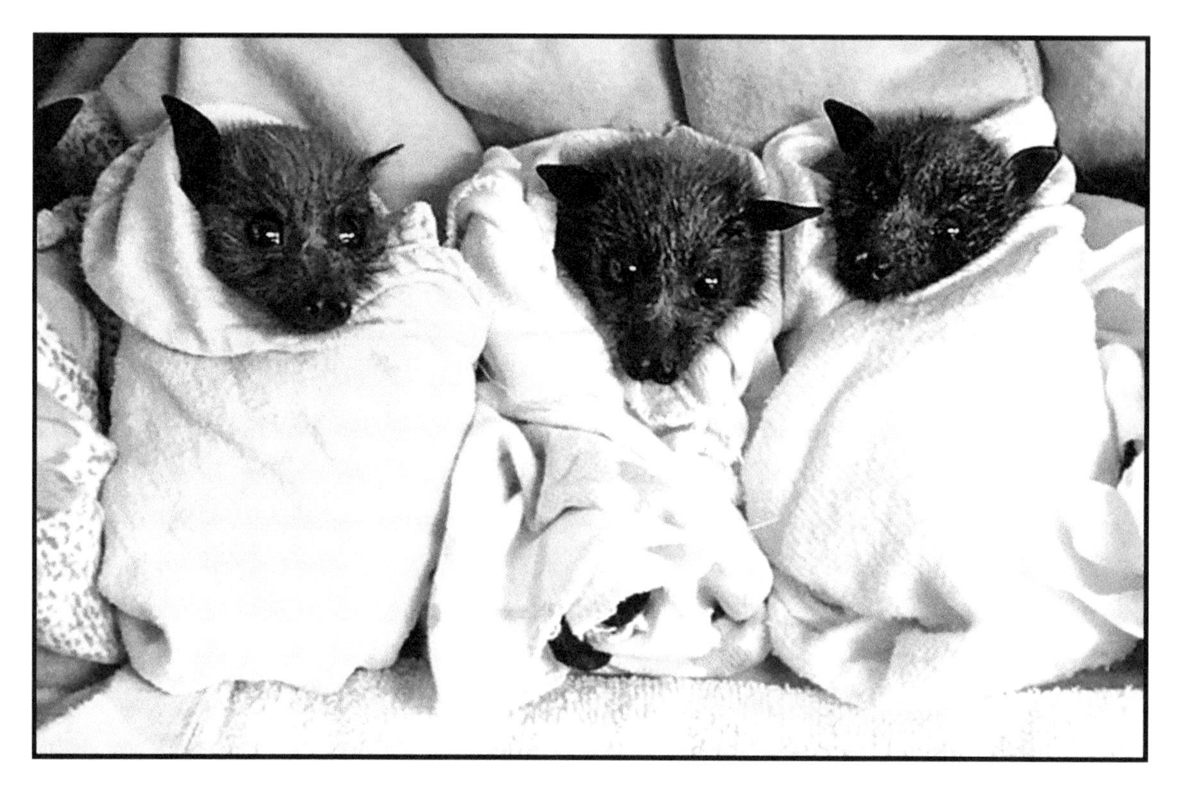

FIGURE 17.11 Gray-headed flying fox (*Pteropus poliocephalus*) pups, 2–3 weeks old, in a crib in the Bat Hospital of Wildlife Australia. Photo by Wcawikinfo (CC BY SA 3.0).

FIGURE 17.12 Large bat houses. (a) A bat house east of Comfort, Texas, built in 1918 and designated a Texas Historic Landmark. The 9 m tall structure was originally intended to attract bats to consume mosquitoes. Photo by Larry D. Moore (CC BY SA 3.0). (b) A bat house at the Norikura Nature Conservation Center in Nagano Prefecture, Norikura Heights, Azumi, Matsumoto City, Nagano Prefecture. This was built in 1996, and is used as a conservation and research facility. Photo by Mountainlife (CC BY SA 3.0).

which keep people out while allowing bats to enter and leave freely are now often installed at the entrance of caves housing large numbers of bats (Tobin and Chambers 2017). Disturbance of sleeping bats by people can be a substantial threat to them, as mothers may abandon their young, and rousing hibernating bats forces them to burn energy that they may need to survive winter. In addition to preserving natural roosting sites, there are efforts to promote bat welfare by providing artificial roosts (Figure 17.12).

HOUSE MICE

INTRODUCTION

Numerous small rodents are known as mice, for example 'dormice,' which are small rodents, sometimes employed as pets. However, by far the most prominent is the common house mouse, *Mus musculus*. This is sometimes called the 'field mouse' although many other species also have this name, especially any of the other approximately 40 species of the genus *Mus*. The house mouse originated in the grasslands of Asia (northern India according to Boursot et al. 1996) and has spread around the world. It may have a more extensive range than any other mammal except *Homo sapiens* (MacKay and Invasive Species Specialist Group 2010). This species has become intimately associated with people (Boursot et al. 1993), negatively as a pest, and usefully as a domesticated pet and much more importantly as the laboratory mouse, serving as an experimental animal in biology, medicine, industry, and education.

The evolution and classification of the house mouse have been extensively studied (Fox et al. 2007; Macholán et al. 2012). Its considerable degree of variation is indicated by Hardouin et al. (2015): 'the more widely distributed populations are grouped into three different subspecies: *Mus musculus musculus* in Eastern Europe, Central and North East Asia, *Mus musculus domesticus* in Northern Africa and Western Europe, and *Mus musculus castaneus* in South East Asia. These last two subspecies have further expanded in modern times to the Americas, Australia, and Oceania. In addition, *Mus musculus molossinus*, a hybrid between *M. m. musculus* and *M. m. castaneus* found in Japan, is often considered as a subspecies on its own. Closer to the center of the distribution, *Mus musculus gentilulus* has been identified in the eastern part of the Arabic peninsula on the basis of its mitochondrial DNA lineage while from the same type of data it has been shown that certain populations considered as *M. m. castaneus* in Iran, Pakistan and Afghanistan should probably be considered as belonging to further sub-specific groups. Moreover, another completely independent lineage has recently been identified on this basis in Nepal.'

Adult house mice (Figure 17.13) usually have bodies 6.5–10 cm long, with tails 5–10 cm long, and they generally weigh less than 20 g (range: about 12–30 g). Wild forms are usually dark (gray, light brown, or black) while domesticated mice are available in a variety of colors. The house mouse superficially resembles the brown rat but is much smaller (Figure 17.14). House mice breed prolifically, often reaching sexual maturity at 35 days, and bearing up to 10 litters of 4–7 in a year. Wild house mice usually die within a year, but domesticated mice can live

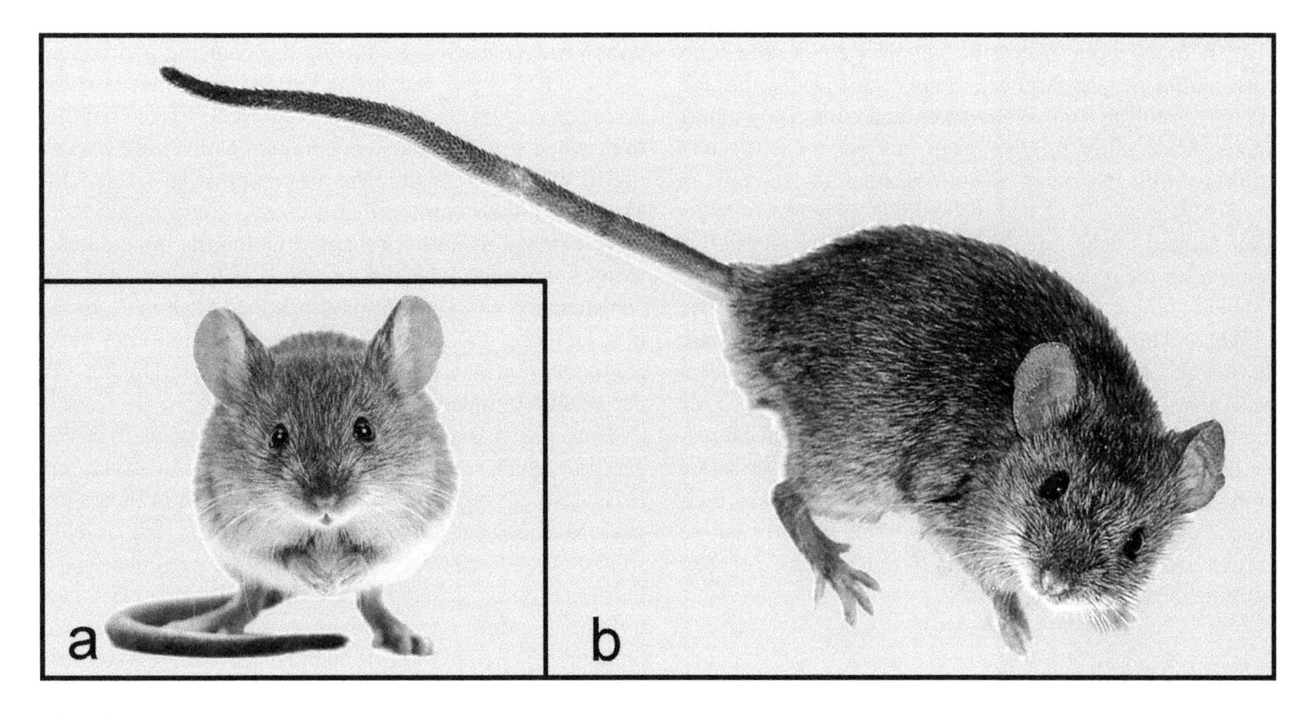

FIGURE 17.13 House mouse (*Mus musculus*). (a) Source: Shutterstock, contributor: Szasz-Fabian Jozsef. (b) Photo by George Shuklin (CC BY 2.0).

FIGURE 17.14 Comparison of house mouse (*Mus musculus*, bottom right) and brown rat (*Rattus norvegicus*, top). Source: Nelson, E. W. 1918. Wild Animals of North America. Washington, DC, The National Geographical Society.

for 2 or several years (rarely as long as 6 years). A baby mouse is called a 'pinky,' a male is a 'buck,' and a female is a 'doe.'

The ecology of house mice has been extensively studied (Berry 1970). They thrive in and near buildings, in open fields, and in agricultural lands. Indeed, human-associated mice are much more successful than they are in wild areas remote from people. A house mouse can squeeze through a hole as small as 1 cm in diameter, and so it easily enters homes. Mice often migrate from outdoor locations into buildings with the onset of cold weather in late fall, in search of warmth, food, and shelter. (Other species of mice, often termed 'field mice,' often also behave similarly.) House mice are quick runners (up to 13 km per hour), good climbers, jumpers, and swimmers. Mice prefer to be active at night and mostly consume plant matter although they are omnivorous. Mice tend to nibble, eating small but frequent meals that can amount to as much as 20% of their body weight daily, and as a result, they pass 50 or more fecal pellets per day (Witmer and Jojola 2006). Although mice are rarely observed in buildings, the droppings give away their presence. Mice obtain much of their water from their food and can live on limited availability of water because of their ability to concentrate their urine.

Harmful Aspects

Indoors, house mice feed on food left unprotected by humans (Figure 17.15), degrading and contaminating domestic, agricultural, and commercial edible materials along with their packaging. Feces deposited by mice in buildings are a potential source of disease and also cause psychological stress. Mouse odor, especially from the urine of male mice, is remarkably persistent (Kwak et al. 2016). The teeth of mice grow continuously, facilitating gnawing and damaging wood, textiles, and insulation. Of particular concern, mice chewing on electrical wires represent a risk of building fires. Mice similarly damage communication wires, resulting in widespread shutdowns of computer-based systems. Mice are more common than rats and cause significantly more damage to indoor property. Outdoors, mice damage many field crops especially grains and legumes, and also consume and contaminate livestock feed at animal production facilities. Occasionally in some regions, mice reach plague proportions, numbers sometimes exceeding 200,000 per hectare (Singleton and Redhead 1990).

House mice and the fleas and other parasites that they carry transmit several diseases to humans, livestock, and pets, although other rodents are often more significant carriers. Mice can be sources of human diseases, such as leptospirosis, cryptosporidiosis, salmonellosis, streptobacillosis, murine typhus, rickettsial pox, and tularemia. However, unlike rats, they are not major carriers of the plague or of hemorrhagic fevers like hantavirus (Phifer-Rixey and Nachman 2015). For homeowners, a particular concern is that mouse feces can contain *Salmonella* bacteria causing food poisoning. Generally, the house mouse is not an

FIGURE 17.15 Mice damaging food. Photo (public domain) from pxhere.

important source of infectious diseases affecting humans (Blackwell 1981).

Rodent control with toxins (especially rodenticides, designed to be consumed, but also fumigants and repellents; note Figure 17.16) is widely practiced, with unintended harm to innocent species. Most rodenticides sold over the counter are anticoagulants, which kill by interfering with normal blood clotting, causing the animals to die from internal bleeding. Anticoagulants are less effective against mice than against Norway rats, and house mice have evolved resistance against anticoagulants (Prescott et al. 2018). Mice baits are usually poisoned seeds or grains, and when discarded, they are a threat to wild animals consuming them.

In some regions of the world, invasive house mice have seriously endangered native wild species. Introduced house mice are a particular threat to the native flora and fauna of islands, where the biota has not adapted to them, especially to seabird nestlings (Cuthbert and Hilton 2004; Wanless et al. 2007; Angel et al. 2009). Mice have been eliminated from some islands, but they are harder to remove than rats (MacKay et al. 2007).

Beneficial Aspects

The house mouse is the world's most frequently used vertebrate laboratory animal, due in large part to low cost, ease of maintenance and handling, and high reproductive rate (Berry 1981). In Britain (which compiles the best national statistics on animal usage) in 2018 about 1 million of 1.8 million experimental vertebrate animals were mice (Understanding Animal Research 2019). (Dogs, cats, horses, and primates have special protection under U.K. law and can only be used with special justification.) Because of opposition to animal experimentation, most countries and organizations are reluctant to report accurate numbers of laboratory animals employed. Estimates of world annual usage of all vertebrate animals tend to range up to over 100 million. However, if one were to extrapolate world usage just for mice, about 100 million experimental mice may be sacrificed annually.

Many standardized inbred and mutant mouse strains are used in medical research (Figure 17.17), and genetically engineered strains are being produced to study and potentially cure inherited diseases of humans. The inevitable harm and pain inflicted on mice has been argued to be morally justified by the benefits to people (e.g., Baertschi and Gyger 2011). Scientific and medical experimentation on animals (especially mammals) is subject to national guidelines governing their welfare. The Association for Assessment and Accreditation of Laboratory Animal Care International (AAALAC; https://www.aaalac....org/) is a private, non-profit organization promoting humane treatment of animals employed in science. People for the Ethical Treatment of Animals (PETA; https://www.peta.org/) is the largest animal rights organization in the world

FIGURE 17.16 Early 20th-century lithograph showing a farmer killing mice by piping poison gas into an underground passage. Credit: Wellcome Collection (CC BY 4.0).

and promotes the abandonment of animal usage. Ethical treatment of animals and ethical treatment of biodiversity are different areas, but both require consideration of appropriate relationships of people and the rest of the living world.

'Fancy mice' (Figure 17.18), bred for docility and attractive colors, have been employed as pets for centuries, and most laboratory mice strains arose from fancy mice (Royer 2015). Domesticated mice termed 'feeder mice' are often sacrificed as a food source for carnivorous pets, especially snakes. Local animal cruelty laws may forbid feeding live mice (or other rodents) to snakes, and many but by no means all suppliers furnish only dead and frozen mice.

Conservation Aspects

Mice are such a ubiquitous pest that it is odd to consider that some unique wild kinds are in danger of extinction (Textbox 17.3; Figure 17.19). Nevertheless, conservation of rodents, which make up 40% of all mammal species, is a difficult undertaking (Textbox 17.4).

Conservation of mice is complicated by distinctly ambivalent attitudes toward them. Of course, the harm that they do as pests and their potential to carry diseases generate hostility. As well, some people seem to be born with, or develop early in their lives, a pathological fear of mice, a psychological condition termed musophobia (murophobia is fear of rats and mice;

FIGURE 17.17 An albino laboratory mouse strain (BALB/c) widely used in animal experimentation. Photo by Aaron Logan (CC BY 1.0).

FIGURE 17.18 'Fancy mice' (domesticated strains bred as pets). Public domain illustration from Haacke (1893).

Figure 17.20). Fear of mice tends to be associated with fear of rats. Animal phobias tend to be considerably more common in women than in men (see discussions in Chapters 15 and 16).

In a more favorable vein, mice have been associated with respected religious and mythological figures (Powell 1929; Festing and Lovell 1981; Figure 17.21) and have often been treated kindly in literature for children (Figure 17.22), notably

TEXTBOX 17.3 SAVING MICE FROM EXTINCTION

It seems strange to worry about the disappearance of animals many people consider pests. Nevertheless, dozens of mouse subspecies are going extinct around the world. For example, the Pacific pocket mouse is sitting on some of the most desirable coastal real estate in California. Fortunately, this little guy is protected by conservation regulations strong enough to deter developers from pursuing building projects in coastal lands worth millions of dollars, causing projects to be put on hold or completely shut down to insure the health and safety of its habitat.

—*Thornton (2013)*

FIGURE 17.19 The St Kilda house mouse (*Mus musculus muralis*), an extinct subspecies of the house mouse that was found only on the islands of the St Kilda archipelago of northwest Scotland, and vanished before World War II. These mice were as much as twice as heavy as house mice on the mainland of Scotland. Painting (public domain) from Barrett-Hamilton (1899).

TEXTBOX 17.4 DIFFICULTIES WITH RODENT CONSERVATION

Rodents are a hard sell when it comes to conservation. In most human cultures, rodents are generally viewed as vermin... Admittedly, rodents are in fact often pests, being guilty of damaging crops, pilfering stored grains, invading households, and spreading zoonoses... As a group, rodents face the same suite of threats to their welfare and existence as do other organisms. Paramount are habitat destruction, over-exploitation by humans, and disease. A new threat of unknown but potentially major influence is rapid global warming. Additionally, rodent conservation is hampered by indifference and complacency. The former is the result of a conservation focus on larger, more charismatic and generally familiar species, as well as residual negative attitudes toward rodents. The latter stems from the view that rodents are abundant, have high reproductive rates, widespread distributions, and are adaptable. Therefore, they can take care of themselves, or so it is thought.

—*Lidicker (2007)*

FIGURE 17.20 Musophobia (fear of mice), shown in a children's book illustration (1898, public domain) titled 'Three Blind Mice.' Credit: New York Public Library Digital Collections.

in Beatrix Potter's books (Figure 17.23). The conflict between mice and cats has been a favorite theme in stories for young people, most memorably between the comic characters Tom (the cat) and Jerry (the mouse) (note Figure 17.24). Micky Mouse and Mighty Mouse are familiar, heroic cartoon characters, and indeed mice are often the subjects of popular art (Figure 17.25).

The public's attitude toward mice is reflected in how they are treated in the traditional fairy tale Cinderella. Just as she was elevated from a low-class servant girl into a princess, lowly small mice were transformed into a team of big, beautiful horses to pull her magically transformed pumpkin-coach (Figure 17.26). At least, the mice received honorable mention.

Despite the generally negative attitude toward mice, there is recognition of their entertainment and scientific values (Figure 17.27).

FIGURE 17.21 Lord Ganesha, elephant-headed Hindu deity, mounted on his giant mouse (also often said to be a rat). Subduing the mouse by riding it symbolizes control over the rodent's destructive tendencies. Bazaar art (public domain) from India, 1910.

FIGURE 17.22 Anthropomorphic illustrations (public domain) in a children's book. Source: Coloma et al. (1914).

FIGURE 17.23 Examples of Beatrix Potter's illustrations (public domain) for children's books.

FIGURE 17.24 Early 20th-century postcard (public domain) showing cat dining with mice. The French artist, Maurice Boulanger, was known for cute and funny anthropomorphic cats.

FIGURE 17.25 Street art featuring mice. (a) In Albania. Photo by Quinn Dombrowski (CC BY 2.0). (b) A mouse-eared Frankenstein in Montreal. Photo by Photographymontreal (public domain). (c) In New Orleans (attributed to Banksy). Photo by Mark Gstohl (2.0).

FIGURE 17.26 Cinderella: a tale about human preferences and prejudices. (a) Cinderella's mouse-powered pumpkin-wagon. Source (public domain): Ames, E. 1901. The Bedtime Book. London, Grant Richards. Credit: Special Collections Toronto Public Library. (b) Cinderella's fairy godmother transforming the pumpkin into a magnificent enchanted carriage and changing the humble mice into a much more attractive team of beautiful horses. Source: Shutterstock, contributor: Arbit.

FIGURE 17.27 Tributes to mice. (a) Statue in Oakdale, Wisconsin of a huge mouse on cheese, for which the state is famous. Photo by Amy Meredith (CC BY ND 2.0). (b) Monument to the laboratory mouse, in Novosibirsk, Siberia, Russia. The sculpture is located at the Institute of Cytology and Genetics of the Russian Academy of Sciences. The mouse is knitting a double helix of DNA. Photo by Irina Gelbukh (CC BY SA 3.0). (c) Statue of Walt Disney and Mickey Mouse at Disneyland, Anaheim, California. Photo by Wood26 (CC BY SA 3.0).

HYENAS

Introduction

Living hyenas (occasionally spelled hyaenas) include four species constituting the family Hyaenidae. The spotted hyena is the dominant species and is genetically separated from the others. The brown and striped hyenas are somewhat related. The aardwolf is not a wolf and is considered to be a 'hyena' despite its divergent lifestyle. Although they are dog-like in appearance, hyenas are more closely related to cats (Smith and Holekamp 2019). Hyenas are among the leading carnivores of Africa (Figure 17.28).

The spotted hyena (*Crocuta crocuta*; Figures 17.29 and 17.30) is the largest living hyena species. It is also known as laughing hyena and indeed is the hyena that 'laughs' (Brottman 2012). The females on average are about 10% larger and much more aggressive than the males, frequently dominate them, and lead the clans (Watts and Holekamp 2007). The biggest specimens weigh over 70 kg and can be almost 1 m tall at the shoulder. The species is widespread but fragmented over Sub-Saharan Africa, especially East and southern Africa. It was once common in Eurasia. Spotted hyenas are the most social of hyenas, occurring in groups sometimes exceeding 100, and they are the most successful and frequent of the large African carnivores. Although they scavenge, at least half of their food is obtained by hunting, contrary to the depiction in the popular and influential Disney film *Lion King*. Indeed, spotted hyenas compete fairly successfully against the African lion (Kruuk 1972). Spotted hyenas have powerfully muscled jaws (possibly the

FIGURE 17.28 The dominant large predators of Africa. Artwork by M. Antón in Turner and Antón 2006 (CC BY 4.0).

FIGURE 17.29 Spotted hyena (*Crocuta crocuta*). (a) Photo by Charles J. Sharp (CC BY SA 4.0). (b) Female and cubs. Photo by Bernard Dupont (CC BY SA 2.0).

strongest of any living mammal) with enlarged premolars enabling them to shear and crush even the largest bones. Their highly acidic stomachs facilitate digestion of bones, horns, and teeth that competing carnivores cannot use. The same is true for striped and brown hyenas (Smith and Holekamp 2019). 'By virtue of their size and abundance, spotted hyenas are among the most significant predators on the African Savannah. In terms of tonnage of meat consumed, they are, perhaps, the most significant terrestrial carnivore on the planet' (Glickman 1995). Spotted hyenas are keystone predators in many African ecosystems, i.e., they have a predominant influence on the balance of species (Trinkela 2009). In much of Africa where lions are not present, they are the apex predator.

The striped hyena (*Hyaena hyaena*, also known as the Barbary hyena; Figure 17.31) is the smallest 'hyena' species (if one excludes the aardwolf), adults averaging about 35 kg, but sometimes weighing as much as 55 kg. Striped hyenas are indigenous to North and East Africa, the Middle East, the Caucasus, Central Asia, and the Indian subcontinent. They rarely form small packs, but mated pairs cooperate to raise their cubs. The species is mostly a nocturnal scavenger. Like skunks, when threatened they spray a noxious liquid from their anal glands.

The brown hyena (*Parahyaena brunnea*, Figure 17.32; it is also called Strandwolf, a reference to the frequency of the species in the Strand of the Western Cape of South Africa) is the rarest species (Eaton 1976), occurring only in southernmost Africa (Namibia, Botswana, Zimbabwe, Mozambique, and South Africa). It is adapted to desert and semi-desert areas and open woodland savannahs and survives near urban areas by scavenging. Indeed, brown hyenas are mostly scavengers. Adults weigh about 40 kg, the males slightly larger than the females. Like wolves, they live in small clans made up of a mated pair and their offspring.

The aardwolf (*Proteles cristata*; Figure 17.33) looks like a small hyena, usually weighing 7–10 kg and having

FIGURE 17.30 Spotted hyenas consuming their prey. Source: Shutterstock, contributor: EreborMountain.

FIGURE 17.31 Striped hyena (*Hyaena hyaena*). Photo by Rigelus (CC BY SA 4.0).

a height at the shoulder of 40–50cm. The name 'aardwolf' means earth-wolf in Afrikaans. The aardwolf has been placed in its own subfamily, Protelinae, the other three species in subfamily Hyaeninae. Unlike their hyena relatives, they do not kill or scavenge large animals, although they have sometimes been accused of eating corpses. The cheek teeth of aardwolves are specialized for eating harvester termites, especially of the genus *Trinervitermes*. Unlike aardvarks which dig into termite mounds, aardwolves lick termites off the ground with their long sticky tongues and can consume up to 250,000 termites in a night. Aardwolves are found in two regions: Southern Africa, and in East and Northeast Africa. They are restricted to these regions where their food – termites of the family Hodotermitidae – occur. Like the striped hyena, threatened aardwolves can spray a noxious liquid from their anal glands. Also like striped hyenas, they are mostly nocturnal and live as monogamous pairs along with their young.

In prehistoric times, there were many species of hyenas, the largest of which was the giant short-faced hyena

FIGURE 17.32 Brown hyena (*Parahyaena brunnea*). Photo by Bernard Dupont (CC BY SA 2.0).

FIGURE 17.33 Aardwolf (*Proteles cristata*). Photo by Greg Hume (CC BY SA 3.0).

FIGURE 17.34 Drawing of restoration of the extinct giant short-faced hyena (*Pachycrocuta brevirostris*) by Mariomassone (CC BY SA 4.0), in comparison to a human (added by B. Brookes).

(*Pachycrocuta brevirostris*; Figure 17.34), some individuals reaching over 1 m at the shoulder and weighing well over 100 kg (Turner and Antón 1996). The species arose about 3 million years ago, spread across Eurasia and Africa, and became extinct about 400,000 years ago.

Harmful Aspects

During periods of extreme stress (war, famine, disease), there has been extensive predation on weakened and unprotected people by hyenas (Gade 2006). Today, 'Spotted hyenas and humans often come into conflict where they coexist in the landscape. Usually the conflicts involve hyena predation on livestock and retaliatory killings by humans; however, direct attacks on humans by rabid or otherwise healthy hyenas do occur' (Baynes-Rock 2013). Hyenas significantly attack livestock in Africa, but protective measures are usually effective (Gade 2006). In parts of Africa, spotted hyenas have adapted to infiltrating cities, where they scavenge garbage, prey on feral dogs and cats, eat human corpses, and represent a potential danger to children and homeless people (Gade 2006). Hyenas can also vector diseases of humans and livestock, but this is not considered a notable danger.

Beneficial Aspects

Like most carnivores, hyenas play important roles in regulating the balance of their prey in relation to other animals in ecosystems. Carnivores usually catch weaker prey (especially very young, old, or sick animals), which tends to leave the strongest and fittest individuals. Like other scavengers, hyenas also contribute to the recycling of energy in ecosystems and to the prevention of accumulating corpses that can contribute to diseases. Hyenas can consume carrion in advanced decomposition because of their resistant immune system and very strong digestive system, so they can survive the presence of infectious agents like anthrax that could be deadly to other species. Referring to spotted hyenas in the Horn of Africa, Gade (2006) stated that they are 'tolerated as efficient sanitation units' for their roles in removing garbage and carrion from towns, notably reducing rodents, flies, and bad odors.

Hyenas are attractions in zoos (but see Textbox 17.5). Rarely, they have been kept as pets. In ancient Egypt, it appears that hyenas were artificially fattened as food (Zeuner 1963). Under Islamic law, the meat of hyenas is considered halal (acceptable for consumption), and in areas of North Africa, the Middle East, Iran, and Pakistan, hyena meat has become popular (Tubei 2019).

Conservation Aspects

The International Union for Conservation of Nature currently assesses the spotted hyena and aardwolf as 'Least Concern,' although spotted hyenas have become locally extinct in parts of Africa. Habitat loss and widespread hunting have downgraded the assessments of the striped

TEXTBOX 17.5 HYENA CONSERVATION IS PENALIZED BY THEIR BAD REPUTATION

Spotted hyenas, or their immediate relatives, the brown and striped hyenas, are rarely found in zoological parks… this serious miscarriage of biological justice can be traced to the poor public reputation of the hyena. There are no 'save-the-hyenas' committees, and their persistent public relations problems could have very serious consequences for the preservation of hyena habitats and the long-term prospects for these species.

—*Glickman (1995)*

Hyenas inspire horror in people… in people's minds, hyenas are inexorably linked with garbage cans, corpses, feces, bad smells, and hideous cackles. Indeed, in places hyenas subsist on the refuse of human society, and that side of their behavior is, of course, most often noticed.

—*Kruuk (1972)*

We have always imagined the hyena to be involved with dead bodies, and, as a result, we have convinced ourselves that hyenas are vile, horrible creatures. In mythology and magic, they have been associated with putrefaction and the macabre, with waste and disease. The hyena is the totem animal of the outcast and the taboo, lurking in wastelands, laughing and scavenging. We think about hyenas this way because it is easy to do so, and because we need to have our villains, even in the animal world. But these imagined creatures, almost universally feared and reviled, are a product of human culture, not of nature.

—*Brottman (2012)*

and brown hyenas to 'Near-threatened.' Humans and lions cause most adult hyena mortality (Smith and Holekamp 2019). Spotted hyenas are the largest of the hyenas, and when harvested for trophies hunters seek the largest, which will often be the dominant breeding alpha females, thereby removing some of the superior genes and leadership of the clan.

The striped hyena (Figure 17.35) is at risk. 'In some parts of its global distribution range, striped hyenas are considered as critically endangered… the striped hyena is already extinct in many localities… Populations are generally declining throughout their geographical range due to persecution, poisoning, and hunting for meat or medicinal purposes' (Alam et al. 2015).

The brown hyena is particularly at risk. 'Despite its listing as Near Threatened, brown hyenas continue to be persecuted, often considered as problem animals by farmers or killed for trophy hunting. Incidental and often deliberate poisoning, shooting, and trapping of these animals all hamper the survival of this ecologically important species' (Westbury et al. 2018).

Hyenas, particularly the spotted hyena, are mostly viewed negatively in Western culture, where they are considered to be dangerous, vicious, ugly, cowardly, and treacherous (Gould 1981; Glickman 1995). They are invariably depicted in a bad light in movies and cartoons, often as evil, deceitful bullies. Hyenas are also viewed negatively to a considerable extent in African tradition – often characterized as dull-witted despite it being known that hyenas are cunning and dangerous predators (Crandall 2002). In North African folklore the 'werehyena' was equivalent to a werewolf. In Middle Eastern literature and folklore, striped hyenas were often viewed as symbols of treachery and stupidity. The extensive historical and modern vilification of hyenas makes it extremely difficult to enact conservation measures for them (Textbox 17.5).

Illustrative of the bad light in which hyenas have been shown is the story of 'The Beast of Gévaudan,' interpreted as a man-eating animal which terrorized the former province of Gévaudan, France, allegedly killing hundreds between 1764 and 1767. It is not possible today to determine how much of the story is true, and what was the actual cause of the killings at the time, but one popular interpretation was that a hyena, perhaps escaped from a zoo, was responsible (Figure 17.36).

By analogy with humans, scavengers are interpreted as cowards who would rather steal meals from more

FIGURE 17.35 Striped hyena (*Hyaena hyaena*, at left) and spotted hyena (*Crocuta crocuta*, at right), the two species that are least threatened at present. Lithograph (public domain image) designed by Friedrich Specht (1839–1909) in 1878 and printed by Leipziger Schulbilderverlag von F. E. Wachsmuth, Leipzig.

successful predators than hunt or kill their own prey. Scavengers are also often associated with gluttony, uncleanliness, and disease. While the scavenger stereotype is inaccurate for spotted hyenas, it is at least somewhat applicable to striped and brown hyenas. As discussed earlier for vultures (Chapter 16), although scavengers are not admired by humans, they play indispensable roles in ecosystem function.

For over a thousand years, records indicate great concern about the role of hyenas as graverobbers of human corpses. Hyenas are able to locate buried bodies and dig them up (Figure 17.37), and in areas where hyenas are indigenous, care is often taken to keep them from accessing the bodies. In Moslem areas of Africa and the Middle East, walls are often built around cemeteries. In times of war, epidemics, and famine, consumption of bodies by hyenas has been widespread (Gade 2006). Consuming dead bodies is one of the natural ecological roles of hyenas, but when they do so in graveyards they greatly offend humans and make it extremely difficult to sympathize with their welfare.

Spotted hyenas are often disturbingly noisy, sometimes emitting bloodcurdling howls or more often maniacal laughs or demented giggles which can seem remarkably human-like. Although these sounds seem sinister and distressful to many, it should be noted that spotted hyenas have a large repertoire of calls that serve for communication among the numerous members of the clan.

Still another odd feature of hyenas is their stubby hind legs, which seem disproportionately short, and give their bodies a curious sloping appearance. However, while the resulting gait seems awkwardly lumbering, it appears to increase energy efficiency, allowing the animals to lope easily. Unfortunately for hyenas, the movements make them appear as if they are skulking, further adding to their reputation as deceitful. As pointed out by Glickman (1995), 'Bears also have short hindlegs and [this] has not prevented

FIGURE 17.36 Eighteenth-century print (public domain) depicting the Beast of Gévaudan, allegedly the savage killer of several hundred people in 18th-century France. It is often interpreted as a hyena.

FIGURE 17.37 Striped hyena that has dug up human remains in a graveyard. Source (public domain): Buffon, G. L. L. 1855. *Oeuvres complètes*. Paris, Garnier.

FIGURE 17.38 Old French trade card (public domain) showing a striped hyena being hunted in North Africa.

them from being adopted as positive cultural icons in the form of children's toys.'

'Sexual mimicry' is one sex mimicking the opposite sex in behavior, appearance, or chemical signaling. It is frequent in invertebrate species but rare in vertebrates. An extreme form of sexual mimicry occurs in spotted hyenas. Female spotted hyenas are notorious for having a 'pseudopenis' – a clitoris that is large, elongated, erectile, and looks like the male penis when erect. In place of a separate vagina and urethra, the pseudopenis contains a single urogenital canal used to urinate, have sex, and deliver babies – a unique anatomy among female mammals. (In addition, the outer labia are fused and filled with tissue, taking the form of a 'pseudoscrotum' resembling the male's scrotum.) This odd anatomy led to the false interpretation, persistent to this day, that spotted hyenas are hermaphrodites. The possible adaptive advantages to spotted hyenas are unclear (Gould 1981; Muller and Wrangham 2002), but it is certainly disadvantageous to their reputation, contributing to their image as extremely deviant.

Hyenas have often been accused of smelling bad, but since they often feed on decaying carcasses this might be inevitable. Like many other animals, hyenas also sometimes roll in smelly materials for social interaction motives. Additionally, hyenas produce pungent anal gland secretions called 'hyena butter,' which they employ to mark their territories.

Hyenas have been hunted for body parts employed in traditional medicine (Frembgen 1998), a doubtful usage that unfortunately is paralleled by similar persecution of many other animals. Hyenas have also been hunted for sport (Figure 17.38), but infrequently because they are not considered to make attractive trophies.

In modern times, hyenas are particularly attracted to road-kill, with the result that they themselves become traffic victims (Brottman 2012). In areas of war, many striped hyenas have been killed by land mines (Brottman 2012).

It is ironic that hyenas are viewed as cowardly, much less worthy of conservation than their brave distant cousins, the iconic, charismatic big cats of Africa, with which they compete (Figure 17.39). Lions, especially well-maned specimens, are frequently depicted at rest in noble poses, while hyenas are frequently shown looking quite vicious (Figure 17.40). In fact, lions are at least as savage, and just as willing to scavenge, but are fortunate in being far more appealing to the aesthetic senses of humans.

FIGURE 17.39 Competition of hyenas and African big cats. (a) Brown hyena stealing a springbok from cheetahs, in South Africa. Photo by Derek Keats (CC BY 2.0). (b) Museum diorama showing spotted hyenas attempting to rob lions of their killed zebra. Credit (public domain photo): Bruce Emmerling, Pixabay.com.

FIGURE 17.40 Lithograph showing the four species of hyena. (a) Brown hyena (*Parahyaena brunnea*). (b) Aardwolf (*Proteles cristata*). (c) Striped hyena (*Hyaena hyaena*). (d) Spotted hyena (*Crocuta crocuta*).

Source (public domain): Craig (1897, plate 13).

RATS

Introduction

Although many species (mostly rodents) have 'rat' in their name, 'true rats' are members of the genus *Rattus*. Musser and Carleton (2005) recognized 66 species in the genus. *Rattus* is the largest genus of mammals (Feng and Himsworth 2014) and is arguably among the most complex and least understood of all mammalian genera (Pagès et al. 2010). True rats are native to southeast Asia, Australia, and nearby islands. Most true rats live in forests or near water, often constructing nests or burrows. Many species live in groups. In nature, rats feed on plants (especially seeds), insects, and sometimes other small animals. Male rats are termed bucks, virgin females are does, pregnant and mother females are dams, infants are kittens or pups, and a group of rats is a mischief.

Some rats, the so-called 'common rats,' have become associated with humans, and indeed are considered to be the world's most successful mammals, with the house mouse in second place. Several books have been written detailing the historical and economic associations of rats and humans (Hendrickson 1983; Hodgson 1997; Barnett 2001; Sullivan 2004; Twigg 1975; Langton 2006). The common rats are omnivorous, have a very high reproductive rate, and are extremely adaptable, and these characteristics assist them to colonize new locations and habitats made available by people. In terms of their importance to humans, the most significant species are the black rat and the brown rat (Figure 17.41), discussed in detail here. Several other species are also noteworthy pests. The Asian black rat (*R. tanezumi*), like the black and brown rats, has colonized urban ecosystems globally for centuries. Another invasive species, the Pacific or Polynesian rat (*R. exulans*), is limited to tropical Asia–Pacific areas (Invasive Species Specialist Group. 2014a; Kosoy et al. 2015). It is a major agricultural and environmental pest in parts of Southeast Asia and the Pacific (Invasive Species Specialist Group 2014a). The Himalayan field rat (*R. nitidus*) and Turkestan rat (*R. turkestanicus*) are pests in southern and central Asia.

The word 'commensal' (a noun and an adjective) is used in a general sense to refer to relationships in which one species benefits without harming or helping its host. An often cited example is remora fish attaching to sharks, the former benefitting by eating scraps from the latter's kills, while not significantly harming the sharks. The common rats are frequently called commensals. However, since rats often cause considerable harm to humans, the word commensal is usually inappropriate.

The brown rat (*R. norvegicus*; Figure 17.42) has also been called barn rat, common rat, gray rat, Hanover rat, house rat, Norway rat (the most frequent alterative name), Norwegian rat, Parisian rat, sewer rat, street rat, water

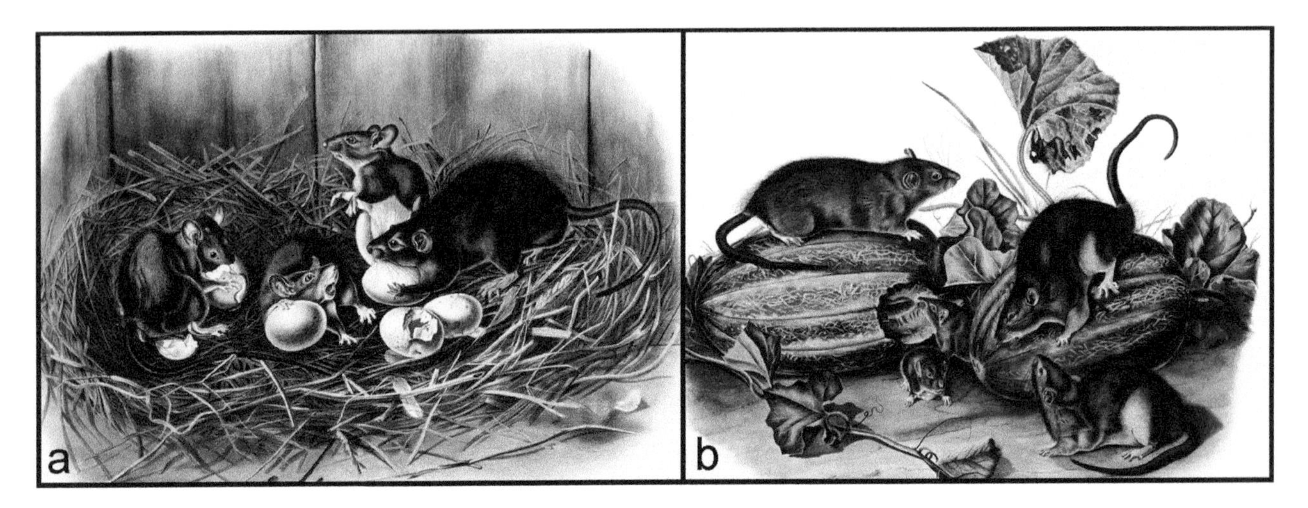

FIGURE 17.41 Common rats illustrated by the Audubon family (John Woodhouse Audubon, 1812–1862, and his father John James Audubon, 1785–1851). (a) Black rat (*Rattus rattus*) by J. W. Audubon. (b) Brown rat (*R. norvegicus*), by J. J. Audubon. Source: Audubon et al. (1851–1854). Digitally enhanced by Rawpixel (CC BY 4.0).

FIGURE 17.42 Brown rat (*Rattus norvegicus*), side and face views. Photos by Ian Kirk (CC BY 2.0).

rat, and wharf rat. (When there are so many names for just one species, it is a clear indication of its importance.) The brown rat is believed to have originated in northern China. 'Norway rats (*Rattus norvegicus*) possess the most extensive geographic range of any terrestrial mammal... Wherever they have been introduced, their extraordinary capacity for adaptation and procreation distinguish them as one of, if not the most, successful of vertebrate invasive species' (Porter et al. 2015). The species is brown or gray, up to 28 cm long excluding the tail which is slightly shorter. Wild rats weigh 140–500 g, while some domesticated strains range up to 1 kg.

The black rat (*R. rattus*; Figure 17.43) has also been called blue rat, European house rat, gray rat, Old English rat, roof rat, and ship rat. It is a native of the Indian subcontinent. The species may be gray-brown on the back with either a similarly colored or creamy-white belly, or it may be black all over. It is slender and has large hairless ears. The animal usually weights 120–160 g, occasionally over 200 g. There are genetically distinctive forms of black rat, which sometimes are recognized as different species (Musser and Carleton 2005; Robins et al. 2007; Pagès et al. 2010). The black rat is lighter and has a shorter head-body length (excluding tail) than the brown rat. Its tail is much longer than its head-body length and is uniformly colored. The tail of the brown rat is clearly shorter than its head-body length, and it has a pale underside. The upper side of the hind foot of the black rat is usually dark, whereas it is always completely pale in the brown rat. The black rat is a very agile and frequent climber, rarely burrows, nests mainly in trees and shrubs, and swims infrequently. The brown rat burrows extensively, nests underground, and is a strong swimmer. It also climbs much less frequently. On continents and large islands (such as the U.K.), the brown rat is more closely associated with humans, and it is dominant over other introduced rats. However, on oceanic islands (such as New Zealand) where it is free-living in forests and wetlands, black rats are dominant. Black rats are also most abundant in coastal areas and in the tropics, while brown rats prefer temperate climates. Although black and brown rats inhabit both rural and urban areas, black rats are more successful in the former, brown rats in the latter. Sanitary sewers, which are necessary for cities, provide superb conditions for rats: continuous supplies of water and food, warmth, and absence of predators. In the U.S., black rats are more common in storm sewers, while Norway rats are more common in sanitary sewers. (Some cities combine both sewer systems.) As feral (i.e., free-living outside of its native area) species, black and brown rats rarely remain alive for as long as 1 year, although under human care they can live for several years (very rarely 6 years).

Harmful Aspects

Rats are agile, capable of impressive jumping, climbing, and swimming. They are intelligent, inquisitive, courageous, adapt rapidly to different situations, and breed prolifically. From a human viewpoint, they are talented, indeed dangerous competitors (Textbox 17.6). Very extensive efforts have been made to eliminate rats from locations but rarely with complete success (Figure 17.44).

Most rat species are of limited concern in their indigenous areas but will take advantage of grain crops grown locally. Some rats are agricultural pests, especially of grain crops (Figure 17.45), and they consume and contaminate considerable stored foods by depositing saliva, urine, and feces on them. 'The primary economic impact of *R. rattus* relates to agricultural and horticultural damage. It is capable of destroying up to 30% of crops annually. Of the 60 or more species in the genus *Rattus*, *R. rattus* is likely to be the most damaging to agricultural crops globally' (Invasive Species Specialist Group 2013). The brown rat is also a major pest of both rural and urban areas (Invasive Species Specialist Group 2014b).

Rats are serious vectors of infections such as typhus, salmonella, Lyme disease, leptospirosis, Hantavirus, and bubonic plague (Himsworth et al. 2013; Kosoy et al. 2015; Strand and Lundkvist 2019), and they transfer diseases to

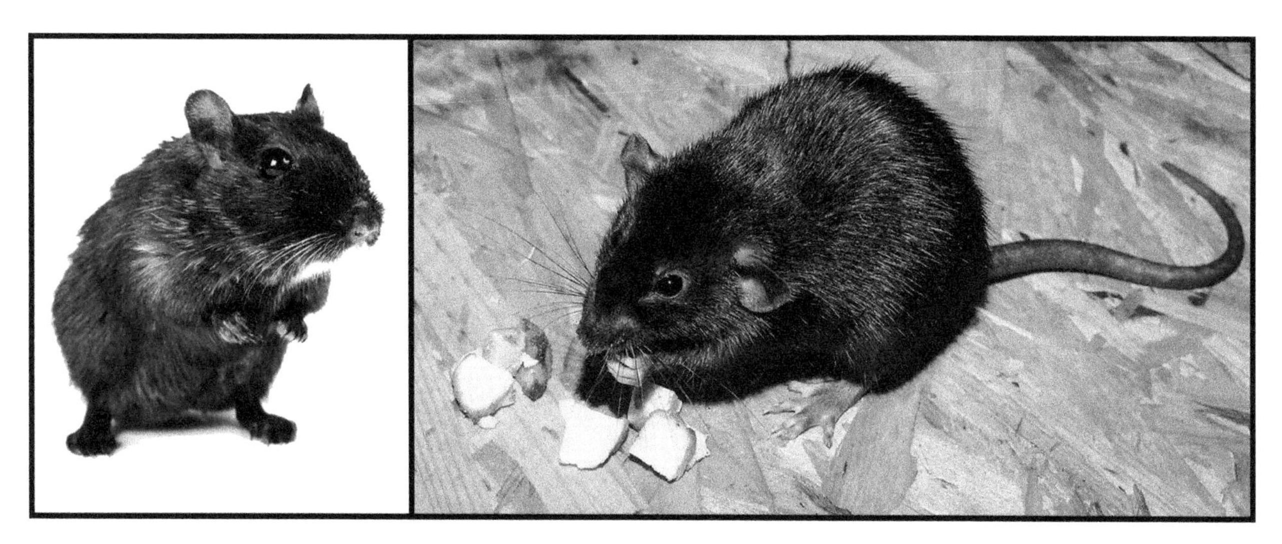

FIGURE 17.43 Black rat (*Rattus rattus*). Photos (public domain) from Pikrepo.

TEXTBOX 17.6 THE DANGERS FROM RATS

The presence of urban Norway and black rats (*Rattus norvegicus* and *Rattus rattus*) is an important and growing issue in cities globally due to their associated health and economic impacts. For example, rats pose a risk to public health as they are the source of a variety of zoonotic pathogens (disease-causing microbes transmissible from rats to people, e.g., *Leptospira interrogans*) responsible for significant human morbidity and mortality. Infestations can also serve as a chronic stressor, impacting both the mental and physical health of residents. Rats also damage urban infrastructure (due to chewing and burrowing activities) and contaminate foodstuffs. Finally, infestations can result in substantial economic losses, both directly (i.e., costs associated with rat control) and indirectly (i.e., costs associated with mitigating and repairing rat-associated damage). Given rapid urbanization, these issues are likely to increase in future; 55% of the world's population resides in cities, with a projected increase to 68% by 2050. Much of this growth will occur in developing regions where rat-associated risks are higher due to issues of inadequate housing, infrastructure, and sanitation.

—*Byers et al. (2019)*

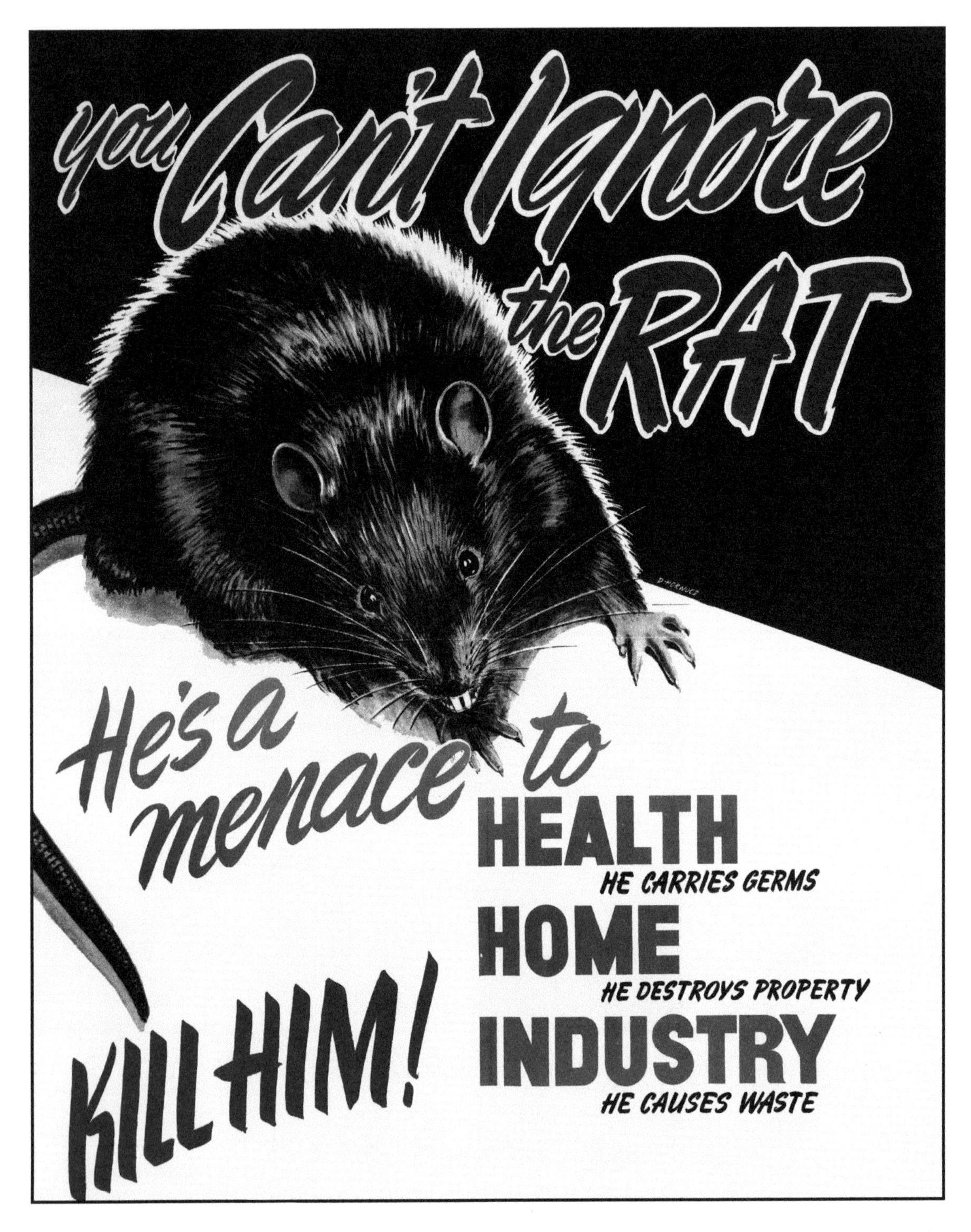

FIGURE 17.44 Anti-rat poster (public domain) circulated in 1948 by the Province of Alberta, which has managed to remain rat-free by diligently preventing entry of rats.

wildlife, livestock, and pets as well as humans. The black rat is generally credited with having carried fleas that vectored the bacterium *Yersinia pestis* which caused the devastating Black Death in Europe (1346–1353), but there is evidence that other disease carriers were responsible (Dean et al. 2018). Nevertheless, it is disquieting that both black and brown rats are capable of transmitting *Yersinia pestis*, which is commonly present in rat populations of Africa, southeast Asia, and South America. Bacterial infections can spread from rats to humans via rat bites, contact with the animal's urine, and by fleas. Rats are also an important source of producing bacteria resistant to antimicrobial medicines, which accordingly make humans more susceptible. Brown rats are the primary

FIGURE 17.45 Agricultural damage by rat nibbling on grain on a farm. Colored wood engraving published in 1843 by J. W. Whimper. (Identified as a brown rat [and by the old synonym '*Mus decumanus*'], but appears to be a black rat.) Credit: Wellcome Collection (CC BY 4.0).

reservoir of Seoul virus, which causes a hemorrhagic fever in humans.

In houses, rats cause fires and electrical interruptions by gnawing electrical wires. By cutting cables, rats interfere with communications. Using their constantly growing rodent teeth, they can even burst metal water pipes. Rats also cause structural damage to premises and ruin household contents by chewing. Outdoors, their burrowing impairs sewers, damages the banks of irrigation canals and levees, undermines building foundations and slabs, causes settling in roads and railroad track beds, and even endangers dams.

Rats threatened by humans or pets defend themselves by biting, and rats sometimes attack sleeping people, especially very young children. Reports of mutilation of babies by rats receive sensational media coverage, reminiscent of how rare shark attacks on people generate morbid interest.

In cities, rat control relies heavily on the use of poisons (Parsons et al. 2017), which are potentially harmful to humans and biodiversity. Unfortunately, resistance to rodenticides has become substantial (Buckle 2013; Prescott et al. 2018). The term 'super rat' has been applied to rats that have become immune to conventional anticoagulant poisons.

Starting in the early 2000s, car manufacturers began using soy-based materials for insulation for wiring in cars, to reduce dependency on petroleum and to save money. This led to complaints that rodents, mainly rats, were being attracted to consume the insulation and so to damage the cars. In recent years, a half-dozen class-action lawsuits were filed against auto manufacturers. However, the evidence isn't conclusive, and to date, judges have sided with the companies (Brulliard 2020).

Beneficial Aspects

Rats have been employed for entertainment for centuries, often in circuses in which they were trained to conduct acrobatic acts (Hedrich 2006). More cruelly, rat-baiting contests in which dogs killed large numbers were widespread in Europe and North America until the 19th century. Some strains of brown rats (often called 'fancy rats') have been bred to be raised as pets (Figure 17.46). The brown rat is also the principal laboratory rat (Modlinska and Pisula 2020). Wild and black rats are also occasionally employed in research, but not infrequently scientific papers from Asia have incorrectly identified their research subject as the black rat (Hulin and Quinn 2006). About 90% of vertebrate animals employed in laboratory procedures are mice and rats. After mice, the rat is the most common laboratory animal sacrificed in biomedical research and testing. Several strains of brown rats have been bred specifically for medical research into human diseases (Figure 17.47). Carbone (2004) estimated that in the United States alone, 80,000,000 mice

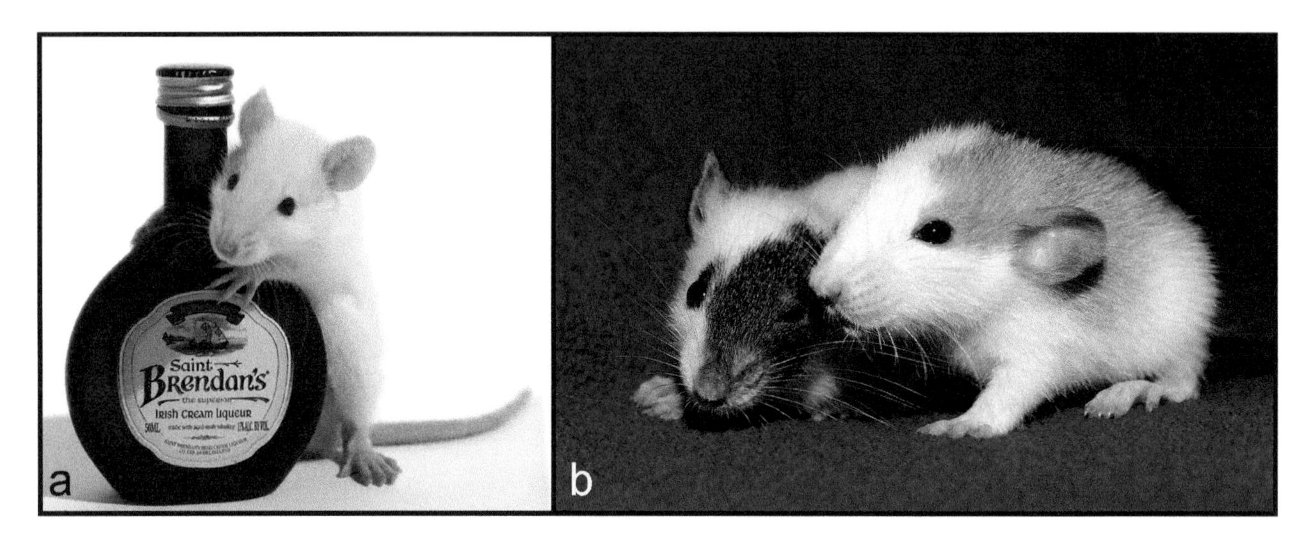

FIGURE 17.46 Pet (fancy) rats. (a) A pet rat posed beside a bottle of liqueur. Photo by Alexey Krasavin (CC BY SA 2.0). (b) Young rats. Photo (public domain) by Karsten Paulick.

FIGURE 17.47 A strain of laboratory rat (Rowett nude rat). Photo by Armin Kübelbec (CC BY SA 3.0).

and rats (collectively) were used for experimentation in 2001. The concern in this review is primarily about conservation of the world's biota, not moral aspects of animal welfare. Nevertheless, physically harming enormous numbers of animals to benefit humans is an issue that, even when judged to be necessary and ethical, should be re-examined regularly in the hope of reducing or even eliminating the practice. For a sympathetic guide to the care of lab animals, see Liss et al. (2015).

Numerous mammals, birds, and reptiles are predators of rats, and for many, they are an important staple. In addition, rats are a significant source of nutrition for people in many Third-World nations, where they are hunted and consumed widely. In times of famine in Europe, rats were widely employed as food (Hedrich 2006). Nevertheless, some religions (notably Judaism and Islam) forbid the consumption of rat meat, and in most of the Western World, rats are not considered a socially acceptable food. Rat fur is employed in garments to a small degree, although so-called rat fur is likely to be from nutria (*Myocastor coypus*), muskrat (*Ondatra zibethicus*), or some other rodent. Although rats are generally harmful to plants, sometimes they serve as distributors of seeds so that the plants can establish in new areas.

Conservation Aspects

As pointed out in the treatment of the house mouse, it is difficult to find support for rodent conservation in general, although they make up 40% of all mammal species. The generally unsavory reputation of rats makes it hard to sympathize with numerous other rodent species that are not significantly harmful to people, especially the many species with 'rat' in their name.

In their native habitats, rats serve useful ecological roles, but when introduced in foreign locations, they can devastate wildlife by competing for the same resources, introducing diseases, and simply killing native species. Rats introduced around the world by seagoing vessels have been extremely harmful to biodiversity. Over 80% of the world's oceanic islands have been invaded by rats, which have devastated some of the endemic species (Russell et al. 2008). Indeed, they can greatly degrade and modify island ecosystems (Banks and Hughes 2012). 'Removing invasive rats from islands is a powerful conservation tool' (Howald et al. 2010). Rats have been especially harmful to birds on islands (Textbox 17.7), where they have preyed on eggs and contributed to seabird extinction and endangerment (Jones et al. 2008). Invasive rats are not only harmful to animals but also to native plants, which they reduce by eating seeds and seedlings. Ironically, foreign rats particularly endanger the survival of their rodent relatives when they invade their indigenous areas. While certain species of *Rattus* are a threat to biodiversity, the IUCN Red List of Threatened Species shows that several other species of the genus are threatened, and two have become extinct (https://www.iucnredlist.org/search?query=rattus&searchType=species; Figure 17.48).

TEXTBOX 17.7 RAT THREAT TO SEABIRDS

The current crisis of global biodiversity loss and species extinctions requires the urgent implementation of long-term conservation actions. Among birds, colonial breeding seabirds are most threatened. Indeed, out of the 346 species of seabirds in the world, 28% are listed as threatened and 10% as Near-Threatened. Seabirds are threatened by a combination of interlinked factors, most notably competition with the fishing industry, climate change, degradation of breeding sites, and egg/chick predation by introduced species. On oceanic islands and archipelagos, the most serious threat to colonial seabirds is often the introduction of non-native species. Invasive rodents have probably had the largest impact on seabird populations and occur on over 90% of all islands worldwide. Seabird breeding colonies are particularly vulnerable to rats because most species nest on the ground or in burrows, and chicks are poorly adapted to escape from predators. Rats have been observed to prey on seabird eggs, chicks, and adults, and are estimated to be directly or indirectly responsible for 42% of bird extinctions on islands.

—Sarmento et al. (2014)

FIGURE 17.48 The extinct bulldog rat (*Rattus nativitatis*), once endemic to Christmas Island (Australia) in the Indian Ocean. It has been speculated that a chief cause of extinction was disease introduced by the black rat. Public domain painting from Andrews (1900).

If there ever was an animal with a reputation in desperate need or rehabilitation it is the rat. The word rat is often used as a slang pejorative, indicating an unscrupulous, despicable, contemptible character, a liar, double-crosser, or hateful person. Among criminals, a rat is a person who betrays fellow criminals by providing information to the police (less kindly expressed: the lowest form of human being – a backstabbing snitch). Strikebreakers are sometimes called rats. Another informal meaning is a person who deserts his party, side, or cause. 'To smell a rat' is to suspect something is wrong in a particular situation. The expression 'rats' is an exclamation of disappointment, disgust, or disbelief.

Rat stories are also often rather negative in tone. 'The Pied Piper of Hamelin' is an ancient, well-known tale featuring a rat-catcher who, after being refused payment for leading away a town's rats using enchanted music, leads away the town's children (Figure 17.49). Cartoon and movie depictions of rats are also usually uncomplimentary, especially compared to mice, which are often shown in attractive poses. Nevertheless, rats are sometimes shown sympathetically (Figure 17.50).

Murophobia is fear of rats and mice (musophobia is fear of mice). Fear of rats tends to be associated with fear of mice. Animal phobias tend to be considerably more common in women than in men (see discussion in the sections dealing with mice; and for frogs and snakes in Chapter 16). Phobic fear of classes of animals are not only harmful to people with such fears, they also harm the target organisms by reducing sympathy for their survival.

Despite the very negative regard for rats by most people, they are respected or considered useful in some places. Rats play a very large role in the religious traditions of Asia (Figure 17.51). The Karni Mata Temple in Rajasthan, India, also known as the 'Temple of Rats,' is a Hindu shrine dedicated to worship of up to 25,000 black rats that are maintained there (Figure 17.52). The Chinese zodiac is based a sequence of 12 animals, which are sequentially associated with a Chinese calendar, the first year of which is the rat. 2020 was a year of the rat (Figure 17.53). Clearly, the culture in which one is born has a determining effect on attitudes toward particular animals. As discussed in Chapter 10, once prejudices against particular species are acquired, they are extremely difficult to modify, and the most efficient way of doing so is to educate the young about their admirable qualities. As noted in Figure 17.54, children are quite capable of establishing respectful, indeed even loving relationships with rats.

FIGURE 17.49 Illustrations (public domain) of the tale 'The Pied Piper of Hamelin.' (a) The piper enticing rats to leave town: a very early 20th-century postcard. (b) The piper enticing children to leave town: painting by James Elder Christie (1847–1914) housed in the Scottish National Gallery of Modern Art.

FIGURE 17.50 Vintage early 20th-century postcards (public domain) showing rats, by Luciano Achille Mauzan (1883–1952), who prepared over 1000 cards. (a) The proverbial visit of a country rat to a city rat. (b) A rat enjoying his retirement.

FIGURE 17.51 Rat religious iconography. Source of illustrations: Shutterstock. (a) A golden sculpture at a Buddhist temple in Thailand. Contributor: Anant Kasetsinsombut. (b) Statue in a temple in India. Contributor: Tommywang. (c) Painting on a temple wall in Thailand. Contributor: Prapann. (d) A gilded statue of a praying rat at a Hindu monastery in India. Contributor: Milkovasa.

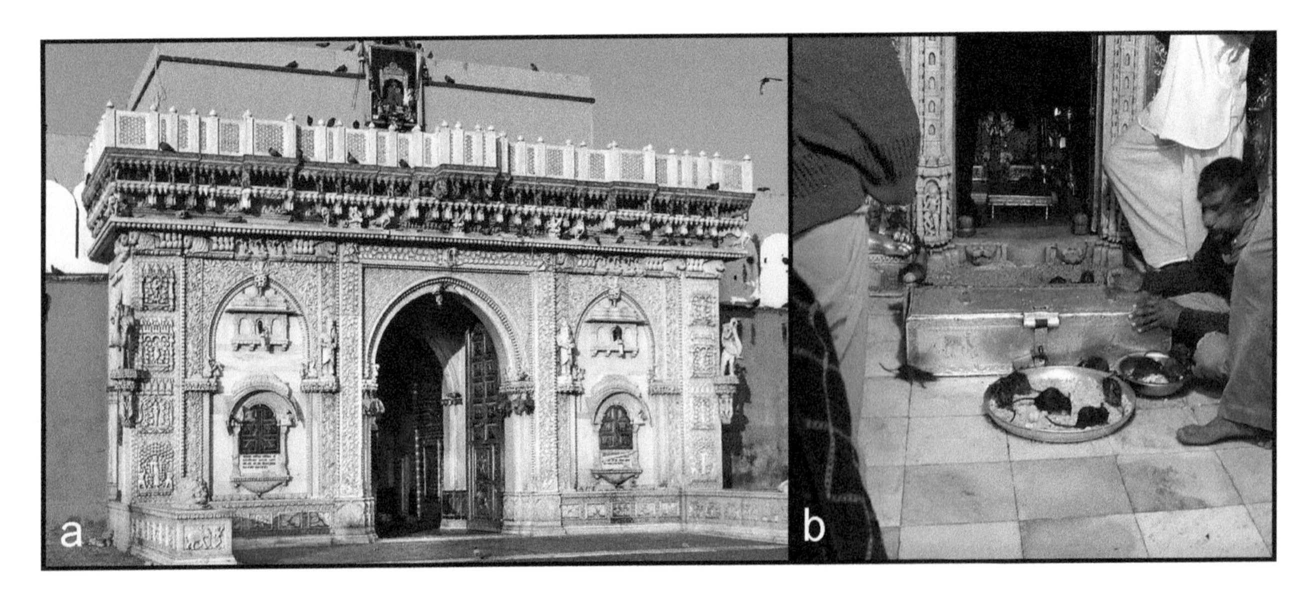

FIGURE 17.52 The Rat Temple in Rajasthan India. (a) Facade of the temple. Photo by Jean-Pierre Dalbéra (CC BY 2.0). (b) Feeding rats inside the temple. Photo by Amanderson2 (CC BY 2.0).

FIGURE 17.53 Year-of-the-Rat celebrations of the 2020 Chinese New Year. (a) Display in Singapore. Photo by GT#2 (public domain). (b) Costumed performance in Manchester, UK. Photo by Author Medowduk (CC BY SA 4.0).

FIGURE 17.54 Painting (public domain), entitled 'The White Rats' by French artist Jean de Francqueville (1840–1939). The young girls are clearly comfortable playing with rats.

SKUNKS

Introduction

Numerous plants and animals produce chemicals that attract (e.g., sex hormones often cause animals of one sex to seek the opposite sex; floral odors draw in pollinators). Conversely, many species produce repellent chemicals for self-defense. A few have perfected the ability to project a harmful liquid to protect themselves. For example, the hundreds of species of bombardier beetles eject a hot noxious chemical spray from the tip of their abdomen, which can kill attacking insects. Spitting cobras (mostly species of *Naja*), true to their name, project toxic chemicals into the eyes of creatures that threaten them. The most well-known animal that launches noxious repulsive chemicals is the skunk. The awful odor associated with skunks is primarily responsible for their demonization by humans (Miller 2015).

There are ten living species of skunks, placed in three genera: *Conepatus* (hog-nosed skunks, four species); *Mephitis* (the hooded and striped skunks, two species); and *Spilogale* (spotted skunks, four species). These are indigenous to the Americas, from Canada through central South America.

Related to the skunks (and sometimes also called skunks) are two species of the genus *Mydaus*, known as stink badgers (optionally hyphenated as stink-badgers), which are found in Indonesia and the Philippines. The stink badgers resemble the

American skunks and also have the ability to spray extremely potent anal secretions, but before resorting to spraying they often first feign death. The most obvious external difference between stink badgers and skunks is that the former have short, pointed tails while skunks have bushy tails. The Palawan stink badger (*M. marchei*) reportedly can spray noxious liquid up to a meter, while the Sunda stink badger (*M. javanensis*) Figure 17.55) is limited to about 15 cm. As noted later, the North American skunks have longer ranges.

The skunks and stink badgers constitute the mammal family Mephitidae. Skunks were previously considered to be in the weasel family (Mustelidae, which includes weasels, otters, badgers, and others), but DNA research indicated that they warrant separate recognition, and as a result, they were placed in their own family (Dragoo and Honeycutt 1997).

Most skunks are the size of house cats. The species vary in total length (including tails) from about 40–95 cm. The smallest, the spotted skunks, may be only 0.5 kg in weight, while the largest, the hog-nosed skunks, may weigh over 8 kg. Their short legs do not allow skunks to run away rapidly from predators, and they move at a deliberate waddle-like walk, slow trot, or clumsy gallop. Skunks seem placid, sluggish, and unconcerned about predators, which is understandable given their formidable defense system. Most skunks are black and white; so are numerous other mammals, including giant pandas, killer whales, zebras, and hundreds more – a phenomenon which may have several explanations, such as serving as camouflage, as a warning, or for recognition of fellow members of a species (Caro 2011).

The claws of a skunk's forefeet are long, curved, and sharp (Figure 17.56), well-adapted to digging, and skunks often unearth soil-dwelling creatures. While they sometimes excavate their own tunnels, skunks generally den in burrows of other animals, rock crevices, brush piles, or spaces under buildings. Skunks occur in agricultural areas, deserts, grasslands, open fields, rocky or mountainous areas, and woodlands. They are primarily insectivores, but they eat a variety of other invertebrates such as spiders, crustaceans, and molluscs, vertebrates smaller than themselves such as toads, frogs, lizards, snakes, mice, chipmunks, rabbits, and eggs of birds and turtles, as well as fruits, nuts, roots, and leaves. Skunks are opportunistic feeders, strongly attracted to edible garbage, carrion, and food left for wild birds and pets by humans. They are adept at stealing honey from beehives as well as eating the bees. Skunks are nocturnal, and during the winter, they may rest periodically or enter a phase of lethargic sleep (torpor) without falling into a complete state of hibernation (skunks do not hibernate). For many skunks, sex must be rough to induce ovulation; male skunks may bite the female on the back of her neck, often drawing blood. A male skunk is called a buck, the female a doe, and the baby a kit or kitten. A group of skunks is called a surfeit or a huddle, but most skunks (especially the males) are solitary.

The striped skunk (*Mephitis mephitis*; Figure 17.57a) is by far the most common species in North America (Fergus 2010). It occurs in all 48 contiguous states, southern Canada, and northern Mexico, from sea level to timberline. The fur characteristically has a white stripe that starts at the forehead and splits into a V shape along the skunk's back. However, other patterns also occur (Verts 1967; Rosatte et al. 2010), and as in other skunk species, albino mutants arise. The hooded skunk (*Mephitis macroura*; Figure 17.57b) is similar to the striped skunk but has longer and softer fur and a longer tail. Some hooded skunks have a single, solid white stripe down their back from their forehead to the tip of their tail, some have two thin white stripes running down the side of the body from shoulder to

FIGURE 17.55 Sunda stink badger (Malayan stink badger; *Mydaus javanensis*). Pre-20th century print (public domain). Source: Iconographia Zoologica (University of Amsterdam).

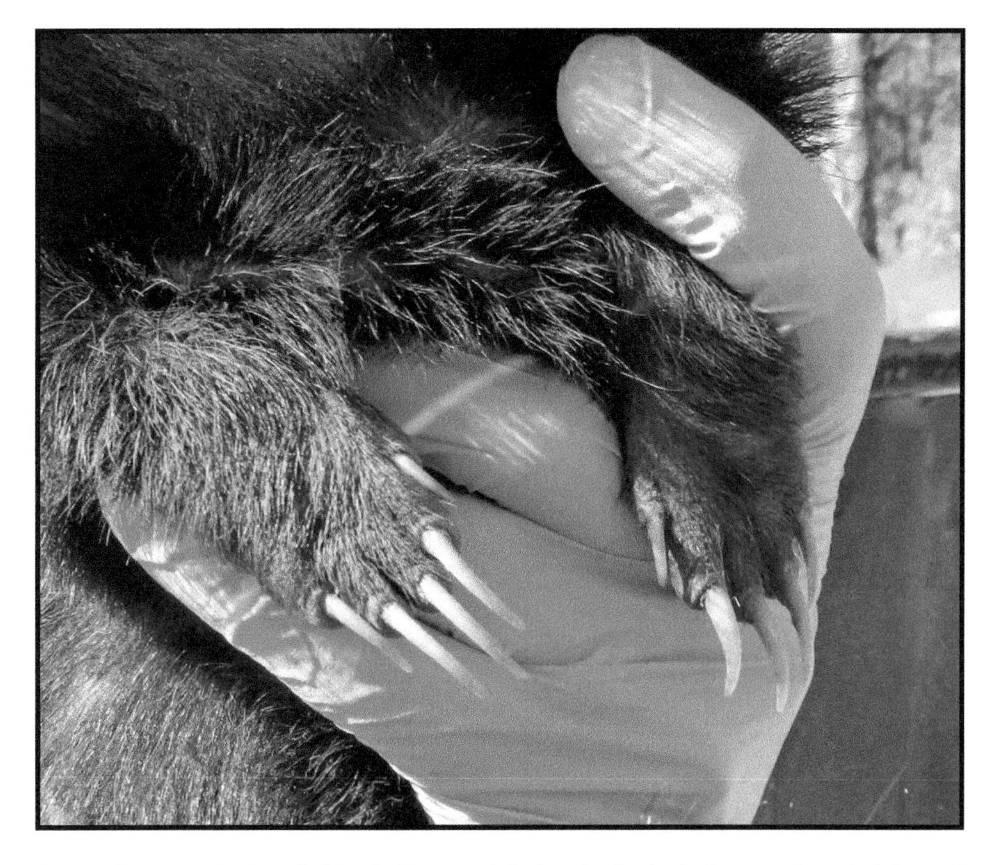

FIGURE 17.56 Claws of a young skunk in California. Photo (public domain) by Andy Greenawalt, United States Geological Survey.

FIGURE 17.57 Skunks of the genus *Mephitis*. (a) Striped skunk (*M. mephitis*), from Audubon and Bachman (1851), plate XLII. (b) Hooded skunk (*M. macroura*, from Audubon and Bachman (1854), plate CII. These paintings are in the public domain.

stomach, and some have a combination of these two patterns. Both species stomp with their front feet to warn animals they consider threatening before spraying; sometimes they try to spray fast-approaching cars and are killed.

The spotted skunks (*Spilogale* species; Figure 17.58) do not actually have 'spots' on their back, but this appearance is due to a series of interrupted white lines. These are the smallest members of the skunk family and the only ones that can climb trees (Figure 17.66), which they do frequently to hunt or avoid predators.

The hog-nosed skunks (*Conepatus* species; Figure 17.59) have elongated snouts which serve to extract bugs and grubs from the ground. These superb diggers have long, specialized claws and powerful forearms which they employ to dig dens and climb up rocky slopes. Hog-nosed skunks occur in the U.S., Mexico, and Central and South America.

Skunks are infamous for their ability to protect themselves from predators by projecting a foul-smelling thick oily liquid from their musk glands, one on either side of the end of the rectum, just inside the anus (the glands are sometimes referred to as rectal glands, but usually as anal glands). These scent glands have nozzle-like ducts ending in nipples which protrude through the anus and can be aimed. The anal glands are operated by muscular action, and baby skunks as young as a week after birth may have some ability to expel smelly liquid. The irritating spray contains sulfur compounds – thiols – sometimes called mercaptans. (The simplest thiol is hydrogen sulfide, H_2S, the sulfur analog of water, H_2O.) Thiols, which are responsible for the offensive long-lasting stench, have a rotten egg or garlic smell. Thiols in raw onions cause people to cry, and thiols are added to natural gas supplied to homes, so that a leak will have a noticeable odor. Skunk thiols are responsible for their horrible smell. Skunks can target accurately up to 3 m, less accurately up to 6 m, and the highly persistent spray can cause temporary blindness and nausea. Skunk spray can be detected by humans at concentrations as low as 10 parts per billion, a level that may occur by wind transport over 5 km away from the release point. All of the species have fur with

FIGURE 17.58 Some spotted skunks (*Spilogale* species). (a) Eastern spotted skunk (*S. putorius*). This species occurs from southern Manitoba and Ontario in Canada through the eastern U.S. to northeastern Mexico. Public domain painting from Nelson (1918). (b) Island spotted skunk (*S. gracilis amphiala*), endemic to the two largest California Channel Islands, Santa Cruz and Santa Rosa. Photo by Brian Kentosh (released into the public domain).

FIGURE 17.59 Some hog-nosed skunks (*Conepatus* species). (a) The American hog-nosed skunk (*C. leuconotus*) from Central and North America. This large skunk grows to 85 cm in length. Painting (public domain) from Nelson (1918). (b) Molina's hog-nosed skunk (*C. chinga*) from western South America. This specimen is brown with white stripes, but most animals are black with white stripes. Painting (public domain) from Jardine (1843), plate 13.

warning coloration, most often bold white stripes, streaks or spots over black, which they have from birth (some are all-black, some have brown instead of black areas, and a few are all-white). Skunks are usually shy and reclusive, avoiding confrontation. They will first attempt to ward off attackers by hissing, foot-stamping, or other threat postures, because their store of noxious liquid (typically about 15 cc) is sufficient only for five or six uses, and regenerating the supply may take more than a week. Very few animals feast on skunk meat, although skunks are sometimes killed by dogs, foxes, coyotes, bobcats, and mountain lions. However, great horned owls, which have a dull sense of smell and razor sharp talons, frequently prey on skunks. Skunks have been shown to be more fearful of horned owl shrieks than calls from mammalian predators (Fisher and Stankowich 2018).

Harmful Aspects

Skunks are a significant vector of the rabies virus, one of the most lethal of infectious diseases. It has been estimated that about 20% of reported cases in animals in the U.S. and Canada are due to skunks (Rosatte et al. 2010). Often over 20,000 people die annually from rabies, over 95% in Africa and Asia (but not acquired from skunks). However, in developed countries rabies in humans and domesticated animals is rare, due to surveillance and vaccination programs. Vaccines to treat humans are available but need to be employed promptly following potential infection. Globally, dogs are overwhelmingly the principal source of infection of humans, but in the Americas bats are the most common source. Rodents are very rarely infected, and non-mammalian species are immune (some birds can be temporarily weakened). Rabies virus is transmitted by direct contact, such as a bite, or by contact with mucosal surfaces of an infected animal. Many wild New World mammals, including bats, foxes, raccoons, and skunks, are reservoirs of infection in the Americas, posing a danger to domestic mammals including cats, dogs, and livestock. In some parts of the world, culling (i.e., killing) of species is carried out when it is discovered that the risk of rabies has increased. In the New World, for most potential carriers, provision of vaccines in targeted food ('bait') is considered the most efficient way of controlling the risk, and this has been applied to skunks (Rosatte et al. 2009; Mainguy et al. 2012). Different host species often are infected by particular viral strains (a species may have a characteristic viral strain, or several strains, and a given viral strain may occur in more than one species). In the U.S., skunk rabies has the broadest geographical distribution of all terrestrial (i.e., excluding bats) rabies virus strains (Brown et al. 2014). Because of the difficulty of creating combinations of vaccines that will immunize skunks against all virus strains, trap-vaccinate-release programs are also employed, although they are more costly and labor-intensive (Jojola et al. 2007; Wohlers et al. 2018). Skunks suffer from an unwarranted reputation as being rabies 'carriers' that transmit but do not succumb to the virus (such resistance occurs in bats, as noted earlier); in fact, skunks, die like almost all other infected mammals once symptoms occur.

Skunks are common in suburban areas, where they often raid garbage containers and annoy residents (Rosatte et al. 2010). They often establish their homes in crawl spaces under porches, and in barns, garages, sheds, junk cars, and woodpiles. They also search for rodents in such places. Sometimes skunks become trapped in window wells. Discovering that a skunk family has moved onto one's property can be disturbing. Skunks also dig holes in lawns, golf courses, and gardens to capture insect grubs. They occasionally kill poultry (Figure 17.60) and eat their eggs. Unfortunately for skunks, European colonizers of the New World confused the European polecat (*Mustela putorius*) with skunks and indeed employed the word polecat to label skunks. They also confused the considerable reputation of European polecats for destroying game and poultry with the behavior of skunks (Miller 2015). Although most wild animals that skunks consume are not admired by humans, skunks will eat eggs of wild ground-nesting birds, upland game birds, and waterfowl, sometimes resulting in bird enthusiasts and hunters lobbying for skunk control and removal (Larivière and Messier 1998). Skunks may also damage beehives when they try to feed on bees or obtain honey (Storer and Vansell 1935). Skunks occasionally consume vegetables and grains from gardens and commercial crops. However, 'An economic evaluation of the feeding habits of skunks shows that only 5% of the diet is made up of items that are economically valuable to people' (Knight 1994).

Irritating skunk odor is regularly generated in neighborhoods from skunk encounters with other animals, and especially when a skunk is run over on roadways. The relatively poor vision of skunks and their slow movements make them vulnerable to vehicle collisions.

Dog fur and skunks rarely coexist peacefully (Figure 17.61). Dogs are often sprayed, causing them and their owners considerable annoyance. Occasionally, people are also sprayed, but skunk scent gets on humans most often from sprayed dogs. Regardless of the advice offered below, when skunks spray humans, it is a good idea to seek expert medical advice, and when they spray dogs, it is also advisable to consult a veterinarian. Contrary to folklore, bathing in tomato sauce or juice to remove the stench is not recommended because it merely masks the skunk smell, without eliminating it. The odor eventually may become less evident to the victim simply because of olfactory fatigue (Wood 1999). If sprayed in the eyes, immediately flush with cool water (there are commercial eyewash products made specifically for dogs). Dog fur can be treated with a mixture of 1 L of 3% hydrogen peroxide, 80 mL of baking soda, and 5 mL of liquid soap (a quart of 3% hydrogen peroxide, a quarter cup of baking soda, and a teaspoon or two of liquid detergent). The mix should be made in a large, open container, and used while it is still bubbling. Wear rubber gloves and lather the mixture on, leaving it for 30 minutes, and then rinse thoroughly with tap water. The mix should not be stored because the oxygen build-up could blow off the top of the container, and it might bleach the dog's fur (Cosier 2006). Commercial mixtures are also often available from veterinarians, but their clinics are usually closed at night when dogs are sprayed. If an odor lingers

FIGURE 17.60 The common striped skunk (*Mephitis mephitis*) viewed in early times as a serious poultry killer. (a) Source (public domain): John Doughty (1832). (b) Source (public domain): Bryson (1911).

FIGURE 17.61 Peaceful co-existence between dogs and skunks: A skunk at a dogs' dinner party. Based on *The Dogs Dinner Party* by Harrison Weir (1824–1906), modified by Brenda Brookes.

after washing clothes exposed to the spray, a dash of ammonia can be added to the wash cycle. To get rid of the odor on non-living objects, use dilute chlorine bleach, ammonia, or vinegar, after conducting a test spot. Although skunk odor can last for weeks, it will eventually naturally dissipate. Although the smell of skunks is highly objectionable, being sprayed by skunks almost never leaves lasting damage.

Beneficial Aspects

Skunks are highly beneficial to farmers, gardeners, and landowners because they feed on large numbers of agricultural and garden pests. They help control pests such as rodents (especially mice and rats), moles, rabbits (occasionally), insects (notably beetles, crickets, grasshoppers, wasps, and a variety of grubs), and other harmful animals such as snails (Kelker 1937; Lantz 1917). Without the pest-control services of skunks, health problems from mice and rats would increase. Although skunks damage lawns by digging up grubs, they benefit lawns by reducing grubs and other insect pests, as well as moles.

In North America, skunks were once widely trapped for the fur trade (Figure 17.62). Skunks which are shot or caught in leg traps are likely to discharge their musk and may contaminate their fur, so that trappers preferred to catch the animals in box traps and drown them, or kill them with a paralyzing blow. Trappers sometimes ate young skunks and found the meat reasonably palatable. For much of the 20th-century skunks were farmed for their pelts, with selective breeding for tameness and dark coats. The much more marketable names 'Alaska sable' and 'American sable' were substituted for skunk by furriers during the late 19th century. On a global basis, about 80% of fur (all species collectively) is produced from captive animals today, and skunk fur is insignificant.

During the 19th century, farmers sometimes tamed striped skunks and maintained them in barns to kill rodents, much like domesticated cats.

'Skunk oil' is a viscous preparation made as a rendered byproduct of killed skunks. It is obtained by slowly heating the fat from a skunk. This use of the phrase skunk oil does not refer to the foul-smelling musk oil of the two anal glands, although many publications and some scientists refer to the anal gland liquid as skunk oil. Neither of these skunk oils is the same as cannabis (marijuana) oil, which is a third meaning of the phrase skunk oil. Skunk oil (rendered fat) was employed in indigenous North American medicine and occasionally is still used in folk medicine, although effectiveness has not been demonstrated. This kind of skunk oil is also used as a hunting lure (some preparations are odorous, others not).

FIGURE 17.62 A skunk fur coat. Source (public domain): Larisch and Schmid (1902).

Skunks are sometimes kept as pets (Hume 1958), especially the striped skunk *Mephitis mephitis*, which is relatively social. The species generally lives no longer than 3 years in the wild but can reach 15 years in captivity. Wild skunks are sometimes adopted, especially as babies, but this usually represents a violation of local wildlife regulations. Breeds have been selected that are relative tame and have various fur patterning and coloration (popular categories include albino, apricot, blonde, champagne, chocolate chip, chocolate swirl, cream, lavender, smoke, and violet). In the U.S., skunks cannot be kept as pets in many states. The scent glands of skunks are often surgically removed, but there is controversy in some jurisdictions about the ethics and legality of this. Skunks have been imported to England as pets, and there have been allegations that this has resulted in some escapes becoming feral. Skunks are high-maintenance pets requiring a major commitment. They can be kept responsibly as pets but are unsuitable except for a small minority of enthusiasts (Dragoo 2009; Cipriani 2011; note Figure 17.63).

Conservation Aspects

Skunks are vilified primarily because of their smell (e.g., Figures 17.64 and 17.65). A colloquial French Canadian phrase for skunk, *enfant du diable*, means 'child of the devil.' The word skunk is rarely used as a compliment and does not serve to improve the image of skunks for conservation purposes. Pejorative meanings related to skunk include: skunk: a contemptible ill-mannered person; to skunk: to deprive by cheating; skunked: being overwhelmingly defeated; drunk as a skunk: highly inebriated and stinking of alcohol. 'Skunked beer' has been exposed to too much light, which transforms certain components to skunky-smelling compounds. Several plant species have 'skunk' in their English names, reflecting their bad odor. Examples of 'skunky' plants include skunk cabbage (*Symplocarpus foetidus*), skunk currant (*Ribes glandulosum*), and skunk sumac (*Rhus trilobata*). *Cannabis sativa* often has a distinctly skunky odor, and in the U.S., the word 'skunk' is often employed to designate certain strains

FIGURE 17.63 Pet skunk rescuer Deborah Cipriani with domesticated breeds from people who no longer wanted to care for them. Photo (in 2002) by Paul M. Walsh (CC BY 2.0).

FIGURE 17.64 'Reddy Fox' is repelled by the prospect of being sprayed by 'Jimmy the Skunk' in a children's story, *The Adventures of Jimmy the Skunk*. Public domain illustration by Harrison Cady, from Burgess (1918). T. W. Burgess (1874–1965), an American naturalist and conservationist, wrote over 100 children's stories.

FIGURE 17.65 Early 20th-century postcards (public domain) showing unwelcome visits from skunks. The beach scene is printed on linen, an early practice in some countries.

of marijuana with an especially skunky smell. In Great Britain, the word skunk often designates premium quality, indoor-grown cannabis. Because of the derogatory nature of the word, its use for medicinal forms of cannabis is sometimes not appreciated (Potter and Chatwin 2012).

The curious phrase 'skunk works,' unlike most uses of the word skunk, is employed in a respectable sense. The phrase was first employed by the aeronautics industry, having originated from cartoonist Al Capp's comic strip Li'l Abner, where the 'skonk works' was a secret laboratory that operated a backwoods still. 'When an established firm aspires to experiment in a radical direction, management gurus recommend opening a skunk works – a home for high-priority original thinking and projects. It is housed away from the organization's main operations, sometimes in secret or with access restrictions. Typically the projects involve something of value to the future but are not directly connected to the present operational or service missions. Sometimes a skunk works has the approval of senior management, and sometimes it does not' (Greenstein 2016).

No species of the skunk family (Mephitidae) is currently evaluated as 'Endangered,' according to the International Union for Conservation of Nature. The pygmy spotted skunk is listed as 'Vulnerable' because of recent reductions in populations. Concern has been expressed about the eastern spotted skunk, *Spilogale putorius* which has become rare in large parts of its distribution range (Gompper 2017; Eastern Spotted Skunk Cooperative Study Group 2018), particularly the subspecies Plains spotted skunk (Dowler et al. 2017; Figure 17.66).

Skunks, as archetypical despised animals, need to be better understood in order to generate more respect for nature in general. For those with a religious bent, there are spiritual reasons why skunks should be respected (Textbox 17.8; Figure 17.67).

FIGURE 17.66 Plains spotted skunk (*Spilogale putorius interrupta*). *Spilogale* species are the only skunk that climbs trees. Photo courtesy of the Missouri Department of Conservation.

FIGURE 17.67 A skunk from heaven. Prepared by B. Brookes.

TEXTBOX 17.8 WHY DID GOD MAKE SKUNKS? SOME ANSWERS FROM THE WORLD WIDE WEB.

Why did God make skunks? They have no right to live. I hate them.

Why did God makes skunks? Honestly? Why not leave it at bunny rabbits?

Some people ask, 'Why did God make skunks?' I don't know why God made skunks, but I do know that skunks are part of God's awesome plan of nature.

Why did God make skunks? I think perhaps for the same reason He created lightning bugs and platypus. He has a sense of humor.

Why God made skunks? I guess to keep us on our toes.

God made skunks because of their beauty and so they will enjoy life. As for skunks and humans, God knew that humans could learn a lot in their relations with wild animals, including skunks.

Skunks help keep the ecosystem in place, just like every other living thing on earth. Every life has a part in this world.

18 In Defense of the World's Most Despised Toxic Plants

INTRODUCTION

Photosynthetic organisms or 'plants' in the most general sense capture solar energy and accordingly are the ultimate indispensable source of energy that sustains all life, including all humans. Nevertheless, some plants are harmful to people, at least under some circumstances. The five species discussed in detail here are arguably the most harmful. Three of these – opium poppy, tobacco, and sugar cane – are deleterious because they contain extremely addictive substances, which when consumed to excess are poisons. Two others – poison ivy and ragweed – are environmental contact toxins, which are more or less unavoidable.

This and the remaining chapters deal with plants. Human attitudes toward plants (both love and hatred) are much less intense than for animals (Parsley 2020; Textbox 18.1), a phenomenon that has been termed 'zoochauvinism' and 'zoocentrism.' Nevertheless, both the benefits from the plants we depend on and the harm from those that injure us, our crops, and our livestock exceed the effects of animals on human welfare, and so need to be addressed.

Addictive Plant Poisons

Most natural drugs (in contrast to those synthesized chemically) come from plants. Numerous drugs are beneficial when employed under the supervision of competent medical professionals, but some are employed non-medicinally merely for pleasure, with substantial resulting harm from excessive consumption. Several commonly misused, dangerous, and regulated (often illegal) drugs are known as 'narcotics,' a word that is used in different senses. Etymologically, narcotic refers to substances that when ingested or otherwise administered induce sleep. In practice, however, narcotics are widely understood to refer to substances that affect the nervous system and may be (1) addictive, (2) harmful, and/or (3) used illegally. Narcotics can have a variety of effects, for example they may be psychotomimetic (mood-altering), psychotropic (mind-altering), and/or hallucinogenic. The most disreputable of so-called narcotic plants is opium poppy, presented here as a representative of this class of plants. Plants are not only the primary sources of illicit narcotics, they are also the sources of two other widely abused (hence regulated) addictive, harmful substances: tobacco and alcohol. The tobacco plant is presented here as a second example of plants that have been condemned for their evil effects on people. Because of the horribly destructive effects of addictive substances, they are widely condemned (Figure 18.1).

Notorious plants such as those mentioned above pose three principal problems in terms of commercial exploitation for legitimate, beneficial purposes: (1) they are illegal or at least controlled, so security issues make production and trade both expensive and complicated; (2) considerable stigma is attached to the plants, making it difficult to attract commercial, development interest; and (3) they are highly politicized, with the result that research support is readily available only for projects that are consistent with the often-prevailing view that these plants are just too dangerous to examine their possible redeeming values.

As horrible as the effects have been historically, and as damaging as these species continue to be, there are reasons for defending them. As explained by Small (2004), such plants have actual or potential values:

1. All plant species have some unique biochemical characteristics, one never knows which one will be the basis of important scientific and technological developments, and so poisonous addictive plants should not be eliminated a priori from consideration as candidates for new uses.
2. Toxic addictive plants have a long history of use in medicine, there is no doubt that some chemicals from some species are useful, and the search

TEXTBOX 18.1 COMPARISON OF EMOTIONAL REACTIONS TO PLANTS AND TO ANIMALS

Humans are interested in plants much less than in animals. Consequently, plants receive lower conservation support than animals, although it should be noted that the factors underlying human's willingness to preserve plants are rarely investigated. Human neglect and ignorance of plants are also known as plant blindness or plant awareness disparity. One of the major proposed causes of plant blindness, that is, the human natural tendency to overlook plants, is their homogeneity and the absence of apparent movement, which leads (among other things) to insensitivity to the aesthetic characteristics of plants. Females are generally more attentive to plants than males.

— Fančovičová et al. (2022)

DOI: 10.1201/9781003412946-18

FIGURE 18.1 Anti-drug art (public domain). (a) Credit: Pxfuel. (b) Credit: U.S. Centre for Drug Control. (c) Scene showing drug addiction being treated. Published in Puck 72(854) in 1912. Prepared by artist S. D. Ehrhart (ca. 1862–1937). Source: Library of Congress.

for additional useful constituents could well be fruitful.

3. Considerable research has been completed on the biochemistry of toxic addictive plants, and this knowledge is useful as a basis for further research and development.
4. The most commonly used toxic addictive plants are very well understood in terms of their agronomy and processing characteristics. By and large, these species are easy to grow and very productive.
5. In some cases, industries and people have become dependent on income generated from addictive plants, and alternative income-generating schemes have not proven successful. Growing the same plants but for legitimate purposes seems like a tactic at least worth considering.
6. In some cases, varieties of otherwise toxic, addictive species are available that are lacking in the harmful constituent, or at least are so low that additional breeding could produce 'harmless' forms of the plant.
7. Genetic engineering offers the possibility of complete inactivation of the enzymes responsible for the production of harmful constituents. Moreover, morphological markers could be added to facilitate the identification of such relatively innocuous forms of otherwise harmful plants.

Addictive Plant Foods

In nature, animals almost never have access to more food than they need, except for brief periods, and so compulsive 'overeating' when the opportunity arises is normal. Indeed some animals must binge, when opportunities to eat occur, in order to accumulate energy reserves to survive periods of famine. As omnivores, people have flexibility in diet, and indeed we humans as a species consume a wider variety of foods than any other animal (Ungar and Teaford 2002). But we are especially partial to the taste and/or texture of fats and sweets. Most plants produce some form of sugar (carbohydrates), but sugar is especially present in fleshy fruits, which sustain many primates, probably including our simian ancestors, which may account for our sweet tooth. In ancient times, honey was the primary source of concentrated sweetness, but collecting honey from wild bee hives is difficult, and so was in short supply. While purified sugar has been available for millennia, efficient extraction of table sugar (sucrose) is only several centuries of age. Progressively, sugar consumption has increased along with serious correlated health issues (Figure 18.2). As discussed later, it is probable that in much of the world, sucrose consumption, derived mainly from sugar cane, is the leading cause of avoidable human illness.

FIGURE 18.2 Excessive consumption of sugar and sweet confections is a major health hazard, contributing to the worldwide obesity epidemic and associated diseases. Prepared by Jessica Hsiung.

Environmental Allergens

Numerous animals and microorganisms attack humans, as a source of 'food' (energy). However, plants get their energy from the sun, so they have no reason to deliberately harm people. Nevertheless, there are several dangerous ways that plants repel animals (including spines, thorns, surface irritants, and poisons) to prevent being eaten or even contacted, and pollen is a major irritant for many (Figure 18.3).

Poison ivy and ragweed, which are examined in this chapter, are the leading causes of medical problems for humans associated with wild plants in North America (Rostenberg 1955), and ragweed has blossomed into a health threat for

FIGURE 18.3 Ways that plants harm humans. Top left: Livestock have been poisoned by consuming larkspur (*Delphinium*), and a woman in a bathing suit is suffering dermatitis from contact with another plant species. Top right: people who consumed toxic fruit are experiencing nausea. Center: Victims of hayfever. Bottom left: A girl is pricked by a rose thorn. Prepared by Jessica Hsiung.

much of the world. However, numerous species of plants produce chemicals that upon contact by skin results in toxic reactions (Weston and Chen 2020). There are also numerous poisonous plants that when ingested produce sickness or even death. Consuming plant materials is arguably a voluntary exercise (debatably so in the case of addictive substances), whereas contacting allergenic plants is either unavoidable or at least very difficult to avoid.

Poison ivy and ragweed are serious allergenic plants, but by no means produce the most serious reactions. Many other plants contain much more dangerous substances. Moreover, possibly every individual is capable of having an extremely serious reaction to some plants. The situation is analogous to venomous snakes. Some, such as sea snakes, have highly potent venom but rarely kill people because they are docile, whereas others with less potent venom kill many more people because they are aggressive and widespread where humans live (Chapter 16). Poison ivy and ragweed may cause more cumulative pain to humans from physical contact than any other plants.

Wind-borne pollen breathed into the lungs (or even contacting eyes and nose tissues) may produce 'hay fever' (allergic rhinitis; Figure 18.4a). This is due to reactions to the protein constituents of the pollen exine (outer surface) contacting 'internal' (mucosal) nasal tissues. The purpose of such pollen is simply to pollinate distant individuals of the same species, and negatively affecting humans and other animals is of no benefit to the plants (indeed, it represents a waste of material). Rarely, pollen landing directly on bare skin causes dermatitis.

On the other hand, causing an animal to suffer when it contacts a plant is an obvious adaptation for discouraging the latter from consuming the former. In particular, many mammals, including humans, eat plants, and so it is not surprising that plants have evolved protective mechanisms that include toxins transferred to the animals simply from topical contact (physically touching the skin; Figure 18.4b). Evans and Schmidt (1980) classified allergenic plants that exert their effect by such direct transfer of toxins into five categories (for alternative classifications, see Mitchell 1975; Rozas-Muñoz et al. 2012):

- Mechanical irritants
  Such plants have small, easily detachable rough hairs or bristles, or needle-like calcium oxalate crystals that penetrate skin when touched. For example, the needle-like crystals in the dry outer scales of daffodil (*Narcissus* species) and hyacinth (*Hyacinthus* species) bulbs cause rashes. Thorns, which are larger, serve a similar purpose.
- Stinging nettles
  The nettles (*Urtica* species and relatives) have a highly specialized weapon: tiny syringe-like hypodermic needles that penetrate skin and inject toxin. A number of other species have much simpler hairs topped by toxins that achieve the same result.
- Phototoxic compounds
  Various plant species have compounds that when they contact skin and are subsequently exposed to light are 'photoactivated,' transforming into an irritant or allergen. A class of chemicals known as furocoumarins is notorious for this phenomenon. The result may resemble sunburn, and in some cases, the wounds can last for months. Giant hogweed (*Heracleum mantegazzianum*) and wild parsnip (*Pastinaca sativa*) are particularly widespread species causing such dermatitis.

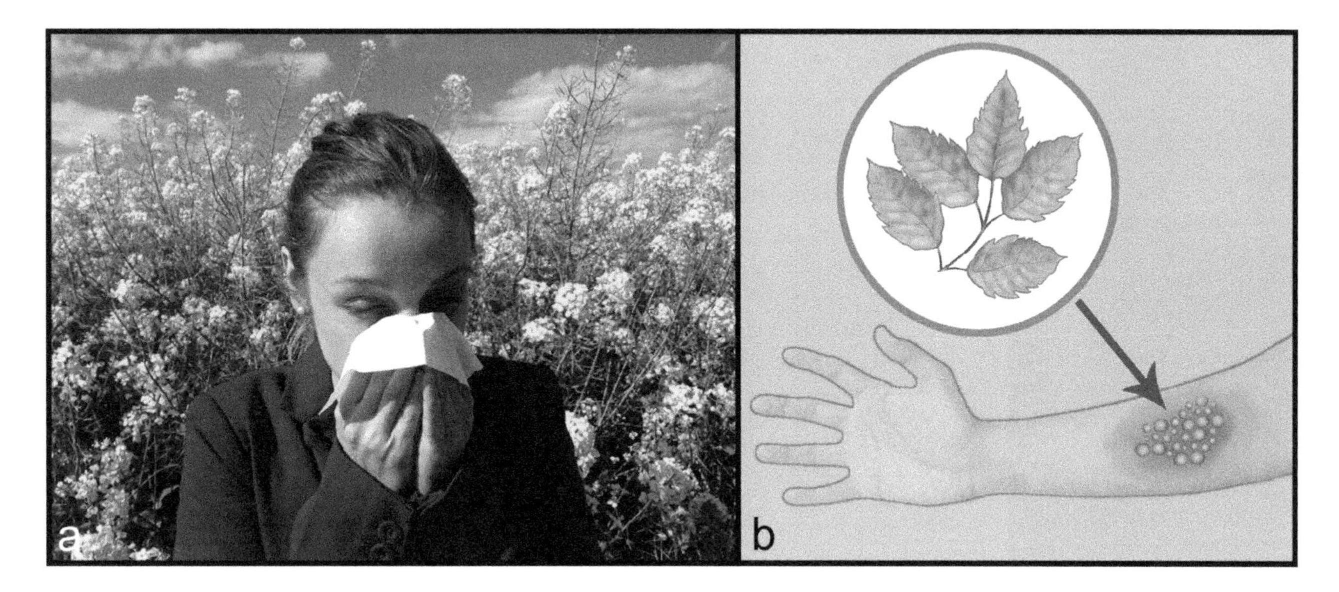

FIGURE 18.4 Allergic reactions to plants. (a) A woman experiencing a hayfever (allergic rhinitis) reaction to pollen in a field of canola (oilseed type *Brassica napus*). Photo (public domain) from Pixabay. (b) A contact allergic reaction. Prepared by B. Brookes.

- Allergens

  The most common type of inflammatory skin condition induced by plants is 'allergic eczematous contact dermatitis,' and this can take many forms. Susceptibility often varies by race and sex, and sometimes by age. Often an initial exposure or several exposures are required to sensitize an individual, who upon subsequent exposure develops dermatitis. A single exposure may suffice, or sometimes a person may not react for many years before finally becoming susceptible. Poison ivy is the most widespread example. Unfortunately, some humans also become sensitized to common ornamental or food plants.
- Primary irritants

  The skin reaction to chemicals of this class occurs within a short period after contact. Unlike allergenic substances, primary irritants affect everyone upon first exposure if the dosage is sufficient. The most familiar example is capsaicin, which is responsible for the pungent taste of chili peppers (*Capsicum* species) but is also a powerful irritant.

Dangerous Spiny Seeds and Fruits

Not included in this review are plants that distribute their seeds or fruits by hooks or spines which attach to the flesh (not just the fur) of animals, sometimes inflicting serious wounds. One plant of this category – puncture vine (*Tribulus terrestris*) – is a noxious invasive weed of warm and hot regions, where it is often one of the world's most hated plants. The spines of its 'burs' (small woody fruits; Figure 18.5) penetrate bare feet, thin-soled shoes, and bicycle tires.

Numerous plants are too poisonous to consume safely, but most people have learned to avoid them, although there are regular cases of children eating toxic berries. There are many extremely poisonous plants, but on the whole, they infrequently harm people. Illustrative of this is castor bean (*Riccinus communis*), allegedly the world's most poisonous plant, but which is nevertheless widely cultivated as a garden ornamental (Small 2011a). In famine situations, people may be forced to consume dangerously toxic plants (Small and Catling 2004b) or even grain seeds treated with toxic chemicals. Perhaps the most infamous example of plant toxicity is ergot, a fungal disease of the fruiting heads of cereals and other grasses (Wegulo and Carlson 2011). The fungus, *Claviceps purpurea*, produces toxic alkaloids, and

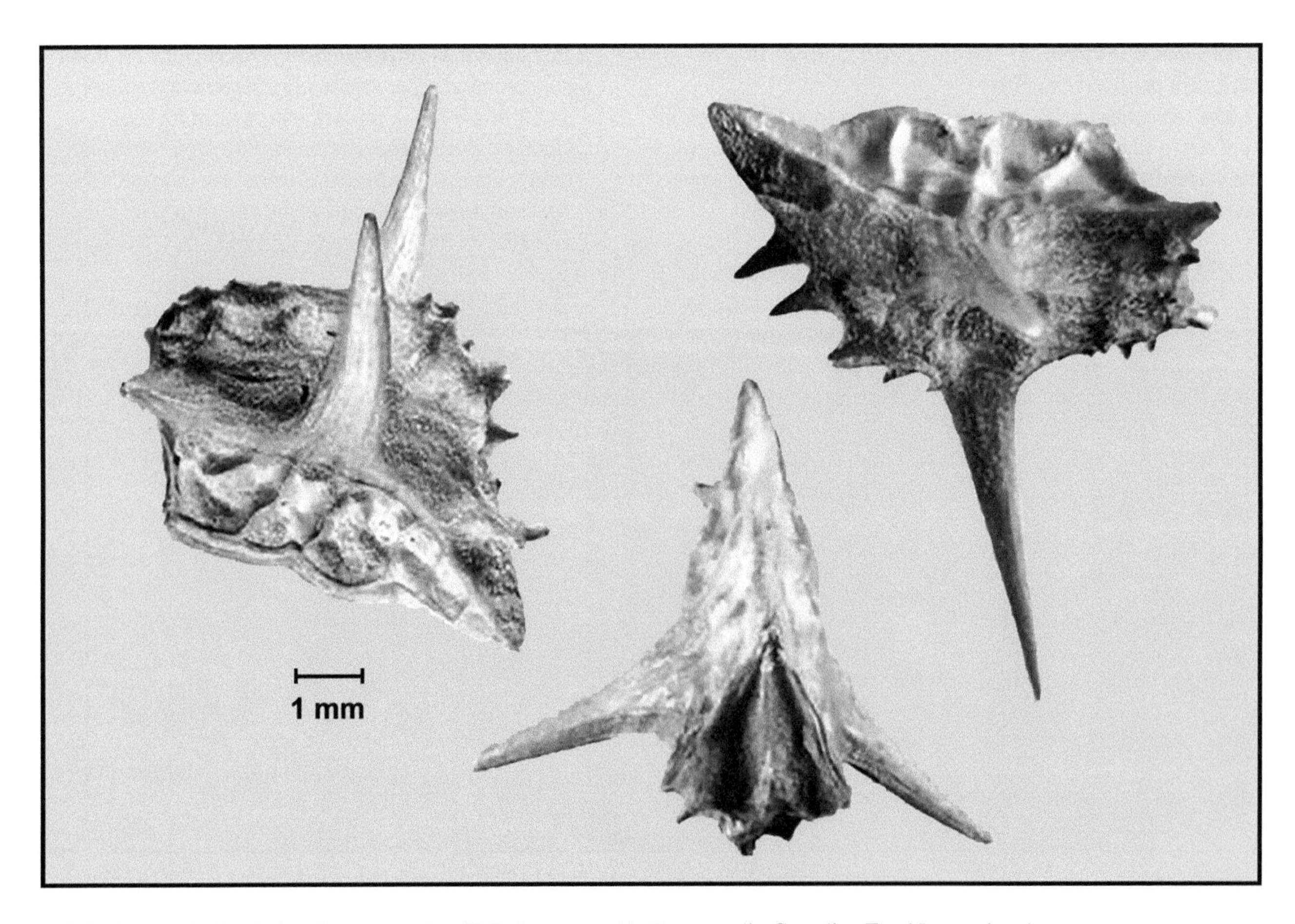

FIGURE 18.5 Spiny fruits of puncture vine (*Tribulus terrestris*). Photo credit: Canadian Food Inspection Agency.

consuming infected grain has resulted through the centuries of epidemics of ergotism, a disease marked by terrible extensive gangrene. In medieval times, this was called St. Anthony's Fire, the 'fire' a reference to the burning sensation sufferers of the disease felt. Associated symptoms of mental degeneration have been hypothesized as responsible for odd behavior which may have led to the Salem witch trials. In modern times, the most common occurrence of human toxicity from the consumption of plants is food poisoning related to accidental bacterial contamination of commercial vegetables, particularly leafy greens. Also notable is the presence of pesticide residues in popular fruits and vegetables (Park 2019). However, these problems are relatively easily controlled with appropriate care. Although eating toxic plants is not a major health issue for people, ingestion of many species of poisonous plants of pastures and rangelands is a significant cause of livestock pathology (Burrows and Tyrl 2013).

OPIUM POPPY

Introduction

True poppies belong to *Papaver*, a genus of up to 100 species which are native mostly to temperate regions of the Northern Hemisphere. Many unrelated plants have 'poppy' in their names – for example, the California poppy (*Eschscholzia californica*) and the Mexican poppy (*Argemone mexicana*). Numerous species of *Papaver* are grown as ornamentals, and a few are used in medicine because of their content of alkaloids. Although many of the true poppy species contain alkaloids that can be employed for illicit drug purposes, only the opium poppy, *P. somniferum*, is significant as a source of drugs of abuse.

Alkaloids are plant-produced compounds that contain nitrogen. They are generally bitter and often very toxic (hence they protect plants against herbivores). There are thousands of alkaloids, some of which are very important in foods. For example, piperine is responsible for the burning taste of black pepper, and caffeine is the stimulant in coffee and tea. The first alkaloid ever isolated chemically was morphine, the basic compound of opium. This was carried out in 1803 by F. W. A. Sertürner (1793–1841), a pharmacist's assistant in the German town of Paderbom.

Papaver somniferum (Figure 18.6) is an annual, growing to as much as 2 m in height, with attractive white, pink, red, bluish, or purple flowers ('doubled' in some garden forms, i.e., with extra petals) up to 10 cm across. The species is believed to grow wild in the Mediterranean region, from the Canary Isles eastward. It is found as an escape from cultivation in fields, roadsides, and waste places in scattered localities throughout North America and in other regions of the world. *Papaver somniferum* has been treated taxonomically in different ways. According to a popular interpretation, the cultivated phase is placed in subspecies *somniferum* while the wild plants are assigned to subsp. *setigerum* (Dittbrenner et al. 2008, 2009). Others divide the cultivated phase itself into various subspecies. For example, the oilseed varieties have been placed in subsp. *hortensis* (Tétényi 1995, 1997). There is appreciable genetic variation within *P. somniferum* (Labanca et al. 2018; Tamiru-Oli et al. 2019).

Crude opium is the hardened latex (milky sap, as shown in Figure 18.7a) of the unripe fruit ('capsule') of the opium poppy. Dried latex is obtained from unripe capsules and is used medicinally as well as for illicit narcotics. The drug called opium is a mixture of many constituents, including the alkaloids morphine and codeine (Yadav et al. 2006). Morphine is normally the most abundant alkaloid present in opium. Opium was traditionally obtained by making incisions into the nearly ripe poppy capsules 10–20 days after flowering (Figure 18.7b). In cooler climates, incisions do not seem to result in good exudation of latex, and mature capsules are simply collected for chemical extraction.

The opium poppy has been selected to produce three kinds of plants: drug varieties, oilseed/condiment varieties, and ornamentals. Most people are unaware that the source of poppy seeds commonly used on bakery products is the opium poppy, exactly the same species that produces opiate drugs such as heroin. Indeed, the very same plants are often the source of both condiments and narcotics. Most people are also unaware that ornamental poppies, often cultivated on their property, are also frequently opium poppies. While the content of opium is much higher in drug varieties, condiment and ornamental forms of the opium poppy may have significant content of opium. Morphine is often extracted from the capsules of oilseed cultivars after the seeds have been harvested, although drug cultivars are more productive (Frick et al. 2005). The capsules of some ornamental forms of opium poppy have less than 1% opiate alkaloids, while the morphine content of some selected pharmacological varieties is more than 25% by dry weight.

Harmful Aspects

Opium poppy has been employed as a source of drugs since ancient times (Merlin 1984; Kapoor 1995; Norn et al. 2005; Damania 2011). The narcotic effects may have been known to the ancient Sumerians, about 4000 B.C.E., as they had a symbol for it which has been translated as 'joy plant.' The species was well established as a crop in classical times. It was familiar to ancient Egyptian, Greek, and Roman civilizations, which seem to have used it effectively as a therapeutic drug. By the time of Mohammed (A.D. 570–632), the drug qualities were also appreciated in Arabia. Islamic traders and missionaries spread the cultivation of the opium poppy to Persia, India, China, and Southeast Asia, where it was used to relieve pain, but also began to be used excessively as a habit-forming, destructive narcotic, first in India, then in China. The drug was initially eaten in the Orient, but was smoked in the 17th century, which resulted in a tremendous upsurge of use and production. Crude opium has a disagreeable smell and a hot biting taste. The Swiss

FIGURE 18.6 Painting (public domain) of opium poppy (*Papaver somniferum*), from Thomé (1885, Plate 25.9). (a) Cross-section of a fruit capsule, showing fruit sectors filled with seeds. (b) Fruit capsule. (c) Seed. (d) Red flower. (e) White-flowered plant.

alchemist Paracelsus (1493–1541) overcame these objectionable features by making up a solution of opium in alcohol. This deadly mixture, known as tincture of opium or laudanum, enslaved numerous people during the following centuries (Figure 18.8). Portuguese, Dutch, and British merchants trafficked extensively in opium. The British brought about the Opium Wars of 1840 and 1855 to prevent the Chinese from outlawing the opium trade (Figure 18.9). It has been estimated that in 1886 about a fourth of the Chinese (approximately 15 million people) were addicted to opium. During Victorian times in England, tincture of opium was readily available and was often administered to teething or upset babies to make them sleep (Figure 18.10a and b). In the U.S., opium preparations became widely available in the 19th century in patent medicines and were considered to be benign (Figure 18.10c). Morphine was used so commonly as a painkiller for wounded soldiers during the American Civil War that opium addiction became known as 'the army disease' and 'soldier's disease.'

As a source of drugs of abuse, the opium poppy may have caused more human pain than any other plant. While morphine is the premier medication for agonizing pain and suffering, it has high potential for addiction, and so it is strongly regulated. Heroin, which is produced by chemical conversion of morphine, acts relatively rapidly to produce euphoria and is a chief illegal drug of abuse.

FIGURE 18.7 Traditional opium production. (a) White latex exuding from a freshly cut opium poppy seed pod that has been slit to cause release of the milky substance. Photo by KGM007 (released into the public domain). (b) An opium poppy field in Afghanistan. Notice that the seed capsules have been slit longitudinally, and the exuded latex has turned brown. Photo by Davric, released into the public domain.

FIGURE 18.8 Artists' conceptions (public domain) of Chinese opium dens. (a) In San Francisco. Source: Detroit Publishing Company Postcard series number 7339, 1903–1904. (b) In France. Source: cover of Le Petit Journal, July 5, 1903.

Illicit opium is produced chiefly in Asia. The major illegal growing areas are: the highlands of mainland Southeast Asia (especially Burma, Laos, Thailand, and adjacent southern China and northwestern Vietnam); Southwest Asia (notably Pakistan, Iran, and Afghanistan); Mexico; and to a lesser extent Lebanon, Guatemala, and Columbia. In recent times, Afghanistan became the world's predominant supplier of illicit opium (Parenti 2015).

Opium poppy tea (also misleadingly known as poppy seed tea) is a (usually) weak narcotic preparation made by soaking opium poppy capsules in water (the poppies are usually first ground to a powder, and often served in tea). Called 'poor man's opium' and 'doda,' it has been popular among taxi and truck drivers in Afghanistan, Pakistan, and India for boosting alertness. However, in recent times opium poppies have been illicitly cultivated for the production of doda in North America, particularly by Asian immigrants. While usually notably lower in the opiate chemicals that are found in opium, doda is now considered to be a significant health and law enforcement issue. Some deaths have resulted from concentrated preparations (Centre for Science in the Public Interest 2019).

FIGURE 18.9 Scenes of the First Anglo-Chinese War (1839–1842) popularly known as the First Opium War or simply the Opium War. (a) Painting (public domain) by British military artist Richard Simkin (1840–1926), entitled 'The 98th Regiment of Foot at the attack on Chin-Kiang-Foo.' (b) The British ship Nemesis (a steam-driven vessel made entirely out of iron) easily destroying Chinese war junks in Anson's Bay in 1841. Painting (public domain) by English artist Edward Duncan (1803–1882).

By United Nations international conventions, the opium poppy can only be cultivated and sold for authorized medicinal purposes in a restricted number of countries by special permits. Major opium poppy cultivating nations are Turkey, India, Australia, France, Spain, Hungary, Czech Republic, and China (Baser and Arslan 2014), and occasionally in other countries (e.g., Figure 18.11). Countries differ in their laws governing opium poppy, including capital punishment in some jurisdictions. Generally, opium poppy seeds are exempted from criminal penalties, and ornamental poppies rarely attract the attention of the authorities.

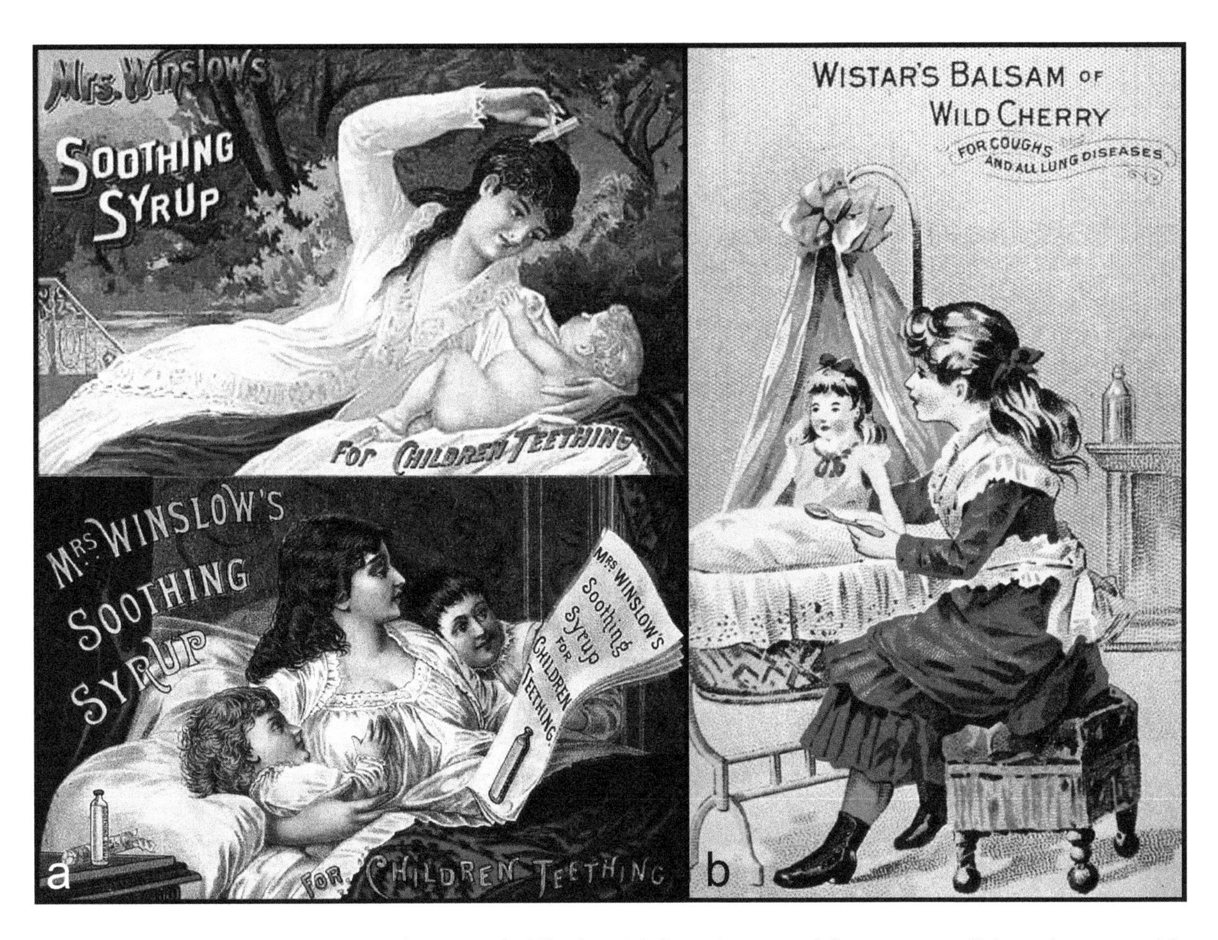

FIGURE 18.10 Nineteenth-century advertisements (public domain) for opium-containing patent medicines. (a – top and bottom figures) Mrs. Winslow's Soothing Syrup, which contained one grain (65 mg) of morphine sulfate per American fluid ounce (29.6 mL). This preparation was marketed as an indispensable mother's aid for relief of babies' discomfort while teething. The administration of opium to infants dates to ancient times: An Egyptian text dated at about 1300 B.C.E. described the custom of giving poppy extracts to children to stop them from crying. While effective at quieting restless infants and small children, such medications became known as 'baby killers.' (b) Advertisement for Wistar's Balsam of Wild Cherry, which contained cherry bark, alcohol, and opiates. This 1986 (public domain) trading card features a young girl giving medicine to a doll. Credit: U.S. National Library of Medicine.

FIGURE 18.11 A field of opium poppies cultivated for pharmaceuticals in North Dorset, England. Photo by Marilyn Peddle (CC BY 2.0).

Beneficial Aspects

- Drugs

Although opium poppy may be the source of more human suffering than any other plant (Figure 18.12), as a source of legitimate pharmaceuticals it has relieved more human pain than any other plant (Figure 18.12). Morphine is considered to be the most important analgesic used for severe pain. At least 80 additional alkaloids are produced by opium poppy, and several are important medicinally (Duke 1973). One of these is codeine, but, in the opium poppy, codeine is usually a minor constituent. Most legally produced morphine is chemically converted to codeine. Codeine is a widely prescribed analgesic and cough-suppressing drug. It is the most widely used opiate alkaloid and indeed is probably the most commonly used drug in the world. Codeine is a constituent of over 22% of analgesic preparations in the U.S. Other poppy alkaloids are also useful: noscapine suppresses coughs and tumors; papaverine dilates veins and relaxes smooth muscle tissue; and sanguinarine is antimicrobial and anti-inflammatory. Thebaine is not narcotic and is used as an antagonist to curb heroin addiction. By chemical conversion of opioid alkaloids, a host of additional compounds are produced and employed in medicine. In recent times, new varieties of opium poppy, and different species such as the thebaine poppy (*P. bracteatum*), are being employed as sources of useful medicinal alkaloids with negligible narcotic potential (Nyman and Hall 1974; Palevitch and Levy 1992; Sharma et al. 1999; Small 2010). Much of the poppy grown for the extraction of opium for medicinal purposes is cultivated in government-regulated farms in Europe, India, Turkey, and Tasmania (Australia).

- Edible seeds

Although the seeds of some other poppy species are quite edible, the opium poppy remains the exclusive commercial source of poppy seed products (Lim 2013; Figure 18.13). The kidney-shaped seeds range from white and yellow to slate-blue or black and are very small, 1.0–1.5 mm long (Figure 18.13e), with more than 2 million in 1 kilogram. In mature wild plants, the wind shakes the seeds out of the fruit through pores at the apex (Figure 18.13c). The culinary use of the opium poppy is as ancient as the drug usage. Many seeds have been found at the sites of Neolithic domiciles, including Swiss Lake Dwellings (at least 4,000 years of age). The opium poppy has been employed as a food and spice since ancient times (Andrews 1952). In the

FIGURE 18.12 The good and evil sides of opium poppy. As a source of pharmaceuticals, the opium poppy has relieved more human pain than any other plant. As a source of drugs of abuse, it may have caused more human pain than any other plant. Prepared by B. Brookes.

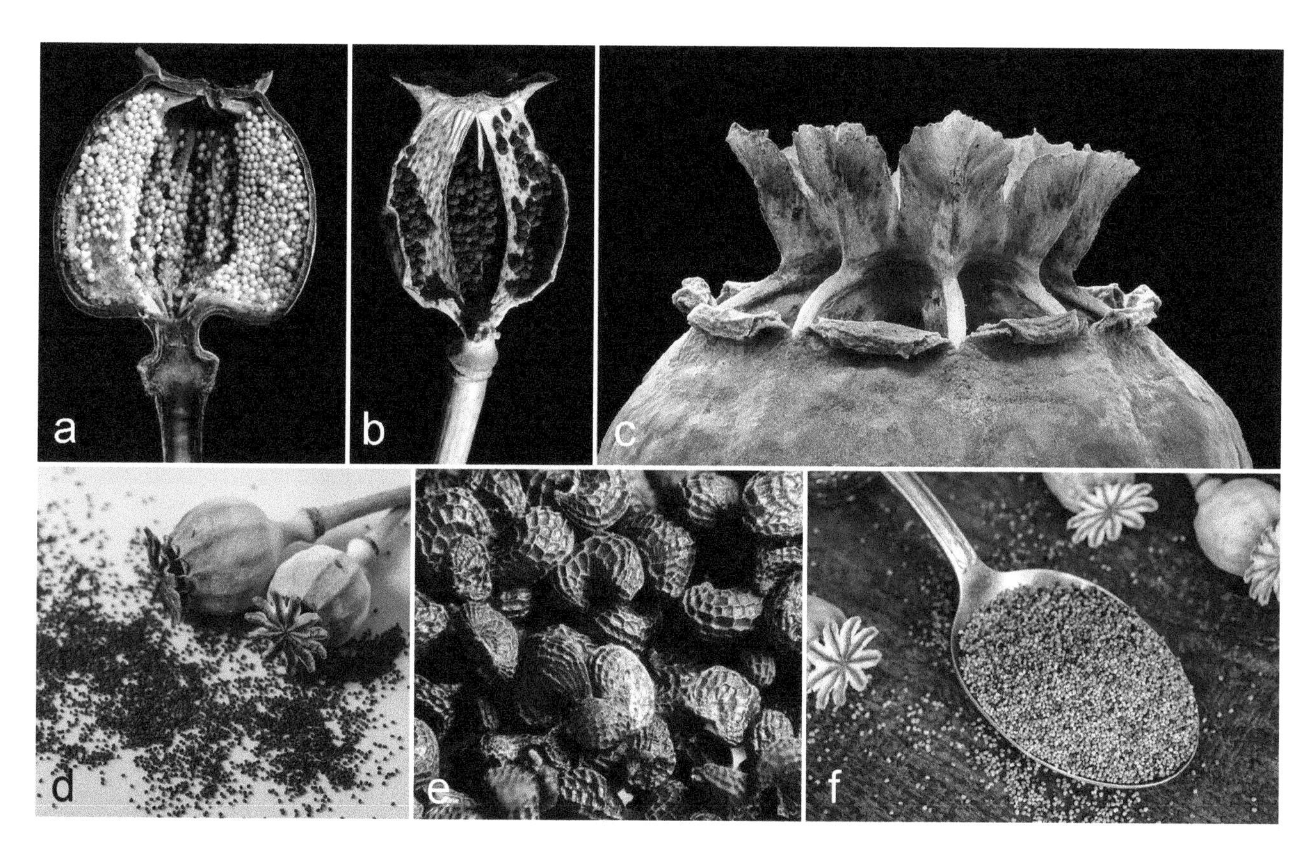

FIGURE 18.13 Seeds of opium poppy (*Papaver somniferum*). (a) Long section of young capsule. Photo credit: William Copeland McCalla family fonds, Provincial Archives of Alberta. (b) Mature capsule broken open. Photo (public domain) by Andy Faeth from Pixabay. (c) Top of dried capsule showing pores that allow seeds to be knocked out and blown away by wind currents. Photo (public domain) by Ian Lindsay from Pixabay. (d) Mature capsules spilling seeds. Photo by Keith Ellwood (CC BY 2.0). (e) Close-up of seeds (which are 1.0–1.5 mm long). Photo by Thomas Bresson (CC BY 3.0). (f) A spoonful of seeds. Source: Shutterstock, contributor: Madeleine Steinbach.

2nd millennium B.C.E., the Egyptians cultivated poppy to obtain the edible oil in the seeds. The classical Greeks also used poppy seeds as a food. Both the ancient Greeks and Romans added the seeds to cakes and bread as a flavoring. By the Middle Ages, the use of poppy seeds as a spice on bread was well established. Today, poppy seeds are still a popular condiment. Poppy seed for culinary use is produced in several countries, particularly the Netherlands, Poland, Romania, the Czech Republic, Russia, India, Iran, Turkey, Argentina, and many Asian and Central and South American countries.

The ripe seeds (sometimes called maw seed) have a mildly spicy, oily, agreeable nutty flavor, and are used as a condiment on baked goods, rolls, and pastry (Figure 18.14). The flavor and crunchy texture are brought out by toasting or baking. Poppy seeds are used as a topping on rolls, breads, cookies, cakes, and other baked goods, or as a garnish. The seeds can also be used in a variety of cooked vegetable dishes and stews. Crushed seeds are mixed with other ingredients and used as a filling for cakes and pastries, cream cheese, and sour cream dips. Poppy seeds are additionally employed to flavor butter and various food preparations.

Since poppy seed is a widespread culinary item, but is obtained from a dangerously 'narcotic' plant, there has been interest in establishing whether or not the seeds contain any of the alkaloids found in opium that are considered to be dangerous (Bonicamp and Santana 1998; Lachenmeier et al. 2010). The statement is frequently recorded that the seeds and the oil expressed from them contain no drug properties. However, there is some evidence that minute quantities of the legislatively controlled alkaloids occur in the embryos and endosperm (nutritive tissue) of the seeds. Nevertheless, these trace amounts are generally considered insignificant, and it has been speculated that findings of opiates in the seeds is the result of contamination by latex from the capsules. 'Poppy seeds do not contain the opium alkaloids, but can become contaminated with alkaloids as a result of pest damage and during harvesting' (Knudsen et al. 2018). Regardless, it is clear that minute amounts of opiates are sometimes associated with the seeds.

FIGURE 18.14 Examples of bakery uses of opium poppy seeds. (a) Poppyseed bagel. Photo by Stas Ovsky (released into the public domain). (b) The 'kolache' is a traditional Czech sweet bun, made with an egg and yeast dough and filled with preserves and poppy seeds. Photo by Petr Brož (CC BY SA 3.0).

It has been demonstrated that, in at least some cases, subjects who have consumed poppy seeds can test positive by drug exams for the illicit consumption of narcotics. Because of concern that poppy seeds may contain opiates, efforts have been made to breed oilseed cultivars with very limited alkaloids (Németh and Bernáth 2009).

- Oil

  Opium poppy is also commonly grown as a source of oil, and there is interest in breeding improved crop varieties (Singh et al. 1995). Forty percent of the seed weight is a pale yellow, fixed, tasteless, oil, useful as a salad oil as it is less liable to become rancid than olive oil. The oil is also used to make margarine, salad dressing, and as a cooking oil. Poppy oil is rich in the unsaturated oleic and linoleic acids, and therefore is potentially useful as a source of nutritional supplements. Artists also use poppy seed oil as a drying oil, useful in paints and varnishes. The oil is also a component of some soaps. As an animal feed, poppy seed is commonly used as birdfeed, and after oil has been expressed from seeds, the protein-rich oil cake has been fed to cattle.
- Ornamental poppies

  Opium poppy is extensively grown as an ornamental (so are other species of *Papaver*, and some varieties are hybrids). Ornamental strains often have double or fringed petals (Figure 18.15). In Western countries, the widespread cultivation of opium poppies as ornamentals is almost always overlooked by the authorities, except in rare cases where individuals appear to have deliberately attempted to use the plants for illicit drugs. Seeds for ornamental opium poppies are very widely available from catalogs and stores, and the plants are widely cultivated.

Conservation Aspects

Considerable damage to biodiversity often results from the illegal cultivation of addictive drug plants. Criminals associated with the illicit production and trade have limited respect not only for the laws of society, but often also show no regard for the environment. Illicit plants are often cultivated in wild areas to avoid detection, with consequent damage to ecosystems. A very large proportion of the illegally grown plants are confiscated by authorities, necessitating the wasteful production of much material to compensate for the losses. Irresponsible and wasteful agricultural practices – diversion of water, clearcutting, heavy use of herbicides and pesticides, poisoning of birds and mammals that might consume the plants – harm the environment and its biota, often contributing to erosion and pollution of water bodies. Irresponsible use of chemicals may result in the deposition of toxic substances in the soil. An associated problem with illegal cultivation is the eradication tactics often adopted by law enforcement when illegal plantations are discovered. The broadcasting of herbicides has often occurred, with consequent damage to the plants and other species in the area.

One of the great tragedies associated with illicit plants is that impoverished people have often become dependent on income generated from growing and harvesting the 'black market' crops. There have been frequent attempts to persuade such poor people to convert to 'white market' plants, i.e., legitimate crops, but with very limited success because alternatives are not as profitable. Ninety percent of the world's illegal poppy production in recent times has occurred in Afghanistan, with the income from the trade having contributed to the war effort of the Taliban against the West. NATO forces (Figure 18.16) were very reluctant to engage in widespread destruction of the poppy crop because of fear of antagonizing much of the population whose income depended on it. For several years, there were proposals to initiate a program in which Afghanistan

FIGURE 18.15 Ornamental variety of opium poppy with doubled flowers. Photo by Yewchan (CC BY SA 2.0).

FIGURE 18.16 A U.S. marine in a poppy field in Afghanistan. Source: U.S. Government public domain photo.

poppy farmers would grow legitimate poppy crops (both narcotic and non-narcotic), but this seemingly noble initiative was not realized. The Afghanistan situation, in which illicit drugs were entangled with war, is not only particularly damaging to people but also to biodiversity. In a combat theater, concern for biodiversity is of virtually no consequence to military forces. Good farmers everywhere have learned the value of ecologically based agricultural practices, but in times of military conflict concern for the environment becomes less of a priority.

Most crops are harmful to biodiversity, if for no other reasons than they take up space that was formerly used by wild plants and animals. However, very high-value crops that can be grown on limited acreages are, in reality, 'biodiversity-friendly' because they sacrifice less of nature. Opium poppy crops, whether grown for (1) legitimate narcotic pharmaceuticals, (2) non-narcotic pharmaceuticals, (3) edible seeds, or (4) seed oil, are very high-value crops that need relatively limited areas. Moreover, opium poppy crops are fairly 'tough' (tolerant of climatic and edaphic stresses). Compared to the major crops, they have significantly smaller needs for agricultural inputs (water, biocides, fertilizers) and energy-consuming cultivation. Thus their cultivation tends to limit the consumption of fossil fuels; decrease the generation of greenhouse gases; decrease environmental pollution due to runoff of fertilizers, pesticides, and herbicides; and decrease pressure to obtain water from a world that is running out of it. Because they occupy acreages very much smaller than required for conventional crops, they save wildlands from being converted to agricultural purposes. Still another consideration is that the harvested seeds or extracted chemicals constitute an extremely small part of the crop's biomass. The remaining biomass serves two useful ecological purposes: it ties up CO_2 ('sequesters carbon') for a period, thus decreasing the greenhouse gas problem; and it constitutes natural compost that contributes to health of the soil.

TOBACCO

Introduction

Tobacco comes from species of the genus *Nicotiana*, a name which commemorates Jean Nicot (1530–1600), French ambassador to Portugal. He obtained tobacco from sailors returning to Lisbon from America and introduced it into France in 1560. Eighty-two species of *Nicotiana* have been recognized, distributed in tropical and warm temperate America, Australasia, and Polynesia. Most are native to South America, with three dozen in Australia (Goodspeed 1954; Knapp 2020). The most frequently grown tobacco plant is common tobacco, *N. tabacum* (Figures 18.17 and 18.18). The species is thought to be a hybrid, whose ancestors are distributed in the eastern Andes Mountains of South America. Tobacco from *N. tabacum* is produced in most temperate and tropical regions of the world and is a major crop of many nations. The common tobacco is a herbaceous annual (or less commonly a biennial), growing 1–3 m in height. The stem tends to become woody at its base. The flowers can be white, yellow, pink, purple, or red, 3–6 cm long. Leaves vary in size depending on variety, and the lower ones are often longer than 50 cm. The leaves of cigar-wrapper varieties are thin, fine-textured, and small-veined, while those of plug and pipe tobacco varieties are usually coarser, tougher, and thicker. Growth of the plant is modified by removing the top, usually when 8–12 leaves have developed, and also by removing branches. This causes the remaining leaves to increase in size by as much as 50%, increases nicotine content, and encourages uniform ripening of the leaves.

Nicotiana rustica (Figure 18.19a), known as wild tobacco and Aztec tobacco, was also domesticated for tobacco production, although it is not much grown commercially in North America today. Like *N. tabacum*, it is thought to have arisen as a hybrid, probably from ancestral species in Peru, and it does not have a natural native distribution. The name 'wild tobacco' is a misnomer, since the species arose in cultivation and has no natural wild distribution area. When Europeans came to North America in the 15th and 16th centuries, both *N. tabacum* and *N. rustica* were widely cultivated by the Indians. *Nicotiana rustica* was the first species to be brought back to the Old Word, but *N. tabacum* was imported soon afterward and quickly became much more popular. *Nicotiana rustica* is a course South American annual, hardier than *N. tabacum*, and is naturalized in eastern North America, occurring as far north as southern Ontario. *Nicotiana rustica* was widely grown by North America Indians and was cultivated for tobacco by them as recently as the middle of the 20th century. Tobacco was first cultivated by European colonists in Virginia in 1612, and the species first grown was *N. rustica*. It is now very rarely cultivated in North America, although it is grown in Europe and Asia (Figure 18.19b) for smoking tobacco and as a source of the insecticide nicotine. The dried leaves of *N. rustica* contain up to 10% nicotine, whereas those of regular tobacco usually have 1.5%–4% (although up to 8% is possible).

Tobacco has a long history (Gurstel and Sisson 1995; Doll 1999; Gately 2002; Goodman 2004; Fox 2015). It was employed 10,000 years according to Castaldelli-Maia et al. (2016). Tobacco was chewed by New World Indians and became a traditional ceremonial item (Figure 18.20). In 1492, Christopher Columbus was given a gift from the native Arawaks, of what he recorded as 'certain dried leaves,' when he and his men set foot on the New World for the first time. Columbus politely had the tobacco brought back to his ship, but afterward, not realizing what he had been given, ordered it thrown away. That same year in Cuba, Columbus's crewmen Rodrigo de Jerez and Luis de Torres were the first Europeans to observe smoking, including the use of smoldering rolls of leaves inserted into the nostrils. Jerez became the first European to smoke tobacco and indeed became addicted. When he returned to Spain and

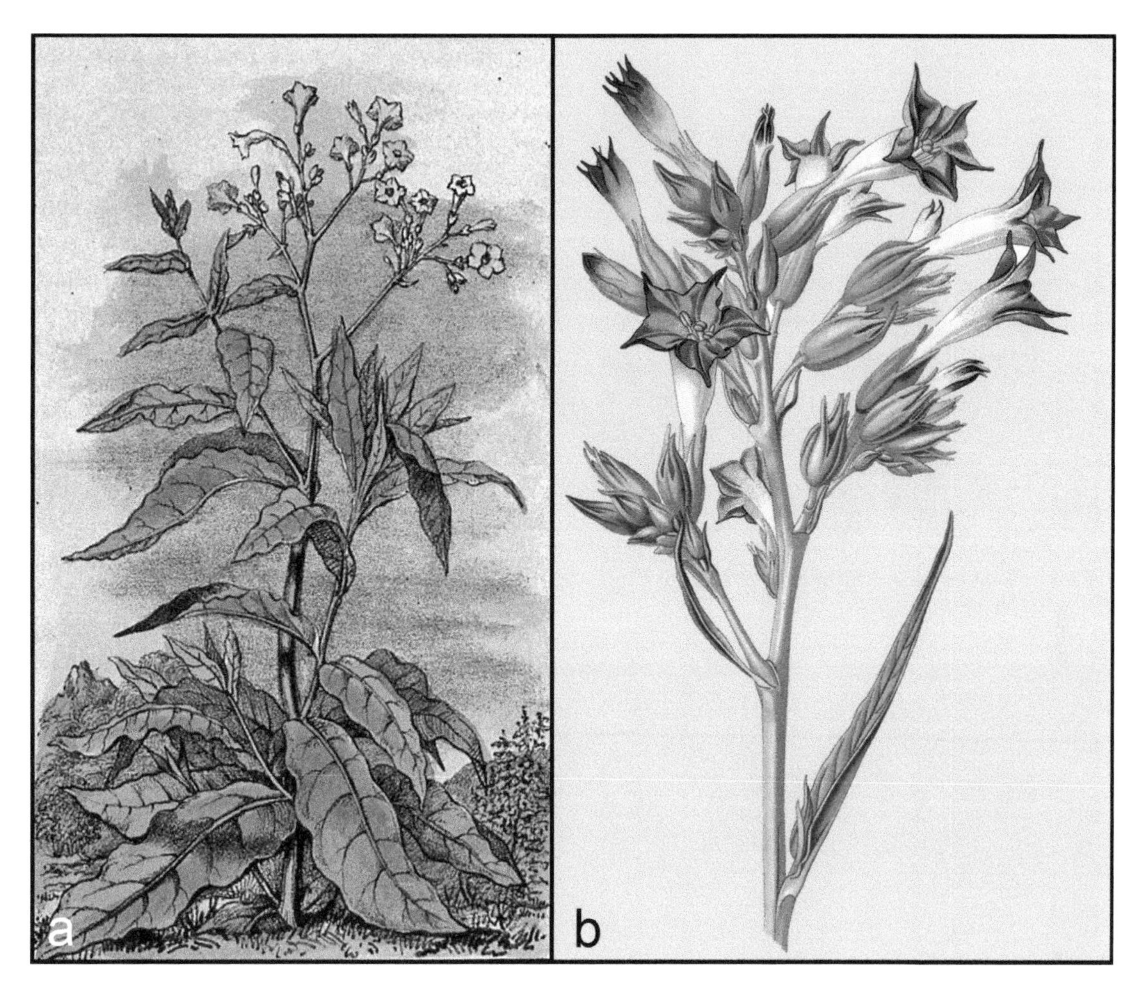

FIGURE 18.17 Common tobacco (*Nicotiana tabacum*). (a) Colored lithograph, ca. 1840. Source: Wellcome Collection (CC BY 4.0). (b) Flowering top. Source (public domain) Thomé (1885).

FIGURE 18.18 A tobacco field in Cuba. The flowers are removed in order to promote improved growth of the top leaves. Photo by Henryk Kotowski (CC BY SA 3.0).

FIGURE 18.19 'Wild tobacco' (*Nicotiana rustica*). (a) Flowering top. Source (public domain): Köhler (1883–1914). (b) Field in Vietnam. Photo by Hungda (CC BY SA 3.0).

FIGURE 18.20 Artwork entitled 'Peace Pipe,' representing traditional ceremonial use of tobacco. Painting by contemporary Russian artist Marianna Ochyra. Photo by Maxx9035 (CC BY SA 4.0).

FIGURE 18.21 Humorous cartoons (public domain) of 'smoking clubs' illustrating early European misgivings about tobacco. (a) Prepared by James Gilray in 1793. (b) Source: Fairholt (1859).

smoke was observed billowing from his mouth and nose, the Holy Inquisition imprisoned him for 7 years. In Europe, attitudes about the use of tobacco varied. Tobacco smoking was often ridiculed (Figure 18.21), but the upper classes and nobility became enamored of extremely expensive tobacco containers (Figure 18.22), and commercial packages of tobacco were decorated with the finest of art (Figure 18.23). Indeed, smoking became so common in the Western world that artists and cartoonists commonly showed their subjects smoking (Figure 18.24).

FIGURE 18.22 Luxurious tobacco containers owned by European nobles. (a) Tobacco jar made in Britain in 1879, housed in the Science Museum, London. Photo from the Wellcome collection (CC BY 4.0). (b) Snuffbox decorated with gold and diamonds, manufactured about 1765 for Frederick II of Prussia, now in the collection of the Victoria and Albert Museum. Photo by Vassil (public domain). (c) Jeweled snuffbox from 18th-century Russia. Photo by Shakko (CC BY 3.0).

FIGURE 18.23 Cigar box label (1905, public domain) illustrating the high quality of art that often decorated tobacco containers.

FIGURE 18.24 Examples of smoking in cartoons. (a) Victorian clip art showing a puppy smoking. Photo by Egg Studio (CC BY SA 2.0). (b) Gentleman smoking a cigarette (public domain). (c) Late 19th-century American print (public domain) depicting a frog smoking a pipe.

In the United States, chewing tobacco was so prevalent in the late 1800s that over 90% of tobacco factories only made chewing tobacco, and spittoons were far more common than ash trays. Chewing tobacco was so popular during the latter 19th and early 20th centuries that 12,000 brands were registered in the United States. Today, the practice of chewing tobacco has remained most popular in the United States, where it has been associated with cowboys and sports, notably baseball, where the stimulant value of nicotine may have been the motivation. At the end of the 20th century, smokeless tobacco (spit tobacco) was used by about 15 million Americans, and in the United States, about 6% of men aged 18 and older (chewing tobacco is primarily a male practice) were using some form of spit tobacco.

Smoking tobacco is not entirely an American (i.e., New World) invention. Australian aborigines also used several Australian species of *Nicotiana* for the purpose, and it appears that people native to the southwestern Pacific may have also used local species of the genus. Perhaps ten species of *Nicotiana* containing the alkaloid nicotine were used by native peoples for religious and medicinal purposes. The distribution ranges of these species were extended by such usage before European colonization.

Tobacco is used as a fumitory (i.e., for smoking) in pipes (Figure 18.25), cigars, and cigarettes, inhaled as a powder (snuff is simply finely ground tobacco; Figure 18.26a), chewed as plugs (Figure 18.26b), and savored by covering the gums with a fine powder. 'Smokeless' tobacco includes snuff that is inhaled, and all material that is taken orally – so-called 'spit tobacco.' Chewing tobacco is available in several forms. Finely shredded tobacco is dipping tobacco (really not much different from snuff); pressed bricks and cakes are called plugs; and rope-like strands are called twists.

Harmful Aspects

Nicotine, the primary addictive chemical in tobacco, was first chemically isolated in 1807. This alkaloid is extremely toxic. One or two drops (60–120 mg) of pure nicotine placed on the skin can kill an adult human. A typical cigar contains enough nicotine to kill two people, if injected into their bodies. Species of *Nicotiana* have been used as sources of dart-poison ingredients in South America. Nicotine was once employed as a pesticide (Figure 18.27). Because of extreme toxicity to humans, the use to control pests was discontinued. In *Nicotiana* plants, nicotine serves as a defense against herbivores. Curiously, nicotine in the tobacco plant is produced in the roots and translocated to the leaves (Zenkner et al. 2019).

Neonicotinoids are synthetic analogues of nicotine, with much lower toxicity to humans and greater field persistence. However, Furlan et al. (2019) reported that the use of neonicotinoids over the past 20 years has inflicted serious damage to birds, pollinators, and other insects without generally increasing yields.

Tobacco has extremely serious health hazards (U.S. Department of Health and Human Services. 2020; Textbox 18.2) and has killed more humans in history than any other plant substance. Worldwide, over 1 billion people

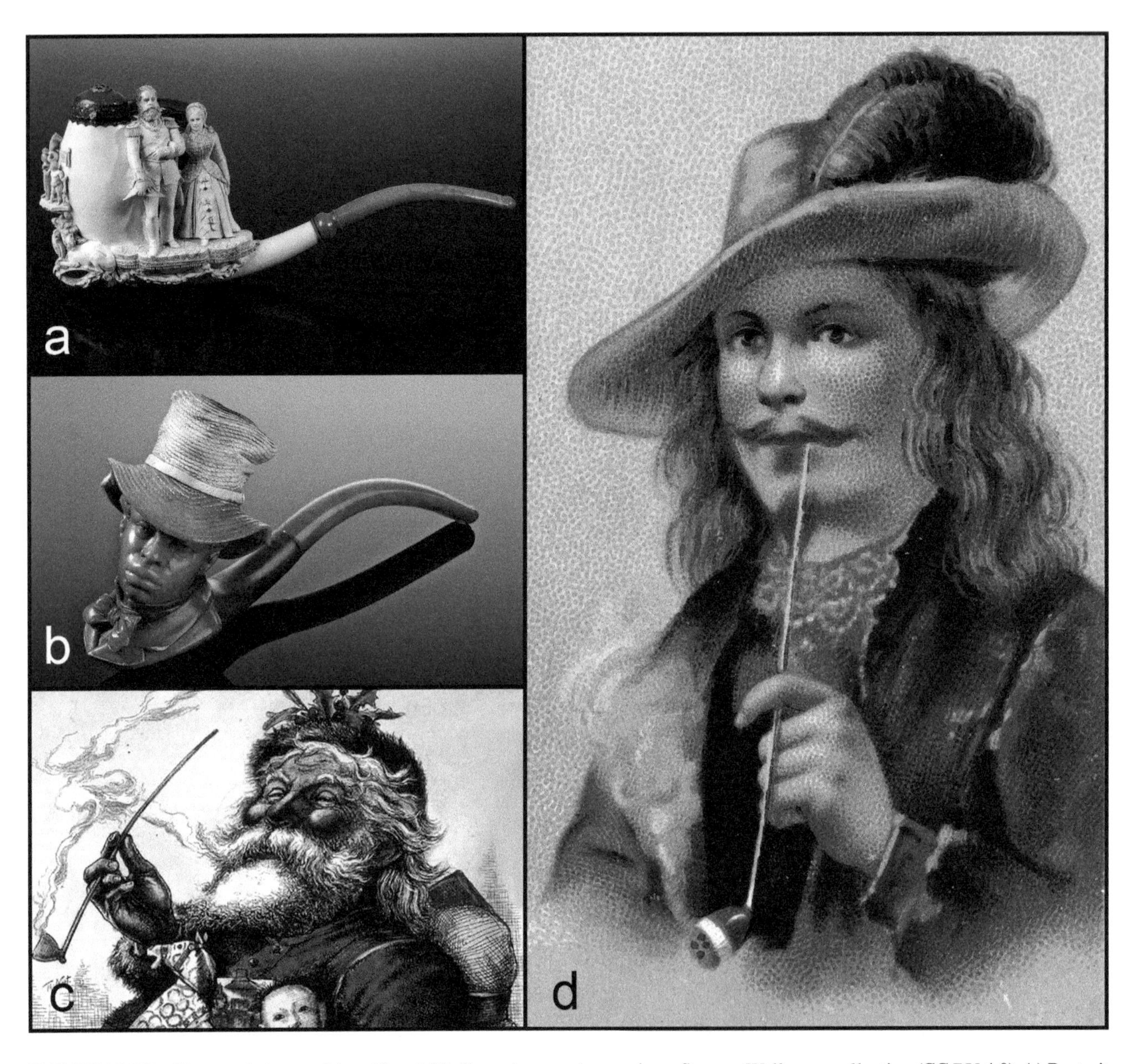

FIGURE 18.25 Pipes and pipe smoking. (a) and (b). Carved meerschaum pipes. Source: Wellcome collection (CC BY 4.0). (c) Portrait (public domain) of Santa Claus smoking a pipe, by Thomas Nast, Published in *Harper's Weekly*, 1881. (d) Seventeenth-century portrait of a Dutch man smoking a pipe. Credit: Jefferson R. Burdick Collection, Metropolitan Museum of Art.

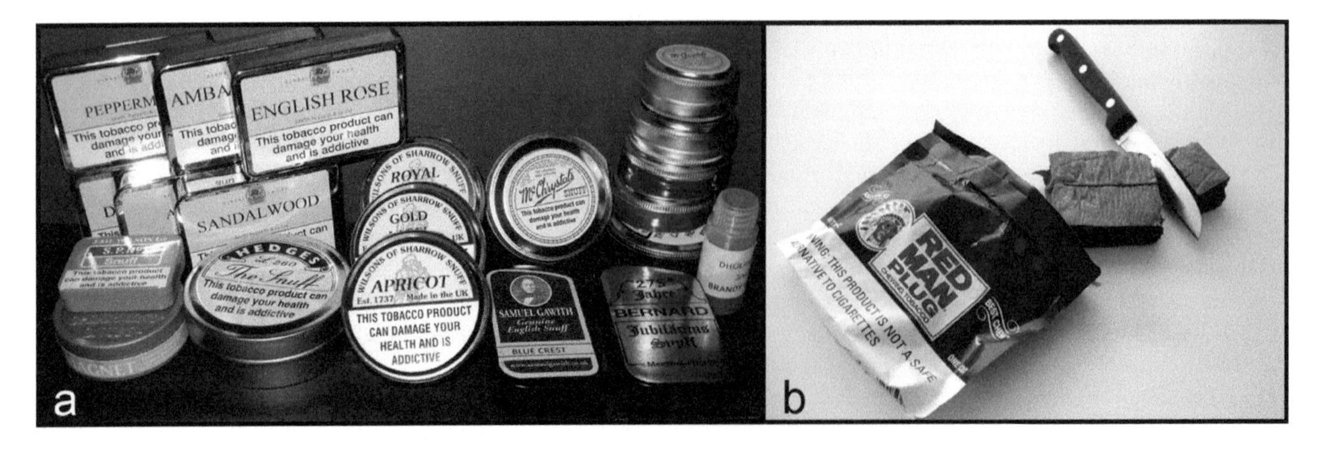

FIGURE 18.26 'Smokeless tobacco.' (a) An assortment of nasal snuff. Photo by Hellahulla (CC BY SA 3.0). (b) Plug chewing tobacco. Photo by Fredderik (CC BY SA 3.0).

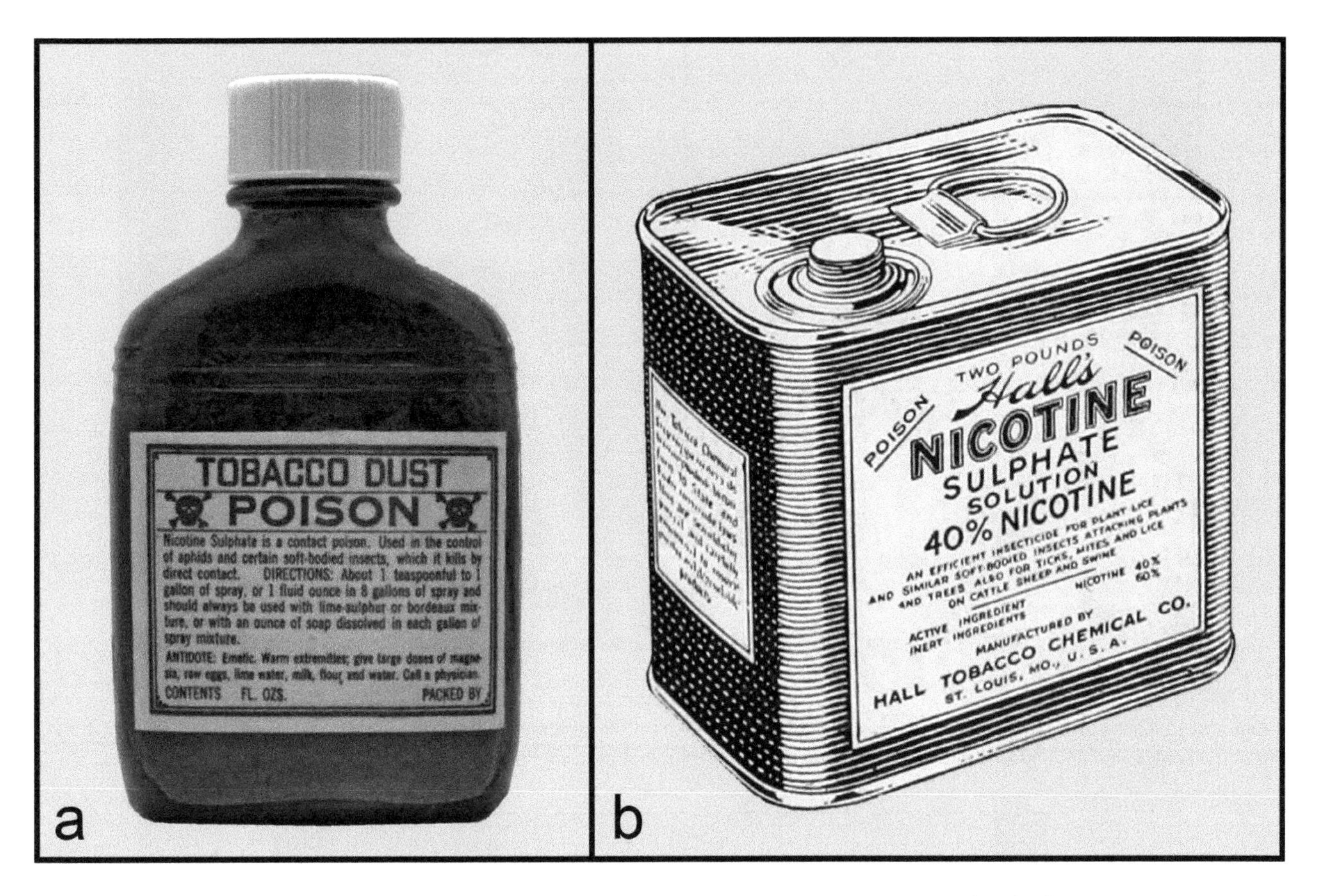

FIGURE 18.27 Early 20th-century nicotine pesticides (public domain figures). (a) Nicotine dust. (b) Nicotine spray.

TEXTBOX 18.2 NEGATIVE HEALTH EFFECTS OF SMOKING TOBACCO

Nicotine contained in tobacco is highly addictive and tobacco use is a major risk factor for cardiovascular and respiratory diseases, over 20 different types or subtypes of cancer, and many other debilitating health conditions. Every year, more than 8 million people die from tobacco use. Most tobacco-related deaths occur in low- and middle-income countries, which are often targets of intensive tobacco industry interference and marketing. Tobacco can also be deadly for non-smokers. Second-hand smoke exposure has also been implicated in adverse health outcomes, causing 1.2 million deaths annually. Nearly half of all children breathe air polluted by tobacco smoke and 65,000 children die each year due to illnesses related to second-hand smoke. Smoking while pregnant can lead to several life-long health conditions for babies. Heated tobacco products (HTPs) contain tobacco and expose users to toxic emissions, many of which cause cancer and are harmful to health. Electronic nicotine delivery systems and electronic non-nicotine delivery systems, commonly known as e-cigarettes, do not contain tobacco and may or may not contain nicotine but are harmful to health and undoubtedly unsafe. However, it is too early to provide a clear answer on the long-term impact of HTPs and/or e-cigarette use.

— *World Health Organization (2020)*

are smokers and more than 7 million die from the negative effects of smoking every year (World Health Organization 2017). The World Health Organization (2011) also notes that tobacco dependence is the leading cause of preventable death in the world. The tobacco industry ('Big Tobacco') is notorious for its insidious techniques of promoting the use of tobacco despite its known harmful effects (Robinson 1998; Klein et al. 2007; Brandt 2012; Figure 18.28). About 500 million people alive today will eventually be killed by tobacco use. Smoking tobacco is a cause of emphysema and lung cancer, other cancers (note Figure 18.29), heart disease, and numerous other disorders. It has been estimated

FIGURE 18.28 Some of the deceitful tactics employed by 'Big Tobacco' to promote smoking. (a) A painting of a young woman with a cigarette, making smoking seem glamorous, healthy, and sexy. Public domain image from PublicDomainPictures.net. (b) Candy cigarettes, employed to condition children that smoking is acceptable. Photo by Phillip Stewart (CC BY SA 2.0).

FIGURE 18.29 Sigmund Freud (1856–1939), founder of psychoanalysis, whose doctor assisted his suicide because of oral cancer caused by smoking. Photo (colorized; public domain) by Max Halberstadt of Christie's, in 1921.

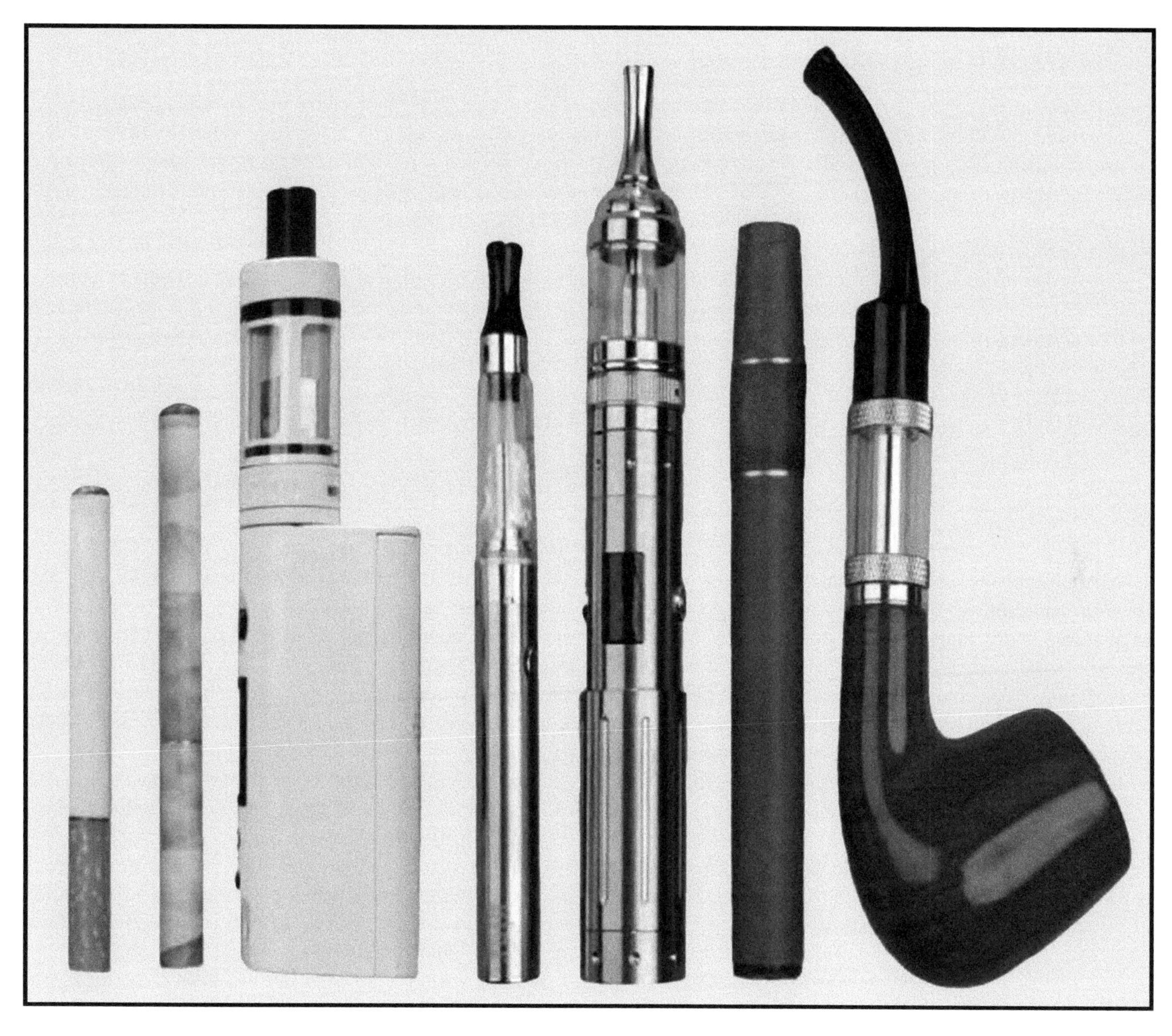

FIGURE 18.30 Various types of electronic cigarettes. Public domain photo, U.S. Centers for Disease Control and Prevention.

that each cigarette smoked takes 5 minutes off one's life, a figure that indicates that smoking a standard pack of 20 cigarettes a day shortens an expected year of life by 1 month. Smoking is one of the most common causes of 'accidental' fires. It is also a cause of social unrest, with the addictive qualities of tobacco causing otherwise thoughtful people to subject innocent passers-by to the deleterious effects of the disgusting habit.

Chewing tobacco is addictive and is not, contrary to the belief of many, safer than smoking tobacco, although the set of health hazards are somewhat different. The nicotine in chewing tobacco is absorbed through the oral mucous membranes and has similar adverse effects as that in cigarettes. Tobacco placed in the mouth is associated particularly with cancers and other pathological conditions of the mouth, tongue, lips, and throat. Studies have found that half to three quarters of spit tobacco users have oral lesions. Chewing tobacco users are four times more likely than non-users to develop tooth decay.

Electronic cigarettes are battery-powered devices that provide users nicotine to inhale following its aerosolization from a liquid solution. E-cigarettes (Figure 18.30) and other methods of vaping have been touted as safer alternatives to continuing to satisfy nicotine addiction than traditional smoking and also as a means to assist in breaking the habit of smoking. Much fewer harmful chemicals are released in vaping, since nicotine is vaporized at relatively low temperatures (by contrast, 'burning' or combustion of materials occurs at higher temperatures and releases many more harmful chemicals). However, the safety of some of the carrier chemicals in which nicotine is dissolved is an unsettled issue, there is clear evidence that youth have taken up vaping in alarmingly large numbers, and there is a widespread conviction in the medical community that the harm associated with e-cigarettes is very significant although not sufficiently evaluated at present (Polosa et al. 2019; Simonavicius et al. 2019; Bozier et al. 2020; Cao et al. 2020; Winnicka and Shenoy 2020).

TEXTBOX 18.3 ENVIRONMENTAL AND BIODIVERSITY COSTS OF TOBACCO PRODUCTION

Tobacco cultivation has serious environmental consequences, such as loss of biodiversity, erosion and soil degradation, water pollution, and a significant increase of carbon dioxide in the atmosphere. Tobacco production involves the use of chemicals such as pesticides, fertilizers, and growth regulators that can pollute sources of drinking water and groundwater; it also impoverishes the soil by assimilating more nitrogen, phosphorus, and potassium than other cultures. The soil will collapse in a few years with this massive and constant pollution, so it is necessary to move crops elsewhere or to rejuvenate the soil by rotating the crops. Although the literature available on the subject is poor, it has shown that tobacco plantations cause environmental degradation and disruption of the ecosystem not only because of the intensive use of pesticides but also because of deforestation needed to increase the availability of fertile land. Every year in the world, 4.3 million hectares of land are used for tobacco plantations, resulting in a global deforestation between 2% and 4%. Moreover, the entire tobacco industry (cultivation, processing, and distribution) produces more than 2 million tons of solid waste. It is estimated that from 1995 to 2015, global tobacco production will deposit a total of 45 million tons of solid waste, 6 million tons of nicotine waste, and over 4 million tons of chemical waste.

— *Masanotti et al. (2019)*

All major crops have detrimental effects on the environment and biodiversity, if for no other reason than they necessarily usurp land that was once employed by wild species. However, from a societal perspective, it is difficult to condemn crops that provide essential material such as food, fiber, and medicine. On the other hand, the cultivation of tobacco simply for the purpose of producing a harmful addictive chemical is difficult to justify (albeit, the tobacco crop provides income and employment for many). Tobacco cultivation has been criticized for its harmful effects on the environment and biodiversity (Montford and Small 1999a, b; Novotny and Zhao 1999; Lecours et al. 2012; Lecours 2014; Masanotti et al. 2019; Textbox 18.3).

Beneficial Aspects

Curiously, tobacco was employed medicinally in past times (Charlton 2004). From the 1500s to the 1700s, tobacco was sometimes prescribed by European doctors to treat ailments such as headaches, toothaches, arthritis, and bad breath. The idea that tobacco smoke is beneficial persisted down to modern times among New World peoples. It is now known that second-hand smoke can be deadly. Nevertheless, New World medicine men blew the smoke of cigars or pipes over the skin of the sick in order to cure them. In the Antilles (West Indies excluding the Bahamas), medicine men placed the lighted end of a cigar that was 60 cm long in their mouth, and forcefully blew the smoke out the other end of the cigar, over the patient (a procedure that has been observed in recent times in South America). While these practices seem peculiar, it is conceivable that smoke with nicotine could kill infectious organisms. Today, the chief medicinal usage of tobacco is simply to furnish nicotine for the purpose of curing people from nicotine addiction.

The World Health Organization's Farm and Agriculture division in 1981 raised the possibility of extracting protein from tobacco plants as a means of alleviating world hunger (American Council on Science and Health 1992). The project has not subsequently received significant funding, in considerable part because tobacco has become a pariah crop in the eyes of many people. The proposal was the subject of research at North Carolina State University. The possibility remains of producing nicotine-free 'tobaccoburgers,' either as simulated meat, or actual meat of animals raised and fed on tobacco.

Tobacco has been used as a source of rutin. This chemical has been considered to exhibit 'vitamin P' (supposedly anti-viral) activity. Rutin is also an antioxidant, i.e., a substance that counteracts the free radicals produced by metabolism in the body, which are thought to be damaging. Rutin can be obtained from members of the Rutaceae family, notably lemons and oranges.

Several species of *Nicotiana* are grown for ornamental purposes. They typically resemble common tobacco in appearance and height (about 1.5 m) but lack sufficient nicotine to make them suitable for smoking. However, they may have enough nicotine to be considered as potentially toxic to people and pets (whether they can poison wildlife is unclear). The name 'flowering tobacco' is predominantly applied to *N. alata* (Figure 18.31b), which is the most popular species grown in gardens. This perennial from the Andes region of Argentina and Bolivia is grown as a tender annual in cool temperate areas. Another popular species is *N. sylvestris* (Figure 18.31a), which also sometimes called flowering tobacco, and is notable for its large leaves and masses of trumpet shaped white flowers. This is also a tender perennial. Both species are known for producing a fragrance in the evenings to attract moths. Increasingly, hybrids, including dwarf forms (under 50 cm in height), are becoming popular.

It is as a 'factory' for molecular farming that tobacco has the greatest significance for both nutraceuticals (nutritional additives) and pharmaceuticals (Small 2004). Tobacco has been described as an ideal platform for producing recombinant proteins (Daniell et al. 2001; Rymerson, Menassa, and Brandle 2002; Twyman et al. 2003; Gadani et al. 1995).

FIGURE 18.31 Ornamental species of *Nicotiana* (often termed 'flowering tobacco'). (a) *Nicotiana sylvestris.* Photo by Magnus Manske (CC BY SA 3.0). (b) *Nicotiana alata*. Photo by Carl Lewis (CC BY 2.0).

Recombinant proteins are being produced using genetically modified tobacco, and therefore, there is an excellent possibility that health-promoting tobacco-based protein additives could become very important as functional food components (compounds added to prepared foods to fortify health). A recent COVID vaccine was manufactured using *Nicotiana benthamiana*, an Australian species, as the production base, but because Big Tobacco was one of the sponsors, the vaccine has had difficulty being marketed (Khandekar and Roy 2022).

Conservation Aspects

Several *Nicotiana* species are useful as ornamentals, but genetic resource collections of the genus have been of interest primarily for improving tobacco cultivars. About 60 species and over a thousand tobacco accessions are represented in the United States Department of Agriculture *Nicotiana* germplasm collection. This was originally established to serve the needs of tobacco breeders but is now advertised as a resource for molecular farming and plant-made pharmaceuticals. Currently available tobacco cultivars have been bred for smoking quality, not protein production, a chief requirement of 'molecular farming' (growing genetically modified plants as sources of valued extracts), but available germplasm collections are nevertheless of great value for breeding new cultivars suitable for the purpose. It is also possible that new kinds of plant can be bred to eliminate threats to biodiversity and the environment (e.g., varieties could be created that cannot produce seeds capable of escape from containment; or with lowered needs for pesticides). This demonstrates how germplasm collections can serve future needs that have not been anticipated and emphasizes the need to conserve a broad range of genetic resources for the future, even when they seem to have limited present importance.

The relationship of 'genetic engineering' to the welfare of the environment and biodiversity remains controversial. Small and Catling (2006b) examined the risks and benefits, from the point of view of biodiversity, of employing tobacco to produce pharmaceuticals. Since pharmaceutical plants have relatively high concentrations of potentially toxic chemicals, there is uncertainty regarding whether the chemicals could affect wildlife that feed on the plants, or could leach into the soil and affect soil ecology, and the ecology of streams and lakes.

SUGAR CANE

Introduction

Sugar cane (often spelled as one word, sugarcane) is the species *Saccharum officinarum* (Figure 18.32). Modern cultivated varieties are the products of hybridization with other species of *Saccharum*, of which there are at least five. All species of the genus are native to the Old World tropics. *Saccharum barberi* and *S. sinense* are also cultivated. The native land of sugar cane is uncertain, but the plant is thought to have originated in Southeast Asia and neighboring islands. Sugar cane was known in China and India 2,500 years ago, reached the Mediterranean countries in the 8th century and the Americas in early colonial times. As for many other extremely labor-intensive, major crops, for many centuries Europeans and European colonists employed slaves imported from Africa to grow and process the crop, especially in the Caribbean area, usually under

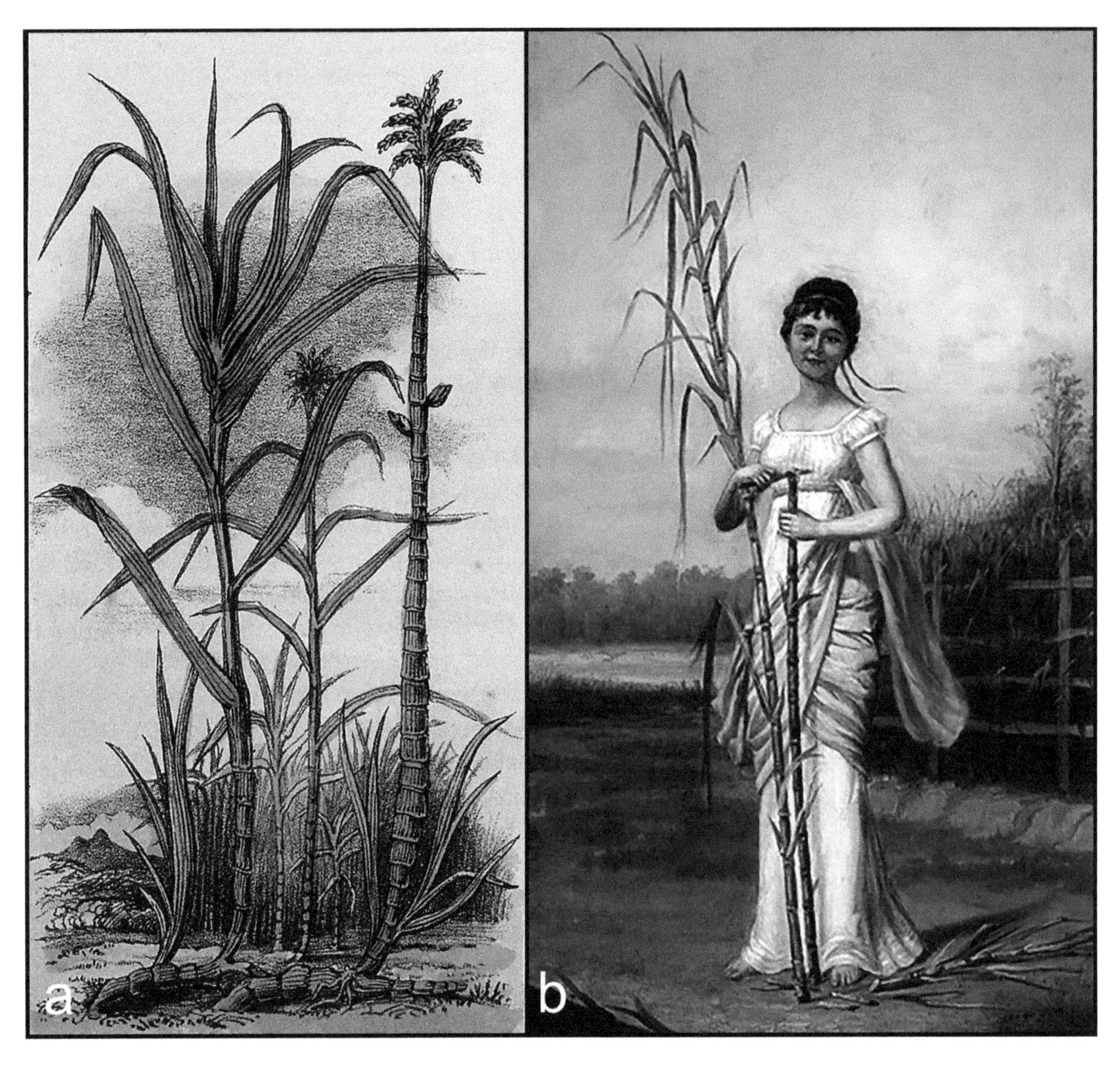

FIGURE 18.32 Public domain paintings of sugar cane (*Saccharum officinarum*). (a) Lithograph, ca. 1849. Courtesy of Wellcome Collection (CC BY 4.0). (b) 'Allegory of sugar cane,' oil on canvas painting by John Genin (1830–1895).

appalling conditions (Figure 18.33). Today, sugar cane is grown in every tropical and semi-tropical country. Brazil and India are the largest growers, and other major producers are Pakistan, China, Thailand, Mexico, and Cuba. In the United States, sugar cane is cultivated from Florida to Texas (Louisiana is sometimes called the 'Sugar State' because of its large crop). The world devotes about 24 million ha to growing sugar cane, resulting in about 100 million metric tons of sugar (sucrose) annually. Brazil alone produces about 40 million tons yearly. The sugar cane crop ranks third highest in quantity of plant calories in the human diet after rice and wheat (Singh and Rai 2018). 'About 110 countries produce sugar from either cane or beet, and eight countries produce sugar from both cane and beet. Sugarcane, on average, accounts for nearly 80% of global sugar production… the top ten producing countries (India, Brazil, Thailand, China, the US, Mexico, Russia, Pakistan, France, and Australia) accounted for nearly 70% of global output' (International Sugar Organization 2020). Chow (2017) noted 'The habitual consumption of added sugar in food and drinks in human history is actually very recent, beginning only about 200 years ago and steadily increasing until today, where the industry annually produces up to 25 kg of refined sugar for every man, woman and child. This works out to 180 million tons of sugar annually, which is equivalent to 17 teaspoons of sugar per person per day in the world.'

Sugar cane is a tall (3–5 m, occasionally up to 7 m) perennial grass (family Poaceae) with arching stems, long broad leaves, and prominent whitish feathery flower clusters (Figure 18.34). The canes (stems) are unbranched, solid,

FIGURE 18.33 Paintings of slaves processing sugar cane. A tragic side of the history of sugar cane is the fact that slavery was once a major component of the industry. Harvesting the crop by hand, which is still the predominant practice, is labor-intensive and back-breaking work. Source (public domain): Clark, W. 1823. *Ten Views in the Island of Antigua*. London, Thomas Clay. (a) Harvesting the sugar cane. (b) Extracting the sweet juice by grinding the stems in a windmill.

yellowish green, 2–5 cm thick, with prominent nodes and internodes, reminiscent of bamboo. The stems of different varieties may be variegated and strikingly colored in different shades of green, purple and yellow (Figure 18.35a). Several stems often arise close together from the rhizomes (underground stems), as shown in Figure 18.32a. The pith (soft central tissue) of the stems between the nodes contains 12%–15% sugar (Figure 18.35b). The canes become tough and turn pale yellow when ready for cutting, 12–20 months from when they were planted. They are cut close to the ground, as the root end of the cane is richest in sugar. The plant is propagated commercially by stem cuttings and, unlike most major crops, is generally grown for many years in the same ground, without crop rotation. The rhizomes continue to produce harvestable crops for at least 3 years, sometimes more than 8 years.

FIGURE 18.34 Sugar cane (*Saccharum officinarum*) in flower. Depending on cultivar and season, flowering may increase or decrease yield of sugar (Ekstein 2014). (a) Photo by Ton Rulkens (CC BY SA 2.0). (b) Photo by Sektordua (CC BY 2.0).

FIGURE 18.35 Sugar cane (*Saccharum officinarum*) stems. (a) The stems of different cultivars, showing that they can be quite ornamental, colored in different shades and patterns of green, purple, and yellow. Photo by Forest and Kim Starr (CC BY 2.0). (b) A stalk of sugar cane broken open to reveal the sugar-rich pithy interior, from which sugar, shown below, is extracted and refined. Photo by Carl Davies, CSIRO (CC BY 3.0).

Sugar cane stalks are pressed to produce a syrup that is used to make sugar (i.e., table sugar, sucrose). The cane yields about 10% of its weight in raw sugar, and this is converted to refined sugar with 96% efficiency. Cane syrup, molasses, wax, rum, and alcohol are other products of sugarcane. Refuse cane, called bagasse, is used in the manufacture of paper, cardboard, and fuel. White sugar is made by removing all the molasses during preparation, while some is left in brown sugar. Icing sugar is very fine-textured and is used particularly for decorating cakes. Table sugar is also produced from sugar beet (*Beta vulgaris*), which accounts for about 20% of the world's sugar intake. Sugar is cheap, easily transported, easily stored, and relatively imperishable, characteristics which have contributed to its importance as a food. For the past two centuries, sugar has been treated as a staple, often accompanying hot beverages, dispensed from the finest China for those with the means (Figure 18.36).

Harmful Aspects

Earlier in this chapter, opium poppy and tobacco were discussed as examples of plants providing extremely harmful addictive substances that have ruined the health of untold millions. Is it fair to include sugar cane in this disreputable category? DiNicolantonio et al. (2018) noted: 'In animal studies, sugar has been found to produce more symptoms than is required to be considered an addictive substance. Animal data has shown significant overlap between the consumption of added sugars and drug-like effects,

FIGURE 18.36 Pre-20th-century luxurious European porcelain sugar bowls, at the Metropolitan Museum of Art. Public domain photos.

FIGURE 18.37 Illustrations of sugar addiction. Source Shutterstock. (a) Contributor: Tijana Moraca. (b) Contributor: Sabelskaya. (c) Contributor: Irina Ashpina. (d) Contributor: Jacky Brown. (e) Contributor: Igor Zakowski. (f) Contributor: NoraVector.

including bingeing, craving, tolerance, withdrawal, cross-sensitisation, cross-tolerance, cross-dependence, reward and opioid effects. Sugar addiction seems to be dependence to the natural endogenous opioids that get released upon sugar intake. In both animals and humans, the evidence in the literature shows substantial parallels and overlap between drugs of abuse and sugar, from the standpoint of brain neurochemistry as well as behaviour.' As shown in Figure 18.37, most people are all too familiar with the irresistible nature of sweets prepared with sugar.

Refined sugar has been claimed to be associated with health problems such as tooth decay, obesity, diabetes, cardiovascular disease, and behavioral changes in children (Textbox 18.4), and as a result has become rather disreputable (Figure 18.38). Although it is often unclear whether sugar consumption alone can be blamed for ill health, the majority of medical researchers currently consider sugar a potential hazard (e.g., Imamura et al. 2015; Misra et al. 2016; Temple and Alp 2016; Chow 2017; Kristen 2018; Chen et al. 2019; Ebrahimpour-koujan et al. 2020;

TEXTBOX 18.4 HEALTH PROBLEMS ASSOCIATED WITH EXCESSIVE SUGAR CONSUMPTION

The prevalence of overweight and obesity among children and adolescents has increased worldwide and has reached alarming proportions. Currently, sugar-sweetened beverages (SSBs) are the primary source of added sugar in the diet of children and adolescents... The majority of reviews concluded that there was a direct association between SSB consumption and weight gain, overweight, and obesity in children and adolescents.

—Keller and Torre (2015)

The various noncommunicable diseases that are closely related to excessive sugar consumption are obesity, cardiovascular disease, gout, diabetes, peptic ulcers, hiatus hernia, gallstones, Crohn's disease, irritable bowel syndrome, dental caries, dermatitis, joint disease, liver disease, cancer, negative effects on growth, maturation, and longevity. Worldwide, it is estimated that 180,000 deaths every year are attributed to the consumption of SSBs, including 133,000 from diabetes, 44,000 from cardiovascular disease, and 6,000 from cancer.

—Chow (2017)

SSBs are potential contributors to weight gain and increase the risk for elevations in blood pressure, type 2 diabetes, coronary heart disease, and stroke.

—Monnard and Grasser (2018)

Sugar consumption from foods and beverages high in added sugars has been associated with an increased risk of type 2 diabetes, cardiovascular diseases, some cancers as well as non-alcoholic fatty liver disease. In addition, emerging research suggest that diet high in sugars may increase the risk of developing dementia such as Alzheimer disease.

— Breda et al. (2019)

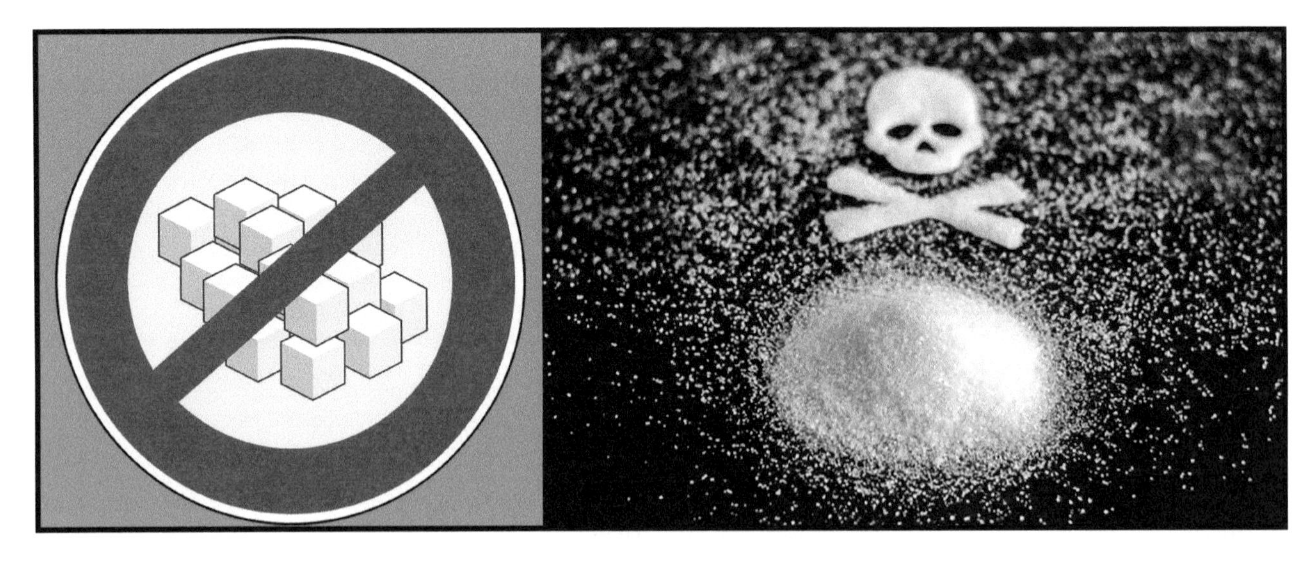

FIGURE 18.38 Anti-sugar signs (public domain).

Haque et al. 2020; Yin et al. 2020). Aside from simply supplying calories, sugar has very little nutritional value and is often said to contain 'empty calories.' Dentists have urged brushing teeth after eating sweet substances that stick to tooth enamel. It has been widely recommended that sugar consumption be reduced, which is not easy since human taste buds crave sweetness.

Reminiscent of the tobacco industry's efforts at concealing the harm of their product, 'When it comes to added sugars, there are clear conflicts between public health interests

FIGURE 18.39 Children are especially attracted to sugary foods (public domain images from needpix.com).

and the interests of the food and beverage (F&B) industry. Studies are more likely to conclude there is no relationship between sugar consumption and health outcomes when investigators receive financial support from F&B companies. Industry documents show that the F&B industry has manipulated research on sugars for public relations purposes' (Schillinger and Kearns 2017). Similarly Litman et al. (2018) concluded that 'Industry-related research… appears biased to underestimate the adverse health effects of SSBs, potentially delaying corrective public health action.' The sugar industry has long been interested in increasing consumption by children, who are naturally strongly attracted to sweetness (Figure 18.39). More generally, Erickson et al. (2017) concluded that 'Guidelines on dietary sugar do not meet criteria for trustworthy recommendations and are based on low-quality evidence.' The conservative guidelines advanced by the World Health Organization (2015; Textbox 18.5) are widely accepted. The key recommendation that the consumption of added sugars should be reduced to less than 10% of total daily energy intake has also been advocated in the 2015–2020 Dietary Guidelines for Americans (USDHHS 2015). Practically, this means that a person should not consume more than ten teaspoons of added sugar daily (a can of soft drink typically contains 10–15 teaspoons of sugar).

Beneficial Aspects

Sugar is the world's most widely used sweetener and probably the cheapest form of food produced. It represents between 15% and 30% of the total calorific intake of the world, and 24% of the calories consumed by the average American. Huge amounts of sugar are used in the manufacture of alcoholic beverages, soft drinks, candy (the word has been traced to the Arabic *quand*, cane sugar), and prepared cereals. A cup of sugar-coated cereal contains almost two tablespoons of sugar. A can of cola has over three tablespoons. The canning industry adds sugar to canned fruits, jams, marmalades, jellies, and evaporated milk. Meat packers use sugar to cure ham, bacon, and other meat products.

TEXTBOX 18.5 WORLD HEALTH ORGANIZATION SUGAR GUIDELINES FOR DENTAL HEALTH AND OBESITY PREVENTION

There is increasing concern that intake of free sugars – particularly in the form of SSBs – increases overall energy intake and may reduce the intake of foods containing more nutritionally adequate calories, leading to an unhealthy diet, weight gain, and increased risk of noncommunicable diseases (NCDs). Another concern is the association between intake of free sugars and dental caries. Dental diseases are the most prevalent NCDs globally... Based on the entire body of evidence, WHO generated the following recommendations for free sugars (i.e., not naturally present) intake in adults and children:

- WHO recommends a reduced intake of free sugars throughout the lifecourse (strong recommendation).
- In both adults and children, WHO recommends reducing the intake of free sugars to less than 10% of total energy intake (strong recommendation).
- WHO suggests a further reduction of the intake of free sugars to below 5% of total energy intake (conditional recommendation).

—*World Health Organization (2015)*

Sugar is present in condiments, canned soups, vegetables, and hundreds of other food products. Fresh cane stems are often chewed for their sweet taste. Sugar sweetens without changing the flavor of food and drink, unlike such traditional sweeteners as fruit syrups, honey, and maple syrup. As bad as sugar may be for health, undeniably it is the basis of countless pleasurable foods (Figure 18.40).

Sugar has a number of additional, very significant properties in prepared foods. It is a preservative agent in many foods. For example, in jams and jellies, sugar binds water, thereby inhibiting the growth of microorganisms. Sugar helps to maintain texture in fruit, and fruit frozen after being coated with sugar, or frozen in 40% sugar solution tends to remain intact. Sugar promotes elasticity of gluten, and cakes rise better; the protein of eggs in cakes and custards similarly remains more elastic. Sugar combines with pectin in fruit, when fruit acids are present, to form jelly. Sugar adds bulk and texture to cakes and ice cream. Sugar makes meat more tender. Saltpeter (potassium nitrate), which is used to cure ham and bacon, has less of a hardening effect in the presence of sugar.

Molasses is a thick syrup produced as the uncrystallized residue from sugar refining. The color ranges from bright amber to dark brown, the lighter color corresponding to finer grades. Molasses is used in breads, cookies, pastries, in various confectionery, and for the manufacture of rum.

Rum is a liquor distilled mainly from fermented molasses and other products of sugar refining, but also sometimes from sugarcane juice. Rum is generally manufactured wherever sugar cane is grown. The natural color is clear, but often caramel is added to produce a darker color, ranging from amber to deep mahogany. The Daiquiri, a combination of rum, sugar, and lime juice, is the best known rum cocktail, while rum and cola is a classic mixed rum drink.

The sugar industry of the world generates billions of dollars of commerce annually (Chow 2017), and employs millions (Figure 18.41), often in low-income countries.

Conservation Aspects

Sugar cane is the world's main source of bioethanol, used as a substitute for gasoline made from fossil petroleum. Indeed, sugarcane biofuel has shown more promise for replacing fossil fuels than any other source (Small and Catling 2006a). In Brazil, more than half of the sugar crop is turned into ethanol used as fuel for cars (Figure 18.41b), and the country leads the world in running vehicles on ethanol. Demafelis et al. (2017) examined the carbon budget of sugar cane production and concluded that agricultural production of raw sugar results in more carbon being fixed than released into the atmosphere. Demafelis et al. (2020) indicate that bioethanol production in the Philippines had the potential of reducing greenhouse gas emissions from comparable use of fossil fuels by as much as 90%.

Saranraj and Stella (2014) examined the deleterious effects of sugar mill effluent on the environment in India, where 30,000–40,000 L of effluent are generated for every ton of sugar produced. While the sugar industry is very important for the country, the effluents result in organic and heavy metal pollution of aquatic environments and soils, and affect the flora and fauna. Sugar mill effluent is widely used in third-world countries as fertilizer, but this can result in contamination of soil and deterioration of soil ecosystems.

Sugar cane is commonly burned at maturity to remove the outer, dried leaves around the stalks, leaving only the sugar-containing stalk in place, before harvesting (Figure 18.42a).

The result is that huge amounts of hazardous pollutants are released into the air. In some parts of Brazil and Australia, sugar growers have shifted to greener approaches. The entire stalks can be harvested (Figure 18.42b) and the foliage separated afterward, or the leafy parts can be separated from the standing crop (which is labor-intensive). The foliage can be used to mulch fields. Sugar growers could also use the leafy cuttings as biofuel in processing plants that (unlike the open fields) have pollution control equipment.

FIGURE 18.40 Sugar and sugary foods (public domain figures from needpix.com).

FIGURE 18.41 Sugar cane production is a major industry, employing millions and generating billions of dollars of commerce. (a) Sugar cane truck in Philippines. Photo by Brian Evans (CC BY ND 2.0). (b) Sugar and ethanol plant in São Paulo, Brazil. Photo by Marco Aurelio Esparz (CC BY SA 3.0).

FIGURE 18.42 A contrast of unsustainable and sustainable practices for harvesting sugar cane. (a) A field being burned in Australia prior to harvest to remove the dried foliage. Photo by Willem van Aken, CSIRO (CC BY 3.0). (b) A mechanized harvest operation (without burning) in Brazil. Photo by Mariordo Mario Roberto Duran Ortiz (CC BY 3.0).

TABLE 18.1
The North American Species of *Toxicodendron*

Toxicodendron Scientific Name	*Rhus* Scientific Name	Common Names	Distribution
T. radicans	*R. toxicodendron*	Poison ivy, common poison ivy, eastern poison ivy	Common in eastern North America, also Central North America
T. rydbergii	*R. rydbergii*	Western poison ivy, northern poison ivy	Northern & western U.S., occasionally in eastern U.S., in Canada
T. toxicarium (*T. pubescens*)	*R. toxicarium*	Eastern poison oak, Atlantic poison ivy	Eastern U.S.
T. diversilobum	*R. diversiloba*	Western poison oak, Pacific poison oak	Common in western North America
T. vernix	*R. vernix*	Poison sumac	Eastern North America

POISON IVY AND RELATIVES

Introduction

There are numerous plants that cause dermatitis (Juckett 1996; Rohde 2012), but the genus *Toxicodendron* is exceptional. It is a group of about 20 species of shrubs, small trees, and woody vines, the majority in eastern Asia, with several in North America (Qi 2008). Some botanists maintain a traditional view that these species belong to the genus *Rhus*, and so there are scientific names for the species in both *Toxicodendron* and *Rhus*. However conceptualized, both *Toxicodendron* and *Rhus* belong to the Anacardiaceae (cashew or pistachio family). Depending on authority, the genus *Rhus* is defined as large, with over 200 species (including *Toxicodendron*), or small, with only about ten species (excluding *Toxicodendron*), notable for shrubs known as sumac (Miller et al. 2001). Regardless, *Toxicodendron* does comprise a distinct group, notorious for its resin containing urushiol, which causes severe dermatitis (Nie et al. 2009). All species of *Toxicodendron* produce this allergen.

Five North American species of *Toxicodendron* (Table 18.1) are especially responsible for dermatitis (Frankel 1991). Of these, common or eastern poison ivy (*T. radicans*) causes more cases of dermatitis than any of the other species, mostly in eastern North America where it is ubiquitous. The closely related, similar western poison ivy (*T. rydbergii*; Figure 18.43), found in the western and northeastern U.S. as well as in Canada, also is a significant problem. Western poison oak (*T. diversilobum*; Figure 18.44) of western North America, eastern poison oak (T. *toxicarium* or *T. pubescens*; Figure 18.45) of Eastern North America, and poison sumac (*T. vernix*; Figure 18.46) of eastern North America can cause especially severe skin reactions but are less common. A guide to the American species is provided by Guin et al. (1981). Unfortunately, many harmless unrelated plants mimic poison ivy, and many people become frightened by these similar plants (McGovern et al. 2000).

Because it is such a major problem, Eastern poison ivy (often just called poison ivy), *T. radicans* (Figure 18.47), deserves special mention. It is native to the New World and is abundant in North America and Central America, Bermuda, and the Bahama Islands (Gillis 1971; Francis 2004). This species has been introduced to Europe (where poison ivy dermatitis is rare), South Africa, Australia, and New Zealand (Mohan et al. 2006). Asian poison ivy (*T. orientale*, *R. orientale*) of eastern Asia is very similar and is sometimes considered to be part of *T. radicans*. Like all species of *Toxicodendron*, *T. radicans* is a woody perennial,

FIGURE 18.43 Western poison ivy (*Toxicodendron rydbergii*). (a) In fall coloration. Photo by Mamdy G, Botanic Garden, Brussels (CC BY SA 4.0). (b) In summer coloration. Photo by Dave Powell, USDA Forest Service; credit: Bugwood.org (CC BY 3.0).

FIGURE 18.44 Western poison oak (*Toxicodendron diversilobum*). (a) and (b) Shrub growing in California. Credit: labradorite_luster (public domain). (c) Vine in fall coloration, growing in Oregon. Photo credit: Gary Halvorson, Oregon State Archives.

spreading by seeds. Most populations also spread by rhizomes (laterally spreading stems). Male and female flowers occur on different plants. The species takes several forms: a trailing vine, a subshrub (low shrub; Figure 18.47d), a shrub over 1 m in height, and (especially in subtropical areas) a high-climbing vine (Figure 18.47c) capable of reaching over 20 m in height. One plant in Florida in the form of a tree reached a height of 7 m (Gillis 1975). Some poison ivy vines have achieved a diameter of 15 cm (Gillis 1975). Aerial or adventitious roots (Figure 18.48) assist the vine forms to attach to bark and other rough surfaces. [The botanical term 'adventitious' refers to an organ developing in an unusual place; roots normally develop at the base of plants, so roots from various parts of the stem are termed adventitious. When these roots develop in the air, they are termed 'aerial.'] Older stems may be covered with slim aerial roots. The species has compound leaves that consist of three leaflets (Figure 18.49), giving rise to the saying 'Leaves of three, let it be,' although as pictured in this chapter, several species of *Toxicodendron* have leaves split into more than three leaflets. The leaves are dark green in the summer, becoming yellow to red in the fall. Clusters of small, round, shiny, whitish, greenish-white, or yellowish fruits 3–6 mm in diameter develop in the autumn and remain on the plant well after the leaves fall (Figure 18.45b).

Harmful Aspects

Urushiols (collectively called urushiol), the allergenic compounds of *Toxicodendron*, are colorless or slightly yellow in their natural state but turn black when exposed to air. The oil is found in all parts of the plants. As little as 2 mg can cause a reaction in sensitive individuals. Usually, damage to the plant is necessary for release of the oil, and light contact with uninjured parts may not cause harm. The nonvolatile oil solidifies quickly on contacted objects, which in turn

FIGURE 18.45 Eastern poison oak (*Toxicodendron toxicarium* or *T. pubescens*). Source (public domain): De Chazelles (1796).

when touched can result in an allergenic reaction. Under hot, humid conditions, urushiol may become inert in about a week. However, under dry conditions, it can remain active for a very long period. Contaminated gloves stored at room temperature for 16 months can still cause poison ivy dermatitis (Mulligan and Junkins 1977). Contact with clothing, shoes, tools, pets, and even smoke from burning plants can cause a reaction. Vinyl (PVC) gloves are useful for protection, but rubber gloves allow penetration of urushiol (Fisher 1996). Washing clothing with detergent is usually effective in eliminating the plant toxin, and rubbing alcohol will remove it from tools. As much as 80% of humans develop dermatitis upon exposure to urushiol (Mohan et al. 2006), but a minority of people are immune. The chemical structure of urushiol differs slightly in the different species of *Toxicodendron*, but exposure to any one of these usually increases sensitivity to the others. Similar chemicals occur in other species of the cashew family, and in some unrelated species, and rarely these compounds elicit similar responses to the urushiol of *Toxicodendron* (Gladman 2006).

FIGURE 18.46 Poison sumac (*Toxicodendron vernix*). (a) Leaf and reproductive parts. Source (public domain): Ypey (1813). (b) Young tree. Photo credit: James H. Miller & Ted Bodner, Southern Weed Science Society, and Bugwood.org (CC BY 3.0). (c) An old trunk. Photo credit: Keith Kanoti, Maine Forest Service, and Bugwood.org (CC BY 3.0).

First contact does not usually provoke a rash but sensitizes the individual so that the next contact produces an allergic dermatitis, typically within 24–48 hours. Within a few minutes of exposure, washing with soap or use of a solvent that absorbs fats may remove at least some of the urushiol from the skin and lessen an allergic response; even just water can be helpful (Guin 2001). (Some authorities state that washing with soap and water is ineffective, others recommend that washing must be within 5 minutes of contact.) Rubbing alcohol or one of the commercial products formulated for poison ivy prevention are alternatives. The dermatitis is characterized by intense pruritus and redness, followed by the appearance of small skin eruptions. The fluid contained in these does not produce reactions if contacted. In most cases the irritation, inflammation, and blistering of the skin persist 1–3 weeks and leave no permanent damage. However, sometimes secondary bacterial infections or, rarely, longer-lasting symptoms may develop. Medical management may include cool compresses, calamine lotion, and anti-inflammatory agents, especially topical corticosteroids, or any of several commercial preparations (Lee and Arriola 1999; Guin 2001). Traditional treatment of the dermatitis with jewelweed (*Impatiens biflora*, *I. pallida*) has been found to be ineffective (Zink et al. 1991; Long et al. 1997). (Some are of the opinion that jewelweed at least provides temporary relief from the itching. The hypothesis has been advanced that saponins in jewelweed act like soap, so some relief might be expected.) 'Although extensive efforts have been made to develop therapies that prevent and treat contact dermatitis to these plants, there lacks an entirely effective method besides complete avoidance' (Kim et al. 2019).

Aside from causing enormous misery, *Toxicodendron* dermatitis has economic consequences. People in outdoor occupations where contact with the plants is frequent, including firefighters, forestry workers, highway construction and maintenance workers, and farmers, are often incapacitated, and the costs of medical treatment generally are extensive. The considerable use of herbicides to control poison ivy (Figure 18.50) also has potential for harming all living things, including humans.

The Anacardiaceae family is well known for several edible plants: cashews, pistachios, and mangos. The perilous *Toxicodendron* is also a prominent constituent, and so are other species that are quite toxic. Particularly dangerous is *Metopium toxiferum* (poisonwood, Florida poisontree; Figure 18.51), a beautiful tree widely found in southern Florida, the Bahamas, and other areas of the Caribbean. It also has urushiol, but exposure produces a much more serious rash than does *Toxicodendron*. Urushiol is found in the shells of cashews and pistachios, and in the skin of mangoes as well as where the stalk meets the mango fruit. To destroy the urushiol (and avoid lawsuits), cashews and pistachios are marketed after they are roasted or steamed (so-called 'raw cashews' in stores will in fact have been steamed).

FIGURE 18.47 Poison ivy (*Toxicodendron radicans*). (a) Flowering branch. Source (public domain) Thomé (1885). (b) Fruiting branch. Photo by Sam Fraser-Smith (CC BY 2.0). (c) An old vine on a tree trunk. Photo by X099008 (public domain). (d) Shrub form. Photo by Jaknouse (CC BY SA 3.0).

Despite these treatments, individuals who are very sensitive to urushiol may experience allergies from eating these nuts. It has been recommended that mangoes be brushed in warm water to remove residual urushiol beads, and peeled (Pradhan 2017).

Beneficial Aspects

Toxicodendron vernicifluum, the lacquer tree (Figure 18.52b), is the main source of 'oriental lacquer' (oriental varnish, Asian lacquer) in several Asian countries including China, Japan, and Korea, which has been

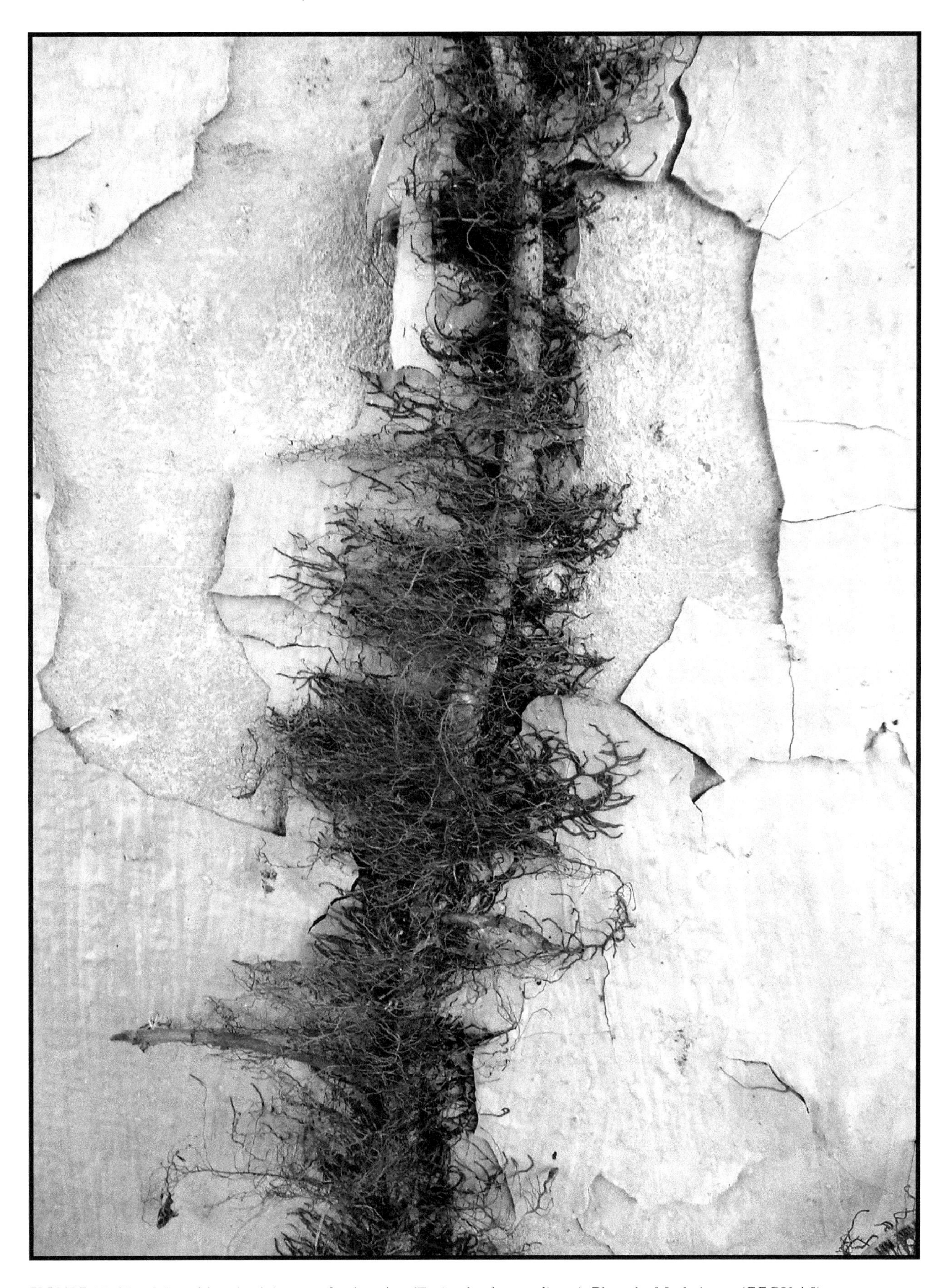

FIGURE 18.48 Adventitious/aerial roots of poison ivy (*Toxicodendron radicans*). Photo by Mark Apgar (CC BY 4.0).

FIGURE 18.49 Eastern poison ivy (*Toxicodendron radicans*), showing leaves with three leaflets. Photo by Hardyplants (public domain).

FIGURE 18.50 Spraying herbicide on poison ivy (*Rhus radicans*) in Maine. Photo (public domain) by Beth Goettel, United States Fish and Wildlife Service.

FIGURE 18.51 Florida poisontree (*Metopium toxiferum*), one of the world's most dangerous rash-producing plants. The bird is the Greater Antillean bullfinch, *Coccothraustes purpurea* (originally identified as 'violet grossbeak,' *Loxigilla violacea*). Public domain watercolor from Catesby, M. *The Natural History of Carolina, Florida, and Bahama Islands* (early 1700s).

FIGURE 18.52 Asian Toxicodendron trees employed as sources of lacquer and wax. (a) Wax tree (*T. succedaneum*) growing in Australia. Photo by Tatters (CC BY 2.0). (b) Lacquer tree (*T. vernicifluum*) growing in Japan. Photo by Takami Torao (public domain).

FIGURE 18.53 Chinese lacquer (from the lacquer tree, *Toxicodendron vernicifluum*). (a) Commode constructed in 1763, covered with Chinese lacquer. Photo by Wmpearl (public domain). (b) Lacquer set. Photo by Mk2010 (CC BY SA 3.0).

employed for centuries as a glossy protective for furniture and objets-d'art (Figure 18.53). After 10 years of growth, the stem of the tree is mature enough to be cut, and a sap oozes out and is harvested, much as rubber is collected from the rubber tree. In Vietnam and Taiwan, lacquer is obtained from *T. succedaneum* (Figure 18.52a), and some other species are employed in other countries. The lacquer from all toxicodendrons causes dermatitis (Vogl 2000). *Toxicodendron succedaneum* and *T. vernicifluum* are also the source of a fat by-product known as Japan wax and sumac wax which is an alternative to wax. This is also employed in cosmetic products including hair and skin creams.

Most wildlife appear immune to *Toxicodendron* (only humans and a few primates have been found to be susceptible; Figure 18.54). Many birds eat the fruits (Figure 18.55a), and many mammals consume both the fruits and foliage (Senhina 2008; Innes 2012). Bees produce a non-poisonous honey from the nectar of poison ivy (Mulligan and Junkins 1977), and insects commonly visit the flowers (Figure 18.55b). Because several *Toxicodendron* species are widespread, they provide significant cover for many animals and represent a valuable resource for biodiversity (Textbox 18.6).

In the north of the Netherlands, poison ivy has been used since the early 19th century to stabilize dikes (Gillis 1975). Historical accounts describe a number of unusual uses of *Toxicodendron* in North America (many described by Gillis 1975), but the accuracy of this information is difficult to evaluate (Senchina 2006; Shah et al. 2019).

Occasionally poison ivy (with fall colors ranging from yellow through orange to dark red) and poison sumac (with its bronze-red or cherry-red leaves in autumn) have been employed in decorative indoor arrangements because of their autumn coloration (Gillis 1975; Figure 18.56). The plants have also been cultivated as ornamental shrubs for their attractive foliage.

Allegedly therapeutic preparations of *Toxicodendron* termed 'Rhus toxicodendron' ('Rhus tox'; Figure 18.57) are sometimes employed as a homeopathic anti-inflammatory remedy for arthritis pain (Signore 2017). Classical homeopathy (the application of extremely small doses of substances that seem to elicit symptoms of a condition requiring treatment: the principle of similia) is disreputable, and its allegations should not be taken at face value.

Conservation Aspects

Poison ivy and its cousins are part of the natural biodiversity of much of the world and contribute to the orderly functioning of ecosystems. Unfortunately for these species, they are also naturally harmful to people. There is very limited sympathy for the toxicodendrons (Figure 18.58).

It has been suggested that the higher temperatures and CO_2 levels being generated by climate change may increase the frequency of *Toxicodendron* (Mohan et al. 2006, 2008), but this has been disputed (Schnitzer et al. 2008). In any event, with the urbanization of North America, *Toxicodendron* species have become more widespread, because they are adapted to the edges of woods and forests, and land clearing has greatly increased this habitat (Gillis 1975). The control method that is most benign to nature is mechanical removal, which is very labor-intensive (Figure 18.59).

TEXTBOX 18.6 VALUES OF POISON IVY

Poison ivy is a plant of immense ecological value. Wild mammals from mice to moose, livestock like goats and cattle, hundreds of species of insects – all defy poison ivy's nasty nature and feast on its leaves. Bees and butterflies suck its sweet nectar. (Yes, it has flowers.) Woodpeckers, wild turkeys, robins, and bluebirds all feed with gusto on its fruit. Cardinals are even known to line their nests with fuzzy poison ivy rootlets, in which the young birds nestle comfortably… Poison ivy was for centuries a popular – and very expensive – garden plant. I'm serious. There was a time, centuries ago, when seeds of poison ivy were almost literally worth their weight in gold. Planted in the gardens of emperors, presidents, and kings, poison ivy was displayed like a captive tiger equally prized for its beauty and its deadliness. Alas for poison ivy, that's no longer the case… poison ivy is drenched with herbicides in backyards and schoolyards, hiking trails and campgrounds across the nation. I would guess that few plants on Earth undergo such a barrage of deadly chemicals… Poison ivy… is a lens through which to take a broader look at the whole green world around us. The human love/hate relationship with this plant is a microcosm of our changing attitudes to nature over the centuries.

—*Sanchez (2016)*

FIGURE 18.54 Poison ivy is harmless to most animals, but dogs allowed to contact the plant acquire the irritating oils on their fur and often transfer it to people. Prepared by B. Brookes.

FIGURE 18.55 Wildlife consuming *Toxicodendron* species. (a) Japanese tits (*Parus minor minor*), eating fruits of wax tree (*T. succedaneum*) in Japan. Photo by Alpsdake (CC BY SA 3.0). (b) Halictid bees harvesting pollen and nectar from flowers of poison ivy (*T. radicans*) in southern Florida. Photo by Bob Peterson (CC BY 2.0).

FIGURE 18.56 Beautiful fall coloration of *Toxicodendron*. (a) Poison ivy (*T. radicans*) in Illinois. Photo by Amos Oliver Doyle (CC BY SA 3.0). (b) Poison sumac (*T. vernix*) in Massachusetts. Photo by Doug_McGrady (CC BY 2.0).

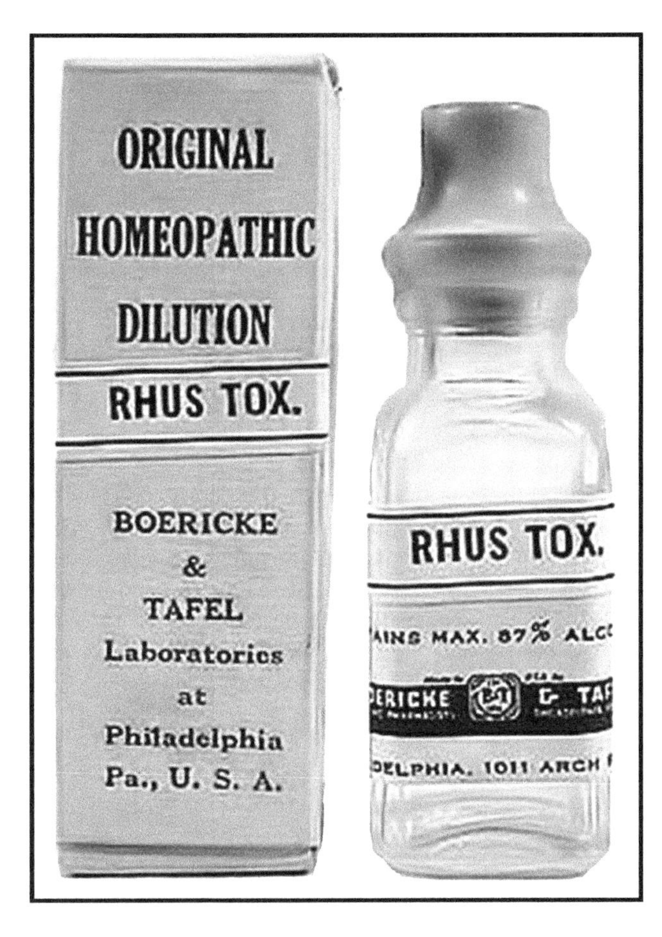

FIGURE 18.57 Questionable homeopathic remedy known as 'Rhus Toxicodendron,' derived from poison ivy. Photo by Wikidudeman (public domain).

FIGURE 18.58 Humorous drawing (public domain) of a 'poison ivy fairy' from Gordon and Ross (1910).

FIGURE 18.59 Cutting out poison ivy (*Toxicodendron radicans*) in eastern U.S. national parks. Photos by U.S. National Park Service (public domain).

RAGWEED

INTRODUCTION

The serious allergenic plants known as ragweeds belong to the genus *Ambrosia*. It consists of about 40 species (for classification information, see Payne 1964; Peterson and Payne 1973; Strother and Baldwin 2002; Strother 2006; Hrabovský and Mičieta 2013). The species are native to tropical, subtropical, and temperate parts of the New World, with slightly more than half of them in North America. Several New World species have become established in the Old World (Montagnani et al. 2017). The plants are annuals, perennials, or shrubs. The principal species of interest are common ragweed (*A. artemisiifolia*) and great ragweed (*A. trifida*). Why the ragweeds have this name is unclear. It may be because their leaves are quite divided and look 'ragged,' but perhaps the name was coined by European settlers in North America because they concluded that the uncomplimentary name reflects a lack of value of the plants. The genus name *Ambrosia*, meaning food of the gods or an ointment conferring immortality, was assigned to these plants by the Father of Biological Nomenclature, Carl Linnaeus, in Sweden in the middle 1700s. Why he applied this name to such unheavenly plants is unknown.

The vernacular name ambrosia is often applied to an unrelated species, *Chenopodium botrys*, also known as Mexican tea (confusingly, this is sometimes invalidly termed '*Ambrosia mexicana*' in the horticultural trade). This plant has become a worldwide weed. It is sometimes used for tea and medicine and has a superficial resemblance to the ragweeds.

Common ragweed (also known as annual ragweed, low ragweed, short ragweed, and by many other names; Figure 18.60) is a strongly branched annual herb typically growing up to 70 cm in height but varying from 0.1 to 2.5 m depending on environment (Essl et al. 2015). It is widely distributed in the Americas where it is indigenous and has spread to much of the world (Bassett and Crompton 1975). The plant reproduces by seeds. A single plant can produce as many as 18,000 seeds, and they can remain dormant in the soil for over 40 years (Milakovic and Karrer 2016; CABI 2019a). The plant produces harmful chemicals in its roots, a form of chemical warfare (allelopathy) against competing plants (Hall et al. 2020).

FIGURE 18.60 Common ragweed (*Ambrosia artemisiifolia*). (a) Photo by James H. Miller & Ted Bodner, Southern Weed Science Society, Bugwood.org (CC BY 3.0). (b) Photo by Steve Dewey, Utah State University, Bugwood.org (CC BY 3.0).

Giant ragweed (also known as great ragweed, buffalo weed, and by other names; Figure 18.61) is also an annual herb. It typically grows to a height of over 2 m and is known to have grown over 6 m (Figure 18.61a). This native of North America has also spread to Eurasia. A single large plant can produce over 1 billion pollen grains in a season (CABI 2019b) and over 10,000 seeds, although less than 2,000 is typical (Goplen et al. 2016).

Both male and female flowers occur on the same plant. The true flowers are tiny, and in small clusters called 'heads,' which in common and giant ragweed have either a group of male flowers or a single female flower. The male heads – of course responsible for pollen production – are found on elongated branches at the top of the flowering part of the plant, high up where the wind can pick up the pollen (Figure 18.62). The female flowers are located singly, tight against branches. The pollen grains may be transported by wind for hundreds of kilometers. The major allergenic ragweeds flower in late summer and early autumn.

Harmful Aspects

Ragweed pollen is a leading cause of allergic rhinitis ('hayfever') in humans. A variety of proteins in the pollen are the causative agents (Bordas-Le Floch et al. 2015; Figure 18.63). For those who are susceptible, special cells (mast cells) in the body react to pollen by producing antibodies, which stimulate histamines to flood into the bloodstream (Figure 18.64). Histamines cause the uncomfortable allergy symptoms that hay fever sufferers experience. Accordingly, the typical treatment for allergies is anti-histamines, which block the body's excessive histamine production. In the past, a variety of questionable cures have been marketed (Figure 18.65). An estimated half of all pollen-related allergic rhinitis in North America is due to ragweeds. Often the rhinitis is accompanied by asthma (Déchamp 2013), bronchitis, and/or eye irritation. In addition, sensitization to ragweed pollen often leads to allergies against foods like celery and spices. Also, for some individuals simply touching the plant can result in contact dermatitis. Climate warming has been shown to be increasing the length of season and the severity of allergenic ragweed pollen in North America (Ziska et al. 2011; Zhang et al. 2015; Case and Stinson 2018) and in Europe (Hamaoui-Laguel et al. 2015; Rasmussen et al. 2017; Mang et al. 2018).

While the principal harm from ragweeds is due to the medical consequences of their huge pollen production, they are also very harmful as agricultural weeds and as invaders of natural ecosystems (Figure 18.66; Textbox 18.7).

Beneficial Aspects

There are few known uses of the ragweeds featured here that would redeem them in the eyes of most people. In past times, Native Americans apparently consumed the seeds, which are rich in protein and oil. Common ragweed

FIGURE 18.61 Giant ragweed (*Ambrosia trifida*). (a) Tall plant in Texas. Photo by Forrest M. Mims III (CC BY SA 3.0). (b) Painting (public domain) from Lindman (1901–1926). 1, Male head with several flowers. 2, Female head with one flower. 3, Fruit ('seed'). (c) A large plant in Russia. Photo by Le.Loup.Gris (CC BY SA 3.0).

TEXTBOX 18.7 SUMMARIES OF THE HARM CAUSED BY COMMON AND GIANT RAGWEEDS

Ambrosia artemisiifolia [common ragweed] is an annual herb native to Central and Northern America. It has been accidentally introduced into a large number of countries as a contaminant of seed and grains. *A. artemisiifolia* typically colonizes disturbed land where it produces a large number of seeds which can remain viable in the soil for 40 years or more. The pollen produced by species of *Ambrosia* is highly allergenic and can induce allergic rhinitis, fever, or dermatitis. As a result, high medical costs have been reported in areas with large infestations in both its native and introduced range. *A. artemisiifolia* can also invade agricultural land where it acts as a weed in a number of crops (in particular in sunflower, maize, soybean, and cereals) and can cause significant decreases in yields.

—*CABI (2019a)*

Ambrosia trifida [giant ragweed] is an annual herb native to temperate North America which is now present in a number of countries in Europe and Asia. The primary means of spread of *A. trifida* occurs accidentally as a contaminant of seed or agricultural equipment. This species readily colonizes disturbed areas and is often one of the first plants to emerge in early spring. As a result, it has an initial competitive advantage and can therefore behave as a dominant species throughout the entire growing season. *A. trifida* is a particular problem for cultivated agricultural and horticultural crops where it can significantly decrease yields. Like many species of *Ambrosia*, *A. trifida* produces pollen which is allergenic and can induce allergic rhinitis, fever, or dermatitis. A. *trifida* is extremely competitive and can also decrease native biodiversity.

—*CABI (2019b)*

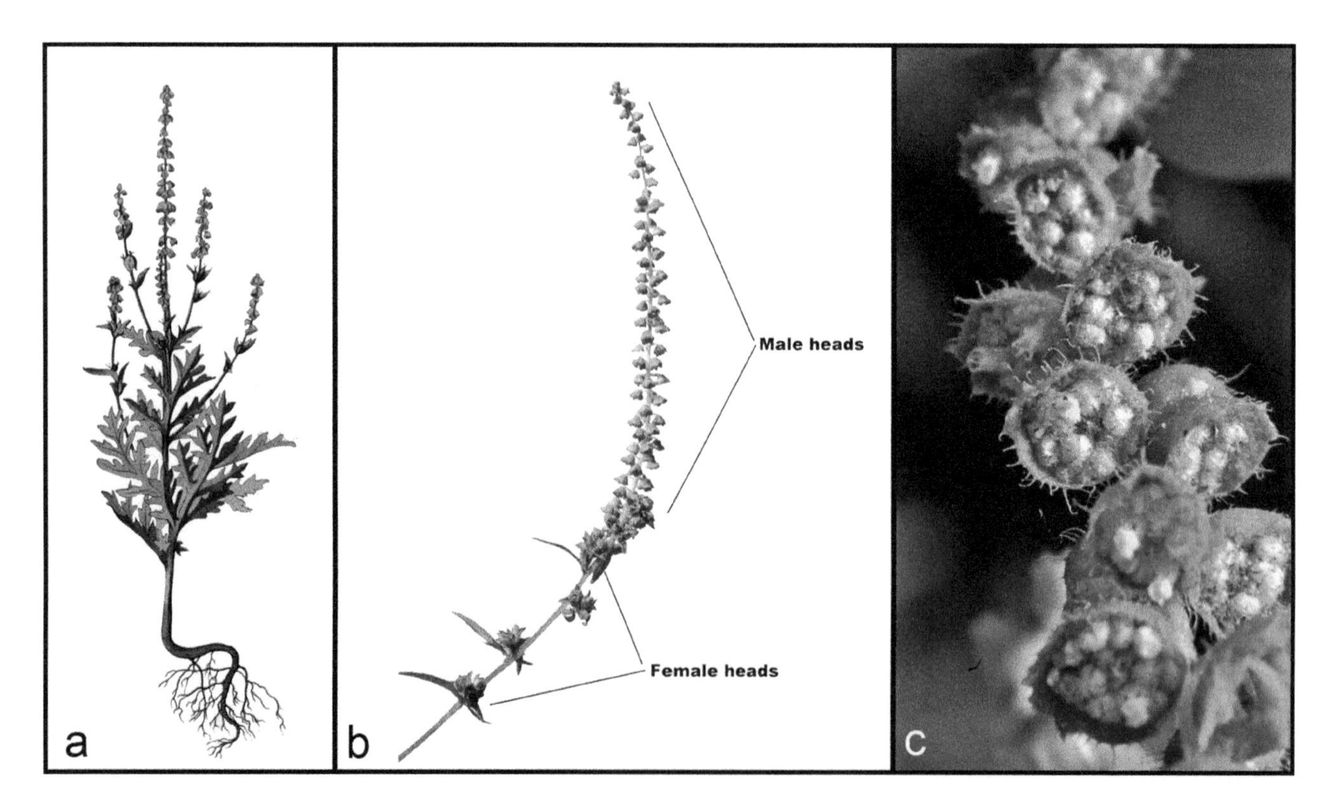

FIGURE 18.62 Disposition of flowers and fruits in common ragweed (*Ambrosia artemisiifolia*). (a) Drawing of a plant showing flowers on branches. Source (public domain): Millspaugh (1892). (b) Numerous cream to pale-green male flower heads occur in branches above the much fewer female heads (with just one flower each). Photo by Harry Rose (CC BY 2.0). (c) Close-up of heads (clusters) of male flowers. Photo by Frank Mayfield (CC BY SA 2.0).

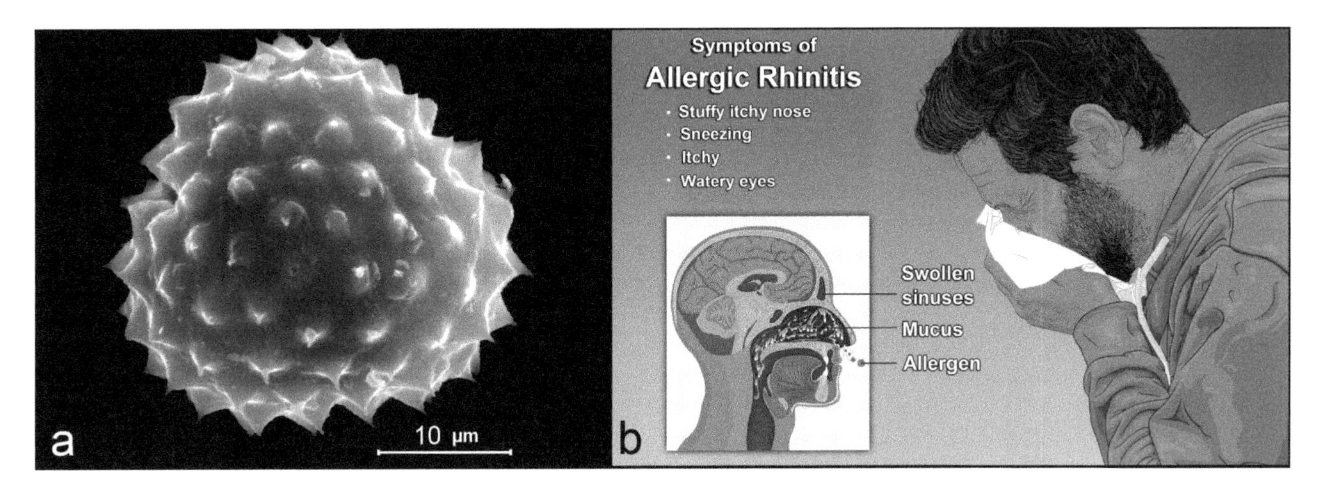

FIGURE 18.63 Ragweed pollen and its harmful effects. (a) A pollen grain of common ragweed (*Ambrosia artemisiifolia*). The round, prickly grains are 20–25 μm (μ = micron; 1 μ = 0.001 mm) in diameter. (The average cross-section of a human hair is 50 μm in width. The human eye cannot distinguish anything smaller than 40 μm.) Photo by Marie Majaura (CC BY SA 3.0). (b) Depiction (public domain) of a person suffering from allergic rhinitis.

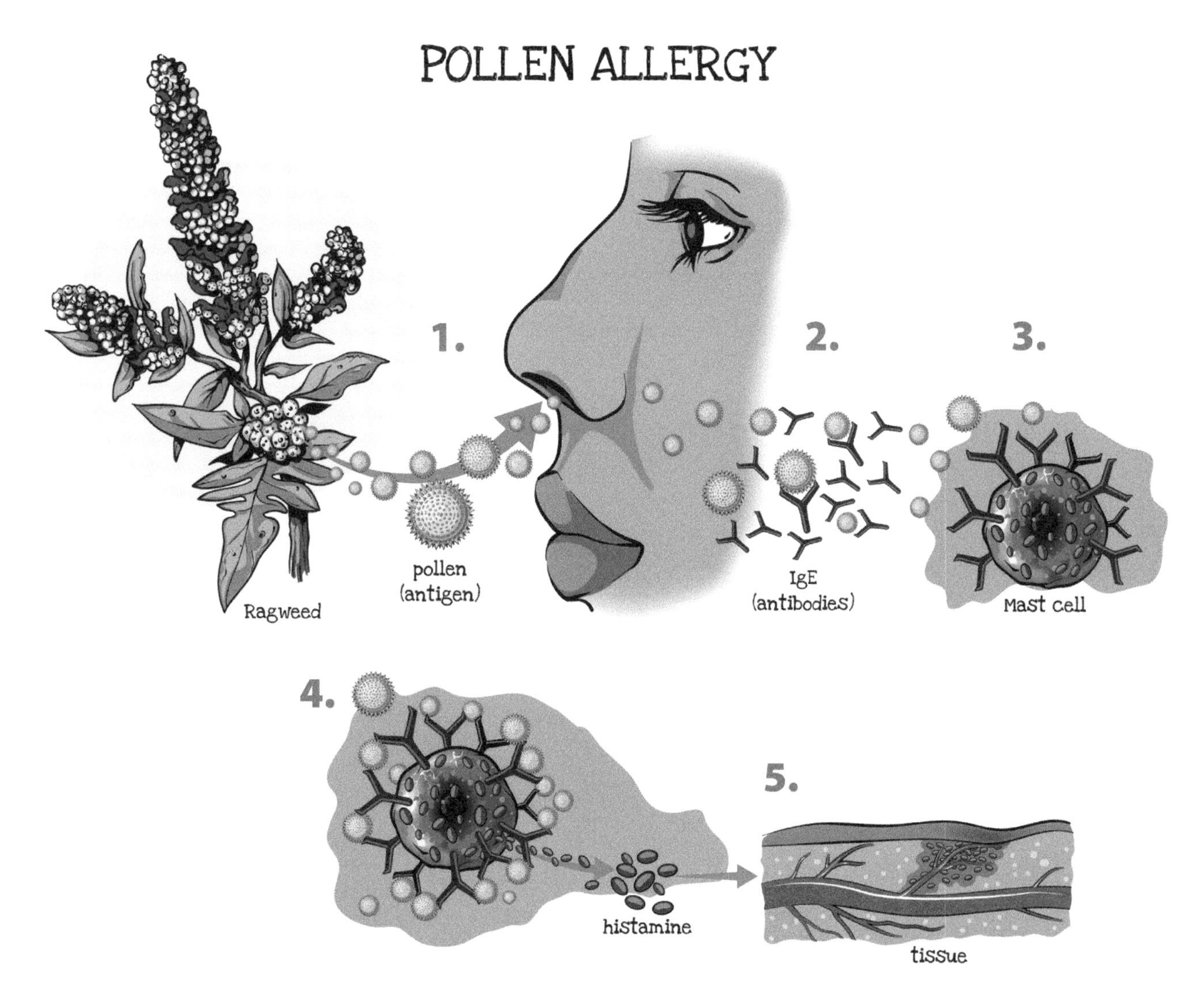

FIGURE 18.64 Immune system allergic response to pollen. Source: Shutterstock, contributor: EreborMountain.

FIGURE 18.65 Nineteenth-century trade cards (public domain) advertising pollen allergy treatments by showing healthy young people.

FIGURE 18.66 A field in France, covered with invasive common ragweed (*Ambrosia artemisiifolia*). Photo by Amage9 (CC BY SA 4.0).

can be employed for phytoremediation in soils contaminated with heavy metals (Bassett and Crompton 1975; Kang et al. 1998). It can remove soil lead and cadmium (Pichtel et al. 2000; Makra et al. 2015). The plant has been fed to pigs and sheep (Crockett 1977) but is unsuitable for cattle (Stubbendieck et al. 1995). The species also has some medical potential as an anti-inflammatory agent (Stubbendieck et al. 1995) and as an antibacterial agent (Kim et al. 1993). Giant ragweed has, like common ragweed, been employed for a number of obsolete folk usages. It has been suggested that it might have some value as an ornamental (CABI 2019b).

Plants like the ragweeds that produce considerable biomass are candidates for collecting and employing the material for useful purposes. It has been proposed that introduced giant ragweed in Korea could be employed as an absorbent for removing contaminants from water (Yakkala et al. 2013; Ahmad et al. 2014).

Conservation Aspects

In Western nations at least 20% of people suffer from hay fever – so much so that sufferers are very common (Figure 18.67). Perhaps 10% of dogs and 20% of cats also seem to be susceptible. Aside from the medical consequences, the fact that the natural world is providing the pollen that results in so much suffering does not endear nature to afflicted people, and this has the unfortunate consequence of lessening support for biodiversity conservation. At least, it should be explained to society what are the best ways to address the worst source of hay fever – the ragweeds.

Biodiversity-friendly, pesticide-free options to control the ragweeds have been employed. Several insects, especially beetles, have been enlisted as biological control agents (Palmer and McFadyen 2012; Watson and Teshler 2013; Sun et al. 2017; Mouttet et al. 2018; Schaffner et al. 2020), although the plants remain a serious problem, increasing in many areas. Regular, sustained mowing of the plants can be effective, but occasional mowing is not (Katz et al. 2014; Milakovic and Karrer 2016; Karrer 2016). Gentili et al. (2017) recommended that in pastures in Italy, no control measures be taken to control common ragweed, as it would naturally be reduced by competition with native species.

In areas where they are native, the fruits of ragweeds are a significant source of food for small birds and mammals (Stubbendieck et al. 1995). However, the ragweeds highlighted here are invasive weeds that compete with the native flora in much of the world, and because of their harmful effects on people and crops they invite the use of pesticides, which also represent threats to natural habitats and ecosystems (Smith et al. 2013). Herbicide-resistant forms of giant ragweed now occur in the United States (Goplen et al.

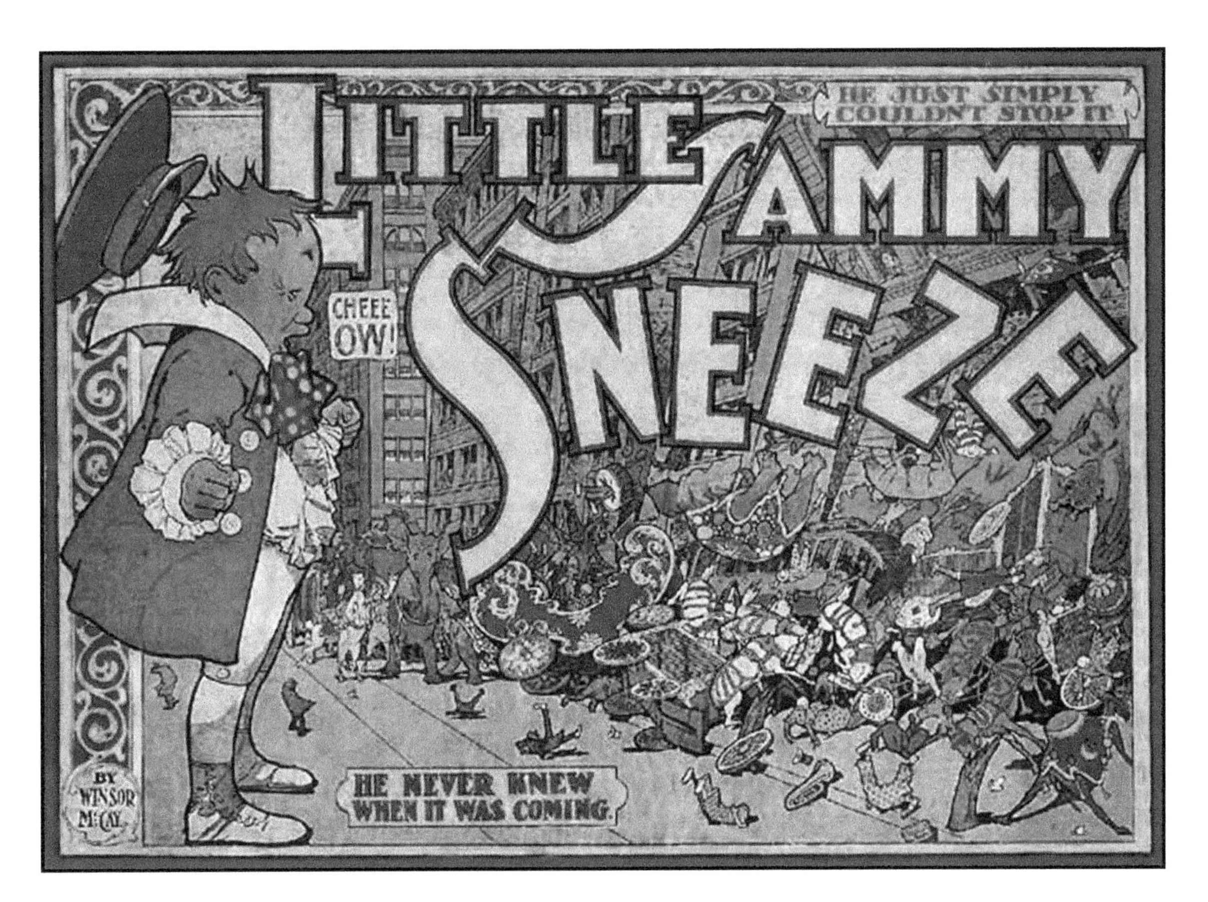

FIGURE 18.67 Illustration of the perils of allergic rhinitis ('hayfever'). Cover (public domain) of a 1905 book, *Little Sammy Sneeze*, by American cartoonist Winsor McCay.

2018). In Europe, while the detrimental effects of invading common ragweed on health and crops is clear, the extent of damage to biodiversity has not yet been determined (Alberternst et al. 2016).

Mostly or entirely harmless plants that are related to pernicious weeds are often identified as the harmful plants, and eliminated. When such relatives are rare and endangered, they are frequently in the unfortunate position of finding little support for conservation. South Texas ambrosia (*A. cheiranthifolia*) of southern Texas and adjacent Mexico, and San Diego ambrosia (*A. pumila*) of southern California and adjacent Mexico are protected species in the U.S. They are far safer when they are called 'ambrosia' than when referred to as 'ragweed.'

One of the most common cases of mistaken identity is the confusion of goldenrods with ragweeds. Goldenrods are any of over 100 perennial species of *Solidago*, which like *Ambrosia* are members of the daisy family (Asteraceae), and include common weeds. They often flower at the same time as ragweeds, and there is a superficial resemblance (Figure 18.68). Goldenrods have foliage that is not divided or dissected like ragweed leaves. Advertisements concerning hay fever often mistakenly show pictures of goldenrods, reflecting the common belief that they cause hay fever in people. However, unlike ragweed pollen, which is carried by wind for pollination, goldenrod pollen is transported by insects and is too heavy and sticky to be blown a significant distance from the flowers. Goldenrods are basically harmless, although there are varieties that are employed as ornamentals (Williamson 2018) and some florists handling these have experienced serious dermatitis (De Jong et al. 1998).

FIGURE 18.68 Canadian goldenrod (*Solidago canadensis*). Photo by Harry Rose (CC BY 2.0).

19 In Defense of the World's Most Despised Agricultural Weeds

INTRODUCTION

This and the next two chapters deal with weeds. Weeds are the villains of the plant kingdom, causing enormous damage to the welfare of both humans and biodiversity (Textbox 19.1). Before people appeared on the planet, there were no weeds. As humans came to dominate Earth, they unintentionally facilitated certain plants to become menacing invaders of places where they are unwelcome (which is the standard definition of weeds). Weeds are a grave threat to people, most importantly by injuring the crops that sustain human life. However, they are also a grave threat to biodiversity: directly, by altering natural ecosystems and eliminating wild species, and indirectly by provoking environmentally unsustainable agricultural practices to control the weeds. This chapter stresses the beneficial as well as the harmful economic and ecological aspects, and emphasizing that a better understanding of the biology of weeds is key to managing both their adverse and advantageous potentials.

Expanding agricultural use of land is the leading cause of habitat and ecosystem degradation and consequently of biodiversity reduction (Small 2011c, d). Contributing to the need for expansion of agriculture are weeds specialized to thrive in crop fields. These reduce yield and consequently exacerbate the need for more intensive agriculture. Inevitably, unintended harm results to the environment by usurpation of wildlands, drainage of scarce water sources, burning of fossil fuels to generate greenhouse gases, and eutrophication of aquatic habitats from runoff of synthetic fertilizers. In particular, the primary method employed in Western agriculture to control agricultural weeds – herbicides – is a threat to wildlife and hence to ecosystems (Prosser et al. 2016; Figure 19.1). In parts of the world, hoeing remains an important method of removing weeds but is highly labor-intensive (Figure 19.2). As pointed out in Figure 19.3, sometimes excessive efforts to kill weeds are counterproductive. Itodo (2019) noted 'The intensive use of herbicides and other classes of chemical products in agricultural practices has resulted in serious environmental impact causing increasing levels of herbicide residues in natural water, soil, and foodstuffs.' Alarmingly, the use of herbicides is expected to increase in the future (Hossain 2015), despite the fact that weeds are evolving resistance to herbicides (Kumar et al. 2018). This is so, despite the development of relatively benign ways of reducing weeds (Nazarko et al. 2005; Chicouene 2007; Petit et al. 2011; Chauhan et al. 2012; Harding and Raizada 2015; Pandey et al. 2020). By definition, the worst agricultural weeds are those that have proven to be most difficult to control by any means. In this presentation, the leading agricultural weeds are examined.

Inevitably, the ranking of the harmfulness of weeds is at least partly a subjective exercise, because of knowledge limitations, the fact that the harm of particular weeds varies geographically, and by what gets damaged (metaphorically, it depends on whose ox is being gored). An additional consideration is that the invasions of many weeds are not complete, so some weeds are becoming more significant, and accordingly their comparative seriousness is changing with time. Holm (1978) stated that about 200 plant species are responsible for about 95% of the weed problems of food, feed, and fiber production in the world. 'The World's Worst Weeds' by Holm et al. (1977) recognized 18 weeds (Table 19.1) as very distinctly the worst of the 76 species that they examined, on the basis 'that workers have not only cited these species more often than other world weeds, but have also ranked them as the greatest troublemakers in the

TEXTBOX 19.1 HARM AND BENEFIT OF WEEDS

Weeds are plants that are unwanted in a given situation and may be harmful, dangerous, or economically detrimental. They are responsible for substantial losses of farm production and extensive damage to the environment. Weeds, through competition with other plants, would almost always have deleterious effects on them and can have a lethal effect on livestock.... Weed invasion has become the most dreaded and deleterious impact... in nature; it adversely affects agriculture, alters the balance of ecological communities, disrupts the natural diversity, and interferes in the aesthetic value of the environment. Weeds can interfere in water management, thereby reducing the economic value of water. Weeds, however, besides their deleterious impacts in nature, have many beneficial properties, which include, but not limited to... ethnomedical and ethnopharmaceutical uses... and the use of weeds as feed for livestock.

—*Ekwealor et al. (2019)*

DOI: 10.1201/9781003412946-19

FIGURE 19.1 Excessive use of herbicides. Prepared by Jessica Hsiung.

largest number of crops.' Holm et al. restricted their analysis to agricultural weeds, whereas weeds are also devastatingly harmful to the natural world and its biodiversity. While therefore restricted in scope, their analysis remains invaluable. Weber (2017) is a guide to environmental weeds. Lowe et al. (2000) provided a list of '100 of the world's worst invasive alien species' (including plants, animals, and microorganisms), noting that 'Species were selected for the list according to two criteria: their serious impact on biological diversity and/or human activities and their illustration of important issues surrounding biological invasion.'

A few categories of weeds that are harmful to agriculture are not discussed in this review but deserve mention. Many plants considered to be agricultural or environmental weeds primarily exert their harm by vectoring (i.e., acting as a source of) diseases caused by pathogenic microorganisms (bacteria, fungi, viruses) or harmful herbivores (particularly insects). Stem rust of wheat and some other cereals

FIGURE 19.2 Hoeing away weeds in crops. Although herbicides are employed predominantly in Western agriculture to remove weeds, in much of the world hoeing by hand is widely practiced. Although extremely labor-intensive, it is much friendlier to the environment. Source of photos: Shutterstock. (a) Indonesian farmers hoeing potatoes. Contributor: Dany Kurniawan. (b) Farmer with hoe in a sugarcane farm in Thailand. Contributor: Au_Uhoo. (c) Farmer hoeing soybeans in Vietnam. Contributor: John Bill. (d) Farmer hoeing rice field in Thailand. Contributor: Ukrit Yasuwan.

is caused by the fungus *Puccinia graminis*. The life cycle of the fungus requires alternate infection of barberry (the genus *Berberis*) plants and the cereal. Some wheat varieties and some barberry varieties are resistant to the fungus, but stem rust has been a very serious problem historically and remains dangerous to wheat farmers. Just as common barberry (*Berberis vulgaris*) is an alternate host of wheat rust (Figure 19.4), similarly currants and gooseberries (*Ribes* species) are alternate hosts of white pine blister rust (Malloch 1995). A few plants are parasitic on other plants, and some (notably dodders – species of *Cuscuta*) are quite harmful to crops (Kroschel 2001). Some crops and weeds accumulate nitrate (KNO_3), especially when stressed by drought. Plants containing more than 1.5% dry weight nitrate may be lethal to livestock. Excessive fertilization can result in high content of nitrate in some vegetables, which can be toxic to humans (Anjana and Iqbal 2007). The edibility of plants grown over septic fields and chemically contaminated soils is not clear, but consumption is considered inadvisable.

The grass family (Poaceae) is the most important of all economic families, because it includes the true cereals (rice, wheat, barley, etc., as well as sugarcane examined in Chapter 18, and numerous species grown as forage). It also contains some of the world's most damaging weeds. In this chapter, the agricultural weeds barnyard grass, Bermuda grass, and weedy rice are discussed. Crabgrass is reviewed in Chapter 21. Many other grasses are pre-adapted to becoming weeds because they can not only spread by numerous seeds but also by laterally growing (horizontal) stems (both by rhizomes, which are underground spreading stems and stolons, which are above-ground spreading stems), and by 'tillers' (branches which create new shoots growing from the base of the plant), as shown in Figure 19.5. Moreover, since grasses produce much of their growth from the base of the plant, cutting the above-ground stems often merely slows their growth rather than eliminating them.

BARNYARD GRASS

Introduction

The genus *Echinochloa* contains 40–50 species (Michael 2003) of moist habitats, mostly in tropical and warm-

FIGURE 19.3 Biblical condemnation of overenthusiastic elimination of agricultural weeds, as reflected by 'The Parable of the Weeds' in Matthew 13:24–43. Jesus advised servants eagerly pulling up weeds that they would root out their wheat as well, and so it was better to let both grow together until the harvest. (a) Religious icon from a church based on the parable. Having allowed both the wheat and the weeds to grow until harvest, the wheat (top) is being harvested, while the weeds (bottom) are being removed. Photo by Ted (CC BY SA 2.0). (b) 'Parable of the Weeds among the Wheat,' painting (public domain) by Dutch artist Isaac Claesz van Swanenburg (1537–1614) in the Artsrijksmuseum. Note how the wheat harvest has some weeds. (c) Painting (public domain) of the devil sowing weeds, by artist Heinrich Füllmaurer (1526–1546), in Kunsthistorisches Museum (Vienna). The medallion at top states 'Math. 13.' in reference to Matthew 13, the Parable of the Weeds. (d) Artist impression of a farmer in biblical times cautioning his workers that some tolerance of weeds is best. Source: Shutterstock (contributor Askib).

temperate locations of all continents except Antarctica. The most important species, *E. crus-galli*, is known by numerous names, including: barn grass, barnyard grass (barnyardgrass), barnyard millet, billion dollar grass, chicken panic grass, cocksfoot panicum, cockspur, cockspur grass, common barnyard grass, German grass, Japanese millet, panic grass, water grass, and wild millet. As will be noted, several of these vernacular names are also applied to different species of *Echinochloa*. Although the name barnyard grass is employed for other species of *Echinochloa*, it usually refers to *E. crus-galli* and is the most frequent name used for it.

Occasionally, the hyphen in *crus-galli* is omitted, in the mistaken belief that the Code of Botanical Nomenclature

TABLE 19.1
The world's worst (agricultural) weeds according to Holm et al. (1977).

1. *Cyperus rotundus* (purple nutsedge)
2. *Cynodon dactylon* (Bermuda grass)
3. *Echinochloa crus-galli* (barnyard grass)
4. *Echinochloa colonum* (jungle rice)
5. *Eleusine indica* (goose grass)
6. *Sorghum halepense* (johnson grass)
7. *Imperata cylindrica* (cogon grass)
8. *Eichhornia crassipes* (water hyacinth)
9. *Portulaca oleracea* (purslane)
10. *Chenopodium album* (lambsquarters)
11. *Digitaria sanguinalis* (crabgrass)
12. *Convolvulus arvensis* (field bindweed)
13. *Avena fatua* (wild oat)
14. *Amaranthus hybridus* (green amaranth)
15. *Amaranthus spinosus* (spiny amaranth)
16. *Cyperus esculentus* (yellow nutsedge)
17. *Paspalum conjugatum* (buffalo grass)
18. *Rottboellia exaltata* (itch grass)

requires this. As noted by Liu et al. (2013), '*Echinochloa crus-galli* gets its epithet from the phrase "crus galli" (meaning "rooster's shank"), and it is necessary to join the two words by a hyphen... More incorrect is to write it as *E. "Crusgallii,"* taking it as derived from a personal same [sic], as in some other published works.'

Barnyard grass originated in Eurasia (Maun and Barrett 1986) and has become established in tropical to temperate regions throughout the world as a result of having been introduced for cultivation as a fodder and forage crop. This annual grass can grow to over 1.5 m in height but is often shorter depending on environment. It frequently grows in tufts but does not spread by rhizomes (underground stems) like many other weedy grasses. Some plants spread on top of the ground (Figure 19.6a) but most are upright (Figure 19.6b). Several stems usually develop from the base. The stems frequently grow along the ground before turning upward. The long, flat leaves are often purplish at the base, and the stem bases may also be reddish or purplish (Figure 19.7). The seed heads are also often purplish (Figure 19.8a). The seeds (technically grass 'seeds' are fruits called caryopses) develop in branching seed heads (Figures 19.8b and c) and, characteristic of the grass family, each seed is enveloped within bracts (tiny leaves) one of which often has a long pointed extension (awn).

The species occurs in many different forms, and some authorities have subdivided it into different varieties, while others have split it into different species. Plants identified as weedy *E. crus-galli* may in fact be very similar related species, or hybrids (Darbyshire et al. 2009), and many specialists speak of *E. crus-galli* and closely related plants as a 'species complex.'

Different kinds of barnyard grass vary in physiological and habitat adaptation (Norris 1996). The plants tend to be in damp or wet habitats (Figure 19.9), including saline soils, but can survive some drought. They are not tolerant of shade or strong drought. The species is a C4 plant, i.e., it is among the 3% of flowering plants that conduct photosynthesis by a metabolic pathway that is advantageous under high light, high temperature, and drought. Under suitable environmental conditions, barnyard grass produces large amounts of seeds (up to 1,000,000 by a single plant according to Mitich 1990), which may remain dormant in the soil for several years. The seeds can be dispersed by wind, water, agricultural machinery, the feet of humans and other animals, and as a contaminant in commercial seeds. Additional adaptations include allelopathy (the production of chemicals that are toxic to competing plants; Guo et al. 2017) and the evolution of resistance to herbicides (Bajwa et al. 2015).

'Vavilovian mimicry' (named after Nikolai Vavilov, a prominent Russian plant geneticist and plant explorer) is evolution by weeds to resemble domesticated crop plants. This is the result of unintentional selection by humans, who remove the weeds in cultivated crops when they recognize

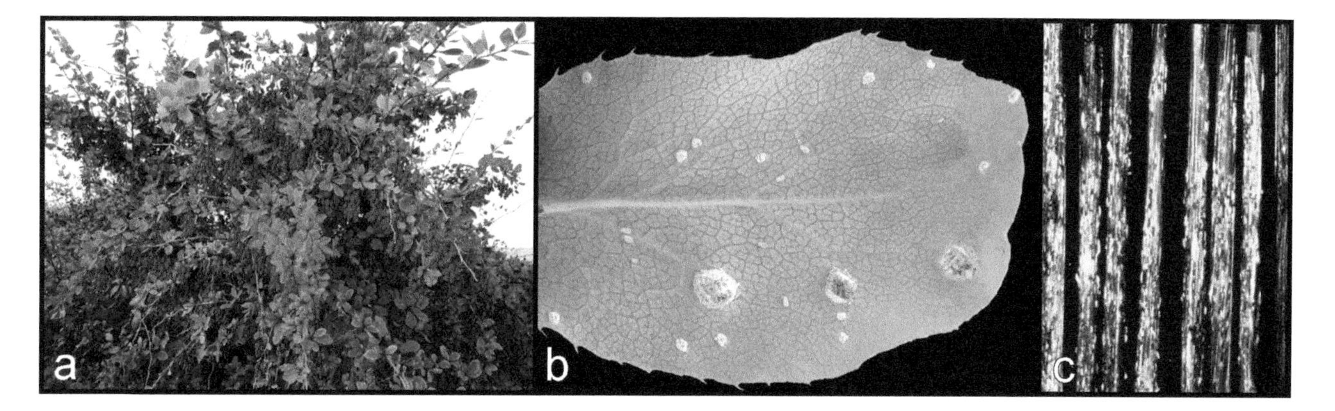

FIGURE 19.4 Barberry as a source of fungus infection of wheat. (a) European barberry (*Berberis vulgaris*), a native to the Old World but widely invasive in North America. The species is a source of fungus spores that infect nearby wheat. Photo by Arnstein Rønning (CC BY SA 3.0). (b) Fungus infection of a leaf of European barberry by wheat stem rust. Photo by Haruta Ovidiu, University of Oradea, Bugwood.org (CC BY 3.0). (c) Wheat stems affected by stem rust acquired from barberry. Photo (public domain) by Scot Nelson.

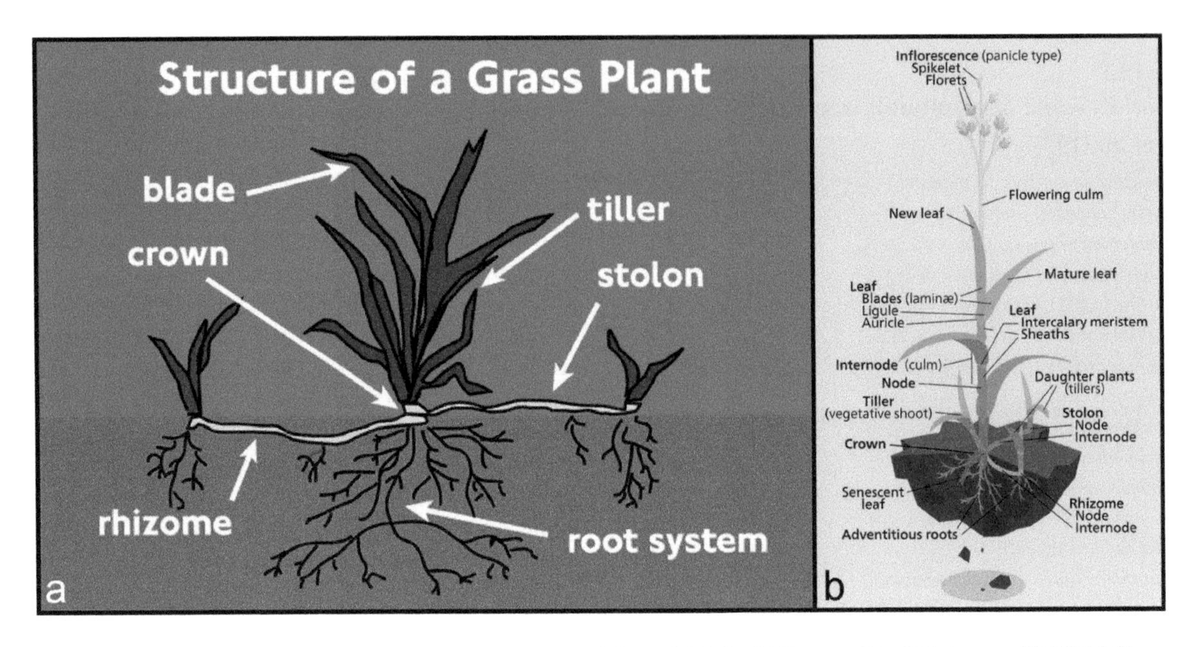

FIGURE 19.5 Anatomy of a grass. (a) Prepared by Wackymacs (CC BY SA 3.0). (b) Prepared by Kelvinsong (CC BY 3.0).

FIGURE 19.6 Divergent growth forms of barnyard grass (*Echinochloa crus-galli*). (a) Erect form. Photo by AnRo0002 (CC0 1.0). (b) Prostrate form. Photo by Howard F. Schwartz, Colorado State University, Bugwood.org (CC BY 3.0).

FIGURE 19.7 Base of barnyard grass (*Echinochloa crus-galli*). Note reddish areas. (a) Overhead view of prostrate plant. Photo by Michael Becker (CC BY SA 3.0). (b) Excavated plant showing fibrous roots. Photo by Steve Dewey, Utah State University, Bugwood.org (CC BY 3.0).

FIGURE 19.8 Development of fruits ('seeds') of barnyard grass (*Echinochloa crus-galli*). (a) Young fruiting branches. This form has prominent 'awns' (stiff pointed extensions) on one of the bracts covering the seeds. Photo by Matt Lavin (CC BY SA 3.0). (b) Seed head with young fruits. Photo by Steve Dewey, Utah State University, Bugwood.org (CC BY 3.0). (c) Seeds covered by awned bracts. Photo by Stefan.lefnaer (CC BY SA 4.0).

FIGURE 19.9 Barnyard grass (*Echinochloa crus-galli*) thriving in a wet habitat. Photo by Ewen Cameron, Auckland Museum (CC BY 4.0).

them, allowing weeds that become more similar to the crop to survive. In China, populations of barnyard grass (sometimes recognized as *E. crus-galli* var. *oryzicola*) that closely mimic rice plants have evolved (Barrett 1983; Ye et al. 2019).

Harmful Aspects

Barnyard grass is one of the world's worst weeds of crops, most notably of rice paddies, which is notable because more people depend on rice than any other crop. It also infests maize (corn), soybean, alfalfa (lucerne), and many other leading crops, including vegetables, orchards, and vineyards (CABI 2019h). Barnyard grass is capable of dominating fields, replacing whatever crops were planted (Figure 19.10). In addition to being an extremely competitive agricultural weed, in some soils, barnyard grass accumulates high levels of nitrates which can poison livestock (Marten and Andersen 1975), and it is also a host for several insect pests and virus diseases of crops. Many grasses produce seeds enveloped by an awned bract (a miniature leaf with a slender bristle-like extension), and this is the case for some forms of barnyard grass. These can pierce and so cause mechanical damage to the mouth of horses (Equine News 2007). In addition to being an agricultural weed, barnyard grass is also a frequent occupant of human-disturbed sites, such as roadsides and waste areas, and its elimination can be costly.

Beneficial Aspects

There is evidence that barnyard grass (or a near-relative) was harvested for grain in ancient China (Yang et al.

FIGURE 19.10 A field that has been taken over by barnyard grass. Photo by Harry Rose (CC BY 2.0).

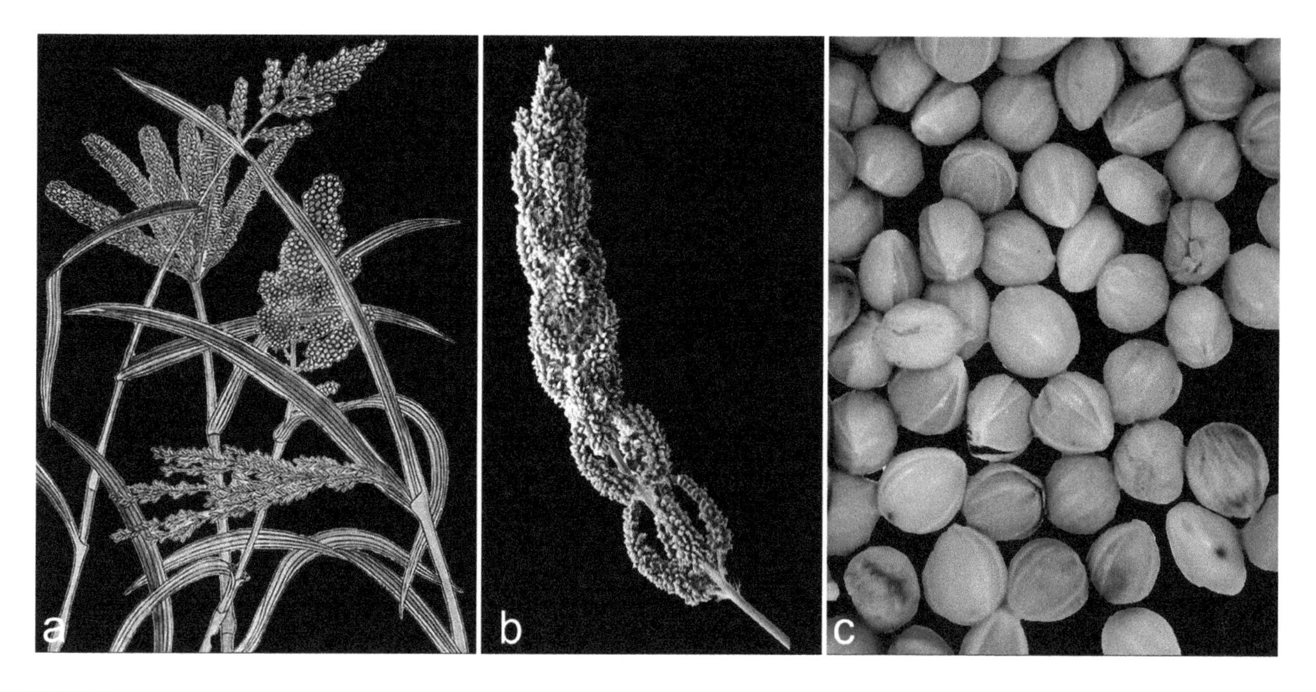

FIGURE 19.11 Japanese barnyard millet (*Echinochloa esculenta*). (a) Illustration from Seikei Zusetsu Agricultural Encyclopedia: Japanese Agriculture in the Early 19th Century. Photo credit: University of Leiden Digital Collections (CC BY 4.0). (b) Seed head. Photo by Jschnable (CC BY SA 4.0). (c) Grain. Photo (public domain) by Kochfutter2.

2015). Barnyard grass is currently grown in the tropics and subtropics for grain (Holm et al. 1977; Jun 2019), although the plant is a very minor cereal, and references to its cultivation actually are based on the *Echinochloa* cultigens mentioned below. *Echinochloa esculenta* is also sometimes cultivated as a livestock feed, with 'fair to poor value' (Heuzé et al. 2020). It can be grazed in the field, harvested fresh for feeding in a green state, and made into silage. It is also grown for hay (Esser 1994) although its high water content makes haymaking difficult (Heuzé et al. 2020).

'Millets' are a miscellaneous group of small-seeded grasses grown as cereal crops or grains for human food and livestock feed. The millets are very important in developing countries, where they are primarily grown. Barnyard grass is considered to be a millet, but the following millet species are more important. *Echinochloa esculenta* (*E. utilis*; Japanese barnyard millet, Japanese millet; Figure 19.11) and *E. frumantacea* (*E. crus-galli* var. *frumantacea*; billion dollar grass, Indian barnyard millet, Japanese millet, sawa millet, Siberian millet, white millet, white panicum, white-panic; Figure 19.12) are cultivated species related to *E. crus-galli* (and sometimes considered as part of it). They are grown in Asia as food for humans and livestock (Yabuno 1962; Hilu 1994). *Echinochloa crus-galli* is thought to be the main ancestor of *E. utilis* (Hilu 1994). *Echinochloa frumantacea* is employed for various preparations, including 'barnyard millet cookies' (Ballolli 2010; Figure 19.12d).

Conservation Aspects

In addition to being a devastating agricultural weed, and significant urban weed, barnyard grass is also an environmental weed which invades natural grasslands, wetlands, and coastal forests. It outcompetes many native plants, altering ecosystems.

On the positive side, barnyard grass has been employed to some extent for erosion control and soil reclamation, and may have potential for habitat rehabilitation (U.S.D.A. 2002; Heuzé et al. 2020). It has been shown to reduce arsenic in contaminated soil (Sultana and Kobayashi 2011) and to also have potential for removing cadmium and lead (Xu et al. 2017). The species provides cover for birds (Esser 1994), and 'Barnyard grass is commonly planted on wildlife refuges as the seeds are utilized by waterfowl and other birds' (North Carolina Government n.d.).

FIGURE 19.12 Indian barnyard millet (*Echinochloa frumantacea*). (a) Young seed head. Photo by James Schnable (CC BY SA 4.0). (b) Harvested seeds (covered by bracts). Photo by Omar hoftun (CC BY SA 4.0). (c) Grain on sale in a market in India. Photo by Thamizhpparithi Maari (CC BY SA 3.0). (d) Cookies made from Indian barnyard millet. Photo by Kalaiselvi Murugesan (CC BY SA 4.0).

BERMUDA GRASS

Introduction

Bermuda grass (*Cynodon dactylon*; Figure 19.13) is a perennial warm-season grass that probably originated in sub-Saharan Africa and/or Indian Ocean islands but is now distributed to much of the tropical and warm-temperate world (Mitich 1989b; CABI 2021b). Winter-dormant forms are sometimes mislabelled as 'annual' (the above-ground plants perish, and new shoots are generated). The plant is mostly 10–50 cm tall. Sometimes the name 'couch grass' is applied to Bermuda grass, but more often it designates another weedy grass, *Elymus repens*.

Harmful Aspects

Bermuda grass is a worldwide weed of crops, lawns, and native ecosystems between 45 degrees north and 45 degrees south latitude. Nicknamed 'devil grass,' the species has been characterized as one of the most damaging agricultural and environmental weeds in the world and indeed has been ranked as the world's second-most serious weed (Holm et al. 1977). The species has also been called 'the world's weediest grass' (Newman and Global Invasive Species Team 2014; Figure 19.14). Bermuda grass is a host for several insect, fungal, viral, and nematodal pests that spread to various crops. In addition to the harm it causes to agriculture and the environment, the large amount of pollen produced is a serious allergen (Rawls et al. 2020).

Beneficial Aspects

It highly ironic that many of the world's worst weeds, such as Bermuda grass, have a benign side. The species is widely planted for hay and pasturage (Harlan 1970; Figure 19.15) and is also employed as a turf grass for lawns (Figure 19.16),

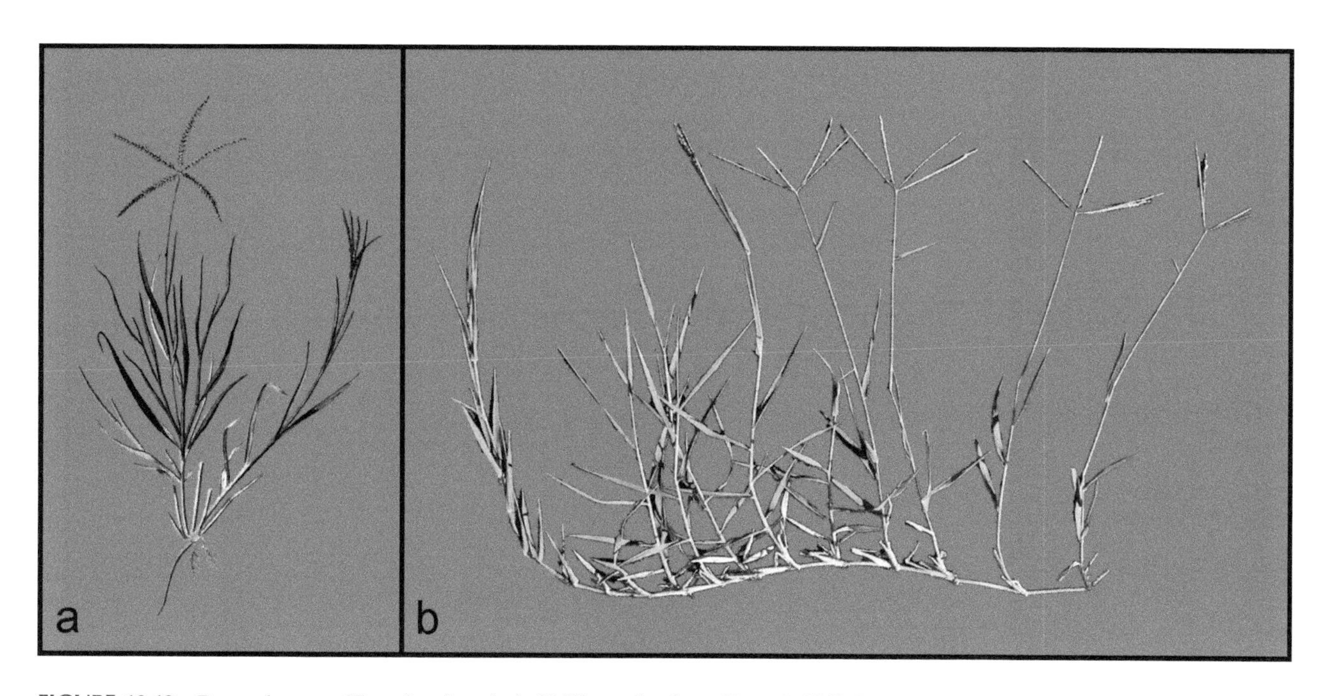

FIGURE 19.13 Bermuda grass (*Cynodon dactylon*). (a) Illustration from Kops, J. 1849. Flora Batava of Afbeelding en Beschrijving van Nederlandsche Gewassen, X Deel. (Illustrated by Christiaan Sepp.) (b) View showing stolons (above-ground horizontal stems) which creep along the ground and root wherever a node touches the ground. Photo by Steve Dewey, Utah State University, Bugwood.org (CC BY 3.0).

FIGURE 19.14 Bermuda grass (*Cynodon dactylon*) growing as an invasive weed. (a) In Australia. Photo by Harry Rose (CC BY 2.0). (b) In Hawaii. Photo by Forest and Kim Starr, Starr Environmental, Bugwood.org (CC BY 3.0). (c) Extensive network of spreading rhizomes (underground stems). Photo by Shawn Wright, University of Kentucky, Bugwood.org (CC BY 3.0).

FIGURE 19.15 Bermuda grass (*Cynodon dactylon*) as a source of nutrition for animals. (a) A miniature horse in a Bermuda grass pasture in Hawaii. Photo by Forest and Kim Starr (CC BY 2.0). (b) A cat in a Bermuda grass pasture in Hawaii, benefitting from the large supply of rodents which consume the numerous seeds. Photo by Forest & Kim Starr (CC BY 3.0). (c) A herd of crossbred beef cattle in a bermudagrass pasture. Source: Shutterstock, contributor: JNix.

FIGURE 19.16 Bermuda grass (*Cynodon dactylon*) as a lawn. (a) Bermudagrass sod for preparing a new lawn. Source: Shutterstock, contributor: MarinaDphotography. (b) Expansive planted Bermuda grass lawn at Vicksburg, Mississippi (the site of a key Civil War battle). Photo by U.S. National Parks Conservation (CC BY ND 2.0).

FIGURE 19.17 Audi Soccer Stadium in Washington DC, in 2018, with a Bermuda grass field surface. Photo by Rainclaw7 (CC BY SA 4.0).

athletic fields (Figure 19.17), and golf courses. It is a leading lawn choice in warm and hot climates. It does not survive cold winters and is difficult to maintain as a lawn grass in extreme northern areas. Its roots can penetrate the soil to depths of 2 m, providing considerable drought resistance. There are hundreds of cultivated varieties, reflecting the value of the species.

Bermuda grass is respected in Hinduism, and employed in ceremonies honoring Lord Ganesha, and in various religious ceremonies (Figure 19.18).

Bermuda grass is employed in traditional herbal medicine in Asia and is under investigation for its potential pharmacological values (Ashokkumar et al. 2013; Al-Snafi 2016b; Das et al. 2021). Like other plants that produce considerable plant material, Bermuda grass also has potential as biomass for various purposes (Xu et al. 2011).

Conservation Aspects

Bermudagrass is a worldwide aggressive weed that is so competitive in grasslands that it represents a threat to the rare species in the same habitats. Contrarily, it is very useful as a forage and turfgrass and is widely employed in agriculture, so that often decision-makers are ambivalent about control measures. It is represented in germplasm banks and is so common that at least the more common forms of species are not in need of conservation.

LAMBSQUARTERS

Introduction

Lambsquarters (lamb's quarters; *Chenopodium album*) is an annual herb (Figures 19.19 and 19.20) growing as tall as 3 m, usually much less, often less than 30 cm in unfavorable sites. The flowers are inconspicuous individually but are in fluffy clusters. Well-grown plants produce tens of thousands of small fruits ('seeds') 1.0–1.5 mm long (Figure 19.20b). Very large plants could produce 50,000 seeds (hence the species can be utilized for grain). Parts of the plant are often purple-tinged, and a whitish, 'mealy' powder is present on young leaves (Figure 19.21b). The plant is known by dozens of other names, including common lambsquarters, fat hen, pigweed, and white goosefoot. It is found throughout much of the world as a weed of disturbed natural habitats and in crop fields. Lambsquarters has been so widely distributed that its original distribution range is difficult to identify. Various sources suggest that Europe and/or Asia is the indigenous area, and some even

FIGURE 19.18 A garland made of Bermuda grass to be worn by the bride and groom in their Hindu marriage ceremony. Source: Shutterstock, contributor: Nabaraj Regmi.

suggest that there are indigenous North American races. The species is quite variable and has been treated taxonomically in different ways (Ohri 2015; Krak et al. 2016; Habibi et al. 2018). '*Chenopodium album*, one of the worst weeds and most widespread synanthropic plants on the Earth, in its broad circumscription is also among the most polymorphic plant species. It is a loosely arranged aggregate of still insufficiently understood races... It should be also kept in mind that many enigmatic and deviant forms of the *Chenopodium album* aggregate are in fact hybrids' (Clemants and Mosyakin 2003).

Harmful Aspects

Lambsquarters is one of the world's most serious agricultural weeds, occurring in many different crops (Williams 1963; Mitich 1988c; CABI 2021a; Eslami and Ward 2021; Figure 19.22b) as well as in uncared areas (Figure 19.22a). Holm et al. (1977) ranked it as the tenth most serious weed. Brijačak et al. (2018) ranked it as the fifth worst weed. The Weed Science Society of America (2017) published a list of 'most common' weeds in broadleaf crops, and lambsquarters ranked first. The plant acts as an alternate host of several crop pests (Bajwa et al. 2019). The pollen can cause hay fever (Bassett and Crompton 1978; Amini et al. 2011).

Beneficial Aspects

Lambsquarters may have been cultivated as a grain crop in prehistorical times in Europe (Stokes and Rowley-Conwy, 2002), and domesticated forms are still grown for grain in the Himalayan region. It is consumed as a garden-grown leafy vegetable in India (Jansen 2004). It is considered to have great potential to alleviate nutritional deficiencies among poor people in Africa (Gqaza et al. 2013) and to have healthful properties meriting consumption by people generally (Saini and Saini 2020; Figure 19.23). The seeds are also consumed (as a grain) to a lesser extent. Antinutritional substances (saponins and oxalic acid) are present in the foliage, although not in large concentrations,

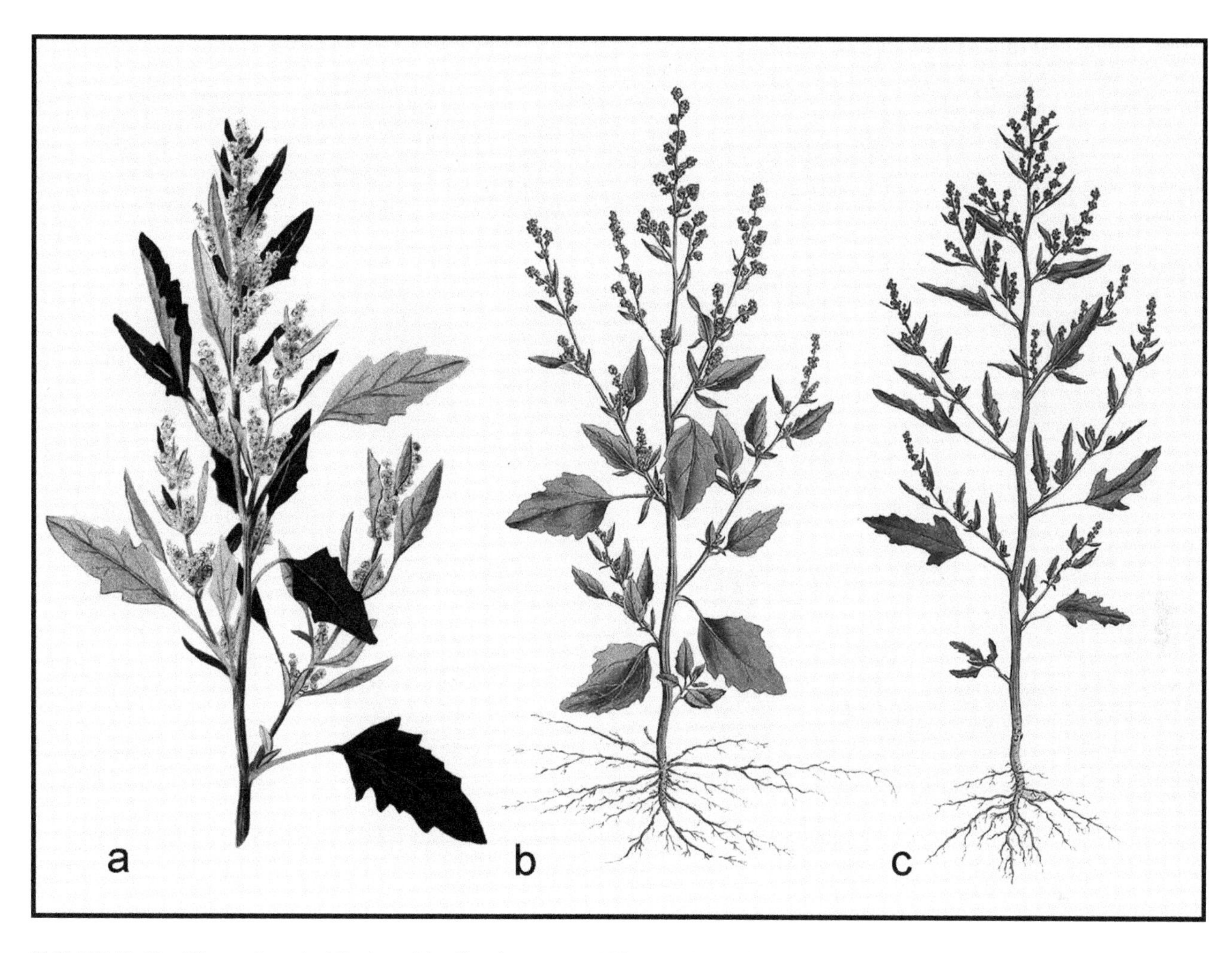

FIGURE 19.19 Illustrations (public domain) of lambsquarters (*Chenopodium album*). (a) Source: Clark, G. H. and N. Criddle. 1906. Farm Weeds of Canada. Ottawa, Canada Dept. of Agriculture. (b and c) Source: Curtis, W. 1777. *Flora Londinensis*, vol. 2.

suggesting that quantities consumed should be limited. Like many other plants, nitrates can be taken up from nitrogen-rich soils and cause illness in humans and livestock, once again indicating that consumption should be moderate (Plants for a Future 2022). In Asia, lambsquarters has been employed in traditional herbal medicine (Singh et al. 2010; Poonia and Upadhayay 2015) and has potential for development of new pharmaceuticals (Aman et al. 2016). Aside from the plant's values for people, it is invaluable for wildlife. Many wild birds and rodents consume the seeds.

Conservation Aspects

Lambsquarters is a worldwide weed that is not threatened by the possibility of extinction. Like other invasive weeds, it is a threat to natural environments. However, it does have appreciable agricultural potential and deserves to be considered for development as a crop. For this purpose, promising plant material should be preserved in germplasm banks (Partap et al 1998).

FIGURE 19.20 Reproductive parts of lambsquarters (*Chenopodium album*). (a) Inflorescence. Photo by NY State IPM Program at Cornell University (CC BY 2.0). (b) Seeds. Photo by Stefan Lefnaer (CC BY SA 4.0).

FIGURE 19.21 Young plants of lambsquarters (*Chenopodium album*), the most suitable stage for consumption of the foliage. (a) Photo by Babij (CC BY SA 2.0). (b) Close-up of young leaves (note whitish powder on the surfaces). Photo by Wendell Smith (CC BY 2.0).

FIGURE 19.22 Infestations of lambsquarters (*Chenopodium album*). (a) Roadside stand. Photo by Joost J. Bakker Ijmuiden (CC BY 2.0). (b) Growth as a weed on a farm. Photo by Harry Rose (CC BY 2.0).

FIGURE 19.23 Culinary preparations employing lambsquarters (*Chenopodium album*) foliage. (a) A curry preparation of rice, onions, potatoes, and lambsquarters. Photo by Xufanc (CC BY SA 3.0). (b) 'Bathua Raita,' an Indian dish made with lambsquarters, yogurt or curd, spice powders, other herbs, and vegetables. Source: Shutterstock, contributor: Pravruti.

NUTSEDGES

Introduction

The genus *Cyperus* contains about 700 species of grass-like plants, most often referred to as 'sedges,' but this term is also applied to thousands of other species in the sedge family (Cyperaceae), especially the 'true sedges' (species of *Carex*). Only a few species of *Cyperus* provide products of much interest. The best known is the papyrus sedge (*C. papyrus*) of Africa, which historically was the source of papyrus, the ancient equivalent of writing paper. Purple nutsedge (*C. rotundus*) and yellow nutsedge (*C. esculentus*) are the black sheep of the sedge family, infamous for the harm they cause as weeds (Bryson and Carter 2008). They are extremely damaging to crops and are very difficult to eradicate from lawns, vegetable and flower gardens, and urban landscapes.

Just what should be included in the name 'nutsedge' (sometimes spelled 'nut sedge') is open to interpretation. The New Oxford American Dictionary (Oxford University Press 2010) defines the word as follows: 'an invasive sedge with small edible nutlike tubers. Also called nutgrass. Genus *Cyperus*, family Cyperaceae: two species, purple nutsedge (*C. rotundus*) and yellow nutsedge (*C. esculentus*),' and defines nutgrass as: 'another term for nutsedge.' Webster's Third New International Dictionary (1993) states: that 'nut sedge' is a synonym for 'nut grass' and defines nut grass as: '1: any of several aggressively weedy sedges of the genus *Cyperus*; esp.: a perennial sedge (*C. rotundus*) of wide distribution having slender rootstocks that bear small edible nutlike tubers. 2: any of several American sedges of the genus *Scleria*.' The word 'nutgrass' was more popular in the past than the word 'nutsedge' and has been applied to more species than the two highlighted here. Occasionally, other species of *Cyperus* are referred to as nutsedge, but for practical purposes, nutsedge refers to purple and/or yellow nutsedge.

Nutsedges superficially resemble grasses (members of the Poaceae family), which also have narrow relatively long leaves and small inconspicuous flowers in clusters. However, nutsedge stems are triangular and solid in cross section (Figure 19.24), while grass stems are round and hollow in the internodes. The leaves of nutsedges are thicker and stiffer than the foliage of most grasses and are arranged in groups of three at the base rather than sets of two as occurs in most grasses. Purple nutsedge flowering/fruiting stems can grow to 50 cm in height (or as high as 1 m according to some authorities), and yellow nutsedge stems grow up to 1 m. Purple nutsedge develops a characteristic reddish-purple seedhead, while that of yellow nutsedge is yellowish.

Purple nutsedge (also known as coco sedge; Figure 19.25) originated in India according to some authors, while others think that it is native to the tropical and subtropical old world, mainly Eurasia and Africa, and still others believe that the native area is even more widespread, including Australia (CABI 2019e; Saha et al. 2019). Several genetically separated variants have been found to have been distributed to several areas of the world, reflecting the fact that the plant has usually been introduced as clones (Molin et al. 2019). As a weed, purple nutsedge is most troublesome in tropical and warm-temperate climates, where it is often much more difficult to eradicate than yellow nutsedge.

The indigenous area of yellow nutsedge (also known as chufa, chufa flatsedge, earth almond, rush nut, and yellow nutgrass; Figure 19.26) has been disputed (CABI 2019d), but the species appears to be native to the warmer parts of the Northern Hemisphere (including Europe, Asia, and North America) and parts of Africa and South America (De Castro et al. 2015). It has become widely distributed in tropical, subtropical, and temperate regions of the world (Bendixen and Nandihalli 1987).

Both nutsedge species are very competitive in irrigated croplands and also thrive in moist soils such as on the margins of lakes, rivers, streams, and marshes (hence the occasional name 'watergrass'). However, they can also survive well in drier areas and are tolerant of heat, drought, and flooding. The two species often grow together, but in hotter, drier areas purple nutsedge is more vigorous (Figure 19.27).

The nutsedges produce rhizomes (laterally growing underground stems), tubers (swollen structures attached to the rhizomes), and corm-like bulbs (at the base of the plant) – types of underground organs from which the species produce shoots that grow into new plants (Figure 19.28). Purple nutsedge produces tubers in chains connected by rhizomes, whereas yellow nutsedge produces tubers only at the tips of rhizomes. Purple nutsedge rarely sets viable seeds but produces many tubers (Figure 19.28b). Yellow nutsedge produces seeds more frequently than purple nutsedge but nevertheless also reproduces mainly from tubers (Figure 19.28a). Pieces of rhizome, the tubers, and the bulbs can be dispersed by wind and water to new locations, where they can establish new growths. Usually, the plants are spread by unintentional movement of soils due to agricultural activities. The starchy tubers (often called 'nutlets') can remain dormant to survive periods of extreme environmental conditions. While dormant, they are very resistant to herbicides, making both species very difficult to control in crops and lawns. Yellow nutsedge is able to tolerate colder winter conditions than purple nutsedge, and accordingly is more widespread worldwide. In Canada, where only yellow nutsedge occurs, the plant overwinters as tubers (Mulligan and Junkins 1976). The tubers and rhizomes of both species also produce allelopathic compounds that reduce the growth of some crops and competing plants. Another adaptation of the nutsedges is their ability to resume growth in the spring and grow through thick mulches, and they have even been known to penetrate through asphalt. Still another adaption is concerned with mode of photosynthesis. About 3% of flowering plants, including the nutsedges, process carbon dioxide by the C4 fixation pathway, which is advantageous under high light, high temperature, and drought. The result of all these adaptations is that both nutsedges often produce large dense colonies where they are not wanted.

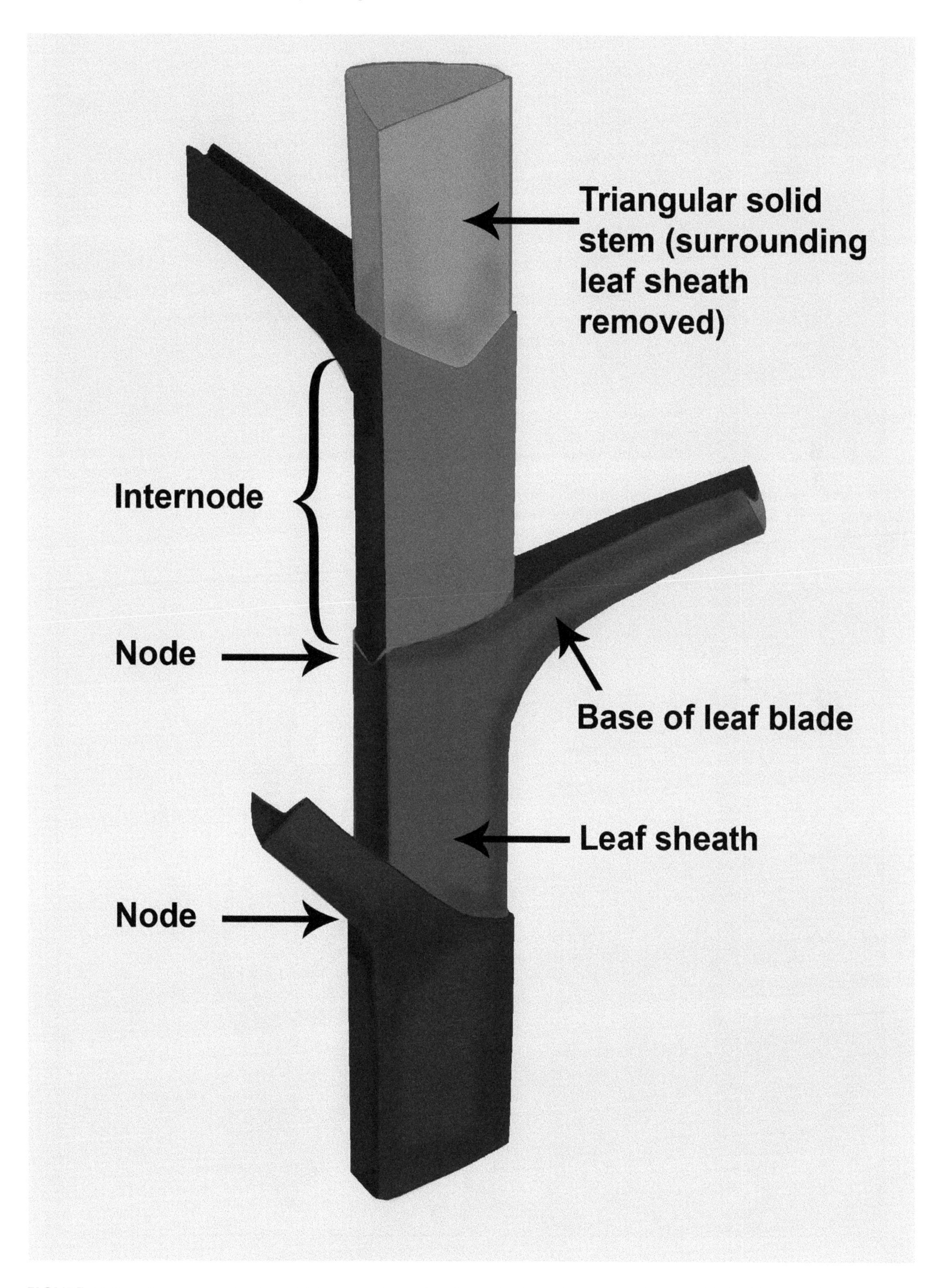

FIGURE 19.24 Diagrammatic view of triangular stem of yellow nutsedge (*Cyperus esculentus*) with leaf bases arising alternately from corners of the triangle. Drawn by B. Brookes, after Alex and Switzer (1977).

FIGURE 19.25 Purple nutsedge (*Cyperus rotundus*). Note variation of height of flowering/fruiting stems. (a) Tall plants. Photo by Rickjpelleg (CC BY SA 3.0). (b) Short plants. Photo by Javier Martin (released into the public domain).

FIGURE 19.26 Yellow nutsedge (*Cyperus esculentus*). (a) Plant in fruit. Photo credit: New York State IPM Program at Cornell University (CC BY 2.0). (b) Painting (public domain) of plant in flower, with underground tubers. Source: Host, N.T. 1805. Icones et Descriptions Graminum Austriacorum, Vindobonae, Austria. Vol. 3, plate 75.

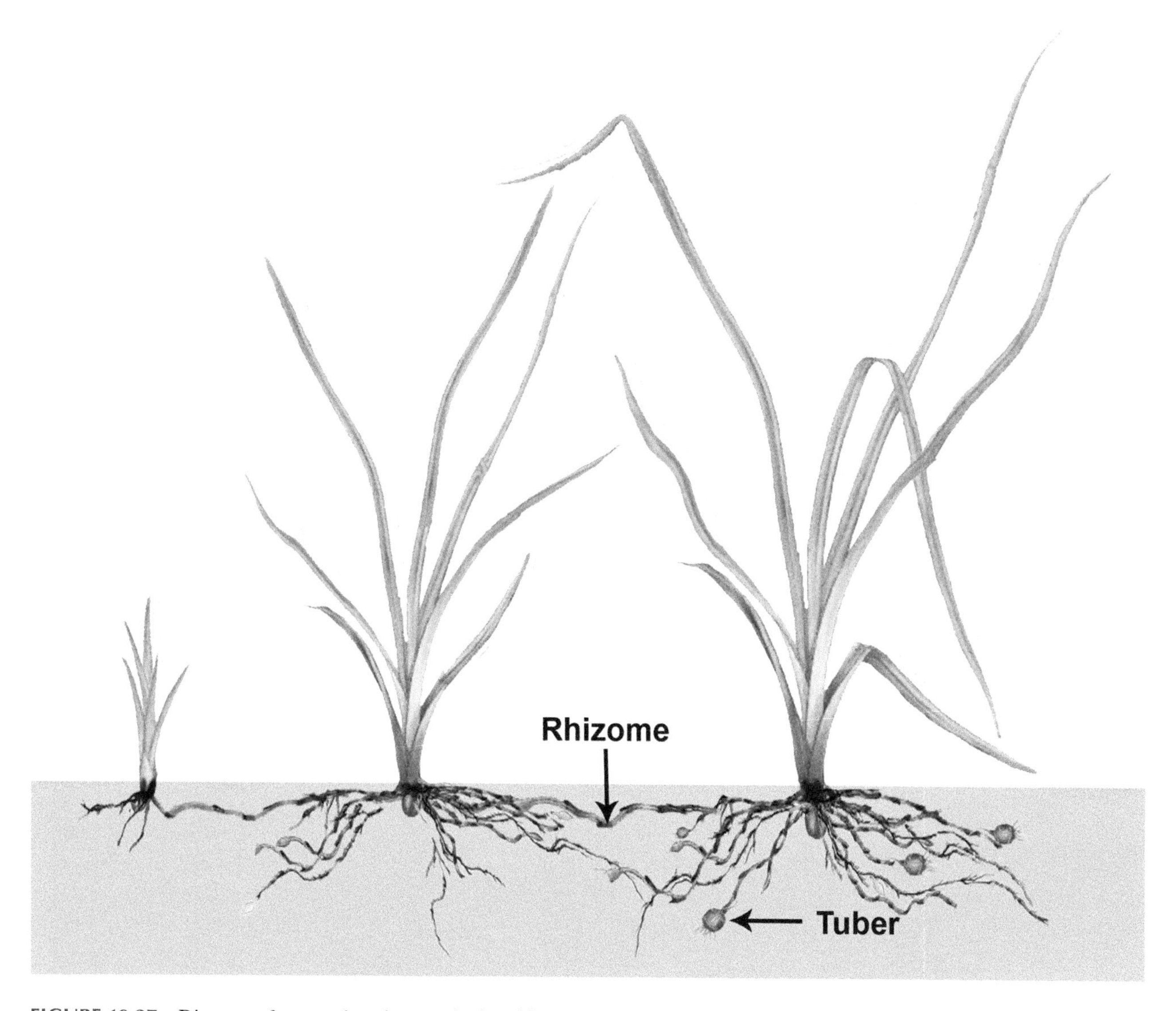

FIGURE 19.27 Diagram of a nutsedge plant producing rhizomes and tubers. Prepared by B. Brookes.

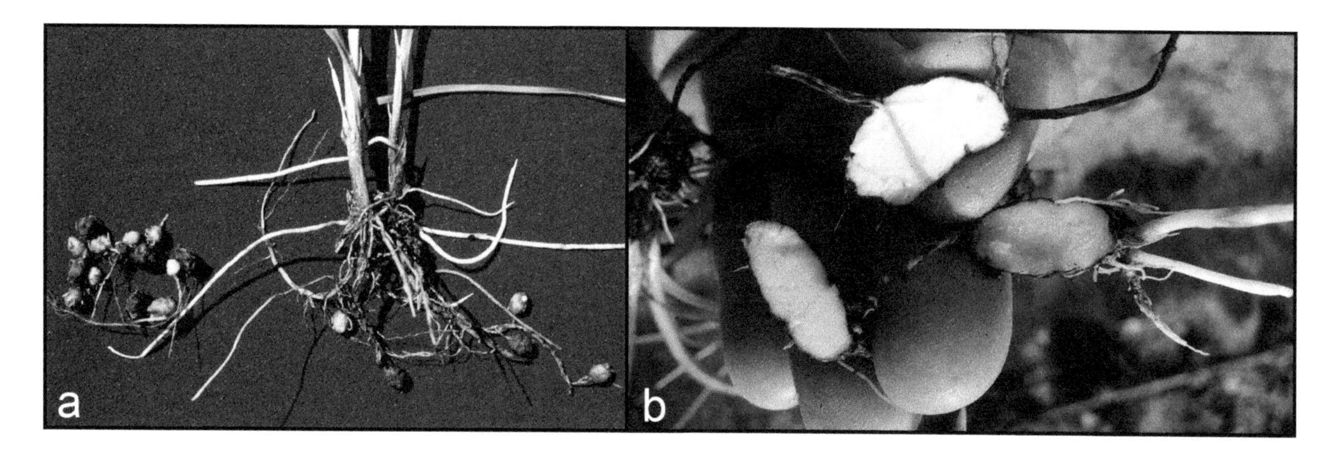

FIGURE 19.28 Tubers of the nutsedges. (a) Yellow nutsedge (*Cyperus esculentus*). Photo by Steve Dewey, Utah State University, Bugwood.org (CC BY 3.0). (b) Purple nutsedge (*Cyperus rotundus*), showing sectioned tubers. Photo by Iftahm (CC BY SA 4.0).

Harmful Aspects

Extremely extensive research has been conducted on the harmful aspects of the nutsedges and their control (Mero 1996; Textbox 19.2). Purple nutsedge has been ranked as the world's worst weed because of its ability to survive, spread, and compete, especially in agricultural areas (Holm et al. 1977). The species is extremely damaging to agriculture (Peerzada 2017). It is also a significant weed of urban areas and natural communities. Yellow nutsedge has been ranked as the 16th worst weed (Holm et al. 1977). It reduces crop yield and is a significant urban weed (Follak et al. 2016). Yellow nutsedge is a principal weed of potato (*Solanum tuberosum*), sugarbeet (*Beta vulgaris*) and many other cool-season crops.

Beneficial Aspects

Purple nutsedge tubers were a staple food of ancient Egyptians of the late Paleolithic era, ca. 1600 B.C.E. (Negbi 1992). The species has also been employed since ancient times as a minor or famine food plant (Dwyer 2016), and occasional food uses are still sometimes encountered (Lim 2016a). Purple nutsedge has also been used since ancient times as a medicinal plant and is being investigated for modern therapeutic applications (Al-Snafi 2016a; Lim 2016a).

Yellow nutsedge has often been deliberately established in natural areas so that deer, turkey, and other game animals will be attracted by the tubers (Stribling 1990; Kelley and Fredrickson 1991), which are much sweeter than the bitter tubers of purple nutsedge. Such planting may contribute to spreading of the plant as a weed. The cultivated kind of yellow nutsedge (discussed in the following) poses much less of an invasive weed problem than the wild kinds of plant (Sams 1999).

Livestock are occasionally used to control the weedy nutsedges, simultaneously providing them with fodder. Hens and geese confined to an area will consume the plants. Pigs are especially talented at rooting up the tubers.

A domesticated form of yellow nutsedge is cultivated for its large tubers, called chufas, tiger nuts, and rush nuts, employed as food (De Vries 1991; Negbi 1992; Pascual et al. 2000; Ejoh et al. 2006; Sánchez-Zapata et al. 2012; Akabassi et al. 2022; Figures 19.29 and 19.30). Rather than being referred to by the pejorative name yellow nutsedge, the crop form of the species is called chufa, tiger nut, rush nut, or by several other names. 'Chufa' is a Spanish dialect name for the plant. The 'tiger' in the name 'tigernut' is based on the striping often evident on the tubers of domesticated forms of the plant (Figure 19.30a). The name 'rush nut' is based on the similarity of chufa to plants of the rush family (Juncaceae). People also employ these tubers as food for domesticated animals such as chickens and pigs. The tubers are used particularly to prepare a beverage known as 'horchata de chufas' or just horchata (Figure 19.30c), which originated in the Valencia area of Spain. The refreshing milky drink has the flavor of vanilla and almonds, and is served cold or at room temperature. The Zulus of South Africa have long used chufa, hence the name 'Zulu nut' as well as using the tubers for food, the Zulus chew them to relieve indigestion and to sweeten breath.

Conservation Aspects

As noted earlier, there are uncertainties regarding the native ranges of both purple and yellow nutsedges, so it is also uncertain where the plants can be considered invasive. Nevertheless, while both species are highly unwelcome in most of the world, they appear to be primarily of concern as weeds of agricultural and urban environments, which are artificial habitats created by humans. In natural areas, they are much less competitive. Indeed, in North America where yellow nutsedge is probably indigenous, it is considered to be a valuable food source for wildlife, meriting maintenance (Taylor and Smith 2003). The tubers are eaten by some songbirds and many species of mammals, waterfowl, and upland game birds (Mitchell and Martin 1986).

As for most agricultural weeds and lawn weeds, herbicides are the chief method of control (USDA 1965; Webster 2003). Unfortunately, because the nutsedges are so difficult to remove, repeated applications are needed, with the inevitable result that adjacent natural ecosystems are exposed to chemical runoff and/or drift.

TEXTBOX 19.2 NUTSEDGES: AMONG THE WORST AGRICULTURAL WEEDS

Weeds cost billions of dollars in agriculture, forestry, and urban areas and threaten diversity in natural communities worldwide. Of an estimated 8,000 species of weeds worldwide, only about 200 species cause approximately 95% of the problems in production of food, feed, fiber, and livestock... Of these, sedges are among the most troublesome and difficult to control. The most important cyperaceous weeds in terms of their adverse effect on agriculture include *Cyperus rotundus* L., *C. esculentus* L... Economic losses result from interference or competition with crops and forests and the costs of pest-control chemicals, fuel, equipment, labor, cultural-control practices, and additional irrigation and fertilizer.

—*Bryson and Carter (2008)*

FIGURE 19.29 Chufa form of yellow nutsedge (*Cyperus esculentus*). (a) A chufa farm in Spain. Photo by Joanbanjo (CC BY SA 4.0). (b) A clump of plants. Source (public domain): Vilmorin-Andrieux. M. M. 1885. *The Vegetable Garden*. London, John Murray (colored by B. Brookes).

FIGURE 19.30 Commercial chufa products. (a) Tuber (notice stripes, the basis of the name 'tigernut'). Source: same as Figure 19.29b. (b) Plate of chufa tubers. Photo by Tamorlan (CC BY 3.0). (c) Horchata, the traditional chufa beverage of Spain. Photo by Heather Cowper (CC BY 2.0).

WEEDY RICE

Introduction

The subjects of this subchapter are forms of Asian rice (*Oryza sativa*), usually called 'weedy rice' and sometimes classified as *Oryza sativa* forma *spontanea* (cf. Roma-Burgos et al. 2021 for nomenclatural discussion). These weeds are very unusual in that they arise from the crop itself and act mainly as weeds of the crop. It may be noted that essentially all domesticated crops tend to revert (mutate) to characteristics that are not desirable from the point of view of farmers, requiring vigilance to maintain the genetic purity of the variety. However, the problem seems to be much more serious in rice.

Rice and wheat contend for the title of the world's most important crop in terms of economic value, but more people depend on rice for survival. About half of the world's population relies on rice as the primary source of calories. Because weedy rice is the most serious weed of rice, it is sometimes called the world's worst weed. Representative reviews of the subject are Delouche et al. (2007), Ziska et al. (2015), Nadir et al. (2017), Ajaykumar and Sharmili (2020), Fogliatto et al. (2020), and Svizzero (2021).

Weedy rice is often called 'red rice' because the fruits (grains) are frequently red, but as noted below this phrase is also employed for African rice. Moreover, some cultivated varieties of Asian rice also can have red grains, so the name red rice is ambiguous and is best avoided.

Weedy rices arise repeatedly and independently in countries throughout the world (Qi et al. 2015; Kanapeckas et al. 2016; Figure 19.31). Sometimes the plants originate simply as mutations, at other times they are hybrids. They closely resemble their parents but often have traits well-known in weeds, such as seed shattering (spontaneous disengagement at maturity from the plant) and seed dormancy (germination only after a prolonged period). They also often are taller and have reddish seeds, and sometimes mature earlier. They strongly tend to outcompete the normal crop but produce much less seed, so they are agriculturally harmful.

Asian rice is an annual native grass of the tropics and subtropics of Southeast Asia and grows best in a hot, moist climate. The plants are typically 60–180 cm tall. Domestication of rice occurred perhaps as far back as 10,000 years, in the river valleys of South and Southeast Asia and China. Rice is commercially cultivated in over 100 countries, representing all continents of the world except Antarctica. 'Upland rice' is grown without submersion, usually on terraced hillsides; 'lowland rice,' the predominant type, is grown in beds that are flooded during much of the growing season, with a water level of 10–15 cm with the water drained off as the plants approach maturity. In Asia, the monsoon (a season of torrential rains followed by a dry spell) naturally provides the necessary cycle of wetness and dryness.

More than 90% of the world's rice production is in Asia (Figure 19.32). China, India, Indonesia, Bangladesh, and Thailand are the largest producers. While rice is responsible for 80%–90% of the dietary intake of Asians, at the latitude north of Beijing called the 'Rice Line,' wheat replaces rice as the staple food. India tends to be divided into two food areas: wheat and meat in the north, rice and vegetables in the tropical south. Italy is the leading rice-producing nation in Europe. In the United States, rice is grown in appreciable quantities in the 'Southern Rice Belt,' Arkansas, Louisiana, Mississippi, Missouri, and Texas, as well as in California. Rice is the primary staple of more than 2 billion people in Asia and hundreds of millions of people in Africa and Latin America. Throughout history failures of the rice crop have caused widespread famine and death; conversely, in human history, there is a close relationship between expansion of rice cultivation and rapid rises in population growth.

This subchapter is concerned with Asian rice, not other species referred to as rice. African rice (*O. glaberrima*, also called red rice) is thought to have been domesticated as early as 3,500 years ago in the delta of Niger, from which it was introduced into other parts of West Africa, where diverse cultivars were selected. Asian rice is more easily produced, harvested, and milled. African rice continues to be produced mostly by small-scale farmers, mainly for local consumption, often in remote areas of Africa. It is cultivated particularly in Nigeria, Mali, Niger's inland delta, and to a lesser extent on the hills near the Ghana–Togo border and in Sierra Leone. Asian rice should also not be confused with so-called 'wild rice' (*Zizania* species). Indigenous wild rice species of North America have been traditionally collected from natural stands by Native Americans, but now the commercial supply is mostly generated by cultivation of specially bred varieties.

FIGURE 19.31 Weedy rice. (a) Along a ditch. (b) Along the edge of a rice field. Photos by Nilda Burgos, Arkansas Agricultural Experiment Station (CC BY 2.0).

FIGURE 19.32 Asian rice. (a) Farmers transplanting rice plantlets into a paddy in Thailand. Photo by Torikai Yukihiro (CC BY SA 2.5). (b) Toddler in a rice field in Viet Nam. Photo by Thái Văn Trà/Pixabay. (c) Rice field in China. Photo by JACKSON FK/Pixabay. (d) Harvesting upland rice in Laos. Photo by Basile Morin (CC BY SA 4.0).

HARMFUL ASPECTS

Like some other weedy grasses (Figure 19.33), weedy rice is a serious weed of rice fields. Weedy rice is very difficult to control, and yield losses due to infestations can be as high as 80% (Abraham and Nimmy 2015). Only specialists can reliably distinguish weedy rice from normal cultivated rice, so while agronomists hate it, most people are simply unaware of the harm caused. Accordingly, although weedy rice may be the world's most harmful weed, it does not suffer from the kinds of bias and prejudice described in this book for most serious pests. Weedy rice is the plant equivalent of a wolf in sheep's clothing.

BENEFICIAL ASPECTS

The agricultural community views weedy rice very negatively, but there are some benefits. Weedy rices are often physiologically superior to the domesticated varieties from which they arose (Figure 19.34), and their gene

FIGURE 19.33 A farmer in Thailand removing weedy grass plants from a rice field. Weedy grasses, including weedy rice, seriously compromise production. Source: Shutterstock, contributor: 1EYEman.

FIGURE 19.34 Experimental study near Stuttgart, Arkansas, comparing regular cultivated rice and weedy rice. Weedy rice (the taller plants) proved to be more competitive. Photo (public domain) by David Gely, U.S. Department of Agriculture (https://www.ars.usda.gov/oc/images/photos/nov14/d3344-1/).

mutations are potentially useful for breeding improved cultivars (Rutger et al. 1983; Swati et al. 2020; Wu et al. 2022). As with other very damaging pests, weedy rices provide employment for a vast array of researchers and control specialists.

CONSERVATION ASPECTS

Because of the economic harm caused by weedy rice, there are very extensive studies being conducted of the complex ways the plants arise genetically. This knowledge is useful in addressing the subject of crop germplasm conservation.

20 In Defense of the World's Most Despised Environmental Weeds

INTRODUCTION

This chapter reviews the world's worst environmental weeds, a phrase that designates weeds of natural environments, which are almost always plants introduced from foreign locations that are outcompeting native plants and altering ecosystems. There are numerous weeds considered to be environmental weeds (Weber 2017). Many are specialists of natural habitats and directly threaten indigenous species that are exclusively or primarily confined to fragile ecosystems.

Freshwater Environmental Weed Specialists

In the natural world, it happens that freshwater habitats are more susceptible to invading weeds than any other class of ecosystem. The reason is simple: it is almost impossible to control water weeds once they have become established, because we humans are terrestrial creatures and can't prevent aggressive water organisms from migrating in water as they choose. Aquatic plants have 'the highest risk of becoming natural area invaders' (Daehler 1998). Examples include giant salvinia (*Salvinia molesta*, Figure 20.1), waterthyme (*Hydrilla verticillata*), and alligator weed (*Alternanthera phyloeroides*). This chapter discusses common reed and purple loosestrife, which are specialists of shallow freshwaters and damp locations, and water hyacinth, a specialist of the surfaces of canals, rivers, and lakes. Adding to the difficulties posed by aquatic weeds, spraying herbicides on them as they grow in freshwater systems (the main method of control) is much more likely to poison innocent creatures and contaminate drinking water than treating weeds on land (Textbox 20.1).

Desert Environmental Weed Specialists

Deserts have been considered to be rather unaffected by plant invaders (Holt and Barrows n.d.), but native species specialized to survive in water-deficient substrates are highly stressed and can hardly tolerate additional pressure from unfamiliar invasive competitors. Relatively few plant species can live in extremely dry places, so biodiversity is low (Figure 20.2). Endangering plants specialized to survive in extreme environments can drastically affect their ecosystems. For example, Sahara mustard (*Brassica tournefortii*), indigenous to North Africa and parts of Asia, has recently become a source of danger to native plants of much of the drylands of the southwestern United States by fueling wildfires that they are unable to survive (Barows et al. 2009; Marushia et al. 2010; Figure 20.3). In this chapter, the most serious invasive plant of dry environments, cactus pear, is examined in detail.

Vine Environmental Weed Specialists

In some parts of the world, such as the UK, the term 'vine' often refers specifically to the grapevine (especially the wine grape, *Vitis vinifera*). In general, however, 'vines' are any of numerous plants that have long but relatively weak and flexible stems which either climb upon self-supporting tall plants, natural physical objects (such as rocks and cliffs) or human-made structures (such as fences, power posts, and houses), or simply sprawl over the ground. Unlike 'epiphytes' (plants which entirely or mostly live on trunks or canopies of trees, but rarely grow in the ground), vines root permanently in soil. 'Climber' is a general term for vines that grow from lower to higher levels, often employing twining, tendrils, or special attachments (which may be modified roots, leaves, or hairs). 'Creeper' is applied to plants that trail along the ground, but some so-called creepers also climb on objects, including trees, if they encounter them. Ecologists often restrict the word 'liana' to woody, large vines that grow in the Tropics, applying 'vine' to herbaceous plants (Rowe 2018). For a more detailed classification of vines, see Gentry (1991).

There are thousands of vine species because there are innumerable trees providing habitat upon which they can prosper. The more than 60,000 tree species (Beech et al. 2017) represent 20% of all 'vascular plants' (plants with internal conducting tissues, including angiosperms, gymnosperms, ferns, and allies). The world currently has over 3 trillion trees, although 12,000 years ago, before the advent of agriculture, there were twice as many (Crowther et al. 2015). Trees invest considerable energy into producing woody trunks to elevate their foliage so that they can outcompete smaller plants for sunlight. Rather than employing their own resources to produce sturdy stems for mechanical support, vines are 'structural parasites' which simply grow up and over trees to reach sunlight, an exploitive relationship pointed out long ago by Charles Darwin (1865). Vines harm trees by outcompeting them for sunlight (they have also been termed 'light parasites'). Since they root beside trees, they also compete with them for water and nutrients from the soil. (Numerous parasitic animals and a few parasitic plants directly penetrate host tissues to extract food, but almost all vines rob nourishing chemicals from their hosts indirectly.) Sometimes vines physically damage trees by abrasion, and they can even topple a tree by weighing it down.

DOI: 10.1201/9781003412946-20

FIGURE 20.1 Giant salvinia (*Salvinia molesta*), an invasive aquatic fern native to southeastern Brazil, that floats on water and produces mats sometimes over 50 cm thick. It spreads laterally by rhizomes and reproduces by budded off portions. (a) A small portion of growth. Photo by Mokkie (CC BY SA 3.0). (b) A young growth, showing leaf-like (fronds) above, root-like extensions below, and branch-like extensions below with (infertile) spore sacs. (c) Female and grown-up ducklings of rosy-billed pochard (*Netta peposaca*) in a pond overgrown with giant salvinia in a park in Buenos Aires Argentina. Source: Shutterstock, contributor: Nick Pecker. (d) Infestation of a river in Australia. Photo by Harry Rosen (CC BY 2.0). (e) Infestation of a farm pond in Florida. Photo by Ted D. Center, USDA Agricultural Research Service, Bugwood.org (CC BY 3.0).

TEXTBOX 20.1 HARMFUL EFFECTS OF HERBICIDES ON AQUATIC SPECIES

Reducing invasive species populations can mitigate the adverse effects on native species and ecosystems; however, control actions also have the potential for unintended and harmful effects on native species and ecosystems… Given the potential for undesirable consequences, herbicide treatments are often used as a management tool… Yet several studies have also found that large-scale herbicide treatments can cause significant declines in native aquatic plants, in addition to the target invasive species.

—*Mikulyuk et al. (2020)*

Pesticides play a very important role in reducing losses and maintaining quality in crop production. Although positive effects of pesticide use are undeniable, adverse effects are frequent. This has led to a comprehensive re-evaluation of the benefits of pesticide use and potential adverse effects on human health and the environment before placing them on the market. The fact that pesticides are designed to be toxic and are deliberately introduced into the environment makes them a very important and strictly regulated group of pollutants. The most commonly used group of pesticides are herbicides, and their detection in surface water bodies has been repeatedly reported. In spite of being designed to be toxic to target species, adverse effects on other inhabitants of aquatic environments have also been observed.

—*Marija and Slavica (2019)*

> Glyphosate-based herbicides (GBHs) are chemicals developed to control unwanted plants such as weeds or algae... Despite the target use, GBHs have been related to toxic effects on non-plant organisms, such as invertebrates, fishes, amphibians, reptiles, birds, and mammals, including humans... The most used herbicide worldwide is GBH... glyphosate per se has low toxicity when compared to its commercial formulation containing surfactants. However, those formulations are toxic to a large number of organisms due mainly to products added to the formulae... GBH can reach the aquatic environment through many ways. It can be applied directly on water bodies for algae control... GBH can also reach the aquatic environment through leaching, runoff, and contaminated food sources.
>
> —*Gonçalves et al. (2019)*

FIGURE 20.2 Representative dryland landscapes (Mojave Desert). Note the very low biodiversity typical of the more extreme drylands. (a) Joshua tree (*Yucca brevifolia*). Photo by Marshal Hedin (CC BY 2.0). (b) Creosote (*Larrea tridentata*). Photo (public domain) by Wilson44691.

FIGURE 20.3 Invasive Sahara mustard (*Brassica tournefortii*) in Joshua Tree National Park, southern California (public domain photos). (a) Photo (public domain) by Robb Hannawacker. (b) Volunteers removing the plant. Photo by Lian Law, National Park Service.

FIGURE 20.4 Trumpet vine (*Campsis radicans*) at the Chicago Botanical Garden. This extremely attractive vine, much used as an ornamental, is significantly invasive in some regions. Photos by Cultivar413 (CC BY 2.0).

Vines play critical roles in regulating forest biodiversity, affecting different species differently, especially trees (Schnitzer and Bongers 2002; Schnitzer 2015). Vines are a common component of forests worldwide, and they contribute to forest ecology, diversity, and dynamics (Schnitzer et al. 2014). They can have both positive and negative effects in forests. Vines are an important resource for some animals, as food (in the form of nectar, pollen, fruits, leaves, or sap), nesting sites, and shelter. By climbing among many tree crowns, vines can also provide aerial highways for many arboreal animal species. By contrast, vines also compete intensively with trees, reducing tree recruitment, growth, reproduction, and survival, as well as tree diversity and forest-level carbon sequestration. Many ornamental vine species have been introduced into foreign lands and have proven to be so invasive in some locations that they are now major pests, often endangering native biodiversity and degrading ecosystems (Niemiera and Von Holle 2008).

North America in particular appears to be the victim of invasive vines (Leicht-Young and Pavlovic 2014). Notable invasive vines in North America include ivies (*Hedera* species), kudzu (*Pueraria montana* var. *lobata*), Japanese honeysuckle (*Lonicera japonica*), mile-a-minute (*Polygonum perfoliatum*), Oriental bittersweet (*Celastrus orbiculatus*), porcelain berry (*Ampelopsis brevipedunculata*), bigleaf periwinkle (*Vinca major*), and dog-strangling vines (swallowworts; *Vincetoxicum* species).

Many ornamental vines are extremely attractive, and this has led to their use in regions foreign to their native lands. Often, vines have escaped and caused ecological problems in their new lands. Trumpet vine (*Campsis radicans*, Figure 20.4), a native of the southeastern U.S., has proven to be invasive when introduced in some other areas. English ivy, another widely employed ornamental vine, is examined in this chapter as an example of how devastating vines can be in forests, and kudzu is discussed in Chapter 21 as an urban vine threat.

Human Hypocrisy: We Despise Weeds, but Not When They Are Gorgeous Like Garden Lupins

Aside from edible plants (which are feasts for the stomach), the most beloved plants are garden ornamentals (which are feasts for the eye). However, numerous ornamental plants escape from cultivation and become harmful or 'invasive.' Many ornamentals are adapted to hot climates and easily survive when they escape to places that meet their requirements. For example, in the hot climates of Australia, two-thirds of that continent's invasive plants originated from introduced ornamentals (Small (2012b). Even in the cold winters of Canada, which frequently prevent plants from overwintering, a quarter of the country's invasive plants originated from garden ornamentals (Small 2012b).

Garden lupins (*Lupinus polyphyllus*) exemplify the dangers associated with escaped ornamentals (Small 2012c, 2022). They are widely cultivated beautiful plants (Figure 20.5a). In parts of the world (particularly in Europe, New Zealand, and eastern North America), they have escaped and established wild populations (Figure 20.5b). Exacerbating the problem, people have deliberately scattered lupin seeds in the wild, sometimes on the theory that beautifying the wilderness will make an area more attractive to tourists and therefore increase business.

Introduced garden lupins can change the ecology of a region, harming plants that are adapted to particular habitats. The plants significantly alter soils supporting wild plants and their associated fauna, tending to enrich the ground with nitrogen, and leaching toxic alkaloids into the soils. In New Zealand, weedy lupins have stabilized river gravels, forcing rivers to form deep, rapid channels unsuitable for wading birds such as the wrybill (*Anarhynchus frontalis*, Figure 20.6), a plover endemic to New Zealand, which has lost about half of its habitat.

Invasive garden lupins have proven to be harmful to local vegetation. For example, in New Zealand, the woodrush (*Luzula celata*) has been shaded out in areas by lupins.

FIGURE 20.5 Garden lupins (*Lupinus polyphyllus*) in cultivation and escaped to the wild. (a) A display of garden lupin varieties at a flower show. Source: Shutterstock, contributor: Devonpaul. (b) Weedy garden lupin growing rampant in New Zealand, where the highly invasive plant has caused extensive ecological damage. Photo (public domain) by SONY DSC.

Introduced garden lupins can even replace other *Lupinus* species. This has endangered several butterfly species in North America, the larvae of which cannot live without them. Larvae of the mission blue butterfly (*Aricia icarioides missionensis*, Figure 20.7a), a native of the San Francisco Bay area, feed only on the leaves of several lupin plants native to the area. Much of the original habitat of this butterfly has been destroyed. The xerces blue butterfly (*Glaucopsyche xerces*) also was once a native of the San Francisco area and depended on certain lupins; it is thought to be the first American butterfly species to become extinct because of loss of habitat caused by urban development. The Karner blue butterfly (*Lycaeides melissa samuelis*, Figure 20.7b) is another endangered species dependent on certain lupins. It occurs in the eastern U.S. and has suffered dramatic population declines in recent decades. The Karner blue is especially

FIGURE 20.6 The wrybill (*Anarhynchus frontalis*) of New Zealand, a narrowly distributed bird being threatened by invasive garden lupins. Source (public domain): Buller, W. L. 1888. *A History of the Birds of New Zealand*. 2nd edition. London, UK, W.L. Buller.

interesting for several reasons: it was described as a separate subspecies of the melissa blue in the 1940s by Vladimir Nabokov, the author of the controversial novel *Lolita*; it is the state butterfly of New Hampshire; and glands on the rear ends of the caterpillars secrete a nourishing liquid consumed by ants which in turn tend and protect them. Fender's blue butterfly (*Icaricia icarioides fenderi*, Figure 20.7c and d) was thought to be extinct until rediscovered in 1989 in the Willamette Valley of Oregon. It depends on the threatened Kincaid's lupine (*L. sulphureus* subsp. *kincaidii*) as a larval food plant. To assist in the preservation of both plant species, dogs have been trained to locate them by their odor.

CACTUS PEAR

Introduction

The cactus family (Cactaceae) is made up of over 1900 species placed in over 120 genera, notable of which is *Opuntia* with about 300 species. (Classification and identification of opuntias are very complex and require more study.) Nearly all cacti have succulent stems and are spiny. The plants characteristically grow in hot, dry, and hostile desert areas. Cacti are native to North and South America and the West Indies. The only exception seems to be mistletoe cactus (*Rhipsalis baccifera*), which apparently originated in tropical Americas, and was dispersed across the Atlantic Ocean by birds, reaching southern Africa, Madagascar, and Sri Lanka (Novoa et al. 2014). Cacti are favorite ornamentals, and many people specialize in collecting and cultivating these unusual plants (Figure 20.8). It has been claimed that there are more species of cacti growing in homes (especially in cold regions) than species of any other plant family (about 300 species of cacti are available as ornamentals). Unfortunately because cacti are highly desirable as ornamental plants, many wild species have been badly overcollected from nature. There is legislation that now protects cacti. CITES (Convention on International Trade in Endangered species of Wild Fauna and Flora) requires appropriate permits and documentation for moving plants across international boundaries, and this applies to all cacti except certain species used in commerce.

FIGURE 20.7 Endangered butterflies dependent for survival on particular wild lupin species that are being eliminated by introduced garden lupins. (a) Male mission blue butterfly, photo by P. Kobernus. (b) Female Karner blue butterfly, photo by J. & G. Hollingsworth. (c) View of upper side of female Fender's blue butterfly, photo by J. Dillon. (d) View of underwing of female Fender's blue butterfly, photo by G. Gentry. These photos are public domain, from the U.S. Fish and Wildlife Service.

Cacti have several adaptations to drought that allow them to survive in dry environments, but the most significant is special photosynthetic machinery, crassulacean acid metabolism, or CAM for short, which permits the plants to keep their stomates closed during daylight (Winter and Smith 1996). Plants acquire carbon dioxide through pores (the stomates), simultaneously losing water through them. Most plants keep their stomates open during the day, when sunlight is available because the usual photosynthetic machinery requires a continuous influx of carbon dioxide while the sun is shining. Plants regulate the opening of the pores, and most plants close them at night to prevent unnecessary loss of water (keeping the stomates open during the day is the price paid to acquire carbon dioxide). In arid environments, so much water can be lost when pores are open during the day that plants may die. Six to 7% of the world's plant species, including cacti, have CAM. During the much cooler nights, when water loss is much lower, the pores open, taking in carbon dioxide and storing it (like a battery) in a fixed chemical form (as 'organic acids') until light is once again available during the day to process the organic acids further by photosynthesis. Cacti have

FIGURE 20.8 A collection of cacti suitable for home cultivation. Photo by Phuket@photographer.net (CC BY 2.0).

additional physiological mechanisms that result in astonishing conservation of water. They typically lose 60% of their water before showing signs of stress, while most plants wilt and die when they lose only 20% of their water. It has been calculated that cacti lose 1/6,000th as much water as a typical plant with the same surface area.

The cactus pear (*Opuntia ficus-indica*) or prickly pear has also been called pear apple, nopal cactus, and tuna or tuna cactus (in Latin America; the word tuna is a much-used Mexican term for cactus fruit), and Barbary fig and Indian fig (in many European countries; the taste of some varieties is reminiscent of figs). *Ficus-indica* in the scientific name means 'fig of India.' The names with 'Indian' (for India) and 'Barbary' (for the Barbary Coast, extending from the Egyptian border to the Atlantic, and including Morocco, Algiers, Tunis, and Tripoli) reflect the past tendency of Europeans to give exotic names to unfamiliar fruit (inappropriate names in this case, as the fruit comes from the New, not the Old World). The cactus pear has been called the prickly pear for many years, and there is a tendency to use the phrase to describe the spiny, wild-growing, and weedy types of *O. ficus-indica* and to reserve 'cactus pear' for the cultivated selections (CABI 2019j). Unfortunately, some 20 different species of *Opuntia* are called prickly pear, so the name can be misleading. In Israel, the cactus pear fruit is called sabra (from the Hebrew *tzabar*), a term that was coined after the cactus was brought to the Middle East for use as a living fence. Sabra is also used to denote the cactus pear in Arabic, and in northern Africa and southwestern Asia. The word has also come to be applied to native-born Israelis (sometimes to Arab women as well), by analogy meaning tough on the outside, sweet on the inside. Cactus pear plants are sometimes called mission pears because they were often grown about missions in Mexico and the southwestern United States. However, the phrase usually refers to very old pear varieties brought over from Europe by missionaries, especially from France, and often by Jesuits.

The cactus pear has the appearance of a fleshy bush or small tree 3–7 m in height (Figure 20.9), with a trunk up to 45 cm in diameter (Pinkava 2003) or even 1 m (CABI 2019j). The root system is very shallow, mostly about 30 cm below the soil. When rain falls, numerous rootlets called rain roots develop to absorb the water. In *Opuntia* species, the flattened branches or stem sections, reminiscent of beaver tails, are technically called cladodes and are popularly called pads (or sometimes paddles) and joints (although cladodes are often joined together by what appear to be joints). These have the function of leaves (some anatomists have mistakenly equated the spines on the stems with the leaves of conventional plants). The cladodes are up to 60 cm long and 40 cm wide, generally a half to two-thirds as broad

FIGURE 20.9 Large specimens of cactus pear (*Opuntia ficus-indica*). (a) Plants growing in South Africa. Photo by JMK (CC BY SA 3.0). (b) Plants in the southern U.S. Photo (public domain) from U.S. Department of Agriculture.

FIGURE 20.10 Flowers of cactus pear (*Opuntia ficus-indica*). (a) Photo (public domain) by G. M. Stolz, U.S. Fish and Wildlife Service. (b) Photo by Forest and Kim Starr (CC BY 3.0).

as long, and the older ones become woody. On the edges of the youngest cladodes, new cladodes are budded off (usually after 1 year of growth). Mostly on the edges of 2-year-old cladodes, one or several flowers (occasionally more than 30) are produced. The attractive flowers (Figure 20.10) are yellow, orange or reddish (often combinations of these colors within a flower), sometimes white or pinkish, and are up to 7 cm long and 7 cm wide. Fertilized flowers develop into fruits (Figure 20.11), with less than 100 to more than 400 seeds. Some plants have been known to be parthenocarpic, i.e., to develop fruits without fertilization (Weiss et al. 1993). Commercial cactus pear fruit weigh from 120 to 200 g and are up to 10 cm long, depending on cultivar. The skin and pulp of the fruits on ripening may be green, yellow, red, or creamy white, and the color of the rind and interior may differ. The sugar content may be as high as 20%. Without controlled storage, the shelf life of the fruits is 3–4 days (Inglese 2009).

Growth (cell multiplication) of the organs (such as branches, leaves, and flowers) of flowering plants occurs in concentrated areas called meristems. In the cactus family, the growth areas that produce the spines are often recognizable as distinct small zones called areoles (singular: areole), which are often somewhat raised. Two kinds of spines may be present in a tuft, produced from the areole (Figure 20.12). The larger, more conspicuous, menacing

FIGURE 20.11 Fruits of cactus pear (*Opuntia ficus-indica*). (a) Shrub in fruit. Photo by Valérie Heuzé, AFZ (CC BY 3.0). (b) Close-up of fruits. Photo by Ken Bosma (CC BY 2.0). (c) Painting of a cladode bearing fruits. Source (public domain): Bessler, B. 1640. Hortus Eystettensis, Quintus ordo collectarum plantarum aestivalium (d) Cladodes topped by fruits. Photo by Pietro (CC BY SA 2.0).

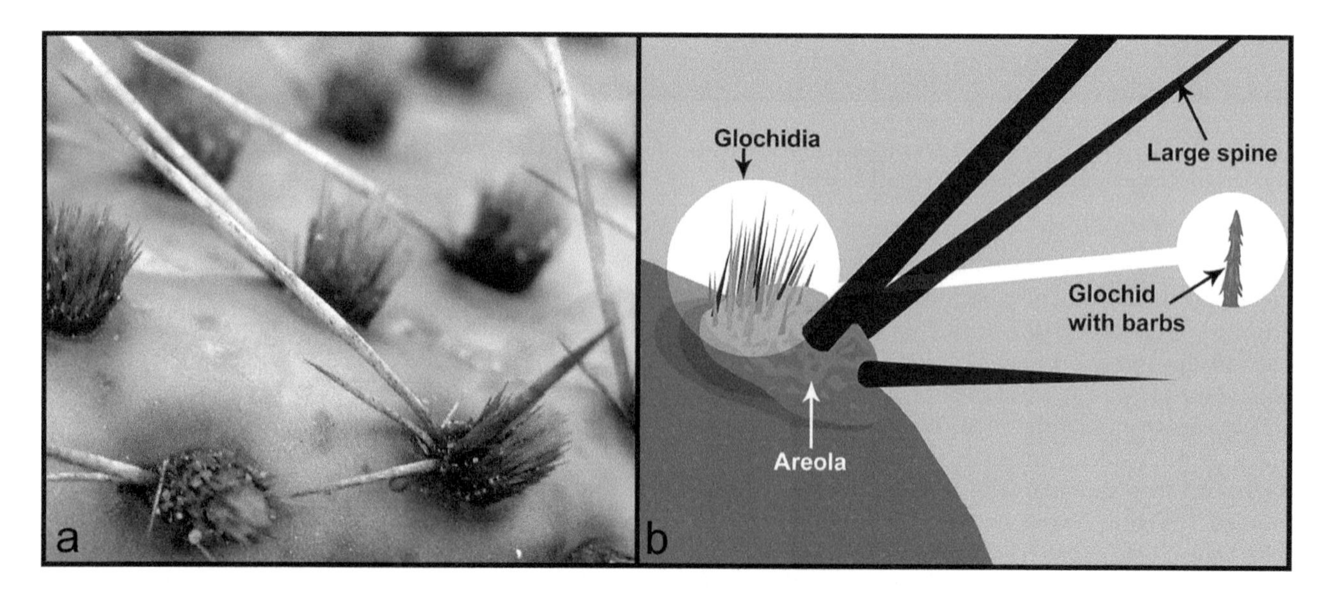

FIGURE 20.12 Large spines and glochids (hair-like spines) arising on an areole (raised area on the surface) of a cactus fruit. (a) Close-up of *Opuntia howeyi*. Photo by RoRo (CC BY 3.0). (b) Diagram by Morbid Mumbles (CC BY 4.0).

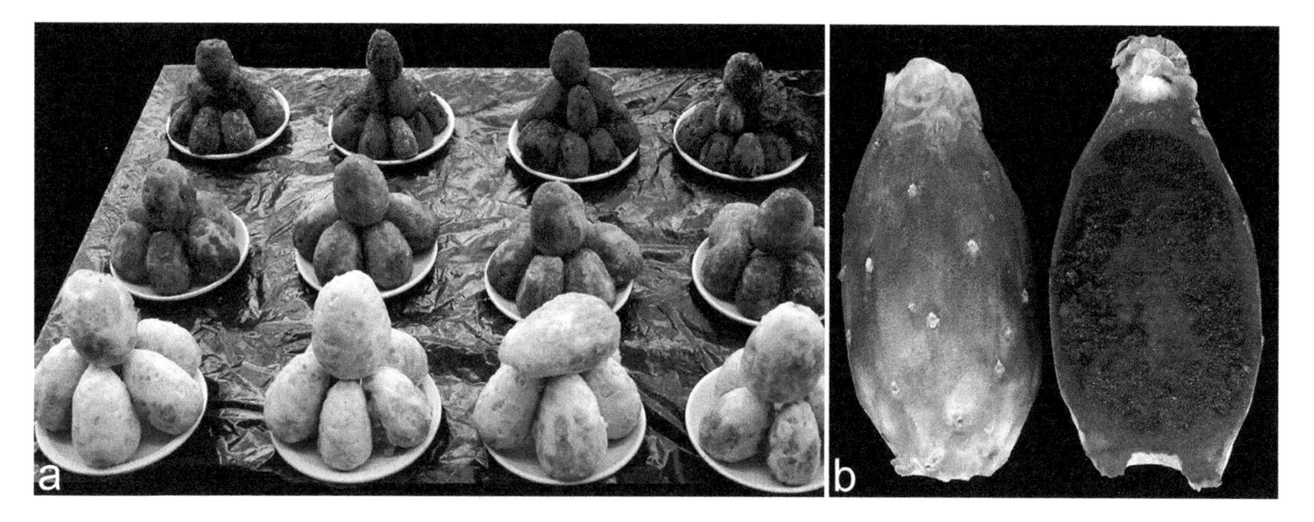

FIGURE 20.13 Fruits of commercial varieties of cactus pear (*Opuntia ficus-indica*). (a) Peeled fruit showing flesh color of different varieties. Photo by Tomás Castelazo (CC BY SA 2.5). (b) An intact and a sectioned fruit. Photo by Willamette Biology (CC BY SA 2.0).

spines (when present) are up to 2.5 cm in length. They can cause serious wounds and serve to repel large herbivores. Much smaller, shorter, slimmer, hair-like spines called glochids or glochidia (singular glochid or glochidium) may be present on an areola in addition to or instead of the large spines, often occurring in the hundreds, although very hard to see. These easily pierce skin, detach readily, and their backward barbs make them almost impossible to withdraw. They can cause extreme irritation for days. Spines occur on the fruits and cladodes, so that anyone handling cacti is strongly advised to wear gloves and protective clothing. Because the spines need to be removed for sale of the fruits and cladodes, and they can harm livestock that eat them, spineless forms have been bred. These are generally reproduced by cladodes (which need merely be placed on the ground, where they will develop roots), not by seeds, because plants grown from the seeds usually revert back to producing spines (Novoa et al. 2019).

Cactus pear is native to the desert zones of northwestern Mexico and possibly also the southwestern United States. It is extremely well adapted to arid and semi-arid climates of tropical and subtropical regions. Today, 'there is probably hardly a country with a Mediterranean or subtropical to tropical climate that is without this species' (CABI 2019j). It seems that cactus pear originated as a cultivated crop in Mexico by hybridization of various species (Griffith 2004; Caruso et al. 2010; Ervin 2012), and that its wild-growing range in the New World is based on escapes from ancient cultivation. Accordingly, ascertaining its original indigenous range is probably impossible. In the United States, it has been collected as a wild (i.e., uncultivated) plant in many places, particularly in California and Arizona. Apparently, wild-growing plants have been reported in many other locations in the southern U.S. and adjacent Mexico, but these are likely to have originated as introduced plants. In the United States, cactus pear has been collected from coastal chaparral, sage scrub, arid uplands, washes, canyons, and disturbed sites, at altitudes of sea level to about 300 m. The plant was brought to Europe by the first Spanish colonists from Mexico and has been cultivated along the Mediterranean coast since the late 17th century.

The cactus pear fruit is generally pear-shaped and except in spineless forms has a number of small spines. The skin of commercial cactus pear fruits can be red, pink, orange, purple, green, or yellow. The fruits vary in weight from 100 to 200 g and are generally 5–10 cm in length. The skin or rind is thick and fleshy (and represents 30%–40% of the weight of the fruit). The center of the fruit is filled with a soft juicy pulp (making up 60%–70% of the total fruit weight; about 85% of the pulp is water). The flesh may be green, light yellow-green, orange-yellow, deep golden, purple, or dark red depending on variety (Figure 20.13). The sugar content of the fruits is about 15%. There are many hard-coated, small, black seeds in the pulp (composing 5%–10% of the pulp weight). These seeds are similar in size and texture to those in grapes, and some new consumers find them objectionable to swallow, so there are efforts underway to breed seedless varieties.

While *Opuntia* species reproduce sexually by seeds, they also reproduce vegetatively when cladodes fall off the plant to the ground and produce roots and new daughter cladodes (indeed, commercial plantations are established entirely with cladodes or portions of cladodes). Tender young cactus pear cladodes, called nopales (and usually referred to as nopalitos when cut up for culinary use), have been consumed as a vegetable in central Mexico since pre-Hispanic times. Nopalitos are mostly water (over 90%). In Mexico, most nopalitos come from *O. ficus-indica*, but in the United States, mostly in southern California and Texas, considerable amounts of nopalitos are also obtained from *Opuntia cochenillifera* (*Nopalea cochenillifera*), called the cochineal cactus and nopal cactus.

FIGURE 20.14 Invasive *Opuntia ficus-indica* in southern Portugal. Photo by Peter Broster (CC BY 2.0).

Harmful Aspects

Cactus pear is the most widely distributed invasive species of cactus, not surprising since it has been more extensively imported and utilized for various purposes than any other cactus (Novoa et al. 2014; Figure 20.14). In Australia, where introduced *Opuntia* species are pernicious weeds, they have been called 'pest pears' and are legislatively controlled in some regions (Landscape South Australia undated Meadley 1958). Similarly in other hot-climate countries, *Opuntia* species are thought to be detrimental to livestock pastures (the spiny plants wound domestic animals who try to consume them and outcompete more desirable plants). Invasive opuntias are also claimed to degrade native ecosystems. Similarly, in African nations where grazing lands are valued, *Opuntia* species are considered very objectionable (Textbox 20.2).

Beneficial Aspects

Native Americans dried wild cactus fruit in the sun, and these could be stored for at least a year. Today, the edible fruits of more than 40 species of cacti can be purchased in Latin America, although only a few of these are grown

TEXTBOX 20.2 NEGATIVE IMPACTS OF OPUNTIA IN KENYAN GRAZING LANDS

Despite that *Opuntia* plants are appreciated for both direct and indirect uses by many livelihoods in the world, the species is hardly used in Kenya due to the presence of spines and glochids that cause injuries to both humans and animals. They are only valued for their use as ornamental plants or as a live hedge. However, these uses cannot compensate for the overall negative impacts they portray. Once they escape into the environment, they usually adapt to a broad range of environmental conditions where they are widely established. *Opuntia* species mainly invade the natural pasture which forms about 80% of the Kenyan land mass. Pastoralists are the main inhabitants of these lands and contribute more than 50% to agricultural Gross Domestic Product. However, the loss of productivity, stability, and ecological functioning of these lands by invasive alien species hinder the economic development. In many African countries, forage losses to unpalatable invasive plants have reduced rangeland capacities to three-folds. The potential costs associated with *Opuntia* invasion in the rangelands are based on reduced grazing land, replacement of native forage species as well as negative impacts on human and livestock health.

—*Githae (2018)*

FIGURE 20.15 Cactus pears (*Opuntia ficus-indica*) on sale in various countries. (a) In Sicily, Italy. Source: Shutterstock, contributor: Nyker. (b) In Morocco. Photo by Claude Renault (CC BY 2.0).

on a major scale. Numerous species of *Oputnia* (commonly called prickly pear) in Mexico are cultivated and harvested for food. *Opuntia tuna*, often called tuna and also known as elephant ear prickly pear, is particularly popular in Mexico, for although it produces rather spiny, small fruit, the plants make an ideal hedge. ('Tuna' is also a frequent term in Mexico for cactus fruit.) It has been estimated that the average Mexican citizen eats as much cactus as the average American eats cauliflower, and both the fruits and nopales are staple culinary items. Cactus pear is the chief species of interest, and the main cactus commercially available outside of Mexico. A mature plant can yield 100–200 fruits in a season. There are dozens of cultivated varieties (Modragon-Jacobo 2001). In Europe, especially in Sicily, there are very tasty varieties of cactus pears with whitish or yellowish fruits. In Sicily, these are called bastardi or bastadoni. Cactus pear fruits are now often sold in markets throughout much of the world (Savio 1987; Figure 20.15).

The flavor of ripe fruit depends on the variety but may resemble strawberries, watermelons, honeydew melons, figs, bananas, or citrus. Cactus pears may be eaten raw, alone, or with lemon or lime juice. The fruit pulp can be dried and ground into flour for baking into small sweet cakes, or stored for future use. The fruit can be made into jams and preserves. Cactus pears have also become staples in ethnic food preparations. Cactus syrup can be reduced into a dark red or black paste that is fermented into a potent alcoholic drink called 'coloncha.' Queso de tuna pronounced KEH-soh day too-nah) is a sweet paste prepared from fermented cactus pear juice, used in Mexican confections. It often serves as a base for jelly and candy – the 'cactus candy' in some Mexican food stores. Several alcoholic beverages are produced from cactus pear in the Old World. Ficodi is a cactus-pear-flavored liqueur produced in the village of Gagliano Castelferrato in central Sicily. Bajtra (the Maltese name for cactus pear) is the name of a liqueur made from cactus pear growing wild in Malta. Tungi Spirit is a locally distilled liqueur produced using cactus pear on the island of Saint Helena.

The cladodes are used as vegetables (Figure 20.16), although their main use is as a source of flour, which can partly substitute for wheat or corn flour in bread, cookies, and cakes (Barba et al. 2020). Some Mexican companies use nopalitos to make candy or process them for export as pickles, sauce, or jam. The pads are said to taste something like green beans, or between that of green pepper and asparagus, although when preserved in jars they have a tart, pickle-like taste. Fresh nopales are sometimes available in supermarkets as 'cactus leaves.' Cactus pie is prepared from nopalitos, and tastes like apple pie, possibly because both the cactus and apples contain high levels of malic acid. Huevos rancheros con nopalitos (fried eggs with onions and diced cactus) is a popular meal in Mexico.

Cactus pear has developed into an important crop in semi-tropical climates that are arid or semi-arid (Russell and Felker 1987a, b; Figure 20.17). The species has become the most widely cultivated cactus grown for crops in the world. The plants are cultivated for fruit production on all continents except Antarctica, with at least 140,000 ha of orchards (Yahia and Sáenz 2011). Losada et al. (2017) stated that the area in Mexico alone dedicated to 'prickly pears' (all *Opuntia* species) is 230,000 ha, of which 67,000 is for orchard fruits. In addition, considerable fruit is harvested from wild plants and home gardens. Cactus pear is grown in about 30 countries, mostly in Mexico, but with notable crops also in Chile, Bolivia, South Africa, Italy, Argentina, the United States, and Israel.

Cactus pear is rich in polyphenols, vitamins, polyunsaturated fatty acids, and amino acids, claimed to have nutritional and health benefits (Lim 2012; El-Mostafa et al. 2014; Isaac 2016; Yahia and Sáenz 2017; Aragona et al. 2018; Tilahun and Welegerima 2018). Cota-Sánchez (2016) characterized the nutritional value of the fruit as 'relatively modest.' However, Beccaro et al. (2015) stated that 'The nutritional composition of cactus pear is comparable to apple, cherry,

FIGURE 20.16 Cladodes (nopales) in Mexico. (a) & (c): On sale in markets. Photo (a) by Pro bug_catcher (released into the public domain). Photo (c) by Gary Stevens (CC BY 2.0). (b) & (d): Culinary preparations. Photo (b), a traditional Mexican meat- and nopal dish, by Gzzz (CC BY SA 4.0). Photo (d), a salad of prickly pear, cheese, and oil, by goodiesfirst (CC BY 2.0).

FIGURE 20.17 Cactus pear orchards. (a) In Sicily. Photo by Jos Dielis (CC BY 2.0). (b) In Mexico. Photo by AlejandroLinaresGarcia (CC BY SA 4.0).

FIGURE 20.18 An impenetrable hedge of cactus pear in Spain. Photo by Grez (CC BY SA 3.0).

kiwifruit and apricot.' Commercial health products made with cactus pear include body lotions, shampoos, and creams (Kaur et al. 2012), and seed oil (Krist 2020).

Large cacti such as cactus pear are frequently grown into hedges and fences by planting them 30 cm or so apart. Within several years, the plants grow together to form a spiny barrier that will repel any intruder larger than a rabbit (Figure 20.18).

In some countries, cattle are allowed to forage on cactus pear plants that are growing wild (Figure 20.19). In Texas and Mexico, ranchers commonly use propane torches to singe the spines of wild *Opuntia* species so that cattle can forage easily on them. Cattle readily eat the singed plants, consuming cactus up to 10% of their body weight daily. In many areas of the world where drought is severe and frequent, notably semi-arid central Mexico, southern Texas, and northeastern Brazil, cactus pear is valued as an emergency cattle food. Because the cacti typically contain about 90% water, the cattle can be sustained for many months of drought by relying entirely on the water in the plants. In many nations, cladodes are fed to livestock, often after being chopped up.

Before cheap synthetic aniline dyes were developed from coal tar in 1856, cactus plantations (including the cactus pear) were established for the production of cochineal (carminic acid), a dye (Donkin 1977; Figure 20.20). This is obtained by extracting the dye that cochineal insects (Figure 20.20b) produce (from the females of the genus *Dactylopius*, which are much larger than the males). About 32,000 female insects weigh 1 kg. The scale insect, which is native to Central America and Peru, feeds on the pads and fruit and develops a large quantity of cochineal stain within its body. The dye was once used to color the robes of Aztec emperors, including those of Montezuma (1466?–1520), a deep royal red. The dye was used for the robes of European royalty, the red jackets of British soldiers (Figure 20.21b; made famous by American patriot Paul Revere when he warned in 1775 that 'the redcoats are coming'), and the crimson jackets of the Northwest Mounted Police (Figure 20.21a; now the Royal Canadian Mounted Police). To supply cochineal for its overseas military forces, the British imported *Opuntia* species, which often subsequently became invasive. Cochineal is used to make a modern dye called carmine. Cochineal or carmine is still used

FIGURE 20.19 Cows grazing in a field of cactus pear in a nature reserve near Palermo Sicily, Italy. Source: Shutterstock, contributor: Poludziber.

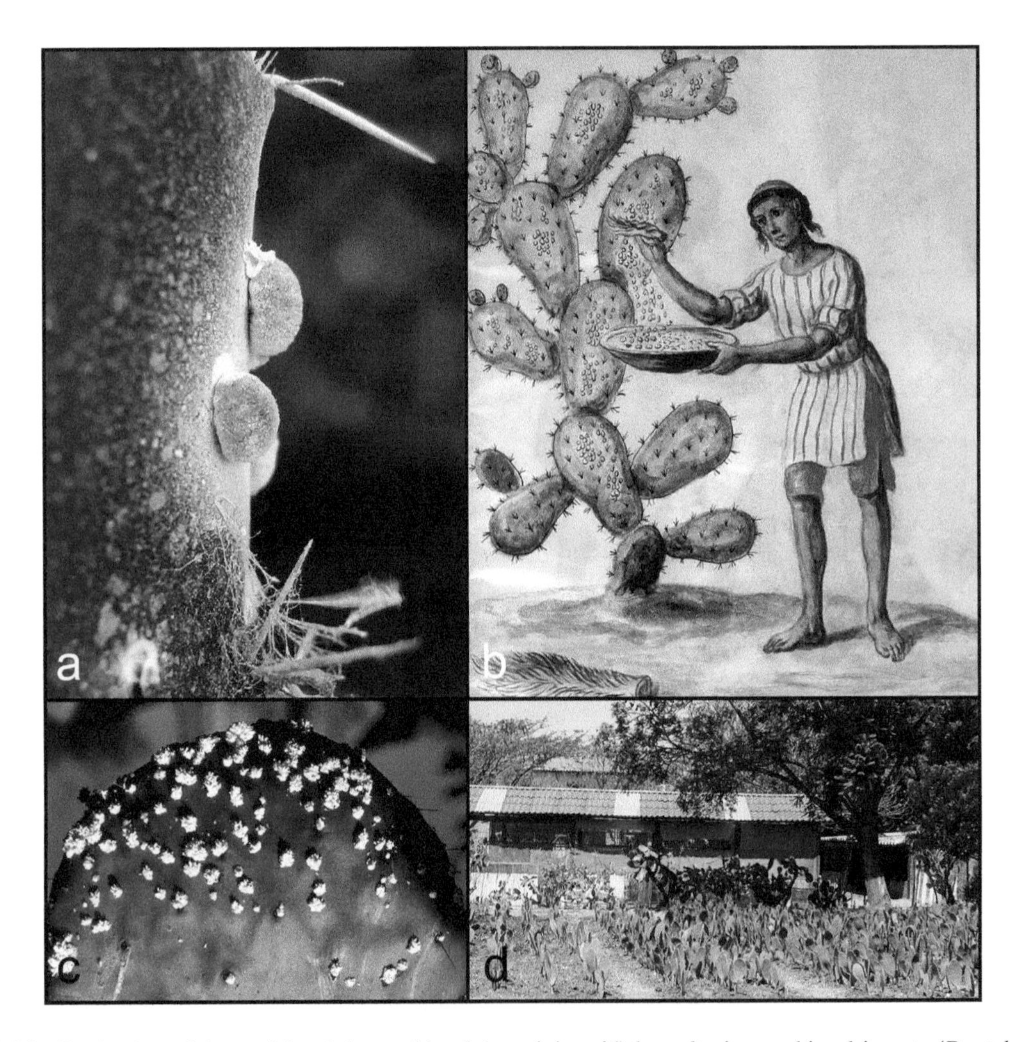

FIGURE 20.20 Production of the traditional dye cochineal (carminic acid) by culturing cochineal insects (*Dactylopius coccus*) on cactus pear (*Opuntia ficus-indica*). (a) Close-up of the insects. Photo by IT-management, German Wikipedia (CC BY SA 3.0). (b) 'Indian Collecting Cochineal with a Deer Tail,' a 1777 illustration from Memoria sobre la naturaleza, cultivo, y beneficio de la grana by José Antonio de Alzate y Ramírez. (c) A cladode infested by the insect. (The white appearance is due to a wax secreted by the insect as protection against drying out. Crushing the insect by hand stains the skin red.) Photo by James Denny Ward, USDA Forest Service, Bugwood.org (CC BY 3.0). (d) A small cultivated patch of cactus pear cultivated in Mexico for the production of cochineal. Photo by Whitney Cranshaw, Colorado State University, Bugwood.org (CC BY 3.0).

FIGURE 20.21 Traditional cochineal-red uniforms. (a) Postcard (public domain) of red-jacketed Royal Canadian Mounted Police officer (circa 1920). (b) '1775 Battle of Bunker Hill' showing red-coated British troops. Painted (1909) by E. Percy Moran (1862–1935), from the Library of Congress Prints and Photographs (public domain).

in botanical stains and as a textile dye, and has some use as an edible (but tasteless and odorless) dye for foods like maraschino cherries, icings, creams, jellies, cakes, candies, wines, and liqueurs, as well as in lipsticks. The dye has re-acquired importance since coal tar (aniline) dyes used for these purposes have been linked to cancer in laboratory animals.

There have been several international conferences dedicated to cactus pear, for the most part dealing more with the useful than its invasive aspects (Inglese and Brutsch 1997; Nefzaoui and Inglese 2002; Nefzaoui et al. 2010). There are also books dedicated to the species (Barbera et al. 1995; Pareek et al. 2001; Inglese et al. 2017) and a beautifully illustrated online guide to cultivation (Louhaichi et al. 2020).

Conservation Aspects

The introduction of *Opuntia* cacti to Australia in 1832 as hedging plants caused an ecological disaster. By 1925, several species were spreading through Australia like wildfire. A variety of control measures proved ineffective until 1925, when the cactus moth *Cactoblastis cactorum* was introduced, the larvae of which feed on the cacti (Figure 20.22).

FIGURE 20.22 The cactus moth (*Cactoblastis cactorum*), the chief biocontrol agent for invasive *Opuntia*, which ironically has become an invasive pest of cultivated cactus pear. (a) Caterpillars (larvae). Photo (public domain) by Peggy Greb, United States Department of Agriculture. (b) Adult (moth). Photo by Susan Ellis, USDA APHIS PPQ, Bugwood.org (CC BY 3.0).

By 1933, 90% of the cacti had been eradicated. A similar situation occurred in South Africa with *O. ficus-indica* covering a huge territory before suitable biological control agents (a weevil and a beetle, as well as the moth used in Australia) were found in the 1930s. The cactus moth originated in South America but has been moved all around the globe to control invasive prickly pear cacti. Unfortunately, it is harming the native *Opuntia* of southern North America. Moreover, now that cactus pear has become an important crop in many nations where the insect was employed in the past as a biological control agent, the insect, not the plant, is often considered to be the pest that requires control. While the use of biological control agents is generally considered to be worthwhile, the undesirable consequences of the cactus pear–cactus moth interaction indicate the need for careful planning.

In much of Africa, particularly in Ethiopia, people have come to rely on introduced cactus pear as a seasonal staple source of food (Shackleton et al. 2011; Lennox et al. 2012; Iqbal et al. 2020). This is not a unique situation: other invasives have become valued additions to the flora or fauna of some regions, albeit they may still retain damaging aspects (Novoa et al. 2016). Philosophically, this poses a challenge, since attempting to control the invasive plant on behalf of the natural world could endanger the survival of people (Textbox 20.3). Yes, the plant is invasive, but it provides significant food to people. Mdweshu and Maroyi (2020) commented 'The positive socio-economic contributions of *O. ficus-indica* need to be taken into account when evaluating the costs and benefits resulting from invasions caused by alien plant species.' In a similar vein, Illoldi-Rangel et al. (2012) stated 'any policy that seriously threatens (human) livelihoods only in the interest of preserving biota is not ethically defensible.' Moreover, wild animals (which are often victimized by invasive species) also find the plant to be a wonderful source of food (Figure 20.23).

In their indigenous locations, cacti provide food and shelter for wildlife species in arid environments (Figure 20.24). In Mexico, deer and other mammals forage on *Opuntia* species. White-tailed deer may utilize opuntias as the source of over 20% of their annual diet. Javelinas or collared peccaries (a kind of wild pig), various rodents, and some turtles and tortoises forage frequently on opuntias. The spines of *Opuntia* species can be quite damaging to vertebrates if consumed, but some have learned to scrape them off before consumption. Dessert cottontails, for example, discard the spines, often leaving piles of them at the base of the plant on which they are feeding. Cross pollination is carried out primarily by bees, but flowers of some species may be pollinated by grasshoppers which feed on the stamens. Over 100 insect species have been found to feed regularly on *Opuntia* species in Mexico. The developing cactus fruit is naturally protected by tufts of glochids, but as the fruit ripens the glochids tend to drop off, so that birds and mammals will not be discouraged from eating the fruits and dispersing the seeds (either by transporting whole fruits or by releasing viable seeds in their droppings).

Aside from its invasive tendencies, the cactus pear is one of the most environmentally-friendly of agricultural crops, because it is an ideal fruit plant to raise in arid regions (Small and Catling 2004a; Bautista-Cruz et al. 2018; Iqbal et al. 2020). Fresh water constitutes only about 2.5% of the water on Earth and is becoming so scarce that it has been described as 'the world's next gold.' Agriculture uses 70% of the world's freshwater. The welfare of both human societies and ecosystems, especially in the arid and semiarid zones of the world, depends on choosing crops that are less wasteful of water. Cactus pear requires far less water than conventional crops, and so is much less demanding of scarce freshwater supplies than virtually any other cultivated fruits. Given the dramatically decreasing supplies of water in the world, and cactus pear's natural adaptation to dry, hot climates, the species is likely to become more important in the future.

Coal, oil, and wood are examples of fuels containing chemically stored energy that is extracted simply by burning. Crop residues also contain energy. Burning the residues is one of several methods of recycling agricultural waste. Other methods include the production of animal feed, fertilizer, building and industrial materials, and soil conditioners, but fuel use is attracting increasing attention. By fermentation of agricultural wastes, alcohol or methane fuels can be produced. The high productivity of cactus pear has led to its consideration as a possible source of biomass

TEXTBOX 20.3 THE COST-BENEFIT DILEMMA OF HOW TO TREAT CACTUS PEAR

Opuntia ficus-indica reduces the availability of pasture grasses and invades crop land in some countries, though the actual economic loses have not been quantified. Also, where livestock are forced to feed almost exclusively on *O. ficus-indica*, this can lead to a loss of condition or even result in death. However, *O. ficus-indica* is regarded as both an aggressive invader and as a source of food in the areas where the plant proliferates. That *O. ficus-indica* is a commercial crop will mean that there are often substantial economic gains from presence of this species leading to a requirement for balancing economic calculations to evaluate whether there are gains or losses in any given situation. This conflict of interest also restricts the control options available. The more *O. ficus-indica* invades an area the more people become dependent on it because it will also reduce the availability of other options or resources.

—CABI (2019j)

FIGURE 20.23 Wild animals eating invasive cactus pear. (a) Elephant in Africa eating plant. Source: Shutterstock, contributor: Four Oaks. (b) Olive baboon (*Papio anubis*) eating fruit in Kenya. Photo by Amanda Lea (CC BY SA 3.0). (c) Gopher tortoise (*Gopherus polyphemus*) snacking on cladode in Florida. Photo by Andrea Westmoreland (CC BY SA 2.0). (d) African blue tit (*Cyanistes teneriffae*) eating fruit in Las Palmas de Gran Canaria, Spain. Photo by Juan Emilio (CC BY SA 2.0). (e) Canary Islands chiffchaff (*Phylloscopus canariensis*), an endemic of the Canary Islands, eating a cactus pear flower. Source: Shutterstock, contributor: Pacotoscano.

for energy production as bioethanol (Beccaro et al. 2015). When the plant is used as fodder, both the remains of the plant and the resulting animal waste can be used for biogas (methane) production. Although not yet practical, research has suggested that biogas production from cactus pear may be a good method for increasing the sustainability of agricultural systems in arid zones.

Opuntia species often play a useful ecological role in conservation projects for arid zones. They are excellent for remediation of eroded, overgrazed, or salinized areas. Desertification (the conversion of lands to deserts), typically caused by degradation and destruction of vegetation cover, is a major concern today. Cactus pear has been employed to revegetate areas in Tunisia and Algeria in order to retard

FIGURE 20.24 Animals consuming cactus pear plants in the southwestern U.S., where the plant is probably native. (a) Deer mouse (*Peromyscus maniculatus*) eating fruit in New Mexico. Photo (public domain) by Sally King, U.S. National Park Service. (b) Round-tailed ground squirrel (*Xerospermophilus tereticaudus*), eating ripe fruit. (c) Cottontail rabbit eating a fruit. (d) Cottontail rabbit (*Sylvilagus audubonii*) eating a cladode. (b–d) From Shutterstock, photos taken in the Sonoran Desert, near Tucson, Arizona, contributor: Charles T. Peden.

sand movement, since the roots keep considerable soil in place, and the species is an excellent wind-break. The plant also stabilizes land terraces that prevent water erosion during times of heavy rainfall.

COMMON REED

Introduction

'Reeds' include several tall, slim, hollow-stemmed, grass-like plants of shallow wetlands. In north-temperate areas, reeds are especially species of the grass genus *Phragmites*, of which four or five are recognized (Packer et al. 2017). However, these are often confused with cattails (of the genus *Typha*), which also grow in shallow water. In Europe, cattail species are often called reed-mace (reed mace, reedmace), contributing to confusion between *Phragmites* and *Typha* species. Common reed (*Phragmites australis*, formerly called *P. communis*), the subject of this discussion, is also sometimes misidentified as giant reed, *Arundo donax* L., an exotic Asian introduction to North America, where it is a troublesome aquatic weed in the southern United States. Occasionally, common reed is also called giant reed, further confusing identification. In the Southern Hemisphere, different but superficially similar species are also called reeds.

There are many plant species, termed 'emergent aquatics,' that root in shallow water but grow out of the water. Common reed can grow in water as deep as 2 m but can also be found on the relatively dry parts of river banks and in damp fields. It is a tall, robust, perennial grass (the above-ground parts dying with freeze-up), usually 1–4 m in height, sometimes as high as 7 m (Figure 20.25). The species is clonal, producing large stands by extension of horizontal basal stems. Although the plant is perennial, no portion of the clone lives more than 8 years (Gucker 2008). However, a clone can persist in a site for a thousand years (Haslam 1972). The jointed stems are slim but stout and can be so close together in a reed bed (over 600 stems/m^2 have been recorded) that they prevent movement of large animals. The

FIGURE 20.25 Tall specimens of common reed (*Phragmites australis*). (a) Photo by James H. Miller, USDA Forest Service, Bugwood.org (CC BY 3.0). (b) Photo by Elizabeth Banda/U.S. National Aeronautics and Space Administration (CC BY 2.0).

FIGURE 20.26 Common reed (*Phragmites australis*) at the reproductive stage. (a) Head of flowers. Photo by Harry Rose (CC BY 2.0). (b) Plants with fruiting heads. Photo by R. A. Nonenmacher (CC BY SA 4.0).

leaves are flat, pennant-shaped, gray-green, 20–50 cm long. The plant produces a large plume-like, whitish (often purplish when young) flowering/fruiting head 20–40 cm long (Figure 20.26). The head appears to be feathery because of the presence of long, silky hairs attached to the seeds, which serve as sails for dispersion by wind, although the seeds are also distributed by water. The edible grains are small, but the seed head may contain over 2,000 seeds.

FIGURE 20.27 A large bed of common reed (*Phragmites australis*) found at Bolata Cove on the Black Sea coast of Bulgaria. Photo by MrPanyGoff (CCBY SA 3.0).

Common reed has been alleged to be the most broadly distributed flowering plant in the world (Good 1974). The species occurs on every continent except Antarctica and is especially widespread in the temperate zones. There are genetically distinctive races indigenous to different areas of the world, and some have been introduced elsewhere. The classification of these is controversial (Saltonstall 2016), and additional information is given later. Common reed is a 'semi-aquatic,' usually growing with its roots in water or at least in soil that is periodically wet. It inhabits fresh water, and alkaline wetlands, typically in marshes, swamps, wet meadows, and fens. It also very frequently is in salt marshes and similar habitats on the coasts of oceans. Populations differ in tolerance to 'brackish' water (saline, but less so than seawater), some plants growing despite inundation in water with over 3% salt. The species roots in a wide variety of soil types. It is usually the dominant plant where it occurs, often forming very large, continuous patches (Figure 20.27). The Danube Delta (mostly in Romania) probably has the world's greatest stretch of reed beds (ca. 2,700 km^2). In European lakes, extremely large floating or loosely rooted mats of vegetation are often developed, dominated by common reed.

Although seed reproduction is comparatively rare, the seeds are sometimes numerous. They are normally dispersed by the wind or water, but sometimes by birds. Seeds will germinate in water depths up to 5 cm, and salinities up to 2% (but germination is reduced in water with over 1% salt). The plant is mainly dispersed by fragments of the underground stems (rhizomes) or stolons (above-ground laterally spreading stems) – sometimes by adhering to the feet of animals or floating away after being dug up – but more often unintentionally by people. The fragments can be carried to newly opened sites by machinery, floodwaters, birds gathering nesting material, and even by wind. Rhizome fragments are generated not only by human activity (e.g., plowing and cleaning out ditches) but also naturally by rodents. Common reed also spreads by stolons that trail horizontally along the ground for as long as 15 m (Figure 20.28). These very long runners typically develop out of water, serve to rapidly spread the plant, and produce shoots initially with abnormally small leaves.

Harmful Aspects

Common reed has been listed as one of the 'world's worst weeds' (Holm et al. 1977) and is legislated as a noxious weed in several jurisdictions in North America. It is a major pest of irrigation and flood control channels, where it invades the wet banks and sometimes extends into water 2 m deep, reducing flow. In addition, it causes water loss and stagnation in reservoirs and impedes navigation and recreation in rivers and canals. Common reed is a particularly troublesome weed in marshes that have been disturbed by excess sediment from flooding, pollution, and cultivation. It does well in disturbed areas, especially along roadside ditches, railroad tracks, and on dredged soils and mine spoils. Common reed often becomes very aggressive in aquatic areas where the water table is lowered. The ability of the plant to tolerate salinity has allowed it to thrive along roadsides that have been treated with de-icing salt (Mal and Narine 2005). The species is also sometimes a significant weed of vegetables and grains (CABI 2019k). When the above-ground parts of the plant die in the autumn, they become a source of fuel for fires and are a fire hazard where they are close to buildings. Although herbicides are the main control method, the plant is also often mowed or burned. These methods need to be repeated for several years at considerable cost.

Beneficial Aspects

The most significant modern commercial use of common reed in Europe is as thatching for roofing and construction of small buildings (Haslam 1969; Figure 20.29). The most important usage in North America is for phytoremediation (Kiviat 2013), i.e., planted to improve degraded habitats. Common reed has often been used for weaving mats and for cordage. Pens for writing on parchment were once cut and fashioned from the plant. Other miscellaneous uses include animal bedding, upholstery insulation, fencing, musical instruments, fishing rods, basket weaving, and crafts (Figure 20.30). North American Indians used the stems for a very wide variety of purposes, including shafts of arrows, prayer sticks, weaving rods, and pipe stems (Kiviat and Hamilton 2001). The plant has also been employed as

FIGURE 20.28 Rhizomes and stolons (laterally spreading stems that produce new shoots) of common reed. (a) Rhizome. Photo by Kenraiz (CC BY SA 4.0). (b) Portion of a stolon (above-ground stem, capable of spreading very long distances) Photo credit: Dan Engel, Five Rivers Services (public domain).

a source of cellulose for making paper and construction board, especially in Europe. Because of its short fibers, common reed is unsuitable for high-quality paper, but up to 20% can be added to paper products. Common reed is also a potential source of furfural, a major industrial chemical widely used in the chemical, petrochemical, and pharmaceutical industries. The rhizomes have served as raw material for manufacturing alcohol (Cotana et al. 2015). The flowers heads are often a component of dried flower arrangements and have been used as an upholstery

FIGURE 20.29 Use of common reed for thatching. (a) Bundles of harvested stems collected for roof thatching. Photo by Rasbak (CC BY SA 3.0). (b) Ornamental garden building with a thatched common reed roof. Source (public domain): Papworth, J. B. 1823. *Hints on Ornamental Gardening*. London, R. Ackermann. (c) Workers using the stems to thatch a large building in the Netherlands. Photo by E. Dronkert (CC BY 2.0). (d) Thatched-roofed house from the 17th century in Freesenort, Germany. Photo by Global Fish (CC BY SA 4.0).

FIGURE 20.30 Handcrafted (mid-20th century) traditional Norwegian children's shoes, top with braided reeds, bottom made from canvas (cotton). Photo by Anne-Lise Reinsfelt, Norwegian Museum of Cultural History (CC BY SA 3.0).

FIGURE 20.31 Livestock feeding on common reed. (a) Highland cattle in the Isle of Mull, Scotland. Photo by Alex Dixon (CC BY SA 2.0). (b) Horses in the Camargue, southern France. Photo by Hedwig Storch (released into the public domain).

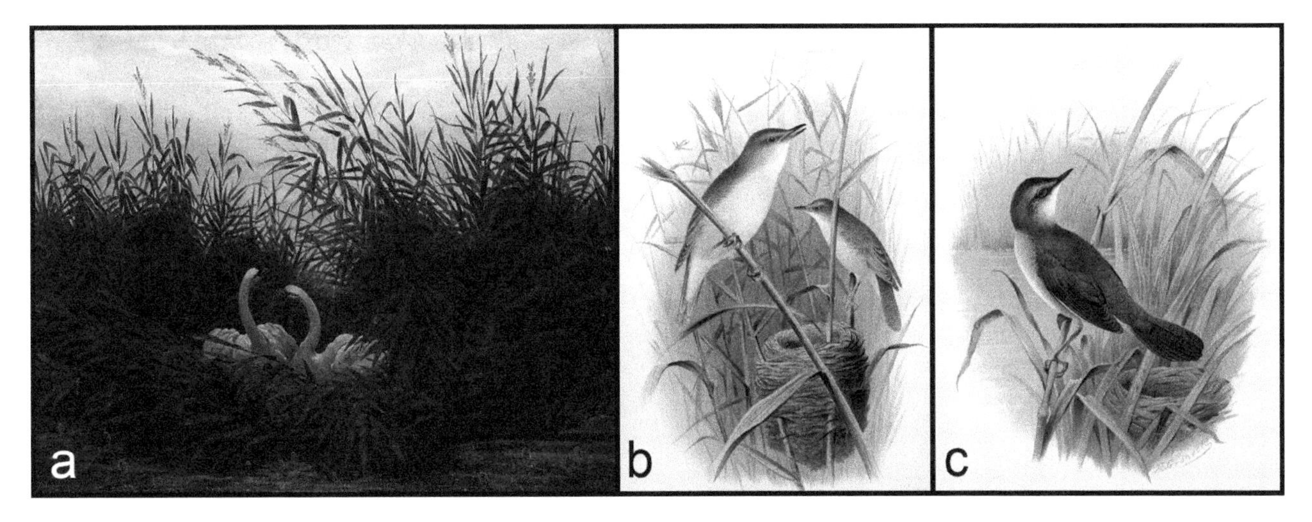

FIGURE 20.32 Birds taking advantage of cover from common reed. Indigenous races of common reed provide important habitat for many birds, although invasive races often reduce needed habitat and harm some birds. (a) 'Swans in the reed with first morning-red' by German landscape painter Caspar David Friedrich (1774–1840; public domain image of the original, housed in Goethe house, Frankfurt am Main, Germany). (b) Reed warblers at their nest, from A. G. Butler, 1907, *Birds of Great Britain and Ireland*, vol. I, plate 13. (c) Savi's Warbler, published in same source as (b).

stuffing. Variegated ornamental cultivars of common reed, with thin gold and green stripes, are sometimes grown as garden plants. Common reed in early growth is palatable to cattle, sheep, goats, and other livestock and natural stands are employed as forage (Duncan et al. 2019; Figure 20.31).

Conservation Aspects

Indigenous forms of common reed provide useful services to wildlife, including hiding cover for deer, rabbits, pheasants and other animals, nesting concealment for wide variety of waterfowl and shoreline birds (Figure 20.32), edible seeds for waterfowl, and nutritious rhizomes for muskrats and nutrias (Tilley and St. John 2012). In Africa, common reed is frequent on river banks and is consumed by large herbivores (Figure 20.33).

In recent centuries, invasive races of common reed have been introduced to North America and Australia from Eurasia and are of considerable concern because they are displacing native forms of the species as well as other aquatic plants and reducing habitat for wildlife (Hocking et al. 1983; Catling and Mitrow 2011; Textbox 20.4). The superior salt tolerance of invasive forms has allowed them to thrive and displace indigenous populations in coastal salt

FIGURE 20.33 Wild mammals grazing reeds in South Africa. Source: Shutterstock. (a) Young hippopotamus. Contributor: Madelein Wolfaardt. (b) African elephant. Contributor: Gaston Piccinetti.

TEXTBOX 20.4 A DANGEROUS WOLF IN AN OTHERWISE SHEEP-LIKE SPECIES

There's a war being waged in the wetlands... The enemy is an aggressive line of the common reed, *Phragmites australis*, with European origins. It is practically identical to a native strain that has existed in North America's wetland plant communities for more than 3,000 years. The major difference between the two is the non-native plant's aggressive proliferation. *Phragmites* can spread at rates of up to 15 feet per year, choking out native marsh plants... Across its invasive range, numerous negative effects of *Phragmites* on marsh ecosystems have been reported, including reductions in plant diversity and in habitat used by threatened and endangered birds... Furthermore, *Phragmites* management generally entails the eradication of existing stands within marshes using fire or herbicides, methods that often require control techniques to be employed for many years after. These efforts are costly, labor-intensive, and potentially dangerous.

—*Theuerkauf (2015)*

The non-native *Phragmites* is a very adaptable plant form with many characteristics that give it a competitive edge over our native plants. It sheds its broad lower leaves creating a thick litter that deprives native vegetation of sunlight and prevents germination by native plants. It conducts chemical warfare by secreting a substance that suppresses other plants from growing in a process similar to the allelochemicals produced by tomatoes, walnuts, and many other plants.

—*Simmons (2013)*

Phragmites invasions are often associated with decreases in plant biodiversity, declines in habitat quality for fish and wildlife, and disruptions to biogeochemical cycles and other ecosystem services.

—*Hazelton et al. (2014)*

marshes (Figure 20.34). The species grows so tall that few plants can outshade it. Hybridization between the indigenous and foreign races is occurring and threatens to alter the genetics of the native populations and generate especially aggressive races (Meyerson et al. 2009, 2010; Paul et al. 2010; Saltonstall et al. 2014; Williams et al. 2019). While some bird species thrive in reed beds, others do not (Parsons 2003). Several studies have demonstrated the harmful effects of invasive forms of common reed on animals (Able and Hagan 2003; Benoit and Askins 1999; Bolton and Brooks 2010). Invasive reed is particularly common in eastern North America, whereas native North American kinds of reed are relatively predominant in western North America. Native North American plants tend to look different from the Eurasian races (in the northeastern U.S. and adjacent Canada, native *Phragmites* usually has a reddish stem, often with black spots, and is smaller in stature with a less prominent seed head). However, botanical expertise is required as the differences are difficult to discern (for identification keys, see Saltonstall et al. 2005,

FIGURE 20.34 A huge salt marsh bed of common reed in the Parker River National Wildlife Refuge, in eastern Massachusetts, seriously invaded by the invasive form of the species. Photo (public domain) by Matt Poole, U.S. Fish and Wildlife Service.

and Tilley and St. John 2012). Most native North American plants have been placed in *P. australis* subsp. *americanus*; most invasive common reed plants (which come from Eurasia) have been labeled *P. australis* subsp. *australis*.

Several uses of the plant contribute to improving the environment, either directly by phytoremediation (restoring degraded habitats by growing appropriate plants) or indirectly by reducing dependence on fossil fuels and industrial agriculture. Common reed is a source of biomass, for production of building materials such as fiberboard, insulation, and similar products, and as fuel, fodder, and agricultural compost (Vaičekonytė et al. 2013/2014; Rezania et al. 2019; Baibagyssov et al. 2020). The species is efficient at removing heavy metals from contaminated water (Vymazal and Březinová 2016; Milke et al. 2020). It is also useful to reduce excess nutrients from water (Nikolić et al. 2014). Common reed is occasionally planted to dry out excessively wet areas, especially near water courses. The cultivar 'Shoreline' is used as a land stabilizer in the southern Central Great Plains of North America, and 'Southwind' has been recommended for streambank and shoreline stabilization, rehabilitation of polluted waters, and for treatment of sewage and sludge in constructed wetlands (Tilley and St. John 2012).

Invasive populations are mainly controlled by herbicides in North America (Packer et al. 2017), which when applied over water can easily endanger biodiversity. Mechanical harvesting (Figure 20.35a) and fire (Figure 20.35b) are also employed (Marks et al. 2014). Plants can be dug up, but this causes severe destruction of wetland soils. Biological control agents to destroy invasive common reed is a tricky issue, because of the danger that these same agents will harm native reedbed populations of the species, about which there is already concern that they are being reduced. The pro-biological agent position is represented by Blossey et al. (2020); the contra-biological agent position is presented by Kiviat et al. (2019).

ENGLISH IVY

Introduction

There are many plant species known as 'ivy' (e.g., Boston ivy, German ivy, poison ivy, Swedish ivy), but the name is most widely understood to apply to species of the genus *Hedera*, especially English ivy (*H. helix*; Figure 20.36). *Hedera* consists of a dozen or more species, but its classification is problematical, and the taxonomy of the genus

FIGURE 20.35 Some control methods for eliminating introduced common reed. (a) A specially built tractor designed to remove common reed from the vicinity of houses in Staten Island, New York, where the late-season dead plants provide tinder for fires. Photo (public domain) by U.S. National Park Service. (b) A controlled burn. Photo (public domain) by Jeremy Smith, U.S. Department of Agriculture.

FIGURE 20.36 English Ivy. (a) Vines climbing on a tree. Public domain photo from Pixabay. (b) An old vine growing against the ruins (bombed during the Second World War) of the Tempelherrenhaus (House of the Templers) in Weimar, Germany. Notice the impressive thickness of the woody stem at its base. Photo by ArtMechanic (CC BY SA 3.0). (c) Adventitious rootlets covering stems. These special small roots serve to anchor the vines to objects. Photo by Bialowieza 2005 (CC BY SA 3.0). d) Variation in leaf shape. The variegated (white/green) leaves are cultivated forms. Photo by Magpie Ilya and Kenraiz (CC BY SA 3.0). (e) Flowering branches. Photo by Jan Samanek, Bugwood.org (CC BY 3.0). (f) Fruits. Photo by NatureServe (CC BY 2.0).

requires considerable study (Green et al. 2011; Strelau et al. 2018). Some of the recognized species are difficult to distinguish (Ackerfield and Wen 2002). *Hedera helix* is widely considered to be the most important species, but it is poorly separable by appearance from *H. hibernica*, and the two are widely confused in the commercial trade. Both (and sometimes other species) are commonly known as 'English ivy,' but *H. hibernica* is also called Irish ivy and Atlantic ivy. On the West Coast of North America, *H. hibernica* is the most common invasive species (Murai 1999; Green et al. 2013), while in eastern North America *H. helix* is more frequent (Clarke et al. 2006). *Hedera helix* is also called 'true ivy' and 'common ivy.' Algerian ivy (*H. algeriensis*) and Canary Island ivy (*H. canariensis*) are additional closely related species believed to also appear in cultivation. When the name 'ivy' is used alone, it usually refers to *H. helix*. Most cultivated varieties of *Hedera* are identified as *H. helix*. *Hedera* species are native to Asia, Europe, North Africa, and Macaronesia (Ackerfield and Wen 2003). In nature, ivies are common in forests, starting growth on the ground and climbing up trees. Ivies also scramble up rocks and cliffs, and if they do not encounter objects on which to grow, they simply scatter laterally to produce a prostrate dense cover on the ground. Whether deliberately planted or not, ivies are also found on the walls of buildings, fences, posts, hedges, shrubs, and trees. The plants may be very long-lived, sometimes reaching an age of over 400 years (CABI 2018). The seeds are spread particularly by birds, but also by other animals.

Hedera species are woody, perennial, branching vines, sometimes growing to lengths (or heights) of over 30 m. Old stems can attain diameters greater than 10 cm (rarely, more than 30 cm) and may develop short trunks (Figure 20.36b). In initial growth, germinated seeds produce stems which scramble laterally over the soil surface, and roots develop at the stem nodes (where leaves originate), anchoring the horizontally growing stems to the ground. These roots absorb minerals and water. When the horizontally growing stems contact above-ground objects like trees, rocks, or man-made objects, they climb vertically. Numerous 'adventitious rootlets' (Figure 20.36c) develop on the vertically growing stems. (In botany, 'adventitious' means growing in an unusual place; in this case, the rootlets [small roots] grow out of stems that are remote from the soil where normal roots are located.) Because the adventitious rootlets develop from aerial stems, they are also termed 'aerial rootlets.' Unlike underground roots, these are unbranched, permanently slim, and do not absorb water and nutrients. Instead, the aerial rootlets anchor the vines to surfaces, clinging with the aid of an adhesive that they produce from a dense covering of very short, tiny root hairs (Xia et al. 2011; Melzer et al. 2012). The evergreen foliage is deep green, leathery, glossy, and often lobed. The stems trailing on the ground are non-flowering and tend to produce 'juvenile' leaves that are noticeably lobed, while climbing and flowering stems produce less-lobed or unlobed leaves. Most cultivated selections are based on the juvenile form, and these may not develop flowers even when climbing upward (Pennisi et al. 2009). The hundreds of cultivars include forms with variegated foliage (with whitish areas) and different leaf sizes and shapes have been selected (Figures 20.36d and 20.37). The veins of the leaves are often whitish, contributing to their attractiveness. Pierot (1974) classified ivy cultivars into eight categories based simply on leaf shape and variegation. After about 10 years of age, upright vines may produce small, yellowish-green flowers

FIGURE 20.37 Paintings (public domain) of ornamental variegated leaf varieties of English ivy. Source (public domain): Hibberd, S. 1871. *The Floral World and Garden Guide*, vol. 14. Figures digitally enhanced by Rawpixel.com.

FIGURE 20.38 Harmful ecological effects of ivy (*Hedera* species) vines smothering trees. (a) In Jungfernheide, a former forest and heathland located in Berlin, Germany. Photo by Rolf Dietrich Brecher (CC BY SA 2.0). (b) On a leafless tree in Connecticut. Photo by Leslie J. Behrhoff (CC BY 3.0). (c) Removing invasive ivy by hand labor (identified as 'probably *Hedera hibernica*') near Seattle, Washington. Photo by Everyguy (released into the public domain).

in clusters (Figure 20.36e), and these mature into small purplish or black (occasionally yellowish or orange) berries 5–10 mm wide (Figure 20.36f). The degree to which given species and cultivated forms develop physical and physiological characteristics differs appreciably. For example, some kinds of *Hedera* are non-climbing.

Two acrimonious debates have developed regarding *Hedera* ivies: Are they bad for biodiversity, even in their native areas, or do they contribute significantly to ecosystem functions? Second, do ivies planted on buildings and other structures physically deteriorate them, or are there insulation benefits that contribute to energy efficiencies and planetary climate control?

Harmful Aspects

Ivies have escaped from cultivation in widespread areas of the world, where they have become significant alien invasives, condemned by many ecologists (Small 2019a). *Hedera* species are considered detrimental to forests, especially in non-native areas (Waggy 2010; Castagneri et al. 2013; CABI 2018; Strelau et al. 2018). They 'smother' trees (Figure 20.38a and b), reducing their survival, and are especially harmful to deciduous trees as the vines can carry on photosynthesis while the trees are dormant, thereby growing faster (in areas with cold winters, ivies may simply be killed by frost). When vines develop mainly on one side of trees, the latter become susceptible to wind fall from the uneven distribution of weight (ivy vines can weigh as much as a ton). Ivies invade urban natural areas, woodlands, the sides of stream corridors, and semi-open and deeply shaded forests (Strelau et al. 2018). It has been suggested that in North America the weight of water and/or ice on the evergreen ivy leaves may increase storm damage (Reichard 2000). The dense ivy cover sometimes formed either over open ground or in forests can reduce native biodiversity, and such areas have been termed 'ivy deserts' (Okerman 2000; Strelau et al. 2018). In California, invasive ivy harms the northern spotted owl (*Strix occidentalis caurina*, 'endangered' or 'threatened' in some regions; Armanino 2017) by so densely covering the ground that prey are hard to find. In North America, invasive ivy serves as a reservoir for bacterial leaf scorch (*Xylella fastidiosa*), a pathogen that harms elms, oaks, maples, and other native plants. Rodents commonly nest in dense ivy vines on the ground, which is fine for the rodents, but considered objectionable when this occurs near buildings.

Destruction of ivy that has become established in areas as an invasive is difficult. The most common eradication technique is simply removal by hand (Figure 20.38c), which is effective with small patches, but is very labor-intensive, expensive, and difficult in large infestations. Recommendations for removing invasive ivy are provided by Soll (2005) and McQueeney (2018).

Ivy-covered buildings are icons of upper-class status and quality (Figure 20.39; note that species other than *Hedera* ivies are often employed on the walls of buildings). Nevertheless, *Hedera* ivies are one of the vines thought to endanger buildings, especially those of historical value (Mishra et al. 1995; Bartoli et al. 2017; Figure 20.40a).

FIGURE 20.39 Impressive ornamental use of ivies. Note that the species have not been identified, and may include vines other than *Hedera*. *Hedera* ivies usually cannot be reliably identified just from photographs of buildings. Boston ivy (*Parthenocissus tricuspidata*) is frequently grown, and its foliage is reminiscent of English ivy leaves. (a) University library, Lund, Sweden. Source: Shutterstock, contributor: Canbedone. (b) Wrigley (baseball) Field, Chicago, showing center field wall with 400-foot home-run marker. Boston ivy is the predominant plant. Photo by Jimcchou (CC BY 2.0). (c) The Empress Hotel in Victoria, British Columbia. Photo by TMAB2003 (CC BY ND 2.0). (d) Beautifully ivy-covered building in Buenos Aires. Photo by Sergejf (CC BY 2.0).

FIGURE 20.40 Weedy ivy (*Hedera* species) growing on deteriorating structures. As discussed in the text, ivy has been described as both good and bad for preservation of buildings and monuments. (a) Ivy-clad ruin of ancient church at Freneystown, United Kingdom. Photo by Kevin Higgins (CC BY SA 2.0). (b) Ivy running rampant in a neglected graveyard. Public domain photo from Pixabay.

FIGURE 20.41 Outdoor ornamental uses of English Ivy (*Hedera helix*). (a) As a ground cover in Germany. Photo by Presse03 (CC BY SA 3.0). (b) As a building cover, of a house in Weener, Germany. Photo by Frank Vincentz (CC BY SA 3.0).

The adventitious roots that are produced to anchor them to objects can penetrate cracks in brick and stone walls, expanding openings to allow penetration by water (and ice), insects and molds. High-climbing vines can block gutters and lift roof tiles. Potentially serious structural faults can be hidden by ivy, delaying needed repairs. When ivy vines are removed, they can leave unsightly remains on exterior walls, necessitating expensive resurfacing. Neglected graveyards can become quite unsightly when overrun by ivy (Figure 20.40b). Unwelcome pests can nest in ivy.

All parts of *Hedera* ivies are toxic to humans when consumed and contact with skin can produce dermatitis in sensitive individuals (De Smet 1993; Paulsen et al. 2010), but reports of negative reactions by people are uncommon (Burrows and Tyrl 2013). Cows, sheep, and dogs have been poisoned by consuming the stems and/or foliage, although ivies are not considered to be a significant danger to domesticated animals. Some deer species in Europe graze on indigenous ivy vines, although in North America ivies are considered deer- and rabbit-resistant. Most domesticated mammals (including cattle and sheep) avoid ivies, although goats have been observed to consume the foliage and have even been employed as a control measure (Ingham and Borman 2010).

Beneficial Aspects

Hedera ivy species were universally admired ornamental vines until several decades ago when the problems discussed above came to light. Nevertheless, most horticulturalists still defend their use. The attractive, evergreen, lustrous foliage (turning red in the autumn in some selections), and the ability to climb smooth surfaces to considerable heights have resulted in ivies becoming very popular outdoor ornamentals (McAllister and Marshall 2017; Figure 20.41). Although not particularly cold-tolerant (Castagneri et al. 2013), *Hedera* species are hardy in warm temperate areas, very vigorous, easily propagated, drought-tolerant, and shade-tolerant. They grow best in full sun but can survive in as little as 3% of sunlight (Metcalfe 2005). Ivies are very attractive, cheap, and require little maintenance, but if not kept in check they can become invasive. The invasive potential is a major impediment to the industry, and in some jurisdictions, sales are regulated.

Ivies are also popular houseplants (Figure 20.42), often grown in hanging baskets (Figure 20.42c) and in planters with mixed species. Cultivated forms with short internodes (the portion of stem between nodes, where leaves are attached) are particularly suitable for indoor cultivation. Ivies are reputed to improve air quality by removing volatile organic compounds (Yang et al. 2009). Formaldehyde is the most common volatile organic compound emitted in households, and potted ivy has been demonstrated to significantly decrease its concentration (Lin et al. 2017).

Ivies are also used by the florist industry as greenery in floral arrangements. Fake (i.e., plastic) ivy vines for indoor display are often encountered, although true lovers of plants should regard artificial plants as abominations. Ivy wood has been used to fashion curios but is too limited to be significant. Ivies can be trained to grow on structures and are often used in topiary (Figure 20.43). Ivy growing over wire frames in the shape of animals is particularly attractive (Figure 20.43b).

Some authors have alleged that ivy (and other vines) growing on walls protects buildings from excessive sun, rain, temperature fluctuations, chemicals, and other environmental stresses (Rose 1996; Sternberg et al. 2010). Ivy-covered walls of residences can be cooler in the summer

FIGURE 20.42 Indoor ornamental uses of English Ivy (*Hedera helix*). (a) Cultivated in a pot as a houseplant. Public domain photo from Pixabay. (b) Bonsai. Photo by Jerry Norbury (CC BY ND 2.0). (c) Hanging pot. Source: Shutterstock, contributor: Amnuay Kaewkatmanee.

FIGURE 20.43 Topiary employing English ivy. (a) Fireplace made entirely of living plants, including English ivy, at the Centennial Park Christmas Show in Toronto in 2010. Photo (public domain) by Torontofiredancer. (b) Ivy dinosaurs in Park Terra Nostra in Azores. Source: Shutterstock, contributor: Will eye.

because of intercepted sunlight (Cuce 2017), and warmer in the winter because of the insulating effect of the vegetation (Sternberg et al. 2011). It seems that the insulating effect of ivy can protect masonry from damaging frosts (Coombes et al. 2018). Sulgrove (2002) provided a list of publications contending that ivies are either good or bad for structures, and it appears that there is by no means a unanimous viewpoint on this subject.

Despite the toxicity, ivy preparations have been employed in herbal medicine, both internally (especially for respiratory conditions) and externally (especially for skin conditions), to treat a wide variety of disorders (Lutsenko et al. 2010; Al-Snafi 2018). Ivy continues to be used as a natural remedy for respiratory illness (Rehman et al. 2017), although the efficacy has been challenged (Holzinger and Chenot 2011). Bees visiting ivy flowers produce an excellent honey which is apparently non-toxic (CABI 2018).

Conservation Aspects

As noted above, invasive ivies are seriously detrimental to forests. Nevertheless, they provide some support for wildlife. A wide variety of insects visit the flowers for nectar and pollen (Figure 20.44a and b). The berries are bitter and poisonous to humans but are extensively consumed by birds (Figure 20.44c). Indeed, the seeds usually need to be abraded by passing through a bird's digestive system before they will germinate (Reichard 2000). Well-developed leafy vines can provide nesting habitat for birds (Figure 20.44d) and rodents.

Secondary compounds in ivy foliage are potential natural product pesticides for insects, which might be of benefit in reducing the use of toxic synthetic pesticides. Ivy has been employed to control soil erosion along highway embankments and medians (Strelau et al. 2018), but while

FIGURE 20.44 English ivy (*Hedera helix*) as food for wildlife. (a) Red admiral butterfly (*Vanessa atalanta*) feeding on ivy flowers in Great Britain. Photo by Andrew Curtis (CC BY SA 2.0). (b) The ivy bee (*Colletes hederae*) feeding on ivy flowers in Germany. This European solitary bee forages mainly on nectar and pollen from ivy flowers. Photo by Pjt56 (CC BY SA 4.0). (c) Woodpigeon (*Columba palumbus*) feeding on ivy berries. Photo by MPF (CC BY SA 3.0). (d) A blackbird (*Turdus merula*) nest in ivy vines. Photo (public domain) by Max Pixel.

it quickly provides cover, the shallow roots are not efficient at retaining the ground. Moreover, invasive ivy may actually increase erosion in some circumstances when native species with superior soil-binding properties are displaced.

PURPLE LOOSESTRIFE

Introduction

Purple loosestrife (*Lythrum salicaria*) is a perennial herb, usually 0.5–1.5 m tall but occasionally exceeding 2 or even 3 m in height. The stems become somewhat woody and the plant may appear shrub-like (Figure 20.45). The very attractive, purplish flowers occur in elongated spikes (Figure 20.46). Charles Darwin (1864) described three forms of the flowers differing in height of the stigma (pollen-receptive top of the pistil; the pistil is the female part of the flower), and this 'sexual polymorphism' has since generated hundreds of research papers. The plants are 'tristylous' (Figure 20.46a), that is the flowers' styles (the stalk between the ovary and the stigma) occur in a given plant in one of three different lengths (long, medium, or short), and the

FIGURE 20.45 Clump of purple loosestrife. Source: Shutterstock, contributor: Antero Topp.

same is true of the stamens (the male parts of the flower), and pollination is more successful when the sexual organs are of the same height. A single plant often produces more than 1 million seeds (Figure 20.47). This native of Eurasia has been distributed to much of the world. It thrives in low-lying coastal areas, wet marshy places, moist pastures, stream banks, ditches, and flood plains (Figures 20.48 and 20.49). As many as 50 (but usually less than 15), aboveground stems arise from the perennial rootstock. These are killed by frost in cold climates, new stems growing from the base in the spring. Where fireweed (*Epilobium angustifolium*) occurs, it is frequently confused with purple loosestrife (both species produce conspicuous massed displays of blue or purplish blooms).

FIGURE 20.46 Flowers of purple loosestrife. (a) Illustration of flowering loosestrife, showing the three kinds of flower. Source (public domain): Thomé, O. W. 1885. Flora von Deutschland, Österreich und der Schweiz, Gera, Germany, Kohler. (b) Close-up of flowering stem. Photo by Christian Fischer (CC BY SA 3.0). (c) Close-up of flowers. Photo by H. Zell (CC BY SA 3.0).

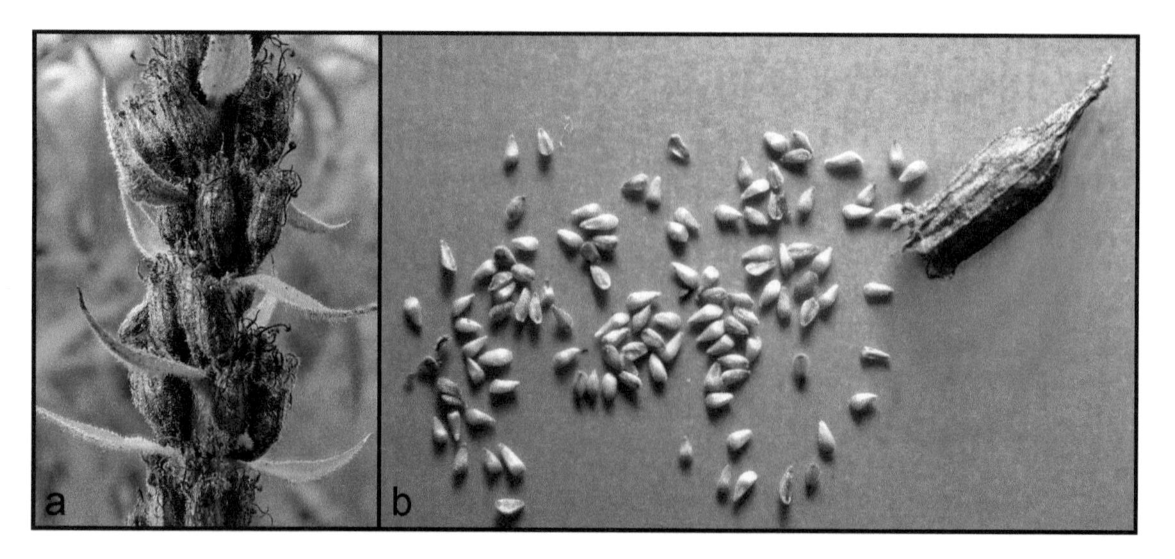

FIGURE 20.47 Seeds of purple loosestrife. (a) Stem bearing seed pods. Photo by Krzysztof Ziarnek, Kenraiz (CC BY SA 4.0). (b) Pod broken open spilling some of the numerous seeds. Photo by Gary L. Piper, Washington State University, Bugwood.org.

FIGURE 20.48 Purple loosestrife in shallow water bordering ponds and streams in Massachusetts, USA. Photos by Liz West (CC BY 2.0).

FIGURE 20.49 Purple loosestrife in flood plains. (a) In Oregon. Photos by Eric Coombs, Oregon Department of Agriculture, Bugwood.org. (b) In Massachusetts. Photos by Liz West (CC BY 2.0).

Harmful Aspects

Purple loosestrife outcompetes many native plants in wetlands, such as cattails, sedges, rushes, and other shallow water plant species. It also degrades habitats for wildlife, notably amphibians and birds (Mitich 1999; CABI 2021c). Lavoie (2010) argued that the harm caused by purple loosestrife has been greatly exaggerated. Rogers et al. (2022) concluded that 'Overall, the consensus is that purple loosestrife has limited value in the ecosystem as a source of nutrients or habitat for wildlife and disrupts a functional ecosystem.' Most reviews have concluded that the species is one of the worst invasive plants in the world. It reduces nesting habitat for some waterfowl, although Tavernia and Reed (2012) found that while the plant harms some bird species, it benefits others. Large stands can clog irrigation canals, reduce water flow, inhibit transportation, and degrade hunting and fishing areas. As noted next, while it is somewhat palatable, it reduces the forage value of pastures. Several insects have been investigated as biological control agents (Blossey et al. 2015). In 1992, the Canadian and American governments approved the release of two European specialist leaf-eating beetles, *Galerucella calmariensis* and *G. pusilla*, to reduce the populations of invasive weedy purple loosestrife. The insects have been somewhat effective in controlling the plant (St. Louis et al. 2020).

Beneficial Aspects

Purple loosestrife was considered to be an excellent garden ornamental in the past, and some of the horticultural varieties are still used for this purpose (Figure 20.50) and are sold in some nurseries (sale is prohibited in many jurisdictions). It is unclear whether cultivated varieties significantly escape to the wild and so contribute to the damaging environmental effects of the species. The huge flower production provides considerable nectar and pollen for honeybees and other insects (Figure 20.51), and so purple loosestrife has frequently been described as an excellent honey plant, although the honey is

FIGURE 20.50 Cultivated purple loosestrife. Source: Shutterstock. (a) Purple loosestrife in an ornamental artificial pond. Contributor: Nadezda Verbenko. (b) A garden cultivar. Contributor: Dan Gabriel Atanasie.

FIGURE 20.51 Insects visiting purple loosestrife flowers. (a) Syrphid fly, in Germany. Photo by Dirk Paehr-Heine (CC BY ND 2.0). (b) Cabbage white (*Pieris rapae*), an invasive butterfly, in New York State. Photo by Manjith Kainickara (CC BY SA 2.0). (c) Shrill carder bee (*Bombus sylvarum*), in Northwestern Estonia. Photo by Ivar Leidus (CC BY SA 4.0).

not considered to be of high quality (Mal et al. 1992). The young foliage is consumed by wild and domesticated grazing animals, but the mature plant is much less frequently browsed. The plant has a long history of use in traditional herbal medicine and has potential modern therapeutic usages (Piwowarski et al. 2015; Al-Snafi 2019).

Conservation Aspects

Purple loosestrife is an extremely competitive invasive plant that reduces the presence of natural plants and animals in many areas of the world. Its benefits to some species are outweighed by harm to many others (Munger 2002b), and the extremely extensive and costly efforts that have been made to eradicate or at least control its presence reduced funds from environmental organizations that could have been otherwise employed for conservation.

WATER HYACINTH

Introduction

The often-misspelled genus *Eichhornia* was named in honor of J. A. F. Eichhorn (1779–1856), a Prussian statesman. The genus contains about six perennial aquatic species which are indigenous to tropical and subtropical South America, and one species native to Africa. *Eichhornia crassipes* (common water hyacinth or just water hyacinth; Figure 20.52), the most significant species, is native to the Amazon Basin. The key monograph on water hyacinth by Gopal (1987) listed thousands of scientific papers on the plant, and since then, thousands more have been published. For this review, only representative recent publications are cited. The species is highly invasive and has become established in southern North America, Africa, Asia, Australia, New Zealand, and elsewhere. Water hyacinth has been found to be substantially uniform genetically, and indeed most of the world's plants are genetically identical (Zhang et al. 2020b). As much as 95% of a water hyacinth plant is water (Degaga 2018), and the species uses water as its habitat. The plant is mostly free-floating by means of buoyant swollen inflations at the base of the leaf stalks (Figure 20.53). The feathery roots hang down in the water, but in shallow areas

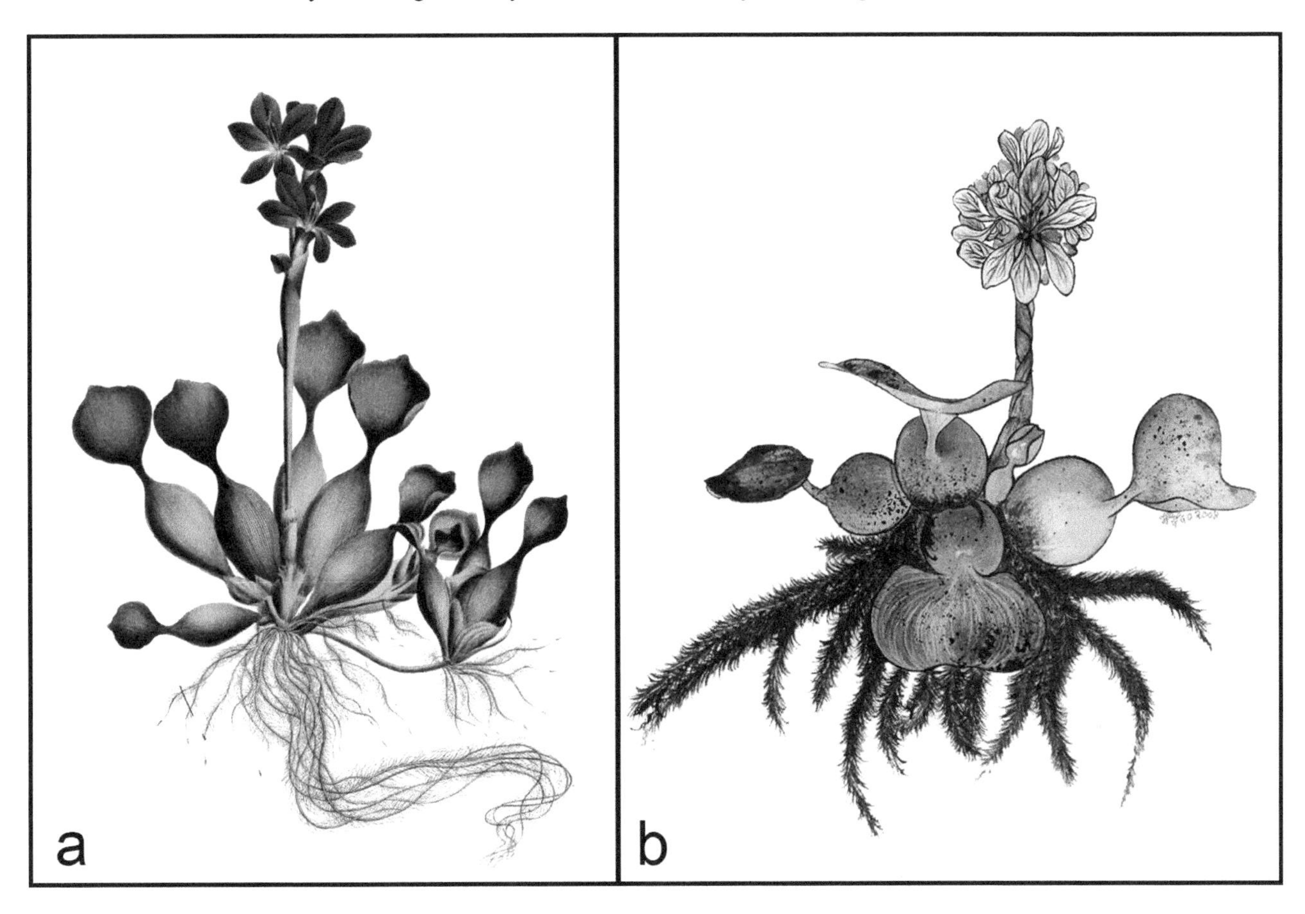

FIGURE 20.52 Paintings of water hyacinth (*Eichhornia crassipes*). (a) Image from Martius, C. P. F. von. 1823. Nova Genera et Species Plantarum Brasiliensium, 1: 9, plate 4, Munich, Lindaueri. (b) Painting by Ozhe (CC BY SA 3.0).

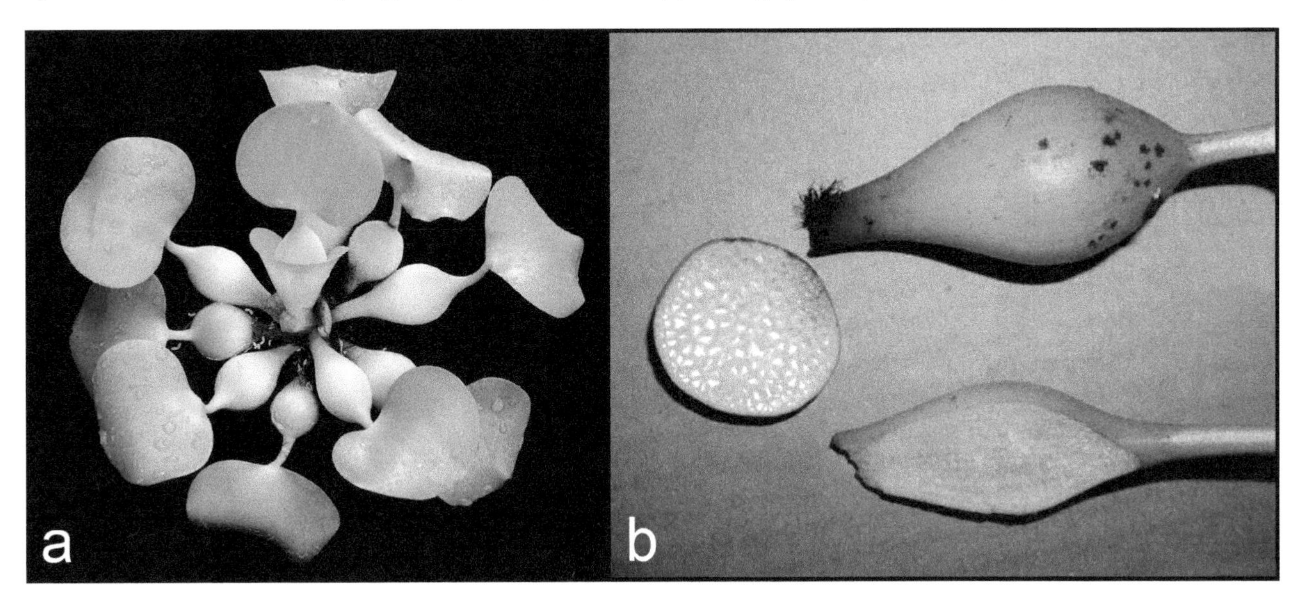

FIGURE 20.53 Buoyant swollen inflations at base of leaf stalks of water hyacinth. (a) Rosette of leaves. Photo by Amada44 (CC BY 3.0). (b) Views of an unsectioned and two sectioned inflations (CC BY 3.0).

FIGURE 20.54 Researcher with several separated shoots of water hyacinth connected by stolons (laterally spreading stems; these produces shoots, which can eventually develop into independent plants). Photo by Leslie J. Mehrhoff, University of Connecticut, and Bugwood.org (CC BY 3.0).

they frequently anchor in mud. The plant produces a flowering stalk with up to 15 attractive flowers, which are usually lavender or pink, with a yellow patch at the center. Barrett (1988) noted that considerable variation of floral size and structure is known in the species. The top of the stalks can rise to a height of more than 1 m above water level. Water hyacinth produces stolons (also called runners, and sometimes also termed rhizomes) that develop new shoots at regular intervals of 30 cm or less (Figure 20.54). These generate attached small plants which eventually become independent large plants. The species also develops numerous long-lived seeds that germinate under water. Water hyacinth requires a minimum water temperature of 12°C to grow well and can just survive minimal frost (Bowles and Bowles 2013). It tolerates slightly salty (i.e., brackish) water.

Harmful Aspects

During the 19th century, water hyacinth was brought to numerous countries around the world to be grown as an ornamental in aquaria and ponds (Williams 2005; Figure 20.55). It escaped in many locations and now is widely considered to be the most damaging aquatic weed of the tropics (many consider it to be the world's worst aquatic plant; note Figure 20.56). Water hyacinth is now banned in most countries where it is not native, so that it is no longer planted deliberately (except occasionally in quite northern areas where it cannot survive outdoors). Its ability to cover water surfaces and clog rivers and ponds (Figures 20.57 and 20.58) as well as drainage and irrigation channels seriously compromises water transportation, fishing, farmland irrigation, hydro-electric

FIGURE 20.55 Nineteenth-century garden catalogue illustrations (public domain) of water hyacinth cultivated in containers of water. (a) Painting from John Lewis Childs, Inc. catalogue (1892). (b) A widely reproduced drawing of water hyacinth in garden catalogues of the 19th century, a time when the plant was viewed as a harmless ornamental for aquaria and ponds.

operations, and recreation (CABI 2019i; Yigermal and Assefa 2019; Ayanda et al. 2020). Rice production in paddies is also significantly reduced when water hyacinth invades. In some places, clogged waterways result in flooding that is harmful to human activities and may also be bad for ecosystems. Water hyacinth can also reduce water quality. The stagnant water that often result can affect human health by promoting the breeding of disease-transmitting flies, malarial mosquitoes, snails that are associated with schistosomiasis, and other organisms linked to human illnesses such as encephalitis, filariasis, and cholera (Degaga 2018). In Africa, increased attacks by crocodiles have been attributed to the heavy cover provided by water hyacinth, which also hides dangerous poisonous snakes and hippos (Patel 2012; Figure 20.59), the last-mentioned considered to be Africa's leading wild large killer of humans.

Beneficial Aspects

As water hyacinth has invaded various countries, people have learned to use it for a variety of purposes and are also exploring how to harvest it for new purposes. The species is employed as animal feed (Gunnarsson and Mattsson-Petersen 2007). It contains needle-shaped crystals of calcium carbonate which can damage intestinal tracts, although the plants are nevertheless fed to a wide variety of livestock (Su et al. 2018). The dried stalks and extracted fibers are useful for making handicrafts (Rakotoarisoa et al. 2016; Figure 20.60). The plants are sometimes employed as a medium to cultivate mushrooms (Mukhopadhyay 2019) and earthworms (Ogutu 2019). Other suggested uses include fertilizer or compost (Malik 2007; Montoya et al. 2013), charcoal and fuel briquettes (Rodrigues et al. 2014), pulp

FIGURE 20.56 A water hyacinth invasion in Africa, harming wildlife and people. Prepared by Jessica Hsiung.

FIGURE 20.57 Water bodies clogged with water hyacinth. (a) A ferry wading through water hyacinth in a canal in Alppuzha, India. Photo by P. K. Niyogi (released into the public domain). (b) Hartbeespoort Dam, South Africa, covered by floating water hyacinth. Photo by Olga Ernst (CC BY SA 4.0).

FIGURE 20.58 Extensive infestations of water hyacinth on waterways was reported to prevent these schoolchildren from going to school by canoe in Benin, Africa. Photo by Degan Gabin (CC BY SA 4.0).

FIGURE 20.59 Hippo (*Hippopotamus amphibious*) emerging from a pool covered by water hyacinth in Mikumi National Park, Tanzania. Photo by Muhammad Mahdi Karim (GFDL).

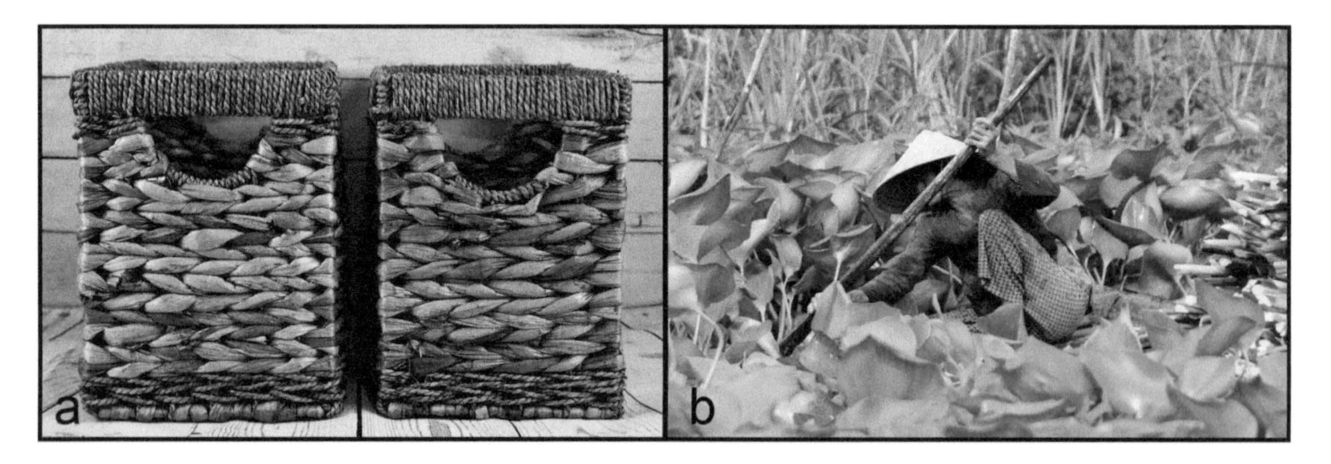

FIGURE 20.60 Use of the fibrous stems of water hyacinth to prepare handicrafts. (a) Baskets woven from water hyacinth. Source (public domain): Pixabay. (b) Harvesting water hyacinth in Vietnam for handicrafts. Notice the bundle of harvested plants at right. Photo from www.all-free-photos.com (CC BY SA 2.5).

and paper, construction materials (Ajithram et al. 2020), and extraction of industrial chemicals (Sindhu et al. 2017).

Inle Lake, the second largest lake in Myanmar (Burma), has become seriously infested by water hyacinth. In recent decades, farmers cleverly harvested the mats and employed them to build and anchor the so-called 'floating gardens' of the lake, on which they grow vegetables (Figure 20.61).

FIGURE 20.61 Very young farm workers in a boat with their harvest of vegetables beside one of the 'floating gardens' constructed with water hyacinth on Inle Lake, Myanmar. The big bamboo poles anchor the floating mats to the bottom mud, while allowing the mats to bob up and down as the water level changes. Photo credit: Piktour UK (CC BY 2.0).

Conservation Aspects

Water hyacinth is widely found as floating mats that are monotypic (with only one species), or as anchored extensive colonial growths (Figure 20.62) – wild counterparts of monocultured crops in agriculture, which by their nature strongly tend to exclude most other species. Indeed, water hyacinth can reduce aquatic biodiversity (Villamagna and Murphy 2009). Fishing can be made very difficult, which is good for the fish, but other changes to the environment can be quite detrimental. The thick floating mats can drastically reduce oxygen in the water to levels that some fish species cannot tolerate. When the plants die, they decay and enrich the water (the process known as eutrophication), but excessively, and the resulting increases in algal growth, water turbidity, and sedimentation make life difficult for other species. In some situations, evapotranspiration from the plants increases greatly (Textbox 20.5), sometimes reducing aquatic habitats. When herbicides are used to control water hyacinth, non-targeted species are also often affected. A wide variety of biological control agents have been tested, with variable success (Harley 1990). To date, the two known species of *Neochetina*, which are South American beetles know as water hyacinth weevils because both the adults and larvae feed almost exclusively on the plants, have been the most successful controlling organisms (CABI 2019d;

FIGURE 20.62 Extensive monocultural growth of water hyacinth (notice that all other plant species are excluded). (a) A colony on a riverbank in Lumbini, Nepal. Photo by Alexey Komarov (CC BY SA 4.0). (b) A floating mat in Japan. Photo by Gyoda_Suijo_Park (GFDL).

TEXTBOX 20.5 ECOLOGICAL AND BIODIVERSITY DAMAGE CAUSED BY WATER HYACINTH

Eichhornia crassipes forms dense mats that reduce light transmission to submerged plants and competes with other plants, often displacing wildlife forage and habitats, depletes oxygen in aquatic communities resulting in a lack of phytoplankton and an alteration of the composition of invertebrate communities, ultimately affecting fisheries. It reduces the area available for water birds and harbors mosquitoes and other animal disease vectors. *Eichhornia crassipes* increases evapotranspirative water losses, with estimates varying from 2.67 to 3.2 times more from a mat of *E. crassipes* compared to open water. This effect is of particular relevance in regions that suffer chronic or seasonal droughts (e.g., Mediterranean or Wet-Dry tropics). In Spain and Portugal as in other parts of the world, impacts have been noted in fisheries, recreational water sport, and boat navigation.

—*Kriticos and Brunel (2016)*

FIGURE 20.63 The two known *Neochetina* species, *N. eichhorniae* (left) and *N. bruchi* (right), feeding on water hyacinth foliage. These two weevils (which are beetles) are among the leading biological control agents that have significantly reduced water hyacinth populations. Photo by Willey Durden, USDA Agricultural Research Service, Bugwood.org (CC BY 3.0).

Figure 20.63). Mechanical removal is also widely practiced (Figure 20.64), but eradication is extremely difficult once water hyacinth has become established. At least mechanical harvesting provides considerable material, which as noted previously does have potential uses.

On the positive side, water hyacinth has the potential to remove pollutants from water contaminated by various activities, such as domestic wastewater, agricultural runoff, livestock wastes, and industrial effluent (Mishra and Maiti 2017; Dahake and Hedaoo 2018). It also has potential for the production of biodiesel fuel (Venu et al. 2019), bioethanol fuel (Das et al. 2016), and biogas (Rozy et al. 2017), all of which can reduce the harvest and burning of fossil fuels.

FIGURE 20.64 Removal of water hyacinth by hand labor in Sacramento, California. Source: California Department of Fish and Wildlife (CC BY 2.0).

21 In Defense of the World's Most Despised Urban Weeds

INTRODUCTION

The Urban War between People and Plants

Our simian ancestors lived in trees, so wild plants literally constituted their domiciles. However, the first priority wherever humans move is to clear the land of vegetation so homes can be built. Literally, we evict plants from their natural habitats. Most plants, indeed most species, are incapable of fighting back, but many do. On the outskirts of urban areas, plants are constantly sending seeds in to try to re-establish. Some (as discussed in this chapter) manage to thrive in the urban theater despite the best efforts of mankind to eradicate them. Literally, there is a constant war being waged by plants to re-occupy areas usurped by people, and without regular 'weeding,' the plants at least partly succeed (Figure 21.1).

Most wild plants that grow in urban settings are more or less harmless, and many are useful in various respects – as wild foods, soil stabilizers, erosion control, sustenance for honey bees and other pollinators, carbon storage to lower atmospheric pollution, and for temperature moderation (particularly by trees). The main objection to most weeds growing in cities is that 'spontaneous urban vegetation often leaves something to be desired from the aesthetic point of view, and most people perceive it as ugly or messy and an indication of neglect' (Del Tredici 2020). The appearance of residential properties is a key determinant of their financial value, most businesses view weeds on their property as an admission of low quality, and city officials know that both taxpayers and tourists will view weedy parks and streets as indications of failing management.

The Biased Human Perspective of Urban Weeds

Long before there were humans, the plant world occupied every habitable part of the planet. We humans have progressively uprooted and displaced them from their natural homes, and we continue to destroy them when they attempt to return. It seems to be part of human nature that we easily find philosophical justifications for killing other species, but using pejorative terms like 'weed' and 'pest' reflects our biases. The case has been made that, ironically, humans are the most successful of weeds (Ketcham 2010).

Urban weeds are simply wild plants that people dislike when they grow in their yards, gardens, parks, and cities. 'Weeds' are undesirable plants (Textbox 21.1), but there are few species that do not have at least some redeeming virtues (Mabey 2010). Philosopher/essayist Ralph Waldo Emerson (1803–1882) famously wrote 'What is a weed? A plant whose virtues have not yet been discovered.'

Although weeds of commercial crops and natural wildlands cause much more economic and environmental damage, urban weeds are far more visible to most people, and therefore are responsible for generating negative attitudes toward nature. In most cases, urban weeds are merely an aesthetic issue, but since the majority of society values 'clean' landscapes, huge expenditures and resources are dedicated to their elimination. Urban weeds are often also 'environmental weeds' (degrading habitats, ecosystems, and biodiversity) and/or 'agricultural weeds' (harming crops and livestock), so placing them in a given category depends on where they are most significant. There are numerous species that are particularly significant weeds in urban areas (Orlando 2018).

Lawns and Lawn Weeds

Lawns are the most popular form of gardening in the Western world. Lawn weeds are a subdivision of urban weeds. The most hated of lawn weeds – crabgrass and dandelion – generate fanatical efforts at eradication. Unfortunately, conventional turf lawns are a significant source of pollution and harm for the natural world. However, much more sustainable and eco-friendly groundcovers are available, and it is gratifying that social acceptance of their cultivation is growing. As will be noted, even the worst urban weeds have redeeming properties, and where control is necessary, management strategies need to be based on ecological principles.

For most people, owning a house is a priority, and in Western culture, houses are usually accompanied by lawns. Many homeowners have acquired a fanatical hatred of lawn weeds, and this is the basis of a critical environmental issue. Lawns represent a significant source of pollution and associated harm for the natural world and its living inhabitants, particularly because of efforts to eliminate the weeds by chemical agents. Especially at risk are beneficial soil organisms, pollinators, birds, and aquatic animals (not to mention pets and children).

A green, attractive lawn is widely viewed as an aesthetic achievement reflecting social status and taste (Jenkins 1994; Figure 21.2). At least, this was true in past decades, although in recent times lawns have become the subject of attack by ecologically conscious environmentalists, concerned with unintended harmful consequences. The chief source of harm associated with lawns is the use of pesticides (including insecticides, herbicides, and fungicides). Lawn care agencies typically apply herbicides (both pre-emergence and

DOI: 10.1201/9781003412946-21

FIGURE 21.1 Urban war waged by plants to occupy urban areas. Source of Photos: Shutterstock. (a) Weeds growing over an abandoned bicycle. Contributor: Jeanette Dietl. (b) Weeds growing over an abandoned car. Contributor: Sorapong Chaipanya. (c) Strangler fig growing over ruin of Prasat Pram in Koh Ker, the lost capital city of Cambodia, built in the 10th century. Contributor: Haklao. (d) Ruined overgrown apartment house, the consequence of war in the Republic of Abkhazia (Georgia). Contributor: Vladimir Mulder. (e) Houses overgrown with bushes and trees in an abandoned village in the Chernobyl exclusion zone in Belarus (the result of the nuclear disaster in 1986). Contributor: Intreegue Photography.

TEXTBOX 21.1 'WEEDS': AN ARBITRARY PEJORATIVE TERM

From a utilitarian perspective, a weed is any plant that grows by itself in a place where people do not want it to grow. The term is a value judgment that humans apply to plants we do not like, not a biological characteristic. Calling a plant a weed gives us license to eradicate it. In a similar vein, calling a plant invasive allows us to blame it for ruining the environment when really it is humans who are actually to blame. From the biological perspective, weeds are plants that are adapted to disturbance in all its myriad forms, from bulldozers to acid rain. Their pervasiveness in the urban environment is simply a reflection of the continual disruption that characterizes this habitat. Weeds are the symptoms of environmental degradation, not its cause, and as such they are poised to become increasingly abundant within our lifetimes.

—*Del Tredici (2020)*

FIGURE 21.2 Spectacular expansive lawns. (a) The White House in Washington, D.C. Photo by Daniel Schwen (CC BY SA 3.0). (b) The Palace of Versailles, about 19 km west of Paris. Source: Shutterstock, contributor: Elena Elisseeva. (c) State capitol, Saint Paul, Minnesota. Source: Shutterstock, contributor: Randy Runtsch.

post-emergence) on turf to eliminate weeds. Many companies also care for ornamentals that are strategically placed to complement lawns. Flowering plants are treated primarily with different insecticides than employed on lawns, and the result is that a wide variety of different chemicals are applied in a generally small residential area. Fungicides are also often applied to both grass and flowering plants. The second class of chemical for which lawn care companies are usually responsible is fertilizers, and since nitrogen is the principal needed element to make grass grow vegetatively (i.e., to produce more grass), high-nitrogen synthetic fertilizers are favored for ease of application. In many jurisdictions, only licensed personnel are authorized to employ the more toxic chemicals. Professional lawn care personnel are trained to employ minimal amounts of pesticides and fertilizers (both as an ethical practice and simply not to waste expensive materials). Just the same, urban homeowners still have access to a very wide range of commercial chemicals and are motivated to use them by ceaseless advertising campaigns. Rarely does the average homeowner have either the training to use potentially toxic chemicals properly, or the equipment to apply them precisely, and even when the attempt is made to follow label instructions, excess material is frequently dispersed. The inevitable result of application of lawn chemicals is harm to unintended target plants and animals. Indeed, based on equivalent areas, lawns use considerably more chemicals than does farmland (Coates 2004). In addition, lawns need water, often in very large amounts. It has been estimated that in some nations (such as the U.S.) lawns use much more water than any crop in the entire country (Milesi et al. 2005). This exacerbates toxicity by promoting run-off of the applied chemicals to affect species that are sometimes in far-away aquatic systems (Overmyer et al. 2005). Finally, whether done by hired specialists or the homeowner, grass lawns need mowing, and even this simple practice can exacerbate the ecological harm associated with lawns. Gas-powered lawnmowers produce considerable greenhouse gases, and sloppy handling of gas containers is a frequent occurrence, contaminating the soil. Removing clippings (a natural form of fertilizer) and excessively low cutting (harmful to the plants as well as wasting energy) are simply bad practices.

In recent years, a movement to maintain lawns without the use of pesticides (including herbicides) has developed. The most common recommendations in this regard is to keep the grass growing very vigorously by irrigation, fertilization, and mowing, so that the plants can simply outcompete dandelions and other weeds (Busey 2003; Textbox 21.2). Unfortunately, the resulting grass monocultures or near-monocultures, while attractive, mostly exclude natural biodiversity. The use of alternative, biodiversity-friendly ground covers and hardy local shrubs is preferable, but often the plants available are just not as attractive or do not meet the desire of many to have a green outdoor carpet that serves some of the same functions as indoor carpets.

Lawns that most people consider to be 'ideal' are often monocultures, i.e., expanses of just one grass species. As with crop monocultures, to establish and maintain just one species it is necessary to eliminate all other plants that try to establish in the same area – frequently accomplished with herbicides. Similarly, there is also a need to prevent herbivorous insects, fungi, and other microorganisms – leading to the application of pesticides. Also, the presence of burrowing small mammals, such as shrews and moles and other mammals like skunks that dig up lawns to eat the grubs, encourages the use of poisons. When successful, the lawn industry refers to the product as a 'healthy lawn,' but from a biodiversity perspective, such an unnatural, artificial habitat is produced only by extensive 'specicide' – the elimination of hundreds of native species. Ironically, among the casualties are several weedy species of the pea family (Fabaceae), which are associated with nitrogen-fixing bacteria located in the roots, which take nitrogen out of the air and provide it to the plants, and eventually to the soil, thereby reducing the need for synthetic fertilizers. These include bird's-foot trefoil (*Lotus corniculatus*), black medic (*Medicago lupulina*), and clovers (*Trifolium* species), all of which are rather inconspicuous and do little to harm the appearance of lawns. People need to eat, so crop monocultures have considerable moral justification, but there are alternative, ecologically friendly plants that can better serve the purpose of providing aesthetic groundcovers (Textbox 21.3).

The current obsession with lawns traces to the invention of mowing machines in the early 19th century in England. Before that, large expanses of low vegetation around residences were maintained simply as pastures, clipped by livestock. High humidity and regular rainfall in countries like England limit the need for water, one of the primary wasteful needs of most lawns. Livestock, of course, provide natural fertilizer by way of their droppings and also eliminate the need for lawnmowers. Even with today's sophisticated technologies, grazing animals can still provide invaluable lawn maintenance services (Figure 21.3).

The human love affair with lawns may not be reversible. Certainly the many sports that are played on grass are unlikely to lose their popularity (the alternative prospect of playing on artificial grass is not appealing even to environmentalists, in view of associated production energy costs and pollution). One can at least hope that lawn grasses will be more sustainably cultivated in the future. Dandelions and crabgrass, which are highlighted in this review, are considered to be the worst lawn weeds, and the species which are most responsible for excessive and unnecessary ecologically unsustainable practices.

Rogue Ornamentals

Many of the most annoying city-dwelling plant species were imported from foreign lands for the purpose of being grown as ornamentals in gardens and parks. Unexpectedly, they have proven to be so aggressive, uncontrollable, and harmful that they have become as hated as mosquitoes and as despised as cockroaches. Unlike the most despised animals, the most disliked plants are often very attractive. This is the case for two of the species highlighted here: Japanese knotweed (*Reynoutria japonica*) and kudzu

TEXTBOX 21.2 LAWN CARE: AN ECOSYSTEM APPROACH

Like forests or prairie grasslands, lawns are dynamic ecosystems: communities of plants, soil, and microbes; insects and earthworms and the birds that feed on them; and humans who mow, water, fertilize, and play on the lawn. The interactions of all these community members shape the dynamic equilibrium we see as a lawn. Understanding and working within the natural processes that shape the lawn and its soil community can yield a durable, beautiful lawn that is easier to care for. As it turns out, these ecologically sound methods will also help reduce water use, waste generation, and water pollution.

—McDonald (1999)

TEXTBOX 21.3 PERMACULTURAL GROUNDCOVERS: AN ECOLOGICALLY SUSTAINABLE ALTERNATIVE TO TURF LAWNS

In order to end the constant cycle of environmental degradation that turf lawns promote, people must adopt landscaping schemes that promote a healthy ecosystem... The best ethical alternative to turf lawns is permaculture landscaping. The term 'permaculture' was coined by Bill Mollison and David Holmgren in 1978. Their goal was to develop 'a new, self-sustaining alternative to conventional agriculture' that works with, rather than against the environment... permaculture uses domestic species and configures them in a way that maximizes energy usage and minimizes waste.' A momentous permaculture landscaping movement could significantly benefit the environment and help create a more sustainable society. It is evident that maintaining a turf lawn is wasteful and harmful for the environment. The damage it does cannot be justified by the simple, artificial, aesthetic that people have been conditioned to believe is beautiful. Adopting a permaculture yard allows individuals to break free from the status quo and live harmoniously with nature.

—*Louis-Charles (2017)*

FIGURE 21.3 Maintaining lawns sustainably with 'organic lawnmowers.' (a) Horses on the estate of Balmoral Castle, Scotland, owned by the monarch of the UK. Source: Shutterstock, contributor: Byunau Konstantin. (b) A goat 'mowing' the grass of a football stadium. Source: Shutterstock, contributor: Volodymyr Nikitenko. (c) Sheep grazing a lawn in Paris, France. Photo by Guilhem Vellut (CC BY 2.0). (d) Sheep grazing a golf course. Photo by MrMark (CC BY 2.0).

(*Pueraria montana*), which are so beautiful that they have until recently been considered to be superb garden ornamentals. They may, nevertheless, represent the very worst of urban weeds.

Another infamous urban weed at least deserves special mention: the city specialist, *Ailanthus altissima*, usually called tree of heaven but also termed tree from hell. The species is native to China, where it is known as stink tree (the odor of the male plants is claimed by many to be like semen). Reflecting its disdain in its homeland China, the pejorative Chinese phrase 'good for nothing ailanthus stump-root' refers to a spoiled or irresponsible child who fails to meet parental expectations. In parts of the U.S., the species is called ghetto palm because it is common in the inhospitable conditions of rundown urban areas (Figure 21.4a). This toughest of trees (Textbox 21.4) easily tolerates stresses that would kill most other plants, including extreme shade, drought, road salt, air pollution, and contaminated soils, and when cut down, it simply regrows from the roots. Tree of heaven is now considered to be one of the world's most invasive weeds of temperate countries. Although it is somewhat harmful to natural ecosystems, like most weeds in urban environments, it is usually intensely disliked simply because of its undesirable appearance and the consequent perceived need to invest resources to eliminate it. In the Old World, it is often cultivated as one of the world's most attractive ornamental trees (Figure 21.4b).

Prickly Plants

Animals that bite people are the most hated of all species. Plants with thorns or prickles are usually similarly hated (thorny roses are an exception). Thistles are notorious for unexpectedly scratching people (Figure 21.5). In this chapter, Canada thistle, the leading urban prickly plant, is examined.

CRABGRASS

Introduction

Digitaria is a genus of grasses with over 200 species (Wipff 2003). The name is based on the Latin digitus, meaning 'finger,' a reference to flowering and fruiting branches of many of the species, which somewhat look like fingers arising from the top part of the stem or branch on which they are located. The origin of the name 'crabgrass' is obscure. One interpretation is that the name is based on a fancied resemblance of the creeping stems to the legs of a crab.

FIGURE 21.4 Tree of heaven (*Ailanthus altissima*). Public domain photos. (a) A weedy clump of young trees, growing characteristically in a neglected urban environment. Photo by AnRo0002. (b) A huge cultivated ornamental tree on display at Praterstern, one of Vienna's main railway stations. Photo by Peter Gugerell.

TEXTBOX 21.4 THE WORLD'S TOUGHEST TREE

There's a tree that grows in Brooklyn. Some people call it the tree of heaven. No matter where its seed falls, it makes a tree which struggles to reach the sky. It grows in boarded up lots and out of neglected rubbish heaps. It grows up out of cellar gratings. It is the only tree that grows out of cement. It grows lushly... survives without sun, water, and seemingly earth. It would be considered beautiful except that there are too many of it.

—Betty Smith (1943)

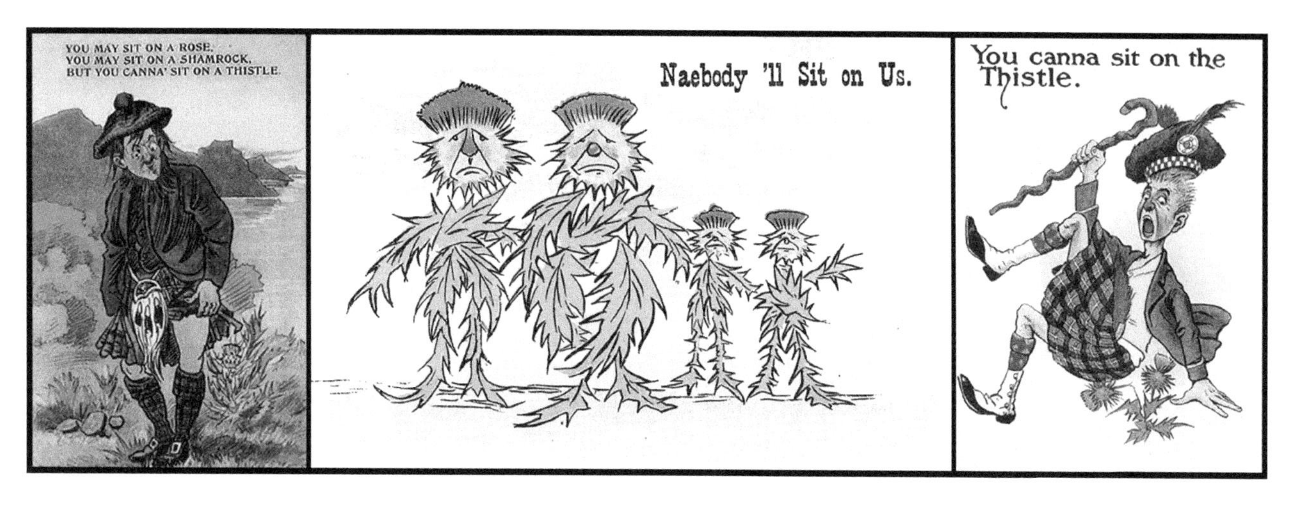

FIGURE 21.5 Early 20th-century postcards (public domain) indicating the dangerous nature of thistles.

FIGURE 21.6 Crabgrass (*Digitaria sanguinalis*) growing aggressively as a streetside weed. (a) A Yorkshire terrier visiting a weedy crabgrass mat at the base of a hydro pole in Alaska. Photo by Qwen Wan (CC BY 2.0). (b) Crabgrass growing between paving blocks and a house in Germany. Photo by Andreas Rockstein (CC BY SA 2.0). (c) Crabgrass growing between paving blocks in Australia. Photo by Harry Rose (CC BY 2.0).

The genus *Digitaria* is infamous for crabgrass (occasionally spelled 'crab grass') – species that are serious annual weeds in lawns and agricultural fields but are also frequently encountered as a streetside weed (Figure 21.6). (Some so-called 'crabgrass' species are in other genera of grasses, and people often use the word crabgrass as a label for any undesirable weed in their lawn that looks like a grass.) The most significant lawn weed species of *Digitaria* in North America and indeed in most temperate and tropical regions of the world are large crabgrass (*D. sanguinalis*; Figures 21.6 and 21.7), also called crab finger grass, hairy crabgrass, hairy finger grass, and purple crabgrass, and the closely related smooth crabgrass (*D. ischaemum*; Figure 21.8), also called small crabgrass. Large crabgrass is more upright and has larger stems, leaves, inflorescences, and seeds than small crabgrass. Large crabgrass has hairs on its leaves and stems, while the leaves and stems of smooth crabgrass have very few hairs except on the collar region (the intersection of the leaf and stem of the plant). Large crabgrass roots at the stem nodes while smooth crabgrass does so uncommonly. Because it can root at the nodes and so keep growing, large crabgrass is sometimes a perennial in some areas (Holm et al. 1977). *Digitaria ciliaris*, known by many names including southern crabgrass in the U.S. where it is cultivated, is very closely related to *D. sanguinalis* and indeed is debatably separated from it (CABI 2019f).

Large crabgrass (*D. sanguinalis*), the most prevalent crabgrass, is native to Eurasia and is distributed worldwide as a weed. It can grow to over a meter in height (Figure 21.7a) although it often spreads in a decumbent (reclining) fashion (Figure 21.7b). Parallel variation occurs in smooth crabgrass (*D. ischaemum*), which is often smaller (Figure 21.8). A single plant can produce as many as 150,000 seeds and 700 tillers (branches from the base). Large crabgrass thrives in full or partial sun, moist to slightly dry conditions, and heavy clay-loam soils. Outside of cultivated areas, it occurs in disturbed places, weedy fields, meadows, vacant lots, and along roads, paths, and railroads. It is also prolific in cultivated situations, including lawns, pastures, golf courses, athletic fields, gardens, orchards, crop fields, vineyards,

FIGURE 21.7 Large crabgrass (*Digitaria sanguinalis*). (a) Growing erect. Photo by Andreas Rockstein (CC BY SA 2.0). (b) Growing prostrately. Photo (public domain) by AnRo0002 (CC BY 2.0).

FIGURE 21.8 Smooth crabgrass (*Digitaria ischaemum*). (a) Erect plant. Watercolor over pencil (public domain), by an unknown artist. Source: Naturalis Biodiversity Center, Leiden, Netherlands. (b) Prostrate plant. Photo (public domain) by AnRo0002 (CC BY 2.0).

sod farms, nurseries, and landscaped areas. Smooth crabgrass similarly occurs in disturbed places such as roadsides and ditches, and in cultivated situations such as crop fields, orchards, vineyards, gardens, landscaped areas, turf, nurseries, and pastures.

There are two widespread classes of plants that differ in the way that they carry out photosynthesis. C_3 plants are so named because the first stable compound formed when carbon dioxide is processed is a three-carbon compound, i.e., C_3. C_4 plants are so named because the first organic compound incorporating CO_2 is a four-carbon compound. The C_4 system works more efficiently at high temperatures and high light intensity, and so many tropical plants belong to the C_4 category. Conversely, the C_3 system works best in the more moderate temperatures and light intensity of the temperate zone, where most plants are in the C_3 category. Some C_4 plants native to the hotter areas of the world, most notably corn (maize), when imported into northern regions, grow relatively slowly during cool springs, but when hot temperatures and intense sunlight arrive in the summer their C_4 system allows them to grow much faster than most other plants. The C_4 system of crabgrass allows it to grow much faster than the C_3 turfgrasses cultivated in temperate regions. The C_4 system is another of the weapons that makes crabgrasses possibly the world's most hated lawn weeds.

Harmful Aspects

As a weed, crabgrass is responsible for harmful effects worldwide (Holm et al. 1977; Mitich 1988a; CABI 2019g). Because of their high production of seeds, rapid growth, and relatively high tolerance to heat and drought, both large and smooth crabgrass can often outcompete domestic turf grasses (Turner et al. 2012). Crabgrass especially outcompetes turf grasses in lawns that are insufficiently watered and fertilized, as well as in poorly drained soils. The seeds grow quickly, typically producing a circle of stems up to 30 cm in diameter. The following year, the new seeds germinate in the same location where the parental plant died, and this cycle can continue for years. Crabgrass also spreads by the prostrate stems sprouting roots where they touch the ground (Figure 21.9). The fibrous roots of crabgrass can extend down 2 m in soil, making the plant more drought tolerant than most turf grasses. Crabgrass is also hard to remove from lawns because its low growth escapes the mower (it can survive when lawns are cut at a height of only 12 mm, which is much too low for lawn grasses). There is also evidence for the production of allelochemical compounds that serve to harm competing plants (Zhou et al. 2013; Hesammi et al. 2014). It is often said that, practically speaking, crabgrass can never be completely eradicated.

Tools and mechanical inventions (Figure 21.10a and b) have been employed for many years in attempts to eradicate crabgrass, generally with mediocre success. Herbicides are also traditionally employed to eliminate crabgrass on lawns (Figure 21.10c). Nevertheless, few herbicides can control crabgrass without damaging the lawn. Pre-emergence herbicides work by killing the young seedlings as they germinate. Unfortunately, pre-emergence herbicides also damage germinating desirable grass seed that might be planted at the same time. Also unfortunately, dormant seeds are not affected by pre-emergence herbicides, and so some seeds may survive the treatment. (A common misconception is that pre-emergence herbicides act by preventing weed seeds from germinating.) Post-emergent herbicides are employed to control crabgrass after it has already germinated and is growing. However, several countries and regions have banned the use of lawn herbicides for cosmetic purposes (Turner et al. 2012), and there are prospects for additional regions to also do so in respect to environmental and human health concerns.

With the recent trend against employing synthetic herbicides on ornamental plants, biological control agents are being investigated. Thaxtomin A, produced by the bacterium *Streptomyces scabies*, has been shown to have some value as a pre-emergent biological control agent against crabgrass (Wolfe et al. 2016). Krupska (2012) examined the

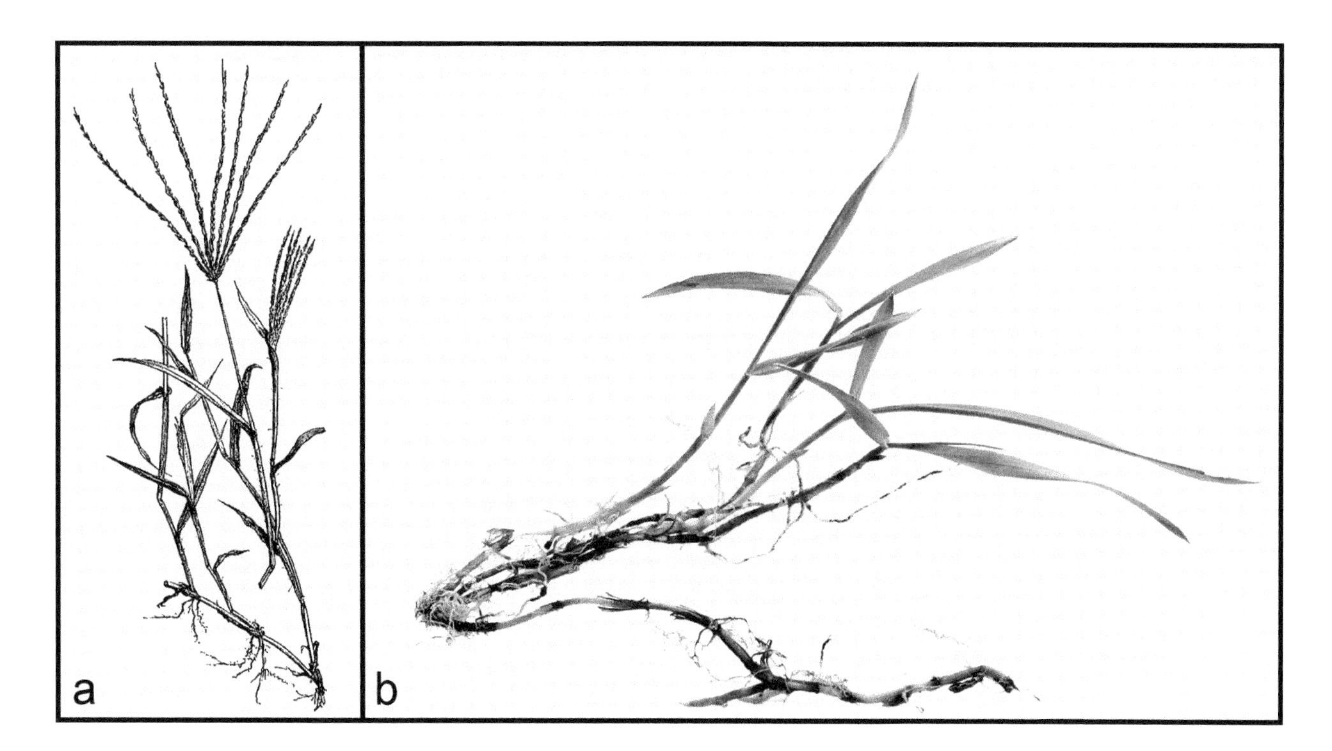

FIGURE 21.9 (a) Large crabgrass (*Digitaria sanguinalis*) showing a portion of a laterally spreading rhizome that has rooted to produce new shoots. Source (public domain): Hitchcock and Chase (1951). (b) Excavated crabgrass rhizome. Source: Shutterstock, contributor: Marekuliasz.

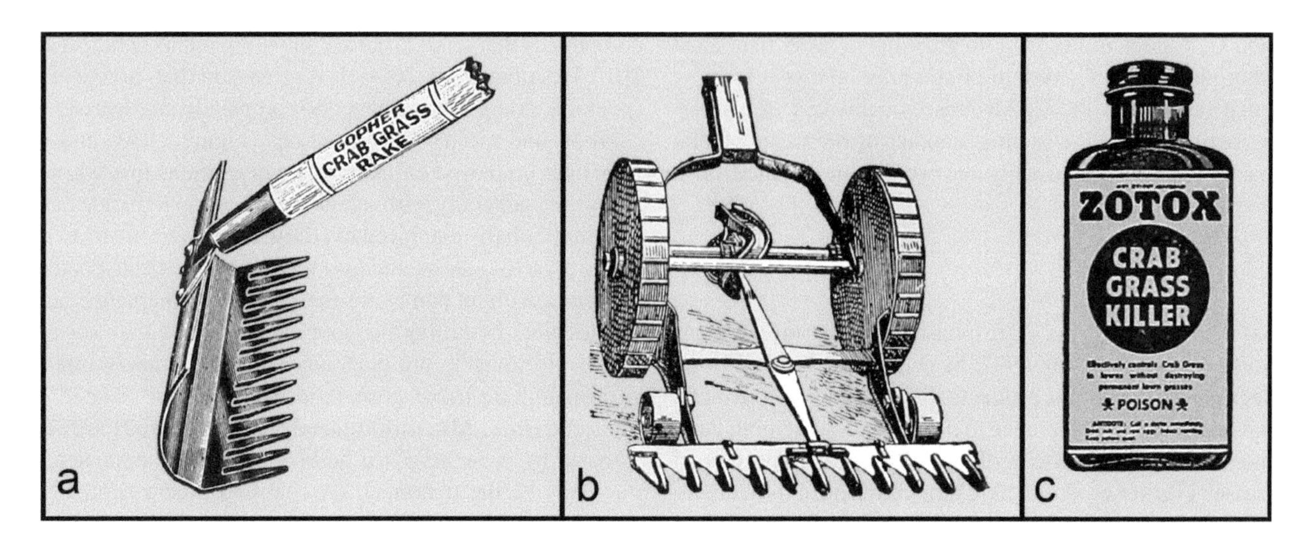

FIGURE 21.10 Traditional materials employed to eradicate crabgrass. (a) Crabgrass rake. 'This rake is formed so that the teeth pull flat on the sod; thus they get under the branches of the crab grass, shear off the entire head of the plants and keep them from further seeding.' Source (public domain): Vaughan's Seed Company and Gilbert Nursery (1942). (b) A wheeled contraption for digging up crabgrass. Image (public domain) from *The American Florist*, 1885. (c) An early crabgrass herbicide. Source (public domain): Vaughan's Seed Company and Gilbert Nursery (1942).

effects of several pathogenic fungi on crabgrass, but these agents had only limited success. On the whole, there are not yet good biological control agents for crabgrass.

To eliminate crabgrass infestation, lawn specialists widely recommend improving lawn health by appropriate irrigation, fertilization, and soil treatment, since this makes the desired grasses more competitive. Appropriate cultural practices might include core aeration to reduce compaction, topdressing with organic matter, overseeding to increase grass plant density, carefully controlling irrigation, managing thatch, and mowing at the proper height (usually about 6 cm). Inevitably, even well-cared lawns are attacked by a variety of species (particularly insects), requiring additional care (most frequently, chemicals). All of these measures can amount to considerable financial expenditures, to say nothing of the associated environmental costs.

Crabgrass in lawns is merely an aesthetic issue in most situations: the somewhat yellowish blades contrast starkly with the dark green of most turfgrasses, producing an appearance which is intolerable to many lawn enthusiasts. However, crabgrass actually provides a reasonable lawn for those who aren't fussy (Figure 21.11). Indeed, it may be argued that the lazy homeowner who neglects caring for his lawn is providing a far superior service to the welfare of the world than those who mindlessly nurture grass plants too weak to survive without excessive use of water, chemicals, and energy that damages the ecology of the planet.

Beneficial Aspects

Crabgrass is the motivation for considerable economic activity and employment. As with most major pests, there are associated industries to research and apply methods for controlling crabgrass. There are many weed and lawn specialists conducting research, and numerous retail products being sold to the public. Crabgrass is probably the most important reason why there are so many lawn service companies.

Some species of *Digitaria* have been employed as turf grasses (Figures 21.11 and 21.12). Throughout the world, a number of *Digitaria* species are employed as forages (Pitman et al. 2004). In temperate regions, large crabgrass and smooth crabgrass are occasionally used as animal forage and fodder, particularly in mixed pastures (Andrae 2002; Blount et al. 2013). Large crabgrass allegedly was introduced into the U.S. in 1849 by the U.S. Patent Office (predecessor of the United States Department of Agriculture) as a potential forage crop (Foy et al. 1983; Dalrymple 2010 could not find proof of this). Because it grows well in poor soils, it is sometimes grown as a forage grass for cows, sheep, and horses. Several cultivars of *D. ciliaris* are available: 'Red River' and 'Quick-N-Big' (Dalrymple 1975, 1994, 2001, 2010) and 'Impact' (Bouton et al. 2019). Jennings et al. (2014) provide a guide to the forage use of large crabgrass. Crabgrass makes good forage (Dalrymple 1980; Figure 21.13) and good hay (Gotcher et al. 2014). The seeds of large crabgrass are edible and have been used as a grain especially in Poland, where it is occasionally cultivated, and as a result, it has become known as Polish millet. The species has also been cultivated for its seeds in Germany. Wild food collectors sometimes harvest the seeds of crabgrass, a quite labor-intensive activity.

Two species of *Digitaria* – *D. exilis* (white fonio, the more popular plant; Figure 21.14) and *D. iburua* (black fonio) – are important African cereals (Small 2009). Both are annuals with small but abundant seeds, and both are

FIGURE 21.11 A crabgrass lawn. Public domain photo from Peakpx.

FIGURE 21.12 Queensland blue couch (*Digitaria didactyla*), an invasive species from Madagascar, cultivated in Australia on sports fields (photographed growing at the Royal Botanic Gardens Sydney, Australia, in January). Photo by Raffi Kojian, Gardenology.org (CC BY SA 3.0).

FIGURE 21.13 Black angus cow in a mixed crabgrass (*Digitaria ciliaris*) pasture in Ardmore, Oklahoma. Photo by Rob Mattson, Noble Research Institute, reproduced with permission.

indigenous to West Africa. Fonio is said to be Africa's oldest cereal, and it has been cultivated for millennia. It is also claimed to be the world's fastest maturing cereal, some varieties producing grain in just 6 weeks after planting. Fonio was once the major food of West Africa and is still a staple of the diet of certain regions of western Africa (such as Mali, Burkina Faso, Guinea, and Nigeria). Fonio is grown mostly on small farms in Africa for home consumption, but there is increasing interest in producing it in Western countries as a specialty product for niche markets.

Like most plant species, *Digitaria* species have been employed in traditional folk herbal medicine. Uses include treatment of venereal diseases, cataracts, parasites, digestive difficulties, and heart problems (Quattrocchi 2012). Also as with numerous other plants, modern pharmaceutical research is exploring the constituents of several *Digitaria* species for their medicinal values (e.g., Park et al. 2020).

Conservation Aspects

Most people have acquired an extremely negative opinion of crabgrass (Textbox 21.5, Figure 21.15) and have a greatly exaggerated opinion of its harmfulness, which is regrettable because the species can play positive roles. Crabgrass has potential for disposal of livestock waste, as it has been demonstrated to employ manure efficiently as a fertilizer (Dalrymple 1994). Gotcher et al. (2014) demonstrated that crabgrass effectively removes excess P from soils. This is important because nutrient buildup in pastures from repeated animal manure application may increase soil P and contribute to eutrophication and water quality

FIGURE 21.14 Fonio (*Digitaria exilis*), an important African cereal sold in some Western country stores. (a) A bowl of grain. Photo by Gabbri (CC BY SA 4.0). (b) A village chief in Senegal hold up sheaves of fonio. Photo (public domain) by Richard Nyberg, United States Agency for International Development.

TEXTBOX 21.5 CRABGRASS QUOTATIONS

Big sisters are the crabgrass in the lawn of life.
—*Charles M. Schulz*

Inflation is the crabgrass in your savings.
—*Robert Orben*

The Democrats are the party that says government will make you smarter, taller, richer, and remove the crabgrass on your lawn. The Republicans are the party that says government doesn't work and then they get elected and prove it.
—*P. J. O'Rourke*

Crabgrass can grow on bowling balls in airless rooms, and there is no known way to kill it that does not involve nuclear weapons.
—*Dave Barry*

Writing is like crabgrass: You have to keep weeding it.
—*Elyse Sommer*

I also have bills, a mortgage, and crabgrass like everybody else.
—*Jack Irion*

If your neighbor makes no attempt to control his crabgrass, you are fighting a losing battle.
—*Mark Foley*

If you don't immediately kill errant bulls**t, no matter how ridiculous, it can grow and thrive and eventually take over like crabgrass.
—*Bill Maher*

Crabgrass is like a boomerang: it keeps coming back.
—*Anonymous*

Stupidity, like crabgrass, need no tending to survive or thrive.
—*Anonymous*

FIGURE 21.15 The stereotypical view of crabgrass: A dangerous enemy requiring extermination. Prepared by B. Brookes.

deterioration. Crabgrass also has potential to be used to stabilize sites subject to erosion, as well as denuded areas requiring rapid cover (Aleshire 2005).

DANDELION

Introduction

Species of *Taraxacum* are perennial herbs, mostly native to North Temperate and Arctic regions of the Northern Hemisphere. The genus occurs in Europe, Asia, and the Arctic, Antarctic, and montane regions of the Americas. *Taraxacum* is an extremely complex genus of herbs. Many of the species reproduce mostly apomictically, i.e., by a pseudo-sexual process that produces seeds without fertilization. Mutations in species of such genera tend to be preserved because apomixis resembles vegetative reproduction, so that numerous 'microspecies' are generated, i.e., races differing only slightly. Taxonomists outside North America have recognized many of the races as different species (over 2,500 species have been recognized), but others submerge most of the races into only a few species (Richards 1973, 1985, 1996; Kirschner and Štěpánek 1997). Away from boreal and temperate regions, native kinds of *Taraxacum* are usually sexual, and these (about 100 species) are comparable in distinctiveness to the species recognized by most botanists. The widespread familiar weedy dandelions (which are apomictic) are usually collectively called the *Taraxacum officinale* complex or just *Taraxacum officinale*. Not all plants called 'dandelion' are species of *Taraxacum*. For example, 'Italian dandelion' is a usually large-rooted variety of the vegetable chicory (*Cichorium intybus*).

The dandelion (frequently termed common dandelion and French dandelion; Figure 21.16) is normally simply called dandelion in English but is known by dozens of other local names, including: bitterwort, blow ball, blowballs, bum-pipes, cankerwort, clock flower, crow parsnip, dindle, dumble-dor, face clock, fortune teller, golden milk, grunsel, heart-fever grass, horse gowan, Irish daisy, lion-tooth, lion's tooth, love's oracle, milk gowan, milk witch, monk's head (monk's-head), peasant's cloak, piss-a-bed (pissabed), piss-the-beds, pissenlit, priest's crown (priest's-crown, pries' crown), puffball, radicchiello, schoolboy clock, swine's snout, stink davie, telltime, time-table, wet-a-bed, white endive, wild endive, wishes, witch gowan, witches' milk, and yellow gowan. Such a wide variety of names reflects the importance and prevalence of the dandelion.

Dandelion is a herbaceous perennial with jagged, irregularly lobed leaves in a rosette (circularly arranged leaves at ground level; Figure 21.17). The rosette permits dandelions to survive mowing, grazing, and competition with

FIGURE 21.16 Dandelion (*Taraxacum officinale*). Public domain illustration from Köhler (1883–1914).

grasses. The leaf margins may be nearly smooth-margined, saw-toothed, or deeply cut. The taproot (Figure 21.18) is long (up to 2 m), thick (up to 3 cm), fleshy, and branched. The root is contractile, i.e., at the end of the season it contracts, pulling the crown of the plant downward so it is protected from the cold, herbivores, and lawnmowers. Cut roots can quickly produce new shoots (Naylor 1941; Mann and Cavers 1979), so that mechanical removal of the plant is difficult (repeated cutting, which progressively weakens the plants, is usually necessary). The single flowering stalk, sometimes over 50 cm tall, is hollow and bears a head (technically a 'capitulum;' Figure 21.19a) of up to 250 tiny yellow florets, the whole head referred to as a flower by non-botanists. The flowering stalk elongates with age. When the sun shines, the flower heads are open, and when the weather turns dull, the flower heads close up. Although dandelion produces flowers particularly in mid-spring under moist conditions, plants often continue flowering right up to frost. Fruiting heads produce tiny (3–5 mm) brown 'seeds' (technically, fruits called achenes), each carried by a 'parachute' of white, fluffy hairs on a stalk (Figure 21.19b–d). White, bitter, milky juice exudes from all parts the plant where it is cut or broken; this stains hands brown and is difficult to remove. Cultivated selections differ in various respects from wild plants, some tending to have broader, more deeply-lobed leaves, others with a very high production of leaves, often semi-erect. Most weedy forms of dandelion in the U.S., and probably most if not all cultivated selections, originated from Europe and Asia.

FIGURE 21.17 Dandelion rosette (foliage in a whorl of leaves at ground level). Photo (public domain) by Hans, Pixabay.com.

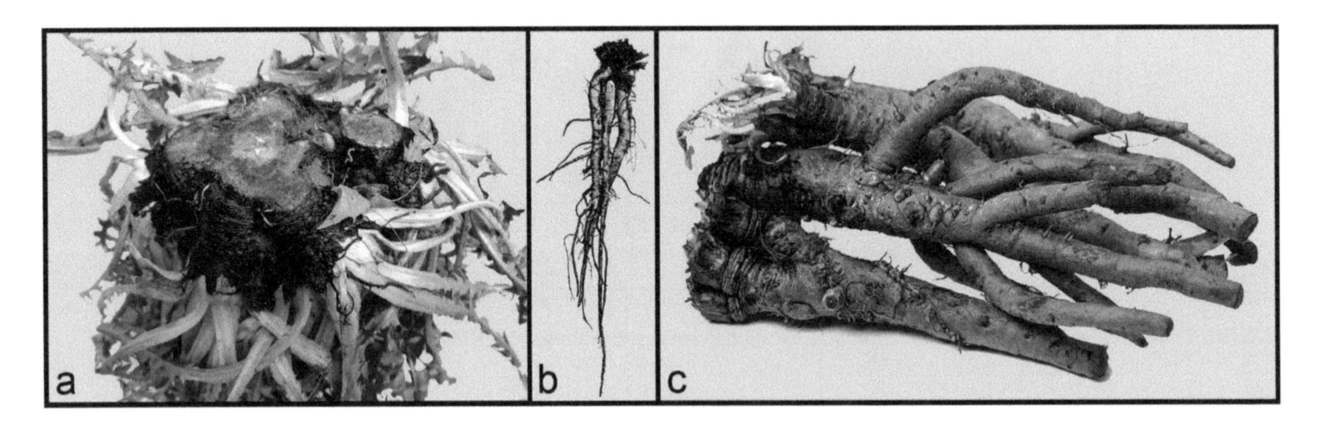

FIGURE 21.18 Taproots of dandelion (*Taraxacum officinale*). (a) Bottom view of the laterally cut root of a mature plant. Photo by MHM55 (CC BY SA 4.0). (b) A clump of roots extending to a depth of 40cm. Drawing (public domain) from Henkel (1904). (c) Large roots harvested for medicinal use. Source: Shutterstock, contributor: Zhengchengbao.

Harmful Aspects

Dandelions thrive under a broad range of climatic and soil conditions. The plant reproduces quickly and rapidly, and does not need pollinators. It disperses widely by wind, water, animals, and humans. It is very difficult to eliminate. These characteristics make the plant a leading weed, often blanketing large regions (Figure 21.20). The dandelion is now a common weed throughout the world in temperate regions, often in pastures, meadows, orchards, gardens, and waste ground, and along roadsides. Dandelion is a significant weed of temperate region crops and of lawns. In some orchards where the trees require insect pollination, dandelions can draw away pollinators. However, the species is not a detriment in most pastures as it increases the nutritious value of forages and fodders for most livestock. For a particularly extensive review of the harm caused by dandelion, see CABI (2019l).

Dandelions in residential lawns frequently enrage homeowners, who often spend considerable effort in trying to eradicate them (Figure 21.21). Several control measures are employed to eradicate them (Mann 1981; Mitich

FIGURE 21.19 Reproductive parts of dandelion (*Taraxacum officinale*). (a) Flower head in full bloom. Photo by Daniel Villafruela (CC BY SA 4.0). (b) Dandelion head with mature 'seeds' (achenes, a type of dry fruit). Photo by Michael Gäbler (CC BY 3.0). (c) Seeds being blown off a dandelion flower head by wind (photoshopped for appearance). Photo by Louise Docker (CC BY 2.0). (d) Seeds. Photo (public domain) by Steve Hurst, United States Department of Agriculture, Agricultural Research Service Systematic Botany and Mycology Laboratory.

FIGURE 21.20 Fields contaminated with dandelions. (a) Dandelions in flower. Photo by Vadim Indeikin (CC BY SA 30). (b) Dandelions gone to seed. Photo by Mike Mozart (CC BY 2.0).

FIGURE 21.21 Cartoon illustrating the misperception that dandelions are a major problem justifying the excessive use of pesticides. Prepared by B. Brookes.

1989a; Stewart-Wade et al. 2002). Mechanical removal (by hand, tools, and machines) can reduce infestations, but this requires considerable human effort and/or resources and is rarely practiced in Western countries for cropland, parks, and golf courses. Herbicides (especially 2,4-D) are commonly employed to control dandelions (and other weeds) on farms, and urban governments also spend considerable amounts on pesticides simply to remove these weeds from public property. However, there is increasing sentiment against the use of pesticides simply to improve the aesthetic appearance of public and private landscapes. Applying pesticides to areas where people spend considerable time, such as athletic fields, requires particular care to avoid possible contact with chemical residues.

Beneficial Aspects

- Human food usage

  Extensive information on the food usage of dandelion is given in Small (2006). The plant is an excellent source of vitamins, minerals, and protein (Schmidt 1979; Kuusi et al. 1982; Escudero et al. 2003; Ghaly et al. 2012; Pădureţ et al. 2016). Dandelion has been consumed for centuries in Eurasia. Anglo-Saxons, Celts, Gauls, and Romans used the dandelion as food. In Europe, it was widely used as a pot herb by the poor, who gathered it from nature. It was also grown in a blanched form and used in salads, and the roots were used as a coffee substitute. In Victorian England, dandelion was widely grown in kitchen gardens for use as a salad herb. Dandelion was brought to North America by early colonists as a pot herb. During times of war and famine, it was a much-valued emergency food. In modern times, dandelions are a favorite of exotic food enthusiasts (Figure 21.22).

  More than a century ago in Europe, domesticated, relatively tasty, succulent dandelion varieties were grown (Figure 21.23). Today, cultivars selected for edibility are marketed (Figure 21.24a), although compared to the common vegetables, the market is relatively small. Almost all parts of the plant can be eaten (Figure 21.24c). The leaves are consumed primarily as a cooked potherb, although young leaves are sometimes eaten raw in salads. When cultivated in a blanched form (i.e., protected from light to produce succulent, tender tissue; Figure 21.24b), the leaf stalks and crown (the central white portion of the rosette) can be eaten raw or as a cooked vegetable. If wild dandelions are harvested, the leaves are best cooked in one or two changes of water to reduce the bitterness. Dandelion roots are sometimes eaten raw but are best after two changes of water. They can be

FIGURE 21.22 'Please don't eat the dandelions,' prepared by B. Brookes.

FIGURE 21.23 Nineteenth-century domesticated salad varieties of dandelion (*Taraxacum officinale*) with succulent leaves. Source (public domain): Vilmorin-Andrieux et Cie (1883).

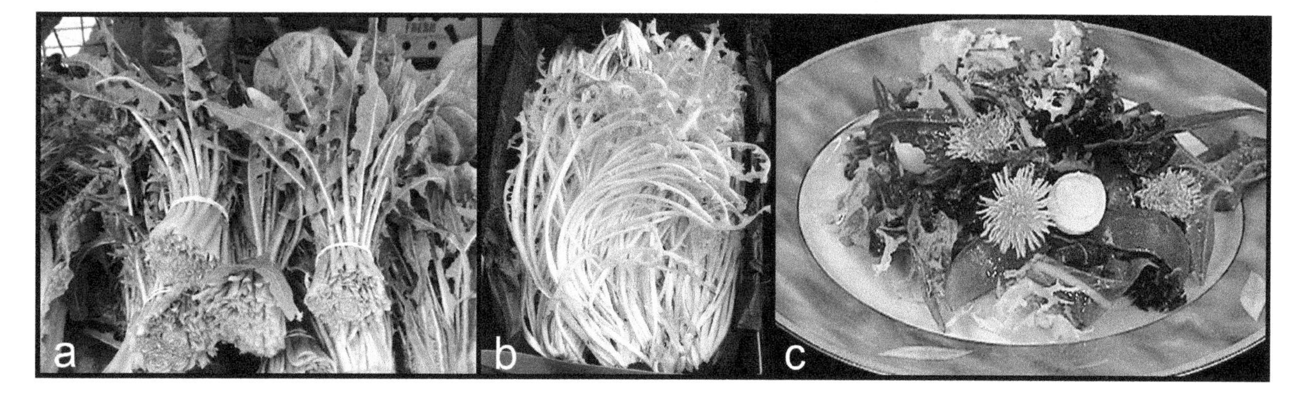

FIGURE 21.24 Food uses of dandelion leaves. (a) Dandelion greens for sale. Photo by Darya Pino (CC BY 2.0). (b) Blanched dandelion greens (grown in darkness to promote succulence). Note that the leaf blades remain stunted. Photo by Usien (CC BY SA 4.0). (c) Tossed salad with dandelion leaves and flowers. Photo by Burkhard Mücke (CC BY SA 3.0).

consumed roasted or fried. The roots can also be dried, roasted, and ground as a caffeine-free coffee substitute (Figure 21.25). Some of the carbohydrates in the root are in the form of inulin, which is better than sucrose for diabetics. Alcoholic beverages are made from dandelion, using flowers for wine and whole plants for beer. In Scandinavia, flowers are used to flavor schnapps. The flowers can be eaten outright, although they are slightly bitter and turn one's saliva a startling yellow. Unopened flower-buds are sometime eaten in pancakes, omelets, and fritters, and the flowers are used in the Arabic cake *yublo*. However ingested, dandelion is invariably a marked diuretic, promoting urination.

- Animal food usage

  Dandelion is very rich in nutrients (Neamtu et al. 1992) and has high feed value for livestock (Figure 21.26). It is considered to be an excellent, highly nutritious pasture plant for cattle (Jackson 1982; Bergen et al. 1990). The plants have also been found to be beneficial for poultry (Qureshi et al. 2017). Despite its nutritional value, Falkowski et al. (1990) reported that it was not eaten readily by most domestic animals because of its bitterness (reflecting the adaptive value of the bitter chemicals present). Dandelion is a valuable bee plant because the flowers bloom early in the spring, and at that time may be the only major source of nourishment for bees (Dalby 1999; Figure 21.27).

- Medicinal usage

  As a medicinal plant, dandelion has been used at least since the time of the Arabian physicians of the 10th and 11th centuries. The flowers have been used in folk medicine to treat jaundice and other liver ailments (Small and Catling 1996, 1999; Schütz et al. 2006; Grauso et al. 2019; Figure 21.28). Dandelion has also served as a tonic. A tradition developed in Europe of taking dandelion as a 'cleansing cure' in the spring. Other ailments treated include fever, insomnia, jaundice, rheumatism, eczema, other skin diseases, constipation, warts, cancers, and tumors. These folk usages are considered obsolete today, although as with many other herbals, a niche market exists in many countries for dandelion extracts. There are also many recent experimental studies (mostly on laboratory rodents) demonstrating that extracts of dandelion (especially of roots) can alter metabolism, with possible relevance to human diseases (including cancer), but the plant is not considered to be commercially significant in modern Western medicine (Martinez et al. 2015).

  The most common medicinal usage of dandelion has been as a diuretic (to promote urination). Root extracts were once used extensively for this purpose and are still sometimes so employed. In the 18th-century dandelion was termed Herba Urinaria, and in Holland, it is still sometimes called beddepissers. The French pissenlit, literally 'wet-a-bed,' is still another allusion to the reputation of dandelion greens for stimulating the kidneys. Such names as wet-a-bed, piss-a-bed, and piss-the-beds are old English names for dandelion. In parts of the U.S., the dandelion was once called pissabed, and a dandelion-based salad was called 'pissabed salad.'

FIGURE 21.25 Dandelion root coffee. Photo by Gavin Bloemen (CC BY SA 3.0).

FIGURE 21.26 Dandelions being consumed by livestock. Source of photos: Shutterstock. (a) Cattle. Contributor: Ben Schonewille. (b) Vietnamese pig. Contributor: Kuttelvaserova Stuchelova. (c) Horse. Contributor: Ben Schonewille. (d) Girl feeding a horse. Contributor: Boitano. (e) Pygmy goats. Contributor: Slowmotiongli.

FIGURE 21.27 Honey bee on a dandelion flower head. Photo by Juan Carlos Fonseca Mata (CC BY 4.0).

FIGURE 21.28 Nineteenth-century American advertising posters (public domain) for 'dandelion bitters,' which were popularly used to treat disorders of the liver, kidneys, and blood.

The bitter taste of dandelions is due to 11.13-dihydrotaraxine acid-1'O-B-D-glucopyranoside and several other awesomely-named chemicals. Because food cultivars of dandelion have been selected to have reduced amounts of the bitter chemicals thought to be responsible for pharmacological properties, food cultivars are likely less effective medicinally than wild forms.

- Industrial extracts

  Hybrids of *T. kok-saghyz*, Russian dandelion (from Turkmenistan), have been grown in Russia and elsewhere as a source of rubber, derived from the plant latex. A golden yellow dye that colors wool and cotton can be obtained from the flowers, and a purple dye from the roots. These uses are considered obsolete.
- Ornamental usage

  Several species of *Taraxacum* are occasionally grown as ornamentals, particularly a cultivated form called 'Pink Dandelion', debatably assigned to the species *T. pseudoroseum* (Kirschner and Štěpánek 1993). Frequently, different species of the daisy family are misleadingly sold as Pink Dandelion.

Conservation Aspects

Dandelion is a serious invasive weed in some areas, threatening biodiversity. In some regions, notably Alaska, introduced weedy dandelion is hybridizing with rare local species of *Taraxacum*, and thereby threatening the very existence of their own close relatives. Efforts are underway to remove such invading dandelions (Figure 21.29). However, most people are prejudiced against the dandelion because it is a bothersome weed of gardens and lawns, and seems improbable as a useful plant. In partial compensation for their deleterious effect on biodiversity, it may be noted that dandelions and other flowering weeds in turfgrass lawns provide significant food resources for declining pollinator populations in urbanized landscapes (Larson et al. 2014). Although weedy dandelions are generally undesirable in natural ecosystems, they do provide food for a wide variety of wildlife (Figure 21.30). Oxlade and Clifford (1999) proposed many ways of using the plant as experimental material for teaching students about biology, which would be a good way of improving the reputation of dandelions. Curiously, the dandelion was once declared an endangered wildflower in England (Smith 1995). Dandelions almost became extinct during the 1950s in the British Isles, and a 'Save the Dandelion' club was formed. Similar societies were started in the U.S.

JAPANESE KNOTWEED

Introduction

Japanese knotweed (also called Asian knotweed, donkey rhubarb, Japanese bamboo, Mexican bamboo, and many other names) is the perennial semi-woody species *Reynoutria japonica* (also known in older literature as *Fallopia japonica* and *Polygonum cuspidatum*). The species is a native of eastern Asia, including China, Japan, Korea, and Taiwan. In the 19th century, the plant was introduced to Europe (Thiébaut et al 2020) and North America (Del Tredici 2017) for use as an ornamental, and it is indeed an impressive, attractive shrub (Figure 21.31). It has become a

FIGURE 21.29 Members of the Southeast Alaska Guidance Association, removing invasive dandelions. Photo (public domain) by U.S. National Park Service.

FIGURE 21.30 Wild animals consuming dandelions. (a) Black bear (cinnamon-coated) in Waterton Lakes National Park, Alberta, Canada. Photo by Traveler100 (CC BY SA 3.0). (b) European rabbit eating dandelions in the UK. Photo by Gidzy (CC BY 2.0). (c) Swan eating dandelion in Germany. Photo by FHgitarre (CC BY 2.0).

FIGURE 21.31 Japanese knotweed (*Reynoutria japonica*). (a) Photo by Gerald@volp.com (CC BY SA 3.0). The photographer's daughter in the foreground is 90 cm tall. (b) A portion of a flowering plant. Photo by W. Carter (CCO 1.0).

FIGURE 21.32 Giant knotweed (*Reynoutria sachalinensis*). (a) Pre-flowering plant. Photo by Leslie J. Mehrhoff, University of Connecticut, Bugwood.org (CC BY 3.0). (b) Plant in fruit. Photo by Barbara Tokarska-Guzik, University of Silesia, Bugwood.org (CC BY 3.0).

very aggressive introduced weed in Australia, Europe, New Zealand, and North America.

The genera *Reynoutria*, *Fallopia*, and *Muehlenbeckia* are closely related (Schuster et al. 2011) and are known to have invasive potential. Giant knotweed (*R. sachalinensis*, also called Sakhalin knotweed; Figure 21.32) is similar to Japanese knotweed and has also become invasive in many locations. Hybrids among the species are especially vigorous. The hybrid between Japanese knotweed and giant knotweed, known as Bohemian knotweed *Reynoutria* × *bohemica*; Figure 21.33), is very invasive, particularly in North America (Stone 2010).

The green to purple stems of Japanese knotweed are reminiscent of bamboo, with well-demarcated nodes or 'joints' (Figure 21.34a), and stems that are hollow between the nodes (Figure 21.34b). Moreover, the plant grows in bamboo-like clumps. The knotweed family (Polygonaceae) has other genera with swollen or at least well-demarcated nodes, also referred to as knotweeds, and indeed the 'knots' in knotweeds is likely a reference to these nodes.

The plants arise in the spring from overwintering rhizomes (underground stems), looking initially like asparagus spears (Figure 21.35). The stems become 1–3 m in height, sometimes exceeding 4 m, and often have reddish-purple spots or a reddish caste. Extra-floral nectaries (i.e., nectar-producing organs that are not in flowers) are located on stem nodes or at the petiole bases and seem to represent a means of attracting ants that chase away other insects that can harm the plants (Kawano et al. 1999; Giuliani et al. 2019). Small, creamy, white, or greenish-white flowers are produced in late summer and early autumn. Some populations

FIGURE 21.33 Bohemian knotweed (*Reynoutria* × *bohemica*), the hybrid of Japanese and giant knotweeds. Photo by AfroBrazilian (CC BY SA 3.0).

FIGURE 21.34 Structure of stem of Japanese knotweed (*Reynoutria japonica*). (a) Prominent nodes (location on stem where leaf stalks are attached). Photo by Frank Vincentz (CC BY SA 3.0). (b) Stem split open longitudinally showing hollow internodes between the solid nodes (which appear as partitions). Photo by Lamiot (CC BY SA 3.0).

FIGURE 21.35 Young shoots of Japanese knotweed in the spring. Photo by Rob Routledge, Sault College, Bugwood.org (CC BY 3.0).

generate many viable seeds, which are winged, triangular, shiny, and very small (Forman and Kesseli 2003; Barney et al. 2006; Figure 21.36). The wings facilitate wind and water dispersal. The stems grow up to 2.5 cm in diameter and die back in the autumn, the dead stalks standing over the winter.

Many plants of Japanese knotweed bear only female flowers or only male flowers, but some populations contain plants with mostly flowers of one sex but also some flowers of the opposite sex. In Great Britain and much of the U.S., the majority of plants only bear flowers that are functionally female. Indeed, in Great Britain and the U.S., most Japanese knotweed belongs to one genetically identical female clone that has been divided and transported to different regions (Hollingsworth and Bailey 2000; Grimsby et al. 2007; Gammon and Kesseli 2010). However, giant knotweed has been found to represent different genotypes (Krebs et al. 2010). Of course, both sexes are necessary in a location to produce seeds, and since this is often not the case, Japanese knotweed spreads predominantly by rhizomes in most places (often unintentionally taken to other locations by people).

There are several garden cultivars of Japanese knotweed (Figure 21.37). 'Crimson Beauty' has dark red flowers, 'Variegata' has leaves with white margins, and 'Compactum' is a dwarf form, usually less than 70 cm tall (Beerling et al. 1994). These selections can also be invasive. Although they typically reproduce vegetatively and do not produce seeds, if pollinated by other plants of *Reynoutria* the resulting seeds can also spread the plants. Although planting material is legally available in some jurisdictions, in view of the harm potential, Japanese knotweed and its hybrids should not be grown as ornamentals. While the plants are quite attractive when growing, the overwintering stems, typically very crowded together, are unsightly (Figure 21.38).

Japanese knotweed is a remarkably tough and harmful weed of roadsides, right-of-ways, waste places, the banks of streams and rivers, and various urban spaces. The species is often on sites of old homesteads where it was planted as an ornamental. In Poland, Sołtysiak and Brej (2014) observed that the invasive knotweeds occurred particularly in urban areas. Fragments of rhizomes and pieces of stems are spread easily by moving water in ditches, canals, and streams. Pieces of rhizome are also spread in

FIGURE 21.36 Winged fruits ('seeds') of Japanese knotweed (*Reynoutria japonica*). (a) A fruiting branch, (b) Close-up of fruits. Photos by Udo Schmidt (CC BY SA 2.0). (c) Branches with very heavy production of fruits. Photo by Leslie J. Mehrhoff, University of Connecticut, Bugwood.org (CC BY 3.0).

FIGURE 21.37 The cultivar 'Spectabile' of Japanese knotweed (*Reynoutria japonica*). The young leaves are whitish (a), and turn green splotched with white during the summer (b). Photos by Cultivar 413 (CC BY 2.0).

FIGURE 21.38 Unsightly persisting overwintering dead stems of Japanese knotweed (*Reynoutria japonica*). (a) Plants in a neglected field. Photo by Leslie J. Mehrhoff, University of Connecticut, Bugwood.org (CC BY 3.0). (b) Plants in the front yard of a house. Photo by Steve Brewer (CC BY SA 2.0).

FIGURE 21.39 A spreading rhizome of Japanese knotweed (*Reynoutria japonica*) being pulled out. The rhizomes are capable of growing laterally 20 m from the point of origin. Photo by Klarerwiki (CC BY SA 2.0).

contaminated soil, which may be carried by machinery. Seeds are sometimes produced and are spread especially by wind. Once a plant establishes in a given location, it spreads by rhizomes forming clones (genetically uniform growths) that mature into extremely dense colonies which crowd out most other species. The plant grows well in various soils, including those that are salty or contaminated with heavy metals, and is resistant to atmospheric pollution. This tolerance traces to its adaptation of growing on volcanic soils in its homeland, Asia. Although it prefers open areas, Japanese knotweed tolerates shade. The rhizomes can grow laterally as much as 20 m from the parent shoot (Figure 21.39) and 3 m down, and can survive winter temperatures of −35°C.

Harmful Aspects

Japanese knotweed is very difficult to remove once it has become established, and time-consuming, expensive procedures are usually required (Shaw 2008; Anderson 2012; Textbox 21.6). The rhizome can reproduce when buried 1 m deep (allegedly much deeper). The plant is extremely difficult to excavate, and sprouts vigorously from pieces of rhizome that have not been removed. A rhizome portion weighing less than 1 g can grow into a new clone. Pieces of stem can also regenerate the plant. Soil in which the plant has grown needs to be treated as hazardous waste, adding to costs. Growth of the underground system has been known to damage concrete foundations, buildings, roads, and other paved surfaces (it can grow through 5 cm of asphalt), as well as retaining walls (Figures 21.40 and 21.41). Japanese knotweed has been listed as one of the 100 world's worst alien species (Lowe et al. 2000; a rather arbitrary list). In the U.K., residences put up for sale need to be accompanied by a statement whether Japanese knotweed is on the property (if present, property values are lowered). Giant knotweed and hybrids with Japanese knotweed are expected to pose greater problems in the future, which will add to the already very significant damage (Gillies et al. 2016).

Beneficial Aspects

Japanese knotweed was once considered to be desirable for planting as an ornamental, forage, and for erosion control, but its ability to escape cultivation precludes it for these purposes.

Honey bees are attracted to Japanese knotweed flowers (Ferrazzi and Marletto 1990; Figure 21.42), and the resultant honey (sometimes called bamboo honey) is excellent (Bobis et al. 2019). This is not surprising since the plant is in the buckwheat family (Polygonaceae), and buckwheat honey is outstanding.

The young shoots which are reminiscent of asparagus shoots are harvested as a wild food (Figure 21.43) and are particularly popular in Japan (Shimoda and Yamasaki 2016). The taste is like very sour rhubarb. To improve palatability, the

TEXTBOX 21.6 THE PLANT THAT CONQUERED BRITAIN

Britain does indeed have it worst. A single stalk of knotweed found on a property, or even on a neighbor's lot, devastates a house's value and makes it near impossible to obtain a mortgage or insurance. It has led to financial ruin, depression, even a murder-suicide. A botched attempt to remove the weed may only drive it underground, where it can remain dormant for a decade or more. Once the coast is clear, once you've built your nice, new conservatory, up it pops again…

The slightest fragment of root or stalk dropped on disturbed ground will colonize. It advances as much as 8 cm a day and can reach 5 m in height or more. It is hardy. Subsurface, its roots can extend 3 m or more deep, and 20 m across, in a constant search for water and the tiniest cracks or seams in barriers blocking its quest for light. Just over a century and a half after the plant arrived in Kew, there is not a single 1,500-ha patch of ground in the entire U.K. that is not rooted with at least one Japanese knotweed…The U.K. government estimates that the cost of controlling knotweed has hit the equivalent of $3 billion… it is Brits without deep pockets who have it worst. An estimated 220,000 homes are infested, fertile ground for a lucrative new branch of the legal industry. Reports abound of it ripping through foundations, infesting floor and wall cavities, and poking out of baseboards and electrical sockets.

—Ken MacQueen (2015)

FIGURE 21.40 Japanese knotweed damaging pavements, concrete, and brickwork. (a) Shoots growing through an asphalt driveway. Photo by Lamiot (CC BY SA 3.0). (b) Shoots penetrating between pavement slabs. Photo by Gordon Joly (CC BY SA 2.0). (c) Plant establishing in a crack between a brick wall and a concrete sidewalk. Photo by Gordon Joly (CC BY SA 2.0).

fibrous outer skin is peeled off, and the shoots are subjected to prolonged soaking or parboiling in water, and cooked.

Japanese knotweed is used in traditional Asian herbal medicine to treat various conditions (Peng et al. 2013; Zhang et al. 2013). The roots of Japanese knotweed and giant knotweed contain relatively high levels of resveratrol, a controversial anti-cancer drug, shown to have anti-tumor effects in mice (Kimura and Okuda 2001). Resveratrol, also extracted from grapes, soy, and peanuts (Tian and Liu 2020), is also alleged to have other medicinal properties (Ahmadi and Ebrahimzadeh 2020). There are numerous other constituents in *Reynoutria* under examination for possible medicinal applications (Patocka et al. 2017).

Japanese Knotweed grows so aggressively that is has potential for harvest of biomass (Strašil and Kára 2010; Gregorczyk et al. 2012).

Conservation Aspects

Reynoutria is naturally adapted to spreading along watercourses (Figure 21.44), and when colonies develop, they inhibit other plants and animals, negatively affecting natural ecosystems. Colonies also sometimes occur in natural terrestrial environments, in which case they tend strongly to become expanding monocultures which displace native plant species and drastically reduce habitat for animals (Gerber et al. 2008; Aguilera et al. 2010; Maurel et al. 2010; Hajzlerová and Reif 2014; Lavoie 2017). Although the *Reynoutria* knotweeds have some soil stabilizing ability, they actually increase erosion by outcompeting deep-rooted plants which more efficiently hold soil. The roots release compounds into the soil that are detrimental to some soil organisms and other plants.

FIGURE 21.41 A residence under attack by Japanese knotweed. Painting by Jessica Hsiung.

Japanese and giant knotweed and their hybrid have become major pests in urban situations. Because they are so difficult (many say impossible) to remove, people commonly turn to herbicides as the cheapest alternative. Because the plants are so tenacious, overuse of herbicides is common, and harm to the environment occurs. Unfortunately, herbicides are only partly effective on the knotweeds and require repeated application for several years (Bashtanova et al. 2009). There are more sustainable practices, including grazing animals (Figure 21.45a), mechanical removal (Figure 21.45b), and application of opaque coverings (Figure 21.45c), and indeed an industry has developed devoted to the removal of knotweed in urban areas (Figure 21.45d).

FIGURE 21.42 Bee on Japanese knotweed flowers. Photo by Stanze (CC BY SA 2.0).

FIGURE 21.43 Young, edible shoots of Japanese knotweed, reminiscent of asparagus spears. (a) Photo by Lamiot (CC BY SA 3.0). (b) Sprouts on sale in a market. Photo by Foodista (www.foodista.com; CC BY 2.0).

FIGURE 21.44 Infestation of Japanese knotweed (*Reynoutria japonica*) on a riverbank, a frequent habitat colonized by the species. Photo by Leslie J. Mehrhoff, University of Connecticut, Bugwood.org (CC BY 3.0).

FIGURE 21.45 Eradicating Japanese knotweed. (a) U.S. Forest Service staff eliminating an infestation of giant knotweed in Michigan (public domain photo). (b) Killing Japanese knotweed by covering it with a sheet of opaque plastic to prevent sunlight from reaching the young shoots in the spring. Photo by Kiu77 (CC BY SA 3.0). (c) Piglets eliminating Japanese knotweed. Photo by Ryan (CC BY ND 2.0). (d) Car from a company specializing in Japanese knotweed removal in Great Britain. Photo by Sludge G (CC BY SA 2.0).

KUDZU

Introduction

The genus *Pueraria* (named for the Swiss botanist Marc Nicolas Puerari, 1766–1845) has been recognized as containing as many as 20 species, but some of these appear to belong other genera (Egan and Pan 2015; Egan et al. 2016). *Pueraria* is infamous for kudzu (a word designating certain *Pueraria* species, especially *P. montana* var. *lobata*, sometimes called *P. lobata*). The English name is based on the Japanese common name for the plant, kuzu. Among the many local names that the plant has acquired are porch vine, telephone vine, and wonder vine. The classification of kudzu is somewhat unsettled, as it is very variable genetically (Pappert et al. 2000; Sun et al. 2005), with groupings such as variety *lobata* distinguishable by DNA fingerprinting (Zhang et al. 2020a). Kudzu is native mainly to Southeast Asia. It is naturally distributed in eastern and southern China, the Korean peninsula, and Japan, and southward to northern Myanmar, northern Laos, and northern Vietnam (Li et al. 2011). It has become widely introduced and naturalized elsewhere in Asia, Africa, Oceania, the tropical Americas, and the United States, and to a lesser extent in southern Europe (Pasiecznik 2007). In the United States, kudzu occurs from New York south to Florida, and west to Texas, with occasional infestations elsewhere. The species has been collected in southern Ontario in Canada, the northernmost location in North America.

Kudzu is a perennial, climbing, semi-woody vine that can exceed 30 m in length, re-growing from the woody root system each spring in areas with cold or cool winters, which includes most of North America. The vine climbs by twining, typically on trees and shrubs (van der Maesen 2002). The leaves are made up of three leaflets, somewhat reminiscent of poison ivy foliage (Figure 21.46). The root system is extensive, and in sandy soil, the roots can extend 3.6 m deep (Munger 2002a). Massive tuber-like swellings termed tuberous roots occur on the roots. These can be over 2 m long and over 50 cm in diameter, and weigh as much as 180 kg (van der Maesen 1985). The flowers are reddish-purple or blue (Figure 21.47), with a pleasant fragrance. The fruit pods are greenish when young, brown when mature (Figure 21.48),

FIGURE 21.46 Foliage of kudzu. Photo by James H. Miller and Ted Bodner, Southern Weed Science Society, Bugwood.org (CC BY 3.0).

FIGURE 21.47 Flowering stems of kudzu. (a) Photo by Matt Lavin (CC BY SA 2.0). (b) Photo by Leslie J. Mehrhoff, University of Connecticut, Bugwood.org (CC BY 3.0). (c) Painting (public domain image) from the Yokohama Nursery Company, 1909–1910. Descriptive catalogue of the Yokohama Nursery Co., Bulbs, Plants, Seeds. Yokohama Nursery Co., Yokohama, Japan. Biodiversitylibrary.org (CC BY 2.0).

FIGURE 21.48 Mature fruits (pods) of kudzu. Photo by James H. Miller & Ted Bodner, Southern Weed Science Society, Bugwood.org (CC BY 3.0).

up to 13 cm long, with seeds 3–5 mm long. Extra-floral nectaries (i.e., glands secreting sweet liquid but not located in the flowers) occur on the stems just under the flower clusters (Harvey 2009). These are thought to protect the seeds by attracting ants, which repay the food service from the plant by chasing away other insects.

Like most other species of the pea family (Fabaceae), kudzu has a symbiotic relationship with nitrogen-fixing bacteria which are in the roots. In most natural habitats, nitrogen is in very short supply, so the relatively few plant species that have associated with bacteria that make nitrogen available have a distinct advantage. In agriculture, such plants are greatly valued for improving soil fertility, although today synthetic nitrogen fertilizers are routinely employed to make crops grow faster and bigger.

Harmful Aspects

Kudzu has been listed as one of the 100 world's worst alien species (Lowe et al. 2000; a rather arbitrary list), and it has been a frequent subject of books about harmful plants (Hoots and Baldwin 1996; Lembke 2001; Peters 2018). Kudzu has some potential for invading various areas of the world (European and Mediterranean Plant Protection Organization 2007; Follak 2011; Figure 21.49). However, by a considerable margin, the species has been most harmful in the southeastern U.S. (Britton et al. 2002). In the U.S., kudzu has become known as 'the vine that ate the South,' a reflection of its common occurrence in the southern states of Mississippi, Alabama, and Georgia. The first introduction to the U.S. is often stated to have been from Japan in 1876, although there is evidence of earlier ornamental introductions at least by 1855 (Lindgren et al. 2013). Especially in the first half of the 20th century, the plant was employed as an ornamental vine, for erosion control, and as fodder for livestock, but after about 1950 it became evident that kudzu is extremely detrimental (Lowney and Best 1998; Mitich 2000; Lowney 2002; Forseth and Innis 2004; Simberloff 2011). In the Southeastern U.S., kudzu has proven to be particularly damaging to forests employed for lumber (Harron et al. 2020; Figure 21.50), as well as harming orchards and some perennial crops. Kudzu kills trees and other plants by smothering (preventing access to sunlight). It also overgrows and damages power transmission lines, communication lines, railways, and abandoned buildings. While kudzu can make an attractive ornamental on buildings (Figure 21.51a), it can easily grow uncontrollably (Figures 21.51b and 21.52).

In the southeastern U.S., kudzu has been most successful in natural habitats disturbed by human activities, including roadsides, logged forests, and abandoned agricultural lands. However, the plant also invades natural areas, parks, and urban regions. U.S. military land managers have complained that kudzu has compromised some of the wildlands employed for training purposes (Guertin et al. 2008).

FIGURE 21.49 Kudzu spreading in Hawaii. Photo by Forest and Kim Starr, Starr Environmental, Bugwood.org (CC BY 3.0).

FIGURE 21.50 Kudzu smothering trees in Southeastern U.S. (a) At a golf course. Photo by Bill Sutton (CC BY 2.0). (b) Beside a field. Photo by James H. Miller, USDA Forest Service, Bugwood.org (CC BY 3.0).

FIGURE 21.51 Use of kudzu as an ornamental vine to cover buildings. (a) A postcard (public domain) showing a kudzu-covered Catholic church in Camden, South Carolina in the 1930s, published by Asheville Post Card Company. (b) An out-of-control vine that has become rampant. Photo by Florida Division of Plant Industry, Florida Department of Agriculture and Consumer Services, Bugwood.org (CC BY 3.0).

Kudzu outcompetes most other plants in the Southeastern U.S. and is harmful to them. It is responsible for producing considerable nitrogen in the soil, which is normally welcome by other plants, but this alters natural competitive relationships, and when the nitrogen leaches into streams the increased growth of algae harms some fish. Some chemicals produced in the roots inhibit growth of other plants (Rashid et al. 2010).

FIGURE 21.52 A residence under attack by kudzu. Painting by Jessica Hsiung.

Beneficial Aspects

Pueraria species have been employed for thousands of years in China and Japan as sources of medicines, textiles, food, starch, fodder, forage, paper, and building materials (van der Maesen 1985; Bodner and Hymowitz 2002). Tropical kudzu (*P. phaseoloides*) is employed in hot countries as a cover crop (one intended to maintain or improve soil between plantings), and it is admired for this purpose (Tian et al. 2001). However, the principal species of *Pueraria* of economic value is kudzu, which has been at least experimentally grown in numerous countries for its use as forage and fodder crops, soil conservation, and green manure (Tsugawa 1986a). In the early part of the 20th century, kudzu was an admired agricultural crop (McKee and Stephens 1948; Winberry and Jones 1973). Kudzu indeed has admirable qualities as a livestock feed (Tsugawa 1986b; Corley et al. 1997; Glass and Al-Hamdani 2016; Gulizia and Downs 2019) and for other purposes (Tanner et al. 1979; Textbox 21.7). In China and Japan, kudzu roots have been a valued source of edible starch (Figure 21.53b), although because the roots are difficult to collect and extract, production is limited (Pasiecznik 2007). Kudzu foliage, shoots, and flowers can be steamed or pickled as a vegetable (van der Maesen 1985). Kudzu can be employed in numerous recipes (Shurtleff and Aoyagi 1985). In Asia, the root is an ingredient in many herbal medicinal remedies (Wong et al. 2011; Maji et al. 2014; Lim 2016b;

TEXTBOX 21.7 IN PRAISE OF KUDZU

Alongside almost any southern road during hot summer days, a seething, undulating sea of green dominates the landscape. A sea whose fearsome waves have inundated houses, barns, trees, utility poles, and hillsides. If you stop to view the many-shaped, self-sculptured topiary, you may hear... the gnashing of teeth from despairing property owners trying to stem the onslaught. Many have abandoned the fight for ownership and ceded victory to a voracious enemy – kudzu... In reality, it has many good and profitable attributes. For instance, it is used for erosion control (does it ever control!). It is an ideal source for livestock grazing, fodder, and hay. The leguminous roots enrich the soil by providing nitrogen-producing bacteria. The vines, which have the tensile strength of sisal, are made to order for strong sturdy wicker basketry, and the long, silky fibers can be turned into cloth. The large roots and vines furnish the basic raw material for fine paper making. The purple and magenta flowers, with their distinct grape odor, are attractive to bees and thus ensure a good honey crop. Young kudzu leaves may be eaten raw and in salads, or cooked for a green vegetable. However, it is the root which is most widely used. Dried and diced, it makes a medicinal tea highly prized in the Orient. The powdered root is available in capsule form in health food stores.

—*Trimble (1996)*

FIGURE 21.53 Harvesting and processing of kudzu in 19th-century Japan. Source (public domain): Dai Nippon Bussan Zue [Products of Greater Japan]. 1877. Ōkura Magobei, Tokyo. Drawings by Utagawa Hiroshige III (1842–1894). (a) Use of young shoots to prepare a textile fiber. (b) Use of roots to prepare an edible starch-based powder called 'Japanese arrowroot.'

FIGURE 21.54 Kudzu roots in a market, in London, England. Photo by Emőke Dénes (CC BY SA 4.0).

Figures 21.54 and 21.55), most notably for the treatment of alcohol-induced hangovers (Keung and Vallee 1998). In Asia, kudzu stems have been the source of stem fibers (Figure 21.53a), woven into cloth, fishing lines, and baskets, and employed for paper (van der Maesen 1985). The stems are tough and can be used for wicker basketwork (Figure 21.56). Although the stems are the usual source, fiber can also be obtained from the roots (Lee et al. 2018a). Because it produces so much biomass, kudzu has been considered to be a potential source of biofuel (Sage et al. 2009). Kudzu accumulates lead in its roots, and so may be useful to remediate soils contaminated by this element (Schwarzauer-Rockett et al. 2013).

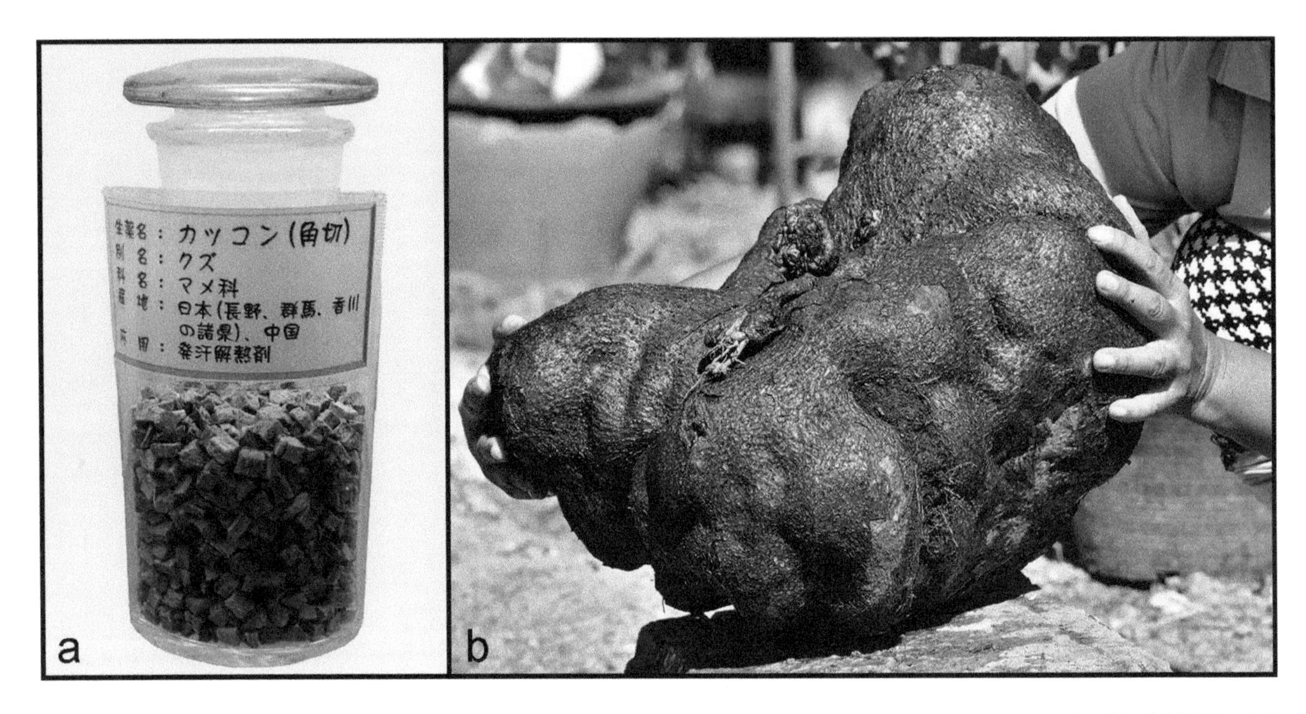

FIGURE 21.55 Medicinal use of *Pueraria*. (a) Japanese medicinal preparation of *P. lobata*. Photo by Matsuoka Akari (CC BY 3.0). (b) An exceptionally large root of *Pueraria candollei* var. *mirifica*, a Thai medicinal plant used traditionally as a rejuvenating herb. It is a rich source of phytoestrogens and is advertised as a breast enlarger. Source: Shutterstock, contributor: Pisitpong2017.

FIGURE 21.56 Basket made out of kudzu stems by basket-maker Matt Tommey. Photo by Mijtommey (CC BY SA 3.0).

Conservation Aspects

Probably the most important environmental benefit of kudzu is its value for controlling erosion (Figure 21.57). Although the plant has very desirable agricultural properties, on balance its introduction into the Southeastern U.S. has been ecologically disastrous (Textbox 21.8). Aside from its threat to terrestrial environments, it also appears to have potential for atmospheric harm. Kudzu seems to have the potential to raise ozone levels by increasing nitric oxide emissions from soils as a consequence of its ability to enrich soil nitrogen. Ozone is often regarded as the most important air pollutant in terms of its impacts on human health and agriculture (Hickman et al. 2010).

Controlling kudzu has been a frustrating challenge. Mechanical removal is effective but is slow and expensive. Livestock, especially goats and sheep, are useful but can only consume a small amount of the plant (Figure 21.58). Herbicides are widely employed (Miller and Edwards 1983; Miller 1996; Figure 21.59), but such usage usually represents a threat to the environment. There has been an intensive search for biological control agents, particularly fungi and insects. To date, kudzu has resisted effects to eliminate it, and it may be that unless a better biological control agent is found, this invader may remain as a permanent pest. The kudzu bug (*Megacopta cribraria*; Figure 21.60), a natural herbivore on kudzu from Asia, was first

FIGURE 21.57 A steep slope in Tennessee, stabilized from erosion by kudzu. Photo (public domain) by Norm Stephens.

TEXTBOX 21.8 KUDZU: AN UNINTENDED ECOLOGICAL DISASTER

Humans have had a profound impact on the environment. We have destroyed species for food. We have destroyed homes and habitats of numerous species because of our needs for space. Some of the most interesting and also devastating impacts have been unintended outcomes of our desire to enhance or control our environment. Numerous species of plants and animals have been introduced in many, if not all, countries around the world in an attempt to secure a better life for humans. To provide a familiar food supply to those moving to a new place, domestic plants and animals were imported from one country into another country. Many of these domestic organisms were deliberately released into the wild while others escaped. Many introduced species have had a major impact on their new environments and have eliminated or now threaten native plants and animals... These species include... kudzu in the eastern United States.

—*Matthews and Cummo (1999)*

FIGURE 21.58 Hair sheep grazing on kudzu. ('Hair sheep' are breeds with sparse fur, which does not require shearing. They are better suited for brush control and meat than wool sheep.) Photo by James H. Miller, USDA Forest Service, Bugwood.org (CC BY 3.0).

FIGURE 21.59 Eliminating kudzu by using herbicides. (a) Spraying by hand. Photo by James H. Miller, USDA Forest Service, Bugwood.org (CC BY 3.0) (b) Aerial application. Photo by John D. Byrd, Mississippi State University, Bugwood.org (CC BY 3.0).

discovered in the United States in 2009 and has spread rapidly in the Southeastern U.S. (Britt and Taylor 2019). Unfortunately, it not only likes kudzu (which it can reduce by over 30%) but also is fond of soybean, which it can reduce by over 60%. The insect also consumes other crops in the pea family, and has become a household pest.

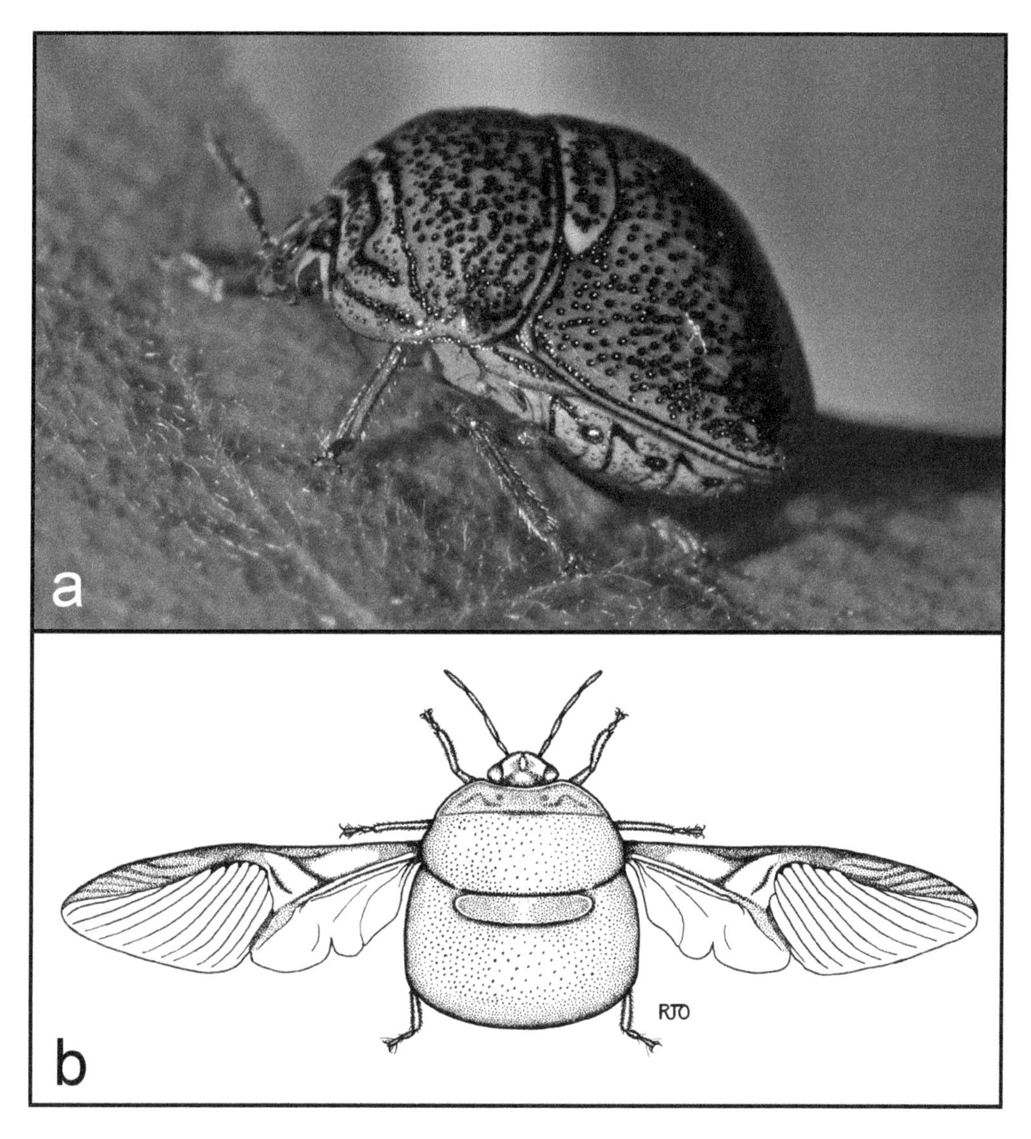

FIGURE 21.60 Kudzu bug (*Megacopta cribraria*), an Asian insect first discovered on kudzu in Georgia in 2009, it has spread throughout the Southeastern U.S. Adults are 3.5–6mm long and have a small plate in the middle of the back, above the elytra (wing coverings). This species would make a good biocontrol agent but harms crops, particularly soybean. (a) Photo by Robert Webster/xpda.com (CC BY SA 4.0). (b) Line drawing showing wings extended. Prepared by Russ Ottens, University of Georgia, Bugwood.org (CC BY 3.0).

CANADA THISTLE

Introduction

In its general sense, a 'thistle' is a prickly plant. Such plants are extremely widespread in the world. Several are pernicious weeds, and of these, Canada thistle (*Cirsium arvense*) is considered to be the most invasive and the most harmful. Thistles and other prickly plants, like biting insects, have the unfortunate effect of indicating to the public that creatures capable of inflicting pain are widespread in nature. Thistles provide the opportunity to point out that even these objectionable plants have a good side. Unfortunately, Canada thistle has a very bad side. It is not merely an urban weed that people regularly encounter in cities, it is also one of the world's worst agricultural weeds and is a vicious invader of natural ecosystems.

Thistles include many plants in different families which have acquired prickles or spines, normally on the leaf edges, sometimes additionally on the stems, and occasionally on the flowering parts. These protect the plants against herbivores. Numerous plants in the daisy family (Asteraceae) are thistles, particularly tribe Cardueae, which includes the genera *Carduus*, *Cirsium*, and *Onopordum*. Few thistles are popular with people. Globe or French artichoke (*Cynara cardunculus*) is the source of artichokes, and so is liked. The thistle (which species is not clear) has been adopted as a heraldic symbol, particularly in Scotland, France, and countries with colonial relationships with them, where in fact the plant has been respected historically (Figures 21.61 and 21.62). The blessed thistle (*Cnicus benedictus*) has a long history as a folk medicinal plant. Just which species was the subject of Genesis 3:18, 'It [the cursed ground] will produce thorns and thistles for you, and you will eat the plants of the field' is open to debate. Of all thistles, Canada thistle is the most widespread and the most troublesome weed.

Canada thistle (Figure 21.63) is one of about 200 species of the genus *Cirsium* (Keil 2006). It is also known as Canadian thistle, California thistle (Californian thistle), creeping thistle, field thistle, and by many other names, including lettuce from Hell thistle and cursed thistle. Despite the predominant name Canada thistle, the species is a native of southeast Europe, North Africa, and central Asia. It appears to have been introduced into North

FIGURE 21.61 Examples of thistles in heraldry. (a) Greater coat of arms of the city of Nancy, France. Photo by Oie blanche (CC BY SA 4.0). (b) Thistle of Scotland in the Royal Badge of the Kingdom of Scotland. Photo by Sodacan (CC BY SA 3.0). (c) Arms of Nova Scotia. Photo by Fenn-O-maniC (CC BY SA 4.0).

FIGURE 21.62 Humorous depictions (public domain) of thistles in reference to France and Scotland, where the plants have in fact been respected for centuries. (a) Source: Grandville (1867). (b and c) Source: Early 20th century postcards.

FIGURE 21.63 Illustrations of Canada thistle (*Cirsium arvense*) in old plant guides. (a) Source: Thomé (1885). (b) Source: Schmeil and Fischen (1913). (c) Source: Clark et al. (1906).

America from different locations in Europe (Guggisberg et al. 2012). Canada thistle occurs in many habitats, including gardens, roadsides, pastures, and croplands.

The branching stems (Figure 21.64) usually range from 30 to 150 cm in height (plants sometimes grow to 2 m) and rarely have spiny wings, unlike many other thistles. The leaves are lobed, with the margins developing very spiny points, beginning with the seedling stage (Figure 21.65), although occasionally, plants with spineless leaves occur.

Canada thistle spreads by deep, creeping roots, forming colonies (Figure 21.66). The roots can reach down 5 m. Mowing or pulling out plants is not effective because they can grow back from the remaining roots. Cultivation of the soil can break the roots into sections, but each of the pieces can grow into a new plant, making the infestation worse. Ploughing can spread the pieces over a field.

For the most part, Canada thistle plants develop either male or female flowers. Often, because a given plant grows into a genetically uniform colony, some colonies produce only pollen and some produce only seeds. In the daisy family (Asteraceae), what appear to be flowers are assemblages ('flower heads') of many tiny flowers ('florets'). Most flower heads are purple or lavender, occasionally white. Canada thistle flower heads typically contain more than 100 florets (Figure 21.67). Each fertilized floret produces a tiny one-seeded fruit (generally called a seed), which has a tuft of hairs attached to the top (Figure 21.68c). As in dandelion seeds, the tuft of hairs is a sail serving for dispersion by wind, although water and animals can also distribute the seeds. The seeds can remain viable in the soil for 20 years. Each flower head of female plants can produce as many as 100 small seeds (2–4 mm long), and a given plant can produce thousands of seeds (Figure 21.68).

FIGURE 21.64 Canada thistle (*Cirsium arvense*) plants showing branching pattern. Photo by Steve Dewey, Utah State University, Bugwood.org (CC bY 3.0).

FIGURE 21.65 Foliage of young plants of Canada thistle (*Cirsium arvense*), showing spiny leaf margins. (a) Seedling. Photo by Judith (CC BY SA 4.0). (b) Young rosette. Photo by Phil Westra, Colorado State University, Bugwood.org (CC BY 3.0).

FIGURE 21.66 Extensive root system of Canada thistle (*Cirsium arvense*) producing a colony of plants. Illustration redrawn and colorized by B. Brookes based on a figure in Hill (1915).

FIGURE 21.67 Flower heads of Canada thistle (*Cirsium arvense*). (a) Intact flower head. Photo by Photogramma (CC BY SA 2.0). (b) Flower head split open to show the tiny florets. Photo by Andrey Zharkikh (CC BY 2.0).

FIGURE 21.68 Fruit (seed) production by Canada thistle (*Cirsium arvense*). (a) Plants at the fruiting stage, with evident floss ('thistle-down'). Photo by Pierre Henri Giraud (CC BY SA 2.5). (b) A single head of fruiting florets. Photo by Matt Lavin (CC BY 2.0). (c) A single fruit (seed) with attached pappus (floss). Photo by Rasbak (CC BY SA 3.0).

Harmful Aspects

Canada thistle is a very serious weed of crops, pastures, rangelands, urbanized areas, and natural ecosystems (Moore 1975; Mitich 1988b; Donald 1994; Tiley 2010; Kurtz and Hansen 2019; Textbox 21.9). It often forms large stands in fields (Figure 21.69). In summing up a review of Canada thistle, Hill (1983) commented 'It lacks most virtuous qualities and is rightly considered noxious.' Control measures, normally involving repeatedly applying a burdensome combination of elimination procedures, are discussed by Jacobs et al. (2006), Drahota (2010), Orloff et al. (2018), and Favrelière et al. (2020). Eckberg et al. (2017) commented 'Canada thistle is one of the most studied invasive plants in the world and is the target of extensive biological control and weed management programs. A growing body of research continues to focus on Canada thistle resulting in approximately 3000 publications on the management of this weed. Several strategies have been employed to manage Canada thistle including herbicides, mechanical destruction, biological control agents, selection of competitive native plants, and pathogens.'

TEXTBOX 21.9 CANADA THISTLE'S HORRIBLE PERFORMANCE EVALUATION

Cirsium arvense is a major pest and is considered one of the world's worst weeds, ranked as the third most important weed in Europe. Yield losses due to *C. arvense* occur in horticultural crops, field crops, pastures, rangelands, lawns, vineyards, and orchards... *C. arvense* causes greater crop losses than any other broadleaf weed in its growth range... *C. arvense* is among the ten most frequently listed noxious weeds in North America... In pastures, *C. arvense* reduces forage consumption and cattle will not graze near the plants because of the sharp spines on its leaves...

The natural communities where *C. arvense* has an impact include various non-forested plant communities such as prairies, barrens, savannas, glades, sand dunes, fields, and open meadows that have been impacted by disturbance. It negatively impacts natural environments by crowding out and replacing native grasses and forbs, changing the structure and species composition of natural plant communities, and reducing species diversity. Although primarily seen as a weed of field and horticultural crops or of natural areas, other environments are also affected such as turf, landscape, and nurseries. As various countries put in place ecological compensation areas or set aside land formerly in agriculture, these lands become vulnerable to invading *C. arvense* which is often listed as the most common weed in these areas... *C. arvense* produces allelochemicals that are released into the soil that can be toxic to surrounding vegetation, and the phytotoxicity of soil incorporated plant parts can persist... Negative impacts of *C. arvense* invasions on biodiversity include crowding out and replacing native grasses and forbs, decreasing the species diversity of an area, and altering ecosystem structure and composition.

—*CABI (2019c)*

FIGURE 21.69 Fields heavily infested by Canada thistle (*Cirsium arvense*). (a) Plants in flower. Photo by Steve Dewey, Utah State University, Bugwood.org (CC BY 3.0). (b) Plant in fruit. Photo by Huhu Uet (CC BY 3.0).

Beneficial Aspects

Thistles are generally avoided by livestock. However goats eat thistles at both the rosette and flowering stage as well as post-flowering (Figure 21.70a), and sheep, horses, and cattle may engage in limited grazing (Figure 21.70b–d), especially when the plants are very young, or they may simply consume the flowers. Cattle can be trained to eat thistles (Vickers 2016). Like several other plants, Canada thistle can accumulate high levels of nitrates, depending on the soil, which can be toxic to livestock, so careful monitoring of consumption is necessary (Vickers et al. 2017). Thistle flowers are very attractive to bees (Figure 21.71), and so contribute to honey production. Thistle flowers are also extremely attractive to butterflies (Figure 21.72), and the seeds are loved by wild birds (Figure 21.73). People have consumed thistles, but mostly as a famine food. Like most plants, Canada thistle has been utilized in folk medicine for a variety of uses. Foster and Duke (2000) note its employment

FIGURE 21.70 Livestock consuming Canada thistle (*Cirsium arvense*). (a) A goat eating thistles. Photo by Floodllama (CC BY 2.0). (b) Cows in a pasture infested with Canada thistle. Photo by Odd Wellies (CC BY 2.0). (c) A horse eating thistle. Source: Shutterstock, contributor: Nigel Baker photography. (d) A sheep eating thistle. Source: Shutterstock, contributor: Howard Marsh.

FIGURE 21.71 A honey bee (*Apis mellifera*) collecting nectar from Canada thistle (*Cirsium arvense*). Photo by Ivar Leidus (CC BY SA 4.0).

FIGURE 21.72 Butterflies on flowers of Canada thistle (*Cirsium arvense*). (a) Small tortoiseshell (*Aglais urticae*). Photo by Rea (CC BY ND 2.0). (b) European peacock (*Inachis io*). Photo by Jörg Hempel (CC BY SA 2.0). (c) Painted lady (*Vanessa cardui*). Photo by hedera.baltica (CC BY SA 2.0).

FIGURE 21.73 Goldfinches feeding on seeds of thistles (*Cirsium* species). (a) American goldfinch (also known as the eastern goldfinch and wild canary; *Carduelis tristis*) on Canada thistle (*C. arvense*). Photo by Will Sweet (CC BY 2.0). (b) European goldfinch (*Carduelis carduelis*) on spear thistle (*Cirsium vulgare*). Photo by Andreas Trepte (CC BY SA 2.5).

for dysentery, diarrhea, stimulating bowels, deworming, skin eruptions and ulcers, poison ivy rash, and tuberculosis. Admittedly, the harmful aspects of Canada thistle greatly outweigh the beneficial aspects.

Conservation Aspects

Some insects and fungi have been employed as biological control agents, but with limited success (Cripps et al. 2011). There is concern that some of these agents have threatened protected endangered species (Textbox 21.10; for a defense of biological control, see Sheppard et al. 2006). Moreover, there is also concern that because thistle species provide extensive support to pollinators, laws against weedy thistles actually endanger the survival of some bees (Vray et al. 2017). Herbicides remain widely employed and need to be applied repeatedly, which is undesirable for biodiversity. There is considerable worry about the survival of some endangered species of *Cirsium*, but the poor image of thistles makes their conservation challenging. Attempts are underway to save the 'Mingan thistle' (*Cirsium scariosum* var. *scariosum*), a threatened species in Quebec (Parks Canada 2020). Eckberg et al. (2017) provide a detailed guide to conserving native *Cirsium* species.

TEXTBOX 21.10 RISKING THE WELFARE OF NATIVE SPECIES BY RELEASING BIOCONTROL AGENTS AGAINST INVASIVE THISTLES

Some weeds are so noxious, their crimes so heinous, and their control so challenging that desperation leads us to introduce other non-native organisms to contain them. Alien vs. alien duking it out in a novel environment. It seems counterintuitive – if an introduced species has reached the status of invasive, is it worth the risk of bringing in yet another foreign species in attempt to defeat it? We all know what happened to the old lady who swallowed the spider to catch the fly, yet for decades now we have been doing just this. It's something we call classical biological control – introducing pathogens, insects, or other organisms to help control the spread of problematic ones. Such attempts mostly fail, but we keep trying. The attempts made on *Cirsium arvense* exemplify this… The trouble is that even when such efforts fail, they aren't always benign… Unfortunately, and perhaps not surprisingly… biocontrol currently provides little or no control of Canada thistle populations [and]… quite a few native thistles, several of which are rare or threatened… are at risk of such agents.

—*Awkward Botany (2018)*

Thistles (*Cirsium* spp.) are prickly plants native to North America that are numerically minor and are often considered unattractive or undesirable. So, thistles might be considered expendable… Few individuals are known to admire thistles (although the Scots are a notable exception). Most express a disregard for thistles, native or not, without knowing much about them or their ecological interactions and ecosystem functions. The impact on native North American thistles was not a major factor in the decision to release the biocontrol weevil *Rhinocyllus conicus* in 1969, in spite of evidence that it would use *Cirsium* species. Nor is the potential impact of *Larinus planus* on thistle populations, exotic or native, a factor influencing its redistribution within North America. Distribution continues, in spite of evidence that each of these weevils accepts *Cirsium* species into its diet and that each can have a major non-target effect on native species. Perhaps a disregard for noncharismatic species, especially prickly ones, is understandable. However, neither a lack of charisma nor relative rarity provides an adequate scientific basis for deciding whether a species has a significant ecological function or indirect economic value.

—*Louda and Rand (2003)*

Epilogue

Tolerant Co-existence vs. Justifiable Biocide

'CAN'T LIVE WITH THEM, CAN'T LIVE WITHOUT THEM'

This familiar saying is frequently cited in reference to the frustration experienced by members of one sex with those of the opposite sex. It is also appropriate in addressing the relationship between humans and numerous pests, especially the most despised species featured in this book (Figure 1). Like most proverbs, it oversimplifies. The basic need is to find a reasonable way of co-existing that minimizes harm, and this is possible in most circumstances (Figure 2). Nevertheless, some species are so dangerous that they represent existential threats to our existence. Just how drastically they need to be controlled is the key issue that must be addressed.

FIGURE 1 Creatures (some hypothetical) that seem so dangerous under some circumstances that their elimination is essential for human welfare. All figures are from Shutterstock. (a) Carnivorous plant (contributor: User friendly). (b) Snake (contributor: Satori.artwork). (c) Mouse chewing through electrical cable (contributor: Teguh Mujiono). (d) Coronavirus (contributor: Vector bucket). (e) Giant leech (contributor: Crystal Eye Studio). (f) Tick on dog (contributor: Decter). (g) Bedbug biting a human's foot (contributor: Refluo). (h) Girl being bitten by mosquitos (contributor: BlueRingMedia). (i) Rats consuming garbage (contributor: GraphicsRF.com).

FIGURE 2 Peaceful co-existence among people and disliked species. Prepared by Jessica Hsiu.

MAKING DIFFICULT DECISIONS TO ELIMINATE HARMFUL SPECIES

Many species have gone totally or at least regionally extinct because of the activities of people, but usually not deliberately. Humankind has, as a result, significantly changed the world although these changes have been relatively slow. As discussed in Chapter 14, advancing technologies offer the prospects of efficiently controlling, even eliminating species. The rapid changes that are now conceivable are both promising and terrifying.

In the case of infectious diseases, a distinction is often made between 'eradication' (regional but not world elimination) and 'elimination' (reduction of world prevalence to zero). In this sense, smallpox in humans and rinderpest in ruminants have been 'eradicated' although the causal organisms still exist (basically in secure laboratories). Regardless, deliberately driving species to extinction in nature is rapidly becoming an option for society.

In the human sphere, degrees of culpability for legally terminating the lives of other humans are widely recognized. In particular, circumstances matter: the extent of provocation and the possibility of non-lethal alternatives are key considerations. Self-defense (including war) can be a defense for killing. Often, killing other people constitutes 'Justifiable Homicide,' freeing one from legal (if not necessarily moral) responsibility (see, for example, Bedau and Cassell 2005; Mandery 2005; Marzilli 2008). One of the signs of the state of morality of a nation is the issue of capital punishment: nations that still apply the death penalty include the most brutalized, anti-democratic countries, while societies that value all individuals have entirely or mostly abandoned state-sanctioned killing of the imprisoned. Indeed, the history of civilization over the millennia is marked by continuously decreasing savagery, albeit our innate tendencies to be cruel to each other probably is no different from that of our ancestors who lived in the very distant and much more dangerous past.

Remarkably, scholars have examined analogies between the ethics of executing humans and killing non-human animals (Beirnes 1994; Ebury 2021). Today, very few would apply morality and associated legislation from the human world to the non-human species, but most reasonable people are sufficiently empathetic to appreciate that wanton killing of living things without a reasonable purpose is unjustified. However, the creatures highlighted in this book provoke considerable hostility, and indeed the majority can be extremely dangerous to human survival. Probably most of these despised species would be totally exterminated if it were possible. This book argues that this would be unwise and unjustified, as such actions endanger ourselves. Nevertheless, protecting oneself from attack can be a question of survival. A well-known scriptural recommendation is to turn the other cheek in response to injury – a noble perspective, but one that most people are incapable of following literally. Creatures that are genuinely harmful to the health and welfare of ourselves, our domesticates, and our inanimate property need to be controlled, and frequently, this means killing them. The decision of how much deadly force to use on any form of life should require careful consideration although currently there is very little sympathy for most of the world's wild species. One of the wisest of credos is the Alcoholics Anonymous Serenity Prayer, written by American theologian Rheinhold Niebuhr ((1892–1971): 'God, grant me the serenity to accept the things I cannot change, the courage to change the things I can, and the wisdom to know the difference.' Serenity is advisable to accept the minor annoyances caused by many pests, courage is needed to tolerate the many ugly species that look dangerous but aren't, and wisdom is essential to take sufficient but not excessive actions to limit significant threats.

Literature Cited

Able, K. W., and S. M. Hagan. 2003. "Impact of Common Reed, *Phragmites australis*, on Essential Fish Habitat: Influence on Reproduction, Embryological Development, and Larval Abundance of Mummichog (*Fundulus heteroclitus*). *Estuaries* 26: 40–50.

Abraham, C. T., and J. Nimmy. 2015. "Weedy Rice Invasion and Its Management." *Indian Journal of Weed Science* 47: 216–223.

Abrol D. P. 2012. *Pollination Biology*. Dordrecht, Springer.

Ackerfield, J., and J. Wen. 2002. "A Morphometric Analysis of *Hedera* L. (the Ivy Genus, Araliaceae) and Its Taxonomic Implications." *Adansonia Série 3* 24: 197–212.

Ackerfield, J., and J. Wen. 2003. "Evolution of *Hedera* (the Ivy Genus, Araliaceae): Insights from Chloroplast DNA Data." *International Journal of Plant Sciences* 164: 593–602.

Ackerman, J. D. 2000. "Abiotic Pollen and Pollination: Ecological, Functional, and Evolutionary Perspectives." *Plant Systematics and Evolution* 222: 167–185.

Adams, J. 2009. *Species Richness. Patterns in the Diversity of Life*. Berlin, Springer. https://doi.org/10.1007/978-3-540-74278-4_5

Adams, J. C., E. McBride, C. Anne, A. Carr, and K. Carnelly. 2010. "The Human-Animal Bond: Role of Anthropomorphism in Diversity and Variation." *Journal of Veterinary Behaviour: Clinical Applications and Research* 5: 41–42.

Adams, R. A., and S. C. Pedersen. Eds. 2013. Bat Evolution, Ecology, and Conservation. New York, Springer-Verlag.

Adams, W. M. 2017. "Geographies of Conservation I: De-extinction and Precision Conservation." *Progress in Human Geography* 41: 534–545.

Adelaja, A. O., Y. G. Hailu, and A. T. Tekle. 2008. "Conservation Spending in the United States: A Political Economy Analysis." *Journal of Environmental Monitoring and Restoration* (Conference Paper, December 2008). https://www.researchgate.net/profile/Adesoji-Adelaja/publication/250362011_The_political_economy_of_state_conservation_spending_-_the_case_of_Michigan/links/5850e03308ae4bc8993b7353/The-political-economy-of-state-conservation-spending-the-case-of-Michigan.pdf

Aguilera, A. G., P. Alpert, J. S. Dukes, and R. Harrington. 2010. "Impacts of the Invasive Plant *Fallopia japonica* (Houtt.) on Plant Communities and Ecosystem Processes." *Biological Invasions* 12: 1243–1252.

Ahmad, M., D. H. Moon, M. Vithanage, A. Koutsospyros, S. S. Lee, J. E. Yang, et al. 2014. "Production and Use of Biochar from Buffalo-weed (*Ambrosia trifida* L.) for Trichloroethylene Removal from Water." *Chemical Technology and Biotechnology* 89: 150–157.

Ahmadi, R., and M. A. Ebrahimzadeh. 2020. "Resveratrol – A Comprehensive Review of Recent Advances in Anticancer Drug Design and Development." *European Journal of Medicinal Chemistry* 200: 112356. https://doi.org/10.1016/j.ejmech.2020.112356.

Ajaykumar, R., and K. Sharmili. 2020. "Weedy Rice: An Emerging Threat to Rice Cultivation." In *Research Trends in Crop and Weed, Volume 3*, edited by V. Nand, 107–130. Uttar Pradesh, India, AkiNik.

Ajithram A., J. T. W. Jappes, T. S. M. Kumar, N. Rajini, A. V. Rajulu, S. M. Rangappa, et al. 2020. "Water Hyacinth for Biocomposites – An Overview." In *Biofibers and Biopolymers for Biocomposites*, edited by A. Khan, S. M. Rangappa, S. Siengchin, and A. Asiri, 171–179. Cham, Springer.

Akabassi, G. C., K. K. Palanga, E. A. Padonou, E. A. Y. D. Dagnon, K. Tozo, and A. E. Assogbadjo. 2022. "Biology, Production Constraints and Uses of *Cyperus esculentus* L. (Neglected and Underutilized Crop Species), Implication for Valorization: A Review." *Genetic Resources and Crop Evolution* 69: 1979–1992.

Alam, M. S., J. A. Khan, and B. J. Pathak. 2015. "Striped Hyena (*Hyaena hyaena*) Status and Factors Affecting Its Distribution in the Gir National Park and Sanctuary, India." *Journal of Vertebrate Biology* 64: 32–39.

Albert, C., G. M. Luque, and F. Courchamp. 2018. "The Twenty Most Charismatic Species." *PLoS One*. https://doi.org/10.1371/journal.pone.0199149

Alberternst, B., S. Nawrath, and U. Starfinger. 2016. "Biodiversity Impacts of Common Ragweed." *Julius-Kühn-Archiv* 455: 188–226. https://doi.org/10.5073/jka.2016.455.45; https://www.researchgate.net/profile/Uwe_Starfinger2/publication/310484408_Biodiversity_impacts_of_common_ragweed/links/584fe96a08aed95c250b4db9/Biodiversity-impacts-of-common-ragweed.pdf

Albo, M. J., L. Montes De Oca, and I. Estevan. 2021. "Fearless and Positive Children after Hands-On Educational Experience with Spiders in South America." *Journal of Biological Education* 55: 395–405.

Aleshire, E. B. 2005. *Forage Systems for the Southeastern United States: Crabgrass and Crabgrass-Lespedeza Mixtures*. M.Sc. Thesis, Blacksburg, Virginia, Virginia Polytechnic Institute and State University. https://pdfs.semanticscholar.org/47c0/0a5ddc72ffc13e7a1b7d55f3cba34b7e629e.pdf

Alex, J. F., and C. M. Switzer. 1977. *Ontario Weeds*. Publication 505, Toronto, Ontario Ministry of Agriculture and Food.

Aligica, P. D. 2010. "Elinor Ostrom - Nobel Prize in Economics 2009." *Economic Affairs* 30: 95–96.

Alker, S., V. Joy, P. Roberts, and N. Smith. 2000. "The Definition of Brownfield." *Journal of Environmental Planning and Management* 43: 49–69.

Al-Khleif, A., M. Roth, C. Menge, J. Heuser, G. Baljer, and W. Herbst. 2011. "Tenacity of Mammalian Viruses in the Gut of Leeches Fed with Porcine Blood." *Journal of Medical Microbiology* 60: 787–792.

Allen, B. 2016. *Animals in Religion: Devotion, Symbol and Ritual*. London, Reaktion Books.

Almeling, R. 2007. "Selling Genes, Selling Gender: Egg Agencies, Sperm Banks, and the Medical Market in Genetic Material." *American Sociological Review* 72: 319–340.

Aloto, D., and E. Eticha. 2018. "Leeches: A Review on Their Pathogenic and Beneficial Effects." *Journal of Veterinary Science & Technology* 19(1): 511. https://doi.org/10.4172/2157-7579.1000511.

Al-Snafi, A. E. 2016a. "A review on *Cyperus rotundus*: A Potential Medicinal Plant." *IOSR Journal of Pharmacy* 6: 3–48.

Al-Snafi, A. E. 2016b. "Chemical Constituents and Pharmacological Effects of *Cynodon dactylon* – A Review." *IOSR Journal of Pharmacy* 6: 17–31.

Al-Snafi, A. E. 2018. "Pharmacological and Therapeutic Activities of *Hedera helix* – A Review." *IOSR Journal of Pharmacy* 8: 41–53.

Al-Snafi, A. E. 2019. "Chemical Constituents and Pharmacological Effects of *Lythrum salicaria* – A Review." *IOSR Journal of Pharmacy* 9: 51–59.

Altringham, J. D. 2011. *Bats: From Evolution to Conservation*. Second Edition. Oxford, Oxford University Press.

Alves, R. R. N., and U. P. Albuquerque. Eds. 2017. *Ethnozoology: Animals in Our Lives*. London, Academic Press.

Alves, R. R. N., and H. N. Alves. 2011. "The Faunal Drugstore: Animal-based Remedies Used in Traditional Medicines in Latin America." *Journal of Ethnobiology Ethnomedicine* 7: 9. https://doi.org/10.1186/1746-4269-7-9

Alves, R. R. N., and G. A. P. Filho. 2007. "Commercialization and Use of Snakes in North and Northeastern Brazil: Implications for Conservation and Management." *Biodiversity and Conservation* 16: 969–985.

Alves, R. R. N., V. N. Silva, D. M. Trovão, J. V. Oliveira, J. S. Mourão, T. L. Dias, et al. 2014. "Students' Attitudes toward and Knowledge about Snakes in the Semiarid Region of Northeastern Brazil." *Journal of Ethnobiology and Ethnomedicine* 10: 1. https://fr.art1lib.org/book/62493448/a74a56

Alvey, A. A. 2006. "Promoting and Preserving Biodiversity in the Urban Forest." *Urban Forestry & Urban Greening* 5: 195–201.

Aman, S., A. Mazumder, U. K. Gupta, and A. Nayak. 2016. "Pharmacological Activities of *Chenopodium album* Linn – A Review." *World Journal of Pharmaceutical Research* 5: 361–371.

Amarello, M. 2017. "Teaching People to Hate Snakes Is a Disaster for Ecology. Our Negative Attitude toward Snakes Is Their Biggest Hurdle." *AlterNet (Blog)*. https://www.salon.com/2017/07/29/teaching-people-to-hate-snakes-is-a-disaster-for-ecology_partner/

American Council on Science and Health. 1992. "Food from Tobacco. A Well Kept Secret." https://www.acsh.org/news/1992/01/01/food-from-tobacco-a-well-kept-secret

American Society of Plastic Surgeons. 2019. *Plastic Surgery Statistics Report*. https://www.plasticsurgery.org/documents/News/Statistics/2019/plastic-surgery-statistics-full-report-2019.pdf

Amini, A. M. Sankian, M. A. Assarehzedegan, F. Vahedi, and A. Varasteh. 2011. "*Chenopodium album* Pollen Profilin (Che A 2): Homology Modeling and Evaluation of Cross-Reactivity with Allergenic Profilins Based on Predicted Potential Ige Epitopes and Ige Reactivity Analysis." *Molecular Biology Reports* 38: 2578–2587.

Amiot, C. E., and B. Bastian. 2015. "Toward a Psychology of Human–Animal Relations." *Psychological Bulletin* 141: 6–47.

Amri, F. 2022. "House Passes Bill to Prohibit Discrimination Based on Hair." *The Associated Press*, March 22. https://www.mymcmurray.com/2022/03/18/house-passes-bill-to-prohibit-discrimination-based-on-hair/

Andelman, S. J. and W. F. Fagan. 2000. "Umbrellas and Flagships: Efficient Conservation Surrogates or Expensive Mistakes?" *Proceedings of the National Academy of Sciences* 97: 5954–5959.

Anderl, C. T. Hahn, A.-K. Schmidt, H. Moldenhauer, K. Notebaert, C. C. Clément, et al. 2016. "Facial Width-to-Height Ratio Predicts Psychopathic Traits in Males." *Personality and Individual Differences* 88: 99–101.

Anderson, H. 2012. *Invasive Japanese Knotweed (*Fallopia japonica *(Houtt.)) Best Management Practices in Ontario*. Peterborough, Ontario Invasive Plant Council. https://www.ontarioinvasiveplants.ca/wp-content/uploads/2016/06/OIPC_BMP_JapaneseKnotweed.pdf

Andrae, J. 2002. "Crabgrass: Friend or Foe?" *Georgia Cattleman* 2002(August). https://athenaeum.libs.uga.edu/bitstream/handle/10724/33521/Aug02.pdf?sequence=1

Andrews, A. C. 1952. "The Opium Poppy as a Food and Spice in the Classical Period." A*gricultural History* 26: 152–155.

Andrews, C. W. 1900. *A monograph of Christmas Island (Indian Ocean)*. London, Longmans and Company.

Angel, A., R. M. Wanless, and J. Cooper. 2009. "Review of Impacts of the Introduced House Mouse on Islands in the Southern Ocean: Are Mice Equivalent to Rats?" *Biological Invasions* 11: 1743–1754.

Angelici, F. M., and L. Ross. Eds. 2020. *Problematic Wildlife II: New Conservation and Management Challenges in the Human-Wildlife Interactions*. Switzerland, Springer.

Angold, P. G., J. P. Sadler, M. O. Hill, A. Pullin, S. Rushton, K. Austin, et al. 2006. "Biodiversity in Urban Habitat Patches." *Science of the Total Environment* 360: 196–204.

Anjana, S. U., and M. Iqbal. 2007. "Nitrate Accumulation in Plants, Factors Affecting the Process, and Human Health Implications. A Review." *Agronomy for Sustainable Development* 27: 45–57.

Apfelbeck, B., R. P. H. Snep, T. E. Hauck, J. Ferguson, M. Holy, C. Jakoby, et al. 2020. "Designing Wildlife-inclusive Cities That Support Human-Animal Co-existence." *Landscape and Urban Planning* 200: 103817. https://doi.org/10.1016/j.landurbplan.2020.103817

Aragona, M., E. R. Lauriano, S. Pergolizzi, and C. Faggio. 2018. "*Opuntia ficus-indica* (L.) Miller as a Source of Bioactivity Compounds for Health and Nutrition." *Natural Product Research* 32: 2037–2049.

Archer, J. 1997. "Why Do People Love Their Pets?" *Evolution and Human Behavior* 18: 237–259.

Archer, J., and S. Monton. 2011. "Preference for Infant Facial Features in Pet Dogs and Cats." *Ethology* 117: 217–226.

Archibald, K. 2017. "Arctic Capital: Managing Polar Bears in Churchill, Manitoba." In *Animal Metropolis: Histories of Human-Animal Relations in Urban Canada* (e-book), edited by J. Dean, D. Ingram, and C. Sethna, 255–283. Calgary, University of Calgary Press. https://prism.ucalgary.ca/bitstream/handle/1880/51826/9781552388655_chapter09.pdf;jsessionid=FCD297B4AE23135AE4F035377B2C46F7?sequence=12

Arena, C., T. Graham, V. Legué, and R. Paradiso. 2021. "Editorial: Higher Plants, Algae and Cyanobacteria in Space Environments." *Frontiers in Plant Science* 12: 629014. https://doi.org/10.3389/fpls.2021.629014.

Arluke, A., and C. R. Sanders. 1996. *Regarding Animals*. Philadelphia, Temple University Press.

Armanino, S. 2017. "Attack of the English Ivy." *North Coast Journal of Politics, People and Art*, February 16. https://www.northcoastjournal.com/NewsBlog/archives/2017/02/16/attack-of-the-english-ivy

Arroz, A. M., R. Gabriel, I. R. Amorim, R. S. Marcos, and P. A. V. Borges. 2016. "Bugs and Society I: Raising Awareness about Endemic Biodiversity." In *Biodiversity and Education*

for Sustainable Development, edited by P. Castro, U. M. Azeiteiro, P. Bacelar-Nicolau, W. L. Filho, and A. M. Azul, 69–89. Switzerland, Springer International.
Arruda, K., and A. Pomés. 2013. "Every Cockroach Is Beautiful to Its Mother." *International Archives of Allergy and Immunology* 161: 289–292.
Asebe, G., Y. Hailu, and A. K. Basu. 2016. "Overview of the Biology, Epidemiology and Control Methods against Hard Ticks: A Review." *Global Journal of Science Frontier Research: C Biological Science* 16: 33–45. https://journalofscience.org/index.php/GJSFR/article/view/1775
Ashokkumar, K., S. Kumarakurubaran, and S. Muthukrishnan. 2013. "Review *Cynodon dactylon* (L.) Pers.: An updated Review of its Phytochemistry and Pharmacology." *Journal of Medicinal Plant Research* 7: 3477–3483.
Astudillo-Scalia, Y., and F. S. de Albuquerque. 2020. "The Geography of High-Priority Conservation Areas for Marine Mammals." *Global Ecology and Biogeography* 29: 2097–2106.
Athreya, V, M. Odden, J. D. C. Linnell, J. Krishnaswamy, and U. Karanth. 2013. "Big Cats in Our Backyards: Persistence of Large Carnivores in a Human Dominated Landscape in India." *PLoS One* 8(3), https://doi.org/10.1371/journal.pone.0057872
Atitwa, S. C. 2018. "How Many Dogs Are There in The World?" *WorldAtlas*. https://www.worldatlas.com/articles/how-many-dogs-are-there-in-the-world.html
Audubon, J. J. 1827–1838. *The Birds of North America*, Vol. 4. London, Robert Havell Jr.
Audubon, J. J., J. W. Audubon, and J. Bachman. 1851–1854. *The Quadrupeds of North America*. New York, J. J. Audubon.
Aukema, J. E., N. G. Pricope, G. J. Husak, and D. Lopez-Carr. 2017. "Biodiversity Areas under Threat: Overlap of Climate Change and Population Pressures on the World's Biodiversity Priorities." *PLoS One* 12(1): e0170615. https://doi.org/10.1371/journal.pone.0170615
Aust, P. W., N. Van Tri, D. J. D. Natusch, and G. J. Alexander. 2017. "Asian Snake Farms: Conservation Curse or Sustainable Enterprise?" *Oryx* 51: 498–505.
Avery, M. L., and M. Lowney. 2016. "Vultures." Ft. Collins, CO, Wildlife Damage Management Technical Series. USDA, APHIS, WS National Wildlife Research Center. https://www.aphis.usda.gov/wildlife_damage/reports/Wildlife%20Damage%20Management%20Technical%20Series/FINAL_MS%20Publisher%20Layout_Vultures_October%202016.pdf
Avise, J. C. 2015. "Evolutionary Perspectives on Clonal Reproduction in Vertebrate Animals." *Proceedings of the National Academy of Sciences of the United States* 112(29): 8867–8873.
Awkward Botany. 2018. "Attempts to Avenge the Acts of *Cirsium arvense* – A Biocontrol Story." *Awkward Botany: Citizen Botany for the Phytocurious*. https://awkwardbotany.com/2018/11/14/attempts-to-avenge-the-acts-of-cirsium-arvense-a-biocontrol-story/
Ayanda, O. I., T. Ajayi, and F. P. Asuwaju. 2020. "*Eichhornia crassipes* (Mart.) Solms: Uses, Challenges, Threats, and Prospects." *Scientific World Journal* Article ID 3452172. https://doi.org/10.1155/2020/3452172
Ayensu, J., R. A. Annan, A. Edusei, and H. Lutterodt. 2019. "Beyond Nutrients, Health Effects of Entomophagy: A Systematic Review." *Nutrition & Food Science* 49: 2–17.
Bacigalupe, L. D., C. Soto-Azat, C. García-Vera, I. Barría-Oyarzo, and E. L. Rezende. 2017. "Effects of Amphibian Phylogeny, Climate and Human Impact on the Occurrence of the Amphibian-Killing Chytrid Fungus." *Global Change Biology* 23: 3543–3553.
Badaracco, R. J. 1973. "Scorpions, Squirrels, or Sunflowers?" *American Biology Teacher* 35: 528–538.
Baertschi, B., and M. Gyger. 2011. "Ethical Considerations in Mouse Experiments." *Current Protocols in Mouse Biology* 1: 155–167.
Baerwald, E. F., G. H. D'Amours, B. J. Klug, and R. M. R. Barclay. 2008. "Barotrauma Is a Significant Cause of Bat Fatalities at Wind Turbines." *Current Biology* 18: R695–R696.
Baibagyssov, A., N. Thevs, S. Nurtazin, R. Waldhardt, V. Beckmann, and R. Salmurzauly. 2020. "Biomass Resources of *Phragmites australis* in Kazakhstan: Historical Developments, Utilization, and Prospects." *Resources* 9(6), 74. https://doi.org/10.3390/resources9060074
Bajwa, A. A., K. Jabran, M. Shahid, H. H. Ali, B. S. Chauhan, Ehsanullah. 2015. "Eco-biology and Management of *Echinochloa crus-galli*." *Crop Protection* 75: 151–162.
Bajwa, A. A, U. Zulfiqar, S. Sadia, P. Bhowmik, and B. S. Chauhan. 2019. "A Global Perspective on the Biology, Impact and Management of *Chenopodium album* and *Chenopodium murale*: Two Troublesome Agricultural and Environmental Weeds." *Environmental Science and Pollution Research* 26: 5357–5371.
Bakker, I., T. van der Voordt, P. Vink, J. de Boon, and C. Bazley. 2015. "Color Preferences for Different Topics in Connection to Personal Characteristics." *Color Research and Application* 40: 62–71.
Bakmazian, A. 2014. "The Man Behind the Beard: Perception of Men's Trustworthiness as a Function of Facial Hair." *Psychology* 5: 185–191.
Ballolli, U. 2010. *Development and Value Addition to Barnyard Millet (*Echinochloa frumantacea *Link) Cookies*. M.Sc. Thesis. Dharwad, India, University of Agricultural Sciences. https://krishikosh.egranth.ac.in/displaybitstream?handle=1/84544
Balmford, A. 2013. "Pollution, Politics, and Vultures." *Science* 339: 653–654.
Balmford, A., G. M. Mace, and N. Leader-Williams. 1996. "Designing the Ark: Setting Priorities for Captive Breeding." *Conservation Biology* 10: 719–727.
Banks, P. B., and D. F. Hochuli. 2017. "Extinction, De-extinction and Conservation: A Dangerous Mix of Ideas." *Australian Zoologist* 38: 390–394.
Banks, P. B., and N. Hughes. 2012. "A Review of the Evidence for Potential Impacts of Black Rats (*Rattus rattus*) on Wildlife and Humans in Australia." *Wildlife Research* 39: 78–88.
Barba, F. C., C. Garcia, A. Fessard, P. E. S. Munekata, J. M. Lorenzo, A. Aboudia, et al. 2020. "*Opuntia ficus-indica* Edible Parts: A Food and Nutritional Security Perspective." *Food Reviews International*. https://doi.org/10.1080/87559129.2020.1756844
Barbera, G., P. Inglese, and E. Pimienta-Barrios, E. Eds. 1995. *Agro-ecology, Cultivation and Uses of Cactus Pear*. FAO Plant Production and Protection Paper 132. Rome, Food and Agriculture Organization of the United Nations.
Barker, S. C. 1994. "Phylogeny and Classification, Origins, and Evolution of Host Associations of Lice." *International Journal for Parasitology* 24: 1285–1291.
Barnett, S. A. 2001. *The Story of Rats: Their Impact on Us, and Our Impact on Them*. Crows Nest, Australia, Allen & Unwin.
Barney J. N., N. Tharayil, A. DiTommaso, and P. C. Bhowmik. 2006. "The Biology of Invasive Alien Plants in Canada. 5. *Polygonum cuspidatum* Sieb. & Zucc. [= *Fallopia japonica* (Houtt.) Ronse Decr.]." *Canadian Journal of Plant Science* 86: 887–905.

Barnosky, A. D., N. Matzke, S. Tomiya, G. O. U. Wogan, B. Swartz, T. B. Quental, et al. 2011. "Has the Earth's Sixth Mass Extinction Already Arrived?" *Nature* 471: 51–57.

Bar-On, Y. M., and R. Phillips. 2018. "The Biomass Distribution on Earth." *Proceedings of the National Academy of Sciences of the United States of America* 115: 6506–6511.

Barows, C. W., E. B. Allen, M. L. Brooks, and M. F. Allen. 2009. "Effects of an Invasive Plant on a Desert Sand Dune Landscape." *Biological Invasions* 11: 673–686.

Barras, V. and G. Greub. 2014. "History of Biological Warfare and Bioterrorism." *Clinical Microbiology and Infection* 20: 497–502.

Barrett, S. C. H. 1983. "Crop Mimicry in Weeds." *Economic Botany* 37: 255–282.

Barrett, S. C. H. 1988. "Evolution of Breeding Systems in *Eichhornia* (Pontederiaceae): A Review." *Annals of the Missouri Botanical Garden* 75: 741–760.

Barrett-Hamilton, G. E. H. 1899. "On the Species of the Genus *Mus* Inhabiting St. Kilda." *Proceedings of the Zoological Society of London for the Year 1899*: 77–88. https://archive.org/stream/proceedingsofgen99scie#page/n6/mode/1up

Barrowclough, G. F., J. Cracraft, J. Klicka, and R. M. Zink. 2016. "How Many Kinds of Birds Are There and Why Does It Matter?" *PLoS One* 11(11): e0166307. https://doi.org/10.1371/journal.pone.0166307

Bartoli, F., F. Romiti, and G. Caneva. 2017. "Aggressiveness of *Hedera helix* L. Growing on Monuments: Evaluation in Roman Archaeological Sites and Guidelines for a General Methodological Approach." *Plant Biosystems* 151: 866–877.

Barua, M. 2011. "Mobilizing Metaphors: The Popular Use of Keystone, Flagship and Umbrella Species Concepts." *Biodiversity Conservation* 20: 1427–1440.

Barua, M., D. J. Gurdak, R. A. Ahmed, and J. Tamuly. 2012. "Selecting Flagships for Invertebrate Conservation." *Biodiversity and Conservation* 21: 1457–1476.

Baser K. H. C., and N. Arslan. 2014. "Opium Poppy (*Papaver somniferum*)." In *Medicinal and Aromatic Plants of the Middle-East. Medicinal and Aromatic Plants of the World*, Vol. 2, edited by Z. Yaniv and N. Dudai, 305–332. Dordrecht, Springer.

Bashtanova, U. B., K. P. Beckett, and T. J. Flowers. 2009. "Review: Physiological Approaches to the Improvement of Chemical Control of Japanese Knotweed (*Fallopia japonica*)." *Weed Science* 57: 584–592.

Bassett, I. J., and C. W. Crompton. 1975. "The Biology of Canadian Weeds. 11. *Ambrosia artemisiifolia* L. and *A. psilostachya* DC." *Canadian Journal of Plant Science* 55: 463– 476.

Bassett, I. J. and C. W. Crompton. 1978. "The Biology of Canadian weeds. 32. *Chenopodium album* L." *Canadian Journal of Plant Science* 58: 1061–1072.

Basu, S. K., and W. Cetzal-Ix. 2018. "Call of the Wild: Conservation of Natural Insect Pollinators Should Be a Priority." *Biodiversity* 19(3–4): 240–243.

Bates, C. 2016. "Would It Be Wrong to Eradicate Mosquitoes?" *BBC News Magazine*. http://www.bbc.com/news/magazine-35408835

Bates, M. 1967. *Gluttons and Libertines*. New York, Vintage.

Batson, C. D. 2011. *Altruism in Humans*. New York, Oxford University Press.

Bautista-Cruz, A., T. Leyva-Pablo, F. de León-González, R. Zornoza, V. Martínez-Gallegos, M. Fuentes-Ponce, et al. 2018. "Cultivation of *Opuntia ficus-indica* under Different Soil Management Practices: A Possible Sustainable Agricultural System to Promote Soil Carbon Sequestration and Increase Soil Microbial Biomass and Activity." *Land Degradation & Development* 29: 38–46.

Baynes-Rock, M. 2013. "Local Tolerance of Hyena Attacks in East Hararge Region, Ethiopia." *Anthrozoös* 26: 421–423.

Baynes-Rock, M. 2017. "Human Perceptual and Phobic Biases for Snakes: A Review of the Experimental Evidence." *Anthrozoös* 30: 5–18.

Beattie, M. H. 1995. "The Endangered Species Act: Myths and Realities." *U.S. Fish & Wildlife Service National Conservation Training Center*, https://nctc.fws.gov/History/ConservationHeroes/BeattieSpeech1995.html

Beazley, K., and N. Cardinal. 2004. "A Systematic Approach for Selecting Focal Species for Conservation in the Forests of Nova Scotia and Maine." *Environmental Conservation* 31: 91–101.

Beccaro, G. L., L. Bonvegna, D. Donno, M. G. Mellano, A. K. Cerutti, G. Nieddu, et al. 2015. "*Opuntia* spp. Biodiversity Conservation and Utilization on the Cape Verde Islands." *Genetic Resources and Crop Evolution* 62: 21–33.

Bedau, H. A. and P. G. Cassell. Eds. 2005. *Debating the Death Penalty: Should America Have Capital Punishment? The Experts on Both Sides Make Their Case*. Oxford, Oxford University Press.

Beech, E., M. Rivers, S. Oldfield, and P. P. Smith. 2017. "GlobalTreeSearch: The First Complete Global Database of Tree Species and Country Distributions." *Journal of Sustainable Forestry* 36: 454–489.

Beerling, D. J., J. P. Bailey, and A. P. Conolly. 1994. "*Fallopia japonica* (Houtt.) Ronse Decraene." *Journal of Ecology* 82: 959–979.

Beever, E. A. 2003. "Management Implications of the Ecology of Free-Roaming Horses in Semi-arid Ecosystems of the Western United States." *Wildlife Society Bulletin* 31: 887– 895.

Beirnes, P. 1994. "The Law Is an Ass: Reading EP Evans' the Medieval Prosecution and Capital Punishment of Animals." *Society & Animals* 2(1): 27–46.

Bélanger, J., and D. Pilling. 2019. *The State of the World's Biodiversity for Food and Agriculture*. Rome, FAO Commission on Genetic Resources for Food and Agriculture Assessment.

Bell, M. J., R. J. Eckard, and J. E. Pryce. 2012. "Breeding Dairy Cows to Reduce Greenhouse Gas Emissions." In *Livestock Production*, edited by K. Javed, 47–58. London, Intech Open.

Bell, W. J., L. M. Roth, and C. A. Nalepa. 2007. Cockroaches: Ecology, Behavior, and Natural History. Baltimore, MD, John Hopkins University Press.

Bellon, A. M. 2019. "Does Animal Charisma Influence Conservation Funding for Vertebrate Species under the US Endangered Species Act?" *Environmental Economics and Policy Studies* 21: 399–411.

Beltz, L. 2018. *Bats and Human Health: Ebola, SARS, Rabies and Beyond*. Hoboken, Wiley. (e-book)

Bender, B. 1975. *Farming in Prehistory: From Hunter-Gatherer to Food-Producer*. London, Hutchinson.

Bendixen, L. E., and U. B. Nandihalli. 1987. "Worldwide Distribution of Purple and Yellow Nutsedge (*Cyperus rotundus* and *C. esculentus*)." *Weed Technology* 1: 61–65.

Bennett, J. R., R. Maloney, and H. P. Possingham. 2015. "Biodiversity Gains from Efficient Use of Private Sponsorship for Flagship Species Conservation." *Proceedings of the Royal Society B. Biological Science*. 282: 20142693. https://doi.org/10.1098/rspb.2014.2693

Benoit, K. K., and R. A. Askins. 1999. "Impact of the Spread of *Phragmites* on the Distribution of Birds in Connecticut Tidal Marshes." *Wetlands* 19: 194–208.

Berenbaum, N, 2010. "This Bedbug's Life." *The New York Times*. https://www.nytimes.com/2010/08/08/opinion/08berenbaum.html

Bergen, P., J. R. Moyer, and G. C. Kozub. 1990. "Dandelion (*Taraxacum officinale*) Use by Cattle Grazing on Irrigated Pasture." *Weed Technology* 4: 258–263.

Berger, K. M. 2006. "Carnivore-Livestock Conflicts: Effects of Subsidized Predator Control and Economic Correlates on the Sheep Industry." *Conservation Biology* 20: 751–761.

Berry, R. J. 1970. "The Natural History of the House Mouse." *Field Studies* 3: 219–262.

Berry, R. J. Ed. 1981. *Biology of the House Mouse*. Proceedings of a Symposium Held at the Zoological Society of London on 22 and 23 November 1979. London, Academic Press.

Beseris, E. A., S. E. Naleway, and D. R. Carrier. 2020. "Impact Protection Potential of Mammalian Hair: Testing the Pugilism Hypothesis for the Evolution of Human Facial Hair." *Integrative Organismal Biology* 2: 1. https://doi.org/10.1093/iob/obaa005

Bezeredi, S. 2014. "Shark Ecotourism: Time to Attack the Unknown." https://biol420eres525.wordpress.com/2014/05/04/shark-ecotourism-time-to-attack-the-unknown/

Bhagwat, S. A., N. Dudley, and S. R. Harrop. 2011. "Religious Following in Biodiversity Hotspots: Challenges and Opportunities for Conservation and Development." *Conservation Letters* 4: 234–240.

Bhat, Z. F., S. Kumar, and H. Fa. Bhat. 2017. "In Vitro Meat: A Future Animal-free Harvest." *Critical Reviews in Food Science and Nutrition* 57: 782–789.

Bianca, H., M. Haward, J. McGee, and A. Fleming. 2021. "Regional Fisheries Management Organizations and the New Biodiversity Agreement: Challenge or Opportunity?" *Fish and Fisheries* 22: 226–231.

Biehler, D. D. 2013. *Pests in the City. Flies, Bedbugs, Cockroaches, and Rats*. Seattle, University of Washington Press.

Billé, R., R. Lapeire, and R. Purard. 2012. "Biodiversity Conservation and Poverty Alleviation: A Way Out of the Deadlock?" *Sapiens* 5(2). https://sapiens.revues.org/1452

Bilska, B., M. Tomaszewska, D. Kołożyn-Krajewska, K. Szczepański, R. Łaba, and S. Łaba. 2020. "Environmental Aspects of Food Wastage in Trade – A Case Study." *Environmental Protection and Natural Resources* 31: 24–34.

Binkley, S. K. 2001. "Color On, Color Off." *Minnesota Conservation Volunteer*, 29–39. http://files.dnr.state.mn.us/mcvmagazine/young_naturalists/young-naturalists-article/albino_animals/albino_animals.pdf

Bisht, I. S. 2013 "Biodiversity Conservation, Sustainable Agriculture and Climate Change: A Complex Interrelationship." In *Knowledge Systems of Societies for Adaptation and Mitigation of Impacts of Climate Change*, edited by S. Nautiyal, K. Rao, H. Kaechele, K. Raju, and R. Schaldach, 119–142. Berlin, Springer.

Bjerke, T., B. P. Kaltenborn, and T. S. Odegardstuen. 2001. "Animal-related Activities and Appreciation of Animals among Children and Adolescents." *Anthrozoös* 14: 86–94.

Blackwell, J. M. 1981. "The Role of the House Mouse in Disease and Zoonoses." In *Biology of the House Mouse* (Proceedings of a Symposium Held at the Zoological Society of London on 22 and 23 November 1979), edited by R. J. Berry, 591–616. London, Academic Press.

Blanchard, K. D., and K. J. O'Brien. 2014. *An Introduction to Christian Environmentalism: Ecology, Virtue, and Ethics*. Waco, TX, Baylor University Press.

Blancher, P. 2013. "Estimated Number of Birds Killed by House Cats (*Felis catus*) in Canada." *Avian Conservation and Ecology* 8(2): 3. https://doi.org/10.5751/ACE-00557-080203

Bland, L. M., D. A. Keith, R. M. Miller, N. J. Murray, and J. P. Rodríguez. Eds. 2017. *Guidelines for the Application of IUCN Red List of Ecosystems Categories and Criteria. Version 1.1*. Gland, Switzerland, International Union for Conservation of Nature and Natural Resources.

Bland, L. M., E. Nicholson, R. M. Miller, A. Andrade, A. Carré, A. Etter, et al. 2019. "Impacts of the IUCN Red List of Ecosystems on Conservation Policy and Practice." *Conservation Letters* 12(5) e12666. https://doi.org/10.1111/conl.12666

Blechman, A. 2007. *Pigeons: The Fascinating Saga of the World's Most Revered and Reviled Bird*. St Lucia, Queensland, University of Queensland Press.

Blossey, B., S. B. Endriss, R. Casagrande, P. Häfliger, H. Hinz, A. Dávalos, et al. 2020. "When Misconceptions Impede Best Practices: Evidence Supports Biological Control of Invasive *Phragmites*." *Biological Invasions* 22: 873–883.

Blossey, B., C. B. Randall, and M. Schwarzländer. 2015. *Biology and Biological Control of Purple Loosestrife*. Second Edition. United States Department of Agriculture, Forest Service. USDA Forest Service. FHTET-2015-3.

Blount, A. R., D. M. Ball, R. K. Sprenkel, R. O. Myer, and T. D. Hewitt. 2013. "Crabgrass as a Forage and Hay Crop." Agronomy Department, University of Florida, Publication SS-AGR-193. http://www.kingsagriseeds.com/wp-content/uploads/2014/12/Crabgrass.pdf

Blumberg, H. H. 2015. "The Relevance of Psychology to Conservation Conflicts." In *Conflicts in Conservation: Navigating towards Solution*, edited by S. M. Redpath, R. J. Gutiérrez, K. A. Wood, and J. C. Young, 122–133. Cambridge, Cambridge University Press.

Bobis, O., D. S. Dezmirean, V. Bonta, A. Moise, C. Pasca, T. E. Domokos, et al. 2019. "Japanese Knotweed (*Fallopia japonica*): Landscape Invasive Plant versus High Quality Honey Source." *Scientific Papers. Series D. Animal Science* 52: 2393–2260. http://animalsciencejournal.usamv.ro/pdf/2019/issue_1/Art34.pdf

Bodine, M. C., and D. N. Ueckert. 1975. "Effect of Desert Termites on Herbage and Litter in a Shortgrass Ecosystem in West Texas." *Journal of Range Management* 28: 353–358.

Bodner, C. C., and Hymowitz, T. 2002. "Ethnobotany of *Pueraria* Species." In *The Genus Pueraria*, edited by W. M. Keung, 29–58. New York, Taylor & Francis.

Boesch, C. 2020. "The Human Challenge in Understanding Animal Cognition." In *Neuroethics and Nonhuman Animals. Advances in Neuroethics*, edited by L. Johnson, A. Fenton, and A. Shriver, 33–51. Cham, Springer.

Bogin, B., and M. I. Varela-Silva. 2010. "Leg Length, Body Proportion, and Health: A Review with a Note on Beauty." *International Journal of Environmental Research and Public Health* 7: 1047–1075.

Boitani, L., G. M. Mace, and C. Rondinini. 2015. "Challenging the Scientific Foundations for an IUCN Red List of Ecosystems." *Conservation Letters* 8: 125–131.

Bokota, S. 2021. "Defining Human-Animal Chimeras and Hybrids: A Comparison of Legal Systems and Natural Sciences." *Ethics & Bioethics* 11: 101–114.

Bologna, C. 2018. "Here's Why People Say 'Don't Let the Bedbugs Bite'. The Origin of the Creepy Rhyme." *Huffpost US*. https://www.huffpost.com/entry/heres-why-people-say-dont-let-the-bedbugs-bite_n_5a5eb9e6e4b00a7f171b947c

Bolton, R. M., and R. J. Brooks. 2010. "Impact of the Seasonal Invasion of *Phragmites australis* (Common Reed) on Turtle Reproductive Success." *Chelonian Conservation and Biology* 9: 238–254.

Bonachela, J. A., R. M. Pringle, E. Sheffer, T. C. Coverdale, J. A. Guyton, K. K. Caylor, et al. 2015. "Termite Mounds Can Increase the Robustness of Dryland Ecosystems to Climatic Change." *Science* 347: 651–655.

Bond, W. 2001. "Keystone Species – Hunting the Snark?" *Science* 292(5514): 63–64.

Bonicamp, J. M., and L. L. Santana. 1998. "Can a Poppy Seed Food Addict Pass a Drug Test?" *Micro-chemical Journal* 58: 73–79.

Bonilla, D. L., L. A. Durden, M. E. Eremeeva, and G. A. Dasch. 2013. "The Biology and Taxonomy of Head and Body Lice – Implications for Louse-Borne Disease Prevention." *PLoS Pathogens* 9(11):e1003724. https://doi.org/10.1371/journal.ppat.1003724

Bonthoux, S., M. Brun, F. Di Pietro, S. Greulich, and S. Bouché-Pillon. 2014. "How Can Wastelands Promote Biodiversity in Cities? A Review." *Landscape and Urban Planning* 132: 79–88.

Bordas-Le Floch, V., R. Groeme, H. Chabre, V. Baron-Bodo, E. Nony, L. Mascarell, et al. 2015. "New Insights into Ragweed Pollen Allergens." *Current Allergy and Asthma Reports* 15(63). https://doi.org/10.1007/s11882-015-0565-6

Borel, B. 2016. *Infested: How the Bed Bug Infiltrated Our Bedrooms and Took Over the World.* Chicago, University of Chicago Press.

Boren, Z. D. 2015. "Women Like Cuddly Creatures Whilst Men Like Creepy-Crawlies, Says YouGrow." *The Independent*, March 2. https://www.independent.co.uk/news/weird-news/the-most-popular-animals-in-the-uk-men-like-creepy-crawlies-women-like-miniature-pigs-10080109.html

Borgi, M., and F. Cirulli. 2015. "Attitudes toward Animals among Kindergarten Children: Species Preferences." *Anthrozoös* 28: 45–59.

Borgi, M., I. Cogliati-Dezza, V. Brelsford, K. Meints, and F. Cirulli. 2014. "Baby Schema in Human and Animal Faces Induces Cuteness Perception and Gaze Allocation in Children." *Frontiers in Psychology* 5: 411. https://doi.org/10.3389/fpsyg.2014.00411

Borgi, M., and B. Majolo. 2016. "Facial Width-to-Height Ratio Relates to Dominance Style in the Genus *Macaca*." *PeerJ* 4:e1775. https://doi.org/10.7717/peerj.1775

Bostrom, N. 2014. *Superintelligence: Paths, Dangers, Strategies.* Oxford, Oxford University Press.

Bottrill, M. C., L. N. Joseph, J. Carwardine, M. Bode, C., Cook, E. T. Game, et al. 2008. "Is Conservation Triage Just Smart Decision Making?" *Trends in Ecology and Evolution* 23: 649–654.

Bouchard, S., B. K. Wiederhold, and J. Bossé. 2014. "Arachnophobia and Fear of Other Insects: Efficacy and Lessons Learned from Treatment Process." In *Advances in Virtual Reality and Anxiety Disorders*, edited by B. K. Wiederhold and S. Bouchard, 91–117. New York, Springer.

Bourne, J. B. 1998. "Norway Rat Exclusion in Alberta." In *Proceedings 18th Vertebrate Pest Conference*, edited by R. O. Baker and A. C. Crabb. Davis, University of California. https://digitalcommons.unl.edu/vpc18/32/

Boursot, P., J. C. Auffray, J. Britton-Davidian, and F. Bonhomme. 1993. "The Evolution of House Mice." *Annual Review of Ecology and Systematics* 24: 119–152.

Boursot, P., W. Din, R. Anand, D. Darviche, B. Dod, F. Von Deimling, et al. 1996. "Origin and Radiation of the House Mouse: Mitochondrial DNA Phylogeny." *Evolutionary Biology* 9: 391–415.

Bouton, J. H., B. Motes, M. A. Trammell, and T. J. Butler. 2019. "Registration of 'Impact' Crabgrass." *Journal of Plant Registrations* 13: 19–23.

Bouyer, J., N. H. Carter, C. Batavia, and M. P. Nelson. 2019. "The Ethics of Eliminating Harmful Species: The Case of the Tsetse Fly." *BioScience* 69: 125–135.

Bowen-Jones, E., and A. Entwistle. 2002. "Identifying Appropriate Flagship Species: The Importance of Culture and Local Contexts." *Oryx* 36: 189–195.

Bower, B. 2009. "Humans: Girls but Not Boys May Be Primed for Arachnophobia, Ophidiophobia: Fear of Crawly, Slithery Things Could Begin Before First Birthday." *Science News* 176: 11. http://onlinelibrary.wiley.com/doi/10.1002/scin.5591760712/epdf

Bowles, D. E., and B. D. Bowles. 2013. "Evidence of Overwintering in Water Hyacinth, *Eichhornia crassipes* Pontederiaceae in Southwestern Missouri, USA." *Rhodora* 115: 112–114.

Boyce, J. K. 2006. "A Future for Small Farms? Biodiversity and Sustainable Agriculture." In *Human Development in the Era of Globalization: Essays in Honor of Keith B. Griffin*, edited by J. K. Boyce, S. Cullenberg, P. K. Pattanaik, and R. Pollin, 83–104. Northampton, MA, Edward Elgar Publishing.

Boyles, J. G., P. M. Cryan, G. F. McCracken, and T. H. Kunz. 2011. "Economic Importance of Bats in Agriculture." *Science* 332(6025): 41–42.

Bozier, J., E. K. Chivers, D. G. Chapman, A. N. Larcombe, N. A. Bastian, J. A. Masso-Silva, et al. 2020. "The Evolving Landscape of E-Cigarettes: A Systematic Review of Recent Evidence." *Thoracic Oncology: CHEST Reviews* 157: 1362–1390.

Bradshaw, C. J. A., P. R. Ehrlich, A. Beattie, G. Ceballos, E. Crist, J. Diamond, et al. 2021. "Underestimating the Challenges of Avoiding a Ghastly Future." *Frontiers in Conservation Science* 1:615419. https://doi.org/10.3389/fcosc.2020.615419

Bradshaw, J. 2017a. *The Animals among Us: How Pets Make Us Human.* New York, Basic Books.

Bradshaw, J. 2017b. "The Science Behind Why Some People Love Animals and Others Couldn't Care Less." *The Conversation* (online newsletter). https://theconversation.com/the-science-behind-why-some-people-love-animals-and-others-couldnt-care-less-84138

Bramble, D., and D. Lieberman. 2004. "Endurance Running and the Evolution of *Homo*." *Nature* 432: 345–352.

Brandt, A. M. 2012. "Inventing Conflicts of Interest: A History of Tobacco Industry Tactics." *American Journal of Public Health* 102: 63–71.

Brandt, J. S., and R. C. Buckley. 2018. "A Global Systematic Review of Empirical Evidence of Ecotourism Impacts on Forests in Biodiversity Hotspots." *Current Opinion in Environmental Sustainability* 32: 112–118.

Branton, M., and J. S. Richardson. 2010. "Assessing the Value of the Umbrella-Species Concept for Conservation Planning with Meta-Analysis." *Conservation Biology* 25: 9–20.

Braun, S., C. Peus, and D. Frey. 2012. "Is Beauty Beastly? Gender-Specific Effects of Leader Attractiveness and Leadership Style on Followers' Trust and Loyalty." *Zeitschrift für Psychologie* 220: 98–108.

Breda, J., J. Jewella, and A. Keller. 2019. "The Importance of the World Health Organization Sugar Guidelines for Dental Health and Obesity Prevention." *Caries Research* 53: 149–152.

Brenner, R. J., and R. D. Kramer. 2018. "Chapter 6: Cockroaches (*Blattaria*)." In *Medical and Veterinary Entomology*, Third Edition, edited by G. R. Mullen and L. A. Durden, 61–77. Cambridge, Academic Press.

Breyer, T. 2000. "Symmetry and Beauty of Human Faces." In *Bridges: Mathematical Connections in Art, Music, and Science*, 339–346. Bridges Conference, 2000. http://archive.bridgesmathart.org/2000/bridges2000-339.pdf

Brickell, C. D., C. Alexander, J. J. Cubey, J. C. David, M. H. A. Hoffman, A. C. Leslie, et al. 2016. *The International Code of Nomenclature for Cultivated Plants*. Leuven, Belgium, International Society for Horticultural Science.

Brijačak, E., V. Šoštarčić, and M. Šćepanović. 20118. "Biology and Ecology of Common Lambsquarters (*Chenopodium album* L.)." *Agronomski glasnik: Glasilo Hrvatskog agronomskog društva* 80: 19–34.

Brinton, L. A., J. D. Figueroa, D. Ansong, K. M. Nyarko, S. Wiafe, J. Yarney, et al. 2018. "Skin Lighteners and Hair Relaxers as Risk Factors for Breast Cancer: Results from the Ghana Breast Health Study." *Carcinogenesis* 39: 4. https://doi.org/10.1093/carcin/bgy002; https://core.ac.uk/download/pdf/200748111.pdf

Britt, K., and S. Taylor. 2019. "Kudzu Bug, *Megacopta cribraria*, a Pest of Soybeans." St. Petersburg, VA, Virginia Tech, Virginia Cooperative Extension Publication ENTO-303NP. https://vtechworks.lib.vt.edu/bitstream/handle/10919/88377/ENTO-303NP.pdf?sequence=1

Britton, K. O., D. Orr, and J. Sun. 2002. "Kudzu." In *Biological Control of Invasive Plants in Eastern United States*, edited by R. Van Driesche, S.Lyon, B. Blossey, M. Hoddle, and R. Reardon, 413 p. USDA Forest Service Publication FHTET-2002-04. https://www.invasive.org/biocontrol/25Kudzu.cfm

Brockhaus, F. A. 1892. *Brockhaus' Konversations-Lexikon*, Vol. 7. Berlin, F.A. Brockhaus.

Brooks, T. M., R. A. Mittermeier, G. A. B. da Fonseca, J. Gerlach, M. Hoffmann, J. F. Lamoreux, et al. 2006. "Global Biodiversity Priorities." *Science* 313(5783): 58–61.

Broom, D. M. 1999. "The Welfare of Vertebrate Pests in Relation to Their Management." In *Advances in Vertebrate Pest Management*, edited by P. D. Cowan and C. J. Feare, 309–329. Fürth, Filander Verlag.

Brottman, M. 2012. *Hyena*. London, Reaktion Books.

Brown, L. J., R. C. Rosatte, C. Fehlner-Gardiner, J. A. Ellison, F. R. Jackson, P. Bachmann, et al. 2014. "Oral Vaccination and Protection of Striped Skunks (*Mephitis mephitis*) against Rabies Using ONRAB®." *Vaccine* 32: 3675–3679.

Brown, S., and S. Ponsonby-McCabe. 2014. *Brand Mascots and Other Marketing Animals*. London, Routledge.

Browning, W., C. Ryan, and J. Clancy. 2014. "14 Patterns of Biophilic Design. Improving Health & Well-Being in the Built Environment." New York, Terrapin Bright Green. https://www.terrapinbrightgreen.com/reports/14-patterns/#what-is-good-biophilic-design

Brulliard, K. 2020. "Rats Will Devour Your Car." *The Washington Post*. https://www.washingtonpost.com/science/2020/02/13/rats-will-devour-your-car/

Brunsdon, S. 2019. "Fairy Trees of Ireland." *Irish Culture, University of Alberta, 2020 (Blog)*. https://irishculture.home.blog/2019/04/04/fairy-trees-of-ireland/

Bryson, C. and R. Carter. 2008. "The significance of Cyperaceae as Weeds." In *Sedges: Uses, Diversity, and Systematics of the Cyperaceae*, edited by R. F. C. Naczi and B. A. Ford, 15–101. St. Louis, Missouri Botanical Garden Press.

Bryson, C. L. 1911. *Woodsy Neighbours of Tan and Teckle*. New York, Fleming H. Revell.

Buckle, A. 2013. "Anticoagulant Resistance in the United Kingdom and a New Guideline for the Management of Resistant Infestations of Norway Rats (*Rattus norvegicus* Berk.)." *Pest Management Science* 69: 334–341.

Buckley, M. T., F. Racimo, M. E. Allentoft, M. K. Jensen, A. Jonsson, H. Huang, et al. 2017. "Selection in Europeans on Fatty Acid Desaturases Associated with Dietary Changes." *Molecular Biology and Evolution* 34: 1307–1318.

Buckner, W. 2021. "Disguises and the Origins of Clothing." *Human Nature* 32: 706–728.

Buechley, E. R., and C. H. Sekercioglu. 2016a. "How Are Vultures Doing?" *Current Biology* 26: R560–R561. https://www.cell.com/current-biology/pdf/S0960-9822(16)00134-2.pdf

Buechley, E. R., and Ç. H. Sekercioglu. 2016b. "The Avian Scavenger Crisis: Looming Extinctions, Trophic Cascades, and Loss of Critical Ecosystem Functions." *Biological Conservation* 198: 220–228.

Buglife. n.d. "Introduction to Brownfields." https://cdn.buglife.org.uk/2020/01/Introduction-to-brownfields.pdf

Bulte, E. H., and G. C. Van Kooten. 1999. "Marginal Valuation of Charismatic Species: Implications for Conservation." *Environmental and Resource Economics* 14: 119–130.

Bungay, N. 2018. "Pestwatch: Cat and Dog Fleas." British Pest Control Association Publication PPC3. https://bpca.org.uk/News-and-Blog/pestwatch-cat-and-dog-fleas

Bure, C. M., and M. S. Moore. 2019. "White-nose Syndrome: A Fungal Disease of North American Hibernating Bats." In *Encyclopedia of Caves*, Third Edition, edited by W. B. White, D. C. Culver, and T. Pipan, 1165–1174. Cambridge, Academic Press.

Burgess, T. W. 1918. *The Adventures of Jimmy the Skunk*. Toronto, Little Brown and Company. https://www.gutenberg.org/files/21015/21015-h/21015-h.htm

Burgin, C. J., J. P. Colella, P. L. Kahn, and N. S. Upham. 2018. "How Many Species of Mammals Are There?" *Journal of Mammalogy* 99: 1–14.

Burrows, A. M., J. Kaminski, B. M. Waller, K. M. Omstead, C. Rogers-Vizena, and B. Mendelson. 2021. "Dog Faces Exhibit Anatomical Differences in Comparison to Other Domestic Animals." *The Anatomical Record* 304: 231–241.

Burrows, A. M., and K. M. Omstead. 2022. "Dog Faces Are Faster than Wolf Faces." [Abstract] *The FASEB Journal* https://doi.org/10.1096/fasebj.2022.36.S1.R5001

Burrows, G. E., and R. J. Tyrl. 2013. *Toxic Plants of North America*. Second Edition. Hoboken, NJ, Wiley-Blackwell.

Buschinger, A. 2009. "Social Parasitism among Ants: A Review (Hymenoptera: Formicidae)." *Myrmecological News* 12: 219–235.

Busey, P. 2003. "Cultural Management of Weeds in Turfgrass: A Review." *Crop Science* 43: 1899–1911.

Butterfield, M. E., S. E. Hill, and C. G. Lord. 2012. "Mangy Mutt or Furry Friend? Anthropomorphism Promotes Animal Welfare." *Journal of Experimental Social Psychology* 48: 957–960.

Butterworth, A. Ed. 2018. *Animal Welfare in a Changing World*. Wallingford, Oxfordshire, CABI.

Byers, K. A., M. J. Lee, D. M. Patrick, and C. G. Himsworth. 2019. "Rats about Town: A Systematic Review of Rat Movement in Urban Ecosystems." *Frontiers in Ecology and Evolution* 7. https://doi.org/10.3389/fevo.2019.00013

CABI. 2018. "*Hedera helix* L." *Invasive Species Compendium*, Oxfordshire, UK, CAB International. https://www.cabi.org/isc/datasheet/26694

CABI. 2019a. "*Ambrosia artemisiifolia* (Common Ragweed)." *Invasive Species Compendium*, Oxfordshire, UK, CAB International. https://www.cabi.org/isc/datasheet/4691

CABI. 2019b. "*Ambrosia trifida* (Giant Ragweed)." *Invasive Species Compendium*, Oxfordshire, UK, CAB International. https://www.cabi.org/isc/datasheet/4693

CABI. 2019c. "*Cirsium arvense* (Creeping Thistle)." *CABI Invasive Species Compendium*, Oxfordshire, CAB International. https://www.cabi.org/isc/datasheet/13628
CABI. 2019d. "*Cyperus esculentus* (Yellow Nutsedge)." *Invasive Species Compendium*, Oxfordshire, UK, CAB International. https://www.cabi.org/isc/datasheet/17496
CABI. 2019e. "*Cyperus rotundus* (Purple Nutsedge)." *Invasive Species Compendium*, Oxfordshire, UK, CAB International. https://www.cabi.org/isc/datasheet/17506
CABI. 2019f. "*Digitaria ciliaris* (Southern Crabgrass)." *CABI Invasive Species Compendium*, Oxfordshire, UK, CAB International. https://www.cabi.org/isc/datasheet/18912
CABI. 2019g. "*Digitaria sanguinalis* ((Large Crabgrass)." *CABI Invasive Species Compendium*, Oxfordshire, UK, CAB International. https://www.cabi.org/isc/datasheet/18916
CABI. 2019h. "*Echinochloa crus-galli* (Barnyard Grass)." *Invasive Species Compendium*, Oxfordshire, UK, CAB International. https://www.cabi.org/isc/datasheet/20367
CABI. 2019i. "*Eichhornia crassipes* (Water Hyacinth)." *Invasive Species Compendium* (online). Oxfordshire, UK, CAB International. https://www.cabi.org/isc/datasheet/20544
CABI. 2019j. "*Opuntia ficus-indica* (Prickly Pear)." *Invasive Species Compendium*, Oxfordshire, UK, CAB International. https://www.cabi.org/isc/datasheet/37714
CABI. 2019k. "*Phragmites australis* (Common Reed)." *CABI Invasive Species Compendium*, Oxfordshire, UK, CAB International. https://www.cabi.org/isc/datasheet/40514#D0C9AD4F-BEA4-4F9C-9C53-87955E75EB67
CABI. 2019l. "*Taraxacum officinale* Complex (Dandelion)." *CABI Invasive Species Compendium*, Oxfordshire, UK, CAB International. https://www.cabi.org/isc/datasheet/52773#tosummaryOfInvasiveness
CABI. 2021a. "*Chenopodium album* (Fat Hen)." *CABI Invasive Species Compendium*, Oxfordshire, UK, CAB International. https://www.cabi.org/isc/datasheet/12648
CABI. 2021b. "*Cynodon dactylon* (Bermuda Grass)." *CABI Invasive Species Compendium*, Oxfordshire, UK, CAB International. https://www.cabi.org/isc/datasheet/17463
CABI. 2021c. "*Lythrum salicaria* (Purple Loosestrife)." *CABI Invasive Species Compendium*, Oxfordshire, UK, CAB International. https://www.cabi.org/isc/datasheet/31890
Cafaro, P., and E. Crist. Eds. 2012. *Life on the Brink: Environmentalists Confront Overpopulation*. Athens, University of Georgia Press.
Cailly Arnulphi, V. B., S. A. Lambertucci, and C. E. Borghi. 2017. "Education Can Improve the Negative Perception of a Threatened Long-Lived Scavenging Bird, the Andean Condor." *PLoS One* 12(9): e0185278, https://doi.org/10.1371/journal.pone.0185278
Calder, W. A. 1996. *Size, Function, and Life History*. Mineola, NY, Dover.
Call, E. 2006. *Mending the Web of Life: Chinese Medicine and Species Conservation*. IFAW/AHPA-ERB Foundation [International Fund for Animal Welfare/American Herbal Products Association Foundation for Education and Research on Botanicals].
Callahan, M. M., T. Satterfield, and J. Zhao. 2021. "Into the Animal Mind: Perceptions of Emotive and Cognitive Traits in Animals." *Anthrozoös*. https://doi.org/10.1080/08927936.2021.1914439
Callegari, B., and O. Mikhailova. 2021. "RRI and Corporate Stakeholder Engagement: The Aquadvantage Salmon Case." *Sustainability* 13(4): 1820. https://doi.org/10.3390/su13041820
Camhi, M., S. Fowler, J. Musick, A. Bräutigam, and S. Fordham. 1998. *Sharks and their Relatives – Ecology and Conservation*. Gland, Switzerland, International Union for Conservation of Nature and Natural Resources. http://citeseerx.ist.psu.edu/viewdoc/download?doi=10.1.1.232.2135&rep=rep1&type=pdf
Campbell, K. R., and T. S. Campbell. 2001. "The Accumulation and Effects of Environmental Contaminants on Snakes: A Review." *Environmental Monitoring and Assessment* 70: 253–301.
Campbell, M. O. 2015. *Vultures: Their Evolution, Ecology and Conservation*. Boca Raton, FL, Taylor & Francis.
Campbell, M. O. 2019. "Conservation Psychology, Social Media and the Conservation of Large Carnivores." In *Advances in Psychology Research*, edited by A. M. Columbus, 1–47. Hauppauge, NY, Science Publishers Inc.
Campbell-Staton, S. C., B. J. Arnold, D. Gonçalves, P. Granli, J. Poole, R. A. Long, et al. 2021. "Ivory Poaching and the Rapid Evolution of Tusklessness in African Elephants." *Science* 374(6566): 483–487.
Caniglia, R., E. Fabbri, L. Mastrogiuseppe, and E. Randi. 2013. "Who Is Who? Identification of Livestock Predators Using Forensic Genetic Approaches." *Forensic Science International: Genetics* 7: 397–404.
Cano-Terriza, D., R. Guerra, S. Lecollinet, M. Cerdà-Cuéllar, O. Cabezón, S. Almería, et al. 2015. "Epidemiological Survey of Zoonotic Pathogens in Feral Pigeons (*Columba livia* var. *domestica*) and Sympatric Zoo Species in Southern Spain." *Comparative Immunology, Microbiology and Infectious Diseases* 43: 22–27.
Cao, D. J., K. Aldy, S. Hsu, S. M. McGetrick, G. Verbeck, I. De Silva, et al. 2020. "Review of Health Consequences of Electronic Cigarettes and the Outbreak of Electronic Cigarette, or Vaping, Product Use-Associated Lung Injury." *Journal of Medical Toxicology*. https://doi.org/10.1007/s13181-020-00772-w
Capoccia, S., C. Boyle, and T. Darnell. 2018. "Loved or Loathed, Feral Pigeons as Subjects in Ecological and Social research." *Journal of Urban Ecology* 4(1). https://doi.org/10.1093/jue/juy024
Capron, M. 2011. "The Effect of Parasite Infection on Allergic Disease." *Allergy* 66(Suppl. 95): 16–18.
Carbone, L. 2004. *What Animals Want*. Oxford, Oxford University Press.
Cardoso, P., P. A. V. Borges, K. A. Triantis, M. A. Ferrández, and J. L. Martín. 2012. "The Underrepresentation and Misrepresentation of Invertebrates in the IUCN Red List." *Biological Conservation* 149: 147–148.
Cardoso, P., T. L. Erwin, P. A. V. Borges, and T. R. New. 2011. "The Seven Impediments in Invertebrate Conservation and How to Overcome Them." *Biological Conservation* 144: 2647–2655.
Carlile, N., D. Priddel, and T. O'Dwyer. 2018. "Preliminary Surveys of the Endangered Lord Howe Island Cockroach *Panesthia lata* (Blattodea: Blaberidae) on two Islands within the Lord Howe Group, Australia." *Austral Entomology* 57: 207–213.
Carlson, C. J., K. R. Burgio, E. R. Dougherty, A. J. Phillips, V. M. Bueno, C. F. Clements, et al. 2017a. "Parasite Biodiversity Faces Extinction and Redistribution in a Changing Climate." *Science Advances* 3(9): e1602422. https://doi.org/10.1126/sciadv.1602422
Carlson, C. J., K. R. Burgio, E. R. Dougherty, A. J. Phillips, V. M. Bueno, C. F. Clements, Brenner, R. J., and R. D. Kramer. 2018. "Chapter 6: Cockroaches (*Blattaria*)." In *Medical and Veterinary Entomology*, Third Edition, edited by G. R. Mullen and L. A. Durden, 61–77. Cambridge: Academic Press.

Carlson, C. J., S. Hopkins, K. C. Bell, J. Doña, S. S. Godfrey, M. L. Kwak, et al. 2020. "A Global Parasite Conservation Plan." *Biological Conservation* 250. https://doi.org/10.1016/j.biocon.2020.108596

Carlson, C. J., O. C. Muellerklein, A. J. Phillips, K. R. Burgio, G. Castaldo, C. A. Cizauskas, et al. 2017b. "The Parasite Extinction Assessment & Red List: An Open-Source, Online Biodiversity Database for Neglected Symbionts." *BioRxiv*, 192351. https://doi.org/10.1101/192351

Caro, T. 2011. "The Functions of Black-and-White Coloration in Mammals." In *Animal Camouflage: Mechanisms and Function*, edited by M. Stevens and S. Merilaita, 298–328. Cambridge: Cambridge University Press.

Caro, T., A. Engilis Jr., E. Fitzherbert, and T. Gardner. 2004. "Preliminary Assessment of the Flagship Species Concept at a Small Scale." *Animal Conservation* 7: 63–70.

Caro, T., H. Walker, Z. Rossman, M. Hendrix, and T. Stankowich. 2017. "Why Is the Giant Panda Black and White?" *Behavioral Ecology* 28: 657–667.

Caro, T. M., and S. Girling. 2010. *Conservation by Proxy: Indicator, Umbrella, Keystone, Flagship, and Other Surrogate Species*. Washington, DC, Island press.

Caro, T. M., and G. O'Doherty. 1999. "On the Use of Surrogate Species in Conservation Biology." *Conservation Biology* 13: 805–814.

Carrier, J. C., J. A. Musick, and M. R. Heithaus. Editors. 2012. *Biology of Sharks and Their Relatives*. Second Edition. Boca Raton, FL, CRC Press.

Carrito, M. L., I. M. Santos, L. Alho, J. Ferreira, S. C. Soares, P. Bem-Haja, et al. 2017. "Do Masculine Men Smell Better? An Association between Skin Color Masculinity and Female Preferences for Body Odor." *Chemical Senses* 42: 269–275.

Carson, C. 1951. *The Sea around Us*. Oxford, Oxford University Press.

Caruso, M., S. Currò, G. Las Casas, S. La Malfa, and A. Gentile. 2010. "Microsatellite Markers Help to Assess Genetic Diversity among *Opuntia ficus-indica* Cultivated Genotypes and Their Relation with Related Species." *Plant Systematics and Evolution* 290: 85–97.

Cascio, C. J., D. Moore, and F. McGlone. 2019. "Social Touch and Human Development." *Developmental Cognitive Neuroscience* 35: 5–11.

Case, M. J., and K. A. Stinson. 2018. "Climate Change Impacts on the Distribution of the Allergenic Plant, Common Ragweed (*Ambrosia artemisiifolia*) in the Eastern United States." *PLoS One* 13(10): e0205677. https://doi.org/10.1371/journal.pone.0205677

Castagneri, D., M. Garbarino, and P. Nola. 2013. "Host Preference and Growth Patterns of Ivy (*Hedera helix* L.) in a Temperate Alluvial Forest." *Plant Ecology* 214: 1–9.

Castaldelli-Maia, J. M., A. Ventriglio, and D. Bhugra. 2016. "Tobacco Smoking: From 'Glamour' to 'Stigma'. A Comprehensive Review." *Psychiatry and Clinical Neurosciences* 70: 24–33.

Castillo-Huitrón, N. M., E. J. Naranjo, D. Santos-Fita, and E. Estrada-Lugo. 2020. "The Importance of Human Emotions for Wildlife Conservation." *Frontiers in Psychology* 11: 1277. https://doi.org/10.3389/fpsyg.2020.01277

Catling, P. M., and G. Mitrow 2011. "Major Invasive Alien Plants of Natural Habitats in Canada. 1. European Common Reed (Often Just Called Phragmites), *Phragmites australis* (Cav.) Trin. ex Steud. subsp. *australis*." *Canadian Botanical Association Bulletin* 44(2): 52–61.

Cavico, F. J., S. C. Muffler, and B. G. Mujtaba. 2013. "Appearance Discrimination in Employment: Legal and Ethical Implications of 'Lookism' and 'Lookphobia'." *Equality, Diversity and Inclusion* 32: 83–119.

CDC. 2018. Tickborne *Diseases of the United States. A Reference Manual for Healthcare Providers*. Fifth Edition. Atlanta, GA, U.S. Department of Health and Human Services, Centers for Disease Control and Prevention. https://www.cdc.gov/ticks/tickbornediseases/index.html

Ceballos, G., P. R. Ehrlich, and R. Dirzo. 2017. "Biological Annihilation via the Ongoing Sixth Mass Extinction Signaled by Vertebrate Population Losses and Declines." *Proceedings of the National Academy of Sciences of the United States* 114(30): E6089–E6096. https://doi.org/10.1073/pnas.1704949114

Ceballos, G., P. R. Ehrlich, and P. H. Raven. 2020. "Vertebrates on the Brink as Indicators of Biological Annihilation and the Sixth Mass Extinction." *Proceedings of the National Academy of Sciences of the United States of America* 117(24): 13596–13602.

Cela-Conde, C. J., G. Marty, F. Maestu, T. Ortiz, E. Munar, A. Fernández, et al. 2004. "Activation of the Prefrontal Cortex in the Human Visual Aesthetic Perception." *Proceedings of the National Academy of Sciences of the United States of America* 101: 6321–6325.

Cellan-Jones, R. 2014. "Stephen Hawking Warns Artificial Intelligence Could End Mankind." *BBC News*. https://www.bbc.com/news/technology-30290540

Centre for science in the Public Interest. 2019. "CSPI Urges FDA, DEA to Crack Down on Sale of Contaminated Poppy Seeds. At Least 12 U.S. Deaths Are Associated with 'Tea' Made with Poppy Seeds or Pods." https://cspinet.org/news/cspi-urges-fda-dea-crack-down-unwashedl-poppy-seeds

Chandran, M. D. S., and J. D. Hughes. 2000. "Sacred Groves and Conservation: The Comparative History of Traditional Reserves in the Mediterranean Area and in South India." *Environment and History* 6: 169–186.

Chapman, A. D. 2009. *Numbers of Living Species in Australia and the World*. Second Edition. Toowoomba, Australia, Australian Biodiversity Information Services. http://www.environment.gov.au/biodiversity/abrs/publications/other/species-numbers/2009/03-exec-summary.html

Chapman, B. 2017. *Shark Attacks: Myths, Misunderstandings and Human Fear*. Clayton South, Victoria, CSIRO Publishing.

Chapman, B. K., and D. McPhee. 2016. "Global Shark Attack Hotspots: Identifying Underlying Factors Behind Increased Unprovoked Shark Bite Incidence." *Ocean & Coastal Management* 133: 72–84.

Charlton, A. 2004. "Medicinal Uses of Tobacco in History." *Journal of the Royal Society of Medicine* 97: 292–296

Chauhan, B. S., R. G. Singh, and G. Mahajan. 2012. "Ecology and Management of Weeds under Conservation Agriculture: A Review." *Crop Protection* 38: 57–65.

Chaurasia, A., D. L. Hawksworth, and M. Pessoa de Miranda. Eds. 2020. *GMOs. Topics in Biodiversity and Conservation, vol 19*. Cham, Springer.

Chen, H., J. Wang, L. Zheng, C. W. K. Lam, Y. Xiao, W. Qibiao, et al. 2019. "Consumption of Sugar-Sweetened Beverages Has a Dose-Dependent Effect on the Risk of Non-Alcoholic Fatty Liver Disease: An Updated Systematic Review and Dose-Response Meta-Analysis." *International Journal of Environmental Research and Public Health* 16(12): 2192. https://doi.org/10.3390/ijerph16122192

Chen, J., B. Zhang, L. Luo, F. Zhang, Y. Yi, Y. Shan, et al. 2021. "Review on Recycling Techniques for Bioethanol Production from Lignocellulosic Biomass." *Renewable and Sustainable Energy Reviews* 149 (2021): 111370. https://doi.org/10.1016/j.rser.2021.111370

Chen, J. M., and A. Francis-Tan. 2021. "Setting the Tone: An Investigation of Skin Color Bias in Asia." *Race and Social Problems*. https://doi.org/10.1007/s12552-021-09329-0

Chen, S.-L., H. Yu, H.-M. Luo, Q. Wu, C.-F, Li, and A. Steinmetz. 2016. "Conservation and Sustainable Use of Medicinal Plants: Problems, Progress, and Prospects." *Chinese Medicine* 11: 37. https://doi.org/10.1186/s13020-016-0108-7

Chen, Y., B. Becker, Y. Zhang, H. Cui, J. Du, J. Wernicke, et al. 2020. "Oxytocin Increases the Pleasantness of Affective Touch and Orbitofrontal Cortex Activity Independent of Valence." *European Neuropsychopharmacology* 39: 99–110.

Cherniack, E. P. 2010. "Bugs as Drugs, Part One: Insects. The 'New' Alternative Medicine for the 21st Century?" *Alternative Medicine Review* 15: 124–135.

Cherniack, E. P. 2011. "Bugs as Drugs, Part Two: Worms, Leeches, Scorpions, Snails, Ticks, Centipedes, and Spiders." *Alternative Medicine Review* 16: 50–58.

Chiarelli, B., and M. Annese. 2009. "The Carrying Capacity of the Environment as It Relates to Human Consumerism." *Global Bioethics* 22(1–4): 3–18.

Chicouene, D. 2007. "Mechanical Destruction of Weeds. A Review." *Agronomy for Sustainable Development* 27: 19–27.

Child, K. and C. Darimont. 2015. "Hunting for Trophies: Online Hunting Photographs Reveal Achievement Satisfaction with Large and Dangerous Prey." *Human Dimensions of Wildlife* 20: 531–541.

Chole, A. S., A. R. Jadhav, and V. N. Shinde. 2021. "Vertical Farming: Controlled Environment Agriculture." *Just Agriculture* 1: 249–256. https://justagriculture.in/files/newsletter/2021/jan/055.pdf

Chow, K. F. 2017. "A Review of Excessive Sugar Metabolism on Oral and General Health." *Chinese Journal of Dental Research* 201: 193–198. https://cjdr.quintessenz.de/cjdr_2017_04_s0193.pdf

Christenhusz, J. M., and J. W. Byng. 2019. "The Number of Known Plants Species in the World and its Annual Increase." *Phytotaxa* 261(3). https://doi.org/10.11646/phytotaxa.261.3.1

Christian, K. A. 2011. "Dog Bites and Dangerous Pets." In *Animals, Diseases, and Human Health: Shaping Our Lives Now and in the Future*, edited by R. G. Davis, 33–50. Santa Barbara, CA, Praeger.

Cipriani, D. 2011. "Skunks Are Affectionate, Intelligent Pets for Owners Who Offer the Proper Care." *Critters USA* 2011: 2–6.

Cisneros-Montemayor, A. M., and U. R. Sumaila. 2014. "Economic Rationale for Shark Conservation." In *Sharks: Conservation, Governance and Management*, edited by E. J. Techera and N. Klein, 197–212. New York, Routledge.

Clark, G., J. Fletcher, and N. Criddle. 1906. *Farm Weeds of Canada*. Ottawa, Canada Department of Agriculture.

Clarke, M. M., S. H. Reichard, and C. W. Hamilton. 2006. "Prevalence of Different Horticultural Taxa of Ivy (*Hedera* spp., Araliaceae) in Invading Populations." *Biological Invasions* 8: 149–157.

Clayton, S., and G. Myers. 2015. *Conservation Psychology: Understanding and Promoting Human Care for Nature*. Hoboken, NJ, Wiley.

Clemants, S. E., and S. L. Mosyakin. 2003. "*Chenopodium*. Linnaeus Sp. Pl. 1: 218." In *Flora of North America, Vol. 4*, edited by Flora of North America Editorial Committee, 275–299. New York, Oxford University Press.

Clucas, B., K. McHugh, and T. Caro. 2008. "Flagship Species on Covers of US Conservation and Nature Magazines." *Biodiversity and Conservation* 17: 1517–1528.

Clucas, B., I. D. Parker, and A. M. Feldpausch-Parker. 2018. "A Systematic Review of the Relationship between Urban Agriculture and Biodiversity." *Urban Ecosystems* 21: 635–643.

Clutton-Brock, T. H. 1986. "Sex Ratio Variation in Birds." *Ibis* 128: 317–329.

Clutton-Brock, T. H., and G. Iason. 1986. "Sex Ratio Variation in Mammals." *Quarterly Review of Biology* 61: 339–374.

Coates, P. 2004. "Emerging from the Wilderness: (or, From Redwoods to Bananas): Recent Environmental History in the United States and the Rest of the Americas." *Environment and History* 10: 407–438.

Cock, M. J. W., J. C. Biesmeijer, J. C. Cannon, P. J. Gerard, D. Gillespie, J. J. Jiménez, et al. 2012. "The Positive Contribution of Invertebrates to Sustainable Agriculture and Food Security." *CAB Reviews* 7: 043. http://www.cabi.org/cabreviews

Coddington, J. A. 1991. "Systematics and Evolution of Spiders (Araneae)." *Annual Review of Ecology and Systematics* 22: 565–592.

Cohen, E. 2014. "Recreational Hunting: Ethics, Experiences and Commoditization." *Tourism Recreation Research* 39: 3–17.

Cohen, E. B., L. D. Auckland, P. P. Marra, S. A. Hamer, and H. L. Drake. 2015. "Avian Migrants Facilitate Invasions of Neotropical Ticks and Tick-Borne Pathogens into the United States." *Applied and Environmental Biology* 81: 8366–8378.

Cohen, M. N. 1977. *The Food Crisis in Pre-History. Overpopulation and the Origins of Agriculture*. New Haven, Yale University Press.

Colding, J., and C. Folke. 1997. "The Relations among Threatened Species, Their Protection, and Taboos." *Conservation Ecology* 1(1): 1–19. https://www.ecologyandsociety.org/vol1/iss1/art6/

Colléony, A., S. Clayton, D. Couvet, M. Saint Jaimie, and A.-C. Prévot. 2017. "Human Preferences for Species Conservation: Animal Charisma Trumps Endangered Status." *Biological Conservation* 206: 263–269.

Coloma, L., A. M. S. Moreton, and G. H. Vyse. 1914. *Perez the Mouse*. London, J. Lane.

Conte, F., E. Voslarova, V. Vecerek, R. W. Elwood, P. Coluccio, M. Pugliese, et al. 2021. "Humane Slaughter of Edible Decapod Crustaceans." *Animals* 11(4): 1089. https://doi.org/10.3390/ani11041089

Conway Morris, S. 2005. *Life's Solution: Inevitable Humans in a Lonely Universe*. Cambridge, UK, Cambridge University Press.

Cook, M., and S. Mineka. 1990. "Selective Associations in the Observational Conditioning of Fear in Rhesus Monkeys." *Journal of Experimental Psychology Animal Behavior Processes* 16: 372–389.

Coombes, M. A., H. A. Viles, and H. Zhang. 2018. "Thermal Blanketing by Ivy (*Hedera helix* L.) Can Protect Building Stone from Damaging Frosts." *Scientific Reports* 8: 9834. https://doi.org/10.1038/s41598-018-28276-2.

Cooper, J. E. 1977. "Diseases of Lower Vertebrates and Biomedical Research." *Laboratory Animals* 11: 119–123.

COPD. 2018. "Synthetic Biology." Draft decision submitted by the Chair of Working Group II Conference of the Parties to the Convention on Biological Diversity, Fourteenth Meeting, Sharm El-Sheikh, Egypt. https://www.cbd.int/doc/c/2c62/5569/004e9c7a6b2a00641c3af0eb/cop-14-l-31-en.pdf

Coren, S. 1999. The *Intelligence of Dogs: A Guide to the Thoughts, Emotions, and Inner Lives of Our Canine Companions*. New York, Bantam Books.

Corley, R. N., A. Woldeghebriel, and M. R. Murphy. 1997. "Evaluation of the Nutritive Value of Kudzu (*Pueraria lobata*) as a Feed for Ruminants." *Animal Feed Science and Technology* 68: 183–188.

Cornelius, R. R., and J. R. Averill. 1983. "Sex Differences in Fear of Spiders." *Journal of Personality and Social Psychology* 45: 377–383.

Corvalán, C., S. Hales, and A. J. McMichael. 2005. *Ecosystems and Human Well-Being: Health Synthesis*. Geneva, World Health Organization.

Corvellec, H., A. F. Stowell, and N. Johansson. 2022. "Critiques of the Circular Economy." *Journal of Industrial Ecology* 26: 421–432.

Cosier, S. 2006. "Is It True That Tomato Sauce Will Get Rid of the Smell of a Skunk?" *ScienceLine*. https://scienceline.org/2006/07/ask-cosier-skunk/

Cotana, F., G. Cavalaglio, A. L. Pisello, M. Gelosia, D. Ingles, and E. Pompili. 2015. "Sustainable Ethanol Production from Common Reed (*Phragmites australis*) through Simultaneous Saccharification and Fermentation." *Sustainability* 7: 12149–12163.

Cota-Sánchez, J. H. 2016. "Nutritional Composition of the Prickly Pear (*Opuntia ficus-indica*) Fruit. In *Nutritional Composition of Fruit Cultivars*, edited by M. S. J. Simmonds and V. R. Preedy, 691–712. Cambridge, MA, Academic Press.

Cotter, J., K. Kawall, and C. Then. 2020. "New Genetic Engineering Technologies." Risk *Assessment of Genetically Engineered Organisms in the EU and Switzerland*. https://www.testbiotech.org/sites/default/files/RAGES_report-%20new%20genetic%20engineering%20techniques.pdf

Cowie, R., C. Régnier, B. Fontaine, and P. Bouchet. 2017. "Measuring the Sixth Extinction: What Do Mollusks Tell Us?" *The Nautilus* 131: 3–41.

Cox, P. D., and D. B. Pinniger. 2007. "Biology, Behaviour and Environmentally Sustainable Control of *Tineola bisselliella* (Hummel) (Lepidoptera: Tineidae)." *Journal of Stored Products Research* 43: 2–32.

Craig, H. Ed. 1897. *The Animal Kingdom; Based upon the Writings of the Eminent Naturalists, Audubon, Wallace, Brehm, Wood and Others*, Vol. 1. New York, Johnson & Bailey.

Crandall, D. P. 2002. "Himba Animal Classification and the Strange Case of the Hyena." *Africa* 72: 293–311.

Cripps, M. G., A. Gassmann, S. V. Fowler, G. W. Bourdôt, A. S. McClay, and G. R. Edwards. 2011. "Classical Biological Control of *Cirsium arvense*: Lessons from the Past." *Biological Control* 57: 165–174.

Crist, E., C. Mora, and R. Engelman. 2017. "The Interaction of Human Population, Food Production, and Biodiversity Protection." *Science* 356(6335): 260–264.

Crivens, M. A. 1999. *What Color Is Attractive? A Look at the Relationships between Skin Tone Preference, Self-Esteem, and Ethnic Identity Attitudes*. ProQuest Dissertations Publishing, 9951938. https://www.proquest.com/docview/304522437?pq-origsite=gscholar&fromopenview=true

Crockett, L. J. 1977. *Wildly Successful Plants: A Handbook of North American Weeds*. New York, Macmillan.

Cronon, W. 1995. "The Trouble with Wilderness; or, Getting Back to the Wrong Nature." In *Uncommon Ground: Rethinking the Human Place in Nature*, edited by W. Cronon, 69–90. New York, W. W. Norton & Co. https://www.williamcronon.net/writing/Trouble_with_Wilderness_Main.html

Crook, J. H., and J. D. Goss-Custard. 1972. "Social Ethology." *Annual Review of Psychology* 23: 277–312.

Crook, R. J. 2013. "The Welfare of Invertebrate Animals in Research: Can Science's Next Generation Improve Their Lot?" *Journal of Postdoctoral Research* 1(2): 1–20.

Cross, L. 2016. *Ogres*. New York, Cavendish Square Publishing.

Crowther, T. W., H. B. Glick, K. R. Covey, C. Bettigole, D. S. Maynard, S. M. Thomas, et al. 2015. "Mapping Tree Density at a Global Scale." *Nature* 525: 201–205.

Cuce, E. 2017. "Thermal Regulation Impact of Green Walls: An Experimental and Numerical Investigation." *Applied Energy* 194: 24–254.

Culet, P. 2003. "The International Treaty on Plant Genetic Resources for Food and Agriculture." International Environmental Law Research Centre, Briefing Paper 2003-2. http://www.ielrc.org/content/f0302.htm

Curtin, P., and S. Papworth. 2020. "Coloring and Size Influence Preferences for Imaginary Animals, and Can Predict Actual Donations to Species-Specific Conservation Charities." *Conservation Letters* 13(4): e12723. https://doi.org/10.1111/conl.12723

Curtis, G. C., W. J. Magee, W. W. Eaton, H.-U. Wittchen, and R. C. Kessler. 1998. "Specific Fears and Phobias: Epidemiology and Classification." *British Journal of Psychiatry* 173: 212–217.

Curtis, T. H., B. D. Bruce, G. Cliff, S. F. J. Dudley, A. P. Klimley, A. A. Kock, et al. 2012. "Responding to the Risk of White Shark Attack Updated Statistics, Prevention, Control Methods, and Recommendations." In *Global Perspectives on the Biology and Life History of the Great White Shark*, edited by M. L. Domeier, 477–509. Boca Raton, FL, CRC Press.

Curtis, V., M. De Barra, and R. Aunger. 2011. "Disgust as an Adaptive System for Disease Avoidance Behaviour." *Philosophical Transactions of the Royal Society B: Biological Sciences*, 366: 389–401.

Cuthbert, R., and G. Hilton. 2004. "Introduced House Mice *Mus musculus*: A Significant Predator of Threatened and Endemic Birds on Gough Island, South Atlantic Ocean?" *Biological Conservation* 117: 483–489.

Cuthill, I. C., W. L. Allen, K. Arbuckle, B. Caspers, G. Chaplin, M. E. Hauber, et al. 2017. "The Biology of Color." *Science* 357(6350). https://doi.org/10.1126/science.aan0221

Cyranoski, D. 2015. "Gene-edited 'Micropigs' to Be Sold as Pets at Chinese institute." *Nature* 526: 18. https://doi.org/10.1038/nature.2015.18448

D'Silva, J., and C. McKenna. 2018. *Farming, Food and Nature: Respecting Animals, People and the Environment*. London, Routledge.

Daehler, C. D. 1998. "The Taxonomic Distribution of Invasive Angiosperm Plants: Ecological Insights and Comparison to Agricultural Weeds." *Biological Conservation* 84: 167–180.

Dafni, A. 2006. "On the Typology and the Worship Status of Sacred Trees with a Special Reference to the Middle East." *Journal of Ethnobiology and Ethnomedicine* 2: 26. https://doi.org/10.1186/1746-4269-2-26

Dahake, A. S., and M. N. Hedaoo. 2018. "Application of Water Hyacinth (*Eichhornia crassipes*) in Wastewater Treatment – A Review." *International Research Journal of Engineering and Technology* 5: 1573–1577.

Dalby R, 1999. "The Delightful Dandelion." *American Bee Journal* 139: 300–301.

Dalrymple, R. L. 1975. *Crabgrass as a Forage*. Publication No. CG-75. Ardmore, OK, Noble Foundation.

Dalrymple, R. L. 1980. "Crabgrass Pasture Produces Good Calf Gains." *Rangelands* 2:107–109.

Dalrymple, R. L. 1994. *Crabgrass for Forage: Management from the 1990s*. Publication No. NFFO-99-18. Ardmore, OK, Noble Foundation.

Dalrymple, R. L. 2001. "Registration of 'Red River' Crabgrass." *Crop Science* 41: 1998–1999.

Dalrymple, R. L. 2010. "'Quick-n-Big': A New Forage Crabgrass Variety/Cultivar." In *Proceedings of the American Forage and Grassland Council*, June 21–23, 2010. Springfield, MO, American Forage and Grassland Council. CD-Rom.

Daly, B., and L. L. Morton. 2003. "Children with Pets Do Not Show Higher Empathy: A Challenge to Current Views." *Anthrozoös* 16: 298–314.

Daly, J. W. 1995. "The Chemistry of Poisons in Amphibian Skin." *Proceedings of the National Academy of Sciences of the United States of America* 92: 9–13.

Daly, M., and M. Wilson. 1988. *Homicide*. Hawthorne, NY, Aldine.

Daly, N. 2017. "The Scientific Reason So Many Animals Are Black and White." *National Geographic*. https://www.rd.com/culture/black-and-white-animals/

Damania, A. B. 2011. "The Origin, History, and Commerce of the Opium Poppy (*Papaver somniferum*) in Asia and the United States." *Asian Agri-History* 15: 109–123.

Daniell, H., S. J. Streatfield, and K. Wycoff. 2001. "Medical Molecular Farming: Production of Antibodies, Biopharmaceuticals and Edible Vaccines in Plants." *Trends in Plant Science* 6: 219–226.

Dantas-Torres, F. 2015. "Climate Change, Biodiversity, Ticks and Tick-Borne Diseases: The Butterfly Effect." *International Journal for Parasitology: Parasites and Wildlife* 4: 452–461.

Danvir, R. E. 2018. "Multiple-use Management of Western U.S. Rangelands: Wild Horses, Wildlife, and Livestock." *Human–Wildlife Interactions* 12(1): 4. https://digitalcommons.usu.edu/hwi/vol12/iss1/4

Darbyshire, S. J., J. Y. Leeson, and A. G. Thomas. 2009. "Identification and Distribution of Barnyard Grass (*Echinochloa crus-galli* and *E. muricata*)." In *Soils and Crops Workshop*, University of Saskatchewan. https://harvest.usask.ca/bitstream/handle/10388/9212/S.J.%20Darbyshire%20et%20al.%2c%202009.pdf?sequence=1&isAllowed=y

Darlington, S. M. 2019. "The Potential of Buddhist Environmentalism." *The Ecological Citizen* 3: 25–26. https://www.ecologicalcitizen.net/pdfs/v03n1-04.pdf

Darwin, C. 1864. "On the Sexual Relations of the Three Forms of *Lythrum salicaria*." *Botanical Journal of the Linnean Society* 8: 169–196.

Darwin, C. 1865. *On the Movements and Habits of Climbing Plants*. London, John Murray.

Darwin, C. 1871. *The Descent of Man and Selection in Relation to Sex*. London, John Murray.

Darwin, C. 1872. *The Expression of the Emotions in Man and Animals*. London, John Murray.

Das, A., P. Ghosh, T. Paul, U. Ghosh, B. R. Pati, and K. C. Mondal. 2016. "Production of Bioethanol as Useful Biofuel through the Bioconversion of Water Hyacinth (*Eichhornia crassipes*). *3 Biotech* 6: 70. https://doi.org/10.1007/s13205-016-0385-y

Das, S., S. Morya, A. Neumann, and V. K. Chattu. 2021. "A Review of the Pharmacological and Nutraceutical Properties of *Cynodon dactylon*." *Pharmacognosy Research* 13: 104–112.

Dasgupta, P. 2021. *The Economics of Biodiversity: The Dasgupta Review*. London, HM Treasury. www.gov.uk/official-documents

Dato, V. M., E. R. Campagnolo, J. Long, and C. E. Rupprecht. 2016. "A Systematic Review of Human Bat Rabies Virus Variant Cases: Evaluating Unprotected Physical Contact with Claws and Teeth in Support of Accurate Risk Assessments." *PLoS One* 11(7): e0159443. https://doi.org/10.1371/journal.pone.0159443

Davies, T., A. Cowley, J. Bennie, C. Leyshon, R. Inger, H. Carter, et al. 2018. "Popular Interest in Vertebrates Does Not Reflect Extinction Risk and Is Associated with Bias in Conservation Investment." *PLoS One* 14(2): e0212101. https://doi.org/10.1371/journal.pone.0212101

Dawkins, R. 1989. *The Selfish Gene*. Oxford, Oxford University Press.

Dawson, J., E. J. Stewart, H. Lemelin, and D. Scott. 2010. "The Carbon Cost of Polar Bear Viewing Tourism in Churchill, Canada." *Journal of Sustainable Tourism* 18: 319–336.

De Castro, O., R. Gargiulo, E. Del Guacchio, P. Caputo, and P. De Luca. 2015. "A Molecular Survey Concerning the Origin of *Cyperus esculentus* (Cyperaceae, Poales): Two Sides of the Same Coin (Weed vs. Crop)." *Annals of Botany* 115: 733–745.

De Chazelles, L. 1796. *Arbres, arbrisseaux, plantes, fleurs et fruits peints par le Pt de Chazelles, pour être joints au Dictionnaire des jardiniers, et Supplément qui en donne la description*. [Trees, Shrubs, Plants, Flowers and Fruits Painted by Pt de Chazelles, to Be Included in the Dictionary of Gardeners, and Supplement Which Gives the Description.] Vol. 3. Paris. https://galeries.limedia.fr/ark:/79345/d60wf7nl66xsws83/p1

De Cleene, M., and M. C. Lejeune. 2003. *Compendium of Symbolic and Ritual Plants in Europe*. 2 vols. Ghent, Man and Culture.

De Freitas Netto, S. V., M. F. F. Sobral, A. R. B. Ribeiro, and G. R. da Luz Soares. 2020. "Concepts and Forms of Greenwashing: A Systematic Review." *Environmental Sciences Europe* 32: 19. https://doi.org/10.1186/s12302-020-0300-3

De Jong, N. W., A. M. Vermeulen, R. Gerth van Wijk, and H. de Groot. 1998. "Occupational Allergy Caused by Flowers." *Allergy* 53: 204–209.

DeLeeuw, J. L., L. W. Galen, C. Aebersold, and V. Stanton. 2007. "Support for Animal Rights as a Function of Belief in Evolution, Religious Fundamentalism, and Religious Denomination." *Society & Animals: Journal of Human-Animal Studies* 15: 353–363.

Delouche, J. C., R. Labrada, N. R. Burgos, and D. R. Gealy. 2007. *Weedy Rices: Origin, Biology, Ecology and Control*. Food and Agriculture Organization Plant Production and Protection Paper No. 188. Geneva, Food & Agriculture Organization of the United Nations.

De Paiva, S. R., C. M. McManus, and H. Blackburn. 2016. "Conservation of Animal Genetic Resources – A New Tact." *Livestock Science* 193: 32–38.

De Pinho, J. R., C. Grilo, R. B. Boone, K. A. Galvin, and J. G. Snodgrass. 2014. "Influence of Aesthetic Appreciation of Wildlife Species on Attitudes towards Their Conservation in Kenyan Agropastoralist Communities." *PLoS One* 9: e88842. https://doi.org/10.1371/journal.pone.0088842.

De Smet, P. A. G. M. 1993. "*Hedera helix*." In *Adverse Effects of Herbal Drugs, Vol. 2*, edited by P. A. G. M. Smet, K. Keller, R. Hänsel, and R. Chandler, 209–215. Berlin, Springer.

De Vries, F. T. 1991. "Chufa (*Cyperus esculentus*, Cyperaceae): A Weedy Cultivar or a Cultivated Weed?" *Economic Botany* 45: 27–37.

De Vries, M., and de Boer, I. J. 2010. "Comparing Environmental Impacts for Livestock Products: A Review of Life Cycle Assessments." *Livestock Science* 128: 1–11.

De Waal, F., R. Wright, C. Korsgaard, P. Kitcher, and P. Singer. 2006. *Primates and Philosophers: How Morality Evolved*. Princeton, NJ, Princeton University Press.

Dean, K. R., F. Krauer, L. Walløe, O. C. Lingjærde, B. Bramanti, N. C. Stenseth, et al. 2018. "Human Ectoparasites and the Spread of Plague in Europe during the Second Pandemic." *Proceedings of the National Academy of Sciences of the United States of America*. https://doi.org/10.1073/pnas.1715640115

Déchamp, C. 2013. "Ambrosia Pollinosis." *Revue des Maladies Respiratoires* 30: 316–327.

Dedek, A., J. Xu, L.-É. Lorenzo, A. G. Godin, C. M. Kandegedara, G. Glavina, et al. 2022. "Sexual Dimorphism in a Neuronal Mechanism of Spinal Hyperexcitability across Rodent and Human Models of Pathological Pain." *Brain* awab408. https://doi.org/10.1093/brain/awab408

Degaga, A. H. 2018. "Water Hyacinth (*Eichhornia crassipes*) Biology and its Impacts on Ecosystem, Biodiversity, Economy and Human Wellbeing." *Journal of Life Science and Biomedicine* 8: 94–100.

Dejean, A., B. Bolton, and J. L. Durand. 1997. "*Cubitermes subarquatus* Termitaries as Shelters for Soil Fauna in African Rainforests." *Journal of Natural History* 31: 1289–1302.

Dejean, A., and J. L. Durand. 1996. "Ants Inhabiting *Cubitermes* Termitaries in African Rain Forests." *Biotropica* 28: 701–713.

Del Tredici, P. 2017. "The Introduction of Japanese Knotweed, *Reynoutria japonica*, into North America." *Journal of the Torrey Botanical Society* 144: 406–416.

Del Tredici, P. 2020. *Wild Urban Plants of the Northeast: A Field Guide*. Second Edition. Ithaca, NY, Comstock Publishing Associates.

Delaunay, P., V. Blanc, P. Del Giudice, A. Levy-Bencheton, O. Chosidow, P. Marty, et al. 2011. "Bedbugs and Infectious Diseases." *Clinical Infectious Diseases* 52: 200–210.

DeLoache, J. S., and V. LoBue. 2009. "The Narrow Fellow in the Grass: Human Infants Associate Snakes and Fear." *Developmental Science* 12: 201–207.

DeLong-Bas, N. 2018. "Islam, Nature, and the Environment." *Oxford Bibliographies*. https://www.oxfordbibliographies.com/view/document/obo-9780195390155/obo-9780195390155-0258.xml

Demafelis, R. B., T. C. Mendoza, A. E. D. Matanguihan, J. A. S. Malabuyoc, V. M. R. Magadia Jr., A. A. Pector, et al. 2017. "Is Raw Sugar Produced from Sugarcane (*Saccharum officinarum* L.) Carbon Positive or Negative?" *International Journal of Agricultural Technology* 13: 565–581.

Demafelis, R. B., J. L. Movillon, C. D. Predo, D. S. Maligalig, P. J. M. Eleazar, and B. Tongko-Magadia. 2020. "Socioeconomic and Environmental Impacts of Bioethanol Production from Sugarcane (*Saccharum officinarum*) and Molasses in the Philippines." *Journal of Environmental Science and Management* 23: 96–110.

Demirbaş, N. 2019. "Vertical Agriculture: A Review of Developments. In *Proceedings International Balkan and Near Eastern Social Sciences Congress Series XI. Ibaness Congress Series-Tekirdağ / Turkey*, edited by D. K. Dimitrov, D. Nikoloski, and R. Yilmaz, 620–626. Tekirdağ, Turkey. https://www.researchgate.net/profile/Ipek-Okkay/publication/331981135_Tuketiciyi_Hem_Eglendirerek_Hem_de_Hissettirmeden_Ulasmanin_Yolu_Olarak_Viral_Pazarlama/links/5c98abaa92851cf0ae9606b9/Tueketiciyi-Hem-Eglendirerek-Hem-de-Hissettirmeden-Ulasmanin-Yolu-Olarak-Viral-Pazarlama.pdf#page=633

Deryabina, T. G., S. V. Kuchmel, L. L. Nagorskaya, T. G. Hinton, J. C. Beasley, A. Lerbours, et al. 2015. "Long-Term Census Data Reveal Abundant Wildlife Populations at Chernobyl." *Current Biology* 25: R811–R826.

Devault, T. L., J. C. Beasley, Z. H. Olson, M. Moleón, M. Carrete, A. Margalida, et al. 2016. "Ecosystem Services Provided by Avian Scavengers." In *Why Birds Matter: Avian Ecological Function and Ecosystem Services*, edited by C. H. Şekercioglu, D. G. Wenny, and C. J. Whelan, 235–270. Chicago, IL, University of Chicago Press.

Dhingra, N., S. Sharma, and L. K. Dwivedi. 2013. "Protective Role of Parasites against the Autoimmune Diseases: A Benefit of Being Host." *South Asian Journal of Experimental Biology* 3: 335–339.

Dhont, K., and G. Hodson. Eds. 2019. *Why We Love and Exploit Animals. Bridging Insights from Academia and Advocacy*. London, Routledge.

Di Minin, E., H. S. Clements, R. A. Correia, G. Cortés-Capano, C. Fink, A. Haukka, et al. 2021a. "Consequences of Recreational Hunting for Biodiversity Conservation and Livelihoods." *One Earth* 4: 238–253.

Di Minin, E., R. A. Correia, and T. Toivonen. 2021b. "Conservation Geography." *Trends in Ecology & Evolution* 36(10). https://doi.org/10.1016/j.tree.2021.08.009

Dias De Oliveira, M. E., B. E. Vaughan, and E. J. Rykiel. 2005. "Ethanol as Fuel: Energy, Carbon Dioxide Balances, and Ecological Footprint." *BioScience* 55: 593–602.

Diaz, J. H. 2004. "The Global Epidemiology, Syndromic Classification, Management, and Prevention of Spider Bites." *American Journal of Tropical Medicine and Hygiene* 71: 239–250.

Dickson, B. 2009. "The Ethics of Recreational Hunting." In *Recreational Hunting, Conservation and Rural Livelihoods: Science and Practice*, edited by B. Dickson, J. Hutton, and W. A. Adams, 59–72. Chichester, UK, Blackwell.

DiNicolantonio, J. J., J. H. O'Keefe, and W. L. Wilson. 2018. "Sugar Addiction: Is It Real? A Narrative Review." *British Journal of Sports Medicine* 52: 910–913.

Dion, K., E. Berscheid, and E. Walster. 1972. "What Is Beautiful Is Good." *Journal of Personality and Social Psychology* 24: 285–290.

Dittbrenner, A., U. Lohwasser, H.-P. Mock, and A. Börner. 2008. "Molecular and Phytochemical Studies of *Papaver somniferum* in the Context of Infraspecific Classification." *Acta Horticulturae* 799: 81–88.

Dittbrenner, A., H.-P. Mock, A. Börner, and U. Lohwasser. 2009. "Variability of Alkaloid Content in *Papaver somniferum* L." *Journal of Applied Botany and Food Quality* 82: 103–107.

Diverio, S., B. Boccini, L. Menchetti, and P. C. Bennett. 2016. "The Italian Perception of the Ideal Companion Dog." *Journal of Veterinary Behavior* 12: 27–35.

Dixson, A., B. Dixson, and M. Anderson. 2005. "Sexual Selection and the Evolution of Visually Conspicuous Sexually Dimorphic Traits in Male Monkeys, Apes, and Human Beings." *Annual Review of Sex Research* 16: 1–19.

Dixson, J. W., and M. J. Rantala. 2016. "The Role of Facial and Body Hair Distribution in Women's Judgments of Men's Sexual Attractiveness." *Archives of Sexual Behavior* 45: 877–889.

Dixson, J. W. B., J. M. Sherlock, W. K. Cornwell, and M. M. Kasumovic. 2018. "Contest Competition and Men's Facial Hair: Beards May Not Provide Advantages in Combat." *Evolution and Human Behavior* 39: 147–153.
Djernæs, M. 2018. "Biodiversity of Blattodea – The Cockroaches and Termites." In *Insect Biodiversity: Science and Society*, edited by R. G Foottit and P. H. Adler, 359–387. Oxford, Wiley, Blackwell.
Doctor, J. 2013. "*Musca domestica*." *Animal Diversity Web*. https://animaldiversity.org/accounts/Musca_domestica/
Doepler, E., and W. Ranisch. 1903. *Walhall, die Götterwelt der Germanen* [Valhall, the Gods of the Teutons]. Berlin, Martin Oldenbourg.
Doggett, S. L., D. E. Dwyer, P. F. Peñas, and R. C. Russell. 2012. "Bed Bugs: Clinical Relevance and Control Options." *Clinical Microbiology Reviews* 25: 164–192.
Doi, H., Gałęcki, R., and R. N. Mulia. 2021. "The Merits of Entomophagy in the Post COVID-19 World." *Trends in Food Science & Technology* 110: 849–854.
Dolbeer, R. A. 1999. "Overview and Management of Vertebrate Pests." In *Handbook of Pest Management*, edited by J. R. Ruberson, 663 –691. New York, Marcel Dekker.
Doll, R. 1999. *Tobacco: A medical history. Journal of Urban Health* 76: 289–313.
Donald, W. W. 1994. "The Biology of Canada Thistle (*Cirsium arvense*)." *Reviews of Weed Science* 6: 77–101.
Donaldson, L. J., J. Cavanagh, and J. Rankin. 1997. "The Dancing Plague: A Public Health Conundrum." *Public Health* 111: 201–204.
Donaldson, M. R., N. J. Burnett, D. C. Braun, C. D. Suski, S. G. Hinch, S. J. Cooke, et al. 2016. "Taxonomic Bias and International Biodiversity Conservation Research." *Facets*. https://doi.org/10.1139/facets-2016-0011; http://www.facetsjournal.com/article/facets-2016-0011/
Donkin, R. 1977. "Spanish Red: An Ethnogeographical Study of Cochineal and the Opuntia Cactus." *Transactions of the American Philosphical Society* 67: 1–77.
Donovan, S. E., P. Eggleton, W. E. Dubbin, M. Batchelder, and L. Dibog. 2001. "The Effect of a Soil-Feeding Termite, *Cubitermes fungifaber* (Isoptera: Termitidae) on Soil Properties: Termites May Be an Important Source of Soil Microhabitat Heterogeneity in Tropical Forests." *Pedobiologia* 45: 1–11.
Dorcas, M. E., S. E. Pittman, and J. D. Willson. 2017. "Burmese Pythons." In *Ecology and Management of Terrestrial Vertebrate Invasive Species in the United States*, edited by W. C. Pitt, J. Beasley, and G. W. Witmer, 136–162. Boca Raton, FL, CRC Press.
Dossey, A. T., J. A. Morales-Ramos, and M. G. Rojas. Eds. 2016. *Insects as Sustainable Food Ingredients: Production, Processing and Food Applications*. Amsterdam, Elsevier.
Dougherty, E. R., C. J. Carlson, V. M. Bueno, K. R. Burgio, C. A. Cizauskas, C. F. Clements, et al. 2016. "Paradigms for Parasite Conservation." *Conservation Biology* 30: 724–733.
Douvere, F. 2016. "Conservation: The Seas Cannot Be Saved on a Budget of Breadcrumbs." *Nature* 534: 30–32.
Dowler, R. C., J. C. Perkins, A. A. Shaffer, B. D. Wolaver, B. J. Labay, J. P. Pierre, et al. 2017. "Conservation Status of the Plains Spotted Skunk, *Spilogale putorius interrupta*, in Texas, with an Assessment of Genetic Variability in the Species." The University of Texas at Austin Texas Scholarworks. https://repositories.lib.utexas.edu/handle/2152/62348
Dragoo, J. W. 2009. "Nutrition and Behavior of Striped Skunks." *Veterinary Clinics of North America: Exotic Animal Practice* 12: 313–326.
Dragoo, J. W., and R. L. Honeycutt. 1997. "Systematics of Mustelid-like Carnivores." *Journal of Mammalogy* 78: 426–443.
Drahota, J. 2010. "Canada Thistle: A Literature Review of Treatment Success Specific to Current Land Management on Waterfowl Production Areas in the Rainwater Basin." Washington, DC, U.S. Fish and Wildlife Service.
Ducatez, S., and R. Shine. 2017. "Drivers of Extinction Risk in Terrestrial Vertebrates." *Conservation Letters* 10: 186–194.
Duckitt, J., C. Wall, and B. Pokroy. 1999. "Color Bias and Racial Preference in White South African Preschool Children." *Journal of Genetic Psychology* 160: 143–154.
Dudley, J., and M. Woodford. 2002. "Bioweapons, Biodiversity, and Ecocide: Potential Effects of Biological Weapons on Biological Diversity." *American Institute of Biological Sciences* 52: 583–592.
Dudley, N., and S. Alexander. 2017. "Agriculture and Biodiversity: A Review." *Biodiversity* 18(2–3): 45–49.
Duke, J. A. 1973. "Utilization of *Papaver*." *Economic Botany* 27: 390–400.
Dulvy, N. K., S. L. Fowler, J. A. Musick, R. D. Cavanagh, P. M. Kyne, L. R. Harrison, et al. 2014. "Extinction Risk and Conservation of the World's Sharks and Rays." *eLife* 3, e00590. http://doi.org/10.7554/eLife.00590
Duncan, B. L., R. Hansen, C. Cranney, J. F. Shaw, K. Veblen, and K. M. Kettenring. 2019. "Cattle Grazing for Invasive *Phragmites australis* (Common Reed) Management in Northern Utah Wetlands." Utah State University Extension Fact Sheet NR/Wildlands/2019-01pr. https://digitalcommons.usu.edu/cgi/viewcontent.cgi?article=3038&context=extension_curall
Dunham, R. A. 2009. "Transgenic Fish Resistant to Infectious Diseases, Their Risk and Prevention of Escape into the Environment and Future Candidate Genes for Disease Transgene Manipulation." *Comparative Immunology, Microbiology and Infectious Diseases* 32: 139–161.
Dunn, R. R. 2005. "Modern Insect Extinctions, the Neglected Majority." *Conservation Biology* 19: 1030–1036.
Dunn, R. R. 2009. "Coextinction: Anecdotes, Models, and Speculation." In *Holocene Extinctions*, edited by S. T. Turvey, 167–180. Oxford, Oxford University Press.
Durden, L. A. 2018. "Chapter 7: Lice (Phthiraptera)." In *Medical and Veterinary Entomology*, Third Edition, edited by G. R. Mullen and L. A. Durden, 79–106. Cambridge, Academic Press.
Durden, L. A., and N. C. Hinkle. 2018. "Chapter 10: Fleas (Siphonaptera)." In *Medical and Veterinary Entomology*, Third Edition, edited by G. R. Mullen and L. A. Durden, 145–169. Cambridge, Academic Press.
Durden, L. A., and Keirans, J. E. 1996. "Host–Parasite Coextinction and the Plight of Tick Conservation." *American Entomologist* 42: 87–91.
Dwyer, J. 2016. "*Cyperus rotundus* L.: An Ancient Food Staple but Now Designated the World's Worst Weed." In *Twentieth Australasian Weeds Conference, Perth*, edited by R. Randall, S. Lloyd, and C. Borger, 251–254. Perth, Weeds Society of Western Australia. http://caws.org.nz/old-site/awc/2016/awc201612511.pdf
Eads, D. A., T. M. Livieri, P. Dobesh, E. Childers, L. E. Noble, M. C., Vasquez, et al. 2021. "Fipronil Pellets Reduce Flea Abundance on Black-tailed Prairie Dogs: Potential Tool for Plague Management and Black-footed Ferret Conservation." *Journal of Wildlife Diseases* 57: 434–438.
Eastern Spotted Skunk Cooperative Study Group. 2018. "Eastern Spotted Skunk Conservation Plan." https://easternspottedskunk.weebly.com/uploads/3/9/7/0/39709790/ess_conservation_plan_5july18_final.pdf

Eaton, R. L. 1976. "The Brown Hyena: A Review of Biology, Status and Conservation." *Mammalia* 40: 377–400.

Eaton, W. W., O. J. Bienvenu, and B. Miloyan. 2018. "Specific Phobias." *The Lancet, Psychiatry* 5: 678–686.

Ebel, R., F. Menalled, S. Ahmed, S. Gingrich, G. M. Baldinelli, and G. F. Félix. 2021. "How Biodiversity Loss Affects Society." In *Handbook on the Human Impact of Agriculture*, edited by H. S. James, Jr., 352–376. Cheltenham, UK, Edward Elgar.

Eberle, C. E., D. P. Sandler, K. W. Taylor, and A. J. White. 2020. "Hair Dye and Chemical Straightener Use and Breast Cancer Risk in a Large US Population of Black and White Women." *International Journal of Cancer* 147: 383–391.

Ebrahimpour-koujan, S., P. Saneei, B. Larijani, and A. Esmaillzadeh. 2020. "Consumption of Sugar-sweetened Beverages and Serum Uric Acid Concentrations: A Systematic Review and Meta-analysis." *Journal of Human Nutrition and Dietetics*. https://doi.org/10.1111/jhn.12796

Ebury, K. 2021. "Animals and Modern Capital Punishment." In *Modern Literature and the Death Penalty, 1890-1950*. Palgrave Studies in Literature, Culture and Human Rights, 117–139. Cham, Palgrave Macmillan. https://doi.org/10.1007/978-3-030-52750-1_5

Echeverri, A., M. M. Callahan, K. M. A. Chan, T. Satterfield, and J. Zhao. 2017. "Explicit Not Implicit Preferences Predict Conservation Intentions for Endangered Species and Biomes." *PLoS One*. https://doi.org/10.1371/journal.pone.0170973

Eckberg, J., E. Lee-Mäder, J. Hopwood, S. F. Jordan, and B. Borders. 2017. *Native Thistles: A Conservation Practitioner's Guide. Plant Ecology, Seed Production Methods, and Habitat Restoration Opportunities*. Portland, OR, Xerces Society for Invertebrate Conservation. http://xerces.org/sites/default/files/2018-05/16-029_01_XercesSoc_Native-Thistles-Conservation-Guide_web.pdf

Eco, U. 2005. *History of Beauty*. London, Secker & Warburg.

Egan, A. N., and B. Pan. 2015. "Resolution of Polyphyly in *Pueraria* (Leguminosae, Papilionoideae): The Creation of Two New Genera, *Haymondia* and *Toxicopueraria*, the Resurrection of *Neustanthus*, and a New Combination in *Teyleria*." *Phytotaxa* 218: 201–226.

Egan, A. N., M. Vatanparast, and W. Cagle. 2016. "Parsing Polyphyletic *Pueraria*: Delimiting Distinct Evolutionary Lineages through Phylogeny." *Molecular Phylogenetics and Evolution* 104: 44–59.

Egan, E. 2021. "What Our Use of Animal-based Slurs and Endearments Says about Us." *Psyche* (online newsletter), May 12. https://psyche.co/ideas/what-our-use-of-animal-based-slurs-and-endearments-says-about-us

Egerer, M., and S. Buchholz. 2021. "Reframing Urban 'Wildlife' to Promote Inclusive Conservation Science and Practice." *Biodiversity and Conservation* 30: 2255–2266.

Eggleton, P. 2011. "An Introduction to Termites: Biology, Taxonomy, and Functional Morphology." In *Biology of Termites: A Modern Synthesis*, edited by D. E. Bignell, Y. Roisin, and N. Lo, 1–26. New York, Springer.

Ejoh, R. A., Djomdi, and R. Ndjouenkeu. 2006. "Characteristics of Tigernut (*Cyperus esculentus*) Tubers and Their Performance in the Production of a Milky Drink." *Journal of Food Processing and Preservation* 30: 145–163.

Eklund, A., J. López-Bao, M. Tourani, G. Chapron, and J. Frank. 2017. "Limited Evidence on the Effectiveness of Interventions to Reduce Livestock Predation by Large Carnivores." *Scientific Reports* 7: 2097. https://doi.org/10.1038/s41598-017-02323-w

Eklund, J., A. Arponen, P. Visconti, and M. Cabeza. 2011. "Governance Factors in the Identification of Global Conservation Priorities for Mammals." *Philosophical Transactions of the Royal Society B: Biological Sciences* 366: 2661–2669.

Ekstein, A. 2014. "Flowering and Pithing in Sugarcane." Information Sheet 11, South African Sugarcane Research Institute. https://sasri.org.za/wp-content/uploads/Information_Sheets/IS_11.5-Flowering-in-sugarcane.pdf

Ekwealor, K. U., C. B. Echereme, T. N. Ofobeze, and C. N. Okereke. 2019. "Economic Importance of Weeds: A Review." *Asian Plant Research Journal* 3(2): 1–11. https://doi.org/10.9734/aprj/2019/v3i230063

Elfes, C. T., S. R. Livingstone, A. Lane, V. Lukoschek, K. L. Sanders, A. J. Courtney, et al. 2013. "Fascinating and Forgotten: The Conservation Status of Marine Elapid Snakes." *Herpetological Conservation and Biology* 8: 37–52.

Elia, I. E. 2013. "A Foxy View of Human Beauty: Implications of the Farm Fox Experiment for Understanding the Origins of Structural and Experiential Aspects of Facial Attractiveness." *The Quarterly Review of Biology* 88: 163–183.

Elliot, A. J. 2015. "Color and Psychological Functioning: A Review of Theoretical and Empirical Work." *Frontiers in Psychology* 6: 368. https://www.frontiersin.org/articles/10.3389/fpsyg.2015.00368/full

Elliot, A. J., and M. A. Maier. 2014. "Color Psychology: Effects of Perceiving Color on Psychological Functioning in Humans." *Annual Review of Psychology* 65: 95–120.

Elliott, J. M., and U. Kutschera. 2011. "Medicinal Leeches: Historical Use, Ecology, Genetics and Conservation." *Freshwater Reviews* 4: 21–41.

El-Mostafa, K., Y. El Kharrassi, A. Badreddine, P. Andreoletti, J. Vamecq, M. S. El Kebbaj, et al. 2014. "Nopal Cactus (*Opuntia ficus-indica*) as a Source of Bioactive Compounds for Nutrition, Health and Disease." *Molecules* 19: 14879–14901.

Elmer, E. M., and J. Houran. 2020. "Physical Attractiveness in the Workplace: Customers Do Judge Books by Their Covers." *Academia* (website). https://www.academia.edu/58096778/Physical_Attractiveness_in_the_Workplace_Customers_Do_Judge_Books_by_Their_Covers

Elmqvist, T., M. Fragkias, J. Goodness, B. Güneralp, P. Marcotulli, R. I. McDonald, et al. 2013. *Urbanization, Biodiversity and Ecosystem Services: Challenges and Opportunities: A Global Assessment*. New York, Springer.

Elmqvist, T., W. C. Zipperer, and B. Güneralp. 2016. "Urbanization, Habitat Loss, Biodiversity Decline: Solution Pathways to Break the Cycle." In *Routledge Handbook of Urbanization and Global Environmental Change*, edited by K. Seta, W. D. Solecki, and C. A. Griffith, 139–151. London, Routledge.

Enoch, J., L. McDonald, L. Jones, P. R. Jones, and D P. Crabb. 2019. "Evaluating Whether Sight Is the Most Valued Sense." *JAMA Ophthalmology* 137: 1317–1320.

Enquist, M., and A. Arak. 1994. "Symmetry, Beauty and Evolution." *Nature* 372: 169–172.

Entin, P. 2007. "Do Racehorses and Greyhound Dogs Exhibit a Gender Difference in Running Speed?" *Equine and Comparative Exercise Physiology* 4: 135–140.

Equine News. 2007. "Dangerous 'Awn Grasses' May Cause Horses to Stop Eating." *Equine News* (Washington State University College of Veterinary Medicine newsletter) Fall, 2007. http://sbbch.org/wp-content/uploads/2012/04/dangerous-grass2.pdf

Erickson, J., B. Sadeghirad, L. Lytvyn, J. Slavin, and B. C. Johnston. 2017. "The Scientific Basis of Guideline Recommendations on Sugar Intake: A Systematic Review." *Annals of Internal Medicine* 166: 257–267.

Erlich, N., O. V. Lipp, and V. Slaughter. 2013. "Of Hissing Snakes and Angry Voices: Human Infants Are Differentially Responsive to Evolutionary Fear-relevant Sounds." *Developmental Science* 16: 894–904.

Ervin, G. N. 2012. "Indian Fig Cactus (*Opuntia ficus-indica* (L.) Miller) in the Americas: An Uncertain History." *Haseltonia* 17: 70–81.

Escudero, N., M. De Arellano, S. Fernández, G. Albarracín, and S. Mucciarelli. 2003. *Taraxacum officinale* as a Food Source." *Plant Foods for Human Nutrition* 58: 1–10.

Eslami, S. V., and S. Ward. 2021. "*Chenopodium album* and *Chenopodium murale*." In *Biology and Management of Problematic Crop Weed Species*, edited by B. Chauhan, 89–112. Cambridge, MA, Academic Press.

Esser, H. J., E. A. Herre, R. Kays, Y. Liefting, and P. A. Jansen. 2019. "Local Host-Tick Coextinction in Neotropical Forest Fragments." *International Journal for Parasitology* 49: 225–233.

Esser, L. L. 1994. "*Echinochloa crus-galli*." In *Fire Effects Information System [Online]*. U.S. Department of Agriculture, Forest Service, Rocky Mountain Research Station, Fire Sciences Laboratory (Producer). https://www.fs.usda.gov/database/feis/plants/graminoid/echcru/all.html

Essl, F., K. Biró, D. Brandes, O. Broennimann, J. M. Bullock, D. S. Chapman, et al. 2015. "Biological Flora of the British Isles: *Ambrosia artemisiifolia*." *Journal of Ecology* 103: 1069–1098.

Estevão-Costa, M. I., R. Sanz-Soler, B. Johanningmeier, and J. A. Eble. 2018. "Snake Venom Components in Medicine: From the Symbolic Rod of Asclepius to Tangible Medical Research and Application." *International Journal of Biochemistry & Cell Biology* 104: 94–113.

Estéve, A. T. 2007. "The Genetic Creation of Bioluminescent Plants for Urban and Domestic Use." *Leonardo* 40(1): 18.

Estren, M. J. 2012. "The Neoteny Barrier: Seeking Respect for the Non-Cute." *Journal of Animal Ethics* 2: 6–11.

ETC Group and Heinrich Böll Foundation. 2018. "Forcing the Farm: How Gene Drive Organisms Could Entrench Industrial Agriculture and Threaten Food Sovereignty." https://www.boell.de/sites/default/files/etc_forcingthefarm-report_8.pdf

Etcoff, N. 1999. *Survival of the Prettiest: The Science of Beauty*. New York, Doubleday.

European and Mediterranean Plant Protection Organization. 2007. "Data Sheets on Quarantine Pests: *Pueraria lobata*." *European and Mediterranean Plant Protection Organization Bulletin* 37: 230–235. http://citeseerx.ist.psu.edu/viewdoc/download?doi=10.1.1.619.242&rep=rep1&type=pdf

European Commission. 2016. *Science for Environment Policy, Synthetic Biology and Biodiversity. Future Brief 15*. Bristol, European Commission Science Communication Unit. http://ec.europa.eu/science-environment-policy

Evans, F. F., and R. J. Schmidt. 1980. "Plants and Plant Products That Induce Contact Dermatitis." *Planta Medica* 38: 289–316.

Evans, J., M. H. Alemu, R. Flore, M. B. Frøst, A. Halloran, A. B. Jensen, et al. 2015. "'Entomophagy:' An Evolving Terminology in Need of Review." *Journal of Insects as Food and Feed* 1: 293–305.

Evans, T. A., B. T. Forschler, and J. K. Grace. 2013. "Biology of Invasive Termites: A Worldwide Review." *Annual Review of Entomology* 58: 455–474.

Everett, J. A. C., L. Caviola, J. Savulescu, and N. S. Faber. 2019. "Speciesism, Generalized Prejudice, and Perceptions of Prejudiced Others." *Group Processes & Intergroup Relations* 22: 785–803.

Everts J. W., and L. Ba. 1997. "Environmental Effects of Locust Control: State of the Art and Perspectives." In *New Strategies in Locust Control*, edited by S. Krall, R. Peveling, and D. B. Diallo, 331–336. Basel, Birkhäuser.

Fairholt, F. W. 1859. *Tobacco: Its History and Associations; Including an Account of the Plant and Its Manufacture; with Its Modes of Use in All Ages and Countries*. London, Chapman & Hall.

Faith, D. P. 2008. "Threatened Species and the Potential Loss of Phylogenetic Diversity: Conservation Scenarios Based on Estimated Extinction Probabilities and Phylogenetic Risk Analysis." *Conservation Biology* 22: 1461–1470.

Falkowski, M., I. Kukulka, and S. Kozlowski. 1990. "Biological Properties and Fodder Value of Dandelion. Soil-Grassland-Animal Relationships." In *Proceedings of 13th General Meeting of the European Grassland Federation*, June 25-29, 1990, edited by N. Gáborčik, V. Krajčovič, and M. Zimková, 2: 208–211. Banská Bystrica, Czechoslovakia, Grassland Research Institute.

Fančovičová, J., P. Prokop, and M. Kubíčková. 2022. "The Effect of Aposematic Signals of Plants on Students' Perception and Willingness to Protect Them." *Sustainability* 14(15): 9121. https://doi.org/10.3390/su14159121

Fang, S. 2010. "A World without Mosquitoes." *Nature* 466: 432–434.

FAO. 2010. *FAO Guidelines for the Cryoconservation of Animal Genetic Resources*. Rome, Italy, Food and Agricultural Organization of the United Nations.

FAO. 2021. "Overview of Global Meat Market Developments in 2020." Food and Agriculural Organization of the United Nations Meat Market Review March 2021. http://www.fao.org/3/cb3700en/cb3700en.pdf

Fardisi, M., A. D. Gondhalekar, A. R. Ashbrook, and M. E. Scharf. 2019. "Rapid Evolutionary Responses to Insecticide Resistance Management Interventions by the German Cockroach (*Blattella germanica* L.)" *Scientific Reports* 9: 8292. https://doi.org/10.1038/s41598-019-44296-y

Favreau, J. M., C. A. Drew, G. R. Hess, M. J. Rubino, F. H. Koch, and K. A. Eschelbach. 2006. "Recommendations for Assessing the Effectiveness of Surrogate Species Approaches." *Biodiversity and Conservation* 15: 3949–3969.

Favrelière, E., A. Ronceux, J. Pernel, and J.-M. Meynard. 2020. "Nonchemical Control of a Perennial Weed, *Cirsium arvense*, in Arable Cropping Systems. A Review." *Agronomy for Sustainable Development* 40: 31. https://doi.org/10.1007/s13593-020-00635-2

Feldman, W., Feldman, E., and Goodman, J. T. 1988 "Culture versus Biology: Children's Attitudes towards Thinness and Fatness." *Pediatrics* 81: 190–194.

Feldstein, S. 2017. "Wasting Biodiversity: Why Food Waste Needs to Be a Conservation Priority." *Biodiversity* 18(2–3): 75–77.

Feldstein, S. 2018. *The Animal Lover's Guide to Changing the World: Practical Advice and Everyday Actions for a More Sustainable, Humane, and Compassionate Planet*. New York, St. Martin's Press.

Feng, A. Y. T., and C. G. Himsworth. 2014. "The Secret Life of the City Rat: A Review of the Ecology of Urban Norway and Black Rats (*Rattus norvegicus* and *Rattus rattus*)." *Urban Ecosystems* 17: 149–162.

Fenoglio, S., G. Boano, and G. B. Delmastro. 2018. “Conservation and Prejudice: Why Adopt Double Standards for Fish and Homoeothermic Vertebrates?” *European Zoological Journal* 85: 227–228.

Fenton, M. B., and N. B. Simmons. 2015. *Bats: A World of Science and Mystery*. Chicago, IL, University of Chicago Press.

Fergus, C. 2010. “Striped Skunk.” *Wildlife Note* 175(23): 1–4. Harrisburg, Pennsylvania Game Commission Bureau of Information and Education.

Ferguson, C. J., and K. M. Beaver. 2009. “Natural Born Killers: The Genetic Origins of Extreme Violence.” *Aggression and Violent Behavior* 14: 286–294.

Ferguson, D., I. Colditz, T. Collins, L. Matthews, and P. Hemsworth. 2013. *Assessing the Welfare of Farm Animals – A Review*. Australia, APL. https://www.awstrategy.net/uploads/1/2/3/2/123202832/assessing_the_welfare_of_farm_animals_-_part_1_v7_final.pdf

Ferl, R., R. Wheeler, H. G. Levine, and A-L. Paul. 2002. “Plants in Space.” *Current Opinion in Plant Biology* 5: 258–263.

Fernau S., M. Braun, and P. Dabrock. 2020. “What Is (Synthetic) Life? Basic Concepts of Life in Synthetic Biology.” *PLoS One* 15(7): e0235808. https://doi.org/10.1371/journal.pone.0235808

Ferrazzi, P., and F. Marletto. 1990. “Bee Value of *Reynoutria japonica* Houtt.” *Apicoltore Moderno* 81: 71–76.

Ferretti, F., B. Worm, G. L. Britten, M. R. Heithaus, and H. K. Lotze. 2010. “Patterns and Ecosystem Consequences of Shark Declines in the Ocean.” *Ecology Letters* 13: 1055–1071.

Festing, M. F. W., and D. P. Lovell. 1981. “Domestication and Development of the Mouse as a Laboratory Animal.” In *Biology of the House Mouse* (Proceedings of a Symposium Held at the Zoological Society of London on 22 and 23 November 1979), edited by R. J. Berry, 43–62. London, Academic Press.

Fidrmuc, J., B. Paphawasit, and Ç. B. Tunalı. 2017. “Nobel Beauty.” Rimini Centre for Economic Analysis paper WP 17-27. http://www.rcea.org/RePEc/pdf/wp17-27.pdf

Figueirêdo, R. E. C. R., A. Vasconcellos, I. S. Policarpo, and R. R. N. Alves. 2015. “Edible and Medicinal Termites: A Global Overview.” *Journal of Ethnobiology and Ethnomedicine* 11: 1–17.

Finn, F. 1909. *The Wild Beasts of the World*, Vol. 1. London, T. C. & E. C. Jack.

Fisher, A. A. 1996. “Poison Ivy/Oak Dermatitis. Part I: Prevention – Soap and Water, Topical Barriers, Hyposensitization.” *Cutis* 57: 384–386.

Fisher, K. A., and T. Stankowich. 2018. “Antipredator Strategies of Striped Skunks in Response to Cues of Aerial and Terrestrial Predators.” *Animal Behaviour* 143: 25–34.

Flather, C. H., G. D. Hayward, S. R. Beissinger, and P. A. Stephens. 2011. “Minimum Viable Populations: Is There a ‘Magic Number’ for Conservation Practitioners?” *Trends in Ecology and Evolution* 26: 307–316.

Fleishman, E., D. D. Murphy, and P. F. Brussard. 2000. “A New Method for Selection of Umbrella Species for Conservation Planning.” *Ecological Applications* 10: 569–579.

Fleiss, A., and K. S. Sarkisyan. 2019. “A Brief Review of Bioluminescent Systems.” *Current Genetics* 65: 877–882.

Focks, D. A. 2007. “*Toxorhynchites* as Biocontrol Agents.” *Journal of the American Mosquito Control Association* 23(2 Suppl): 118–127.

Fogliatto, S, A Ferrero, and F. Vidotto. 2020. “How Can Weedy Rice Stand against Abiotic Stresses? A Review.” *Agronomy* 10(9): 1284. https://doi.org/10.3390/agronomy10091284

Follak, S. 2011. “Potential Distribution and Environmental Threat of *Pueraria lobata*.” *Central European Journal of Biology* 6: 457–469.

Follak, S., R. Belz, C. Bohren, O. De Castro, E. Del Guacchio, N. Pascual-Seva, et al. 2016. “Biological Flora of Central Europe: *Cyperus esculentus* L.” *Perspectives in Plant Ecology, Evolution and Systematics* 23: 33–51.

Fontecha, A. F., and R. M. J. Catalán. 2003. “Semantic Derogation in Animal Metaphor: A Contrastive-Cognitive Analysis of Two Male/Female Examples in English and Spanish.” *Journal of Pragmatics* 35: 771–797.

Food and Agriculture Organization of the United Nations. 2019a. “The State of the World’s Biodiversity for Food and Agriculture in Brief.” Food and Agriculture Commission on Genetic Resources for Food and Agriculture. https://upload.wikimedia.org/wikipedia/commons/a/ab/The_State_of_the_World%27s_Biodiversity_for_Food_and_Agriculture_%E2%88%92_In_Brief_%28FAO%29.pdf

Food and Agriculture Organization of the United Nations. 2019b. *The State of the World’s Biodiversity for Food and Agriculture*, edited by J. Bélanger and D. Pilling. Rome, FAO Commission on Genetic Resources for Food and Agriculture Assessments. http://www.fao.org/3/CA3129EN/CA3129EN.pdf

Forister, M. L., E. M. Pelton, and S. H. Black. 2019. “Declines in Insect Abundance and Diversity: We Know Enough to Act Now.” *Conservation Science and Practice* 1(8): e80 https://doi.org/10.1111/csp2.80

Forman, J., and R. V. Kesseli. 2003. “Sexual Reproduction in the Invasive Species *Fallopia japonica* (Polygonaceae).” *American Journal of Botany* 90: 586–592.

Forseth, I. N., and A. F. Innis. 2004. “Kudzu (*Pueraria montana*): History, Physiology, and Ecology Combine to Make a Major Ecosystem Threat.” *Critical Reviews in Plant Sciences* 23: 401–413.

Foster, S., and J. A. Duke. 2000. *A Field Guide to Medicinal Plants and Herbs of Eastern and Central North America*. Second Edition. New York, Houghton Mifflin.

Foster, W. A., and E. D. Walker. 2018. “Chapter 15: Mosquitoes (Culicidae).” In *Medical and Veterinary Entomology*, Third Edition, edited by G. R. Mullen and L. A. Durden, 261–325. Cambridge, Academic Press.

Foubister, L. 2003. *Goddess in the Grass: Serpentine Mythology and the Great Goddess*. Victoria, BC, Ecce Nova.

Fourcade, Y., A. G. Besnard, and J. Secondi. 2017. “Evaluating Interspecific Niche Overlaps in Environmental and Geographic Spaces to Assess the Value of Umbrella Species.” *Journal of Avian Biology* 48: 1563–1574.

Fox, G. L. 2015. *The Archaeology of Smoking and Tobacco*. Gainesville, University Press of Florida.

Fox, J. G., S. W. Barthold, M. T. Davisson, C. E. Newcomer, F. W. Quimby, and S. L. Smith. Eds. 2007. *The Mouse in Biomedical Research, Vol. 1. History, Genetics and Wild Mice*. Second Edition. San Diego, CA, Academic Press.

Foy, C. L., D. R. Forney, and W. E. Cooley. 1983. “History of Weed Introductions.” In *Exotic Plant Pests and North American Agriculture*, edited by C. L Wilson and C. L. Graham, 65–92. New York, Academic Press.

Francis, B. 2012. “Before and after ‘Jaws’: Changing Representations of Shark Attacks.” *The Great Circle* 34: 44–64.

Francis, J. K. 2004. “*Toxicodendron radicans* (L.) Kuntze.” In *Wildland Shrubs of the United States and Its Territories: Thamnic Descriptions: Volume 1*, edited by J. K. Francis, 769–771. Fort Collins, CO, Rocky Mountain Research

Station, United States Department of Agriculture, Forest Service. https://data.fs.usda.gov/research/pubs/iitf/iitf_gtr026.pdf#page=779

Frankel, E. 1991. *Poison Ivy, Poison Oak, Poison Sumac and Their Relatives; Pistachios, Mangoes and Cashews*. Pacific Grove, CA, The Boxwood Press.

Franz, V., G. Grimpe, F. Hempelmann, L. Nick, H. Simroth, and E. Wagner. 1918. *Brehms Tierleben. Allgemeine kunde des Tierreichs* [Brehms' Animal Life. General Kinds of the Animal Kingdom]. Leipzig, Bibliographisches Institut.

Frederick, D. A., M. Hadji-Michael, A. Furnham, and V. Swami. 2010. "The Influence of Leg-to-Body Ratio (LBR) on Judgments of Female Physical Attractiveness: Assessments of Computer-Generated Images Varying in LBR." *Body Image* 7: 51–55.

Fredrickson, M., P. Annas, H. Fischer, and G. Wik. 1996. "Gender and Age Differences in the Prevalence of Specific Fears and Phobias." *Behaviour Research & Therapy* 26: 241–244.

Freeman, C., E. Leane, and Y. Watt. Eds. 2011. *Considering Animals: Contemporary Studies in Human–Animal Relations*. Oxfordshire, UK, Routledge.

Frembgen, J. W. 1998. "The Magicality of the Hyena: Beliefs and Practices in West and South Asia." *Asian Folklore Studies* 57: 331–344.

Fressoz, J.-B., and C. Bonneuil. 2017. "Growth Unlimited: The Idea of Infinite Growth from Fossil Capitalism to Green Capitalism." In *History of the Future of Economic Growth*, edited by I. Borowy and M. Schmelzer, 52–68. New York, Routledge.

Frey, N. and E. Thacke. 2018. "Wild Horses and Burros: An Overview." *NR/Wildlife* 2018-01pr. https://digitalcommons.usu.edu/cgi/viewcontent.cgi?article=2852&context=extension_curall

Frick, S., R. Kramell, J. Schmidt, A. J. Fist, and T. M. Kutchan. 2005. "Comparative Qualitative and Quantitative Determination of Alkaloids in Narcotic and Condiment *Papaver somniferum* Cultivars." *Journal of Natural Products* 68: 666–673.

Friedrich, L. A., R. Jefferson, and G. Glegg. 2014. "Public Perceptions of Sharks: Gathering Support for Shark Conservation." *Marine Policy* 47: 1–7.

Frost, P. 2005. *Fair Women, Dark Men*. Christchurch, New Zealand, Cybereditions Corporation.

Frost, P. 2015. "Evolution of Long Head Hair in Humans." *Advances in Anthropology* 5: 274–281.

Frynta, D., O. Šimková, S. Lišková, and E. Landová. 2013. "Mammalian Collection on Noah's Ark: The Effects of Beauty, Brain and Body Size." *PLoS One* 8(5): 1–12.

Fukano, Y., and M. Soga. 2021. "Why Do So Many Modern People Hate Insects? The Urbanization–Disgust Hypothesis." *Science of the Total Environment* 777: 146229. https://doi.org/10.1016/j.scitotenv.2021.146229

Furlan, L., A. Pozzebon, C. Duso, C. N. Simon-Delso, F. Sánchez-Bayo, P. A. Marchand, et al. 2019. "An Update of the Worldwide Integrated Assessment (WIA) on Systemic Insecticides. Part 3: Alternatives to Systemic Insecticides." *Environmental Science and Pollution Research* https://doi.org/10.1007/s11356-017-1052-5

Gabbatiss, J. 2017. "Nasty, Brutish and Short: Are Humans DNA-Wired to Kill?" *Scientific American Newletter*. https://www.scientificamerican.com/article/nasty-brutish-and-short-are-humans-dna-wired-to-kill/

Gadani, F, D. Ayers, and W. Hempfling. 1995. "Tobacco: A Tool for Plant Genetic Engineering Research and Molecular Farming. Part II." *AgroFood Industry High Technology* 6(2): 3–6.

Gade, D. W. 2006. "Hyenas and Humans in the Horn of Africa." *Geographical Review* 96: 609–632.

Gallagher, A., and N. Hammerschlag. 2011. "Global Shark Currency: The Distribution, Frequency, and Economic Value of Shark Ecotourism." *Current Issues in Tourism* 14: 797–812.

Gallagher, A. J., G. M. S. Vianna, Y. P. Papastamatiou, C. Macdonald, T. L. Guttridge, and N. Hammerschlag. 2015. "Biological Effects, Conservation Potential, and Research Priorities of Shark Diving Tourism." *Biological Conservation* 184: 365–379.

Gammon, M. A., and R. Kesseli. 2010. "Haplotypes of *Fallopia* Introduced into the US." *Biological Invasions* 12: 421–427.

Gardner, M. 2007. "Is Beauty Truth and Truth Beauty? How Keats's Famous Line Applies to Math and Science." *Scientific American Newsletter*, April 1. https://www.scientificamerican.com/article/is-beauty-truth-and-truth/

Gaskett, A. C., C. G. Winnick, and M. E. Herberstein. 2008. "Orchid Sexual Deceit Provokes Ejaculation." *American Naturalist* 171(6): E206–E212.

Gately, I. 2002. *Tobacco: A Cultural History of How an Exotic Plant Seduced Civilization*. New York, Grove Press.

Geniole, S. N., T. F. Denson, B. J. Dixson, J. M. Carré, and C. M. McCormick. 2015. "Evidence from Meta-Analyses of the Facial Width-to-Height Ratio as an Evolved Cue of Threat." *PLoS One* 10(7): e0132726. https://doi.org/10.1371/journal.pone.0132726

Gentili, R., C. Montagnani, F. Gilardelli, M. F.Guarino, and S. Citterio. 2017. "Let Native Species Take Their Course: *Ambrosia artemisiifolia* Replacement during Natural or 'Artificial' Succession." *Acta Oecologica* 82: 32–40.

Gentry, A. H. 1991. "The Distribution and Evolution of Climbing Plants." In *The Biology of Vines*, edited by F. E. Putz and H. A. Mooney, 3–42+unpaginated appendix. Cambridge, Cambridge University Press.

Georgiev, A. V., A. C. E. Klimczuk, D. M. Traficonte, and D. Maestripieri. 2013. "When Violence Pays: A Cost-Benefit Analysis of Aggressive Behavior in Animals and Humans." *Evolutionary Psychology* 11: 678–699.

Gerber, E., C. Krebs, C. Murrell, M. Moretti, R. Rocklin, and U. Schaffner. 2008. "Exotic Invasive Knotweeds (*Fallopia* spp.) Negatively Affect Native Plant and Invertebrate Assemblages in European Riparian Habitats." *Biological Conservation* 141: 646–654.

Gerdes, A. B. M., G. Uhl, and G. W. Alpers. 2009. "Spiders Are Special: Fear and Disgust Evoked by Pictures of Arthropods." *Evolution & Human Behavior* 30: 66–73.

Gerhardt, R. R., and L. J. Hribar. 2018. "Chapter 11: Flies (Diptera)." In *Medical and Veterinary Entomology*, Third Edition, edited by G. R. Mullen and L. A. Durden, 171–190. Cambridge, Academic Press.

Geries-Johnson, B., and J. H. Kennedy. 1995. "Influence of Animals on Perceived Likability of People." *Perceptual and Motor Skills* 80: 432–434.

Ghaly, A. E., N. Mahmoud, and D. Dave. 2012. "Nutrient Composition of Dandelions and Its Potential as Human Food." *American Journal of Biochemistry and Biotechnology* 8: 118–127.

Gielis, J. 2017. *The Geometrical Beauty of Plants*. Paris, Atlantis Press.

Gillies, S., D. R. Clements, and J. Grenz. 2016. "Knotweed (*Fallopia* spp.) Invasion of North America Utilizes Hybridization, Epigenetics, Seed Dispersal (Unexpectedly), and an Arsenal of Physiological Tactics." *Invasive Plant Species and Management* 9: 71–80.

Gillis W. T. 1971. "The Systematics and Ecology of Poison-Ivy and the Poison-Oaks (*Toxicodendron*, Anacardiaceae)." *Rhodora* 73: 72–159, 161–237, 370–443, 465–540.

Gillis W. T. 1975. "Poison Ivy and Its Kin." *Arnoldia* 35: 93–123.

Githae, E. W. 2018. "Status of *Opuntia* Invasions in the Arid and Semi-arid Lands of Kenya." *CAB Reviews* 13, No. 003. https://www.cabi.org/bni/FullTextPDF/2018/20183117948.pdf

Giuliani, C., L. Lastrucci, L. Cresti, G. Santini, B. Foggi, and M. M. Lippi. 2019. "The Morphology and Activity of the Extrafloral Nectaries in *Reynoutria* × *bohemica* (Polygonaceae)." *Plant Biology* 21: 975–985.

Gladman, A. C. 2006. "*Toxicodendron* Dermatitis: Poison Ivy, Oak, and Sumac." *Wilderness & Environmental Medicine* 17: 120–128.

Glass, D., and S. Al-Hamdani. 2016. "Kudzu Forage Quality Evaluation as an Animal Feed." *American Journal of Plant Sciences* 7: 702–707.

Glickman, S. E. 1995. "The Spotted Hyena from Aristotle to the Lion King: Reputation Is Everything." *Social Research* 62: 501–537.

Global Footprint Network. 2011. "What Happens When Infinite-Growth Economy Runs into a Finite Planet." In *Global Footprint Network 2011 Annual Report*. Oakland, CA, Footprint Network. https://www.footprintnetwork.org/content/images/article_uploads/Annual_Report_2011.pdf

Goddard, J., and R. deShazo. 2009. "Bed Bugs (*Cimex lectularius*) and Clinical Consequences of Their Bites." *Journal of the American Medical Association* 301: 1358–1366.

Goddard, M. A., A. J. Dougill, and T. G. Benton. 2010. "Scaling Up from Gardens: Biodiversity Conservation in Urban Environments." *Trends in Ecology & Evolution* 25: 90–98.

Golle, J., F. Probst, F. W. Mast, and J. S. Lobmaier. 2015. "Preference for Cute Infants Does Not Depend on Their Ethnicity or Species: Evidence from Hypothetical Adoption and Donation Paradigms." *PLoS One* 10(4): e0121554. https://doi.org/10.1371/journal.pone.0121554

Gómez, J., M. Verdú, A. González-Megías, and M. Méndez. 2016. "The Phylogenetic Roots of Human Lethal Violence." *Nature* 538: 233–237.

Gómez-Betancur, I., V. Gogineni, A. Salazar-Ospina, and F. León. 2019. "Perspective on the Therapeutics of Anti-Snake Venom." *Molecules* 24(18): 3276. https://doi.org/10.3390/molecules24183276

Gompper, M. E. 2017. "Range Decline and Landscape Ecology of the Eastern Spotted Skunk." In *Biology and Conservation of Musteloids*, edited by D. W. Macdonald, C. Newman, and L. A. Harrington, 478–492. Oxford, Oxford University Press.

Gonçalves, B. B., P. C. Giaquinto, D. dos Santos Silva, C. de Melo e Silva Neto, A. A. de Lima, A. A. B. Darosci, et al. 2019. "Ecotoxicology of Glyphosate-Based Herbicides on Aquatic Environment." IntechOpen. https://doi.org/10.5772/intechopen.85157. https://www.intechopen.com/books/biochemical-toxicology-heavy-metals-and-nanomaterials/ecotoxicology-of-glyphosate-based-herbicides-on-aquatic-environment

Gonzalez, W. Ed. 2015. *New Perspectives on Technology, Values, and Ethics*. Cham, Springer.

Good, R. 1974. *Geography of Flowering Plants*. Harlow, UK, Longman Group.

Goodall, J. 1999. *Reason for Hope: A Spiritual Journey*. New York, Grand Central Publishing.

Goodman, J. 2004. *Tobacco in History and Culture: An Encyclopedia*. 2 vols. Detroit, Charles Scribner's Sons.

Goodspeed, T. H. 1954. *The genus Nicotiana; Origins, Relationships, and Evolution of Its Species in the Light of Their Distribution, Morphology, and Cytogenetics*. Waltham, MA, Chronica Botanica.

Gopal, B. 1987. *Water Hyacinth*. Amsterdam, Elsevier.

Goplen, J. J., J. A. Coulter, C. C. Sheaffer, R. L. Becker, F. R. Breitenbach, L. M. Behnken, et al. 2018. "Economic Performance of Crop Rotations in the Presence of Herbicide-Resistant Giant Ragweed." *Agronomy Journal* 110: 260–268.

Goplen, J. L., C. C. Sheaffer, R. L. Becker, J. A. Coulter, F. R. Breitenbach, L. M. Behnken, et al. 2016. "Giant Ragweed (*Ambrosia trifida*) Seed Production and Retention in Soybean and Field Margins." *Weed Technology* 30: 246–253.

Gordon, E., and M. T. Ross. 1910. *Flower Children: The Little Cousins of the Field and Garden*. Chicago, IL, P. F. Volland.

Gordon, E. A., O. E. Franco, and M. L. Tyrrell. 2005. "Protecting Biodiversity: A Guide to Criteria Used by Global Conservation Organizations." *Forestry & Environmental Studies Publications Series* 26. https://elischolar.library.yale.edu/fes-pubs/26

Gosling, S. D., J. S. Carson, and J. Potter. 2010. "Personalities of Self-identified 'Dog People' and 'Cat People.'" *Anthrozoös* 23: 213–222.

Gotcher, M. J., H. Zhang, J. L. Schroder, and M. E. Payton. 2014. "Phytoremediation of Soil Phosphorus with Crabgrass." *Agronomy Journal* 106: 528–536.

Goud Collins, M. 2016. "International Organizations and Biodiversity." In *Encyclopedia of Biodiversity (Second Edition)*, edited by S. A. Levin, 324–331.Waltham, MA, Elsevier/Academic.

Gould, S. J. 1979. "Mickey Mouse Meets Konrad Lorenz." *Natural History* 88: 30–36.

Gould, S. J. 1980. *The Panda's Thumb: More Reflections in Natural History*. New York, W. W. Norton.

Gould, S. J. 1981. "Hyena Myths and Realities." *Natural History* 90(2): 16–24.

Gould, S. J. 1989. *Wonderful Life: The Burgess Shale and the Nature of History*. New York, W. W. Norton.

Gouveia, S. S. Ed. 2020. *The Age of Artificial Intelligence: An Exploration*. Wilmington, DE, Vernon Press.

Govorushko, G. 2019. "Economic and Ecological Importance of Termites: A Global Review." *Entomological Science* 22: 21–35.

Gowdy, J. M. 1973. "The Value of Biodiversity: Markets, Society, and Ecosystems." *Land Economics* 73: 25–41.

Gqaza, B. M., C. Njume, N. I. Goduka, and G. George. 2013. "Nutritional Assessment of *Chenopodium album* L. (Imbikicane) Young Shoots and Mature Plant-Leaves Consumed in the Eastern Cape Province of South Africa." *International Proceedings of Chemical, Biological and Environmental Engineering* 53: 97–102.

Grammer, K., B. Fink, A. P. Møller, and R. Thornhill. 2003. "Darwinian Aesthetics: Sexual Selection and the Biology of Beauty." *Biological Reviews* 78: 385–407.

Grandville, J. J. 1867. *Les Fleurs Animées (The Flowers Personified), Volume 1*. Paris, Garnier Frères/Librairie Philosophique de Lagrange.

Grant, D. 2011. "Political and Practical Problems with Dangerous Dogs." *The Veterinary Record* 168: 133.

Grassmann, M., B. McNeil, and J. Wharton. 2017. "The Role of Husbandry, Breeding, Education, and Citizen Science in Shark Conservation." *Advances in Marine Biology* 78: 89–119.

Grauso, L., S. Emrick, B. de Falco, V. Lanzotti, and G. Bonanomi. 2019. "Common Dandelion: A Review of Its Botanical, Phytochemical and Pharmacological Profiles." *Phytochemistry Reviews* 18: 1115–1132.

Grebowicz, M. 2010. "When Species Meat: Confronting Bestiality Pornography." *Humanimalia* 1(2): 1–17.

Greely, H. T. 2019. "CRISPR'd Babies: Human Germline Genome Editing in the 'He Jiankui Affair.'" *Journal of Law and the Biosciences* 6: 111–183.

Green, A. F., T. S. Ramsey, and J. Ramsey. 2011. "Phylogeny and Biogeography of Ivies (*Hedera* spp., Araliaceae), a Polyploid Complex of Woody Vines." *Systematic Botany* 36: 1114–1127.

Green, A. F., T. S. Ramsey, and J. Ramsey. 2013. "Polyploidy and Invasion of English Ivy (*Hedera* spp., Araliaceae) in North American Forests." *Biological Invasions* 15: 2219–2241.

Greenaway, K. 1884. *Language of Flowers*. London, F. Wane.

Greenstein, S. 2016. "What Does a Skunk Works Do?" *IEEE Micro* 36(2): 70–71.

Gregorczyk, A., J. Wereszczaka, and S. Stankowski. 2012. "Utilisation of Biomass of Japanese Knotweed (*Polygonum cuspidatum* Siebold & Zucc.) for Energy Purposes." *Folia Pomeranae Universitatis Technologiae Stetinensis, Agricultura, Alimentaria, Piscaria et Zootechnica* 293: 35–40. (In Polish, English abstract.)

Greitemeyer, T. 2020. "Unattractive People are Unaware of Their (Un)attractiveness." *Scandinavian Journal of Psychology* 61: 471–483.

Griffin, A. 2015. "Genetically-engineered, Extra-Muscular Dogs Created by Chinese Scientists." *The Independent*. https://www.independent.co.uk/news/science/mutant-extramuscular-dogs-created-by-chinese-scientists-a6701156.html

Griffith, M. P. 2004. "The Origins of an Important Cactus Crop, *Opuntia ficus-indica* (Cactaceae): New Molecular Evidence." *American Journal of Botany* 91: 1915–1921.

Grimsby, J. L., D. Tsirelson, M. A. Gammon, and R. Kesseli. 2007. "Genetic Diversity and Clonal vs. Sexual Reproduction in *Fallopia* spp. (Polygonaceae)." *American Journal of Botany* 94: 957–964.

Gross, M. 2020. "Virus Outbreak Crosses Boundaries." *Current Biology* 30: R191–R194.

Grossman, M. R. 2016. "Genetically Engineered Animals in the United States: The AquAdvantage Salmon." *European Food and Feed Law Review* 11: 190–200.

Grosso, M., and L. Falasconi. 2018. "Addressing Food Wastage in the Framework of the UN Sustainable Development Goals." *Waste Management and Research* 36: 97–98.

Gucker, C. L. 2008. "*Phragmites australis*." In *Fire Effects Information System* [Online]. U.S. Department of Agriculture, Forest Service, Rocky Mountain Research Station, Fire Sciences Laboratory. https://www.fs.usda.gov/database/feis/plants/graminoid/phraus/all.html

Guertin, P. J., Denight, M. L., Gebhart, D. L. and Nelson, L. 2008. *Invasive Species Biology, Control, and Research. Part 1. Kudzu (Pueraria montana)*. Washington, DC, US Army Corps of Engineers, Engineer Research and Development Center. ERDC TR-08-10.

Guggisberg, A., E. Welk, R. Sforza, D. P. Horvath, J. V. Anderson, M. E. Foley, et al. 2012. "Invasion History of North American Canada Thistle, *Cirsium arvense*." *Journal of Biogeography* 39: 1919–1931.

Guin, J. D. 2001. "Treatment of *Toxicodendron* Dermatitis (Poison Ivy and Poison Oak)." *Skin Therapy Letter* 6(7). http://www.skintherapyletter.com/eczema/toxicodendron-dermatitis/

Guin, J. D., W. T. Gillis, and J. H. Beaman. 1981. "Recognizing the Toxicodendrons (Poison Ivy, Poison Oak, and Poison Sumac)." *Journal of the American Academy of Dermatology* 4: 99–114.

Guiney, M. S., and K. S. Oberhauser. 2008. "Insects as Flagship Conservation Species." *Terrestrial Arthropod Reviews* 1: 111–123.

Gulizia, J. P., and K. M. Downs. 2019. "A Review of Kudzu's Use and Characteristics as Potential Feedstock." *Agriculture* 9(10): 220. https://doi.org/10.3390/agriculture9100220

Gunn, A. S. 2001. "Environmental Ethics and Trophy Hunting." *Ethics and the Environment* 6: 68–95.

Gunnarsson, C., and C. Mattsson-Petersen. 2007. "Water Hyacinths as a Resource in Agriculture and Energy production: A Literature Review." *Waste Management* 27: 117–129.

Guo, L., J. Qiu, C. Ye, G. Jin, L. Mao, H. Zhang, et al. 2017. "*Echinochloa crus-galli* Genome Analysis Provides Insight into Its Adaptation and Invasiveness as a Weed." *Nature Communications* 8: 1031. https://doi.org/10.1038/s41467-017-01067-5

Gurstel, D. U., and V. A. Sisson. 1995. "Tobacco." In *Evolution of Crop Plants*, Second Edition, edited by J. Smartt and N. W. Simmonds, 458–463. Burnt Mill, Harlow, Essex, UK, Longman Scientific & Technical.

Gustafson, J. P., P. H. Raven, and P. R. Ehrlich. Eds. 2020. *Population, Agriculture, and Biodiversity: Problems and Prospects*. Colombia, MO, University of Missouri Press.

Haacke, W. 1893. *Die Schöpfung der Tierwelt* [The Creation of the Animal World]. Leipzig, Bibliographisches Institut.

Habel, J. C., L. Rasche, U. A. Schneider, J. O. Engler, E. Schmid, D. Rödder, et al. 2019. "Final Countdown for Biodiversity Hotspots." *Conservation Newsletter* 12(6): e12668. https://doi.org/10.1111/conl.12668

Haberzettl, J., P. Hilgert, and M. von Cossel. 2021. "A Critical Review on Lignocellulosic Biomass Yield Modeling and the Bioenergy Potential from Marginal Land." *Agronomy* 11(12): 2397. https://doi.org/10.3390/agronomy11122397

Habibi, F., P. Vít, M. Rahiminejad, and B. Mandák. 2018. "Towards a Better Understanding of the *Chenopodium album* Aggregate (Amaranthaceae) in The Middle East: A Karyological, Cytometric and Morphometric Investigation." *Journal of Systematics and Evolution* 56: 231–242.

Haeckel, E. 1899–1904. *Kunstformen der Natur (Artforms in Nature)*. Leipzig, Bibliographisches Institut. (Originally published in ten installments. Widely reproduced, for example at https://archive.org/details/KunstformenDerNaturErnstHaeckel).

Haemig, P. D. 2012. "Ecology of Condors." *ECOLOGY.INFO*: 25. http://www.ecology.info/condors.htm

Hajzlerová, L., and J. Reif. 2014. "Bird Species Richness and Abundance in Riparian Vegetation Invaded by Exotic *Reynoutria* Spp." *Biologia* 69: 247–253.

Haldane, J. B. S. 1942. "The Selective Elimination of Silver Foxes in Eastern Canada." *Journal of Genetics* 44: 296–304.

Hall, R. M., H. Bein, B. Bein-Lobmaier, G. Karrer, H.-P. Kaul, and J. Novak. 2020. "Know Your Enemy: Are Biochemical Substances the Secret Weapon of Common Ragweed (*Ambrosia artemisiifolia* L.) in the Fierce Competition with Crops and Native Weeds?" *Julius-Kühn-Archiv* 464: 121–126. https://core.ac.uk/download/pdf/288219861.pdf

Hamaoui-Laguel, L., R. Vautard, L. Liu, F. Solmon, N. Viovy, D. Khvorostyanov, et al. 2015. "Effects of Climate Change and Seed Dispersal on Airborne Ragweed Pollen Loads in Europe." *Nature Climate Change* 5: 766–771.

Hamermesh, D. S. 2011. *Beauty Pays: Why Attractive People Are More Successful*. Princeton, NJ, Princeton University Press.

Handlin, L., A. Nilsson, M. Ejdebäck, E. Hydbring-Sandberg, and K. Uvnäs-Moberg. 2012. "Associations between the Psychological Characteristics of the Human-Dog Relationship and Oxytocin and Cortisol Levels." *Anthrozoös* 25: 215–228.

Hanson, T. 2018. "Biodiversity Conservation and Armed Conflict: A Warfare Ecology Perspective." *Annals of the New York Academy of Sciences* 1429: 50–65.

Haque, M., J. McKimm, M. Sartelli, N. Samad, S. Z. Haque, and M. A. Bakar. 2020. "A Narrative Review of the Effects of Sugar-sweetened Beverages on Human health: A Key Global Health Issue." *Journal of Population Therapeutics & Clinical Pharmacology* 27(1). https://doi.org/10.15586/jptcp.v27i1.666

Hardin, G. 1968. "The Tragedy of the Commons." *Science* 162(3859): 1243–1248.

Harding, D. P., and M. N. Raizada. 2015. "Controlling Weeds with Fungi, Bacteria and Viruses: A Review." *Frontiers in Plant Science* 28. https://doi.org/10.3389/fpls.2015.00659

Hardouin, E. A., A. Orth, M. Teschke, J. Darvish, D. Tautz, and F. Bonhomme. 2015. "Eurasian House Mouse (*Mus musculus* L.) Differentiation at Microsatellite Loci Identifies the Iranian Plateau as a Phylogeographic Hotspot." *BMC Evolutionary Biology* 15: 26. https://doi.org/10.1186/s12862-015-0306-4.

Hardy, T. 1988. "Entomophobia: The Case for Miss Muffet." *Bulletin of the Entomological Society of America* 34: 64–69.

Harlan, J. 1970. "*Cynodon* species and their value for grazing and hay." *Herbage Abstract* 40: 233–238.

Harlan, J. R., P. Gepts, T. R. Famula, R. L. Bettinger, S. B. Brush, A. B. Damania, et al. 2012. *Biodiversity in Agriculture: Domestication, Evolution, and Sustainability*. Cambridge, Cambridge University Press.

Harley, K. L. S. 1990. "The Role of Biological Control in the Management of Water Hyacinth, *Eichhornia crassipes*." *Biocontrol News and Information* 11: 11–22.

Harr, S. 2016. "Insects That Get a Bad Reputation." https://foleylionsroar.com/2016/10/insects-that-get-a-bad-reputation/

Harris, C. 2014. "The Conservation Beauty Pageant." *The Gist: Glasgow Institute into Science and Technology*. https://thegist.org/2014/11/the-conservation-beauty-pageant/

Harris, J. M. 2000. "Cats and Birds." *Journal of Avian Medicine and Surgery* 14(1): 1. https://doi.org/10.1647/1082-6742(2000)014[0001:CAB]2.0.CO;2

Harrison, I. J., M. F. Laverty, and E. J. Sterling. 2021. "Setting Priorities for Biodiversity Conservation." *Ecology Center* (private blog). https://www.ecologycenter.us/natural-history-2/setting-priorities-for-biodiversity-conservation.html

Harron, P., O. Joshi, C. B. Edgar, S. Paudel, and A. Adhikari. 2020. "Predicting Kudzu (*Pueraria montana*) Spread and its Economic Impacts in Timber Industry: A Case Study from Oklahoma." *PLoS One* 15(3): e0229835. https://doi.org/10.1371/journal.pone.0229835

Hart, D., and R. W. Sussman, R. W. 2008. *Man the Hunted: Primates, Predators, and Human Evolution*. Boulder, CO, Westview Press.

Harvey, A. W. 2009. "Extrafloral Nectaries in Kudzu, *Pueraria montana* (Lour.) Merr., and Groundnut, *Apios americana* Medicus (Fabaceae)." *Castanea* 74: 360–371.

Harvey, D., P. Bardelang, S. L. Goodacre, A. Cockayne, and N. R. Thomas. 2016a. "Antibiotic Spider Silk: Site-Specific Functionalization of Recombinant Spider Silk Using 'Click' Chemistry." *Advanced Materials* 1604245. https://doi.org/10.1002/adma.201604245

Harvey, M., N. Gasz, and S. Voss. 2016b. "Entomology-based Methods for Estimation of Postmortem Interval." *Research and Reports in Forensic Medical Science* 6: 1–9.

Haslam, S. M. 1969. *The Reed: A Study of* Phragmites communis *Trin., in Relation to Its Cultivation and Harvesting in East Anglia for the Thatching Industry*. Norwich, UK, Norfolk Reed Growers Association.

Haslam, S. M. 1972. "Biological flora of the British Isles. No. 128. *Phragmites communis* Trin. *Journal of Ecology* 60: 585–610.

Havelock, E. 1927. *Studies in the Psychology of Sex, Vol. 1*. Philadelphia, PA, Davis.

Hazell, P., and A. Rahman. 2014. *New Directions for Smallholder Agriculture*. Oxford, Oxford University Press.

Hazelton, E. L. G., T. J. Mozdzer, D. M. Burdick, K. M. Kettenring, and D. F. Whigham. 2014. "*Phragmites australis* Management in the United States: 40 Years of Methods and Outcomes." *AoB Plants* 6(2014): plu001. https://doi.org/10.1093/aobpla/plu001

He, F., and S. P. Hubbell. 2011. Species–Area Relationships Always Overestimate Extinction Rates from Habitat Loss." *Nature* 473: 368–371.

He, J., N. M. Evans, H. Liu, and S. Shao. 2020. "A Review of Research on Plant-based Meat Alternatives: Driving Forces, History, Manufacturing, and Consumer Attitudes." *Comprehensive Reviews in Food Science and Food Safety* 19: 2639–2656.

Heatwole, H., and J. W. Wilkinson Eds. 2012. *Amphibian Biology, Volume 10, Conservation and Decline of Amphibians: Ecological Aspects, Effects of Humans, and Management*. Baulkham Hills, Australia, Surrey Beatty & Sons.

Hedrich, H. J. 2006. "Taxonomy and Stocks and Strains." In *The Laboratory Rat (2nd edition)*, edited by M. A. Suckow, S. H. Weisbroth, and C. L. Franklin, 71–92. Amsterdam, Academic Press.

Heerwagen, J. H., and G. H. Orians. 1995. "Humans, Habitats, and Aesthetics." In *The Biophilia Hypothesis*, edited by S. R. Kellert, 138–172. Washington, DC, Island Press.

Hendrickson, R. 1983. *More Cunning than Man: A Complete History of the Rat and Its Role in Civilization*. New York, Kensington Books.

Henkel, A. 1904. *Weeds Used in Medicine*. Washington, DC, U.S. Government Printing Office.

Herzog, H. A. 2007. "Gender Differences in Human–Animal Interactions: A Review." *Anthrozoös* 20: 7–21.

Herzog, H. A., N. S. Betchart, and R. B. Pittman. 1991. "Gender, Sex Role Orientation, and Attitudes toward Animals." *Anthrozoös* 4: 184–191.

Hesammi, E., A. B. J. Talebi, and F. Sadatebrahimi. 2014. "Effect of Hairy Crabgrass (*Digitaria sanguinalis*) Aqueous Extract on Germination, Alpha Amylase, Antioxidant Enzymes Activity, and Destruction of Cellular Membrane in Canada Thistle Seedling (*Cirsium arvense*)." *Advances in Environmental Biology* 8: 3088–3090.

Hessel, D. T., and R. D. Ruether. Eds. 2000. *Christianity and Ecology: Seeking the Well-Being of Earth and Humans*. Cambridge, MA, Harvard University Press.

Heupel, M. R., D. M. Knip, C. A. Simpfendorfer, and N. K. Dulvy. 2014. "Sizing Up the Ecological Role of Sharks as Predators." *Marine Ecology Progress Series* 495: 291–298.

Heuzé, V., H. Thiollet, G. Tran, and F. Lebas. 2020. "Cockspur Grass (*Echinochloa crus-galli*) Forage." *Feedipedia*, a programme by INRA, CIRAD, AFZ and FAO. https://feedipedia.org/node/451

Hewitson, W. C. 1856–1876. *Illustrations of New Species of Exotic Butterflies: Selected Chiefly from the Collections of W. Wilson Saunders and William C. Hewitson*. London, John Van Voorst.

Hickman, J., S. Wu, L. Mickley, and M. Lerdau. 2010. "Kudzu (*Pueraria montana*) Invasion Doubles Emissions of Nitric Oxide and Increases Ozone Pollution." *Proceedings of the National Academy of Sciences of the USA* 107: 10115–10119.

Hildebrandt, K. A., and H. E. Fitzgerald. 1978. "Adults' Responses to Infants Varying in Perceived Cuteness." *Behavioral Processes* 3: 159–172.

Hildebrandt, K. A., and H. E. Fitzgerald. 1979. "Adults' Perceptions of Infant Sex and Cuteness." *Sex Roles* 5: 471–481.

Hill, J., E. Nelson, D. Tilman, S. Polasky, and D. Tiffany. 2006. "Environmental, Economic, and Energetic Costs and Benefits of Biodiesel and Ethanol Biofuels." *Proceedings of the National Academy of Sciences of the United States of America* 103(30): 11206–11210.

Hill, J. E., and J. D. Smith. 1984. *Bats: A Natural History*. London, British Museum.

Hill, R. J. 1983. "Canada Thistle, *Cirsium arvense* (L.) Scop. Compositae." Regulatory Horticulture 9(1–2), Weed Circular No. 2, Pennsylvania Department of Agriculture Bureau of Plant Industry. https://www.agriculture.pa.gov/Plants_Land_Water/PlantIndustry/NIPPP/Documents/canada_thistle%20article.pdf

Hill, T. E. 1915. *The Open Door to Independence; Making Money from the Soil*. Chicago, IL, Hill.

Hilu, K. W. 1994. "Evidence from RAPD Markers in the Evolution of *Echinochloa* Millets (Poaceae)." *Plant Systematics and Evolution* 189: 247–257.

Himsworth, C. G., K. L. Parsons, C. Jardine, and D. M. Patrick. 2013. "Rats, Cities, People, and Pathogens: A Systematic Review and Narrative Synthesis of Literature Regarding the Epidemiology of Rat-Associated Zoonoses in Urban Centers." *Vector Borne Zoonotic Diseases* 13: 349–359.

Hitchcock, A. S., and A. Chase. 1951. *Manual of the Grasses of the United States*. Second Edition. U.S.D.A. Miscellaneous Publication 200. Washington, DC, United States Department of Agriculture.

Hocking, P. J., C. M. Finlayson, and A. J. Chick. 1983. "The Biology of Australian Weeds. 12. *Phragmites australis* (Cav.) Trin. ex Steud." *Journal of the Australian Institute of Agricultural Science* 49: 123–132.

Hodgson, B. 1997. *The Rat: A Perverse Miscellany*. Berkeley, CA, Ten Speed Press.

Hodgson, E. W., J. L. Trina, and A. H. Roe. 2008. "Clothes Moths." Utah State University Extension Paper 877. https://digitalcommons.usu.edu/extension_curall/877

Hoffmann, M., C. Hilton-Taylor, A. Angulo, M. Böhm, T. M. Brooks, S. H. M. Butchart, et al. 2010. "The Impact of Conservation on the Status of the World's Vertebrates." *Science* 330(6010): 1503–1509.

Hogue, J. M. 2009. "Cultural Entomology." In *Encyclopedia of Insects*, Second Edition, edited by V. H. Resh and R. T. Cardé, 239–245. Cambridge, MA, Academic Press.

Holder, C. F., and D. S. Jordan. 1909. *Fish Stories: Alleged and Experienced, with a Little History Natural and Unnatural*. New York, Henry Holt and Company.

Holland, K. K., L. R. Larson, and R. B. Powell. 2018. "Characterizing Conflict between Humans and Big Cats *Panthera* spp: A Systematic Review of Research Trends and Management Opportunities." *PLoS One*. https://doi.org/10.1371/journal.pone.0203877

Hollingsworth, M. L., and J. P. Bailey. 2000. "Evidence for Massive Clonal Growth in the Invasive Weed *Fallopia japonica* (Japanese Knotweed)." *Botanical Journal of the Linnean Society* 133: 463–472.

Holm, L. 1978. "Some Characteristics of Weed Problems in Two Worlds." *Proceedings of the Western Society of Weed Science* 31: 3–12.

Holm, L. G., D. L. Plucknett, J. V. Pancho, and J. P. Herberger. 1977. *The World's Worst Weeds: Distribution and Biology*. Honolulu, University Press of Hawaii.

Holm, S. 2019. "Deciding in the Dark: The Precautionary Principle and the Regulation of Synthetic Biology." *Ethics, Policy & Environment* 22: 61–71.

Holmern, T., J. Nyahongo, and E. Røskaft. 2007. "Livestock Loss Caused by Predators Outside the Serengeti National Park, Tanzania." *Biological Conservation* 135: 518–526.

Holmes, G., T. A. Smith, and C. Ward. 2018. "Fantastic Beasts and Why to Conserve Them: Animals, Magic and Biodiversity Conservation." *Oryx* 52: 231–239.

Holoyda, B., R. Sorrentino, S. H. Friedman, and J. Allgire. 2018. "Bestiality: An Introduction for Legal and Mental Health Professionals." *Behavioral Sciences & the Law* 36: 687–697.

Holt, J. S. and C. W. Barrows. n.d. "Sahara Mustard." Center for Invasive Species Research, University of California, Riverside (online report). https://cisr.ucr.edu/invasive-species/sahara-mustard

Holzinger, F., and J.-F. Chenot. 2011. "Systematic Review of Clinical Trials Assessing the Effectiveness of Ivy Leaf (*Hedera helix*) for Acute Upper Respiratory Tract Infections." *Evidence-Based Complementary and Alternative Medicine* Article ID 382789. https://doi.org/10.1155/2011/382789.

Home, R., C. Keller, P. Nagel, N. Bauer, and M. Hunziker. 2009. "Selection Criteria for Flagship Species by Conservation Organizations." *Environmental Conservation* 36: 139–148.

Hoots, D., and J. Baldwin. 1996. *Kudzu, the Vine to Love or Hate*. Kodak, TN, Suntop Press.

Hornok, S., K. Szőke, S. A. Boldogh, A. D. Sándor, J. Kontschán, V. T. Tu, et al. 2017. "Phylogenetic Analyses of Bat-Associated Bugs (Hemiptera: Cimicidae: Cimicinae and Cacodminae) Indicate Two New Species Close to *Cimex lectularius*." *Parasites & Vectors*. https://doi.org/10.1186/s13071-017-2376-1

Horta, O. 2010. "What Is Speciesism?" *Journal of Agricultural and Environmental Ethics* 23: 243–266.

Horta, O., and F. Albersmeier. 2020. "Defining Speciesism." *Philosophy Compass* 15: 1–9.

Horvath, K., D. Angeletti, G. Nascetti, and C. Carere. 2013. "Invertebrate Welfare: An Overlooked Issue." *Annali dell'Istituto Superiori di Sanità* 49: 9–17.

Hossain, M. M. 2015. "Recent Perspective of Herbicide: Review of Demand and Adoption in World Agriculture." *Journal of the Bangladesh Agricultural University* 13(1): 13–24.

Hovardas, T. Ed. 2018. *Large Carnivore Conservation and Management: Human Dimensions*. London, Routledge.

Howald, G., C. Donlan, K. Faulkner, S. Ortega, H. Gellerman, D. Croll, et al. 2010. "Eradication of Black Rats *Rattus rattus* from Anacapa Island." *Oryx* 44: 30–40.

Hoyt, J. R., K. E. Langwig, K. Sun, K. L. Parise, A. Li, Y. Wang, et al. 2020. "Environmental Reservoir Dynamics Predict Global Infection Patterns and Population Impacts for the Fungal Disease White-Nose Syndrome." *Proceedings of the National Academy of Sciences* 201914794. https://doi.org/10.1073/pnas.1914794117.

Hrabovský, M., and K. Mičieta. 2013. "Review of the Taxonomic Concepts of the Invasive Species *Ambrosia artemisiifolia* L. as the Basis for the Evaluation of Variability of This Species in Europe." *Acta Botanica Universitatis Comenianae* 48: 9–13.

Hsu, J. 2017. "The Hard Truth about the Rhino Horn 'Aphrodisiac' Market. Media Coverage Hyping the Supposed use of Rhino Horn to Pump up Sex Drive Does No Favors for Conservation Efforts." *Scientific American Online Newsletter*. https://www.scientificamerican.com/article/the-hard-truth-about-the-rhino-horn-aphrodisiac-market/

Huang, W.-H., C.-M. Yang, and W.-J. Chou. 2019. "Get Away! The Effects of Pest Anthropomorphism on Consumer Willingness to Purchase Healthcare Products." *Advances in Consumer Research* 47: 635–636.

Hudson, M. J., M. Aoyama, T. Kawashima, and T. Gunji. 2008. "Possible Steatopygia in Prehistoric Central Japan: Evidence from Clay Figurines." *Anthropological Science* 16: 87–92.

Hugo, K. 2017. "Remember the Lab Mouse with a Human Ear on Its Back? The Scientist Accused of 'Playing God' Explains His Work." *Newsweek* (online). https://www.newsweek.com/tissue-surgeon-ear-mouse-human-organs-transplant-cell-phones-666082

Hůla, M., and J. Flegr. 2016. "What Flowers Do We Like? The Influence of Shape and Color on the Rating of Flower Beauty." *PeerJ* 4: e2106. https://doi.org/10.7717/peerj.2106

Hůla, M., and J. Flegr. 2021. "Habitat Selection and Human Aesthetic Responses to Flowers." *Evolutionary Human Sciences* 3: e5. https://doi.org/10.1017/ehs.2020.66

Hulin, M. S., and R. Quinn. 2006. "Wild and Black Rats." In *The Laboratory Rat*, Second Edition, edited by M. A. Suckow, S. H. Weisbroth, and C. L. Franklin, 865–882. Amsterdam, Academic Press.

Hume, C. 1958. *Skunks as Pets*. Fond du Lac, WI, All-Pets Books.

Hund, A. 2012. "Charismatic Megafauna." In *Encyclopedia of Global Warming & Climate Change*, Second Edition, edited by S. G. Philander, 237–241. Thousand Oaks, CA, Sage Publications.

Hunt, K. 2020. "Bats, the Source of So Many Viruses, Could Be the Origin of Wuhan Coronavirus, Say Experts." *CNN Health*. https://www.cnn.com/2020/01/29/health/bats-viruses-coronavirus-scn/index.html

Hunter, P. 2014 "Brown Is the New Green: Brownfield Sites Often Harbour a Surprisingly Large Amount of Biodiversity." *Embo Reports* 15: 1238–1242. https://doi.org/10.15252/embr.201439736

Hurricane Matthew. 2015. "Blog Discussion: How Can We Drive Cockroaches to Extinction?" *Personality Café*. http://personalitycafe.com/science-technology/542370-how-can-we-drive-cockroaches-extinction.html

Hurwitt, M. C. 2018. "Freedom versus Forage: Balancing Wild Horses and Livestock Grazing on the Public Lands." *Idaho Law Review* 53(2): 425–463.

Hutchinson, G. E. and R. H. MacArthur. 1959. "A Theoretical Ecological Model of Size Distributions among Species of Animals." *American Naturalist* 93(869): 117–125.

Huynh, L. T. M., A. Gasparatos, J, Su, R. Dam Lam, E. I. Grant, and K. Fukushi. 2022. "Linking the Nonmaterial Dimensions of Human-Nature Relations and Human Well-Being through Cultural Ecosystem Services." *Science Advances* 8: 31. https://doi.org/10.1126/sciadv.abn8042; https://www.science.org/doi/10.1126/sciadv.abn8042

Hyde, N. 2014. *Glow-in-the-Dark Creatures*. Markham, ON, Fitzhenry and Whiteside.

Iacona, G., R. F. Maloney, I. Chadès, J. R. Bennett, P. J. Seddon, and H. P. Possingham. 2017. "Prioritizing Revived Species: What Are the Conservation Management Implications of De-extinction? *Functional Ecology* 31: 1041–1048.

Ibrahim, B. U., and D. A. Adebote. 2012. "Appraisal of the Economic Activities of Termites: A Review." *Bayero Journal of Pure and Applied Sciences* 5: 84–89.

Ikonomopoulou, M., and G. King. 2013. "Natural Born Insect Killers: Spider-Venom Peptides and Their Potential for Managing Arthropod Pests." *Outlooks on Pest Management* 24: 16–19.

Illoldi-Rangel, P., M. Ciarleglio, L. Sheinvar, M. Linaje, V. Sánchez-Cordero, and S. Sarkar. 2012. "*Opuntia* in México: Identifying Priority Areas for Conserving Biodiversity in a Multi-Use Landscape." *PLoS One* 7(5): e36650. https://doi.org/10.1371/journal.pone.0036650

Imamura, F. L. O'Connor, Z. Ye, J. Mursu, Y. Hayashino, S. N. Bhupathiraju, et al. 2015. "Consumption of Sugar Sweetened Beverages, Artificially Sweetened Beverages, and Fruit Juice and Incidence of Type 2 Diabetes: Systematic Review, Meta-Analysis, and Estimation of Population Attributable Fraction." *British Medical Journal* 351: h3576. https://doi.org/10.1136/bmj.h3576

Ingham, C. S., and M. M. Borman. 2010. "English Ivy (*Hedera* spp., Araliaceae) Response to Goat Browsing." *Invasive Plant Science and Management* 3: 178–181.

Inglese, P. 2009. "Cactus Pear: Gift of the New World." *Chronica Horticulturae* 49: 15–18.

Inglese, P. and M. O. Brutsch, M. O. Eds. 1997. *Third International Congress on Cactus Pear and Cochenille*. Wageningen, The Netherlands, International Society for Horticultural Science.

Inglese, P., C. Mondragon, A. Nefzaoui, and C. Saenz. Eds. 2017. *Crop Ecology, Cultivation and Uses of Cactus Pear*. Rome, Food and Agriculture Organization of the United Nations.

Ingram, V., B. Vinceti, and N. van Vliet. 2017. Wild Plant and Animal Genetic Resources." In *Routledge Handbook of Agricultural Biodiversity*, edited by D. Hunter, L. Guarino, C. Spillane, and P. C. McKeown, 65–85. Oxfordshire, UK, Routledge.

Innes, R. J. 2012. "*Toxicodendron radicans*, *T. rydbergii*." In *Fire Effects Information System* [Online]. U.S. Department of Agriculture, Forest Service, Rocky Mountain Research Station, Fire Sciences Laboratory. https://www.fs.usda.gov/database/feis/plants/shrub/toxspp/all.html

International Sugar Organization. 2020. "About Sugar." https://www.isosugar.org/sugarsector/sugar

Invasive Species Specialist Group. 2013. "*Rattus rattus* (Black Rat)." *CABI Invasive Species Compendium*. https://www.cabi.org/isc/datasheet/46831

Invasive Species Specialist Group. 2014a. "*Rattus exulans* (Pacific Rat)." *CABI Invasive Species Compendium*. https://www.cabi.org/isc/datasheet/46834

Invasive Species Specialist Group. 2014b. "*Rattus norvegicus* (Brown Rat)." *CABI Invasive Species Compendium*. https://www.cabi.org/isc/datasheet/46829

Inward, D., G. Beccaloni, and P. Eggleton. 2007. "Death of an Order: A Comprehensive Molecular Phylogenetic Study Confirms That Termites Are Eusocial Cockroaches." *Biology Letters* 3(3): 331–335. https://doi.org/10.1098/rsbl.2007.0102

Ioannis, K., F. F. Brunel, and R. A. Coulter. 2014. "Judgment Is Not Color Blind: The Impact of Automatic Color Preference on Product and Advertising Preferences." *Journal of Consumer Psychology* 24: 87–95.

IPBES (Intergovernmental Science-Policy Platform on Biodiversity and Ecosystem Services). 2019. *Global Assessment Report on Biodiversity and Ecosystem Services of the Intergovernmental Science-Policy Platform on Biodiversity and Ecosystem Services*, edited by E. S. Brondizio, J. Settele, S. Díaz, and H. T. Ngo. Bonn, Germany, IPBES Secretariat. https://doi.org/10.5281/zenodo.3831673

Iqbal, M. A., A. Hamid, H. Imtiaz, M. Rizwan, M. Imran, U. A. A. Sheikh, et al. 2020. "Cactus Pear: A Weed of Dry-Lands for Supplementing Food Security under Changing Climate." *Planta Daninha* 38. https://doi.org/10.1590/s0100-83582020380100040

Iqbal, W., M. F. Malik, M. K. Sarwar, I. Azam, N. Iram, and A. Rashda. 2014. "Role of Housefly (*Musca domestica*, Diptera; Muscidae) as a Disease Vector; a Review." *Journal of Entomology and Zoology Studies* 2: 159–163.

Isaac, A. A. 2016. "Overview of Cactus (*Opuntia ficus-indica* L): A Myriad of Alternatives." *Studies on Ethno-Medicine* 10: 195–205.

Isaac, N. J. B., S. T. Turvey, B. Collen, C. Waterman, J. E. Baillie, E. M. Jonathan, et al. 2007. "Mammals on the EDGE: Conservation Priorities Based on Threat and Phylogeny." *PLoS One* 2(3): e296. http://www.plosone.org/article/info:doi/10.1371/journal.pone.0000296

Isasi-Catala, E. 2011. "Indicator, Umbrellas, Flagships and Keystone Species Concepts: Use and Abuse in Conservation Ecology." *Interciencia* 36: 31–38. (In Spanish, English abstract)

Itodo, H. U. 2019. "Controlled Release of Herbicides Using Nano-Formulation: A Review." *Journal of Chemical Reviews* 1: 130–138.

IUCN. 2017. *The IUCN Red List of Threatened Species*. http://www.iucnredlist.org/search

IUCN. 2021. *The IUCN Red List of Threatened Species. Version 2021-1*. https://www.iucnredlist.org

Jabr, F. 2018. "It's Official: Fish Feel Pain. But Will Our Oceanic Friends Ever Get the Same Legal Protections as Land Animals?" *Hakai Magazine*. https://www.smithsonianmag.com/science-nature/fish-feel-pain-180967764/

Jach, L., and M. Moroń. 2020. "I Can Wear a Beard, but You Should Shave… Preferences for Men's Facial Hair from the Perspective of Both Sexes." *Evolutionary Psychology* 18: 4. https://doi.org/10.1177/1474704920961728

Jackson, B. S. 1982. "The Lowly Dandelion Deserves More Respect." *Canadian Geographic* 102: 54–59.

Jackson, J. A., I. M. Friberg, L. Bolch, A. Lowe, C. Ralli, P. D. Harris, et al. 2009. "Immunomodulatory Parasites and Toll-Like Receptor-mediated Tumour Necrosis Factor Alpha Responsiveness in Wild Mammals." *BMC Biology* 7(1): 16. https://doi.org/10.1186/1741-7007-7-16.

Jackson, J. B. C., and A. G. Coates. 1986. "Life Cycles and Evolution of Clonal (Modular) Animals." *Philosophical Transactions of the Royal Society of London. Series B, Biological Sciences* 313(1159): 7–22.

Jackson, T. N., and B. G. Fry. 2016. "A Tricky Trait: Applying the Fruits of the 'Function Debate' in the Philosophy of Biology to the 'Venom Debate' in the Science of Toxinology." *Toxins* 8: 263. https://doi.org/10.3390/toxins8090263

Jackson, T. N. W., H. Jouanne, and N. Vidal. 2019. "Snake Venom in Context: Neglected Clades and Concepts." *Frontiers in Ecology and Evolution*. https://doi.org/10.3389/fevo.2019.00332

Jacobs, J., J, Sciegienka, and F. Menalled. 2006. "Ecology and Management of Canada Thistle [*Cirsium arvense* (L.) Scop.]." United States Department of Agriculture, Natural Resources Conservation Service, Invasive Species Technical Note No. MT-5. https://www.researchgate.net/publication/242497779_Ecology_and_Management_of_Canada_thistle_Cirsium_arvense_L_Scop

Jacot, J., A. S. Williams, and J. R. Kiniry. 2021. "Biofuel Benefit or Bummer? A Review Comparing Environmental Effects, Economics, and Feasibility of North American Native Perennial Grass and Traditional Annual Row Crops When Used for Biofuel." *Agronomy* 11(7): 1440. https://doi.org/10.3390/agronomy11071440

Jaeggi, A. V.; and C. P. van Schaik. 2011. "The Evolution of Food Sharing in Primates." *Behavioral Ecology and Sociobiology* 65: 2125. https://doi.org/10.1007/s00265-011-1221-3

Janeiro-Otero, A., T. M. Newsome, L. M. Van Eeden, W. J. Ripple, and C. F. Dormann. 2020. "Grey Wolf (*Canis lupus*) Predation on Livestock in Relation to Prey Availability." *Biological Conservation* 243. https://doi.org/10.1016/j.biocon.2020.108433

Jansen P. C. M. 2004. "*Chenopodium album* L." In *Vegetables. Plant Resources of Tropical Africa (PROTA), volume 2*, edited by G. J. H. Grubben and O. A. Denton, 178–180. Wageningen, ROTA Foundation / Backhuys Publishers / CTA. http://edepot.wur.nl/417517

Jardine, W. H. 1843. *Naturalist's Library*. Edinburgh, W. H. Lizars.

Jena, M. 2012. "Worms, Termites, Microbes Offer Food Security." https://ourworld.unu.edu/en/worms-termites-microbes-offer-food-security

Jenkins, V. S. 1994. *The Lawn: A History of an American Obsession*. Washington, DC, Smithsonian Institution.

Jennings, J., P. Beck, D. Philipp, and K. Simon. 2014. "Crabgrass for Forage." Arkansas, University of Arkansas, Division of Agriculture Research & Extension, FSA3138. https://www.uaex.uada.edu/publications/pdf/FSA-3138.pdf

Jennings, K. 2019. "A Literature Review of the Function of Bats as Disease Vectors across Fragmented Habitats." *Minnesota Undergraduate Research & Academic Journal* 2(4). https://pubs.lib.umn.edu/index.php/muraj/article/view/1563

Jepson, P., and M. Barua. 2015. "A Theory of Flagship Species Action." *Conservation & Society* 13: 95–104.

Jerolmack, C. 2008. "How Pigeons Became Rats: The Cultural-Spatial Logica of Problem Animals." *Social Problems* 55: 72–94.

Jeronen, E., I. Palmberg, and E. Yli-Panula. 2017. "Teaching Methods in Biology Education and Sustainability Education Including Outdoor Education for Promoting Sustainability – A Literature Review." *Education Sciences* 7: 1. https://doi.org/10.3390/educsci7010001

Jiang, C., and M. Galm. 2014. "The Economic Benefit of Being Blonde: A Study of Waitress Tip Earnings Based on Their Hair Color in a Prominent Restaurant Chain." *Journal of Behavioral Studies in Business* 7(September). http://www.aabri.com/manuscripts/141934.pdf

Jobling, I. 2001. "The Psychological Foundations of the Hero-Ogre Story." *Human Nature* 12: 247–272.

John Doughty (Publisher). 1832. *The Cabinet of Natural History and American Rural Sports*, Vol. 2. Philadelphia, PA, John Doughty.

Johnson, C., G. Schreer, and K. J. Bao. 2021. "Effect of Anthropomorphizing Food Animals on Intentions to Eat Meat." *Anthrozoös*. https://doi.org/10.1080/08927936.2021.1914442

Johnson, D. W. 2010. "Physical Appearance and Wages: Do Blondes Have More Fun?" *Economics Letters* 108: 10–12.
Johnson, M. T. J., and J. Munshi-South. 2017. "Evolution of Life in Urban Environments." *Science* 358(6363): eaam8327. https://doi.org/10.1126/science.aam8327
Jojola, S. M., S. J. Robinson, and K. C. Vercauteren. 2007. "Oral Rabies Vaccine (ORV) Bait Uptake by Captive Striped Skunks." *Journal of Wildlife Diseases* 43: 97–106.
Jokela, M., 2009. "Physical Attractiveness and Reproductive Success in Humans: Evidence from the Late 20th Century United States." *Evolution and Human Behavior* 30: 342–350.
Jones, A. L., R. Russell, and R. Ward. 2015. "Cosmetics Alter Biologically-Based Factors of Beauty: Evidence from Facial Contrast." *Evolutionary Psychology* 13: 21–229.
Jones, H. P., B. R. Tershy, E. S. Zavaleta, D. A. Croll, B. S. Keitt, M. E. Finkelstein, et al. 2008. "Severity of the Effects of Invasive Rats on Seabirds: A Global Review." *Conservation Biology* 22: 16–26.
Jouquet, P., E. Chaudhary, and A. R. V. Kumar. 2018. "Sustainable Use of Termite Activity in Agro-Ecosystems with Reference to Earthworms. A Review." *Agronomy for Sustainable Development* 38: 1–11.
Joy, M. 2001. "From Carnivore to Carnist: Liberating the Language of Meat." *Satya* 8(2): 26. http://www.satyamag.com/sept01/joy.html
Joy, M. 2009. *Why We Love Dogs, Eat Pigs, and Wear Cows.* Newburyport, MA, Conari Press.
Joye, J., and A. De Block. 2011. "'Nature and I Are Two': A Critical Examination of the Biophilia Hypothesis." *Environmental Values* 20: 189–215.
Juckett, G. 1996. "Plant Dermatitis." *Postgraduate Medicine* 100: 159–171.
Jun, S. H. 2019. "Cultivating Barnyard Grass." In *Agriculture and Korean Economic History*, edited by S. H. Jun, 101–102. Singapore, Palgrave Macmillan. https://doi.org/10.1007/978-981-32-9319-9_14
Jung, C. 1959. *The Archetypes and the Collective Unconscious, Collected Works, Volume 9, Part 1.* Princeton, NJ, Princeton University Press.
Kaltenborn, B. P., T. Bjerke, and J. Nyahongo. 2006. "Living with Problem Animals - Self-Reported Fear of Potentially Dangerous Species in the Serengeti Region, Tanzania." *Human Dimensions of Wildlife* 11: 397–409.
Kampmann, M. L., I. B. Schnell, R. H. Jensen, J. Axtner, A. F. Sander, A. J. Hansen, et al. 2017. "Leeches as a Source of Mammalian Viral DNA and RNA – A Study in Medicinal Leeches. *European Journal of Wildlife Research* 63: 36. https://doi.org/10.1007/s10344-017-1093-6.
Kanapeckas, K. L., C. C. Vigueira, A. Ortiz, K. A. Gettler, N. R. Burgos, A. J. Fischer, et al. 2016. "Escape to Ferality: The Endoferal Origin of Weedy Rice from Crop Rice through De-Domestication." *PLoS One* 11(9): e0162676. https://doi.org/10.1371/journal.pone.0162676
Kang, B.-H., S.-I. Shim, S.-G. Lee, K.-H. Kim, and I.-M. Chung. 1998. "Evaluation of *Ambrosia artemisiifolia* var. *elatior*, *Ambrosia trifida*, *Rumex crispus* for Phytoremediation of Cu and Cd Contaminated Soil." *Korean Journal of Weed Science* 18: 262–267.
Kapoor, L. D. 1995. *Opium Poppy: Botany, Chemistry, and Pharmacology.* New York, Food Products Press.
Kareiva, P., and I. Kareiva. 2017. "Biodiversity Hotspots and Conservation Priorities." Oxford, Oxford University Press, Environmental Science, https://doi.org/10.1093/acrefore/9780199389414.013.95 (online article).
Kareiva, P., and M. Marvier. 2003. "Conserving Biodiversity Coldspots." *American Scientist* 91: 344–351.
Karraker, K., and M. Stern. 1990. "Infant Physical Attractiveness and Facial Expression: Effects on Adult Perceptions." *Basic and Applied Social Psychology* 11: 371–385.
Karrer, G. 2016. "Control of Common Ragweed by Mowing and Hoeing." *Julius-Kühn-Archiv* 455: 118–124. https://pdfs.semanticscholar.org/969f/274836dd3a8236d0697164b946e1b57dee14.pdf
Kasterine, A., R. Arbeid, O. Caillabet, and D. Natusch. 2012. "The Trade in South-East Asian Python Skins." Geneva, International Trade Centre. https://papers.ssrn.com/sol3/papers.cfm?abstract_id=2362381
Kattwinkel, M., R. Biedermann, and M. Kleyer. 2011. "Temporary Conservation for Urban Biodiversity." *Biological Conservation* 144: 2335–2343.
Katz, D. S. W., B. T. C. Barrie, and T. S. Carey. 2014. "Urban Ragweed Populations in Vacant Lots: An Ecological Perspective on Management." *Urban Forestry & Urban Greening* 13: 756–760.
Kaufman, S. B., D. B. Yaden, E. Hyde, and E. Tsukayama. 2019. "The Light vs. Dark Triad of Personality: Contrasting Two Very Different Profiles of Human Nature." *Frontiers in Psychology* 10: 467. https://doi.org/10.3389/fpsyg.2019.00467
Kaur, M., A. Kaur, and R. Sharma. 2012. "Pharmacological actions of *Opuntia ficus-indica*: A Review." *Journal of Applied Pharmaceutical Science* 2(7): 15–18.
Kawai, N. 2019. *The Fear of Snakes: Evolutionary and Psychobiological Perspectives on Our Innate Fear.* Singapore, Springer.
Kawai, Y., S. Tohyama, H. Shimizu, K. Fukuda, and E. Kobayashi. 2019. "Pigs as Models of Preclinical Studies and In Vivo Bioreactors for Generation of Human Organs, Xenotransplantation - Comprehensive Study." In *Xenotransplantation*, edited by S Miyagawa. IntechOpen. https://doi.org/10.5772/intechopen.90202; https://www.intechopen.com/chapters/70174
Kawano, S., H. Azuma, M. Ito, and K. Suzuki. 1999. "Extrafloral Nectaries and Chemical Signals of *Fallopia japonica* and *Fallopia sachalinensis* (Polygonaceae), and Their Roles as Defense Systems against Insect Herbivory." *Plant Species Biology* 14: 167–178.
Keil, D. J. 2006. "*Cirsium*." In *Flora of North America North of Mexico, Volume 19*, edited by Flora of North America Editorial Committee, 95–164. New York, Oxford University Press.
Keith, D. A., J. P. Rodríguez, K. M. Rodríguez-Clark, E. Nicholson, K. Aapala, A. Alonso, et al. 2013. "Scientific Foundations for an IUCN Red List of Ecosystems." *PLoS One* 8(5): e62111. https://doi.org/10.1371/journal.pone.0062111
Keith, R. J., L. M. Given, J. M. Martin, and D. F. Hochuli. 2022. "Urban Children and Adolescents' Perspectives on the Importance of Nature." *Environmental Education Research.* https://doi.org/10.1080/13504622.2022.2080810
Kelker, G. H. 1937. "Insect Food of Skunks." *Journal of Mammalogy* 18: 164–170.
Keller, A., and S. B. D. Torre. 2015. "Sugar-Sweetened Beverages and Obesity among Children and Adolescents: A Review of Systematic Literature Reviews." *Childhood Obesity* 11: 338–346.
Keller, V., and K. Bollmann. 2004. "From Red Lists to Species of Conservation Concern." *Conservation Biology* 18: 1636–1644.

Kellert, S. R. 1993. "Values and Perceptions of Invertebrates." *Conservation Biology* 7: 845–855.

Kellert, S. R. 1996. *The Value of Life: Biological Diversity and Human Society*. Washington, DC, Island Press.

Kellert, S. R., and J. K. Berry. 1987. "Attitudes, Knowledge, and Behaviors toward Wildlife as Affected by Gender." *Wildlife Society Bulletin* 15: 363–371.

Kelley, J. R., and L. H. Fredrickson. 1991. "Chufa Biology and Management." Washington, DC, U.S. Department of the Interior, Fish and Wildlife Service. https://digitalcommons.unl.edu/cgi/viewcontent.cgi?article=1033&context=icwdmwfm

Kelly, D. 2011. Yuck! *The Nature and Moral Significance of Disgust*. Cambridge, MA, MIT Press.

Kelly, M. 2015. "Tiny Termites Can Hold Back Deserts by Creating Oases of Plant Life." https://www.princeton.edu/news/2015/02/05/tiny-termites-can-hold-back-deserts-creating-oases-plant-life

Kemp, D. J., M. Hilbert, and M. R. Gillings. 2016. "Information in the Biosphere: Biological and Digital Worlds." *Trends in Ecology & Evolution* 31: 180–189.

Kendrick, H. M. N. 2018. "Autonomy, Slavery, and Companion Animals." *Between the Species: A Journal for the Study of Philosophy and Animals* 22(1): 7. https://digitalcommons.calpoly.edu/bts/vol22/iss1/7

Kerksiek, K. 2008. "Parasites and the Hygiene Hypothesis." *Infection Research: News and Perspectives*. https://pdfs.semanticscholar.org/a190/371b0ebe04fd605a9077d55f1a200ed1d9bf.pdf

Ketcham, C. 2010. "New Dog in Town." *Orion Magazine* (online blog). https://orionmagazine.org/article/new-dog-in-town/

Keung, W. M., and B. L. Vallee. 1998. "Kudzu Root: An Ancient Chinese Source of Modern Antidipsotropic Agents." *Phytochemistry* 47: 499–506.

Khamesipour, F., K. B. Lankarani, B. Honarvar, and T. E. Kwenti. 2018. "A Systematic Review of Human Pathogens Carried by the Housefly (*Musca domestica* L.)." *BMC Public Health* 18(1): 1049. https://doi.org/10.1186/s12889-018-5934-3.

Khandekar, A. and M. Roy. 2022. "Medicago's Tobacco Ties Jeopardize Growth of Its COVID Shot." *Reuters (Blog)*, March 28. https://www.reuters.com/business/healthcare-pharmaceuticals/medicagos-tobacco-ties-jeopardize-growth-its-covid-shot-2022-03-27/

Kidd, A. H., H. T. Kelley, and R. M. Kidd. 1984. "Personality Characteristics of Horse, Turtle, Snake, and Bird Owners." In *The Pet Connection: Its Influence on Our Health and Quality of Life*, edited by R. K. Anderson, B. L. Hart, and L. A. Hart, 200–206. Minneapolis, University of Minnesota Press.

Kiire, S. 2016. "Effect of Leg-to-Body Ratio on Body Shape Attractiveness." *Archives of Sexual Behavior* 45: 901–910.

Kikel, S. K. 2018. "Ticks and Tick-Borne Infections: Complex Ecology, Agents, and Host Interactions." *Veterinary Sciences* 5(2): 60. https://doi.org/10.3390/vetsci5020060

Kim, C.-J., B.-H. Kang, I.-K. Lee, I.-J. Ryoo, D.-J. Park, K. H. Lee, et al. 1993. "Screening of Biologically Active Compounds from Weeds." *Korean Journal of Weed Science* 14: 16–22.

Kim, Y., A. Flamm, M. A. ElSohly, D. H. Kaplan, R. J. Hage, Jr., C. P. Hamann, et al. 2019. "Poison Ivy, Oak, and Sumac Dermatitis: What Is Known and What Is New?" *Dermatitis* 30: 183–190.

Kimura, Y., and H. Okuda. 2001. "Resveratrol Isolated from *Polygonum cuspidatum* Root Prevents Tumor Growth and Metastasis to Lung and Tumor-Induced Neovascularization in Lewis Lung Carcinoma-Bearing Mice." *Journal of Nutrition* 131: 1844–1849.

Kincaid, A. T. 2008. "Wild or Feral? Historical and Biological Consideration of Free Roaming Horses (FRH) in Alberta." Contributed paper for the Canadian Parks for Tomorrow: 40th Anniversary Conference, May 8–11, 2008, Calgary, AB, University of Calgary. http://hdl.handle.net/1880/46932

King, G. F., and M. C. Hardy. 2013. "Spider-Venom Peptides: Structure, Pharmacology, and Potential for Control of Insect Pests." *Annual Review of Entomology* 58: 475–496.

King, T., L. C. Marston, and P. C. Bennett. 2009. "Describing the Ideal Australian Companion Dog." *Applied Animal Behaviour Science* 120: 84–93.

Kirby, W. F., and G. H. Schubert. 1889. *Natural History of the Animal Kingdom for the Use of Young People*. Brighton, E. & J. B. Young and Co.

Kirkness, E. F., B. J. Haas, W. Sun, H. R. Braig, M. A. Perotti, J. M. Clark, et al. 2010. "Genome Sequences of the Human Body Louse and Its Primary Endosymbiont Provide Insights into the Permanent Parasitic Lifestyle." *Proceedings of the National Academy of Sciences of the United States* 107: 12168–12173.

Kirschner, J., and J. Štěpánek. 1993. "The genus *Taraxacum* in the Caucasus 1. Introduction, 2. The Section *Porphyrantha*." *Folia Geobotanica et Phytotaxonomica* 28: 295–320.

Kirschner, J., and J. Štěpánek. 1997. "A Nomenclatural Checklist of Supraspecific Names in *Taraxacum*." *Taxon* 46: 87–98.

Kiviat, E., and E. Hamilton. 2001. "*Phragmites* Use by Native North Americans." *Aquatic Botany* 69: 341–357.

Kiviat, E., L. E. Meyerson, T. J. Mozdzer, W. J. Allen, A. H. Baldwin, G. P. Bhattarai, et al. 2019. "Evidence Does Not Support the Targeting of Cryptic Invaders at the Subspecies Level Using Classical Biological Control: The Example of *Phragmites*." *Biological Invasions* 21: 2529–2541.

Kiviat, K. 2013. "Ecosystem Services of *Phragmites* in North America with Emphasis on Habitat Functions." *AoB Plants* 5: plt008. https://doi.org/10.1093/aobpla/plt008; https://www.ncbi.nlm.nih.gov/pmc/articles/PMC4104640/

Klein, J. D., R. K. Thomas, and E. J. Suttera. 2007. History of Childhood Candy Cigarette Use Is Associated with Tobacco Smoking by Adults." *Preventive Medicine* 45: 26–30.

Klimley, A. P., and T. H. Curtis. 2006. "Shark Attack versus Ecotourism: Negative and Positive Interactions." In *Proceedings 22nd Vertebrate Pest Conference*, edited by R. M. Timm and J. M. O'Brien, 33–44. Davis, University of California. https://escholarship.org/uc/item/52h8t628

Klimpel, S., and H. Melhorn. Eds. 2013. *Bats (Chiroptera) as Vectors of Diseases and Parasites. Facts and Myths*. New York, Springer.

Knapp, S. 2020. "Biodiversity of *Nicotiana* (Solanaceae)." In *The Tobacco Plant Genome. Compendium of Plant Genomes*, edited by N. Ivanov, N. Sierro, and M. Peitsch, 21–41. Cham, Springer.

Knight, J. E. 1994. "Skunks." *The Handbook: Prevention and Control of Wildlife Damage*, 42. https://digitalcommons.unl.edu/icwdmhandbook/42

Knudsen, H. K., J. Alexander, L. Barregård, M. Bignami, B. Brüschweiler, S. Ceccatelli, et al. 2018. "Update of the Scientific Opinion on Opium Alkaloids in Poppy Seeds." European Food Safety Authority Journal 16(5): e05243. https://doi.org/10.2903/j.efsa.2018.5243

Köhler, E. 1883–1914. *Köhler's Medizinal-Pflanzen*. 4 vols. Gera-Untermhaus, Germany, F.E. Köhler.

Konur, O. 2021. "Algal Biomass Production for Biodiesel Production: A Review of the Research." In *Biodiesel Fuels Based on Edible and Nonedible Feedstocks, Wastes, and Algae*, edited by O. Konur, 695–717. Boca Raton, FL, CRC Press.

Kopecký, O., A. Bílková, V. Hamatová, D. Kňazovická, L. Konrádová, B. Kunzová, et al. 2019. "Potential Invasion Risk of Pet Traded Lizards, Snakes, Crocodiles, and Tuatara in the EU on the Basis of a Risk Assessment Model (RAM) and Aquatic Species Invasiveness Screening Kit (AS-ISK)." *Diversity* 11(9): 164. https://doi.org/10.3390/d11090164

Kordsmeyer, T. L., D. Freund, S. R. Pita, J. Jünger, and L. Penke. 2019. "Further Evidence that Facial Width-to-Height Ratio and Global Facial Masculinity Are Not Positively Associated with Testosterone Levels." *Adaptive Human Behavior and Physiology* 5: 117–130.

Kos, M., J. Jerman, and G. Torkar. 2021. "Preschool Children's Attitude toward Some Unpopular Animals and Formation of a Positive Attitude toward Them through Hands-On Activities." *Journal of Biological Education*. https://doi.org/10.1080/00219266.2021.1877779.

Kosch, T.A., A. W. Waddle, C. A. Cooper, K. R. Zenger, D. J. Garrick, L. Berger, et al. 2022. "Genetic Approaches for Increasing Fitness in Endangered Species." *Trends in Ecology & Evolution* 37: 332–345.

Kosoy, M., L. Khlyap, J.-F. Cosson, and S. Morand. 2015. "Aboriginal and Invasive Rats of Genus *Rattus* as Hosts of Infectious Agents." *Vector-Borne and Zoonotic Diseases* 15(1): 3–12.

Kotze, M. 2018. "The Theological Ethics of Human Enhancement: Genetic Engineering, Robotics and Nanotechnology." *Die Skriflig* 52(3): 1–8.

Koyama, R., Y. Takahashi, and K. Mori. 2006. "Assessing the Cuteness of Children: Significant Factors and Gender Differences." *Social Behavior and Personality* 34: 1087–1100.

Krak, K., P. Vít, A. Belyayev, J. Douda, L. Hreusová, and B. Mandák. 2016. "Allopolyploid Origin of *Chenopodium album* s. str. (Chenopodiaceae): A Molecular and Cytogenetic Insight." *PLoS One* 11(8): e0161063. https://doi.org/10.1371/journal.pone.0161063

Kramer, G. R., S. M. Peterson, K. O. Daly, H. M. Streby, and D. E. Andersen. 2019. "Left Out in the Rain: Comparing Productivity of Two Associated Species Exposes a Leak in the Umbrella Species Concept." *Biological Conservation* 233: 276–288.

Kramer, P. 2008.*Totem Poles*. Custer, WA, Heritage House.

Kraus, F. 2015. "Impacts from Invasive Reptiles and Amphibians." *Annual Review of Ecology, Evolution, and Systematics* 46: 75–97.

Krebs, C., G. Mahy, D. Matthies, U. Schaffner, M.-S. Tiébré, and J.-P. Bizoux. 2010. "Taxa Distribution and RAPD Markers Indicate Different Origin and Regional Differentiation of Hybrids in the Invasive *Fallopia* Complex in Central-Western Europe." *Plant Biology* 12: 215–223.

Krist, S. 2020. "Cactus Pear Seed Oil." In *Vegetable Fats and Oils*, edited by S. Krist, 165–169. Cham, Springer.

Kristen L. P. 2018. "The Detrimental Health Effects of Sugar." *Journal of Obesity and Diabetes* 1: 21–22.

Kriticos, D. J., and S. Brunel. 2016. "Assessing and Managing the Current and Future Pest Risk from Water Hyacinth, (*Eichhornia crassipes*), an Invasive Aquatic Plant Threatening the Environment and Water Security." *PLoS One* 11(8). https://doi.org/10.1371/journal.pone.0120054

Kroschel, J. 2001. "Introduction." In *A Technical Manual for Parasitic Weed Research and Extension*, edited by J. Kroschel, 1–6. Dordrecht, Springer.

Krupska, I. 2012. "Fungal Pathogens for Biological Control of Crabgrass '*Digitaria* spp.' in Canada." M.Sc. thesis. Montreal, McGill University. https://escholarship.mcgill.ca/concern/theses/tx31qn451

Kruuk, H. 1972. *The Spotted Hyena: A Study of Predation and Social Behavior*. Chicago, IL, University of Chicago Press.

Kuehn, B. M. 2013. "CDC Estimates 300,000 US Cases of Lyme Disease Annually." *Journal of the American Medical Association* 310(11): 1110. https://doi.org/10.1001/jama.2013.278331

Kularatne, S. A. M., and N. Senanayake. 2014. "Chapter 66: Venomous Snake Bites, Scorpions, and Spiders." In *Handbook of Clinical Neurology*, Volume 120, edited by J. Biller and J. M. Ferro, 987–1001. Amsterdam, Elsevier.

Kumar, A., and C. Jnanesha. 2016. "Conservation of Rare and Endangered Plant Species for Medicinal Use." *International Journal of Science and Research* 5: 1370–1372.

Kumar, D., D. Jayaswal, A. Jangra, K. K. Mishra, and S. Yadav. 2018. "Recent Approaches for Herbicide Resistance Management in Weeds: A Review." *International Journal of Chemical Studies* 6: 2844–2850.

Kuo, D.-H., and Y.-T. La. 2019. "On the Origin of Leeches by Evolution of Development." *Ecology, Evolution and Development* 61: 43–57.

Kurtz, C. M., and M. H. Hansen. 2019. "An Assessment of Canada Thistle in Northern U.S. Forests." Research Note NRS252. Newtown Square, PA, U.S. Department of Agriculture, Forest Service, Northern Research Station. https://doi.org/10.2737/NRS-RN-252

Kushwaha, S. 2016. "Vultures in the Cultures of the World." *Asian Journal of Agriculture & Life Sciences* 1(2): 34–40.

Kutschera, U., I. Pfeiffer, and E. Ebermann. 2007. "The European Land Leech: Biology and DNA-Based Taxonomy of a Rare Species That Is Threatened by Climate Warming." *Naturwissenschaften* 94: 967–974.

Kuusi, T., K. Hardh, and H. Kanon. 1982. "The Nutritive Value of Dandelion Leaves." *Alternative/Appropriate Technologies in Agriculture* 3: 53–60.

Kwak, J., M. Jackson, A. Faranda, K. Osada, T. Tashiro, K. Mori, et al. 2016. "On the Persistence of Mouse Urine Odour to Human Observers: A Review." *Flavour and Fragrance Journal* 31: 267–282.

Kwak, M. 2018. "Australia's Vanishing Fleas (Insecta: Siphonaptera): A Case Study in Methods for the Assessment and Conservation of Threatened Flea Species." *Journal of Insect Conservation* 22: 545–550. https://doi.org/10.1007/s10841-018-0083-7.

Kyrou, K., A. M. Hammond, R. Galizi, N. Kranjc, A. Burt, A. K. Beaghton, et al. 2018. "A CRISPR–Cas9 Gene Drive Targeting Doublesex Causes Complete Population Suppression in Caged *Anopheles gambiae* Mosquitoes." *Nature Biotechnology* 36: 1062–1066.

Labanca, F., J. Ovesnà, and L. Milella. 2018. "*Papaver somniferum* L. Taxonomy, Uses and New Insight in Poppy Alkaloid Pathways." *Phytochemistry Reviews* 17: 853–871.

LaBarbera, M. 1989. "Analyzing Body Size as a Factor in Ecology and Evolution." *Annual Review of Ecology and Systematics* 20: 97–117.

Lachenmeier, D. W., C. Sproll, and F. Musshoff. 2010. "Poppy Seed Foods and Opiate Drug Testing – Where Are We Today?" *Therapeutic Drug Monitoring* 32: 11–18.

LaFrance, A. 2016. "Genetically Modified Mosquitoes: What Could Possibly Go Wrong?" *The Atlantic*. https://www.theatlantic.com/technology/archive/2016/04/genetically-modified-mosquitoes-zika/479793/

Lahart, B., L. Shalloo, J. Herron, D. O'Brien, R. Fitzgerald, T. M. Boland, et al. 2021. "Greenhouse Gas Emissions and Nitrogen Efficiency of Dairy Cows of Divergent Economic Breeding Index under Seasonal Pasture-Based Management." *Journal of Dairy Science* 104: 8039–8049.

Lai, K. M. 2017. "Are Cockroaches an Important Source of Indoor Endotoxins?" *International Journal of Environmental Research and Public Health* 14(1): 91. https://doi.org/10.3390/ijerph14010091.

Lai, O., D. Ho, S. Glick, and J. Jagdeo. 2016. "Bed Bugs and Possible Transmission of Human Pathogens: A Systematic Review." *Archives of Dermatological Research* 308: 531–538.

Lamba, N., D. Holsgrove, and M. L. Broekman. 2016. "The History of Head Transplantation: A Review." *Acta Neurochirurgica* 158: 2239–2247.

Lambeck, R. J. 1997. "Focal Species: A Multi-species Umbrella for Nature Conservation." *Conservation Biology* 11: 849–856.

Lamoreux, J., J. Morrison, T. Ricketts, D. M. Olson, E. Dinerstein, M. W. McKnight, et al. 2006. "Global Tests of Biodiversity Concordance and the Importance of Endemism." *Nature* 440: 212–214.

Landová, E., P. Poláková, S. Rádlová, M. Janovcová, M. Bobek, and D. Frynta. 2018. "Beauty Ranking of Mammalian Species Kept in the Prague Zoo: Does Beauty of Animals Increase the Respondents' Willingness to Protect Them?" *Science of Nature* 105: 69. https://doi.org/10.1007/s00114-018-1596-3

Meadly, G. R. W. 1958. "Weeds of Western Australia – Prickly Pear (*Opuntia* spp.)." *Journal of the Department of Agriculture, Western Australia*, Series 3, 7: 297–302.

Langton, J. 2006. *Rat: How the World's Most Notorious Rodent Clawed Its Way to the Top*. Toronto, Key Porter.

Lantz, D. E. 1917. *Economic Value of North American Skunks*. Revised Edition. Farmers Bulletin 587. Washington, DC, United States Department of Agriculture.

Lanz, B., S. Dietz, and T. Swanson. 2018. "The Expansion of Modern Agriculture and Global Biodiversity Decline: An Integrated Assessment." *Ecological Economics* 144: 260–277.

Larisch, P., and J. Schmid. 1902. *Das-Kürschner-Handwerk. Eine gewerbliche Monographie 1. Jahrgang [The furrier-craft. A commercial monograph. 1st year]*. Paris, P. Larisch.

Larivière, S., and F. Messier. 1998. "Denning Ecology of the Striped Skunk in the Canadian Prairies: Implications for Waterfowl Nest Predation." *Journal of Applied Ecology* 35: 207–213.

Larson, J. L., A. J. Kesheimer, and D. A. Potter. 2014. "Pollinator Assemblages on Dandelions and White Clover in Urban and Suburban Lawns." *Journal of Insect Conservation* 18: 863–873.

Latchininsky, A. V. 2013. "Locusts and Remote Sensing: A Review." *Journal of Applied Remote Sensing* 7(1): 075099. https://doi.org/10.1117/1.JRS.7.075099

Latchininsky, A. V., G. Sword, M. Sergeev, M. M. Cigliano, and M. Lecoq. 2011. "Locusts and Grasshoppers: Behavior, Ecology, and Biogeography." *Psyche: A Journal of Entomology* 2011: 578327. https://doi.org/10.1155/2011/578327

Lavoie, C. 2010. "Should We Care about Purple Loosestrife? The History of an Invasive Plant in North America." *Biological Invasions* 12: 1967–1999.

Lavoie, C. 2017. "The Impact of Invasive Knotweed Species (*Reynoutria* spp.) on the Environment: Review and Research Perspectives." *Biological Invasions* 19: 2319–2337.

Lawrence, M. J., H L. J. Stemberger, A. J. Zolderdo, D. P. Struthers, and S. J. Cooke. 2015. "The Effects of Modern War and Military Activities on Biodiversity and the Environment." *Environmental Reviews* 23: 443–460.

Leandro, C., P. Jay-Robert, and A. Vergnes. 2017. "Bias and Perspectives in Insect Conservation: A European Scale Analysis." *Biological Conservation* 215: 213–224.

Leary, C. 2020. "10 Curly Animals That Aren't Poodles or Sheep." *Treehugger: Sustainabiliy for All* (online newsletter). https://www.treehugger.com/curly-animals-arent-poodles-sheep-4869696

Lecoq, M. 2010. "Integrated Pest Management for Locusts and Grasshoppers: Are Alternatives to Chemical Pesticides Credible?" *Journal of Orthoptera Research* 19: 131–132.

Lecours, N. 2014. "The Harsh Realities of Tobacco Farming: A Review of Socioeconomic, Health and Environmental Impacts." In *Tobacco Control and Tobacco Farming: Separating Myth from Reality*, edited by W. Leppan, N. Lecours, and D. Buckles, 99–138. New York, Anthem Press and the International Development Research Centre.

Lecours, N., G. E. G. Almeida, J. M. Abdallah, and T. E. Novotny. 2012. "Environmental Health Impacts of Tobacco Farming: A Review of the Literature." *Tobacco Control* 21: 191–196.

LeDoux, J. E. 2012. "Evolution of Human Emotion: A View through Fear." *Progress in Brain Research* 195: 431–442.

Lee, G. H. 2019. "Social Exclusion, Raising Companion Animals, and Psychological Well-Being: An Exploratory Study." *Science of Emotion & Sensibility* 22: 3–14.

Lee, J.-Y., C.-H. Kim, J.-E. Lee, S. Kwon, H.-H. Park, H.-T. Yim, et al. 2018a. "Papermaking Characteristics of Kudzu Root Fibers with Different Morphology." *Journal of Korea TAPPI* 50: 22–30. http://www.ktappi.kr/xml/16713/16713.pdf

Lee, N. P., and E. R. Arriola. 1999. "Poison Ivy, Oak, and Sumac Dermatitis." *Western Journal of Medicine* 171: 354–355.

Lee, S., M. Pitesa, M. Pillutla, and S. Thau. 2015. "When Beauty Helps and When It Hurts: An Organizational Context Model of Attractiveness Discrimination in Selection Decisions." *Organizational Behavior and Human Decision Processes* 128: 15–28.

Lee, Y.-T., X. Chen, Y. Zhao, and W. Chen. 2018b. "The Quest for Today's Totemic Psychology: A New Look at Wundt, Freud and Other Scientists." *Journal of Pacific Rim Psychology* 12, https://doi.org/10.1017/prp.2018.13

Lees, A. C., S. Attwood, J. Barlow, and B. Phalan. 2020. "Biodiversity Scientists Must Fight the Creeping Rise of Extinction Denial." *Nature Ecology & Evolution* 4: 1440–1443.

Lefevre, C. E., G. J. Lewis, D. I. Perrett, and L. Penke. 2013. "Telling Facial Metrics: Facial Width Is Associated with Testosterone Levels in Men." *Evolution and Human Behavior* 34: 273–279.

Lefevre, C. E., V. A. D. Wilson, F. B. Morton, S. F. Brosnan, A. Paukner, and T. C. Bates. 2014. "Facial Width-to-Height Ratio Relates to Alpha Status and Assertive Personality in Capuchin Monkeys. *PLoS One* 9(4): e93369. https://doi.org/10.1371/journal.pone.0093369

Leftwich, P. T., M. Bolton, and T. Chapman. 2016. "Evolutionary Biology and Genetic Techniques for Insect Control." *Evolutionary Applications* 9: 212–230.

Legagneux, P., N. Casajus, K. Cazelles, C. Chevallier, M. Chevrinais, L. Guéry, et al. 2018. "Our House Is Burning: Discrepancy in Climate Change vs. Biodiversity Coverage in the Media as Compared to Scientific Literature." *Frontiers in Ecology and Evolution* 5. https://doi.org/10.3389/fevo.2017.00175.

Lehmann, S. 2010. *The Principles of Green Urbanism: Transforming the City for Sustainability*. London, Earthscan.

Lehner, E., and J. Lehner. 1960. *Folklore and Symbolism of Flowers, Plants and Trees*. New York, Tudor.

Leicht-Young, S. A., and N. B. Pavlovic. 2014. "Lianas as Invasive Species in North America." In *Ecology of Lianas*, edited by S. A. Schnitzer, F. Bongers, R. J. Burnham, and F. E. Putz, 427–422. Hoboke, NJ, John Wiley & Sons.

Lembke, J. 2001. *Despicable Species: On Cowbirds, Kudzu, Hornworms, and Other Scourges*. Guilford, CT, Lyons Press.
Lennox, S., R. Mulaudzi, M. Potgieter, and L. Erasmus. 2012. "Not All Invasive Species Are Equal! Communities in Limpopo Rely on the Prickly Pear." *Veld & Flora* 98: 70. https://hdl.handle.net/10520/EJC121431
Leo, N. P., and S. C. Barker. 2005. "Unravelling the Evolution of the Head Lice and Body Lice of Humans." *Parasitology Research* 98: 44–47.
Leroi-Gouram, A. 1982. *The Dawn of European Art – An Introduction to Palaeolithic Cave Painting*. Cambridge, Cambridge University Press.
Leveau, L. M. 2020. "Artificial Light at Night (ALAN) Is the Main Driver of Nocturnal Feral Pigeon (*Columba livia* f. *domestica*) Foraging in Urban Areas." *Animals* 10(4): 554. https://doi.org/10.3390/ani10040554
Lewis, G. J., C. E. Lefevre, and T. C.Bates. 2012. "Facial Width-to-Height Ratio Predicts Achievement Drive in US Presidents." *Personality and Individual Differences* 52: 855–857.
Li, S., F. Hauser, S. Skadborg, S. Nielsen, N. Kirketerp, Cornelis, and C. Grimmelikhuijzen. 2016. "Adipokinetic Hormones and Their G protein-coupled Receptors Emerged in Lophotrochozoa." *Scientific Reports* 6: 32789, https://doi.org/10.1038/srep32789
Li, S., Zhu, S., Jia, Q., Yuan, D., Ren, C., Li, K., et al. 2018. "The Genomic and Functional Landscapes of Developmental Plasticity in the American Cockroach." *Nature Communications* 9(1008): https://www.nature.com/articles/s41467-018-03281-1
Li, Z., Q. Dong, T. P. Albright, and Q. Guo. 2011. "Natural and Human Dimensions of a Quasi-wild Species: The Case of Kudzu." *Biological Invasions* 13: 2167–2179.
Libell, M. 2014. "Seeing Animals. Anthropomorphism between Fact and Function." In *Exploring the Animal Turn. Human-Animal Relations in Science, Society and Culture*, edited by E. A. Cederholm, A. Björck, K. Jennbert, and A.-S. Lönngren, 141–153. Lund, Pufendorf Institute for Advanced Studies.
Lidicker, W. Z. Jr. 2007. "Issues in Rodent Conservation." In *Rodent Societies: An Ecological and Evolutionary Perspective*, edited by J. O. Wolff and P. W. Sherman, 453–462. Chicago, IL, University of Chicago Press.
Lieberman, D. E., D. M. Bramble, D. A. Raichlen, and J. J. Shea. 2009. "Brains, Brawn, and the Evolution of Human Endurance Running Capabilities." In *The First Humans – Origin and Early Evolution of the Genus* Homo. *Vertebrate Paleobiology and Paleoanthropology*, edited by F. E. Grine, J. G. Fleagle, and R. E. Leakey, 77–92. Dordrecht, Springer.
Lieberman, D. E., H. Pontzer, E. Cutright-Smith, and D. A. Raichlen. 2005. "Why Is the Human Gluteus so Maximus?" *American Journal of Physical Anthropology* 126, S40: 138, https://dash.harvard.edu/handle/1/2797436
Liles, M., M. Peterson, K. Stevenson, and M. Peterson. 2021. "Youth Wildlife Preferences and Species-Based Conservation Priorities in a Low-Income Biodiversity Hotspot Region." *Environmental Conservation* 48: 110–117.
Lilywhite, H. B. 2014. *How Snakes Work: Structure, Function and Behavior of the World's Snakes*. New York, Oxford University Press.
Lim, T. K. 2012. "*Opuntia ficus-indica*." In *Edible Medicinal and Non-Medicinal Plants*, edited by T. K. Lim, 660–682. Dordrecht, Springer.
Lim, T. K. 2013. "*Papaver somniferum*." In *Edible Medicinal and Non-Medicinal Plants, Vol. 5, Fruits*, edited by T. K. Lim. 202–217. Dordrecht, Springer.
Lim, T. K. 2016a. "*Cyperus rotundus*." In *Edible Medicinal and Non-Medicinal Plants, Volume 10, Modified Stems, Roots, Bulbs*, edited by T. K. Lim, 178–208. New York, Springer.
Lim, T. K. 2016b. "*Pueraria montana* var. *lobata*." In *Edible Medicinal and Non-Medicinal Plants, Volume 10, Modified Stems, Roots, Bulbs*, edited by T. K. Lim, 482–540. Dordrecht, Springer.
Lin, B. B., S. M. Philpott, and S. Jha. 2015. "The Future of Urban Agriculture and Biodiversity-Ecosystem Services: Challenges and Next Steps." *Basic and Applied Ecology* 16: 189–201.
Lin, M.-W., L.-Y. Chen, and Y.-K. Chuah. 2017. "Investigation of a Potted Plant (*Hedera helix*) with Photo-Regulation to Remove Volatile Formaldehyde for Improving Indoor Air Quality." *Aerosol and Air Quality Research* 17: 2543–2554.
Lindemann-Matthies, P. 2005. "'Loveable' Mammals and 'Lifeless' Plants: How Children's Interest in Common Local Organisms Can Be Enhanced through Observation of Nature." *International Journal of Science Education* 27: 655–677.
Lindenmayer, D. B., S. Cunningham, and A. Young. Eds. 2012. *Land Use Intensification: Effects on Agriculture, Biodiversity and Ecological Processes*. Boca Raton, FL, CRC Press.
Lindenmayer, D. B., A. D. Manning, P. L. Smith, H. P. Possingham, J. Fischer, I. Oliver, et al. 2002. "The Focal-species Approach and Landscape Restoration: A Critique." *Conservation Biology* 15: 338–345.
Lindgren, C. J., K. L. Castro, H. A. Coiner, R. E. Nurse, and S. J. Darbyshire. 2013. "The Biology of Invasive Alien Plants in Canada. 12. *Pueraria montana* var. *lobata* (Willd.) Sanjappa & Predeep." *Canadian Journal of Plant Science* 93: 71–95.
Lindman, C. A. M. 1901–1926. *Bilder ur Nordens Flora*. [Pictures from the Flora of the North.] Stockholm, Wahlstrom & Wildstrand.
Linzey, A. 2009. *Why Animal Suffering Matters: Philosophy, Theology, and Practical Ethics*. Oxford, Oxford University Press.
Liss, C., K. Litwak, D. Tilford, and V. Reinhardt. Eds. 2015. *Guide for Laboratory Animal Care Personnel on the Humane Housing and Handling of Individual Animal Species in Research Facilities*. Tenth Edition. Washington, DC, Animal Welfare Institute. https://awionline.org/sites/default/files/publication/digital_download/AWI-ComfortableQuarters-2015.pdf
Litman, E. A., S. L. Gortmaker, C. B. Ebbeling, and D. S. Ludwig. 2018. "Source of Bias in Sugar-sweetened Beverage Research: A Systematic Review." *Public Health Nutrition* 21: 2345–2350.
Littin, K. E., D. J. Mellor, B. Warburton, and C. T. Eason. 2004. "Animal Welfare and Ethical Issues Relevant to the Humane Control of Vertebrate Pests." *New Zealand Veterinary Journal* 52: 1–10.
Liu, D., F. Wang, and W. Li. 2011, "Cloning and Expression of Spider Dragline Silk Protein Gene in *Escherichia coli* and Eukaryotic Cells." *Genomics and Applied Biology* 30: 16–20.
Liu, S., B. Liu, and X.-Y. Zhu. 2013. "Corrections of Wrongly Spelled Scientific Names in Flora Reipublicae Popularis Sinicae." *Journal of Systematics and Evolution* 51: 231–234.
Lobmaier, J. S. R. Sprengelmeyer, B. Wiffen, and D. I. Perrett. 2010. "Female and Male Responses to Cuteness, Age and Emotion in Infant Faces." *Evolution and Human Behavior* 31: 16–21.
LoBue, V., and J. S. DeLoache. 2008. "Detecting the Snake in the Grass: Attention to Fear-Relevant Stimuli by Adults and Young Children." *Psychological Science* 19: 284–289.

LoBue, V., and D. H. Rakison. 2013. "What We Fear Most: A Developmental Advantage for Threat-Relevant Stimuli." *Developmental Review* 33: 285–303.

Lockwood, J. A. 2004. *Locust: The Devastating Rise and Mysterious Disappearance of the Insect that Shaped the American Frontier*. New York, Basic Books.

Lockwood, J. A. 2013. *The Infested Mind: Why Humans Fear, Loathe, and Love Insects*. Oxford, Oxford University Press.

Loeb, J. 2020. "Keeping Dangerous Pets." *The Veterinary Record* 186: 333.

Logan, C. J., S. Avin, N. Boogert, A. Buskell, F. R. Cross, A. Currie, et al. 2018. "Beyond Brain Size: Uncovering the Neural Correlates of Behavioral and Cognitive Specialization." *Comparative Cognition and Behavior Reviews* 13: 55–90.

Loike, J. D. 2015. "When Does a Smart Mouse Become Human?: Ethical Issues Attend the Creation of Animal-Human Chimeras." *Scientist (Philadelphia, PA)* 29(7): 25–25.

Lomer, C. J., R. P. Bateman, D. L. Johnson, J. Langewald, and M. B. Thomas. 2001. "Biological Control of Locusts and Grasshoppers." *Annual Review of Entomology* 46: 667–702.

Long, D., N. H. Ballentine, and J. G. Marks, Jr. 1997. "Treatment of Poison Ivy/Oak Allergic Contact Dermatitis with an Extract of Jewelweed." *American Journal of Contact Dermatitis* 8: 150–153.

Longbottom, J., F. M. Shearer, M. Devine, G. Alcoba, F. Chappuis, D. J. Weiss, et al. 2018. "Vulnerability to Snakebite Envenoming: A Global Mapping of Hotspots." *The Lancet* 392: 673–684.

López-Baucells, A., R. Rocha, and Á. Fernández-Llamazares. 2018. "When Bats Go Viral: Negative Framings in Virological Research Imperil Bat Conservation?" *Mammal Review* 48: 62–66.

Lorenz, K. 1943. "Die angeborenen Formen möglicher Erfahrung. [The Innate Conditions of the Possibility of Experience.]" *Zeitschrift für Tierpsychologie* 5: 235–409. https://doi.org/10.1111/j.1439-0310.1943.tb00655.x

Lorenz, K. 1971. *Studies in Animal and Human Behaviour, Vol. 2*. (Translated from the 1950 original by R. Martin). Cambridge, MA, Harvard University Press.

Lorimer, J. 2006. "Nonhuman Charisma: Which Species Trigger Our Emotions and Why?" *Ecos (British Association of Nature Conservationists)* 27(1): 20–27.

Losada, H. E., J. E. Vieyra, L. Luna, J. Cortés, and J. M. Vargas. 2017. "Economic Indicators, Capacity of the Ecosystem of Prickly Pear Cactus (*Opuntia megacantha*) and Environmental Services in Teotihuacan, México to Supply Urban Consumption." *Journal of Agriculture and Environmental Sciences* 6: 85–91.

Louda, S. M., and T. A. Rand. 2003. "Native Thistles: Expendable or Integral to Ecosystem Resistance to Invasion." In *The Importance of Species*, edited by P. Kareiva and S. Levin, 5–15. Princeton, NJ, Princeton University Press.

Louhaichi, M., S. Hassan, and G. Liguori. 2020. *Manual: Cactus Pear Agronomic Practices*. International Center for Agricultural Research in the Dry Areas (ICARDA). https://www.researchgate.net/profile/Mounir_Louhaichi/publication/339747118_Manual_Cactus_Pear_Agronomic_Practices/links/5e625e83a6fdcc37dd07befd/Manual-Cactus-Pear-Agronomic-Practices.pdf

Louis-Charles, M. A. 2017. "Turf Lawns vs. Permaculture." https://static1.squarespace.com/static/59b5f9a8d2b85729bbc525bc/t/5a99cc0de2c48334e31b3f42/1520028698161/Turf+Lawns+vs+Permaculture.pdf

Louv, R. 2005. *Last Child in the Woods: Saving Our Children from Nature-Deficit Disorder*. New York, Algonquin Books.

Lowe, S, M. Browne, S. Boudjelas, and M. De Poorter. 2000. "100 of the World's Worst Invasive Alien Species: A Selection from the Global Invasive Species Database." Auckland, NZ, IUCN Species Survival Commission of the World Conservation Union, Invasive Species Specialist Group. https://portals.iucn.org/library/sites/library/files/documents/2000-126.pdf

Lowney, K. S. 2002. "Friend or Foe? Changing Cultural Definitions of Kudzu." In *Pueraria: The genus Pueraria*, edited by W. M. Keung, 273–286. New York, Taylor & Francis.

Lowney, K. S., and J. Best. 1998. "Floral Entrepreneurs: Kudzu as Agricultural Solution and Ecological Problem." *Sociological Spectrum* 18: 93–114.

Lu, H, K. A. McComas, D. E. Buttke, S. Roh, and D. J. Wild. 2016. "A One Health Message about Bats Increases Intentions to Follow Public Health Guidance on Bat Rabies." *PLoS One* 11(5): e0156205. https://doi.org/10.1371/journal.pone.0156205

Lu, H., K. A. McComas, D. E. Buttke, S. Roh, M. A. Wild, and D. J. Decker. 2017. "One Health Messaging about Bats and Rabies: How Framing of Risks, Benefits and Attributions Can Support Public Health and Wildlife Conservation Goals." *Wildlife Research* 44: 200–206.

Lu, T., B. Yang, R. Wang, and C. Qin. 2020. Xenotransplantation: "Current Status in Preclinical Research." *Frontiers in Immunology*. https://doi.org/10.3389/fimmu.2019.03060

Lu, Z-H., L. Ma, and Q.-X. Gou. 2001. "Concepts of Keystone Species and Species Importance in Ecology." *Journal of Forest Research* 12: 250–252.

Lück, M. 2007. "Captive Marine Wildlife: Benefits and Costs of Aquaria and Marine Parks." In *Marine Wildlife and Tourism Management: Insights from the Natural and Social Sciences*, edited by J. E. S. Higham and M Lück, 130–144. Wallingford, CABI.

Lute, M. L., and N. H. Carter. 2020. "Are We Coexisting with Carnivores in the American West?" *Frontiers in Ecology and Evolution* 8: 48. https://doi.org/10.3389/fevo.2020.00048

Lutsenko, Y., W. Bylka, I. Matławska, and R. Darmohray. 2010. "*Hedera helix* as a Medicinal Plant." *Herba Polonica* 56: 83–96.

Lutsenko, Y., W. Bylka, I. Matławska, O. Gavrylyuk, M. Dworacka, A. Krawczyk, et al. 2017. "Anti-inflammatory, Antimicrobial Activity and Influence on the Lungs and Bronchus of *Hedera helix* Leaves Extracts." *Acta Poloniae Pharmaceutica-Drug Research* 74: 1159–1166.

Lydekker, R. 1895. *The New Natural History*. New York, Merrill and Baker.

Lyons, J. A., and D. J. D. Natusch. 2011. "Wildlife Laundering through Breeding Farms: Illegal Harvest, Population Declines and a Means of Regulating the Trade of Green Pythons (*Morelia viridis*) from Indonesia." *Biological Conservation* 144: 3073–3081.

Lyons, P. C., K. Okuda, M. T. Hamilton, T. G. Hinton, and J. C. Beasley. 2020. "Rewilding of Fukushima's Human Evacuation Zone." *Frontiers in Ecology and the Environment* 18: 127–134.

Mabey, R. 2010. *Weeds: In Defense of Nature's Most Unloved Plants*. New York, Harper Collins.

Macadam, C. R., and S. Z. Bairner. 2012. "Urban Biodiversity: Successes and Challenges: Brownfields: Oases of Urban Biodiversity." *Glasgow Naturalist* 25: 4. http://www.gnhs.org.uk/urban_bio/macadam.pdf

Macdonald, E. A., A. Hinks, D. J. Weiss, A. Dickman, D. Burnham, C. J. Sandom, et al. 2017. "Identifying Ambassador Species for Conservation Marketing." *Global Ecology and Conservation* 12: 204–214.

MacFarlane, D., and R. Rocha. 2020. "Guidelines for Communicating about Bats to Prevent Persecution in the Time of COVID-19." *Biological Conservation* 248:108650. https://doi.org/10.1016/j.biocon.2020.108650

Machineni, L. 2020. "Lignocellulosic Biofuel Production: Review of Alternatives." *Biomass Conversion and Biorefinery* 10: 779–791.

Macholán, M., S. J. E. Baird, P. Munclinger, and J. Piálek. Eds. 2012. *Evolution of the House Mouse*. Cambridge, Cambridge University Press.

MacKay, J., and Invasive Species Specialist Group. 2010. "CABI Invasive Compendium: House Mouse (*Mus musculus*)." https://www.cabi.org/isc/datasheet/35218

MacKay, J. W. B., J. C. Russell, and E. C. Murphy. 2007. "Eradicating House Mice from Islands: Successes, Failures and the Way Forward." In *Managing Vertebrate Invasive Species: Proceedings of an International Symposium*, edited by G. W. Witmer, W. C. Pitt, and K. A. Fagerstone, 294–304. Fort Collins, CO, National Wildlife Research Center. http://www.aphis.usda.gov/wildlife_damage/nwrc/symposia/invasive_symposium/content/MacKay294_304_MVIS.pdf

MacQueen, K. 2015. "The Plant that's Eating B.C." *McClean's (Magazine)*, 215, June 12. https://www.macleans.ca/society/science/the-plant-thats-eating-b-c/

Madsen, B., N. Carroll, D. Kandy, and G. Bennett. 2011. *State of Biodiversity Markets Report: Offset and Compensation Programs Worldwide*. Washington, DC, Ecosystem Marketplace.

Maeng, A., and P. Aggarwal. 2016. "Dominant Designs: The role of Product Face-Ratios and Anthropomorphism on Consumer Preferences." In *The Psychology of Design: Creating Consumer Appeal*, edited by R. Batra, C. Seifert, and D. Brei, 133–148. Boca Raton, FL, Routledge/Taylor & Francis.

Magnarelli, L. A. 2009. "Global Importance of Ticks and Associated Infectious Disease Agents." *Clinical Microbiology Newsletter* 31: 33–37.

Mahlke, S. 2014. *Das Machtverhältnis Zwischen Mensch und Tier im Kontext Sprachlicher Distanzierungsmechanismen: Anthropozentrismus, Speziesismus und Karnismus in der Kritischen Diskursanalyse. [The Power Relationship between Humans and Animals in the Context of Linguistic Distancing Mechanisms: Anthropocentrism, Speciesism and Carnism in Critical Discourse Analysis.]* Hamburg, Germany, Diplomica Verlag.

Mainguy, J., E. E. Rees, P. Canac-Marquis, D. Bélanger, C. Fehlner-Gardiner, and G. Séguin. 2012. "Oral Rabies Vaccination of Raccoons and Striped Skunks with ONRAB® Baits: Multiple Factors Influence Field Immunogenicity." *Journal of Wildlife Diseases* 48: 979–990.

Maji, A. K., S. Pandit, P. Banerji, and D. Banerjee. 2014. "*Pueraria tuberosa*: A Review on Its Phytochemical and Therapeutic Potential." *Natural Product Research* 28: 2111–2127.

Makra, L., I. Matyasovszky, L. Hufnagel, and G. Tusnády. 2015. "The History of Ragweed in the World." *Applied Ecology and Environmental Research* 13: 489–512.

Mal, T. K., J. Lovett-Doust, L. Lovett-Doust, and G. A. Mulligan. 1992. "The Biology of Canadian Weeds. 100. *Lythrum salicaria*." *Canadian Journal of Plant Science* 72: 1305–1330.

Mal, T. K., and L. Narine. 2005. "The Biology of Canadian Weeds. 129. *Phragmites australis* (Cav.) Trin. ex Steud." *Canadian Journal of Plant Science* 84: 365–396.

Małecki, W., P. Sorokowski, B. Pawłowski, and M. Cieński. 2020. *Human Minds and Animal Stories: How Narratives Make Us Care about Other Species*. Milton Park, Routledge.

Malik, A. 2007. "Environmental Challenge vis a vis Opportunity: The Case of Water Hyacinth." *Environment International* 33: 122–138.

Malik, A., N. Singh, and S. Satya. 2007. "House Fly (*Musca domestica*): A Review of Control Strategies for a Challenging Pest." *Journal of Environmental Science and Health, Part B* 42: 453–469.

Malloch, D. 1995. "Fungi with Heteroxenous Life Histories." *Canadian Journal of Botany* 73(Suppl. 1): S1334–S1342.

Maloney, D., F. A. Drummond, and R. Alford. 2003. *Spider Predation in Agroecosystems: Can Spiders Effectively Control Pest Populations?* Maine Agricultural and Forest Experiment Station, Technical Bulletin 190. Orono, University of Maine. https://digitalcommons.library.umaine.edu/cgi/viewcontent.cgi?referer=https://www.google.com/&httpsredir=1&article=1018&context=aes_techbulletin

MaMing, R., L. Lee, X. Yang, and P. Buzzard. 2016. "Vultures and Sky Burials on the Qinghai-Tibet Plateau." *Vulture News* 71: 22–35.

Mandery, E. J. 2005. *Capital Punishment: A Balanced Examination*. Sudbury, MA, Jones and Bartlett.

Manfredo, M. J., E. G. Urquiza-Haas, A. W. Don Carlos, J. T. Bruskotter, and A. M. Dietsch. 2020. "How Anthropomorphism Is Changing the Social Context of Modern Wildlife Conservation." *Biological Conservation* 241: 108297. https://doi.org/10.1016/j.biocon.2019.108297

Mang, T., F. Essl, D. Moser, and S. Dullinger. 2018. "Climate Warming Drives Invasion History of *Ambrosia artemisiifolia* in Central Europe." *Preslia* 90: 59–81.

Mann, H. 1981. "Common Dandelion (*Taraxacum officinale*) Control with 2,4-D and Mechanical Treatments." *Weed Science* 29: 704–708.

Mann, H., and P. B. Cavers. 1979. "The Regenerative Capacity of Root Cuttings of *Taraxacum officinale* under Natural Conditions." *Canadian Journal of Botany* 57: 1783–1791.

Mans, B. J., J. Featherston, M. Kvas, K. A. Pillay, D. G. de Klerk, R. Pienaar, et al. 2019. "Argasid and Ixodid Systematics: Implications for Soft Tick Evolution and Systematics, with a New Argasid Species List." *Ticks and Tick-borne Diseases* 10: 219–240.

Marc, P., A. Canard, and F. Ysnel. 1999. "Spiders (Araneae) Useful for Pest Limitation and Bioindication." *Agriculture, Ecosystems & Environment* 74: 229–273.

Marija, S., and G. Slavica. 2019. "Herbicides in Surface Water Bodies – Behaviour, Effects on Aquatic Organisms and Risk Assessment." *Pesticidi i Fitomedicina* 34: 157–172.

Marin, C., M.-D. Palomeque, F. Marco-Jiménez, and S. Vega. 2014. "Wild Griffon Vultures (*Gyps fulvus*) as a Source of *Salmonella* and *Campylobacter* in Eastern Spain." *PLoS One* 9(4): e94191. https://doi.org/10.1371/journal.pone.0094191

Marinelli, J. 2005. "Charismatic Plants: Endangered Plants Need Poster Children Too." *Landscape Architecture* 58(60): 62–63.

Markandya, A., T. Taylor, A. Longo, M. N. Murty, S. Murty, and K. Dhavala. 2008. "Counting the Cost of Vulture Decline – An Appraisal of the Human Health and Other Benefits of Vultures in India." *Ecological Economics* 67: 194–204.

Marks, G. S. 2019. "Glow-in-the-Dark Vampire Bats Could Help Curtail Rabies." *American Scientist* 107(3): 133. https://doi.org/10.1511/2019.107.3.133.

Marks, M., B. Lapin, and J. Randall. 2014. "*Phragmites australis.*" *Bugwood Wiki* https://wiki.bugwood.org/Phragmites_australis

Markwell, K. 2021. "Why Do We Love Koalas So Much? Because They Look Like Baby Humans." *The Conversation (online newsletter).* https://theconversation.com/why-do-we-love-koalas-so-much-because-they-look-like-baby-humans-153619

Maron M., R. J. Hobbs, A. Moilanen, J. W. Matthews, K. Christie, T. A. Gardner, et al. 2012. "Faustian Bargains? Restoration Realities in the Context of Biodiversity Offset Policies." *Biological Conservation* 155: 141–148

Marques, A., I. S. Martins, T. Kastner, C. Plutzar, M. C. Theurl, N. Eisenmenger, et al. 2019. "Increasing Impacts of Land Use on Biodiversity and Carbon Sequestration Driven by Population and Economic Growth." *Nature Ecology & Evolution* 3: 628–637.

Marten, G. C., and R. N. Andersen. 1975. "Forage Nutritive Value and Palatability of 12 Common Annual Weeds." *Crop Science* 15: 821–827.

Martin, J.-L., V. Maris, and D. S. Simberloff. 2016. "The Need to Respect Nature and Its Limits Challenges Society and Conservation Science." *Proceedings of the National Academy of Sciences of the United States* 113(22): 6105–6112.

Martin, J. S., N. Staes, A. Weiss, J. M. G. Stevens, and A. V. Jaeggi. 2019. "Facial Width-to-Height Ratio Is Associated with Agonistic and Affiliative Dominance in Bonobos (*Pan paniscus*)." *Biology Letters* 15: 8. https://doi.org/10.1098/rsbl.2019.0232

Martinez, J. 2015. "Debate over Origin of Long Necks in Giraffes (*Giraffa camelopardalis*)." *Eukaryon (Lake Forest College Review)*, March 11. https://www.lakeforest.edu/live/files/martinezdebatepdf.pdf

Martinez, M., P. Poirrier, R. Chamy, D. Prüfer, C. Schulze-Gronover, L. Jorquera, et al. 2015. "*Taraxacum officinale* and Related Species – An Ethnopharmacological Review and Its Potential as a Commercial Medicinal Plant." *Journal of Ethnopharmacology* 169: 244–262.

Martín-López, B., C. Montes, and J. Benayas. 2007. "The Non-economic Motives behind the Willingness to Pay for Biodiversity Conservation." *Biological Conservation* 139: 67–82.

Marty, P. R., M. A. van Noordwijk, M. Heistermann, E. P. Willems, L. P. Dunkel, M. Cadilek, M. Agil, et al. 2015. "Endocrinological Correlates of Male Bimaturism in Wild Bornean Orangutans." *American Journal of Endocrinology* 77: 1170–1178.

Marushia, R. G., M. W. Cadotte, and J. S. Holt. 2010 "Phenology as a Basis for Management of Exotic Annual Plants in Desert Invasions." *Journal of Applied Ecology* 47: 1290–1299.

Marusik, Y. M., and S. Koponen. 2000. "Circumpolar Diversity of Spiders: Implications for Research, Conservation and Management." *Annales Zoologici Fennici* 37: 265–269.

Marzilli, A. 2008. *Capital Punishment – Point-Counterpoint.* Second Edition. Broomall, PA, Chelsea House.

Masanotti, G. M., E. Abbafati, E. Petrella, S. Vinciguerra, and F. Stracci. 2019. "Intensive Tobacco Cultivations, a Possible Public Health Risk?" *Environmental Science and Pollution Research* 26: 12616–12621.

Mason, A. 2021. "What's Wrong with Everyday Lookism?" *Politics, Philosophy & Economics.* https://doi.org/10.1177/1470594X20982051

Matavel, A., G. Estrada, and F. De Marco Almeida. 2016. "Spider Venom and Drug Discovery: A Review." In *Spider Venoms, Toxicology*, edited by P. Gopalakrishnakone, G. A. Corzo, E. Diego-Garcia, and M. E. de Lima, 273–292. Dordrecht, Springer.

Mathäs, A. 2014. "Defining the Human and the Animal." *Konturen* 6: 1–16. http://journals.oregondigital.org/index.php/konturen/article/view/3538/3242

Mather, J. A. 2019. "Ethics and Care: For Animals, Not Just Mammals." *Animals* 9(12): 1018. https://doi.org/10.3390/ani9121018; https://www.ncbi.nlm.nih.gov/pmc/articles/PMC6941085/

Matthews, C. E., and E. Cummo. 1999. "All Wrapped Up in Kudzu & Other Ecological Disasters." *American Biology Teacher* 61: 42–46.

Mattison, C. 2007. *The New Encyclopedia of Snakes.* Princeton, NJ, Princeton University Press.

Maun, M. A., and S. C. H. Barrett. 1986. "The Biology of Canadian Weeds. 77. *Echinochloa crus-galli* (L.) Beauv." *Canadian Journal of Plant Science* 66: 739–759.

Maurel, N., S. Salmon, J.-F. Ponge, N. Machon, J. Moret, and A. Muratet. 2010. "Does the Invasive Species *Reynoutria japonica* Have an Impact on Soil and Flora in Urban Wastelands?" *Biological Invasions* 12: 1709–1719.

Maxmen, A. and S. Mallapaty. 2021. "The COVID Lab-Leak Hypothesis: What Scientists Do and Don't Know." *Nature News Explainer* (blog), June 8. https://www.nature.com/articles/d41586-021-01529-3

Maxwell, R., and T. Miller. 2013. "How to Rally Liberals and Conservatives to Environmentalism: The Art and Psychology of Green Persuasion." *Psychology Today.* https://www.psychologytoday.com/ca/blog/greening-the-media/201304/how-rally-liberals-and-conservatives-environmentalism

May, T. 2005. "Looking after the Bad Guys: The Conservation of Pathogenic Fungi." *Australasian Plant Conservation: Journal of the Australian Network for Plant Conservation* 13(4): 20–21.

May, R.M. 1976. "Coexistence with Insect Pests." *Nature* 264: 211–212.

Maynard, D. S., T. W. Crowther, J. R. King, R. J. Warren, and M. A. Bradford. 2015. "Temperate Forest Termites: Ecology, Biogeography, and Ecosystem Impacts." *Ecological Entomology* 40: 199–210.

McAllister, H., and R. Marshall. 2017. *Hedera: The Complete Guide.* London, Royal Horticultural Society.

McClanahan, T. R., and P. S. Rankin. 2016. "Geography of Conservation Spending, Biodiversity, and Culture." *Conservation Biology* 30: 1089–1101.

McCleery, R. A., C. E. Moorman, and M. N. Peterson. Eds. 2014. *Urban Wildlife Conservation: Theory and Practice.* New York, Springer.

McDonald, D. K. 1999. "Ecologically Sound Lawn Care for the Pacific Northwest: Findings from the Scientific Literature and Recommendations from Turf Professionals." Seattle, Seattle Public Utilities. http://citeseerx.ist.psu.edu/viewdoc/download;jsessionid=2218263A8270D20BB6B2C334C9C02A97?doi=10.1.1.170.3838&rep=rep1&type=pdf

McFarlane, G. R., C. B. A. Whitelaw, and S. G. Lillico. 2018. "CRISPR-Based Gene Drives for Pest Control." *Trends in Biotechnology* 36: 130–133.

McGeoch, M. A., H. Sithole, M. J. Samways, J. P. Simaika, J. S. Pryke, M. Picker, et al. 2011. "Conservation and Monitoring of Invertebrates in Terrestrial Protected Areas." *Koedoe* 53(2): Art. #1000. https://doi.org/10.4102/koedoe.v53i2.1000

McGovern, T. W., S. R. LaWarre, and C. Brunette. 2000. "Is It, or Isn't It? Poison Ivy Look-a-Likes." *American Journal of Contact Dermatitis* 11: 104–110.

McGowan, J., L. J. Beaumont, R. J. Smith, A. L. M. Chauvenet, R. Harcourt, S. C. Atkinson, et al. 2020. "Conservation Prioritization Can Resolve the Flagship Species Conundrum." *Nature Communications* 11: 994. https://doi.org/10.1038/s41467-020-14554-z

McKee, R., and J. L. Stephens. 1948. *Kudzu as a Farm Crop*. Revised Edition. Washington, DC, U.S. Department of Agriculture.

McKim, R. Ed. 2020. *Laudato Si' and the Environment. Pope Francis' Green Encyclical*. London, Routledge.

McKinney, M. L. 2002. "Urbanization, Biodiversity, and Conservation: The Impacts of Urbanization on Native Species Are Poorly Studied, but Educating a Highly Urbanized Human Population about These Impacts Can Greatly Improve Species Conservation in All Ecosystems." *BioScience* 52: 883–890.

McQueeney, C. 2018. "January's Weed of the Month: English Ivy." https://weedwise.conservationdistrict.org/tag/hedera-canariensis

McWilliams, J. 2017. "The Ethics of Humane Animal Agriculture." In *The Routledge Handbook of Food Ethics*, edited by M. C. Rawlinson and C. Ward, 243–252. London, Routledge.

Mdweshu, L., and A. Maroyi. 2020. "Local Perceptions about Utilization of Invasive Alien Species *Opuntia ficus-indica* in Three Local Municipalities in the Eastern Cape Province, South Africa." *Biodiversitas* 21: 1653–1659.

Measey, J., A. Basson, A. D. Rebelo, A. L. Nunes, G. Vimercati, M. Louw, et al. 2019. "Why Have a Pet Amphibian? Insights from YouTube." *Frontiers in Ecology and Evolution* 7: 52. https://doi.org/10.3389/fevo.2019.00052

Meehan, C. J., E. J. Olson, M. W. Reudink, T. K. Kyser, and R. L. Curry. 2009. "Herbivory in a Spider through Exploitation of an Ant–Plant Mutualism." *Current Biology* 19: R892–R893.

Mehrtens, J. M. 1987. *Living Snakes of the World in Color*. New York, Sterling.

Melchiori, K. J., and R. K. Mallett. 2015. "Using Shrek to Teach About Stigma." *Teaching of Psychology* 42: 260–265.

Melzer, B., R. Seidel, T. Steinbrecher, and T. Speck. 2012. "Structure, Attachment Properties, and Ecological Importance of the Attachment System of English Ivy (*Hedera helix*)." *Journal of Experimental Botany* 63: 191–201.

Meng, X., S. Lu, J. Yang, D. Jin, X. Wang, X. Bai, et al. 2017. "Metataxonomics Reveal Vultures as a Reservoir for *Clostridium perfringens*." *Emerging Microbes & Infections* 6(2): e9. https://doi.org/10.1038/emi.2016.137

Merlin, M. D. 1984. *On the Trail of the Ancient Opium Poppy*. London, Associated University Presses.

Mero, H. 1996. "The Weedy Nutsedges (*Cyperus* spp.) Bibliography." http://agron-www.agron.iastate.edu/~weeds/WeedBiolLibrary/u4nutsg2.html

Metcalfe, D. J. 2005. "Biological Flora of the British Isles No 268 *Hedera helix* L." *Journal of Ecology* 93: 632–648.

Metzl, J. 2019. *Hacking Darwin: Genetic Engineering and the Future of Humanity*. Naperville, IL, Sourcebooks.

Meyer-Rochow, V. B. 2001. "Light of My Life – Messages in the Dark." *Biologist* 48: 163–167.

Meyer-Rochow, V. B. 2004. "Traditional Food Insects and Spiders in Several Ethnic Groups of Northeast India, Papua New Guinea, Australia, and New Zealand." In *Ecological Implications of Minilivestock: Rodents, Frogs, Snails, and Insects for Sustainable Development*, edited by M. G. Paoletti, 385–409, USA, Enfield, NH, Science Publications.

Meyer-Rochow, V. B. 2009 "Food Taboos: Their Origins and Purposes." *Journal of Ethnobiology and Ethnomedicine* 5(18). https://doi.org/10.1186/1746-4269-5-18

Meyerson, L. A., A. M. Lambert, and K. Saltonstall. 2010. "A Tale of Three Lineages: Expansion of Common Reed (*Phragmites australis*) in the U.S. Southwest and Gulf Coast." *Invasive Plant Science and Management* 3: 515–520.

Meyerson, L. A., D. Viola, and R. Brown. 2009. "Hybridization of Invasive *Phragmites australis* with a Native Subspecies in North America." *Biological Invasions* 12: 103–111.

Mia, M. M., M. Hasan, and M. R. Hasnath. 2022. "Global Prevalence of Zoonotic Pathogens from Pigeon Birds: A Systematic Review and Meta-Analysis." *Heliyon* 8(6): e09732. https://doi.org/10.1016/j.heliyon.2022.e09732

Michael, P. W. 2003. "*Echinochloa* P. Beauv." In *Flora of North America North of Mexico, Volume 25*, edited by M. E. Barkworth, K. M. Capels, S. Long, and M. B. Piepet, 390–403. New York, Oxford University Press.

Michel, G. 2010. "When Less Means More: Impact of Myostatin in Animal Breeding." *Immunology, Endocrine & Metabolic Agents in Medicinal Chemistry* 10: 240–248.

Mickleburgh, S., K. Waylen, and P. Racey. 2009. "Bats as Bushmeat: A Global Review." *Oryx* 43: 217–234.

Midway, S. R, T. Wagner, and G. H. Burgess. 2019. "Trends in Global Shark Attacks." *PLoS One* 14(2): e0211049. https://doi.org/10.1371/journal.pone.0211049

Migiro, G. 2018. "How Many Cats Are There in the World?" *WorldAtlas*. https://www.worldatlas.com/articles/how-many-cats-are-there-in-the-world.html

Mihalca, A. D., C. M. Gherman, and V. Cozma. 2011. "Coendangered Hard-Ticks: Threatened or Threatening?" *Parasites & Vectors* 4: 71. https://doi.org/10.1186/1756-3305-4-71

Mikulyuk, A., E. Kujawa, M. E. Nault, S. Van Egeren, K. I. Wagner, M. Barton, et al. 2020. "Is the Cure Worse Than the Disease? Comparing the Ecological Effects of an Invasive Aquatic Plant and the Herbicide Treatments Used to Control it." *Facets*. https://doi.org/10.1139/facets-2020-0002

Milakovic, I., and G. Karrer. 2016. "The Influence of Mowing Regime on the Soil Seed Bank of the Invasive Plant *Ambrosia artemisiifolia* L." *NeoBiota* 28: 39–49.

Mildenstein, T., I. Tanshi, and P. A. Racey. 2016. "Exploitation of Bats for Bushmeat and Medicine." In *Bats in the Anthropocene: Conservation of Bats in a Changing World*, edited by C. C. Voigt and T. Kingston, 325–375. New York, Springer International.

Milesi, C., S. W. Running, C. D. Elvidge, J. B. Dietz, B. T. Tuttle, and R. R. Nemani. 2005. "Mapping and Modeling the Biogeochemical Cycling of Turf Grasses in the United States." *Environmental Management* 36: 426–438.

Milke, J., M. Gałczyńska, and J. Wróbel. 2020. "The Importance of Biological and Ecological Properties of *Phragmites australis* (Cav.) Trin. ex Steud., in Phytoremediation of Aquatic Ecosystems – The Review." *Water* 12(6). https://www.mdpi.com/2073-4441/12/6/1770

Miller, A. 2015. *Skunk*. London, Reaktion Books.

Miller, A., D. A. Young, and J. Wen. 2001. "Phylogeny and Biogeography of *Rhus* (Anacardiaceae) Based on ITS Sequence Data." *International Journal of Plant Sciences* 162: 1401–1407.

Miller, G. F. 1998. "How Mate Choice Shaped Human Nature: A Review of Sexual Selection and Human Evolution." In *Handbook of Evolutionary Psychology: Ideas, Issues, and Applications*, edited by C. B. Crawford and D. L. Krebbs, 87–129. Mahwah, NJ, Lawrence Erlbaum Associates Publishers.

Miller, H., M. Ward, and J. A. Beatty. 2019. "Population Characteristics of Cats Adopted from an Urban Cat Shelter and the Influence of Physical Traits and Reason for Surrender on Length of Stay." *Animals (Basel)* 9(11): 940. https://doi.org/10.3390/ani9110940

Miller, J. H. 1996. "Kudzu Eradication and Management." In *Kudzu the Vine to Love or Hate*, edited by D. Hoots and J. Baldwin, 137–149. Kodak, TN, Suntop Press. https://www.fs.usda.gov/treesearch/pubs/985

Miller, J. H., and B. Edwards, B. 1983. "Kudzu, *Pueraria lobata* Where Did It Come from and How Can We Stop It?" *Southern Journal of Applied Forestry* 7: 165–169.

Miller, R. L., H. Marsh, A. Cottrell, and M. Hamann. 2018. "Protecting Migratory Species in the Australian Marine Environment: A Cross-Jurisdictional Analysis of Policy and Management Plans." *Frontiers in Marine Science*. https://doi.org/10.3389/fmars.2018.00229

Millspaugh, C. F. 1892. *Medicinal plants, Volume 1*. Philadelphia, PA, John C. Yorston & Co.

Minter, D. 2010. "Fungi: The Orphans of Rio. Fighting for the Future of Fungi." *Species (Magazine of the Species Survival Commission)* 52: 5–7.

Miralles, A., M. Raymond, and G. Lecointre. 2019." Empathy and Compassion toward Other Species Decrease with Evolutionary Divergence Time." *Scientific Reports* 9: 19555. https://doi.org/10.1038/s41598-019-56006-9

Mirhoseini, S. Z., and J. Zare. 2012. "The Role of Myostatin on Growth and Carcass Traits and Its Application in Animal Breeding" *Life Science Journal* 9: 2353–2357.

Mishra, A. K., K. K. Jain, and K. L. Garg. 1995. "Role of Higher Plants in the Deterioration of Historic Buildings." *Science of the Total Environment* 167: 375–392.

Mishra, S., and A. Maiti. 2017. "The Efficiency of *Eichhornia crassipes* in the Removal of Organic and Inorganic Pollutants from Wastewater: A Review." *Environmental Science and Pollution Research* 24: 7921–7937.

Misra, V., A. K. Shrivastava, S. P. Shukla, and M. I. Ansari. 2016. "Effect of Sugar Intake towards Human Health." *Saudi Journal of Medicine* 1: 29–36.

Mitchell, B. R., M. M. Jaeger, and R. H. Barrett. 2004. "Coyote Depredation Management: Current Methods and Research Needs." *Wildlife Society Bulletin* 32: 1209–1218.

Mitchell, J. C. 1975. "Contact Allergy from Plants." In *Recent Advances in Phytochemistry*, edited by V. C. Runeckles, 119–138. Boston, MA, Springer.

Mitchell, W. A., and C. O. Martin. 1986. "Chufa (*Cyperus esculentus*): Section 7.4.1, U.S. Army Corps of Engineers Wildlife Resources Management Manual." Springfield, VA, U.S. Army Engineer Waterways Experiment Station. https://apps.dtic.mil/dtic/tr/fulltext/u2/a171202.pdf

Mitich, L. W. 1988a. "Intriguing World of Weeds: Crabgrass." *Weed Technology* 2: 114–115.

Mitich, L. W. 1988b. "Thistles I: *Cirsium* and *Carduus*." *Weed Technology* 2: 228–229.

Mitich, L. W. 1988c. "Common Lambsquarters." *Weed Technology* 2: 550–552.

Mitich, L. W. 1989a. "Common Dandelion - The Lion's Tooth." *Weed Technology* 3: 537–539.

Mitich, L. W. 1989b. "Bermudagrass." *Weed Technology* 3: 433–435.

Mitich, L. W. 1990. "Intriguing World of Weeds: Barnyardgrass." *Weed Technology* 4: 918– 920. http://wssa.net/wssa/weed/intriguing-world-of-weeds/#x

Mitich, L. W. 1999. "Purple Loosestrife, *Lythrum salicaria* L." *Weed Technology* 13: 843–846.

Mitich, L. W. 2000. "Kudzu [*Pueraria lobata* (Willd.) Ohwi]." *Weed Technology* 14: 231–235.

Mitsuhashi, J. 2016. *Edible Insects of the World*. Boca Raton, FL, CRC Press.

Mizik, T., and G. Gyarmati. 2021. "Economic and Sustainability of Biodiesel Production – A Systematic Literature Review." *Clean Technologies* 3: 19–36.

Modesto, S. P, and J. S. Anderson. 2004. "The Phylogenetic Definition of Reptilia." *Systematic Botany* 53: 815–821.

Modlinska, K., and W. Pisula. 2020. "The Norway Rat, from an Obnoxious Pest to a Laboratory Pet." *eLife* 9:e50651. https://doi.org/10.7554/eLife.50651

Modragon-Jacobo, C. 2001. "Cactus Pear Domestication and Breeding." *Plant Breeding Reviews* 20: 135–166.

Mohan, J. E., L. H. Ziska, W. H. Schlesinger, R. B. Thomas, R. C. Sicher, K. George, et al. 2006. "Biomass and Toxicity Responses of Poison Ivy (*Toxicodendron radicans*) to Elevated Atmospheric CO_2." *Proceedings of the National Academy of Sciences* 103: 9086–9089.

Mohan, J. E., L. H. Ziska, R. B. Thomas, R. C. Sischer, K. George, J. S. Clark, et al. 2008. "Biomass and Toxicity Responses of Poison Ivy (*Toxicodendron radicans*) to Elevated Atmospheric CO_2: Reply." *Ecology* 89: 585–587.

Mojoodi, M., G. Alizadeh, and M. Ebrahimian. 2020. "Bioterrorism and Biological Software Warfare: A Classic Review Study from the Past to the Present." *War Studies* 1: 107–130.

Mokossesse, J. A., G. Josens, J. Mboukoulida, and J. F. Ledent. 2012. "Effect and Field Application of *Cubitermes* (Isoptera, Termitidae) Mound Soil on Growth and Yield of Maize in Central African Republic." *Agronomie Africaine* 24: 241–252.

Molin, W., R. Kronfol, J. Ray, B. Scheffler, and C. Bryson. 2019. "Genetic Diversity among Geographically Separated *Cyperus rotundus* Accessions Based on RAPD Markers and Morphological Characteristics." *American Journal of Plant Sciences* 10: 2034–2046.

Moll, R. J., J. D. Cepek, P. D. Lorch, P. M. Dennis, E. Tans, T. Robison, et al. 2019. "What Does Urbanization Actually Mean? A Framework for Urban Metrics in Wildlife Research." *Journal of Applied Ecology* 56: 1289–1300.

Momigliana, P., and R. Hartcourt. 2014. "Shark Conservation, Governance and Management. The Science-Law Disconnect." In *Sharks, Conservation and Management*, edited by E. J. Techera and N. Klein, 89–106. London, Routledge.

Monnard, C. R., and E. K. Grasser. 2018. "Perspective: Cardiovascular Responses to Sugar-Sweetened Beverages in Humans: A Narrative Review with Potential Hemodynamic Mechanisms." *Advances in Nutrition* 9: 70–77.

Montagnani, C., R. Gentili, M. Smith, M. F. Guarino, and S. Citterio. 2017. "The Worldwide Spread, Success, and Impact of Ragweed (*Ambrosia* spp.)." *Critical Reviews in Plant Sciences* 36: 139–178.

Montford, S., and E. Small. 1999a. "Measuring Harm and Benefit: The Biodiversity Friendliness of *Cannabis sativa*." *Global Biodiversity* 8(4): 2–13.

Montford, S., and E. Small. 1999b. "A Comparison of the Biodiversity Friendliness of Crops with Special Reference to Hemp (*Cannabis sativa*)." *Journal of the International Hemp Association* 6: 53–63.

Montoya, J. E., T. M. Waliczek, and M. L. Abbott. 2013. "Large Scale Composting as a Means of Managing Water Hyacinth (*Eichhornia crassipes*)." *Invasive Plant Science and Management* 6: 243–249.

Mookerjee, S., Y. Cornil, and J. Hoegg. 2021. "Express: From Waste to Taste: How 'Ugly' Labels Can Increase Purchase of Unattractive Produce." *Journal of Marketing*. https://journals.sagepub.com/doi/pdf/10.1177/0022242920988656

Moore, D., M. M. Nauta, S. E. Evans, and M. Rotheroe. Eds. 2001. *Fungal Conservation: Issues and Solutions*. Cambridge, UK, Cambridge University Press.

Moore, R. J. 1975. "The Biology of Canadian Weeds: 13. *Cirsium arvense* (L.) Scop." *Canadian Journal of Plant Science* 55: 1033–1048.

Mora, C., and P. Sale. 2011. "Ongoing Global Biodiversity Loss and the Need to Move Beyond Protected Areas: A Review of the Technical and Practical Shortcoming of Protected Areas on Land and Sea." *Marine Ecology Progress Series* 434: 251–266.

Morand, S. 2015. "(Macro-) Evolutionary Ecology of Parasite Diversity: From Determinants of Parasite Species Richness to Host Diversification." *International Journal for Parasitology: Parasites and Wildlife* 4: 80–87.

Moratelli, R., and C. H. Calisher. 2015. "Bats and Zoonotic Viruses: Can We Confidently Link Bats with Emerging Deadly Viruses?" *Memórias do Instituto Oswaldo Cruz* 110: 1–22.

Mordaka, J., and R. Gentle. 2003. "The Biomechanics of Gender Difference and Whiplash Injury: Designing Safer Car Seats for Women." *Acta Polytechnica* 43: 47–54.

Moro, D., M. Byrne, M. Kennedy, S. Campbell, and M. Tizard. 2018. "Identifying Knowledge Gaps for Gene Drive Research to Control Invasive Animal Species: The Next CRISPR Step." *Global Ecology and Conservation* 13: e00363. https://doi.org/10.1016/j.gecco.2017.e00363

Morreall, J. 1991. "Cuteness." *British Journal of Aesthetics* 31: 39–47.

Morris, A. L., E. J. Livengood, and F. A. Chapman. 2010. "Sharks for the Aquarium and Considerations for Their Selection." University of Florida Institute of Food and Agricultural Sciences Publication FA 179. http://citeseerx.ist.psu.edu/viewdoc/download?doi=10.1.1.652.7804&rep=rep1&type=pdf

Morris, D. 1969. *The Naked Ape*. New York, Random House.

Morrison, C., C. Simpkins, J. G. Castley, and R. C. Buckley. 2012. "Tourism and the Conservation of Critically Endangered Frogs." *PLoS One* 7(9): e43757. https://doi.org/10.1371/journal.pone.0043757

Mosaheb, M. U. F. Z., N. A. Khan, and R. Siddiqui. 2018. "Cockroaches, Locusts, and Envenomating Arthropods: A Promising Source of Antimicrobials." *Iranian Journal of Basic Medical Sciences* 21: 873–877.

Mosquito Reviews. 2022. Why Do Mosquitoes Buzz? They're Actually Singing Mating Songs for Each Other. *MosquitoReviews.com* https://mosquitoreviews.com/learn/mosquitoes-buzzing-noise/

Moss, A., and M. Esson. 2010. "Visitor Interest in Zoo Animals and the Implications for Collection Planning and Zoo Education Programmes." *Zoo Biology* 29: 715–731.

Mousseau T. A., and A. P. Møller. 2016. "The Animals of Chernobyl and Fukushima." In *Genetics, Evolution and Radiation*, edited by V. Korogodina, C. Mothersill, S. Inge-Vechtomov, and C. Seymour, 251–266. Cham, Springer.

Mouttet, R. A. Benno, M. Bonini, B. Chauvel, N. Desneux, E. Gachet, et al. 2018. "Estimating Economic Benefits of Biological Control of *Ambrosia artemisiifolia* by *Ophraella communa* in Southeastern France." *Basic and Applied Ecology* 33: 14–24.

Mueller, G. M., A. Dahlberg, and M. Krikorev. 2014. "Bringing Fungi into the Conservation: The Global Fungal Red List Initiative." *Fungal Conservation* 4: 12–16.

Mukhopadhyay, S. B. 2019. "Oyster Mushroom Cultivation on Water Hyacinth Biomass: Assessment of Yield Performances, Nutrient, and Toxic Element Contents of Mushrooms." In *An Introduction to Mushroom*, edited by A. K. Passari and S Sánchez. IntechOpen, https://doi.org/10.5772/intechopen.90290; https://www.intechopen.com/books/an-introduction-to-mushroom/oyster-mushroom-cultivation-on-water-hyacinth-biomass-assessment-of-yield-performances-nutrient-and-

Mullen, G. R., and R. S. Vetter. 2018. "Chapter 25: Spiders (Araneae)." In *Medical and Veterinary Entomology*, Third Edition, edited by G. R. Mullen and L. A. Durden, 507–531. Cambridge, Academic Press.

Mullen, G. R., and J. M. Zaspel. 2018. "Chapter 21: Moths and Butterflies (Lepidoptera)." In *Medical and Veterinary Entomology*, Third Edition, edited by G. R. Mullen and L. A. Durden, 439–458. Cambridge, Academic Press.

Muller, M. N., and R. Wrangham. 2002 "Sexual Mimicry in Hyenas." *Quarterly Review of Biology* 77(1): 3–16.

Mulligan, G. A., and B. E. Junkins. 1976. "The Biology of Canadian Weeds. 17. *Cyperus esculentus* L." *Canadian Journal of Plant Science* 56: 339–350.

Mulligan, G. A., and B. E. Junkins. 1977. "The Biology of Canadian Weeds. 23. *Rhus radicans* L." *Canadian Journal of Plant Science* 57: 515–523.

Mullin, S. J., and R. A. Seigel. Eds. 2009. *Snakes: Ecology and Conservation*. Ithaca, NY, Cornell University Press.

Mumcuoglu, K. Y. 2008. "The Louse Comb: Past and Present." *American Entomologist* 2008(Fall): 164–168.

Munawar, A., S. A. Ali, A. Akrem, and C. Betzel. 2018. "Snake Venom Peptides: Tools of Biodiscovery." *Toxins* 10(11): 474. https://doi.org/10.3390/toxins10110474

Munger, G T. 2002a. "*Pueraria montana* var. *lobata*." *Fire Effects Information System* [Online]. U.S. Department of Agriculture, Forest Service, Rocky Mountain Research Station, Fire Sciences Laboratory (Producer). https://www.fs.fed.us/database/feis/plants/vine/puemonl/all.html

Munger, G. T. 2002b. "*Lythrum salicaria*." *Fire Effects Information System* [Online]. U.S. Department of Agriculture, Forest Service, Rocky Mountain Research Station, Fire Sciences Laboratory (Producer). https://www.fs.usda.gov/database/feis/plants/vine/puemonl/all.html

Mankad, A., E.V. Hobman, and L. Carter. 2021. "Genetically Engineering Coral for Conservation: Psychological Correlates of Public Acceptability." *Frontiers in Marine Science* 8: 710641. https://doi.org/10.3389/fmars.2021.710641

Muñoz-Reyes, J. A., M. Iglesias-Julios, M. Pita, and E. Turiegano. 2015. "Facial Features: What Women Perceive as Attractive and What Men Consider Attractive." *PLoS One* 10(7): e0132979. https://doi.org/10.1371/journal.pone.0132979

Murai, M. 1999. *Understanding the Invasion of Pacific Northwest Forests by English Ivy (*Hedera *spp., Araliaceae).* Seattle, University of Washington.

Murdoch, W. W., F.-I. Chu, A. Stewart-Oaten, and M. Q. Wilber. 2018. "Improving Wellbeing and Reducing Future World Population." *Plos One* 13(9): e0202851. https://doi.org/10.1371/journal.pone.0202851

Murphy, J. B., and B. Gratwicke. 2017. "History of Captive Management and Conservation Amphibian Programs Mostly in Zoos and Aquariums. Part I – Anurans." *Herpetological Review* 48: 241–260.

Murray, M. G., M. J. B. Green, G. C. Bunting, and J. R. Paine. 1996. *Biodiversity Conservation in the Tropics: Gaps in Habitat Protection and Funding Priorities.* Cambridge, UK, U.N. Environment Program, World Conservation Monitoring Centre. https://citeseerx.ist.psu.edu/viewdoc/download?doi=10.1.1.583.6590&rep=rep1&type=pdf

Muscat, A., E. M. De Olde, I. J. M. de Boer, and R. Ripoll-Bosch. 2020. "The Battle for Biomass: A Systematic Review of Food-Feed-Fuel Competition." *Global Food Security* 25: 100330. https://doi.org/10.1016/j.gfs.2019.100330

Musser, G. G., and M. D. Carleton. 2005. "Family Muridae." In *Mammal Species of the World: A Taxonomic and Geographic Reference*, edited by D. E. Wilson and D. M. Reeder, 894–1531. Baltimore, MD, The John Hopkins University Press.

Myers, N., R. A. Mittermeier, C. G. Mittermeier, G. A. B. da Fonseca, and J. Kent. 2000. "Biodiversity Hotspots for Conservation Priorities." *Nature* 403: 853–858.

Nadir, S., H.-B. Xiong, Q. Zhu, X.-L. Zhang, H.-Y. XU, J. Li, et al. 2017. "Weedy Rice in Sustainable Rice Production. A Review." *Agronomy for Sustainable Development* 37: 46. https://doi.org/10.1007/s13593-017-0456-4

Nagar, I. 2018. "The Unfair Selection: A Study on Skin-Color Bias in Arranged Indian Marriages." *Sage Open* 8(2). https://doi.org/10.1177/2158244018773149

Nai, A., and E. Toros. 2020. "The Peculiar Personality of Strongmen: Comparing the Big Five and Dark Triad Traits of Autocrats and Non-autocrats." *Political Research Exchange* 2: 1. https://doi.org/10.1080/2474736X.2019.1707697

Naish, D. 2012. "Dinosauroids revisited." *Scientific American Blog Network* https://blogs.scientificamerican.com/tetrapod-zoology/dinosauroids-revisited-revisited/

Nakajima, S., M. Yamamoto, and N. Yoshimoto. 2009. "Dogs Look Like Their Owners: Replications with Racially Homogenous Owner Portraits." *Anthrozoös* 22: 173–181.

Narladkar, B. W. 2018. "Projected Economic Losses Due to Vector and Vector-Borne Parasitic Diseases in Livestock of India and Its Significance in Implementing the Concept of Integrated Practices for Vector Management." *Veterinary World* 11: 151–160.

National Academies of Sciences, Engineering, and Medicine. 2016. *Gene Drives on the Horizon: Advancing Science, Navigating Uncertainty, and Aligning Research with Public Values.* Washington, DC, The National Academies Press. https://doi.org/10.17226/23405

Naylor, E. 1941. "The Proliferation of Dandelion from Roots." *Bulletin of the Torrey Botanical Club* 68: 351–358.

Naylor, G., and N. Aschliman. 2013. "How Many Species of Living Sharks, Skates and Rays Are There, and How Did They Arise over the Course of Evolution?" *Proceeding of the International Symposium – Reproduction of Marine Life, Birth of New Life! Investigating the Mysteries of Reproduction*, February 21–22, 2009, Okinawa, edited by K. Sato, 5–12. Okinawa, Japan, Okinawa Churashima Research Center. https://churashima.okinawa/userfiles/files/page/ocrc/5-12_naylor_and_aschliman_edit.pdf

Nazarko, O. M., R. C. Van Acker, and M. H. Entz. 2005. "Strategies and Tactics for Herbicide Use Reduction in Field Crops in Canada: A Review." *Canadian Journal of Plant Science* 85: 457–479.

Ndava, J., S. D. Llera, and P. Manyanga. 2018. "The Future of Mosquito Control: The Role of Spiders as Biological Control Agents: A Review." *International Journal of Mosquito Research* 5: 6–11.

Neamtu, G., C. Tabacaru, and C. Sociaciu. 1992. "Phytochemical Research on Higher Plants. V. The Content of Carotenoids, Chlorophylls, Nutrients and Mineral Elements in *Taraxacum officinale* L. (Dandelion)." *Buletinul Universitatii de stiinte Cluj-Napoca. Seria Agricultura si Horticultura* 46: 93–99.

Neff, C., and R. Hueter. 2013. "Science, Policy, and the Public Discourse of Shark 'Attack': A Proposal for Reclassifying Human–Shark Interactions." *Journal of Environmental Studies and Sciences* 3: 65–73.

Nefzaoui, A., and P. Inglese. Eds. 2002. *Fourth International Congress on Cactus Pear and Cochineal.* Wageningen, The Netherlands: International Society for Horticultural Science.

Nefzaoui, A., P. Inglese, and T. Belay. Eds. 2010. *Improved Utilization of Cactus Pear for Food, Feed, Soil and Water Conservation and Other Products in Africa. Proceedings of International Workshop*, Mekelle (Ethiopia), 19–21 October, 2009. Cactus Newsletter Special Issue 12. https://www.researchgate.net/publication/282204744_Improved_utilization_of_cactus_pear_for_food_feed_soil_and_water_conservation_and_other_products_in_Africa

Negbi, M. 1992. "A Sweetmeat Plant, a Perfume Plant and Their Weedy Relatives: A Chapter in the History of *Cyperus esculentus* L. and *C. rotundus* L." *Economic Botany* 46: 64–71.

Negi, C. S. 2005. "Religion and Biodiversity Conservation: Not a Mere Analogy." *International Journal of Biodiversity Science and Management* 1: 85–96.

Nelson, E. W. 1918. *Wild Animals of North America.* Washington, DC, The National Geographical Society.

Nelson, M. M. 1960. "The Classroom as a Wedge in Removing Fear of Our Common Snakes." *The American Biology Teacher* 22: 143–145.

Németh, E., and J. Bernáth. 2009. "Selection of Poppy (*Papaver somniferum* L.) Cultivars for Culinary Purposes." *Acta Horticulturae* 826: 413–420.

Nestor, A., and M. J. Tarr. 2008. "Gender Recognition of Human Faces Using Color." *Psychological Science* 19: 1241–1246.

New, T. R. 1999. "Untangling the Web: Spiders and the Challenges of Invertebrate Conservation." *Journal of Insect Conservation* 3: 251–256.

New, T. R. 2004. "A Special Issue on Moths and Conservation." *Journal of Insect Conservation* 8: 77. https://doi.org/10.1007/s10841-004-1327-2.

New, T. R. 2007. "Broadening Benefits to Insects from Wider Conservation Agendas." In *Insect Conservation Biology: Proceedings of the Royal Entomological Society 23rd Symposium*, edited by A. J. A. Stewart, T. R. New, and O. T. Lewis, 301–348, Oxfordshire, CABI.

New, T. R., and A. L. Yen. 1995. "Ecological Importance and Invertebrate Conservation." *Oryx* 29: 187–191.

Newman, D., and Global Invasive Species Team. 2014. "*Cynodon dactylon.*" The Nature Conservancy, Center for Invasive Species and Ecosystem Health at the University of Georgia. Cynodon dactylon - Bugwoodwiki

Nezerith, C., and C. S. Wareham. 2020. "Ethical Considerations in Xenotransplantation: A Review." *Current Opinion in Organ Transplantation* 25: 483–488.

Nicholson, W. L., D. E. Sonenshine, B. H. Noden, and R. N. Brown. 2018. "Chapter 27: Ticks (Ixodida)." In *Medical and Veterinary Entomology*, Third Edition, edited by G. R. Mullen and L. A. Durden, 604–672. Cambridge, Academic Press.

Nico, D., and R. J. Cooper. 2009. "Impacts of Free-Ranging Domestic Cats (*Felis catus*) on Birds in the United States: A Review of Recent Research with Conservation and Management Recommendations." In *Proceedings of the Fourth International Partners in Flight Conference: Tundra to Tropics*, 205–219. Partners in Flight (publisher). https://www.researchgate.net/publication/281579799_Impacts_of_free-ranging_domestic_cats_Felis_catus_on_birds_in_the_United_States_A_review_of_recent_research_with_conservation_and_management_recommendations

Nie, Z.-L., H. Sun, Y. Meng, and J. Wen. 2009. "Phylogenetic analysis of *Toxicodendron* Anacardiaceae) and Its Biogeographic Implications on the Evolution of North Temperate and Tropical Intercontinental Disjunctions." *Journal of Systematics and Evolution* 47: 416–430.

Nield, C. T. 2016. "In Defence of the Bed Bug." *The Conversation*. https://theconversation.com/in-defence-of-the-bed-bug-54218

Nielsen, J., R. B. Hedeholm, J. Heinemeier, P. G. Bushnell, J. S. Christiansen, J. Olsen, et al. 2016. "Eye Lens Radiocarbon reveals Centuries of Longevity in the Greenland Shark (*Somniosus microcephalus*)." *Science* 353: 702–704.

Niemiera, A. X., and B. Von Holle. 2008. "Invasive Plant Species and the Ornamental Horticulture Industry." In *Management of Invasive Weeds*, edited by Inderjit, 167–188. Berlin, Springer.

Nijhawan, S., I. Mitapo, J. Pulu, C. Carbone, and J. M. Rowcliffe. 2019. "Does Polymorphism Make Asiatic Golden Cat the Most Adaptable Predator in Eastern Himalayas?" *Ecology* 100: 1–4.

Nijwala, D., and A. K. Sandhu. 2021. "Vertical Farming – An Approach to Sustainable Agriculture." *International Journal for Research in Applied Science & Engineering Technology* 9: 145–149.

Nikolić, L., D. Džigurski, and B. Ljevnaić-Mašić. 2014. "Nutrient Removal by *Phragmites australis* (Cav.) Trin. ex Steud. in the Constructed Wetland System." *Contemporary Problems of Ecology* 7: 449–454.

Nimal, C. 2014. "Living with Weeds – A New Paradigm." *Indian Journal of Plant Science* 46: 96–110.

Niroula, G. and N. B. Singh. 2015. "Religion and Conservation: A Review of Use and Protection of Sacred Plants and Animals in Nepal." *Journal of Institute of Science and Technology* 20(2): 61–66.

Nita, M. 2016. *Praying and Campaigning with Environmental Christians*. New York, Palgrave Macmillan.

Nogueira-de-Sá, F., and J. R. Trigo. 2002. "Do Fecal Shields Provide Physical Protection to Larvae of the Tortoise Beetles *Plagiometriona flavescens* and *Stolas chalybea* against Natural Enemies?" *Entomologia Experimentalis et Applicata* 104: 203–206.

Nombre, C. A., G. Sampaio, L. S. Borma, J. C. Castilla-Rubio, J. S. Silva, and M. Cardoso. 2016. "Land-use and Climate Change Risks in the Amazon and the Need of a Novel Sustainable Development Paradigm." *Proceedings of the National Academy of Sciences of the United States of America* 113(39): 10759–10768.

Norn, S., P. R. Kruse, and E. Kruse. 2005. "History of Opium Poppy and Morphine." *Dansk Medicinhistorisk Arbog* 33: 171–184. (In Danish, English abstract.) https://europepmc.org/article/med/17152761

Norris, R. F. 1996. "Morphological and Phenological Variation in Barnyardgrass (*Echinochloa crus-galli*) in California." *Weed Science* 44: 804–814

North Carolina Government. n.d. *Common Wetland Plants of North Carolina*. Raleigh, North Carolina Department of Environment, Health and Natural Resources Division of Water Quality, https://files.nc.gov/ncdeq/Water%20Quality/Surface%20Water%20Protection/401/Policies_Guides_Manuals/Common%20Wetlands%20Plants%20of%20NC.pdf

Norwood, K. J. Ed. 2014. *New Directions in American History. Color Matters: Skin Tone Bias and the Myth of a Postracial America*. Milton, UK, Routledge/Taylor & Francis Group.

Norwood, K. J. 2015. "If You Is White, You's Alright… Stories about Colorism in America." *Global Studies Law Review (Washington University, St. Louis)* 585: https://openscholarship.wustl.edu/law_globalstudies/vol14/iss4/8

Noser, E., J. Schoch, and U. Ehlert. 2018. "The Influence of Income and Testosterone on the Validity of Facial Width-to-Height Ratio as a Biomarker for Dominance." *PLoS One* 13(11): e0207333. https://doi.org/10.1371/journal.pone.0207333

Novak, B. J. 2018. "De-Extinction." *Genes* 9(11): 548. https://doi.org/10.3390/genes9110548

Novoa, A., V. Flepu, and J. S. Boatwright. 2019. "Is Spinelessness a Stable Character in Cactus Pear Cultivars? Implications for Invasiveness." *Journal of Arid Environments* 160: 11–16.

Novoa, A., H. Kaplan, J. R. U. Wilson, and D. M. Richardson. 2016. "Resolving a Prickly Situation: Involving Stakeholders in Invasive Cactus Management in South Africa." *Environmental Management* 57: 998–1008.

Novoa, A., J. J. Le Roux, M. P. Robertson, J. R. Wilson, and D. M. Richardson. 2014. "Introduced and Invasive Cactus Species: A Global Review." *AoB Plants* 7: plu078. https://doi.org/10.1093/aobpla/plu078

Novotny, T. E., and F. Zhao. 1999. "Consumption and Production Waste: Another Externality of Tobacco Use." *Tobacco Control* 8: 75–80.

Ntiamoa-Baidu, Y. 2008. "Indigenous Beliefs and Biodiversity Conservation: The Effectiveness of Sacred Groves, Taboos and Totems in Ghana for Habitat and Species Conservation." *Journal for the Study of Religion, Nature & Culture* 2: 309–326.

Nyffeler, M., E. J. Olson, and W. O. C. Symondson. 2016. "Plant-Eating by Spiders." *Journal of Arachnology* 44: 15–27.

Nyman, U., and O. Hall. 1974. "Breeding Oil Poppy (*Papaver somniferum*) for Low Content of Morphine." *Hereditas* 76: 49–54.

O'Brien, K. J. 2010. *An Ethics of Biodiversity: Christianity, Ecology, and the Variety of Life*. Washington, DC, Georgetown University Press.

O'Bryhim, J. R., and E. C. M. Parsons. 2015. "Increased Knowledge about Sharks Increases Public Concern about Their Conservation." *Marine Policy* 56: 43–47.

O'Neill, O. 1997. "Environmental Values, Anthropocentrism and Speciesism." *Environmental Values* 6: 127–142.

O'Neill, R. V., and J. R. Kahn. 2000. "*Homo economus* as a Keystone Species." *BioScience* 50: 333–337.

O'Shea, M. 2018. *The Book of Snakes: A Life-Size Guide to Six Hundred Species from Around the World*. Chicago, IL, University of Chicago Press.

O'Shea, T. J., P. M. Cryan, D. T. S. Hayman, R. K. Plowright, and D. G. Streicker. 2016. "Multiple Mortality Events in Bats: A Global Review." *Mammal Review* 46: 175–190.

Oda, S. 2003. *Fifty Years of the Law of the Sea*. Leiden, Brill.

Odendaal, J. S. J., and R. A. Meintjes. 2003. "Neurophysiological Correlates of Affiliative Behaviour between Humans and Dogs." *Veterinary Journal* 165: 296–301.

Office of Laboratory Animal Welfare. 2015. *Public Health Service Policy on Humane Care and Use of Laboratory Animals*. Bethesda, MD, National Institutes of Health. https://olaw.nih.gov/policies-laws/phs-policy.htm

Ogada, D., P. Shaw, R. L. Beyers, R. Buij, C. Murn, J. M. Thiollay, et al. 2016. "Another Continental Vulture Crisis: Africa's Vultures Collapsing toward Extinction." *Conservation Letters* 9: 89–97.

Ogada, D. L., F. Keesing, and M. Virani. 2012. "Dropping Dead: Causes and Consequences of Vulture Population Declines Worldwide." *Annals of the New York Academy of Sciences* 1249: 57–71.

Ogutu, P. A. 2019. "Vermicomposting Water Hyacinth: Turning Fisherman's Nightmare into Farmer's Fortune." *International Journal of Research and Innovation in Applied Science* 4: 12–14.

Ohler, A., and V. Nicolas. 2017. "Which Frog's Legs Do Froggies Eat? The Use of DNA Barcoding for Identification of Deep Frozen Frog Legs (Dicroglossidae, Amphibia) Commercialized in France." *European Journal of Taxonomy* 271: 1–19.

Ohman, A. 2005. "Conditioned Fear of a Face: A Prelude to Ethnic Enmity?" *Science* 309(5735): 711–712.

Öhman, A. 2009. "Of snakes and Faces: An Evolutionary Perspective on the Psychology of Fear." *Scandinavian Journal of Psychology* 50: 543–552.

Ohman, A., A. Flykt, and F. Esteves. 2001a. "Emotion Drives Attention: Detecting the Snake in the Grass." *Journal of Experimental Psychology: General* 130: 466–478.

Ohman, A., D. Lundqvist, and F. Esteves. 2001b. "The Face in the Crowd Revisited: A Threat Advantage with Schematic Stimuli." *Journal of Personality and Social Psychology* 80: 381–396.

Ohman, A., and S. Mineka. 2001. "Fears, Phobias, and Preparedness: Toward an Evolved Module of Fear and Fear Learning." *Psychological Review* 108: 483–522.

Ohri, D. 2015. "The Taxonomic Riddle of *Chenopodium album* L. Complex (Amaranthaceae)." *The Nucleus* 58: 131–134.

Okerman, A. 2000. "Combating the 'Ivy Desert': The Invasion of *Hedera helix* (English Ivy) in the Pacific Northwest United States." St. Paul, Department of Horticultural Science, University of Minnesota. http://www.theconservationcompany.co.nz/pdf/Okerman2000.pdf

Oksanen M., and H. Siipi. Eds. 2014. *The Ethics of Animal Re-creation and Modification*. London, Palgrave Macmillan.

Oliveira, S. V., G. S. Gazeta, and R. Gurgel-Gonçalves. 2017. "Climate, Ticks, and Tick-Borne Diseases: Mini Review." *Vector Biology Journal* 2: 1–3.

Olivos-Jara, P., R. Segura-Fernández, C. Rubio-Pérez, and B. Felipe-García. 2020. "Biophilia and Biophobia as Emotional Attribution to Nature in Children of 5 Years Old." *Frontiers in Psychology* 11: 511. https://doi.org/10.3389/fpsyg.2020.00511

Ollerton J., R. Winfree, and S. Tarrant. 2011. "How Many Flowering Plants Are Pollinated by Animals?" *Oikos* 120: 321–326.

Omura, H. 2004. "Trees, Forests and Religion in Japan." *Mountain Research and Development* 24: 179–182.

Orians, G. H., and J. H. Heerwagen, J. H. 1992. "Evolved Responses to Landscape." In *The Adapted Mind: Evolutionary Psychology and the Generation of Culture*, edited by J. H. Barkow, L. Cosmides, and J. Tooby, 555–579. Oxford, Oxford University Press.

Orlando, R. 2018. *Weeds in the Urban Landscape: Where They Come from, Why They're Here, and How to Live with Them*. Berkeley, North Atlantic Books.

Orloff, N., J. Mangold, Z. Miller, and F. Menalleda. 2018. "A Meta-analysis of Field Bindweed (*Convolvulus arvensis* L.) and Canada Thistle (*Cirsium arvense* L.) Management in Organic Agricultural Systems." *Agriculture, Ecosystems & Environment* 254: 264–272.

Orlofske, S. A., R. C. Jadin, and P. T. J. Johnson. 2015. "It's a Predator–Eat–Parasite World: How Characteristics of Predator, Parasite and Environment Affect Consumption." *Oecologia* 178: 537–547.

Orsolini, L., M. Ciccarese, D. Papanti, D. De Berardis, A. Guirguis, J. M. Corkery et al. 2018. "Psychedelic Fauna for Psychonaut Hunters: A Mini-Review." *Frontiers in Psychiatry* 9: 153. https://doi.org/10.3389/fpsyt.2018.00153

Ortiz, A. M. D., C. L. Outhwaite, C. Dalin, T. Newbold. 2021. "A Review of the Interactions between Biodiversity, Agriculture, Climate Change, and International Trade: Research and Policy Priorities." *One Earth* 4: 88–101.

Oudejans, L. 2015. "Why Do We Hate Insects?" (Blog). https://leonoudejans.blogspot.com/2015/03/why-do-we-hate-insects.html

Oven, A. 2019. "If Hunting Is Just a Necessary Duty, Why Do People Enjoy It So Much?" (Blog). https://www.alice-animalwelfare.com/if-hunting-is-simply-a-necessary-moral-duty-why-do-people-enjoy-it-so-much/

Overmyer, J. P., R. Noblet, and K. L. Armbrust. 2005. "Impacts of Lawn-Care Pesticides on Aquatic Ecosystems in Relation to Property Value." *Environmental Pollution* 137: 263–272.

Oxford University Press. 2010. *New Oxford American Dictionary*. Third Edition. New York, Oxford University Press.

Oxlade, E. L., and P. E. Clifford. 1999. "The Versatile Dandelion." *Journal of Biological Education* 33: 125–129.

Pacelle, W. 2016. *The Humane Economy: How Innovators and Enlightened Consumers Are Transforming the Lives of Animals*. New York, William Morrow.

Packer, J. G., L. A. Meyerson, H. Skálová, P. Pyšek, and C. Kueffer. 2017. "Biological Flora of the British Isles: *Phragmites australis*." *Journal of Ecology* 105: 1123–1162.

Pădureţ, S., S. Amariei, G. Gutt, and B. Piscuc. 2016. "The Evaluation of Dandelion (*Taraxacum officinale*) Properties as a Valuable Food Ingredient." *Romanian Biotechnological Letters* 21: 11569–11575.

Pagès, M., Y. Chaval, V. Herbreteau, S. Waengsothorn, J.-F. Cosson, J.-P. Hugot, et al. 2010. "Revisiting the Taxonomy of the Rattini Tribe: A Phylogeny-based Delimitation of Species Boundaries." *BMC Evolutionary Biology* 10: 184. https://doi.org/10.1186/1471-2148-10-184

Pak, E. 2021. "Screening Science: The Curious Genetic Improbabilities of George, Giant Gorilla of 'Rampage.'" *Daily Bruin*, August 19. https://dailybruin.com/2018/04/11/screening-science-the-curious-genetic-improbabilities-of-george-giant-gorilla-of-rampage

Palevitch, D., and A. Levy. 1992. "Domestication of *Papaver bracteatum* as a Source of Thebaine." *Acta Horticulturae* 306: 33–52.

Palmer, B., and R. C. McFadyen. 2012. "*Ambrosia artemisiifolia* – Annual Ragweed." In *Biological Control of Weeds in Australia*, edited by M. H. Julien, R. E. McFadyen, and J. Cullen, 50–59. Melbourne, CSIRO Publishing.

Palmer, S. E., and K. B. Schloss. 2010. "An Ecological Valence Theory of Human Color Preference." *Proceedings of the National Academy of Sciences of the United States of America* 107(19): 8877–8882.

Pandey, P., H. N. Dakshinamurthy, and S. Young. 2020. "A Literature Review of Nonherbicide, Robotic Weeding: A Decade of Progress." Raleigh, North Carolina State University. https://cottoncultivated.cottoninc.com/wp-content/uploads/2020/06/Robotic-Weeding-LitReview-White_Paper_Pandey_Dakshinamurthy_Young_2020.pdf

Panlasigui, S., E. Spotswood, E. Beller, and R. Grossinger. 2021. "Biophilia Beyond the Building: Applying the Tools of Urban Biodiversity Planning to Create Biophilic Cities." *Sustainability* 13(5): 2450. https://doi.org/10.3390/su13052450

Pappert, R. A., J. L. Hamrick, and L. A. Donovan. 2000. "Genetic Variation in *Pueraria lobata* (Fabaceae), an Introduced, Clonal, Invasive Plant of the Southeastern United States." *American Journal of Botany* 87: 1240–245.

Paramita, W., and S. Winahjoe. 2014. "Analyzing Fundamental Factors of Indonesians' Skin Color Preference: A Qualitative Approach to Develop Research Framework." *International Journal of Business Anthropology* 5: 106–117.

Pareek, O. P., R. S. Singh, V. Nath, and B. B. Vashishtha. 2001. *The Prickly Pear (*Opuntia ficus-indica *L. Mill.).* Jodhpur, India, Agrobios.

Parenteau, P. A. 1998. "Rearranging the Deck Chairs: Endangered Species Act Reforms in an Era of Mass Extinction." *William & Mary Environmental Law and Policy Review* 22(2): 227–311. https://scholarship.law.wm.edu/cgi/viewcontent.cgi?article=1276&context=wmelpr&httpsredir=1&referer

Parenti, C. 2015. "Flower of War: An Environmental History of Opium Poppy in Afghanistan." *SAIS Review of International Affairs* 35(1): 183–200. https://muse.jhu.edu/article/582537

Park, A. 2019. "Kale Is One of the Most Contaminated Vegetables You Can Buy. Here's Why." *Time* (online). https://time.com/5554573/kale-dirty-dozen-list-pesticides/#:~:text=Strawberries%20top%20the%20list%2C%20followed, tomatoes%2C%20celery%20and%20potatoes

Park, S. M., K. J. Won, D. I. Hwang, D. Y. Kim, H. B. Kim, Y. Li, et al. 2020. "Potential Beneficial Effects of *Digitaria ciliaris* Flower Absolute on the Wound Healing-Linked Activities of Fibroblasts and Keratinocytes." *Planta Medica* 86: 348–355.

Parks Canada. 2020. "To the Rescue of the Mingan Thistle." https://www.pc.gc.ca/en/pn-np/qc/mingan/nature/conservation/restaurer-restoration/chardon-thistle/education

Parr, L. A., and B. M. Waller. 2006. "Understanding Chimpanzee Facial Expression: Insights into the Evolution of Communication." *Social Cognitive and Affective Neuroscience* 1: 221–228.

Parsley, K. M. 2020. "Plant Awareness Disparity: A Case for Renaming Plant Blindness." *Plants People Planet* 2: 598–601.

Parsons, K. C. 2003. "Reproductive Success of Wading Birds Using *Phragmites* Marsh and Upland Nesting Habitats." *Estuaries* 26: 596–601.

Parsons, M. H., P. B. Banks, M. A. Deutsch, R. F. Corrigan, and J. Munshi-South. 2017. "Trends in Urban Rat Ecology: A Framework to Define the Prevailing Knowledge Gaps and Incentives for Academia, Pest Management Professionals (PMPS) and Public Health Agencies to Participate." *Journal of Urban Ecology* 3(1): jux005. https://doi.org/10.1093/jue/jux005

Partap, T., B. D. Joshi, and N. W. Galwey. 1998. *Chenopods.* Chenopodium spp. *Promoting the Conservation and Use of Underutilized and Neglected Crops*. Rome, International Plant Genetics Research Institute.

Parthasarathy, N., K. Naveen Babu. 2019. "Sacred Groves: Potential for Biodiversity and Bioresource Management." In *Life on Land. Encyclopedia of the UN Sustainable Development Goals*, edited by W. Leal Filho, A. Azul, L. Brandli, P. Özuyar, and T. Wall. Cham, Springer. https://doi.org/10.1007/978-3-319-71065-5_10-1

Pascual, B., J. V. Maroto, S. LopezGalarza, A. Sanbautista, and J. Alagarda. 2000. "Chufa (*Cyperus esculentus* L. var. *sativus* Boeck.): An Unconventional Crop. Studies Related to Applications and Cultivation." *Economic Botany* 54: 439–448.

Pasiecznik, N. 2007. "*Pueraria montana* var. *lobata* (Kudzu)." *Crop Protection Compendium*. Wallingford, UK, CAB International. https://www.cabi.org/isc/datasheet/45903

Patel, S. 2012. "Threats, Management and Envisaged Utilizations of Aquatic Weed *Eichhornia crassipes*: An Overview." *Reviews in Environmental Science and Bio-Technology* 11: 249–259.

Patocka, J., Z. Navratilova, and M. Ovando-Martínez. 2017. "Biologically Active Compounds of Knotweed (*Reynoutria* spp.)." *Military Medical Science Letter* 86: 17–31. https://doi.org/10.31482/mmsl.2017.004

Paul, E., R. S. Sikes, S. J. Beaupre, and J. C. Wingfield. 2016. "Animal Welfare Policy: Implementation in the Context of Wildlife Research – Policy Review and Discussion of Fundamental Issues." *ILAR Journal* 56: 312–334.

Paul, J., N. Vachon, C. J. Garroway, and J. R. Freeland. 2010. "Molecular Data Provide Strong Evidence of Natural Hybridization between Native and Introduced Lineages of *Phragmites australis* in North America." *Biological Invasions* 12: 2967–2973.

Paulsen, E., L. P. Christensen, and K. E. Andersen. 2010. "Dermatitis from Common Ivy (*Hedera helix* L. subsp. *helix*) in Europe: Past, Present, and Future." *Contact Dermatitis* 62: 201–209.

Paulus, D. L., and K. M. Williams. 2002. "The Dark Triad of Personality: Narcissism, Machiavellianism, and Psychopathy." *Journal of Research in Personality* 36: 556–563.

Pawlowski, B., R. I. M. Dunbar, and A. Lipowicz. 2000. Evolutionary Fitness: Tall Men Have More Reproductive Success." *Nature* 403: 156.

Payne, W. W. 1964. "A Re-evaluation of the Genus *Ambrosia* (Compositae)." *Journal of the Arnold Arboretum* 45: 401–430.

Pazda, A. D., C. A. Thorstenson, A. J. Elliot, and D. I. Perrett. 2016. "Women's Facial Redness Increases Their Perceived Attractiveness: Mediation through Perceived Healthiness." *Perception* 45: 739–754.

Pearce, A. 2015. "Shark Attack: A Cultural Approach." *Anthropology Today* 31: 3–7.

Pearce, J. M. 2019. "Animal Intelligence." In *The Routledge Handbook of Animal Ethics*, edited by B. Fischer, 43–54. New York, Routledge.

Peek, S. F., and S. Buczinski. 2017. "Cardiovascular Diseases." In *Rebhun's Diseases of Dairy Cattle*, Third Edition, edited by S. Peek and T. J. Divers, 46–93. Amsterdam, Elsevier.

Peerzada, A. M. 2017. "Biology, Agricultural Impact, and Management of *Cyperus rotundus* L.: The World's Most Tenacious Weed." *Acta Physiologiae Plantarum* 39: 270. https://doi.org/10.1007/s11738-017-2574-7

Peng, L.-H., and L.-C. Chung. 2017. "A Study of Forming Taiwanese Aborigines' Totem and Conversional Products." In *2017 International Conference on Applied System Innovation (Sapporo, Japan)*, 397–400. IEE (online publisher).

Peng, W., R. Qin, X. Li, and H. Zhou. 2013. "Botany, Phytochemistry, Pharmacology, and Potential Application of *Polygonum cuspidatum* Sieb.et Zucc.: A Review." *Journal of Ethnopharmacology* 148: 729–745.

Peng, W., C. Sonnea, S. S. Lam, Y. S. Ok, and A. K. O. Alstrup. 2020. "The Ongoing Cut-down of the Amazon Rainforest Threatens the Climate and Requires Global Tree Planting Projects: A Short Review." *Environmental Research* 181. https://doi.org/10.1016/j.envres.2019.108887

Pennisi, B. V., R. D. Oetting, F. E. Stegelin, P. A. Thomas, and J. L. Woodward. 2009. *Commercial Production of English Ivy (*Hedera helix *L.)*. Bulletin 1206. Athens, University of Georgia. https://athenaeum.libs.uga.edu/bitstream/handle/10724/12400/B1206.pdf?sequence=1

Pennisi, L. A., S. M. Holland, and T. V. Stein. 2004. "Achieving Bat Conservation through Tourism." *Journal of Ecotourism* 3: 195–207.

Perrine, R. M., and H. L. Osbourne. 1998. "Personality Characteristics of Dog and Cat Persons." *Anthrozoös* 11: 33–40.

Perveen, F. K., and A. Khan. 2018. "Moths." In *Pests of Potato, Maize and Sugar Beet*, edited by F. K. Perveen. IntechOpen. https://www.intechopen.com/books/moths-pests-of-potato-maize-and-sugar-beet/introductory-chapter-moths

Perzigian, A. B. 2003. "Detailed Discussion of Genetic Engineering and Animal Rights: The Legal Terrain and Ethical Underpinnings." Michigan State University College of Law. https://www.animallaw.info/article/detailed-discussion-genetic-engineering-and-animal-rights-legal-terrain-and-ethical#III

Peters, C. C. 2018. *Pandora's Garden: Kudzu, Cockroaches, and Other Misfits of Ecology*. Athens, University of Georgia Press.

Peterson, K. M., and W. W. Payne. 1973. "The genus *Hymenoclea* (Compositae: Ambrosieae)." *Brittonia* 25: 243–256.

Petit, S., A. Boursault, M. Le Guilloux, N. Munier-Jolain, and X. Reboud. 2011. "Weeds in Agricultural Landscapes. A Review." *Agronomy for Sustainable Development* 31: 309–317.

Peveling, R. 2001. "Environmental Conservation and Locust Control – Possible Conflicts and Solutions." *Journal of Orthoptera Research* 10: 171–187.

Pfeiffer, M. B. 2018. "In the Battle against Lyme Disease, the Ticks Are Winning." *Scientific American Newsletter*, June 11. https://blogs.scientificamerican.com/observations/in-the-battle-against-lyme-disease-the-ticks-are-winning/

Phifer-Rixey, M., and M. W. Nachman. 2015. "Insights into Mammalian Biology from the Wild House Mouse *Mus musculus*." *Elife*. https://doi.org/10.7554/eLife.05959.

Piazza, E. A., M. C. Iordan, and C. Lew-Williams. 2017. "Mothers Consistently Alter Their Unique Vocal Fingerprints When Communicating with Infants." *Current Biology* 27(20): 3162–3167.

Pichtel, J., K. Kuroiwa, and H. T. Sawyerr. 2000. "Distribution of Pb, Cd and Ba in Soils and Plants of Two Contaminated Sites." *Environmental Pollution* 110: 171–178.

Pierot, S. W. 1974. *The Ivy Book, the Growing and Care of Ivy and Ivy Topiary*. London, Macmillan.

Pigeon, G., M. Festa-Bianchet, D. W. Coltman, and F. Pelletier. 2016. "Intense Selective Hunting Leads to Artificial Evolution in Horn Size." *Evolutionary Applications* 9: 521–530.

Pimentel, D., and M. Pimentel. 2003. "Sustainability of Meat-based and Plant-based Diets and the Environment." *American Journal of Clinical Nutrition* 78: 660S–663S.

Pimm, S. L., C. N. Jenkins, R. Abell, T. M. Brooks, J. L. Gittleman, L. N. Joppa, et al. 2014. "The Biodiversity of Species and Their Rates of Extinction, Distribution, and Protection." *Science* 344(6187). https://www.science.org/lookup/doi/10.1126/science.1246752

Pincheira-Donoso, D., A. M. Bauer, S. Meiri, and P. Uetz. 2013. "Global Taxonomic Diversity of Living Reptiles." *PLoS One* 8(3): e59741. https://doi.org/10.1371/journal.pone.0059741

Pinkava, D. J. 2003. "*Opuntia*." In *Flora of North America North of Mexico, Volume 4*, edited by Flora of North America Editorial Committee, 123–148. Oxford, UK, Oxford University Press.

Pinkerton, E., A. K. Salomon, and F. Dragon. 2019. "Reconciling Social Justice and Ecosystem-Based Management in the Wake of a Successful Predator Reintroduction." *Canadian Journal of Fisheries and Aquatic Sciences* 76: 6. https://doi.org/10.1139/cjfas-2018-0441

Pitman, W. D., C. G. Chambliss, and J. B. Hacker. 2004. "Digitgrass and Other Species of *Digitaria*." In *Warm-Season (C4) Grasses*, edited by L. E. Moser, B. L. Burson, and L. E. Sollenberger, 715–743. Hoboken, NJ, Wiley.

Piwowarski, J. P., S. Granica, and A. K. Kiss. 2015. "*Lythrum salicaria* L. – Underestimated Medicinal Plant from European Traditional Medicine. A Review." *Journal of Ethnopharmacology* 170: 226–250.

Pizzi, R. 2009. "Veterinarians and Taxonomic Chauvinism: The Dilemma of Parasite Conservation." *Journal of Exotic Pet Medicine* 18: 279–282.

Plants for a Future. 2022. "*Chenopodium album*." *Plants for a Future (online database)*. https://pfaf.org/user/plant.aspx?LatinName=Chenopodium+album

Plarre, P., and B. Krüger-Carstensen. 2011. "An Attempt to Reconstruct the Natural and Cultural History of the Webbing Clothes Moth *Tineola bisselliella* Hummel (Lepidoptera: Tineidae)." *Journal of Entomological and Acarological Research* 43: 83–93.

Platt, J. R. 2018. "When This Rat Went Extinct, So Did a Flea. The Extinction of the Christmas Island Flea – and the Current Risk to Other Parasites – Shows a Major Gap in Conservation Efforts." *Scientific American* (blog). https://blogs.scientificamerican.com/extinction-countdown/when-this-rat-went-extinct-so-did-a-flea/?redirect=1

Platt, J. R. 2019a. "Why Don't We Hear about More Species Going Extinct?" *Scientific American Newsletter*. https://blogs.scientificamerican.com/extinction-countdown/why-dont-we-hear-about-more-species-going-extinct/

Platt, J. R. 2019b. "Rise of the Extinction Deniers." *Scientific American Newsletter*. https://blogs.scientificamerican.com/extinction-countdown/rise-of-the-extinction-deniers/

Plaza, P. I., E. Martínez-López, and S. A. Lambertucci. 2019. "The Perfect Threat: Pesticides and Vultures." *Science of the Total Environment* 687: 1207–1218.

Plotz, D. 2006. *The Genius Factory: The Curious History of the Nobel Prize Sperm Bank*. New York, Random House.

Plummer, B. 2016. "'Gene drive.' Learn the Term. Because It Could One Day Transform the World." *Vox*. https://www.vox.com/2016/6/9/11890472/gene-drive-benefits-risk

Podberscek, A. L. and A. M. Beetz. Eds. 2005. *Bestiality and Zoophilia: Sexual Relations with Animals*. West Lafayette, IN, Purdue University Press.

Podberscek, A. L., and S. D. Gosling. 2000. "Personality Research on Pets and Their Owners: Conceptual Issues and Review." In *Companion Animals and Us: Exploring the Relationships between People and Pets*, edited by A. L. Podberscek, E. S. Paul, and J. A. Serpell, 143–167. Cambridge, Cambridge University Press.

Polák, J., S. Rádlová, M. Janovcová, J. Flegr, E. Landová, and D. Frynta. 2020. "Scary and Nasty Beasts: Self-reported Fear and Disgust of Common Phobic Animals." *British Journal of Psychology* 111: 297–321.

Polosa, R., K. Farsalinos, and D. Prisco. 2019. "Health Impact of Electronic Cigarettes and Heated Tobacco Systems." *Internal and Emergency Medicine* 14: 817–820.

Poonia, A. and A. Upadhayay. 2015. "*Chenopodium album* Linn: Review of Nutritive Value and Biological Properties." *Journal of Food Science and Technology (Mysore)* 52: 3977–3985.

Pope, J., A. Morrison-Saunders, A. Bond, and F. Retief. 2021. "When is an Offset Not an Offset? A Framework of Necessary Conditions for Biodiversity Offsets." *Environmental Management* 67: 424–435.

Porshinsky, B. S., S. Saha, M. D. Grossman, P. R. Beery II, and S. P. A. Stawicki. 2011. "Clinical Uses of the Medicinal Leech: A Practical Review." *Journal of Postgraduate Medicine* 57: 65–71.

Porter, K. 2016. "Why Aren't People Adopting Black Pets?" https://www.sittingforacause.com/blog/adopt-dont-shop/why-arent-people-adopting-black-pets/

Porter, F. H., F. Costa, G. Rodrigues, H. Farias, M. Cunha, and G. E. Glass, et al. 2015. Morphometric and Demographic Differences between Tropical and Temperate Norway Rats (*Rattus norvegicus*)." *Journal of Mammalogy* 96 (2): 317–323, https://doi.org/10.1093/jmammal/gyv033

Posadzki, P. 2011. "Shark Cartilage for Cancer Patients: A Mini Systematic Review." *Focus on Alternative and Complementary Remedies* 16: 204–207.

Pospischil, R. 2015. "Cimicidae, Cimicids." In *Encyclopedia of Parasitology*, edited by H. Mehlhorn. Berlin, Springer. https://doi.org/10.1007/978-3-642-27769-6_618-2

Potter, G., and C. Chatwin. 2012. "The Problem with 'Skunk'." *Drugs and Alcohol Today* 12: 232–240.

Pouydebat, E. 2017. *L'intelligence Animale - Cervelle D'oiseaux et Mémoire d'éléphants*. [Animal Intelligence - Bird Brains and Elephant Memory.] Paris, Odile Jacob.

Powell, J. U. 1929. "Rodent-Gods in Ancient and Modern Times." *Folklore* 40: 173–179.

Pradhan, D. 2017. "What Do Mangoes, Cashews, Pistachios, and Poison Ivy Have in Common?" https://medium.com/@TildeCafe/what-do-mangoes-cashews-pistachios-poison-ivy-have-in-common-66d6bfda63cb

Prescott, C., E. Coan, C. Jones, M. Baxter, D. Rymer, and A. Buckle. 2018. "Anticoagulant Resistance in Rats and Mice in the UK – Current Status in 2018." (Technical Report, Campaign for Responsible Rodenticide Use UK). Reading, Vertebrate Pests Unit, University of Reading. https://research.reading.ac.uk/resistant-rats/wp-content/uploads/sites/49/Unorganized/CRRUUniversityofReadingUKResistanceReport2018.pdf

Pressler, M. P., M. L. Kislevitz, J. J. Davis, and B. Amirlak. 2022. "Size and Perception of Facial Features with Selfie Photographs, and Their Implication in Rhinoplasty and Facial Plastic Surgery." *Plastic and Reconstructive Surgery* 149: 859–867.

Preston, C. J. 2021. "De-extinction and Gene Drives: The Engineering of Anthropocene Organisms." In *Animals in Our Midst: The Challenges of Co-existing with Animals in the Anthropocene*, edited by B. Bovenkerk and J. Keulartz, 495–511. Cham, Switzerland, Springer.

Price, M. K. 2008. "Fund-raising Success and a Solicitor's Beauty Capital: Do Blondes Raise More Funds?" *Economic Letters* 100: 351–354.

Prince, A. 2010. *The Politics of Black Women's Hair*. London, ON, Insomniac Press.

Prokop, P., and J. Fančovičová. 2012. "Beautiful Fruits Taste Good: The Aesthetic Influences of Fruit Preferences in Humans." *Anthropologischer Anzeiger* 69: 71–83.

Prokop, P., and J. Fančovičová. 2013. "Self-protection versus Disease Avoidance: The Perceived Physical Condition is Associated with Fear of Predators in Humans." *Journal of Individual Differences* 34: 15–23.

Prokop, P., and J. Fančovičová. 2017. "Animals in Dangerous Postures Enhance Learning, but Decrease Willingness to Protect Animals." *Eurasia Journal of Mathematics Science and Technology Education* 13: 6069–6077.

Prokop, P., and P. Fedor. 2011. "Physical Attractiveness Influences Reproductive Success of Modern Men." *Journal of Ethology* 29: 453–458.

Prokop P., and R. Randler. 2017. "Biological Predispositions and Individual Differences in Human Attitudes toward Animals." In *Ethnozoology: Animals in our Lives*, edited by R. R. N. Alves and U. P. Albuquerque, pp. 447–466. London, Academic Press.

Prokop, P., and S. D. Tunnicliffe. 2010. "Effects of Keeping Pets at Home on Children's Attitudes toward Popular and Unpopular Animals." *Anthrozoös* 23 21–35.

Prokop, P., M. Usak, and J. Fančovičová. 2010. "Risk of Parasite Transmission Influences Perceived Vulnerability to Disease and Perceived Danger of Disease-Relevant Animals." *Behavioural Processes* 85: 52–57.

Prokop, P., M. Zvaríková, M. Zvarík, A. Pazda, and P. Fedor. 2021. "The Effect of Animal Bipedal Posture on Perceived Cuteness, Fear, and Willingness to Protect Them." *Frontiers in Ecology and Evolution* 9: 681241. https://doi.org/10.3389/fevo.2021.681241.

Prosser, R. S., J. C. Anderson, M. L. Hanson, K. R. Solomon, and P. K. Sibley. 2016. "Indirect Effects of Herbicides on Biota in Terrestrial Edge-of-Field Habitats: A Critical Review of the Literature." *Agriculture, Ecosystems & Environment* 232: 59–72.

Prum, R. O. 2017. *The Evolution of Beauty: How Darwin's Forgotten Theory of Mate Choice Shapes the Animal World – and Us*. New York, Doubleday.

Pulver, S., and B. Manski. 2021. "Corporations and the Environment." In *Handbook of Environmental Sociology*, edited by B. Schaefer Caniglia, A. Jorgenson, S. A. Malin, L. Peek, D. N. Pellow, and X. Huang, 89–114. Cham, Springer.

Puts, D. A. 2010. "Beauty and the Beast: Mechanisms of Sexual Selection in Humans." *Evolution and Human Behavior* 31: 157–175.

Pyron, R. A., and J. J. Wiens. 2011. "A Large-Scale Phylogeny of Amphibia Including Over 2800 Species, and a Revised Classification of Extant Frogs, Salamanders, and Caecilians." *Molecular Phylogenetics and Evolution* 61: 543–583.

Qi, S. S. 2008. "*Toxicodendron*." In *Flora of China*, edited by Z. Y. Wu, P. H. Raven, and D. Y. Hong, 11: 348–354. Beijing, Science Press, and St. Louis, Missouri Botanical Garden Press. http://flora.huh.harvard.edu/china/pdf/pdf11/toxicodendron.pdf

Qi, X, Y. Liu, C. C. Vigueira, N. D. Young, A. L. Caicedo, Y. Jia, et al. 2015. "More than One Way to Evolve a Weed: Parallel Evolution of US Weedy Rice through Independent Genetic Mechanisms." *Molecular Ecology* 24: 3329–3344.

Quattrocchi, U. 2012. *CRC World Dictionary of Medicinal and Poisonous Plants: Common Names, Scientific Names, Eponyms, Synonyms, and Etymology*. 5 volumes. Boca Raton, FL, CRC Press.

Qureshi, S., S. Adil, M. E. Abd El-Hack, M. Alagawany, and M. R. Farag. 2017. "Beneficial Uses of Dandelion Herb (*Taraxacum officinale*) in Poultry Nutrition." *World's Poultry Science Journal* 73: 591–602.

Rafles, H. 2010. *Insectopedia*. New York, Pantheon.

Raguso, R. A. 2004. "Why Do Flowers Smell? The Chemical Ecology of Fragrance-Driven Pollination." In *Advances in Insect Chemical Ecology*, edited by R. T. Cardé and J. G. Millar, 151–178. Cambridge, Cambridge University Press.

Raheem, D., A. Raposo, O. B. Oluwole, M. Nieuwland, A. Saraiva, and C. Carrascosa. 2019. "Entomophagy: Nutritional, Ecological, Safety and Legislation Aspects." *Food Research International* 126: 108672. https://doi.org/10.1016/j.foodres.2019.108672

Rakison, D. H. 2009. "Does Women's Greater Fear of Snakes and Spiders Originate in Infancy?" *Evolution & Human Behavior* 30: 438–444.

Rakotoarisoa, T. F., T. Richter, H. Rakotondramanana, and J. Mantilla-Contreras. 2016. "Turning a Problem into Profit: Using Water Hyacinth (*Eichhornia crassipes*) for Making Handicrafts at Lake Alaotra, Madagascar." *Economic Botany* 70: 365–379.

Ralls, K. 1976. "Mammals in Which Females are Larger than Males." *Quarterly Review of Biology* 51: 245–276.

Rand, L. R., P. Anker, D. Fritz, L. Leigh, and S. Rosenheim. 2020. "Biosphere 2: Why an Eccentric Ecological Experiment Still Matters 25 Years Later." https://edgeeffects.net/biosphere-2/

Randimbiharinirina, R. D., T. Richter, B. M. Raharivololona, J. H. Ratsimbazafy, and D. Schüßler. 2021. "To Tell a Different Story: Unexpected Diversity in Local Attitudes towards Endangered Aye-Ayes *Daubentonia madagascariensis* Offers New Opportunities for Conservation." *People and Nature* 3: 484–498.

Rankin, A. 2018. *Jainism and Environmental Philosophy: Karma and the Web of Life*. New York, Routledge.

Rashid, M. H., T. Asaeda, and M. N. Uddin. 2010. "The Allelopathic Potential of Kudzu (*Pueraria montana*)." *Weed Science* 58: 47–55.

Rasmussen, K., J. Thyrring, R. Muscarella, and F. Borchsenius. 2017. "Climate-change-induced Range Shifts of Three Allergenic Ragweeds (*Ambrosia* L.) in Europe and Their Potential Impact on Human Health." *PeerJ* 5: e3104. https://doi.org/10.7717/peerj.3104

Rawls, M., J. Thiele, D. E. Adams, L. M. Steacy, and A. K. Ellis. 2020. "Clinical Symptoms and Biomarkers of Bermuda Grass-Induced Allergic Rhinitis Using the Nasal Allergen Challenge Model." *Annals of Allergy, Asthma & Immunology* 124: 608–615.

Ray, J. C. 2005. "Large Carnivorous Animals as Tools for Conserving Biodiversity: Assumptions and Uncertainties." In *Large Carnivores and the Conservation of Biodiversity*, edited by J. C. Ray, K. H. Redford, R. S. Steneck, and J. Berger, 34–56. Washington, DC, Island Press.

Read, N. 2012. *City Critters: Wildlife in the Urban Jungle*. Custer, WA, Orca.

Redford, K. H., T. M. Brooks, N. B. W. Macfarlane, and J. S. Adams. 2019. *Genetic Frontiers for Conservation: An Assessment of Synthetic Biology and Biodiversity Conservation: Technical Assessment*. Gland, International Union for Conservation of Nature.

Reed, J., J. Oldekop, J. Barlow, R. Carmenta, J. Geldmann, A. Ickowitz, et al. 2020. "The Extent and Distribution of Joint Conservation-Development Funding in the Tropics." *One Earth* 3: 753–762.

Rehman, S. U., I. S. Kim, M. S. Choi, S. H. Kim, Y. Zhang, and H. H. Yoo. 2017. "Time-dependent Inhibition of CYP2C8 and CYP2C19 by *Hedera helix* Extracts, a Traditional Respiratory Herbal Medicine." *Molecules* 22(7): 1241. https://doi.org/10.3390/molecules22071241

Reichard, S. 2000. "*Hedera helix* L." In *Invasive Plants of California's Wildlands*, edited by C. C. Bossard, J. M. Randall, and M. C. Hoshovsky, 212–216. Berkeley, University of California Press.

Reid, W. V., H. A. Mooney, A. Cropper, D. Capistrano, S. R. Carpenter, K. Chopra, et al. 2005. *Millennium Ecosystem Assessment. Ecosystems and Human Well-being: Synthesis*. Washington, DC, Island Press.

Reinhardt, K., and M. T. Siva-Jothy. 2007. "Biology of the Bed Bugs (Cimicidae)." *Annual Review of Entomology* 52: 351–374.

Reynolds, J. L. 2021. "Engineering Biological Diversity: The International Governance of Synthetic Biology, Gene Drives, and De-extinction for Conservation." *Current Opinion in Environmental Sustainability* 49: 1–6.

Rezania, S., J. Park, P. F. Rupani, N. Darajeh, X. Xu, and R. Shahrokhishahraki. 2019. "Phytoremediation Potential and Control of *Phragmites australis* as a Green Phytomass: An Overview." *Environmental Science and Pollution Research* 26: 7428–7441.

Rhodes, G. 2006. "The Evolutionary Psychology of Facial Beauty." *Annual Review of Psychology* 57: 199–266.

Rhodes, G., F. Proffitt, J. M. Grady, and A. Sumich. 1998. "Facial Symmetry and the Perception of Beauty." *Psychonomic Bulletin & Review* 5: 659–669.

Rhodes, G., S. Yoshikawa, A. Clark, K. Lee, R. McKay, and S. Akamatsu.2001. "Attractiveness of Facial Averageness and Symmetry in Non-Western Cultures: In Search of Biologically Based Standards of Beauty." *Perception* 30: 611–625.

Riccucci, M., and L. Benedetto. 2014. "Bats and Insect Pest Control: A Review." *Vespertilio* 17: 161–169.

Richards, A. J. 1973. "The Origin of *Taraxacum* Agamospecies." *Botanical Journal of the Linnean Society* 66: 189–211.

Richards, A. J. 1985. "Sectional Nomenclature in *Taraxacum* (Asteraceae)." *Taxon* 34: 633–644.

Richards, A. J. 1996. "Genetic Variability in Obligate Apomicts of the Genus *Taraxacum*." *Folia Geobotanica et Phytotaxonomica* 31: 405–414.

Ricketts, T., and M. Imhoff. 2003. "Biodiversity, Urban Areas, and Agriculture: Locating Priority Ecoregions for Conservation." *Conservation Ecology* 8(2): 1. http://www.consecol.org/vol8/iss2/art1/

Riechert, S. E., and T. Lockley. 1984. "Spiders as Biological Control Agents." *Annual Review of Entomology* 29: 299–320.

Riedel, S. 2004. "Biological Warfare and Bioterrorism: A Historical Review." *Baylor University Medical Center Proceedings* 17: 400–406.

Riley, S. P. D., J. L. Brown, J. A. Sikich, C. M. Schoonmaker, and E. E. Boydston. 2014. "Wildlife Friendly Roads: The Impacts of Roads on Wildlife in Urban Areas and Potential

Remedies." In *Urban Wildlife Conservation: Theory and Practice*, edited by C. E. Moorman, R. A. McCleery, and M. N. Peterson, 323–360. New York, Springer.

Ripple, W. J., C. Wolf, T. M. Newsome, M. Hoffmann, A. J. Wirsing, and D. J. McCauley. 2017. "Extinction Risk Is Most Acute for the World's Largest and Smallest Vertebrates." *Proceedings of the National Academy of Sciences of the United States* 114: 10678–10683, https://doi.org/10.1073/pnas.1702078114

Ritter, M. 2006. *The Physical Environment: An Introduction to Physical Geography*. University of Wisconsin. https://web.archive.org/web/20070522024431/http://www.uwsp.edu/gEo/faculty/ritter/geog101/textbook/contents.html

Rittmeyer, E. N., A. Allison, M. C. Gründler, D. K. Thompson, and C. C. Austin. 2012. "Ecological Guild Evolution and the Discovery of the World's Smallest Vertebrate." *PLoS One* 7(1): e29797. https://doi.org/10.1371/journal.pone.0029797

Rizos, V., K. Tuokko, and A. Behrens. 2017. "The Circular Economy: A Review of Definitions, Processes and Impacts." *Centre for European Policy Studies* Paper 12440. https://ideas.repec.org/p/eps/cepswp/12440.html

Rizwan, M., M. Hussain, H. Shimelis, M. U. Hameed, R. M. Atif, M. T. Azhar, et al. 2019. "Gene Flow from Major Genetically Modified Crops and Strategies for Containment and Mitigation of Transgene Escape: A Review." *Applied Ecology and Environmental Research* 17(5): 11191–11208.

Robbins, C. R. 2012. *Chemical, Weird and Physical Behavior of Human Hair*. Berlin, Springer-Verlag.

Roberge, J.-M., and P. Angelstam. 2004. "Usefulness of the Umbrella Species Concept as a Conservation Tool." *Conservation Biology* 18: 76–85.

Robertson, I. A. 2015. *Animals, Welfare and the Law. Fundamental Principles for Critical Assessment*. London, Routledge.

Robertson, J., M. Roodbol, M. Bowles, S. Dures, and J. Rowcliffe. 2020. "Environmental Predictors of Livestock Predation: A Lion's Tale." *Oryx* 54: 648–657.

Robins, J. H., M. Hingston, E. Matisoo-Smith, and H. A. Ross. 2007. "Identifying *Rattus* Species Using Mitochondrial DNA." *Molecular Ecology Notes* 7: 717–729.

Robinson, J. G. 2012. "Common and Conflicting Interests in the Engagements between Conservation Organizations and Corporations." *Conservation Biology* 26: 967–977.

Robinson, M. 1998. "Tobacco: The Greatest Crime in World History?" *The Critical Criminologist* 8(3): 20–22. https://divisiononcriticalcriminology.com/wp-content/uploads/Critical-Criminology-08-3.pdf#page=20

Robson, D. 2018. "The Reasons Why Women's Voices Are Deeper Today." *BBC* (Blog), June 12. https://www.bbc.com/worklife/article/20180612-the-reasons-why-womens-voices-are-deeper-today

Rodrigues, A. J., M. O. Oderob, P. O. Hayombe, W. Akunod, D. Kerich, and I. Maobe. 2014. "Converting Water Hyacinth to Briquettes: A Beach Community Based Approach." *International Journal of Sciences: Basic and Applied Research* 15: 358–378.

Roe, E. T., and C. Leonard-Stuart. 1911. *Webster's New Illustrated Dictionary*. New York, Syndicate Publishing Company.

Roe, K., A. McConney, and C. F. Mansfield. 2014. "The Role of Zoos in Modern Society – A Comparison of Zoos' Reported Priorities and What Visitors Believe They Should Be." *Anthrozoös* 27: 529–541.

Rogers, J., K. Humagain, and A. Pearson. 2022. "Mapping the Purple Menace: Spatiotemporal Distribution of Purple Loosestrife (*Lythrum salicaria*) along Roadsides in Northern New York State." *Scientific Reports* 12: 5270. https://doi.org/10.1038/s41598-022-09194-w

Rohde, M. 2012. *Contact-Poisonous Plants of the World* (e-book). mic-ro.com (out of print). https://www.you-books.com/book/M-Rohde/Contactpoisonous-Plants-Of-The-World

Rollin, B. E. 2020. "Ethical and Societal Issues Occasioned by Xenotransplantation." *Animals* 10(9): 1695. https://doi.org/10.3390/ani10091695

Rollins, B. E. 2011. "An Ethicist's Commentary on Animal Intelligence and Animal Welfare." *Canadian Veterinary Journal* 52: 702–704.

Roma-Burgos, N., M. San Sudo, K. Olsen, I. Werle, and B. Song. 2021. "Weedy Rice (*Oryza* spp.): What's in a Name?" *Weed Science* 69: 505–513.

Romero, A., and T. D. Anderson. 2016. "High Levels of Resistance in the Common Bed Bug, *Cimex lectularius* (Hemiptera: Cimicidae), to Neonicotinoid Insecticides." *Journal of Medical Entomology* 53: 727–731.

Romero, A., A. M. Sutherland, D. H. Gouge, H. Spafford, S. Nair, V. Lewis, et al. 2017. "Pest Management Strategies for Bed Bugs (Hemiptera: Cimicidae) in Multiunit Housing: A Literature Review on Field Studies." *Journal of Integrated Pest Management* 8: 1–10.

Rosatte, R., K. Sobey, J. W. Dragoo, and S. D. Gehrt. 2010. "Striped Skunks and Allies." In *Urban Carnivores: Ecology, Conflict, and Conservation*, edited by S. D. Gehrt, S. P. D. Riley, and B. L. Cypher, 97–108. Baltimore, MD, John Hopkins University Press.

Rosatte, R. C., D. Donovan, J. C. Davies, M. Allan, P. Bachmann, and B. Stevenson. 2009. "Aerial Distribution of ONRAB® Baits as a Tactic to Control Rabies in Raccoons and Striped Skunks in Ontario, Canada. *Journal of Wildlife Diseases* 45: 363–374.

Rose, P. Q. 1996. *The Gardener's Guide to Growing Ivies*. Portland, Timber Press.

Rosenthal, M. F., M. Gertler, A. D. Hamilton, S. Prasad, and M. C. B. Andrade. 2017. Taxonomic Bias in Animal Behavior Publications." *Animal Behavior* 127: 83–89.

Røskaft, E. 2004. "Large Carnivores and Human Safety: A Review." *AMBIO: A Journal of the Human Environment* 33: 283–288.

Røskaft, E., T. Bjerke, B. P. Kaltenborn, J. D. C. Linnell, and R. Andersen. 2003. "Patterns of Self-Reported Fear towards Large Carnivores among the Norwegian Public." *Evolution and Human Behavior* 24: 184–198.

Rossbach, K. A., and J. P. Wilson. 1992. "Does a Dog's Presence Make a Person Appear More Likable? Two Studies." *Anthrozoös* 5: 40–51.

Rostenberg, A., Jr. 1955. "An Anecdotal Biographical History of Poison Ivy." *American Medical Association Archives of Dermatology* 72: 438–445.

Roth, S., O. Balvín, M. T. Siva-Jothy, O. Di Iorio, P. Benda, O. Calva, et al. 2019. "Bedbugs Evolved before Their Bat Hosts and Did Not Co-speciate with Ancient Humans." *Current Biology*. https://doi.org/10.1016/j.cub.2019.04.048

Rothmund, T., M. Elson, M. Appel, J. Kneer, J. Pfetsch, F. M. Schneider, and C. Zahn. 2018. "Does Exposure to Violence in Entertainment Media Make People Aggressive?" *VerKörperungen* 23: 267–282.

Rottman, J., C. R. Crimston, and S. Syropoulos. 2021. "Tree-Huggers Versus Human-Lovers: Anthropomorphism and Dehumanization Predict Valuing Nature Over Outgroups." *Cognitive Science* 45: e12967. https://doi.org/10.1111/cogs.12967

Rouland-Lefèvre, C. 2011. "Termites as Pests of Agriculture." In *Biology of Termites: A Modern Synthesis*, edited by D. E. Bignell, Y. Roisin, and N. Lo, 499–517. New York. Springer.

Rowe, N. 2018. "Lianas." *Current Biology* 28: R249–R252.
Roy, M. M. and N. J. S. Christenfeld. 2004. "Do Dogs Resemble Their Owners?" *Psychological Science* 15: 361–363.
Royer, N. 2015. "The History of Fancy Mice." *American Fancy Rat & Mouse Association*. http://www.afrma.org/historymse.htm
Royet, E. 2016. "Vultures Are Revolting. Here's Why We need to Save Them." *National Geographic*. https://www.nationalgeographic.com/magazine/2016/01/vultures-endangered-scavengers
Rozas-Muñoz, E., J. P. Lepoittevin, R. M. Pujol, and A. Giménez-Arnau. 2012. "Allergic Contact Dermatitis to Plants: Understanding the Chemistry Will Help our Diagnostic Approach." *Actas Dermo-Sifiliográficas* (English Edition) 103: 456–477.
Rozsa, L., and P. Apari. 2012. "Why Infest the Loved Ones – Inherent Human Behaviour Indicates Former Mutualism with Head Lice." *Parasitology* 139: 696–700.
Rózsa, L., and Z. Vas. 2015. "Co-extinct and critically Co-endangered species of Parasitic Lice, and Conservation-Induced Extinction: Should Lice Be Reintroduced to Their Hosts?" *Oryx* 49: 107–110.
Rozy, R., R. A. Dar, and U. G. Phutela. 2017. "Optimization of Biogas Production from Water Hyacinth (*Eichhornia crassipes*)." *Journal of Applied and Natural Science* 9: 2062–2067.
Rubio, N. R., N. Xiang, and D. L. Kaplan. 2020. "Plant-based and Cell-based Approaches to Meat Production." *Nature Communications* 11: 6276. https://doi.org/10.1038/s41467-020-20061-y
Ruby, M. B., and S. J. Heine. 2011. "Meat, Morals, and Masculinity." *Appetite* 56: 447–450.
Ruby, M. B., and S. J. Heine 2012. "Too Close to Home. Factors Predicting Meat Avoidance." *Appetite* 59: 47–52.
Ruby, M. B., S. J. Heine, S. Kamble, T. K. Cheng, and M. Waddar. 2013. "Compassion and Contamination. Cultural Differences in Vegetarianism" *Appetite* 71: 340–348.
Rusch, H., and E. Voland. 2013. "Evolutionary Aesthetics: An Introduction to Key Concepts and Current Issues." *Aisthesis. Pratiche, Linguaggi E Saperi dell'estetico* 6: 113–133.
Ruso, B., L. Renninger, and K. Atzwanger. 2003. "Human Habitat Preferences: A generative Territory for Evolutionary Aesthetics Research." In *Evolutionary Aesthetics*, edited by K. Grammer and E. Voland, 279–294. New York, Springer.
Ruspoli, M. 1987. *The Caves of Lascaux, the Final Photographic Record*. London, Thames & Hudson.
Russell, D. A., and R Séguin. 1982. "Reconstruction of the Small Cretaceous Theropod *Stenonychosaurus inequalis* and a Hypothetical Dinosauroid." *Syllogeus* 37: 1–43.
Russell, E. 2001. *War and Nature: Fighting Humans and Insects with Chemicals from World War I to Silent Spring*. Cambridge, Cambridge University Press.
Russell, E. C., and P. Felker. 1987a. "The Prickly Pears (*Opuntia* spp. Cactaceae): A Source of Human and Animal Food in Semiarid Regions." *Economic Botany* 41: 433–445.
Russell, E. C., and P. Felker. 1987b. "Comparative Cold-hardiness of *Opuntia* spp. and cvs. Grown for Fruit, Vegetable and Fodder Production." *Journal of Horticultural Science* 62: 545–550.
Russell, J. C., D. R. Towns, and M. N. Clout. 2008. *Review of Rat Invasion Biology Implications for Island Biosecurity*. Wellington, New Zealand, Science & Technical Publishing Department of Conservation. https://www.doc.govt.nz/documents/science-and-technical/sfc286entire.pdf
Rust, M. K. 2018. "The Biology and Ecology of Cat Fleas and Advancements in Their Pest Management: A Review." *Insects* 8(4): 118. https://doi.org/10.3390/insects8040118
Rutger, J. N., R. K. Webster, and R. A. Figoni. 1983. "Weedy Species of Rice Show Promise for Disease Resistance." *California Agriculture* 1983: 7–9.
Rutherford, S. 2022. *Villain, Vermin, Icon, Kin: Wolves and the Making of Canada*. Montreal & Kingston, McGill-Queens University Press.
Rutkowska, M. 2017. "Dangerous Pets, Misguided Owners: The Pitfalls of Pet-Keeping in TC Boyle's Stories." In *Yearbook of the Polish Association for American Studies, Volume 11*, edited by M. Paryż, 133–145. Warsaw, Polish Association for American Studies.
Ryan, T. 2014 "The Moral Priority of Vulnerability and Dependency: Why Social Work Should Respect Both Humans and Animals." In *Animals in Social work: Why and How They Matter*, edited by T. Ryan, 80–81. London, Palgrave Macmillan.
Ryder, R. D. 2010 [Republication of Ryder 1970]. "Speciesism: The Original Leaflet." *Critical Society* 2: 1–2.
Ryder, R. D. 2011. *Speciesism, Painism and Happiness: A Morality for the Twenty-First Century*. Exeter, UK, Academic Imprint.
Rymerson, R. T., R. Menassa, and J. E. Brandle. 2002. "Tobacco, a Platform for the Production of Recombinant Proteins." In *Molecular Farming of Plants and Animals for Human and Veterinary Medicine*, edited by L. E. Erickson, W.-J. Yu, J. Brandle, and R. Rymerson, 1–31. Dordrecht, The Netherlands, Kluwer Academic.
Saad, G. 2007. *The Evolutionary Bases of Consumption*. Mahwah, NJ, Lawrence Erlbaum Associates.
Safford, R., J. Andevski, A. Botha, C. G. R. Bowden, N. Crockford, R. Garbett, R., et al. 2019. "Vulture Conservation: The Case for Urgent Action." *Bird Conservation International* 29: 1–9.
Sage, R. F., H. A. Coiner, D. A. Way, G. B. Runion, S. A. Prior, H. A. Torbert, et al. 2009. "Kudzu (*Pueraria montana* (Lour.) Merr. variety *lobata*): A New Source of Carbohydrate for Bioethanol Production." *Biomass Bioenergy* 33: 57–61.
Saglam, N. 2018. "The Effects of Environmental Factors on Leeches." *Advances in Agriculture and Environmental Science* 1(1): 1–3.
Saha, D., C. Marble, N. Boyd, and S. Steed. 2019. "Biology and Management of Yellow (*Cyperus esculentus*) and Purple Nutsedge (*C. rotundus*) in Ornamental Crop Production and Landscapes." University of Florida, Environmental Horticulture Department Document ENH1305. https://journals.flvc.org/edis/article/view/107052/117712
Saiki, D., A. D. Adomaitis, and J. Gundlach. 2017. "An Examination of 'Lookism' in Scholarly Literature." *International Textile and Apparel Association Proceedings* 74(1). https://www.iastatedigitalpress.com/itaa/article/1609/galley/1482/view/
Saini, D. C., K. Kulshreshtha, S. Kumar, D. K. Gond, and G. K. Mishra. 2011. "Conserving Biodiversity Based on Cultural and Religious Values." In *National Conference on Earth's Living Treasures*, 22 May, 2011, edited by Uttar Pradesh State Biodiversity Board, 145–152. Lucknow, India, Uttar Pradesh State Biodiversity Board. https://upsbdb.org/pdf/2013/reserch_publications/2.UPBIODIVER_I.pdf
Saini, S., and K. Saini. 2020. "*Chenopodium album* Linn: An Outlook on Weed cum Nutritional Vegetable along with Medicinal Properties." *Emergent Life Sciences Research* 6: 28–33.

Saltonstall, K. 2016. "The Naming of *Phragmites* haplotypes." *Biological Invasions* 18: 2433–2441.

Saltonstall, K., D. Burdick, S. Miller, and B. Smith. 2005. "Native and Non-native *Phragmites*: Challenges in Identification, Research, and Management of the Common Reed." National Estuarine Research Reserve Technical Report Series 2005. https://coast.noaa.gov/data/docs/nerrs/Research_TechSeries_Phrag_Final_2009.pdf

Saltonstall, K., H. E. Castillo, and B. Blossey. 2014. "Confirmed Field Hybridization of Native and Introduced *Phragmites australis* (Poaceae) in North America." *American Journal of Botany* 101: 211–215.

Samaddar, M. 2020. "A Critical Analysis of Ethics of Genetic Engineering." *International Journal for Research in Applied Science & Engineering Technology* 8: 419–421.

Sams, D. J. 1999. "Nutsedge: Weedy Pest or Crop of the Future?" *Ethnobotanical Leaflets*, Southern Illinois University Carbondale. https://opensiuc.lib.siu.edu/cgi/viewcontent.cgi?referer=https://www.google.com/&httpsredir=1&article=1451&context=ebl

Samways, M. J. 2007. "Insect Conservation: A Synthetic Management Approach." *Annual Review of Entomology* 52: 465–487.

Sanchez, A. 2016. *In Praise of Poison Ivy. The Secret Virtues, Astonishing History, and Dangerous Lore of the World's Most Hated Plant.* New York, Taylor Trade.

Sánchez-Bayo, F., and K. A. G. Wyckhuys. 2019. "Worldwide Decline of the Entomofauna: A Review of Its Drivers." *Biological Conservation* 232: 8–27.

Sánchez-Zapata, E., J. Fernández-López, and J. Pérez-Alvarez. 2012. "Tiger Nut (*Cyperus esculentus*) Commercialization: Health Aspects, Composition, Properties, and Food Applications." *Comprehensive Reviews in Food Science and Food Safety* 11: 366–377.

Sanderson, E. W. 2006. "How Many Animals Do We Want to Save? The Many Ways of Setting Population Target Levels for Conservation." *Bioscience* 56: 911–922.

Sandler, R. 2020. "The Ethics of Genetic Engineering and Gene Drives in Conservation." *Conservation Biology* 34: 378–385.

Santangeli, A., M. Girardello, E. Buechley, A. Botha, E. D. Minin, and A. Moilanen. 2019. "Priority Areas for Conservation of Old World Vultures." *Conservation Biology* 33: 1056–1065.

Santillán, M. A., D. L. Carpintero, M. A. Galmes, and J. H. Sarasola. 2009. "Presence of Cimicid Bugs (Hemiptera: Cimicidae) on a Crowned Eagle (*Harpyhaliaetus coronatus*) Nestling." *Journal of Raptor Research* 43: 255–256.

Santos J. C., R. D. Tarvin, and L. A. O'Connell. 2016. "A Review of Chemical Defense in Poison Frogs (Dendrobatidae): Ecology, Pharmacokinetics, and Autoresistance." In *Chemical Signals in Vertebrates 13*, edited by B. Schulte, T. Goodwin, and M. Ferkin, 305–337. Cham, Springer.

Saporito, R. A., M. A. Donnelly, T. F. Spande, and H. M. Garraffo. 2012. "A Review of Chemical Ecology in Poison Frogs." *Chemoecology* 22: 159–168.

Saranraj, P., and D. Stella. 2014. "Impact of Sugar Mill Effluent to Environment and Bioremediation: A Review." *World Applied Sciences Journal* 30: 299–316.

Sarmento, R., D. Brito, R. M. Ladle, G. da Rosa Leal, and M. Amorim Efe. 2014. "Invasive House (*Rattus rattus*) and Brown Rats (*Rattus norvegicus*) Threaten the Viability of Red-Billed Tropicbird (*Phaethon aethereus*) in Abrolhos National Park, Brazil." *Tropical Conservation Science* 7: 614–627.

Sasaki, K., Y. Sasaki, and S. F. Fox. 2010. "Endangered Traditional Beliefs in Japan: Influences on Snake Conservation." *Herpetological Conservation and Biology* 5: 474–485.

Saunders, C. D., A. T. Brook, O. E. Myers Jr. 2006. "Using Psychology to Save Biodiversity and Human Well-Being." *Conservation Biology* 20: 702–705.

Savio, Y. 1987. "Prickly Pear Cactus: The Pads Are 'Nopales,' and the Fruits Are 'Tunas': They're Easy to Grow and Wonderful to Eat." *Cactus Succulent Journal* 59(3): 113–117.

Sawler, R. T. 1981. "Leech Biology and Behavior." In *Neurobiology of the Leech*, edited by J. Muller, J. G. Nichols, and G. Stent, 7–26. New York, Cold Spring Harbor Laboratory.

Scanes, C. G. 2018. "Invertebrates and Their Use by Humans." In *Animals and Human Society*, edited by C. G. Scanes and S. R. Toukhstai, 181–193. Amsterdam, Elsevier.

Scanes, C. G., and S. R. Toukhsati. 2018. "Parasites." In *Animals and Human Society*, edited by C. G. Scanes and S. R. Toukhstai, 383–412. Amsterdam, Elsevier.

Scarpino, S. V., and B. M. Althouse. 2019. "Uncovering the Hidden Cost of Bed Bugs." *Proceedings of the National Academy of Sciences of the USA* 116: 7160–7162.

Schaffner, U., S. Steinbach, Y. Sun, C. A. Skjøth, L. A. de Weger, S. T. Lommen et al. 2020. "Biological Weed Control to Relieve Millions from *Ambrosia* Allergies in Europe." *Nature Communications* 11: 1745. https://doi.org/10.1038/s41467-020-15586-1

Scherf, B. D., and D. Pilling. Eds. 2015. *The Second Report on the State of the World's Animal Genetic Resources for Food and Agriculture.* Rome, Italy, FAO Commission on Genetic Resources for Food and Agriculture Assessments. https://www.fao.org/documents/card/en/c/fea3da3d-d6ed-4a27-8f58-2d83222b29d9

Schillinger, D., and C. Kearns. 2017. "Guidelines to Limit Added Sugar Intake: Junk Science or Junk Food?" *Annals of Internal Medicine*. https://www.acpjournals.org/doi/pdf/10.7326/M16-2754

Schlegel, J., G. Breuer, and R. Rupf. 2015. "Local Insects as Flagship Species to Promote Nature Conservation? A Survey among Primary School Children on Their Attitudes toward Invertebrates." *Anthrozoös* 28: 229–245.

Schloss, K. B., C. S. Goldberger, S. E. Palmer, and C. A. Levitan. 2015. "What's That Smell? An Ecological Approach to Understanding Preferences for Familiar Odors." *Perception* 44: 23–38.

Schmeil, D. and J. Fischen. 1913. *Pflanzen der Heimat. Eine Auswahl der verbreitetsten Pflanzen unserer Fluren. (Plants of the Homeland. A Selection of the Most Common Plants in Our Districts).* Leipzig, Quelle und Meyer.

Schmidt M. 1979. "The Delightful Dandelion." *Organic Gardening* 26: 112–117.

Schneeberger, K., and C. C. Voigt. 2016. "Zoonotic Viruses and Conservation of Bats." In *Bats in the Anthropocene: Conservation of Bats in a Changing World*, edited by C. C. Voigt and T. Kingston, 263–292. Basel, Springer International.

Schnitzer, S. A. 2015. "The Contribution of Lianas to Forest Ecology, Diversity, and Dynamics." In *Biodiversity of Lianas*, edited by N. Parthasarathy, 149–160. Cham, Springer.

Schnitzer, S. A., and F. Bongers. 2002. "The Ecology of Lianas and Their Role in Forests." *Trends in Ecology & Evolution* 17: 223–230.

Schnitzer, S. A., F. Bongers, R. J. Burnham, and F. E. Putz. Eds. 2014. *Ecology of Lianas*. Hoboken, NJ, John Wiley & Sons.

Schnitzer, S. A., R. A. Londré, J. Klironomos, and P. B. Reich. 2008. "Biomass and Toxicity Responses of Poison Ivy (*Toxicodendron radicans*) to Elevated Atmospheric CO_2: Comment." *Ecology* 89: 581–585.

Schuessler, R. 2016. "The Mountain Chicken Frog's First Problem: It Tastes Like… An Ambitious Reintroduction Program Aims to Save This Endangered, Odd, Gigantic Frog." *National Geographic News*. https://www.nationalgeographic.com/news/2016/01/160128-mountain-chicken-frog-endangered-animals/

Schultes, R. E., and A. Hofmann. 1979. *Plants of the Gods: Their Sacred, Healing and Hallucinogenic Powers*. New York, McGraw-Hill.

Schulz, F., M. T. Engel, A. J. Bath, and L. R. Oliveira 2017. "Human-Wildlife Interaction: The Case of Big Cats in Brazil." In *Biological Conservation in the 21st Century: A Conservation Biology of Large Wildlife*, edited by M. O. Campbell, 31–56. New York, Nova.

Schultz, T. R. 2000. "In Search of Ant Ancestors." *Proceedings of the National Academy of Sciences of the United States of America* 97(26): 14028–14029.

Schultze, L., and W. Dönitz. 1910. *Zoologische und anthropologische Ergebnisse einer Forschungsreise im westlichen und zentralen Südafrika, ausgeführt in den Jahren 1903–1905 mit Unterstützung der Kgl. Preussischen Akademie der Wissenschaften zu Berlin. [Zoological and Anthropological Results of a Research Trip in Western and Central South Africa, Carried Out in the Years 1903–1905 with the Support of the Royal Prussian Academy of Sciences in Berlin.]* Jena, Gustav Fischer.

Schuppli, C. A., D. Fraser, and H. J. Bacon. 2014. "Welfare of Non-traditional Pets." *Scientific and Technical Review of the Office International des Epizooties (Paris)* 33: 221–231.

Schuster, T. M., K. L. Wilson, and K. A. Kron. 2011. "Phylogenetic Relationships of *Muehlenbeckia*, *Fallopia* and *Reynoutria* (Polygonaceae) Investigated with Chloroplast and Nuclear Sequence Data." *International Journal of Plant Sciences* 172: 1053–1066.

Schütz, K., R. Carle, and A. Schieber. 2006. "*Taraxacum* – A Review on Its Phytochemical and Pharmacological Profile." *Journal of Ethnopharmacology* 107: 313–323.

Schwarzauer-Rockett, K., S. H. Al-Hamdani, J. R. Rayburn, and N. O. Mwebi. 2013. "Utilization of Kudzu as a Lead Phytoremediator and the Impact of Lead on Selected Physiological Responses." *Canadian Journal of Plant Science* 93: 951–959.

Schweitzer, A. 1959. *The Light within Us*. New York, The Philosophical Library.

Scripps Howard News Service. 1997. "Protect Endangered Flies? No Way." https://www.deseret.com/1997/4/7/19305326/protect-endangered-flies-no-way

Secretariat of the Convention on Biological Diversity. 2011. *Nagoya Protocol on Access to Genetic Resources and the Fair and Equitable Sharing of Benefits Arising from Their Utilization to the Convention on Biological Diversity*. Montreal, Secretariat of the Convention on Biological Diversity (United Nations). http://www.cbd.int/abs/text/; update: https://www.cbd.int/abs/

Secretariat of the Convention on Biological Diversity. 2015. *Synthetic Biology*. Montreal, Secretariat of the Convention on Biological Diversity.

Seddon, P. J., and M. King. 2019. "Creating Proxies of Extinct Species: The Bioethics of De-extinction." *Emerging Topics in Life Sciences* 3: 731–735.

Selbach, C., P. J. Seddon, and R. Poulin. 2018. "Parasites Lost: Neglecting a Crucial Element in De-Extinction." *Trends in Parasitology* 34: 9–11.

Selin, S. 2020. "Symbols of Napoleon: The Violet." In *Shannon Selin, Imagining the Bounds of Nature* (blog). https://shannonselin.com/2020/01/symbols-napoleon-violet/

Selvan, C., D. Dutta, A. Thukral, T. Nargis, M. Kumar, S. Mukhopadhyay, et al. 2016. "Neck Height Ratio Is an Important Predictor of Metabolic Syndrome among Asian Indians." *Indian Journal of Endocrinology and Metabolism* 20: 831–837.

Senchina, D. S. 2006. "Ethnobotany of Poison Ivy, Poison Oak, and Relatives (*Toxicodendron* spp., Anacardiaceae) in America: Veracity of Historical Accounts." *Rhodora* 108: 203–227.

Senchina, D. S. 2008. "Fungal and Animal Associates of *Toxicodendron* spp. (Anacardiaceae) in North America." *Perspectives in Plant Ecology, Evolution and Systematics* 10: 197–216.

Seppälä, S., S. Henriques, M. L. Draney, S. Foord, A. T. Gibbons, L. A. Gomez, S. Kariko, et al. 2018. "Species Conservation Profiles of a Random Sample of World Spiders IV: Scytodidae to Zoropsidae." *Biodiversity Data Journal* 2018 (6): e30842. https://doi.org/10.3897/BDJ.6.e30842

Serpell, J. A., 1995. *The Domestic Dog: Its Evolution, Behaviour, and Interactions with People*. Cambridge, Cambridge University Press.

Serpell, J. A. 1996. *In the Company of Animals: A Study of Human-Animal Relationships*. Revised Edition. Cambridge, Cambridge University Press.

Serpell, J. A., 2000. "Creatures of the Unconscious: Companion Animals as Mediators." In *Companion Animals and Us: Exploring the Relationships between People and Pets*, edited by A. L. Podberscek, E. S. Paul, and J. A. Serpell, 108–121. Cambridge, Cambridge University Press.

Serpell, J. A. 2004. "Factors Influencing Human Attitudes to Animals and Their Welfare." *Animal Welfare* 13: S145–S151.

Serpell, J. A., and E. S. Paul. 1994. "Pets and the Development of Positive Attitudes to Animals." In *Animals and Human Society: Changing Perspectives*, edited by A. Manning, and J. A. Serpell, 127–144. London, Routledge.

Severns, P. M., and A. R. Moldenke. 2010. "Management Tradeoffs between Focal Species and Biodiversity: Endemic Plant Conservation and Solitary Bee Extinction." *Biodiversity and Conservation* 19: 3605–3609.

Shackleton, S., D. Kirby, and J. Gambiza. 2011. "Invasive Plants – Friends or Foes? Contribution of Prickly Pear (*Opuntia ficus-indica*) to Livelihoods in Makana Municipality, Eastern Cape, South Africa." *Development Southern Africa* 28: 177–193.

Shah, J. N., S. P. Peerzada, A. D. Chinche, A. B. Jadhav, and A. D. Patil. 2019. "The Role of Rhus Toxicodendron: A Homoeopathic Remedy in its Various Attenuations on Patients with Rheumatic Disorders." *International Journal of Medical Research & Health Sciences* 8: 103–107.

Shapiro, B. 2017. "Pathways to De-extinction: How Close Can We Get to Resurrection of an Extinct Species?" *Functional Ecology* 31: 996–1002.

Shapiro, D. 2020. "How Did Human Butts Evolve to Look That Way? An Evolutionary Anthropologist Tackles the Mystery of the Butt." *Massive Science*. https://massivesci.com/articles/butts-shape-big-anthropologist-evolution-how-why-explainer/

Sharma, H. O. 2020. "Production of Biodiesel: Industrial, Economic and Energy Aspects: A Review." *Plant Archives* 20: 2058–2066.

Sharma, J. R., R. K. Lal. A. P. Gupta, H. O. Misra, V. Pant, N. K. Singh, et al. 1999. "Development of Non-narcotic (Opiumless and Alkaloid-free) Opium Poppy, *Papaver somniferum*." *Plant Breeding* 118: 449–452.

Sharma, N., N. Dhawale, and M. Venkadesan. 2014. "Limits on the Body Size and Shape of Animals." *Abstracts, 2013 Annual Meeting, Society for Integrative and Comparative Biology*, 53: E369. Oxford, Oxford University Press. https://sicb.burkclients.com/meetings/2013/schedule/abstractdetails.php?id=1038

Shaw, D. 2008. "*Fallopia japonica* (Japanese Knotweed)." *CABI Invasive Species Compendium*. Wallingford, Oxfordshire, CAB International. https://www.cabi.org/isc/datasheet/23875

Shaw, G. 1813. *The Naturalist's Miscellany, or Coloured Figures of Natural Objects*. London, Elizabeth Nodder.

Shelomi, M. 2016a. "What Is the Purpose of Mosquitoes and Flies on Earth?" *Quora (Blog Post)*. https://www.quora.com/What-is-the-purpose-of-mosquitoes-and-flies-on-earth

Shelomi, M. 2016b. "Mosquitoes: Can We Get Rid of Them, and What Would Happen f We Did?" *Quora (Blog Post)*. https://lupbpktfbtldwbuc.quora.com/Mosquitoes-Can-we-get-rid-of-them-and-what-would-happen-if-we-did

Shephard, T. V., S. E. G. Lea, and N. Hempel de Ibarra. 2014. "The Thieving Magpie'? No Evidence for Attraction to Shiny Objects." *Animal Cognition* 18: 393–397.

Sheppard, A. W., R. H. Shaw, and R. Sforza. 2006. "Top 20 Environmental Weeds for Classical Biological Control in Europe: A Review of Opportunities, Regulations and Other Barriers to Adoption." *Weed Research* 46: 93–107.

Sherlock, R., and J. D. Morrey. Eds. 2002. *Ethical Issues in Biotechnology*. Lanham, MD, Rowman & Littlefield.

Shermer, M. 2004. *The Science of Good & Evil*. New York, Time Books.

Shiffman, D. S., and N. Hammerschlag. 2016. "Shark Conservation and Management Policy: A Review and Primer for Non-Specialists." *Animal Conservation* 19: 401–412.

Shimoda, A., and S. Ohsawa. 2017. "A Quantitative Study on Body Modification (Neck Ring Wear) by Kayan Women." *International Journal of Human Culture Studies* 2017(27): 610–620. (In Japanese.)

Shimoda, M., and N. Yamasaki. 2016. "*Fallopia japonica* (Japanese Knotweed) in Japan: Why Is It Not a Pest for Japanese People?" In *Vegetation Structure and Function at Multiple Spatial, Temporal and Conceptual Scales*, edited by E. O. Box, 447–473. Berlin, Springer.

Shinde, K. A., and D. H. Olsen. Eds. 2020. *Religious Tourism and the Environment*. Oxfordshire, CAB International.

Shine, R. 2018. *Cane Toad Wars*. Oakland, University of California Press.

Shivambu, N., C. T. Shivambu, and C. T. Downs. 2020. "Rock Dove (*Columba livia* Gmelin, 1789)." In *Invasive Birds: Global Trends and Impacts*, edited by C. T. Downs and L. A. Hart, 109–117. Wallingford, UK, CABI.

Shivanna K. R., R. Tandon, and M. Koul. 2020. "'Global Pollinator Crisis' and Its Impact on Crop Productivity and Sustenance of Plant Diversity." In *Reproductive Ecology of Flowering Plants: Patterns and Processes*, edited by R. Tandon, K. Shivanna, and M. Koul, 395–413. Singapore, Springer.

Shrestha, N., X. Shen, and Z. Wang. 2019. "Biodiversity Hotspots Are Insufficient in Capturing Range-restricted Species." *Conservation Science and Practice* 1(10): e103. https://doi.org/10.1111/csp2.103

Shrivastava, A., L. Shrestha, S. Prakash, and R. Mehta. 2019. "Transgenic Mosquitoes Fight against Malaria: A Review." *Journal of Universal College of Medical Sciences* 7: 59–65.

Shurtleff, W., and A. Aoyagi. 1985. *The Book of Kudzu: A Culinary and Healing Guide*. Garden, Avery Publishing Group.

Sig, A. K., M. Guney, A. Uskudar Guclu, and E. Ozmen. 2017. "Medicinal Leech Therapy – An Overall Perspective." *Integrative Medicine Research* 6: 337–343.

Signore, R. J. 2017. "Prevention of Poison Ivy Dermatitis with Oral Homeopathic Rhus Toxicodendron." *Dermatology Online Journal* 23(1): 17. https://escholarship.org/content/qt3rm4r9hk/qt3rm4r9hk.pdf

Silva, A. 2013. "Dangerous Snakes, Deadly Snakes and Medically Important Snakes." *Journal of Venomous Animals and Toxins Including Tropical Diseases* 19: 26. https://doi.org/10.1186/1678-9199-19-26

Simaika, J. P., and M. J. Samways. 2018. "Insect Conservation Psychology." *Journal of Insect Conservation* 22: 635–642.

Simberloff, D. 1998. "Flagships, Umbrellas, and Keystones: Is Single-species Management Passé in the Landscape Era?" *Biological Conservation* 83: 247–257.

Simberloff, D. 2011. "Kudzu." In *Encyclopedia of Biological Invasions*, edited by D. Simberloff and D. Rejmánek, 396–399. Berkeley, University of California Press.

Simmons, L. W., and T. J. Ridsdill-Smith. Eds. 2011. *Ecology and Evolution of Dung Beetles*. Hoboken, NJ, Wiley-Blackwell.

Simmons, T. 2013. "Successfully Managing *Phragmites*." *Ecological Landscape Alliance*. https://www.ecolandscaping.org/07/landscape-challenges/invasive-plants/successfully-managing-phragmites/

Simon, S., M. Otto, and M. Engelhard. 2018. "'Gene Drive Organisms' to Combat Invasive Alien Species? – Not Ready for Release." *Natur und Landschaft* 93: 462–464. (In German).

Simonavicius, E., A. McNeill, L. Shahab L, and L. S. Brose. 2019. "Heat-Not-Burn Tobacco Products: A Systematic Literature Review." *Tobacco Control* 28: 582–594.

Simons, E. L., and D. M. Meyers. 2001. "Folklore and Beliefs about the Aye Aye. (*Daubentonia madagascariensis*)." *Lemur News* 6: 11–16.

Simpfendorfer, C. A., M. R. Heupel, W. T. White, and N. K. Dulvy. 2011. "The Importance of Research and Public Opinion to Conservation Management of Sharks and Rays: A Synthesis." *Marine and Freshwater Research* 62: 518–527.

Simpson, C. 2021. "An Ecological Driver for the Macroevolution of Morphological Polymorphism within Colonial Invertebrates." *Journal of Experimental Zoology. Part B, Molecular and Developmental Evolution* 336: 231–238.

Sinatra, G. M., and B. K. Hofe. 2021. *Science Denial: Why It Happens and What to Do about It*. Oxford, Oxford University Press.

Sindhu, R., P. Binod, A. Pandeya, A. Madhavana, J. A. Alphonsa, N. Viveka, et al. 2017. "Water Hyacinth: A Potential Source for Value Addition: An Overview." *Bioresource Technology* 230: 152–162.

Singer, P. 1975. *Animal Liberation: A New Ethics for Our Treatment of Animals*. New York, HarperCollins.

Singh, H. R., and S. A. Rahman. 2012. "An Approach for Environmental Education by Non-Governmental Organizations (NGOs) in Biodiversity Conservation." *Procedia-Social and Behavioral Sciences* 42: 144–152.

Singh, P., and R. K. Rai. 2018. "Tailoring Sugarcane for Smart Canopy Architecture." *Advances in Plants & Agriculture Research* 8: 142–147.

Singh, P., Y. Shivhare, A. K. S, and A. Sharma. 2010. "Pharmacological and Phytochemical Profile of *Chenopodium album* Linn." *Research Journal of Pharmacy and Technology* 3: 960–963.

Singh, R., A. Arora, and V. Singh. 2021. "Biodiesel from Oil Produced in Vegetative Tissues of Biomass – A Review." *Bioresource Technology* 326 (2021): 124772. https://doi.org/10.1016/j.biortech.2021.124772

Singh, S. P.,. K. R. Khanna, S. Schukla, B. S. Dixit, B. S., and R Banerji. 1995. "Prospects of Breeding Opium (*Papaver somniferum* L.) as a High Linoleic Acid Crop." *Plant Breeding* 114: 89–91.

Singleton, G. R., and T. D. Redhead. 1990. "Structure and Biology of House Mouse Populations That Plague Irregularly: An Evolutionary Perspective." *Biological Journal of the Linnean Society* 41: 285–300.

Sinha, R. K. 1995. "Biodiversity Conservation through Faith and Tradition in India: Some Case Studies." *International Journal of Sustainable Development and World Ecology* 2: 278–284.

Sinha, R. K., V. Thakur, and T. N. Lakhanpal. 2005. *Sacred and Magico-Religious Plants of India*. Jodhpur, India, Scientific Publishers.

Sinski, J., R. M. Carini, and J. D. Weber. 2016. "Putting (Big) Black Dog Syndrome to the Test: Evidence from a Large Metropolitan Shelter." *Anthrozoös* 29: 639–652.

Sirwinchester. 2016. "Meet the World's Deadliest Animal – The Mosquito." *Steemit (Blog)*. https://steemit.com/life/@sirwinchester/meet-the-world-s-deadliest-animal-the-mosquito

Sitas, N., J. E. M. Baillie, and N. J. B. Isaac. 2009. "What Are We saving? Developing a Standardized Approach for Conservation Action." *Animal Conservation* 12: 231–237.

Skamel, U. 2003. "Beauty and Sex Appeal: Sexual Selection of Aesthetic Preferences." In *Evolutionary Aesthetics*, edited by V. Eckhard, 173–183. New York, Springer.

Skerl, K. L. 1999. "Spiders in Conservation Planning: A Survey of US Natural Heritage Programs." *Journal of Insect Conservation* 3: 341–347.

Skerratt, L. F., L. Berger, R. Speare, S. Cashins, K. R. McDonald, A. D. Phillott, et al. 2007. "Spread of Chytridiomycosis Has Caused the Rapid Global Decline and Extinction of Frogs." *EcoHealth* 4: 125. https://doi.org/10.1007/s10393-007-0093-5

Sket, B., and P. Trontelj. 2008. "Global Diversity of Leeches (Hirudinea) in Freshwater." *Hydrobiologia* 595: 129–137.

Skibins, J. C., E. Dunstan, and K. Pahlow. 2017. "Exploring the Influence of Charismatic Characteristics on Flagship Outcomes in Zoo Visitors." *Human Dimensions of Wildlife* 22: 157–171.

Skov, M., and O. Vartanian. 2009. "Introduction: What Is Neuroaesthetics?" In *Neuroaesthetics*, edited by M. Skov and O. Vartanian, 1–7. New York, Routledge.

Smaers, J. B., R. S. Rothman, D. R. Hudson, A. M. Balanoff, B. Beatty, D. K. N. Dechmann, et al. 2021. "The Evolution of Mammalian Brain Size." *Science Advances* 7(18): eabe2101. https://doi.org/10.1126/sciadv.abe2101

Small, E. 2004. "Narcotic Plants as Sources of Medicinals, Nutraceuticals, and Functional Foods." In *Proceedings of the International Symposium on the Development of Medicinal Plants*, Taiwan, August 24–25, 2004, edited by F.-F. Hou, H.-S. Lin, M.-H. Chou, and T.-W. Chang, 11–67. Hualien, Taiwan, Hualien District Agricultural Research and Extension Station.

Small, E. 2006. *Culinary Herbs*. Second Edition. Ottawa, NRC Research Press.

Small, E. 2009. *Top 100 Food Plants: The World's Most Important Culinary Crops*. Ottawa, NRC Research Press.

Small, E. 2010. "Blossoming Treasures of Biodiversity: 33. Non-narcotic Drug Poppies – Benefits for People and Biodiversity." *Biodiversity* 11(3&4): 73–80.

Small, E. 2011a. "Blossoming Treasures of Biodiversity – 36. Castor Bean – Taming the World's Most Poisonous Plant." *Biodiversity* 12(3): 186–195.

Small, E. 2011b. "Blossoming Treasures of Biodiversity – 37. *Spirulina* – Food for the Universe." *Biodiversity* 12(4): 255–265.

Small, E. 2011c. "The New Noah's Ark: Beautiful and Useful Species Only. Part 1: Biodiversity Conservation Issues and Priorities." *Biodiversity* 12(4): 232–247.

Small, E. 2012a. "The New Noah's Ark: Beautiful and Useful Species Only. Part 2: The Chosen Species." *Biodiversity* 13(1): 37–53.

Small, E. 2012b. "Top Canadian Ornamental Plants. 1. Overview (Part 3)." *Canadian Botanical Association Bulletin* 45(1): 33–44.

Small, E. 2012c. "Blossoming Treasures of Biodiversity – 38. Lupins – Benefit and Harm Potentials." *Biodiversity* 13(1): 54–64.

Small, E. 2016a. "The Value of Cartoons for Biodiversity Conservation." *Biodiversity* 17(3): 106–114.

Small, E. 2016b. *Cannabis: A Complete Guide*. Boca Raton, FL, Taylor & Francis/CRC Press.

Small, E. 2019a. "Blossoming Treasures of Biodiversity – 58. Ivy (*Hedera species*) – Virtues and Vices of the World's Most Popular Ornamental Vine." *Biodiversity* 20(1): 62–74.

Small, E. 2019b. "In Defence of the World's Most Reviled Invertebrate 'Bugs'." *Biodiversity* 20(4): 168–221.

Small, E. 2021a. "In Defence of the World's Most Reviled Vertebrate Animals: Part 1: 'Lower' Species (Sharks, Snakes, Vultures, Frogs & Toads)." *Biodiversity* 22(1, 2) 159–193.

Small, E. 2021b. "In Defence of the World's Most Reviled Vertebrate Animals: Part 2: Mammals (Bats, Hyenas, Mice, Rats, and Skunks)." *Biodiversity* 22(3, 4) 194–244.

Small, E. 2021c. "Top Canadian Ornamental Plants. 28. Pansies and Other Violets." *Canadian Botanical Association Bulletin* 54(1): 20–29.

Small, E. 2022. "Top Canadian Ornamental Plants. 32. Garden Lupins (*Lupinus polyphyllus* Hybrids)." *Bulletin of the Canadian Botanical Association* 55(2): 51–59.

Small, E., and P. M. Catling. 1996. "Poorly Known Economic Plants of Canada – 9. Dandelion Species, *Taraxacum* spp." *Bulletin of the Canadian Botanical Association* 29(2): 24–26.

Small, E., and P. M. Catling. 1999. *Canadian Medicinal Crops*. Ottawa, National Research Council of Canada.

Small, E., and P. M. Catling. 2003. "Blossoming Treasures of Biodiversity: 8. Coco de Mer (*Lodoicea maldivica*) – What Do You Do with a 70 pound Coconut?" *Biodiversity* 4(2): 25–28 + back cover.

Small, E., and P. M. Catling. 2004a. "Blossoming Treasures of Biodiversity: 11. Cactus Pear (*Opuntia ficus-indica*): Miracle of Water Conservation." *Biodiversity* 5(1): 27–31.

Small, E., and P. M. Catling. 2004b. "Blossoming Treasures of Biodiversity 14. Grass Pea (*Lathyrus sativus*). Can a Last Resort Food Become a First Choice?" *Biodiversity* 5(4): 29–32.

Small, E., and P. M. Catling. 2006a. "Blossoming Treasures of Biodiversity: 21. Sugarcane: An Old Star with a New Act." *Biodiversity* 7(3/4): 37–46.

Small, E., and P. M. Catling. 2006b. “Blossoming Treasures of Biodiversity: 22. Tobacco: Another Old Star with a New Act.” *Biodiversity* 7(3/4): 47–54.

Small, E., and J. Cayouette. 1992. “Biodiversity Diamonds – The Example of Wild Corn.” *Canadian Biodiversity* 2(3): 24–28.

Smith, A. Ed. 1995. “A Dandy Plant.” *Cornell Plant* 50(2): 22.

Smith, B. 1943. *A Tree Grows in Brooklyn*. New York, Harper.

Smith, J. E., and K. E. Holekamp. 2019. “Spotted Hyenas.” In *Encyclopedia of Animal Behavior*, Second Edition, Vol. 3, edited by M. D. Breed and J. Moore, 190–208. Oxford, Academic Press.

Smith, M., L. Cecchi, C. A.Skjøth, G. Karrer, and B. Šikoparija. 2013. “Common Ragweed: A Threat to Environmental Health in Europe.” *Environment International* 61: 115–126.

Smith, R. J., D. Veríssimo, N. J. B. Isaac, and K. E. Jones. 2012. “Identifying Cinderella Species: Uncovering Mammals with Conservation Flagship Appeal.” *Conservation Letters* 5: 205–212.

Smithsonian Institution Press. 1987. “A Mouse Like a House? Pocket Elephant? How Size Shapes Animals, and What the Limits Are.” *Art to Z*. http://www.smithsonianeducation.org/educators/lesson_plans/size_shapes_animals/ATZ_AMouseLikeAHouse_Winter1987.pdf

Sneath, J., and J. W. Toole. 2010. “Head Lice: A Review of Topical Therapies and Rising Pediculicidal Resistance.” *STL Pharmacist* 5(2). https://www.skintherapyletter.com/pharmacist-edition/head-lice-pediculicidal-resistance-pharm/

Sneddon, J. N., U. Evers, and J. A. Lee. 2021. “Giving to Animal Charities: A Nine-Country Study.” *Anthrozoös*. https://doi.org/10.1080/08927936.2021.1938409

Soave, O. and C. D. Brand. 1991. “Coprophagy in Animals: A Review.” *The Cornell Veterinarian* 81: 357–364.

Sobolevskaya, O. 2015. “Young People are Obsessed with Slim Body.” *HSE (Newsletter)* January 23. https://iq.hse.ru/en/news/177665931.html

Sobotková, K., W. Parker, J. Levá, J. Růžková, J. Lukeš, and K. J. Pomajbíková. 2019. “Helminth Therapy – From the Parasite Perspective.” *Trends in Parasitology* 35: 501–515.

Sogari, G, C. Mora, and D. Menozzi Eds. 2019. *Edible Insects in the Food Sector: Methods, Current Applications and Perspectives*. Switzerland, Springer.

Soll, J. 2005. “Controlling English Ivy (*Hedera helix*) in the Pacific Northwest.” *The Nature Conservancy*. https://www.invasive.org/gist/moredocs/hedhel02.pdf

Sołtysiak, J., and T. Brej. 2014. “Invasion of *Fallopia* Genus Plants in Urban Environment.” *Polish Journal of Environmental Studies* 23: 449–458.

Sommer, R. 1988. “The Personality of Vegetables: Botanical Metaphors for Human Characteristics.” *Journal of Personality* 56: 665–683.

Sonenshine, D. E., and M. R. Roe. Eds. 2014. *Biology of Ticks*. Second Edition. 2 vols. New York, Oxford University Press.

Sood, S. K., V. Thakur, and T. N. Lakhanpal. 2005. *Sacred and Magico-religious Plants of India*. Jodhpur, India, Scientific Publishers.

Sorokowski, P. 2010. “Did Venus Have Long Legs? Beauty Standards from Various Historical Periods Reflected in Works of Art.” *Perception* 39: 1427–1430.

Soucacos, P. N., A. E. Beris, K. N. Malizos, C. T. Kabani, and S. Pakos. 1994. “The Use of Medicinal Leeches, *Hirudo medicinalis*, to Restore Venous Circulation in Trauma and Reconstructive Microsurgery.” *International Angiology* 13: 251–258.

Souchet, J., and F. Aubret. 2016. “Revisiting the Fear of Snakes in Children: The Role of Aposematic Signalling.” *Scientific Reports* 6: 37619. https://doi.org/10.1038/srep37619

Soulé, M., and R. Noss. 1998. “Rewilding and Biodiversity: Complementary Goals for Continental Conservation.” *Wild Earth* 8: 19–28.

Soulsbury, C. D., G. Iossa, S. Kennell, and S. Harris. 2009. “The Welfare and Suitability of Primates Kept as Pets.” *Journal of Applied Animal Welfare Science* 12: 1–20.

South, R. 1907. *The Moths of the British Isles*. London, Frederick Warne.

Spence, N., L. Hill, and J. Morris. 2020. “How the Global Threat of Pests and Diseases Impacts Plants, People, and the Planet.” *New Phytologist* 2: 5–13.

Spennemann, D. H., and M. J. Watson. 2017. “Dietary Habits of Urban Pigeons (*Columba livia*) and Implications of Excreta pH – A review.” *European Journal of Ecology* 3: 27–41.

Spitzer, M. 2005. “Spiders, Snakes and Humans – The Amydaloid Nucleus and the Fear of Strangers.” *Nervenheilkunde* 24: 736–739.

Sponsel, L. E., and P. Natadecha-Sponsel. 2017. “Buddhist Environmentalism.” In *Teaching Buddhism: New Insights on Understanding and Presenting the Traditions*, edited by T. Lewis and G. DeAngelis, 318–343. New York, Oxford University Press.

Sprengelmeyer, R., J. Lewis, A. Hahn, and D. Perrett. 2013. “Aesthetic and Incentive Salience of Cute Infant Faces: Studies of Observer Sex, Oral Contraception and Menstrual Cycle.” *PLoS One* 8: 1–7.

Sprengelmeyer, R., D. Perrett, E. Fagan, R. Cornwell, J. Lobmaier, A. Sprengelmeyer, et al. 2009. “The Cutest Little Baby Face: A Hormonal Link to Sensitivity to Cuteness in Infant Faces.” *Psychological Science* 20: 149–154

St John, F. A. V., G. Edwards-Jones, and J. P. G. Jones. 2010. “Conservation and Human Behaviour: Lessons from Social Psychology.” *Wildlife Research* 37: 658–667.

St. Louis, E., M. Stastny, and R. D. Sargent. 2020. “Impacts of Biological Control on the Performance of *Lythrum salicaria* 20 Years Post-Release.” *Biological Control* 140: 104123. https://doi.org/10.1016/j.biocontrol.2019.104123

Staňková, H., M. Janovcová, Š. Peléšková, K. Sedláčková, E. Landová, and D. Frynta. 2021. “The Ultimate List of the Most Frightening and Disgusting Animals: Negative Emotions Elicited by Animals in Central European Respondents.” *Animals* 11: 747. https://doi.org/10.3390/ani11030747

Stanley, J. W. 2008. “Snakes: Objects of Religion, Fear, and Myth.” *Journal of Integrative Biology* 2: 42–58.

Steel, Z., C. Marnane, C. Iranpour, T. Chey, J. W. Jackson, V. Patel, et al. 2014. “The Global Prevalence of Common Mental Disorders: A Systematic Review and Meta-analysis 1980–2013.” *International Journal of Epidemiology* 43: 476–493.

Steg, L., and J. I. M. de Groot. Eds. 2018. *Environmental Psychology: An Introduction*. Second Edition. Hoboken, NJ, John Wiley & Sons.

Stejskal, V., and P. Horák. 1999. “Webbing Clothes Moth, *Tineola bisselliella* (Hum.), Causing Serious Feeding Damage to *Lactuca sativa* and Other Plant Seeds.” *Anzeiger für Schädlingskunde* (*Journal of Pest Science*) 72: 87–88.

Sternberg, T., H. Viles, and A. Cathersides. 2011. “Evaluating the Role of Ivy (*Hedera helix*) in Moderating Wall Surface Microclimates and Contributing to the Bioprotection of Historic Buildings.” *Building and Environment* 46: 293–297.

Sternberg, T., H. Viles, A. Cathersides, and M. Edwards. 2010. "Dust Particulate Absorption by Ivy (*Hedera helix* L.) on Historic Walls in Urban Environments." *Science of the Total Environment* 409: 162–268.

Sternglanz, S. H., J. L. Gray, and M. Murakami. 1977. "Adult Preference for Infantile Facial Features: An Ethological Approach." *Animal Behaviour* 25: 108–115.

Stevens, J. R. 2004. "The Selfish Nature of Generosity: Harassment and Food Sharing in Primates." *Proceedings of the Royal Society of London B: Biological Sciences* 271: 451–456.

Stewart, A. J. A., and T. R. New. 2007. "Insect Conservation in Temperate Biomes: Issues, Progress and Prospects." In *Insect Conservation Biology. Proceedings of the Royal Entomological Society's 23rd Symposium*, edited by A. J. A. Stewart, T. R. New, and O. T. Lewis, 1–33. Wallingford, Oxfordshire, UK, CABI.

Stewart, C., and N. Coverstone. n.d. "Beneficial Insects and Spiders in Your Maine Backyard." University of Maine Cooperative Extension Bulletin 7150. https://extension.umaine.edu/publications/wp-content/uploads/sites/82/2015/04/7150.pdf

Stewart, H. 2009. *Looking at Totem Poles*. Vancouver, Douglas & McIntyre.

Stewart-Wade, S. M., S. Neumann, L. L. Collins, and G. J. Boland. 2002. "The Biology of Canadian Weeds. 117. *Taraxacum officinale* G. H. Weber ex Wiggers." *Canadian Journal of Plant Science* 82: 825–853.

Stokes, P. and P. Rowley-Conwy. 2002. "Iron Age Cultigen? Experimental Return Rates for Fat Hen (*Chenopodium album* L.)." *Environmental Archaeology* 7: 95–99.

Stolton, S., N. Maxted, B. V. Ford-Lloyd, S. P. Kell, and D. Dudley. 2006. *Food stores: Using Protected Areas to Secure Crop Genetic Diversity*. Gland, Switzerland: World Wildlife Fund.

Stone, K. R. 2010. "*Polygonum sachalinense, P. cuspidatum, P. × bohemicum*." In *Fire Effects Information System* [Online]. U.S. Department of Agriculture, Forest Service, Rocky Mountain Research Station, Fire Sciences Laboratory (Producer). https://www.firescience.gov/projects/08-1-2-04/project/08-1-2-04_polspp.pdf

Storer, T. I, and G. H. Vansell. 1935. "Bee-Eating Proclivities of the Striped Skunk." *Journal of Mammalogy* 16: 118–121.

Strachey, J. Ed. 1953–1974. *The Standard Edition of the Complete Psychological Works of Sigmund Freud*. 24 vols. London, Hogarth Press.

Strand, J., R. T. Carson, S. Navrud, A. Ortiz-Bobea, and J. R. Vincent. 2017. "Using the Delphi Method to Value Protection of the Amazon Rainforest." *Ecological Economics* 131: 475–484.

Strand, T. M., and Å. Lundkvist. 2019. "Rat-borne Diseases at the Horizon. A Systematic Review on Infectious Agents Carried by Rats in Europe 1995–2016." *Infection Ecology & Epidemiology* 9:1. https://doi.org/10.1080/20008686.2018.1553461

Strašil, Z., and J. Kára. 2010. "Study of Knotweed (*Reynoutria*) as Possible Phytomass Resource for Energy and Industrial Utilization." *Research in Agricultural Engineering* 56: 85–91.

Strelau, M., D. R. Clements, J. Benner, and R. Prasad. 2018. "The Biology of Canadian Weeds. *Hedera helix* L. and *Hedera hibernica* (G. Kirchn.) Bean." *Canadian Journal of Plant Science* 98(5). https://doi.org/10.1139/CJPS-2018-0009

Stribling, H. L. 1990. "Growing Chufa for Wild Turkeys." Auburn, AL: Auburn University, Alabama Cooperative Extension Service Circular ANR.

Strother, J. L. 2006. "*Ambrosia*." In *Flora of North America North of Mexico, Volume 19*, edited by Flora of North America Editorial Committee, 10–18. New York, Oxford University Press.

Strother, J. L., and B. G. Baldwin. 2002. "Hymenocleas are Ambrosias (Compositae)." Madroño 49: 143–144.

Stubbendieck, J., G. Y. Friisoe, and M. R. Bolick. 1995. *Weeds of Nebraska and the Great Plains*. Nebraska, Nebraska Department of Agriculture.

Su, N. Y., and R. H. Scheffrahn. 2000. "Termites as Pests of Buildings." In *Termites: Evolution, Sociality, Symbioses, Ecology*, edited by T. Abe, D. E. Bignell, and M. Higashi, 437–453. Dordrecht, Springer.

Su, W., Q. Sun, M. Xia, Z. Wen, and Z. Yao. 2018. "The Resource Utilization of Water Hyacinth (*Eichhornia crassipes* (Mart.) Solms) and Its Challenges." *Resources* 7(3): 46. https://doi.org/10.3390/resources7030046

Suetsugu, K. 2019. "Social Wasps, Crickets and Cockroaches Contribute to Pollination of the Holoparasitic Plant *Mitrastemon yamamotoi* (Mitrastemonaceae) in Southern Japan." *Plant Biology* 21: 176–182.

Sulgrove, S. M. 2002. "The Consequences of Ivies and Other Vines on Walls and Trees: Annotated Reference List." http://www.ivy.org/pdf%20files/WALL-TRS.pdf

Sullivan, R. 2004. *Rats: A Year with New York's Most Unwanted Inhabitants*. New York, Bloomsbury. [= Sullivan, R. 2004. *Rats: Observations on the History and Habitat of the City's Most Unwanted Inhabitants*. New York, Bloomsbury.]

Sultana, R., and K. Kobayashi. 2011. "Potential of Barnyard Grass to Remediate Arsenic-Contaminated Soil." *Weed Biology and Management* 11: 12–17.

Sun, J. H., Z.-C. Li, D. K. Jewett, K. O. Britton, W. H. Ye, and X.-J Ge. 2005. "Genetic Diversity of *Pueraria lobata* (Kudzu) and Closely Related Taxa as Revealed by Intersimple Sequence Repeat Analysis." *Weed Research* 45: 255–260.

Sun, Y., O. Brönnimann, G. K. Roderick, A. Poltavsky, S. T. E. Lommen, and H. Müller-Schärer. 2017. "Climatic Suitability Ranking of Biological Control Candidates: A Biogeographic Approach for Ragweed Management in Europe." *Ecosphere* 8(4): e01731. https://doi.org/10.1002/ecs2.1731

Sunderland, K. 1999. "Mechanisms Underlying the Effects of Spiders on Pest Populations." *Journal of Arachnology* 27: 308–316.

Suomi, S. J. 1995. "Touch and the Immune System in Rhesus Monkeys." In *Touch in Early Development*, edited by T. M. Field, 67–79. Hillsdale, NJ, Lawrence Erlbaum Assoc.

Suryawanshi, K. R., Y. V. Bhatnagar, S. Redpath, and C. Mishra. 2013. "People, Predators and Perceptions: Patterns of Livestock Depredation by Snow Leopards and Wolves." *Journal of Applied Ecology* 50: 550–560.

Sutherland, W. J., W. M. Adams, R. B. Aronson, R. Aveling, T. M. Blackburn, S. Broad, et al. 2009. "One Hundred Questions of Importance to the Conservation of Global Biological Diversity." *Conservation Biology* 23: 557–567.

Svizzero, S. 2021. "Agronomic Practices and Biotechnological Methods Dealing with the Occurrence, Dispersion, Proliferation and Adaptation of Weedy Rice (*Oryza sativa* f. *spontanea*)." *International Journal of Pest Management*. https://doi.org/10.1080/09670874.2021.1944697

Swami, V., D. Einon, and A. Furnham. 2006. "The Leg-to-Body Ratio as a Human Aesthetic Criterion." *Body Image* 3: 317–323.

Swami, V., D. Einon, and A. Furnham. 2007 "Cultural Significance of Leg-to-Body Ratio Preferences? Evidence from British and Rural Malaysia." *Asian Journal of Social Psychology* 10: 265–269.

Swati, S., G. Sharma, N. R. Burgos, and T.-M. Tseng. 2020. "Competitive Ability of Weedy Rice: Toward Breeding Weed-Suppressive Rice Cultivars." *Journal of Crop Improvement* 34: 455–469.

Talbot, B., N. Keyghobadi, and B. Fenton. 2019. "Bed Bugs: The Move to Humans as Hosts." *Facets*. https://www.facetsjournal.com/doi/full/10.1139/facets-2018-0038

Tam, K.-P. 2019. "Anthropomorphism of Nature, Environmental Guilt, and Pro-Environmental Behavior." *Sustainability* 11(19): 5430. https://doi.org/10.3390/su11195430

Tamiru-Oli, M., S. D. Premaratna, A. R. Gendall, and M. G. Lewsey. 2019. "Biochemistry, Genetics, and Genomics of Opium Poppy (*Papaver somniferum*) for Crop Improvement." *Annual Plant Reviews Online* 2(4). https://doi.org/10.1002/9781119312994.apr0711

Tan, T., J. Wu, C. Si, S. Dai, Y. Zhang, N. Sun, et al. 2021. "Chimeric Contribution of Human Extended Pluripotent Stem Cells to Monkey Embryos Ex Vivo." *Cell* 184: 2020–2032.

Tang, C., D. Yang, H. Liao, H. Sun, C. Liu, L. Wei, et al. 2019. "Edible Insects as a Food Source: A Review." *Food Production Processing and Nutrition* 1: 8. https://doi.org/10.1186/s43014-019-0008-1

Tanner, R. D., S. S. Hussain, L. A. Hamilton, and F. T. Wolf. 1979. "Kudzu (*Pueraria lobata*): Potential Agricultural and Industrial Resource." *Economic Botany* 33: 400–412.

Tapley, B., K. S. Bradfield, C. Michaels, and M. Bungard. 2015. "Amphibians and Conservation Breeding Programmes: Do All Threatened Amphibians Belong on the Ark?" *Biodiversity and Conservation* 24: 2625–2646.

Tavernia, B. G., and J. M. Reed. 2012. "The Impact of Exotic Purple Loosestrife (*Lythrum salicaria*) on Wetland Bird Abundances." *The American Midland Naturalist* 168: 352–363.

Taylor, J. P., and L. M. Smith. 2003. "Chufa Management in the Middle Rio Grande Valley, New Mexico." *Wildlife Society Bulletin* 31: 156–162.

Taylor, M., and M. D. Tuttle. 2018. *Bats: An Illustrated Guide to All Species*. Washington, DC, Smithsonian Books.

Taylor, M. P., and M. J. Wedel. 2013. "Why Sauropods Had Long Necks; And Why Giraffes Have Short Necks." *PeerJ* 1: e36. https://doi.org/10.7717/peerj.36

Taylor, P. J., I. Grass, A. J. Alberts, E. Joubert, and T. Tscharntke. 2018. "Economic Value of Bat Predation Services – A Review and New Estimates from Macadamia Orchards." *Ecosystem Services* 30: 372–381.

Team Ocean Portal. 2018. "Sharks." Ocean (Smithsonian National Museum of Natural History). https://ocean.si.edu/ocean-life/sharks-rays/sharks

Temple N. J., and K. Alp. 2016. "Sugar in Beverages: Effects on Human Health." In *Beverage Impacts on Health and Nutrition. Nutrition and Health*, edited by T. Wilson and N. Temple, 277–283. Cham, Humana Press. https://doi.org/10.1007/978-3-319-23672-8_19

Tepper, B. J., and N. V. Ullrich. 2002. "Taste, Smell, and the Genetics of Food Preferences." *Topics in Clinical Nutrition* 17(4): 1–14.

Tessler, M., S. R. Weiskopf, L. Berniker, R. Hersch, K. P. Mccarthy, D. W. Yu, et al. 2018. "Bloodlines: Mammals, Leeches, and Conservation in Southern Asia." *Systematics and Biodiversity* 16: 488–496.

Tétényi, P. 1995. "Biodiversity of *Papaver somniferum* L. (Opium Poppy)." *Horticulturae* 390: 191–201.

Tétényi, P. 1997. "Opium (*Papaver somniferum*): Botany and Horticulture." *Horticultural Reviews* 19: 373–408.

Thakur, R. K., N. Hooda, and V. Jeeva. 2003. "Termites and Global Warming – A Review." *The Indian Forester* 129: 923–930.

Thanikachalam, S., N. Sanchez, and A. J. Maddy. 2019. "Long Hair throughout the Ages." *Dermatology* 235: 260–262.

The Guardian. 2013. "Pests That Bug Us Have Their Own Ecological Importance." https://www.theguardian.com/environment/2013/may/21/insects-cockroach-bed-bugs-environment

Theuerkauf, S. 2015. "Phragmites: Always a Foe, or Sometimes a Friend?" *Sea Grant North Carolina Currents*. https://ncseagrant.ncsu.edu/currents/2015/09/phragmites-always-a-foe-or-sometimes-a-friend/

Thiébaut, M., N. Sébastien, and F. Piola. 2020. "'The Fad for *Polygonum* will fade away!': Historic Aspects of the Propagation and Success in France of the *Reynoutria* Complex Based on Archives." *Botany Letters*. https://doi.org/10.1080/23818107.2020.1750478

Thomas-Walters, L., and N. J. Raihan. 2017. "Supporting Conservation: The Roles of Flagship Species and Identifiable Victims." *Conservation Letters* 10: 581–587.

Thomé, O. W. 1885. *Flora von Deutschland,* Österreich und der Schweiz [*Flora of Germany, Austria and Switzerland*.] Gera, Germany, Verlag von Fr. Eugen Köhler.

Thompson, P. B. 2020. *Food and Agricultural Biotechnology in Ethical Perspective*. Third Edition. Cham, Springer.

Thompson, S. L., and Wildlife Conservation Society 2006. *Amazing Snakes!* New York, HarperCollins.

Thornhill, R. 2003. "Darwinian Aesthetics Informs Traditional Aesthetics." In *Evolutionary Aesthetics*, edited by K. Grammer and E. Voland, 9–35. New York, Springer.

Thornhill, R., and S. W. Gangestad. 1999. "Facial Attractiveness." *Trends in Cognitive Sciences* 3: 452–460.

Thornton, A. 2019. "This Is How Many Animals We Eat Each Year." https://www.weforum.org/agenda/2019/02/chart-of-the-day-this-is-how-many-animals-we-eat-each-year/

Thornton, S. 2013. "Twenty Animals You Might Not Know Are Going Extinct." *VetStreet*, March 18. http://www.vetstreet.com/our-pet-experts/20-animals-you-might-not-know-are-going-extinct

Thrasher, C., and V. LoBue. 2016. "Do Infants Find Snakes Aversive? Infants' Physiological Responses to 'Fear-relevant' Stimuli." *Journal of Experimental Child Psychology* 142: 382–390.

Thresher, R. E. 2007. "Genetic Options for the Control of Invasive Vertebrate Pests: Prospects and Constraints." In *Managing Vertebrate Invasive Species: Proceedings of an International Symposium*, edited by G. W. Witmer, W. C. Pitt, and K. A. Fagerstone, 318–331. Fort Collins, CO, USDA/APHIS Wildlife Services, National Wildlife Research Center.

Tian, B., and J. Liu. 2020. "Resveratrol: A Review of Plant Sources, Synthesis, Stability, Modification and Food Application." *Journal of the Science of Food and Agriculture* 100: 1392–1404.

Tian, G., S. Hauser, L.-S. Koutika, F. Ishida, and J. N. Chianu. 2001. "*Pueraria* Cover Crop Fallow Systems: Benefits and Applicability." In *Sustaining Soil Fertility in West Africa*, edited by G. Tian, F. Ishida, D. Keatinge, R. Carsky, and J. Wendt, 137–155. Madison, WI, Soil Science Society of America.

Tierney, D. A., K. D. Sommerville, K. E. Tierney, M. Fatemi, and C. L. Gross. 2017. "Trading Populations – Can Biodiversity Offsets Effectively Compensate for Population Losses?" *Biodiversity and Conservation* 26: 2115–2131.

Tilahun, Y., and G. Welegerima. 2018. “Pharmacological Potential of Cactus Pear (*Opuntia ficus-indica*): A Review.” *Journal of Pharmacognosy and Phytochemistry* 7: 1360–1363.

Tiley, G. E. D. 2010. “Biological Flora of the British Isles: *Cirsium arvense* (L.) Scop.” *Journal of Ecology* 98: 938–983.

Tilley, D. J., and L. St. John. 2012. “Plant Guide for Common Reed (*Phragmites australis*).” Aberdeen, Idaho, USDA-Natural Resources Conservation Service. https://plants.usda.gov/plantguide/pdf/pg_phau7.pdf

Tilman, D., and M. Clark. 2014. “Global Diets Link Environmental Sustainability and Human Health.” *Nature* 515(7528): 518–522.

Ting, W., W. Meng, W. Wenfang, L. Qianxuan, J. Liping, T. Huai, et al. 2019. “Spider Venom Peptides as Potential Drug Candidates Due to Their Anticancer and Antinociceptive Activities.” *Journal of Venomous Animals and Toxins including Tropical Diseases* 25: e146318. https://doi.org/10.1590/1678-9199-jvatitd-14-63-18

Tingley, R., G. Ward-Fear, L. Schwarzkopf, M. J. Greenlees, B. L. Phillips, G. Brown, et al. 2017. “New Weapons in the Toad Toolkit: A Review of Methods to Control and Mitigate the Biodiversity Impacts of Invasive Cane Toads (*Rhinella marina*).” *Quarterly Review of Biology* 92: 123–149.

Tirosh-Samuelson, H. 2015. “Judaism and the Environment.” *Oxford Bibliographies* https://www.oxfordbibliographies.com/view/document/obo-9780199840731/obo-9780199840731-0118.xml

Tisdell, C., and H. S. Nantha. 2006. “Comparison of Funding and Demand for the Conservation of the Charismatic Koala with Those for the Critically Endangered Wombat *Lasiorhinus krefftii*.” In *Vertebrate Conservation and Biodiversity*, edited by D. L. Hawksworth and A. T. Bull, 435–455. Dordrecht, Springer.

Tisdell, C., C. Wilson, and H. S. Nantha. 2006. “Public Choice of Species for the ‘Ark’: Phylogenetic Similarity and Preferred Wildlife Species for Survival.” *Journal for Nature Conservation* 14: 97–105.

Titeux, N., K. Henle, J.-B. Mihoub, and L. Brotons. 2016. “Climate Change Distracts Us from Other Threats to Biodiversity.” *Frontiers in Ecology and the Environment* 14: 291. https://esajournals.onlinelibrary.wiley.com/doi/pdf/10.1002/fee.1303

Titley, M. A., J. L. Snaddon, and E. C. Turner. 2017. “Scientific Research on Animal Biodiversity Is Systematically Biased Towards Vertebrates and Temperate Regions.” *PLoS One* 12(12): e0189577. https://doi.org/10.1371/journal.pone.0189577

Tobin, A., and C. L. Chambers. 2017. “Mixed Effects of Gating Subterranean Habitat on Bats: A Review.” *Wildlife Management* 81: 1149–1160.

Todorov, A. 2017. *Face Value: The Irresistible Influence of First Impressions*. Princeton, NJ, Princeton University Press.

Toon, A., and J. M. Hughes. 2008. “Are Lice Good Proxies for Host History? A Comparative Analysis of the Australian Magpie, *Gymnorhina Tibicen*, and Two Species of Feather Louse.” *Heredity* 101: 127–135.

Toups, M. A., A. Kitchen, J. E. Light, and D. L. Reed. 2011. “Origin of Clothing Lice Indicates Early Clothing Use by Anatomically Modern Humans in Africa.” *Molecular Biology and Evolution* 28: 29–32.

Tramper, J. 2019. “Can We Engineer Life? 5.1a Animal Gene Technology, AquAdvantage GM Salmon: The Gene Construct.” https://www.biobasedpress.eu/2019/08/can-we-engineer-life-5-1a-animal-gene-technology-aquadvantage-gm-salmon-the-gene-construct/

Třebický, V., J. Fialová, K. Kleisner, S. C. Roberts, A. C. Little, and J. Havlíček. 2015. “Further Evidence for Links between Facial Width-to-Height Ratio and Fighting Success: Commentary on Zilioli et al. (2014).” *Aggressive Behavior* 41: 331–334.

Treves, A., and L. Naughton-Treves. 1999. “Risk and Opportunity for Humans Coexisting with Large Carnivores.” *Journal of Human Evolution* 36: 275–282.

Treves, A., and P. Palmqvist. 2007. “Reconstructing Hominin Interactions with Mammalian Carnivores (6.0–1.8 Ma).” In *Primate Anti-Predator Strategies. Developments in Primatology: Progress and Prospects*, edited by S. L. Gursky and K. A. I. Nekaris, 355–382. Boston, MA, Springer.

Trew, B. T., and I. M. D. Maclean. 2021. “Vulnerability of Global Biodiversity Hotspots to Climate Change.” *Global Ecology and Biogeography* 30: 768–783.

Trimble, R. 1996. “Kudos to Kudzu.” *Appalachian Heritage* 24(4): 59–60.

Trinkela, M. 2009. “A Keystone Predator at Risk? Density and Distribution of the Spotted Hyena (*Crocuta crocuta*) in the Etosha National Park, Namibia.” *Canadian Journal of Zoology* 87: 941–947.

Trombulak, S. C., K. S. Omland, J. A. Robinson, J. J. Lusk, T. L. Fleischner, G. Brown, et al. 2004. “Principles of Conservation Biology: Recommended Guidelines for Conservation Literacy from the Education Committee of the Society for Conservation Biology.” *Conservation Biology* 18: 1180–1190.

Tronson, D. 2001. “The Odour, the Animal and the Plant.” *Molecules* 6: 104–116.

Tsugawa, H. 1986a. “Cultivation and Utilization of Kudzu-vine (*Pueraria lobata* Ohwi). Taxonomy, Geographical Distribution, Use, Breeding and Propagation.” *Japanese Journal of Grassland Science* 31: 435–443.

Tsugawa, H. 1986b. “Cultivation and Utilization of Kudzu-vine (*Pueraria lobata* Ohwi). Adaptability, Cultivation Method, Cutting Frequency, Yield, Grazing and Feeding Value.” *Japanese Journal of Grassland Science* 32: 173–183.

Tu, T. N., and T. D. Thien. Eds. 2019. *Buddhist Approach to Responsible Consumption and Sustainable Development*. VietNam, Hong Duc.

Tuan, Y.-F. 1984. *Dominance and Affection: The Making of Pets*. New Haven, Yale University Press.

Tubei, G. 2019. “The Spotted Hyena, One of the Most Intelligent, Powerful and Unique Wild Animals Habiting North and East Africa, Is Being Eaten to Extinction by Humans.” *Business Insider*, February 15. https://www.pulselive.co.ke/bi/lifestyle/the-spotted-hyena-is-being-eaten-to-extinction-by-humans/mq2vs7h

Tucker, C. M., M. W. Cadotte, T. J. Davies, and T. G. Rebelo. 2012. “Incorporating Geographical and Evolutionary Rarity into Conservation Prioritization.” *Conservation Biology* 26: 593–601.

Tucker, M., and N. W. Bond. 1997. “The Roles of Gender, Sex Role, and Disgust in Fear of Animals.” *Personality and Individual Differences* 22: 135–138.

Tunnessen, N., and M. Hsieh. 2018. “Eating Worms to Treat Autoimmune Diseases?” *Frontiers for Young Minds* 6: 32. https://doi.org/10.3389/frym.2018.00032

Turner, A., and M. Antón. 1996. “The Giant Hyaena, *Pachycrocuta brevirostris* (Mammalia, Carnivora, Hyaenidae).” *Geobios* 29: 455–468.

Turner, A., and M. Antón. 2006. “Africa – The Evolution of a Continent and Its Large Mammal Fauna.” *Cranium* 23: 17–40.

Turner, D. D. 2017. "Biases in the Selection of Candidate Species for De-extinction." *Ethics, Policy & Environment* 20: 21–24.

Turner, F. A., K. S. Jordan, and R. C. Van Acker. 2012. "Review: The Recruitment Biology and Ecology of Large and Small Crabgrass in Turfgrass: Implications for Management in the Context of a Cosmetic Pesticide Ban." *Canadian Journal of Plant Science* 92: 829–845.

Turvey, S. T., M. M. Marr, I. Barnes, S. Brace, B. Tapley, R. W. Murphy, et al. 2019. "Historical Museum Collections Clarify the Evolutionary History of Cryptic Species Radiation in the World's Largest Amphibians." *Ecology and Evolution* 18: 10070–10084.

Tuttle, M. D. 2017. "Fear of Bats and Its Consequences." *Journal of Bat Research & Conservation* 10(1). https://doi.org/10.14709/ BarbJ.10.1.2017.09; http://secemu.org/wp-content/uploads/2018/05/Tuttle_et_al_2017.pdf

Tuttle, M. D. 2020. "Bat Flash! COVID-19 Coronavirus Leads to More Premature Scapegoating of Bats." *Merlin Tuttle's Bat Conservation*, January 30. https://www.merlintuttle.org/2020/01/30/wuhan-coronavirus-leads-to-more-premature-scapegoating-of-bats/

Twigg, G. 1975. *The Brown Rat*. London, David & Charles.

Twyman, R. M., E. Stoger, S. Schillberg., P. Christou, and R. Fischer. 2003. "Molecular Farming in Plants: Host Systems and Expression Technology." *Trends in Biotechnology* 21: 570–578.

Tyack, N., H. Dempewolf, and C. K. Khoury. 2020. "The Potential of Payment for Ecosystem Services for Crop Wild Relative Conservation." *Plants* 9(10): 1305. https://doi.org/10.3390/plants9101305

Tyler, M. J., T. Burton, and A. M. Bauer. 2001. "Parotoid or Parotid? On the Nomenclature of an Amphibian Skin Gland." *Herpetological Review* 32: 79–81.

U.S. Department of Health and Human Services. 2020. *Smoking Cessation: A Report of the Surgeon General*. Rockville, MD, U.S. Department of Health and Human Services, Public Health Service, Office of the Surgeon General. https://www.cdc.gov/tobacco/data_statistics/sgr/2020-smoking-cessation/index.html?s_cid=OSH_misc_M181&CDC_AA_refVal=https%3A%2F%2Fwww.cdc.gov%2Fcessationsgr%2Findex.html

U.S.D.A. 1965. *Controlling Lawn Weeds with Herbicides*. Revised Edition. U.S. Department of Agriculture, Home and Garden Bulletin No. 79. Washington, DC, U.S. Government Printing Office.

U.S.D.A. 2002. "Billion-dollar Grass, *Echinochloa frumentacea* Link." United States Department of Agriculture, Natural Resources Conservation Service. https://webarchive.library.unt.edu/eot2008/20080920044550/http://plants.usda.gov/factsheet/pdf/fs_ecfr.pdf

Ulinski, S. 2018. *Amazing Heights: How Short Guys Stand Tall*. Chicago, IL, Publisher Services.

Ulrich, R. 1986. "Human Responses to Vegetation and Landscapes." *Landscape and Urban Planning* 13: 29–44.

Ulrich, W. 2008. "Body Size and the Relative Abundance of Species." *Ecological Questions* 10: 19–29.

Understanding Animal Research. 2019. "Animal Research Numbers in 2018." http://www.understandinganimalresearch.org.uk/news/communications-media/animal-research-numbers-in-2018/

Ungar, P. S., and M. F. Teaford. Eds. 2002. *Human Diet: Its Origin and Evolution*. Westport, CT, Greenwood Publishing Group.

United Nations. 2012. *Resilient People, Resilient Planet. A Future Worth Choosing, The Report of the United Nations Secretary-General's High Level Panel on Global Sustainability*. New York, United Nations. https://earthcharter.org/library/resilient-people-resilient-planet-a-future-worth-choosing/?gclid=Cj0KCQjwmtGjBhDhARIsAEqfDEcyKD22sgeP5r0EqDw9id5w-0i_HE9S1otxylxioIl65kavCsTqvP8aApAaEALw_wcB

United Nations. 2018. *World Urbanization Prospects 2018*. https://population.un.org/wup/

United Nations. 2019. *World Population Prospects 2019: Ten Key Findings*. New York, United Nations, Department of Economic and Social Affairs, Population Division.

United Nations. 2020. "UN Report: Nature's Dangerous Decline 'Unprecedented'; Species Extinction Rates 'Accelerating'." https://www.un.org/sustainabledevelopment/blog/2019/05/nature-decline-unprecedented-report/

United States Department of Agriculture Forest Service. 1984. *Wildlife, Fish and Sensitive Plant Habitat Management. Title 2600. Amendment 48*. Washington, DC, U.S. Forest Service.

United States Environmental Protection Agency. 2021. "Overview of EPA's Brownfields Program." https://www.epa.gov/brownfields/overview-epas-brownfields-program

University of Utah. 2016. "Why Vultures Matter – And What We Lose If They're Gone." https://unews.utah.edu/why-vultures-matter-and-what-we-lose-if-theyre-gone/

Uniyal, V. P. 2004. "Spiders as Conservation Monitoring Tools." http://vpuniyal.com/images/spiders_as_conservation_monitoring_tools.pdf

U.S. Department of Health and Human Services and U.S. Department of Agriculture. 2015. *2015–2020 Dietary Guidelines for Americans. 8th Edition*. http://health.gov/dietaryguidelines/2015/guidelines/

Uspensky, I. 1992. "General Principles of Protecting People from Arthropod Pests." *Annals of the New York Academy of Sciences* 661: 229–235.

Vaičekonytė, R., E. Kiviat, F. Nsenga, and A. Ostfeld. 2013/2014. "An Exploration of Common Reed (*Phragmites australis*) Bioenergy Potential in North America." *Mires and Peat* 13: 1–9.

Value of Nature to Canadians Study Taskforce. 2017. *Completing and Using Ecosystem Service Assessment for Decision-Making: An Interdisciplinary Toolkit for Managers and Analysts*. Ottawa, Federal, Provincial, and Territorial Governments of Canada. https://publications.gc.ca/collections/collection_2017/eccc/En4-295-2016-eng.pdf

Van der Maesen, L. J. G. 1985. *Revision of the Genus* Pueraria *with Some Notes on* Teyleria *(Leguminosae)*. Wageningen, Agricultural University.

Van der Maesen, L. J. G. 2002. "*Pueraria*: Botanical Characteristics." In *The Genus* Pueraria, edited by W. M. Keung, 1–28. New York, Taylor & Francis.

Van der Sluijs, J. P., S. Foucart, and J. Casas. 2021. "Editorial Overview: Halting the Pollinator Crisis Requires Entomologists to Step up and Assume Their Societal Responsibilities." *Current Opinion in Insect Science* 46: vi–xiii.

Van Eeden, L. M., K. Slagle, M. S. Crowther, C. R. Dickman, and T. M. Newsome. 2020. "Linking Social Identity, Risk Perception, and Behavioral Psychology to Understand Predator Management by Livestock Producers." *Restoration Ecology* 28: 902–910.

Van Huis, A., B. Rumpold, C. Maya, and N. Roos. 2021. "Nutritional Qualities and Enhancement of Edible Insects." *Annual Review of Nutrition* 41: 1. https://doi.org/10.1146/annurev-nutr-041520-010856

Van Schaik, C. 2006. "Why Are Some Animals So Smart?" *Scientific American* 294: 64–71.

Van Slyck, K. M. 2017. "Salmon with a Side of Genetic Modification: The FDA's Approval of AquAdvantage Salmon and Why the Precautionary Principle Is Essential for Biotechnology Regulation." *Seattle University Law Review* 41: 311. https://digitalcommons.law.seattleu.edu/sulr/vol41/iss1/10/

Van Valen, L. 1972. "Body Size and Numbers of Plants and Animals." *Evolution* 27: 27–35.

Vashi, N. A. Ed. 2015. *Beauty and Body Dysmorphic Disorder. A Clinician's Guide*. Cham, Springer.

Vaughan's Seed Company, and Henry G. Gilbert Nursery. 1942. *Book for Florists*. Chicago, IL, Vaughan's Seed Store.

Venu, H., D. Venkataraman, P. Purushothaman, and D. R. Vallapudi. 2019. "*Eichhornia crassipes* Biodiesel as a Renewable Green Fuel for Diesel Engine Applications: Performance, Combustion, and Emission Characteristics." *Environmental Science and Pollution Research* 26: 18084–18097.

Veríssimo, D., G. Vaughan, M. Ridout, C. Waterman, D. MacMillan, and R. J. Smith. 2017. "Increased Conservation Marketing Effort Has Major Fundraising Benefits for Even the Least Popular Species." *Biological Conservation* 211(Part A): 95–101.

Verma, V., S. Sharma, and R. Prasad. 2009. "Biological Alternatives for Termite Control: A review." *International Biodeterioration & Biodegradation* 63: 959–972.

Vermeersch, E. 2000. "Ethical Aspects of Genetic Engineering." In *Biotechnology, Patents and Morality*, edited by S. Sterckx, 165–171. London, Routledge.

Versluys, T. M. M., and W. J. Skylark. 2017. "The effect of Leg-to-Body Ratio on Male Attractiveness Depends on the Ecological Validity of the Figures." *Royal Society Open Science* 4: 10. https://doi.org/10.1098/rsos.170399

Verts, B. J. 1967. *The Biology of the Striped Skunk*. Urbana, University of Illinois Press.

Vickers, L. 2016. "Using Cattle to Manage Invasive Plants." https://www.cattlemen.bc.ca/docs/presentation_cattle_managing_weeds_bcca_2016635930694904173555.pdf

Vickers, L., K. Bondaroff, and S. Burton. 2017. "Nutritional Value of Thistle." Peace River Forage Association of British Columbia, https://peaceforage.bc.ca/wp-content/uploads/2021/06/FF_104_Grazing_Weeds_Nutritional_Value.pdf

Videbæk, P. N., and K. G. Grunert. 2020. "Disgusting or Delicious? Examining Attitudinal Ambivalence towards Entomophagy among Danish consumers." *Food Quality and Preference* 83. https://doi.org/10.1016/j.foodqual.2020.103913

Villamagna, A., and B. Murphy. 2009. "Ecological and Socio-economic Impacts of Invasive Water Hyacinth (*Eichhornia crassipes*): A Review." *Freshwater Biology* 55: 282–298.

Vilmorin-Andrieux et Cie (Vilmorin, H. de, et Vilmorin, M. de). 1883. Les Plantes Potagères. Paris, Vilmorin-Andreux et Cie. [Translated by W. Robinson. 1885. The Vegetable Garden. London, John Murray].

Vitt, L. J., and J. P. Caldwell. 2013. *Herpetology: An Introductory Biology of Amphibians and Reptiles*. Fourth Edition. Cambridge, MA, Academic Press.

Vlasáková, B., J. Pinc, F. Jůna, and Z. Kotyková Varadínová. 2019. "Pollination Efficiency of Cockroaches and Other Floral Visitors of *Clusia blattophila*." *Plant Biology* 21: 753–761.

Vodičková, B., V. Večerek, and E. Voslarova. 2019. "The Effect of Adopter's Gender on Shelter Dog Selection Preferences." *Acta Veterinaria Brno* 88: 93–101.

Vogel, E. R., C. D. Knott, B. E. Crowley, M. D. Blakely, M. D. Larsen, and N. J. Dominy. 2012. "Bornean Orangutans on the Brink of Protein Bankruptcy." *Biology Letters* 8: 333–336.

Vogl, O. 2000. "Oriental Lacquer, Poison Ivy, and Drying Oils." *Journal of Polymer Science, Part A, Polymer Chemistry* 38: 4327–4335.

Voigt, C. C., and T. Kingston. Eds. 2016. *Bats in the Anthropocene: Conservation of Bats in a Changing World*. Geneva, Springer International.

Voland, E., and K. Grammer. 2003. *Evolutionary Aesthetics*. Berlin, Springer-Verlag.

Volk, A. A., A. V. Dane, and Z. A. Marini. 2014. "What Is Bullying? A Theoretical Redefinition." *Developmental Review* 34: 327–343.

Völker, T., Z. Kovacic, and R. Strand. 2020. "Indicator Development as a Site of Collective Imagination? The Case of European Commission Policies on the Circular Economy." *Culture and Organization* 26: 103–120.

Von Essen, E., and H. P. Hansen. 2016. "Sport Hunting and Food Procurement Ethics." In *Encyclopedia of Food and Agricultural Ethics*, edited by P. B. Thompson, and D. M. Kaplan, 1–7. Dordrecht, Springer.

Vray, S., T. Lecocq, S. P. M. Roberts, and P. Rasmont. 2017. "Endangered by Laws: Potential Consequences of Regulations against Thistles on Bumblebee Conservation." *Annales de la Société entomologique de France (N.S.)* 53: 33–41.

Vymazal, J., and T. Březinová. 2016. "Accumulation of Heavy Metals in Aboveground Biomass of *Phragmites australis* in Horizontal Flow Constructed Wetlands for Wastewater Treatment: A Review." *Chemical Engineering Journal* 290: 232–242.

Wacker, S., T. Aronsen, S. Karlsson, O. Ugedal, O. H. Diserud, E. M. Ulvan, et al. 2021. "Selection against Individuals from Genetic Introgression of Escaped Farmed Salmon in a Natural Population of Atlantic Salmon." *Evolutionary Applications* 14: 1450–1460.

Waggy, M. A. 2010. "*Hedera helix*." *Fire Effects Information System*. U.S. Department of Agriculture, Forest Service, Rocky Mountain Research Station, https://www.fs.usda.gov/database/feis/plants/vine/hedhel/all.html

Wakahara, M. 1995. "Cannibalism and the Resulting Dimorphism in Larvae of a Salamander *Hynobius retardatus*, Inhabited in Hokkaido, Japan." *Zoological Science* 12: 467–473.

Walker, T. I. 1998. "Can Shark Resources Be Harvested Sustainably? A Question Revisited with a Review of Shark Fisheries." *Marine and Freshwater Research* 49: 552–572.

Waller, B., and J. Micheletta. 2013. "Facial Expression in Nonhuman Animals." *Emotion Review* 5: 54–59.

Walls, J. G. 2013. "The World's Deadliest Snakes." *Reptiles (Magazine)*. http://www.reptilesmagazine.com/Snakes/Wild-Snakes/The-Worlds-Deadliest-Snakes/

Walpole, M. J., and N. Leader-Williams 2002. "Tourism and Flagship Species in Conservation." *Biodiversity and Conservation* 11: 543–547.

Walsh, N. P., and V. Cotovio. 2020. "Bats Are Not to Blame for Coronavirus. Humans Are." *CNN*. https://www.cnn.com/2020/03/19/health/coronavirus-human-actions-intl/index.html

Waltz, E. 2021. "First Genetically Modified Mosquitoes Released in the United States: Biotech Firm Oxitec Launches Controversial Field Test of Its Insects in Florida after Years of Push-Back from Residents and Regulatory Complications." *Nature* (online) 593: 175–176. https://doi.org/10.1038/d41586-021-01186-6

Wang, D., K. Nair, M. Kouchaki, E. J. Zajac, and X. Zhao. 2019. "A Case of Evolutionary Mismatch? Why Facial Width-to-Height Ratio May Not Predict Behavioral Tendencies." *Psychological Science* 30: 1074–1081.

Wang, F., and F. Basso. 2019. "'Animals Are Friends, Not Food': Anthropomorphism Leads to Less Favorable Attitudes Toward Meat Consumption by Inducing Feelings of Anticipatory Guilt." *Appetite* 138: 153–173.

Wang, L.-F., and C. Cowled. 2015. *Bats and Viruses: A New Frontier of Emerging Infectious Diseases*. Hoboken, NJ, Wiley.

Wang, M., and J. Cao. 2020. "Chinese Skin Colour Preference of Different Genders." In *Advanced Graphic Communication, Printing and Packaging Technology. Lecture Notes in Electrical Engineering, Vol. 600*, edited by P. Zhao, Z. Ye, M. Xu, and L. Yang, 92–97. Singapore, Springer. https://doi.org/10.1007/978-981-15-1864-5_13

Wanless, R., A. Angel, R. Cuthbert, G. Hilton, and P. Ryan. 2007. "Can Predation by Invasive House Mice Drive Seabird Extinctions?" *Biological Letters* 3: 241–244.

Ward, J., D. J. McCafferty, D. C. Houston, and G. D. Ruxton. 2008. "Why Do Vultures Have Bald Heads? The Role of Postural Adjustment and Bare Skin Areas in Thermoregulation." *Journal of Thermal Biology* 33: 168–173.

Ward, P. I., N. Mosberger, C. Kistler, and O. Fischer. 1998. "The Relationship between Popularity and Body Size in Zoo Animals." *Conservation Biology* 12: 1408–1411.

Warkentin, I. G., D. Bickford, N. S. Sodhi, and C. J. A. Bradshaw. 2009. "Eating Frogs to Extinction." *Conservation Biology* 23: 1056–1059.

Warner, D. 2011. "Frog Leg Feast Has Scientist Shopping Mad." https://www.irishexaminer.com/lifestyle/outdoors/dick-warner/frog-leg-feast-has-scientist-shopping-mad-163475.html

Watson, A. K., and M. Teshler. 2013. "*Ambrosia artemisiifolia* L., Common Ragweed." In *Biological Control Programs in Canada 2001–2012*, edited by P. G. Mason and D. R. Gillespie, 296–302. Wallingford, Oxfordshire, UK, CAB International.

Watts, D. P. 2020. "Meat Eating by Nonhuman Primates: A Review and Synthesis." *Journal of Human Evolution* 149. https://doi.org/10.1016/j.jhevol.2020.102882

Watts, H. E., and K. E. Holekamp. 2007. "Hyena Societies." *Current Biology* 17(16): R657–660.

Wauters, E., K. D'Haene, and L. Lauwers. 2017. "The Social Psychology of Biodiversity Conservation in Agriculture." *Journal of Environmental Planning and Management* 60: 1464–1484.

Weaver, A. J. 2011. "A Meta-Analytical Review of Selective Exposure to and the Enjoyment of Media Violence." *Journal of Broadcasting & Electronic Media* 55: 232–250.

Weber, E. 2017. *Invasive Plant Species of the World: A Reference Guide to Environmental Weeds*. Wallingford, Oxfordshire, CABI.

Webster, J. 2011. "Zoomorphism and Anthropomorphism: Fruitful Fallacies?" *Animal Welfare* 20: 29–36.

Webster, T. M. 2003. "Nutsedge (*Cyperus* spp.) Eradication: Impossible Dream?" In *Southern Forest Nursery Association and the Northeastern Forest and Conservation Nursery Association Conference*, Gainseville, FL, July 15–18, 2002, Coordinated by L. R. Riley, R. K. Dumroese, and T. D. Landis, 21–25. Ogden, Utah, U.S. Department of Agriculture Forest Service, Rocky Mountain Research Station.

Webster's Third New International Dictionary. 1993. *Webster's Third New International Dictionary*, edited by P. B. Gove and the Merriam-Webster Editorial Staff. Springfield, MA, Merriam-Webster Inc.

Weebly.com. 2015. "Mammals: Ecological Importance." https://mammal2015.weebly.com/ecological-importance.html

Weed Science Society of America. 2017. "WSSA Survey Ranks Most Common and Most Troublesome Weeds in Broadleaf Crops, Fruits and Vegetables." *Weed Science Society of America*. https://wssa.net/2017/05/wssa-survey-ranks-most-common-and-most-troublesome-weeds-in-broadleaf-crops-fruits-and-vegetables/

Wegulo, S. N., and M. O. Carlson. 2011. "Ergot of Small Grain Cereals and Grasses and Its Health Effects on Humans and Livestock." University of Nebraska–Lincoln Extension EC1880. https://extensionpublications.unl.edu/assets/pdf/ec1880.pdf

Weiss, J., A. Nerd, and Y. Mizrahi. 1993. "Vegetative Parthenocarpy in the Cactus Pear *Opuntia ficus-indica* (L.) Mill." *Annals of Botany* 72: 521–526.

Weitzman, M. L. 1998. "The Noah's Ark problem." *Econometrica* 66: 1279–1298.

Welch, W. R. 1967. "Sedentary Bottom Animals." *American Biology Teacher* 29: 465–467.

Wendell, L. 1977. *The Pigeon*. Sumter, SC, Levi Publishing.

Westbury, M. V., S. Hartmann, A. Barlow, I. Wiesel, V. Leo, R. Welch, D. M., et al. 2018. "Extended and Continuous Decline in Effective Population Size Results in Low Genomic Diversity in the World's Rarest Hyena Species, the Brown Hyena." *Molecular Biology and Evolution* 35: 1225–1237.

Weston, G., and A. Y. Y. Chen. 2020. "Allergic Contact Dermatitis Due to Plants." In *Dermatological Manual of Outdoor Hazards*, edited by J. Trevino and A. Y. Chen, 57–72. Cham, Springer.

Whitaker, I. S., D. Izadi, D. W Oliver, G. Monteath, and P. E. Butler. 2004. "*Hirudo medicinalis* and the Plastic Surgeon." *British Journal of Plastic Surgery* 57: 348–353.

Whitehead, M. L. and C. Vaughn-Jones. 2015. "Suitability of Species Kept as Pets." *Veterinary Record* 177: 573–574.

Whittle, P. M., E. J. Stewart, and D. Fisher. 2015. "Re-creation Tourism: De-extinction and Its Implications for Nature-Based Recreation." *Current Issues in Tourism* 18: 908–912.

Whyte, S. 2015. *Human Mating in the Informal Market for Sperm Donation: Preferences and Decision Making*. Master's Thesis, Brisbane, Australia, Queensland University of Technology, https://eprints.qut.edu.au/90658/

Wiktelius, S., J. Ardö, and T. Fransson. 2003. "Desert Locust Control in Ecologically Sensitive Areas: Need for Guidelines." *Ambio: A Journal of the Human Environment* 3: 463–468.

Wilkinson, D. M., and G. D. Ruxton. 2012. "Understanding Selection for Long Necks in Different Taxa." *Biological Reviews* 87: 616–630.

Williams, A. E. 2005. "Water Hyacinth." In *Van Nostrand's Scientific Encyclopedia*, Ninth Edition, edited by G. D. Considine. New York, John Wiley & Sons. https://www.researchgate.net/profile/Adrian_Williams6/publication/278306432_Water_Hyacinth/links/59edc9e44585158fe53404b7/Water-Hyacinth.pdf

Williams, D. J., M. A. Faiz, B. Abela-Ridder, S. Ainsworth, T. C. Bulfone, A. D. Nickerson, et al. 2019a. "Strategy for a Globally Coordinated Response to a Priority Neglected Tropical Disease: Snakebite Envenoming." *PLoS Neglected Tropical Diseases* 13(2): e0007059. https://doi.org/10.1371/journal

Williams, J., A. M. Lambert, R. Long, and K. Saltonstall. 2019b. "Does Hybrid *Phragmites australis* Differ from Native and Introduced Lineages in Reproductive, Genetic, and Morphological Traits?" *American Journal of Botany* 106: 29–41.

Williams, J. T. 1963. "Biological Flora of the British Isles: *Chenopodium album* L. *Journal of Ecology* 51: 711–725.

Williams, M. O., L. Whitmarsh, and D. M. G. Chríosta. 2021. "The Association between Anthropomorphism of Nature and Pro-Environmental Variables: A Systematic Review." *Biological Conservation* 255: 109022. https://doi.org/10.1016/j.biocon.2021.109022

Williams, N. S. G., J. Lundholm, and J. S. MacIvor. 2014. "Do Green Roofs Help Urban Biodiversity Conservation?" *Journal of Applied Ecology* 51: 1643–1649.

Williamson, J. 2018. "Goldenrod & Ragweed." Clemson Cooperative Extension Factsheet HGIC 2326. https://hgic.clemson.edu/factsheet/goldenrod-ragweed/

Willson, J. D. 2017. "Indirect Effects of Invasive Burmese Pythons on Ecosystems in Southern Florida." *Journal of Applied Ecology* 54: 1251–1258.

Wilson, E. E., and E. M. Wolkovich. 2011. "Scavenging: How Carnivores and Carrion Structure Communities." *Trends in Ecology and Evolution* 26: 129–135.

Wilson, E. O. 1975. "Slavery in Ants." *Scientific American* 232(6): 32–40.

Wilson, E. O. 1984. *Biophilia*. Cambridge, MA, Harvard University Press.

Wilson, E. O. 1993. "Biophilia and the Conservation Ethic." In *The Biophilia Hypothesis*, edited by S. R. Kellert and E. O. Wilson, 31–41. Washington, DC, Island Press.

Wilson, E. O. 2000. *Sociobiology: the New Synthesis*. Cambridge, MA, Harvard University Press.

Wilson, E. O., and B. Hölldobler. 2005. "Eusociality: Origin and Consequences." *Proceedings of the National Academy of Sciences of the United States of America* 102: 13367–13371.

Wilson, M. L., C. Boesch, B. Fruth, T. Furuichi, I. C. Gilby, C. Hashimoto, et al. 2014. "Lethal Aggression in *Pan* Is Better Explained by Adaptive Strategies Than Human Impacts." *Nature* 513(7518): 414. https://doi.org/10.1038/nature13727

Wilson, R. P., A. Gómez-Laich, J.-E. Sala, G. Dell'Omo, M. D. Holton, and F. Quintana. 2017. "Long Necks Enhance and Constrain Foraging Capacity in Aquatic Vertebrates." *Proceedings Royal Society B* 284: 20172072. https://doi.org/10.1098/rspb.2017.2072

Wilson, V., A. Weiss, C. E. Lefevre, T. Ochiai, T. Matsuzawa, M. I. Murayama, et al. 2020. "Facial Width-to-Height Ratio in Chimpanzees: Links to Age, Sex and Personality." *Evolution and Human Behavior* 41: 226–234.

Winberry, J. J., and D. M. Jones. 1973. "Rise and Decline of the 'Miracle Vine': Kudzu in the Southern landscape." *Southeastern Geographer* 13: 61–70.

Windsor, D. A. 1995. "Equal Rights for Parasites." *Conservation Biology* 9: 1–2.

Windsor, D. A. 1998. "Most of the Species on Earth are Parasites." *International Journal of Parasitology* 28: 1939–1941.

Windsor, D. A. 2021. "Differences between Conservation of Parasites and Hosts." *SciAesthetics Essays* March 13: 1–4. https://www.researchgate.net/profile/Donald-Windsor-3/publication/350090655_Differences_between_conservation_of_parasites_and_of_hosts/links/605088e2458515e8344ab5f6/Differences-between-conservation-of-parasites-and-of-hosts.pdf

Winegard, T. C. 2019. *The Mosquito: A Human History of Our Deadliest Predator*. New York, E. P. Dutton.

Winnicka, L., and M. A. Shenoy. 2020. "EVALI and the Pulmonary Toxicity of Electronic Cigarettes: A Review." *Journal of General Internal Medicine*. https://doi.org/10.1007/s11606-020-05813-2

Winston, M. L. 1997. *Nature Wars: People vs. Pests*. Cambridge, MA, Harvard University Press.

Winter, K., and J. A. C. Smith. Eds. 1996. *Crassulacean Acid Metabolism: Biochemistry, Ecophysiology and Evolution*. Berlin, Springer.

Wipff, J. K. 2003. "*Digitaria* Haller." In *Flora of North America. Volume 25. Magnoliophyta: Commelinidae (in part): Poaceae, part 2*, edited by M. E. Barkworth, K. M. Capels, S. Long, and M. B. Piep, 358–383. New York, Oxford University Press.

Witmer, G. 2007. "The Ecology of Vertebrate Pests and Integrated Pest Management (IPM)." In *Perspectives in Ecological Theory and Integrated Pest Management*, edited by M. Kogan and P. Jepson, 393–441. Cambridge, Cambridge University Press.

Witmer, G. W., and S. M. Jojola. 2006. "What's Up with House Mice? A Review." In *Proceedings 22nd Vertebrate Pest Conference*, edited by R. M. Timm and J. M. O'Brien, 124–130. Davis, University of California. https://escholarship.org/uc/item/64r826p6

Wohlers, A., E., W. Lankau, E. H. Oertli, and J. Maki. 2018. "Challenges to Controlling Rabies in Skunk Populations Using Oral Rabies Vaccination: A Review." *Zoonoses Public Health* 65: 373–385.

Woinarski, J., B. Murphy, L. A. Woolley, S. Legge, S. Garnett, and T. Doherty. 2017. "For Whom the Bell Tolls: Cats Kill More Than a Million Australian Birds Every Day." *The Conversation* (Blog), October 4. https://openresearch-repository.anu.edu.au/bitstream/1885/206357/1/01_Woinarski_For_whom_the_bell_tolls%253A_cats_2017.pdf

Wolf, E., E. Kemter, N. Klymiuk, and B. Reichart. 2019. "Genetically Modified Pigs as Donors of Cells, Tissues, and Organs for Xenotransplantation." *Animal Frontiers* 9: 13–20.

Wolfe, J. C., J. C. Neal, D. Harlow, and T. W. Gannon. 2016. "Efficacy of the Bioherbicide Thaxtomin A on Smooth Crabgrass and Annual Bluegrass and Safety in Cool-Season Turfgrasses." *Weed Technology* 30: 733–742.

Wolff, L.-A., and T. H. Skarstein. 2020. "Species Learning and Biodiversity in Early Childhood Teacher Education." *Sustainability* 12(9): 3698. https://doi.org/10.3390/su12093698

Wong, K. H., G. Q. Li, K. M. Li, V. Razmovski-Naumovskia, and K. Chan. 2011. "Kudzu Root: Traditional Uses and Potential Medicinal Benefits in Diabetes and Cardiovascular Diseases." *Journal of Ethnopharmacology* 134: 584–607.

Wong, S., S. Lau, P. Woo, and K. Y. Yuen. 2007. "Bats as a Continuing Source of Emerging Infections in Humans." *Reviews in Medical Virology* 17: 67–91.

Woo, P., S. Lau, and K.-Y. Yuen. 2006. "Infectious Diseases Emerging from Chinese Wet-Markets: Zoonotic Origins of Severe Respiratory Viral Infections." *Current Opinion in Infectious Diseases* 19(5): 401–407.

Wood, W. F. 1999. "The History of Skunk Defensive Secretion Research." *Chemical Educator* 4: 44–50.

Wood, C.L. and P.T.J. Johnson. 2015. "A World without Parasites: Exploring the Hidden Ecology of Infection." *Frontiers in Ecology and the Environment* 13: 425–434.

Wookey, O. A. 2022. "Human-Wildlife Coexistence in the Urban Domain: Promoting Welfare through Effective Management, Responsibility and the Recognition of Mutual Interest." In

Human/Animal Relationships in Transformation, edited by A. Vitale and S. Pollo, Cham, Palgrave Macmillan. https://doi.org/10.1007/978-3-030-85277-1_15
World Book. 1998. "A World Book Science Year Interview with Peter H. Raven." *World Book Online*. http://www.well.com/~davidu/raven.html
World Health Organization. 2011. "WHO Report on the Global Tobacco Epidemic, 2011: Warning About the Dangers of Tobacco: Executive Summary." Geneva, World Health Organization.
World Health Organization. 2014. *The Guidance Framework for Testing Genetically Modified Mosquitoes*. Geneva, World Health Organization. http://apps.who.int/iris/bitstream/10665/127889/1/9789241507486_eng.pdf?ua=1
World Health Organization. 2015. *Guideline: Sugars Intake for Adults and Children*. Geneva, World Health Organization. https://www.ages.at/download/0/0/d5caf21e89583a1d8da9e383569a272905750dcc/fileadmin/AGES2015/Themen/Ernaehrung_Dateien/9789241549028_eng.pdf
World Health Organization. 2017. *WHO Report on the Global Tobacco Epidemic 2017: Monitoring Tobacco Use and Prevention Policies*. Geneva, World Health Organization.
World Health Organization. 2018. *WHO Guidelines for the Production, Control and Regulation of Snake Antivenom Immunoglobulins*. Geneva, World Health Organization.
World Health Organization. 2020. "Tobacco." Geneva, World Health Organization. https://www.who.int/health-topics/tobacco#tab=tab_1
World Wildlife Fund. 2014. "Black-Footed Ferret Facts: The Masked Bandits of the Northern Great Plains." https://www.worldwildlife.org/stories/black-footed-ferret-facts-the-masked-bandits-of-the-northern-great-plains
Wrangham, R. W., and D. Peterson. 1996. *Demonic Males: Apes and the Origins of Human Violence*. Boston, MA, Houghton Mifflin Harcourt.
Wu, D., J. Qiu, J. Sun, B. K. Song, K. M. Olsen, and L. Fan. 2022. "Weedy Rice, a Hidden Gold Mine in the Paddy Field." *Molecular Plant* 15: 566–568.
Xia, L., S. C. Lenaghan, M. Zhang, Y. Wu, X. Zhao, J. N. Burris, et al. 2011. "Characterization of English Ivy (*Hedera helix*) Adhesion Force and Imaging Using Atomic Force Microscopy." *Journal of Nanoparticle Research* 13: 1029–1037.
Xu, J., Q. Cai, H. Wang, L. Xuejun, J. Lv, D. Yao, et al. 2017. "Study of the Potential of Barnyard Grass for the Remediation of Cd- and Pb-contaminated Soil." *Environmental Monitoring and Assessment* 189: 224. https://doi.org/10.1007/s10661-017-5923-5
Xu, J., Z. Wang, and J. J. Cheng. 2011. "Bermuda Grass as Feedstock for Biofuel Production: A Review." *Bioresource Technology* 102: 7613–7620.
Xu, R., A. J. Boreland, X. Li, A. Posyton, K. Kwan, and R. P. Hart. 2019. "Functional Mature Human Microglia Developed in Human iPSC Microglial Chimeric Mouse Brain." *Nature Communications*. https://doi.org/10.1038/s41467-020-15411-9
Xu, Y., E. H. Jeong, S. C. S. Jang, and X. Shao. 2021. "Would You Bring Home Ugly Produce? Motivators and Demotivators for Ugly Food Consumption." *Journal of Retailing and Consumer Services* 59: 102376. https://doi.org/10.1016/j.jretconser.2020.102376
Yabuno, T. 1962. "Cytotaxonomic Studies on the Two Cultivated Species and the Wild Relatives in the Genus *Echinochloa*." *Cytologia* 27: 296–305.
Yadav, H. K., S. Shukla, and S. P. Singh 2006. "Genetic Variability and Interrelationship among Opium and Its Alkaloids in Opium Poppy (*Papaver somniferum* L.)." *Euphytica* 150: 207–214.
Yahia, E. M., and C. Sáenz. 2011. "Cactus pear (*Opuntia* species)." In *Postharvest Biology and Technology of Tropical and Subtropical Fruits. Açai to Citrus*, edited by E. M. Yahia, 290–329, 330e–331e. Cambridge, UK, Woodhead.
Yahia, E. M., and C. Sáenz. 2017. "Cactus Pear Fruit and Cladodes." In *Fruit and Vegetable Phytochemicals: Chemistry and Human Health*, Second Edition, edited by E. M. Yahia, 941–956. Hoboken, NJ, Wiley.
Yakkala, K., M.-R. Yu, H. Roh, J.-K. Yang, and Y.-Y. Chang. 2013. "Buffalo Weed (*Ambrosia trifida* L. var. *trifida*) Biochar for Cadmium (II) and Lead (II) Adsorption in Single and Mixed System." *Desalination and Water Treatment* 51: 7732–7745.
Yamaura Y., M. Higa, M. Senzaki, and I. Koizumi. 2018. "Can Charismatic Megafauna Be Surrogate Species for Biodiversity Conservation? Mechanisms and a Test Using Citizen Data and a Hierarchical Community Model." In *Biodiversity Conservation Using Umbrella Species*, edited by F. Nakamura, 151–179. Singapore, Springer.
Yang, D. S., S. V. Pennisi, K.-C. Son, and S. J. Kays. 2009. "Screening Indoor Plants for Volatile Organic Pollutant Removal Efficiency." *HortScience* 44: 1377–1381.
Yang, X., D. Q. Fuller, X. Huan, L. Perry, Q. Li, Z. Li, et al. 2015. "Barnyard Grasses Were Processed with Rice Around 10000 Years Ago." *Science Reports* 5: 16251. https://doi.org/10.1038/srep16251; https://www.ncbi.nlm.nih.gov/pmc/articles/PMC4633675/
Ye, C.-Y., W. Tang, D. Wu, L. Jia, J. Qiu, M. Chen, et al. 2019. "Genomic Evidence of Human Selection on Vavilovian Mimicry." *Nature Ecology & Evolution* 3: 1474–1482.
Yen, A. L. 2009. "Entomophagy and Insect Conservation: Some Thoughts for Digestion." *Journal of Insect Conservation* 13: 667. https://doi.org/10.1007/s10841-008-9208-8
Yigermal, H., and F. Assefa. 2019. "Impact of the Invasive Water Hyacinth (*Eichhornia crassipes*) on Socio-Economic Attributes: A Review." *Journal of Agriculture and Environmental Sciences* 4: 46–56.
Yin, S. 2018. "In a Cockroach Genome, 'Little Mighty' Secrets'." *The New York Times* [Online]. https://www.nytimes.com/2018/03/20/science/american-cockroach-genome.html
Yin, J., Y. Zhu, V. Malik, X. Li, X. Peng, F. F. Zhang, et al. 2020. "Intake of Sugar-Sweetened and Low-Calorie Sweetened Beverages and Risk of Cardiovascular Disease: A Meta-Analysis and Systematic Review." *Advances in Nutrition* 12: 89–101. https://doi.org/10.1093/advances/nmaa084
Yli-Panula, E., E. Jeronen, P. Lemmetty, and A. Pauna. 2018. "Teaching Methods in Biology Promoting Biodiversity Education." *Sustainability* 10(10): 3812. https://doi.org/10.3390/su10103812
Young, J., R. Pritchard, C. Nottle, and H. Banwell. 2020. "Pets, Touch, and COVID-19: Health Benefits from Non-Human Touch through Times of Stress." *Journal of Behavioral Economics for Policy* 4(Special Issue 2): 25–33.
Young, O. P., and G. B. Edwards. 1990. "Spiders in United States Field Crops and Their Potential Effect on Crop Pests." *Journal of Arachnology* 18: 1–29.
Ypey, A. 1813. *Vervolg ob de Avbeeldingen der artseny-gewassen met derzelver Nederduitsche en Latynsche beschryvingen* [Continued from the Images of the Doctors' Crops with Their Dutch and Latin descriptions]. Amsterdam, J. C. Sepp en zoon.

Yuan, J. J., S. Yi, H. A. Williams, and O.-H. Park. 2019. "US Consumers' Perceptions of Imperfect 'Ugly' Produce." *British Food Journal* 121: 2666–2682.

Yue, C., and B. Behe. 2010. "Consumer Color Preferences for Single-Stem Cut Flowers on Calendar Holidays and Noncalendar Occasions." *HortScience* 45: 78–82.

Zacchariah, T. T. 2012. "Invertebrate Animal Welfare." In *Invertebrate Medicine*, Second Edition, edited by G. A. Lewbart, 445–450. Chichester, UK, John Wiley & Sons.

Zaidel, D. W., S. M. Aarde, and K. Baig. 2005. "Appearance of Symmetry, Beauty, and Health in Human Faces." *Brain and Cognition* 57: 261–263.

Zaidel, D. W., and M. Hessamian. 2010. "Asymmetry and Symmetry in the Beauty of Human Faces." *Symmetry* 2: 136–149.

Zaidi, S. M., S. S. Jameel, F. Zaman, S. Jilani, A. Sultana, and S. A. Khan. 2011. "A Systematic Overview of the Medicinal Importance of Sanguivorous Leeches." *Alternative Medicine Review* 16: 59–65.

Zenkner, F. F., M. Margis-Pinheiro, and A. Cagliari. 2019. "Nicotine Biosynthesis in *Nicotiana*: A Metabolic Overview." *Tobacco Science* 56: 1–9.

Zeuner, F. E. 1963. *A History of Domesticated Animals*. New York, Harper and Row.

Zhang, G., J. Liu, M. Gao, W. Kong, Q. Zhao, L. Shi, et al. 2020a. "Tracing the Edible and Medicinal Plant *Pueraria montana* and Its Products in the Marketplace Yields Subspecies Level Distinction Using DNA Barcoding and DNA Metabarcoding." *Frontiers in Pharmacology*. https://doi.org/10.3389/fphar.2020.00336

Zhang, H., C. Li, S.-T. Kwok, Q.-W. Zhang, and S.-W. Chan. 2013. "A Review of the Pharmacological Effects of the Dried Root of *Polygonum cuspidatum* (Hu Zhang) and Its Constituents." *Evidence-Based Complementary and Alternative Medicine* 2013: Article 208349. https://doi.org/10.1155/2013/208349

Zhang, L., M. Lecoq, A. Latchininsky, and D. Hunter. 2019. "Locust and Grasshopper Management." *Annual Review of Entomology* 64: 15–34.

Zhang, Y., L. Bielory, Z. Mi, T. Cai, A. Robock, and P. Georgopoulos. 2015. "Allergenic Pollen Season Variations in the Past Two Decades under Changing Climate in the United States." *Global Change Biology* 21: 1581–1589.

Zhang, Y.-Y., D.-Y. Zhang, and S. C. H. Barrett. 2020b. "Genetic Uniformity Characterizes the Invasive Spread of Water Hyacinth (*Eichhornia crassipes*), a Clonal Aquatic Plant." *Molecular Ecology* 19: 1774–1786.

Zhou, B, C.-H. Kong, Y.-H. Li, P. Wang, and X.-H. Xu. 2013. "Crabgrass (*Digitaria sanguinalis*) Allelochemicals That Interfere with Crop Growth and the Soil Microbial Community." *Journal of Agricultural and Food Chemistry* 61: 5310–5317.

Zhou, P, X.-L. Yang, X.-G. Wang, B. Hu, L. L. Zhang, W. Zhang, et al. 2020. "A Pneumonia Outbreak Associated with a New Coronavirus of Probable Bat Origin." *Nature* 579: 270–273. https://www.nature.com/articles/s41586-020-2012-7

Ziegler, M. 2016. "The Promiscuous Human Flea." https://contagions.wordpress.com/2016/09/27/the-promiscuous-human-flea/

Zilcha-Mano, S., M. Mikulincer, and P. R. Shaver. 2012. "Pets as Safe Havens and Secure Bases: The Moderating Role of Pet Attachment Orientations." *Journal of Research in Personality* 46: 571–580.

Zimmerman, P. R., J. P. Greenberg, S. O. Wandiga, and P. J. Crutzen. 1982. "Termites: A Potentially Large Source of Atmospheric Methane, Carbon-Dioxide, and Molecular-Hydrogen." *Science* 218: 563–565.

Zink, B. J., E. J. Otten, M. Rosenthal, and B. Singal. 1991. "The Effect of Jewel Weed in Preventing Poison Ivy Dermatitis." *Journal of Wilderness Medicine* 2: 178–182.

Ziska, L., K. Knowlton, C. Rogers, D. Dalan, N. Tierney, M. A. Elder, et al. 2011. "Recent Warming by Latitude Associated with Increased Length of Ragweed Pollen Season in Central North America." *Proceedings of the National Academy of Sciences of the United States of America* 108: 4248–4251.

Ziska, L. H., D. R. Gealy, N. Burgos, A. L. Caicedo, J. Gressel, A. L. Lawton-Rauh, et al. 2015. "Weedy (Red) Rice: An Emerging Constraint to Global Rice Production." *Advances in Agronomy* 129: 181–228.

Zucker, I., and A. Beery. 2010. "Males Still Dominate Animal Studies." *Nature* 465: 690. https://doi.org/10.1038/465690

Zvaríková, M., P. Prokop, M. Zvarík, Z. Ježová, W. Medina-Jerez, and P. Fedor. 2021 "What Makes Spiders Frightening and Disgusting to People?" *Frontiers in Ecology and Evolution* 9: 694569. https://doi.org/10.3389/fevo.2021.694569

Zwarun, L., and G. R. Camilo. 2021. "Facts Aren't Enough: Addressing Communication Challenges in the Pollinator Crisis and Beyond." In *The Palgrave Handbook of International Communication and Sustainable Development*, edited by M. J. Yusha'u and J. Servaes, 393–423. Cham, Palgrave Macmillan.

General Index

Note: Bold page numbers refer to tables and *italic* page numbers refer to figures.

Scientific Name Index

Note: This index is based on mention of scientific names in the book, such as for families of species, genera, species, and varieties. Usually, the pages indicated provide the vernacular (common, English) names, and the Vernacular Name Index, as well as the Table of Contents, often indicate where additional information is located.

Note: *Italic* page numbers refer to figures.

Vernacular Name Index

Note: *Italic* page numbers refer to figures.